2008 13th International Power Electronics and Motion Control Conference

Poznan, Poland
1-3 September 2008

Pages 1-514

IEEE Catalog Number: CFP0834A-PRT
ISBN 13:   978-1-4244-1741-4

Copyright © 2008 by The Institute of Electrical and Electronics Engineers, Inc.
All Rights Reserved

Copyright and Reprint Permissions: Abstracting is permitted with credit to the source. Libraries are permitted to photocopy beyond the limit of U.S. copyright law for private use of patrons those articles in this volume that carry a code at the bottom of the first page, provided the per-copy fee indicated in the code is paid through Copyright Clearance Center, 222 Rosewood Drive, Danvers, MA 01923.

For other copying, reprint or republications permission, write to IEEE Copyrights Manager, IEEE Operations Center, 445 Hoes Lane, Piscataway, New Jersey USA 08854. All rights reserved.

IEEE Catalog Number: CFP0834A-PRT

ISBN 13:   978-1-4244-1741-4

ISSN: 2007906910

Additional Copies of This Publication Are Available from:

IEEE Service Center
445 Hoes Lane
Piscataway, NJ 08854
Phone: (800) 678-IEEE
                       (732) 981-1393
Fax: (732) 981-9667
E-mail: customer-service@ieee.org

Table of Contents

Electric Drive System for Automatic Guided Vehicles Using Contact-free Energy Transmission 1
Marcel Jufer

State-of-the-Art High Power Density and High Efficiency DC-DC Chopper Circuits for HEV and FCEV Applications 7
Atsuo Kawamura, Martin Pavlovsky, Yukinori Tsuruta

Current-Based Condition Monitoring of Electrical Machines in Safety Critical Applications 21
Thomas G. Habetler

The Essence of Three-Phase AC/AC Converter Systems 27
J. W. Kolar, T. Friedli, F. Krismer, S. D. Round

An Analysis on Turn-off Behaviour of 1.2kV NPT-CIGBT under Clamped Inductive Load Switching 43
S.T. Kong, L.Ngwendson, M. Sweet, E.M. Sankara Narayanan

Turn-off behaviour of high voltage NPT- and FS-IGBT 48
Hans-Guenter Eckel, Karl Fleisch

Exact Circuit Power Loss Design Method for High Power Density Converters Utilizing Si-IGBT/SiC-Diode Hybrid Pairs 54
Kazuto Takao, Hiromichi Ohashi

A Forward Converter with a Monolithic Cascode Device: Design and Experimental Investigation 61
F. Chimento, S. Musumeci, A. Raciti, L. Abbatelli, S. Buonomo, R. Scollo

Switching and conducting performance of SiC-JFET and ESBT against MOSFET and IGBT 69
André Knop, W. Toke Franke, Friedrich W. Fuchs

In-Service Life Consumption Estimation in Power Modules 76
Mahera Musallam, C Mark Johnson, Chunyan Yin, Hua Lu, Chris Bailey

Measurement Of Temperature Sensitive Parameter Characteristics Of Semiconductor Silicon And Silicon-Carbide Power Devices 84
Mietek Nowak, Jacek Rabkowski, Roman Barlik

Unsymmetrical Gate Voltage Drive for High Power 1200V IGBT4 Modules Based on Coreless Transformer Technology Driver 88
Piotr Luniewski, Uwe Jansen

A Novel RESURFed Double Gates IGBT with Superior Performance 97
Dongming Wu, Kaihang Li, Lingling Yang

An Empiric Approach to Establishing MOSFET Failure Rate Induced by Single-Event Burnout 102
Jeroen van Duivenbode, Bart Smet

Comparative Study on Paralleled vs. Scaled Dc-dc Converters in High Voltage Gain Applications 108
Pawel Klimczak, Stig Munk-Nielsen

A Low-Loss Dc-Dc Converter For A Renewable Energy Converter 114
David S. Thompson, Otu A. Eno

A Single Active Edge-Resonant Snubber Cell-assisted ZCS Half-Bridge DC-DC Converter with Constant Frequency Asymmetrical PWM Scheme 119
Tomokazu Mishima, Mutsuo Nakaoka, Eiji Hiraki

A New Approach to High Efficiency in Isolated Boost Converters for High-Power Low-Voltage Fuel Cell Applications 127
Morten Nymand, Michael A.E. Andersen

New Modulation Strategy with Low Switching Frequency and Minimum Baseband Distortion 132
N. E. Ruger, O. Schnick, W. Mathis, A. Mertens

Table of Contents

A Bit-Stream Based PWM Technique for Variable Frequency Sinewave Generation 139
N. D. Patel, U. K. Madawala

Control Strategies of the Quasi-Resonant DC-Link Inverter 144
Slawomir Mandrek, Piotr J. Chrzan

Consideration for Input Current-Ripple of Pulselink DC-AC Converter for Fuel Cells 148
Kentaro Fukushima, Tamotsu Ninomiya, Masahito Shoyama, Isami Norigoe, Yosuke Harada, Kenta Tsukakoshi

New Practical Approach to Input Current Shaping in AC-DC Power Converters 154
Kuno Janson, Viktor Bolgov, Lauri Kütt, Ants Kallaste, Heigo Mõlder

LLCC-PWM Inverter for Driving High-Power Piezoelectric Actuators 159
Rongyuan Li, Norbert Fröhleke, Joachim Böcker

Modelling and Analysis of a Matrix-Reactance Frequency Converter Based on Buck-Boost Topology by DQ0 Transformation 165
Pawel Szczeniak, Zbigniew Fedyczak, Marius Klytta

A Modular AC/DC Rectifier Based on Cascaded H-bridge Rectifier 173
H. Iman-Eini, Sh. Farhangi, JL. Schanen

Low Loss Soft Switching Boost Converter 181
So-Ri Park, Sang-Hoon Park, Chung-Yuen Won, Yong-Chae Jung

Methods for Experimental Assessment of Component Losses to Validate the Converter Loss Model 187
Yi Wang, Sjoerd de Haan, Jan Abraham Ferreira

Modified multistage semiconductor-Fitch generator topology with magnetic compression 195
Stanislaw Kalisiak, Marcin Holub

Modeling and Measuring Results of a Shunt Current Source Active Power Filter with Series Capacitor 201
P. Parkatti, M. Salo, H. Tuusa

A Multi-Drive System Based on a Two-stage Matrix Converter 207
Dinesh Kumar, Patrick W Wheeler, Jon C Clare, Lee Empringham

Characteristics of the Single Active Bridge Converter with Voltage Doubler 213
Andreas Averberg, Axel Mertens

Analysis of Capacitor Dividers for Multilevel Inverter 221
Oleg Sivkov, Jiri Pavelka

Space Vector Modulation for a Capacitor Clamped Multi-level Matrix Converter 229
Xu Lie, Jon C. Clare, Patrick W. Wheeler, Lee Empringham

New Family of Matrix-Reactance Frequency Converters Based on Unipolar PWM AC Matrix-Reactance Choppers 236
Zbigniew Fedyczak, Pawel Szczesniak, Igor Korotyeyev

Consideration of Conduction Losses for the Series Resonant Converter by Means of a Simple Extension to the SPA Approach 244
Alexander Bucher, Thomas Duerbaum, Daniel Kuebrich, Markus Schmid

Validation and Comparison of different PWM Converter Small Signal Models 250
Alexander Bucher, Markus Schmid, Lukas Bendkowski, Thomas Duerbaum

Dynamic Behaviour of a Series - Connected Multilevel Converter with Interleaved Switching 256
C. Fahrni, A. Rufer

Simple Analysis of a Flying Capacitor Converter Voltage Balance Dynamics for DC Modulation 260
A. Ruderman, B. Reznikov, M. Margaliot

Simulation of Simplified Seven Level Multilevel Converter Circuit 268
Gerardo Ceglia, Víctor Guzmán, Carlos Sánchez, Fernando Ibáñez, Julio Walter, María Giménez

Table of Contents

SEPP High-Frequency Inverter Incorporating an Auxiliary Switch and Its Performance Evaluation......................275
H.Ogiwara,Y.Fujita, R.Urabe, M.Itoi, T.Sugai, M. kuwata, M.Nakaoka

Multiphase coupled converter models dedicated to transient response and output voltage regulation studies 281
Nadia Bouhalli, Marc Cousineau, Emmanuel Sarraute, Thierry Meynard

A 13.56 MHz Current-output-type Inverter Utilizing An Immittance Conversion Element......................288
Yosei Sakamoto, Keiji Wada, Toshihisa Shimizu

Voltage Fed Zero-Voltage Zero-Current Switching PWM DC-DC Converter......................295
Jaroslav Dudrik, Vladimír Rulscin

PWM Spectrum Evaluation and Over-Modulation Phenomena in a Three-Phase Inverters - Analytical Approach 301
Miro Milanovic

Experimental Study of a Matrix Converter Excited Doubly-Fed Induction Machine in Generation and Motoring......................307
Ivan Shapoval, Jon Clare, Eduard Chekhet

Effect of Type and Interconnection of DG Units in the Fault Current Level of Distribution Networks 313
H.R. Baghaee, M. Mirsalim, M. J. Sanjari, G.B. Gharehpetian

An Isolated Full-Bridge DC/DC Converter..with Bidirectional Communication Capability 320
Lon-Kou Chang, Ru-Shiuan Yang

Efficiency and Power Losses in PM BLDC Motor with Variable Bridge/half-bridge Structure Electronic Commutator......................326
K. Krykowski, A. Bodora

Analysis of a device for converting a unipolar input voltage into two symmetric bidirectional output voltages with a magnetically coupled coil 331
Felix. A. Himmelstoss, Wilhelm Kraeftner

Invariant Modulation Strategy for Two-stage Direct Power Converter......................337
Radiy Bekbudov

Experimental Study of A Multicell ac/ac Converter Balancing Circuit......................345
Robert Stala, Andrzej Mondzik

A Comparison and Optimum Design of Reluctance-Controlled Classical Load-Resonant Converters 350
Stefan V. Mollov, Michael P. Theodoridis

Capacitor Clamped Multilevel Matrix Converter Controlled with Venturini Method......................357
Janina Rzasa

Reliability Consideration for a High Power Zero-Voltage-Switching Flyback Power Supply 365
Arash Rahnamaee, Jafar Milimonfared, Kaveh Malekian, Mohammad Abroushan

The Traction Drive Topology Using the Matrix Converter with Middle-Frequency Transformer 372
Martin Pittermann, Pavel Drábek, Marek Cédl

Analysis of Multipulse Rectifiers with Modulation in DC Circuit in Vector Space Approach......................377
Andrzej KAPLON and Jaroslaw ROLEK

High Efficiency Soft Switching Boost Converter for Photovoltaic System 383
Gil-Ro Cha, Sang-Hoon Park, Chung-Yuen Won, Yong-Chae Jung, Sang-Hoon Song

A Power Converter For Fault Tolerant Machine Development In Aerospace Applications 388
Liliana de Lillo, Patrick Wheeler, Lee Empringham, Chris Gerada, XiaoyanHuang

Optimal Bus Capacitance Design for System Stability in On-Board Distributed Power Architecture 393
Seiya Abe, Masahiko Hirokawa, Masahito Shoyama, Tamotsu Ninomiya

Table of Contents

Steady State Analysis of Hysteretic Control Buck Converters 400
L.K. Wong, T.K. Man

A Novel Control Method for IGBT Current Source Rectifier 405
Longcheng Tan, Yaohua Li, Ping Wang, Congwei Liu, Zixin Li, Yonggang Chen, Wei Xu

A procedure to optimize the inductor design in boost PFC applications 409
Florent Liffran

Electric Vehicle Drive Inverters Simulation Considering Parasitic Parameters 417
Wen Huiqing, Liu Jun, Zhang Xuhui, Wen Xuhui

DC-DC Converters with FPGA Control for Photovoltaic System 422
Jan Leuchter, Pavel Bauer, Vladimir Rerucha, Petr Bojda

Control of a Converter with Superconductive Energy Storage Inductor 428
Rozanov Yurie Konstantinovich, Lepanov Michail Gennadevich, Kiselev Michail Gennadevich

FPGA-based Controllers for Switching Converters 432
Karel Jezernik

Gamesa DAC converter: the way for REE grid code certification 437
Itziar Martinez, Daniel Navarro

Flatness-Based Voltage-Oriented Control of Three-Phase PWM Rectifiers 444
J. Dannehl, F.W. Fuchs

Control of a single phase H-Bridge multilevel inverter for grid-connected PV applications 451
Elena Villanueva, Pablo Correa, Jose Rodriguez

Switching and Voltage Controls for a Flyback Switch-Mode Rectifier 456
Yuan-Chih Chang, Chang-Ming Liaw

Method Of Designing ZVS Boost Converter 463
Miroslaw Luft, Elzbieta Szychta, Leszek Szychta

A New DC-DC Converter with Multi Output: Topology and Control Strategies 468
Arash A Boora, Firuz Zare, Gerard Ledwich, Arindam Ghosh

Maximum Frequency for Hysteretic Control COT Buck Converters 475
L.K. Wong, T.K. Man

Current Control Method Based on Hysteresis Control Suitable for Single Phase Active Filter with LC Output Filter 479
Yukinori Kobayashi, Hirohito Funato

Optimal Slope Compensation for step load in peak current controlled dc-dc Buck Converter 485
Susovon Samanta, Pradipta Patra, Siddhartha Mukhopadhyay, Amit Patra"

Performances of a PLL Based Digital Filter for double-conversion UPS 490
Armando Bellini, Stefano Bifaretti

10A 12V 1 chip digitally-controlled DC/DC converter IC with high resolution and high frequency DPWM 498
Kazutoshi Nakamura, Toshiyuki Naka*, Yuki Kamata*, Toyoki Taguchi, Takaaki Shimizu, Yoshiko Ikeda, Akio Nakagawa, Dragan Maksimovic*

Modelling and Modulation of Voltage Source Converter 504
Grzegorz Radomski

Sliding Mode Control of DC/DC Multiphase Power Converters 512
Vadim Utkin

A New Digital Control Method for High Performance 400 Hz Ground Power Unit 515
Zixin Li, Ping Wang, Haibin Zhu, Yaohua Li, Longcheng Tan, Yonggang Chen, Fanqiang Gao

vi

Table of Contents

Single-phase 50-kW 16.7-Hz Four-Quadrant Line-Side Converter for Railway Traction Application 521
C. Heising, R. Bartelt, V. Staudt, A. Steimel

Technique to Improve IGBT Converter Efficiency and Transient Response 528
Robert W. Turner, Simon Walton, Richard Duke

The control of voltage converter rectifiers 536
Krzysztof Szubert

Load Voltage Regulation and Line Loss Minimization of Loop Distribution Systems Using UPFC 542
Mahmoud A. Sayed, Takaharu Takeshita

Control of Traction Single-Phase Current-Source Active Rectifier under Distorted Power Supply Voltage 550
Jan Michalík, Jan Molnár, Zdenck Peroutka

Simulation Model Of Neural Network Based Synchronous Generator Excitation Control 556
Damir Sumina, Neven Bulic, Gorislav Erceg

**Predictive Current Control of a 7-level AC-DC back-to-back Converter for Universal and Flexible Power
Management System** 561
Stefano Bifaretti, Pericle Zanchetta, Florin Iov, Jon C. Clare

Predictive Stator Current Control For Three-Level Voltage-Source Inverters With Output LC-Filters 569
Tomasz Laczynski, Axel Mertens

Research on Dimming Control Method of Electronic Ballast for the Automotive HID Headlight 576
P. Dong, K.W.E.Cheng, S.L.Ho

**Control Method for a Three-Port Interface Converter Using an Indirect Matrix Converter with an Active
Snubber Circuit** 581
Koji Kato, Jun-ichi Itoh

**Precise Digital Control Method with Multi-rate deadbeat control for Single Phase Utility Interactive
Inverter with FPGA based Hardware Controller** 589
Kenta Hayashi, Tomoki Yokoyama

A Digital Current Controller for Zero-Current Transition Bidirectional Converter 595
Nobuyuki Kasa, Takahiko Iida

Control Method for a Single Phase Arbitrary Waveform-output Inverter 600
Satoshi Taniguchi, Keiji Wada, Toshihisa Shimizu

Elimination of Harmonics in Multilevel Inverters with Non-Equal DC Sources Using PSO 606
A. K. Al-Othman, Tamer H. Abdelhamid

Improved PFC Circuit Having Ladder Type Filter with Only Passive Devices 614
Kenji Ando, Keiju Matsui, Nobuhito Takeuchi, Masaru Hasegawa

Fuel Cell Current Ripple Minimization using a bi-Buck Power Interface 621
Nicu Bizon, Marian Raducu, Mihai Oproescu

Power Control Strategy of Parallel Inverter Interfaced DG Units 629
H.R. Baghaee, M.Mirsalim, M. J. Sanjari, G.B. Gharehpetian

Implementation of Nonlinear power flow controllers to control a VSC 637
Nelson L. Díaz, Fabián H. Barbosa, Cesar L. Trujillo

**Harmonic Distortion Reduction Technique for Uninterruptible Power Supplies with DC Voltage Boost
Technique** 643
Juei Lung Shyu

**Energy-based Modulation Error Control for High-Power Drives with Output LC-Filters and
Synchronous Optimal Pulse Width Modulation** 649
Tomasz Laczynski, Timur Werner, Axel Mertens

Table of Contents

Voltage Harmonic Control of Z-source Inverter for UPS Applications 657
Arkadiusz Kulka, Tore Undeland

A Method of Optimal Control for Switched-Mode Power Converters 663
Anatoly Bekishev, Albert Iskhakov, Leonid Klyachko, Vladimir Pospelov, Sergey Skovpen

Experiment results with modified Hybrid PWM method for three phase induction motor 669
Daniel Lewandowski, Grzegorz Lisowski

Optimized Design of a Delay line based Analog to Digital Converter for Digital Power Management Applications 674
Mukti Barai, Sabyasachi Sengupta, Jayanta Biswas

Overmodulation Region of Multi-Phase Inverters 682
S. Halasz

Optimal Control of Induction Motor Using High Performance Frequency Converter 690
Jerkovic Vedrana, Spoljaric Zeljko, Valter Zdravko

Power Electronic Converter for the Reluctance Pump Drive 695
B. J. Szymanski, K. Kompa, N. Michalke, H. Kuß, U. Schuffenhauer

A Predictive Control Scheme for Current Source Rectifiers 699
Pablo Correa, Jose Rodriguez

Analysis and Design of New Switching Table for Direct Power Control of Three-Phase PWM Rectifier 703
Abdelouahab Bouafia, Jean-Paul Gaubert, Fateh Krim

Improvement of the performance for DC-DC Converter 710
X..She, Yun She

A Drive System With High-Speed Single-Phase Supplied Three-Phase Induction Motor 714
T. Binkowski, M. Grad, M. Latka, W. Malska, D. Sobczynski

A Pulse Width Modulation Technique for a Multilevel Converter in High Voltage High Frequency Applications 718
Jafar Adabi, Hamid Soltani, Firuz Zare

Bidirectional Positive Buck-Boost Converter 723
Arash A Boora, Firuz Zare, Gerard Ledwich, Arindam Ghosh

Control system of power electronics current modulator utilized in diode rectifier with sinusoidal source current 728
Michal Gwózdz, Michal Krystkowiak

Design and control of a half-bridge converter to drive piezoelectric actuators 731
Oriol Gomis-Bellmunt, Josep Rafecas-Sabate, Daniel Montesinos-Miracle, Josep-Maria Fernandez-Mola, Joan Bergas-Jane

Online Diagnosis of PEM Fuel Cell 734
Abdellah Narjiss, Daniel Depernet, Denis Candusso, Frederic Gustin, Daniel Hissel

Application of Kalman filters to the control of independent power electronic voltage sources 740
Ryszard Porada, Lukasz Nyczkowski

Verification of the load sharing characteristics in Autonomous Decentralized UPS system using FPGA based Hardware Controller 744
Nobuaki Doi, Tsuyoshi Saito, Tomoki Yokoyama

Fault Current Reduction in Distribution Systems with Distributed Generation Units by a New Dual Functional Series Compensator 750
H.R. Baghaee, M. Mirsalim, M. J. Sanjari, G.B. Gharehpetian

Dynamic Simulation of PM Motor Drive System based on Reluctance Network Analysis 758
Kenji Nakamura, Osamu Ichinokura

viii

Table of Contents

Performance Improvement of Direct Torque Controlled Interior Permanent Magnet Synchronous Motor Drive by Considering Magnetic Saturation 763
Behrooz Majidi, Jafar Milimonfared, Kaveh Malekian

Condition Monitoring for Mechanical Faults in Fully Integrated Servo Drive Systems 769
Jesus Arellano-Padilla, Mark Sumner, Chris Gerada

Feed-forward Compensation of Load and Parameter Variations of Electric Drive 776
Alon Kuperman, Yoram Horen, Saad Tapuchi, Uri Suissa

Thermal Effect of Short-Circuit Current in Low Power Induction Motors 782
Leo.s Beran

Generalized Model for a Class of Switched Reluctance Motors 787
Constantin Pavlitov, Yassen Gorbounov, Radoslav Rusinov, Alexandar Alexandrov, Kliment Hadjov, Dimitar Dontchev

Neural Network based Fault Detection of PMSM Stator Winding Short under Load Fluctuation 793
J. Quiroga, D.A. Cartes, C.S. Edrington, Li Liu

Review of Electrical Machine in Downhole Applications and the Advantages 799
Anyuan Chen, Ravindra. B. Ummaneni, Robert Nilssen, Arne Nysveen

Broken Rotor Bar Impact on the Closed Loop and Sensorless Control of Induction Machine 804
Piotr Kotodziejek, Elzbieta Bogalecka

Coupled Magnetic Circuit Method and Permeance Network Method Modeling of Stator Faults in Induction Machines 810
Amin Mahyob, Mohamed Y. Ould Elmoctar, Pascal Reghem, Georges Barakat

Explosion Protected Electrical Drives - Risk Assessment and Technical Diagnostics 818
Ivica Gavranic, Drago Ban, Damirarko Zarko

The effect of subharmonics on induction machine heating 826
Piotr Gnacinski, Marcin Peplinski, Mariusz Szweda

Influence of Saturation Effects in a Transverse Flux Machine 830
M. Siatkowski, B. Orlik

A Model of Semiconductor Converter-Fed Asynchronous Machines Taking into Account Energy Losses and Thermal Processes 837
M. Pronin, O. Shonin, Y. Koskin, A. Vorontsov, P. Kalatchikov

Use of an AC Self-excited Switched Reluctance Generator as a Battery Charger 845
Abelardo Martínez, Estanislao Oyarbide, Javier Vicuña, Francisco Perez, Eduardo Laloya, Bonifacio Martín-del-Brío, Tomás Pollán, Beatriz Sánchez, Juan Lladó

Direct Thrust Controlled Linear Induction Motor Including End Effect 850
Berrin Susluoglu, Vedat M. Karsli

Analysis of Short-Circuit Forces at the Top of the Low Voltage U-Type and I-Type Winding in a Power Transformer 855
Leonardo Strac, Franjo Kelemen, Damir Zarko, Josipa Mokrovica

Anisotropy Comparison of Reluctance and PM synchronous Machines for Low Speed Position Sensorless Applications 859
H.W. de Kock, M.J. Kamper, R.M. Kennel

Analysis of VSI-DTC Fed 6-phase Synchronous Machines 867
Ibrahim Abuishmais, Waqas M. Arshad, Sami Kanerva

Optimal Rotor Flux Shape for Multi-phase Permanent Magnet Synchronous Motors 874
Roberto Zanasi, Federica Grossi

Table of Contents

Modelling of Electrical Machines Using the Modelica Bond-Graph Library 880
Mieczyslaw Ronkowski

Induction Motor Parameters Identification using Genetic Algorithms for Varying Flux Levels 887
Konstantinos Kampisios, Pericle Zanchetta, Chris Gerada, Andrew Trentin, Omar Jasim

Study of the sudden symmetrical short-circuit using the mathematical models of the synchronous machine and the numerical methods 893
Petropol Serb Gabriela, Petropol Serb Ion, Campeanu Aurel, Sonia Degeratu, Anca Petrisor

Analytical Method of Calculation of the Current and Torque of a Reluctance Stepper Motor Using Fourier Complex Series 899
Pavel Zaskalicky, Maria Zaskalicka

Bearing Damage Analysis by Calculation of Capacitive Coupling between Inner and Outer Races of a Ball Bearing 903
Jafar Adabi, Firuz Zare, Gerard Ledwich, Arindam Ghosh, Robert D.Lorenz

The Model of the Squirrel Cage AC Motor including Rotor Slot Harmonics 908
Eleonora Darie, Costin Cepisca, Emanuel Darie

Identification of mathematical model induction motor's parameters with using evolutionary algorithm and multiple criteria of quality 912
Hudy Wiktor, Jaracz Kazimierz

Simulation Study on Control of Ultrahigh Speed Drives in Waste Energy Recovery Systems 916
Péter Stumpf, Miklós G. Simon, Rafael K. Járdán, István Nagy

Adaptive Back EMF Parameter Adjustment of Simplified Vector Control for Position Sensorless Permanent Magnet Synchronous Motors 924
Kiyoshi Sakamoto, Yoshitaka Iwaji, Daigo Kaneko, Toshihiro Takeuchi, Tsunehiro Endo, Atsuo Kawamura

Identification and Control of Precision XY Stages with Active Vibration Suppression System 932
Mayumi Nitta, Seiji Hashimoto

Sensitivity of the Currents Input-Output Decoupling Vector Control of the DFIM versus Current Sensors Fault 938
Meriem Abdellatif, Maria Pietrzak-David, Ilhem Slama-Belkhodja

Extended Back EMF model for PM synchronous machines with different inductances in d- and q-axis 945
Andreas Eilenberger, Manfred Schroedl

Gait generation of a two-legged robot by using adaptive network based fuzzy logic control 949
Umit Onen, Mete Kalyoncu, Mustafa Tinkir, Fatihm. Botsali

Walking robot HEXOR® II - a versatile platform for engineering education 956
M. Sajkowski, T. Stenzel, B. Grzesik

Motion Control of Steel Sheet Shears with Rocking Knife Mechanism 961
Jan Fetyko, Frantisek Durovsky, Viliam Fedak

Intelligent Adaptive Control and Monitoring Of Band Sawing 967
Ilhan Asiltürk, Ali Ünüvar

Hierarchical adaptive network based fuzzy logic controller design for a single flexible link robot manipulator 974
Mete Kalyoncu, Mustafa Tinkir

Digital Controlled High Speed Synchronous Motor 982
Zdenk Cerovský, Jaroslav Novák, Martin Novák, Marek Cambál

Analysis of combustion engine - electric Linear generator set operation 988
Jirí Pavelka

Table of Contents

Closed Loop Control of AC Drive with LC Filter...994
Jaroslaw Guzinski

Sensorless IPMSM based drive for reciprocating compressor...................................1002
Anton Dianov, Kim Young-Kwan, Lee Sang-Joon, Lee Sang-Taek, Yoon Tae-Ho

Controlling system of electrodynamic drive..1009
Josef Cernohorský

Expert System for Electric Drive Design...1017
Juhan Laugis, Valery Vodovozo

Improvement of Moving Characteristics of Cableless Micro-actuator and Consideration of Reversible Motion..1020
Hiroyuki Yaguchi, Kazumi Ishikawa, Toshihiro Zamma, Koichi Funayama

Sensorless Control of AC Machines using High-Frequency Excitation..................1024
Heiko Zatocil

Adaptive PF Speed Control of SRAM Drives...1033
Laszlo Szamel

A Very Simple Fuzzy Control System for Inverter Fed Synhronous Motor..........1040
Pawel Fabijanski, Ryszard Lagoda

Distributed control system of DC servomotors for six legged walking robot.........1044
D. Belter, K. Walas, A. Kasinski

Optimization of Starting Process of the Frequency Controlled Induction Motor....1050
I.Ya. Braslavsky, A.V. Kostylev, D.P. Stepanyuk

3-Axes Satellite Attitude Control Based on Biased Angular Momentum.................1054
Azam Ghaedi, Mohammad Ali Nekoui

Modelling and simulation of a signal injection self-sensored drive.........................1058
Alen Poljugan, Mark Sumner, Chris Gerada, Qiang Gao

Robust PI Cascade Control for a Multi-Mass System Optimized by Evolutionary Algorithms.................1064
M. Joost, K. Zielinski, B. Orlik, R. Laur

Permanent Magnet Synchronous Servo-Drive with State Position Controller........1071
Lech M. Grzesiak, Tomasz Tarczewski, Slawomir Mandra

Closed-Loop Control of Virtual FPGA-Coded Permanent Magnet Synchronous Motor Drives using a Rapidly Prototyped Controller..1077
Christian Dufour, Vincent Lapointe, Jean Bélanger, Simon Abourida

Speed Sensorless Nonlinear Control Of Induction Motor In The Field Weakening Region.............1084
MiroslawWlas, Haithem Abu-Rub, Joachim Holtz

Comparison of Dynamic Performances of Speed Control System Containing Time - Minimal Speed Controller with Control System Containing PI Speed Controller..1090
Andrzej Andrzejewski, Marian Roch Dubowski

Optimisation of Real-Time Complex Path Generation in Constrained Intelligent Motion Applications Based on IPM Motor Drives...1097
Silverio Bolognani, Roberto Petrella, Fabio Stefanutti, Piero Stocco

PMSM Sliding Mode Observer for Speed and Position Estimation Using Modified Back EMF............1105
Ilioudis Vasilios C., Margaris Nikolaos I.

Optimal Control of Electrical Drives with Induction Motors for Variable Torques...1111
Corneliu Botan, Marcel Ratoi, Vasile Horga

Table of Contents

An Optimal Control for Saturated Interior Permanent Magnet Linear Synchronous Motors Incorporating Field Weakening 1117
Mohammad Abroshan, Jafar Milimonfared, Kaveh Malekian, Arash Rahnamaee

Improved Direct Torque Control for Induction Machine Drives using Fuzzy Logic and Particle Swarm Optimization 1123
Mohammad Mehdi Rezaei, Mojtaba Mirsalim, Kaveh Malekian

Design and Implementation of High Performance Full-Digital Spindle Drives 1128
Liu Yang, Zhao Jin

Semi hierarchical adaptive network based fuzzy logic controller design for a multi-straight-line path tracing flexible robot manipulator with rotating-prismatic joint 1132
Mete Kalyoncu, Mustafa Tinkir

Control System with the Set Point Observation 1140
Algirdas Baskys, Vitoldas Gobis, Valerijus Zlosnikas

Electropneumatic Servo System with Adaptive Force Controller 1144
Arunas Grigaitis, Vilius Antanas Gele~evicius

New fault tolerant DTC control for induction machine drives 1149
A.Ben Abdelghani Bennani, M. Ghodbane Cherif, I. Slama Belkhodja

Stability Analysis of the Natural Field Orientation Controlled Induction Machine Drive 1155
G. Mirzaeva, A. Rojas

Control of SR motor EV by instantaneous torque control using flux based commutation and phase torque distribution technique 1163
Ayumu Nishimiya, Hiroki Goto, Hai-Jiao Guo, Osamu Ichinokura

Simulation of IPM Motor by Nonlinear Magnetic Circuit Model for Comparing Direct Torque Control with Current Vector Control 1168
Hiroki Goto, Kensuke Kimura, Hai-Jiao Guo, Osamu Ichinokura

A Simplified Model for Induction Machines with Faults to Aid the Development of Fault Tolerant Drives 1173
O. Jasim, C. Gerada, M. Sumner, J. Arellano-Padilla

About the Experimental Results of an Electric Driving System Based on Asynchronous Motor and PWM Converter 1181
Petre-Marian Nicolae, Dan-Gabriel Stanescu, Ioana-Gabriela Sîrbu

Real-World Force Feedback Control for Mobile-Hapto 1187
Wataru Yamanouchi, Yuki Yokokura, Seiichiro Katsura, Kiyoshi Ohishi

The new numerical integration routine applied in sensorless drives 1193
Arkadiusz Gardecki, Krystyna Macek-Kaminska

Application of Fuzzy Logic Techniques To Robust Speed Control of PMSM 1198
Tomasz Pajchrowski, Krzysztof Zawirski

Optimal control of current commutation of high speed SRM drive 1204
Jan Deskur, Tomasz Pajchrowski, Krzysztof Zawirski

Comparison Between Direct Torque Control and Vector Control of a Permanent Magnet Synchronous Motor Drive 1209
Rafa Souad, Houcine Zeroug

Detection and self-tuning compensation of periodic disturbances by the control of DC motor 1215
Michael Ruderman, Frank Hoffmann, Johannes Krettek, Torsten Bertram

A Linear Switched Reluctance Motor Based Position Tracking System 1221
S. W. Zhao, N. C. Cheung, Y. Lu, W. C. Gan, Z. G. Sun

xii

Table of Contents

Mobile Robot Navigation with Obstacle Avoidance Capability 1225
Anca Sorana Popa, Mircea Popa, Ioan Silea

Requirements for Power Electronics in Solid Oxide Fuel Cell System 1233
T. Riipinen, V. Väisänen, M. Kuisma, L. Seppä, P. Mustonen, P. Silventoinen

Power Supply for a IGBT-Driver with High Insulation Voltage based on a Printed Planar Transformers 1239
Günter Schmitt, Wolf Kusserow, Ralph Kennel

Variable Motor Operating Point by Integration of Power Electronic Device into Rotor 1243
Adrian Tulbure, Hans-Peter Beck, Mircea Risteiu

Magnetic Material Comparisons for High-Current Gapped and Gapless Foil Wound Inductors in High Frequency DC-DC Converters 1249
Marek S. Rylko, Brendan J. Lyons, Kevin J. Hartnett, John G. Hayes, Michael G. Egan

Feasibility Study of Half- and Full-Bridge Isolated DC/DC Converters in High-Voltage High-Power Applications 1257
Dmitri Vinnikov, Tanel Jalakas, Mikhail Egorov

Evaluation of Different Loss Calculation Methods for High-voltage IGBT-s Under Small Load Conditions 1263
T. Jalakas, D. Vinnikov, J. Laugis

Control of Power Supply Unit for Military Vehicles Based on Four-Leg Three-Phase VSI with Proportional-Resonant Controllers 1268
Tomál Glasberger, Zdenek Peroutka

Optimal Design of a Half Wave Cockroft-Walton Voltage Multiplier with Different Capacitances per Stage 1274
Ioannis C. Kobougias, Emmanuel C. Tatakis

Calculation of Leakage Inductance of Core-Type Transformers for Power Electronic Circuits 1280
Reinhard Doebbelin, Marcel Benecke, Andreas Lindemann

Enhanced Current Pulsation Smoothing Parallel Active Filter for Single Stage Grid-connected AC-PV Modules 1287
A.C. Kyritsis, N.P. Papanikolaou, E.C. Tatakis

Outline of the Design of a Cascaded H-bridge Medium Voltage STATCOM 1293
R.E. Betz, B.J. Cook, T.J. Summers, R. Fisher, A. Bastiani, S. Shao, P. Stepien, K. Willis

Investigation of High Frequency Effects on Layered Coils 1301
Georgios S. Dimitrakakis, Emmanuel C. Tatakis

Soft Switching PWM Inverter for Induction Heating Applied to Heating of Ferromagnetic Metal 1309
Sachio Kubota, Muneo Sato, Fumio Ito, Yoshihiro Shimaoka, Kunihiro Nishioka

Corona Treatment System with Resonant Inverter - Selected Proprieties 1316
Mucko Jan

Power supply unit for an electric discharge machine 1321
Wojciech Mysinski

High Power, High Voltage, High Frequency Transformer / Rectifier for HV Industrial Applications 1326
T. Filchev, D. Cook, P. Wheeler, A. Van den Bossche, J. Clare, V. Valchev

Small Power Laboratory Model and High Power Prototype of the Four-Level VSI 1332
Ryszard Michal Strzelecki, Pawel Szczepankowski, Andrzej Kasprowicz, Genady Stepanovic Zinoviev, Krzysztof Zymmer, Zbigniew Zakrzewski

AC Voltage Regulator Using PWM Technique and magnetic flux distribution 1337
A.M. Dabroom

Minimum Reactive Power Filter Design for High Power Converters 1345
Alex-Sander Amavel Luiz, Braz Jesus Cardoso Filho

xiii

Table of Contents

Injection of a carrier with higher than the PWM frequency for sensorless position detection in PM synchronous motors..1353
Roberto Leidhold, Peter Mutschler

Parallel Fixed Point FPGA Implementation of Sensorless Induction Motor Torque Control................1359
Jacek D. Lis, Czeslaw T. Kowalski

Design of an FPGA-Based Real-Time Simulator for Electrical System..1365
I. Bahri, M-W. Naouar, E. Monmasson, I. Slama-Belkhodja, L.Charaabi

A New, Ultra-low-cost Power Quality and Energy Measurement Technology................................1371
Alex McEachern, Andreas Eberhard

Rotor Time Constant Adaptation Using Radial Basis Function Network..1375
Pavel Brandltetter, Ondfej Skuta

Application of Speed and Load Torque Observers in High Speed Train..1382
Jaroslaw Guzinski, Marc Diguet, Zbigniew Krzeminski, Arka diusz Lewicki, Haithem Abu-Rub

Position Estimator including Saturation and Iron Losses for Encoder Fault Detection of Doubly-Fed Induction Machine..1390
Kai Rothenhagen, Friedrich W. Fuchs

Wide Range Low Noise Current Sensor..1398
F. Richter, C. Sourkounis

Transducerless Speed Control with Initial Position Detection for Low Cost PMSM Drives................1402
Roman Filka, Peter Balazovic, Branislav Dobrucky

Study About the Possibility of Electrodes Motion Control in the EAF Based on Adaptive Impedance Control..1409
Manuela Panoiu, Caius Panoiu, Sorin Deaconu

Asynchronous machine stator resistance estimation using integrated PWM modulator and sampler unit as FPGA application..1416
Dag Samuelsen, Waldemar Sulkowski

Development of Monitoring System for Series HEV Bus with Touch Panel................................1421
Tae-Won Chun, Quang-Vinh Tran, Uk-Don Choi, Heung-Gun Kim

A Development System for Testing Integrated Circuits Used for Power and Energy Measurements................1426
Vladimir Cuk, Aleksandar Nikolic, Aleksandar Zigic

State and parameter estimation in a hydraulic system - moving horizon approach................................1432
Jerzy Baranowski, Andrzej Tutaj

Technologies of Current Sensors Suitable for Hot High Density Power Electronics................................1440
Filip Grecki, Grzegorz Iwanski, Wlodzimierz Koczara, Jozef Lastowiecki

Nonlinear dynamical feedback for motion control of magnetic levitation system................................1446
Jerzy Baranowski, Pawel Piatek

Speed and position estimation of SRM..1454
Konrad Urbanski, Krzysztof Zawirski

Potential of Digital Gate Units in High Power Appliations..1458
Harald Kuhn, Thies Koneke, Axel Mertens

Disturbance Currents of Inverters..1465
Petr Vrana, Jiri Javurek

Improvement of the Energy Recovery of Traction Electrical Drives using Supercapacitors................1469
Diego Iannuzzi

Table of Contents

A Multi-Core PC-based Simulator for the Hardware-In-the-Loop Testing of Modern Train and Ship Traction Systems..1475
Christian Dufour, Guillaume Dumur, Jean-Nicolas Paquin, Jean Bélanger

Energy Saving Control of Tram Motors Taking Light Signalling and City Disturbances into Account.....................1481
Stanislaw Rawicki

Characterization and Improved Control of a Brushless DC Drive with In-Wheel Motor.............................1491
Manuele Bertoluzzo, Giuseppe Buja, Alessandro Pavoni

Supply of Electric Vehicles via Magnetically Coupled Air Coils..1497
Slawomir Judek, Krzysztof Karwowski

Sliding-Mode Approach to Control Design for Induction Motor Drive fed by a Three-Level Voltage-Source Inverter 1505
Sergey Ryvkin, Richard Schmidt-Obermoeller, Andreas Steimel

Analysis and configuration of supercapacitor based energy storage system on-board light rail vehicles 1512
R. Barrero, X. Tackoen, J. Van Mierlo

Design of High Power Electronic Building Block based on Parallel of IGBTs for Electric Vehicle............................1518
Wen Huiqing, Liu Jun, Zhang Xuhui, Wen Xuhui

Stability Analysis on the DC Power Distribution System of More Electric Aircraft...1523
H. Zhang, C. Saudemont, B. Robyns, N. Huttin, R. Meuret

Design Considerations for Control of Traction Drive with Permanent Magnet Synchronous Machine 1529
Zden..k Peroutka, Karel Zeman

Control of Primary Voltage Source Active Rectifiers for Traction Converter with Medium-Frequency Transformer 1535
Vojtech Blahník, Zdenek Peroutka, Jan Molnár, Jan Michalík

Energy management strategy for Coupling Supercapacitors and Batteries with DC-DC converters for hybrid vehicle applications 1542
M.B. Camara, F. Gustin, H. Gualous, A. Berthon

Dual-Source Fed Multiphase Traction System with Standard and Non-Standard Control Regimes Based on Synchronized PWM..1548
Valentin Oleschuk, Marian P. Kazmierkowski

Analysis of a H-NPC topology for an AC Traction Front-End Converter 1555
I. Etxeberria-Otadui, A. Lopez-de-Heredia, J. San-Sebastian, H. Gaztañaga, U. Viscarret, M. Caballero

Hybrid - type system of power supply for a trolleybus with an asynchronous motor 1562
Zygmunt Gizinski, Marcin Gasiewski, Ireneusz Mascibrodzki, Michal Zych, Krzysztof Zymmer, Marcin Zulawnik

Control of rotor flux in AC tram drive during sudden braking operation..1568
Andrzej Debowski, Piotr Chudzik

A New Novel Power Electronic Circuit to Reduce Stray Current and Rail Potential in DC Railway 1575
Reza Fotouhi, Siamak Farshad

Slip Control Upgrades for Light-Rail Electric Traction Drives 1581
Madis Lehtla, Hardi Hõimoja

Practical Aspects on the Improved DC Driving System Used in Electric Urban Traction 1585
Petre Marian Nicolae, Ioana-Gabriela Sîrbu, Ileana-Diana Nicolae, Lucian Mandache

The study of using the traction drive topology with the middle-frequency transformer 1593
Martin Pittermann, Pavel Drábek, Marek Cédl, Jiří Fořt

Control of a Linear Switched Reluctance Motor as a Propulsion System for Autonomous Railway Vehicles 1598
L. Kolomeitsev, D. Kraynov, S. Pakhomin, F. Rednov, E. Kallenbach, V. Kireev, T. Schneider, J. Böcker

Table of Contents

Motion Copying System Based on Real-World Haptics in Variable Speed..1604
Yuki Yokokura, Seiichiro Katsura, Kiyoshi Ohishi

Adaptive Fuzzy Control of magnetically suspended Rotary Table 1610
Thomas Schallschmidt, Denis Draganov, Frank Palis

Wideband Force Sensing for Haptic Energy Transmission Utilizing FPGA..1614
Seiichiro Katsura, Masaki Kondo, Kiyoshi Ohishi

On the development of BLDC motor control run-up algorithms for aerospace application 1620
Vladimir Hubik, Martin Sveda, Vladislav Singule

Rotor Levitation by Active Magnetic Bearing Using Digital State Controller 1625
Chip Rinaldi Sabirin, Andreas Binder

Dynamical Torque-Speed-Curve Adaption To Damp Load Peaks Occuring In Drive Trains Of Shredding Plants..1633
Constantinos Sourkounis

Traction vehicle distributed control computer system architecture with auto reconfiguration features and extended DMA support 1638
Jiri Zdenek

Analysis and Position Control of a Linear Switched Reluctance Actuator Based on Sliding Mode Control 1646
António Espírito Santo, Maria R. A. Calado, Carlos M. P. Cabrita

Development and Control for a Reaction Wheel System Driven by Permanent Magnet Synchronous Motor 1652
Ming-Chang Chou, Chang-Ming Liaw, Sywe-Bin Chien, Fa-Hwa Shieh, Jih-Run Tsai, Hao-Chi Chang

Nonlinear control design for magnetic bearings via automatic differentiation..1660
Stefan Palis, Mario Stamann, Thomas Schallschmidt

Design of Energy Harvesting Generator Base on Rapid Prototyping Parts..1665
Zdenek Hadas, Jan Zouhar, Vladislav Singule, Cestmir Ondrusek

Control of Bouc-Wen hysteretic systems: Application to a piezoelectric actuator 1670
Oriol Gomis-Bellmunt, Faycal Ikhouane, Daniel Montesinos-Miracle

Electric drive for carding machine draft device..1676
Martin Diblík

Two-level and Multilevel Converters for Wind Energy Systems: A Comparative Study 1682
R. Melício, V. M. F. Mendes, J. P. S. Catalão

A Stand-alone Photovoltaic Supercapacitor Battery Hybrid Energy Storage System 1688
M.E. Glavin, Paul K.W. Chan, S. Armstrong, W.G Hurley

Integrated contactless power transmission systems with high positioning flexibility 1696
Daniel Kürschner, Christian Rathge

A Transformerless Interface Converter for a Distributed Generation System..1704
Tzung-Lin Lee, Zong-Jie Chen

A Comprehensive Analysis and Comparison Between Multilevel Space-Vector Modulation and Multilevel Carrier-Based PWM 1710
Constantinos Sourkounis, Ahmad Al-Diab

Identification of Electrical Parameters in a Power Network Using Genetic Algorithms and Transient Measurements 1716
Wei. Dong, Pericle Zanchetta, David W.P. Thomas

On Acoustic Noise Reduction Procedure for Inverter-Fed Induction Machines 1722
Weiss Helmut, Zaucher Peter, Xiao Jian

Table of Contents

Cascaded Doubly Fed Induction Generator for Mini and Micro Power Plants Connected to Grid 1729
Marek Adamowicz, Ryszard Strzelecki

Contactless power transmission with new secondary converter topology 1734
Matthias Dockhorn, Daniel Kürschner, Rudolf Mecke

Modeling Approach of a Generator with Non-linear Load in Embedded Electrical Network 1740
Nicolas Amelon, Mourad Ait-Ahmed, Mohamed-Fouad Benkhoris

Optimal Use of the 14 V Alternator in 42 V Automotive Supply Systems 1748
Vasile Comnac, Mihai Cernat, Adrian Mailat

New Dual Channel Quasi Resonant DC-DC Converter Topologies for Distributed Energy Utilization 1755
J. Hamar, I. Nagy, P. Stumpf, H. Ohsaki, E. Masada

Output Filtering of the Customer-end Inverter in a Low-Voltage DC Distribution Network 1763
Pasi Peltoniemi, Pasi Nuutinen, Pasi Salonen, Markku Niemelä, Juha Pyrhönen

**Power Flow Control through a Multi-Level H-Bridge based Power Converter for Universal and Flexible
Power Management in Future Electrical Grids** 1771
Stefano Bifaretti, Pericle Zanchetta, Yue Fan, Florin Iov, Jon Clare

Energy Storage Systems The Flywheel Energy Storage 1779
Tomasz Siostrzonek, Stanislaw Piróg, Marcin Baszynski

Analysis of Wide Area Integration of Dispersed Wind Farms Using Multiple VSC-HVDC Links 1784
S. González-Hernández, E. Moreno-Goytia, O. Anaya-Lara

Generator Selection for Offshore Oscillating Water Column Wave Energy Converters 1790
D.L. O' Sullivan, A.W. Lewis

A Novel Approach To Photovoltaic Powered Water Pumping Design 1798
Michael James Case, Ernest Edward Denny

Direct Controls in Voltage-Source Converters - Generalizations and Deep Study 1803
Karoly Veszpremi, Istvan Schmidt

Multipolar double fed induction wind generator with a single phase secondary winding 1811
Leonids Ribickis, Guntis Dilevs, Nikolajs Levins, Vladislavs Pugachevs

The measurement on the solar cells in Liberec city 1815
Jiri Kubin

**Rotor Turn-to-Turn Faults of doubly-fed Induction Generators in Wind Energy Plants - Modelling,
Simulation and Detection** 1819
Vincenz Dinkhauser, Friedrich W. Fuchs

Static and Dynamic Response of a Photovoltaic Characteristics Simulator 1827
Anastasios Ch. Nanakos, Emmanuel C. Tatakis

Modeling and Optimal Sizing of Hybrid Renewable Energy System 1834
Rachid Belfkira, Cristian Nichita, Pascal Reghem, Georges Barakat

**Photovoltaic System MPPTracker Investigation and Implementation using DSP engine and Buck- Boost
DC-DC converter** 1840
Dimosthenis Peftitsis, Georgios Adamidis, Panagiotis Bakas, Anastasios Balouktsis

Multi Objective Distributed Generation Planning Using NSGA-II 1847
Muhammad Ahmadi, Ashkan Yousefi, Alireza Soroudi, Mehdi Ehsan

Testing of the Grid-connected Photovoltaic Systems Using FPGA-based Real-Time Model 1852
Robert Stala

xvii

Table of Contents

Output Maximization Using Direct Torque Control for Sensorless Variable Wind Generation System Employing IPMSG 1859
Yukinori Inoue, Shigeo Morimoto, Masayuki Sanada

Improving Connection and Disconnection of a Small Scale Distributed Generator Using Solid-State Controller 1866
M.M.R. Ahmed

Research control of electric systems in wind generator systems 1872
Stefan Winternheimer, Artem Kolesnikov, Evgeny Glushkin, Alexander Bukatov

Stand-alone Photovoltaic Generation System with Combined Storage using lead Battery and EDLC 1877
Hiroaki Nakayama, Eiji Hiraki, Toshihiko Tanaka, Noriaki Koda, Nobuo Takahashi, Shuji Noda

Active Filter Action of Inverter Exciting Induction Generator for Wind Power Generation 1884
Noriyuki Kimura, Tomoyuki Hamada, Katsunori Taniguchi, Toshimitsu Morizane

The Operation of Power Electronic Converters in Photovoltaic Drive Systems 1890
Marek Niechaj

Experimental results of a hybrid wind/hydro power system connected to isolated loads 1896
Mehdi Nasser, Stefan Breban, Vincent Courtecuisse, Arnaud Vergnol, Benoît Robyns, Mircea M. Radulescu

Grid Connection of Multi-Megawatt Clean Wave Energy Power Plant under Weak Grid Condition 1904
Kai Rothenhagen, Marek Jasinski, Marian P. Kazmierkowski

Improved sizing method of storage units for hybrid wind-diesel powered system 1911
A.M. Tankari, B. Dakyo, C. Nichita

A Research Platform for a Smart-Blade Wind Generation System 1918
J. Davey, Udaya K. Madawala, R. Sharma

Soft Switching Multi-Phase Boost Converter for Photovoltaic System 1924
Joo-Hyuk Lee, Jae-Hyung Kim, Chung-Yuen Won, Su-Jin Jang, Yong-Chae Jung

Soft Switching Boost Converter for Photovoltaic Power Generation System 1929
Doo-Yong Jung, Young-Hyok Ji, Jae-Hyung Kim, Chung-Yuen Won, Yong-Chae Jung

Optimisation Of Wind Power Pmsm To Grid Conversion System 1934
Ince Kayhan, Weiss Helmut

Analysis of Wind Farm and Multilevel Converter Interactions in Medium Voltage Networks Under Steady-State and Transient Conditions 1941
J. Sosa-Ruiz, E. Moreno-Goytia, O. Anaya-Lara

A Simple, Low Cost Design Using Current Feedback to Improve the Efficiency of a MPPT-PV System for Isolated Locations 1947
Herman Fernández, Abelardo Martínez, Víctor Guzmán, María Isabel Gímenez

A Single-Phase Active Power Filter Based in a Two Stages Grid-Connected PV System 1951
Kleber C.A. De Souza, Denizar C. Martins

Wide Bandwidth Power Flow Control Algorithm of the Grid Connected VSI under Unbalanced Grid Voltages 1957
Zoran Ivanovic, Marko Vekic, Stevan Grabic, Evgenije Adzic, Vladimir Katic

The use of Switched Reluctance Generator in wind energy applications 1963
Eleonora Darie, Costin Cepisca, Emanuel Darie

Active Line Shaping of a Single Phase Rectifier using the Switching Function Technique 1967
Christos Marouchos

Control of Reactive Power in Double-Fed Machine Based Wind Park 1975
Elzbieta Bogalecka, Michal Kosmecki

xviii

Table of Contents

A Novel Hybrid Modulation Method for Cascaded H-bridge Active Power Filter 1981
Yonggang Chen, Ping Wang, Yaohua Li, Zixin Li, Longcheng Tan

Apparent Power Ratio of the Shunt Active Power Filter 1987
A. Kouzou, B.S Khaldi, S. Saadi, M.O. Mahmoudi, M.S. Boucherit

Shunt Active Power Filter with Improved Dynamic Performance 1995
Krzysztof Piotr Sozanski

The Research on the Active Power Filter Based on the Cascaded H-bridge Converter 2000
Yonggang Chen, Junling Chen, Ping Wang, Yaohua Li, Longcheng Tan, Zixin Li, Wei Xu

E-laboratory in the Field of Electrical Drives 2005
H.Hõimoja, A.Rosin, T.Möller, M. Müür

Laboratory Setup for Studying Ultracapacitors in Industrial Applications 2011
I. Roasto, D. Vinnikov, T. Lehtla

Synchronous machine direct axis parameters estimation module from an iterative strategy 2015
Emile Mouni, Slim Tnani, Gérard Champenois

Determination of the Characteristic Life Time of Paper-insulated MV-Cables based on a Partial Discharge and tan(..) Diagnosis 2022
I. Mladenovic, Ch. Weindl

Elimination of Increased Excitation of Common- Mode Oscillations in Electrical Drive Systems with Active Front End and Long Motor Cables 2028
Thomas Weidinger

Internal Short Circuit in a Tooth Wound PMSM with Stranded Conductors 2037
Damien Birolleau, Christian Chillet, Laurent Albert

Implementation of a Virtual Laboratory for Low Power Electrical Drives 2043
Gh. BALUTA, V. HORGA, C. LAZAR

DQ-Transformation Approach for Modelling and Stability Analysis of AC-DC Power System with Controlled PWM Rectifier and Constant Power Loads 2049
K-N Areerak, S.V. Bozhko, G.M. Asher, D.W.P. Thomas

Genetic Identification of Parameters the Sandwich Piezoelectric Ceramic Transducers for Ultrasonic Systems 2055
Pawel Fabijanski, Ryszard Lagoda

The Impact of Higher-Order Harmonics on Tripping of Residual Current Devices 2059
Stanislaw Czapp

Estimation of the Untapped Regenerative Braking Energy in Urban Electric Transportation Network 2066
Leonards Latkovskis, Linards Grigans

Performance Evaluation of Electric Power Steering with IPM Motor and Drive System 2071
Hamidreza Akhondi, Jafar Milimonfared, Kaveh Malekian

Optimal Control: Load Frequency Control of a Large Power System 2076
Sílvio José Pinto Simões Mariano*, Luís António Fialho Marcelino Ferreira

LCL-Load Modular Converter For Induction Heating 2082
Maciej A. Dzieniakowski, Jan Fabianowski, Robert Ibach

On-line PID Controller Tuning Using Genetic Algorithm and DSP PC Board 2087
Pawel Fabijanski, Ryszard Lagoda

Regulation Properties of Pumping Station Control System In The Highest Efficiency Range 2091
Szychta Leszek

xix

Table of Contents

Inner Gas Pressure Measurement Based Life-span Estimation of Electrolytic Capacitors......................................**2096**
A. Riz, D. Fodor, O. Klug, Z. Karaffy

Robust Control Methodologies for Optical Micro Electro Mechanical System - New approaches and Comparison **2102**
Alireza Izadbakhsh, S.M.R. Rafiei

Modeling a Buck-Based Switching Amplifier for Sinusoid Wide Band Tracking by Using a Nonlinear Time Varying Map **2108**
A. El Aroudi, E. Alarcón, E. Rodriguez, R. Leyva

Single Inductor Multiple Outputs Interleaved Converters Operating in CCM......................................**2115**
Luis Benadero, Vanessa Moreno-Font, Abdelali El Aroudi, Roberto Giral

Control of a two-cell dc/dc converter in presence of saturating duty cycle **2120**
Moez Feki, Abdelali El Aroudi, Bruno Gerard Michel Robert, Nabil Derbel

Bifurcations and Chaotic Dynamics in a Linear Switched Reluctance Motor **2126**
M.R. De Castro, B.G.M. Robert, C. Goeldel

Modular Architecture for Decentralized Hybrid Power Systems **2134**
E. Ortjohann, M. Lingemann, O.Omari, A. Schmelter, N. Hmasic, A. Mohd, W. Sinsukthavorn, D. Morton

Design of a power management system for an active PV station including various storage technologies **2142**
Di Lu, Tao Zhou, Hicham Fakham, Bruno Francois

Energy Management and Power Flow of Decoupled Generation System for Power Conditioning of Renewable Energy Sources **2150**
Wlodzimierz Koczara, Zdzislaw Chlodnicki, Nazar Al-Khayat, Neil L.Brown

Inversion Based Control of a Diesel Fed Low Temperature Fuel Cell System **2156**
Daniela Chrenko, Marie-Cecile Pera, Daniel Hissel

Power Management in an Autonomous Adjustable Speed Large Power Diesel Gensets **2164**
Grzegorz Iwanski, Wlodzimierz Koczara

Cost evaluation of Generator-set with Energy Storage for 4Q-load **2170**
Freek J.F.Baalbergen, Pavol Bauer

Integrating renewable energy sources and storage into isolated diesel generator supplied electric power systems **2178**
Chad Abbey, Jonathan Robinson, Géza Joós

Performance comparison of different wind generator based hybrid systems **2184**
Vincent Courtecuisse, Benoit Robyns, Marc Petit, Bruno Francois, Jacques Deuse

First Approach for a Fault Tolerant Power Converter Interface for Multi-Stack PEM Fuel Cell Generator in Transportation Systems **2192**
Alexandre De Bernardinis, Gérard Coquery

Development of Electrical System for Hybrid Vehicles Using the Free-swinging Piston Engine and Oscillating Rotating Generator **2200**
Sigitas Kudarauskas

Power flow control in different time scales for a wind/hydrogen/super-capacitors based active hybrid power system **2205**
ZHOU Tao, LU Di, FAKHAM Hicham, FRANCOIS Bruno

Neuro-Fuzzy Adaptive Control of the IM Drive with Elastic Coupling **2211**
Teresa Orlowska-Kowalska, Krzysztof Szabat, Mateusz Dybkowski

Control of Flexible Drive with PMSM employing Forced Dynamics......................................**2219**
Vittek Ján, Bris Peter, Makys Pavol, Stulrajter Marek, Vavrus Vladimír

Table of Contents

The problems of high dynamic drive control under circumstances of elastic transmission..................2227
Jan Deskur, Roman Muszynski

Protective Predictive Control of Electrical Drives with Elastic Transmission 2235
Mario Vasak, Nedjeljko Peric

Low-Cost High-Performance Predictive Control of Drive Systems with Elastic Coupling 2241
Marcin Cychowski, Kieran Delaney, Krzysztof Szabat

Development of an Expert System for Identification, Commissioning and Monitoring of Drives 2248
Mario Pacas, Sebastian Villwock

Control of Axial Flux Permanent Magnet Motor by the PIPCRM Method at Standstill and at Low Speed 2254
Janusz Wisniewski, Wlodzimierz Koczara

Zero Speed Position Estimation of a Matrix Converter Fed AC PM Machine using PWM Excitation 2261
Q. Gao, G. M. Asher, M. Sumner

Sensorless Direct Torque and Flux Control of an IPM Synchronous Motor at Low Speed and Standstill........2269
Gilbert Foo, S. Sayeef, M.F. Rahman

Sensorless Control of PM Synchronous Motors Using a Predictive Current Controller with Integrated INFORM and EMF Evaluation..2275
Manfred Schrödl, Christian Simetzberger

Torque Sensorless Control of Induction Motor 2283
Karel Jezernik, Miran Rodic

Application of the induction motor torque - observer to the control of turbo - machines 2289
Andrzej Debowski, Daniel Lewandowski

Observer of induction motor speed based on exact disturbance model....................................2294
Zbigniew Krzeminski

Experimental Performance Evaluation for Low Speed and Regenerating Operation of Sensor-less Vector Control System of Induction Motor Using Observer Gain Tuning....................................2300
Kazuhiro Ohyama, Greg Asher, Mark Sumner

Application of the Stator Current-based MRAS Speed Estimator in the Sensorless Induction Motor Drive............2306
Mateusz Dybkowski, Teresa Orlowska-Kowalska

State and Parameter Estimation in Induction Motors using Sliding Modes 2312
Sachit Rao, Martin Buss, Vadim Utkin

Torque Transient Alleviation in Fixed Speed Wind Generators by Indirect Torque Control with STATCOM..2318
Marta Molinas, Jon Are Suul and Tore Undeland

Flicker Study on Variable Speed Wind Turbines with Permanent Magnet Synchronous Generator 2325
Weihao Hu, Zhe Chen, Yue Wang, Zhaoan Wang

Power Output Characteristics Analysis of Wind Energy Converter Control Methods 2331
Bingchang Ni, Constantinos Sourkounis

A Cooperative Control Method for Output Power Smoothing and Hydrogen Production by Using Variable Speed Wind Generator 2337
Rion Takahashi, Hirotaka Kinoshita, Toshiaki Murata, Junji Tamura Masatoshi Sugimasa, Akiyoshi Komura, Motoo Futami, Masaya Ichinose, Kazumasa Ide

A new interconnecting method for wind turbine/generators in a wind farm and basic characteristics of the integrated system 2343
Shoji Nishikata, Fujio Tatsuta

Educational aspects of mechatronic control course design for collaborative remote laboratory 2349
Andreja Rojko, Darko Hercog, Karel Jezernik

xxi

Table of Contents

PEMCWebLab - Distance and Virtual Laboratories in Electrical Engineering: Development and Trends..............2354
Pavol Bauer, Viliam Fedák, Otto Rompelman

Integrated multimedia educational program of a DC servo system for distant learning.............................2360
Gabor Sziebig, Istvan Nagy, Rafael Kalman Jardan, Peter Korondi

Electromechanical Actuators WEB-lab...2368
Dusan Maga, Jan Sitar, Juraj Dudak, Rene Hartansky, Peter Siroky, Jan Halgos, Pavol Bauer

Power Quality and Active Filters as Web-Controlled Experiment in the frame of PEMC WebLab........................2371
Volker Staudt, Andreas Steimel, Pavol Bauer, Vítezslav Hájek

Distant learning of Pulse Width Modulation Techniques for Voltage Source Converters............................2378
Bartlomiej Kamiski, Dariusz Sobczuk

Modern design optimisation exploiting field simulation 2383
Jan K. Sykulski

Transmission-Line Modelling of Wave Propagation Effects in Machine Windings.....................................2385
Herbert De Gersem, Olaf Henze, Thomas Weiland, Andreas Binder

An efficient field-circuit coupling method by a dynamic lumped parameter reduction of the FE model...................2393
F. Henrotte, E. Lange, K. Hameyer

Coupled field-circuit-mechanical model of an electromagnetic actuator operating in error actuated control system..2400
Lech Nowak

Simulation and Investigation of Magnetorheological Fluid Brake..2406
Wieslaw Lyskawinski, Wojciech Szelag, Cezary Jedryczka

Field and Field-Circuit Description of Electrical Machines..2412
Andrzej Demenko, Kay Hameyer

Interaction between Thermal Impedance and Parasitics in Power Sections...2420
Stefan Forster, Andreas Lindemann

Discussion of Internal and External High Frequency Common Mode Noise Current on a Chopper Circuit.............2428
Tetsuya Mitani, Keiji Wada, Toshihisa Shimizu, Hiromichi Ohashi

A Novel Digital Control Method for DC-DC Converter 2434
Fujio Kurokawa, Masashi Okamatsu, Yuichi Sumida, Yasuhiro Mimura, Masahiro Sasaki

A Novel Single/Three-phase Matrix Converter For High Power Integration 2439
Makoto Saito

An Effective Design Method for High Power Density Converters 2445
Yusuke Hayashi, Kazuto Takao, Toshihisa Shimizu, Hiromichi Ohashi

Power Devices in Polish National Silicon Carbide Program...2452
Mariusz Sochacki, Andrzej Kubiak, Zbigniew Lisik, Jan Szmidt

SiC Power Semiconductor Devices for new Applications in Power Electronics 2457
Dominique Planson, Dominique Tournier, Pascal Bevilacqua, Nicolas Dheilly, Herve Morel, Christophe Raynaud, Mihai Lazar, Dominique Bergogne, Bruno Allard, Jean-Pierre Chante

Silicon carbide Schottky diodes and MOSFETs: solutions to performance problems 2464
Owen J. Guy, Michal Lodzinski, Ambroise Castaing, P. M. Igic, Amador Perez-Tomas, Michael R. Jennings, Philip A. Mawby

Characterization of the Static and Dynamic Behavior of a SiC BJT 2472
M.M.R. Ahmed, N-A.Parker-Allotey, P.A. Mawby, Muhammed Nawaz, Carina Zaring

An active network control method using distributed energy resources in microgrids....................................2478
Takayuki Tanabe, Yoshinobu Ueda, Toshihisa Funabashi, Shigeo Numata, Kimio Morino, Eisuke Shimoda

Table of Contents

Energy Management in Solar Photovoltaic Plants based on ESS 2481
M. Lafoz, L. García-Tabarés, M. Blanco

A Method of Three-Phase Balancing in Microgrid by Photovoltaic Generation Systems 2487
Masahide Hojo, Yuta Iwase, Toshihisa Funabashi, Yoshinobu Ueda

Development of HILS(Hardware In-Loop Simulation) System for MMS(Microgrid Management System) by using RTDS 2492
Jin-Hong Jeon, Jong-Yul Kim, Seul-Ki Kim, ong-Bo Ahn, JuneHo Park

Power Quality Analysis of Jeju Island Power System with Wind Farm and HVDC System 2498
Jae-Hong Kim, Eel-Hwan Kim, Se-Ho Kim, Jaeho Choi, Gil-Soo Jang, Seung-Ho Song

A New Control Method for Power Turbine Generators Using an Accurate Ship Plant System Model 2504
Nobumasa Matsui, Fujio Kurokawa, Keiichi Shiraishi

Voltage profile support in distribution networks - influence of the network R/X ratio 2510
B. Bla~ic, I. Papic

Modeling, Simulation and Analysis of Conducted Common-Mode EMI in Matrix Converters for Wind Turbine Generators 2516
S. Zhang, K.J. Tseng

Design of Frequency Shift Acceleration Contol for Anti-islanding of an Inverter-based DG 2524
Seul-Ki Kim, Jin-Hong Jeon, Heung-Kwan Choi, Jonng-Bo Ahn

Integrated Power Converter for Photovoltaic and Fuel Cell Systems in Home 2530
Yasuyuki Nishida, Shinichiro Sumiyoshi, Hideki Omori

A Comparison of Position Control Structures for Ironless Linear Synchronous Motor 2538
Martin Hrasko, Pavol Makys, Marek Franko, Jozef Kuchta

A Comparison of Sliding Mode Approaches to a Nanometre Position Control Application 2543
Paul Andreas Stadler, Stephen James Dodds

Sliding Mode Control of PMSM Drives Subject to Torsion Oscillations in the Mechanical Load 2551
Stephen J. Dodds, Jan Vittek

Sliding Mode Vector Control of PMSM Drives with Minimum Energy Position Following 2559
Stephen J. Dodds

xxiii

Electric Drive System for Automatic Guided Vehicles Using Contact-free Energy Transmission

Marcel Jufer

EPFL, Lausanne, Switzerland, e-mail: *marcel.jufer@epfl.ch*

Abstract—**Four technologies, combining mainly innovative solutions, offer the possibility of clean and flexible vehicles, all using only electricity. The main common components are contact-free energy transmission, storage on super-capacitors, holonome axles integrating wheel-motors and automatic guiding. Moreover, the complete energetic chain is managed by power electronics. Their applications are mainly in the field of public and industrial transportation. Several applications are described: electric busses, automatic guided vehicles for container handling, automatic people movers and automatic surveillance vehicles. For more information please visit** *http://www.numexia.com/*

Keywords—**Wireless power transmission, automotive application, energy storage, energy system management, soft switching**

I. INTRODUCTION

A new generation of vehicles, mainly based on 4 different technologies, is presently in development for good or passenger transportation or special applications.

These technologies are :

- Contact-free energy transmission
- Wheel-motor with an external rotor
- Energy storage on super-capacitances
- The use of an energy chain based on power electronics and control, from the ground to the motors.

The main applications are :

- An automatic guided vehicle (AGV) for container transportation and handling
- An automatic people mover at 15 km/h
- An automatic safety and control vehicle
- An electric city bus

II. AIM

Starting from different technologies developed for the high speed Maglev Swissmetro (Fig 1) [1], the approach of technology transfer has been to apply one or several components to innovative vehicles, responding to specific constraints and criteria:

- High flexibility of exploitation ;
- Clean, quiet and sustainable ;
- High automation and control with or without driver;
- Low investment in infrastructure;
- Low maintenance costs.

In order to reach these goals, specifically in the range of speeds between 10 and 100 km/h, one more device had to be implemented : the direct drive wheel-motor with external rotor, also developed at EPFL.

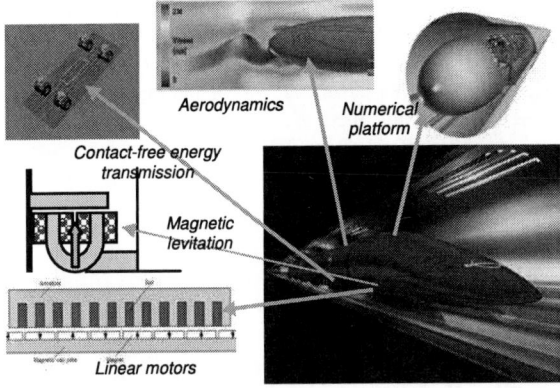

Fig. 1. From Swissmetro project to specific technologies

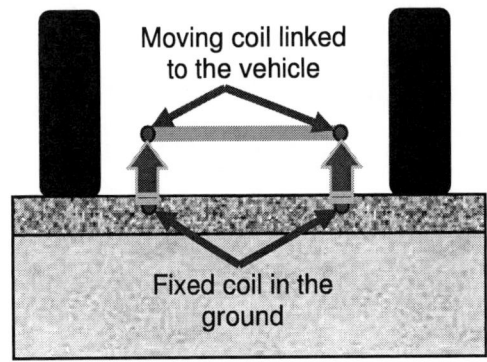

Fig. 2. Contact-free energy transmission - Principle

III. CONTACT-FREE ELECTRICAL ENERGY TRANSFER

A. Principle

This technique [2-7] offers the possibility to transfer electrical energy from a fixed coil on or in the ground to an other coil placed under a vehicle. The process is realized without iron magnetic circuit, in the air. On Figure 2 the principle is illustrated.

In order to reach this ironless magnetic coupling, a high frequency is necessary (>>50 Hz). Effectively, from the back-EMF equation it is possible to write:

$$u_i = d\psi / dt = \text{back-EMF}$$

$$\psi = \text{total flux} = N\phi = NBS_{air}$$

N = coil number of turns

B = magnetic flux density

S_{air} = magnetic air section

So the back-EMF equation becomes :

$$U_i \approx N.f.B.S_{air} \qquad (1)$$

By increasing the frequency and the magnetic section, it is possible to reduce the flux density B and the copper volume by reducing the number of turns N. Thus, it is possible to suppress the iron magnetic circuit, using air instead. An optimisation software, aiming to reduce the copper volume, to reach a transmission efficiency > 98 % and to minimise the electromagnetic radiation has been developed and applied. A system for a total power of 120 kW (Fig 3) has been designed and built with the corresponding power electronic to generate the high frequency and a controlled rectifier on the vehicle . A global maximum efficiency of 95 % has been measured at 110 kW.

Fig.3. Contact-free energy transmission – 120 kW test facility

B. Magnetic field

A specific problem of inductive energy transfer is the magnetic field radiation, limited by normalization, according to the frequency.

The solution to reduce the intensity of radiation for passengers in the vehicle and nearby is to design the system using 3 safety elements:

- To switch on the primary coil only when the secondary coil (i.e. the vehicle) is correctly positioned.
- To place a magnetic shield on the vehicle floor (a thin lamination)
- To impose the same current volume for the primary and the secondary coils

This last condition has been applied to an energy transfer system corresponding to the following conditions:

- Power transferred : 108 kW
- Coil sizes : 4*2 m
- Distance d : 0.115 m
- Primary voltage: 500 V
- Floor level above primary coil : 0.3 m

With current volume opposition, the magnetic field levels without shielding are represented on graphics, in relative values, referred to the earth magnetic field. On Fig. 4, the relative flux density is calculated above the middle of the coils, on a vertical axis (yy), from the vehicle floor level (0.3 m) up to 2 m. On the vehicle floor, the relative amplitude is 0.3.

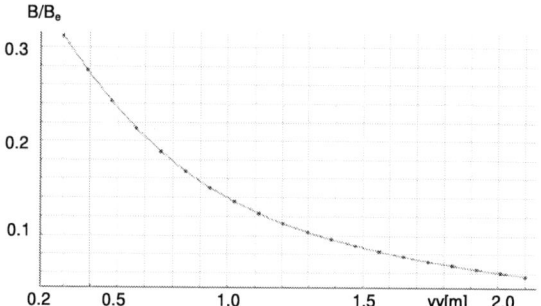

Fig.4. Relative flux density distribution above the coil centre, on a vertical axis – 108 kW power – 4*2 m coils

On Fig.5, the same relative flux density is represented on the floor level (yy= 0.3m) on an horizontal axis, from the coil centre laterally up to 2m; xx = 1 m corresponds to the coil side level. The peak value is 0.75 above the coils.

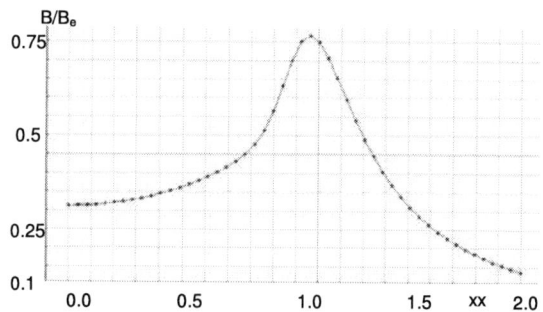

Fig.5. Relative flux density distribution on the vehicle floor level, from the centre to 2 m – 108 kW power – 4*2 m coils

On Fig.6, the same relative flux density is represented on the on a level of yy=1.2m, on an horizontal axis, from the coil centre laterally up to 2m; xx = 1 m corresponds to the coil side level. The peak value is 0.1 m above the coil centre.

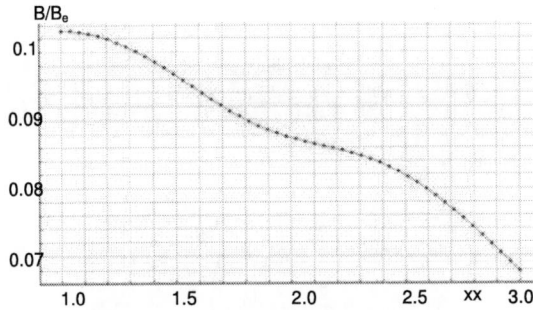

Fig.6. Relative flux density distribution on a level of 1.2 m above the primary coil, from the centre to 2 m – 108 kW power – 4*2 m coils

IV. ENERGY STORAGE ON SUPERCAPACITORS

Energy transfer can be done continuously from the ground to a vehicle, using a continuous succession of coils (Fig.7) . This presents two drawbacks: the investment cost and the absence of track flexibility.

Fig.7.Continuous coil track – Test facility with 5 passengers, 15km/h vehicles

As a variant, contact-free energy transmission with intermediate fast storage on ultra-capacitors [8,9] is a very interesting solution. It allows an autonomy of track with a reduced investment. This solution requires an autonomy in the range of 2.5 to three times the distance between two loading stations.

On Fig.8, the global electrical scheme of such a solution is presented, including the energy transmission system and ultra-capacitor loading control.

The rapid loading operation requires an important peak power. Thus, an interesting solution is to equip a station also with an ultra-capacitor intermediate storage, such as to have the possibility to smooth down the main power (see Fig. 18).

V. WHEEL-MOTOR

For the different categories of vehicles presented hereafter, the electric drive solution has been focused on direct wheel motor drive characterized by:
- No gear transmission
- Brushless DC motors
- External rotor

The absence of transmission leads to a heavier motor. But the solution with external rotor [10] allows a better integration directly in the wheel and not laterally to the wheel (Fig 9) as sometimes realized.

Fig 10 shows a wheel motor according to the proposed solution. An optimized design software leads to an acceptable mass with an important free volume at the inside.

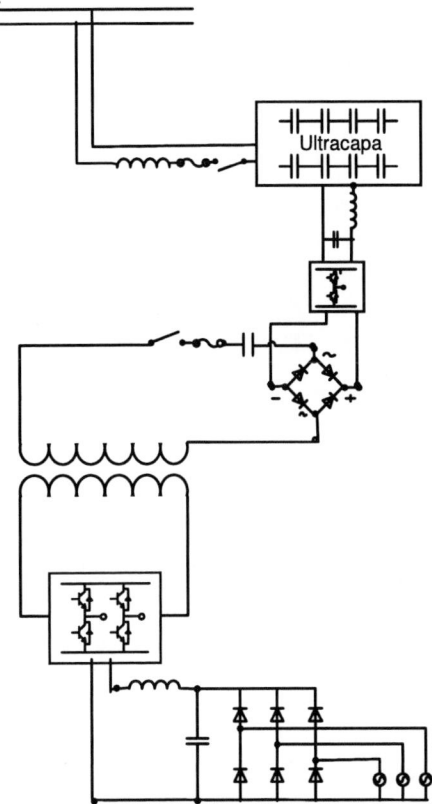

Fig.8.Electric scheme of energy transfer system and storage on ultra-capacitors

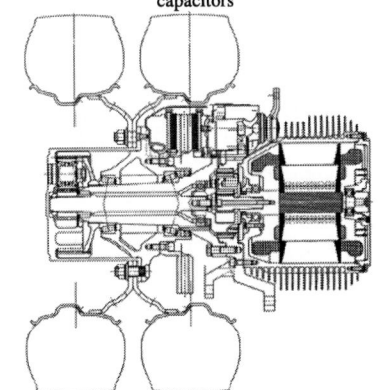

Fig.9.Electric drive for wheel with classical gear transmission

Fig 11 shows the stator (left) and the rotor (right) of such a motor for a rating torque of 4000 Nm.

Based on such motors, vehicle axles with 2 wheels have been developed, with integrated steering motor allowing a very important angle of rotation until ± 90° (Fig 12). In this case, the damping devices have been integrated in the free space internal to the motors. Such an axle has been realized for the vehicle of Fig.14.

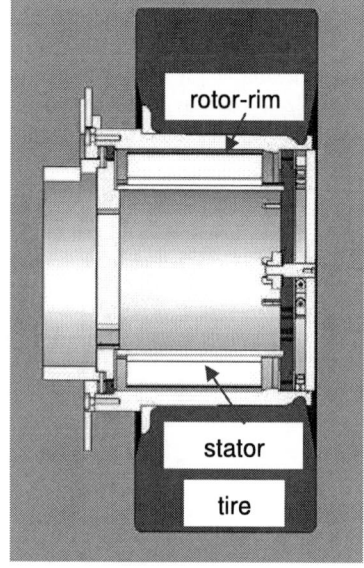

Fig.10.wheel motor with external rotor

Fig.11.Stator (left) and rotor of a wheel motor of 4000 Nm

Fig.13.Automatic Guided Vehicle for 40 T containers -120 kW power at 25km/h.

Fig.12.Vehicle axle with 2 wheel motors and a steering motor. The damping elements are integrated in the motor itself.

VI. VEHICLES INTEGRATING THE DIFFERENT TECHNOLOGIES

A. Automatic guided vehicles for container handling

An automatic guided vehicle for container handling has been developed integrating the following technologies:

- Contact-free energy transmission
- Motor wheel axles
- Energy storage on super-capacitors
- Automatic guiding

The transportation capacity is designed for 40 T containers. The maximum power is 120 kW with a maximum speed of 25km/h.

On Fig.13, a picture of the final vehicle is presented. On Fig 14, the first prototype vehicle is represented. It is supplied by the coil system of Fig.3.

Fig.14.Automatic Guided Vehicle for 40 T containers – First prototype

A first series of such vehicles will be built and tested in a harbor at the end of 2008.

B. Low floor electric bus

The same technology of energy transmission system and axles with wheel motors can be applied to busses with a very low floor, according to the integrated wheel realization . The possibility to move the wheels independently (but in coordination) offers the advantage to reduce the stop area length. A test realization is foreseen for 2009.

Fig.15.Bus with independent wheel axles and very low floor

C. Automatic people mover

Automatic people mover (APM) are small vehicles with a capacity of 8 to 10 people (1200 kg) with a speed of 15 km/h (Fig 16). The rating power is 14 kW with a maximum slope of 16%. The peak power for the energy transfer is 50 kW. This allows a loading time of maximum 10 seconds for an autonomy of 1 km (Fig 17). In order to avoid such peak energy consumption on the main, an intermediate energy storage at each loading station is introduced (Fig 18). So the power request on the main is smoothed.

Such transportation system can be applied to pedestrian streets, airports, large parking areas, exhibition centers, university campus, large factories, etc. Two test vehicles are in construction and a first pilot track is foreseen for 2009, in Lausanne.

Fig.16.Automatic people mover vehicle

Fig.17.Automatic people mover system with loading stations every 400 m

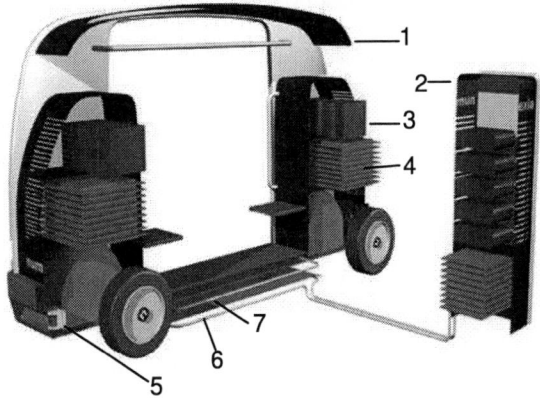

Fig.18.Automatic people mover with energetic chain and loading station
1-Vehicle structure 2-Loading station with intermediate storage
3-Power electronics 4-Super-capacitors 5-Obstacle detection
6-Primary coil 7-Secondary coil

D. Automatic surveillance vehicle

Many different installations or situations require safety surveillance and reconnaissance. They are generally known under the name of *Mobile Detection Assessment Response System (MDARS).* Automatic systems based on the same energetic chain as the APM system have the advantage of reliability, absence of noise, insensitivity to any type of pollution and a smaller size. Such a vehicle has been designed and will be equipped with devices such as radar, NBC sensors, IR camera, etc.

Among the characteristic applications: airports, nuclear power plants, gas and fuel production, high or low temperature environment, catastrophe evaluation and rescue, military investigation, etc. Such a vehicle, based on similar technologies as AGV and APM, have the advantage of no noise, no pollution and high flexibility. A first prototype will be built soon. On Fig 19, the vehicle design is represented.

Fig.17.Automatic people mover system with loading stations every 400 m

VII. ENERGETIC CHAIN

All these different vehicles use basically the same energetic chain, with powers from 10 kW to 250 kW. For these developments, power electronics is a key factor, mainly for the following functions:

- High frequency supply generator for energy transmission, with a high efficiency
- Super-capacitors load and control
- Propulsion and steering motor control and drive

Fig 20 describes the complete energetic chain.

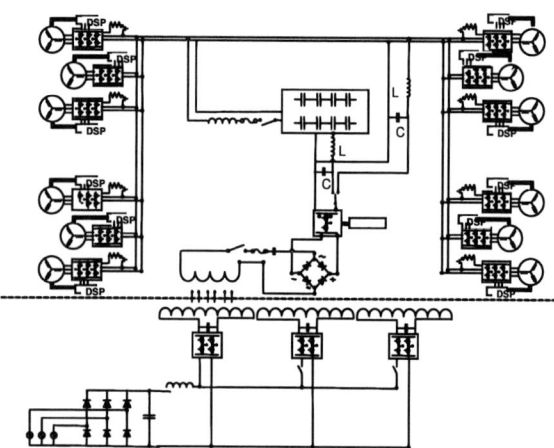

Fig.20.Energetic chain including energy contactless transmission, energy storage and motor drive and control

ACKNOWLEDGMENT

The author thanks particularly the Swiss Federal Commission for Technology and Innovation (CTI http://www.bbt.admin.ch/kti/index.html?lang=en) which supported financially these projects, the Foundation Numexia and the Company Numexia S.A. (http://www.numexia.com/index.php?page=accueil&hl=e n_GB) which designed the different vehicles and realized the prototypes.

REFERENCES

[1] M. Jufer, V. Bourquin, M. Sawley, *"Global Modelisation of the Swissmetro Maglev using a numerical Platform"*, Proceedings MAGLEV 2006, Dresden 13-15 sept 2006

[2] N. Macabrey, "Alimentation et guidage sans contact" *PHD thesis*, EPFL, No 1840, 1998.

[3] M. Jufer, N. Macabrey, M. Perrottet, *"Modeling and Test of Contactless Inductive Energy Transmission"*, 5th International Conference Electrimacs, 1996, Saint-Nazaire (France), Vol 3/3, pp 1199-1204.

[4] M. Jufer, L.Cardoletti, B. Arnet, N. Macabrey, M. Perrottet, *"Inductive powered vehicles for semi-personal transport - The Serpentine project"*, EVS 15, Brussels, October 1998 (CD-ROM).

[5] M. Jufer, N. Macabrey, P.Germano, M. Perrottet, *"Contactless Energy Transmission for Moving Drive,"*Proccedings 27th Annual Symposium on Incremental Motion Control Systems and Drives (IMCSD), San Jose (Ca), July 1998, pp 47-53.

[6] M. Jufer, M. Perrottet, N. Macabrey, *"Contactless Energy Transmission for Electric Drives,"* EPE Chapter Symposium on Electric Drive Design and Applications, Nancy, June 1996, pp 7-12

[7] M. Jufer, R. Perey, *"Contact-less Energy Transmission for Maglev"*, Proceedings MAGLEV 2006, Dresden 13-15 sept 2006

[8] B. Destraz, P. Barrade, A. Rufer, and M. Klohr, *"Study and Simulation of the Energy Balance of an Urban Transportation Network"*, EPE Conference Aalborg, 2-5 September 2007

[9] A. Rufer, *"Benefits of short and long term energy storage in the context of renewable energies and sustainable energy consumption"*, Grimaldi Forum, Monaco, March 29 – April 1 2007.

[10] M. Jufer, *"Limit Performances of direct Electric Drives"*, International Symposium on Advanced Electromechanical Motion Systems, Electromotion'99, Patras, July 1999, pp 1-6.

State-of-the-Art High Power Density and High Efficiency DC-DC Chopper Circuits for HEV and FCEV Applications

Atsuo Kawamura, Martin Pavlovsky, Yukinori Tsuruta

Yokohama National University/ Electrical and Computer Engineering, Yokohama, Japan, e-mail: *kawamura@ynu.ac.jp*

Abstract— recent environmental issues have accelerated the use of more efficient and energy saving technologies in any area of our daily life. One of the major energy consumptions is in the transportation area, especially in the automobile field. DC/DC chopper circuits for use in hybrid electric vehicles (HEV), fuel cell electric vehicles (FCEV) and so on will be discussed in this paper from the view point of power density and efficiency. A typical power range of such converters can be in order of kWs up to over 100 kW with a short term overload requirement of often more than 200 %. Considering the state of the art, switching frequency of these converters is in the range from 50 kHz with IGBTs to 200 kHz with power MOSFETs, the power density peaks at about 25 kW/l, and the highest efficiency is close to 98 [%] depending on the load conditions. As can be seen from the brief introduction, the design of such converter presents multiple challenges from power density as well as efficiency point of view and these are discussed further in the paper.

I. INTRODUCTION

Recent global environmental issues have accelerated the use of more efficient and energy saving technologies in many areas of daily life. Major energy consumptions is the transportation, especially the automobile field. The need of more efficient use of internal combustion engine (ICE) and for an improvement of total system efficiency has created a combination of ICE and electric motors i.e. hybrid electric vehicles (HEV). Even pure electric vehicles (EV) and fuel cell electric vehicles (FCEV) could be commercially available soon. The need of DC/DC choppers is discussed for these electric power trains [1]-[9]. The advantage of DC/DC chopper for HEV is: 1) higher power output in a high speed range [1], and 2) total efficiency increase. FCEV and Pure EV may need a very high efficiency chopper for the higher performance of power-train characteristics [2],[3]-[4].

The power rating of DC/DC choppers for HEV lies in the range between 30 and 90 kW, and it may be beyond 100 kW for EV and FCEV. The main design target points are often high power density and high efficiency. There is few literature available considering this power range and a new performance criterion should be applied. Ohashi proposed the new concept of "Power density" [5] and [6], and he showed the trend of the continuous growth of power density in various applications of power electronics. Kolar in [7] listed four key items which strongly influence final converter design, which are:

- efficiency (loss)
- volume
- weight
- cost

These items are closely coupled together and the total optimisation procedure is discussed in [7].

In this paper, several examples of high power density and high efficiency DC/DC choppers for FCEV and EV will be introduced. They are based on so called SAZZ topology (Snubber assisted zero-voltage-zero-current transition switch) which offers soft switching at turn-on as well as at turn-off in order to reduce the switching losses [23], [24]. The considered power range is 8 to 25 kW with the switching frequency up to 100 kHz with IGBTs and 200 kHz with MOSFETs respectively. The main goal of the presented designs is reaching as high power density and efficiency as possible. The presented converter prototypes exhibit power density as high as 25 kW/l and efficiency close to 98 %.

In the following paragraphs, analyses of several SAZZ based circuit topologies with the main goal being the highest efficiency will be discussed followed by review of various issues related to reaching high power density.

II. HIGH EFFICIENCY

A. Review of Highly Efficient converter topologies

Switching power supplies and regulators have come into widespread use since 1970s. Basic high switching frequency converter design is shown in Figure 1(a) as published in [11], [12]. This circuit based on basic hard-switching buck/boost topology reached the operating frequency limit in the 1980s. In order to increase the frequency further and hence improve the power density, soft switching with one resonant switch as shown in Figure 1(b) was proposed in [13]. However, excessive current and voltage spikes in the switching devices are pointed out as the drawback. A solution to this problem was proposed in [12], [15] as ZVT (Zero Voltage Transition) circuit topology with two switches as shown in Figure 1(c) This partial resonance topology became the basic soft switching topology and various applications were proposed in the literature. Our literature survey showed that boost choppers are mostly based on one resonant switch [13] as shown in Figure 1(b), passive auxiliary resonant circuit [16] (with added passive snubber) as shown in Figure 1(d) and ZVT [15]. In case of high booster ratio, the most used topology is based on coupled inductors [17] as shown in Figure 1(e). However, most of the surveyed papers were in the power range below 1 kW. There are very few examples of papers regarding 100 kW range for EV application with the exception being the non-isolated bidirectional converter discussed in [18] and shown in Figure 1(d).

978-1-4244-1741-4/08/$25.00 ©2008 IEEE

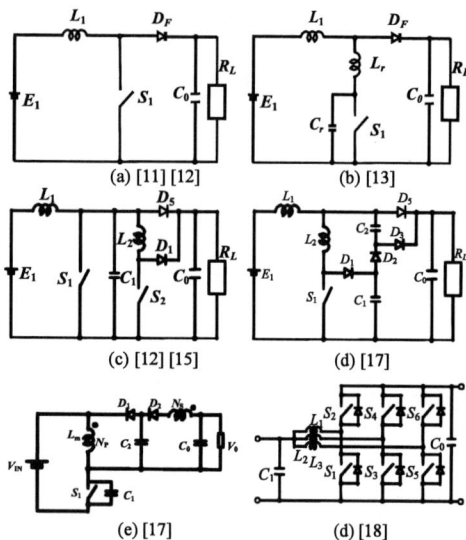

(a) [11] [12]   (b) [13]

(c) [12] [15]   (d) [17]

(e) [17]   (d) [18]

Figure 1: High efficiency switching topologies

B. Proposal of New High Efficiency Chopper Circuits

This paragraph discusses the recent author's effort and research activities in improving the efficiency of high power bi-directional chopper circuits.

C-Bridge switch shown in Figure 2(a) was proposed in [19] for high power application (8 kW) and 96.0 % efficiency at 25 kHz was obtained. Later on, Quasi-resonant Regenerating Active Snubber (QRAS) was proposed in [20] to achieve ZCS turn-on and ZVS turn-off and regenerating the snubber energy. The efficiency of 97.5 % was obtained for an 8 kW prototype operating at 25 kHz. The efficiency was further improved by using new semiconductor devices. This was published in [21] [22] where efficiency improvement from 97.5 % to 98.5 % operating again at 25 kHz was reported. The improvement was achieved by using SiC schottky diodes in 8 kW QRAS chopper. SAZZ (Snubber Assisted Zero Voltage and Zero Current Transition) topology was proposed after reconsidering the QRAS soft switching operation. SAZZ implements ZVZCT turn-on and ZVS turn-off while using fewer circuit components, than in QRAS topology [23]. 97.8 % efficiency was measured on SAZZ converter prototype operating at 100 kHz with the power output of 8 kW [24]. The SAZZ circuit operational principle was extended to a bidirectional buck and boost SAZZ chopper circuit in [25]. The bi-directional converter was tested as a 25 kW converter prototype operating at 50 kHz. The topology reached 96.6 % efficiency in boost mode and 97.4 % in buck mode under nominal power output of 25 kW. The presented topologies are discussed in more details in the following paragraphs with the main goal being high efficiency. As shown further, the proposed topologies offer high-efficiency for high-power applications. The prototypes discussed below were mostly constructed as scaled down models with further power increase kept in mind. The prototypes were tested at rated conditions, under light load, discontinuous current mode and out of resonance operation.

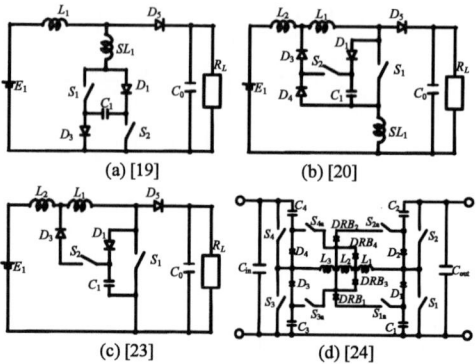

(a) [19]   (b) [20]

(c) [23]   (d) [24]

Figure 2: New high efficiency switching topologies

C. QRAS Chopper Circuit (25 kHz, 8 kW) [20]

Figure 3 depicts the circuit diagram of QRAS DC-DC chopper. The main parts of the proposed topology are: two switches (S_1 and S_2), three diodes (D_1, D_3 and D_4) and a capacitor C_2. This configuration retains the desirable properties of the low turn on loss and low turn off loss by so called "soft switching". Figure 4 shows basic operating waveforms of the QRAS chopper where as Figure 5 depicts the corresponding operational modes. When the switch S_1 turns on, the current rise through the switch is limited by the additional inductor S_{L1} (Mode2 in Figure 4 and in Figure 5(b)). On the other hand, when switch S_1 is turned off, the voltage rise across the switch is limited by the snubber capacitor C_2 (Mode5 in Figure 4 and in Figure 5(e)). Thus, the converter uses a loss less snubber and high efficiency operation can be achieved.

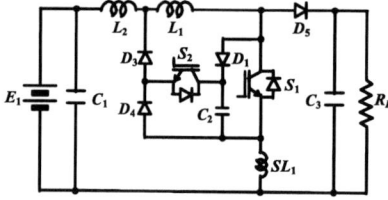

Figure 3: Circuit configuration of the QRAS chopper

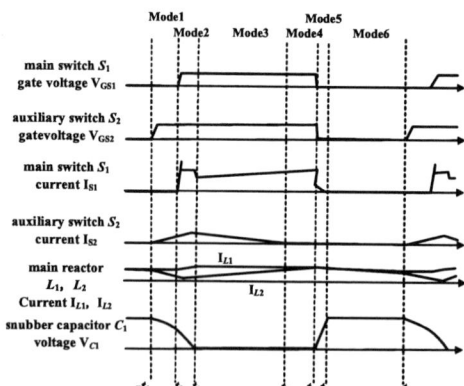

Figure 4: Operating waveforms of the QRAS chopper

8

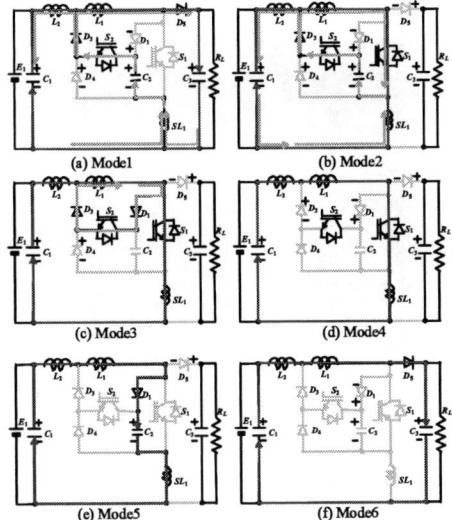

(a) Mode1   (b) Mode2

(c) Mode3   (d) Mode4

(e) Mode5   (f) Mode6

Figure 5: Six operational modes of QRAS chopper.

The soft switching area of QRAS chopper is shown in Figure 6. In almost whole boost operational region, ZCS turn-on and ZVS turn-off can be achieved. 8 kW prototype-model using IGBTs was made as shown in Figure 7. The prototype specifications are listed in Table I. Fig.8 shows the main switch voltage and current waveforms.

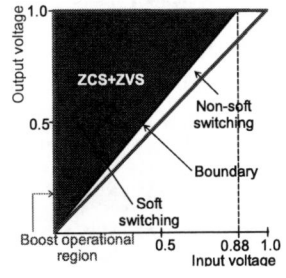

Figure 6: Soft switching area of QRAS topology boost operation

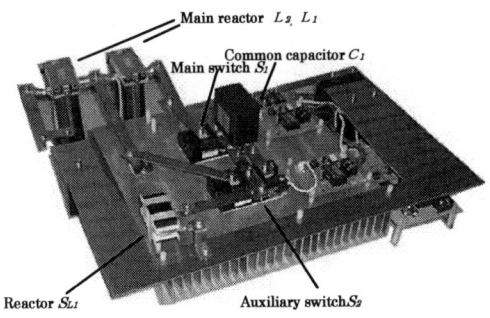

Figure 7: 25 kHz-8 kW QRAS prototype

TABLE I.

SPECIFICATIONS OF 8 KW QRAS PROTOTYPE

Rating	V_0=400 V, P_{OUT}=8 kW
Switch	PDMB100B12C, 1200 V, 100 A
Frequency	25 kHz
Circuit type	QRAS
Control	Pulse width control with open loop
Cooling	Natural air cooling

In order to eliminate the reverse recovery of the output diode D_5, Si diode was replaced by four paralleled SiC schottky diodes (CSD1012D,1200 V-10 A). The efficiency improvement was evaluated by practical measurements. Figure 9 shows the comparison of converter efficiencies with SiC and with conventional Si diodes for the power output ranging from 1 kW to 8 kW. As can be seen, the observed efficiency improvement is in order of 1 %. The efficiency of 98.5 % was measured at full power output of 8 kW. Measurement of power loss break down showed that the S_1 turn on loss and D_5 turn off loss was reduced by using SiC diodes.

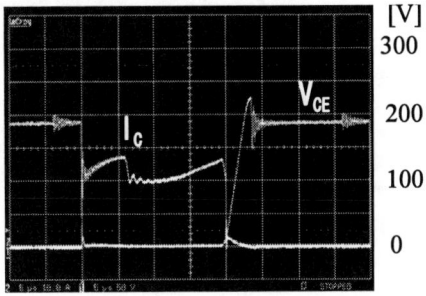

Figure 8: Main switch S_1 voltage and current waveforms; S_1 (PDMB100B12C,1200V,100A)

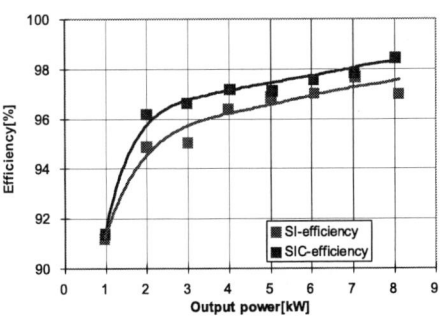

Figure 9: QRAS efficiency measurement results at 400 V output

D. SAZZ Chopper Circuit (100 kHz, 8 kW) [23]

The SAZZ chopper circuit is shown in Figure 10. The main part of the proposed topology includes: two switches (S_1 and S_2), two diodes (D_1 and D_5) and the capacitor C_2.

The topology offers zero voltage and zero current transition (ZVZCT) turn-on and zero voltage switching (ZVS) turn-off of the main switch S_1. Figure 11 depicts the basic steady-state voltage and current waveforms where as operating modes are shown in Figure 12.

Prior to turning on the main switch S_1 (at t_0 in Figure 11), the voltage across capacitor C_2 is discharged by the action of the auxiliary switch S_2 (Mode2 in Figure 12 (b)), resulting in zero voltage and zero current in the main switch S_1. As can be seen, capacitor C_2 is discharged by a resonance with L_2 when switch S_2 turns on (Mode2 in Figure 12). When the switch S_1 is turned off (at t_2 in Figure 11), the voltage across S_1 is snubbed by the capacitor C_2 (Mode5 in Figure 12). The main soft switching area is shown in Figure 13. In almost whole of the boost operational region ZVZCT turn on and ZVS turn off can be achieved. Non-soft switching area depends on the output power. If full discharge condition of the capacitor C_2 is not satisfied with respect to the output power, partial hard switching turn-on and ZVS turn-off are obtained in non-soft switching area [27].

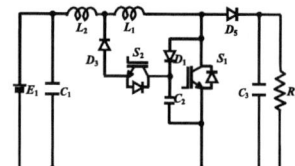

Figure 10: Circuit configuration of the SAZZ chopper

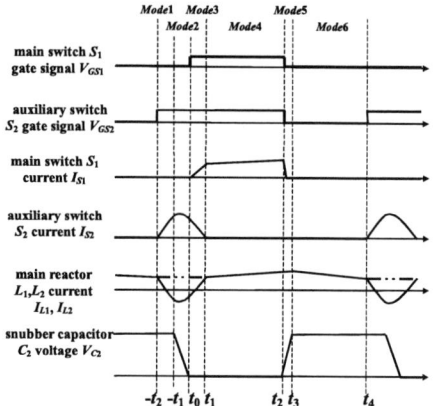

Figure 11: Operating waveforms of the SAZZ chopper.

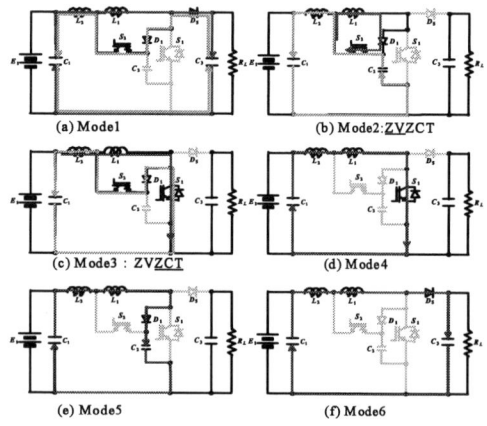

Figure 12: Six operational modes of SAZZ chopper.

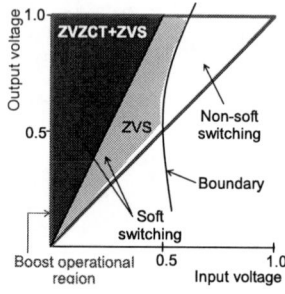

Figure 13: Soft switching area of SAZZ topology, boost operation

100 kHz-8 kW prototype using IGBTs was fabricated as shown in Fig.14. The specification of this prototype-model is listed in Table II. The waveforms measured on the SAZZ prototype are shown in Figure 15. The waveforms confirm that the main switch turns on in ZVZCT mode. The converter efficiency was measured by using HIOKI 3193 POWER HiTESTER. The measured efficiencies for the complete load range are shown in Figure 16. As can be seen, the efficiency of 97.8 % was measured at the full output power of 8 kW and operating frequency of 100 kHz. Table III shows the comparison between SPICE, SiC-SAZZ and SiC-QRAS chopper. As can be seen in Table III, SAZZ topology retains high efficiency of 97.0 % even at the increased operating frequency of 100 kHz. Total reactor loss was reduced from 49.96 W to 15.74 W due to the difference of the topology. However, D_3 loss increased due to the high resonant current (13.9 W to 63.2 W). S_1 turn off loss increased due to the increased higher operating frequency (30 W to 48.8W).

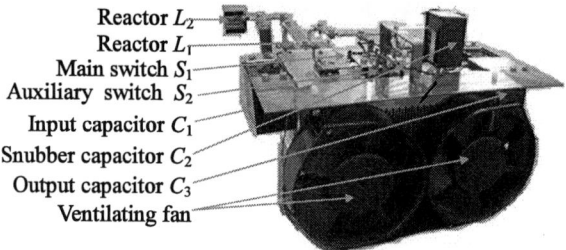

Figure 14: 100 kHz-8 kW SAZZ prototype

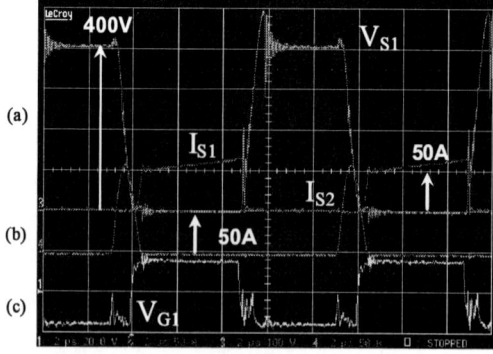

Figure 15: Measured waveforms of SAZZ operation (output=400 V,100 kHz, 8 kW); (a) Main switch S_1 voltage, current (50MT060WH, 600V, 50A), (b) Auxiliary switch S_2 current, (c) Main switch S_1 gate V_{G1}

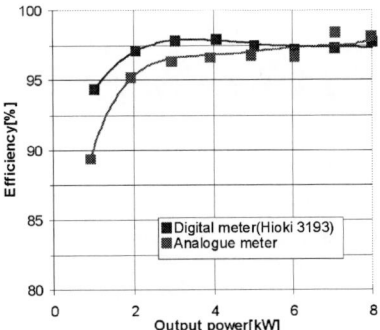

Figure 16: Efficiency curve of SAZZ prototype, 400 V output, 100 kHz.

TABLE II.

SPECIFICATIONS OF 8 KW SAZZ PROTOTYPE

Rating	V_0=400 V, P_{OUT}=8 kW
Switch	50MT060WH, 600 V, 50 A
Frequency	100 kHz
Circuit type	SAZZ
Control	Pulse width control with open loop
Cooling	With ventilating fans

TABLE III.

100 KHZ-8 KW SIC-SAZZ V.S.25 KHZ-8 KW SIC-QRAS LOSS BREAKDOWN.

loss factor	SPICE[W] SAZZ breakdown	100 kHz SiC*-SAZZ[W] Measured by meter	Measured by breakdown	25 kHz SiC*-QRAS [W] Measured by meter	Measured by breakdown
S_1 on-state	60.8		28.9		25.5
S_1 turn-on	0		6.2		5
S_1 turn-off	152.7		48.8		30
S_2 on-state	24.1		24.0		6.5
S_2 turn-on	0		0		0
S_2 turn-off	1.7		0		0
D_1 on-state	14.9		11.5		12.5
D_1 turn-on	0		0		0
D_1 turn-off	0		12.2		0
D_3 on-state	16.2		31.6		13.9
D_3 turn-on	0		0		0
D_3 turn-off	5.2		31.6		0
D_4 on-state					0.2
D_4 turn-on					0
D_4 turn-off					0
D_5 on-state	35		34.9		39
D_5 turn-on	0		0		9
D_5 turn-off	0		0		0
Reactor L_1			13.5		24.5
Reactor L_2			2.24		24.5
Reactor SL_1					0.96
Input power[W]	8109.4	8319		8182	
Output power[W]	7872.5	8079		8049	
Efficiency[%]	97.0	97.1	97.0	98.5	97.6
Total loss[W]	310.6	240	245.4	133	191.6

* Output diode D_5 is SiC schottky diode (CSD1012D,1200V-10A×4P) for QRAS, and SiC schottky diode (CSD20060,600V-20A×3P) for SAZZ

E. Bidirectional SAZZ Chopper Circuit (50 kHz, 25 kW) [25]

SAZZ circuit operational principle can be extended to a bidirectional buck and boost SAZZ circuit. Such topology created by combining two basic SAZZ circuits is shown in Figure 17. Waveforms in the boost-operating mode in forward direction are shown in Figure 18. The six operational modes of the forward buck and boost are shown in Figure 19 and Figure 20. The main soft switching areas are shown in Figure 21 and Figure 22. In almost whole boost operational region and buck operational region as well, ZVZCT turn on and ZVS turn off can be achieved. Non-soft switching area depends on the output power. In both directions, non-soft switching areas appear due to the boundary condition unsatisfying the snubber capacitor full discharge condition [27]. If full discharge condition is satisfied with respect to the output power, partial hard switching turn on and ZVS turn off are obtained in both non-soft switching areas.

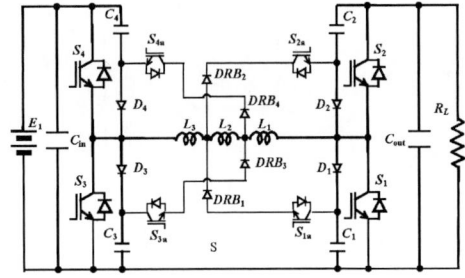

Figure 17: Bidirectional boost and buck SAZZ chopper

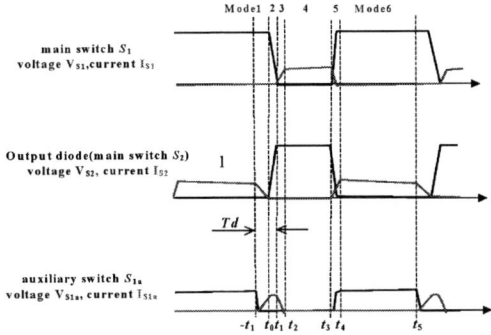

Figure 18: Boost mode operating waveforms of bidirectional SAZZ topology, forward direction.

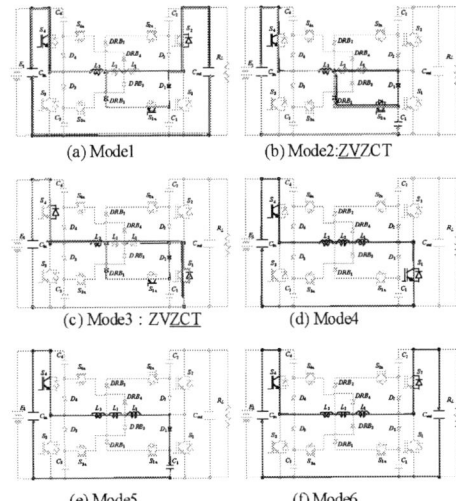

(a) Mode1

(b) Mode2:ZVZCT

(c) Mode3 : ZVZCT

(d) Mode4

(e) Mode5

(f) Mode6

Figure 19: Six operational modes of the forward boost.

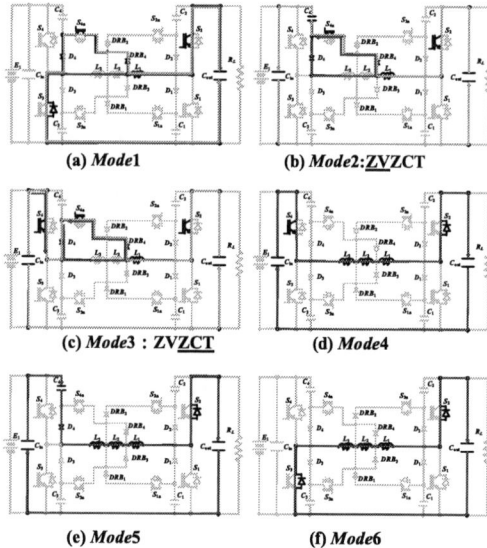

(a) Mode1   **(b) Mode2:ZVZCT**

(c) Mode3 : ZVZCT   **(d) Mode4**

(e) Mode5   **(f) Mode6**

Figure 20: Six operational modes of the forward buck.

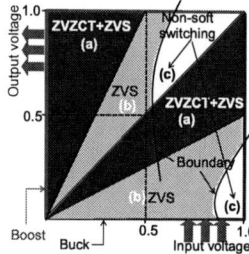

Figure 21: Soft switching area of SAZZ Bidirectional topology (Forward buck and boost operational region).

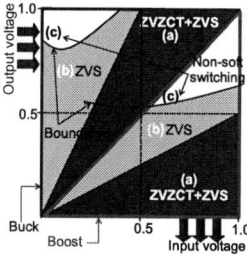

Figure 22: Soft switching area of SAZZ Bidirectional topology (Reverse buck and boost operational region)

Figure 23 and Figure 24 show the efficiency for forward boost operation respectively forward buck operation. Test facilities in the laboratory limited the load condition for boost mode and buck mode operation. Therefore, the efficiency could not be measured under the same input/output voltage condition. The measurements were performed at the load points that were allowed by the load-resistance. It can be concluded that in general the efficiency is higher for buck operation than boost. It is considered that input/output voltage ratio was more suitable for soft switching in buck operation as shown by the comparison of turn-on waveforms between Figure 25 and Figure 26.

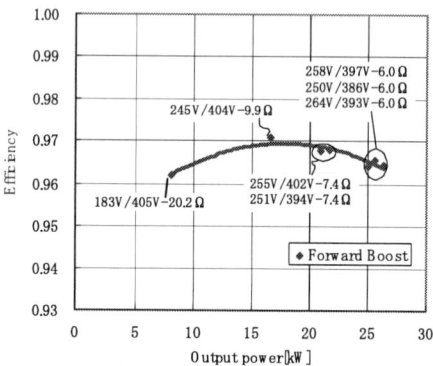

Figure 23: Efficiency of bidirectional SAZZ for forward boost operation

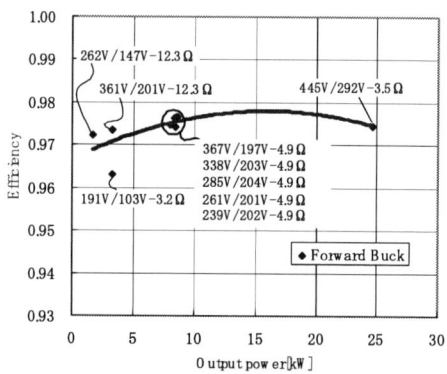

Figure 24: Efficiency of bidirectional SAZZ for forward buck operation

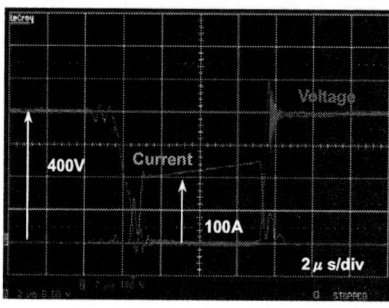

Figure 25: Voltage and current waveforms of the main switch, S_1 (CM200DU-12NFH, 600V, 200A), forward boost mode at 50 kHz-25 kW, 250 V/400 V

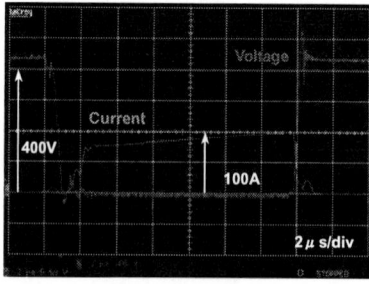

Figure 26: Voltage and current waveforms of the main switch, S_1 (CM200DU-12NFH, 600V, 200A), forward buck mode at 50 kHz-25 kW, 420 V/300 V.

F. High Power 3-Phase Interleaved Boost Chopper Circuit (90kW, 30 kHz)

High power motor drives of FCEV application require high power in the range of 100 kW. Three-phase interleaved boost chopper which operates at 90 kW-30 kHz was developed. Operating frequency of 30 kHz could be reached by implementing SAZZ topology where as conventional hard switching topologies have difficulties to operate over 10 kHz. Figure 27 shows the circuit diagram of the converter. Each phase is phase shifted by 120° with respect to each other. The specifications are defined as follows:

- Operating frequency : 30 kHz
- Output capacity : 90 kW
- Switch : CM600DU24NFH, 1200V, 600A
- Cooling : water cooling

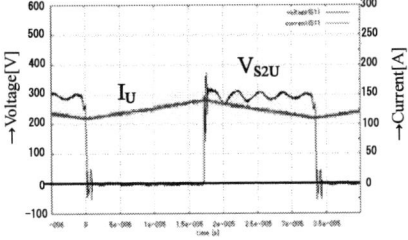

Figure 27: 3-phase interleaved boost chopper.

Figure 28 to Figure 30 show the examples of the waveforms measured on a single phase at 17 kW. The measurements were conducted on per-phase basis due to limit power available in the laboratory. Figure 31 depicts the efficiency characteristics vs. output power. As can be seen, the DC to DC conversion efficiency is more than 95 % and for power levels lower than 8kW it even exceeds 97 %. Figure 32 shows the converter loss breakdown. Total power dissipation consists of main power devices, S1 and D5, loss of about 36 %, auxiliary power devices, S2, D2 and D3, loss of about 14 %, DC reactor loss 22 % and the rest about 28 %.

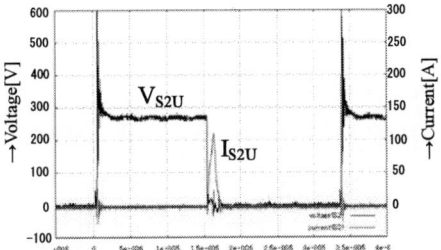

Figure 29: Voltage and current waveforms of the auxiliary switch S_{2U} (CM600DU-24NFH, 1200V, 600A), boost operation at 30 kHz-17 kW, 145 V/291 V

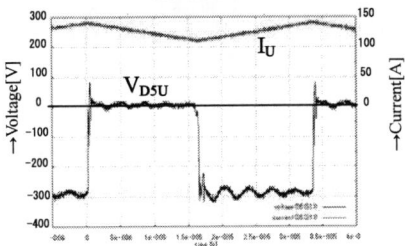

Figure 30: Voltage and current waveforms of the output diode D_{5U} (CM600DU-24NFH, 1200V, 600A), boost operation at 30 kHz-17 kW, 145 V/291 V

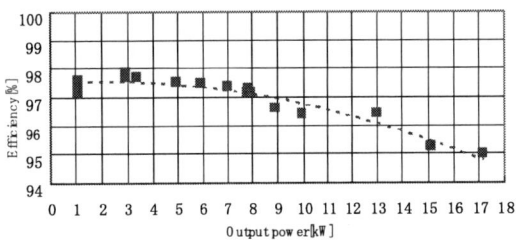

Figure 31: Efficiency of single phase in boost operation.

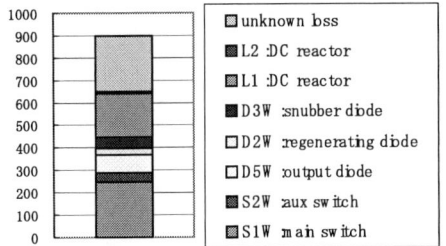

Figure 32: Loss breakdown for a single phase of the 3 phase SAZZ converter

III. HIGH POWER DENSITY

A power electronic converter is typically formed from the following main elements:

- **Power Semiconductors**
- **Power Passives (Filters/Transformers)**
- **Interconnections**
- **Packaging elements**
- **Thermal management**
- **Control and auxiliary circuits**

Figure 28: Voltage and current waveforms of the main switch S_{1U} (CM600DU-24NFH, 1200V, 600A), boost mode at 30 kHz-17kW, 145 V/291 V

All of the elements contribute to the final converter power density and each of them must be considered in order to reach the highest power density possible. Because of the many elements, reaching a high power density is a complex process.

The root of the high power density is usually in high frequency operation and other factors can either support or invalidate its effects. As discussed in the paragraphs above, the high frequency operation can be attained via reducing losses in power semiconductors by employing soft switching converter topologies and advanced semiconductor devices. In the same time, passive components must be designed considering the high frequency operation with related power loss and heat production inside of the components. High power interconnections, packaging elements as well as thermal management assist the main power processing components in performing well close to their performance boundaries. Finally, control and auxiliary circuits control and monitor the converter performance and make sure that it operates as required by design specifications. Operation of all the elements close to the boundary is required in order to achieve the highest power density possible.

A. Contribution of Highly Efficient High Frequency Operation to Power Density

Highly efficient energy conversion has two main effects on the converter power density. The first effect relates to the possibility of increasing the operating frequency and hence reducing the volume of passive components. The second effect relates to the size of cooling system. High conversion efficiency reduces demands on the cooling capacity and hence the volume of the cooling system. Therefore, the high power density converter design should strive for reaching a balance between the volume of the cooling system and volume of the passive components. This is demonstrated in [7] for several components. As shown in Figure 33 for power transformer, power density curves usually exhibit an optimal operating frequency where power density is the highest. This relation is substantially altered by the addition of heatsinks and other components and therefore the final design must be the result of a complete system optimisation.

In addition to direct increase of operating frequency, the frequency can be increased also virtually by using interleaving of multiple phase shifted converter modules [28]. This approach splits the total output power into several portions which are individually processed by separate converter modules/phases. The phases operate phase-shifted with respect to each other which increases the operating frequency from the filtering point of view.

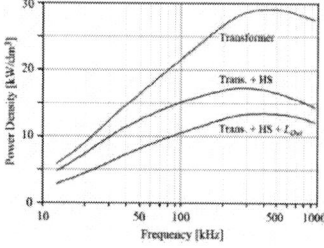

Figure 33: Example of power density of transformer with natural cooling (top line) with volume of semiconductors' heat sink added (middle) and volume of the output inductors added (bottom) [7]

In addition to this, the interleaving also reduces rms currents in the filter capacitors which results in further reduction of the filter volume. An optimisation of dc-dc converter from the number of phases point of view is discussed in [29]. As shown, interleaving has a positive effect on power density as well as on conversion efficiency. In the presented design, four interleaved phases offered the best results and were used in the prototype construction. Interleaved converters can use much more than four phases. A 1 kW converter with as many as 36 interleaved phases is presented in [30]. The presented designs resulted in improved efficiency and in the same time heatsinks could be completely omitted and the whole 1 kW converter could be constructed using solely SMD components.

B. Packaging for High Power Density

Packaging in power electronics has recently attracted a lot of attention. It has been recognised that proper classification of packaging terms [31] and quantification of packaging efficiency [32] can yield improvements in converter performance with respect to various design objectives. As demonstrated in [33], different approaches to packaging yield different results if evaluated from point of view of efficiency, power density, technological complexity, number of components, etc. From the power density point of view, all evaluated advanced concepts performed better than a design based on discrete components with higher than fourfold improvement in two presented cases.

According to [33], integration of components and increasing the functionality of packaging elements have large influence on power density. Increasing the functionality of packaging elements can be demonstrated on so called heat conductor converter [33] shown in Figure 34. In this design, copper bus-bar element fulfils functions of bus-bar, mechanical support, inductor winding and heat removal element in the same time. This substantially reduces the complexity of the converter design as well as the total converter volume due to using single element instead of four.

Figure 34: Heat conductor converter [33]

In integrated components, several devices utilise the same packaging as well as functional elements which reduces the total number of elements in the converter as well as their volume. The effect of this approach can be demonstrated for example on LLCT structure as presented in [34] where two inductors, a capacitor and a transformer are integrated into a single component. The resulting component requires only one set of heat removal and mechanically supporting elements which reduces the total volume. In the same time, air that would be present between the three components is largely rejected which further improves the power density.

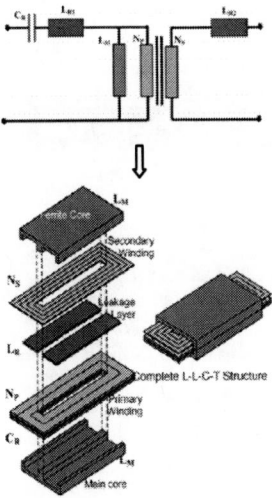

Figure 35: LLCT circuit diagram and internal structure [34]

C. Optimisation of Component Shapes and Layout

Addressing packaging in power converter design often results in using advanced technologies and materials. This is accepted in some applications but in some others a more conventional approach might be preferable due to cost limitations. As demonstrated in [35], truly high power density can be reached even by using solely conventional components by optimising the component shape and layout as well as paying close attention to the thermal management.

In conventional converter design, the largest space is usually wasted between the components due to their poor fitting or not paying attention to the geometrical component layout at all. In reality, any converter design offers certain flexibility to the component layout and increased effort can result in minimising the empty space inside the converter structure and hence in high power density. The flexibility in the design of passive components can help to mitigate the empty space as demonstrated in [36]. In the presented 50 kW converter design, transformer and inductor dimensions were chosen such that they would fit the dimensions of implemented IGBT modules. This resulted in little empty space inside the converter structure and in power density as high as 13 kW/l. Similarly in [35], diodes and capacitors are placed in the cavities next to the inductor winding (as shown in Figure 37 for diodes) which reduces the empty space and saves the extra space where these components would be placed otherwise.

D. Thermal Management and Power Density

A conventional way of removing heat from power components is by convection. In case of active components it often means individually cooled devices placed on heatsinks. In passive components it means airflow directly over components' surfaces. Convection has a big disadvantage with respect to power density which is the requirement of usually large empty spaces to allow for the required airflow. Therefore, in recent years conduction of heat away from passive as well as active components was introduced in power converters aiming for high power density. The use of conduction can be demonstrated on examples shown in Figure 34, Figure 36 to Figure 38. The presented solutions range from using lead frame in relatively low power converters [38], thermal bus-bar [33], heatsink embedded components [39] to individually thermally solved components in high power converter [36]. All the presented solutions use relative simplicity and low volume of heat conducting elements. In the case of the solution presented in [36], the heat conduction is supported by using heat-pipes at critical places like for heat removal from transformer and inductor windings for example.

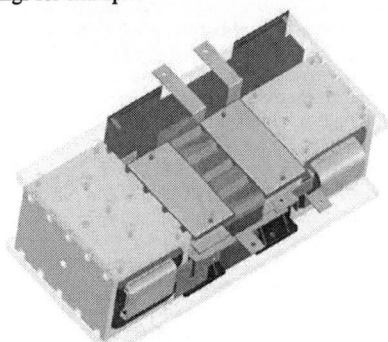

Figure 36: 50 kW converter with shape optimised components [36], [37]

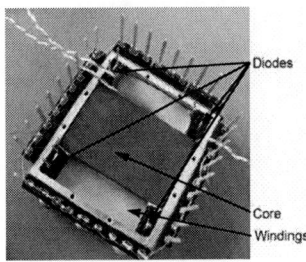

Figure 37: Placing components in winding cavities [35]

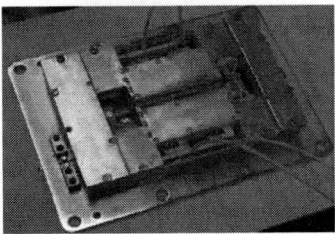

Figure 38: Heat removal based on lead frame [38] (top) and heatsink embedded components [39] (bottom)

E. Influence of Control and Auxiliary Circuits on Power Density

Little attention is usually paid to power density of control, gate drive and auxiliary circuits in conventional high power density converter designs. Their volume is considered small and "negligible" with respect to power processing circuits. The effectiveness of their design starts to play important role as the power density increases and at a certain point it becomes very critical. This can be demonstrated on the case study design as presented below. The power density of the complete converter can be as high as 25 kW/l for liquid-cooling but the power density of the gate drive circuit is only in the range of 0.3 – 0.5 kW/l. Because the gate drive circuit occupies almost 30 % of the converter volume, the power density of the power processing part is as high as 34 kW/l. In spite of the gate drives' low power density, the gate drive design is rather advanced with the printed circuit board almost completely populated by components with vertical boards added to fit more components onto the restricted board area as can be seen in Figure 39. This means that the technology used for gate drive and control circuits becomes very important at really high power densities and improving power density of control and auxiliary circuits could substantially improve the power density of the complete converter.

Figure 39: Gate drive unit of the case study converter

F. Overload Requirement and Power Density

Overload requirement in a converter design typically demands designing all power processing components for the overload power level and thermal management for the overload time duration. In the consequence, both these aspects result in a reduced power density if nominal power is considered due to the over-dimensioning of all components and thermal management parts. This is demonstrated on a simple example presented in Figure 40. The figure depicts the power density of the integrated inductor used in the design presented below at various power levels without considering the required thermal management. The two curves represent the design cases with and without the overload and the arrows indicate the power density reduction if the overload is required. The considered overload is 220 % in this case. The inductor volumes and consequently power densities are calculated by considering the saturation flux density as the core size limiting factor and the core winding gap as the winding size limit; thermal management is not considered. The cores used for this illustration are standard planar E cores as available in the core manufacturers' catalogue.

As can be seen, power density of the inductor decreases as the output power increases. Even more substantial power density reduction is observed if the overload operation is required. The 220 % overload requirement reduces the component power density to approximately 40 %. Further power density reduction can be expected in case if thermal management would be considered.

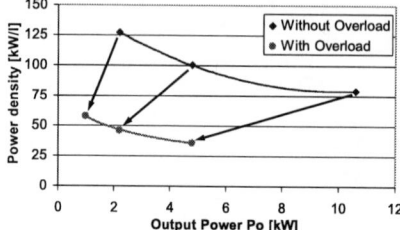

Figure 40. Power density for integrated SAZZ inductor with respect to converter output power, with and without overload requirement (overload 220 % considered)

The power density typically serves as a universal parameter for comparing different design approaches. The fact that it varies strongly with the overload requirement complicates the power density indication for such designs because the power density can be indicated without or with considering the overload. However, in both cases the resulting power density figures are not objective and therefore not comparable with the designs which do not require the overload operation. This occurs because in case of neglecting the overload requirement and considering only the nominal power, the resulting power density figure is very low and its reduction can be compared to what is indicated in Figure 40. Power density calculated in this way underestimates the converter design because all the power processing components must be designed for a higher power level required by the overload operation. On the other hand, if the overload is considered, the resulting power density is usually very high. This is unfair with respect to designs without the overload because converters designed for overload operation can not operate under overload conditions for an extended period of time (the thermal management is usually not designed to withstand such long overload operation). Therefore, in order to compare converter designs with and without the overload operation requirement, a new power density calculation methodology is needed.

The new methodology, first proposed in [40], is based on using maximal continuous power level to calculate the power density. This power level is typically higher than nominal because a thermal overhead is required in order to allow the overload operation. The new method removes the unfavourable conditions related to considering or not considering the overload operation. The resulting power density is higher than in case of using the nominal power level and lower than in the case of using the overload power level. Therefore, it is more realistic and fairer for comparison with designs which do not require the overload operation. Power density calculated in this way includes also the thermal management the design of which varies depending on the required overload time duration.

G. Converter Case Study with High Power Density, High Efficiency and Overload Capability

The main objectives of the presented design concept are high efficiency and power density. These objectives are followed by careful consideration of the issues addressed in the preceding discussion. As the result, following design choices are implemented in the converter design:

- Bidirectional SAZZ converter topology [41]
- Parallel CoolMOS devices – two parallel devices for main switches
- Integrated inductors
- Interleaved converter modules – four modules
- Conduction as the main heat removal method
- Layered converter structure with optimised space utilisation

More details on each design aspect can be found in [35].

1) Case Study Specifications

Main design specifications which are also considered as an example of specifications for traction automotive dc-dc converters are listed in Table IV. As can be seen, automotive converters must typically operate in a broad range of input and output voltages and they must also sustain a considerable overload (almost 240 % overload for the time period of 20 s in the presented case). Especially, the overload requirement has a great impact on the overall converter power density due to the necessity of designing all the components for the overload conditions.

TABLE IV.
MAIN CONVERTER SPECIFICATIONS

Input Voltage V_i [V]	120 – 166 (V_{iN}=150V)
Output Voltage V_o [V]	300 – 500 (V_{oN}=300V)
Output Power P_o [kW]	8.4 (20 kW overload for 20 s)
Operating Frequency f_s [kHz]	200

2) Converter Design

3D converter models are shown in Figure 41 though to Figure 44. Figure 41 and Figure 42 depict the single converter module in collapsed respectively exploded view. As can be seen from Figure 42, the structure is built around the integrated inductor positioned in the modules' centre. The interconnect PCB and Gate drive unit are positioned above the main converter body.

Figure 41. Single module collapsed view

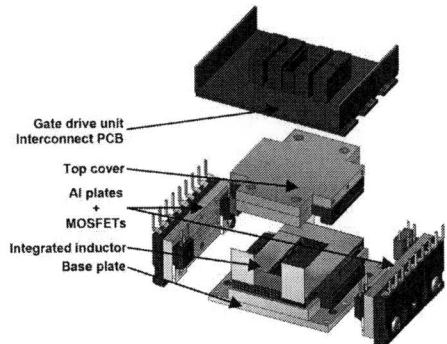

Figure 42. Single module exploded view

Figure 43 and Figure 44 show 3D models of the complete converter structure with air-cooling respectively liquid-cooling. The two structures are almost identical with the only difference being liquid cooled cold plate replaced by heat-pipes attached to forced air-cooled heatsinks in the case of the air-cooled assembly. Both structures are very compact and there is little empty space inside of them.

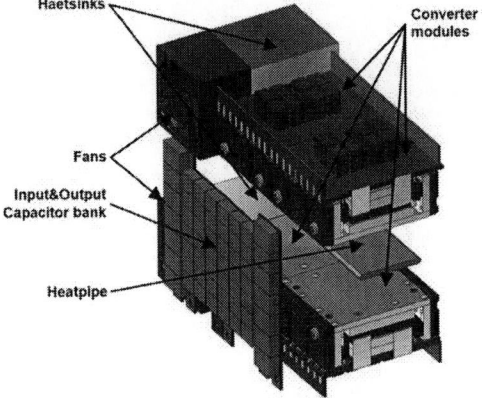

Figure 43. Complete air-cooled assembly

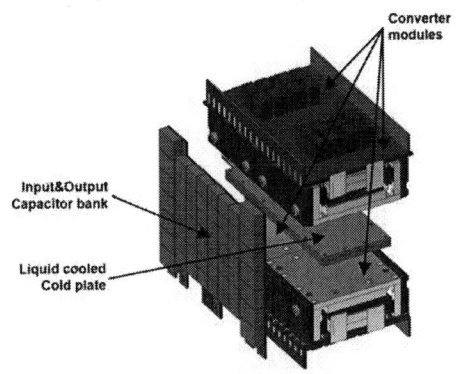

Figure 44. Complete liquid-cooled assembly

Figure 45 shows the prototype of the single 2.1 kW converter module next to an AA battery cell. As can be seen the converter module is very small and compact. The complete assembly of the liquid-cooled converter is then shown in Figure 46. Notice, there is little empty space available inside the converter structure.

Figure 45: Single converter module prototype

Figure 46: Complete liquid-cooled converter prototype

3) Experimental results

Tests performed on the single module converter prototype cover the complete range as required by the design specifications. Converter performance under continuous overload conditions was also tested in order to evaluate the converter power density using the approach as discussed in paragraph III.F. Test results performed on the complete converter were not available when writing this paper and therefore they are not included.

The complete efficiency curve for the power range from 600 W to approx 5 kW is shown in Figure 47. As can be seen the converter exhibits peak efficiency slightly higher than 95 % and this efficiency is held through the whole continuous overload power range up to 3.35 kW. The limit of continuous overload was determined by observing the temperature rise of the converters' top surface. Placing the thermocouples inside the converter structure was also considered but this could not be done due to the limited space. For the continuous power output of 3.35 kW, the measured temperature rise was 60 °C which was considered as the maximal allowed temperature rise. This value was chosen because it corresponds to a hot-spot in the inductor of approx. 150 °C for the heatsink temperature of 70 °C. Short-term overload was also tested under the temperature rise constrain of 60 °C. The time interval of 50 s was measured at the power output of 5 kW from the single module as the time required for the converters' top surface to heat-up by the specified temperature rise of 60 °C. This is much longer than the 20 s required by the design specifications. Notice that the converter efficiency is reduced to approximately 94 % under full overload of 5 kW which is still very high considering the overload of 240 %.

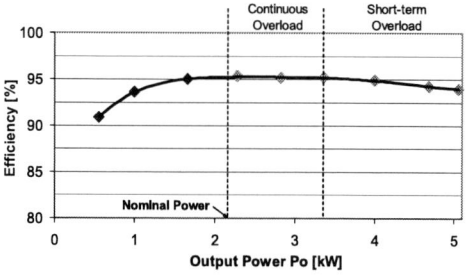

Figure 47. Efficiency curve of single module converter prototype with continuous as well as short-term overload, nominal voltage conversion 150 V to 300 V

Power density figures calculated by using different available power levels are listed in Table V. As discussed in paragraph III.F, the total power density of converters with overload operation should be calculated using the continuous overload power. Based on this assumption, the power density of the complete converter prototype was calculated as 25 kW/l for liquid cooling and 16.4 kW/l for air-cooling. These power densities are among the highest in the class.

TABLE V.
POWER DENSITIES CALCULATED FOR CONVERTER DESIGN II

	Nominal Power 8.4 kW	*Continuous Overload* 13.5 kW	*Short-term Overload* 20 kW
Power Density [kW/l]	Liquid-cooling		
	15.5	*25*	*36*
Power Density [kW/l]	Air-cooling		
	10.2	*16.4*	*24.3*

In case that bi-directional operation is not required, the output MosFETs can be replaced by conventional or SiC diodes. The possibility of using SiC diodes was tested in order to assess their influence on the converter efficiency. The comparison of efficiency curves measured with and without using SiC diodes is shown in Figure 48 for voltage conversion ratio of 120 V to 300 V. As can be seen, the efficiency improvement solely due to using SiC diodes is over 2 % in almost the whole considered power range. The peak efficiency under nominal voltage conversion of 150 V to 300 V was measured as high as 96.2 %. The final efficiency value could be still improved if measured on the complete converter with interleaved converter modules due to loss reduction in the input and output capacitors. Considering that the output MosFETs in the converter structure can be replaced by SiC diodes without any gain in converter volume, doing so would result in a power converter with power density of 25 kW/l and efficiency possibly as high as 97 %. This is a remarkable result considering that the whole converter is based solely on conventional components.

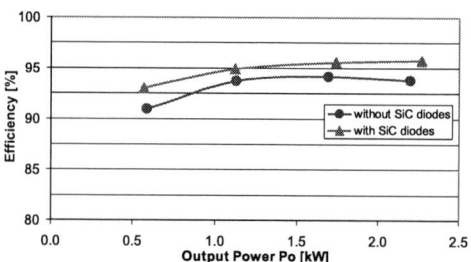

Figure 48. Comparison of efficiency curves of with and without SiC Diode in the rectifier, conversion 120 V to 300 V

IV. CONCLUSIONS

Vehicles with electric propulsion require highly efficient power converters with high power density. In this paper, issues related to reaching high power density and in the same time high efficiency are discussed. At first, several state of the art power converters based on SAZZ soft switching topology are presented. It is shown that efficiencies close to 98 % are attainable in high power converters at operating frequencies as high as 100 kHz using IGBT devices.

In the second part of the paper, issues related to reaching high power density in high power designs are discussed. It is pointed out that high power density is influenced by many design aspects and converters must be designed as close to the operational boundary as possible while most of the empty space inside of the converter structure should be eliminated in order to reaching as high power density as possible. It is also shown that the resulting power density is strongly dependent on the overload requirement and a new method of power density evaluation based on continuous overload is proposed. The discussed approach to high power density designs is concluded by presenting high power converter based on SAZZ topology and CoolMOS devices operating at 200 kHz. The presented prototype reaches power density as high as 25 kW/l while using only conventional components. In the same time, the conversion efficiency of the complete converter is expected to be as high 97 % based on the experimental results of a single converter module.

The presented design studies and converter prototypes are used in this paper as the proof of concept for reaching high power density and high efficiency. The presented results look promising but more studies are needed in order to improve the power density and conversion efficiency even further. The future efforts should concentrate on reaching conversion efficiencies beyond 99 % barrier while not harming the power density.

ACKNOWLEDGMENT

The authors would like to thank NEDO (New Energy Development Organisation, project ID: 05A48701d), JSPS (Japanese Society for Promotion of Science) and KAST(Kanagawa Academy of Science and Technology) for financially supporting the project.

REFERENCES

[1] K.Shingo, K.Kaoru, T.Katsu, Y.Hata: "Development of Electric Motors for the TOYOTA Hybrid Vehicle "PRIUS"", EVS-17, Dec.2000

[2] W.Yu, J.Lai, "Ultra High Efficiency Bidirectional DC-DC Converter With Multi-Frequency Pulse Width Modulation", APEC08, pp.1079-1084, 2008

[3] A. Kawamura, Y.Tsuruta, S. Inasaka, " Proposal of electrical mileage for electric vehicle and discussions", IEEJ, VT-07-13, December 2007

[4] Y. Ito, Y. Tsuruta, M. Bando, and A. Kawamura, "50kHz-25kW Bilateral Chopper Circuit SAZZ-1 for HEV", EVS-22, Oct.23-28, pp.848-857, 2006

[5] H. Ohasi, "Recent Power Devices Trend (in Japanese)", Trans. of IEEJ, vol. 12, no.3, 2002, pp. 168-171.

[6] H. Ohashi, "Power Electronics Inovation with Next Generation Advanced Power Devices", Proceeding of INTELEC'03, Oct. 2003, Yokohama (Japan), pp.9-13

[7] J.W. Kolar, U. Drofenik, J. Biela, M.L. Heldwein, H. Ertl, T. Friedli, S.D. Round, "PWM Converter Power Density Barriers" Proceedings of Power Conversion Conference, PCC 2007, pp. 9 – 29, 2-5 April 2007

[8] T. Teratani,"Vehicle Energy Management and Higher Voltage 42V for New Generation", 2nd International Congress on 42V Power Net(2000)

[9] M.Okamura, E.Sato, S.Sasaki: "Development of Hybrid Electric Drive System Using a Boost Converter", EVS-20, Nov.2003

[10] A.Kawamura, "Future commuter vehicles", KAST seminar on next generation mobility systems, 2007 November

[11] S. Cuk and R.D. Middlebrook : "Advances on Switched-Mode Power Conversion Part I", IEEE Trans. on Industrial Electronics, Vol.IE-30, No.1, pp.10-19 (1983)

[12] Richard G. Hoft : "Semiconductors Power Electronics", Van Nostrand Reinhold Company Inc. New York (1986) (in Japanese)

[13] IEEJ : "Recent development on soft switching", IEEJ Technical Report, No.899, pp. 4-8 (2002-9)(in Japanese)

[14] A. Kawamura : "Modern Power Electronics", SUURIKOUGAKU-SHA Co. Ltd., the First Edition, Tokyo, pp.39 (2005) (in Japanese)

[15] G. Hua, C.S. Leu, and F.C. Lee : "Novel zero-voltage-transition PWM converters", IEEE Trans. on Power Electronics, Vol.9, NO.2, pp.213-219 (1994)

[16] M.Nakamura, T.Myoui, M.Ishitobi,M.Nakaoka:"A Soft-Switching PWM Boost Chopper Comtrolled DC-DC Converter with A Single Passive Auxiliary Resonant Snubber and Its Performance Evaluations.",T.IEE Japan,Vol.122-D,No.10, pp.1006-1016,(2002)

[17] Q. Zhao and F.C.Lee: "High-Efficiency, High Step-Up DC-DC Converters", IEEE Trans. Power Electronics, Vol.18, No.1, pp.65-73 (2003)

[18] J.Zhang, R.-young Kim and J.-Sheng Lai: "High-Power Density Design of a Soft Switching High-Power Bidirectional DC-DC Converter", Proc. of IEEE PESC06, WeA2-1, pp.2119-2125 (2006)

[19] K. Maikawa, Y. Tsuruta, and A. Kawamura: "Soft Switching Chopper Circuit for high power application", IEEJ/JIASC 2003, No.1-101, pp.477-478 (2003-8) (in Japanese)

[20] Y. Tsuruta and A. Kawamura : "Proposed of 98.5% High Efficiency Chopper Circuit QRAS for the Electric Vehicle and the Verification", T.IEE Japan, Vol.125-D, No.11, pp.977-987 (2005) (in Japanese)

[21] Y. Tsuruta, Y. Ito, and A. Kawamura : "The prompt report regarding the efficiency measurement of 8kW QRAS chopper using SiC schottky diode", IEEJ/JIASC 2004, No.1-85, pp.439-442 (2004) (in Japanese)

[22] Y. Tsuruta, Y. Ito, and A. Kawamura : "8kW QRAS Chopper Using SiC Schottky Diode", The 2005 International Power Electronics Conference, IPEC-Niigata 2005, S30-3, pp.1113－1119 (2005)

[23] Y. Tsuruta, Y. Ito, and A. Kawamura : "The proposal of SAZZ chopper circuit and test verification by means of the preceding fabrication", 2005 National Convention Record, No.4-045, pp.71-72 (2005) (in Japanese)

[24] Y. Tsuruta, Y. Ito, and A. Kawamura : "A High Frequency, High Efficiency and High Power Chopper SAZZ and the Test Evaluation at 100kHz-8kW", IEEE 37th Annual Power Electronics Specialists Conference, PESC'06, WeA1-4, pp.1965-1971 (2006)

[25] Y. Tsuruta, Y. Ito, M. Bando, and A. Kawamura : "Proposal of Bilateral Buck and Boost Chopper Circuit SAZZ-1 for EV Drive Application and the Test Evaluation at 25kW", IEEE International Conference on Industrial Technology, ICIT-2006, IF005878, pp.1504-1509 (2006)

[26] Y. Tsuruta and A. Kawamura : "Technical Stream on a High Efficiency and High Frequency Chopper Circuit", IEEJ/JIASC 2006, No. 1-o-6-2, pp. I-139－I-144 (2006) (in Japanese)

[27] Y. Tsuruta, M. Pavlovsky, and A. Kawamura : "Condition Limiting the Formation of the ZVZCT Switching in SAZZ Converter", IEEE Industry Applications Conference 42st Annual Meeting, IAS 2007, IAS56p5, pp.1-6 (2007)

[28] Chin Chang; M.A. Knights, "Interleaving technique in distributed power conversion systems", IEEE Transactions on Circuits and Systems I: Fundamental Theory and Applications, Volume 42 Issue 5, pp. 245 -251

[29] M. Gerber, J.A. Ferreira, I.W. Hofsajer, N. Seliger, "Interleaving optimization in synchronous rectified DC/DC converters", Proceedings of Power Electronics Specialists Conference, PESC 2004, Achen – Germany, pp. 4655 – 4661, Jun 2004

[30] O. Garcia, P. Alou, J.A. Oliver, J.A. Cobos, " A high number of phases enables high frequency techniques and a better thermal management in medium power converters", Proceedings of Conference on Integration of Power Electronics Systems, CIPS 2008, pp. 281 – 284, March 2008

[31] G.R. Blackwell, The electronic packaging handbook, Boca Raton CRC Press 2000

[32] J. Popovic, J.A. Ferreira, "Concepts for high packaging and integration efficiency", Proceedings of Power Electronics Specialists Conference, PESC 2004, Volume 6, June 2004, pp. 4188 – 4194

[33] J. Popovic, J.A. Ferreira, "Design and evaluation of highly integrated dc-dc converters for automotive applications", Proceedings of Industry Applications Conference, IAS 2005, Volume 2, Oct. 2005, pp. 1152 – 1159

[34] J.T. Strydom, J.D. van Wyk, "Electromagnetic modeling for design and loss estimation of resonant integrated spiral planar power passives (ISP/sup 3/)", IEEE Transactions on Power Electronics, Volume 19, Issue 3, May 2004, pp. 603 – 617

[35] M. Pavlovsky, Y. Tsuruta, A. Kawamura, "Automotive DC-DC converter designed for high power-density and high efficiency" Proceedings of Conference on Integration of Power Electronics Systems, CIPS 2008, pp. 191 – 195, March 2008

[36] M. Pavlovsky, S.W.H. de Haan, J.A. Ferreira, "Concept of 50 kW DC/DC converter based on ZVS, quasi-ZCS topology and integrated thermal and electromagnetic design", European Conference on Power Electronics and Applications, EPE 2005, Sept. 2005

[37] M. Pavlovsky, "Electronic DC transformer with High Power Density", PhD thesis Delft University of Technology, The Netherlands

[38] J. Wanes, "A novel integrated packaging technique for high density DC-DC converters providing enhanced efficiency and thermal management", Proceeding of Applied Power Electronics Conference and Exposition, APEC 2004, Volume 2, pp. 1229 – 1235

[39] M. Gerber, J.A. Ferreira, I.W. Hofsajer, N. Seliger, "High density packaging of the passive components in an automotive DC/DC converter", IEEE Transactions on Power Electronics, Volume 20, Issue 2, Mar 2005, pp. 268 – 275

[40] M. Pavlovsky, Y. Tsuruta, A. Kawamura, "Pursuing High Power-Density and High Efficiency in DC-DC Converters for Automotive Application" Proceedings of Power Electronics Specialists Conference, PESC 2008, Jun 2008

[41] Y. Tsuruta, M. Bando, Y. Ito and A. Kawamura, "A new circuit geometry SAZZ for an EV drive application", Proceedings of IEEE Industry Applications Conference, IAS 2006, pp.1-6

Current-Based Condition Monitoring of Electrical Machines in Safety Critical Applications

Thomas G. Habetler

School of Electrical and Computer Engineering
Georgia Institute of Technology
Atlanta, GA 30332 USA
Tel: (404) 894-9829
Email: thabetler@ece.gatech.edu

Abstract-The paper presents a complete summary of state-of-the-art techniques for current-based fault monitoring of low voltage permanent magnet synchronous machines commonly used in safety critical applications such as hybrid vehicles and backup generators. This includes detection of stator winding faults, and bearing faults. In addition, a fault tolerant strategy is presented which can be used to operate the machine in a fault tolerant mode even in the presence of the stator turn fault. Experimental results show the efficacy of these methods which allow for use for PM machines even in automotive applications wherein the machine cannot be stopped.

I. INTRODUCTION

Permanent magnet (PM) synchronous motor drives present a particular safety issue which occurs in the presence of a fault in the stator winding. The ability to sense this, and other faults, greatly improves the reliability, availability and maintainability of PM drives in a variety of critical applications. Taking corrective action as a result of a fault (or fault-tolerant operation) is extremely important in PM drives wherein the rotor cannot be immediately stopped. The addition of fault monitoring and fault tolerant operating schemes can be accomplished very inexpensively by not adding any hardware to the drive. Existing sensors and hardware can provide a vast array of system information that has traditionally not been used.

Stator faults are basically the breakdown of the winding insulation. This may be caused by excess thermal or voltage stress, mechanical vibration, or even an abrasion between the stator and rotor. The weakness of the winding insulation can further result in the turn-turn short circuit, and eventually winding-ground short circuit. It is generally believed that a large portion of stator winding-related failures are initiated by insulation failures in several turns of a stator coil within one phase. This is referred to as a "turn fault."

A major concern with employing interior permanent magnet synchronous machine (IPMSM) drives in safety-critical applications is their unmanageable behavior under fault conditions resulting from the presence of the spinning rotor magnets that cannot be turned off at will. This clear weak point has made engineers hesitant to employ IPMSMs in safety-critical applications, and clearly makes the research on how to increase the fault tolerance of IPMSM drives crucial. A stator turn fault in a symmetrical three-phase AC machine causes a large circulating current to flow and

subsequently generates excessive heat in the shorted turns. Without limiting the heat that is proportional to the square of the circulating current, the fault generally results in complete motor failure. In an automotive application, this could translate into a dangerous situation which may well result in fire or shock hazard

Therefore it is necessary to not only detect stator turn faults with a high degree of reliability and sensitivity, but also to develop a fault tolerant strategy which will allow operation of the machine in the presence of a fault without catastrophic and/or dangerous failure. Highly conservative designs and redundancy are commonly applied in order to improve the fault tolerance of electric machines. The intention of a conservative design is to reduce fault occurrences by over-sizing the system's capacity. The concept of redundancy is well understood: if part of a system fails, there is an extra or spare that is able to operate in place of the failed unit such that the operation of the system is uninterrupted. Although these two approaches are the surest ways to increase the fault tolerance of an electric motor drive, they greatly increase the cost and complexity of the system. Moreover, redundancy may not be practical for an application that has a severe restriction on the installation space, such as in the case of traction drives in electric or hybrid-electric vehicles. As alternatives to these approaches, fault diagnosis and fault-tolerant strategies must be used. The purpose of diagnosis is to detect a certain failure from its point of inception so as to prevent it from developing into the catastrophic phase. Fault-tolerant strategies are based on the concept that a faulty system can maintain its uninterrupted operation with the assistance of a modified topology or control algorithm. To implement such approaches as practical entities, the drives should be designed to perform the following essential tasks:

(1) Fault detection
(2) Fault isolation
(3) Remedial (Emergency) actions after fault detection

The proposed fault-tolerant strategy is applicable in any application that mandates high stator turn fault tolerance. However, the strategy can yield tremendously increased tolerance in specific applications that require the faulty machine to keep rotating in spite of some degradation in the performance. From this perspective, a traction drive in a

978-1-4244-1741-4/08/$25.00 ©2008 IEEE

mild-hybrid vehicle or pure electric vehicle may be most suitable applications for the proposed strategy. The general configuration of the power-train system in a mild-hybrid vehicle is presented in Figure 1. As shown in the figure, since the electric machine is connected directly to the internal combustion engine and the transmission, it is very difficult or even impossible to isolate the electric machine from its mechanical counterparts while the vehicle is moving. Thus, even though the faulty machine is not required to generate any propulsion, the machine should continue to rotate because of the mechanical connections to the internal combustion engine and the transmission.

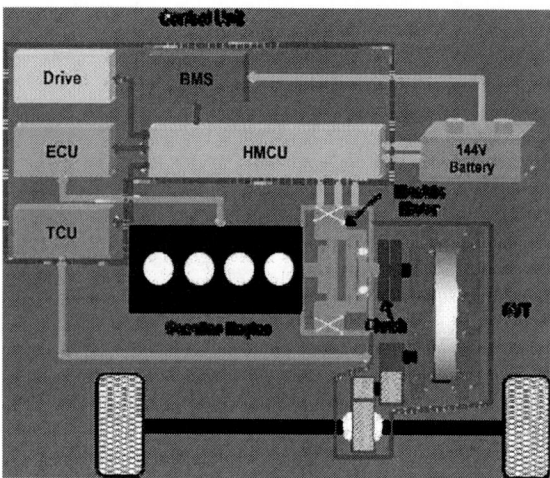

Figure 1. Configuration of the power train system in a mild-type parallel hybrid electric vehicle

This paper presents an overview of work done in solving this problem of safely applying PM synchronous machines in safety critical applications. A method for both detecting the presence of a turn fault and for operating the machine after the turn fault occurs in presented. In order to assure that the machine maintains a safe operating condition, a thermal model is also presented which is used to predict the temperature of the machine in the presence of a fault.

II. WINDING FAULT DETECTION

A. Protection

Many resources show that 35-45% of motor failures are caused by insulation breakdown in the stator winding. The organic materials used for insulation in electric machines are subjected to deterioration from a combination of thermal overloading and cycling, transient voltage stresses on the insulating material, mechanical stresses, and contaminations. Among the possible causes, thermal stresses are the main reason for the degradation of the stator winding insulations. Generally, thermal stresses on the stator winding insulations are categorized into three types: aging, overloading, and

cycling. Even the best insulation will fail quickly if operated above its temperature limit. As a rule of thumb, the life of insulation is reduced by 50% for every $10^\circ C$ increase above the stator winding temperature limit [1]. It is thus necessary to monitor the stator winding temperature so that an electric machine will not operate beyond its thermal capacity. For this purpose, many techniques have been reported in the literature [2, 3]

B. Detection

The next step following protection of the stator is the detection of the early stages of insulation failure. The different types of stator turn faults are depicted in Figure 2. The first stage is small shorts between turns in the same winding. Among the five failure modes, turn-to-turn faults (stator turn fault) have been considered the most challenging one since the other types of failures are usually the consequence of turn faults. Furthermore, turn faults are very

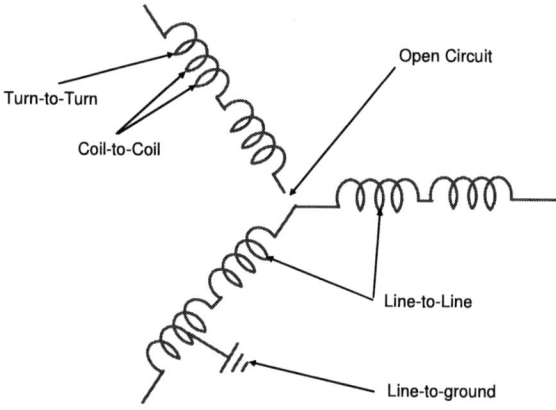

Figure 2. Possible failure modes in wye-connected stator windings.

Conventional turn fault detection schemes merely monitor the negative sequence component of line currents (or the effective negative sequence impedance) and rely on mathematical models for symmetrical induction machines to account for the effect of unbalanced supply voltages on the negative sequence current [5, 6]. However, neglecting inherent asymmetries can lead to misdetection, with catastrophic consequences. The issue of inherent asymmetries can be addressed by using a neural network-based approach [10].

While this technique has been shown to give very good results for line-connected induction machines, it is often not suitable to PM synchronous machine applications since the current controller in the drive attempt to regulate the current to track the reference value. Since the current is regulated, the effect of the asymmetry from the fault is now reflected in the motor voltage. For this reason, voltage-based turn fault detection methods have been proposed, but require additional voltage sensors and cables.

By modeling a machine with a turn fault, it can be concluded that a bolted turn fault reduces the positive sequence components of the machine impedances and back-emf voltages, while increasing the negative sequence and coupling terms in the impedance matrix at the same time. The positive sequence current slightly increases under a stator turn fault condition in a mains-fed application where the power supply is a fixed voltage source [9]. In a CCVSI-driven application, the inverter controls the line currents so as to follow their references by introducing negative sequence voltage and reducing positive sequence voltage under a stator turn fault condition. Since the inverter output voltages are produced according to the voltage references that are generated through the current controllers, the variations in the machine parameters will be reflected into the voltage references. This implies that for a given rotating speed and current references (or alternatively torque reference), the presence of a stator turn fault results in a reduced positive sequence component and an increased negative sequence component of the voltage references as compared to a machine without a turn fault. Thus, it can be concluded that the differences in positive and negative sequence components of the voltage references, for a given torque reference and rotating speed, under a stator turn fault and fault-free conditions can indicate the occurrence of a stator turn fault. The voltage references in the rotating and stationary reference frame until fault and no-fault conditions are shown in Figure 3.

II. FAULT TOLERANT OPERATION

Turn faults are particularly problematic in IPMSM drive in safety-critical applications. This is due to the fact that the rotating magnet can often not be stopped when a fault occurs, and therefore the fault current is allowed to flow until catastrophic or dangerous thermal damage is done to the machine. Therefore, PM machines not only require a reliable turn fault detection method, but also imperatively require a proper remedial action that can maintain the drive's uninterrupted operation.

The most desirable characteristic of a remedial action is to maintain the drive's uninterrupted operation without any degradation in the performance characteristics of the drive in the presence of a stator turn fault. Unfortunately, this is very difficult to achieve, and only redundancy-based approaches can solve this difficulty. But these approaches can be justified in specific applications. In transit applications such as traction drives, an uninterrupted operation during a short period of time, even with a limp operation, can prevent injury or death.

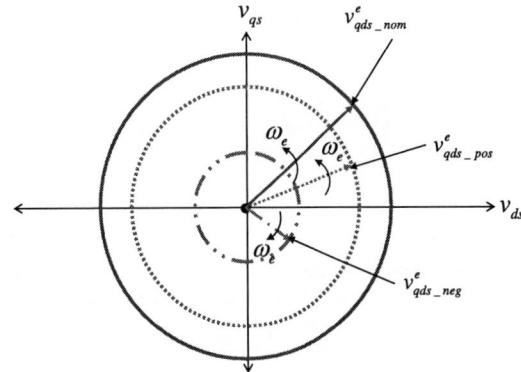

- Blue solid line : Stator voltage vector under fault-free condition
- Red dashed line : Positive sequence voltage with stator turn faults
- Red long-dashed line: Negative sequence voltage with stator turn faults

(a) In the space vector domain

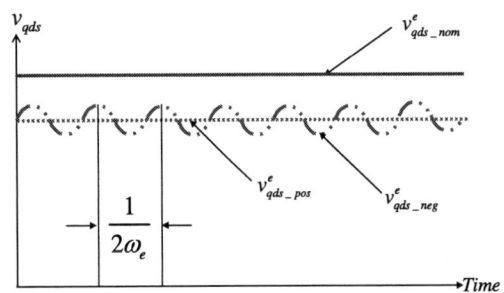

- Blue solid line : Stator voltage under fault-free condition
- Red dashed line : Positive sequence voltage with stator turn faults
- Red long-dashed line: Negative sequence voltage with stator turn faults

(b) In the time domain

Figure 3. Voltage references under fault-free and turn fault conditions.

A simple stator turn fault-tolerant strategy for IPMSM drives that does not require any hardware modification to the standard drive configuration has been proposed [11]. This strategy does not result in the complete loss of availability of the drive. Generally, the asymmetry in the stator voltages resulting from a stator turn fault has only a small effect on the overall stator voltage. Therefore the amplitude of the faulty phase voltage is almost the same as that of the complex stator voltage vector ($\tilde{v}_s^e$) in the synchronous rotating reference frame. Consequently, current in the faulted winding is given by,

$$\left|\tilde{i}_f\right| \approx \frac{\left|\tilde{v}_s^e\right|}{\left|\dfrac{R_f}{\mu} + r_s + j\omega_e \left[L_{ls} \quad \mu\left(L_1 \quad 3\tilde{L}_2\right)\right]\right|} \qquad (1)$$

This implies that an appropriate selection of q- and d-axis current combination for a given operating condition can reduce the stator voltage significantly; consequently, a significant reduction in i_f is achievable while maintaining the given operating condition.

Assuming that a stator turn fault is detected when the motor is working at operating point A in Figure 4, the proposed fault strategy will start to function to minimize the radius of the ellipse of the stator voltage vector while maintaining the given developed torque and rotating speed. If operating point B satisfies this requirement in Figure 4, the operating point of the IPMSM will move from A to B. Figure 5 shows the operating are of the motor under a turn fault condition using the proposed strategy, if the fault current is limited to 3 pu.

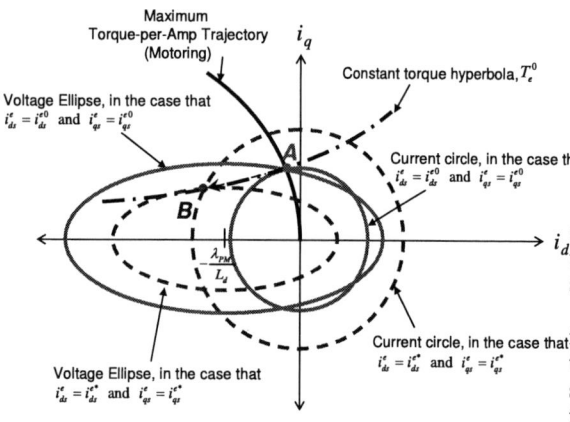

Figure 4. Circle diagram showing how the stator voltage can be reduced in the presence of a turn fault.

V. THERMAL EFFECT OF TURN FAULTS

A simplified, lumped parameter thermal model of a fault-free IPMSM is presented in Figure 6. In the figure, the quantities, θ_S and θ_F are temperature rises [$^{\circ}C$] above ambient temperature (θ_a) at the stator winding and frame, respectively. The current sources, P_{SW} and P_{SC} [W], represent the power losses in the stator winding and core, respectively. The thermal resistance, R_1 [$^{\circ}C/W$], represents the heat dissipation capability of the stator to the

ambient through the combined effects of heat conduction and convection; R_2 [$^{\circ}C/W$] is associated with the heat dissipation capability of the frame to the ambient; the contact thermal resistance, R_3 [$^{\circ}C/W$] is associated with the heat transfer between the stator and frame. The thermal capacitance, C_1 [$J/^{\circ}C$], is the thermal capacity of the stator, while C_2 [$J/^{\circ}C$] is the combined thermal capacity of the frame.

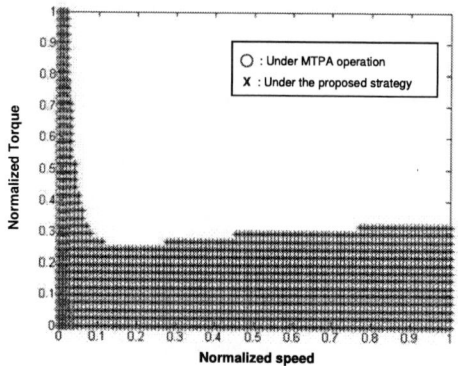

Figure 5. Comparison of the allowable operating areas where is limited to three times the rated coil under MTPA operation and the proposed strategy.

The parameters of the model in Figure 6 are identified strictly from thermal measurements made on the test machine under various operating conditions. The test machine is shown in Figure 8.

A stator turn fault divides the stator windings into the following three parts: (1) the shorted turns, (2) turns adjacent the shorted turns, and (3) turns relatively far away from the shorted turns. For convenience, the turns adjacent and turns far away from the shorted turns are referred to the adjacent turns and healthy turns, respectively. With this in mind, the thermal model of a fault-free IPMSM can be modified into that presented in Figure 7 under a turn fault condition. In the figure, the quantities, θ_T, θ_A, and θ_H are temperature rises [$^{\circ}C$] above ambient temperature (θ_a) at the shorted turns, adjacent turns, and healthy turns, respectively. The current sources, P_{TW}, P_{AW}, and P_{HW} [W], represent the copper losses at the three parts of the stator winding, respectively. The current source, P_{SC}, represents the total stator core loss. However, it should be noted that no change in the stator core losses is caused by the turn fault. The thermal resistances, R_T, R_A, and R_H [$^{\circ}C/W$], represent the heat dissipation capabilities of the three parts of the stator winding to the ambient, respectively. The thermal capacitances, C_T, C_A, and C_H [$J/°C$], represent the thermal capacities of the three parts, respectively. The contact thermal resistances, R_{TA} and R_{AH} [$°C/W$], are associated with the heat transfer between the shorted and adjacent turns and between the adjacent turns and healthy turns, respectively.

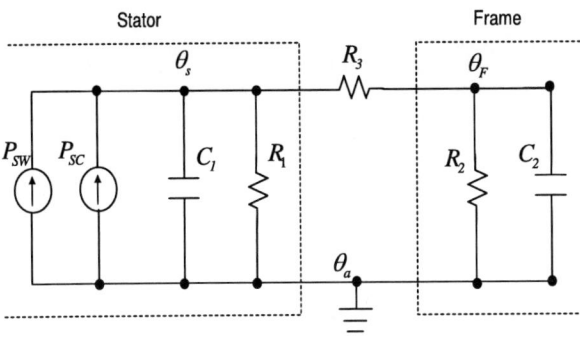

Figure 6. Schematic of the simplified thermal model of a fault-free IPMSM.

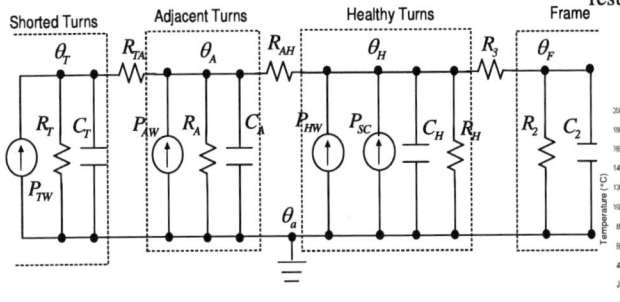

Figure 7. Schematic of the simplified thermal model of an IPMSM with a turn fault.

V. CONCLUSIONS

This paper has presented an overview and summary of work done at Georgia Tech in the field of PM synchronous machine winding fault protection and fault tolerance. condition monitoring. A stator turn fault tolerant strategy for PM drives in safety critical applications was demonstrated to greatly increase the operating range of drives even in the presence of serious turn faults. It was shown that this method can reduce the adverse effects of stator turn faults and maintain the drive operation in the presence of the fault without modification of the hardware found in standard drive configuration. However, the most valuable contribution of the proposed strategy is that the strategy can not only increase

the availability of PM drives, but also can prevent a possible serious accident due to the abrupt shutdown of drive's operation.

Figure 8. Interior PM machine with concentrated stator windings used as a test motor.

Using the thermal model to estimate the motor temperature allows the motor to continue to operate without a thermal failure, even in the presence of a turn fault. The dramatic improvements in operating area are clear from the thermal results shown in Figure 9 with this method.

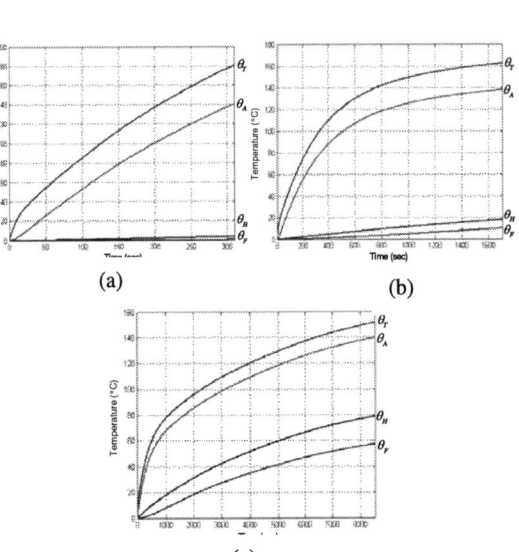

Figure 9. Temperature estimation results showing the durations before θ_A reaches to $140°C$ at three different combinations of i_{ds}^e and i_{qs}^e for 1000rpm rotating speed and 10Nm load operation. (a) Under MTPA operation, (b) under

25% field-weakening, and (c) under 50% field-weakening operation.

VI. REFERENCES

1. T. A. Lipo, *Introduction of AC machine design, Wisconsin Power Electronics Research Center*, 2nd edition, 2004.

2. S. F. Farag, R. G. Bartheld, and W. E. May, "Electronically enhanced low voltage motor protection and control," IEEE Trans. Industry Applications, vol. 29, no. 1, pp. 45-51, Jan./Feb., 1994.

3. J. T. Boys and M. J. Miles, "Empirical thermal model for inverter-driven cage induction machines," IEE Proc., Electr. Power Appl., vol. 141, pp. 360-372, 1994.

4. K. D. Hurst and T. G. Habetler, "A thermal monitoring and parameter tuning scheme for induction machines," *in Conf. Rec. IEEE IAS'97*, pp. 136-142, 1997.

5. G.B. Kliman, W.J. Premerlani, R.A. Koegl, D. Hoeweler, "A New Approach to On-Line Turn Fault Detection in ac Motors," *Conference Record of the IEEE-IAS Annual Meeting,* 1996, pp. 687-693.

6. J.L. Kohler, J. Sottile, F.C. Trutt, "Alternatives for Assessing the Electrical Integrity of Induction Motors," *IEEE Transactions on Industry Applications*, vol. 28, no. 5, Sep/Oct 1992, pp. 1109-1117.

7. B. K. Gupta and W.T. Fink, "A proposed type test for inter-turn insulation in multi-turn coils," *in Conf. Rec. of IEEE International Symposium on Electrical Insulation*, pp. 235-238, 1996.

8. Lee, S.-B. and Habetler, T.G., "An on-line stator winding resistance estimation technique for temperature monitoring of line-connected induction machines," IEEE Transactions on Industry Applications, vol. 39, no. 3 , May/June 2003, Page(s): 685-694.

9. Lee, S.-B., Habetler, T.G.; Harley, R.G.; and Gritter, D.J., "An evaluation of model-based stator resistance estimation for induction motor stator winding temperature," *IEEE Transactions on Energy Conversion*, vol. 17, no. 1, pp. 7-15, March 2002.

10. Tallam, R.M.; Habetler, T.G.; Gritter, D.J.; Burton, B.H.; Harley, R.G., "Neural network based on-line stator winding turn fault detection for induction motors," *Conference Record of the 2000 IEEE Industry Applications Conference*, vol. 1, pp. 375-380, October 2000.

11. Lee, Y-K; and Habetler, T. G.; "An On-Line Stator Turn Fault Detection Method for Interior PM Synchronous Motor Drives," Twenty Second Annual IEEE Applied Power Electronics Conference, APEC 2007, Feb. 2007 Page(s):825 – 831.

12. C. Gerada, K. Bradley, and M. Sumner, "Winding turn-to-turn faults in permanent magnet synchronous machine drives," *in Proc., IEEE 2005 IAS Conf.*, pp. 1029-1036, 2005.

13. R. M. Tallam, T. G. Habetler, and R. G. Harley, "Transient model for induction machines with stator winding turn faults," *IEEE Trans. Industry Application,* vol., 38, no., 3, pp. 632-637, May/June, 2002.

14. P. Milanfar and J. H. Lang, "Monitoring the thermal condition of permanent-magnet synchronous motors," *IEEE Trans. Aerospace and Electronic Systems,* vol. 32, no. 4, pp. 1421-1429, October, 1996.

15. J. F. Moreno, F. P. Hidalgo, and M. D. Martinez, "Realization of tests to determine the parameters of the thermal model of an induction machine," *IEE Proc. Electr. Power Appl.*, vol. 148, no. 5, pp.393-397, September, 2001.

16. M. Kaufhold, G. Borner, M. Eberhardt, and J. Speck, "Failure mechanism of the interturn insulation of low voltage electric machines fed by pulse-controlled inverters," *IEEE Electrical Insulation Magazine* , vol. 12, no. 5, pp. 9-16, Sept./Oct., 1996.

17. W. L. Roux, R. G. Harley, and T. G. Habetler, "Detecting rotor faults in permanent magnet synchronous machines," *in Conf. Rec. SDEMPED'03*, pp. 198-203, 2003.

Thomas G. Habetler (S'83–M'83–SM'92–F'02) received the B.S. and M.S. degrees in electrical engineering from Marquette University, Milwaukee, WI, in 1981 and 1984, respectively, and the Ph.D. degree from the University of Wisconsin, Madison, in 1989. From 1983 to 1985, he was with the Electro-Motive Division, General Motors, as a Project Engineer. Since 1989, he has been with the School of Electrical and Computer Engineering, Georgia Institute of Technology, Atlanta, where he is currently a Professor of electrical engineering. Dr. Habetler was the President of the IEEE Power Electronics Society and the Chair of the Industrial Power Converter Committee of the IEEE Industry Applications Society. He currently serves on the IEEE Board of Directors as Division II Delegate.

The Essence of Three-Phase AC/AC Converter Systems

J. W. Kolar, T. Friedli, F. Krismer, S. D. Round
Power Electronic Systems Laboratory, ETH Zurich, Zurich, Switzerland
email: kolar@lem.ee.ethz.ch

Abstract— In this paper the well-known voltage and current DC-link converter systems, used to implement an AC/AC converter, are initially presented. Using this knowledge and their space vector modulation methods we show their connection to the family of Indirect Matrix Converters and then finally the connection to direct Matrix Converters. A brief discussion of extended Matrix Converter circuits is given and a new unidirectional three-level Matrix Converter topology is proposed. This clearly shows the topological connections of the converter circuits that directly lead to an adaptability of the modulation methods. These allow the reader who is familiar with space vector modulation of voltage and current DC-link converters to simply incorporate and identify new modulation methods. A comparison of the converter concepts, with respect to their fundamental, topology-related characteristics, complexity, control and efficiency, then follows. Furthermore, by taking the example of a converter that covers a typical operation region in the torque-speed plane (incl. holding torque at standstill), the necessary silicon area of the power semiconductors is calculated for a maximum junction temperature. This paper concludes with proposals for subjects of further research in the area of Matrix Converters.

Keywords— AC/AC converter, matrix converter, voltage source converter, current source converter.

I. INTRODUCTION

For conversion from a three-phase mains source to a three-phase voltage load with an arbitrary frequency and amplitude, e.g. variable speed drives, converter systems with either a voltage or current DC-link are mainly used today (Fig. 1). In the case of the voltage DC-link, the mains coupling can, in the simplest case, be implemented by a diode bridge. To accomplish braking operation of the load, a pulse-controlled braking resistor must be placed in the DC-link or an antiparallel thyristor bridge be provided on the mains side. The disadvantages are the relatively high mains distortion and high reactive power requirements, especially during inverter operation.

A mains-friendly AC/AC converter with bidirectional power flow can be implemented by coupling a PWM rectifier and a PWM inverter to the DC-link. The DC-link quantity is then impressed by an energy storage element that is common to both stages, i.e. a capacitor C for the voltage DC-link (U-BBC) or an inductor L for the current DC-link (I-BBC). The PWM rectifier is controlled such that a sinusoidal mains current is drawn that is in phase or anti-phase (energy feedback) with the corresponding mains phase voltage. For the realization of the AC/AC converter, 12 power transistors and 12 diodes are required, or in the future 12 reverse conducting IGBTs (RC-IGBTs) or 12 reverse blocking IGBTs (RB-IGBTs), since for the I-BBC the feedback of energy can be achieved only by the reversal of the DC-link voltage polarity because of the unidirectional current flow set by the power semiconductors.

Due to the DC-link storage element, there is the advantage that both converter stages are to a large extent decoupled for control purposes. Furthermore, a constant, mains independent input quantity exists for the PWM inverter stage, which results in high utilization of the converter's power capability. On the other hand, the DC-link energy storage element has a relatively large physical volume, and when electrolytic capacitors are used, in the case of a voltage DC-link, there is potentially a reduced system lifetime.

With the goal of higher power density and reliability, it is hence obvious to consider the so-called Matrix Converter concepts that achieve three-phase AC/AC conversion without any intermediate energy storage element. The physical basis of these systems is the constant instantaneous power produced by a symmetrical three-phase current-voltage system, which allows, for example, to directly provide the constant power consumption of an electrical machine generating constant torque and running at constant speed (Fig 2). Conventional direct Matrix Converters (CMC, Fig. 3a) carry out voltage and current conversion in one stage. Alternatively, there is the option of indirect conversion by means of an Indirect Matrix Converter (IMC, Fig. 3b). As with the U-BBC and I-BBC, separate stages are again provided for voltage and current conversion, but the DC-link has no intermediate storage element. Realization of

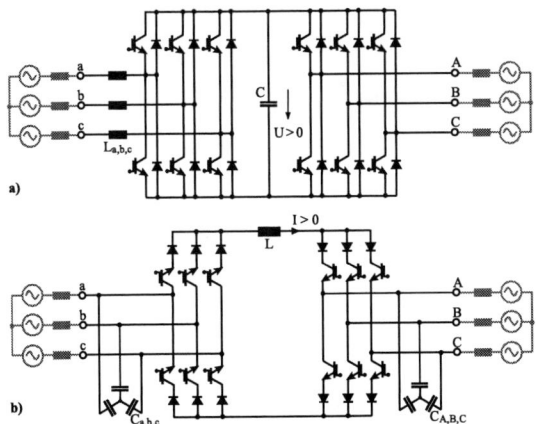

a)

b)

Figure 1: Three-phase AC/AC converter with a) voltage DC-link (U-BBC) and b) current DC-link (I-BBC).

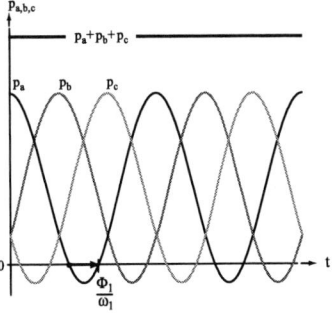

Figure 2: Instantaneous power $p_{a,b,c}(t)$ of the phases a, b, c of a symmetrical three-phase current-voltage system; also shown is the total power of the phases $p(t) = p_a(t) + p_b(t) + p_c(t) = P$.

978-1-4244-1741-4/08/$25.00 ©2008 IEEE

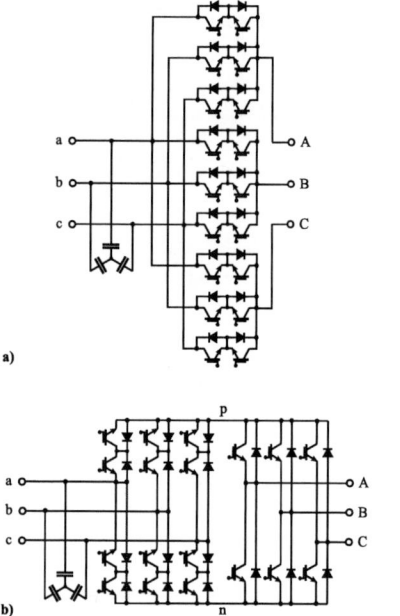

a)

b)

Figure 3: Basic Matrix Converters: a) Conventional (Direct) Matrix Converter (CMC), b) conventional Indirect Matrix Converter (IMC).

both converter systems requires in the basic configuration 18 IGBTs and 18 diodes, or 18 RB-IGBTs (CMC) or 12 RB-IGBTs and 6 RC-IGBTs (IMC): thus the storage element in the DC-link is eliminated at the cost of more semiconductors.

Matrix Converters are frequently seen as a future concept for variable speed drives technology, but despite intensive research over the decades they have until now only achieved low industrial penetration. The reason for this could be, apart from technical aspects, the more complex modulation and dimensioning calculations compared with DC-link converters and the high topological variations, especially with the introduction of the group of the Sparse Matrix Converters (SMC) and the Hybrid Matrix Converters (HMC), which is a mixture between matrix

and DC-link converters. A classification of DC-link and Matrix Converter concepts presented up to now in the literature is shown (limited to forced commutated systems) in Fig. 4.

In this paper, starting from the well-known converter systems with voltage and current DC-links and their space vector modulation methods (Section II), a connection is made to the Indirect Matrix Converters (Section III) and finally to direct Matrix Converters (Section IV). A brief discussion of extended Matrix Converter circuits is given and a new unidirectional three-level Matrix Converter topology is proposed (Section V). This clearly shows the topological connections of the converter circuits that directly lead to an adaption of the modulation methods and allows the readers, who are familiar with space vector modulation of the voltage and current DC-link converters, to easily incorporate and identify the new modulation methods. Then a comparison of the converter concepts with respect to their fundamental, topology-related characteristics, complexity, control requirements and efficiency (Section VI) follows. Furthermore, taking the example of a converter that covers a typical operating region in the torque-speed plane (incl. holding torque at standstill), the necessary silicon area of the power semiconductor is calculated for maintaining a maximum junction temperature. This paper concludes with suggestions of subjects for further research in the field of Matrix Converters (Section VII).

II. AC/AC CONVERTERS WITH DC-LINK

In the following, the fundamentals of space vector modulation of PWM DC-link converters are briefly discussed and their functional equivalent circuit diagrams shown. For the U-BBC, the PWM inverter stage and for the I-BBC the PWM rectifier stage are considered. This allows the development of the IMC topology by simply combining the two subsystems, and the control of the IMC by connection and coordination of the subsystems' modulation methods.

A. Voltage DC-link PWM Inverter

The PWM inverter stage of the voltage DC-link AC/AC converter system shown in Fig. 1a is composed of three bridge legs where each exhibits the function of a switch that connects the output to either the positive or the negative DC bus, p and n.

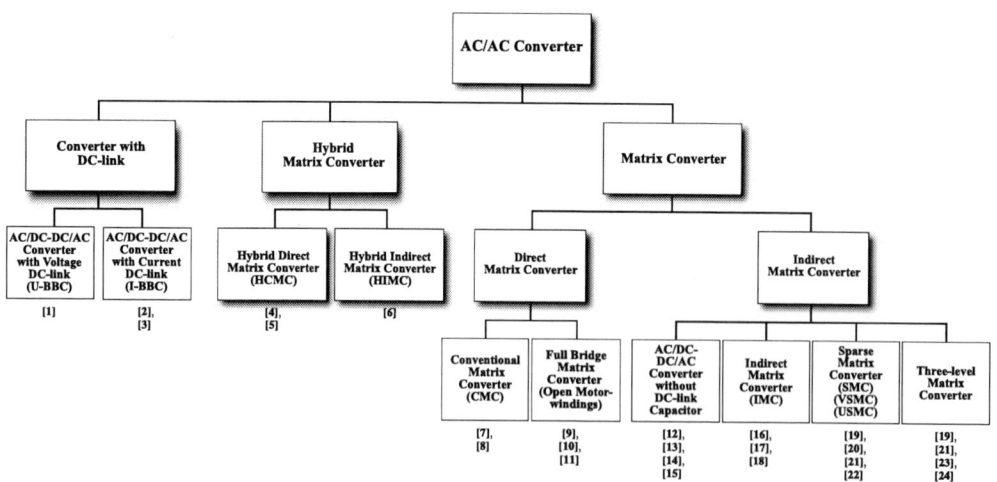

Figure 4: Classification of three-phase AC/AC converter circuits with chronologically ordered references to the technical literature.

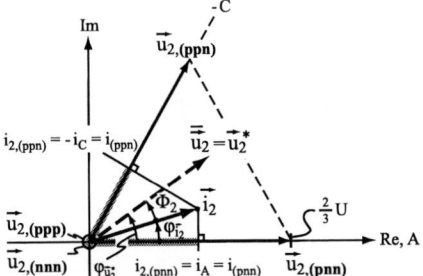

Figure 5: Voltage space vector sector for the U-BBC PWM inverter (output) stage. The DC-link current i (input current of the PWM inverter stage) can be determined by projection of the current space vector $\vec{i}_2$ onto the instantaneous voltage space vector $\vec{u}_{2,j}$.

The switching state of the inverter is defined by (xxx) where x is either p or n, for example (pnn) means that output A is connected to p, and outputs B and C are connected to n. Control of the switch is carried out such that over a pulse period T_P an average voltage space vector $\vec{u}_2 = \vec{u}_2^*$ is formed at the output of the inverter, where $\vec{u}_2^*$ is the output voltage reference value and '$\bar{}$' is the local average value.

To form a steady-state three-phase voltage system

$$\vec{u}_2^* = \hat{U}_2^* e^{j\varphi_{\vec{u}_2^*}} = \hat{U}_2^* e^{j\omega_2^* t} \qquad (1)$$

(with output voltage amplitude $\hat{U}_2^*$ and output frequency ω_2^*) in the simplest case, one must utilize the voltage space vectors closest to $\vec{u}_2^*$, i.e. two active switching states and the free-wheeling state. For the position of $\vec{u}_2^*$, as in Fig. 5, i.e. for $\varphi_{\vec{u}_2^*} \in [0, \pi/3]$, there results the following possible switching state sequences

$$\left. \cdots \right|_{t_\mu=0} (\underline{p}nn) - (\underline{p}pn) - (\underline{p}pp) \left| (p\underline{p}p) - (p\underline{p}n) - (p\underline{n}n) \right|_{t_\mu=T_P/2} \cdots \left|_{t_\mu=T_P} \right. \qquad (2)$$

$$\left. \cdots \right|_{t_\mu=0} (pp\underline{n}) - (pn\underline{n}) - (nn\underline{n}) \left| (nn\underline{n}) - (pn\underline{n}) - (pp\underline{n}) \right|_{t_\mu=T_P/2} \cdots \left|_{t_\mu=T_P} \right. \qquad (3)$$

in order to achieve

$$\begin{aligned}
\vec{\bar{u}}_2 = \frac{1}{T_P} \int_0^{T_P} \vec{u}_{2,j}\, dt_\mu &= d_{(pnn)} \cdot \vec{u}_{2,(pnn)} + d_{(ppn)} \cdot \vec{u}_{2(ppn)} \\
&= d_{(pnn)} \frac{2}{3} U + d_{(ppn)} \frac{2}{3} U e^{j\pi/3} \qquad (4) \\
&= \vec{u}_2^*
\end{aligned}$$

or the associated sequence of the DC-link current shown in Fig. 6. The DC-link current levels may be simply obtained via projection of the output current space vector $\vec{i}_2$ onto the instantaneous active voltage space vector. By restricting operation to a single free-wheeling state (c.f. (2) and (3)), one bridge leg remains clamped to p or n, and hence there are no switching losses in that bridge leg. It is advantageous to change the clamping between the phases in such a way that the phase carrying the highest current is not switched. With an approximately ohmic load (e.g. a permanent magnet synchronous machine), therefore for the angle $\varphi_{\vec{u}_2^*} \in [0, \pi/6]$ Phase A should be permanently connected to p and for $\varphi_{\vec{u}_2^*} \in [\pi/6, \pi/3]$ Phase C should be permanently connected to n.

Independent of the switching state sequence utilized, a fundamental amplitude of the output phase voltage can be formed without overmodulation, where the modulation index M_2 is

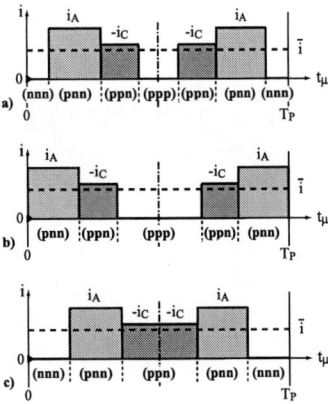

Figure 6: DC-link current i waveform for one pulse period a) switching state sequence (nnn)-(pnn)-(ppn)-(ppp), b) clamping pulse pattern $(\underline{p}nn)$-$(\underline{p}pn)$-$(\underline{p}pp)$, and c) clamping pulse pattern $(nn\underline{n})$-$(pn\underline{n})$-$(pp\underline{n})$.

defined as

$$M_2 = \frac{\hat{U}_2^*}{U/2} = [0, 2/\sqrt{3}]. \qquad (5)$$

The current conversion of the converter, i.e. the transformation of the output phase currents, is determined not only by the switching state, or the modulation index, but also by the phase angle Φ_2 between the output current $\vec{i}_2$ and $\vec{u}_2^*$. As clearly shown by Fig. 7, the average DC-link current, which is formed from the phase current blocks, shifts with increasing Φ_2 from the maximum value to a zero value. The local mean value of the DC-link current is thus given by

$$\bar{i} = I = \frac{3}{4} M_2 \hat{I}_2 \cos\Phi_2. \qquad (6)$$

In summary, the output voltage $\hat{U}_2^*$ (and ω_2^*), is directly formed (i.e. load independent) from the DC-link voltage. In contrast, the backward current conversion is determined not only by the load amplitude $\hat{I}_2$ resulting from $\hat{U}_2^*$ and ω_2^*, but also by the phase angle of the load Φ_2.

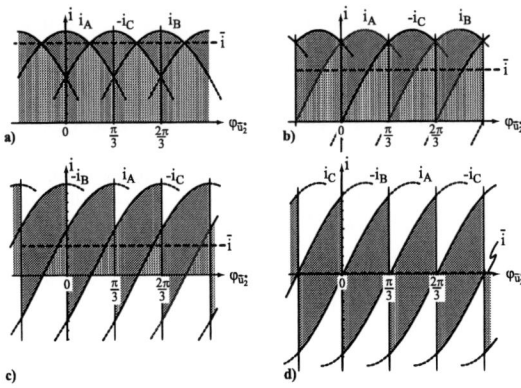

Figure 7: Simulation of the DC-link current waveform for characteristic load current phase angles Φ_2; a) $\Phi_2 = 0$, b) $\Phi_2 = \pi/6$ (ohmic-inductive load), c) $\Phi_2 = \pi/3$ and d) $\Phi_2 = \pi/2$ (purely inductive load). Assumes $M_2 = 2/\sqrt{3}$ and ripple-free phase currents.

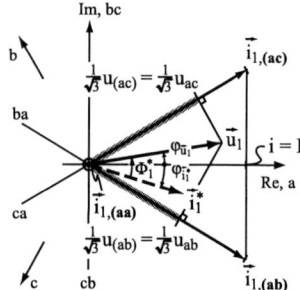

Figure 8: Current space vector sector for the I-BBC PWM rectifier (input) stage. Output voltage u waveform of PWM rectifier can be obtained by projection of the mains voltage space vector $\vec{u}_1$ onto the instantaneous input current space vector $\vec{i}_{1,k}$.

B. Current DC-link PWM Rectifier

The input stage of the current DC-link AC/AC converter has the basic functionality of a diode bridge with regard to the conducting state of the power transistors, which is independent of the filter capacitors' voltage magnitude on the mains side.

Control of the power transistors must be carried out such that a path is always available to the impressed DC-link current. At least one transistor of the positive and one transistor of the negative bridge halves must therefore always be held in the on-state.

If now a sinusoidal, symmetric input current system $\vec{i}_1$ is to be formed after filtering of switching frequency spectral components

$$\vec{i}_1 = \vec{i}_1^* = \hat{I}_1^* e^{j\varphi_{\vec{i}_1^*}} = \hat{I}_1^* e^{j(\omega_1 t - \Phi_1^*)} \qquad (7)$$

with phase current amplitude $\hat{I}_1$, mains frequency ω_1 and mains current phase angle Φ_1^*, then in analogy to the considerations for the U-BBC, the switching states with the current space vectors closest to $\vec{i}_1^*$ must again be utilized. Thus for the sector $\varphi_{\vec{i}_1^*} \in [-\pi/6, +\pi/6]$ shown in Fig. 8, three possible switching state sequences may be employed

$$\dots \left| \begin{matrix} (ab) - (ac) - (aa) \\ t_\mu = 0 \end{matrix} \right| \begin{matrix} (aa) - (ac) - (ab) \\ t_\mu = T_P/2 \end{matrix} \left| \begin{matrix} \dots \\ t_\mu = T_P \end{matrix} \right., \quad (8)$$

$$\dots \left| \begin{matrix} (ac) - (ab) - (aa) \\ t_\mu = 0 \end{matrix} \right| \begin{matrix} (aa) - (ab) - (ac) \\ t_\mu = T_P/2 \end{matrix} \left| \begin{matrix} \dots \\ t_\mu = T_P \end{matrix} \right., \quad (9)$$

$$\dots \left| \begin{matrix} (ac) - (aa) - (ab) \\ t_\mu = 0 \end{matrix} \right| \begin{matrix} (ab) - (aa) - (ac) \\ t_\mu = T_P/2 \end{matrix} \left| \begin{matrix} \dots \\ t_\mu = T_P \end{matrix} \right., \quad (10)$$

which differ regarding the commutation voltages that occur for the individual switching operations. Therefore, for a given Φ_1^* different switching losses result. The details of this can be found in [3], [25]. It is, however, important to note that in contrast to the converter with impressed DC voltage, the free-wheeling state can now also be placed between two active switching states. Starting from one of the two active states, this can be achieved by changing the state of only a single switch. It can be derived from the converter modulation function that a phase current amplitude equal to the magnitude of the DC-link current can be formed without overmodulation, the modulation index M_1 is

$$M_1 = \frac{\hat{I}_1^*}{I} \quad ; \quad M_1 \in [0, 1]. \qquad (11)$$

The output voltage of the input stage for the individual switching states for one pulse period is shown in Fig. 9. As can be seen, the individual voltage levels may be simply obtained

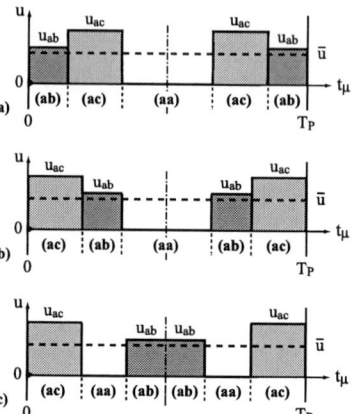

Figure 9: Input stage waveforms of the I-BBC for one pulse period; a) switching state sequence (ab)-(ac)-(aa); b) switching state sequence (ac)-(ab)-(aa); c) switching state sequence (ac)-(aa)-(ab).

by projecting the mains voltage space vector $\vec{u}_1$ onto the instantaneous active current space vector. The waveform of the output voltage is shown in Fig. 10. The current on the input side, i.e. in particular the resulting current amplitude, is determined by the nature of the converter and the output voltage formation is dependent on the (preselectable) mains current phase angle Φ_1^* in order to fulfil the power balance requirement. With increasing Φ_1^*, the mean output voltage

$$\bar{u} = \frac{3}{2} M_1 \hat{U}_1 \cos \Phi_1^* \qquad (12)$$

decreases. Therefore, the instantaneous values of the mains line-to-line voltages become smaller and display positive and negative polarity (Fig. 10c, d), until finally for $\Phi_1^* = \pi/2$ or $\cos \Phi_1^* = 0$, i.e. for impression of a purely reactive current into the mains, $\bar{u} = 0$ results. A further increase of Φ_1^* leads to reversal of the polarity of the output voltage and hence the direction of the power flow. As we shall see later, this is of fundamental importance for the Matrix Converter.

It must be noted that the mains current amplitude $\hat{I}_1^*$ and the mains current phase angle Φ_1^* are predeterminable parameters.

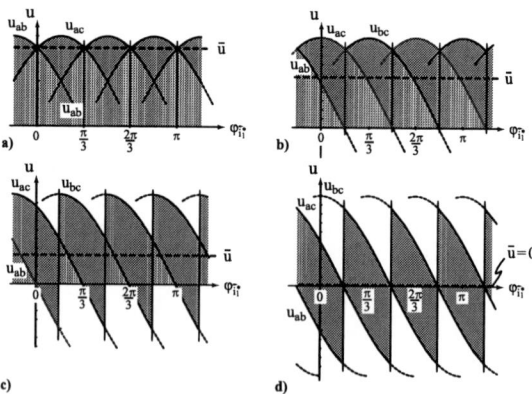

Figure 10: Simulation of the output voltage u waveforms of the I-BBC input stage for characteristic mains current phase angles Φ_1^*; a) $\Phi_1^* = 0$ (ohmic fundamental mains behavior), b) $\Phi_1^* = \pi/6$, c) $\Phi_1^* = \pi/3$ and d) $\Phi_1^* = \pi/2$. Simulation assumes $M_1 = 1$ and ideal input voltages.

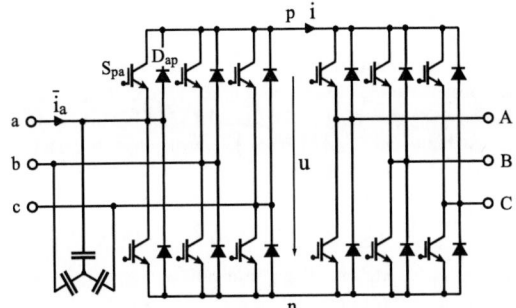

Figure 11: Indirect AC/AC converter with a voltage DC-link without energy storage.

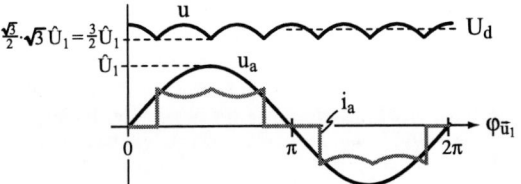

Figure 12: Waveforms of the mains phase voltage u_a, the associated input phase current $\bar{i}_a$ (with filtered switching frequency components) and the DC-link voltage u of the circuit in Fig. 11.

$$u_{\min} = \frac{3}{2}\hat{U}_1 \qquad (13)$$

occurs, so that the amplitude of the output voltage fundamental is limited to

$$\hat{U}_2^* < \frac{2}{\sqrt{3}} \cdot \frac{1}{2} u_{\min} = \frac{\sqrt{3}}{2} \cdot \hat{U}_1 \approx 0.86\,\hat{U}_1 \qquad (14)$$

when operated without overmodulation. We shall find the same limitation of the output voltage later for the IMC and CMC. Furthermore, for a constant power demand of the load, a variation of the local mean value of the DC-link current occurs that is inversely proportional to the variation of the DC-link voltage and appears in the two mains phases of the conducting diodes. The system thus exhibits a relatively high power factor of $\lambda \approx 0.95$, but because of the $\pi/3$-wide gap in the input phase currents, it also has relatively high harmonic current levels.

Hence to obtain a sinusoidal input current, the circuit in Fig. 11 has to be extended according to [26] in such a way that the conductive state of the input stage can be directly defined, i.e. independent of the mains voltage.

B. Indirect Matrix Converter

In order to be able to influence the conductive state of the input stage, it is first necessary to place a power transistor in series with each diode (e.g. S_{ap} and D_{ap} in Fig. 13). However, when this series transistor blocks, a forward voltage can occur that must not appear across the anti-parallel transistor S_{pa}, which is used for the reverse power flow. This is achieved by adding the diode D_{pa} in series with S_{pa}. Overall, the input side now has two mutually anti-parallel current link PWM rectifier stages. Together with the PWM inverter output stage, they form the topology of the Indirect Matrix Converter, IMC. The power transistor and diode combinations between the input phases and the DC-link bus bars form separately controllable four-quadrant switches, which could also be realized by the anti-parallel connection of RB-IGBTs.

The simultaneous turn-on of two four-quadrant switches of the upper or the lower bridge halves would lead to a short circuit

III. INDIRECT MATRIX CONVERTER

By considering the basic functionality and modulation of the AC/DC converter with impressed output current and the DC/AC converter with impressed input voltage, the basis is created for the analysis of Matrix Converter circuits. In the next step we develop the topology of the IMC, starting from the circuit of the U-BBC.

A. AC/AC converter with voltage DC-link without energy storage

In order to derive a Matrix Converter topology from the AC/AC converter with DC-link capacitor, it is obvious to first consider the case where the DC-link capacitor is omitted, or becomes the filter capacitors on the mains side (Fig. 11). Because of the impressed current on the load side, voltage impression must also be assured on the input side. Such a system was suggested in [12] and investigated in more detail in [13]–[15] and is in industrial use.

The input stage of the system in Fig. 11 represents a synchronous three phase rectifier. Its conductive state is directly defined by the mains voltage and cannot be influenced via the control. The load phase current segments in the DC-link, that are generated by the PWM inverter, must be supplied by the input stage through the diodes with the highest instantaneous mains phase-to-phase voltage across them. In order to avoid a mains phase short circuit, only the power transistors that are connected in anti-parallel to a conducting diode may be switched on. The transistors of the input stage thus have no influence on the formation of the DC-link voltage and only permit a reversal of the current flow direction. This is needed since negative components of the DC-link current occur for $\Phi_2 > \pi/6$ (Fig. 7) or for the case of energy feedback into the mains where a negative mean value of the DC-link current exists. The switching frequency of the input stage power transistors is equal to the mains frequency and has the advantage of conducting zero current within a free-wheeling interval of the PWM inverter stage. Thus in contrast to the conventional converter structure with DC-link storage (U-BBC), there are no switching losses of the input stage, hence the system has a higher energy conversion efficiency and still allows bidirectional power flow with the mains. However, the variation of the DC-link voltage with six times mains frequency represents a fundamental disadvantage. According to Fig. 12, a minimum DC-link voltage

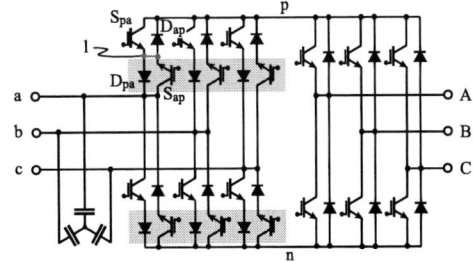

Figure 13: Development of the switching topology of the IMC via extension of the indirect AC/AC converter shown in Fig. 11.

of two mains phases and must therefore be avoided. The basic function of the input stage can thus be abstracted in the same way as for the input section of the I-BBC and can be represented by means of two single-pole, triple-throw switches. With regard to modulation, the IMC represents a combination of the input stage of the I-BBC, with the DC-link inductor shifted to the load side, and the output stage of the U-BBC, with the DC-link capacitor shifted to mains side (Fig. 14a).

The modulation of the system is carried out such that for a given input voltage $\vec{u}_1$ only positive mains line-to-line voltages are switched to the DC-link, and the desired output voltage $\bar{\vec{u}}_2 = \vec{u}_2^*$ is formed along with a sinusoidal mains current $\bar{\vec{i}}_1$ with a defined phase angle Φ_1^* relative to $\vec{u}_1$. On the input side, only the phase angle $\varphi_{\vec{i}_1}^*$ can be predetermined, and not the amplitude of the mains current. The current amplitude adjusts itself via the load current segments, which reach the input via the DC-link, in such a way that the mains provides the real power required by the load.

Starting from, e.g., $\varphi_{\vec{u}_1} \in [0, \pi/6]$ (Fig. 14b), and projecting $\vec{u}_1$ onto the axes ab, ac and bc leads to positive DC-link voltages; correspondingly, the switching states (ab), (ac), (bc) of the input stage or the mains line-to-line voltages u_{ab}, u_{ac}, u_{bc} are permissible within the particular segment of $\varphi_{\vec{u}_1}$. For each of these DC-link voltages, the output stage can form a space vector hexagon. A total of 18 different active voltage vectors and a zero vector are possible, realizable both by the free-wheeling states (nnn) and (ppp) of the output stage and by the free-wheeling states (aa), (bb), (cc) of the input stage (Fig. 15b). The IMC thus exhibits a large variety of output voltage space vectors, similar to a three-level converter.

Considering the current transfer from the load side to the input, we can again limit ourselves to a $\pi/3$-wide interval (of the output period), e.g. $\varphi_{\vec{i}_2} \in [-\pi/6, +\pi/6]$ or $i_A > 0$, $i_B < 0$, $i_C < 0$ (Fig. 14c). For switching states (pnn), (ppn) and (pnp) the respective instantaneous positive DC-link currents $i_{(pnn)} = i_A$, $i_{(ppn)} = -i_C$, $i_{(pnp)} = -i_B$ occur. For the inverse switching states (npp), (nnp) and (npn) DC-link currents of the same absolute value, but of inverse polarity $i_{(npp)} = -i_A$, $i_{(nnp)} = +i_C$ and $i_{(npn)} = +i_B$ result. These DC-link currents are now translated into input current space vectors corresponding to the switching state of the input stage.

As stated above, only the switching states (ab), (ac) and (bc) are permissible in order to provide a positive DC-link voltage.

If, for example, switching state (ac) of the input stage is present, we obtain three input current space vectors pointing in the direction (ac) according to the three DC-link instantaneous values (switching states (pnn), (ppn), (pnp)). Negative DC-link current values result for the inverse switching states and lead to input current space vectors $\vec{i}_{1,(ac)(npp)} = -\vec{i}_{1,(ac)(pnn)}$, $\vec{i}_{1,(ac)(nnp)} = -\vec{i}_{1,(ac)(ppn)}$, $\vec{i}_{1,(ac)(npn)} = -\vec{i}_{1,(ac)(pnp)}$ oriented in the opposite direction to (ac), i.e. in the direction (ca). Input current space vectors of the mutually inverse switching states of the input stage have the same absolute values

$$
\begin{aligned}
|\vec{i}_{1,(ac)(pnn)}| &= |\vec{i}_{1,(ac)(npp)}| = i_A, \\
|\vec{i}_{1,(ac)(ppn)}| &= |\vec{i}_{1,(ac)(nnp)}| = -i_C, \\
|\vec{i}_{1,(ac)(pnp)}| &= |\vec{i}_{1,(ac)(npn)}| = -i_B.
\end{aligned}
\tag{15}
$$

During free-wheeling of the output stage, (nnn) or (ppp), no DC-link current occurs and hence no input current is formed. The same applies for the switching states (aa), (bb) and (cc) of the input stage, which close the DC-link current path without inclusion of the mains.

Through a suitable combination of switching states, a desired target value of the output voltage $\vec{u}_2^*$ and simultaneously desired

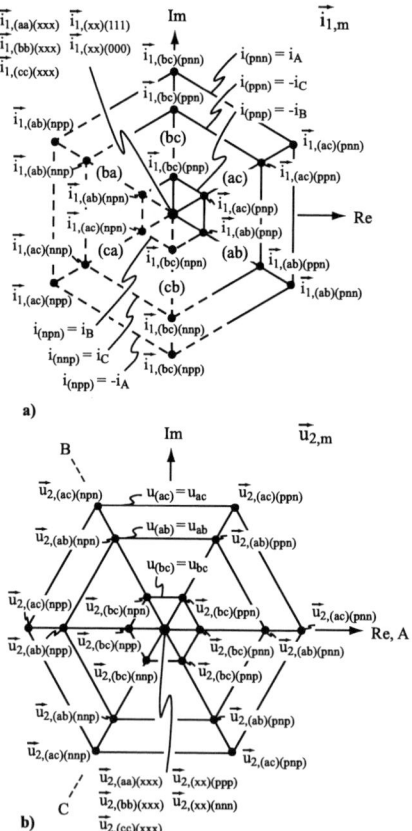

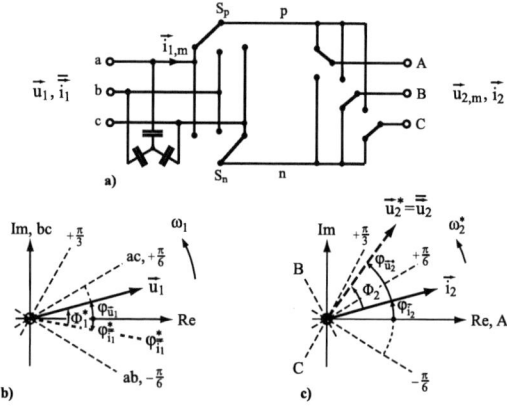

Figure 14: a) Idealized IMC circuit topology, b) mains voltage space vector $\vec{u}_1$ and reference phase angle $\varphi_{\vec{i}_1}^* = \varphi_{\vec{u}_1} - \Phi_1^*$ of input current space vector $\bar{\vec{i}}_1$; c) reference output voltage space vector $\vec{u}_2^*$ and load current space vector $\vec{i}_2$ with angle $\varphi_{\vec{i}_2} = \varphi_{\vec{u}_2^*} - \Phi_2$.

Figure 15: Current conversion a) and voltage conversion b) of the IMC for the input voltage and output current space vector $\vec{u}_1$ and $\vec{i}_2$ according to Fig. 14. To maintain $u > 0$, input current space vectors in the direction of (ca), (ba), (cb) (dotted) can only be formed by inversion of the DC-link current, i.e. inversion of the switching state of the output stage.

phase position $\varphi^*_{\vec{i}_1}$ of the mains current $\vec{i}_1$ must be formed, whereby $\varphi^*_{\vec{i}_1}$ is finally determined by the angular position $\varphi_{\vec{u}_1}$ of the input voltage space vector $\vec{u}_1$ and the desired phase shift Φ^*_1 of $\vec{i}_1$ with reference to $\vec{u}_1$, $\varphi^*_{\vec{i}_1} = \varphi_{\vec{u}_1} - \Phi^*_1$. Further considerations are based upon the condition shown in Fig. 16. For a given $\vec{u}^*_2$, the active switching states (pnn) and (ppn) of the output stage of the IMC should be employed.

The input stage supplies the DC-link voltage, which should be as large as possible in order to obtain the highest possible output voltage magnitude. Thus, operation is limited to the outermost voltage space vector hexagon (Fig. 15b) for the entire pulse period, i.e. the switching state (ac) of the input stage or the DC-link voltage $u_{(ac)} = u_{ac}$. As shown by inspection of the input current space vector diagram (Fig. 16a), this would, however, not allow to achieve the required $\varphi^*_{\vec{i}_1}$ to be set, since only two input phases, a and c, are connected to the DC-link and hence an input current space vector in the direction (ac) is formed over the full pulse period. The absolute value of the vector would be equal to the values $i_{1,(ac)(pnn)} = i_A$, $i_{1,(ac)(ppn)} = -i_C$ and $i_{1,(ac)(nnn)} = i_{1,(ac)(ppp)} = 0$, which is dependent of the switching state of the output stage. Viewed over the mains period, the current space vector would thus retain the direction of a line-to-line axis within $\pi/3$-wide segments, with a block-shaped mains current waveform, as for the circuit in Fig. 11.

In order to always obtain the desired position $\varphi^*_{\vec{i}_1}$, i.e. a continuous rotation of the current space vector $\vec{i}_1$ and hence a sinusoidal input current, it is mandatory to include a second active switching state of the input stage, in this case (ab), in the modulation method. This leads to a second DC-link voltage level $u_{(ab)} = u_{ab}$ and hence to a second output voltage space vector hexagon. For the modulation of the IMC, a switching state sequence must therefore be employed that combines active switching states $(ac)(ppn)$, $(ac)(pnn)$, $(ab)(ppn)$, $(ab)(pnn)$ and the free-wheeling states $(xx)(ppp)$ and $(xx)(nnn)$ or $(aa)(xx)$, $(bb)(xxx)$ and $(cc)(xxx)$.

1) Modulation Method 1: One method consists of leaving the input stage at first in the switching state (ac) while the output side cycles through the switching state sequence $(pnn) - (ppn) - (ppp)$. In this way, voltage space vectors pointing to the corners of the triangular segment of the space vector plane valid for $u = u_{ac}$ are generated, whereby the input current space vector in the direction (ac) is left in place. Next, the input stage changes to (ab) and consequently only input current space vectors oriented in the direction (ab) occur while the output stage repeats the switching state sequence, advantageously in reverse order $(ppp) - (ppn) - (pnn)$. How-

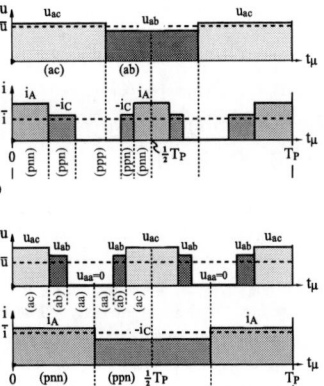

Figure 17: Waveforms of the DC-link voltage u and current i for one pulse period T_P for a) modulation method 1 and b) modulation method 2, assuming the current and voltage conditions in Fig. 14.

ever, the voltage space vectors magnitudes are now equal to $u = u_{ab}$. When the end of the first pulse half-period is reached the sequence is then immediately repeated again in the opposite direction.

$$\cdots \Big|_{t_\mu=0} \; (ac)(pnn) - (ac)(ppn) - (ac)(ppp)$$
$$- (ab)(ppp) - (ab)(ppn) - (ab)(pnn) \Big|_{t_\mu=T_P/2}$$
$$(ab)(pnn) - (ab)(ppn) - (ab)(ppp)$$
$$- (ac)(ppp) - (ac)(ppn) - (ac)(pnn) \Big|_{t_\mu=T_P} \cdots$$
$$(16)$$

The third possible switching state of the input stage, (bc), can be omitted, i.e. the modulation can be limited in this case to those two switching states of the input stage which lead to the higher DC-link voltages and hence to the maximum possible output voltage ($u_{ac} > u_{ab} > u_{bc}$).

For the switching state sequences shown in (16), the switching cycle of the output stage is embedded in the switching cycle of the input stage; thus for a complete cycle of the current space vector triangle, the voltage space vector triangle is run through twice with different DC-link voltages. The switching state sequence basically exhibits the form given in (10) for the current DC-link rectifier. For the presented modulation method, the switching of the input stage occurs during the free-wheeling state of the output stage. It thus has the advantage that the change of the input stage switching state takes place when the DC-link current (Fig. 17a) is absent, so that no switching losses of the input stage occur.

2) Modulation Method 2: As an alternative to modulation method 1, a different switching state of the output stage, e.g. (pnn) can be assumed at the start of a pulse period, where the input stage cycles through the switching state sequence $(ac) - (ab) - (aa)$, i.e. with fixed output voltage space vectors, and a current space vector triangle is run through. After this, the output stage changes from (pnn) to (ppn), whereupon a new position of the discrete output voltage space vector results; the switching state sequence of the input stage is cycled through in reverse sequence, $(aa) - (ab) - (ac)$, and the first pulse half-period ends. During the second half period the entire sequence

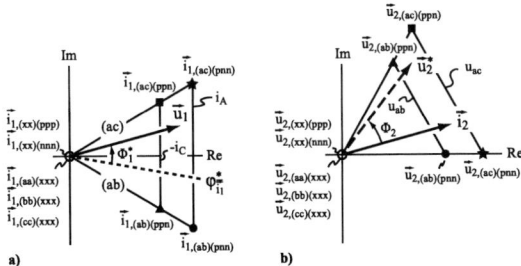

Figure 16: Sector of the space vector diagrams according to Fig. 15 and is relevant for $\varphi^*_{\vec{i}_1} \in [-\pi/6, +\pi/6]$ and $\varphi_{\vec{u}^*_2} \in [0, +\pi/3]$. Switching states for short voltage space vectors are omitted.

is repeated in reverse order

$$
\begin{aligned}
&\dots\big|_{t_\mu=0} \ (ac)(pnn)-(ab)(pnn)-(aa)(pnn)\\
&\quad -(aa)(ppn)-(ab)(ppn)-(ac)(ppn)\big|_{t_\mu=T_P/2}\\
&\quad (ac)(ppn)-(ab)(ppn)-(aa)(ppn)\\
&\quad -(aa)(pnn)-(ab)(pnn)-(ac)(pnn)\big|_{t_\mu=T_P}\ \dots \ .
\end{aligned}
$$
(17)

This switching state sequence is, in principle, known from the AC/AC converter with current DC-link (c.f. (9)), however to date it has not yet been described in the literature for the IMC. The switching sequence of the input stage is embedded in the switching sequence of the output stage. The switching state change of the output stage occurs within the freewheeling interval (aa) of the input stage, i.e. with DC-link voltage absent (Fig. 17b). In this way, switching losses of the output stage are avoided and all of the switching losses occur in the input stage.

Basically, $\vec{u}_2^*$ and $\varphi_{\frac{*}{i_1}}$ may be set by modulation method 1 or 2. However, for modulation method 2, the change of the switching state of the input stage must be carried out at full DC-link current and hence, as later for the CMC (Section IV), a relatively complex multistage commutation sequence must be implemented, which avoids interruption of the DC-link current and a short-circuit of mains phases. In contrast, the conditions are extremely simple for modulation method 1. The four-quadrant switches of the input stage can be switched at zero current and only a small safety time is required to avoid the short circuiting of the mains phases. For a practical realization, therefore, modulation method 1 is clearly preferred to modulation method 2 and is therefore taken as the basis for further considerations.

It is also important to point out here that the voltage transfer ratio of the IMC is independent of the chosen modulation method, as for the AC/AC converter without DC-link capacitor, by

$$
\hat{U}_{2,\max}^* \le \frac{\sqrt{3}}{2}\cdot \hat{U}_1
$$
(18)

(for $\cos \Phi_1^* = 1$). For a general phase angle of the mains current

$$
\hat{U}_{2,\max}^* \le \frac{\sqrt{3}}{2}\cdot \hat{U}_1 \cos \Phi_1^*
$$
(19)

applies.

As shown in Section II, the DC-link voltage formed decreases with increasing Φ_1^*, which results in a corresponding reduction of the achievable maximum output voltage.

C. Sparse Matrix Converter

As shown in Fig. 13, the input stage of the IMC is realized with four-quadrant switches and could in principle also be operated with a negative DC-link voltage. On the other hand, however, for the PWM inverter stage it is mandatory to maintain $u > 0$. It is hence obvious to consider reduction of the number of switches by limiting the operating range of the PWM rectifier stage to a unipolar DC-link voltage that retains the option of bidirectional current flow.

The derivation of a simplified bridge branch structure of the IMC is shown in Fig. 18. If Phase a of the IMC (Fig. 18a) is connected bidirectionally with the positive DC-link bus p, i.e. if S_{pa} and S_{ap} are turned-on and S_{an} and S_{na} turned-off. The positive DC-link voltage will always appears as a blocking voltage over S_{an}. Thus, the function of S_{na} is limited to providing a current path from the negative DC-link bus n (via D_{na}) to the phase input terminal a. For this purpose, no direct connection of the emitter of S_{na} with a is required.

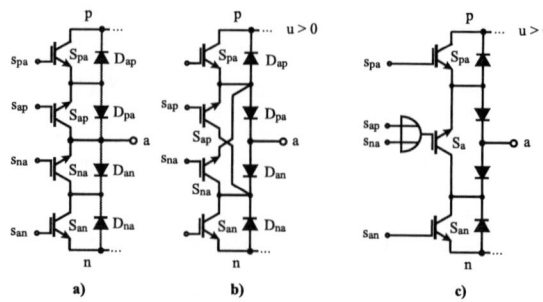

Figure 18: Derivation of the bridge branch topology c) of the SMC, starting from the circuit topology a) of the IMC.

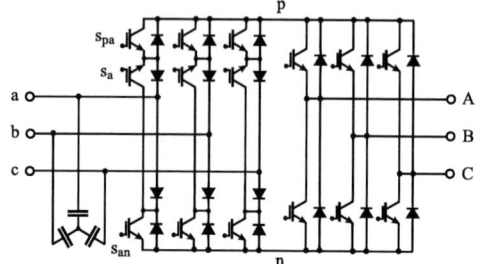

Figure 19: Circuit topology of the Sparse Matrix Converter (SMC, or *Swiss Matrix Converter*).

The return current feedback can be provided via S_{na} and D_{pa}. An analogous consideration for S_{ap} leads to the direct parallel connection of S_{na} and S_{ap} shown in Fig. 18b, which may be replaced by a single switch S_a (Fig. 18c). The effort for the realization of the IMC is thus reduced from 18 IGBTs and (18 diodes) to 15 IGBTs (and 18 diodes). The variant of an IMC shown in Fig. 19 is called a Sparse Matrix Converter (SMC, also known as the Swiss Matrix Converter) [19], [22].

A more comprehensive simplification of the circuit topology is possible by limiting the converter to unidirectional power flow. The transistors S_{pa} and S_{an} (Fig. 18c) only conduct for $i < 0$, i.e. when current flows from the DC-link into the mains, and can thus be omitted when connected to passive (and mostly ohmic) loads. The resulting topology is the Ultra Sparse Matrix Converter (USMC, also known as the Unidirectional Swiss Matrix Converter, Fig. 20a). As is directly obvious with reference to Fig. 7 and Fig. 10, the operation of the converter is limited to

$$
\Phi_1^* = \left[-\frac{\pi}{6}, +\frac{\pi}{6}\right] \quad \text{and} \quad \Phi_2 = \left[-\frac{\pi}{6}, +\frac{\pi}{6}\right],
$$
(20)

i.e. not restricted to a purely ohmic load ($\Phi_2 = 0$). This is explained by the absence of a connection between the mains star point, the DC-link and the load. Apart from the implementation shown in Fig. 20a, the USMC can also be realized with 6 IGBTs in the input stage (Fig. 20b). This circuit variant is obtained by simply omitting the current-carrying power semiconductor of the input stage of the IMC (Fig. 13) for $i < 0$, and exhibits lower conduction losses but a greater realization effort. Finally we must mention a fully bidirectional variant of the IMC, the Very Sparse Matrix Converter (VSMC, [12], [19]) shown in in Fig. 21, whose four-quadrant switches cannot be controlled separately according to the current direction. This controllability is not required when using modulation method 1. As described in Section III-B, the switching of the transistors of the input

34

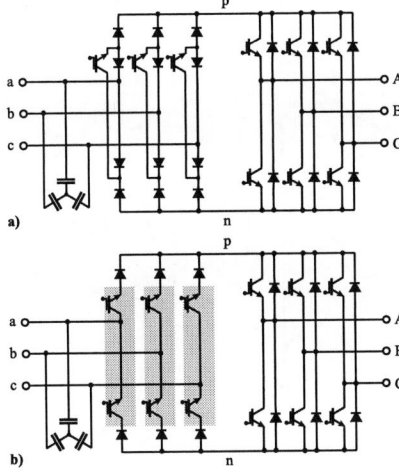

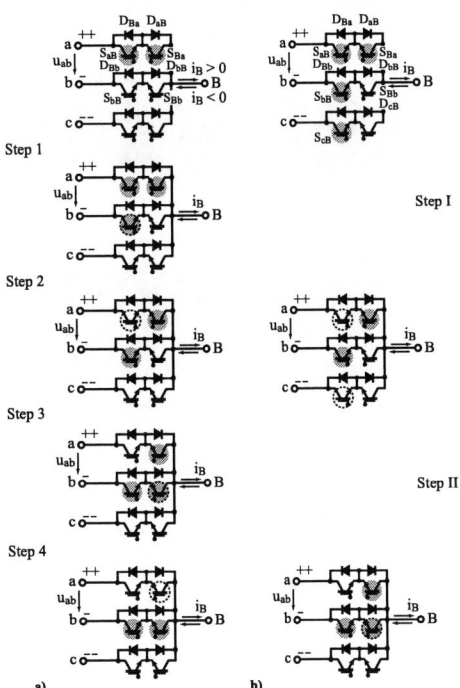

Figure 20: a) Circuit topology of the Unidirectional Ultra-Sparse Matrix Converter (USMC, or Unidirectional Swiss Matrix Converter), b) circuit variant with low conduction losses, which allows the use of PWM inverter half bridge modules (on gray background).

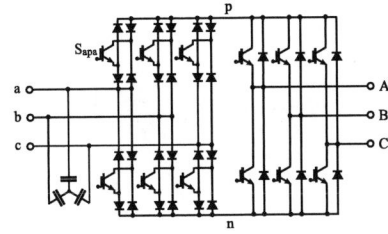

Figure 21: Very-Sparse Matrix Converter (VSMC) circuit topology.

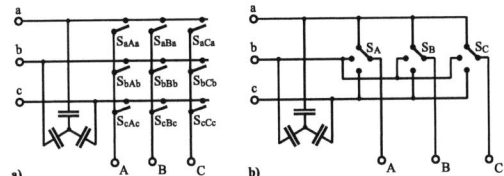

Figure 22: Idealized representation of the CMC as a) a switching matrix and b) in the form of a three-level converter

stage then takes place at zero current within the free-wheeling intervals of the inverter stage and thus only a safety interval is required between switch-off of one four-quadrant switch and the switch-on of the next four-quadrant switch.

IV. DIRECT MATRIX CONVERTER

In contrast to the IMC, for the CMC each of the input phases a, b, c can be directly connected with each output phase A, B, C via a four-quadrant switch and therefore be represented as a matrix (Fig. 22a). The four-quadrant switches, as for the input stage of the IMC, must be realized by the anti-series connection of IGBTs with antiparallel, free-wheeling diodes (an even better solution, with regard to the conduction losses, is by using an anti-series connection of RB-IGBTs).

To take into account an inductive load, it is mandatory to arrange the filter capacitors at the input of the CMC to enable free commutation of the current. The capacitors, in connection with the mains side series inductance, help filter the load current

Figure 23: Switching states of the single-pole, triple-throw switch $B - a$, b, c of the CMC for a) voltage dependent four-step commutation (Steps $1 - 4$) of connection $B - a$ to $B - b$ and b) voltage dependent two-step commutation (Steps I and II); for the connected input phase voltages, $u_a > u_b > u_c$ and $u_a > 0$, u_b, $u_c < 0$ are assumed.

blocks on the input side in order to form a continuous mains current. To avoid a short circuit of the impressed input voltages, simultaneous connection of two input phases to the same output phase must be strictly avoided. On the other hand, because of the impressed load current, one output phase must in any case be connected to one input, whereby it is also permissible to connect all outputs to the same input.

The basic functionality of the Matrix Converter can thus be represented by three single-pole, triple-throw switches whose poles are connected to the outputs (Fig. 22b). The obtained circuit topology clearly shows the strong similarity of the CMC to a three-level voltage DC-link PWM inverter and thus makes it clearer why the Matrix Converter has a greater functionality than the two-level PWM inverter.

The change of the switching state of the CMC must be carried out considering the impressed load current and the impressed input voltage. Suitable multi-step commutation strategies were first described in [27] and [28]. The commutation is thereby carried out depending on the sign of the current in the output phase that is to be connected to another input phase, or, as shown in Fig. 23, with dependence on the sign of the voltage difference of those input phases between which the commutation is carried out. In the case under consideration, u_{ab} must be taken into account.

Fig. 23a depicts the multi-step commutation strategy. There, transistors on a grey background are in the on-state and those surrounded by a dashed line indicate that a change of the switching state has happened in the present commutation step. Prior to transistor turn-off, a transistor of the branch that will take over the current has to be switched on in order to always

35

Table I: Classification of the switching states of the CMC.

Group I	(aaa)	(bbb)	(ccc)	
Group II	(cca)	(ccb)	(aab)	$\left.\right\}\ u_{AB}=0$
	(aac)	(bbc)	(bba)	
	(acc)	(bcc)	(baa)	$\left.\right\}\ u_{BC}=0$
	(caa)	(cbb)	(abb)	
	(cac)	(cbc)	(aba)	$\left.\right\}\ u_{CA}=0$
	(aca)	(bcb)	(bab)	
Group IIIa	(abc)	(cab)	(bca)	
Group IIIb	(acb)	(cba)	(bac)	

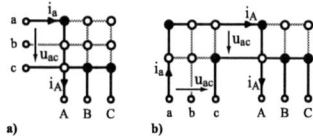

a) b)

Figure 24: a) Conduction state of the CMC for switching state (acc) and b) associated switching state $(ac)(pnn)$ of the IMC.

provide a current path. The transistor to be turned on is selected such that no input phase short circuit occurs for the given commutation voltage, i.e. its series diode blocks the commutation voltage. Consequently, two possible current paths exist for one load current direction, and therefore the remaining transistor of the branch handing over the current can be turned off.

As proposed in [29] and [30], the four-step commutation strategy (Fig. 23a) can also be reduced to two steps (Fig. 23b). For this purpose, as many transistors as possible are always held in the on-state, so that when a commutation is required, fewer steps have to be executed. However, it must be noted that commutation of the CMC in all cases demands that the four-quadrant switches can be separately controlled according to their current direction.

The CMC has a total of $3^3 = 27$ possible switching states, which may be arranged in three groups according to Tab. I. For Group I all output phases are connected to the same input phase, and for Group II two outputs in each case are connected to the same input. The third output is switched to one of the remaining inputs, whereby a total of 18 possibilities exist. Finally, with Group III six further switching states are found where each output phase is connected to a different input phase which generates rotating output voltage or input current space vectors.

The use of rotating space vectors has been widely discussed when the concept of the CMC was first introduced [31] but it has lost considerable importance, therefore we shall in the following limit ourselves exclusively to the switching states of the Groups I and II. This type of output and input connection is also known from the IMC. According to Fig. 3b each connection between the output and the input phases is made via the DC-link rails p and n, whereas only two choices exist for the outputs. The control of the CMC can then be considered simply with reference to a (fictitious) IMC, i.e. obtained by recoding its switching states (Fig. 24). For example, the switching state $(ac)(pnn)$ of the IMC corresponds to the switching state (acc) of the CMC. In both cases the same output and input current space vectors are applied:

$$\vec{u}_{2,(acc)} = \vec{u}_{2,(ac)(pnn)},$$
$$\vec{i}_{1,(acc)} = \vec{i}_{1,(ac)(pnn)}. \tag{21}$$

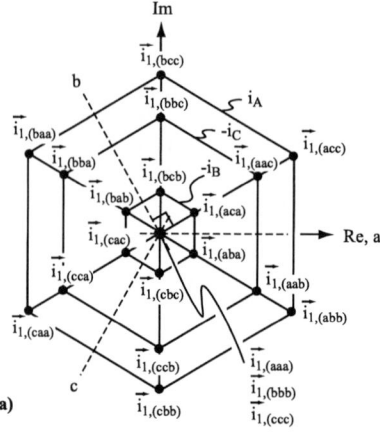

a)

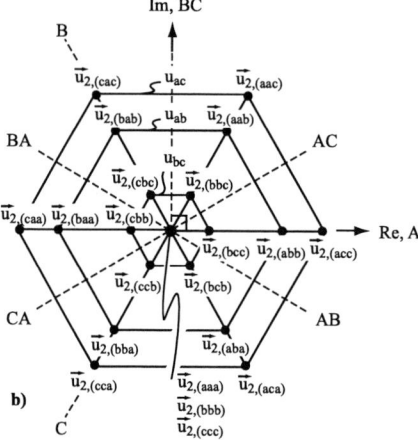

b)

Figure 25: Input current space vector a) and output voltage space vector b) of the CMC for switching states of Groups I and II (Tab. I) assuming the input voltage and output current, $\vec{u}_1$ and $\vec{i}_2$, in Fig. 14.

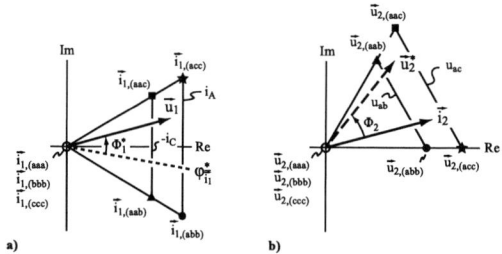

a) b)

Figure 26: Relevant sectors of the space vector diagrams in Fig. 25 for $\varphi^*_{\vec{i}_1} \in [-\pi/6, +\pi/6]$ and $\varphi_{\vec{u}^*_2} \in [0, +\pi/3]$. Switching states leading to short length voltage space vectors are omitted.

For the given input voltage and load current vectors, $\vec{u}_1$ and $\vec{i}_2$, the CMC and IMC show the same space vector diagrams (cf. Fig. 25 and Fig. 15). By considering the change in coding of the switching states or taking into account the equivalence of the space vector diagrams Fig. 16 and Fig. 26, the switching state sequence of the IMC in (16) changes into the switching state sequence

36

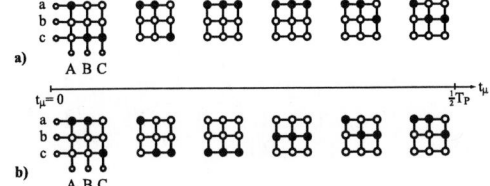

Figure 27: Switching state sequence of the CMC within a half pulse period with a) control in (22) and b) control in (24).

$$
\begin{aligned}
\dots|_{t_\mu=0} \quad &(\underline{a}cc) - (\underline{a}ac) - (\underline{a}aa) \\
&-(\underline{a}aa) - (\underline{a}ab) - (\underline{a}bb)|_{t_\mu=T_P/2} \\
&(\underline{a}bb) - (\underline{a}ab) - (\underline{a}aa) \\
&-(\underline{a}aa) - (\underline{a}ac) - (\underline{a}cc)|_{t_\mu=T_P} \quad \dots
\end{aligned}
\tag{22}
$$

for the CMC. For the IMC in the angular range $\varphi_{\vec{u}_2^*} \in [\pi/6, \pi/3]$ the clamping of Phase C to n brings a greater reduction in the switching losses than the clamping of Phase A to p. This results in a new switching state sequence,

$$
\begin{aligned}
\dots|_{t_\mu=0} \quad &(ac)(ppn) - (ac)(pnn) - (ac)(nnn) \\
&-(ab)(nnn) - (ab)(pnn) - (ab)(ppn)|_{t_\mu=T_P/2} \\
&(ab)(ppn) - (ab)(pnn) - (ab)(nnn) \\
&-(ac)(nnn) - (ac)(pnn) - (ac)(ppn)|_{t_\mu=T_P} \quad \dots
\end{aligned}
\tag{23}
$$

For the associated switching state sequence of the CMC, however, this leads to

$$
\begin{aligned}
\dots|_{t_\mu=0} \quad &(aac) - (acc) - (ccc) \\
&-(bbb) - (abb) - (aab)|_{t_\mu=T_P/2} \\
&(aab) - (abb) - (bbb) \\
&-(ccc) - (acc) - (aac)|_{t_\mu=T_P} \quad \dots,
\end{aligned}
\tag{24}
$$

i.e. in each pulse half-period there is a commutation of all output phases from input c to input b (Fig. 27b). To avoid such commutations with the CMC a clamping of the input phase, which is not switched in the considered angular range, must be implemented. Thus in the present case, the clamping to Phase a or the free-wheeling state (aaa) (free-wheeling state (ppp)) of the IMC must also be retained for $\varphi_{\vec{u}_2^*} \in [\pi/6, \pi/3]$, so the switching state sequence in (22) or Fig. 27a must be used within the entire angular interval $\varphi_{\vec{u}_2^*} \in [0, \pi/3]$. It is an advantage that all commutations always take place between input voltages with large voltage difference, so that high commutation reliability is assured.

It is also advantageous for the IMC to retain the switching state sequence (16) over the entire angular range $\varphi_{\vec{u}_2^*} \in [0, \pi/3]$. Hence, the DC-link bus that is used for clamping the output stage remains permanently connected to an input terminal (via the input stage) during one pulse period. It can be shown that the free-wheeling state used for the output stage is repeated with six times the mains frequency and not with six times the output frequency. In this way, at low output frequencies, a more even loading of the power semiconductors (inherently given by the CMC topology) of the output stage can be attained. This allows for a higher standstill torque being produced for a variable speed drive.

V. EXTENDED MATRIX CONVERTER TOPOLOGIES

Within the last few years especially, numerous extensions of the basic forms of the CMC and IMC have been given in literature that aim to

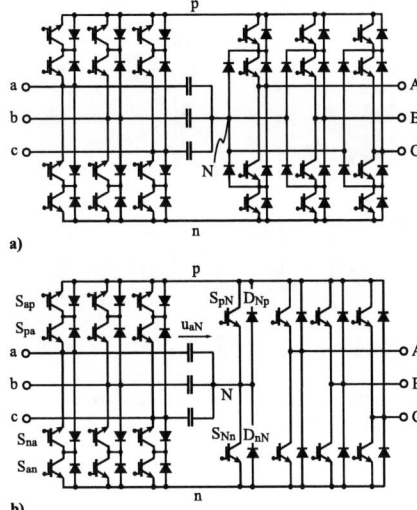

Figure 28: a) IMC with three-level PWM inverter output stage; b) IMC with a further bridge branch that also allows the mains phase voltages to switch directly or inverted into the DC-link and thus attain the functionality of a).

- reduce the load current switching frequency harmonics, or
- operate the IGBTs with low switching losses (e.g. ZCS), or
- realize the converters with conventional power semiconductor modules, or
- increase the voltage control range, or
- reduce or avoid common-mode output voltages.

In the following, only the circuit topologies of the individual systems are briefly discussed; a detailed description of the functionality can be found in the associated literature.

A. Indirect Three-level Matrix Converter

For the IMC, the output stage is a two-level PWM inverter (Fig. 14a). To reduce the switching frequency harmonics of the output voltage or of the load current, it is thus obvious to employ an inverter stage with a three-level characteristic (Fig. 28a). Such a topology with the star point of the mains-side filter capacitors as the voltage center point was proposed in [19] and later designated as SMC3. The space vector modulation for the system is described in [23].

The circuit topology, given in Fig. 28a, enables the line-to-line voltages together with the input phase voltages to form the output voltage. The same functionality can be achieved with Fig. 28b [24], which has the advantage of having a reduced number of switches.

A considerable simplification of the indirect three-level Matrix Converter circuits is possible by restricting the systems to unidirectional power flow (Fig. 29, [21]). The input stage of this new topology exhibits the structure of a Vienna Rectifier [32], [33], i.e. requires only 3 IGBTs, and is thus with regard to complexity of the input stage comparable to a USMC. Hence the abbreviation USMC3 has been agreed for further reference. It should be emphasized that on the input side, the transistors are only switched with twice the mains frequency, i.e. show very low switching losses.

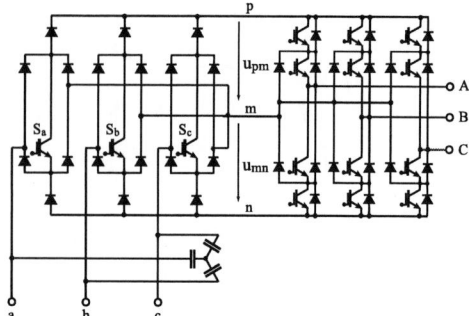

Figure 29: Unidirectional IMC with three-level input stage according to the concept of the Vienna Rectifier and three-level PWM inverter output stage. By switching-on of a transistor in the input stage, the mains phase voltage having the smallest absolute value is connected to the center point of the three-level output stage.

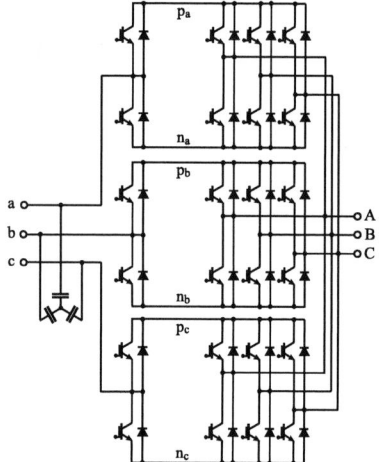

Figure 30: Indirect Matrix Converter with separate sections of a double converter for the individual mains phases.

B. IMC with DC-Links Separated According to Input Phases

In Section III-A, a U-BBC without DC-link storage was examined as an example of a very simple circuit topology of a bidirectional IMC. Because of the lack of controllability of the rectifier stage due to the diodes, zero-current intervals and thus relatively high low-frequency harmonics of the mains current occur.

A sinusoidal shape of the mains current can be achieved with the circuit modification suggested in [34], i.e. by arrangement of an explicit DC-link or a separate input stage for each phase (Fig. 30). However, the system has 24 IGBTs and results in a relatively high component expenditure compared to the IMC with the same functionality. According to [34], a reduction of the switching stress of the IGBTs can be easily achieved. The circuit can be realized with commercially available semiconductor modules.

C. Matrix Converter with Full Bridge Circuit

CMC can be constructed in a half bridge topology (Fig. 3a) or, as with cycloconverter, with a full bridge topology. The latter (Fig. 31) requires open motor windings and 36 instead of 18 IGBTs. With the goal of minimal realization effort, the half bridge topology is often emphasized in technical papers. An

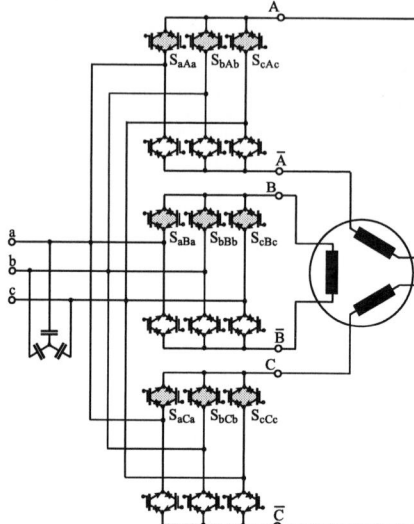

Figure 31: Full bridge CMC topology realized by four-quadrant switches with RB-IGBTs connected in antiparallel. Circuit formed by two CMCs with a half bridge topology, which feed the start A, B, C and end ($\bar{A}$, $\bar{B}$, $\bar{C}$) open windings of an AC machine.

exception is e.g. [9] where in the full bridge topology of the CMC shown in Fig. 31, a control procedure is suggested that achieves a low stress on the insulation of the motor windings, a maximum output voltage of

$$\hat{U}_{2,\max} = \frac{3}{2}\hat{U}_1 \qquad (25)$$

(instead of $\hat{U}_{2,\max} = \sqrt{3}/2\hat{U}_1$), and ohmic fundamental mains behavior. Thus, the topology is of interest especially for high power variable speed drives. However, 6 instead of 3 motor feeds and terminals are necessary for the system.

A reduction of the realization effort of the full bridge topology to 24 power transistors is possible with the variant shown in Fig. 32, i.e. using two PWM inverter output stages I and II in combination with an IMC input stage (Fig. 12 in [9]). When the associated motor terminals are driven in opposition, an output voltage range of

$$\hat{U}_{2,\max} = 2\frac{\sqrt{3}}{2}\hat{U}_1 = \sqrt{3}\hat{U}_1 \qquad (26)$$

is thus available.

D. Hybrid Matrix Converter

The limited voltage control range of the simple Matrix Converters (IMC or CMC) – where an output voltage of approximately only 86% of the mains voltage can be formed without overmodulation – is a significant disadvantage compared to converters with DC-link storage (U-BBC or I-BBC). In the literature, therefore, combinations of CMC and U-BBC, so-called hybrid Matrix Converters have been described that overcome this limitation. However, these converter topologies again have energy storage elements (e.g. electrolytic capacitor) and require a great realization effort.

If the four-quadrant switches in a CMC are replaced by cascaded H-bridge circuits with output capacitors (similar to PWM inverters with cascaded bridge circuits), then the hybrid CMC topology (HCMC, [4]) results as shown in Fig. 33. This enables step-up or step-down converter operation. With an adequate

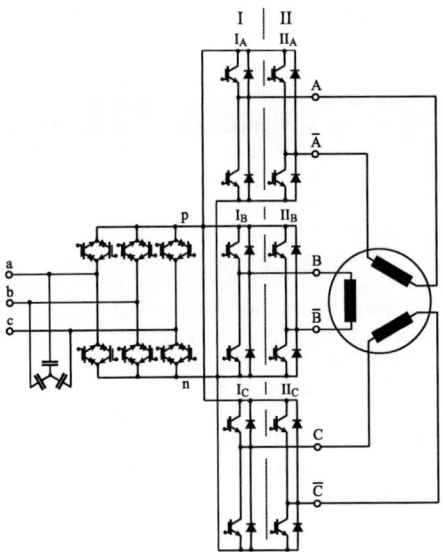

Figure 32: IMC-Topology extended on the output side for the supply of an AC machine with open windings.

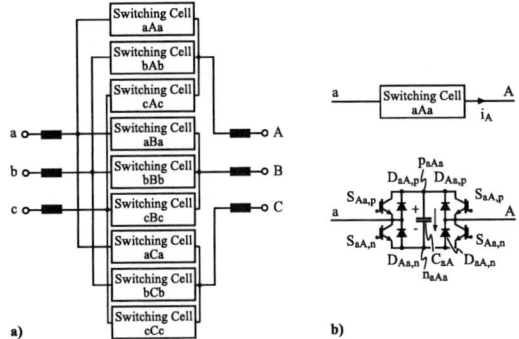

Figure 33: Hybrid CMC (HCMC); also possible is a cascading of several H-bridges in each connection of an input and output.

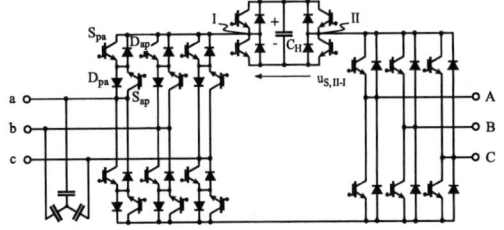

Figure 34: Hybrid IMC with a series voltage source $u_{S,II\text{-}I}$ in the DC-link; this voltage coupling topology is also known from AC/AC converters with an input diode bridge, which simulates a DC-side filter inductor.

modulation scheme, no external supply of the switching cells is required. In contrast to all previously discussed topologies, for the HCMC both the input and the output currents are impressed and can be managed according to [4] via control of always 5 half-bridges.

Transferring the concept of the HCMC to the IMC (HIMC) requires in the simplest case only one H-bridge in the DC-

link [6] (Fig. 34). In the voltage step-up mode, i.e. providing a voltage $u_{S,II\text{-}I}$ with non-zero mean value, a voltage supply of the capacitor C_H is necessary.

In summary, hybrid concepts enable an enlargement of the voltage control range and also offer advantages in the management of mains asymmetries [35]. To obtain these advantages, however, there is a higher complexity of the power stages and of their control.

VI. COMPARISON AND EVALUATION OF DC-LINK AND MATRIX CONVERTERS

Up until now the fundamentals, topologies and control techniques of the IMC and CMC, as well as the modifications and extensions of the circuit topologies have been discussed. We will now conclude by giving a rough qualitative comparison of the converter concepts. All systems have in common

- sinusoidal mains currents (with the exception of the AC/AC converter without energy storage in the DC-link, Section III-A),
- supply of the load with sinusoidal currents, and
- the possibility of bidirectional power transfer (with the exception of the USMC), i.e. in particular feedback of braking energy into the mains.

The converter concepts thus differ not with regard to the basic functionality but in respect of the possible operating range, in particularly the output voltage range, the behavior in characteristic points of the speed-torque plane, of the complexity, of the realization effort, and the physical volume.

To illustrate the differences, we intend in the following to briefly examine the Matrix Converters including CMC and IMC (MC) and standard solution topologies, i.e. provide a comparison of the U-BBC, the CMC, and IMC.

A. Output Voltage Range

The limited output voltage range of the MC represents a clear disadvantage and requires an electrical machine with an adapted nominal voltage. Such a matching of the converter and the machine is possible, especially for niche applications, e.g. for elevator drives. For the U-BBC, on the other hand, the DC-link voltage is freely selectable and thus is highly flexible and a wide control range exists or a broad speed range is covered. Also for the I-BBC, a step-up of the output voltage over the mains voltage can take place via the boost function of the output stage.

Furthermore, mains voltage unbalances directly effect the MC and pulsating load currents are passed through the input filter, with some attenuation, to the mains.

B. DC-link Capacitor and EMC Filter

For the MC, the DC-link capacitor of the U-BBC (an electrolytic capacitor is often used) is omitted but instead filter capacitors (foil capacitor) are required at the mains input. However, a foil capacitor may also be employed in the DC-link of the U-BBC, whereby a similar capacitance value is required to limit the voltage ripple and the voltage change for a step change of the energy flow direction, i.e. on sudden change from motor to generator operation. There is thus no significant difference between the two converter concepts.

For the U-BBC, limiting mains overvoltages or overvoltages caused by the load, e.g. on emergency stop of the drive, may be simply achieved with varistors in the DC-link or pulse-controlled emergency braking resistors integrated in the heat sink. For the CMC, an explicit clamp circuit must be provided.

Both for the U-BBC and for the MC, a multi-stage EMC filter must be provided on the mains side to comply with radio interference regulations. The input inductors of the U-BBC may be regarded as a part of this EMC filter. They cause an increase in apparatus size and hence a reduction in the power density of the converters. However, the input current of the U-BBC is sinusoidal through the use of closed-loop current control. The mains side filter capacitors have values that are significantly lower than those used in the MC. Therefore, the resonance circuit formed between filter capacitors and the input inductors (plus mains impedance) is simpler to damp for the U-BBC.

C. Space Vector Modulation and Commutation

Space vector modulation of the MC is considerable more complex than for the U-BBC. Despite numerous publications by the industry, the often missing knowledge of the modulation details may be one of the reasons that hinder industrial use of the MC today. The space vector modulation of the MC is developed in a stepwise manner, in this paper, from the widely known modulation concepts of the U-BBC and I-BBC in order to provide a better understanding of the MC.

If the modulation of the CMC is chosen such that the commutation always takes place between mains phases with highest voltage difference and two-step commutation is employed, a similar commutation reliability is achieved as for the U-BBC. There are slightly higher switching losses, which can be justified due to the higher system reliability, with this modulation method compared to modulation schemes that are optimized for low switching losses (i.e. schemes that favor commutations between phases of low voltage difference).

For the IMC, the commutation of the input stage takes place during free-wheeling of the output stage. Furthermore, only high mains line-to-line voltages with clearly defined polarity are switched into the DC-link. The commutation strategy of the IMC is thus easier to implement and at least equal to the CMC in respect of reliability.

D. Control

For the MC, the machine currents are impressed by the drive control whereas the mains currents are formed by open-loop control, i.e. not set by a control loop. The system, which is composed of the machine and mains filter, exhibits a relatively high order and has to be controlled by the motor controller. Therefore, especially at low switching frequencies, i.e. close to the mains and output frequencies, the system is difficult to control.

For the U-BBC, the motor currents, DC-link voltage and mains currents are controlled by separate control loops, i.e. the system is broken up into subsystems of lower order. Considering the U-BBC as a virtual MC, regarding mains current and filter capacitor voltage, the input filter can be operated under closed-loop control. Then, generally, the U-BBC offers higher degrees of freedom with its control and therefore the U-BBC offers a more robust control solution.

E. Efficiency and System Volume

The input stage of the IMC, because of the zero-current commutation, exhibits no switching losses. The switching losses of the output stage corresponds, to a first approximation, to those of the output stage of an U-BBC, which works with higher voltage but lower currents. However, the input stage of the U-BBC generates switching losses, so that in total there are higher switching losses and above switching frequencies of approximately 10 kHz a lower efficiency results. With the use of SiC diodes, this limit is shifted to higher frequencies, whereby

for the IMC it is an advantage that SiC diodes are only necessary in the output stage.

The conduction losses of the CMC depend only on the load current amplitude and not on the phase angle of the output current or the voltage modulation index. For the IMC and the U-BBC, the output stage, at low output voltage, will work mainly in freewheeling mode, so that for each phase only one conducting IGBT or diode causes the loss. Furthermore, there exists a low input power and hence a low current through the input stage. Therefore low conduction losses occur both in the input and in the output stages. In respect of conduction losses, there is an advantage for the CMC only at higher speeds.

Apart from power semiconductors or power modules, signal processing and sensors, and the EMC filter, the heat sink occupies a significant fraction of the physical volume of the converter. As shown by a detailed analysis, a $10 - 20\%$ smaller volume of the heat sink and hence a correspondingly higher power density is to be expected on account of the lower losses of the MC as against the U-BBC at medium switching frequencies.

F. Semiconductor Area Based Converter Topology Assessment

The crucial question often raised is which of these converter topologies CMC, IMC, U-BBC, or I-BBC would be the most advantageous for a state-of-the-art motor drive. To select the best converter topology the application's requirements (control dynamics, input and output power quality, overall volume, etc.) and the drive's mission profile must be considered. Therefore a new concept is required to assess and comparing different converter topologies and is introduced in this section.

The core components of three-phase AC/AC converter systems are the power semiconductors, including their packaging. They do not only determine the key performance figures such as efficiency, power density, and allowable switching frequency range, but also contribute to the overall converter costs by approximately 25%. Therefore, any converter comparison should comprise of data for the selected power semiconductors.

The approach is to provide a common basis for comparison by considering the required semiconductor chip area of the individual converter topologies for a given motor drive application profile. For that purpose, in a first step the motor drive requirements defined by its mission profile are mapped to the torque-speed ($M - n$) plane, as depicted in Fig. 35. Then, the motor type (ASM, PMSM) and possibly a gear system need to be selected and suitably matched to the converter output voltage range and the mission profile. Based on that, the required electrical specifications, the semiconductor technology, and the switching frequency are determined. With this information the semiconductor losses can be calculated. Finally, the required semiconductor chip area is determined, according to (27), where the resulting semiconductor junction temperature T_J is equal or less than a selected maximum value $T_{J,max}$ for all power IGBT and diode chips. This automatically leads to minimum semiconductor usage and an optimal partitioning between the IGBT and diode chip area. For this calculation a heat sink temperature T_S, a thermal resistance between junction and sink $R_{th,JS}$, and the overall semiconductor chip losses $P_{L,Chip}$ must be given.

$$T_{J,max} \geq T_J = T_S + R_{th,JS} \cdot P_{L,Chip} \qquad (27)$$

The prerequisite for any semiconductor chip area based comparison is a sufficiently accurate semiconductor model. The individual semiconductor parameters can be derived based on statistical analysis of power module data sheets and should

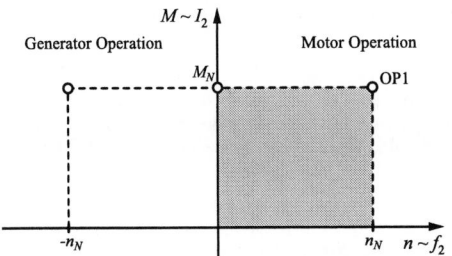

Figure 35: Motor drive requirements represented in the torque-speed plane for an elevator drive example with 2-quadrant operation.

ideally be verified with chip manufacturer data and switching loss measurements of a few selected components.

As an example, the chip area based comparison approach is presented for operating point 1 (OP1 in Fig. 35, nominal motor operation) for the U-BBC, the I-BBC, and the IMC, using latest generation Trench and Field Stop IGBT 4 and diode technology. All of the these converter topologies have in common that they feature a physically dedicated input and output stage but are different regarding their DC-link energy storage elements. The U-BBC, I-BBC, and IMC are designed for a nominal apparent output power $S_{2,N} = 30\,\text{kVA}$, a switching frequency $f_S = 8\,\text{kHz}$, and a heat sink temperature $T_S = 80°\text{C}$ for a RMS phase-to-phase input voltage of 400 V. In this comparison the individual power converters supply a PMSM load machine, optimally matched to the converter output voltage range, such that they can provide a nominal output power of $P_{2,N} = 30\,\text{kW}$ for OP1 with the junction temperature of all IGBT and diode chips limited to $T_J = 150°\text{C}$.

The most economical topology regarding semiconductor usage is the U-BBC with a minimal chip area of roughly $6.5\,\text{cm}^2$, where a DC-link voltage of 750 V has been assumed. The IMC and the I-BBC in contrast typically need a 60% larger chip area in the range of $10.5\,\text{cm}^2$. In terms of efficiency the U-BBC and IMC are similar ($\eta_{OP1} \approx 96\%$), whereas the I-BBC achieves an efficiency of only $\eta_{OP1} \approx 94\%$ due to the comparably high conduction losses.

The new chip area based procedure does not only provide a distinct criterion for assessment and comparison but ultimately enables the determining of the semiconductor costs ($\text{€}/\text{cm}^2$) for a given motor drive application for different converter topologies. Nevertheless, as defined by the drive's requirements, other benefits such as converter compactness may outweigh the higher semiconductor costs resulting from the use of larger chip areas as for MCs.

VII. FUTURE RESEARCH

As shown in Section VI, the CMC and IMC exhibit in contrast to U-BBC a potential for reduction of the losses and the physical volume. An exact statement is only possible with a detailed knowledge of the operating range to be covered and of the motor characteristic of the drive. In each case it makes sense for all converters, MC, U-BBC and I-BBC, to carry out an optimization, taking into account the operating conditions, of the total silicon area distribution used for the semiconductors of the converter topology.

However, MCs are overall in a more difficult competitive position compared to the widespread introduction of a more standard solution, in the form of the U-BBC, and can only offer advantages for special application profiles or ranges, e.g. elevators and escalators. Furthermore, mains feedback or sinusoidal

input current is only necessary for a small fraction of industrial drives.

Fundamentally, there still exists no clear picture in the technical literature, i.e. no comprehensive quantitative comparison of the advantages and disadvantages of MC, U-BBC and I-BBC for specific applications. Future scientific work should therefore pay more attention to the comprehensive comparison of the converter concept, whereby the advantages of new semiconductor technologies, e.g. SiC, should also taken be into account and special attention paid to the obtainable energy conversion efficiency.

REFERENCES

[1] I. Takahashi and Y. Itoh, "Electrolytic Capacitor-Less PWM Inverter," in *Proc. IPEC*, Tokyo, Japan, April 2–6, 1990, pp. 131–138.

[2] K. Kuusela, M. Salo, and H. Tuusa, "A Current Source PWM-Converter Fed Permanent Magnet Synchronous Motor Drive with Adjustable DC-Link Current," in *Proc. NORPIE*, Aalborg, Denmark, June 15–16, 2000, pp. 54–58.

[3] M. H. Bierhoff and F. W. Fuchs, "Pulse Width Modulation for Current Source Converters – A Detailed Concept," in *Proc. 32nd IEEE IECON*, Paris, France, Nov. 7–10, 2006.

[4] R. W. Erickson and O. A. Al-Naseem, "A New Family of Matrix Converters," in *Proc. 27th IEEE IECON*, Denver, CO, Nov. 29–Dec. 2, 2001, vol. 2, pp. 1515–1520.

[5] C. Klumpner and C. I. Pitic, "Hybrid Matrix Converter Topologies: An Exploration of Benefits," in *Proc. 39th IEEE PESC*, Rhodos, Greece, June 15–19, 2008, pp. 2–8.

[6] C. Klumpner, "Hybrid Direct Power Converters with Increased/Higher than Unity Voltage Transfer Ratio and Improved Robustness against Voltage Supply Disturbances," in *Proc. 36th IEEE PESC*, Recife, Brazil, June 12–16, 2005, pp. 2383–2389.

[7] L. Gyugyi, B. R. Pelly, "Static Power Frequency Changers - Theory, Performance, & Application," New York: J. Wiley, 1976.

[8] W. I. Popow, "Der zwangskommutierte Direktumrichter mit sinusförmiger Ausgangsspannung," Elektrie 28, no. 4, pp. 194–196, 1974.

[9] K. K. Mohapatra and N. Mohan, "Open-End Winding Induction Motor Driven with Matrix Converter for Common-Mode Elimination," in *Proc. PEDES*, New Delhi, India, Dec. 12–15, 2006.

[10] M. Braun and K. Hasse, "A Direct Frequency Changer with Control of Input Reactive Power," in *Proc. 3rd IFAC Symp.*, Lausanne, Switzerland, 1983, pp. 187–194.

[11] D. H. Shin, G. H. Cho, and S. B. Park, "Improved PWM Method of Forced Commutated Cycloconverters," in *Proc. IEE*, vol. 136, pt. B, no. 3, pp. 121–126, 1989.

[12] P. D. Ziogas, Y. Kang, and V. R. Stefanovic, "Rectifier-Inverter Frequency Changers with Suppressed DC Link Components," *IEEE Trans. Ind. Appl.*, vol. IA-22, no. 6, pp. 1027–1036, 1986.

[13] S. Kim, S. K. Sul, and T. A. Lipo, "AC/AC Power Conversion Based on Matrix Converter Topology with Unidirectional Switches," *IEEE Trans. Ind. Appl.*, vol. 36, no. 1, pp. 139–145, 2000.

[14] K. Göpfrich, C. Rebbereh, and L. Sack, "Fundamental Frequency Front End Converter (F³E)," in *Proc. PCIM*, Nuremberg, Germany, May 20–22, 2003, pp. 59–64.

[15] B. Piepenbreier and L. Sack, "Regenerative Drive Converter with Line Frequency Switched Rectifier and Without DC Link Components," in *Proc. 35th IEEE PESC*, Aachen, Germany, June 20–25, 2004, pp. 3917–3923.

[16] J. Holtz and U. Boelkens, "Direct Frequency Converter with Sinusoidal Line Currents for Speed-Variable AC Motors," *IEEE Trans. Ind. Electron.*, vol. 36, no. 4, pp. 475–479, 1989.

[17] K. Shinohara, Y. Minari, and T. Irisa, "Analysis and Fundamental Characteristics of Induction Motor Driven by Voltage Source Inverter without DC Link Components (in Japanese)," *IEEJ Trans.*, vol. 109-D, no. 9, pp. 637–644, 1989.

[18] L. Wei and T. A. Lipo, "A Novel Matrix Converter Topology with Simple Commutation," in *Proc. 36th IEEE IAS*, Chicago, IL, Sept. 30–Oct. 4, 2001, vol. 3, pp. 1749–1754.

[19] J. W. Kolar, M. Baumann, F. Stögerer, F. Schafmeister, and H. Ertl, "Novel Three-Phase AC-DC-AC Sparse Matrix Converter, Part I - Derivation, Basic Principle of Operation, Space Vector Modulation, Dimensioning, Part II - Experimental Analysis of the

Very Sparse Matrix Converter," in *Proc. 17th IEEE APEC*, Dallas, TX, March 10–14, 2002, vol. 2, pp. 777–791.

[20] L. Wei, T. A. Lipo, and H. Chan, "Matrix Converter Topologies with Reduced Number of Switches," in *Proc. VPEC*, Blacksburg, VA, April 14–18, 2002, pp. 125–130.

[21] F. Schafmeister, "Sparse und Indirekte Matrix Konverter," PhD thesis no. 17428, ETH Zürich, 2007.

[22] J. W. Kolar, F. Schafmeister, S. D. Round, and H. Ertl, "Novel Three-Phase AC-AC Sparse Matrix Converters," *Trans. Power Electron.*, vol. 22, no. 5, pp. 1649–1661, 2007.

[23] M. Y. Lee, P. Wheeler, and C. Klumpner, "A New Modulation Method for the Three-Level-Output-Stage Matrix Converter," in *Proc. 4th PCC*, Nagoya, Japan, April 2–5, 2007.

[24] C. Klumpner, M. Lee, and P. Wheeler, "A New Three-Level Sparse Indirect Matrix Converter," in *Proc. IEEE IECON*, 2006, pp. 1902–1907.

[25] M. Baumann and J. W. Kolar, "Comparative Evaluation of Modulation Methods for a Three Phase / Switch Buck Power Factor Corrector Concerning the Input Capacitor Voltage Ripple," in *Proc. 32th IEEE PESC*, Vancouver, Canada, June 17–21, 2001, vol. 3, pp. 1327–1333.

[26] J. W. Kolar, H. Ertl, and F. C. Zach, "Power Quality Improvement of Three-Phase AC-DC Power Conversion by Discontinuous-Mode 'Dither'-Rectifier Systems," in *Proc. 6th Int. (2nd European) Power Quality Conf. (PQ)*, Munich, Germany, Oct. 14–15, 1992, pp. 62–78.

[27] J. Oyama, T. Higuchi, E. Yamada, T. Koga, and T. A. Lipo, "New Control Strategy for Matrix Converter," in *Proc. 20th IEEE PESC*, Milwaukee, WI, June 26–29, 1989, vol. 1, pp. 360–367.

[28] Burany, N., "Safe Control of Four-Quadrant Switches," in *Conf. Rec. IEEE IAS*, San Diego, CA, Oct. 1–5, 1989, pp. 1190–1194.

[29] M. Ziegler and W. Hofmann, "A New Two Steps Commutation Policy for Low Cost Matrix Converter," in *Proc. 41st IEEE PCIM*, Nuremberg, Germany, June 6–8, 2000, pp. 445–450.

[30] W. Hofmann and M. Ziegler, "Schaltverhalten und Beanspruchung bidirektionaler Schalter in Matrixumrichtern," ETG/VDE Fachbericht 88 der Fachtagung Bauelemente der Leistungselektronik, Bad Nauheim, Germany, April 23–24, 2002, pp. 173–182.

[31] M. Venturini, "A New Sine Wave In, Sine Wave Out Conversion Technique Eliminates Reactive Elements," in *Proc. Powercon 7*, San Diego, CA, 1980, pp. E3-1–E3-15.

[32] J. W. Kolar and F. C. Zach, "A Novel Three-Phase Utility Interface Minimizing Line Current Harmonics of High-Power Telecommunications Rectifier Modules," *Trans. Ind. Electron.*, vol. 44, no. 4, pp. 456–467, 1997.

[33] J. W. Kolar, U. Drofenik, and F. C. Zach, "VIENNA Rectifier II - A Novel Single-Stage High-Frequency Isolated Three-Phase PWM Rectifier System," *Trans. Ind. Electron.*, vol. 46, no. 4, pp. 674 – 691, 1999.

[34] K. Mino, Y. Okuma, and K. Kuroki, "Direct-Linked-Type Frequency Changer Based on DC-Clamped Bilateral Switching Circuit Topology," *Trans. Ind. Electron.*, vol. 34, no. 6, pp. 1309–1317, 1998.

[35] D. Casadei, G. Serra, G., A. Tani, and P. Nielsen, "Performance of SVM Controlled Matrix Converter with Input and Output Unbalanced Condition." in *Proc. 6th European Conf. on Power Electron. and Appl. (EPE)*, Sevilla, Spain, Sept. 19–21, 1995, vol. 2, pp. 628–633

An Analysis on Turn-off Behaviour of 1.2kV NPT-CIGBT under Clamped Inductive Load Switching

S.T. Kong, L.Ngwendson, M. Sweet and E.M. Sankara Narayanan

Ph/Fax: +44 116 250 6473. The University of Sheffield, UK. e-mail: *S.T.Kong@Sheffield.ac.uk*

Abstract—**For the first time, this paper analyses the turn-off behaviour of the planar 1.2kV/25A NPT-CIGBT under clamped inductive load switching in detail through experiment and simulation. Turn-off behaviour of the CIGBT involves strong interaction between device and circuit parameters. The circuit parameter such as gate resistance was varied, in order to observe the di/dt, dv/dt and turn-off energy loss of the device. Experimental results are shown at 25°C and 125°C. In addition, numerical simulation results are used to enhance understanding of the internal physics of the NPT-CIGBT turn-off process.**

Keywords—**Power Semiconductor Device, Device characteristics.**

I. INTRODUCTION

The Clustered Insulated Gate Bipolar Transistor (CIGBT) is a three terminal device belonging to a new family of MOS controlled power devices with controlled thyristor mode of operation in the on-state. This device uniquely shows current saturation at high gate voltages. In addition it shows enhancements in term of on-state forward voltage drop and switching performance [1-2] over identical IGBT. Furthermore, CIGBT exhibits a wide Safe Operating Area (SOA) and short circuit endurance.

The CIGBT manufacturing technology is simple because it incorporates most of the advancements in IGBT processing. Hence, CIGBT has been experimentally demonstrated at 1.2kV (PT, NPT with planar and trench gates), 1.7kV (NPT), and 3.3kV (NPT) [1-5]. Also Reverse-Blocking CIGBTs (RB-CIGBT) [6] and Reverse-Conducting TCIGBTs (RC-TCIGBT) [7] have been proposed, the latter being useful in AC/AC matrix converter topologies.

As a power semiconductor switch, understanding the turn-off characteristic of CIGBT is crucial to power circuit designers because it enables them to harness the full potential of the device in any given topology. It is shown in this work that the switching performance of the CIGBT under clamped inductive load is influenced by some structural parameters of the device as well as circuit parameters.

II. STRUCTURE AND OPERATION

Fig.1 shows the schematic of the 1.2kV NPT-CIGBT half cell. It can be seen that the IGBT [8-9] – like cathode cells have been created within a floating P-well and N-well, in order to increase the current density per cm^2.

Gate$_1$ and Gate$_2$ are linked together to form a three terminal device. In this structure, Gate$_1$ is used to turn-on the device whereas Gate$_2$ is used for control. All the layers are implanted and diffused. The threshold voltage of Gate$_2$ is higher than that of Gate$_1$, and it determines the Vth of the device. Similarly, the cathode contacts are also connected together and grounded. The P-base, N-well and P-well forms the 'self-clamping' transistor which prevents the potential of N-well and P-well regions increasing excessively when its base (the n-well region) becomes depleted. This action makes possible current saturation at gate voltages and anode voltages by protecting the cathode cells from high anode voltage. In addition the 'self-clamp' action enhances turn-off speed [1-2].

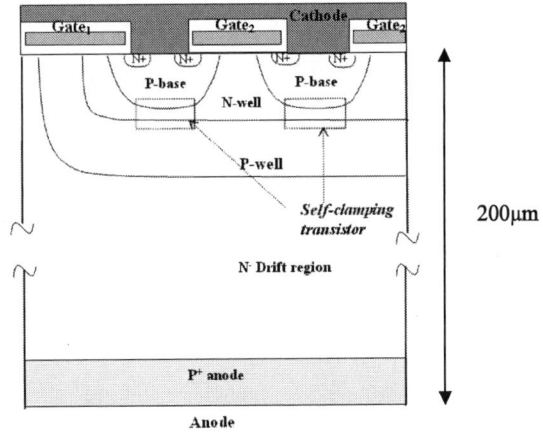

Fig. 1: Cross section of the 1.2kV NPT-CIGBT

The drift region of NPT-CIGBT (~200μm) needs to be wide enough to support the depletion layer when the device is blocking voltage in the off-state. However, the stored charge during forward conduction can generate a long current tail which increases turn-off loss.

III. NPT-CIGBT TURN-OFF BEHAVIOUR

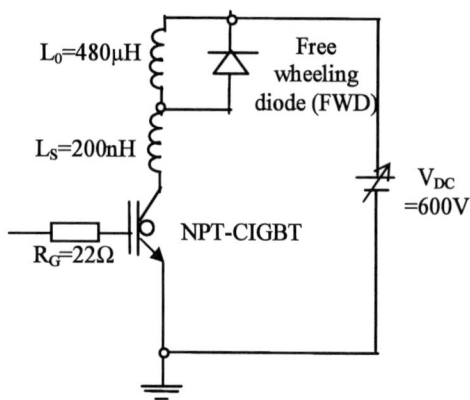

Fig.2: Clamped inductive load switching test in experiment and simulation conditions (L_0= load inductance, Ls=stray inductance, R_G=gate resistance, V_{DC}= DC bus voltage)

In this section, investigation is carried out using Synopsys TCAD Sentaurus device simulator [10], in order to understand the turn-off behaviour of the 1.2kV NPT-CIGBT. The clamped inductive switching circuit directly represents the active component in practical hard-switching applications at the time of commutation as shown in Fig. 2. The load inductor, L_0, starts charging through the device once it is turned on. At the time of turn-off, the device voltage rises while inductor current continues to flow through NPT-CIGBT. When the device voltage exceeds bus voltage, the free wheeling diode becomes forward biased and diverts the inductor current from the device.

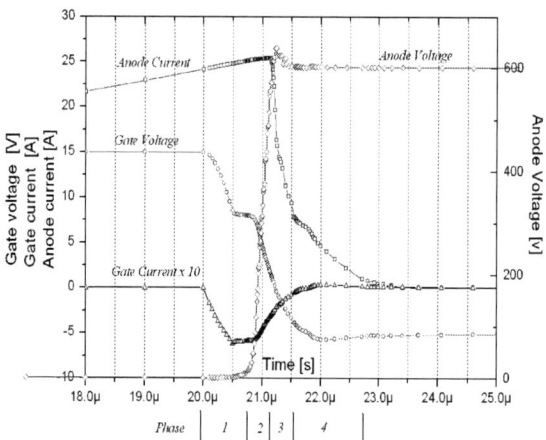

Fig. 3: NPT- CIGBT turn-off waveforms [τ=10µs, R_G=22Ω, V_{DC}=600V, T_j=25°C)

CIGBT turn-off is initiated by shorting the gate to the grounded cathode or a negative bias is applied to shutdown the MOS channels. This results in turning off of electron supply. The turn-off of the device can be divided into 4 phases as shown in Fig.3: Phase 1-Gate

voltage fall, Phase 2-Anode voltage rise, Phase 3-Anode current drop and Phase 4- Tail current decay.

Phase 1-Gate Voltage Fall

The gate voltage begins to decrease from the on-state value of 15V. This causes the gate-cathode capacitance C_{GK} to discharge through R_G, producing a negative gate current I_G. The anode current I_A remains constant, since it is set by the load. The small depletion at P-base/ N-well junction expands as V_{AK} rises until it touches the P-well at the self-clamping voltage as shown in Fig.4. At self-clamping voltage (~10V), the depletion layer from the P-base/N-well junction reaches the P-well region, and clamps the potential in the cathode region. This is known as the "self-clamp" effect .This means the cathode cells and gate oxide are protected from the high anode voltages. The phenomenon also helps in the extraction of holes; resulting in enhanced turn-off speed and low loss in the CIGBT.

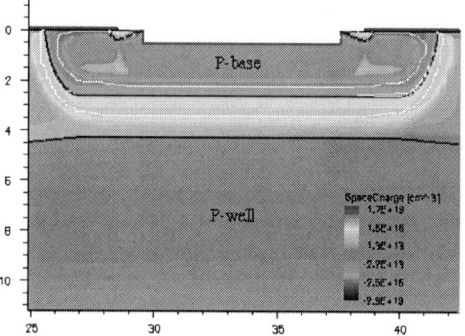

Fig. 4: Tcad Simulation of NPT-CIGBT turn-off at the end of the phase 1 (Time= 20.6µs), showing complete depletion of N-well under P-base regions.

Phase 2: Anode Voltage Rise

After "self-clamp" has occurred additional increase in the anode potential is supported by the P-well/N-drift junction, hence the depletion formed at this junction moves towards the P+ anode as the anode voltage increases (fig.5a). This shrinks the carrier storage region (CSR) as the MOS channel electron current decreases (fig.5b). V_{GK} decreases to around Vth at the end phase 2. Since the charge under the accumulation layer has been removed, C_{GA} (miller capacitance) will starts to reduce at phase 2. Thus, the rate at which V_{AK} increases will accelerate depending on the extraction rate of charge from the CSR.

The depletion layer at P-well/N-well junction will expand as shown in fig. 6b as the anode voltage increases (fig. 6a). The C_{AK} is therefore set by the charge remaining within the CSR. Under high anode voltage and current (t= 21.1µs), the electric field (>1.0E5 V/cm) at drift region/P-well junction is high enough to initiate avalanche (fig.5c and 6b), with the generated electrons flowing into remaining CSR and the holes toward the cathode contact. It should be noted that at this point the anode voltage is still rising towards the DC supply voltage. The ability of avalanche to supply the shrinking CSR results in a change the slope of dE/dx as shown in fig. 6b.

Although the anode current is gradually reducing, the device cannot fully turn-off until the anode voltage reaches the supply voltage and the FWD switches on. Furthermore, fig. 5d represents the Shockley–Read–Hall recombination during phase 2 condition.

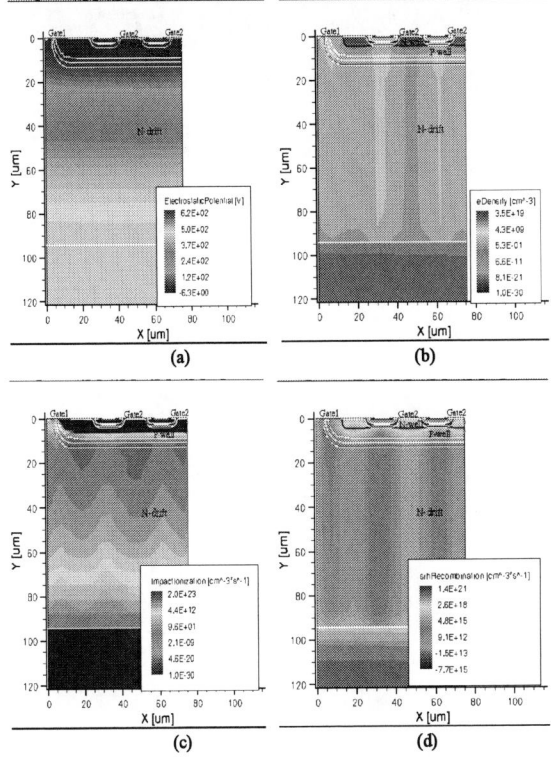

Fig. 5: Tcad Simulated NPT-CIGBT turn-off during phase 2 (Time= 21.1μs), (a) electrostatic potential (b) electron density (c) Impact ionisation rate (d) Shockley–Read–Hall recombination.

Phase 3: Anode Current Drop

Once the anode voltage V_{AK} becomes greater than the supply voltage V_{DC}; the freewheeling diode (FWD) switches on and the space charge expansion is stopped. The freewheeling diode now clamps the device anode voltage to 600V. However the anode voltage will increase beyond 600V by an amount equal to the voltage drop across the stray inductance L_s, or 200nH known as the overshoot voltage (or surge voltage).

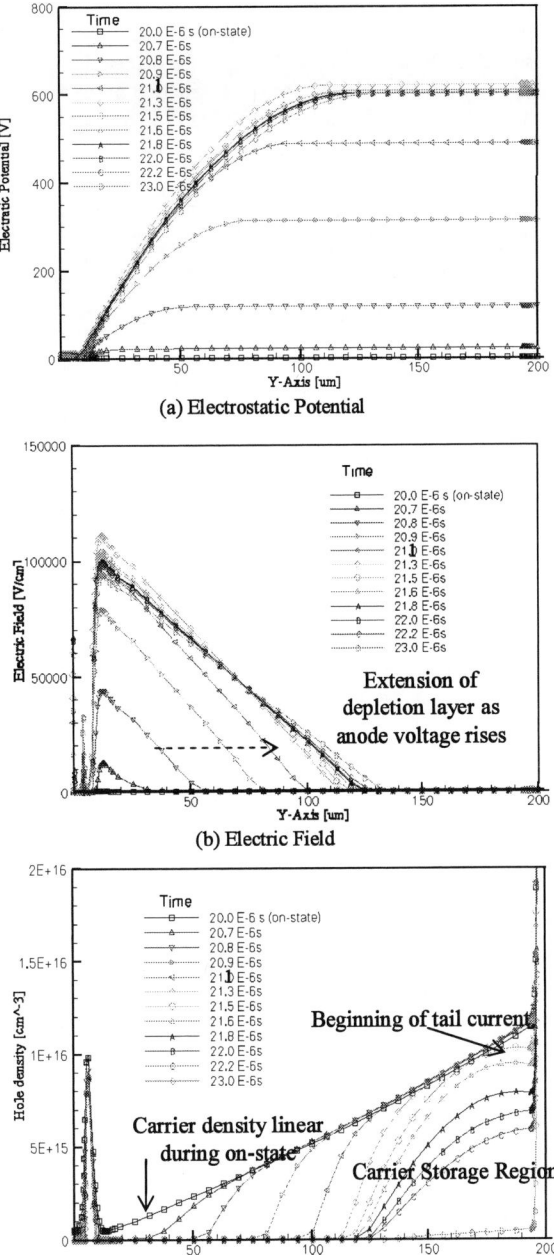

(a) Electrostatic Potential

(b) Electric Field

(c) Hole density

Fig. 6: Synopsys TCAD simulation of NPT-CIGBT showing (a) electrostatic potential, (b) electric field effect and (c) hole density profile during turn-off time.

Before the end of phase 3, the MOS current within the CIGBT falls to zero because V_{GK} has reduced below Vth and inversion layer that maintains the supply of electrons into the N- drift region has been removed. This means the N-well is now disconnected from the grounded cathode terminal. Once the MOS channel is shut we have mostly hole-current flowing through the device and collected at the cathode terminal as shown in fig 6c.

The minority carrier density decays due to three processes **a)** collection of hole through the depleted n-well regions as shown in fig. 7. **b)** recombination process between holes and electrons in the CSR and undepleted N-well region and **c)** electron back injection into the anode.

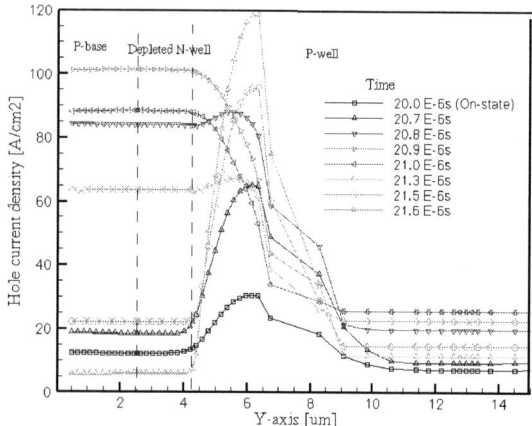

Fig. 7: Magnitude of hole-current density through the depleted n-well regions into cathode electrode at different times.

.Phase 4: Tail Current Decay

In the tail current phase, the charge remaining within the drift region cannot be extracted by the expansion of the depletion layer. This is always the case with NPT devices, unless the n-drift is thin enough (or anode voltage is high enough) to be completely deplete the N-drift region. The tail current (and turn-off energy) resulting from recombination can be reduced with a highly transparent anode at the expense of increased Vce(sat) . The dV_{AK}/dt and dI_A/dt at turn-off and current tail are largely determined by the rate of charge extraction [11], the nature of the anode and lifetime. Turn-off losses are also dependent on the gate driver [12].

The current tail can also be minimised by (i) the FS and PT devices where the depletion layer reaches the buffer. (ii) lifetime control such as electron irradiation.

IV. EXPERIMENTAL RESULTS

A. Gate Resistance

The gate resistance values have a significant impact on the turn-off performance of the NPT-CIGBT. It is known to control dI_A/dt by controlling the gate resistance of the device during turn-off. Fig. 8(a-c) shows the turn-off behaviour of NPT-CIGBT for different gate resistance R_G values such as 4.7, 22, 68Ω. The gate was pulsed from +15V to -2.5V and the DC power supply was set to 600V. It can be seen that a larger gate resistance charges and discharges the CIGBT input capacitance slower and induces an increase of a delay at the turn-off time. This can increases switching times and switching losses during

turn-off. However, a smaller gate resistance improves immunity to dv/dt turn-off.

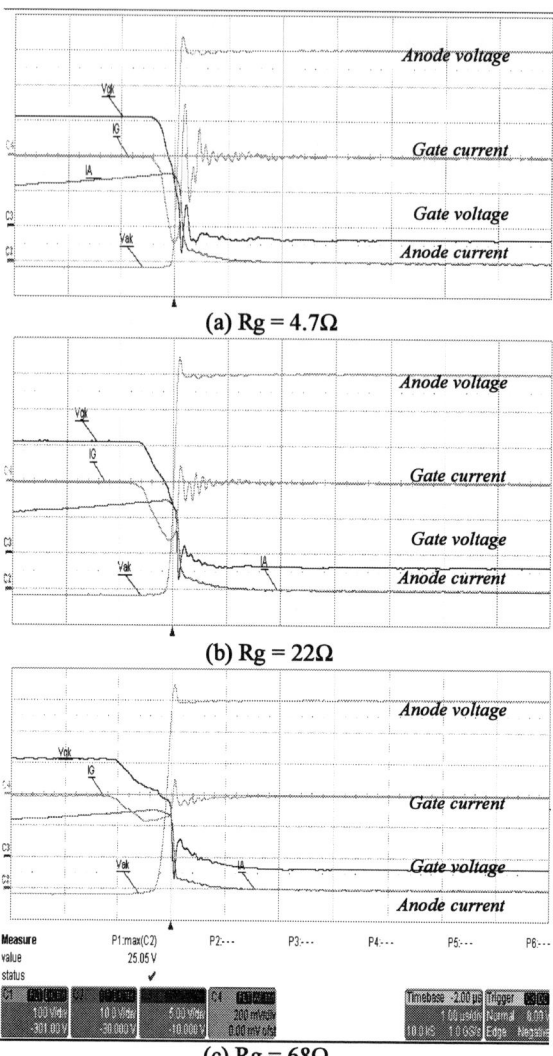

Fig 8: Experimental 1.2kV/25A NPT-CIGBT turn off waveforms under clamped inductive load switching at (a) Rg = 4.7Ω (b) Rg = 22Ω (c) Rg = 68Ω   [V_G= 15V, V_{DC}=600V, I_{off}=25A, Tj=25°C].

Fig.9 shows the turn-off energy loss of the NPT-CIGBT as a function of gate resistance values. It can be seen that the increment of gate resistance, corresponding to an increase in turn-off loss.

Fig. 10(a-b) shows the turn-off waveforms of NPT-CIGBT at 25°C and 125°C. A 600V DC power supply and Rg = 22Ω have been used in the standard DC chopper circuit. The experimental turn-off energy loss E_{off} of NPT-CIGBT increases from 2.42mJ to 3.39mJ from 25°C and 125°C when the junction temperature is increased. This is because of the increase in tail current due to increased lifetime and hole injection at high temperature.

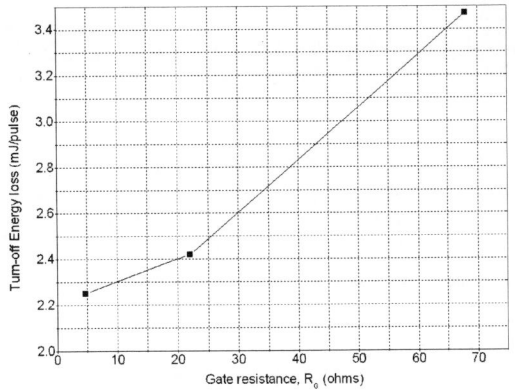

Fig.9: NPT-CIGBT turn-off energy loss as a function of gate resistance at I_{off}=25A

B. *Temperature Effects*

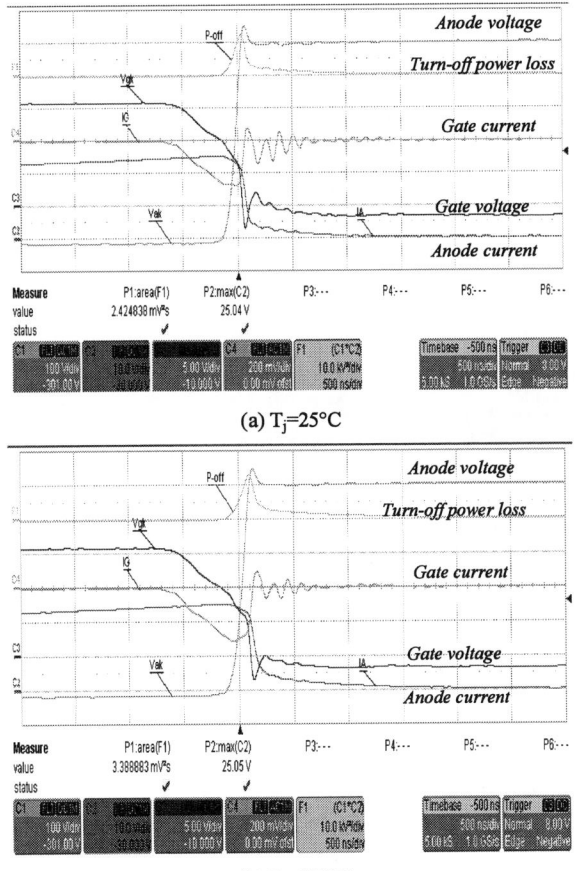

(a) T_j=25°C

(b) T_j=125°C

Fig 10: Typical 1.2kV/25A NPT-CIGBT turn-off waveform under clamped inductive load switching at (a) T_j=25°C and (b) T_j=125°C [R_G=22Ω, V_G= 15V, V_{DC}=600V, I_{off}=25A].

V. CONCLUSION

The turn-off switching behaviour of 1.2kV/25A NPT-CIGBT's has been studied numerically and experimental results presented under clamped inductive load. In the simulation, the device physics and turn-off characteristics of NPT-CIGBT have been fully explained. The Turn-off energy loss of the device is influenced by the gate resistance and temperature. Furthermore, it has been shown that the CIGBT is fully MOS gate controlled, although having a thyristor on-state anode.

REFERENCES

[1] E. M. Sankara Narayanan, M. Sweet, O. Spulber, J. V. Subhas Chandra Bose and M. M. De Souza, "Clustered Insulated Gate Bipolar Transistor (CIGBT) – a New Power Semiconductor Device", in Proceeding of the 10thISPSD1999, pp. 1307-1312.

[2] O. Spulber, M. Sweet, K. Vershinin, N. Luther-King, E. M. Sankara-Narayanan, M. M. De Souza , D. Flores , and J. Millan, "1.7kV NPT V-Groove Clustered IGBT – Fabrication and Experimental Demonstration", ISPSD 2003, pp. 345-348.

[3] K. Vershinin, M. Sweet, L. Ngwendson and E. M. Sankara Narayanan, "Influence of the Device Layout on the performance of 20A 1.2kV CIGBT in PT Technology", SPEEDAM 2006, pp. S38-5 – S38-8.

[4] K. Vershinin, M. Sweet, L. Ngwendson, J.Thomson, P. Waind, J. Bruce and E. M. Sankara Narayanan, "Experimental Demonstration of a 1.2kV Trench Clustered Insulated Gate Bipolar Transistor in Non Punch Through Technology", ISPSD 2006, pp. 221-224.

[5] M. Sweet, L. Ngwendson, S.T.Kong, E. M. Sankara Narayanan, J.Bruce, S.Ray, "Experimental Demonstration of 3.3kV Planar CIGBT in NPT Technology", ISPSD 2008.

[6] L. Ngwendson, M. Sweet, O. Spulber, K. Vershinin, M. M. De Souza, and E. M. Sankara Narayanan, "MOS Control Device Concepts for AC–AC Matrix Converter Applications: The HCD Concept for High-Efficiency Anode-Gated Devices", IEEE Transactions on Electron Devices, 2005, pp.2075-2080.

[7] D.Kumar, M. Sweet, K. Vershinin, L. Ngwendson, E. M. Sankara Narayanan, "RC-TCIGBT: A Reversed conducting Trench clustered IGBT", ISPSD 2007, pp. 161-164.

[8] K.Sheng, B.W.Williams, and S.J.Finney, "A Review of IGBT Models" IEEE Transactions on Electron Devices, 2005, pp.2075-2080.

[9] P.Leturcq, "A Study of Distributed Switching Process in IGBTs and other Power Bipolar Devices", IEEE Transactions on Electron Devices, 1997, pp.139-147

[10] Sysnopsys TCAD Sentaurus Device Package- Reference Manual, 2007.

[11] T.ogura, H.Ninomiya, K.Sugiyama, T.Inoue, "Turn-off Switching Analysis Considering Dynamic Avalanche Effect for Low Turn-off Loss High-Voltage IGBTs", IEEE Transactions on Electron Devices, 2004, pp.629-635.

[12] Y.C. Gerstermater, and M. Stoisiek , "Switching Behaviour of High Voltage IGBTs and Its Dependence on The Gate-Drive", IEEE Transactions on Electron Devices 1997,pp.105-108.

Turn-off behaviour of high voltage NPT- and FS-IGBT

Hans-Guenter Eckel [*], Karl Fleisch [†]

[*] University of Rostock, Institute of Electrical Power Engineering, Germany, e-mail: *hans-guenter.eckel@arcor.de*
[†] Siemens AG, Nuremberg, Germany, e-mail: *karl.fleisch@siemens.com*

Abstract — A simple but physical based one dimensional model is used to characterize the turn-off of high voltage IGBTs. The dependence of the overvoltage and the peak electric field on the gate driving conditions is analyzed. The transition from a triangular to a trapezoidal electric field has a major impact on the turn-off behaviour. If this transition occurs during the voltage slope, the dv/dt increases significantly. If the field-stop layer is reached during the current slope, the current snaps off, which leads to a second voltage spike with a high absolute voltage but only a moderate peak field.

Keywords — IGBT, MOS controlled device.

I. INTRODUCTION

Insulated Gate Bipolar Transistors (IGBTs) are today the standard power semiconductors for inverters from the kW to the MW range. For industrial medium voltage inverters [9] and high power locomotives [10], [14] 3.3kV and 6.5kV IGBTs are used. For these IGBTs, homogeneous silicon with long charge carrier lifetime and a low efficiency of the backside emitter is the preferred technology. To achieve a better product of on-state and turn-off losses, a field-stop layer is inserted (Fig. 1). The higher doped field-stop layer together with a further reduction of the doping concentration of the n-base lead to a trapezoidal electric field in the blocking state. This allows a reduction of the device thickness and thereby a better product of on-state and turn-off losses compared to NPT-IGBT [3], [4], [5]. IGBTs with field stop layer show critical turn-off behaviour, especially in combination with relatively large parasitic inductances of the commutation circuit [7].

Further improvements can be achieved by an optimization of the charge carrier concentration profile in the on-state.

PT-, NPT- and FS-IGBT show a carrier concentration $n(x)$ in the on-state, which is on the back side higher than on the top side. As the on-state voltage is proportional to $1/n(x)$ (1) and the stored charge is proportional to $n(x)$ (2), the lower carrier concentration at the top dominates the on-state voltage while the high concentration at the backside dominates the stored charge and thereby the switching losses.

$$v_{CE(sat)} \sim \int \frac{1}{n(x)} dx \qquad (1)$$

$$Q \sim \int n(x)\, dx \qquad (2)$$

Carrier storage (fig. 2, IGBT with CS layer) or trench technologies allow an increase of the charge carrier concentration at the top of the device, thus improving the ratio between on-state losses and stored charge [2], [8], [11], [12]. The turn-off behaviour of these IGBTs is quite similar to the turn-off behaviour of field-stop devices without carrier enhancement technologies. Due to the higher carrier concentration during the on-state, all effects which depend on the dynamic field steepening are more pronounced.

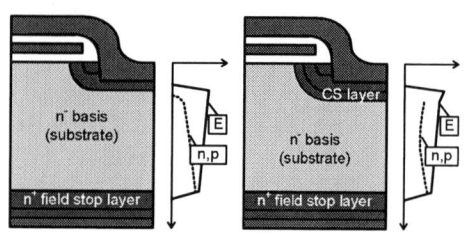

Fig. 2. FS-IGBT and carrier storage IGBT
cross-section of one cell

For low voltage IGBTs, the RBSOA is only limited by the collector-emitter voltage and the collector current. Modern device can turn off currents up to the desaturation current of the IGBT, so there is no current limit. So the gate drive unit has just to limit the collector-emitter voltage to ensure a safe switching of the device. Active clamping is a well proven method to achieve this goal [1]. As the collector-emitter voltage can be measured easily, it is no problem to verify the compliance of the RBSOA limit.

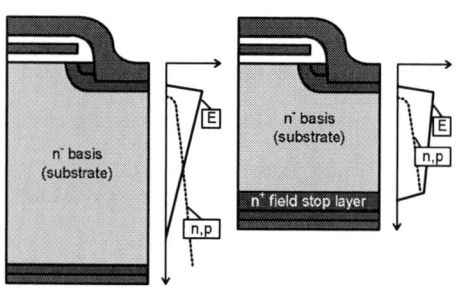

Fig. 1. NPT-IGBT and IGBT with field stop layer
cross-section of one cell

For high voltage IGBTs, the turn-off capability is not only limited by the voltage but also by the maximum electric field. To allow a better understanding of the influence of the gate drive conditions on the electric field during turn-off, a simple one-dimensional but physically based model for the description of the turn-off was introduced in [13]. In this paper, this device model is described in detail and applied on 6.5kV field-stop IGBTs.

II. MODELLING OF THE TURN-OFF OF IGBT

The turn-off behaviour of high voltage IGBTs is dominated by the clear out of the low doped n-basis, which was flooded with charge carriers during the on-state. It is assumed, that the IGBT has a long carrier lifetime, which is the case for most modern high voltage IGBTs. So during the on-state, the recombination of charge carriers in the low doped n-base can be neglected. Consequently, only the drift current in the n-base is considered, the diffusion current is neglected (Fig. 3).

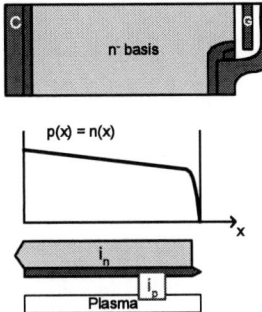

Fig. 3. Current and charge carrier distribution during the on-state

So the ratio between the electron current and the hole current in the plasma is determined by the mobility of the holes and electrons (3).

$$\frac{i_n}{i_p} = \frac{A \cdot e_0 \cdot n \cdot \mu_n \cdot E}{A \cdot e_0 \cdot n \cdot \mu_p \cdot E} = \frac{\mu_n}{\mu_p} \approx 3 \quad (3)$$

When at turn-off the gate is discharged, the conductivity of the channel is reduced until it can no longer carry the electron current, which is necessary for the steady on-state. The difference between the electron current in the plasma and the electron current through the channel is delivered by the charge carriers at the border between the space charge region and the plasma. So the space charge region is enlarged and the IGBT takes up voltage. The electrons flow through the plasma to the backside emitter, the holes flow through the space charge region to the top. If the channel is completely closed, the whole collector current has to flow as hole-current through the space charge region (Fig. 4).

The gradient of the electrical field in the space charge region is proportional to the sum of the doping concentration and the concentration of the free holes (4). The concentration of holes depends on the hole current and the velocity of holes, which again depends on the electric field strength (5).

$$\frac{dE(x)}{dx} = \frac{e_0}{\varepsilon} \cdot \left(N_D + p_{SCR}(x) - n_{SCR}(x) \right) \quad (4)$$

$$p_{SCR}(x) = \frac{i_{pSCR}}{A \cdot e_0 \cdot v_p(E(x))} \quad (5)$$

$$n_{SCR}(x) = \frac{i_{nSCR}}{A \cdot e_0 \cdot v_n(E(x))}$$

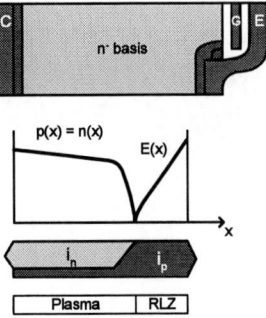

Fig. 4. Current, charge carrier distribution and electrical field during turn-off
RLZ: space charge region

Compared to the off-state, where the hole current and therefore the hole concentration in the space charge region is zero (Fig. 5), the hole current during the turn-off transient leads to a higher gradient of the electrical field. So the same voltage will lead to higher maximum electrical field strength. This can lead to dynamic avalanche, which indeed limits the maximum turn-off current of high voltage IGBTs.

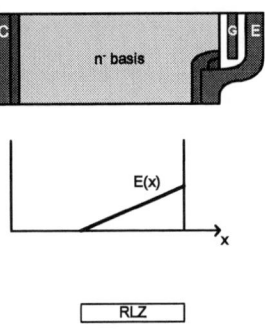

Fig. 5. Electrical field during the off-state

As explained above, the diffusion current in the low doped n-base is neglected. Furthermore, the diffusion current at the boundary between the plasma and the space charge region is neglected. This is leads to an abrupt boundary as shown in Fig. 6. It is assumed, that the charge at the boundary between the plasma and the space charge region is cleared out by the difference between the electron current in the plasma and the space charge region (6). This leads to the same results as the diffusion current, as long as the width of the space charge region is increasing. This is the case during the phases of the turn-off transient, which are critical for the device robustness. Only the tail current cannot be calculated this way, because the tail current is driven by diffusion of the stored charge into the space charge region.

The device losses during the tail phase are calculated using the charge which remains in the low doped base when the space charge region has reached its steady state value.

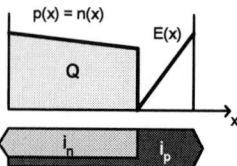

Fig. 6. Current, charge carrier distribution and electrical field during turn-off, neglecting the diffusion current at the boundary between the plasma and the space charge region. MOS channel completely closed.

$$i_{nPlasma} - i_{nSCR} = A \cdot e_0 \cdot n(w_{SCR}) \cdot \frac{dw_{SCR}}{dt} \qquad (6)$$

A further assumption is that the electrons and holes in the space charge region have a constant drift velocity (7), (8). The dependence of the drift velocity on the electrical field strength is neglected. This leads to a constant electron and hole concentration in the space charge region and thereby to a constant gradient of the electric field (9).

$$p_{SCR} = \frac{i_{pSCR}}{A \cdot e_0 \cdot v_{max}} \qquad (7)$$

$$n_{SCR} = \frac{i_{nSCR}}{A \cdot e_0 \cdot v_{max}} \qquad (8)$$

$$\hat{E} = \frac{e_0}{\varepsilon} \cdot (N_D + p_{SCR} - n_{SCR}) \cdot w_{SCR} \qquad (9)$$

The assumption of a triangular electrical field leads to (10) for the collector-emitter voltage.

$$V_{CE} = \frac{1}{2} \cdot \hat{E} \cdot w_{SCR} \qquad (10)$$

III. Turn-off behaviour of NPT-IGBT

The gradient of the electric field and thereby the maximum electric field can be reduced, if there is still an electron current in the space charge region during the turn-off transient. So the peak electrical field can be easily controlled by the gate discharging current. The lower the gate discharging current, the lower is the dv/dt, the dE/dx and thereby the peak electric field.

The switching overvoltage is not as easy controllable as the electric field strength. Starting from high values, a reduction of the gate current leads at first to an increase of the switching overvoltage. The gate current has to be reduced significantly to reduce the overvoltage.

Fig. 7 shows, why the switching overvoltage increases with decreasing gate current. With a gate current of -7.2A, the electron current in the space charge region decreases very fast. As explained before, this leads to a high gradient of the electric field and a small width of the space charge region. When the current decreases, the gradient of the electric field decreases and the width of the space charge region increases. During the increase of the width of the space charge region, the plasma has to be extracted. So a large increase of the width of the space charge region

during the current transient leads to a large amount of charge carriers which have to be extracted.

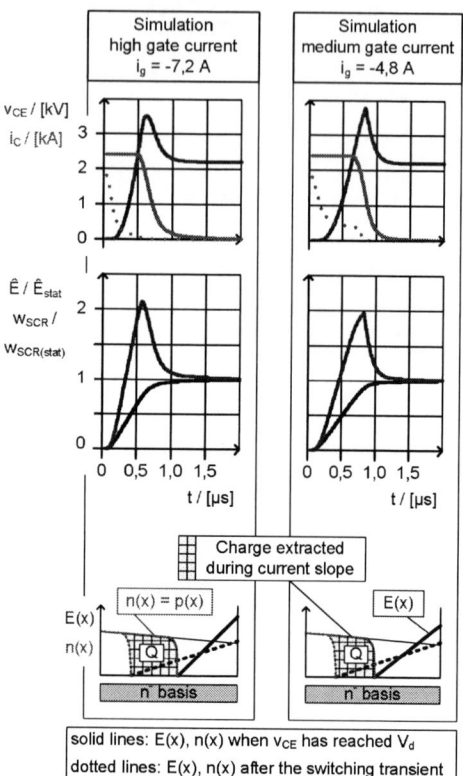

Fig. 7. Simulation of the turn-off behaviour of a NPT-IGBT
Impact of the gate current on the overvoltage

A reduction of the gate current to -4.8A leads to reduction of the dE/dx during the voltage transient, thereby to an increase of the width of the space charge region during the voltage transient, a smaller increase of the width of the space charge region during the current transient, a reduction of the amount of charge which has to be extracted during the current transient and thereby to an increased overvoltage.

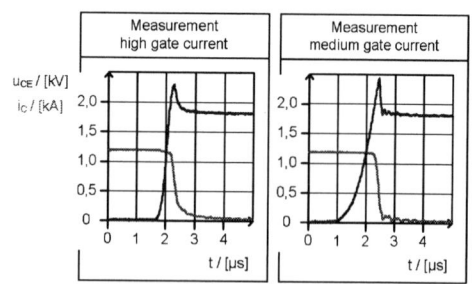

Fig. 8. Measurement of the turn-off behaviour of a NPT-IGBT
Impact of the gate current on the overvoltage

If the gate discharging current is further reduced, there is still a significant electron current in the space charge

50

region which leads to a reduction of the di/dt and the overvoltage.

Due to the poor controllability of the di/dt by the gate current, active clamping circuits are used to limit the overvoltage [1]. A chain of transil diodes between collector and gate is a simple solution for such a circuit, another possibility is a feedback control as described in [6], [10].

IV. TURN-OFF BEHAVIOUR OF FS-IGBT

The main difference of the turn-off behaviour of field-stop IGBT (FS-IGBT) in comparison to NPT-IGBT is the transition of the electric field from a triangle shape to a trapezoidal shape when the space charge region reaches the field-stop layer. At this moment, the differential output capacitance of the IGBT decreases by a factor of 10. If this transition occurs during the voltage slope, the dv/dt increases up to a factor of 10. If the field-stop layer is reached during the current slope, the current snaps off.

A. Influence of the gate-discharging current

Fig. 9 shows the simulation of the turn-off behaviour of a 6.5kV / 600A FS-IGBT in dependence of the gate discharging current. With a high gate current of -2.4 A (fig. 9a), the MOS channel is already closed at the beginning of the voltage transient. As explained on the example of the 3,3 NPT-IGBT, the current through the space charge region flows as hole current, the high dE/dx leads to a small space charge region, there is a lot of charge left, which is swept out during the current transient. The corresponding overvoltage is moderate. Until now, the turn-off behaviour is the same as with the NPT-IGBT. The difference occurs, when the space charge region has reached the field-stop layer. The snap-off of the current leads to a second overvoltage peak, which is higher than the first one. But as the current has already decreased to 50% of the load current, the peak electric field is much lower.

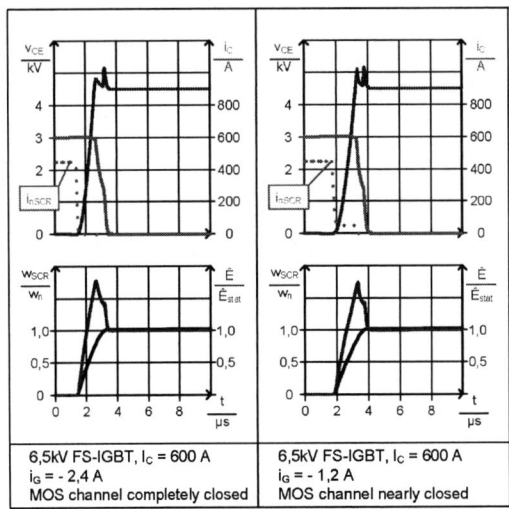

6,5kV FS-IGBT, I_C = 600 A
i_G = - 2,4 A
MOS channel completely closed

6,5kV FS-IGBT, I_C = 600 A
i_G = - 1,2 A
MOS channel nearly closed

Fig. 9 (a, b). Simulation of the turn-off behaviour of a 6.5kV FS-IGBT, impact of the gate current

A reduction of the gate current to -1.2 A (fig. 9b) reduces the dE/dx and, as explained above for the 3.3kV NPT IGBT, leads to a higher first overvoltage peak. If the gate current is further reduced to -0.7 A (fig. 9c), the second voltage peaks melts into the first. This is the most critical operating point in terms of the overvoltage. If the gate current is reduced to -0.4 A (fig. 9d), the dE/dx is so low, that the field-stop layer is reached during the voltage slope. This leads to an increased dv/dt, which is fed back on the gate by the miller capacitance. The gate is pulled up a little bit, the electron current increases, which reduces the dv/dt. So there is a negative feedback loop, as long as the MOS channel is still open. This operating point has the lowest overvoltage and the lowest peak electrical field but the highest turn-off switching losses.

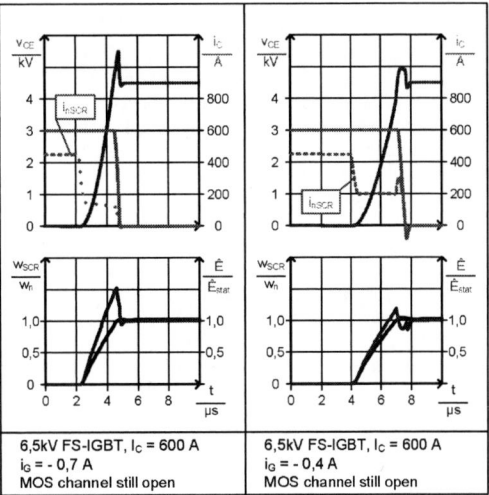

6,5kV FS-IGBT, I_C = 600 A
i_G = - 0,7 A
MOS channel still open

6,5kV FS-IGBT, I_C = 600 A
i_G = - 0,4 A
MOS channel still open

Fig. 9 (c, d). Simulation of the turn-off behaviour of a 6.5kV FS-IGBT, impact of the gate current

Measurements of the turn-off behaviour of a 6.5kV IGBT with field-stop and carrier storage layer are shown in fig. 10. As the simulation has already shown, the reduction of the gate-discharging resistance leads to a higher dE/dx, thereby a higher amount of charge which can be swept out during the current transient and a reduction of the overvoltage. The moment when the space charge region reaches the field-stop layer can be identified by the second spike in the collector-emitter voltage. In comparison to the simulation, the spike in the measurement (fig. 10b) is less pronounced; the transition from a triangular to a trapezoidal electrical field is softer.

With the low gate resistance (fig. 10b), dynamic avalanche leads to a reduction of the dv/dt, when the collector-emitter voltage is higher than 3800 V. So the dv/dt at a collector-emitter voltage above 4 kV is lower with the lower gate-resistance (fig. 10b) than with the higher gate resistance (fig. 10a). This effect also contributes to the lower overvoltage. As dynamic avalanche is not simulated, this effect cannot be seen in fig. 9a and fig. 9b.

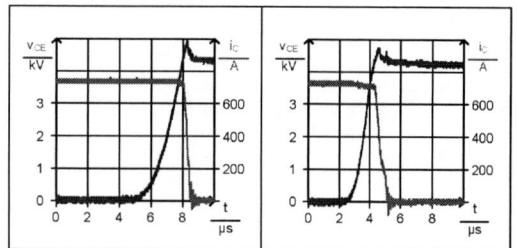

Fig. 10 (a, b). Measurement of the turn-off behaviour of a 6.5kV IGBT
with field-stop and carrier-storage layer at nominal current and 4.2kV
dc-link voltage
left (a): higher gate discharging resistance
right (b): lower gate discharging resistance

B. Influence of the dc-link voltage

Fig. 11a and fig. 11b shows the measurement of the turn-off behaviour for two different dc-link voltages. In both cases, the space-charge region reaches the field-stop layer during the current slope. At this moment, the collector-emitter voltage is the dc-link voltage. As the space-charge region has the same width in both cases and the voltage, which is the integral of the electrical field, is different, the peak electrical field has to be higher with 4.2kV dc-link voltage than with 3.6kV. Therefore, the gradient of the electrical field dE/dx has to be higher in the case of 4.2kV dc-link voltage and, due to equation (4) and equation (5), the current has to be higher in this case. So the current snap-off occurs at a higher instantaneous value of the collector current in the case of 4.2kV dc-link voltage (fig. 11a) than in the case of 3.6 kV (fig. 11b).

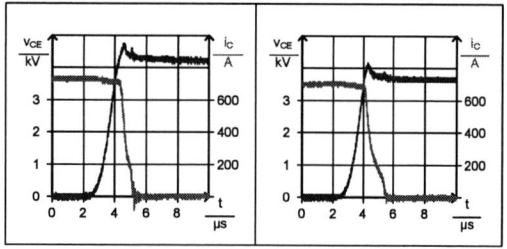

Fig. 11 (a,b). Measurement of the turn-off behaviour of a 6.5kV IGBT
with field-stop and carrier-storage layer at nominal current
left (a): 4.2kV dc-link voltage
right (b): 3.6kV dc-link voltage

C. Influence of the collector current

Fig. 12 shows the influence of the collector current on the switching behaviour (simulation results). Fig. 12a shows the situation at 10% of the nominal current. Due to the low current, the plasma is cleared out slowly, the dv_{CE}/dt is low. Due to the low dv_{CE}/dt, the current through the miller capacitance is lower than the gate current, so the gate current can discharge the gate capacitance of the IGBT. The gate-emitter voltage falls below the threshold voltage, the channel is completely closed. Therefore, the whole current in the space-charge region is hole-current. But as the current is low, the additional hole concentration and the dE/dx remain low. So the space-charge region reaches the field-stop layer at a relatively low voltage of 2.5kV. As soon as the space-charge region has reached the

field-stop layer, the dv_{CE}/dt increases by a factor of 10. The overvoltage is relatively high, the device tends to oscillate but the maximal electrical field is only a little bit higher than in the steady state.

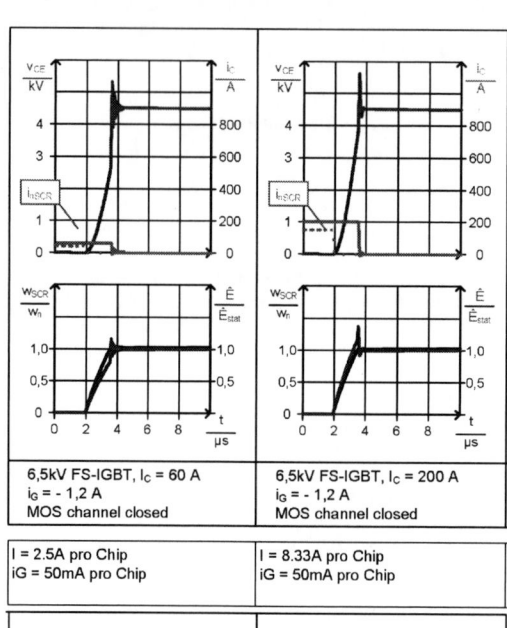

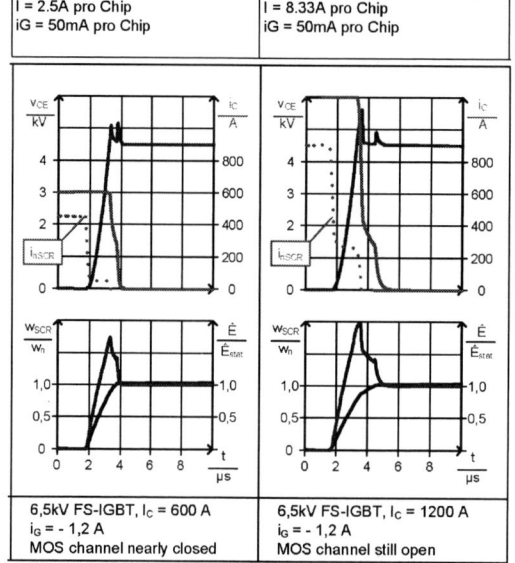

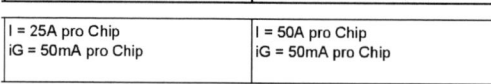

Fig. 12 (a - d). Simulation of the turn-off behaviour of a 6.5kV FS-
IGBT, impact of the collector current

The most critical overvoltage occurs at medium collector current (200A, fig. 12b). The dv_{CE}/dt is still limited by the load current, which has to clear out the plasma. Due to the higher current, the dE/dx is higher and the peak electrical field when the space-charge region reaches the field-stop layer is higher. So the collector-emitter voltage at this moment is higher.

At nominal current (600A, fig. 12c) and twice nominal current (1200A, fig. 12d), the dE/dx is so high, that the space-charge region reaches the field-stop layer during the current slope, there are again two voltage spikes. The dv_{CE}/dt is now so high, that it is limited by the gate current, which has to charge the miller capacitance. The MOS channel is still open, so there is still electron current flowing through the space-charge region.

At twice nominal current (fig. 12d), the overvoltage is not higher than at one third of the nominal current (fig. 12b). But the electrical field is much higher.

The corresponding measurements to these simulations are shown in fig. 13. The high charge carrier concentration of the carrier storage technology leads to a very low dv/dt at low current (fig. 13a) compared to the simulation (fig. 12a). As the current in the measurement is about 50% higher than in the simulation, the gradient of the electrical field is higher and the field stop layer is reached at a higher collector-emitter voltage.

At 1kA load current (fig 13b), the point of dynamic avalanche is already reached at 3.5kV collector-emitter voltage.

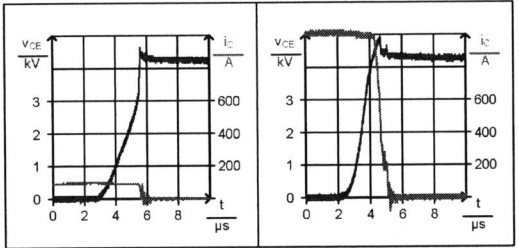

Fig. 13 (a, b). Measurement of the turn-off behaviour of a 6.5kV IGBT with field-stop and carrier-storage layer at 4.2kV dc-link voltage
left (a): 100 A
right (b): 1,000 A

V. CONCLUSION

The paper shows simulation and measurement results of the turn-off behaviour of high voltage IGBT with long carrier lifetime. A simple physical based model helps to understand the characteristic voltage and current waveforms of NPT- and FS-IGBT and shows the internal stress due to high electric fields. The influence of the gate current and the collector current is analyzed.

REFERENCES

[1]. T. Reimann, R. Krümmer, J. Petzoldt, "Active Voltage Clamping Techniques for Overvoltage Protection of MOS-Controlled Power Transistors", *EPE 1997, Trondheim, Norway*, pp. 4.043 – 4.048

[2] M. Mori, et al, "A novel High-conductivity IGBT (HiGT) with a short circuit capability", *ISPSD'98, Japan, 1998*, pp 429-432.

[3] F. Auerbach et al.; "6.5kV IGBT-Modules"; *PCIM 1999, Nürnberg, Germany*, pp. 45-48.

[4] T. Laska, M. Münzer, F. Pfirsch, C. Schaeffer, T. Schmidt; "The Field Stop IGBT (FS IGBT) A New Power Device Concept with a Great Improvement Potential"; *ISPSD 2000, Toulouse, France*, pp. 355-358

[5] J. Bauer, F. Auerbach, A. Porst, R. Roth, H. Ruething, O. Schilling; "6.5 kV-Modules using IGBTs with Field Stop Technology"; *ISPSD 2001, Osaka*, pp. 121-124.

[6] M. M. Bakran, H.-G. Eckel, "Einsatz von IGBTs in Traktionsumrichtern", *ETG-Fachtagung Bauelemente der Leistungselektronik 2002*, S. 163-172

[7] M. M. Bakran, H.-G. Eckel, M. Helsper, A. Nagel, "Challenges in using the latest generation of IGBTs in Traction Converters", *EPE 2003, Toulouse, France*

[8] M. Rahimo, et al, "An assessment of modern IGBT and anti-parallel diode behaviour in hard-switching applications", *EPE2005, Dresden, Germany*.

[9] R.-D. Klug, N. Klaassen, "High Power Medium Voltage Drives – Innovations, Portfolio, Trends", *EPE2005, Dresden, Germany*.

[10] H.-G. Eckel, M. M. Bakran, E. U. Krafft, A. Nagel, "A new Family of Modular IGBT Converters for Traction Applications", *EPE2005, Dresden, Germany*.

[11] Th. Schütze, J. Biermann, R. Spanke, M. Pfaffenlehner, "High power IGBT modules with improved mechanical performance and advanced 3.3kV IGBT3 chip technology", *PCIM 2006, Nuremberg, Germany*

[12] M. Rahmio, A. Kopta, S. Linder, "Novel Enhanced-Planar IGBT Technology rated up to 6.5kV for lower losses and higher SOA Capability", *ISPSD 2006, Naples, Italy*

[13] H.-G. Eckel, M. M. Bakran, "Robustness and turn-off losses of high-voltage IGBT", *EPE2007, Aalborg, Denmark*.

[14] J. Lutz, "Entwicklungstrends bei Halbleiter-Bauelementen für die Traktionstechnik", *Elektrische Bahnen, 3/07*, pp. 130 – 137.

Exact Circuit Power Loss Design Method for High Power Density Converters Utilizing Si-IGBT/SiC-Diode Hybrid Pairs

Kazuto Takao[*] and Hiromichi Ohashi[†]

[*] Corporative Research & Development Center, Toshiba Corporation, 1, Komukai-Toshiba-cho, Saiwaiku, Kawasaki, Japan, e-mail: *kazuto.takao@toshiba.co.jp*

[†] National Institute of Advanced Industrial Science and Technology/Energy Semiconductor Electronics Research Laboratory, 1-1-1, Umezono, Tsukuba, Ibaraki, Japan, e-mail: *h.oohashi@aist.go.jp*

Abstract— An exact design method of circuit power loss is developed. The method is useful for designing high power density converters utilizing Si-IGBT/SiC-Shottky-barrier-diode (SiC-SBD) or high voltage Si-IEGT/SiC-PiN-diode hybrid pairs. For the exact power loss calculation, an empirical method to extract device model parameters is introduced. The calculation results of the power loss are compared with experimental results, and the good agreements are confirmed. By using the method, the power loss of the 4.5 kV Si-IEGT/5 kV SiC-PiN diode hybrid pair is estimated to investigate the possibility of further increase of the switching frequency.

Keywords—SiC-devices, IGBT, Device modeling, High Power Density Systems.

I. INTRODUCTION

High output power density (OPD) converters are key components for future electric vehicles, telecom and data server power supplies and high power industrial systems [1]. To increase the OPD, switching frequency of the converter should be increased to shrink the volume of passive components. An integration design method for the high OPD converters has been extensively studied [2], [3]. The design method is named design platform and its concept is shown in **Fig.1**. The feature of the design platform is that interactions of parameters among the control system, power devices, converter circuit and passive filter are taken into account to calculate exact power losses under high switching frequency conditions. The calculated power loss data are stored in the converter design database. Optimized circuit parameters can be extracted from the database to design the high OPD converters.

In the design platform, novel calculation method for power device loss with respect to unipolar devices such as MOSFETs not only for Si devices but for SiC devices is utilized to calculate exact switching energy data [4]. The method resolves issues of low accuracy and non-convergence, which are discussed in conventional circuit simulators [5]. Although this method is useful to calculate the power loss of the unipolar devices, bipolar devices such as IGBTs and PiN diodes are unsupported option. Si-IGBT/SiC-diode hybrid-pairs are expected to realize the high OPD converters with more than few kW level

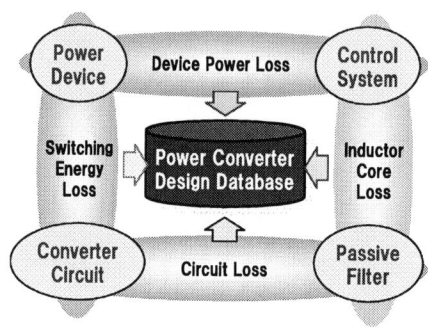

Fig. 1. Concept of a power converter integration design platform.

applications [6]. Therefore, the exact power loss calculation method for the hybrid-pairs is required.

Physics-based analytical models for the IGBTs and PiN diodes have been developed to calculate steady-state and switching transient conditions in circuit simulators [7]-[9]. These models are described with the physical device parameters. In general, the internal structure of the semiconductor devices is not opened to the public. Hence, the extraction of the accurate physical device parameters is difficult for circuit designers.

The purpose of this paper is to present an exact power loss design method for designing the high OPD converters utilizing Si-IGBT/SiC-SBD or high voltage Si-IEGT/SiC-PiN diode hybrid pairs. In the method, switching energies of the hybrid pairs are calculated by analytical switching models derived based on the equivalent circuits for the Si-IGBT, Si-IEGT, SiC-SBD and SiC-PiN diode. To realize exact calculations, an empirical method for the parameter extraction is introduced. Experimental verifications of the method are carried out. By using the proposed method, power losses of the high voltage Si-IEGT/SiC-PiN diode hybrid pair under high switching frequency conditions are calculated to investigate the possibility of further increase of the switching frequency.

978-1-4244-1741-4/08/$25.00 ©2008 IEEE

II. ANALYTICAL SWITCHING MODELS FOR SI-IGBT/SIC-DIODE HYBRID PAIRS

A. Equivalent Circuits of Si-IGBT, Si-IEGT, SiC-SBD and SiC-PiN diode

Power loss of a power device consists of conduction loss and switching loss. The conduction loss is calculated from the static I-V characteristic. In the proposed power loss design method, measured static I-V characteristics are utilized to calculate the conduction loss.

The switching loss is influenced by the circuit voltage, current, stray inductances in the circuit and gate drive conditions. In the proposed method, analytical switching models are utilized to calculate the switching loss.

Fig. 2 shows equivalent circuits for the Si-IGBT, S-IEGT, SiC-SBD and SiC-PiN diode to establish the switching models. The equivalent circuit of the high voltage Si-IEGT is assumed to be same as that of the Si-IGBT. The Si-IGBT, Si-IEGT and SiC-PiN diode are bipolar devices so that they have diffusion capacitances (C_D) that represent minority carrier storage in the drift regions.

To realize the exact power loss calculation, the model parameters in Fig.2 are described with empirically obtained fitting parameters or equations extracted from switching waveforms. The intention of employing the empirical model is to improve the accuracy of the calculation results. In addition, other advantages of the empirical model are simple and ease of implementation.

B. Switching Models and Parameter Extraction Procedures for the Si-IGBT

To analyze the switching behavior of power devices, a chopper circuit with an inductive load is assumed. **Fig. 3** shows the equivalent circuit of the chopper. Stray inductances (L_{s1}, L_{s2}, L_{s3}, L_{s4}, L_{s5}, L_{sg}) are taken into account in the equivalent circuit. Schematic of typical switching waveforms of the Si-IGBT in the chopper is shown in **Fig 4.**

1) Turn-on Phase:

In the period of I, the collector current i_c increases toward the load current I_L with increasing the gate-emitter voltage v_{ge}. The i_c is given as follows [10]:

$$i_c = g_m(v_{ge} - V_{th}) \qquad (1)$$

where g_m is the transconductance and V_{th} is the threshold voltage of the Si-IGBT. The analytical model of v_{ge} considered the influence of the source stray inductance L_{s5} is described in [11].

In the proposed method, i_c is described with an empirical equations as a function of v_{ge}. The empirical equation is extracted from the i_c-v_{ge} curve. Because the Si-IGBT is the bipolar device, the carrier distribution in the N⁻ drift region at the turn on transient is different from that of the steady-state. Therefore, the empirical equation of i_c is extracted from turn on waveforms.

While, the collector-emitter voltage v_{ce} is decreased by inductive voltage of the stray inductance in the main circuit loop L_s ($= L_{s1} + L_{s2} + L_{s3} + L_{s4} + L_{s5}$) and is described as follows:

$$v_{ce} = V_{cc} - L_s \frac{di_c}{dt} \qquad (2)$$

where V_{cc} is the DC link voltage.

In the period of II, the collector-emitter voltage v_{ce} decreases and is described as follow:

$$v_{ce} = V_{ce}' - \left\{ \frac{(V_{GH} + V_{GL}) - V_{GP-on}}{R_g \cdot C_{gc}} \right\} \cdot t \qquad (3)$$

where V_{GH} and V_{GL} are applied gate drive voltages at the on-state and off-state, respectively, V_{GP-on} is the gate-emitter voltage during this period, and V_{ce}' is v_{ce} at $t = t_2$. C_{gc} is the junction capacitance between the gate and collector electrodes and described as follows:

$$C_{gc} = \frac{\varepsilon}{W_{SC}} \cdot A_{gd} = \varepsilon \cdot \left(\frac{2 \cdot \varepsilon \cdot v_{ce}}{q \cdot N_B} \right)^{\frac{1}{2}} \cdot A_{gd} \qquad (4)$$

where ε is the permittivity of silicon, W_{SC} is the width of

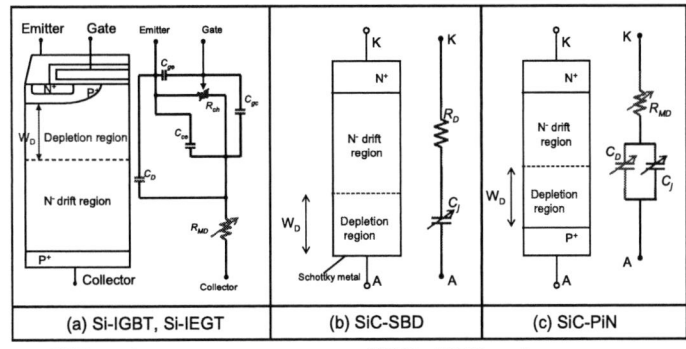

(a) Si-IGBT, Si-IEGT	(b) SiC-SBD	(c) SiC-PiN

Si-IGBT parameters
R_{MD}: conductivity modulated resistance in drift region
Cgc, Cge, Cce: junction capacitances
C_D: diffusion capacitance
Rch: channel resistance

SiC-SBD and SiC-PiN parameters
R_D: resistance in drift region
R_{MD}: conductivity modulated resistance in drift region
C_j: junction capacitance
C_D: diffusion capacitance of SiC-PiN

Fig. 2. Equivalent circuits for the Si-IGBT, Si-IEGT, SiC-SBD and SiC-PiN diode.

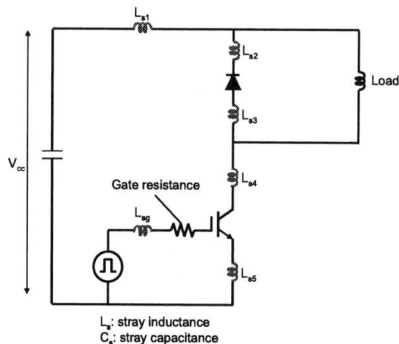

Fig. 3. Equivalent circuit of a chopper.

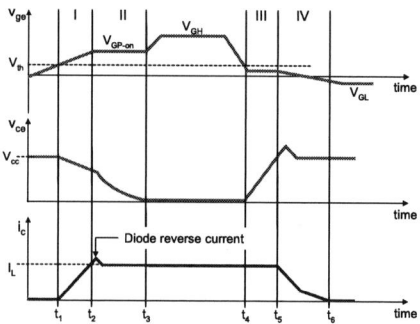

Fig. 4 Schematic of typical switching waveforms under a low gate resistance condition.

the depletion layer in the drift region, A_{gc} is the cross sectional area between the gate and collector electrodes, q is the unit charge, and N_B is the total carrier density in the depletion layer. N_B consists of the doping carrier density in the drift region, the carrier density by the hole-current and the carrier density by the electron-current [12]. By using the v_{ce} and v_{ge} waveforms, C_{gc} is described as follows:

$$C_{gc} = \frac{I_g}{dv_{ce}/dt} = \frac{V_{GP\text{-}on}/R_g}{dv_{ce}/dt} \qquad (5)$$

where I_g is the gate current and $V_{GP\text{-}on}$ is the gate voltage at this period. $V_{GP\text{-}on}$ has constant value because i_c is constant ($= I_L$) in this period. From the waveforms of v_{ge} and v_{ce}, C_{gc} can be extracted.

In this period, the SiC-diode is reverse biased and the reverse diode current i_{diode} is added to the i_c. Therefore, i_c is described by following equation.

$$i_c = I_L + i_{diode} \qquad (6)$$

In the case of the SiC-SBD, i_{diode} is the charging current of the junction capacitance. In the case of the SiC-PiN diode, i_{diode} is the reverse recovery current by discharging the diffusion capacitance. An analytical

model of i_{diode} of the SiC-PiN diode is described in Section II-C.

2) Turn-off Phase:

In the period of III, v_{ge} decreases rapidly and becomes smaller than V_{th} before v_{ce} reaches V_{cc} under low gate resistance conditions [13]. The channel current disappears in this situation. Therefore, i_c is independent of v_{ge}, and v_{ce} is given by following equation [12].

$$v_{ce} = \frac{1}{2} \cdot N_B \cdot \frac{W_B}{Q_0} \cdot \frac{b}{1+b} \cdot I_L \cdot t \qquad (6)$$

where W_B is the width of drift region, Q_0 is the stored charge in the N⁻ drift region at steady on-state, b is the mobility ratio. The equation (6) is re-described into the following form:

$$v_{ce} = a \cdot I_L \cdot t \qquad (7)$$

where a represents $1/2 \cdot N_B \cdot W_B/Q_0 \cdot b/(1+b)$ and is extracted from the v_{ce} waveform. Notice that a is the function of the I_L because Q_0 is the function of the I_L.

In the period of IV, i_c is the discharge current of the diffusion capacitance C_D of the Si-IGBT. This current is expressed as follows under high life time situation [12]:

$$i_c = I_L \cdot \exp\left(-\frac{(1-\alpha_{pnp}) \cdot t}{\tau_c}\right) \qquad (8)$$

where α_{pnp} is the common–base current gain of the pnp transistor in the Si-IGBT at this period and τ_c is the carrier transit time. These parameters can be extracted from the i_c waveform.

Fig. 5 shows the parameter extraction flowchart for the Si-IGBT. First of all, the gate input capacitance C_{iss} is extracted by the static *C-V* characteristic. Then other model parameters are extracted from the switching waveforms. By using the extracted parameters, waveform calculations are implemented. The calculated waveforms are compared with the measured waveforms. If the calculated waveforms are not acceptable results, the model parameters are adjusted to fit the measured

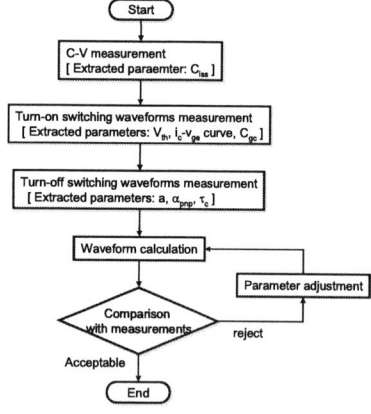

Fig. 5. Flowchart of parameter extraction for the Si-IGBT.

56

waveforms. Acceptable parameters are utilized for the switching waveform calculations.

C. Switching Models and Parameter Extraction Procedures for the SiC-Diodes

Schematic of the turn-off waveforms of the SiC-PiN diode is shown in **Fig. 6**. In the period of I-d, the diode current i_d decreases until the peak reverse recovery current $-I_{RM}$ with the same current slope di/dt of it before time t_{d1}. The reverse recovery current consists of the discharge currents of C_D and C_j. In the case of the PiN diode, C_D is much larger than C_j.

In the period of II-d, i_d is still reverse recovery state. In this period, the diode voltage v_d enters the reverse blocking state. i_d is described as follows [14]:

$$i_d = -I_{RM} \cdot \exp\left(-\frac{t}{\tau_{RR}}\right) \qquad (9)$$

where, I_{RM} is the peak reverse recovery current, τ_{rr} is the diode reverse recovery time constant which can be measured from the current waveform. The diode voltage v_d is described as follows:

$$v_d = V_{cc} - v_{ce} + L_s \frac{di_d}{dt} \qquad (10)$$

At time t_{d3}, v_d corresponds to $-(V_{cc} - v_{ce})$.

In the period of III-d, i_d is constant to its off-state current and v_d reverse biased until V_{cc}.

The turn-off waveforms of the SiC-SBD are shown in **Fig. 7**. The reverse current of the SiC-SBD is caused by the charging of the C_j. At time t_{d1}, the SiC-SBD enters its reverse blocking state. The C_j of the SiC-SBD starts

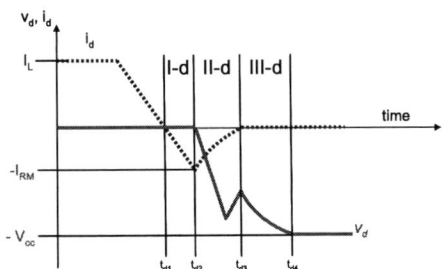

Fig. 6. Schematic of turn-off waveforms of the SiC-PiN diode.

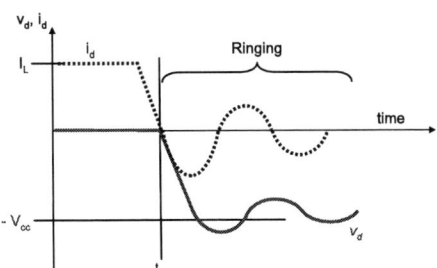

Fig. 7. Schematic of turn-off waveforms of the SiC-SBD.

charging and the charging current flows. i_d and v_d form the ringing waveforms by the resonance of the C_j and L_s. Therefore, the i_d and v_d waveforms after t_{d1} can be formulated by using C_j and L_s.

Fig. 8 shows the parameter extraction flowcharts for the SiC-SBD and SiC-PiN diode. First of all, the static C_j-V characteristics are measured. The C_j is represented as a function of the reverse biased voltage. In the case of the SiC-PiN diode, the reverse recovery charge Q_{rr} and τ_{rr} is extracted from the switching waveforms. By using the all extracted parameters, the switching waveforms are calculated and compared with the measured waveforms. If the calculated waveforms are not acceptable, the parameters will be adjusted until the acceptable results are obtained.

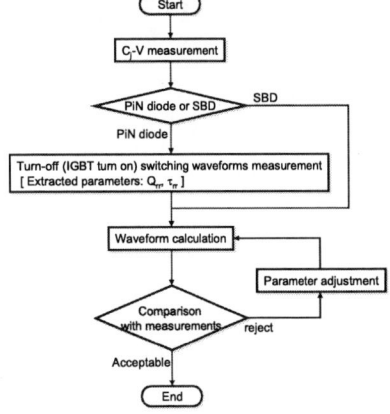

Fig. 8. Flowchart of parameter extraction for the SiC-diodes.

III. EXPERIMENTAL VERIFICATION OF THE ANALYTICAL SWITCHING MODELS

The switching waveforms, which are calculated with the analytical switching models shown in section II, are compared with the measured waveforms to evaluate the accuracy of the models. The 600 V Si-IGBT/SiC-SBD and 4.5 kV Si-IEGT/5 kV SiC-PiN diode hybrid pairs are demonstrated. Furthermore, the switching energies are estimated from the calculated waveforms and compared to the measured values. The model parameters of the 600V/6A Si-IGBT, 600V/6A SiC-SBD both produced by Infineon, 4.5 kV/50A Si-IEGT produced by Toshiba and 5kV/25A SiC-PiN diode fabricated by AIST [15] are extracted. Two 5kV/25A SiC-PiN diode chips are connected in parallel to use with the 4.5kV/50A Si-IEGT. Table I summarizes the extracted parameters of the semiconductor devices. The parameters of the 600V Si-IGBT and SiC-SBD are extracted at the temperature of 150°C, and the parameters of the 4.5kV Si-IEGT and 5kV SiC-PiN diode are extracted at the temperature of 125°C.

57

TABLE I.
MODEL PARAMETERS

Parameters	600V/6A Si-IGBT ($T_j = 150°C$)	4.5 kV/50A Si-IEGT ($T_j = 125°C$)
V_{th}	6 V	7.3 V
i_c-v_{ce}	$i_c = 0.3898 \cdot (v_{ge} - V_{th})^2 - 0.0178 \cdot (v_{ge} - V_{th}) + V_{th}$ (A)	$i_c = 0.3868 \cdot (v_{ge} - V_{th})^2 - 4.78 \cdot (v_{ge} - V_{th}) + V_{th}$ (A)
C_{iss}	368 pF	13 nF
C_{gc}	$2.22 \cdot 10^{-9} \cdot V_{ce}^{-0.834}$ (F) ($V_d \geq 70$V) $4.44 \cdot 10^{-9} \cdot V_{ce}^{-0.834}$ (F) ($V_d < 70$V)	$7.61 \cdot 10^{-9} \cdot V_{ce}^{-0.6271}$ (F) ($V_d \geq 800$V) $2.28 \cdot 10^{-8} \cdot V_{ce}^{-0.6271}$ (F) ($V_d < 800$V)
a	$1.64 \cdot 10^{9} \cdot I_L^{-0.344}$	$7.25 \cdot 10^{8} \cdot I_L^{-0.563}$
α_{pnp}	0.4	0.2
τ_c	$3 \cdot 10^{-8}$ s	$1.5 \cdot 10^{-6}$ s
	600 V SiC-SBD ($T_j = 150°C$)	5 kV SiC-PiN diode ($T_j = 125°C$)
C_j	$216 \cdot V^{-0.4}$ pF	$2281.7 \cdot V^{-0.446}$ pF
τ_{RR}	----------------------	$2 \cdot 10^{-7}$ s

Fig. 9 shows the calculated (dashed) and measured (solid) switching waveforms for the 600V/6A Si-IGBT and 4.5kV/50A Si-IEGT. The switching waveforms are obtained in the chopper shown in **Fig. 3**.

In the case of the 600V/6A Si-IGBT, input DC voltage is 300 V and stray inductance in the main circuit loop L_s is 82 nH. The gate resistance is 10 Ω and gate applied voltage is +15 V. In the case of the 4.5kV/50A Si-IEGT, input DC voltage is 2500 V and stray inductance in the main circuit loop L_s is 3.5 μH. The gate resistance is 100 Ω and gate applied voltage is +15/-10 V. Notice that the stray inductance and gate resistance of the Si-IEGT are scaling values.

The waveform calculations are implemented by using the circuit parameters of the experiment. As seen in **Fig. 9**, good agreement between the measured waveforms and calculated waveforms is obtained. The results indicate that the proposed method can exactly estimate the influence of circuit parameters including the stray inductance. Because the influence of the stray inductance increases with increasing switching frequency, the proposed method is useful for designing the power loss of the future high OPD converters.

Turn-on and -off switching energies are estimated by time integrating the products of v_{ce} and i_c waveforms. The switching energies obtained by measured and calculated waveforms are shown in **Fig. 10**. As seen in the figure, the difference between the measured and calculated switching energies is within 10 %. The error of the Si-IGBT switching energies, which are calculated with the circuit simulator (PSpice), is evaluated in [16]. In that simulation, despite utilizing the optimized device parameters of the physics-based IGBT model, the error in the worst case is 52%. Therefore, the accuracy of the proposed method is significantly improved compared to the conventional circuit simulator.

IV. ESTIMATION OF THE SWITCHING FREQUENCY LIMITATION OF THE HIGH VOLTAGE SI-IEGT/SIC-PIN DIODE HYBRID PAIR

To realize the high OPD high-power-converter, the possibility of the high switching frequency operation of

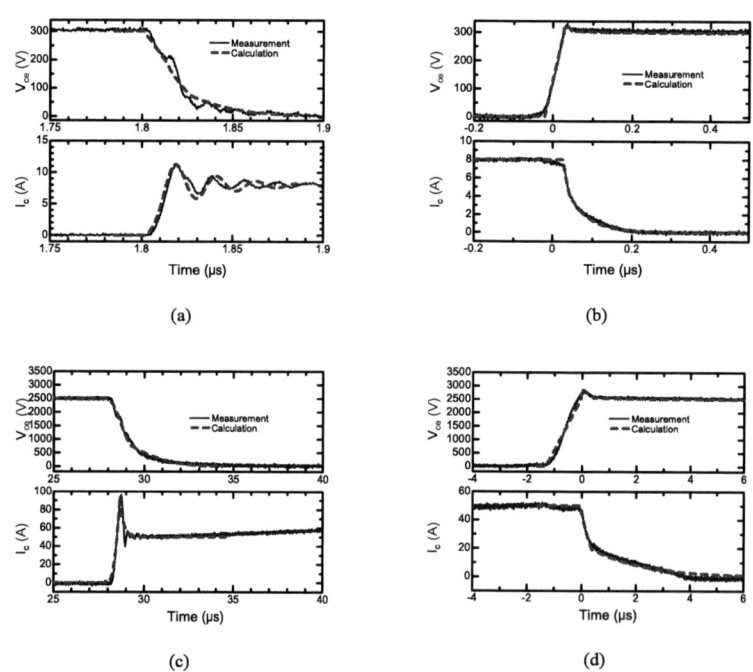

(a)

(b)

(c)

(d)

Fig. 9. Calculated (dashed) and measured (solid) waveforms. (a) Turn-on of the 600 V/6 A Si-IGBT, (b) Turn-off of the 600 V/6 A Si-IGBT, (c) Turn-on of the 4.5 kV Si-IEGT, (d) Turn-off of the 4.5 kV Si-IEGT.

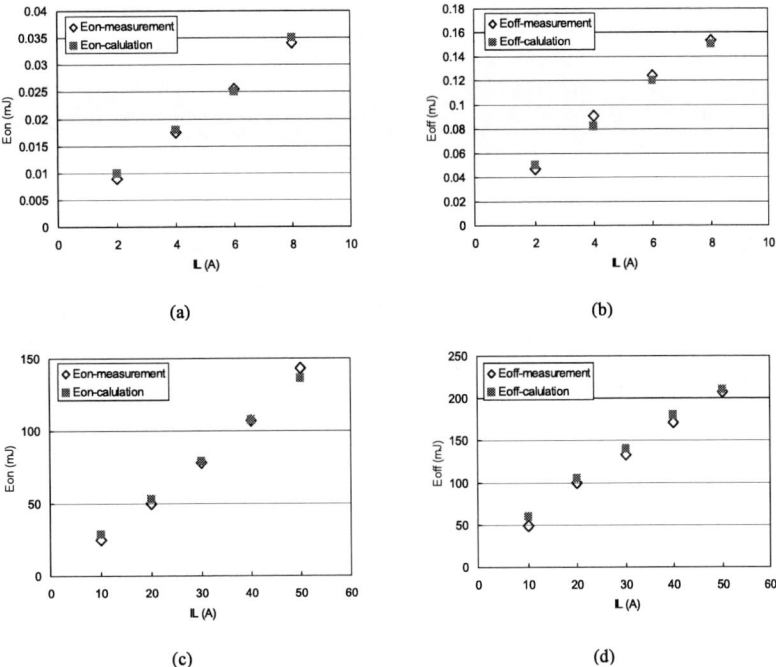

Fig. 10. Dependence on switching energies on load current IL. (a) Turn-on switching energy: E_{on} of the 600 V/6 A Si-IGBT, (b) Turn-off switching energy: E_{off} of the 600 V/6 A Si-IGBT, (c) E_{on} of the 4.5 kV Si-IEGT, (d) E_{off} of the 4.5 kV Si-IEGT.

high voltage Si-IGBT/SiC-PiN diode hybrid pairs have been investigated [15], [17]. In this section, the switching frequency limitation of the high voltage Si-IEGT/SiC-PiN diode hybrid pair is investigated from the point of view of the power loss. The power loss of the Si-IEGT is calculated by the proposed method in this work. The model parameters of 4.5kV/50A Si-IEGT and two parallel 5kV/25A SiC-PiN diodes extracted in previous section are utilized for the power loss calculation.

Fig. 11 shows the turn-on and -off switching energies (E_{on} and E_{off}) dependence on the gate resistance R_g of the Si-IEGT at $T_j = 125°C$. The E_{on} is proportional to the R_g. In contrast, the E_{off} is not influenced by the R_g in the range of that the R_g is smaller than 100 Ω. Both E_{on} and E_{off} increase with increasing the I_L.

Utilizing the E_{on} and E_{off} data shown in **Fig.11**, the dependence of the total power loss of the Si-IEGT P_{total}, which consists of the conduction loss and switching loss, on the switching frequency is estimated. In the power loss calculation, a standard 2-level inverter circuit shown in **Fig. 12** is assumed. The inverter specifications are listed in Table II.

TABLE II.
INVERTER SPECIFICATIONS

DC Voltage V_{DC}	2500V
Modulation method	PWM
Modulation index: M	0.95
Power factor: $\cos\phi$	1
Load current: I_{rms}	35.3A

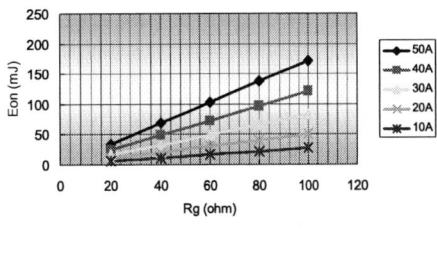

(a)

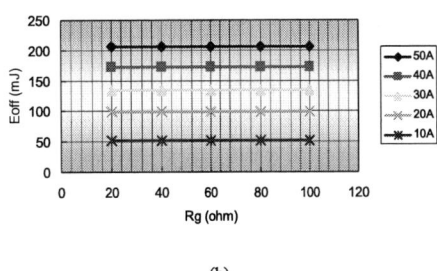

(b)

Fig. 11. Calculated switching energies of the 4.5 kV/50A Si-IEGT. (a) Turn-on switching energy: E_{on}, (b) Turn-off switching energy: E_{off}.

Fig. **13** shows the dependence of the total power loss P_{loss} of the Si-IEGT on the switching frequency f_{sw}. With

59

a conventional forced liquid cooling technique, the maximum allowable power loss per chip of the Si-IEGT is about 180 W. Therefore, the switching frequencies of 1.2 kHz at $R_g = 100\ \Omega$ and 1.6 kHz at $R_g = 20\ \Omega$ can be available. With advanced cooling techniques such as a micro channel structure fin and a double-sided cooling [18], the cooling capacity of more than 1.5 times of the conventional can be available. Assuming the cooling capacity of 270 W (1.5 times of the conventional), the Si-IEGT can operate with the switching frequency more than 2 kHz.

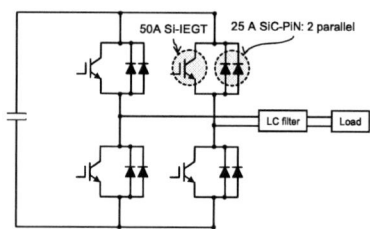

Fig. 12 Equivalent circuit of a 2-level inverter used for power loss calculation of the 4.5 kV/50 A Si-IEGT.

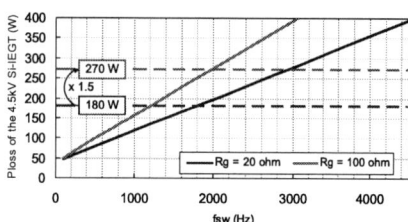

Fig. 13 Dependence of the total power loss P_{loss} of the 4.5 kV/50A Si-IEGT on the switching frequency f_{sw}.

V. CONCLUSION

Switching models and empirical parameter extraction methods for Si-IGBT/SiC-SBD and high voltage Si-IEGT/SiC-PiN diode hybrid pairs are presented. The proposed method is validated by comparing with measurement results. The error of the calculated switching energies is within 10 %. The results indicate that the proposed method could implement an exact power loss design of future converters utilizing hybrid pairs. In addition, the design platform is available for power loss calculations of widely-used power semiconductor devices: MOSFET, super junction MOSFET, SBD, IGBT, IEGT and PiN diode.

By using the proposed method, the switching frequency limitation of the high voltage Si-IEGT/SiC-PiN diode hybrid pair is investigated from the view point of the power loss. The results indicate that more than 2 kHz operations of a 4.5 kV Si-IEGT would be possible with the SiC-PiN diode and the advanced cooling technique.

ACKNOWLEDGMENT

A part of this work was supported by New Energy and Industrial Technology Development Organization (NEDO) project, Development of Inverter System for Power Electronics, Japan.

REFERENCES

[1] H. Ohashi, "Recent Power Devices Trend," *IEEJ*, Vol. 12, No.3, pp. 168-171, 2002 (in Japanease)

[2] Y. Hayashi, K. Takao, T. Shimizu and H. Ohashi, "Power Converter Integration Design based on Evaluation Platform Concept," *in Proc. CD-ROM, 4th International Conference on Integrated Power Systems (CIPS 2006)*, 2006.

[3] Y. Hayashi, K. Takao, T. Shimizu and H. Ohashi, "High Power Density Design Methodology" *in Proc. CD-ROM, 4th Power Conversion Conference (PCC-Nagoya 2007)*, 2007.

[4] K. Takao, Y. Hayashi, S. Harada and H. Ohashi, "Study on advanced power device performance under real circuit conditions with an exact power loss simulator," *in Proc. CD-ROM, 12th European Conference on Power Electronics and Applications (EPE 2007)*, 2007.

[5] X. Wu, S. Wong, C. K. Tse and J. Lu, "Erroneous Results from SPICE Simulations of Switching Converters: A Dynamic System Viewpoint," *in Proc. CD-ROM, 37th IEEE Power Electronics Specialist Conference (PESC 2006)*, 2006.

[6] B. Weis, D. Peters and M. Wölz, "A new 690 VAC Drive with SiC Schottky Freewheeling Diodes," *Power Electronics Europe*, issue 8, pp.33-35, Dec. 2006

[7] A. R. Hefner and D. L. Blackburn, "An analytical model for the steady-state and transient characteristics of the power insulated-gate bipolar ransistor," *Solid-State Electron.*, vol. 31, no. 10, pp. 1513-1532, 1988.

[8] P. R. palmer, E. Santi, J. L. Hudgins, X. Kang, J. C. Joyce, and p. Y. Eng, "Circuit simulator models for the diode and IGBT with full temperature dependent features," *IEEE Trans. Power Electron.*, vol. 18, no. 5, pp.1220-1229, Sep. 2003.

[9] Ty R. McNutt, Allen R. Hefner, Jr., H. Alan Mantooth, Jeff Duliere, David W. Berning and Ranbir Singh, " Silicon Carbide PiN and Merged PiN Schottky Power Diode Models Implemented in the Saber Circuit Simulator", *IEEE Trans. Power Electron.*, vol. 19, no. 19, pp.573-581, May 2004.

[10] B. J. Baliga, "Power Semiconductor devices," PWS publishing company, pp388-395, 1995.

[11] Y. Xiao, H. Shah, T. P. Chow, R. J. Gutmann, "Analytical Modeling and Experimental Evaluation of Interconnect Parasitic Inductance on MOSFET Switching Characteristics," *in Proc. CD-ROM, 19th IEEE Applied Power Electronics Conf. (APEC2004)*, 2004

[12] T. Ogura, H. Ninomiya, K. Sugiyama, and T. Inoue, "Turn-Off Switching Analysis Considering Dynamic Avalanche Effect for Low Turn-Off Loss High-Voltage IGBTs," *IEEE Trans. Electron Devices.*, vol. 51, no. 4, pp.629-635, April 2004.

[13] K. Sheng, F. Udera and G.A.J. Amaratunga, "Optimum Carrier Distribution of the IGBT," Solid-State Electronics, vol. 44, Issue 9, pp.1573-1583, September 2000.

[14] P. O. Lauritzen, and C. L. Ma, "A Simple Diode Model with Reverse Recovery," *IEEE Trans. Power Electron.*, vol. 6, no. 2, pp.188-191, April 1991.

[15] T. Kinjo, K. Takao, Y. Tanaka, K. Sung, T. Ogura and H. Ohashi, "Exact Power Loss Estimation Method for High Voltage Power Converters with 5 kV SiC-PiN Diode," *in Proc. CD-ROM, 23rd IEEE Applied Power Electronics Conf. (APEC2008)*, 2008

[16] A. T. Bryant, X. Kang, E. Santi, P. R. Palmer and J. L. Hudgins, "Two-Step Parameter Extraction Procedure With Formal Optimization for Physics-Based Circuit Simulator IGBT and p-i-n Diode Models," *IEEE Trans. Power Electron.*, vol. 21, no. 2, pp.295-309, March 2006.

[17] W. Bartsch, S. Gediga, H. Koehler, R. Sommer, and G. Zaiser, "Comparison of Si- and SiC-Powerdiodes in 100A-Modules," *in Proc. CD-ROM, 12th European Conference on Power Electronics and Applications (EPE 2007)*, 2007.

[18] C. Gillot, C. Schaeffer, C. Massit and L. Meysenc, " Double-Sided Cooling for High Power IGBT Modules Using Flip Chip Technology," *IEEE Trans. Components and Packaging Technologies.*, vol. 24, no. 4, pp.698-704, Dec. 2001.

A Forward Converter with a Monolithic Cascode Device: Design and Experimental Investigation

F. Chimento[*], S. Musumeci[*], A. Raciti[*], L. Abbatelli[**], S. Buonomo[**], R. Scollo[**]

[*]DIEES - ARIEL, University of Catania, Viale A. Doria, 6 – 95125, Catania, Italy
[**]STMicroelectronics, Stradale Primosole, 50 – 95121, Catania, Italy

Abstract — A forward converter for switched mode power supply (SMPS) applications has been designed and realized, with a monolithic cascode device as active switch, looking for the performance enhancement. The operation of this device in a SMPS application is shown in detail, and results regarding electrical and thermal characteristics in comparison to the power MOSFET device solution are discussed. The experimental tests that have been carried out are targeted to show the suitability of this device when a 1000-1500V operation is required and a single switch topology is the most appropriate. Some remarks about the driving circuit adopted for the cascode are also reported.

Index Terms – Switched mode power supply, forward converter, MOSFET, cascode device.

I. INTRODUCTION

The nowadays growth of the market of switched mode power supply (SMPS) has increased the research efforts of the designers of power devices and has addressed them towards the experimentation of new solutions [1]. Due to the inductive nature of many industrial applications, the devices have to experience heavy switching conditions. In low power applications the MOSFET devices are the switches of the choice, since they have good switching performances, but the conduction losses increase with the increase of the device breakdown voltage. In practice, in applications with voltage levels higher than 900V, the MOSFETs advantages and drawbacks are to be compared with those of other devices in order to verify their convenience.

In the case in which MOSFET devices are used, the high input voltage implies devices with a high on-state resistance (R_{DSon}) with consequent increase of the conduction power losses. The R_{DSon} of the power MOSFET devices, in fact, is roughly proportional to the square of their breakdown voltage. A power MOSFET working in applications with a voltage range in 1000-1500V is to be designed with a large epytaxial layer. Thus, the output resistance is dominated by the epytaxial resistance, being negligible all the other R_{DSon} components [2]. The following relation may be considered in order to comparatively estimate the R_{DSon}:

$$R_{DSON} \propto BV_{DSS}^{2.4 \div 2.5} \qquad (1)$$

For this reason, in the "high voltage" range, alternative devices are welcomed to obtain a suitable reduction of the conduction losses. In particular, in the last years, the cascode device has been considered as a valid alternative to the power MOSFET for applications with the mentioned voltage values at medium range of frequency operations.

This field was in some instance partially covered by the IGBTs that present a low voltage drop due to the bipolar nature of the output stage. IGBT devices have a voltage drop V_{CEs} that increases linearly with the breakdown voltage. i.e.:

$$V_{CE\,igbt} = V_{BE\,bip} + V_{DS\,mosfet} \qquad (2)$$

Unfortunately during the turn-off, the current of the IGBT, which consists of a majority electron carriers current flowing in the MOSFET channel and a minority holes current flowing in the base of the *pnp* bipolar transistor, is impeded to circulate and the charges are trapped in the base. The recombination of this charge gives the tail current phenomenon. Thus, the drawbacks of the utilization of this device are related to the current tail especially in the high frequency range. In the IGBT technology the introduction of life time killers helps the recombination through platinum implantation. Moreover other irradiation techniques are applied in order to create inside the device structure some recombination centers. The devices realized by means of these techniques have the drawback of a trade-off with the *pnp* output transistor gain. Consequently, the increase of the allowed switching frequency causes the reduction of the device static performance.

The cascode technology, which was proposed in order to overcome the drawbacks related to the utilization of MOSFETs and IGBTs in a niche of voltage and frequency, consists in a low voltage MOSFET with the drain directly connected to the emitter of a *npn* high voltage bipolar transistor [3], [4], (Fig. 1). The base of the BJT is not connected with the MOSFET and the turn-off is faster than a conventional BJT since at turn-off the cascode has the base current equal to the collector one. More specifically, during the turn-off at open-emitter condition, a reverse base current equal to the collector one quickly removes the charge in the base. In this way both the storage time and the fall time can be significantly reduced. This fact allows a fast switching of the cascode device that is associated to a low voltage drop at high breakdown voltage. The actual realization of this kind of device goes up to 2200V blocking voltage while maintaining high switching frequency (up to 150 kHz) with a rated current, as a single device, up to tenth of ampere.

In the cross-section of the cascode device shown in Fig. 2 it is possible to recognize an occupied silicon area that has a major dimension in the bipolar part. Moreover there is the vertical MOSFET in which the emitter diffusion is the same of the drain diffusion.

The *p* plant is a thick area of silicon and represents the base of the output bipolar. The collector of the structure is realized with a metal connected to the backslice and

978-1-4244-1741-4/08/$25.00 ©2008 IEEE

corresponds to the epytaxial layer. The choice of a suitable value for the substrate thickness and resistivity, together with an appropriate edge termination, allows the structure to handle high blocking voltage.

The bipolar part gives the low forward voltage drop while the open-emitter turn-off increases the device switching speed. The monolithic device features a deep-base *npn* BJT, and a vertical power MOSFET that is realized inside the emitter of the transistor output stage itself. This solution allows obtaining a silicon area only depending on the BJT size in spite that the whole switch is realized by the series connection of two devices (Fig. 1).

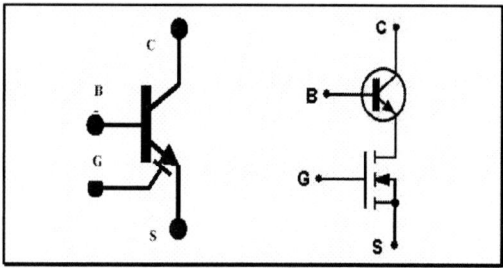

Figure 1 – Cascode device symbol and structure.

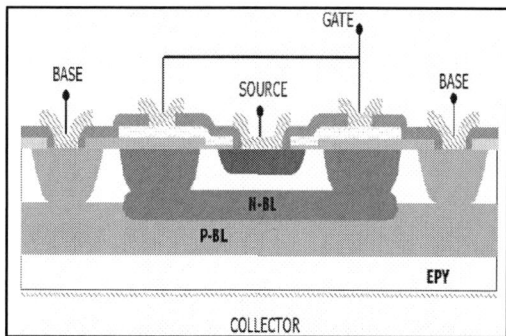

Figure 2 – Cross section of a monolithic cascode device.

II. CONVERTER DESIGN

The forward converter in SMPS applications represents a suitable and cheap solution because it is constituted by a single switch topology and it can reach power values higher than the flyback one [5]. The flyback converter is highly considered as an auxiliary power supply. The blocking voltage of the semiconductor switch used into the topology should account for the dc bus voltage, the reflected output voltage, and the over-voltages due to both the stray inductance of the layout and the transformer leakage inductance. These voltages add up and the breakdown voltage required for the switch may range between 1200 V and 1700 V, thus preventing in some cases the efficient use of MOSFETs. In [5] the usefulness of the cascode devices for such applications due to the good trade off between the high breakdown voltage, the low voltage drop and the high switching frequency capability is deeply analyzed.

In this paper the experimental circuit was realized in a 450W forward topology where the switch is required to work with BV_{CEs} of 1500V and with a switching frequency of 60 kHz. Consequently, in order to evaluate the suitability of the device, the whole system has been realized comparing the global performance in case of both MOSFET and cascode device utilization.

A. Converter operation principle

In the schematic shown in Fig. 3 it is depicted the basic operation principle of the proposed converter solution. The voltage output in the load is kept constant by the control loop which is based on an output voltage control. During the turn-on of the switch, the terminal with the dot in the primary winding N_P has a potential greater than the other terminal, so the diode D_2, is directly biased and the current of the inductor L_1 increases its initial value. In the turn-off phase D_3 is directly biased and a freewheeling current flows in the load, because the inductor discharges the energy stored during the turn-on.

A particular role is carried out by the reset winding N_r and the catch diode D_1. They are used because in the turn-on phase the current on the switch is the sum of the current on the primary and the magnetizing one. In the section D further details will be given about this reset function. The magnetizing current in the turn-off phase does not decrease instantaneously and consequently the diode D_1 has to demagnetize the core bringing it to the initial condition [6], [7], [8].

The main design data for the proposed forward converter are shown in Table I. Moreover, in Table I it is clearly shown that the described SMPS application has requirements of wide input voltage range (240-700V) as it is required in applications such as power solders and industrial welder equipment.

As previously mentioned the analyzed topology will be examined with both the MOSFET and cascode devices which will be driven by a suitable integrated driver.

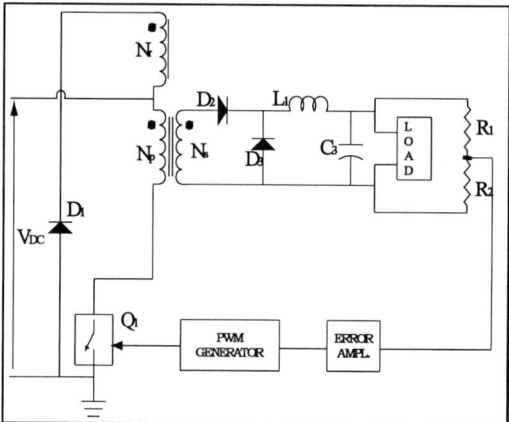

Figure 3 – Schematic of the proposed forward converter.

Table I. Design data for the proposed converter.

V_{DCmin}	Low input voltage	240V
V_{DCmax}	High input voltage	700V
f_{sw}	Switching frequency	60kHz
V_{out}	Output voltage	36V
I_{out_max}	Maximum output current	12.5A
P_{outreg}	Output power	450W

B. Clamping network

A particular attention has been given to the design of the clamping network. In fact, during the off-state the voltage applied to the power device is given by the sum of the input voltage and the demagnetizing one acting on the primary winding. This latter, due to the suitable choice of the turn ratio N_P/N_r has been set to 500V in the worst case. Consequently, according to the design constraints already shown in Table I, the maximum stress on the device is always lower than 1200V. Moreover, it is important to take into account an overvoltage that may occur because of the demagnetization process of the leakage inductance of the transformer. In this circumstance, in fact, a voltage spike is added to the main voltage and consequently, it is necessary to consider a 25-30% more than the theoretical voltage value. This is the reason for the choice of a 1500V device. In order to avoid voltage values over this upper value, a suitable clamping network has been used. The schematic of the clamping circuit is reported in Fig. 4. The Zener use has been preferred in comparison to a traditional RCD (Resistance-Capacitance-Diode) because the latter causes a greater dissipation and consequently a lowering of the whole performance for the converter.

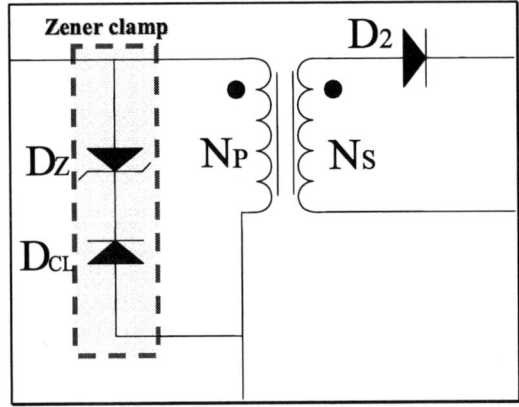

Figure 4 – Zener clamping network.

The operations of this clamping network are the following. During the on-state the diode D_{CL} is unbiased and the clamping network is not working. During the off-state the polarity inversion on the primary winding directly biases the diode and, in the circumstance of a voltage over the limit, the Zener diode conducts and blocks the voltage at the stated value. In the actual application this has been realized by the series connection of two Zener diodes with a Zener breakdown voltage of 300V. Consequently the result is that:

$$\frac{N_P}{N_r} V_{DCmax} + V_{spike} \leq V_{clamp} \qquad (3)$$

According to the diode choice, V_{clamp}, in this occurrence, is 600V. Consequently, the voltage stress on the device, V_{switch}, during the turn off is set by:

$$V_{switch} = V_{DCmax} + V_{clamp} \qquad (4)$$

By considering (3) and the value of V_{DCmax} already expressed in Table I, V_{switch} is equal to 1300V. The choice for a 1500V device, as above mentioned, brings to a 200V

safety margin that allows to expect a good reliability for the designed application.

C. The base driving circuit

The base driving circuit, in case of utilization of the cascode device, needs an accurate analysis [9].

Several solutions have been proposed in the literature for the driving of the cascode device [10]. The driver circuit must supply a suitable current in the base terminal and a timely voltage signal in the gate path.

The current in the base path may be with a constant value or with a value proportional to the collector current. In Fig. 5 a constant current driving circuit is shown. In this circuit the base current at turn on is given by:

$$I_{B,on-state} = \frac{V_{supply} - V_{BS,on-state} - V_{D\_on}}{R_B} \qquad (5)$$

Where V_{supply} is the constant voltage applied to the base terminal of the device, $V_{BS,on-state}$ is the base-source voltage, V_{D_on} is the diode forward voltage drop and R_B is the base resistance of the bipolar transistor.

In this case the design of the driving base current is related to the maximum value of the collector current, and the switching performances are not optimized [9]. Better switching performance is obtained by a proportional base current that is fed by previously sensing the collector current. The proportional base current technique, in fact, is recommended especially in those applications where the current is with a low frequency sinusoidal shape or when the load is variable. The latter is the typical case of a SMPS converter.

A proportional base current driving has been used in the forward application which is described in this paper. In Fig. 6 it is shown that a current transformer was used to detect the collector current. Consequently, during the storage stage (which duration is typically 500ns), when the low voltage MOSFET of the cascode is turned-off, the collector current is equal to the base one thus charging the capacitance C_2 through the diode D_4 until the voltage reaches the value imposed by the Zener diode D_{z1}. During the turn-on the capacitance C_2, charged during the previous phase, supplies the stored energy to the base, while during the on-time the base driving current I_b is proportional to I_c.

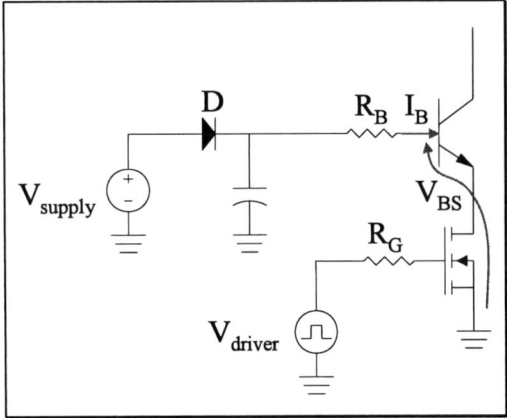

Figure 5 – Cascode device with constant base current driving circuit.

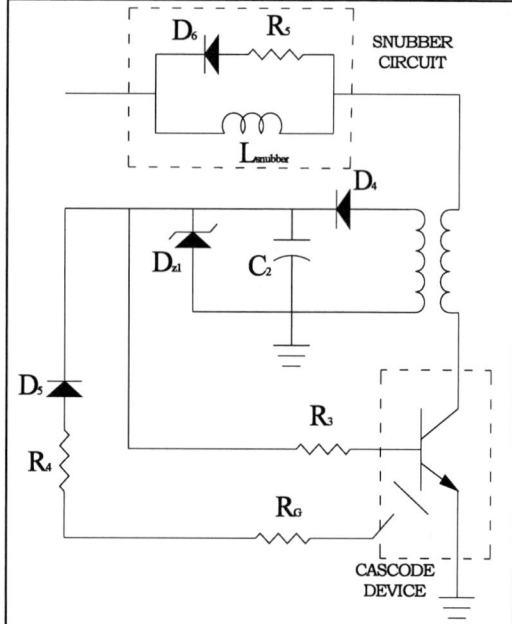

Figure 6 – Cascode device base driving network.

A suitable snubber circuit is connected to the collector to guarantee low losses on the cascode. The utilization of this kind of circuit is very important in the specific case of operation. The higher losses of the converter are related to the turn-on phase rather then to the turn-off transient. This is the reason for the choice for a turn-on snubber circuit [6]. The main role is attributed to the inductance $L_{snubber}$ which is series connected to the primary winding of the transformer. The current that flows through this inductance is the switch main current. During the turn-on there is a high voltage drop on $L_{snubber}$ and, consequently, the collector voltage is lowered in comparison to the case in which the snubber circuit is not used. Moreover the presence of the inductor allows to slow up the current during the turn-on, so that the switch instantaneous power is very low. When the on state is over, the current rate di/dt gets lower and consequently the inductor voltage drops to zero. The positive effect of the snubber circuit is shown in Figures 7 and 8 where the comparison between the turn-on waveforms in case of presence or absence of the snubber circuit is shown. The experimental evaluation allowed to determine an energy consumption E_{ON} of 73µJ in the circumstance of utilization of the snubber circuit versus 136µJ in the case in which this circuit is not used. The experimental traces are referred to a 150W load with a supply voltage of 500V.

During the turn-off the diode D_6 is directly biased and R_5 dissipates the energy previously stored by the inductance.

D. The reset diode

As previously mentioned, the reset winding N_r and the catch diode D_1 play a key role for the demagnetization of the transformer. The design of the catch diode depends on the higher voltage that this device must sustain during the reset operation. In fact, applying a Kirchhoff law to the input circuit, the reset winding and the diode itself yields:

$$V_{reverse} = \left(1 + \frac{N_r}{N_p}\right) V_{DC} \qquad (6)$$

where $V_{reverse}$ is the voltage that the diode must sustain. In the worst case this voltage, according to (6), can reach even 1700V. Consequently, the implementation of the reset function has been done by means of two 1200V diodes, with a large safety margin that takes into account the occurrence of spikes and overvoltage due to inductive nature of the load. The reset winding configuration has been already shown in Fig. 3.

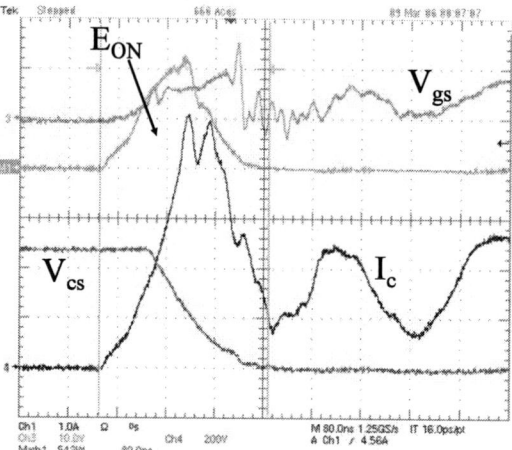

Figure 7 – Turn-on switching waveforms of the cascode device without snubber circuit utilization @P_{out}=150W;
V_{ce} 200/div, I_c 1A/div, V_{gs} 10V/div, power 900W/div, time 80ns/div.

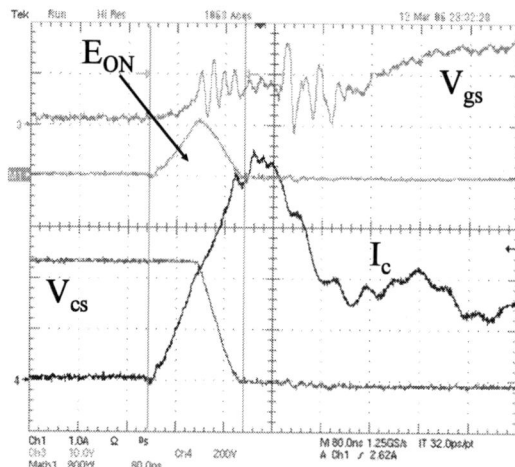

Figure 8 – Turn-on switching waveforms of the cascode device with snubber circuit utilization @P_{out}=150W;
V_{ce} 200/div, I_c 1A/div, V_{gs} 10V/div, power 900W/div, time 80ns/div.

E. The start-up network

The design procedure for the start-up circuit is strictly connected to the choice of the IC driver that has been used in this application. This network is necessary in order to guarantee that the driving circuit (a suitable current mode PWM IC driver) is turned on quickly even in the worst operating conditions, that is, when the input voltage is the lowest admitted for the application. Specifically the integrated PWM driver, requires a start-threshold voltage in the range between 7.8 and 9.0V with a start-up current lower than 0.5mA. The start-up function is guaranteed by two resistances. These two resistances supply a start-up current I_{st} when the converter is switched on and, at the same time, they charge up a capacitance connected to the supply terminal of the IC. Considering a worst-case analysis, taking into account that the lower value for the input voltage, according to Table I, can be 240V and that the higher start-up current, as already mentioned, is set to about 0.5mA the two resistances are chosen as 180kΩ each.

The voltage is supplied to the driver by means of an auxiliary winding of the main transformer. In the meanwhile between the driver switch-on and the steady-state there is a time when this winding is inactive. A capacitance is then connected to the driver circuit in order to ensure the supply voltage even during this time. Finally, the whole driver is protected through a suitable Zener diode which cuts the voltage at the value of 20V, which is an appropriate safety margin for this kind of IC.

III. EXPERIMENTAL RESULTS

The converter realized with the cascode device was compared with the same topology in which the main switch is a MOSFET with similar characteristics. In the first comparison the measurements have been made at the maximum voltage of 700V and with the output power of 150W. In Figures 9 and 10 the curve traces of a turn-off switching are depicted in the case of utilization of the MOSFET device and of the cascode one. In Figures 11 and 12 the turn-on switching is then depicted for the two mentioned devices. As it is shown by the curve traces and as it is reported in Table II, the power dissipated by the cascode device in these experimental tests is near the same of the case the MOSFET one. Specifically, it can be seen that the higher conduction losses for the MOSFET at lower voltage correspond to the higher switching losses for the cascode and vice versa in high voltage operations.

Further experimental tests have been done by increasing the converter output power to 300W. The data of these tests are reported in Table III.

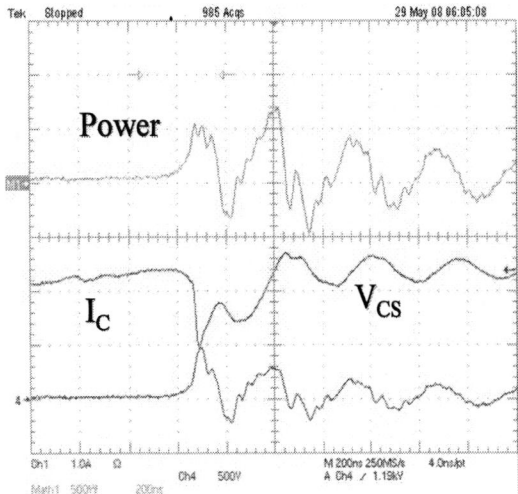

Figure 10 – Cascode turn-off @150W and V_{DC} 700V, I_C 1A/div, V_{CS} 500V/div, power 500W/div, time 200ns/div.

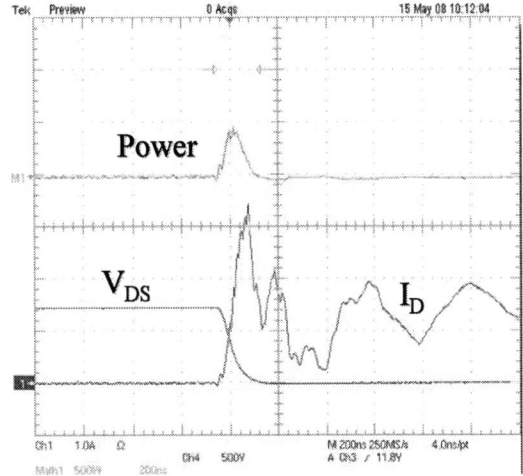

Figure 11 - MOSFET turn-on @150W and V_{DC} 700V, I_D 1A/div, V_{DS} 500V/div, power 500W/div, time 200ns/div.

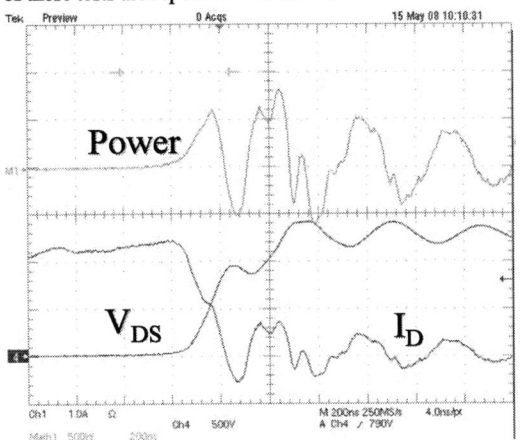

Figure 9 – MOSFET turn-off @150W and V_{DC} 700V, I_D 1A/div, V_{DS} 500V/div, power 500W/div, time 200ns/div.

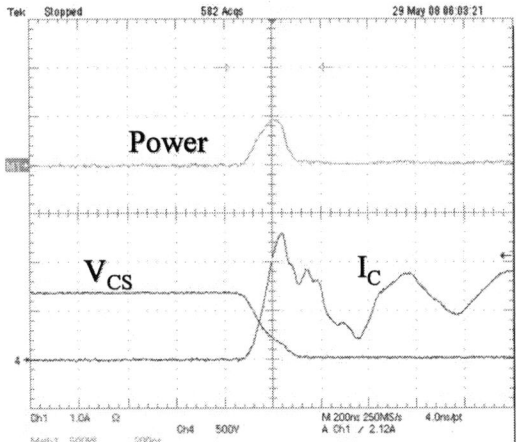

Figure 12 - Cascode turn-on @150W and V_{DC} 700V, I_c 1A/div, V_{CS} 500V/div, power 500W/div time 200ns/div.

Table II. Comparison between the MOSFET and cascode devices power losses at 150W.

DEVICE	V_{DC}	T_{CASE}	P_{TOT}	P_{COND}	P_{ON}	P_{OFF}
MOSFET	300V	35°C	4.3W	1.2W	0.3W	2.8W
CASCODE	300V	37°C	4.3W	0.5W	0.4W	3.4W
MOSFET	500V	39°C	5.5W	0.7W	1.1W	3.7W
CASCODE	500V	42°C	5.2W	0.3W	1.5W	3.4W
MOSFET	700V	44°C	6.9W	0.5W	2.2W	4.2W
CASCODE	700V	50°C	7.9W	0.2W	3.8W	3.9W

Table III. Comparison between the MOSFET and cascode devices power losses at 300W.

DEVICE	V_{DC}	T_{CASE}	P_{TOT}	P_{COND}	P_{ON}	P_{OFF}
MOSFET	300V	73°C	13W	5.4W	0.2W	7.4W
CASCODE	300V	44°C	7.7W	2.0W	0.4W	5.3W
MOSFET	500V	70°C	13.6W	3.2W	0.4W	10W
CASCODE	500V	54°C	9.5W	1.2W	3.3W	5.0W
MOSFET	700V	70°C	15.1W	2.5W	0.7W	11.9W
CASCODE	700V	67°C	14.7W	1.0W	7.8W	5.9W

In Table II and III it is shown the case temperature of the devices that has been reached in the experimental conditions. The devices that have been used for this investigations are two parallel connected 1500V, 8A MOSFET devices in TO247 package with a R_{DSON} 1.8Ω at 25°C, while the cascode devices were two paralleled 1500V 8A in TO247 package too. An interesting fact is that the die size of the MOSFET is 1.75 times the cascode one. For the thermal comparison between the two devices a 6°C/W heatsink has been used. The calculation of the equivalent thermal resistance has been carried out by means of the following expression:

$$R_{th_{eq}} = R_{th\,j\text{-}c} + R_{th\,c\text{-}h} + R_{th\,h\text{-}a} \qquad (7)$$

where:

- $R_{th\,j\text{-}c}$ is the junction-case thermal resistance;
- $R_{th\,c\text{-}h}$ is the case-heatsink thermal resistance;
- $R_{th\,h\text{-}a}$ is the heatsink thermal resistance.

The resulting equivalent thermal resistance is 6.4°C/W for the MOSFET device and 6.6°C/W for the cascode one, due to the device area of the high voltage MOSFET is approximately twice larger than the cascode one. It is important to remark that the on-state voltage of the

MOSFET is always about two times the one of the cascode device. This is the reason because the MOSFET has higher on-state losses at the same working conditions of the cascode one. This kind of behavior, according to the experimental tests carried out at low power level, is sufficient to understand that the MOSFET device is not suitable in this application when used at full load. By simple calculations we can estimate the case temperature at full load, based on the known thermal resistance value.

The conduction power loss will be over 20W and, by considering the other contributes, the whole power dissipation will reach about 40W. This means that a safe operation will require a heatsink with a thermal resistance of about 4°C/W, which will be expensive and too large for this application. Nevertheless further experimental tests have been carried out in order to reach the safety limit for the MOSFET device and to understand if the cascode represents the real alternative at full load and full voltage. Analyses have consequently been done to evaluate the experimental behavior of the converter in case of utilization of the cascode device at 450W. The experimental tests on the MOSFET show that the power losses on the switch bring it to a temperature of 137°C at the minimum applied voltage that is not acceptable for reliability reasons. Moreover the device shows always a worst behavior in comparison with the cascode one for all the conditions that have been tested during the experimental analysis. These considerations led to understand that the MOSFET solution is not suitable in case the requirement is for a single switch topology.

In Table IV the main data extracted for the converter in case of 450W input power are reported. The curve traces of this application are reported in the following figures. In Figures 13 and 14 the overall switching process is shown at full load with the input voltage of 500V when the MOSFET device is used and with 700V when the cascode is used. The next curve traces shown in Figures 15-18 are referred to the turn off and turn on transients.

As it can be seen from the curve traces and as shown in Table IV the cascode device always performs a better thermal behavior than the MOSFET in terms of case temperature at lower power losses, while in the case of full load operation the MOSFET presents a minimum in terms of power losses at the average input voltage level of 500V, and after this boundary the switching losses bring the device to a further increase of dissipation.

Table IV. Comparison between the MOSFET and cascode devices power losses at 450W.

DEVICE	V_{DC}	T_{CASE}	P_{TOT}	P_{COND}	P_{ON}	P_{OFF}
MOSFET	300V	137°C	35.5W	20.2W	TRASC.	15.3W
CASCODE	300V	60°C	13.4W	5.5W	0.5W	7.4W
MOSFET	500V	93°C	25.3W	8.1W	0.3W	16.9W
CASCODE	500V	87°C	21.4W	4.0W	6.5W	10.9W
MOSFET	700V	106°C	27.5W	5.9W	0.7W	20.9W
CASCODE	700V	104°C	27.0W	3.1W	12.3W	11.6W

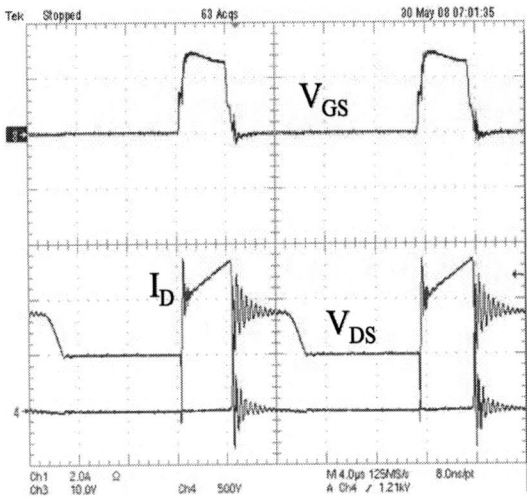

Figure 13 - MOSFET switching transient @450W and V_{DC} 500V, I_D 2A/div, V_{DS} 500V/div, V_{GS} 10V/div, time 4us/div.

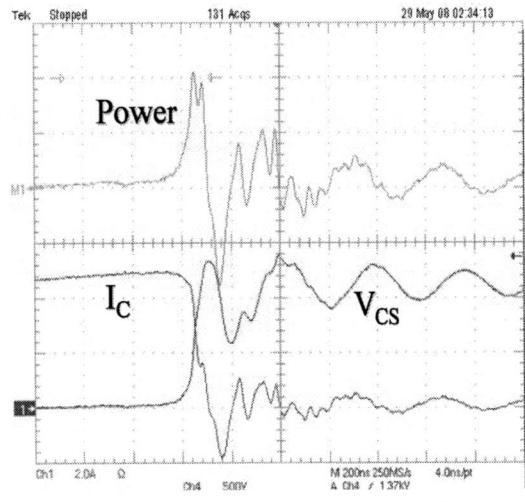

Figure 16 - Cascode Turn-off @450W and V_{DC} 700V, I_C 2A/div, V_{CS} 500V/div, power 1kW/div, time 200ns/div.

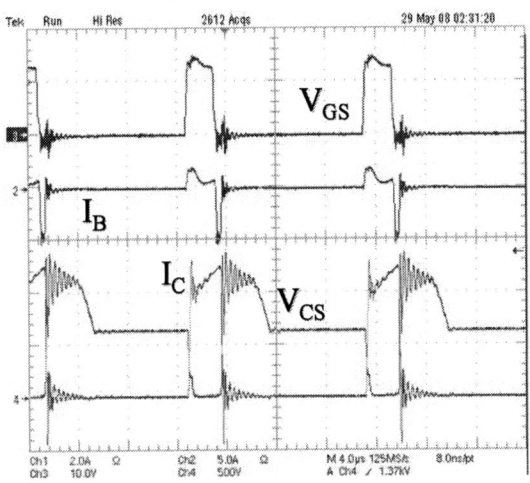

Figure 14 - Cascode switching transient @450W and V_{DC} 700V, I_C 2A/div, V_{CS} 500V/div, V_{GS} 10V/div, time 4us/div.

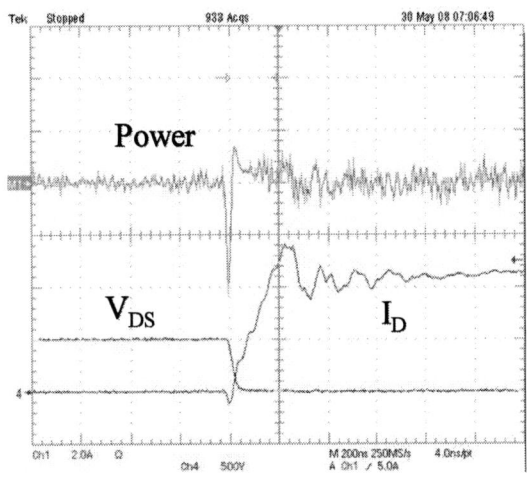

Figure 17 - MOSFET Turn-on @450W and V_{DC} 500V, I_D 2A/div, V_{DS} 500V/div, power 100W/div, time 400ns/div.

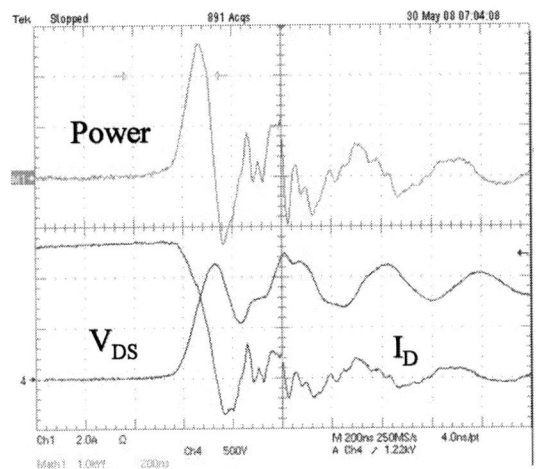

Figure 15 - MOSFET Turn-off @450W and V_{DC} 500V, I_D 2A/div, V_{DS} 500V/div, power 1kW/div, time 200ns/div.

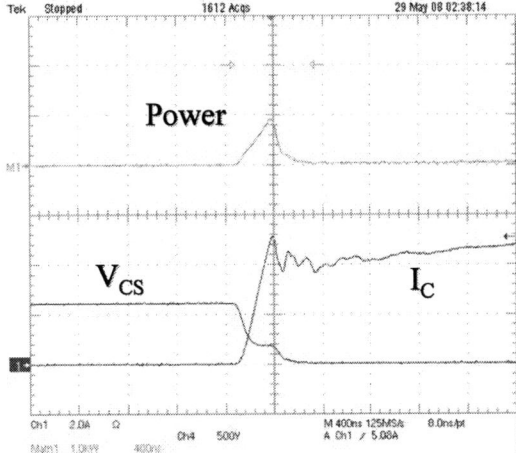

Figure 18 - Cascode Turn-on @450W and V_{DC} 700V, I_C 2A/div, V_{CS} 500V/div, power 1kW/div, time 400ns/div.

Fig. 19 indicates the power losses trend that the devices have in all the tested conditions. Specifically it is valuable that the cascode device offers a better behavior than the MOSFET in terms of total power losses in most of the conditions and, as shown in the previous tables, in all these cases the cascode temperature is maintained lower than the power MOSFET one.

In Fig. 20 a picture of the realized converter is shown.

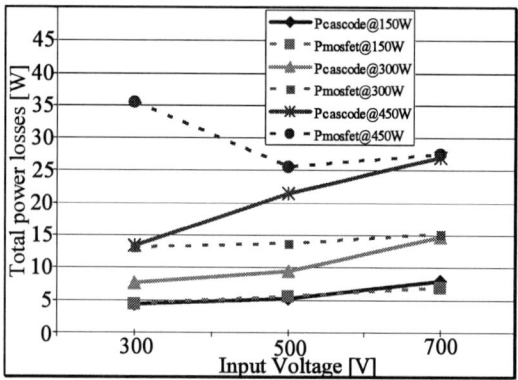

Figure 19 – Comparison between MOSFET and cascode device at different power and different input voltage levels.

IV. CONCLUSIONS

In this work a forward converter for SMPS industrial applications has been investigated and experimentally tested. The converter topology has been considered as the most suitable for the required power level (up to 450W). The main task of the work has been then to consider the most appropriate device accounting for the constraint of utilization of a single device topology. The experimental tests which have been carried out with both MOSFET and cascode devices bring to the result that the MOSFET shows higher power losses in most of the conditions and consequently it is not appropriate in the utilization as single switch as long as the voltage applied to the device is over 1000V. Consequently the cascode has shown its good peculiarities in terms of low power dissipation and operation at high frequency. Many interesting solutions have been also shown regarding the driving circuit for the cascode switch.

The exposed results led to consider the cascode device as a suitable choice when a voltage in the range between 1000-1700V and medium range of switching frequency are required by the application.

The results led to evaluate the suitability of the cascode device especially in case of supplying systems for welding-equipments or very high power solders where peak power values in the range between 1-5kW are achieved by means of series or multiple output converters.

Figure 20 – Photo of the realized converter.

REFERENCES

[1] B. Abdi, J. Milimonfared, S.H. Fatthi, "Modified Interleave Winding of Transformer to Improve SMPS's Performance," *Proceedings of the 12th International Power Electronics and Motion Control Conference*, Aug. 2006, pp. 659 – 662.

[2] B. J. Baliga, Power Semiconductor Devices, PWS Publishing Company, Boston, MA, 1995.

[3] J.L. Duarte, J. Rozenboom, T. Peijnenburg, H. Kemkens, Electromechanics and Power Electronics group, "Simulating a BJT-MOSFET Cascode-Connected Power Switch Suitable for Resonant Converters," *Eindhoven University of Technology*, IEEE Document 1992.

[4] A. Mihaila, F. Udrea, P. Godignon, G. Brezeanu, R.K. Malhan, A. Rusu, J. Millan, G. Amaratunga, "Towards Fully Integrated SiC Cascode Power Switches for High Voltage Applications," *Engineering Department, Cambridge University, Campus Universidad Autonoma de Barcelona, University "Politehnica" of Bucharest, DENSO CORPORATION*, Research Laboratories, IEEE Document 2003.

[5] C. Cavallaro, F. Chimento, S. Musumeci, A. Raciti, S. Buonomo, R. Scollo, "Monolithic Cascode Devices in Low Power Converter Applications with Wide Range Input Voltage," *Proceedings of the 19th International Symposium on Power Electronics, Electrical Drives, Automation and Motion, SPEEDAM 2006*, 23-26 May 2006, Taormina, pp. 646 – 651.

[6] N. Mohan, T. M. Undeland, W. P. Robbins, "Power Electronics: Converters, Applications, and Design", Second edition, John Wiley & Sons, New York, 1995.

[7] A. I. Pressman, "Switching Power Supply Design", Mc Graw-Hill, Inc. New York, 1991.

[8] M. H. Rashid, Power Electronics, Circuits, Devices, and Applications, Second edition, Prentice Hall, Inc., Englewood Cliffs, New Jersey, 1993.

[9] S. Musumeci, R. Pagano, A. Raciti, S. Buonomo, R. Scollo, G. Vitale, "A New Driving Circuit for Cascode Devices Performing Optimal Control of the Storage Time," *Proceedings of the Industry Applications Conference, 2005, Fortieth IAS Annual Meeting*, Conference Record 2005, Vol. 2, 2-6 Ott. 2005, pp. 1130-1137.

[10] F. Chimento, V. Crisafulli, S. Musumeci, A. Raciti, S. Buonomo, R. Scollo, "Experimental Investigation of Monolithic Cascode Devices in Inverter Leg Applications", *Proceedings of the 42st Industrial Application Society Annual Meeting*, IEEE-IAS 2007, 23rd-27th September 2007, New Orleans, Louisiana, USA, pp. 366-373.

Switching and conducting performance of SiC-JFET and ESBT against MOSFET and IGBT

André Knop *, W.-Toke Franke * and Friedrich W. Fuchs *

*Christian-Albrechts-University of Kiel, Institute of Power Electronics and Electrical Drives, Kiel, Germany,
ank@tf.uni-kiel.de, tof@tf.uni-kiel.de, fwf@tf.uni-kiel.de

Abstract—Here the switching and conducting performance of a SiC-JFET, an Emitter-Switching Bipolar Transistor (ESBT) and conventional power semiconductors as MOSFET and IGBT with Si- and SiC-diode is presented. The variety of power semiconductors is growing and there is a need to get rules to select them for the application given. The structure and special characteristics of the new devices are explained. The switching and conducting behavior of the devices is measured and investigated. The test circuit and the measurement method are presented. Based on the measured waveforms the power losses are calculated. The results of the switching and conducting performance of these power semiconductors are discussed.

Keywords—SiC-device, JFET, IGBT, MOSFET, Bipolar device, Power semiconductor device, Device characterization, New switching devices.

I. INTRODUCTION

Power converters as three phase inverters for feeding electrical machines or for application as uninterruptible power supplies are standard for general purpose in industry. In these applications, IGBTs and MOSFETs are the first choice because of their good switching and conducting behavior, their simple driving circuits and their price.

These power devices operate in most applications in the hard switching mode. The variety of power semiconductors is growing.

New devices are introduced into the market with new advantages and disadvantages [1-3]. The aim of the development of power semiconductors is low on-resistance, high operating temperature, high breakdown voltage, and fast switching abilities. These abilities allow higher switching frequencies and that results in a reduction of passive component size. This is very attractive for applications that demand high power density or high operating temperature. Interesting power semiconductors are silicon carbide (SiC) types as SiC junction field effect transistor and SiC-MOSFET.

The intention of this paper is an analysis of the switching and conducting performance of a Junction Field Effect Transistor (JFET) based on silicon carbide and an Emitter-Switching Bipolar Transistor (ESBT) with an IGBT and a 600 V MOSFET. For a fair comparison the IGBT and the MOSFET are tested with their internal Si-freewheeling diode and with an external SiC-diode. The analysis will be on the conducting and switching losses, since there is a trend towards increasing the switching frequency. IGBTs are on behalf of their switching power losses well suited for medium frequency applications [4]. They show the advantage of high blocking voltage. In contrast MOSFETs have lower switching losses but only limited blocking abilities and a poor freewheeling diode. ESBTs and SiC-JFETs seem to combine the advantages and avoid the disadvantages of IGBTs and MOSFETs.

In this paper, the static and switching characteristics of a SiC-JFET and ESBT with a bus voltage of 500 V are measured. The comparative low bus voltage is chosen since the breakdown voltage of the MOSFET is 600 V. Comparisons are carried out with state of the art IGBT and MOSFET. The main emphasis is put on the total losses of the switched device.

Section II describes the specific characteristics of the devices. In section III measurements are presented which show the conducting and switching behavior. After that in section IV the results will be compared and evaluated concerning the power losses and the complexity of their gate drivers. Section V includes the conclusion.

II. DEVICE CHARACTERISTICS AND STRUCTURE

A. ESBT characteristics

The Emitter-Switching Bipolar Transistor is a combination of a NPN bipolar transistor (BJT) and MOSFET. The BJT has an enhanced voltage blocking characteristic. The fast switching low voltage n-channel power MOSFET is realized inside the emitter of the BJT [5]. An equivalent circuit is shown in Fig. 2. In order to drive the BJT and MOSFET independently two separate terminals, gate and base, are required. Thus four terminals are necessary for the cascaded structure. The driving circuit presented in Fig. 1 allows connecting only the gate to the PWM-controller whereas the base is driven automatically by the circuit. The base current is proportional to the collector current. Thus a transformer can couple the collector and the base current [6]. The capacitor in the driving circuit takes the negative base current while the device is switched off. The stored energy of the capacitor provides the positive base current for the following turn-on [7]. Because the capacitor is shortened by the base the turn-on current starts with a peak for a fast opening of the collector-emitter path. The base current is provided by the transformer [6]. For the switch-off firstly the MOSFET is turned off and its drain current falls very fast down to zero. The collector current has to commutate to the base and flows through the capacitor to ground and the transistor starts blocking. With zero bias at the gate node and the zero BJT base current, the device behaves like a reverse biased PN diode in off state. All in all a high voltage device is obtained, with a breakdown voltage equal to the collector-base junction voltage from the BJT [1].

978-1-4244-1741-4/08/$25.00 ©2008 IEEE

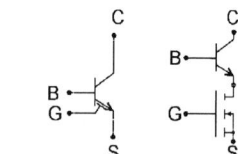

Fig. 2. ESBT symbol and equivalent circuit

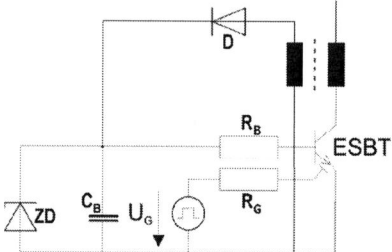

Fig. 1. ESBT driving circuit with transformer for the basis current

The speed of the turn-off switching is limited by the storage time of the devices. Thus, a suitable base current sink is required in order to speed up the turn-off process. So, in an ESBT device, the base charge is removed through a base current equal to the collector current [1]. A switching frequency above 100 kHz is published to be possible [7].

B. SiC-JFET Characteristics

Silicon carbide is a promising material for power semiconductors, since it has a large band gap of 3,26 eV that allows a high blocking voltage and a high junction temperature [8-11]. Besides that SiC has high electron mobility, so that switching frequencies above 100 kHz are published to be possible [12]. After SiC-Schottky-diodes are commercially available since 2001 the first switchable power devices based on SiC will come into the market in 2008. They will be designed as JFETs [3,13]. Characteristic for JFETs is that they are normally on, that means a negative voltage has to be provided to switch the device off. For this reason conventional gate drivers are not suitable. The advantages are that the theoretical operating temperature is above 350°C but momentarily limited by its maximum case temperature. Another advantage is that because of their low switching losses these devices can operate at high switching frequency [12-14]. The principle structure of a JFET is illustrated in Fig. 3. As long as there is no voltage at the gate contacts the n-channel between the p-doped gate zones is conducting. If a negative gate voltage is applied the depletion layer expends until the saturated voltage is reached and the n-channel is totally pinched off so that the JFET is blocking. In the conducting mode there is no pn-junction between drain and source so that the conducting losses are only caused by the $R_{DS,ON}$ of the channel. For power electronics a vertical structure with a buried gate seems to be the most promising structure because the high breakdown field can be fully used. The vertical JFET structure has an internal freewheeling pn-diode. Because of the wide band gap its forward voltage is with 3.5 V very high, so that an external SiC-Schottky-diode is recommended. In addition it is also possible to switch the JFET on while the diode is conducting, so that the conducting losses are just caused by the $R_{DS,ON}$ of the JFET.

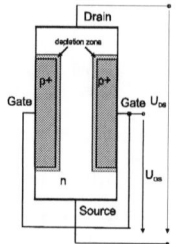

Fig. 3. sketched structure of a JFET

C. IGBT with Si- and SiC-Freewheeling Diode Charcteristics

Since IGBTs are very sensitive to backwards voltages, there is normally an external freewheeling diode placed next to the IGBT chip in the case [15]. For these investigations a non punch through IGBT in combination with an inverse parallel Schottky SiC-diode and a Si-diode to show the benefit of the SiC-diode are used separately. The advantage of the SiC-diode is that its reverse recovery charge is very small compared to Si-diode. That results in a very low reverse recovery current while the IGBT is turned on because there are only a few charge carriers in the diode that has to be removed.

D. MOSFET Characteristics

MOSFETs have an internal freewheeling diode because of their internal pn junctions. Characteristic for this diode is that it has a low forward voltage but a worse switching behavior. Because of the low forward voltage it is not possible to apply an external diode since it will always has a higher forward voltage and the current will flow through the MOS-diode. For a fair comparison the switching behavior is also investigated with a SiC-diode, even if that is not possible in most applications as DC to AC and AC to DC inverters. Also characteristic for the MOSFET is that the typical breakdown is limited to 600 V (however there are some exceptions with $V_{BR} = 800$ V but with an even worse diode).

III. MEASUREMENTS OF SiC-JFET, MOSFET, IGBT AND ESBT

A. Tested Power Semiconductors and Conditions

In this section the measurement results of the semiconductors (B), a SiC-JFET, an ESBT, a CoolMOS and an IGBT with a freewheeling SiC-Schottky diode are presented. For the IGBT and the MOSFET the measurements have been carried out with their internal

TABLE I.
USED SEMICONDUCTORS (*WITH SiC-DIODE)

	Type	Breakdown-voltage (V)	Max Current @ 100 °C (A)
MOSFET	SPW47N60CFD [16]	600 V	27
IGBT	SKW25N120 [17]	1200 V	25
ESBT	STC08IE120HV [18]	1200 V	8
SiC-JFET*	sample SiCED [19]	1200 V	8
SiC-Diode*	C2D20120D [20]	1200 V	2x10

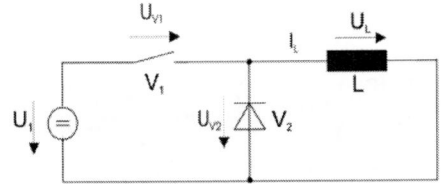

Fig. 4. buck converter as test circuit

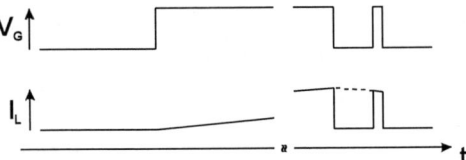

Fig. 5. Test signal (double pulse) for the test circuit: Gate-Signal from V_1 (upper) and the current through the inductor L (lower).

diode and an external SiC-diode, separately. The focus of the measurements is on switching and conducting behavior and their losses. All devices are tested at a bus voltage of 500 V and a current up to 10 A. For the ESBT, MOSFET and IGBT the recommended gate resistors are used. Since there is no datasheet for the JFET available a gate resistor of 20 Ω is applied. This resistor allows safe on- and off-switching without EMC disturbance.

All measurements are done with a very fast gate driver (rise and fall time < 30ns). The signals are measured with a Tektronix DPO4054 scope linked to Matlab for automatic storaging of the data, displaying the waveforms and calculting of the power losses.

B. Test setup for the Switching Characteristics

A buck converter shown in Fig. 4 is chosen as test setup there V_1 is the device under test and V_2 is the freewheeling diode.

For the investigation of the switching characteristics a double pulse gate signal is used (Fig. 5) [15,21]. By the length of the first conducting period of the device under test the current through it is adjusted because of the linear increase of the current through the inductance. Than the device under test (DUT) is switched off for the first time and the current commutates to the freewheeling diode until the DUT is switched on again. With this test setup the switching behavior can be investigated at defined voltage and current. Since the device is operating for very short period of time and only two switching cycles take place its self-heating can be neglected. For the measurements the devices are heated externally by a heating plate to 100°C for the experiments.

C. Calculation of the power losses

The power losses and the switching energy are calculated based on the measured waveforms by the following equations:

$$P_{SW} = u(t) \cdot i(t) \tag{1}$$

$$E_{ON}, E_{OFF} = \int_{t_{ON}, t_{OFF}} u(t) \cdot i(t) dt \tag{2}$$

The integration is realized by summing up all products of the discrete measuring points and multiplying them by the sample rate. The voltage drop across the stray

inductance of the semiconductor L_σ ($L_\sigma = 5...10$ nH) is smaller than one percent because the measuring probe is directly connected to the pins of the device.

D. Measurements

i.)   ESBT

The turn-off switching characteristic of the ESBT for different currents is shown in Fig. 6. When the gate signal of the MOSFET is low, the open emitter condition forces the collector current to flow through the base path of the BJT. Thus, the extraction of the charge stored in the base of the BJT is operated at current equal to collector current. So the storage time and by that the turn-off time varies extremely with the switched current (see TABLE II). The high delay time while turning off the ESBT might cause difficulties regarding driving and controlling of the converter.

The measured current includes the driver current from the base. In Fig. 7 the turn-on and turn-off waveforms with bus voltages of 500V from the ESBT are shown.

It is shown that the ESBT has a very small rise- and fall-time (70 ns, 20 ns) and the reverse recovery effect of the SiC-Diode is very low. The turn-on and turn-off losses of the ESBT are shown in Fig. 8 as function of load current.

ii.)   SiC-JFET

Fig. 9 shows the turn-on and turn-off waveforms of the SiC-JFET. The bus voltage is 500V and the switching current is 8 A. For the turn on Fig. 9 shows that the rise time of the gate voltage is 30ns and of the drain current is 85 ns. Because of the SiC-freewheeling diode the reverse recovery current peak is very low. The drain-source-voltage falls within 80ns to zero after the current reaches the 8 A. The turn on losses are shown in the lower left figure and reach their maximum at the maximum of the reverse recovery current.

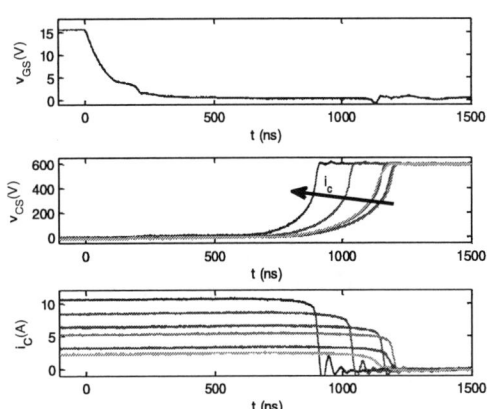

Fig. 6. turn-off switching time versus the current
@ $V_{bus} = 600$ V, $R_G = 47$ Ω, $R_B = 0,33$ Ω, $V_{GS,on} = 15$ V
$I_C = [2,2A; 3,1A; 5,2A; 6,3A; 8,4A; 10,7A]$

TABLE II.
TURN OFF TIME FOR DIFFERENT CURRENTS (ESBT)

I (A)	T_{off} (ns)
5	1200
6	1170
8	1050
10	900

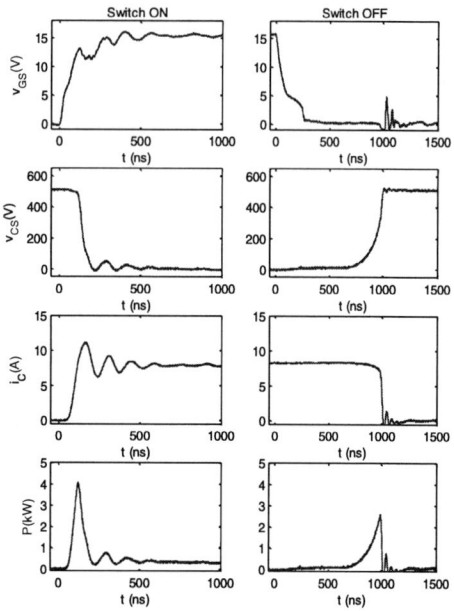

Fig. 7. Switching Waveforms of the ESBT with transformer for the basis current
@ V_{bus} = 500 V, I_C = 8 A, R_G = 47 Ω, R_B = 0,33 Ω, $V_{GS,on}$ = 15 V, T_J = 100°C

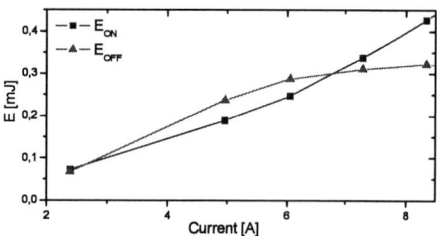

Fig. 8. Turn-on and turn-off loss versus current for the ESBT
@ V_{bus} = 500 V, $V_{GS,on}$ = 15 V, T_J = 100°C

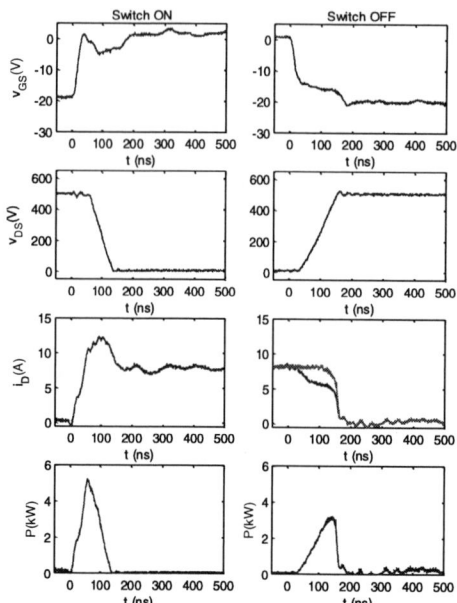

Fig. 9. Switching Waveforms of the SiC-JFET with SiC-Diode with measured (blue) and expected (green) waveform
@ V_{bus} = 500 V, R_G = 20 Ω, $V_{GS,off}$ = -19 V, I_D = 8 A, T_J = 100°C

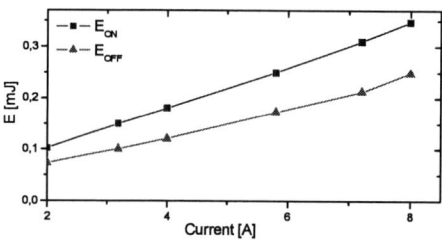

Fig. 10. Turn-on and turn-off loss versus current for the JFET
@ V_{bus} = 500V, $V_{GS,off}$ = -19 V, T_J = 100°C

The turn-off behavior is shown on the right side of Fig. 9. Here the gate source voltage drops initially to 75 % of its pinch-off voltage. However the drain-source-voltage begins to rise at the same time. Since the pinch-off voltage of the JFET is -19.0 V v_{DS} should not rise until the pinch-off voltage is reached. For that reason it can be assumed that this effect (v_{DS} rising before v_{GS} < -19 V) is caused by electromagnetic injection to the measuring probe and that the plateau at 15 V of v_{GS} does not exist in reality. The same effect appears at the drain current that starts slightly to fall at the beginning of the switching event. The green line shows the expected waveform that is also used for calculating the turn off losses. Nevertheless the rise time of the v_{DS} is with 134 ns longer than for the turn on. The turn-off losses are smaller than the one for the turn-on, due to the reverse recovery current of the diode at the transistor turn on.

In Fig. 10 the switching energy versus the current is displayed. It can be seen that the losses increase with the current and that the switching energy is higher for the turn on.

iii.)   IGBT

The main limitation to the turn-off speed of an IGBT is the lifetime of the minority carriers in the base of the BJT. Since the base is not accessible, external drive circuitry cannot be used to improve the switching time. This charge stored in the base causes the characteristic "tail" in the current's waveform of an IGBT. This tail current slightly increases the turn-off losses [4,22]. The switching characteristics are shown in Fig. 11 and the switching losses in Fig. 12.

iv.)   MOSFET

The intrinsic diode of the MOSFET has a very high reverse recovery charge. This charge results from the minority carrier recombination at the turn-off of the diode. This reverse recovery current is added to the MOSFET current and results to a very high current spike at the switch on-time from the MOSFET. This is shown in Fig. 13. A peak reverse recovery current of 74 A at a load current of 10 A is measured and results in very high turn on losses. This current peak results in a voltage drop of v_{ds} across the stray inductance of the source.

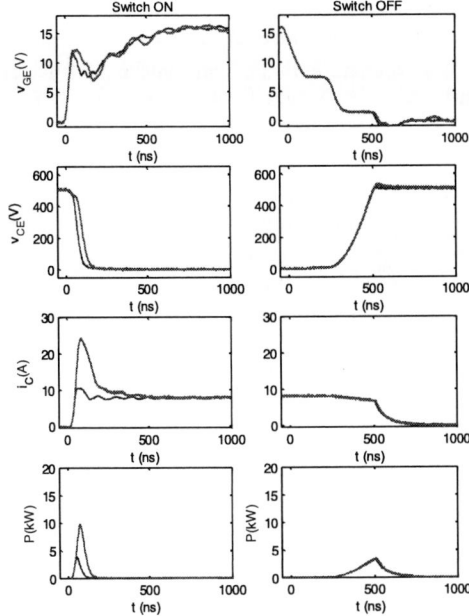

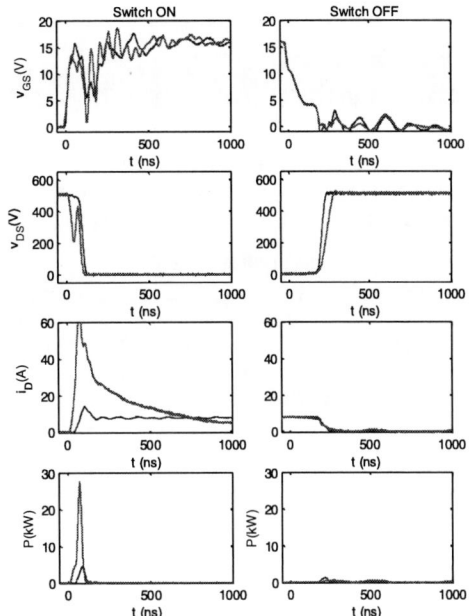

Fig. 11. Switching characteristic from the IGBT with internal Si-Diode (solid green line) and with SiC-Diode (dotted blue line) @ V_{bus} = 500 V, R_G = 22 Ω, $V_{GE,on}$ = 15 V, I_C = 8 A, T_J = 100°C

Fig. 13. Switching characteristic from the MOSFET with internal Si-Diode (solid green line) and with SiC-Diode (dotted blue line) @ V_{bus} = 500 V, R_G = 3,3 Ω, I_D = 8 A, $V_{GS,on}$ = 15 V, T_J = 100°C

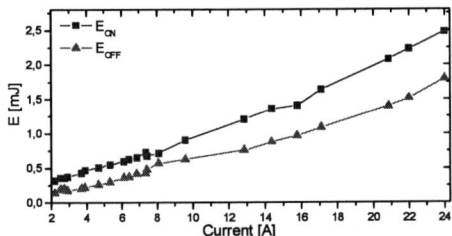

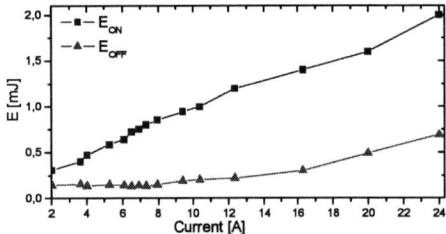

Fig. 12. Turn-on and turn-off loss versus current for the IGBT with Si-Diode @ V_{bus} = 500 V, $V_{GE,on}$ = 15 V, T_J = 100°C

Fig. 14. Turn-on and turn-off loss versus current for the MOSFET with Si-Diode @ V_{bus} = 500 V, $V_{GS,on}$ = 15 V, T_J = 100°C

The very low reverse recovery current of the SiC-Diode is also shown in Fig. 13.

During the turn-off of the MOSFET, the internal capacitances have to be recharged, so that there are no charge carrier influences in the channel area. Thereafter, the neutrality interference in this area will quickly be reduced and the drain current will drop rapidly. This results in very low turn-off losses.

E. Test setup and Measurements for the Conducting Characteristics

For the investigation of the conducting behavior the device under test (DUT) is in conducting state. To prevent self-heating of the device another switched semiconductor is connected in series and is switched on for a short pulse. The load is an adjustable resistor for limiting the current (Fig. 15). As before for the measurement of the switching behavior the DUT is external heated to T_J = 100 °C with a heating plate.

The voltage drop across the ESBT device is given by two contributions: the $V_{CE,SAT}$ of the BJT and the voltage drop across the low voltage MOSFET ($R_{DS,ON}$).

The ESBT has implemented a NPN BJT, that has a lower $V_{CE,SAT}$ in comparison to an PNP BJT that is included in the IGBT. So the saturation voltage of an IGBT is higher.

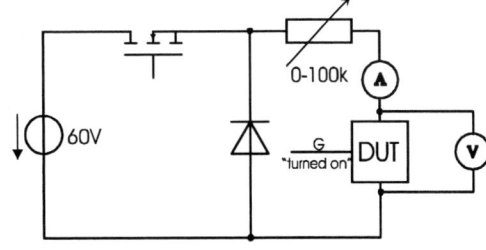

Fig. 15. Test circuit for the conducting behavior

The conducting losses of the JFET and MOSFET are caused by its $R_{DS,ON}$. The conducting losses of ESBT, JFET, MOSFET and IGBT are measured for different input currents in Fig. 16.

IV. COMPARISION OF SiC-JFET, MOSFET, IGBT AND ESBT

The semiconductors are compared with respect to their switching and conducting losses as well as the effort for their driving circuits.

A. Conducting and switching losses

For semiconductor power loss reduction the maximum current of the MOSFET and IGBT has to be derated to 24 A for a switching frequency of $f_{SW} = 50$ kHz. Otherwise the heat cannot be dissipating over the heat sink. Fig. 16 shows the conducting losses of the four devices. It can be seen that the conducting losses of the ESBT and JFET are well below to them of the IGBT and MOSFET.

Fig. 17 shows the switching losses of ESBT, JFET, MOSFET and IGBT. For the ESBT and the JFET it is assumed that three devices are connected in parallel for having the same current carrying ability as the MOSFET and the IGBT. For the MOSFET and the IGBT their internal diodes are used, while for the ESBT and JFET SiC-diodes are connected inverse parallel.

The turn-off and turn-on losses of the ESBT and SiC-JFET are both much smaller than those of the IGBT and MOSFET. Two factors are the main reason of the different switching losses. These are the high switching

speed (rise and fall time) and the reduced reverse recovery current of the freewheeling diode.

B. Total losses in a switching application

To compare the total losses a high switching frequency application with a switching frequency of $f_{SW} = 50$ kHz and a duty cycle of $D = 0.5$ is used.

The switching losses are calculated as

$$P_{SW,ON} = E_{ON} \cdot f_{SW} \tag{3}$$

$$P_{SW,OFF} = E_{OFF} \cdot f_{SW} \tag{4}$$

$$P_{Total} = P_{SW,ON} + P_{SW,OFF} + P_{Cond} \cdot D \tag{5}$$

Fig. 18 and Table III shows a total loss comparison at 24 A and 500 V.

C. Driver efforts

The drivers for the IGBT and MOSFET are comparable simple, since only a constant positive and additional for the IGBT a constant negative voltage has to be provided. However the ESBT demands for a complex circuit providing the base current and for the JFET a high negative voltage is necessary. In practical use the driving circuit for the JFET is even more complex because of the normally on behavior of the device and that each single JFET has a different pinch-off voltage that is close to the maximum negative gate voltage. However it is possible to use the JFET in a cascode circuit for driving it with a MOSFET driver [23]. Therefore the drain of a low voltage MOSFET is connected to the source of the JFET and the gate of JFET is linked to the source of the MOSFET. If the MOSFET is turned off, its drain to source voltages rises and the gate of the JFET gets a negative voltage so that the JFET starts blocking as well. The disadvantages of these circuits are that the conducting losses increase because of the two devices in series

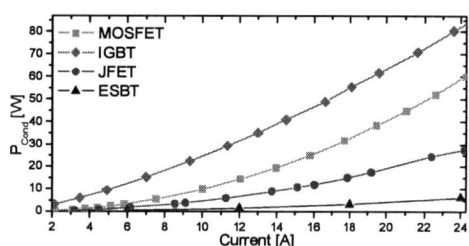

Fig. 16. Conducting losses versus current of ESBT, SiC-JFET, IGBT and MOSFET (For ESBT & JFET three chips in parallel) @ $V_{bus} = 500$ V, $T_J = 100°C$

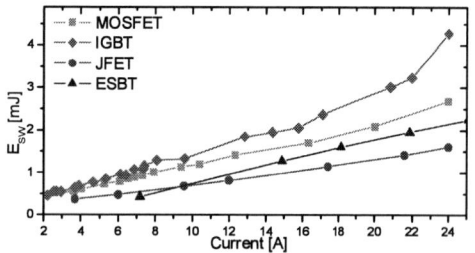

Fig. 17. Comparison of switching loss versus current between ESBT, SiC-JFET, IGBT and MOSFET (For ESBT & JFET three chips in parallel) @ $V_{bus} = 500$ V, $T_J = 100°C$

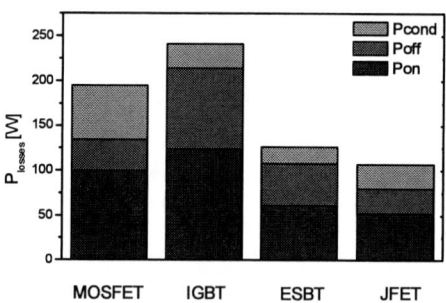

Fig. 18. Loss Comparison of ESBT, SiC-JFET, IGBT and MOSFET @ $V_{bus} = 500$ V, I = 24A $T_J = 100°C$, $f_T = 50$ kHz

TABLE III.
SWITCHING POWER LOSSES $V_{Bus} = 500$ V, I = 24 A, $F_T = 50$ KHZ

	E_{ON} [µJ]	E_{OFF} [µJ]	P_{Cond} [W]	P_{SW} [W]
MOSFET	2000	689	60,2	134,45
IGBT	2480	1800	27	124
3xESBT + SiC-diode	1227	925	18,7	107,61
3xSiC-JFET + SiC-diode	1044	561	27,4	80,25

74

and that the operating temperature is limited to the maximum junction temperature of the MOSFET.

V. CONCLUSION

Four different power semiconductors (ESBT, SiC-JFET, MOSFET and IGBT, last one with SiC-diode and Si-diode) have been investigated concerning their switching and conducting behavior, their power losses and the efforts which have to be done to drive the device.
The measurements show that the new devices JFET and ESBT have a better performance than the well established ones as IGBT and MOSFET. The switching losses of the IGBT can be reduced by replacing the Si-diode by a SiC-freewheeling-diode which has a very low reverse recovery current. However the efforts for the drivers of the JFET and the ESBT are high. Because of the drivers and the SiC substrate of the JFET converters based on these devices are more expensive than converters with traditional power semiconductors. In applications where efficiency, cooling or operation at very low or high temperature is of importance the JFET and the ESBT may be an interesting alternative.

ACKNOWLEDGMENT

This work has been partly financed by the state Schleswig-Holstein (Germany).

REFERENCES

[1] S. Buonomo, S. Musumeci, R. Pagano, C. Porto, A. Raciti, and R. Scollo, "Analysis and performances of a new emitter-switching bipolar transistor device suitable for high-voltage applications," in IECON'03. 29th Annual Conference of the IEEE Industrial Electronics Society, 2 ed, 2003.

[2] AR. Powell and L. Rowland, "SiC materials-progress, status, and potential roadblocks," Proceedings of the IEEE, vol. 90, no. 6, pp. 942-955, 2002.

[3] P. Friedrichs, "SiC power devices - recent and upcoming developments," in IEEE International Symposium on Industrial Electronics, 2006, pp. 993-997.

[4] U. Nicolai, P. R. W. Martin, and SEMIKRON International GmbH & Co.KG, Application manual power modules, 1. ed. Ilmenau: ISLE, 2000.

[5] R. Pagano, "Characterization, parameter identification, and modeling of a new monolithic emitter-switching bipolar transistor," IEEE Transactions on Electron Devices, vol. 53, no. 5, pp. 1235-1244, 2006.

[6] S. Buonomo, F. Saya, and G. Vitale, "ESBT in industrial PFC topologies," in 11th European Conference on Power Electronics and Applications, 2005.

[7] S. Buonomo, C. Ronsisvalle, R. Scollo, S. Musumeci, R. Pagano, and A. Raciti, "A new monolithic emitter-switching bipolar transistor (ESBT) in high-voltage converter applications," in Conference Record of the IEEE Industry Applications Conference, 3 ed, 2003, pp. 1810-1817.

[8] T. Funaki, JC. Balda, J. Junghans, AS. Kashyap, FD. Barlow, HA. Mantooth, T. Kimoto, and T. Hikihara, "Power conversion with SiC devices at extremely high ambient temperatures," IEEE Transactions on Power Electronics, vol. 22, no. 4, pp. 1321-1329, July2007.

[9] PG. Neudeck, RS. Okojie, and Y. C. Liang, "High-temperature electronics - a role for wide bandgap semiconductors?," Proceedings of the IEEE, vol. 90, no. 6, pp. 1065-1076, 2002.

[10] D. Schröder, Leistungselektronische Bauelemente, 2. ed. Berlin: Springer, 2006.

[11] D. Stephani, "Prospects of SiC Power Devices: From the State of the Art to Future Trends," in International Conference and Exhibition on Power Electronics, Intelligent Motion, Power Quality, 2002.

[12] MS. Chinthavali, B. Ozpineci, and L. Tolbert, "High-temperature and high-frequency performance evaluation of 4H-SiC unipolar power devices," in APEC 2005. Twentieth Annual IEEE Applied Power Electronics Conference and Exposition, 1 ed, 2005, pp. 322-328.

[13] P. Friedrichs, "Unipolar SiC devices - latest achievements on the way to a new generation of high voltage power semiconductors," in Conference Proceedings. IPEMC 2006. CES/IEEE 5th International Power Electronics and Motion Control Conference, 1 ed, 2006, pp. 1-5.

[14] A. Elasser and T. Chow, "Silicon carbide benefits and advantages for power electronics circuits and systems," Proceedings of the IEEE, vol. 90, no. 6, pp. 969-986, 2002.

[15] M. Helsper, "Analysis and improvement of the behavior of planar and trench IGBTs modules in hard and soft switching applications (Analyse und Verbesserung des Verhaltens von Planar- und Trench-IGBT-Modulen in hart bzw. weich schaltenden Applikationen)," 136 S Dissertation, Shaker, 2003.

[16] Infineon [Online], "Datasheet SPW47N60CFD," Available: http://www.infineon.com, 2008.

[17] Infineon [Online], "Datasheet SKW25N120," Available: http://www.infineon.com, 2006.

[18] STMicroelectronics [Online], "Datasheet STC08IE120HV," Available: http://www.stmicroelectronics.com, 2007.

[19] SiCED [Online], "SiC-JFET," http://www.siced.de, 2008.

[20] CREE [Online], "Datasheet C2D20120D," Available: http://www.cree.com, 2007.

[21] M. Helsper, F. W. Fuchs, and R. Jakob, "Measurement of Dynamic Characteristics of 1200 A/ 1700 V IGBT-Modules under Worst Case Conditions," in NORPIE Nordic Workshop on Power and Industrial Electronics, 2000.

[22] International Rectifier, "IGBT Characteristics," ONLINE www.irf.com, 2007.

[23] S. Round, M. Heldwein, J. Kolar, I. Hofsajer, and P. Friedrichs, "A SiC JFET driver for a 5 kW, 150 kHz three-phase PWM converter," in Conference Record of the 2005 IEEE Industry Applications Conference Fortieth IAS Annual Meeting, 1 ed, 2005, pp. 410-416.

In-Service Life Consumption Estimation in Power Modules

[1]Mahera Musallam, [2]C Mark Johnson, [3]Chunyan Yin, [4]Hua Lu, [5]Chris Bailey

[1,2]University of Nottingham,School of Electrical and Electronic Engineering

Nottingham, U. K., e-mail: *mahera.musallam@nottingham.ac.uk, Mark.Johnson@nottingham.ac.uk*

[3, 4, 5]School of Computing and Mathematical Sciences, University of Greenwich, UK.

e-mail: *c.yin@gre.ac.uk, h.lu@gre.ac.uk, c.bailey@gre.ac.uk*

Abstract—Health management and reliability form a fundamental part of the design and development cycle of electronic products. In this paper compact real-time thermal models are used to predict temperatures of inaccessible locations within the power module. These models are then combined with physics of failure based reliability analysis to provide in-service predictions of crack propagation in solder layers and at the bond wire joints as a result of thermal cycling. The temperature estimates are combined with lifetime based reliability models to provide a tool for life consumption monitoring. Rainflow counting algorithms are applied to the temperature vs. time data to extract the occurrence frequencies of different thermal cycling ranges. Knowledge of the life consumed for each different cycle then allows the remaining life time to be estimated under arbitrary operational conditions. The technique can be employed to provide functions such as life consumption monitoring and prognostic maintenance scheduling.

Keywords— Reliability, IGBT, Modelling, Prognosis, Pulse Width Modulation (PWM) Thermal stress

I. INTRODUCTION

In this work a typical IGBT half bridge module is studied in which each switch and diode element comprises a number of parallel connected die (in this case 3). The dies are mounted onto electrically insulating DBC substrates using a Sn-Ag eutectic solder and these substrates are then mounted onto a copper baseplate using the same solder, as illustrated in fig. 1. The modules were attached to coolers consisting of an 8 mm thick copper plate which was spray cooled with water. A complete compact real-time thermal model [1] is used to estimate the temperatures of salient features in the power modules such as die junction temperature, substrate, and base plate temperature.

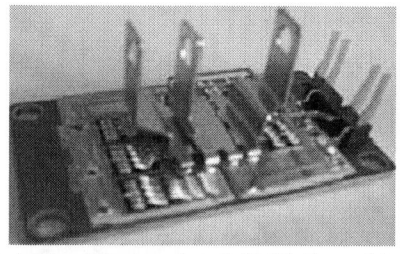

Fig. 1. Internal structure of a typical half bridge module. The IGBT and diode dies with the substrate tiles mounted on the module base-plate with interconnects and bus-bars attached.

An accurate representation of the module's dynamic thermal behavior was created and combined with models that provide estimates of the device losses based on measured values of phase current. The temperature achieved by a particular device is a function of the time-history of power dissipation within the module. In addition, the heat flux from any device dissipating power to the heatsink will affect the temperature of other devices within the module (fig. 2).

The thermal model parameters for each device, including the self-heating and cross-coupled heating effects, were determined over pre-defined ranges of temperature and current through step response measurements. Measurements of device on-state and switching losses were made at a fixed dc-bus voltage for varying current at a range of temperatures to determine the device losses. Fig. 3 shows a schematic diagram showing the half bridge IGBT modules embedded in an inductively-loaded PWM full bridge converter.

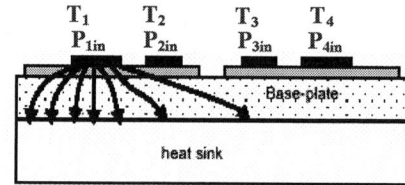

Fig.2. Cross-section of module and heatsink

Previous tests proved the validation of the real time thermal model and example of some test results are shown in fig. 4 and 5 [1].

II. COMPLETE REAL-TIME COMPACT THERMAL MODEL IMPLEMENTATION

In packaged power modules as shown in fig. 1, thermal modelling of the dies and package is generally limited to the die and base plate temperatures [1]. For interfaces away from the surface, such as solder layers, measurement of temperature is difficult and therefore direct determination of the thermal parameters for the solder layers and other hidden layers within the module's package is impractical. To provide a more complete description of the heat transfer path, computational models were developed using the FLOTHERM software packages [2,3] to predict the temperatures of hidden layers within the module. The thermal parameters of the heat transfer path from any heat source (die) to the hidden layers were determined [4]. The FLOTHERM models

978-1-4244-1741-4/08/$25.00 ©2008 IEEE

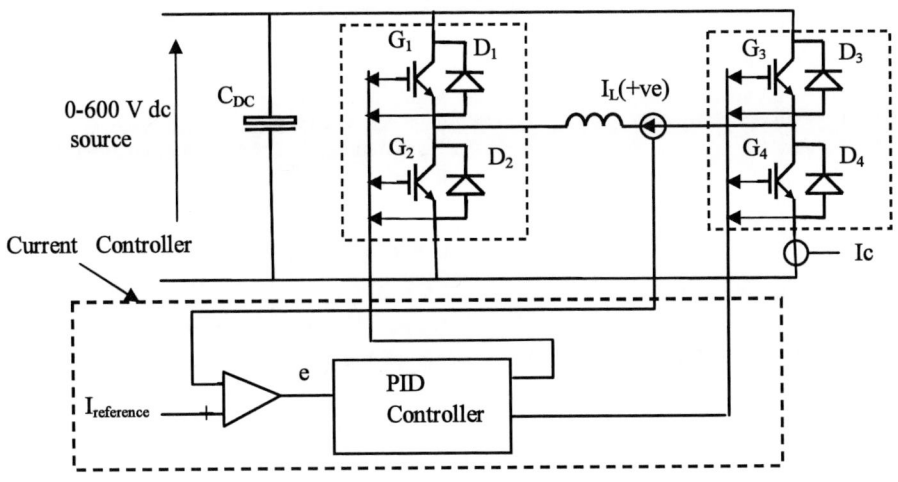

Fig.3. Schematic of real-time thermal model for a half bridge converter with load current controller.

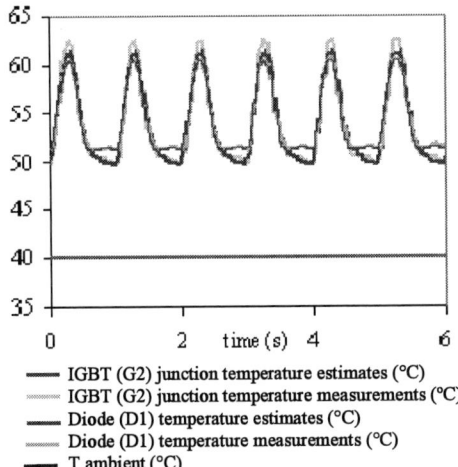

IGBT (G2) junction temperature estimates (°C)
IGBT (G2) junction temperature measurements (°C)
Diode (D1) temperature estimates (°C)
Diode (D1) temperature measurements (°C)
T ambient (°C)

Fig. 4. Comparison between junction temperature estimates and measurements for IGBT (G2) and Diode (D1) at 1 Hz sine wave modulation frequency.

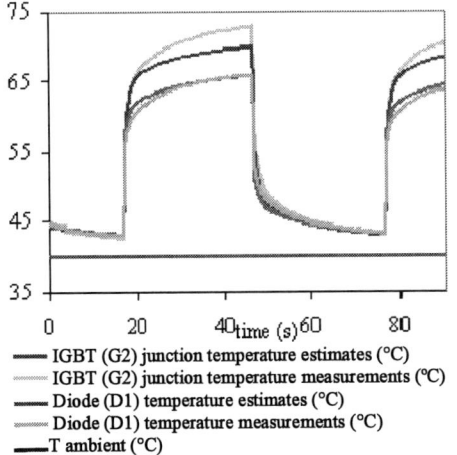

IGBT (G2) junction temperature estimates (°C)
IGBT (G2) junction temperature measurements (°C)
Diode (D1) temperature estimates (°C)
Diode (D1) temperature measurements (°C)
T ambient (°C)

Fig.5. Comparison between junction temperature estimates and measurements for IGBT (G2) and Diode (D1) at 0.017 Hz square wave modulation frequency.

were validated by comparing the results with measurable surface temperatures, for example at the base-plate, substrate and junctions as illustrated in fig.6.

The thermal behavior of the whole module can therefore be represented by an NxM transfer function matrix representing both self-heating, cross-coupling effects and the responses of the hidden layers such as solder, substrate and basplate layers. This matrix representing the combined module thermal impedance is given in (1) where the terms $a_{11}....a_{MN}$ represent the transfer function of each heat transfer path, T_A is the ambient temperature, $P_{1in}...P_{Nin}$ are the heat sources and $T_{J1}....T_{JN}...T_{Baseplate}$ are the temperatures at certain points (such as the device junctions, solder, substrate and basplate) within the module.

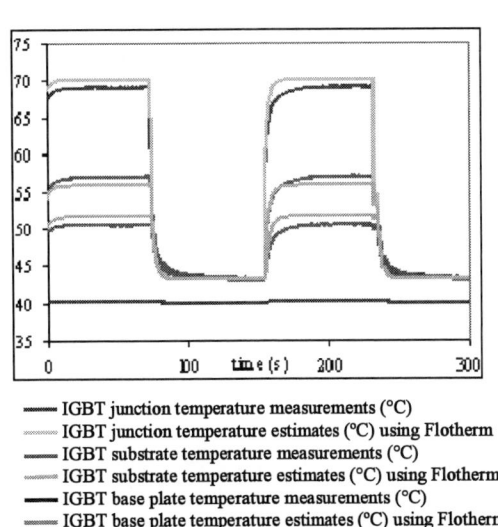

IGBT junction temperature measurements (°C)
IGBT junction temperature estimates (°C) using Flotherm
IGBT substrate temperature measurements (°C)
IGBT substrate temperature estimates (°C) using Flotherm
IGBT base plate temperature measurements (°C)
IGBT base plate temperature estimates (°C) using Flotherm
Heatsink cooling temperature (°C)

Fig.6. Validation of FLOTHERM Models [3].

$$\begin{bmatrix} T_{J1} \\ T_{J2} \\ \vdots \\ T_{JN} \\ T_{Solder\_2} \\ T_{Substrate} \\ T_{Solder\_1} \\ T_{Baseplate} \end{bmatrix} = \begin{bmatrix} a_{11}\,a_{12}\ldots a_{1N} \\ a_{21}\,a_{22}\ldots a_{2N} \\ \vdots \\ a_{N1}\,a_{N2}\ldots a_{NN} \\ \vdots \\ \vdots \\ \vdots \\ a_{M1}\,a_{M2}\ldots a_{MN} \end{bmatrix} \begin{bmatrix} P_{1in} \\ P_{2in} \\ \vdots \\ P_{Nin} \end{bmatrix} + T_A \qquad (1)$$

III. Physics of Failure Models

Thermal stress is generated at various parts within the power module by thermal deformation because of the difference of the coefficients of thermal expansion of power modules' component materials. Fig. 7 shows a cross section of different materials in which power devices consist of. The major causes of failure of power modules occur in the solder layers, bond-wires and interfaces between internal layers due to thermo-mechanical stresses due to CTE mismatch [5]. Environmental and operating conditions such as temperature changes, vibration, etc, can cause degradation within the materials leading to failure [6]. Fig.8 and 9 illustrate two common failures observed in power modules. The physics of failure approach combines mathematical modeling with accelerated life testing to predict the reliability [6,7] of the power module.

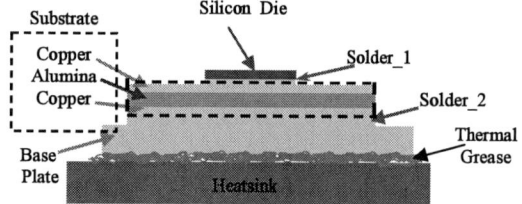

Fig.7. The internal structure of a typical IGBT module.

Fig.8. Ceramic Substrate Failure

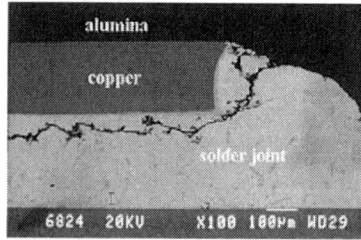

Fig.9. Solder Failure

A given temperature profile such as the IGBT junction temperatures in terms of temperature vs. time data is used as input to the physics of failure models so that these models can predict stress and strain (the physics that generate damage in the module materials) which then is used to provide information about the life expectancy of the power module. In this paper physics of failure models have been developed for the die attach and substrate mount-down solder layers within a typical IGBT power module.

A. Application of life Time Models for Bond Wire Joints

One of the possible modes of failure is bond wire damage. At the bond, large strain is induced on the interface between the wire and the die due to thermal expansion mismatch of the utilized materials under thermal cycling. The fatigue failures are observed at the bond wire joints where the maximum stress could be accumulated as shown in fig.10. As a result a fatigue crack initiates at the edge of the bonded area and propagates parallel to the interface. Fig. 11 shows an example of common wire bond failure.

Compact thermal models combined with life time models are used to provide temperature-time information which can be used to drive a reliability model, for example it can be used to determine wire bond life, which is largely governed by die temperature. On the other hand, these models can provide in-service predictions of crack propagation for the bond wire joints and the substrate mount-down solder layers respectively. Fig. 12 represents the real-time implementation of the compact thermal model with life time model approach.

Thermal fatigue tests of the aluminum bonding wires were studied and a fatigue lifetime model for the wire bond is developed based on in-service real life data. From these results an empirical lifetime model can be derived as shown in fig. 13 and expressed as a function of the cyclic temperature ΔT:

$$N_f = 5.6*10^{11}\Delta T^{-3.597} \qquad (2)$$

where Nf is the material number of cycles to failure

ΔT is temperature variation.

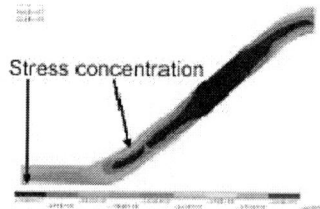

Fig.10. Stress concentration at the wire bond joint

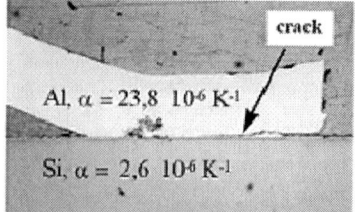

Fig.11. wirebond Failure

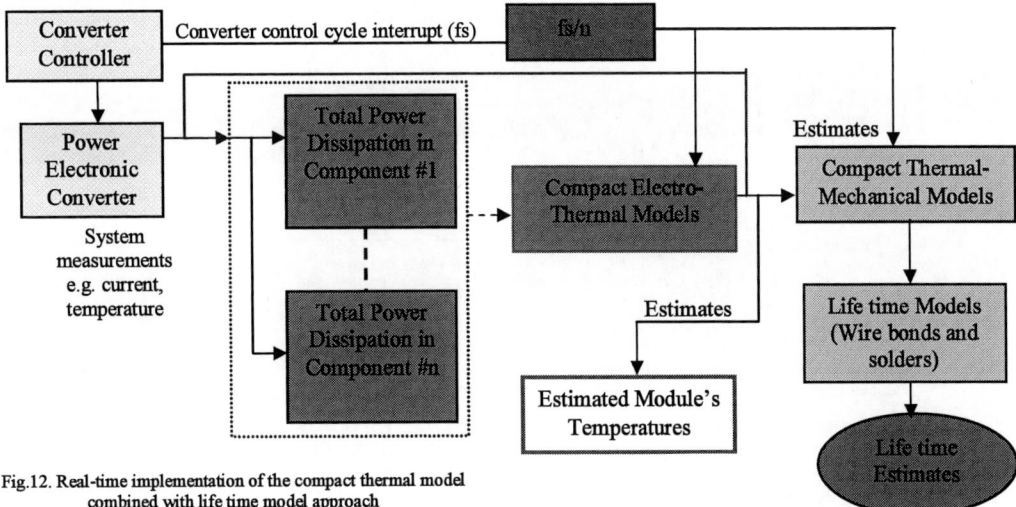

Fig.12. Real-time implementation of the compact thermal model combined with life time model approach

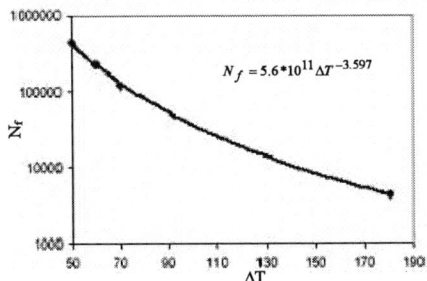

$$N_f = 5.6*10^{11}\Delta T^{-3.597}$$

Fig. 13: life time extraction for the IGBT bond wire interconnect.

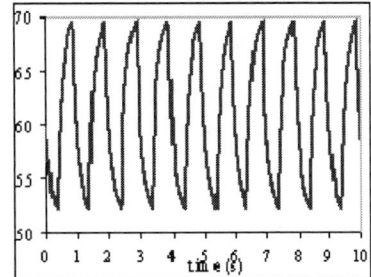

Fig.15: Sample of the real time IGBT junction temperature estimates.

To achieve the purpose of this work and to predict the accurate effect of arbitrary in-service conditions on the device life, an accurate understanding of the physics that drives the principal wear-out mechanisms and a detailed knowledge of the operational profile is required [8]. The real time compact thermal model was applied to a uniform sequence of current demand level over a period of one minute. The load cycling profile has a uniform amplitude (170A) and frequency (1 Hz). An example of this load is shown in fig. 14. Real-time IGBT junction temperature estimates corresponding to this load profile are shown in fig. 15.

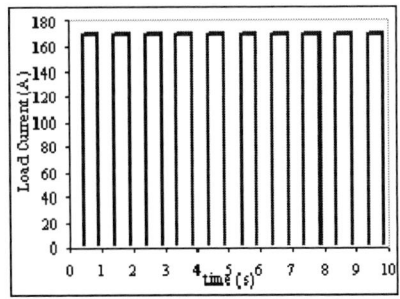

Fig.14. Sample of the load current profile applied to the real time thermal model .

A rainflow counting algorithm was applied to the IGBT junction layer temperature data and used to create a temperature range histogram with a resolution of 5 K. Fig.16 shows the rainflow a sample of the rainflow cycles extracted from the temperature-time data. The temperature values extracted from the rainflow counting method are arranged into histogram bins each of which represents the cycles counted within that bin as shown in fig. 17. To apply the bond wire life time model (2) was recalled. Life estimates may be made by employing Palmgren-Miner rule along with a cycle counting procedure. The life consumption for the bond wire can be obtained as:

$$LC = \sum_{i=1}^{k} \frac{n_i}{N_i}$$

(3)

$$LC = \left[\frac{n_1}{Nf_1}\right]_{k1} + \left[\frac{n_2}{Nf_2}\right]_{k2} + \left[\frac{n_3}{Nf_3}\right]_{k3} + \left[\frac{n_4}{Nf_4}\right]_{k4}$$

Table 1 summarizes the rainflow counting results and the life consumption calculation for the IGBT bond wire.

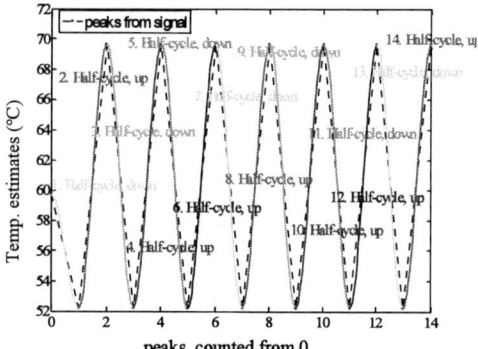

Fig.16. Sample of the rainflow cycles extracted from of IGBT junction temperature data

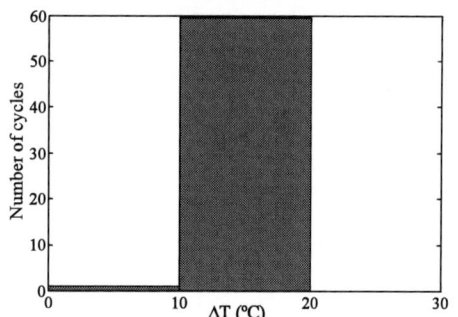

Fig. 17. Rainflow histogram of IGBT junction temperature variations (ΔT) obtained by applying a uniform load profile for one cycle (1 minutes)

Table 1: Calculation of life consumption for the IGBT wire bond with uniform load applied

Number of cycles	$0 \leq \Delta T \leq 10$ (k_1)	$10 < \Delta T \leq 20$ (k_2)	$20 < \Delta T \leq 30$ (k_3)
n (1 cycle)	1	59.5	0
ΔT	10	20	0
Nf	$14.16 * 10^7$	$11.7 * 10^6$	0
LC	7.06E-09	5.08E-06	0

By summing the contributions from each temperature range, the life consumption (LC) for the bond wire can be determined. For this uniform load cycle the total life consumption calculated is 5.09E-06, in other words the expected life of this bond wire joint when exposed to this load cycle would be around 3274 hours.

In real life, such as aircraft, bridges, railroad cars, and railway traction control applications, the cyclic load profiles could be random and expected to last for longer periods e.g. hours to days. For a test case, the real time compact thermal model was applied to a load cycling profile consisting of a series of steps of random amplitude and width. An example of such a loading sequence is shown in fig. 18, which describes a random sequence of current demand levels over a period of three minutes. The corresponding IGBT temperature estimates are shown in

fig. 19. The rainflow counting algorithm was applied again to the IGBT junction layer temperature data and used to create a temperature range histogram as represented in fig. 20 and 21 respectively.

Life estimates were obtained using the bond wire life time model (2) and by applying Palmgren-Miner rule along with a cycle counting method. Summery of these results is in table 2.

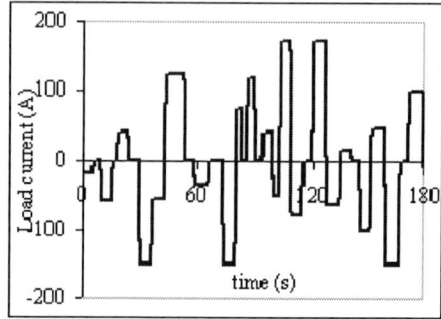

Fig.18. Sample of the random load current profile applied to the real time thermal model.

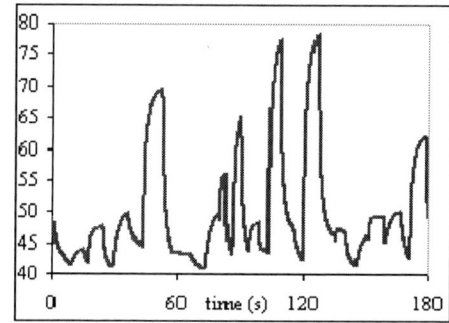

Fig.19: Sample of the real time IGBT junction temperature estimates.

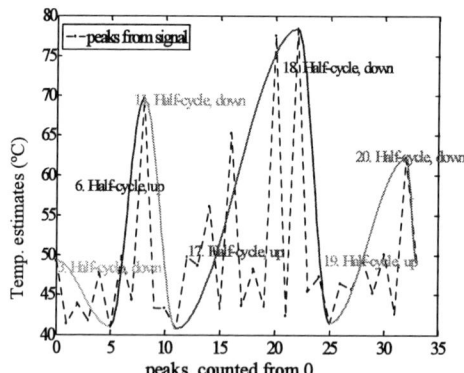

Fig.20. Sample of the rainflow cycles extracted from IGBT junction temperature data for random load profile

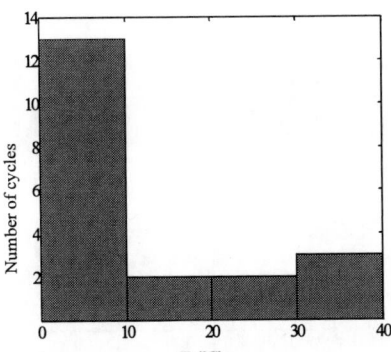

Fig. 21. Rainflow histogram of IGBT junction temperature variations (ΔT) obtained by applying a random load profile for one cycle (3 minutes)

Table 2: Calculation of life consumption for the IGBT wire bond with random load

Number of cycles	$0 \leq \Delta T \leq 10$ (k_1)	$10 < \Delta T \leq 20$ (k_2)	$20 < \Delta T \leq 30$ (k_3)	$30 < \Delta T \leq 40$ (k_4)
n (1 cycle)	13	2	2	3
TΔ	10	20	30	40
Nf	14.16E+7	11.7E+6	27.2E+5	96.7E+4
LC	9.18E-08	1.71E-07	7.35E-07	3.10E-06

The total life consumption (LC) for the bond wire corresponding to this random load is 4.1E-06. In other words the expected life of this bond wire joint when exposed to this load cycle would be around 12199 hours.

B. Application of life Time Models for the substrate mount-down solder layer.

The thermal path from the die to the heatsink includes a number of interfaces between different materials which are subject to failure by thermal cycling. In this section, the crack propagation in the substrate solder layer which is the solder layer connecting the isolation substrate to the baseplate is studied.

Coffin-Manson model can be employed to estimate the time-to-fail of a material in which the number of cycles-to-failure (Nf) of a metal subjected to thermal cycling is given by [9]:

$$N_f = \frac{L}{\alpha(\Delta \varepsilon_p)^b}$$

(4)

This equation applies for a solder interconnect with length L, the damage indicator has an average value of Δεp and

α = a constant, characteristic of the metal

b = another constant, characteristic of the metal.

L = solder interconnect length in millimeters

The application of Coffin-Manson models to in-service data is quiet impractical due to the irregularity of the temperature-time information which makes the identification of regular load cycles problematic. One approach is to generate a temperature range histogram by application of a rainflow counting algorithm to the temperature-time data [10].

Fig. 22a and 22b illustrate the simulated stress distribution arising because of CTE mismatch between the substrate (copper-ceramic-copper) and the adjoining copper baseplate to which it is bonded with solder. The finite element model used includes the effects of plasticity and creep in the solder. This is then translated into a simplified model that can be used to predict the rate of crack propagation [11].

A fatigue lifetime model for substrate mount-down solder layers has been derived by using the experimental lifetime data (Fig.23). The extracted life time solder model can be defined as the number of cycles to fail as in equation (4):

$$N_f = 622860e^{-0.0491\Delta T}$$

(4)

where N_f is the material number of cycles to fail

ΔT is temperature variation.

As a test case, the real time compact model was implemented for same random cycling load signal applied earlier for the wire bond model (fig. 18). Real-time solder layer temperature estimates corresponding to this load profile are shown in fig. 24. The rainflow counting algorithm was applied to the IGBT solder substrate layer temperature data and used to create a temperature range histogram as represented in fig. 25 and 26 respectively.

Life estimates were obtained using the substrate solder life time model (4) and by applying Palmgren-Miner rule along with a cycle counting method. Summery of the results is in table 3.

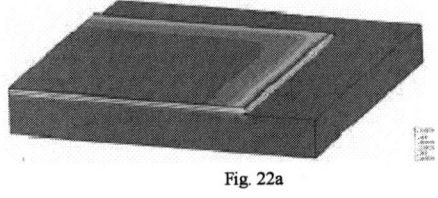

Fig. 22a

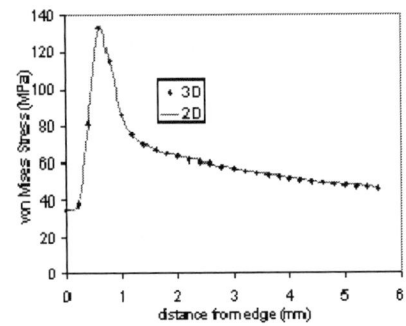

Fig. 22b

Fig. 22a and b. Thermally generated stress distribution for a DBC substrate solder mounted to a Cu baseplate.

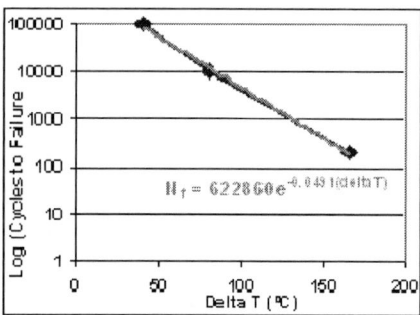

Fig. 23. life time extraction for the IGBT solder interconnect.

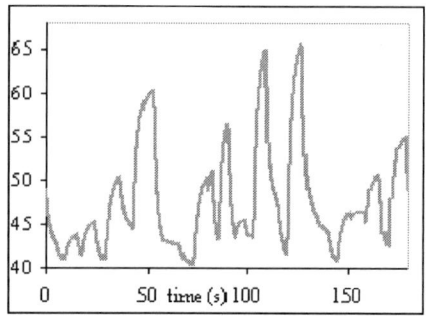

Fig. 24. Sample of the real time IGBT substrate solder temperature estimates.

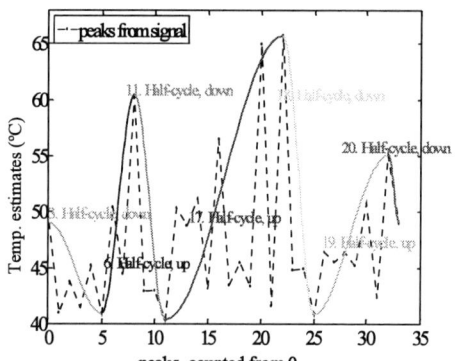

Fig.25. Sample of the rainflow cycles extracted from temperature data for random load profile

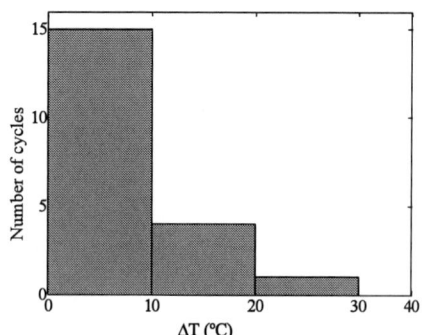

Fig. 26. Rainflow histogram of IGBT substrate solder temperature variations (ΔT) obtained by applying a random load profile for one cycle (3 minutes)

Table 3: Calculation of life consumption for the IGBT substrate solder with random load

Number of cycles	$0 \leq \Delta T \leq 10$ (k_1)	$10 < \Delta T \leq 20$ (k_2)	$20 < \Delta T \leq 30$ (k_3)	$30 < \Delta T \leq 40$ (k_4)
n (1 cycle)	15	4	1	0
TΔ	10	20	30	0
Nf	3.8E+5	2.3E+5	1.4E+5	0
LC	3.93E-05	1.71E-05	6.98E-06	0

The total life consumption (LC) calculated for the substrate solder layer when this random load applied is 6.34E-05. In other words, the expected life of this bond wire joint when exposed to this load cycle would be around 788.5 hours.

The work presented in this paper provides a new real time health assessment for power modules where a real-time model, can be used to estimate the effect of in-service operational conditions on remaining life. Such life consumption monitoring can be used to provide an early indication of imminent wear-out allowing replacement to be undertaken as part of scheduled maintenance. Both the equipment availability and in-service reliability are improved because of the reduced number of unscheduled maintenance events.

IV. CONCLUSIONS

A new technique to predict in-service life consumption in wire bond joints and substrate solder layers under arbitrary operational conditions has been presented. A real time compact model is used to provide temperature-time data estimates at any particular location within the module. A rainflow counting method is applied to the temperature vs. time data to extract the occurrence frequencies of different thermal cycling ranges. Knowledge of the life consumed for each different cycle then allows the remaining life time to be estimated under arbitrary operational conditions. This method has been used successfully to model wire bond joints fatigue and crack growth in solder due to repeated temperature cycling. Consequently, predictions of bond wire fatigue failure and crack length can then be used to give information of the remaining life. This technique can be employed as a tool for prognostic health management of power electronic modules.

ACKNOWLEDGMENT

This work is funded by EPSRC through the Innovative Electronics Manufacturing Research Center (IeMRC) (UK).

REFERENCES

[1] Musallam Mahera, Johnson C Mark, Buttay Cyril "Real-Time Compact Electronic Thermal Modelling for Health Monitoring", the 12th European Conference on Power Electronics and Applications 2 - 5 September 2007, Aalborg, Denmark.

[2] http://www.flomerics.com/

[3] Musallam Mahera, Johnson C Mark, "Extraction of Efficient Thermal Models for Life Limiting Interfaces in Power Modules", CIPS 2008 – the 5th International Conference on Integrated Power Electronics Systems, March, 11-13, 2008, Nuremberg, Germany.

[4] Michael Whitehead, C Mark Johnson, "Determination of Thermal Cross-Coupling Effects in Multi-Device Power Electronic Modules", in *Proc. PEMD 2006 International Conference on Power Electronics Machines and Drives*, 2006, pp. 261-265.

[5] S. Stoyanov, W. Mackay, C. Bailey, D Jibb, C Cregson, "Lifetime Assessment of Electronic Components for High Reliability Aerospace Applications", 6th Electronics Packaging Technology Conference - EPTC 2004, p324 – 329, ISBN 0-7803-8821-6, Pub IEEE, (2004).

[6] Lu H., Tilford T., Bailey C., Newcombe D.R., " Lifetime Prediction for Power Electronics Modules Substrate Mount-down Solder Interconnect", Proceedings of the International Symposium on High Density Packaging and Microsystem Integration (HDP' 07), Shangahai.

[7] Ohring M, 'Reliability and failure of electronic materials and devices', Academic Press, USA, 1998, ISBN 0-12-524985-3

[8] Sergent Jerry E., Al Krum, 'Thermal Management Handbook: For Electronic Assemblies', McGraw-Hill Professional, 1998, ISBN 0070266999.

[9] S. Suresh, Fatigue of Materials, Cambridge University Press (1991), p.133

[10] ASTM E-1049, Standard Practices for Cycle Counting in Fatigue Analysis.

[11] P. Towashiraporn, K. Gall, G. Subbarayan, B. McIlvanie, B. C. Hunter, D. Love and B. Sullivan, "Power cycling thermal fatigue of Sn–Pb solder joints on a chip scale package", International Journal of Fatigue, Volume 26, Issue 5, May 2004, Pages 497-510.

Measurement Of Temperature Sensitive Parameter Characteristics Of Semiconductor Silicon And Silicon –Carbide Power Devices

Mietek Nowak, Jacek Rabkowski, Roman Barlik

Warsaw University of Technology ,Warsaw, Poland mnowak@ee.pw.edu.pl

Abstract— In this paper, results of laboratory investigation of representative samples of semiconductor silicon and silicon carbide power devices, such as PiN diode, Shottky diode, IGBT and JFET, are presented. With use of a thermal chamber the characteristics of temperature sensitive parameters for selected types of devices were determined by measurements within the range 25 – 150°C. The measurement was done using short pulses, which are not able to change measured temperatures. The results have to enable identification of junction temperature in the case of other tests, especially those oriented on energy losses, efficiency, and thermal management.

Keywords—device characterization, IGBT, JFET, Silicon Carbide diode.

I. INTRODUCTION

One of the basic quantities, which must be taken into account in describing properties of manufactured semiconductor power devices is the temperature of the junction structure formed in a silicon chip. All important static and dynamic parameters are strongly influenced by temperature. Nowadays, the data sheets of silicon switches, which absolutely dominate in power electronics applications, describe very precisely all temperature influence related characteristics. Development in modern power electronics is oriented towards miniaturization and integration with the aim to obtain great density of power in converter constructions (over 10kW/dm^3). This leads to serious heat management problems where temperatures of semiconductor structures increase to the permissible limits. In the case of silicon this is not more than 150°C and it is not enough for "hot electronics" applications. This means that the development in power semiconductor switches have concentrated on new materials - especially on the most promising, silicon carbide. This semiconductor withstands working temperatures over 400°C but is very difficult to manufacture. Despite this, some SiC devices have just started to be commercially produced and some other types are widely tested. It is obvious that, in aiming to compare this new solution with traditional silicon devices according to their high temperature properties, it is necessary to use research methods, which permit simple and accurate semiconductor structure temperature testing.

This work is supported as project ordered by Polish Government in the years 2007 - 2010

Accessible methods, without very specialized equipment, permit the measurement of selected parameters, which are temperature dependent according to previously examined characteristics.

In this paper, measurement results of selected temperature sensitive parameters undertaken for some different types of silicon (PiN diode, IGBT) and silicon-carbide (Schottky diode, JFET) devices are presented. The temperature range considered in tests was (25 - 150)°C.

II. MEASUREMENTS OF SILICON PiN DIODE TEMPERATURE SENSITIVE PARAMETERS

In this case, as a temperature sensitive parameter, a PiN diode forward (i.e. conduction) voltage has been selected. The important requirement accordingly for measurement conditions depends on a measuring procedure which doesn't affect the internal temperature despite power dissipation during the measurement interval. The best way to fulfill this condition is to use very short pulse signal setting measuring parameters (e.g. diode current, transistor gate voltage), while a selected temperature sensitive parameter is measured and recorded. This short setting and measuring time interval guarantees that the structure temperature doesn't change significantly. The temperature of the pattern devices under test and used for thermal characterization was controlled by internal thermo-chamber temperature. The basic measuring circuit is schematically presented at Fig.1.a.

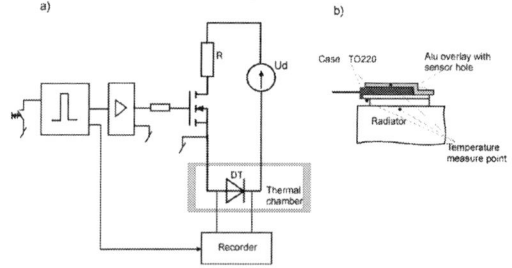

Fig. 1. Equipment used for characterization of temperature dependent parameters of diodes selected for test: a) diode forward voltage test circuit b) temperature measurement arrangement in thermal chamber.

For accurate temperature recognition three sensors have been installed on the tested device as shown in Fig.1.b. Three samples of each type of semiconductor device have been measured, which is too small number for generalization of the obtained results but enough for exclusion of defective test samples. Also, for better confirmation of the measuring procedure each sample was tested twice over the whole temperature range.

Records of measured diode voltage pulse comprised 10000 samples in a 20 μs time interval, as is shown in Fig.2.

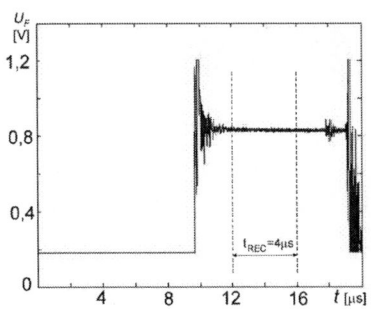

Fig. 2. Example of diode voltage pulse recorded by measurements (t_REC - interval used for calculation of average value)

Only part of these bounded in the interval by the dotted line were used for calculation of mean value of measured voltage. Results obtained for three samples are shown in Fig.3. It is noticeable that there is a very small dispersion of curves measured for randomly selected diodes.

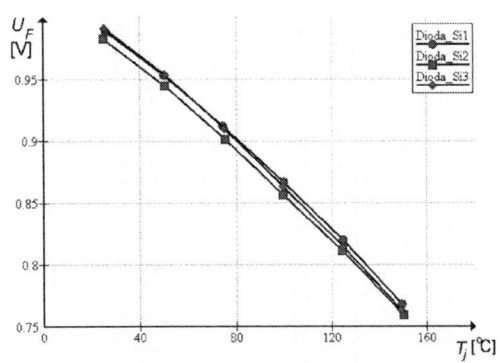

Fig. 3. Examples of measured characteristics of forward voltage of three silicon PiN diodes as junction temperature function

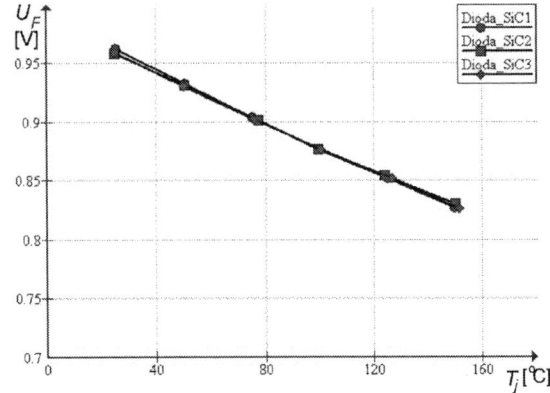

Fig. 4 Measured characteristics of forward voltage of three silicon carbide Schottky diodes as junction temperature function

III. MEASUREMENTS OF SILICON CARBIDE SCHOTTKY DIODE

The measurements were undertaken for 10A/300V diodes. The method applied was the same as in the previous case i.e. a silicon diode (see scheme in Fig.1.), which means that the anode-cathode forward bias voltage was selected as a temperature sensitive parameter. The value of the test current was set at 0,5A, which is significantly lower than in the case of a PiN Si diode. The reason for such settings results from a feature of the voltage temperature coefficient which changes its sign at conducted currents between (0,3 - 0.7) I_N. The following curves were obtained as measurement results

IV. IGBT TEMPERATURE SENSITIVE PARAMETERS

In the case of transistor IGBT as a thermal sensitive parameter, collector current is measured under conditions defined as follows: gate-emitter voltage is only some hundreds of mV higher than the threshold voltage, and emitter-collector voltage is keep constant and equal to 10V. The laboratory test circuit is presented in Fig.4.

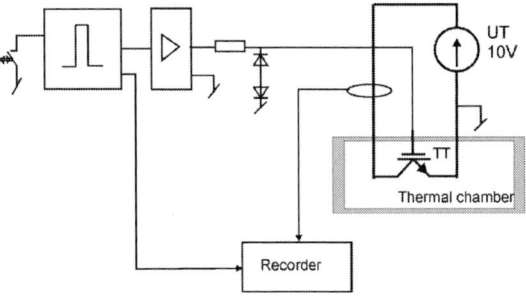

Fig. 5 Scheme of circuit used for characterization of transistors IGBT and JFET collector/drain current as temperature depending parameter.

The same thermal chamber as was applied for diode investigation and the same temperature sensor arrangement as in Fig.1.b. were used for the IGBT test. It is also obvious that measurement must be undertaken in very short time interval if the process is not to influence significantly the junction temperature. An example record of a current test pulse is given in Fig.5.

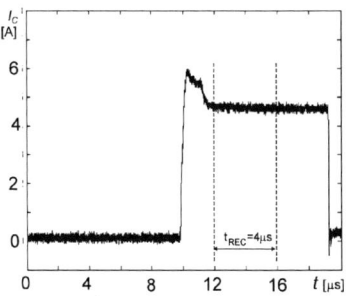

Fig.6 Example of collector/or drain/ current pulse recorded by measurements (t_REC - interval used for calculation of average value)

Only the samples between 6000 and 8000 were used for accurate current determination. Measured characteristics of three tested IGBT's are shown in Fig.7. One of these curves differs apparently from the two others but all have the same shape. Proper adjustment of gate-emitter voltage enables minimizing of characteristic dispersion.

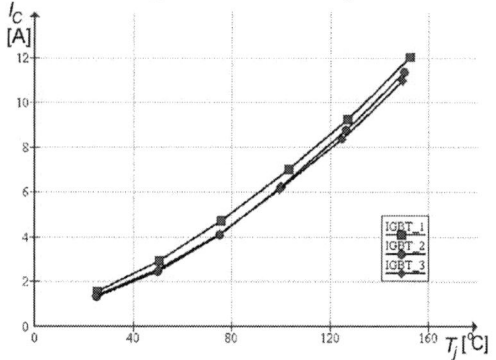

Fig. 7. Examples of measured characteristics of IGBT collector current as junction temperature function

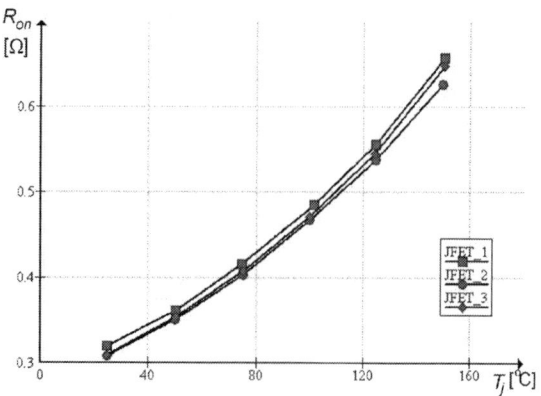

Fig. 9. Examples of measured characteristics of JFET channel resistance as junction temperature function

V. SILICON-CARBIDE JFET MEASUREMENTS

The silicon carbide JFET are new prototypes, which are not detailed as described in accessible data sheets. This involves wider investigation of features of tested samples than in the case of silicon devices. JFET's are "normally ON" switches so it is necessary to polarize the gate with a negative voltage to obtain an OFF state. It is necessary to control the exact value of blocking gate voltage, which is only some volts lower than the gate junction breakdown voltage. The value of gate blocking and breakdown voltages differs for each tested device. The issue of temperature dependent parameter characterization was found in two variants. Firstly was a testing method similar to that applied for IGBTs. During the pulse interval the gate voltage was set higher than the blocking value (ex. -15V) and the value of drain current was measured as a function of temperature, controlled by the thermal chamber. Sample results presented in Fig. 8 shows some problems with repetition of measurement results caused by not accurate enough control of the gate voltage driver.

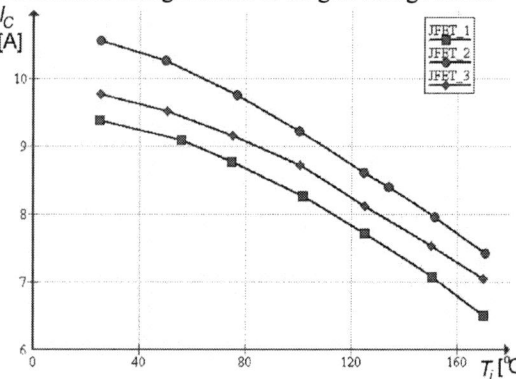

Fig. 8 Examples of measured characteristics of JFET drain current as junction temperature function

This problem can be avoided by selecting as a temperature sensitive parameter a source-drain resistance in the on state of the JFET when the gate is short-circuited with the source electrode. The circuit used for testing is similar to that shown in Fig.1. Results obtained for two tested devices are shown as curves in fig 9.

VI. EXAMPLES OF JUNCTION TEMPERATURE IDENTIFICATION AT DYNAMIC PARAMETER MEASUREMENT

The most convenient experiments useful for design and optimizing of converters are oriented on exact determination of power losses, which depend on features of power semiconductor switches. It is obvious that identification of key parameters of the switches used in the investigation should be done specially in the case when new types of semiconductor devices are tested. During such tests the temperature of the junction structure must be always carefully controlled or estimated. The measurements done in a high temperature environment such as in a thermal chamber may be technically very complicated because, generally, standard voltage and current probes are not fit to use at temperatures over 70℃. The practical solution in this case will be self-heating of the semiconductor tablet containing a junction structure with energy losses arising during current conducting and switching in the application and topology being under test. In the investigations, which have been started with the goal to compare converters built with the use of silicon carbide Schotky diodes with similar converters built using silicon PiN diodes, the measurement of important dynamic parameters i.e. diode recovery current and IGBT on and off energy losses has been undertaken. The simple method of junction structure temperature measurement presented above was applied and useful for direct verifying of laboratory experiments.

In this case a special circuit was prepared, the schema of which is presented in Fig. 10. The step-down converter built with earlier characterized IGBT (T_x) and complementary diode (D_x) represents the main test circuit. Before the proper test a chopper is working during a selected time interval with chosen load, frequency and duty ratio to heat the structure. Then the temperature impulse test procedure is realized using separated voltage sources and additional switches with parameters corresponding exactly to those used in circuits for temperature dependent parameter characterization. After switching off transistor T_L, the test specimens are released from load current. Then two other transistors T_T and T_D are turned on for few microseconds. If the measured temperature sensitive parameter fulfils expectations, the standard dual pulse test is done for measurement of current and voltage waveforms during an "on" and "off" switching process. The main goal of this test was measurement of energy losses characterizing basic switching features of different switches as exactly as possible. As mentioned above, the main subject of investigation in this issue was comparison of converter features - especially its efficiency in the case of replacing fast switching PiN silicon diodes with commercial silicon carbide Schottky diodes.

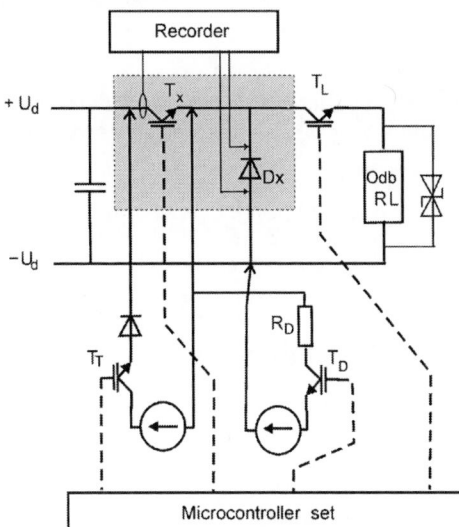

Fig. 10. Test circuit used for measurement of semiconductor power devices parameters including junction temperature measurement of tested diode (Dx) and transistor (Tx)

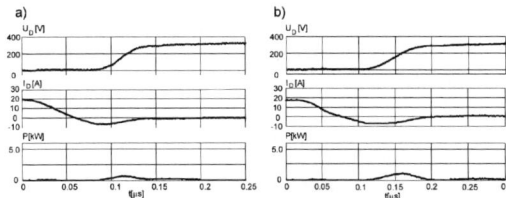

Fig. 11. Recorded PiN silicon ultra-fast diode voltage and current at off process: a) - Tj = 25°C, b) Tj = 125°C (lower trace - calculated power)

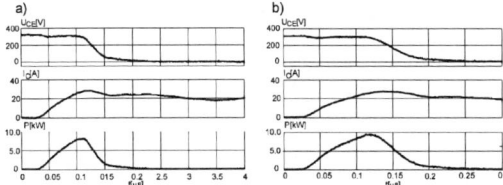

Fig. 12 Recorded IGBT transistor voltage and current at on process in case of PiN silicon diode colaboration: a) - Tj = 25°C, b) Tj = 125°C

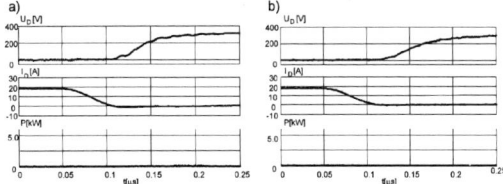

Fig. 13 Recorded Schottky silicon carbide diode voltage and current at off process: a) - Tj = 25°C, b) Tj = 125°C (lower trace - calculated power)

Some examples of recorded measurements give a clear view on the most important advantage of silicon carbide devices. The typical recovery current characterizing the PiN diodes (Fig.11ab) causes a significant increase of transistor turn-on maximal current (Fig.12 a,b) and this also causes much bigger energy losses. The tested silicon carbide diodes don't indicate such a recovery current (Fig.13. a,b) in a wide semiconductor temperature range. As has been measured, the turn-on energy lost in an IGBT transistor for a tested transistor-diode pair may be 40% lower in the case of application of an SiC Schottky diode instead of a silicon PiN diode

VII. . CONCLUSIONS

The main goal of the described investigations was to examine if it is possible for widely used silicon power switches, as well as for newly developed silicon-carbide devices, to select parameters, which are suitable for semiconductor junction structure measurement. Using a thermal chamber the following characteristics were measured for a temperature range of 25-150°C: forward voltage for silicon PiN and silicon-carbide Schottky diode, collector current for IGBT in active state, and on drain-source resistance of silicon-carbide JFET. The obtained curves have a good resolution and could be used for temperature identification in other experiments.

REFERENCES

[1] B. J.Baliga "Silicon Carbide Power Devices" *World Scientific Publishing Company* 2006

[2] W. Janke: "Zjawiska termiczne w elementach i układach półprzewodnikowych". *WNT*, Warszawa 1992,

[3] J. Oleksy , W. Janke.: "SiC and Si Schottky Diodes Thermal Characteristics Comparison". *International Conference Microtechnology and Thermal Problems in Electronics - MicroTherm 2007* , 25.06-27.06.2007 Łódź , pp.117÷122

[4] T. Funaki, J.C. Balada, J. Junghans, A. S. Kashyap, H. A. Mantooth, T. Barlow., T. Kimoto., T. Hikihara "Power Conversion with SiC Devices at Extremaly High Ambient Temperatures". *IEEE Transactions on Power Electronics*. Vol.22. No.4, July 2007, pp.1321-1328

[5] A. Konczakowska, A. Szewczyk, R. Barlik, M. Nowak, J. Rąbkowski: "Silicon Carbide Application Issues" *International Conference Microtechnology and Thermal Problems in Electronics -MicroTherm 2007* , 25.06-27.06.2007 Łódź , pp.229-232.

[6] R. Barlik, J. Rąbkowski, M. Nowak: "Przyrządy półprzewodnikowe z węglika krzemu (SiC) i ich zastosowania w energoelektronice" *Przegląd Elektrotechniczny*, Nr 11/2006, s.1-6,

[7] M. Bakowski: "Status and Prospects of SiC Power Devices", IEEE Trans IA, vol. 126, No. 4,2006

[8] T. Funaki, A.S. Kashyap; H.A. Mantooth,. J.C. Balda,. F.D. Barlow, T. Kimoto,. T. Hikihara. "Characterization of SiC diodes in extremely high temperature ambient" *APEC 2006*

[9] D Marlino, L. E. Seiber, M. B. Scudiere, Chinthavali, F. P. McCluskey " Investigation in to High emperature Components and Packiging *Oak Ridge National Laboratory Raport 2007*

[10] www.siced.de

[11] www.cree.com

[12] www. infineon.com

Unsymmetrical Gate Voltage Drive for High Power 1200V IGBT[4] Modules Based on Coreless Transformer Technology Driver

Piotr Luniewski[*], Uwe Jansen[†]

[*] Infineon Technologies AG/Application Engineering, Warstein, Germany, e-mail: *Piotr.Luniewski@infineon.com*
[†] Infineon Technologies AG/Application Engineering, Warstein, Germany, e-mail: *Uwe.Jansen@infineon.com*

Abstract—The performance of the new IGBT4 chip technology in PrimePACK™ high power module housing is presented here together with the Coreless Transformer technology driver IC for the first time in this paper. These modules usually are driven using symmetrical gate drive voltage of +/-15V. The driver presented here uses unsymmetrical gate drive voltage of −7V and +15V. This alternate approach results in different dynamic module behaviour compared to classical. Thus, this paper discusses differences in both concepts and brings a solution which allows to use the unsymmetrical concept as well as symmetrical.

Keywords—IGBT, Power semiconductor device, High voltage IC's, Thermal design.

I. INTRODUCTION

Modern converters in a power range from 1 kW up to 3 MW and connected to the 380 – 460 VAC power net nowadays, are mostly equipped with 1200V IGBT modules. Although topology and voltage class are the same, requirements on chip performance vary significantly within the large power range. Proper selection of the IGBT and diode chip technology must support the efficiency goal. Hence, in the recent years the great improvement in application oriented silicon took place. Thus, keeping 'Energy Efficiency' target in focus, Infineon has recently introduced a new Trench-/Fieldstop IGBT4 generation. Comparing to third generation the semiconductor family comes as product basically dedicated for three inverter power levels: low (T4), medium (E4) and high power (P4) as in reference [1] and [2]. Further improvements on saturation voltage reduction and switching speed, mean finding a better trade–off between steady and dynamic losses in the final system design, are successfully implemented. However, especially in high power applications usually consisting of high power modules the low dynamic losses are not the main target. Mainly, due to electromagnetic interference, EMI, the appropriate controllability and certain softness during module switching is required and seen as main development target and described in reference [3].

A. High Temperature IGBT Module Operation

Elevated IGBT module operation junction temperature, T_{vjop}, up to 150°C, with the IGBT4 benefits additionally in the system or gives more freedom to engineers during module or heatsink

selection as discussed in [4]. Operating with increased junction temperature up to 150°C offers several possibilities for inverter optimisation:

- higher inverter output power with the same module housing, current rating and same heatsink as in [1], [2], [3], [5]
- same inverter output power with same module housing, current rating but with smaller or less expensive heatsink can be achieved which practically means reduced volume or applying less efficient material as in [1], [2], [3]
- decreased system volume by generally improved heat dissipation as in [4]
- improved system lifetime as prolonged module lifetime. In this case operating with reduced junction temperature $T_{vj\ op} < 150°C$ as in reference [3, Fig. 1] and [4] is mandatory

Power inverter performances in general are governed by power part features, in this case IGBT module, heatsink and the IGBT driver. Based on this, a conclusion can be derived: an inverter property really relies on IGBT module and the IGBT driver. Development of IGBT module housing concept since many years has been pursued in parallel to progressive IGBT and diode silicon evolution. As presented in [5], [6] the IGBT high power module called PrimePACK™ represents a novel high power housing concept equipped with newest 1200V IGBT4 generation. Thus, except increased robustness this module is able to operate with $T_{vjop} = 150°C$ without decrease in power cycling capabilities as pointed out in [3].

B. Driver Selection and Extended Requirements

Every IGBT module needs to be controlled by a dedicated IGBT driver which has to ensure the proper and safe module operation during the entire converter life time. There are many approaches in driving IGBT modules but basically their requirements are split up accordingly to the converter power. For IGBT modules operated with a maximum junction temperature of 125°C, regardless of rated current, the driver selection can be done quite easy. The appropriate selection becomes challenging when IGBT4 is intended to be used. The increased module operation junction temperature may result in increased ambient temperature thus, for drivers placed close to the module, a certain overheating risk or at least life time limitation exists. Drivers which use optocouplers as a method for

978-1-4244-1741-4/08/$25.00 ©2008 IEEE

electric isolation suffer from severe lifetime limitations at ambient operating temperatures close to or exciding 85°C. Their practical use in low and medium power applications is practically excluded. Thanks to Coreless Transformer technology the problem for small as well as medium power IGBTs modules is solved. The 2ED020I12-F driver circuit (IC) is mainly dedicated for small power IGBTs in half bridge configuration where only the high side switch has the basic isolation and the functionality was discussed in [7], [8]. For applications where every IGBT needs functional isolation to the low side control signals and additional security functions the 1ED020I12-F IGBT driver IC comes as best match [9]. Drivers dedicated for high power modules like PrimePACK™ have similar functions as drivers for medium power modules but the drive power and output current are significantly higher. Hence, an approach utilizing powerful PCB driver [10] with dedicated adapter boards is more appropriate as presented in [11], [12]. Such a universal driver system can be used for controlling one IGBT module in 1200V or 1700V class and also several modules in parallel configuration with small modifications [4].

In a converter where modularity in parallel operation as discussed in [4] is not needed the driving system can be optimised resulting in reduction of cost and volume. The PCB driver called 2ED250E12-F, shown in Fig. 1, employing 1ED020I12-F and dedicated especially to PrimePACK™ family in 1200V class is presented in this paper. Thanks to the Coreless Transformer technology and protection functions implemented in the design the 2ED250E12-F connected to PrimePACK™ module makes the building block plug and play solution in half-bridge configuration.

Fig. 1. High Power building block consisting of 2ED250E12-F driver and FF1400R12IP4 IGBT module.

II. THE DIFFERENCES IN DYNAMIC BEHAVIOUR

Characterisation measurements for IGBT modules are usually done with a driver which provides symmetrical and bipolar +/-15V and has sufficient current capability [15]. This voltage is applied to the gate – emitter of an IGBT module through external gate resistor. Dynamic parameters like switching energies and delay times in module datasheets [13] are given for a certain value of R_{Gon} and R_{Goff} resistors, as well as parameters of the external power circuit like stray inductances being precisely specified.

The 2ED250E12-F driver has the negative voltage reduced from -15V to -7V thus, differences on dynamic module performances comparing to symmetrical control should be observed [8], [16]. Due to fixed 2ED250E12-F position to the module effects on changed gate-emitter inductance can be neglected [4]. An assumption that modification on the gate resistor value should result in similar dynamic module behaviour to symmetrical control in this case is justified and practically proven in following chapters. All investigations and measurements presented here are done with a FF900R12IP4 PrimePACK™ module but correctness of the approach has been proven in 1200V PrimePACK™ with high power (P4) as well as with medium power (E4) IGBT versions.

A. Turn-off Behaviour and New Gate Resistor Value

The IGBT as voltage controlled power switch changes its dynamic behaviour by changing the gate-emitter voltage in level and shape [14]. Every module which consists of many IGBT dies besides of external gate resistor, R_{Gext}, where the R_{Gext} represents R_{Gon} and R_{Goff}, has also specified the internal resistance, R_{Gint}, as can be seen in reference [13]. When voltage on the gate and emitter of an IGBT module changes from negative to a positive and returns the power device is forced to switch its collector current with controlled speed. Therefore, IGBT transistor module dynamic behaviour will be different when the driver's negative voltage level changes while keeping the same gate resistances. The influences of the gate resistor value, R_{Goff}, on switching energy and switching speed for turning-off is displayed in Figure 2 separately.

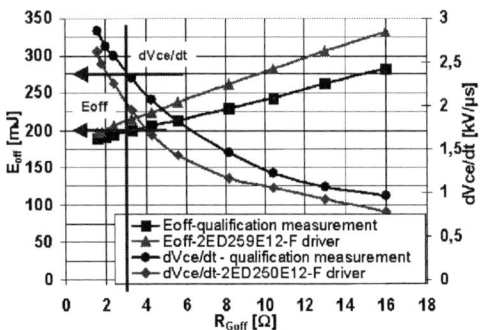

Fig. 2. Eoff losses of an IGBT module and dVce/dt slope versus gate resistor value during switching off where: V_{DC}=600V, I_C=900A, T_{vj}=150°C, Ls=42nH.

The positive slope of a collector-emitter voltage during commutation decreases with increased resistor value and energy losses increase as consequence in both drive concepts. Increased losses elevate the module operation junction temperature when module cooling conditions are the same as before and finally influence the lifetime. Hence, keeping the same module dynamic parameters when driver with unsymmetrical gate voltage is used seems to be appropriate and reduced value of the gate resistor is required.

As can be seen in Fig. 2, decreasing the gate resistor value from 3Ω to approximately 2Ω increases the

dV_{CE}/dt slope and decreases Eoff losses to values as obtained from qualification measurement. Selection of the corrected gate resistor value, R_{Gnew}, in general should allow keeping the same gate discharge current, I_{Goff}, during Miller plateau when the IGBT switches off. By the assumption given in formula (2) where I_{GoffR} is the gate discharge current with reduced negative gate voltage, $I_{Goff(-15V)}$ is the gate discharge current with symmetrical driver the discharge gate current value basically can be adjusted to every negative gate voltage level.

$$I_{GoffR} = I_{Goff(-15V)} \qquad (1)$$

Based on above assumption the formula (2) can be derived where R_{Gnew} is the correct gate resistor value, R_{Goff} is the gate resistor value given in module datasheet, V_M is the Miller plateau voltage, V_{NR} is the reduced negative gate driver voltage and V_N is the negative symmetrical gate voltage driver and typically has -15V value.

$$\frac{V_M - V_{NR}}{R_{Gint} + R_{Gnew}} = \frac{V_M - V_N}{R_{Gint} + R_{Goff}} \qquad (2)$$

Finally, the external gate resistor value for unsymmetrical gate voltage driving is given by formula (3)

$$R_{Gnew} = \frac{V_M - V_{NR}}{V_M - V_N} \cdot (R_{Goff} + R_{Gint}) - R_{Gint} \quad (3)$$

As practical example the new gate resistor value for the FF900R12IP4 high power IGBT module driven by the 2ED250E12-F driver will be calculated in formula (4)

$$R_{Gnew} = \frac{9.8 + 7}{9.8 + 15} \cdot (1.6 + 1.2) - 1.2 \approx 0.7\Omega \quad (4)$$

where the V_M=9.8V for 900A and T_J=25°C taken form the module datasheet – reference [13], V_{NR}=-7V, V_N=-15V, R_{Goff}=1.6Ω and R_{Gint}=1.2Ω.

Suitable waveforms showing an IGBT module dynamic behavior for both driver concepts with matched gate resistors under the same operating conditions are depicted in Fig. 3. The gate current during Miller plateau reaches nearly the same level confirming correctness of the assumption given in formula (1). For practical reasons the calculated gate resistor value in formula (4) for the unsymmetrical drive concept has been increased to 0.75Ω.

One impact of the negative driver voltage supply is a change of energy losses as function of collector current. This dependency is shown in Figure 4. The energy increase with 1.6Ω external gate resistor value between symmetrical and unsymmetrical drive concept measured at Ic=900A is 7%. The small Eoff losses dependency in this case proves that IGBT4 have similar turn-off behavior to the IGBT3 in term of dI_C/dt limitation by applying higher gate resistor value. This phenomenon has been widely described in reference [14].

After the gate resistor value for driver with unsymmetrical gate driver has been reduced according

to formula (4) the turn-off losses are very close to qualification measurement.

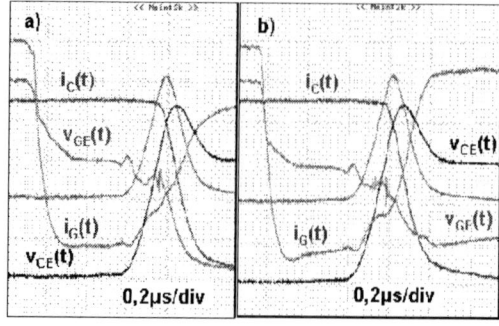

Fig. 3. Waveforms of the collector emitter voltage (200V/div), gate emitter (5V/div) voltage, collector (200A/div) and gate current (2A/div) where: V_{DC}=600V, I_C=900A, T_{vj}=25°C, Ls=42nH, and:
a) symmetrical gate voltage, R_{Goff}=1.6Ω,
b) unsymmetrical gate voltage, R_{Goff}=0.75Ω

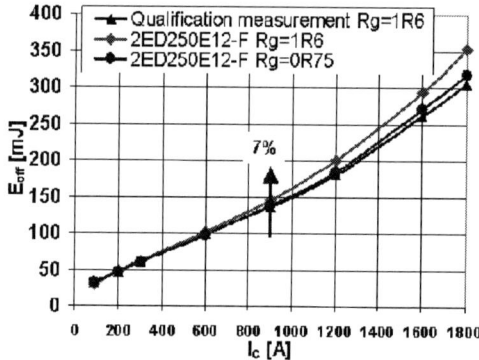

Fig. 4. Turn-off losses of an IGBT module versus switched collector current value during switching-off for symmetrical and unsymmetrical driver concept where:
V_{DC}=600V, T_{vj}=25°C, Ls=42nH.

B. Turn-on Behaviour

Reduction of the negative IGBT driver voltage influences switching energies, Eon, and the positive collector current slope, dI_C/dt, during module turn-on as well. Fig. 5 shows the impact on module turn-on performances between the two drive concepts discussed in this paper. The dI_C/dt, during commutation decreases with increased resistor value and switching losses increase as a consequence in both drive concepts. Considering an example R_{Gon} resistor value of 3Ω with symmetrical +/-15V gate driver voltage the dI_C/dt reaches approximately 4,3kA/μs and Eon=175mJ. The dI_C/dt slope decreases by 12,7% to 3,75kA/μs and Eon consequently increases by 25,7% and reaches 220mJ when the 2ED250E12-F gate driver is used in the same DC-link setup.

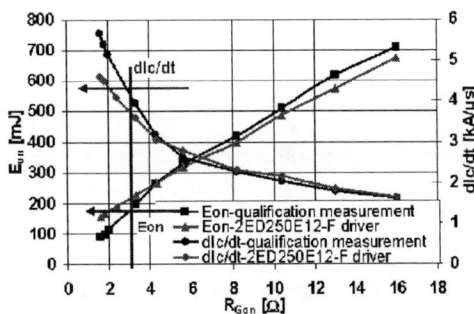

Fig. 5. Turn-on losses of an IGBT module and dIc/dt slope
versus gate resistor value where:
V_{DC}=600V, I_C=900A, T_{vj}=150°C, Ls=42nH.

Differences in losses are also a function of the gate resistor value and are larger when the module is driven with the smaller value. By analogy to turn-off behavior, reduced gate resistor value increases the dI_C/dt slope and decreases Eon losses to values as obtained during qualification measurement. Nevertheless, due to the gate inductance effects on turn-on behavior partly discussed in reference [4] the corrected gate resistor value for unsymmetrical driver can not be calculated using formula (3) in every case. For demonstration purposes the R_{Gon} value in this paper has been chosen using the same criteria as for turning-off. Main target of results presented here is to show that module operation condition can be shifted again into datasheet values when driven with unsymmetrical drive concept.

Figure 6 shows energy losses versus switched collector current when the two driver concepts are applied. As can be noticed at nominal collector current the driver concept with reduced negative voltage increases energy losses by 82% compared to qualification measurement where the gate resistor value is 1.6Ω as in datasheet. The energy increase during switching-on is much higher comparing to 7% during switching-off. After the value of the gate resistor has been corrected according to formula (3) the energy losses up to nominal module current are the same as during qualification. In the range between nominal current and two times nominal current Eon is higher compared to symmetrical gate voltage.

A similar situation shows Figure 7 where Eon is measured as function of stray inductance, Ls, module commutation loop. Higher Ls results in reduced energy for the same module operation conditions as consequence of lover Vce voltage drop measured across the IGBT as can be seen in Figure 8. The energy decrease is true for both driver concepts but the difference between qualification measurement and the 2ED250E12-F driver with R_{Gon}=1.6Ω is higher for small Ls as shown in Fig. 7. Despite of this phenomenon for unsymmetrical drive concept the corrected R_{Gon} brings Eon losses on the same level as for symmetrical concept for all presented stray inductances.

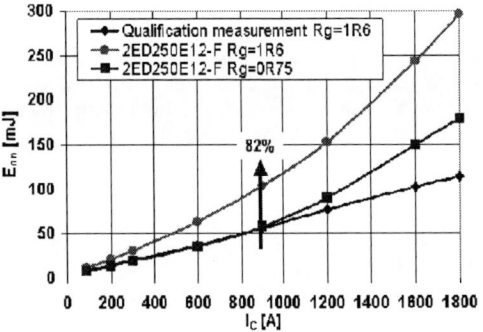

Fig. 6. Turn-on losses of an IGBT module versus switched collector
current for symmetrical and unsymmetrical driver concept where:
V_{DC}=600V, T_{vj}=25°C, Ls=42nH.

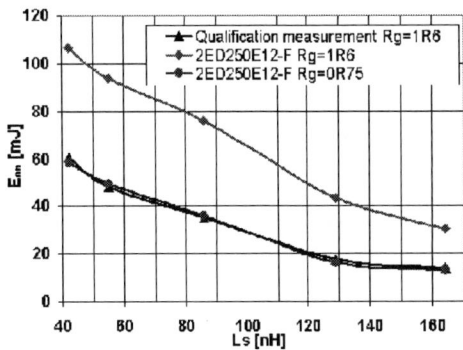

Fig. 7. Turn-on losses of an IGBT module versus stray inductance,
Ls, in commutation loop for symmetrical and unsymmetrical driver
concept where: V_{DC}=600V, I_C=900A, T_{vj}=25°C.

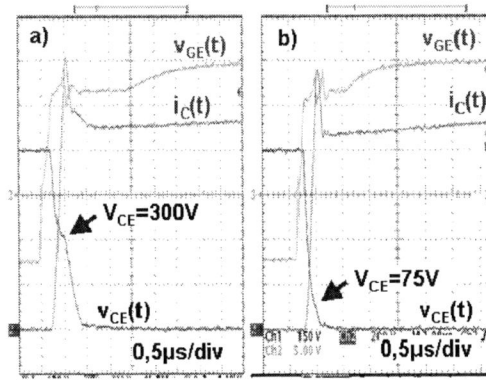

Fig. 8. Waveforms of the collector emitter (150V/div), gate emitter
(5V/div) voltage and collector current (200A/div)
with a 2ED250E12-F driver during switching-on where:
V_{DC}=600V, I_C=900A, R_{Gon}=0.85Ω, T_{vj}=25°C, and
a) Ls=42nH,
b) Ls=164nH

C. FWD Diode Recovery Behaviour

Diode recovery losses for given power setup parameters are mainly determined by IGBT turn-on speed which depends on negative gate driver voltage and gate resistor value. Fig. 9 shows how these two factors influence the diode recovery losses.

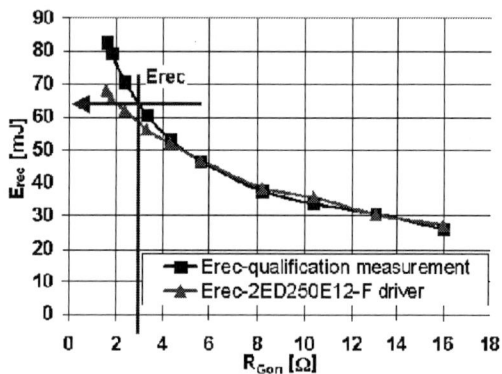

Fig. 9. Recovery losses of an IGBT module versus gate resistor value during diode recovery where: V_{DC}=600V, I_c=900A, T_{vj}=150°C,

In contrast to switching-on (Fig. 5.) the recovery losses decrease with increased R_{Gon} and negative driver power supply voltage. As visible in Fig. 9 and Fig. 5, reduction of the gate resistor value from 3Ω to 2Ω speeds up module commutation process during switching-on process and finally Eon as well as Erec losses are the same as for symmetrical gate driver. Using formula (3) and applying corrected gate resistor value the decreased diode recovery losses are nearly equalized to values obtained during driving with symmetrical gate voltage. Diode recovery losses as function of switched current are depicted Fig. 10.

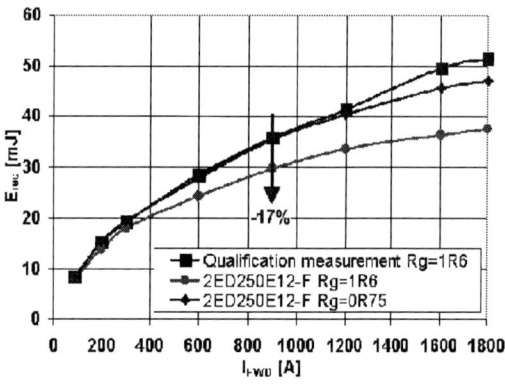

Fig. 10. Recovery losses of an IGBT module versus switched current value for symmetrical and unsymmetrical driver concept where: V_{DC}=600V, T_{vj}=25°C, Ls=42nH.

The negative voltage reduction from -15V to -7V results in Erec reduction of -17% when the same 1,6Ω gate resistor is used. As already mentioned before, changes in gate inductance of the IGBT module influence the switching-on commutation speed and

finally diode recovery losses. Therefore, using formula (3) may not result in Erec equalization when different symmetrical driver is used or interface between driver and module will be different to the one used for qualification measurement.

D. Energy Losses as Function of Ls

One of many parameters which differentiate IGBT operation conditions in different applications is the stray inductance, Ls. Hence, versus this parameter the unsymmetrical driver concept should be compared with the symmetrical concept. Figure 11 shows total energies, E_{TOT}, as sum of Eon, Erec, and Eoff for both driving concepts and Eoff for unsymmetrical and symmetrical drive concept with corrected gate resistor value. A few facts have to be noticed:

- the total energies are stable till Ls reaches approximately 85nH but Eoff is slightly increasing. The Eoff increase is fully compensated by decreased Eon (Fig. 7.) as Erec has the same increasing tendency as Eoff

- in the region were Ls<85nH the total energy increases by 20% when driven with unsymmetrical drive concept keeping datasheet gate resistor value

- after Ls exceeds 85nH the total losses of the unsymmetrical system with corrected gate resistor value increase faster than symmetrical. The 43mJ difference at Ls=168nH is mainly given by increase Eoff by 38mJ compared to symmetrical concept

- increased Eoff at high stray inductances is a result of Active Voltage Clamping circuit operation discussed in reference [4], [12]

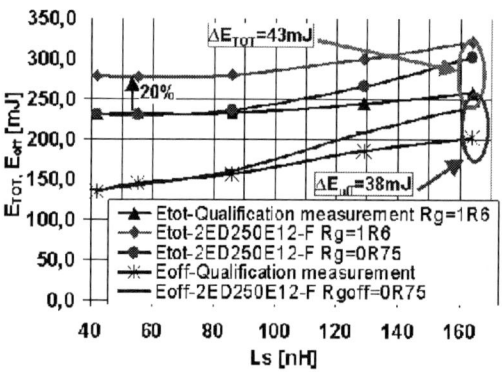

Fig. 11. E_{TOT} and Eoff of an IGBT module versus stray inductance, Ls, in commutation loop for symmetrical and unsymmetrical driver concept where:
V_{DC}=600V, I_c=900A, T_{vj}=25°C.

E. Short-Circuit Performances

Every system properly designed with IGBT modules should survive short circuit event. According to reference [15] tested devices should be able to switch again after short-circuit event with defined short-circuit duration t_{psc}. Figure 12 shows IGBT waveforms under short circuit one, SC1, for both drive concepts.

92

Reduction on the negative voltage does not influence the shape of the waveforms. Differences in values are given mostly by different driver design. Nevertheless, the 2ED250E12-F driver, compared to the symmetric drive concept, reduces the short circuit peak current as a result of better gate-emitter voltage stabilization. It slows down dI_C/dt during switching-on and finally reduces the V_{CE} voltage drop. The Vce overvoltage during switching-off is limited by Active Voltage Clamping circuit.

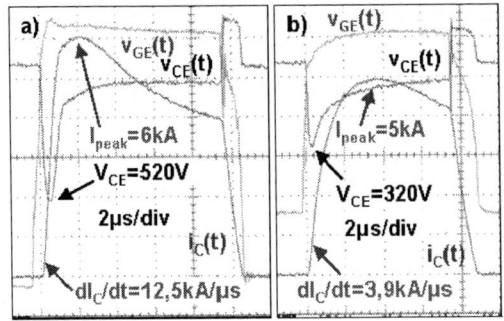

Fig. 12. Waveforms of the collector emitter voltage (150V/div), gate emitter (5V/div) voltage, and collector current (1kA/div) during short circuit one where: V_{DC}=800V, T_{vj}=25°C, Ls=42nH, and:
a) symmetrical gate voltage, R_{Goff}=1.6Ω,
b) unsymmetrical gate voltage, R_{Goff}=0.75Ω.

F. Parasitic Turn-on Phenomena

This effect is defined as IGBT transistors cross conduction if connected in half-bridge configuration. Both shortly conduct current from DC+ to DC-potentials looped into DC-bus capacitors. Due to high dynamics in time and peak value this kind of short circuit current is not considered as safe during module operation and should be avoided by proper inverter design. At least two major reasons determine the phenomena. First is the IGBT module construction and switched current. Second, the value of the negative voltage biased to the gate of a switched-off IGBT transistor. As discussed in reference [17] for modules operating bellow 100A the negative voltage value can be reduced even to zero but for high power modules designed for operating with higher currents having the negative voltage is desired. Nevertheless, the optimum value of the negative voltage in this paper is under discussion. In any case, the off state of an IGBT regardless of the second transistor state in a leg should remain off till is not commanded by driver to be switched-on.

In order to evaluate the impact of driver negative voltage value on the possible parasitic turn-on the real IGBT gate-emitter voltage on the silicon, V_{GE}, has been calculated based on formula (5) where necessary measures have been made based on schematic shown in Fig. 13.

$$V_{GE} = V_{GEaux} + (R_{Gint} \cdot I_G) \qquad (5)$$

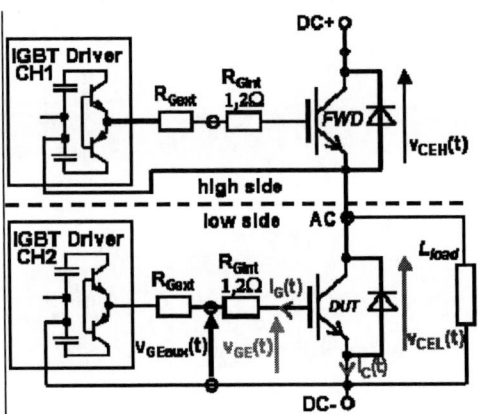

Fig. 13. Block diagram for parasitic turn-on phenomena investigations.

Undesirable switching-on is avoided when the gate-emitter voltage stays bellow the IGBT threshold voltage, V_{GEth}. By this the simple condition where $V_{GE} < V_{GEth}$ is fulfilled. Knowing that the V_{GEth} is a function of junction temperature, switched current, DC-link voltage and time for safety reasons the gate-emitter voltage value should stay bellow zero.

Figure 14 shows the measurement for both driving concepts where the high side IGBT has been switch on the second time in double pulse measurement technique. In this dynamic process the low side antiparallel FWD diode recovers and V_{CEL} across an IGBT is created. At the time where the high side IGBT starts switching-on the positive gate current, I_G, of the low site IGBT flows into the driver and creates a positive voltage drop across the gate resistances. Regardless of the drive concept discussed in this paper the maximum increase on calculated gate-emitter voltage is not more than 4V as shown in Fig. 14a. In the case when unsymmetrical drive concept is considered (Fig. 14b.) the difference between positive V_{GE} peak voltage and negative power supply voltage is 3V. The V_{GE} voltage in this case has been minimized as R_{Goff} was also reduced and the gate current, I_G, determined by high side IGBT turning-on switching speed stays unchanged. The result indicates that there is still at least 4V as safety margin between the maximum V_{GE} voltage and recommended zero volt. Finally, reduction on negative voltage from -15V to -7V did not introduced the parasitic turn-on phenomena and the IGBT can be controlled safely.

Figure 15 shows a case when the low side IGBT has the gate-emitter auxiliary connectors shorted by wire and the high side IGBT is controlled using the unsymmetrical drive concept with reduced gate resistor value. By the case the high side IGBT's commutating speed is unchanged. The V_{GEaux} voltage (Fig. 13) is equal to zero and gate-emitter voltage can be calculated using formula (6).

$$V_{GE} = R_{Gint} \cdot I_G \qquad (6)$$

93

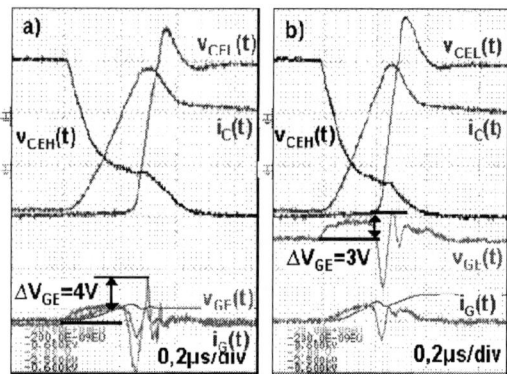

Fig. 14. Waveforms of the collector emitter voltage (200V/div) of low and high side IGBT, gate emitter (5V/div) voltage, collector (500A/div) and gate (2A/div) current during high side IGBT switching-on where: V_{DC}=600V, I_c=900A, T_{vj}=25°C, Ls=42nH, and:
a) symmetrical gate voltage, R_{Gon}=R_{Goff}=1,6Ω,
b) unsymmetrical gate voltage, R_{Gon}= R_{Goff}=0.75Ω

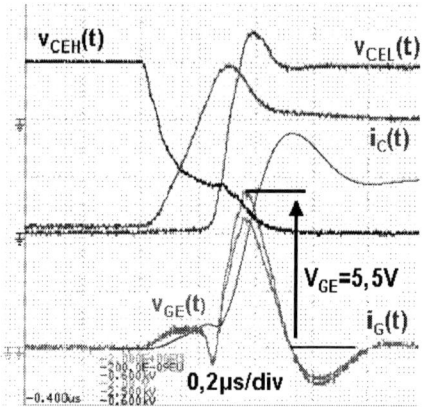

Fig. 15. Waveforms of the collector emitter voltage (200V/div) of low and high site IGBT, gate emitter (2V/div) voltage, collector (500A/div) and gate (2A/div) current during high side IGBT switching-on where: V_{DC}=600V, I_c=900A, T_{vj}=25°C, Ls=42nH, for unsymmetrical gate voltage, R_{Gon}= R_{Goff}=0.75Ω

The VGE maximum voltage during commutation exceeds 5,5V and excides the 5V as the minimum threshold voltage, VGEth, specified in module datasheet. The probability of parasitic turn-on caused by the negative drive voltage reduction till 0V has significantly increased.

III. THERMAL CONSIDERATIONS

Reliability of power electronic systems which would mean lifetime is mainly determined by operating temperature. Ageing processes of electronic components mounted mainly on PCB and operated at high ambient temperature are generally faster compared to operating at lower temperature. The problem is solved for Infineon IGBT modules and based on power cycling curves the usability time can be calculated for operating junction temperatures, T_{vjop}, equal to 150°C. This feature makes the semiconductor fitting to e. g. automotive applications where the coolant used for cooling combustion engine in some cases is intended to

be used for cooling the power electronic as well. As the IGBT driver used to be mounted close to the IGBT module the increased ambient operating temperature for this part of the system has to be considered. Major goal in this case could be to decrease power losses generated in the driver's PCB which would result in decreased temperature keeping the same module switching frequency, baseplate and ambient temperature. In the proper gate drive design the hottest point is in the area of the gate resistors. The temperature can be evaluated based on a simple thermal model consisting of two thermal resistances as shown in Fig. 16.

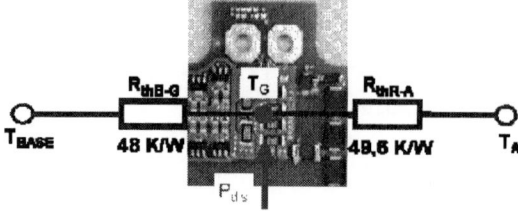

Fig. 16. Simple IGBT driver thermal model used for gate resistors temperature calculation.

One resistance called R_{thB-G} couples the gate resistors to the module baseplate through gate-emitter terminals and a second called R_{thG-A} to the ambient. Based on formula (7) the gate resistor temperature, T_G, can be calculated as function of dissipated power in the external gate resistors, P_{dis}, ambient temperature, T_A, and baseplate temperature T_{BASE}.

$$T_G = R_{thB-G} \cdot \frac{R_{thG-A} \cdot P_{dis} + T_A - T_{BASE}}{R_{thB-G} + R_{thG-A}} + T_{BASE} \quad (7)$$

The maximum value of T_G is usually limited by PCB material in use. When the maximum working temperature for the material is known Pdis can be calculated using formula (8).

$$P_{dis} = \frac{T_G - T_{BASE}}{R_{thB-G}} + \frac{T_G - T_A}{R_{thG-A}} \quad (8)$$

Finally, the maximum IGBT switching frequency, f_s, for a specific driver design can be calculated using formula (9) as function of Pdis and module parameters

$$f_s = \frac{P_{dis} \cdot (R_{Gint} + R_{Gext})}{\Delta V_{GE} \cdot R_{Gext} \cdot Q_q} \quad (9)$$

where Qq (C) is the IGBT gate charge.

Based on the thermal model shown in Fig. 16, Figure 17 shows calculated gate resistor temperature and module switching frequency as function of dissipated power in the gate resistor, Pdis. The temperature T_G increases as a function of dissipated power and ambient temperature. Knowing that the long time maximum operation temperature for FR4 material used in this design is 105°C the maximum allowed power dissipated in gate resistors can be easily determined. For T_A=65°C, T_{BASE}=100°C and T_G=105°C the Pdis=0.95W. The IGBT switching frequency depends on drive concept.

For symmetrical concept fs=8.6kHz can be achieved but when unsymmetrical concept is used the switching frequency can be increased by 190% and reaches 25kHz.

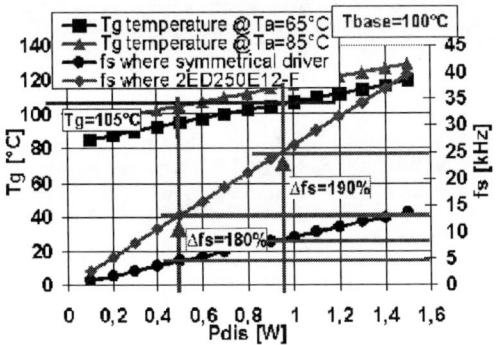

Fig. 17. Calculated gate resistors temperature, T_G, for two ambient temperatures, T_A=65°C, T_A=85°C and IGBT switching frequency, fs, versus gate resistors dissipated power, Pdis.

In order not to exceed the maximum operation temperature for PCB the dissipated power must be reduced to 0.52W when ambient temperature increases to 85°C. This limitation results in decreased switching frequencies of 4.75kHz for symmetrical drive concept. For unsymmetrical drive the switching frequency is higher by 180% and reaches 13.2kHz. Calculated switching frequencies in this chapter for given ambient and baseplate temperatures are only examples. In real operation when T_A and T_B is given by design the module switching frequency is limited by PCB material. Based on formula (9) for specific IGBT driver design can be calculated. For both ambient temperatures the maximum allowed switching frequency for unsymmetrical drive concept is much higher compared to symmetrical.

On one side increased fs can be a design target but often decreased gate resistor temperature for fixed base, ambient temperatures and IGBT switching frequency is the goal. Figure 18 shows the gate resistors temperature and power needed from power supply, P_{sup}, versus module switching frequency for both driver concepts. Driving power need for IGBT switching is calculated according to formula (10)

$$P_{sup} = \Delta V_{GE} \cdot fs \cdot Qq \qquad (10)$$

where ΔV_{GE} is the absolute value of the gate-emitter voltage during IGBT switching. The power, Psup, as well as gate resistor temperature is increasing linearly with the module's switching frequency. For fs=5kHz, T_A=85°C, T_{BASE}=100°C and symmetrical power supply the resistors temperature is around 105°C. Needed driving power from power supply is 0.96W. The 2ED250E12-F driver for the same environmental parameters needs reduced Psup=0.49W which is nearly half and the T_G is reduced by 8.7°C. Hence, the gate resistor temperature is equal to 97.25°C.

Decreased negative voltage from -15V to -7V for the driver design presented in this paper offers the following possibilities:

- the module switching frequency can be increased. In that case temperature of the gate resistors value is not changing as presented in Fig. 17
- for unchanged module switching frequency the gate resistor temperature can be reduced. This results also in lower power needed from driver power supply as presented in Fig. 18
- for unchanged module switching frequency the gate resistor area can be minimized. This results with unchanged higher gate resistor temperature

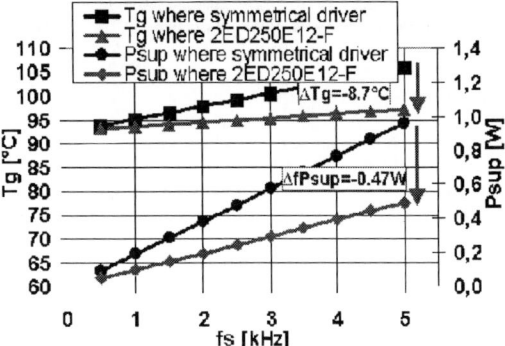

Fig. 18. Calculated gate resistors temperature, T_G, and driving power from driver power supply, P_{sup}, for T_A=85°C, and T_{BASE}=100°C versus IGBT switching frequency, fs.

The 2ED250E12-F design has been practically validated during the 2ED250E12-F operation. Figure 19 shows a thermal camera picture where the hottest point in the driver design is the gate resistors. The temperature T_G reaches 106°C and corresponding to Pdis=0.52W depicted in Fig. 17.

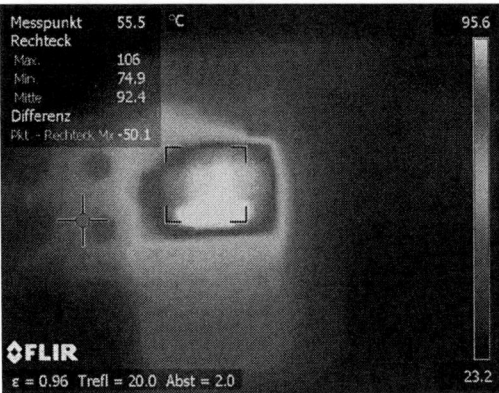

Fig. 19. Thermal picture of the 2ED250E12-F unsymmetrical driver where: T_A=85°C, T_{BASE}=100°C, fs=13.2kHz.

IV. CONCLUSION

Increased IGBT module junction operation temperature requires modified surroundings which can operate a long time with high ambient temperatures. To meet the harsh requirement the IGBT driver technology has been moved from optocoupler devices to Coreless Transformer technology.

The integration of 1ED020I12-F IGBT/MOS medium power driver into PCB for high power modules driver in this paper has been successfully implemented. An appropriate driving with unsymmetrical bipolar gate voltage versus symmetrical bipolar voltage driver has been indicated and proven in all possible module operation conditions. Additional advantages as e.g. improved thermal design are pointed out.

REFERENCES

[1] M. Baessler, et. al.: *1200V IGBT4 Low and Medium Power – Chips designed to the Needs of the Application*, PCIM Europe, 2007

[2] A. Volke, et. al.: *The new power semiconductor generation: 1200V IGBT4 and EmCon4 Diode*, IEEE 2006

[3] M. Baessler, et. al.: *1200V IGBT4 –High Power– a new Technology Generation with Optimized Characteristics for High Current Modules*, PCIM Europe, 2006

[4] P. Luniewski, et. al.: *Benefits of System-oriented IGBT Module Design for High Power Inverters*, EPE 07

[5] O. Schilling, et. al.: *Optimised Integration of PrimePACK™ into Modular Stacks with Increased Power Density*, PCIM Europe 2006

[6] O. Schilling, et. al.: *Properties of a New PrimePACK™ IGBT Module Concept for Optimized Electrical and Thermal Interconnection to a Modern Converter Environment*, PCIM Europe 2005

[7] M. Muenzer, at. al.: *Coreless transformer a new technology for half bridge driver IC's*, PCIM Nuremberg, 2003,

[8] A. Volke, at. al.: *IGBT/MOSFET Applications based on Coreless Transformer Driver IC 2ED020I12-F*, PCIM Nuremberg, 2004

[9] Infineon Technologies AG.: Target datasheet Version 1, EiceDRIVER™, 1ED020I12-F, Single IGBT Driver IC, www.infineon.com

[10] Infineon Technologies AG.: 2ED300C17-S Dual IGBT Driver for Medium and High Power IGBTs, Datasheet and Application Note, Revision 3, www.infineon.com

[11] P. Luniewski, AN-2007-05, 2ED300E17-SFO Evaluation Board for 2ED300C17-S /-ST IGBT Driver, ww.infineon.com

[12] P. Luniewski, AN-2007-06, MA300E12 / MA300E17 Module Adapter Board for PrimePACK™ IGBT Modules, www.infineon.com

[13] Infineon Technologies AG, Datasheets of: FF600R12IE4, FF600R12IP4, FF900R12IP4, FF900R12IP4D, FF1400R12IP4, www.infineon.com

[14] P. Luniewski, at. al.: *Dynamic Voltage Rise Control, the Most Efficient Way to Control Turn-off Switching Behaviour of IGBT Transistors*, PELINCEC2005, paper 80

[15] International Standard, IEC 60747-9

[16] L. Dulau, at. al.: *A New Gate Driver Integrated Circuit for IGBT Devices With Advance Protections*, IEEE Transaction on Power Electronics, VOL.21, NO. 1, January 2006

[17] Baginski, AN-2006-01, Driving IGBTs with unipolar gate voltage, www.infineon.com

A Novel RESURFed Double Gates IGBT with Superior Performance

Dongming Wu, Kaihang Li, Lingling Yang

Xiamen University, Xiamen, China, simon_100@sina.com

Xiamen University, Xiamen, China, khli@xmu.edu.cn

Abstract —In this paper, we proposed a novel RESURFed double channels LIGBT which can achieve high breakdown voltage. The proposed structure, incorporating trench gate and planar gate, can significantly improve the capacity for handling high current and conductance modulation. P-top layer and deep n-drift/p-sub junction have been adopted, which results in a 745V breakdown voltage with the drift length of 44 μ m only. Reduction in drift length can not only reduce on-resistance but also raise current density. Simulation results demonstrate that the forward voltage drop can lower by 12% relative to the conventional one while breakdown voltage can increase by 18%.

Keywords —Power semiconductor device, IGBT, High voltage IC's, IPM.

I. INTRODUCTION

Because of the development of the reduced surface field principle [1], lateral power semiconductor device has attracted much attention as high-voltage output device in power ICs module. As one of application of RESURF technology, the LIGBT has been explored extensively due to its high performance. Many approaches such as multi-channels concept [2] and the variation in lateral doping (VLD) technique [3] have been applied to improving characteristics of LIGBT. Multi-channels enable the reduction in the forward voltage drop due to the additional channel. The MC-IGBT with three cathodes cell can achieve more than 50% area reduction relative to SC-IGBT [3]. Trench-gate injection enhances lateral IECT on SOI, which conducts a drain current twice as large as that of the conventional LIGBT under forward drop of 3V [2]. VLD structure yields BV higher than that of the conventional at a given drift region length about 14% [4-5]. In this paper, a new lateral IGBT structure incorporating a deep n-drift/p-sub junction, a p-top layer and double channels is put forward. Introduction of trench gate and planar gate has decreased forward voltage drop and adoption of deep n-drift/p-sub junction together with p-top layer can raise breakdown voltage obviously. The proposed structure with the drift length of 44 μ m only can reach 745V of breakdown voltage, which increases 18% compared with conventional LIGBT. Because of electron injection of double channels and strong conductivity modulation, significant forward voltage drop reduction and high current carrying capability are obtained. The forward voltage drop at 100A/cm^2 was 1.86V when Vg is 8V and it is reduced 12% relative to the conventional. This paper is arranged as follows, the proposed structure along with the conventional are reported as well as its fabrication processes. Operation principle is presented from aspects of on state and reverse biased case. Detailed simulation results on output and breakdown characteristics comparing with the conventional LIGBT prove the validity of such structure. Dependence of BV varied with p-top length, p-top dose and sub/drift junction depth will be discussed lastly.

II. DEVICE STRUCTURE

Fig.1 illustrates the conventional LIGBT and proposed structure in this paper. In order to assure their comparability, double channels and deep n-drift/p-sub junctions are made in both of them. Furthermore, the spacing of cathode cells is optimized in the conventional and other parameters are the same as the proposed. The new structure introduces a p-top layer and a trench gate extending into bulk but far shallower than P+ isolation. The new approach makes use of junction isolation technology and RESURF principle under blocking conditions to yield a compact device. A vertical channel is formed in the N-drift region on the left side of gate. Although the use of trench gate leads to additional processing complexities, but it maintains the current capability of device after eliminating the wide Lsp which is the distance between two cathodes cell in MC-IGBT [6]. The trench gate not only can reduce layout area, but also remove the JFET effect existing in the pinch-off region [6]. The p-top layer locates underneath poly field plate nearby p-base and its parameters have been optimized. The P-base contacts emitter electrode through the p+ region which is sandwiched between n+ cathodes and served as a collector of holes coming from anode. Fortunately, the p+ region can be finished together with p+ anode and will not add extra manufacturing cost. Of course, this part can be removed at the cost of increasing on-resistance slightly. Related doping and dimensions parameters are showed in fig.1 (a). Above fabrications are accomplished in the epitaxial layer deposited on p substrate. It is well known that the depth of p+isolation junction more than 10 μ m in conventional structure is very difficult. Therefore, Implantation and diffusion have been carried out at the substrate before epitaxy growth of n-drift region material just as buried layer of bipolar

978-1-4244-1741-4/08/$25.00 ©2008 IEEE

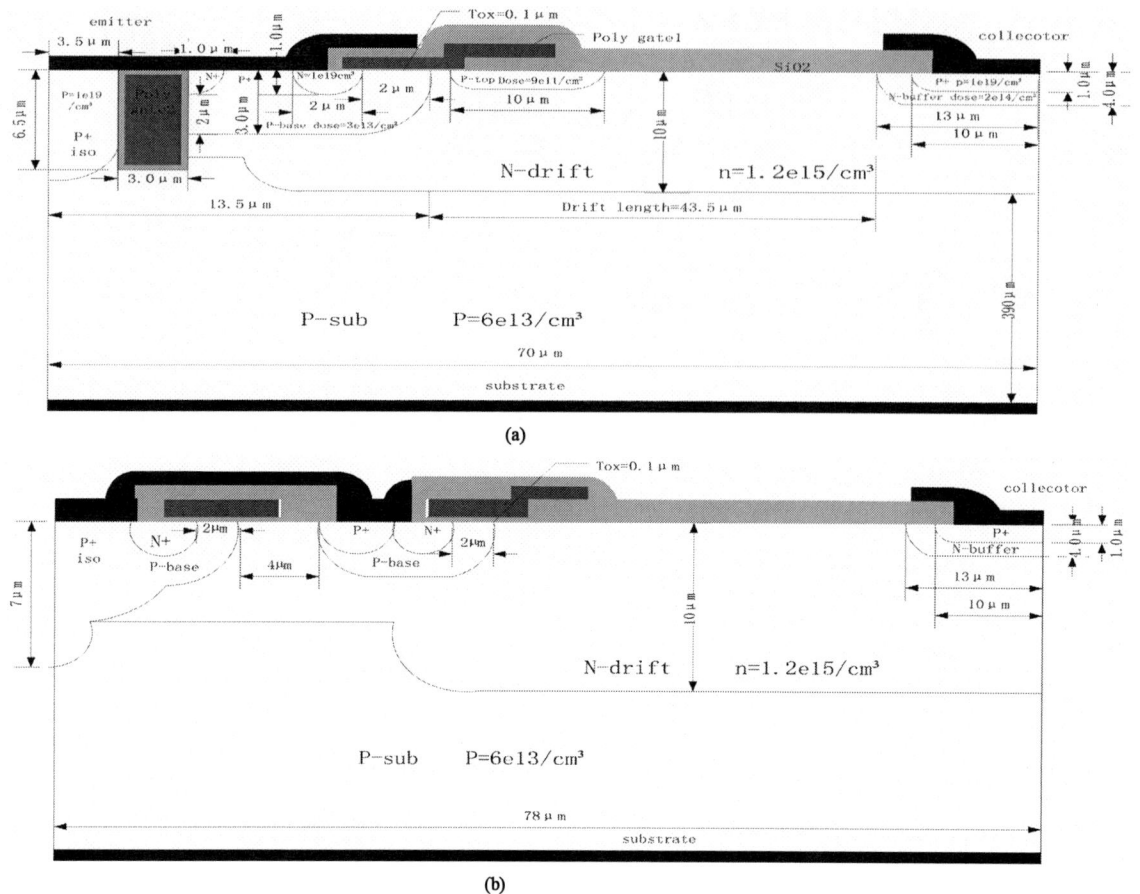

Fig.1 (a) cross section of proposed lateral IGBT and its components   (b) cross section of conventional structure

Technology. By this means, deep drift-sub junction and junction isolation which connects p-substrate with cathode by boron heavy implantation were both realized. The drift region and substrate is optimized to fulfill the RESURF condition.

III. OPERATION PRINCIPLES

It is seen from figure.1 that the proposed structure contains two MOS transistors and two BJTs (P+anode/p-sub and P+anode/N-drift/P-base). When gate voltage Vg increase to threshold voltage, electrons begin to inject into n-drift and the channel from drift region to ground is turned on. Followed the electron injection, the voltage drop of p+-anode/n-drift junction increase quickly so that it turns on the vertical PNP simultaneously. It is observed that the vertical p-n-p transistor is driven into saturation. Because of conductivity modulation, holes from anode can move easily in drift region. Combination takes place there between partial holes and partial electrons and the rest of injected electrons flow into anode. The voltage drop of MOS component ensures the p-base/n-drift junction to be reversed lightly. The holes were drawn to p-base and flow underneath of n+ emitter, and at last were collected by sandwiched p+ region. A significant

number of holes flow towards the p-sub due to holes injection and a little smaller of base width, which make conductivity modulation take place in this region. Current handling capability of device is enhanced due to existence of another channel. Multi-channels strengthen modulation effect in drift region and substrate, resulting in reducing on-state resistance. After removing the Vg, MOS channels were closed and potential of drift region rise to the bus voltage accordingly. Influenced by the RESURF technique, both n-drift/p-sub junction and p-base/n-drift junction reverse to support bias voltage. As long as device drift region have enough length, maximum breakdown voltage of device is chiefly decided by the depth of n-drift/p-sub junction and its bilateral dopant concentration. Deeper junction can make the avalanche take place in the junction depletion region, and keep the strong electrical field peak to locate in this region. With decreasing of depth of n-drift/p-sub junction, the epitaxy layer is fully depleted. Moreover, due to the curvature effect of n-buffer layer, electrical field in crook is stronger than that of any other region. As a result, device could be failure before avalanche breakdown happen at the n-drift/p-sub junction. For given short distance drift, the factors such as p-base/n-drift junction

And the curvature of poly gate terminal can obviously influence BV. Since p-base and n-drift are both medium doping, p-base/n-drift junction is prone to breakdown compared with n-drift/p-sub junction. However, for the thin gate oxide and the curvature effect, the voltage drop between silicon and poly can bring strong electrical field and produce breakdown at the silicon side before other junction failing. For avoiding the breakdown occurs in silicon beneath the place of gate swerving, the p-top is utilized to prevent this kind of failure by the floating p-n space depletion region.

IV SIMULATION RESULTS AND DISCUSSION

To demonstrate the benefits of the proposed structure relative to conventional and analyze its on-state and breakdown performance, numerical simulation was performed in two structures respectively using two-dimension technology Simulator ATHENA and device simulator ATLAS [7-8]. The parameters of simulated structure have been shown in fig.1. The simulated output current-voltage (I-V) is shown in Fig.2. Comparison has been operated on the basis that they have the same threshold voltage and equal Vg are used. So, the output characteristics curve can indicate on-state resistance factually. It can be seen from fig.2 (a) that the dynamic on-resistance of proposed structure is obviously lower than the conventional, which is attributed to the effect of the combination of trench channel and planar channel. Given the collector current density of $100A/cm^2$, forward voltage of the proposed structure and the conventional will be 1.86V and 2.13V respectively under 8V of the gate bias. The on-set voltages are approximately 1V. Compared with planar channel in conventional structure, the advantage of vertical channel in electron injection makes conductivity modulation efficient in drift region and p-substrate and contributes to the drop of forward voltage lastly. We also find the tendency of saturation line current density between them in fig.2 (b) and it shows that the conventional will have worse short circuit capacity. Such less output resistance in conventional IGBT is due to a shortening of the channel due to an increase in the collector voltage. However, shortening of vertical channel in the proposed structure is not obviously relative to the former. For the case that trench extends into substrate and the case not, simulation results show that the former voltage drop is lower and the vertical channel injects electron into substrate to enhance conductivity modulation. Detailed data can be seen in fig.3. Moreover, trench gate extending into substrate would not increase the substrate leakage current which is $0.268A/cm^2$ in the proposed device when the current density is $100A/cm^2$, compared the conventional one of $0.302A/cm^2$. In order to find out the benefit of the p-top Layer and deep drift/sub junction's introduction, the comparison of BV between the conventional and the proposed LIGBT has been carried out. The proposed structure makes BV reach 745V on the condition of drift length less than 44 m

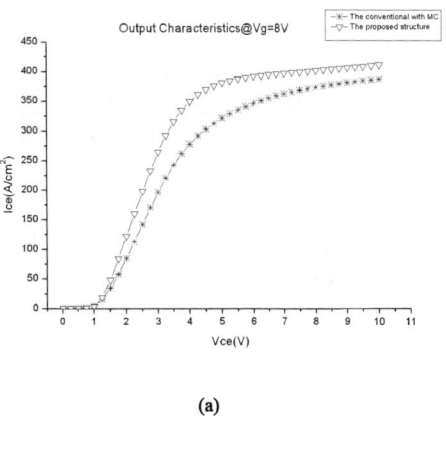

(a)

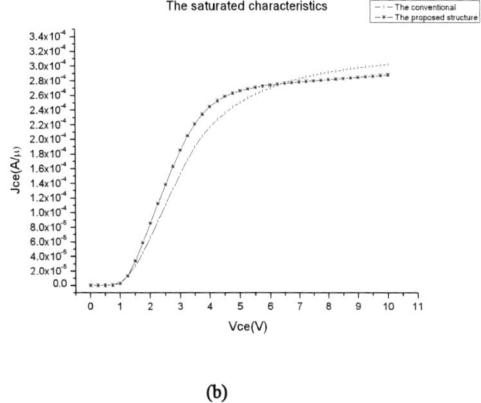

(b)

Fig.2 forward conduction characteristics of the proposed IGBT and conventional one (a) I-V characteristics at Vg=8V, (b) saturation line current density of both structures

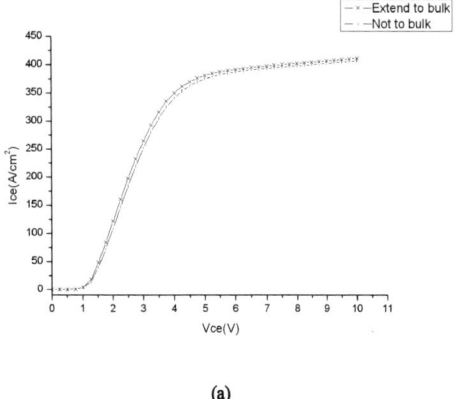

(a)

Fig.3. comparison of forward voltage drop for the case trench extend into bulk and not into bulk at Vg=8V

While the LIGBT only 630V as shown in the Fig.4. It is noted that the BV of proposed structure is higher than that of device made by VLD technique [5]. Considerable difference of BV attribute to conjunct effect of p-top and deep drift/sub junction. When the n-drift/p-sub junction is reverse biased, electrical field lines start from n-drift region and stop at sub or p-base. The potential begins to decrease away from the p+ anode, then becomes zero inside p-base and deeper substrate which is about far away from surface 70 m. Even though p-base/n-drift junction also shares partial reverse bias voltage, the possibility to breakdown is still small. On the contrary, thin gate oxide and curvature effect can easily lead to breakdown at silicon surface underneath the corner of poly prematurely. Structure optimization is the process that we balance the distribution of electrical field in all drift regions to fully make use of the space. The ideal case is that the electrical field gets the maximum of silicon material simultaneously in region where breakdown is prone to take place. P-top layer and deep n-drift/p-sub junction are efficient measures to realize the better electrical field profile. If the p-top layer of structure is depleted, the poly field plate could not stop breakdown efficiently. Influenced by the deep junction, donor in drift region can't be depleted fully. So larger voltage drop will located in gate oxide and it is much more possible to result in breakdown at silicon side of MOS structure. Surface electrical field distribution in above case is displayed as the line 2 in fig.5. Given the existence of p-top, the electrical field line come from the ionization donors of n-drift can be absorbed by acceptor of p-top. If p-top layer is heavily doped, the electrical field will attenuate quickly as go towards the emitter along the p-top, and it leads to a long distance that can not support reversed bias. Above properties is shown by the line3 of fig.5. On the contrary, p-top will lose its function when its doping level approaches that of drift region just like the line4 in fig.5. In that case, it is possible that high voltage drop will be located at p-base/n-drift junction or gate oxide. Local avalanche for strong electrical field will take place in device while other region still far away from breakdown. The ideal surface electrical profile resemble line1 of fig.5, which make sure the electrical field reach silicon limitation in any parts of drift region. Fig.5 also shows that electrical field profile of the proposed structure is closer to the ideal one than others. Fig.6 shows the surface electrical field profile of proposed structure as well as the conventional. It is indicated that the former is more efficient to balance electrical field distribution. Further simulation has been performed to find the dependence of BV with the varied p-top dose on the condition that p-top length was fixed at 10 m. From the fig.7 (a), we can note that breakdown voltage will reach the maximum on the condition of 9e11/cm^2 of the p-top dose, and BV is lightly decreased once the p-top dose is more than 1e12/cm^2. On the other hand, BV decreases quickly with the decreasing of p-top dose. Fig.7(b) illustrates the dependence of BV on p-top length by keeping p-top dose at 9e11/cm^2. Once the length is 10 m, BV arrives at its maximum. A short p-top introduces breakdown at the silicon surface underneath the poly field plate while breakdown takes place at the anode side for curvature

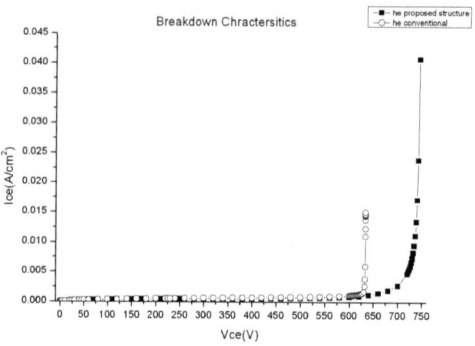

Fig.4 Comparison of BV between the proposed structure and the conventional

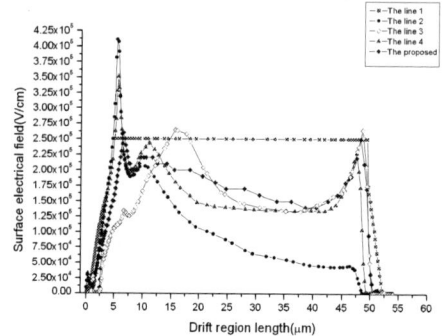

Fig.5 Surface electrical field profile of all kinds of case in the proposed structure

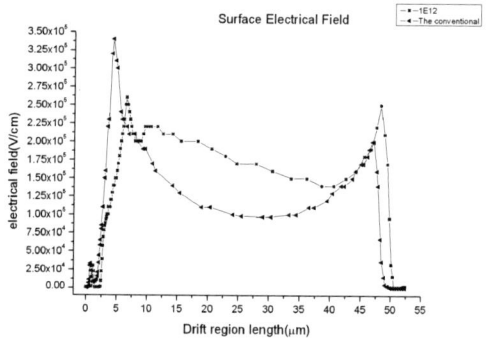

Fig6. Surface electrical field along the surface for two structures at breakdown condition

Effect given usage of long p-top layer. The simulated results also show that n-drift/p-sub junction depth plays an important role in the proposed structure. It seen from Fig.8 that BV increases as the depth of junction increases. But Shallow junction makes depletion region stretch to surface quickly and leads to breakdown in the curvature

Region. The simulated results also indicate that it is important to optimize the concentration of substrate and drift region. Highly doped drift region is preferable in forward voltage drop, but it will lead to breakdown prematurely at poly gate side on the reverse condition. Some cares must be taken to avoid drift region be doped too lightly, because it leads to higher on-resistance and breakdown nearby p$^+$anode without sufficient donor ionization charge. Another curve in Fig.8 verifies the conclusion that high concentration of substrate can reduce the BV significantly.

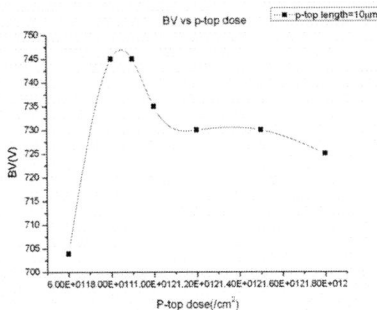

(a) Variation of the BV with respect to p-top dose

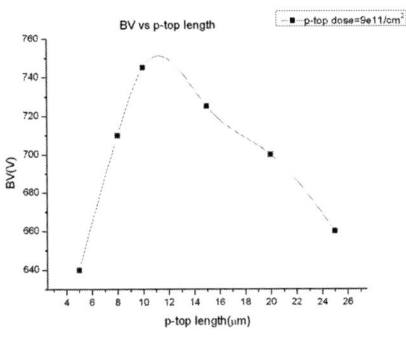

(b)

Fig.7. (b) BV with respect to p-top length

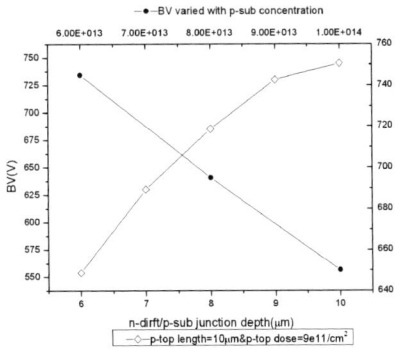

Fig.8 Dependence of BV with n-drift/p-sub junction depth
And dependence of BV with p-substrate concentration

V. CONCLUSION

Double channels are used to enhance conductivity modulation and to reduce on-state resistance in the proposed structure. Forward potential drop at 100A/cm^2 has decreased by 12%. With the technique of combining the p-top layer and deep n-drift/p-sub junction, the proposed structure can support BV more than 745V when the drift region length is less than 44 m. Compared with conventional LIGBT, the BV has increased 18%. The analysis result of surface electrical field profiles in typical cases is given out. At last, the dependence of BV against the p-top length and dose has been discussed in detail.

REFERENCES

[1] J.A Apples and H.M.J. Vaes, "High voltage thin layer devices (RESURF devices)" in IEDM *Tech. Dig.*, 1979, pp, 238-241.

[2] Tomoko Matsudai, Mitsuhiko Kitagawa and Akio Nakagawa, "A trench-gate injection enhanced lateral IECT on SOI" IEEE Proceedings of the 7th ISPSD, May 23-25, 1995, pp.141-145.

[3] Zuxin Qin and E.M.Sankra Narayanan, "A novel multi-channel approach to improve LIGBT performance" IEEE Proceedings of the 10th ISPSD, May 26-29, 1997, pp.313-316

[4] S. Hardikar, R. Tadikonda, D. W. Green, K. Vershinin, and E. M. S.Narayanan, "Realizing high-voltage junction isolated LDMOS transistors with variation in lateral doping," IEEE Trans. Electron Devices, vol. 51,no. 12, pp. 2223–2228, Dec. 2004.

[5] R. Tadikonda, S. Hardikar, D. W. Green, M. Sweet, and E. M. S.Narayanan, "Analysis of Lateral IGBT With a Variation in Lateral Doping Drift Region in Junction Isolation Technology" IEEE Trans. Electron Devices, vol. 53,no. 17, pp. 1740–1744, July. 2006

[6] David W. Green, Shyam Hardikar, Ramakrishna Tadikonda, Mark Sweet, Konstantin V. Vershinin, and E. M. Sankara Narayanan, "Design and Analysis of Multi-channel LIGBTs in Junction Isolation Technology", IEEE Trans. Electron Devices, vol. 52,no. 7, pp. 1672–1676, July. 2005

[7] ATHENA Users manual, SILVACO International Santa Clara, CA 1998

[8] ATLAS Users manual, SILVACO International Santa Clara, CA 1998

An Empiric Approach to Establishing MOSFET Failure Rate Induced by Single-Event Burnout

Jeroen van Duivenbode*, Bart Smet*

* ASML, Veldhoven, Netherlands, e-mail: *jeroen.van.duivenbode@asml.com*

Abstract— **Although the detrimental effect of Single-Event Burnout on semiconductors has been known for over two decades, component manufacturers publish little related data. Through extensive testing, the authors have established trustworthy reliability figures and demonstrate that Single-Event Burnout has a remarkably high impact on power converter failure rate. A standard testing method is proposed for improved power semiconductor qualification testing.**

Keywords—**MOSFET, reliability, cosmic radiation.**

I. INTRODUCTION

ASML provides wafer scanners to microchip manufacturers worldwide. Moore's law has dictated the accuracy roadmap for the last decades [1]. To satisfy it, ASML has successfully applied zero voltage switching technologies in their power amplifiers for minimal switching noise. This technique implies varying switching frequencies which go as high as 300 kHz. Given these high frequencies, MOSFET devices have been the only feasible power switches.

To serve the productivity roadmap, ASML power electronics designers are constantly requested to increase the output power of their amplifiers. Over the past years, current and voltage levels have grown significantly. Recently, the power amplifier supply voltage has risen to 700 V_{DC}, at which level a new failure mechanism has surfaced: Single-Event Burnout. This well documented phenomenon causes MOSFET devices to fail randomly when subjected to a voltage that is high but well within the maximum specified V_{DS} [3]. In this paper, empiric data establish failure rates and show that other failure mechanisms (e.g. voltage or current peaking) are eliminated as possible cause. Consequently, a method is proposed for failure rate testing to be used by MOSFET suppliers, which may be used to quantify SEB-induced failure rates. The ultimate aim is that associated failure rates be adopted in manufacturer's data sheets.

II. SINGLE-EVENT BURNOUT

The detrimental effects of radiation have always been a factor in the design of electronics for space applications. For terrestrial applications, cosmic radiation was also positively identified as the root cause for failure of high voltage power devices when failures were observed to cease when an experiment set up was moved into a salt

In the past, laboratory tests have been carried out showing that MOSFETs and IGBTs are susceptible to proton, neutrons and heavy-ions irradiation [3].

III. ELIMINATION OF DYNAMIC EFFECTS

To eliminate dynamic (voltage/current peaking) effects as possible cause, a "static reverse bias" test was done, as depicted in Fig. 1.

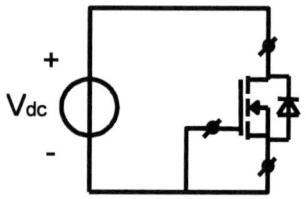

Fig. 1.   Static reverse bias test configuration.

IV. FIRST TEST RESULTS

A preliminary test showed that MOSFETs, subjected to their rated voltage, would fail after less than 70 hours on the average (!).

A test was conducted with 50 FETs in parallel connected to the DC voltage through a fuse (Fig. 2).

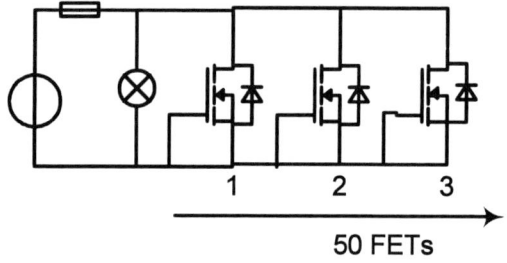

Fig. 2.   Test setup with 50 FETs and fuse.

Each SEB event causes a short circuit tripping the fuse. After manual FET removal, the test is restarted. The test started April 24th 2007 and is still running. FETs with a rated BV_{DSS} of 1000 V that have been cherry picked for a minimal BV_{DSS} of 1075 V are subjected to 700 V_{DC}. There have been 4 failures since the start of the test. The failures occurred after 1176, 1392, 1416 and 4080 test hours. The test is now running for 7000 hours.

978-1-4244-1741-4/08/$25.00 ©2008 IEEE

Failure rate λ is commonly expressed in terms of FIT (failures in time), where 1 FIT is defined as 1 failure per 10^9 hours of operation. It is related to another common reliability term, MTBF (Mean Time Before Failure), by the relation

$$\text{MTBF (h)} = 10^9 \text{ (h/FIT)} / \lambda \text{ (FIT)} \qquad (1)$$

For the 50 FETs, the FIT rate caused by SEB is $10^9 \cdot 4$ / (7000·50) = 11400 FIT. This number should be compared to the target component failure rate of 100 FIT (see below).

V. DESTRUCTIVE OR NON-DESTRUCTIVE?

Note that the effect of the SEB event is greatly influenced by the series resistance and the energy available in the supply.

When applied in a power amplifier, much energy is available in the supply plus close-by buffer capacitors. The series resistance from the other switch in the half bridge is very small. When SEB causes the nominally open MOSFET to conduct, the bus voltage is shorted and the MOSFET typically explodes, often together with its half bridge counterpart. SEM (Scanning-Electron Microscope) inspection of the naked chip showed typical heat damage (Fig. 3 and Fig. 4). This could lead the unprepared observer to suspect electrical overstress, rather than SEB, to be the root cause.

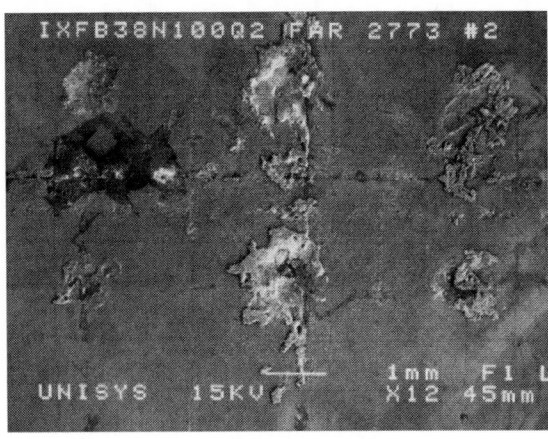

Fig. 3. SEM photograph showing thermal damage under source bonding pads

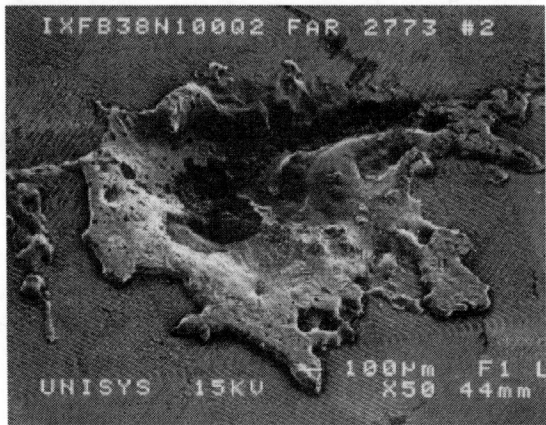

Fig. 4. Detailed SEM photograph of thermal damage under source bond pad.

When the current is limited by fuse, the MOSFET package may remain intact but the junction will be destroyed. SEM inspection showed small failures at a location on the chip that varied randomly from chip to chip, with significantly less massive damage (Fig. 5).

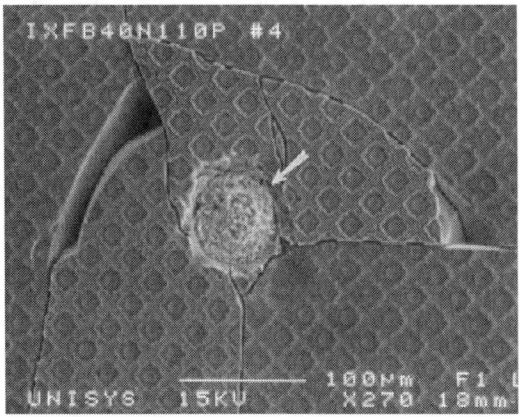

Fig. 5. Detailed SEM photograph of randomly located damage area.

When limiting the current to mere milli-amps by a large series resistor, the MOSFET remains intact with no measurable change in its breakdown voltage.

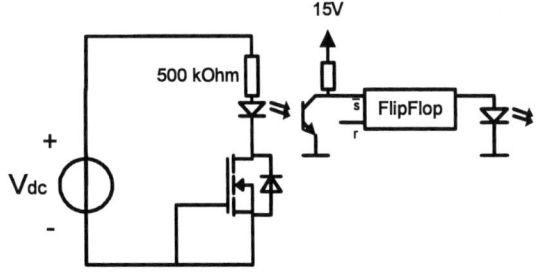

Fig. 6. Non-destructive SEB test setup.

Such a test setup has been used to detect SEB in a non-destructive fashion (Fig. 6).

103

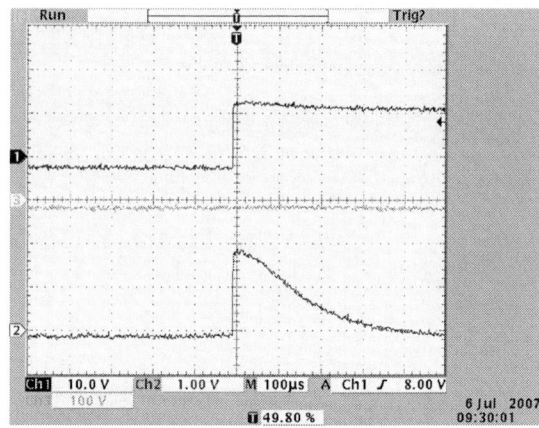

Fig. 7.   SEB current limited by series resistor.

Fig. 7 shows SEB current measured in the non-destructive set up, with $V_{DC} = 1000$ V. The current is presented by Ch2, with 1 mA/division.

VI.   SEB FAILURE RATE TEST SET UP

Acceptable failure rates tend to represent a small fraction over several years of operational use. The challenge of testing for reliability has always been to find a way to get quick results. Temperature often is a factor and ever since Arrhenius has formulated his law, accelerated life testing has been done by raising the test temperature.

Since for SEB the stress factor is not temperature but voltage, testing was started at high voltage stress ($V_{applied}$ = BV_{DSS}) where failure rate was predictably high. Then by stepwise reduction of the voltage stress, more representative results were found at the expense of prolonged test time.

VII.   SEB VOLTAGE DEPENDENCE

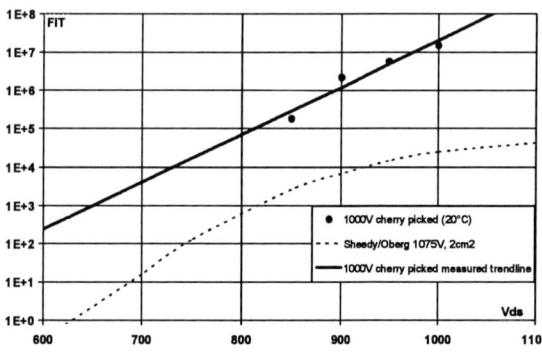

Fig. 8.   SEB voltage dependence.

The observed voltage dependence shows a factor 10 failure rate improvement for each V_{DS} reduction of 80 V. With a rated voltage of 1000 V, this can also be expressed as a factor 10 improvement for each **8%** of derating.

The 4500V rated diode experiment of indicated a factor of 10 improvement per **10%** derating [2]. Three years later, this number is refined to a factor 5 per 5% which corresponds to a factor 10 improvement per **7%** [4]. High voltage MOSFET testing is reported to have a factor 10 improvement per **5.3%** [6]. In Fig. 8 the significant difference between the Sheedy/Oberg prediction [6] and our test results is clearly shown.

The large differences in voltage dependence necessitate testing of each device type, at least until the relation to technology, and voltage rating is fully understood.

VIII.   SEB FIELD RESULTS

To be able to compare the amplifier failure rate in the field with the laboratory measurements the number of failures and the amplifier device hours are needed.

All rejected amplifiers are reported by production and field service in a database. A procedure was started to investigate rejected amplifiers to determine the number of SEB related failures.

The number of device hours was also calculated from the database. With these two numbers the MOSFET FIT rate in the field was calculated, the result is depicted in Fig. 9.

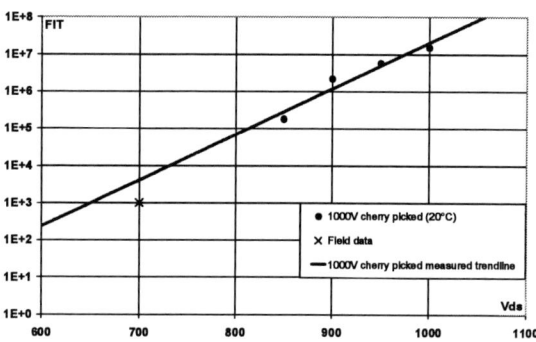

Fig. 9.   Laboratory results compared to field data.

What is remarkable is that the FIT rate in the laboratory is even higher than the FIT rate calculated from the field data. Literature indicates that SEB rate is negatively correlated to junction temperature [6]. The laboratory measurements are performed at an ambient temperature of 20 °C. The junction temperature is higher in a working machine. Therefore the laboratory test was supplemented with a test where the FETs are kept at 80 °C.

IX. SEB TEMPERATURE DEPENDENCE

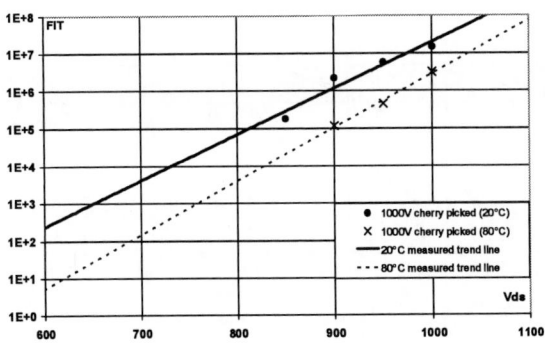

Fig. 10. Temperature dependence of SEB FIT rate.

The measurements show that MOSFETs are *less* susceptible to SEB when the temperature is increased (Fig. 10). A physical explanation may be that the increase in the density of free charge carriers at higher temperatures helps in the recombination of the particles which are generated during an SEB event.

The empiric field number of approximately 1000 FIT per FET exposed to 700 V is situated between the 20 °C and 80 °C lines. Given the fact that the MOSFETs are operated at temperatures in this range, the field data matches the experiment. This confirms that SEB is the major cause of field failures.

X. SEB TECHNOLOGY DEPENDENCE

After consulting the FET supplier it was decided that a good way to improve the amplifier reliability would be to use FETs with a higher BV$_{DSS}$ rating. Before implementing this solution the impact of the new device on SEB FIT rate was checked in the laboratory. The chip size for the new devices has not changed.

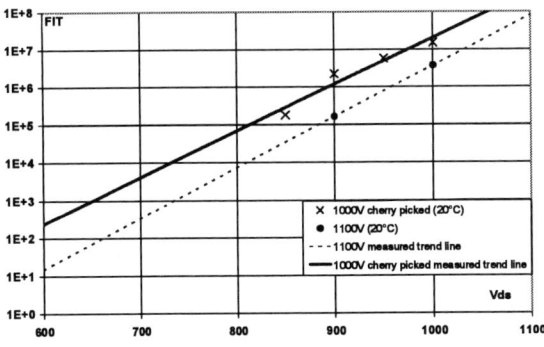

Fig. 11. BV$_{DSS}$ rating dependence of SEB FIT rate.

The results show that, for a given voltage, the 1100V devices perform 10x better than the cherry picked devices (1000 V rated, but selected for BV$_{DSS}$ ≥ 1075 V).

XI. SEB ALTITUDE/LATITUDE DEPENDENCE

The experiments at ASML were conducted at the Veldhoven premises which are located at an elevation of approximately 25 m above mean sea level, and at a latitude of 51.5 degrees North. Literature indicates that the cosmic radiation will be more intense at higher elevations and at higher latitudes (i.e. further from the earth's equator) [7].

XII. ACCEPTABLE FAILURE RATES

Assume a power amplifier (or power supply) contains 3 half bridges with each two switches. Each switch is made up of 3 paralleled MOSFETs to minimize conduction losses. In total, such an amplifier contains 18 MOSFETs. For a population of amplifiers, a yearly failure rate of **1%** is deemed acceptable. This corresponds to 1 failure per 100 years of amplifier operation, so the amplifier failure rate should be better than 10^9 / (365.24 · 100 · 24) equals 1141 FIT.

To calculate the acceptable FIT rate per MOSFET it is important to acknowledge that - due to the switching nature of the amplifier - only half of the switches are open at any time and thereby exposed to the high voltage level. Being forced to accept SEB as the prevailing failure mechanism, we allocate 80% of the 1141 FIT to SEB. Then, the acceptable FIT rate per MOSFET is 1141 (FIT) · 80 (%) / (18 (MOSFETs/amp) · 0.5 (fraction exposed to high voltage level)) which renders approximately **100 FIT per MOSFET.**

Note that with the 80% assumption, *all* other failure mechanisms (both in MOSFETs and in other amplifier parts) will have to make do with the remaining 20% or 228 FIT.

XIII. PROPOSED TEST SETUP

The goal of the test is to establishing failure rate. Therefore, the authors aim for the test to produce 5 failures for acceptable statistical significance. The targeted failure rate *f* being 100 FIT, this implies that a cumulative duration *d* of $50 \cdot 10^6$ device hours is required (!). Using $N = 3000$ devices and an acceleration factor *af* of 10, the required test time *t* is

$$t = \frac{5 \cdot 10^9}{f \cdot N \cdot af}(h) \qquad (2)$$

or 10 weeks test time. The acceleration factor is obtained by deliberately increasing the reverse bias voltage level in accordance with the factor 10 augmented failure rate per 80V voltage increase. The added uncertainty is small as a result of the considerable lab test data that supports the factor 10 per 80 V number.

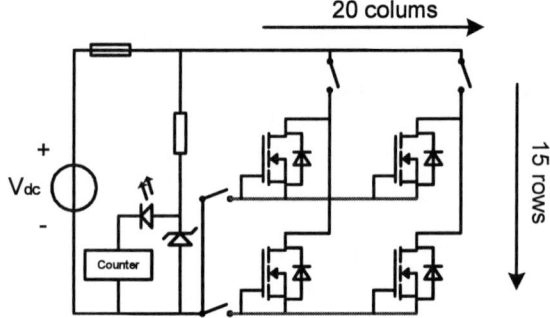

Fig. 12. ASML test set up for N=300, synoptic diagram.

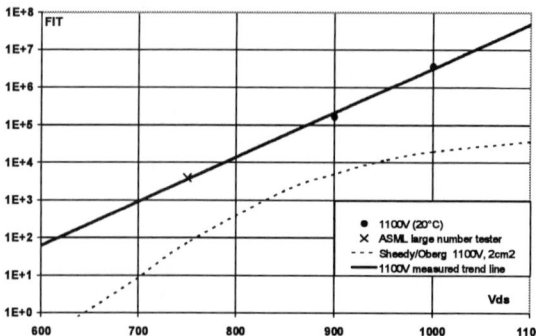

Fig. 15. Result of 3000 MOSFET test at 750 V.

Fig. 13. ASML test set up for N=300, photograph.

Fig. 12 and Fig. 13 show the setup used by ASML. Each setup was implemented on a PCB. Ten such PCBs inserted in a rack to perform the N=3000 experiment (Fig. 14).

Fig. 14. ASML test set up for N=3000, photograph.

The row- and column switches are closed during SEB testing but can be opened to easily identify failed MOSFETs.

Looking at the literature one would expect the line to be steeper for the lower voltages [6]. However, the results show that the measured FIT rate at 750 V is still exactly on the previously measured trend line (Fig. 15). In other words, the target value of 100 FIT is not reached at 750 V (68% derating) but at 620V (56% derating).

XIV. STANDARD MOSFET QUALIFICATION TESTING IS NOT ENOUGH

MOSFET manufacturers generally use a "standard" High Temperature Reverse Bias (HTRB) test for process qualification, during which 30 devices are subjected to an 80% derated reverse bias voltage during 1000 hours. This test is *not* long enough to be significant for SEB. With the acceleration factor 10 per 80 V dependence that the authors have found, an 800 V device normally used at 70% derating, is tested for an effective duration of a mere $10 \cdot 30 \cdot 1000 = 0.3 \cdot 10^6$ device-hours. This number is much too small when compared to the $50 \cdot 10^6$ device-hours required to demonstrate a failure rate of 100 FIT.

XV. PROPOSED DATA SHEET COVERAGE

The authors strongly suggest that MOSFET semiconductor manufacturers start to provide standardized data sheet coverage for "off-state MOSFET failure rate" as suggested in Fig. 16. The data should cover the blocking voltage range from $0.5 \cdot BV_{DSS}$ to $1.0 \cdot BV_{DSS}$, and at least cover junction temperatures 25 °C and 85 °C. Given their influence on the results, altitude and latitude should be normalized to e.g. 0 m above mean sea level and 45° North respectively.

The manufacturer should ensure that the presented data is sufficiently based upon representative measurement data such that actual reliability figures will match within 20%.

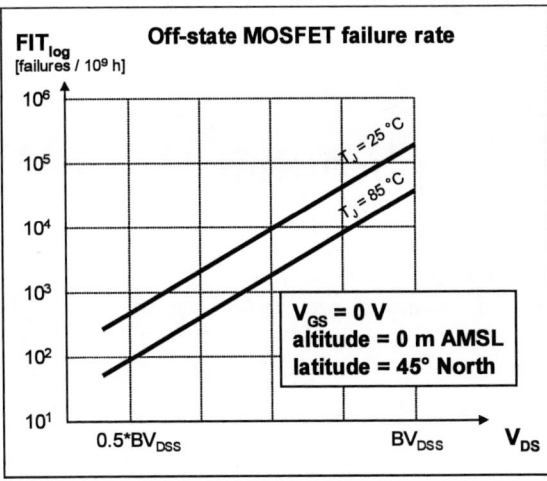

Fig. 16. Off-state MOSFET failure rate example data sheet diagram.

XVI. CONCLUSION

Actual FIT numbers are much worse than previously established models predict.

The authors have shown that testing with a large number of devices can provide representative data within a time span of less than 3 months. MOSFET manufacturers can use this method to provide reliable SEB related FIT numbers in the data sheets.

An alternative test method may be by using accelerated irradiation with an artificial radiation source. However, this will require that the relation to "normal SEB" is well established. Given the solid empiric data presented, the results of this paper may serve as a reliable reference.

XVII. ACKNOWLEDGMENT

The authors wish to thank the companies Microsemi/APT and IXYS for the open-minded discussions on this subject and for the use of the SEM photographs.

XVIII. REFERENCES

[1] G. E. Moore, "Cramming more components onto integrated circuits", *Electronics*, Volume 38, number 8, April 19, 1965

[2] H. Kabza, H.-J. Schulze, Y. Gerstenmaier, P. Voss, J. Wilhelmi, W. Schmid, F. Pirsch, K. Platzöder, "Cosmic Radiation as a Cause for Power Device Failure and Possible Countermeasures", *IEEE Proc. 6th Int. Symp. on Power semiconductor Devices & IC's*, May 1994

[3] D.L. Oberg et al., "First Observations of Power MOSFET Burnout with High Energy Neutrons", *IEEE Tran. Nuc. Sci.*, Vol. 43, Dec 1996

[4] E. Normand, Jerry L. Wert, D. L. Oberg, Peter P. Majewski, P. Voss, S.A. Wender, "Neutron-Induced Single Event Burnout in High Voltage Electronics", *IEEE Tran. on Nuclear Science*, Vol. 44, No. 6, Dec. 1997

[5] Ch. Findeisen, E. Herr, Th. Stiasny, H.R. Zeller, "Cosmic Rays Interacting with Biased High Power Semiconductor Devices", *Int. Foundation HFSJG Activity Report 1999/2000*

[6] R. Sheehy, J. Dekter, N. Machin, "Sea Level Failures of Power MOSFETs Displaying Characteristics of Cosmic Radiation Effects", *IEEE Proc. PESC 2002*

[7] J.F. Ziegler, "Terrestrial Cosmic Ray Intensities", *IBM Journal of R&D*, Vol. 42, number 1, 1998

Comparative Study on Paralleled vs. Scaled Dc-dc Converters in High Voltage Gain Applications

Pawel Klimczak*, Stig Munk-Nielsen*

* Aalborg University / Institute of Energy Technology, Aalborg, Denmark, e-mail: *pak@iet.aau.dk*

Abstract—Today power converters are present in many commercial, medical and industrial applications. A lot of them are high power and high current applications. In order to increase power handling capability several transistors or diodes are paralleled often. However such paralleling may lead to converter's performance degradation or switches quick failure. A parallel modular converter built of many paralleled modules may be an interesting alternative, while a modular converter provides well known advantages like scalability, improved reliability and lower cost. This paper investigates possibility of improving an efficiency by intelligent usage of a modular boost converter in a high voltage gain application.

Keywords— DC power supply, Interleaved converters, Parallel operation, MOSFET

I. INTRODUCTION

If the power level is low it's possible to use only a single power semiconductor as a switch. For higher power levels a single transistor or a single diode may not be enough to handle high current and it is required to use bigger component or few ones in parallel. Usually it's easy to parallel two or three components, but for larger number of paralleled components interconnections become more complex. It's difficult to ensure equal stray inductances in each current path and some components may experience larger stress, especially during fast switching transients [1, 2]. In the same time due to on-state resistance mismatch and temperature rise current sharing problem between transistors may arise. Also there are some applications, e.g. UPS, solar inverters or drive inverters, in which basically the same converter topology is used for different power levels. Often in such case a basic converter is scaled up or down. This approach requires additional design work, more complex production, costly testing facilities and so on. One of disadvantages of a single converter is rather narrow range where high efficiency is achieved – a converter can be optimized for one operating point only. Optimization process is more complex if input voltage varies in a wide range, e.g. fuel cell applications.

A paralleling of whole dc-dc converters may be an alternative for paralleling or scaling of components inside a converter. This approach provides several, well known advantages. It includes expandability, higher reliability, easy maintenance and cheap production due to design standardization and mass production. Additionally, parallel modular converter provides more flexibility for a system designer and a customer. A system designer can

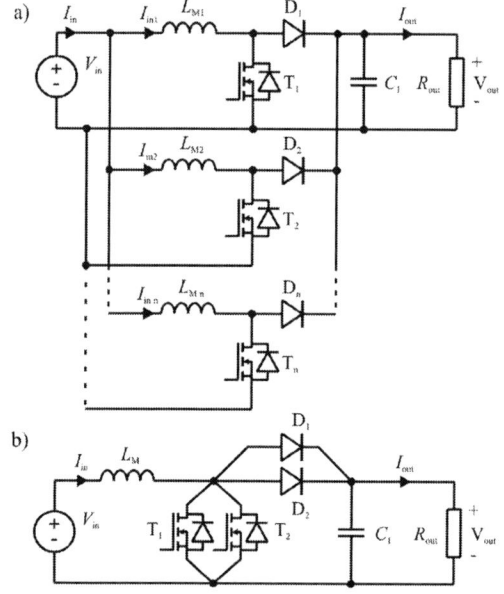

Fig. 1 - Parallel modular boost converter (a) and a boost converter with paralleled/scaled devices (b)

now design and optimize a converter for a power level which ensures the highest efficiency or the best components utilization. Further on, the system can be easily expanded by connecting more modules in parallel. It gives obvious benefits for a customer – instead of replacing whole system, it's necessary to add more modules. However paralleling of many modules creates new challenges. It may lead to unequal load sharing [3-7] or system instability [8-10].

This paper investigates possibility of improving the efficiency by intelligent usage of a modular boost converter in a high voltage gain application.

II. SCALING AND PARALLELING OF A BOOST CONVERTER

Fig. 1 presents a parallel modular converter (a) and a boost converter with paralleled semiconductors (b). Input and output voltages of both converters are the same and the difference is only in rated power of each one. The output power of a single module is limited by a current rating of used transistor – ideally a single module should consist only of one transistor and one diode – a device paralleling should be avoided. To increase power range several modules are paralleled and share the load.

978-1-4244-1741-4/08/$25.00 ©2008 IEEE

Contrary to that approach Fig. 1 b presents a boost converter with paralleled and scaled devices. This converter consists of several paralleled transistors and diodes. Higher power level is achieved by redesigning the converter.

A. Power Semiconductors

In high power and high current applications the main switch is chosen to have sufficiently large current rating. However if such switch is not available or it's performance is not satisfactory then often practice is to parallel few transistors, split current between them and reduce total on-state losses. Paralleling of two or three transistors usually is relatively easy, but putting more devices in parallel may become a challenge – especially in medium/high frequency applications. One, but not the only one, difficulty is to ensure proper layout which provides equal current paths for all transistors – the same stray path length, the same stray inductance etc. During a transistor turn-on period energy is stored in the parasitic stray inductance (leads, connections etc.). When the transistor is being turn-off this energy is transferred between transistor's capacitance and the stray inductance causing oscillations. If there is an exact symmetry in the circuit each transistor absorbs the same amount of energy and experience the same oscillations. But in practice there are small differences e.g. due to component value tolerance or temperature difference. As it is stated in [2] such parameter differences may lead to unexpected oscillations in paralleled MOSFETs. In order to reduce oscillation different snubber circuits are used, but it increases component count and overall converter complexity. Handling of a high current by a switch composed of many devices can push the switching frequency down and result in more bulky passive components.

Also equal current sharing among several transistors is an important issue. As long as MOSFET transistors have a positive temperature coefficient they have an inherent self-balancing capability – up to some extend. Reference [1] analyzes these problems in details and introduces a dynamic current sharing from a gate side as a solution for a current sharing problem during on-state and switching periods. However this method bases on gate current control and may require additional components for a control circuit.

B. Growth of Passive Components

Not only transistor have to be rated for high current. Also all passive components have to be sufficiently large. In the literature one can find at least two approach for power inductor core size selection – one bases on core geometry factor K_g [11] and the other approach uses so called area product A_p [11, 12]. Area product for an inductor core is given by (1). Area product of a particular core can be found by multiplication core cross section are a and window area (2).

$$A_p = \frac{L \cdot I_{dc} \cdot I_{max}}{B_{max} \cdot J \cdot K_u}[cm^4] \qquad (1)$$

$$A_p = A_c \cdot W_a \qquad (2)$$

References [11, 12] give a guidance about core scaling. Now, let's assume that the scaled boost converter has to process twice more power for the same input and output voltages and the same switching frequency. Thus the inductor dc current I_{dc} has to be twice larger. Assuming the same relative ripples $I_{pp\%}$ the peak current I_{max} will be twice larger too (3), but required inductance will be only half according to (4).

$$I_{max} = I_{dc} \cdot \left(1 + \frac{I_{pp\%}}{2}\right) \qquad (3)$$

$$L = \frac{V_{in} \cdot D}{f_s \cdot I_{dc} \cdot I_{pp\%}} \qquad (4)$$

By substituting new I_{dc}, I_{max} and L values to (1) one will find required area product A_p which is exactly twice larger than original one. Extending previous assumptions with constant peak induction B_{max} and constant current density J it's possible to find required number of turns and dc copper losses of the scaled inductor using (5) and (6) respectively. Mean length per turn growth rule is given by (7).

$$n = \frac{L \cdot I_{max}}{B_{max} \cdot A_c} \qquad (5)$$

$$P_{dc} = I_{dc}^2 \cdot \frac{n \cdot MLT \cdot \rho}{A_{Cu}} \qquad (6)$$

$$MLT = K_{MLT} \cdot A_p^{1/4} \qquad (7)$$

It's found that a scaled inductor should have 0.707 number of turns of the original one and dc copper loss will increase by 68% only (under the assumption of optimum design and $P_{fe} = P_{cu}$ [11]). However under the same assumption of optimum design the current density has to decrease when the inductor is scaled up according to (9).

$$surface = K_s \cdot A_p^{1/2} \qquad (8)$$

$$J = \frac{K_j}{A_p^{1/8}} \qquad (9)$$

In the real world such ideal scaling is very rare, because of limited number of different core sizes. In many cases it results in a scaled component which is too bulky or not optimum.

Finally the output capacitor C_1 (Fig. 1) has to be sufficiently larger if the higher power has to be processed. Eq.(10) gives a required capacitance. It can be seen that for constant output voltage V_{out}, voltage peak-peak ripples $V_{pp\%}$, duty cycle D and switching frequency f_s the output filter capacitance C_1 is a directly proportional to the average output current I_{out}.

$$C_1 = \frac{I_{out} \cdot D}{V_{out} \cdot V_{pp\%} \cdot f_s} \qquad (10)$$

Above assumptions will result in sufficiently larger output capacitor and greater capacitor rms current as well.

TABLE I – MODULE INPUT CURRENT DEVIATION UNDER NON-IDENTICAL MODULES OPERATION

	2 modules	4 modules	8 modules
R_s -20%	+4.7%	+7.1%	+8.4%
L_M -20%	0.0%	0.0%	0.0%
R_{DS} -20%	+5.2%	+7.9%	+9.3%
V_f -20%	+1.5%	+2.2%	+2.6%
R_f -20%	+0.8%	+1.1%	+1.3%
Worst case	+13.0%	+20.4%	+24.4%
Duty +0.1%	+32.1%	+40.4%	+53.5%

TABLE II – TRANSISTOR RMS CURRENT DEVIATION UNDER NON-IDENTICAL MODULES OPERATION

	2 modules	4 modules	8 modules
R_s -20%	+4.4%	+6.7%	+7.9%
L_M -20%	+1.7%	+1.7%	+1.7%
R_{DS} -20%	+4.8%	+7.4%	+8.8%
V_f -20%	+1.4%	+2.1%	+2.4%
R_f -20%	+0.7%	+1.1%	+1.2%
Worst case	+12.2%	+19.2%	+23.0%
Duty +0.1%	+30.4%	+44.0%	+50.9%

III. MODULAR POWER CONVERTER APPROACH

Paralleling of dc-dc converters is a well know technique for extending an output power and may be an interesting alternative for a large or scaled up converters with multiple paralleled transistors. This section gives an overview on few common issues and challenges related to paralleling of dc-dc converters.

A. Current Sharing

The most recognized is current sharing problem. If the single dc-dc module has relatively low power it is possible to relay on a simple droop method. Also most of paralleled converters has an inherent self-balancing functionality – up to some extent. If one of modules conducts higher current it heats up, so it's resistances increase and unbalance current is limited in this way. But if safety limits are lower or a single module has higher power then active current sharing method is required. Different paralleling and current sharing methods are presented in [3]. This paper describes several passive and active methods, including Master-Slave and Average Current Programming Methods. More details about active current sharing methods one may find in [4-7].

B. Instability of Paralleled Dc-Dc Converters

Another problem caused by paralleled dc-dc converters is potential system instability. The one reason for this is that dynamic of the parallel system is different that dynamic characteristic of a single converter. It means that a controller tuned for a single module may not work properly with paralleled converters leading to undesired system oscillations. The other reason for instability is presence of multiple current control loops which are introduced by some of current control sharing methods. This problem is widely discussed in [8-10].

IV. SIMULATION OF NON-IDENTICAL MODULES PARALLEL OPERATION

Due to potential current unbalance problem several simulations of non-identical modules are done using Matlab®/PLECS® software. Model parameters and

Fig. 2 - 4-phase modular boost converter

Fig. 3 - Scaled boost converter

components values are selected to fit a breadboard specification described in Section V.

First, simulation of 2, 4 and 8 paralleled identical modules (Fig. 1 a) operating in open loop configuration were performed. Results confirm stable operation and equal current sharing among all modules.

Second step is to simulate behavior of non-identical modules. It's done by changing component values in one of modules. Only one parameter at the time was changed in order to find out which parameter deviation introduces the largest current unbalance. Following parameters (component values) were decreased by 20%: inductor winding resistance R_s, inductance L_M, transistor on-state resistance R_{DS}, diode forward voltage V_f and diode forward resistance R_f.

Next, based on previous observations the 'worst case' scenario was simulated – all components values are increased or decreased by 20% to create the largest unbalance.

Finally, duty cycle value was increased by +0.1% in one of modules to create unbalanced conditions.

Simulation results are summarized in Table I and Table II, where module input current deviation and transistor rms current deviation are presented. It has been found that the inductor series resistance and transistor on-state resistance have big influence on current unbalance. Both resistances are located on low voltage and high current side of the converter. Also, the same disturbance results in larger unbalance when more modules are paralleled. However, the modular converter like this seems to be very

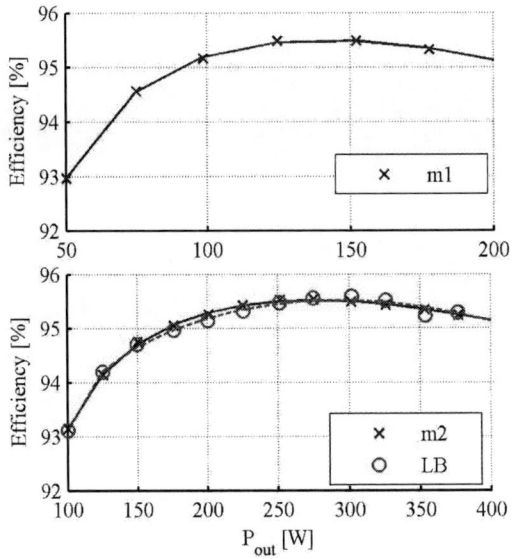

Fig. 4 - Efficiency of a single boost module (m1), two modules in parallel (m2) and the scaled converter (LB) at 30 V input voltage

sensitive to duty cycle unbalance – even a very small deviation in duty cycle may result in huge current unbalance. So, the special attention should be given to PWM gating signals and gate drivers' design.

V. EXPERIMENTAL SETUP

Two demonstrator converters are build and tested in order to compare their performance and efficiency.

A. Description of the Modular Converter

The first one is a modular boost converter built-up from four paralleled boost converter modules (Fig. 1 a, Fig. 2). Desired input voltage range is 30-50 Vdc and output voltage is regulated at 350 Vdc. A single 600 V MOSFET (Fairchild, FCH47N60) and a single 600 V SiC diode (Infineon, SDT05S60) are used. The storage inductor bases on a toroid powder core (Magnetics, High Flux 58071) and it has 68 turns winding. Maximum output power of a single module is 200 W and in fact it's limited

by MOSFET's conduction loss at the lowest input voltage. To achieve higher power level more identical modules are connected in parallel to the first one. Each transistor has its own gate driver and it's controlled by an individual signal. A single Infineon XC167 microcontroller is used for PWM generation. All PWM signals have the same duty cycle and they are equally interleaved – phase shift between PWM signals depends on number of operating modules and it's adjusted continuously.

B. Description of the Scaled Converter

The second converter is scaled up to 400 W version of a single 200 W boost module (Fig. 1 b, Fig. 3). It is designed to process twice larger power like the single module and achieve the same efficiency. This converter consist two paralleled MOSFETs and two SiC diodes – the same type like in the modular converter described above. Each transistor has its own gate driver, but PWM signal is common for both transistors. Both MOSFETs operate synchronously, so effective on-state resistance R_{DSon} is half of a single MOSFET resistance. Also it enables to process twice larger current. Due to higher current the inductor has to be redesign and scaled up – substituting new current and inductance values to (1) it's found that new core should provide twice larger area product value. There are at least three ways to achieve higher power handling capability of the inductor: 1) put two smaller inductors in parallel; 2) redesign an inductor using two stacked smaller cores; 3) redesign an inductor using one larger core. This converter has the storage inductor built on two stacked toroid cores (Magnetics, High Flux 58071). Based on area product approach it's found that such stacked cores have twice larger power handling capability (the same window area W_a, twice larger core cross section area A_c) and it fit's to requirements. Number of turns is decreased by half (34 turns) and, due to doubled core cross section area A_C, inductance of the inductor is decreased by half too. Specific core losses given by Steinmetz equation shall remain on the same level, but total core loss will be twice larger due to larger core volume. The winding wire size is selected in respect to required winding dc resistance, which should be half of dc resistance of the inductor in modular converter. For twice larger current it gives twice larger copper loss.

Using this design approach, the converter is expected to

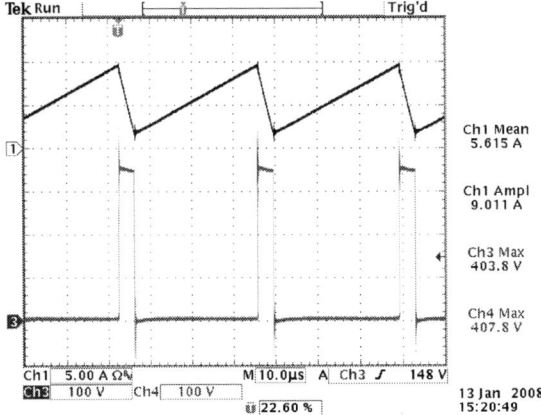

Fig. 5 - Input current (Ch1) and transistor drain-source voltage (Ch3, Ch4)

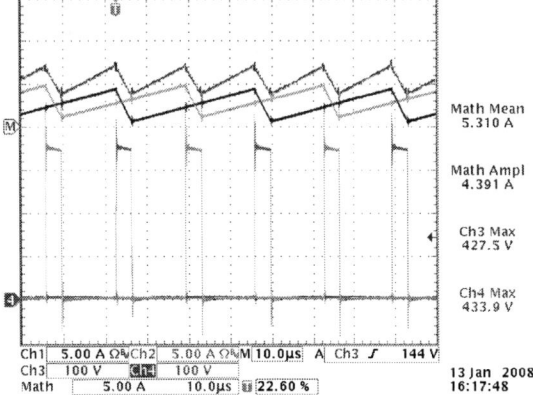

Fig. 6 - Module input currents (Ch1 and Ch2), converter input current (Math) and transistor drain-source voltages (Ch3 and Ch4)

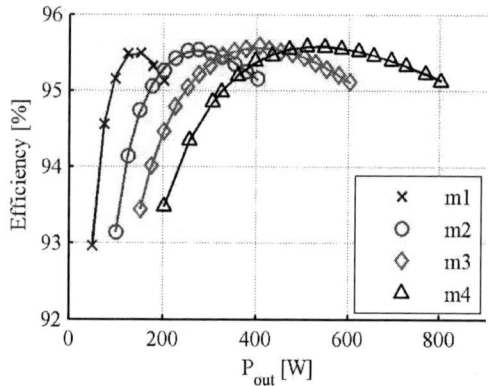

Fig. 7 - Efficiency of a modular converter at 30 V input voltage

have twice larger maximum power and similar efficiency to the single converter module described previously.

VI. EXPERIMENTAL RESULTS

At the beginning three converter configurations were compared in terms of achieved efficiency. At an input voltage of 30 V the single module (curve m1) efficiency was measured in the power range from 50 to 200 W. Next efficiency of a scaled 400 W boost converter (curve LB) was measured and compared with two paralleled modules (curve m2) under the same conditions (input voltage 30 V, output power 100-400 W). Results are presented on Fig. 4. One may observe that efficiency curves m1, m2 and LB are very similar –it means that the scaling process is done correctly. Also, it confirms that paralleling of two switches shall not affect the efficiency significantly.

Fig. 5 and Fig. 6 present input currents and transistor voltages of a scaled 400 W converter and two interleaved modules respectively. One may find that input current ripples are significantly lower for interleaved modular system. Moreover, output voltage ripples have a twice larger frequency, and it results in smaller output capacitor. Also it's should be emphasized that the parallel converter operates in open loop configuration and current sharing is very good in spite of absence of current controller.

Next the input voltage is set to 30 V and efficiency was measured for different number of modules (up to four modules in parallel) for different power levels. Results are presented on Fig. 7. By adjusting number of operating modules it's possible to keep high efficiency over a wide range of output power. By adding more modules it's easy to extend power level of the system.

Finally, measurements under variable input voltage were done. Variable input voltage simulates behavior of a PEM fuel cell – output voltage of the stack is highest at no-load condition and decreases while output power increases. In the lab setup input voltage was set to 50 V for 100 W and decreased linearly to 30 V at 800 W output power. Under such conditions the efficiency was measured for different number of modules. Results are presented on Fig. 8 together with input voltage value. One may observe that efficiency of four modules in parallel (curve m4) drops quickly while output power decreases, e.g. down to 94% at 340 W. In this case it's beneficial to turn-off one or two modules and increase

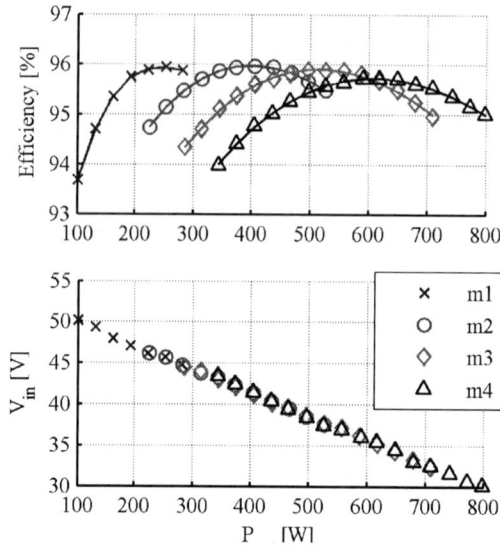

Fig. 8 - Efficiency of a modular converter at variable input voltage

efficiency of the whole system. One may observe that efficiency of such modular system increases while output power decreases – this behavior is opposite to most of traditional converters, where efficiency drops at low load.

VII. CONCLUSION

This paper gives an overview on scaling and paralleling of boost converters. Two demonstrator converters are built, tested and results are presented. Based on these results it's stated that modular converter gives more flexibility and provides additional degree of freedom for a system designer. Adjusting number of operating modules leads to optimum utilization of the converter. In case of a fuel cell application, with variable input voltage, the parallel converter provides high efficiency over a wide range of output power levels.

ACKNOWLEDGMENT

This project is financed by Dansk Energi net under PSO project no. 562/06-14-26808.

REFERENCES

[1] W. Hongfang and F. Wang, "Power MOSFETs Paralleling Operation for High Power High Density Converters," in *Industry Applications Conference, 2006. 41st IAS Annual Meeting. Conference Record of the 2006 IEEE*, 2006, pp. 2284-2289.

[2] J. G. Kassakian and D. Lau, "An analysis and experimental verification of parasitic oscillations in parralleled power MOSFET's," *Electron Devices, IEEE Transactions on*, vol. 31, pp. 959-963, 1984.

[3] L. Shiguo, Y. Zhihong, L. Ray-Lee, and F. C. Lee, "A classification and evaluation of paralleling methods for power supply modules," in *Power Electronics Specialists Conference, 1999. PESC 99. 30th Annual IEEE*, 1999, pp. 901-908 vol.2.

[4] T. Kohama, T. Ninomiya, M. Shoyama, and F. Ihara, "Dynamic analysis of parallel-module converter system with current balance

controllers," in *Telecommunications Energy Conference, 1994. INTELEC '94., 16th International*, 1994, pp. 190-195.

[5] D. J. Perreault, R. L. Selders, and J. G. Kassakian, "Frequency-based current-sharing techniques for paralleled power converters," in *Power Electronics Specialists Conference, 1996. PESC '96 Record., 27th Annual IEEE*, 1996, pp. 1073-1079 vol.2.

[6] N. Hur and N. Kwanghee, "A robust load-sharing control scheme for parallel-connected multisystems," *Industrial Electronics, IEEE Transactions on,* vol. 47, pp. 871-879, 2000.

[7] D. J. Perreault, K. Sato, R. L. Selders, Jr., and J. G. Kassakian, "Switching-ripple-based current sharing for paralleled power converters," *Circuits and Systems I: Fundamental Theory and Applications, IEEE Transactions on [see also Circuits and Systems I: Regular Papers, IEEE Transactions on]*, vol. 46, pp. 1264-1274, 1999.

[8] H. H. C. Iu and V. Pjevalica, "Experimental study of instabilities in two parallel-connected boost converters under current mode control," in *Power Electronics and Applications, 2005 European Conference on*, 2005, p. 8 pp.

[9] S. K. Mazumder, "Stability analysis of parallel DC-DC converters," *Aerospace and Electronic Systems, IEEE Transactions on*, vol. 42, pp. 50-69, 2006.

[10] J. M. Zhang, X. G. Xie, X. K. Wu, and Q. Zhaoming, "Stability study for paralleled DC/DC converters," in *Power Electronics Specialists Conference, 2004. PESC 04. 2004 IEEE 35th Annual*, 2004, pp. 1569-1575 Vol.2.

[11] C. Wm.T.Mclyman, *Transformer and Inductor Design Handbook, Third Edition*. New York: Routledge, 2004.

[12] P. Wallmeier, "Pre-optimization of linear and nonlinear inductors using area-product formulation," in *Industry Applications Conference, 2002. 37th IAS Annual Meeting. Conference Record of the*, 2002, pp. 2445-2450 vol.4.

A LOW-LOSS DC-DC CONVERTER FOR A RENEWABLE ENERGY CONVERTER

David S. Thompson *, Otu A. Eno**
*Department of EEP, University of Dundee, Dundee, U.K, _d.s.thompson@dundee.ac.uk_
**Albacom Ltd, Dundee, U.K, _oaeno@hotmail.com_

Abstract - **The maximization of energy transfer from a low-voltage source to a grid is investigated. The converter has two stages, a DC-DC converter followed by an inverter. Attention to the DC-DC converter losses shows how its energy transfer is maximized and found to be acceptable.**

Keywords – **Resonant Converter, DC-DC converter, Transformer stray load loss.**

I. INTRODUCTION

The maximisation of the energy capture from a renewable source such as PV or Wind for grid connection is dependent on the energy conversion efficiency.

The overall concept is one of a cascade of a bridge-connected DC-DC converter followed by a DC link supplying the inverter whose output voltage is filtered to minimize the adverse effects of harmonics on the grid, generally as shown in Fig. 1 where the inverter's output is linked to the grid by a conventional L-C-L filter. For successful connection to an a.c. grid the converter's output voltage and frequency must match those of the grid at all times; frequency following is achieved with a phase-lock loop, and voltage following at the DC-DC converter. For such an output the inverter is of the 1-phase type; no difficulty is foreseen for adapting the inverter to a 3-phase form.

The regulations for grid connections restrict the level of harmonic content of the injected energy. For this the inverter uses a PWM switching pattern. The modulation frequency of the inverter is high enough to reduce the lowest order of harmonic to not less than 25 times the supply frequency. For this type of converter operation switching loss is unavoidable and so attention is focused on the dc-dc converter.

The converter's input is in the range 45 – 65 V dc and the ac output voltage is 240 V, 1-ph, with a nominal rating of 10 kW.

Galvanic isolation is provided by a high-frequency low-loss transformer at the output of the DC-DC converter. With this concept the transformer's core size is significantly reduced in volume as well as its losses. This gives the complete converter an acceptable performance.

II. OPERATIONAL CONSIDERATIONS

The overall control of this cascade system is by a DSP chip with a range of applications within the converter and described more fully later.

The DC-DC converter is of the resonant type and operates at constant frequency controlled by the DSP controller.

The initial energisation of the DC-DC converter's output capacitor of 7000µF from a constant voltage source of low internal impedance requires care to avoid the use of components, such as Mosfets, etc with ratings far in excess of the converter's nominal energy rating at vast expense. This is achieved by the use of the DSP chip to adjust the DC-DC converter's pulse width to control the rate of rise of capacitor charging current. To achieve such an outcome the pulse width of each half cycle of the dc-dc converter's output voltage is reduced to a value that limits the charging current to an acceptable value and gradually increased as the capacitor voltage increases until it is safe to raise the pulse width to its rated value. The approach to this operational feature is discussed in greater detail elsewhere [1, 4].

The ability of the system to ride through short-term grid voltage sags is also dependent on the dc-link capacitor's energy storage capability.

III. CONVERTER OPTIMIZATION CONSIDERATIONS

For this design the converter is intended to supply energy to a grid irrespective of the grid's operating conditions and no energy storage unit such as fuel cell is included.

978-1-4244-1741-4/08/$25.00 ©2008 IEEE   114

For the system under consideration the main focus of energy-loss control is on the DC-DC converter. Here for minimal losses a series-resonant switching pattern is used to minimize switching losses. Despite the difficulty such resonant systems have in relation to regulation over the load spectrum it would be difficult, without resort to an additional resonant circuit, to arrange a parallel resonant converter with this large DC-link capacitance; an aim of the design of this converter is a minimum component count. With series resonance ZCS is achieved and ZVS is realized by the expedient of a very short interval with complete isolation from the energy source as it transfers from positive to negative output, or vice versa, as shown in Fig. 2.

Successful operation of this type of converter in relation to the supply of energy to its connected grid depends on a "stiff" dc voltage at the output of the dc-dc converter. This in practice means the use of a "large" capacitor for energy storage and of sufficient capacity to enable it to ride out a short term voltage dip arising from a grid short circuit of specified duration, for the converter in question this capacitor is 7000μF and with a voltage rating of not less than 500 V. For this application a low-loss capacitor is required.

Optimal characteristics of the converter's ac output suggest its dc input voltage should be modulated in a PWM format to increase the lowest order harmonic in its output voltage spectrum. The frequency characteristics of the grid can be achieved by the adoption of a phase-lock loop and smooth synchronization to the grid is achieved by ac voltage control of the inverter's PWM switching pattern. For operation with conventional generators the output's power-frequency characteristic is given a drooping shape such that its power-frequency characteristic is compatible with the phase-lock loop's features.

In terms of the above observations the aims of the overall design are to minimize the losses, output harmonic content and respond appropriately to any voltage sags. With reference to the observations about switching loss in the PWM inverter and the filter losses; attention is focused on the DC-DC converter performance.

IV. DC-DC CONVERTER TOPOLOGY

The structure of the converter is a DC-DC converter with an output transformer driving a rectifier supplying a capacitor and finally an inverter with an output filter, generally as shown in Fig. 1. The design of each sub system follows conventional procedures with detail focus on the losses within each stage, particularly the dc-dc converter.

A. DC-DC Optimization Considerations

One advantage of a renewable energy source, such as wind or PV, is its ability to convert the available energy as it comes; when the demand on the grid is low then a suitable form of storage is required to maximize the abstraction of the naturally occurring energy. For this design the converter is intended to supply energy to a grid irrespective of the grid's operating conditions and no energy storage unit is included.

As the design value of the input dc voltage is low a transformer is used to step up the voltage to the value required for direct grid connection.

For the system under consideration the main focus of optimization is on the DC-DC converter. Here for minimal losses a series-resonant switching pattern is used. Despite the difficulty such systems have in relation to regulation over the load spectrum it would be difficult, without resort to an additional resonant circuit, to arrange a parallel resonant converter with the large capacitance in the dc link; an aim of the design of this converter is a minimum component count. With series resonance ZCS is achievable by resonating a capacitor with the transformer's leakage inductance. The achievement of ZVS is realized by the expedient of a very short interval with complete isolation from the energy source as its voltage transfers from positive to negative output. Waveforms of the converter are shown in Fig. 2. These clearly show, in yellow, the resonant current waveform, the Mosfet driver voltages in magenta and the dc voltage in cyan. Both the current and voltage waveforms do show a slight hesitation as they approach zero volts from above. This has a negligible effect on the ac output.

By comparison with some conventional approaches to the optimal design of a transformer it is less easy to optimize a converter of this type because of the erratic nature of the energy input from the basic converter, in this instance a PV or wind-energy converter. To design a transformer, for example, for a conventional power system it is common to design to a specific load pattern with numerical values for the capitalized losses of the transformer. It is then a matter for the designer to arrange the materials and their construction in such a way that the lifetime cost of the transformer, including its prime cost, is a minimum. In practice the operational patterns leading to such specifications are inappropriate to a renewable energy converter where the energy flow is erratic. Optimization is applied to each subsystem of the complete converter.

B. DC-DC Converter

For this design the application of a "conventional" step-up converter is not considered. For this application the effects of source impedance on the system regulation may lead to unrealistically long on times for the main switch and which would be such that it would not operate in an off mode. For this reason a series resonant converter with output transformer is the chosen option. For this converter the switching frequency is 42 kHz and the devices in use are Mosfets.

The dc-dc converter has a full bridge topology and is operated with a unipolar switching arrangement as shown in Fig. 1. A more complete description of its operation may be found in [1,2]. Its output is transferred to the dc-link capacitor by a cascade of a transformer and rectifier. By this particular mode of operation the effects of dc magnetization of the core are minimized without the need for additional switching arrangements at extra cost for its demagnetization. Waveforms of the converter's operation are shown in Fig. 2.

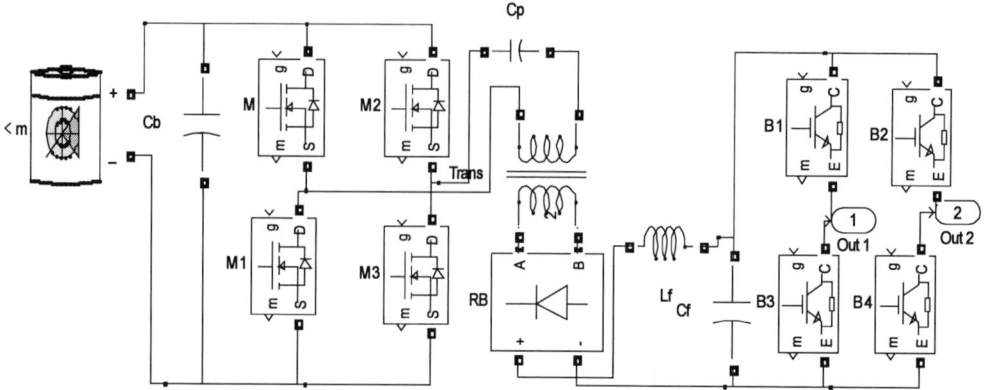

Fig 1 Schematic of the dc-dc converter, transformer and rectifier, dc-link filter, and h-bridge inverter. Out-1 and out-2 are connected to the filter.

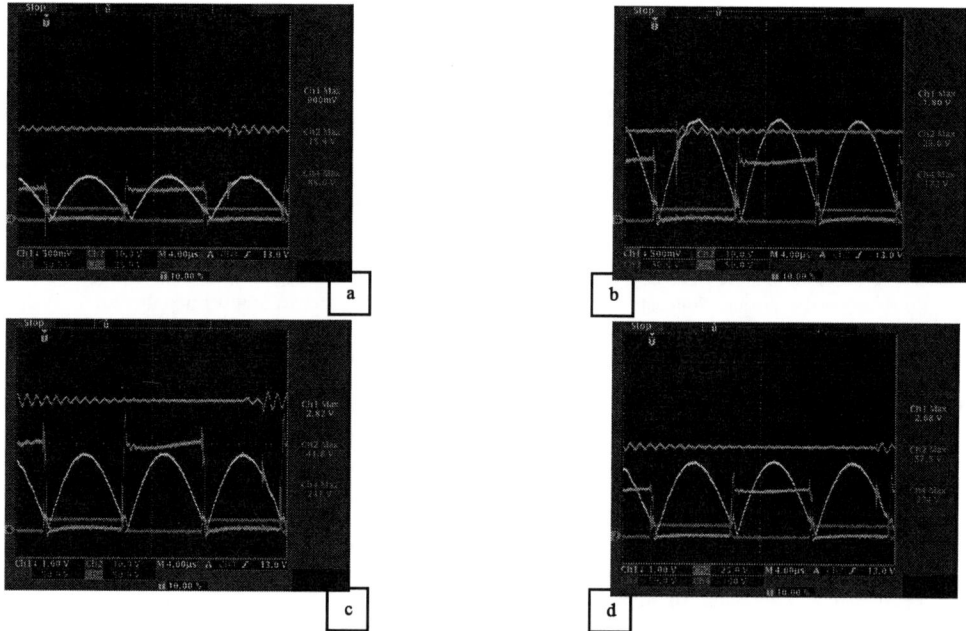

Fig. 2 The figures show the waveforms for the dc-dc converter with increasing input voltage from a to d. Green=Dc-link voltage, Yellow = Mosfet current, Cyan= Mosfet voltage, Magenta=Mosfet gate drive voltage.

V. TRANSFORMER DESIGN FEATURES

The transformer design includes the effects of stray-load loss analysis and minimum overall loss as well as heat dissipation considerations. The converter rating is 10kW. Neglecting losses, at 50 V the corresponding LV, or input current is 200 A. This level of current is such that a helical winding, if selected, would need to be of the multi strand variety to minimize the eddy-current losses. For such a winding with an average current density of 3.5 Amm^{-2} and no subdivision of the turn's cross-sectional area the wire diameter would need to be 17.5 mm, a value greatly in excess of the skin depth at this frequency, hence for minimal winding loss a multi-strand arrangement is essential.

The basic losses within such a winding would comprise the dc resistance loss, skin-effect and proximity effect losses due to the ac current within each strand and all other turns of the winding. The skin-effect loss at 42 kHz is the reason for the need to adopt a multi-strand conductor. Subdivision of the conductor into a set of parallel sub conductors leads to further proximity-effect losses. Also the strands of the subdivided conductor do not coincide geometrically and thus the flux linkage of each sub conductor is different; hence their emfs are unequal. The magnetic coupling between the strands, or sub conductors, is high and thus if each strand is considered as a turn of a transformer winding the leakage reactance of each is low. This low effective leakage reactance per strand leads to circulating currents and further losses which can be significant when all strands are connected in parallel. For this reason it is essential to transpose the sub conductors so that each occupies all possible positions throughout the winding; this approach is to ensure equality of emf and effective leakage reactance per strand and hence reduced losses. The use of Litz wire is one approach to the resolution of this problem. The need for reliable connections of a multi-strand conductor at each end of the winding makes this solution unattractive in relation to the connections. For these reasons a foil-wound winding is adopted for the LV winding. It is such that the radial thickness of each turn is less than the skin depth. Reduction of the turns on this winding minimizes the effects of proximity-effect losses to acceptable values and hence the total winding loss. For the same conditions the equivalent current in the HV winding is 40 A, a value that a single strand can just accommodate without excessive ac losses and heat in a single-layer winding. Further details of these eddy-current problems are covered elsewhere [6, 7].

At this frequency a ferrite core is used and is shaped as an enclosed pot. Its magnetic design is optimized after the manner of Matveev [5]. The losses at 7.0 kVA output are 60 W with an input voltage of 58 V and the dc-link voltage of 442 V; the corresponding efficiency is 98.2% with a standard deviation of 0.0066. The inverter's switching losses are unavoidable and difficult to minimize. The filter components are chosen to achieve minimal losses and THD.

The total losses for both conversion stages are 1253 W. They confirm the difficulty in reducing the losses in a PWM inverter and filter.

An important aspect of the design is the use of the transformer's leakage inductance as one of the resonating components for series resonance. This means adjusting the coefficient of coupling to achieve the desired value of leakage inductance, principally by adjustment of the inter-winding spacing. The relevant parameters of the transformer are: Magnetizing inductance 20.48μH, leakage inductance referred to the LV winding 358 nH, turns ratio 8.017, and inter-winding capacitance 181pF. The turns ratio is chosen to provide sufficient output voltage for reactive energy injection at the point of common coupling. With these values of inductance the value of capacitance for parallel resonance process would be 20 μF. This value is so much less than the referred value of the dc-link capacitance, viz 110 μF, that parallel resonance is not a viable operational mode for this design.

Some details of performance data are given later in Tables 1 and 2; full details are given by Eno [1].

V1. OPERATIONAL OPTIMIZATION

Without control of the overall system the optimization of individual components and systems would be fruitless. For this converter a digital control strategy is adopted with a DSP chip providing the control. The most important feature is its application to the restriction of the inrush current at start up. Thereafter it controls both the dc-dc converter and the inverter, details of their operation may be found elsewhere [1, 2, 4]. The architecture of this class of microcomputer lends itself to the control of more than one system within the overall application. The DSP's internal D/A converter provides the necessary conversion for its effective operation.

The design described focuses on a 1-phase converter. A typical 3-phase converter would only need adjustment to the output stages of inverter and filter for such an extension. The ready availability of sinusoidal PWM inverters for 3-phase systems is not seen as a disadvantage to the process described here, other than to require additional features within the program logic.

The software within the DSP has provision for the inclusion of an MPPT routine for maximising the renewable energy converter's output.

V11. TESTS

Each stage of the converter was tested for functionality. The overall converter was tested for its operation, including successful direct-on-line starts of a 250-W 1-ph reluctance motor with a starting current of 8 times rated current. Loss data are shown in Tables 1 and 2 below.

Table 1

Vin	Iin	Pin	Vout	Iout	Pout	Losses	Effy
Volts	Amps	Watts	Volts	Amps	Watts	Watts	%
48.1	145	6975	353	19.1	6742	233	96.6
50.3	141	7092	372	18.5	6882	210	97.0
56.5	127	7176	426	16.6	7072	104	98.6
58.2	124	7271	442	16.1	7116	101	98.6
62.3	118	7351	476	15.2	7235	116	98.4
66.2	114	7547	509	14.6	7431	116	98.4

Table 1 showing the electrical performance of the DC-DC converter; the test relates to a battery source and the output data refer to the input to the DC link capacitor.

Corresponding data for the inverter stage are shown in Table 2. the input data for Table 2 are identical to the input data in Table 1, thus each row of Table 2 refers directly to the corresponding row of Table 1.

Table 2

Vin	Iin	Pin	Vout	Iout	Pout	Losses	Effy
Volt	Amps	Watts	Volts	Amps	Watts	Watts	%
353	19.1	6742	230	25.0	5691	1151	84.4
372	18.5	6882	233	25.0	5804	1079	84.3
426	16.6	7072	234	25.3	5904	1098	83.5
442	16.1	7116	237	25.4	6010	1106	84.5
476	15.2	7235	239	25.7	6141	1094	84.9
509	14.6	7431	242	26.0	6288	1143	84.6

Table 2 showing the electrical performance of the inverter including the output filter; note that the input data are the output data for the DC-DC converter.

VIII.   CONCLUSIONS

With limited operational experience of renewable energy systems in grid systems it is as yet inappropriate to apply directly the traditional approaches to optimization outlined in [5] to an overall converter. The limited experience with the converter just described seems to justify the approaches described within this contribution. The information in Tables 1 and 2 about the losses confirm the earlier claims for low losses within the

DC-DC converter. The outcome from the attention to the stray-loss aspect of the transformer is particularly pleasing. Although not shown the regulation of the ac output is also pleasingly acceptable, thus justifying the approach to the regulation of the resonant DC-DC converter and the effectiveness of the DSP controller in this aspect of the converter's performance.

Although the discussion focused on the grid connection application, the converter has been used successfully in a stand-alone mode as noted earlier.

The successful outcome is seen to justify the analytical approach described in this contribution. For such success it is essential to focus on each part of the converter stage by stage. For minimal losses this stage-by-stage approach also need a further focus on the overall situation. Attention to the loss distribution and the dissipation of the losses is also essential for a successful outcome. We also acknowledge the many helpful suggestions provided by Julian Coppin of Albacom.

ACKNOWLEDGEMENTS

We wish to thank the Directors of Albacom, the Court of The University of Dundee for the facilities used in the investigation to which this relates and also the support from The UK Government.

REFERENCES

[1] .-"Digital Control of a Two-Stage, High Efficient Converter for Grid Connection of a Renewable Energy Source", Otu Ansa Eno, PhD Thesis accepted by University of Dundee, May 2005.

[2] -"High Power Resonant Topology for DC-DC Converter", O.A. Eno, D. S. Thompson, paper no in Proceedings of EPE 2005, Dresden; on CD.

[3] Minimizing Current Inrush at Resonant Converter Start Up", O. A. Eno, D. S. Thompson, paper no in Proceedings of EPE 2005, Dresden, on CD.

[4] "Keeping a Switch in its SOA with a DSP", O. A. Eno, D. S. Thompson, J. Coppin, Proceedings of EPE-PEMC 2004, Riga, on CD.

[5] "Teaching Magnetic Design Optimization", Alexei Matveev, Robert Nilsson, Tore Undeland, Christian Hartman,   In Proceedings of EPE 2001, Graz, on CD.

[6] "Modelling HF resistance of Parallel Windings of Magnetic Components", Roberto Prieto, Proceedings of EPE2001, Graz, on CD.

[7] Eddy_Current Losses in SMPS Transformers, A Full-Frequency Range Review of 2D Effects Inside Windings", Frederic Robert, Pierre Mathys and Jean-Pierre Schauwers in proceedings EPE 2001, Graz, on CD.

[8] "Inductors and Transformers for Power Electronics" Alex Van den Bosche and Vencislav Valchev. CIRC Press.

A Single Active Edge-Resonant Snubber Cell-assisted ZCS Half-Bridge DC–DC Converter with Constant Frequency Asymmetrical PWM Scheme

Tomokazu Mishima*, Mutsuo Nakaoka[†], and Eiji Hiraki[‡]
*Kure College of Technology, Hiroshima, Japan, e-mail: *mishima@kure-nct.ac.jp*
[†]Kyungnam University, Masan, Republic of Korea
[‡]Yamaguchi University, Yamaguchi, Japan

Abstract— A newly-developed zero current soft-switching (ZCS)-PWM cell-assisted asymmetrical half-bridge (AHB) DC–DC converter topology with a high frequency (HF) link is presented in this paper. The soft-switching half-bridge DC–DC converter consists of a PWM-controlled single-ended half-bridge HF inverter and a center-tapped rectifier linked by a HF transformer. In order to attain the wide range of ZCS commutation in the primary-side HF inverter, the active edge resonant snubber cell composed of a switched capacitor and a lossless inductor is adopted in the half-bridge leg, providing ZCS commutation for a wide range of output power under constant switching frequency. The operation characteristics of the proposed DC–DC converter are described, and its feasibility data is demonstrated and evaluated with simulation and experimental results.

Keywords– Asymmetrical half-bridge DC–DC converter, soft-switching, ZCS, constant frequency PWM, active snubber, edge-resonance.

I. INTRODUCTION

The latest developments of the IGBT-based HF inverter linked DC–DC converters with a large voltage step-down ratio have been playing an important role for establishing more advanced electric power systems in automobiles, telecommunication & information equipments as well as energy storage devices-interfaced power conditioners[1]. In particular, most of PWM DC–DC converters with a center-tapped or a current doubler rectifier have begun to be one of the most promising circuit topologies suitable for this type of power conversion in terms of a high power density, low-profile, and its easy-to-implement control scheme [2].

A series resonant asymmetrical DC–DC converter with Pulse Switching Frequency Modulation (PFM) scheme is utilized widely as an AHB DC–DC converter for the wide range of soft switching performance [3],[4]. The PFM DC–DC converter, however, has a shortcoming of severe audible noise generation in the low output power setting. And, this makes Pulse Width Modulation (PWM) DC–DC converter more attractive to the several electric power systems.

As the soft-switching approach for the IGBT-based PWM DC–DC converters, Zero Voltage Soft-Switching (ZVS) has been popular owing to the less additional passive components [5],[6]. However, the ZVS commutation based on the edge-resonance with lossless capacitors in parallel with switching power devices significantly depends on the load current. This property causes severe limitation of the soft switching operation range under light load regions[8]. On top of that, IGBT with MOS gate-controlled bipolar mode characteristics are generally not suitable for the ZVS scheme because of the tail current transition at its turn-off commutation.

As a solution for the soft switching limitations mentioned above, the counterpart soft-switching scheme, ZCS, is more effective for IGBT-based DC–DC converters. And, the new type of ZCS-PWM circuit topology for the HF-link AHB DC–DC converter have been proposed by the authors as presented in [9].

In this paper, the newly-developed Active Edge-Resonant Snubber (AERS)-assisted ZCS-PWM AHB DC–DC converter are evaluated in more details with experimental analysis performed by the prototype circuit. In particular, the converter characteristics on the actual conversion efficiency as well as the ZCS operation range achieved by the asymmetrical PWM scheme are discussed. In addition, the practical circuit design strategy for the soft-switching DC–DC converter are introduced, and the validity of the circuit design method is demonstrated. And, finally the feasibility of the ZCS-PWM AHB DC–DC converter are discussed from the practical point of view.

II. CIRCUIT DESCRIPTION

Fig. 1 shows the proposed DC–DC converter configuration. Here, E_d represents the input supply DC voltage. This circuit topology is based on a LLC resonant PFM half-bridge DC–DC converter, and derives from the single ended push-pull inverter introduced in [10] and [11].

Besides the main switch $Q_1(S_1/D_1)$, the high side part of the half-bridge leg is comprised of the ZCS-assisted inductor L_{r1}, the resonant lossless capacitor C_r actively switched by the auxiliary switch $Q_3(S_3/D_3)$, all of which operate as an active snubber for Q_1. In the low side part of the leg, the other ZCS-assisted inductor L_{r2} is inserted in series with the other main switch $Q_2(S_2/D_2)$.

978-1-4244-1741-4/08/$25.00 ©2008 IEEE

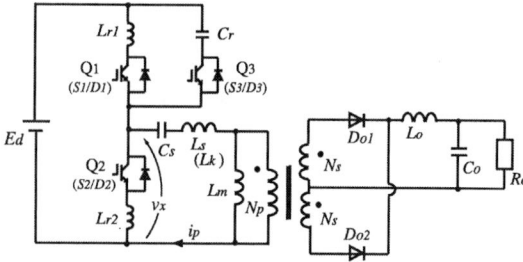

Fig. 1. Proposed AERS-assisted ZCS–PWM half–bridge DC–DC converter with center-tapped rectifier.

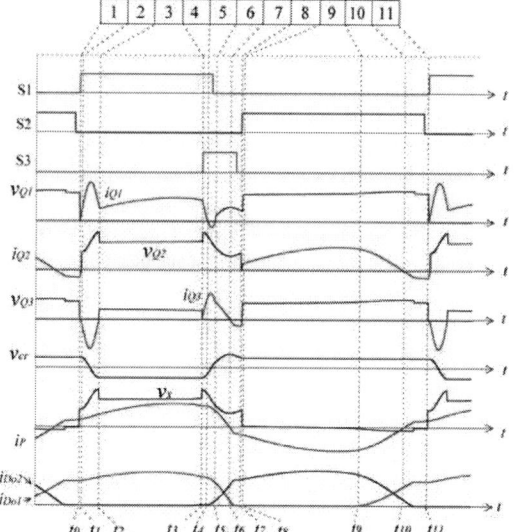

Fig. 2. Voltage and current waveforms of ZCS-PWM DC–DC converter during switching one cycle with 45 % duty cycle.

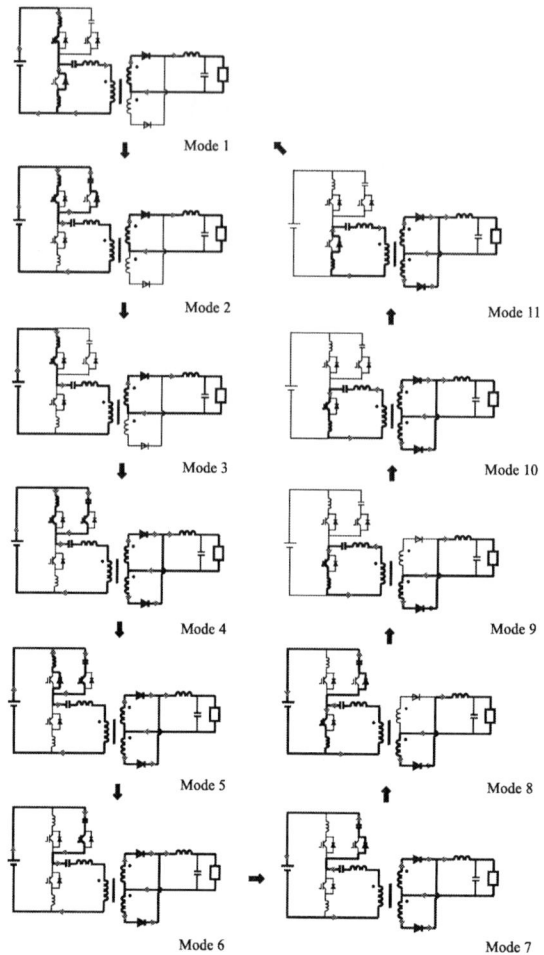

Fig. 3. Switching mode-transitions and equivalent circuits during switching one-cycle under $L_m \gg L_s$.

The resonant and power factor-correcting capacitor C_s, which is effective for block the DC component of the half-bridge inverter current and inhibits the saturation of the HF transformer, is inserted in series with L_s that includes the equivalent leakage inductance L_k of the HF transformer. By utilizing the transformer-parasitic parameters(L_k, L_m)-based series resonance with L_s and C_s, ZCS operation can be performed in Q_2 as well. Since the ZCS turn-on commutation in Q_2 is essentially assisted by L_s, the inductive snubber L_{r2} in series with Q_2 can be small compared to L_{r1}.

The AERS-assisted ZCS-PWM DC–DC converter proposed here has some remarkable advantageous points over the conventional soft-switching schemes such as LLC resonant ZCS-PFM and ZVS-PWM[9]:

i) constant high frequency switching PWM operation.
ii) wide zero current soft switching region.
iii) optimized high frequency transformer and output filter.
iv) current overlapping mode soft commutation between the high- and low-side switching power de-

vices due to the ZCS-assisted inductors L_{r1} and L_{r2}.

III. OPERATION PRINCIPLE

A. Switching Mode Operation

The key waveforms of the proposed DC–DC converter are depicted in Fig. 2. In addition, its switching mode transitions and equivalent circuits during switching one cycle are illustrated in Fig. 3 under the condition of $L_m \gg L_s$.

The operation circuit modes are divided into the eleven steps during the switching one-cycle. While D_2 of Q_2 is conducting after turn-off commutation of D_2, S_1 of Q_1 is turned on. Then, the switch current i_{Q_1} softly increases with the aid of L_{r1}, hereby the ZCS turn-on of S_1 can be achieved. Edge-resonance with L_{r1} and C_r begins at t_1, and then D_3 of the auxiliary switch Q_3 delivers the resonant current circulating in the active snubber.

Prior to the turn-off of S_1, S_3 of the auxiliary switch Q_3 in the AERS cell is turned on. Then, the current i_{Q_3}

through Q_3 softly increases with edge-resonance by L_{r1} and C_r, thereby soft-commutation turn-on in Q_3 can be performed under ZCS mode. Commutation of i_{Q_1} from S_1 to D_1 naturally completes with the edge resonance continuing from the previous operation mode. During this interval, S_1 is turned off, then Zero Current and Zero Voltage Switching (ZCZVS) turn-off can be completely achieved in Q_1. Commutation of i_{Q_3} from S_3 to D_3 can attain owing to the series resonance due to C_r, C_s and L_s. The gate signal to S_3 is removed during this interval, then ZCZVS turn-off can be achieved in Q_3. In addition, the transformer secondary current reverses its direction, consequently soft commutation from D_{o1} to D_{o2} completes in the secondary-side rectifier during Mode 7 due to the leakage inductance of the HF transformer.

S_2 of Q_2 is turned on at t_7. Then, the current i_{Q_2} through Q_2 softly increases with the aid of L_s and L_{r2}, thereby ZCS turn-on can be achieved in Q_2. In this mode, the current commutation from Q_3 to Q_2 completes. The series resonance due to C_s, L_{r2} and L_s with the transformer parasitic parameters takes place. So, the transformer primary current i_p decreases gradually with the series resonance sustaining from the previous mode. As a result, D_{o1} begins to be forward biased, so that the secondary winding of the HF transformer is shorted. Commutation of i_{Q_2} from S_2 to D_2 naturally completes in Mode 11. The gate signal to S_2 is removed during this interval, thereby ZCZVS turn-off can be achieved in Q_2.

The transformer secondary current reverses its direction, and the rectifier diode D_{o2} is reversely biased and cut off. At t_{11}, S_1 is turned on by ZCS as well as at t_0, then this converter operation returns to Mode 1. Note that soft commutation between Q_1 and Q_2 can be performed by the current overlapping mode due to L_{r1} and L_{r2}, and then the no severe diode-recovery appears in Q_1 and Q_2.

B. Gate Pulse Pattern by Asymmetrical PWM-based Duty Cycle

The gate pulse timing sequences for all the switches and the circuit diagram of the pulse generator are illustrated in Figs. 4(a) and (b) respectively[11],[12]. Here, duty cycle D $(= T_{on}/T_s)$ is adjusted for output voltage or power regulation:

In order to achieve the ZCS operation properly, the controllable range of the asymmetrical duty cycle D is determined by:

$$\frac{\pi\sqrt{L_{r1}C_r}}{T_s} < D < D_{max}, \qquad (1)$$

where D_{max} denotes the maximum duty cycle, and it is smaller than 0.5.

The on-period T_{on3} of S_3 can be fixed in a constant value. In this interval, the overlapped on-period T_a of S_1 and S_3 can be estimated from the resonance period during Mode 4 and Mode 5 as follows:

$$T_a \approx \frac{\pi\sqrt{L_{r1}C_r}}{2}. \qquad (2)$$

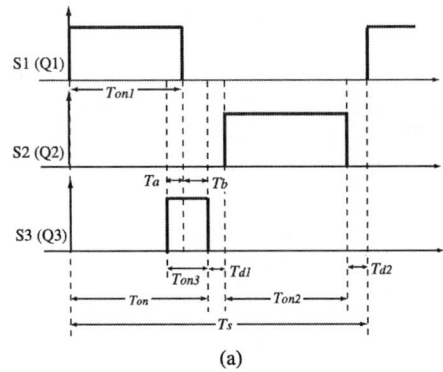

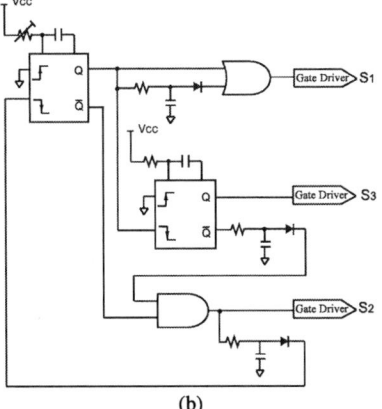

Fig. 4. Asymmetrical PWM duty-cycle control scheme: (a) gate pulse pattern, (b) circuit diagram of gate pulse generator.

Furthermore, the remaining time interval T_b, utilized for the ZCZVS turn-off commutation of S_3, can be specified by

$$T_b \approx \pi\sqrt{L_s \cdot \frac{C_r C_s}{C_r + C_s}}. \qquad (3)$$

IV. CONVERTER DESIGN CONSIDERATION

A. Circuit Parameters of AERS Cell

In order to ensure the ZCZVS mode turn-off commutation of the main switch Q_1 in the AERS cell, the peak value of the edge-resonant current through Q_1 during its turn-on commutation should be larger than the transformer primary current prior to its turn-off commutation. From Fig. 2, this condition can be expressed by:

$$\frac{v_{cr}(t_3)}{\sqrt{L_{r1}/C_r}} > i_p(t_3) \approx I_{pp}, \qquad (4)$$

where I_{pp} denotes the peak value of i_p. Then, the resonant characteristics impedance Z_{q1} of the AERS cell is designed by

$$Z_{q1} = \sqrt{\frac{L_{r1}}{C_r}} < \frac{v_{cr}(t_3)}{I_{pp}}, \qquad (5)$$

where I_{pp} is simply obtained from the HF transformer current at $D = D_{max}$. Accordingly, C_r can be determined

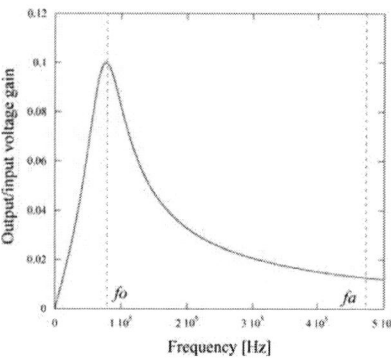

Fig. 5. Converter characteristics on output/input voltage gain vs. converter switching frequency.

by

$$C_r > \frac{L_{r1}I_{pp}{}^2}{v_{cr}(t_3)^2}. \tag{6}$$

The steady-state resonant capacitor voltage $v_{cr}(t_3)$ is smaller than input voltage E_d due to the existence of the capacitor voltage across C_s. In addition, circulating current within the active snubber network increases while the HF transformer current decreases. Therefore, $v_{cr}(t_3)$ increases in accordance with reduction of the duty cycle. Thus, the required capacitance of the resonant snubbing capacitor C_r can be determined in eq.(6) by setting the minimum voltage of $v_{cr} < E_d$.

In addition, the peak current I_{pp} can be determined under condition of $L_m \gg L_s$ by

$$I_{pp} \approx \frac{E_d}{2}\sqrt{\frac{C_s}{L_s}} - N_T V_o \sqrt{\frac{C_s}{L_s}} + \frac{\pi N_T f_s}{2N_T f_o}, \tag{7}$$

where f_o denotes the series resonant frequency of the HF-link half-bridge DC–DC converter.

B. Resonant and Switching Frequency

In Fig. 5, the voltage gain – switching frequency characteristics of the ZCS–PWM DC–DC converter is depicted under a Quality Factor Q determined by the series resonant parameters. The output/input voltage gain $G(=V_o/E_d)$ of the asymmetrical half-bridge DC–DC converter can be defined from the switching frequency f_s and the $L_s C_s$ series resonant frequency f_o by using Fundamental Element Simplification method: [4]:

$$G = \frac{1}{2N_T} \cdot \frac{1}{\sqrt{\left[1 + \frac{1}{k}\left(1 - \frac{f_o{}^2}{f_s{}^2}\right)\right]^2 + \left[\left(\frac{f_s}{f_o} - \frac{f_o}{f_s}\right)Q\right]^2}} \tag{8}$$

$$Q = \sqrt{\frac{L_s}{C_s}} \cdot \frac{\pi^2}{8N_T{}^2 R_o}, \tag{9}$$

where k $(=L_m/L_s)$ is the ratio of parallel to series inductance, and $N_T(=N_p/N_s)$ is the winding turn-ration of the HF transformer.

While the active snubber resonant frequency f_a is defined by $1/\left(2\pi\sqrt{L_{r1}C_r}\right)$, the series resonant frequency f_o is determined by

$$fo = \frac{1}{2\pi\sqrt{\left(L_r + L_s + N_T{}^2 L_o\right)\cdot C_s}}, \tag{10}$$

where $L_r = L_{r1} \simeq L_{r2}$.

In the ZCS-PWM scheme treated here, the switching frequency f_s is fixed to a value in the region which is lower than f_o;

$$f_s < f_o \ll f_a. \tag{11}$$

Moreover, owing to the asymmetrical PWM scheme, the switching frequency of the PWM-based DC–DC converter can be set higher than that of the PFM-based counterpart[9].

V. DESIGN AND SPECIFICATIONS OF PROTOTYPE CIRCUIT

The prototype of the ZCS-PWM half-bridge DC–DC converter has been built in order to evaluate the operation characteristics of the proposed soft-switching circuit topology and control schemes.

The input voltage of the prototype DC–DC converter is set in 200 V, and the DC output voltage rating is designed to 15 V in consideration that the DC–DC converter is suitable as a Point-Of-Load (POL) power converter for the advanced automotive electric power-train architectures[1]. As a consequence, the power rating is selected to 0.8 kW, and the winding turn ratio $N_T=N_p/N_s$ of the secondary-side center-tapped HF transformer is designed to 5/1.

As one of the examples, the series resonant frequency f_o in eq.(10) and the quality factor Q in eq.(9) are selected to be 80 kHz and 5 respectively. Accordingly, the series resonant capacitor C_s and the series inductance L_s with the leakage inductance L_k of the HF transformer are turned in 260 nF and 16.7 µH.

The switching frequency f_s is fixed to 55 kHz, taking into account of the electric performance of the IGBT gate driver using Photocoupler TLP250 under condition of eq.(11). Provided that the edge-resonant period defined by $1/f_a$ is around 10 % of the switching one-cycle time $1/f_s$, the edge-resonant frequency f_a can be provided with 480 kHz.

The resonant capacitor is required to have a capacitance large enough for reversing and decaying the current through the main switch Q_1 in the active snubber prior to its turn-off commutation as mentioned above. Now, we have a the ZCS-assisted inductor L_{r1} of 1.6 µH. Prior to the determination of C_s, the peak current I_{pp} can be approximately calculated by eq.(7)[4]:

$$I_{pp} \approx \frac{E_d}{2}\sqrt{\frac{C_s}{L_s}} - N_T V_o \sqrt{\frac{C_s}{L_s}} + \frac{\pi N_T f_o}{2N_T f_s} = 13\,\text{A}. \tag{12}$$

Therefore, when L_{r1} is decided to 1.6 µH, C_r is given from eq.(10) by:

$$C_r > \frac{L_{r1}I_{pp}{}^2}{v_{cr}(t_3)^2} = 27\,\mu\text{H}, \tag{13}$$

TABLE I
DESIGN SPECIFICATIONS AND CIRCUIT PARAMETERS OF
EXPERIMENTAL PROTOTYPE CIRCUIT.

Parameter & symbol	value & unit
Power rating P_o	0.8 kw
Switching frequency f_s	55 kHz
Input supply DC voltage E_d	200 V
Output voltage V_o	15 V
Rated output current I_o	50 A
Transformer turn ratio $N_T (= N_p/N_s)$	5/1
Transformer magnetizing inductance L_m	2.0 mH
Transformer leakage inductance L_k	5.9 μH
Additional series inductor	10.8 μH
Series resonant inductor L_s	16.7 μH
ZCS-assisted lossless inductors L_{r1}/L_{r2}	1.6 μH/1.6 μH
Series resonant capacitor C_s	260 nF
Snubbing resonant capacitor C_r	85 nF
Output filter inductor L_o	10 μH
Output filter capacitor C_o	100 μF

·Q_1/Q_2, Q_3: MITSUBISHI CM100DU-24NFH (two sets)
·D_{o1}, D_{o2}: SanRex DBA 200AA 01A (Two-in-one)

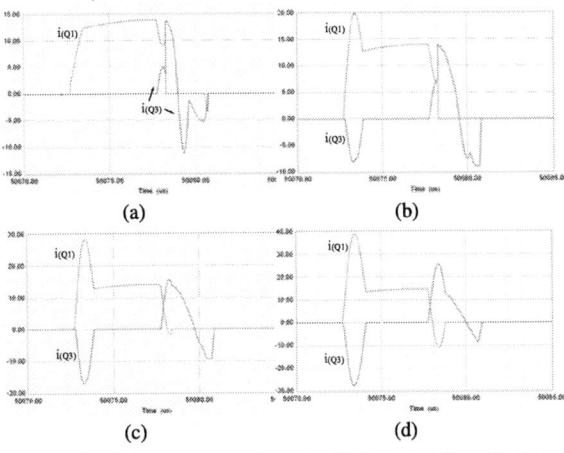

Fig. 6. Simulation current waveforms in AERS cell: (a)$C_r = 30$ nF,
(b)$C_r = 55$ nF, (c)$C_r = 70$ nF, (d)$C_r = 100$ nF

where v_{cr} is selected to be half a DC input voltage (100 V).

Based on the calculated value as a minimum capacitance, the resonant snubber capacitor C_r are designed in accordance with the converter soft-switching performance demonstrated by simulation results in Fig. 6. Thus, in order to ensure the ZCS commutation in Q_1 without significant circulating energy in the active snubber, C_r is selected to be 70 nF in the prototype circuit.

The controlled range of the duty cycle D is designed in accordance with eq.(1) by

$$0.25 < D < 0.45, \qquad (14)$$

where the maximum value D_{max} is set to 0.45.

The prototype design specifications and the experimental set-up are summarized in Table I.

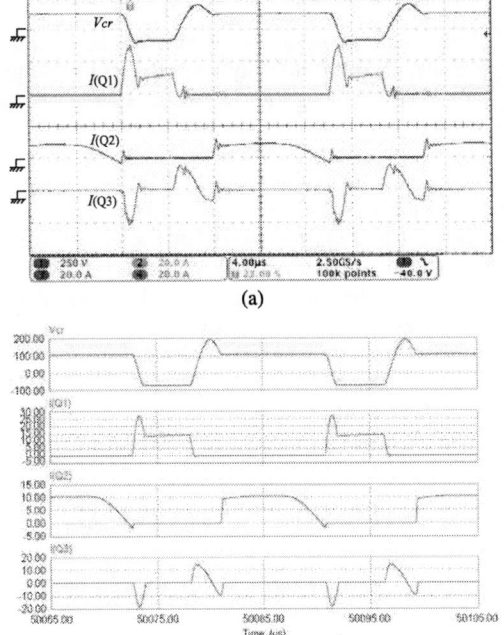

Fig. 7. Steady-state operating waveforms in active snubber at $D = 0.43$: (a) measured waveforms(V_{cr}:250 V/div, I_{Q1}, I_{Q2}, I_{Q3}: 20 A/div, time: 4.0 μs/div), (b) simulation waveforms.

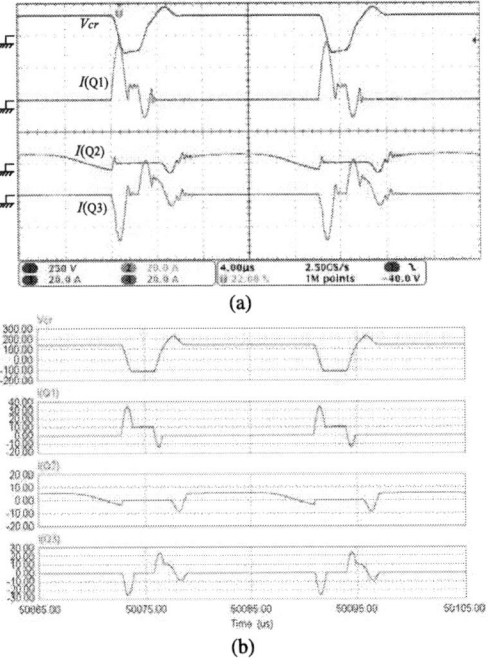

Fig. 8. Steady-state operating waveforms in active snubber at $D = 0.3$: (a) measured waveforms(V_{cr}:250 V/div, I_{Q1}, I_{Q2}, I_{Q3}: 20 A/div, time: 4.0 μs/div), (b) simulation waveforms.

VI. EXPERIMENTAL RESULTS AND EVALUATION

A. Converter Steady-State Operation

The steady-state operating waveforms of the active snubber operation under the duty cycle $D = 0.43$ and $D = 0.3$ are shown in Figs. 7 and 8 respectively, compared with the simulated waveforms. From the measured waveforms, the ZCS commutation between the main and auxiliary switches can be confirmed as well as the snubbing operation of the resonant snubber capacitor C_r. In accordance with reduction of the duty cycle D, the peak currents in the main switch Q_1 and the auxiliary switch Q_3 get to be more outstanding. However, the edge resonant periods related to the peak currents are relatively short especially in the large duty cycle. Thus, the conduction loss due to the peak current are not so significant, and no profound reduction in converter efficiency arises.

B. ZCS Commutations in Switching Power Devices

Switching voltage and current waveforms and their V–I traces of Q_1–Q_3 under the duty cycle $D = 0.4$ are shown in Figs. 9–11, respectively. The complete ZCS turn-on and the ZCZVS turn-off commutation of each switch are observed in the respective result. Thus, the feasibility of the proposed DC–DC converter topology and the soft switching scheme are verified.

Fig. 12 indicates the voltage and current waveforms of the secondary-side diode D_{o1} and D_{o2}. The measured waveforms depict that the ZCS soft commutation between the two diodes can be attained in current overlapping mode due to the leakage inductances of the secondary-side windings in the HF transformer.

C. Output Power Control Characteristics

The converter performance on output power regulation in the ZCS–PWM AHB DC–DC converter are investigated with open and closed loop control schemes. Fig. 13 depicts the converter characteristics for the open loop control scheme. In those results, it can be confirmed that a wide range of output power regulation can be achieved by the constant frequency asymmetrical PWM strategy. The complete ZCS operations can be attained in the output range of $400\,\mathrm{W} - 800\,\mathrm{W}$. And, under $P_o = 400\,\mathrm{W}$, semi-ZCS commutations where only turn-off of Q_2 becomes a hard-switching mode as shown in Fig. 14 are observed. Incidentally, the parameters indicated in the graphs are the resistance value of the electric load (KIKUSUI PLZ 1004).

The converter characteristics for the the closed loop control scheme are shown in Fig. 15, where the output voltage V_o is regulated to $15\,\mathrm{V}$. The experimental results indicate that a wide range of output power regulation can be achieved by the constant frequency asymmetrical PWM. The complete ZCS operations are attained in $P_o = 460\,\mathrm{W} - 810\,\mathrm{W}$, and the semi-ZCS operations described above are also confirmed under $P_o = 400\,\mathrm{W}$. Although there is the region out of the complete ZCS commutation in all switches, the actual conversion efficiencies as high

as 96.2 % at maximum attain under the constant output voltage condition.

VII. CONCLUSION

In this paper, the HF-link asymmetrical half-bridge soft switching DC–DC converter with the single active edge-resonant snubber for zero current switching commutations by the constant frequency PWM scheme has been newly developed, and its performance on high efficiency power conversion has been evaluated by experiments using its 800 W–55 kHz prototype. The ZCS operation of the DC–DC converter are confirmed, and the static characteristics are demonstrated in order to clarify the feasibility as a DC–DC converter. From the results, it has been proved that the ZCS-PWM DC–DC converter topology is effective especially for the large output current type of power conversion with a large voltage step-down ration such as POL power converters. Furthermore, the design guideline for the circuit parameters is indicated, and its validity is also demonstrated by simulation and experimental results.

The next challenges in this research include enhancement of the power density and conversion efficiencies in the low output power settings by improving both switching gate pulse modulation scheme and the circuit topology of AERS cell.

REFERENCES

[1] A. Emadi, Sheldom. S. Williamson, and A. Khalight, "Power Electronics Intensive Solutions for Advanced Electric, Hybrid Electric, and Fuel Cell Vehicular Power Systems", *IEEE Trans. Power Electron.*, Vol.21, No.3, pp.567–577 (May-2006).

[2] R. Chen, J. Tjeerd, and J. D. Wyk, "Design of Planner Integrated Passive Module for Zero–Voltage–Switched Asymmetrical Half–Bridge PWM Converter", *IEEE Trans. on Ind. Appl.*, Vol.39, No.6, pp.1648–1954 (Nov./Dec.-2003).

[3] G. C. Hsieh, C. Y. Tsai, and S. H. Hsieh, "Design Considerations for LLC Series-Resonant Converter in Two-Resonant Regions", *Records of IEEE Power Electronics Specialists Conference 2007*, pp.731–736 (Jun.-2007).

[4] Y. Zang, D. Xu, K. Mino, and K. Sasagawa, "1MHz-1kW LLC Resonant Converter with Integrated Magnetics", *Records of IEEE Applied Power Electronics Conferences and Expositions 2007*, pp.955–961 (Feb.-2007).

[5] P. K. Jain, A. S. Martin, and G. Edwards, "Asymmetrical Pulse-Width-Modulated Resonant DC/DC Converter Topologies", *IEEE Trans. on Power Electron.*, Vol.39, No.6, pp.1775–1782 (Nov./Dec.-2003).

[6] Y. Zhang and P. C. Sen, "A New Soft-Switching Technique for Buck, Boost, and Buck–Boost Converters", *IEEE Trans. Ind. Appl.*, Vol., No.3, pp.413–422 (May.-1996).

[7] N. H. Kutkut, D. S. Divan, and R. W. Gascoigne, "An Improved Full-Bridge Zero-Voltage Switching PWM Converter Using a Two-Inductor Rectifier", *IEEE Trans. Ind. Appl.* , Vol.31, No.1, pp.119–126 (Jan.-1995).

[8] J. L. Russi, M. L. Martins, H. A. Grundling, H. Pinheiro, et al, "A Unified Design Criterion for ZVT DC–DC PWM Converters With Constant Auxiliary Voltage Source", *IEEE Trans. Ind. Appl.*, Vol.52, No.5, pp.1261–1270 (2005-Oct.).

[9] T. Mishima and M. Nakaoka, "Higher constant switching frequency-based HF-link half-bridge PWM DC-DC converter topology with ZCS commutations", *IET Letter 5th June 2008*, Vol. 44, issue 12, pp.769-770 (Jun. 2008).

[10] M. H. Hashem, N. A. Ahmed, E. Hiraki, T. Ahmed, K. Fathy, H. W. Lee, and M. Nakaoka, "Switched-Capacitor Snubber-Assisted Zero Current Switching PWM High Frequency Inverter with Two-Lossless Inductive Snubber", *Records of IEEE Power Electronics and Drive Systems 2005*, vol.1, pp.198–204 (Dec.-2005).

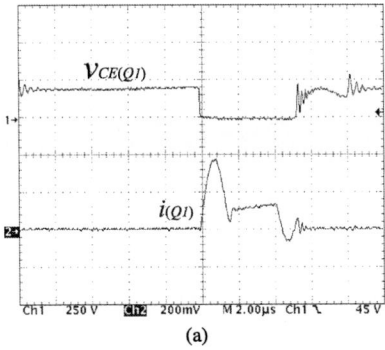
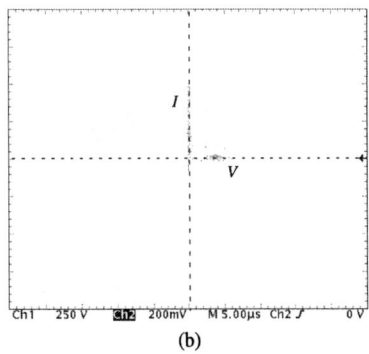

(a)   (b)

Fig. 9. ZCS commutations in Q_1: (a) voltage and current waveforms, (b) V-I trace ($v_{CE(Q1)}$: 250 V/div, i_{Q1}: 20 A/div, time: 2.0 μs/div).

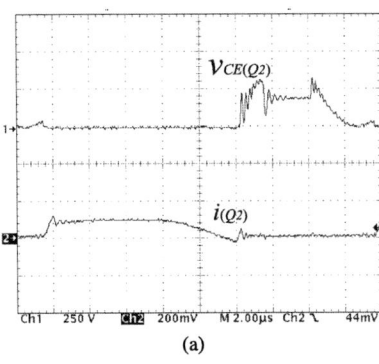
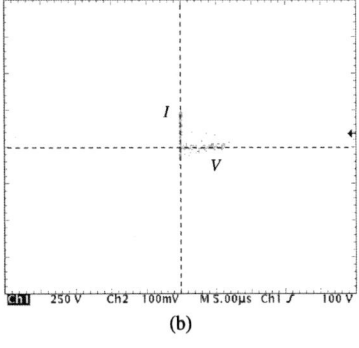

(a)   (b)

Fig. 10. ZCS commutations in Q_2: (a) voltage and current waveforms, (b) V-I trace ($v_{CE(Q2)}$: 250 V/div, i_{Q2}: 20 A/div, time: 2.0 μs/div).

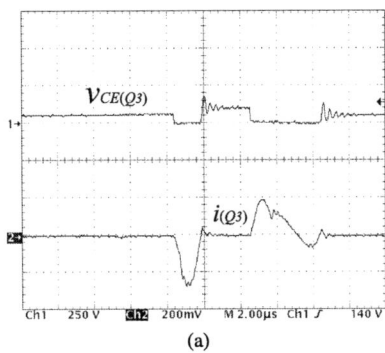
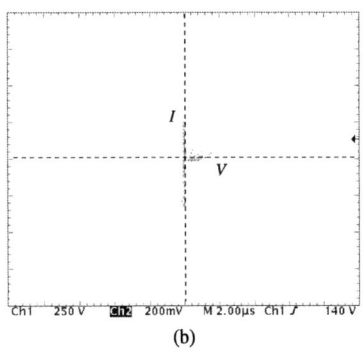

(a)   (b)

Fig. 11. ZCS commutations in Q_3: (a) voltage and current waveforms, (b) V-I trace ($v_{CE(Q3)}$: 250 V/div, i_{Q3}: 20 A/div, time: 2.0 μs/div).

[11] N. A. Ahmed, Ahmad Eid, H. W. Lee, M. Nakaoka, Y. Miura, and E. Hiraki, "Quasi-Resonant Dual Mode Soft Switching PWM and PDM High-Frequency Inverter with IH Load Resonant Tank", *Records of IEEE Power Electronics Specialists Conference*, vol.3, pp.2830–2835.

[12] T. Mishima, E. Hiraki, and T. Tanaka, "A ZCS Lossless Snubber Cells-Applied Half-Bridge Bidirectional DC–DC Converter for Automotive Electric Power Systems", *Records of IEEE Power Electronics Specialists Conference*, pp.2011–2016 (Jun.-2006.)

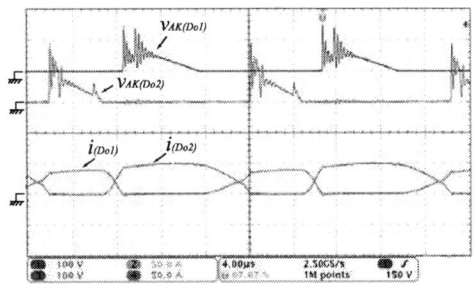

Fig. 12. Voltage and current waveforms in D_{o1} and D_{o2}: (v_{AK}: 100 V/div, i: 50 A/div, time: 4.0 μs/div).

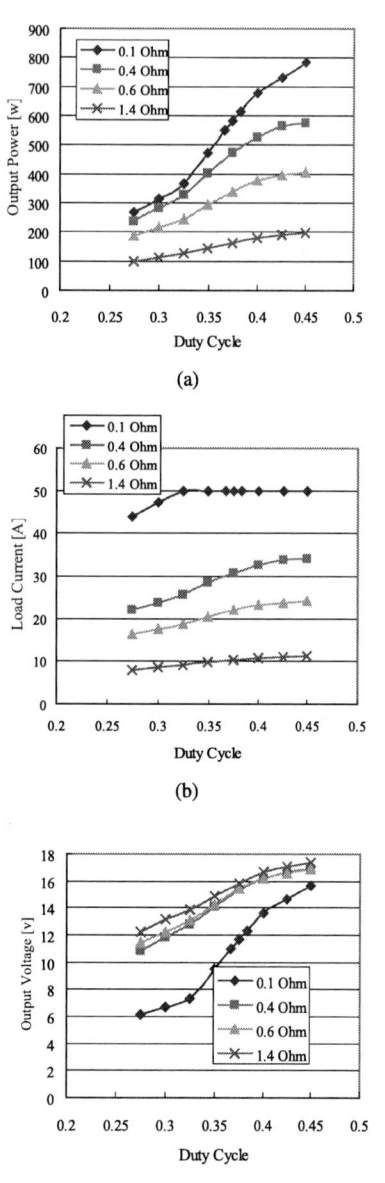

(a)

(b)

(c)

Fig. 13. Open loop control charcateristics: (a) output power – duty cycle, (b) load current – duty cycle, (c) output voltage – duty cycle.

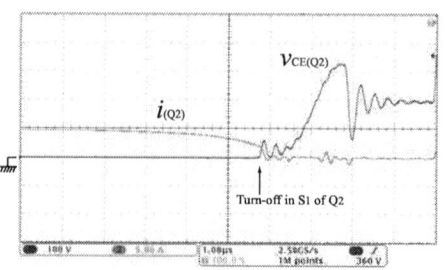

Fig. 14. Semi ZCS turn-off commutation in Q_2 ($v_{CE(Q2)}$: 100 V/div, i_{Q2}: 5 A/div, time: 1 μs/div).

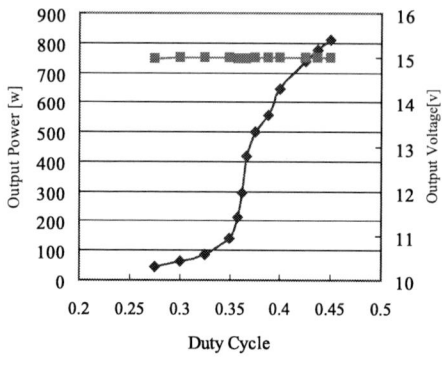

(a)

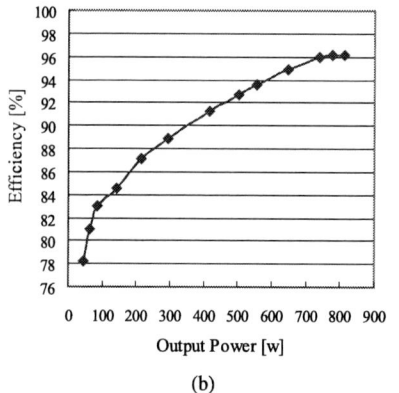

(b)

Fig. 15. Closed loop control characteristics: (a) output power and voltage – duty cycle, (b) actual efficiency.

A New Approach to High Efficiency in Isolated Boost Converters for High-Power Low-Voltage Fuel Cell Applications

Morten Nymand*, Michael A.E. Andersen[†]

* University of Southern Denmark/Dept. of Sensors, Signals and Electrotechnics, Odense, Denmark, *mny@sense.sdu.dk*
[†] Technical University of Denmark/Dept. of Electrical Engineering, Lyngby, Denmark, *ma@elektro.dtu.dk*

Abstract—A new low-leakage-inductance low-resistance design approach to low-voltage high-power isolated boost converters is presented. Very low levels of parasitic circuit inductances are achieved by optimizing transformer design and circuit lay-out. Primary side voltage clamp circuits can be eliminated by the use of power MOSFETs fully rated for repetitive avalanche. Voltage rating of primary switches can now be reduced, significantly reducing switch on-state losses. Finally, silicon carbide rectifying diodes allow fast diode turn-off, further reducing losses. Test results from a 1.5 kW full-bridge boost converter verify theoretical analysis and demonstrate very high efficiency. Worst case efficiency, at minimum input voltage maximum power, is 96.8 percent and maximum efficiency reaches 98 percent.

Keywords—Switched-mode power supply, fuel cell system, efficiency, transformer, SiC-device.

I. INTRODUCTION

High-power fuel cell or battery powered applications such as for transportation, forklift trucks or distributed generation, are often faced with the need for boosting the low input voltage (30-60V) to the much higher voltage (360-400V) required for interfacing to the utility grid fig. 1, [1].

For safety as well as for EMC reasons, galvanic isolation between source and utility grid is often desirable or required.

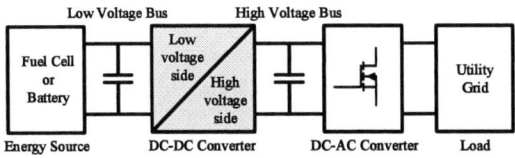

Fig.1. Fuel cell power system with isolated high gain DC-DC converter.

In particular fuel cells, exhibit significant output impedance reducing output voltage as output power is increased. System peak power is therefore reached at converter minimum input voltage. Drop in converter efficiency at minimum input voltage and maximum output power therefore directly reduces available system peak power. While the converter is required to operate over a wide input voltage range, typically up to a factor 1:2, high converter efficiency becomes particular important at minimum input voltage maximum power [1].

Isolated boost converters has some inherent advantages when used in fuel cell applications. With the storage inductor placed at the input side, ripple current is inherently low, only requiring limited extra filtering at the input side.

Output rectifying diodes are placed directly across output capacitors, ensuring minimum voltage stress and effective voltage clamping.

In a 400 V output application, 600 V rated diodes will be sufficient in boost type topologies, whereas buck type topologies would require 1200 V diodes or stacking of multiple outputs. Voltage stress on boost topology diodes are thus less than half of the corresponding voltage stress on a buck derived topology. Buck type topologies will therefore have significantly larger rectifying losses than boost type topologies.

The draw back of the boost type topologies is the need for clamping voltage spikes on primary switches caused by transformer leakage inductance and parasitic circuit inductances. Clamping is typically performed by some sort of voltage clamp circuit or by implementing active reset circuits [2-7]. This however requires significantly increased voltage rating on primary switches severely penalising conduction losses.

A large number of isolated boost converters for fuel cell applications have been presented, among these [2-7]. Ref. [2-4] have input voltage range and power level that are comparable to the converter presented in this paper. Ref. [5-7] are isolated bi-directional full-bridge boost converters intended for electrical vehicles. Input voltage is 12 V (8-15V), output voltage is typically 250-420 V and power range, in boost mode, is from 1.5 kW [5,6] to 3 kW [7].

Even though vastly different designs are represented (hard switched push-pull boost [2], actively clamped, two-inductor boost [3], bi-directional, actively clamped, two-inductor boost [4], and bi-directional soft switching full-bridge boost converters [5-7]), a general efficiency trend is clear. All converters achieve high efficiencies in the medium to high input voltage range, typically peaking at 94-96 percent at medium power. At low input voltage, high power, efficiency however reduces significantly to approx. 90 percent or below.

In this paper, the design of a simple, wide input voltage range, isolated full-bridge boost converter with very high efficiency at low input voltage is presented.

Test results from a 1.5 kW demonstration model achieve peak efficiency of 98 percent. Worst case efficiency at minimum input voltage and maximum power is 96.8 percent.

II. ISOLATED FULL-BRIDGE BOOST CONVERTER

The proposed full-bridge boost converter is presented in fig. 2. Timing diagrams and basic operating waveforms are presented in fig. 3.

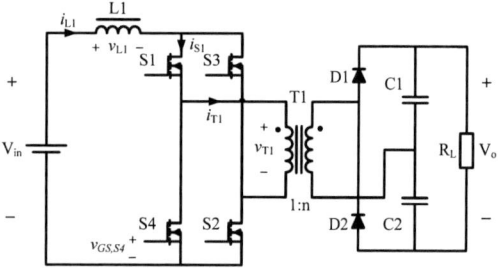

Fig.2. Isolated full-bridge boost converter with voltage doubling rectifier.

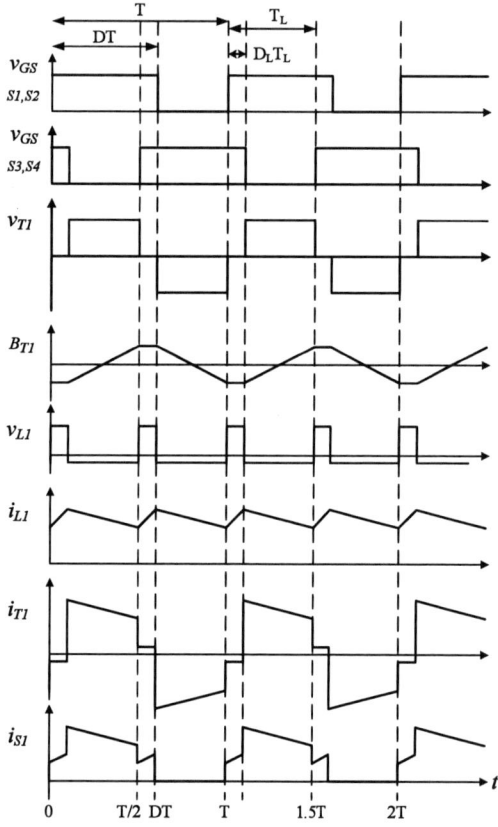

Fig.3. Basic operating waveforms of isolated full-bridge boost converter.

A. Basic converter operation

Primary switches S1-S4, are hard switched and operated in pairs S1-S2, and S3-S4, respectively. Drive signals are 180 degrees phase shifted. Switch transistor duty cycle D, is above 50 percent to ensure switch overlap and thus a continuous current path for the inductor L1, current.

Energy transfer to output starts when switches S3 and S4 are turned off. Inductor current i_{L1}, flows through primary switch S1, transformer T1, rectifier diode D1, output capacitor C1 and returns to input through primary switch S2. Inductor current i_{L1}, discharges. The period ends when primary switches S3 and S4 are turned on again.

During switch overlap, when all switches S1-S4, are turned on, inductor current i_{L1}, is charged. Current in the transformer secondary winding is zero and diodes D1 & D2 are off. Transformer magnetizing current circulates in the transformer primary winding through switches S2-S4 and/or S1-S3. Capacitors C1, C2, supply the load current. The period ends when primary switches S1 and S2 are turned off.

A second energy transfer cycle starts when switches S1 and S2, are turned off and ends when S1 and S2 are turned on again. Inductor current i_{L1}, flows through switches S3, T1, D2, C2, and returns to input through S4.

Finally, a second inductor charging interval similar to the first follows.

The converter transfer function in continuous steady state is:

$$\frac{V_o}{V_{in}} = \frac{n}{1-D} \qquad (1)$$

Where n=Ns/Np is the transformer turns ratio, and D is the switch duty cycle (0.5≤ D<1).

The corresponding inductor duty cycle D_L, and period time T_L, is defined as:

$$D_L \equiv 2D\text{-}1 \qquad (2)$$

$$T_L \equiv {}^T/_2 \qquad (3)$$

Where T=1/f$_S$ is the period time for switches, diodes and the transformer.

III. CONVERTER DESIGN

The four primary switches S1-S4, are 75 V, 2.8 mΩ International Rectifier IRFB 3077 Power MOSFET which are fully repetitive avalanche rated [8]. The two rectifier diodes D1-D2, are 600 V Infineon IDT 10S60C SiC Schottky diodes. The inductor L1, core is a Magnetics Kool Mμ 77439. The transformer core is an EE55/21 ferrite core in 3F3 material. Switching frequency is 45 kHz.

Low current switching times increase efficiency since less charge is being diverted from output into primary side clamp circuits. Current switching times are limited by transformer leakage inductance and primary side stray inductances as well as MOSFET common source inductance whichever is worst case [9].

Transformer leakage inductance can be reduced by extensive interleaving of primary and secondary windings.

Careful primary side lay-out is required to reduce primary side stray inductances. MOSFET common source inductance is a function of package internal wiring (bonding wire length) as well as source external lead length [9].

B. Transformer design

The low input voltage, high power in fuel cell converters, causes high currents to flow in transformer primary windings requiring large copper cross sections in primary windings.

Foil windings are very efficient in providing large copper cross sections with a minimum conductor thickness. However, as power levels increases, even foil winding thicknesses quickly approach or exceed penetration depths in copper. Proximity effect can thereby cause very significant increases in winding AC resistances and thus lead to significantly increased power losses [10,11].

At 45 kHz, penetration depth in copper is only 0.34 mm. A primary winding with 4 turns on an EE55/21 core would allow up to approximately 0.6 mm cobber thickness for each of the 4 primary turns. Winding these 4 turns in a single 4 layer section, would increase AC resistance approximately 13 times compared to winding DC resistance ($F_R = R_{AC}/R_{DC} = 13$) [10,11].

The more frequent case of a single interleaving (primary winding interleaved between two sections of secondary windings), increases AC resistance by a factor 3.5 compared to winding DC resistance.

To avoid severe proximity effect, penetration from both sides of each primary turn can be obtained by interleaving each primary turn between sections of secondary windings. The corresponding increase in AC resistance is now only 5 percent ($F_R = 1.05$).

Keeping AC resistances low in high frequency high-current transformers therefore requires extensive interleaving of windings.

Since interleaving of windings also has the well known effect of reducing transformer leakage inductance [snelling], these transformers will not only have small AC resistances but also extremely low leakage inductances.

AC resistance and leakage inductance of the transformer used in a 1.5 kW isolated boost converter with a turn ratio of 4, are presented in fig. 4. Transferred to primary side, the AC resistance is only 1.9 mΩ and leakage inductance is only 11 nH. The leakage inductance in percent of primary inductance is only 0.01 percent.

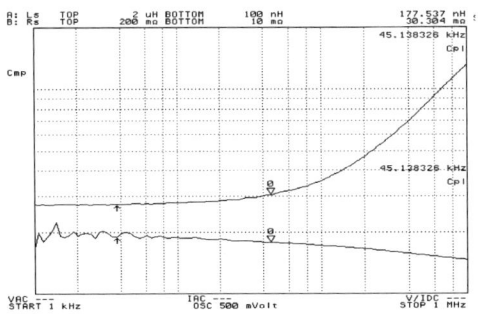

Fig.4. Measured secondary side AC resistance (upper curve) and leakage inductance (lower curve) of 1.5 kW transformer.

C. Primary switch voltage clamping

At low input voltage and high power levels, conduction losses in primary switches are a dominant loss factor. In the voltage range 60-200V, MOSFET on-resistance $R_{DS,ON}$, typically increases quadratic with increasing drain-to-source breakdown voltage $V_{(BR)DSS}$. Voltage rating of primary switches therefore has very significant impact on converter conduction loss and thereby on converter efficiency.

To allow clamp circuits to clamp voltage spikes caused by transformer leakage inductance and circuit stray inductances, voltage rating of primary switches, in isolated boost converters, is typically rated at 2-3 times the maximum input voltage [6].

For primary clamp circuits to be effective, they need to present significantly lower impedance at the clamping point than the circuit which is being clamped. However, with the very low levels of transformer leakage inductances that are achieved in low voltage high power transformers, it becomes indeed very difficult for clamp circuits to present lower impedances than that of the transformer itself.

This result in clamp circuits only taking (small) fractions of the clamp energy, since the major part is being clamped by the transformer which even has lower reflected voltage and thus higher driving voltage across the leakage inductance.

Fortunately, due to the very low leakage inductance in the transformer, the leakage energy in the transformer is very small.

Some new low voltage power MOSFETs are rated for repetitive avalanche and are very robust to unclamped inductive switching [8,12].

With careful primary side lay-out and low leakage design of transformers, converter leakage energy is very small. Using avalanche rated MOSFETs, primary side clamp circuits can be eliminated and switch voltage rating reduced, significantly reducing MOSFET conduction losses.

Since silicon carbide Schottky diodes do not suffer from reverse recovery, they can work at much higher switching frequencies and in particular at much faster current switching speed (turn-off di/dt) without excessive losses.

IV. EXPERIMENTAL RESULTS

Experimental results from 1.5 kW demonstration model are presented in the following.

Fig. 4, is a plot of transformer leakage inductance and AC-resistance measured on secondary side. Transferred to primary side (dividing by transformer turns-ratio squared), the transformer leakage inductance is only 11 nH and the AC-resistance at 45 kHz is only 1.9 mΩ.

Fig. 5, is a plot of voltages and currents in the converter. Transformer current is measured on secondary side in order to avoid adding extensive stray inductance to the primary circuit.

Fig. 6, is an expanded view of fig. 5, showing that at turn off output current rise is only delayed approximately 30 ns from transistor voltage rise. Total avalanche loss, shared between 4 primary switches, can be estimated to:

$$P_{Aval,S1-S4} \approx i_{L1,peak} v_{S1,av} \Delta t f_s = 3.75 \, W \qquad (4)$$

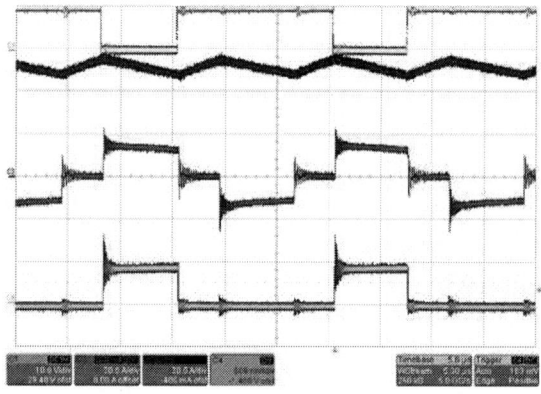

Fig.5. Measured converter waveforms at 30V input and 1.5 kW output power. From top: control signal for transistor S4, inductor L1 current (20A/div), transformer secondary current (20A/div) and bottom trace is transistor S4 drain-source voltage (50V/div). Time base is 5μs/div.

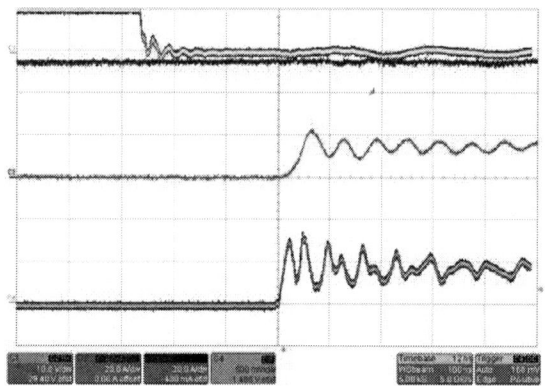

Fig. 6. Expanded view of fig. 5. Time base is 100ns/div.

D. Efficiency measurement

Measuring efficiencies in the 97-98 percent range are particularly critical and extensive care has been taken to ensure very high precision and stability of the efficiency test set-up.

In particular the current measurements are very critical. Measurements were made using 0.1 percent sense resistors with very high temperature stability (<10 ppm) mounted on heat sinks. Sense cables were shielded and supplied with common mode attenuating coils. Agilent 34410A high precision multimeters were used.

Since efficiency of single-input, single-output DC-DC converters is equal to the product of voltage ratio and current ratio (2), the critical current measurement ratio can be checked by simply passing the same current through both current sensors in series and verify that the measured voltage ratios correspond to the ratio of the sense resistor resistances.

$$\eta = \frac{V_o}{V_{in}} \frac{I_o}{I_{in}} \qquad (5)$$

Stability of test set-up was furthermore ensured by using high stability power sources and electronic loads.

Converter heat sink temperature was measured and limited to 40 degree Celsius for repeatability of test.

The measured deviation of the sense resistor voltage ratio compared with the specified ratio (measured at 10 amperes) was less than 0.01 percent.

According to the specification of the Agilent 34410A, the output/input voltage ratio can be measured with a precision of better than +/- 0.012 percent.

The combined precision of the efficiency measurements are better than +/- 0.1 percent.

Measured converter efficiency, including transistor drive losses, is presented in fig. 7.

Maximum efficiency at low input voltage is 97.5 percent. Efficiency at full load is between 96.8 and 97.9 percent. Maximum efficiency is 98 percent.

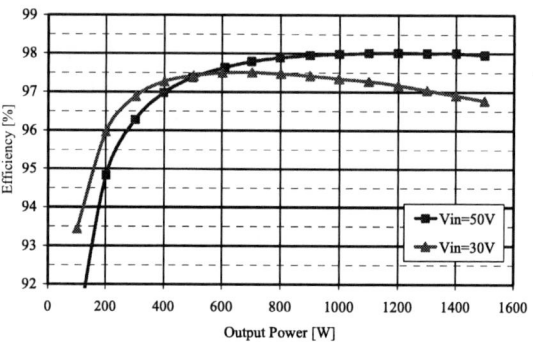

Fig. 7. Converter efficiency including drive power.

A detailed break down of converter power loss at minimum input voltage, maximum power (30 V/1.5 kW), is presented in table 1.

As would be expected, losses are dominated by conduction losses which constitute 83 percent of all calculated losses. Notice also that even though hard switching is used, switching losses are very small. Total inductive clamp losses are only 3.75 W corresponding to a quarter of a percent loss of efficiency.

Transformer efficiency is above 99.6 percent at maximum power.

TABLE I.
CONVERTER POWER LOSS BREAK DOWN AT 1.5 KW / 30 V

Component	Loss type	Loss [W]	Total [W]
MOSFET	Conductive	14.1	18.8
	Capacitive	0.27	
	Drive	0.72	
	Inductive clamp	3.75	
Diodes	Conductive	14.7	15.1
	Capacitive	0.41	
Transformer	Conductive	3.77	5.5
	Core	1.75	
Inductor	Conductive	5.2	6.2
	Core	1.0	
Other	Misc.		4.5
Converter	Measured efficiency		50.1

A photo of the 1.5 kW demonstration model is shown in fig. 8.

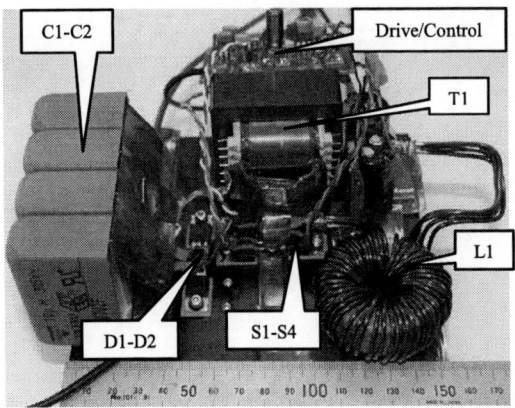

Fig. 7. Photo of 1.5 kW isolated full-bridge boost converter.

V. CONCLUSION

This paper, has presented a design approach to achieve very high efficiency in low-voltage, high-power, isolated boost converters. The design approach is demonstrated on an isolated full-bridge boost converter. Converter operation has been analysed and design details for a 1.5 kW converter has been presented.

Fast current switching is achieved by careful design of transformer and converter layout. Transformer proximity effect losses are reduced by extensive interleaving of primary and secondary windings. Test results on a 1.5 kW demonstration model confirm the achievement of fast current switching, low parasitic circuit inductance and very high efficiency.

Worst case efficiency, at maximum load and minimum input voltage, is 96.8 percent. Maximum efficiency is 98 percent.

References

[1] F. Profumo, A Tenconi, M. Cerchio, R. Bojoi, G. Gianolio, "Fuel cells for electric power generation: Peculiarities and dedicated solutions for power electronic conditioning systems", *EPE Journal* vol. 16, no. 1 February 2006, pp. 44-50.

[2] G. K. Andersen, C. Klumpner, S. Kjær, F. Blaabjerg, "A new power converter for fuel cells with high system efficiency", *International Journal of Electronics*, 2003, vol. 90, pp. 737-750.

[3] J. T. Kim, B. K. Lee, T. W. Lee, S. J. Jang, S. S. Kim, C. Y. Won, "An active clamping current-fed half-bridge converter for fuel-cell generation systems", *Conf. Proc. IEEE PESC 2004*, Aachen, Germany, pp. 4709-4714.

[4] H. Xiao, L. Guo, S. Xie, "A new ZVS bidirectional DC-DC converter with phase-shift plus PWM control scheme", *Conf. Proc. IEEE APEC 2007*, pp. 943-948.

[5] K. Wang, C.Y. Lin, L. Zhu, D. Qu, F.C. Lee, J.S. Lai, "Bi-directional DC to DC converter for fuel cell systems", *Conf. Proc. IEEE Power Electronics in Transportation 1998*, pp. 47-51.

[6] K. Wang, L. Zhu, D. Qu, H. Odendaal, J. Lai, F.C. Lee, "Design, implementation, and experimental results of bi-directional full-bridge DC/DC converter with unified soft-switching scheme and soft-starting capability", *Conf. Proc. IEEE PESC 2000*, pp. 1058-1063.

[7] L. Zhu, "A novel soft-commutating isolated boost full-bridge ZVS-PWM DC-DC converter for bidirectional high power applications", *IEEE Transactions on Power Electronics*, vol. 21, no. 2, March 2006, pp. 422-429.

[8] International Rectifier IRFB3077 Datasheet PD-97047A, 020806.

[9] S. Havanur, "Quasi-clamped inductive switching behaviour of power MOSFETs", *Conf. Proc. IEEE PESC 2008*, Rhodes, Greece, pp. 4349-4354.

[10] P. L. Dowell, "Effects of eddy currents in transformer windings", *Proc. IEE*, vol. 113, no. 8, August 1966, pp. 1387-1394.

[11] E. C. Snelling, *Soft Ferrites – Properties and Applications.* Butterworths, 2nd. Ed., 1988.

[12] T. McDonald, M. Soldano, A. Murray, T. Avram, "Power MOSFET avalanche design guidelines", *Application note AN-1005*, International Rectifier.

New Modulation Strategy with Low Switching Frequency and Minimum Baseband Distortion

N. E. Rüger*, O. Schnick[†], W. Mathis[†], A. Mertens*

*Institute for Drive Systems and Power Electronics, Leibniz Universität Hannover, Hannover, Germany,
e-mail: *rueger@ial.uni-hannover.de*
[†]Institute of Electromagnetic Theory, Leibniz Universität Hannover, Hannover, Germany,
e-mail: *schnick@tet.uni-hannover.de*

Abstract—This paper reports on a novel modulation strategy for generation of band limited signals with binary switches, which was originally developed for audio power amplifiers, and on applying it to energy conversion. This modulation strategy has the extraordinary advantage that the signal band is free of harmonics or carrier intermodulation products. This so called "Zero-Position-Coding with Separated Baseband" (SB-ZePoC) operates at low switching frequencies and thus combines the advantages of PWM and offline-optimised pulse patterns. The advantages and drawbacks are discussed in this paper, considering a single-phase converter application.

Keywords—Modulation Strategy, Pulse Width Modulation (PWM), Harmonics, Power Quality, Signal Processing

I. INTRODUCTION

The objective of modulation in power converters is to generate voltage and current waveforms which usually need to be sinusoidal in steady-state operation. The high frequency components which are introduced by switching are quite pronounced. In voltage source inverters, the unfiltered output voltages with basically rectangular waveforms are fed to the load via simple filters, in most cases a first-order inductive filter (often realised as the motor inductance) or a second-order LC filter with a weakly dampened resonance.

The modulation strategies used so far have been deeply analysed, and their advantages and drawbacks are quite well known. The regular sampled PWM (RPWM) is most commonly used in combination with digital control. With analogue modulation circuits, naturally sampled PWM (NPWM) is a popular choice. The switching frequencies are chosen at about 15 times the maximum fundamental frequency or more, in order to achieve satisfactory spectral performance of the output waveforms. If the waveforms are not adequate, either the switching frequency can be increased, which in turn generates higher losses in the converter and thus reduces its output power rating, or filters of higher order can be used, with the risk of exciting the filter resonance.

Often both ways are not satisfactory, and therefore modulation strategies for very low switching frequency have been developed, the offline-optimised pulse patterns. Here it is possible to minimise some of the harmonics, depending on the number of switchings per fundamental period. This strategy needs complicated trigger sets, since

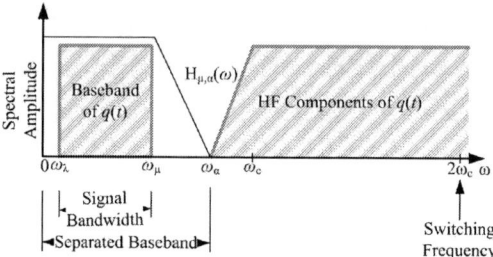

Fig. 1.   Separated signal and high-frequency band

the time regime of RPWM (used at lower output frequency) and synchronous optimised pulse patterns (used at higher frequency or close to overmodulation) is very different.

In this paper, a modulation method which was originally developed for a class-D-audio-amplifier (switched-mode audio amplifier) is proposed as an alternative. It is called "Zero Position Coding with Seperated Baseband" (SB-ZePoC), highlighting its special property that the baseband (range between minimum and maximum fundamental frequency) can be kept free of harmonics of the fundamental frequency, and of intermodulation products between fundamental and switching frequency. This results in a gap in the spectrum between the baseband and the distortion components that are introduced by the modulation (see Fig. 1). Due to this property, it should now be possible to reduce the ratio between switching frequency and fundamental frequency. In case that filters are used, excitation of the resonance by the modulator can be avoided, even if the switching frequency is reduced. Ideally, the switching frequency can be reduced to twice the maximum fundamental frequency (according to the Shannon-Nyquist theorem). For practical applications a reduction of four to five times is achievable, thus resulting in lower switching losses.

In section II of this paper, the modulation strategies are discussed and their properties when used in energy conversion applications are compared. Section III gives results of simulations for a single-phase application.

II. PULSE LENGTH MODULATION (PLM)

A novel general denotation of signal modulation strategies is introduced: Pulse Length Modulation (PLM).

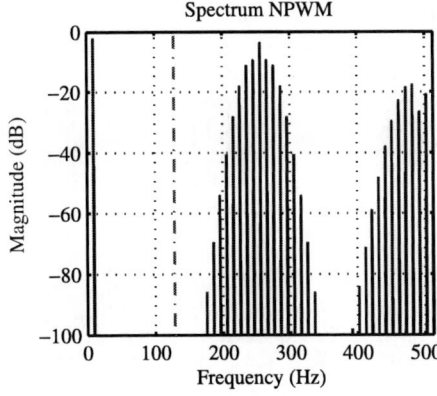

(a) NPWM spectrum with $f_{\text{sig}} = 10\,\text{Hz}$, $f_s = 254\,\text{Hz}$ and $M = 0.75$

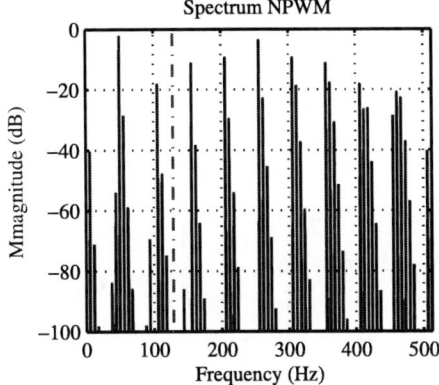

(b) NPWM spectrum with $f_{\text{sig}} = 50\,\text{Hz}$, $f_s = 254\,\text{Hz}$ and $M = 0.75$

Fig. 2.   Spectrum of NPWM generated output

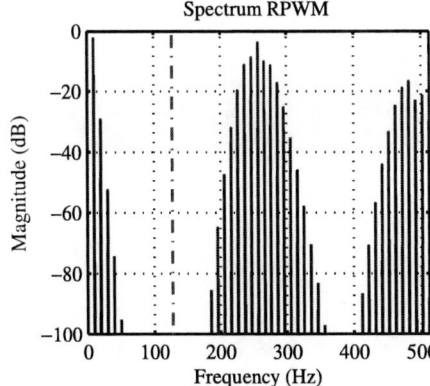

(a) RPWM spectrum with $f_{\text{sig}} = 10\,\text{Hz}$, $f_s = 254\,\text{Hz}$ and $M = 0.75$

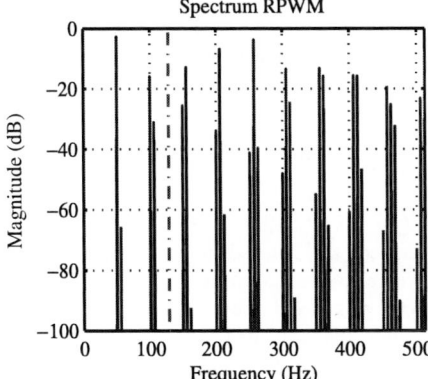

(b) RPWM spectrum with $f_{\text{sig}} = 50\,\text{Hz}$, $f_s = 254\,\text{Hz}$ and $M = 0.75$

Fig. 3.   Spectrum of RPWM generated output

All sorts of Pulse Width Modulation (PWM), like Naturally Sampled Pulse Width Modulation (NPWM) or Regular Sampled Pulse Width Modulation (RPWM), are a subset of PLM and use amplitude samples to represent the bandlimited input signal.

Another subset of PLM are such methods, where the input signal is mainly represented by the zero crossings of the signal. This includes the novel modulation scheme SB-ZePoC (Zero Position Coding with Separated Baseband). SB-ZePoC is introduced for power electronic applications in this paper.

In the following investigation of modulation strategies, the effect of dead time is not considered. If a switching carrier frequency is used, then the carrier is a sawtooth signal.

A. Naturally Sampled Pulse Width Modulation (NPWM)

Historically seen, NPWM is the first PLM method. The output signal is easily generated in analog domain by comparing an input signal with a triangular or sawtooth signal. In digital hardware, NPWM is hard to implement. The spectrum of NPWM shows no harmonic distortion at multiples of the signal frequency, but intermodulation distortion inside the baseband caused by the switching frequency [1]. The distortion is low, if the ratio of switching to signal frequency is large enough ($f_s/f_{\text{sig}} > 10 \cdots 20$) (Fig. 2(a)). For lower ratios of switching to signal frequency, the intermodulation is unacceptable high (Fig. 2(b)). For $f_{\text{sig}} = 50\,\text{Hz}$, the switching frequency should be at least in the range of $500\,\text{Hz}$ to $1000\,\text{Hz}$.

B. Regular Sampled Pulse Width Modulation (RPWM)

RPWM, also known as UPWM (Uniform Sampled Pulse Width Modulation), can be generated in different ways. In the analog domain, a sample-and-hold block is added to the input path of the NPWM strategy. RPWM can be generated easily in digital domain with counters and binary comparison. RPWM has less intermodulation products inside the baseband than NPWM, but harmonic distortions occur (see [1]). The magnitudes of harmonics again depend on the relation f_s/f_{sig}. For $f_s/f_{\text{sig}} > 10 \cdots 20$, the harmonics are low enough, so they can be tolerated. A switching frequency of $f_s = 254\,\text{Hz}$ is much too low for adequate signal quality used for an output signal frequency of $50\,\text{Hz}$ (see (Fig. 3)).

The ratio f_s/f_{sig} needed for good signal quality is

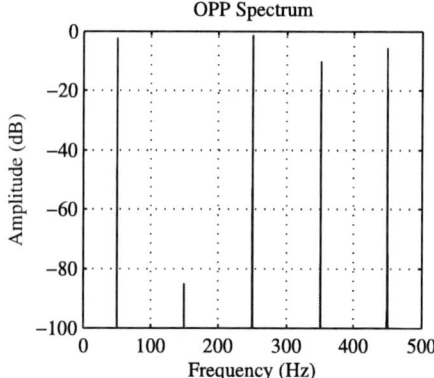

Fig. 4. Optimised Pulse Pattern spectrum with $f_{sig} = 50\,\mathrm{Hz}$, $f_s = 250\,\mathrm{Hz}$ and $M = 0.75$

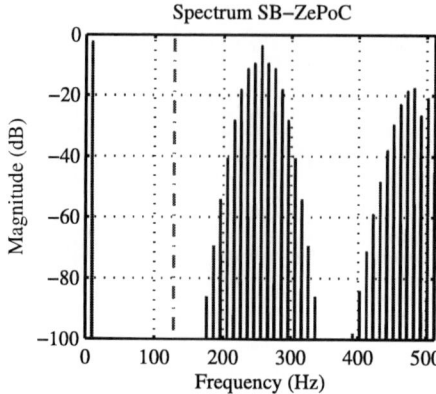

(a) ZePoC Spectrum with $f_{sig} = 10\,\mathrm{Hz}$, $f_s = 254\,\mathrm{Hz}$ and $M = 0.75$

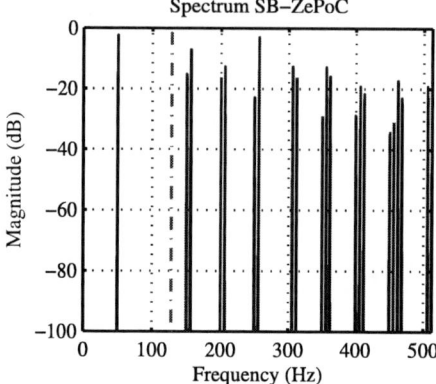

(b) ZePoC Spectrum with $f_{sig} = 50\,\mathrm{Hz}$, $f_s = 254\,\mathrm{Hz}$ and $M = 0.75$

Fig. 5. Spectrum of SB-ZePoC generated output

nearly the same for RPWM and NPWM. So both methods lead to the same minimum switching frequency $f_{s,min}$. This is why RPWM dominates in power electronics. The higher implementation effort for NPWM does not pay off in signal quality.

C. Optimised Pulse Patterns (OPP)

One of the commonly used signal generation methods for power electronic converters with low switching frequency are optimised pulse patterns (for details see e. g. [2]). Here, the switching angles have to be calculated in advance because of the complex optimisation algorithm. Useful optimisation criteria include harmonic cancellation, minimisation of RMS distortion current, or combination of both. One of the disadvantages is the calculation of the switching angles in advance and storing them in look-up tables. Another disadvantage is the fixed integer ratio between fundamental and switching frequency. If the fundamental frequency is changed over a wider range, also the pulse pattern must be changed because of the increasing switching losses. All this leads to a structure of the modulator, which is much different to that of RPWM. Often, both OPP and RPWM have to be implemented in a power electronic converter. For low signal frequencies the RPWM method is used; for higher frequencies the OPP method is used. This results in a complex software implementation. The change of pulse patterns due to changes of the output frequency or amplitude leads to transients, that can be eliminated only with a complex control strategy. An example of the spectrum for a pulse pattern with a signal frequency of $f_{sig} = 50\,\mathrm{Hz}$ and a resulting switching frequency of $f_s = 250\,\mathrm{Hz}$ is shown in Fig. 4. Here, amplitude of the fundamental and 3rd harmonic are optimised. The reduction of the harmonics depends on the number of switchings in one period of the fundamental frequency.

D. Zero Position Coding with Separated Baseband (SB-ZePoC)

The novel modulation method SB-ZePoC (Zero Position Coding with Separated Baseband) was introduced

by Mathis, Streitenberger and Bresch ([3] ,[4], [5]). SB-ZePoC is based on the fundamental work of B. F. Logan between 1977 and 1984 [6]. Logan's theory of impulse or "click" modulation is based on the complex theory of functions, where he constructs analytic signals. These signals do have only positive or only negative frequency content (see [7]). For this purpose, a Hilbert transform was incorporated. The Hilbert transform is a linear convolution operation, that effects the frequency domain of a signal in the following manner ($F(\omega)$ represents the Fourier transform of $f(t)$):

$$\mathcal{H}(F(\omega)) = -\mathrm{i}\,\mathrm{sign}(\omega) \cdot F(\omega).$$

Figure 6 shows the block diagram of SB-ZePoC. A given bandpass input signal $f(t)$ is transformed into a square-waved signal $q(t)$ that has the excellent property of a separated baseband. For a detailed derivation refer to [3]. The input signal $f(t)$ has to be band limited and free of a zero frequency component. First the analytical signal $F(t) = f(t) + i\hat{f}(t)$ is generated, where $\hat{f}(t)$ represents the Hilbert transform of $f(t)$. Then an Analytic Exponential Modulation (AEM) is performed. The result is that negative frequency components are cancelled from

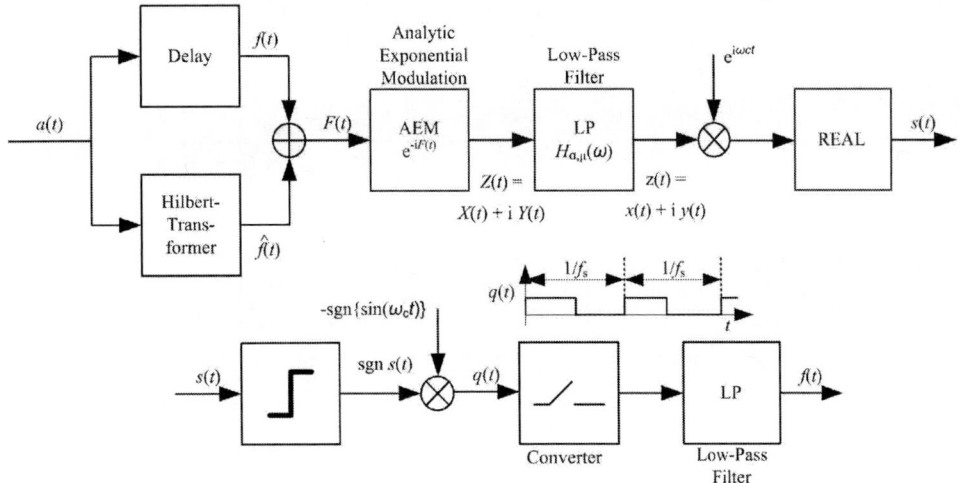

Fig. 6. Block diagram of SB-ZePoC signal processing

the signal (for details see [6]). The low-pass filter behind the AEM is a key component in SB-ZePoC. Next the single side band (SSB) signal $s(t)$ is generated. The binary signal $q(t)$ can finally be constructed by finding the zero crossings of $s(t)$ and of the modulation carrier $\sin \omega_c t$. The resulting switching signal $q(t)$ consist of pulses of constant switching frequency f_s, The pulses are left aligned (turn on occurs at the beginning of each switching period). The signal has similarity with a PWM generated from a sawtooth carrier, but the pulse durations are different.

It should be noted that Logan's concept applies to continuous-time signals. However, it can be implemented in discrete time (e. g. digital). The internally generated AEM signal $Z(t) = X(t) + iY(t)$ requires a higher bandwidth than the input signal. Therefore, appropriate oversampling is necessary in order to avoid aliasing. The fact that $q(t)$ has a separated baseband has been proven by Logan; this means that no harmonic distortions and no intermodulation components in the baseband arise (see Fig.5). Additional filter components might be needed to reduce the distortion above baseband, but this is true for all the modulation strategies.

There are two disadvantages when SB-ZePoC concept is used without change for power electronic applications. First, several blocks in the SB-ZePoC system lead to latency time in their real implementation. The total latency time can be larger than half the period of the input signal. So an implementation in feedback controlled systems would be almost impossible. Second, for the input signal $f(t) = M \cos(\omega_{sig}t + \varphi_0)$, a limit of the signal amplitude M applies at low ratios of switching frequency to signal frequency. For high ratios between switching and signal frequency, the amplitude can be chosen of nearly 0.99. For a low ratio, the limit of the amplitude M has to be decreased down to 0.75. It is a challenge to find a solution for these two problems.

The latency problem is addressed first. When we start

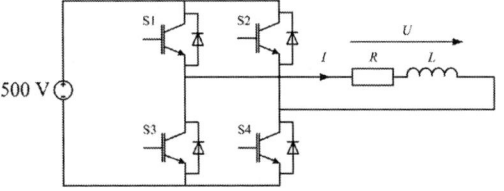

Fig. 7. Circuit used for Simulations (Indication of the measured Voltage and Current)

with an analytic input signal $F(t)$, the Hilbert transform block can be avoided. This can easily be done with periodic input signals. In case of monofrequent input signals the analytic signal is (see [7]):

$$F(t) = A\cos(\omega_{sig}t + \varphi_0) - iA\sin(\omega_{sig}t + \varphi_0). \quad (1)$$

This reduces one of the main reasons for the latency. Further work will address the restriction of the maximum amplitude.

E. Comparison of the Discussed Modulation Strategies

In TABLE I, different PLM modulation strategies are compared. The above mentioned advantages and disadvantages of the SB-ZePoC can been seen.

III. SIMULATION RESULTS

In the following section, simulation results are shown (the figures can be found at the end of the paper). To show the advantages of SB-ZePoC, this modulation strategy is compared with RPWM, NPWM and OPP. The simulation results were produced in a MatLab/Simulink environment. As load a combination of inductive and resistive components ($L = 39.6\,\text{mH}$ and $R = 100\,\text{m}\Omega$) was chosen. The back-emf is neglected, because only the distortions were of interest. For all simulations the same modulation index $M = 0.75$ was chosen. The supply voltage is $500\,\text{V}$. All simulations were done without a

TABLE I

COMPARISON OF DIFFERENT MODULATION STRATEGIES USED IN POWER ELECTRONIC CONVERTERS

	RPWM	NPWM	SB-ZePoC	OPP
range of modulation index	$M < 1.15$	$M < 1.15$	$M < 1.15 \cdot \gamma$ $\gamma = 0.75 \cdots 0.99$	$M < 1.15$
minimum switching frequency $f_{s,min}$	$10 \cdots 20 \cdot f_{sig}$	$10 \cdots 20 \cdot f_{sig}$	$4 \cdots 5 \cdot f_{sig}$	$n \cdot f_{sig}\ n \in \mathbb{N}$
harmonic distortion in the baseband caused by higher order harmonics of fundamental frequency	higher	none	none	selectable
harmonic distortion in the baseband caused by intermodulation of f_s and f_{sig}	low	higher	none	none
switching losses	high	high	low	low
on state losses	nearly equal			

filter between the converter and the load. The simulated circuit is displayed in Fig. 7. A maximum simulation sampling time of $100\,\text{ns}$ was chosen. The noise of the simulation depends on the analysed window, and on the ratio of the switching frequency and simulation sampling time. In our results, the noise of the generated gate signal $q(t)$ is about $-120\,\text{dB}$. Due to the chosen load impedance, the spectrum of all analysed currents is frequency dependent (e.g. see Fig. 8). This behaviour results from the predominantly inductive load. Due to this, the noise is weighted with $1/f$ in the current spectra.

In TABLE II to TABLE IV, several currents of the simulation are compared (fundamental current, RMS of the distortion currents, and RMS of the harmonic currents). The results are listed for a small and a high ratio of switching to signal frequency ($254\,\text{Hz}$ to $50\,\text{Hz}$ and $254\,\text{Hz}$ to $10\,\text{Hz}$).

The time domain signals and frequency spectra of RPWM (Fig. 8) and SB-ZePoC (Fig. 9) at $10\,\text{Hz}$ look similar. However, when looking at TABLE III, it is seen, that distortion current is a little better for SB-ZePoC. In fact, the theory shows, that SB-ZePoC passes into NPWM for low signal frequencies.

By comparing the simulation results of RPWM (Fig. 10) and SB-ZePoC (Fig. 11) for higher frequencies (here $50\,\text{Hz}$, that means a low ratio of switching to signal frequency) SB-ZePoC has some advantages. It can be seen that a band gap exist between the fundamental frequency and the distortions when using SB-ZePoC. RPWM shows intermodulations of signal and switching frequency inside the band gap of SB-ZePoC. However, the distortion current from SB-ZePoC in TABLE III is a little higher than that of RPWM. Looking at the fundamental currents reveals an amplitude error for RPWM, which is not present in SB-ZePoC.

By comparing the spectra of SB-ZePoC (Fig. 11) and of the optimised pulse patterns (Fig. 12), the conclusion can be drawn that the band gap of the OPP is wider than the one of SB-ZePoc. The currents in TABLE II up to TABLE IV show the same results. The OPP has the disadvantage of the complex calculation and implementation. By SB-ZePoC the switching between different

modulation strategies and different pulse patterns can be omitted. This results in an easier implementation and less transient disturbance.

By comparing the spectrum of NPWM (Fig. 13) and the spectrum of SB-ZePoC (Fig. 9), it can be seen that they look almost the same. This underlines the mentioned similarity of SB-ZePoC and NPWM (for details see [3]). The currents of NPWM and of SB-ZePoC (TABLE II to TABLE IV) show the same behaviour for a high ratio of switching to signal frequency. For lower ratios of switching to signal frequency, the NPWM strategy has more distortions and no band gap (see fIG: 2(b)).

TABLE II

COMPARISON OF THE FUNDAMENTAL CURRENT ($M = 0.75$)

Signal Frequency	10 Hz	50 Hz
RPWM	150.41 A	29.34 A
NPWM	150.59 A	30.14 A
SB-ZePoC	150.59 A	30.14 A
OPP	-	30.14 A

TABLE III

COMPARISON OF THE DISTORTION CURRENT (RMS) ($M = 0.75$)

Signal Frequency	10 Hz	50 Hz
RPWM	8.99 A	8.83 A
NPWM	8.26 A	9.72 A
SB-ZePoC	8.26 A	9.30 A
OPP	-	7.47 A

TABLE IV

COMPARISON OF THE HARMONIC CURRENT (RMS) ($M = 0.75$)

Signal Frequency	10 Hz	50 Hz
RPWM	3.49 A	3.34 A
NPWM	0.31 A	0.01 A
SB-ZePoC	0.31 A	0.01 A
OPP	-	7.46 A

IV. CONCLUSION

It was shown that SB-ZePoC is interesting for use in power electronics because of its signal characteristics. SB-ZePoC has advantages over OPP because of the constant sampling time and switching frequency and natural transition to NPWM behaviour at lower signal frequencies. For high ratios of switching to signal frequency, it also has advantages against RPWM. It seems to be very useful for power converters with a low ratio of switching to signal frequency especially when a gap between signal and distortion frequencies is desired; for instance when LC-filters are used. Further investigations will be done on three-phase systems, on reducing the calculation complexity, and on an adapted form of SB-ZePoC to a double side mode (both turn-on and turn-off instants vary in each switching period). The next step is to implement the modulation strategy on a hardware testing environment.

ACKNOWLEDGMENT

This work is funded by Deutsche Forschungsgemeinschaft (DFG).

REFERENCES

[1] D. G. Holmes and T. A. Lipo, *Pulse Width Modulation for Power Converters – Principles and Practice.* IEEE Press, 2003.

[2] J. Sun, *Optimal Pulsewidth Modulation Techniques for High-Power Voltage-Source Inverters.* Düsseldorf: VDI Verlag, 1995.

[3] M. Streitenberger, *Zur Theorie digitaler Klasse-D-Audioleistungsverstärker und deren Implementierung.* Berlin: VDE Verlag, 2005.

[4] M. Streitenberger, M. Streitenberger, H. Bresch, and L. Mathis, "Theory and implementation of a new type of digital power amplifier for audio applications," in *Proc. ISCAS 2000 Geneva Circuits and Systems The 2000 IEEE International Symposium on,* vol. 1, 2000, pp. 511–514.

[5] H. Bresch, W. Mathis, F. Felgenhauer, and M. Streitenberger, "Zero position coding (zepoc) - a generalised concept of pulse-length modulated signals and its application to class-d audio power amplfiers," in *Audio Engineering Society convention preprints, 110th Convention of AES May 12–15,* vol. 5351–5408, no. 5365, Amsterdam, 2001, p. 9.

[6] B. F. Logan, Jr., "Click modulation," *AT&T Bell Lab. Tech. Journal,* vol. 63, no. 3, pp. 401–423, 1984.

[7] E. Bedrosian, "The analytic signal representation of modulated waveforms," *Proceedings of the IRE,* vol. 50, no. 10, pp. 2071–2076, Oct. 1962.

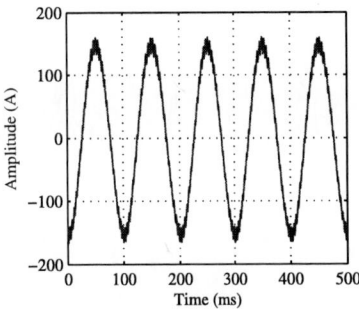

(a) Simulated RPWM current with $f_{sig} = 10\,\text{Hz}$, $f_s = 254\,\text{Hz}$ and $M = 0.75$

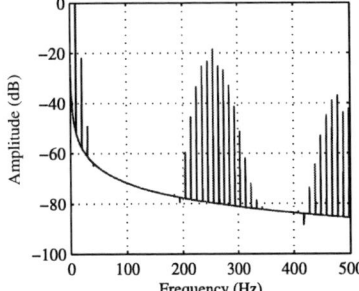

(b) Simulated RPWM spectrum of current with $f_{sig} = 10\,\text{Hz}$, $f_s = 254\,\text{Hz}$ and $M = 0.75$

Fig. 8. Time signal and spectrum of 10 Hz RPWM simulation

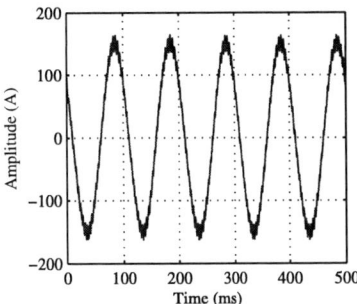

(a) Simulated SB-ZePoC current with $f_{sig} = 10\,\text{Hz}$, $f_s = 254\,\text{Hz}$ and $M = 0.75$

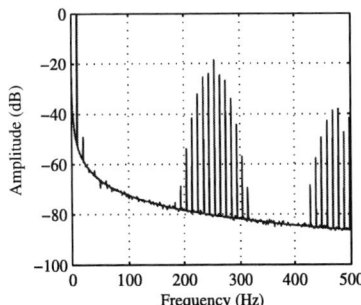

(b) Simulated SB-ZePoC spectrum of current with $f_{sig} = 10\,\text{Hz}$, $f_s = 254\,\text{Hz}$ and $M = 0.75$

Fig. 9. Time signal and spectrum of 10 Hz SB-ZePoC simulation

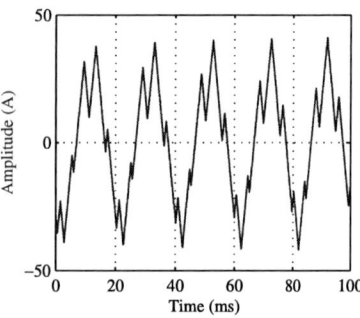

(a) Simulated RPWM current with $f_{sig} = 50\,\text{Hz}$, $f_s = 254\,\text{Hz}$ and $M = 0.75$

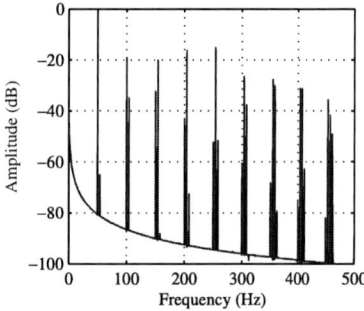

(b) Simulated RPWM spectrum of current with $f_{sig} = 50\,\text{Hz}$, $f_s = 254\,\text{Hz}$ and $M = 0.75$

Fig. 10.   Time signal and spectrum of 50 Hz RPWM simulation

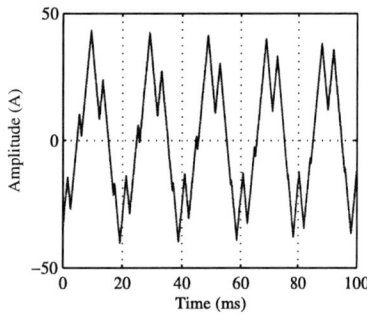

(a) Simulated SB-ZePoC current with $f_{sig} = 50\,\text{Hz}$, $f_s = 254\,\text{Hz}$ and $M = 0.75$

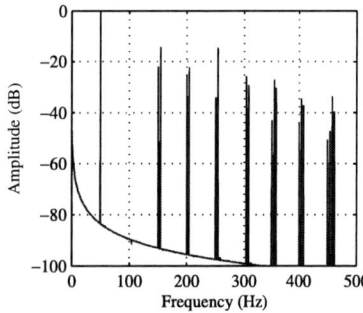

(b) Simulated SB-ZePoC spectrum of current with $f_{sig} = 50\,\text{Hz}$, $f_s = 254\,\text{Hz}$ and $M = 0.75$

Fig. 11.   Time signal and spectrum of 50 Hz SB-ZePoC simulation

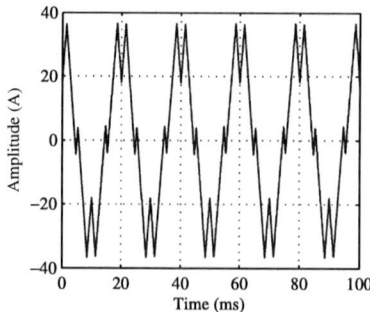

(a) Simulated OPP current with $f_{sig} = 50\,\text{Hz}$, $f_s = 254\,\text{Hz}$ and $M = 0.75$

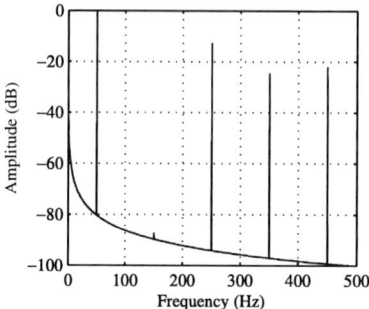

(b) Simulated OPP spectrum of current with $f_{sig} = 50\,\text{Hz}$, $f_s = 250\,\text{Hz}$ and $M = 0.75$

Fig. 12.   Time signal and spectrum of 50 Hz OPP simulation

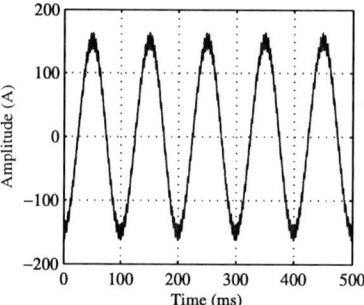

(a) NPWM current with $f_{sig} = 10\,\text{Hz}$, $f_s = 254\,\text{Hz}$ and $M = 0.75$

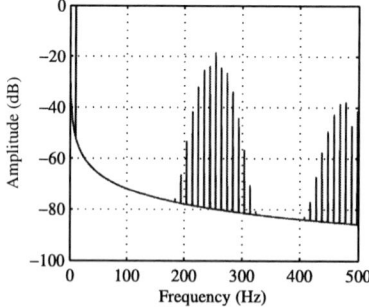

(b) NPWM spectrum of current with $f_{sig} = 10\,\text{Hz}$, $f_s = 254\,\text{Hz}$ and $M = 0.75$

Fig. 13.   Time signal and spectrum of 10 Hz NPWM simulation

A Bit-Stream Based PWM Technique for Variable Frequency Sinewave Generation

N. D. Patel and U. K. Madawala

Dept. of Electrical and Computer Engineering
University of Auckland, Auckland, New Zealand
Email: nd.patel@auckland.ac.nz, u.madawala@auckland.ac.nz

Abstract—**This paper presents a unique technique suitable for the generation of variable frequency sinusoidal waveforms with high quality. Arithmetic operations on the bit streams are performed through digital blocks. The proposed technique is simple and can be implemented on an field programmable gate array (FPGA). Results of a prototype single-phase inverter module with custom built power stage, are presented with simulations. Experimental results indicate that the technique can generate sinusoidal waveforms of frequencies from 20Hz-60Hz with total harmonic distortion (THD) 1.4-4%, respectively, and therefore the technique would be an ideal candidate for variable speed motor applications.**

Index Terms—**Adjustable speed drive, Modulation strategy, Pulse Width Modulation, Variable speed drive**

I. INTRODUCTION

The global market for motor drives and power sources that supply power for various AC loads is also on the rise. This trend, which is expected to continue, can be attributed to the ever increasing consumer demand for an improved standard of living. However, the awareness of both global warming and rapid depletion of fossil reserves has slightly changed the market so that these power sources are now made with stringent specifications and emphasis on improved power quality and minimum carbon footprint. As such there is a revival in research on power supply technology to produce clean-green power, which is environmentally friendly [1]–[3].

Majority of AC loads require power at both variable voltage and frequency, and applications of induction motors operated by variable speed drive (VSD) can be considered as prime examples [4]–[9]. However, irrespective of the level of sophistication of the VSD, the conversion of DC power into a single or three phase AC power at variable frequency is a fundamental requirement and usually achieved electronically with the use of an inverter. An essential requirement of this electronic DC-AC power conversion process is the generation of the appropriate switching instants of the inverter, which is accomplished through the well established PWM techniques.

A comprehensive review on existing PWM techniques can be found in [10]–[12]. The on-line techniques, as classified in [13], such as sinusoidal PWM and space-vector, have been developed for both voltage-sourced and current-sourced inverter systems [14]–[16]. In contrast, the off-line techniques such as Selective Harmonic Elimination (SHE) and fundamental magnitude control etc, are extensively used in current-sourced inverter systems [17]–[19]. They are suitable for many applications with low switching frequency, where low order harmonic elimination is required. However, the generalized SHE method reported in [20] provides low order harmonic elimination for both voltage-sourced and current-sourced inverter systems. The performance of PWM techniques is usually evaluated on the basis of current harmonics, harmonic spectrum, maximum modulation index, torque harmonics, switching frequency, switching loss and dynamic performance [11].

The technique presented here uses uniformly weighted digital bit-streams to produce variable frequency sinusoids. The bit-streams are inherently PWM-like and hence the signal can be directly interfaced to gate drivers. The bit-stream elements are implemented on an field programmable gate array (FPGA). The nature of this implementation is concurrent or parallel and hence multiple instances of sinusoidal generators have no impact on each other. If required, the multiple generators can be synchronized to produce multi-phase sinusoids with user specified phase angle relationships.

II. BIT-STREAM CONCEPTS

The bit-stream technique uses a sequence of two uniformly weighted quanta, $+Q$ and $-Q$ to represent any analogue or binary signal. A logical value of '1' is assigned a $+Q$ while a logic '0' is assigned a $-Q$. Thus a digital square wave, which

978-1-4244-1741-4/08/$25.00 ©2008 IEEE

is an alternating stream of logic 1's and logic 0's, will encode a zero value over several logic transitions. A more complete introduction to bit-streams can be found in [21].

The bit-stream signal can be very easily split to drive an H-bridge. If the DC link voltage is V_{DC} then, a logic '1' and a logic '0' would impress opposite voltages to the load. If $a(i)$ and $b(i)$ represent the positive $(+Q)$ and negative $(-Q)$ quantum, respectively, and $A_N(k)$ and $B_N(k)$ are the summations, at time instant k, of positive and negative quanta over the past N states then, in a normalized analysis where $Q = 1$, the average voltage applied to the load can be expressed by

$$V_{LOAD}(k) = V_{DC} \frac{A_N(k) - B_N(k)}{N} \qquad (1)$$

The digital circuits have been simulated using very high speed integrated circuits hardware description language (VHDL) within Modelsim™ and synthesized on a Stratix FPGA using Altera's Quartus™ tool chain. The power circuits have been designed and assembled in-house.

A. Scaled Fractions

If the resolution is 1 part in N where $N = 2^R$ and if a 2's complement R bit wide register is used to capture the binary value then the minimum and maximum values are $2^{R/2} - 1$ and $-2^{R/2}$ respectively. To make the subsequent explanations more general in nature, scaled fractions are used and define by $a^{sR} = a \times 2^{R-1}$. Here the superscript sR denotes the scaled fraction of a binary word of width R. Thus if $R = 8$ then $0.5^{s8} = 0.5 \times 2^{R-1} = 64$.

B. Binary to Bit-Stream

The schematic shown in Fig. 1 will convert a multi-bit (2's complement) binary word, $B_R(k)$ into a bit-stream, $S_o(k)$. The circuit is driven by a clock such that the bit-rate is f_B. The thick solid lines show the multi-bit (two's complement) data paths. The input, B_R is summed with a value based on the sign of the Sum and Accumulate (SAC). M_1 is a positive value and for a symmetrical operation, $M_2 = -M_1$. The feedback process tries to maintain v_s at zero by accumulating differing proportions of M_1 and M_2. If $B_R(k)$ is assumed to be time invariant then in steady state,

$$\lim_{k \to \infty} \left[k B_R(k) - \left[M_1 \sum^{\infty} a(i) + M_2 \sum^{\infty} b(i) \right] \right] = 0 \quad (2)$$

If a resolution of one part in N is required, the word width, R, of the binary input B_R must satisfy $N \leq 2^R$. Considering the limiting condition, $N = 2^R$, the positive and negative

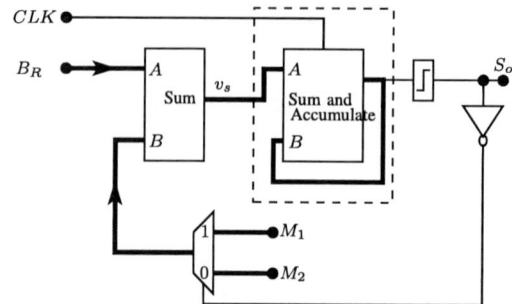

Fig. 1. Digital bit-stream generator

quanta can be written as $\sum_{k-N}^{k} a(i) = A_N$ and $\sum_{k-N}^{k} b(i) = B_N$ respectively. Thus (2) can be written in terms of A_N as

$$A_N(k) = B_R(k) \frac{N}{M_1 - M_2} - M_2 \frac{N}{M_1 - M_2} \qquad (3)$$

The value of the output bit-stream can be obtained by taking the difference $D_N = A_N - B_N$. However, to normalize the numeric value we define $S_o = \frac{D_N}{N} \frac{A_N - B_N}{N}$. Since $A_N + B_N = N$ and using (3)

$$S_o(k) = B_r(k) \frac{2}{M_1 - M_2} - \frac{M_1 + M_2}{M_1 - M_2} \qquad (4)$$

In (4), the first term is a gain term while the second term is an offset. Suitable values of M_1 and M_2 can be calculated depending on requirements. In this application we require a gain of unity with no offset and hence $M_2 = -M_1$ and for a gain of one, $M_1 = N/2$.

C. Bit-Stream Integration

Fig. 2(a) can be shown to be an integrator with a transfer function, I_{BS}, given by

$$I_{BS}(s) = \frac{f_B}{N} \frac{1}{s} \qquad (5)$$

The integrator has been experimentally characterized using standard time domain measurements. With $R = 8$ (i.e. $N = 2^8 = 256$) and $f_B = 512$kHz the frequency response is shown in Fig. 2(b). Also shown on the plot is the theoretical frequency response. The small offsets at the input together with a poor signal to noise ratio makes the measurement of the open loop frequency response of an integrator relatively difficult at high frequencies. The characteristics of this integrator have been verified using frequency and time domain simulations and for brevity, these results have not been presented.

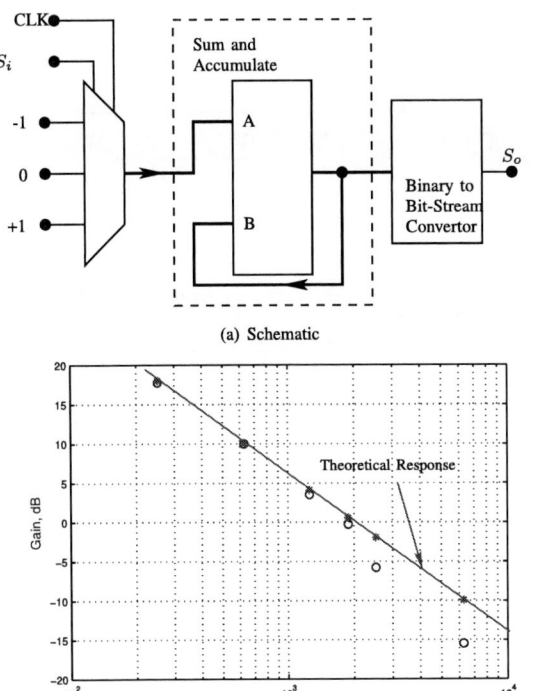

(a) Schematic

(b) Measured Frequency Response

Fig. 2. Bit-Stream Integrator

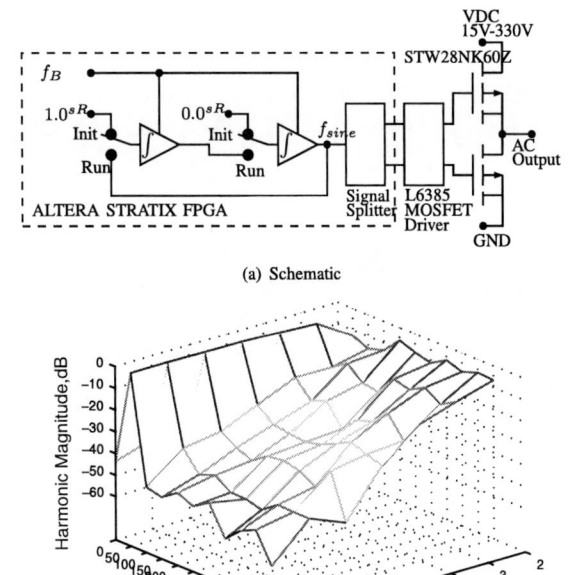

(a) Schematic

(b) Harmonic Content with Different Register Widths

Fig. 3. Sinusoidal Oscillator

III. SINUSOIDAL DRIVES

A. Sinusoidal Oscillator

Fig. 3(a) shows a conceptual schematic of the proposed sinusoidal oscillator. In principle it consists of two cascaded integrators in a negative feedback loop. This system is critically stable since the phase margin is zero and consequently the closed loop system will oscillate at the crossover frequency. In the bit-stream environment this is precisely f_B/N where f_B is the bit-rate, $N = 2^R$ and R is the word width of the SACs in the integrators. The actual implementation requires one of the integrators to be initialized to full-scale while the other one to zero which has been diagrammatically shown using two-pole switches.

The f_{sine} output in Fig. 3(a) is a bit-stream signal which is ideally suited to drive a two state power stage provided the switching frequency is low enough to keep the switching losses within specifications. According to (5), the crossover frequency depends on the bit-rate and the word width of integrator. For a given f_{sine}, if f_B is to be reduced then R must also be reduced. The impact of reducing f_B and

R on the quality of the sinusoidal output has been assessed using a VHDL simulation in Modelsim™. In this experiment a testbench has been created to produce a 50Hz sinusoidal bit-stream. The time domain signals for $2 \leq R \leq 8$ have been sampled at $1\mu s$ and normalized to ± 1. The first 10 harmonics have been extracted and plotted in Fig. 3(b). With $R = 8$ all components other than the 50Hz component are less than -45dB. As R is reduced from 8 to 2, the harmonic magnitudes increase steadily with the worst performance at $R = 2$.

B. Variable Frequency Drive

If a variable frequency sinusoidal generator is required then f_B should be made variable and ideally, the ratio of the f_B to input control signal should be given by $f_B/V_c =$ constant. The 74 (TTL) medium scale integration logic series had a device (now obsolete) called a Bit Rate Multiplier. This IC takes a clock, f_{in} and a 6 bit binary word, M, as inputs. A general expression describing its output bit-rate is $f_o = f_{in}\frac{dec(M)}{2^6}$ where $dec(M)$ is the decimal equivalent of the binary input M. To facilitate the description, a simplified diagram of a 4 bit rate multiplier has been shown in Fig. 4. This circuit has two inputs, a clock and a multi-bit input and one output. Since this is a 4 bit rate multiplier, the *rate* input, $RATE[3..0]$, is also 4 bits wide. The binary counter spans its full range and hence has 16 states. For each of the AND gates in Fig. 4, the following equations can be noted

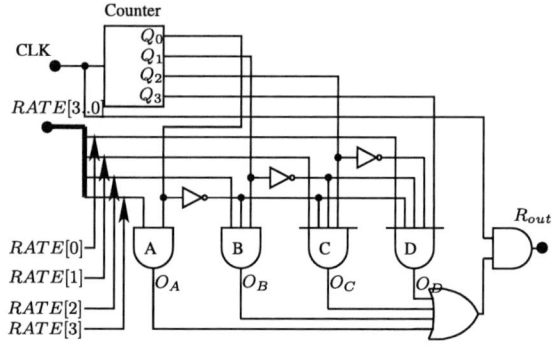

Fig. 4. Rate Multiplier

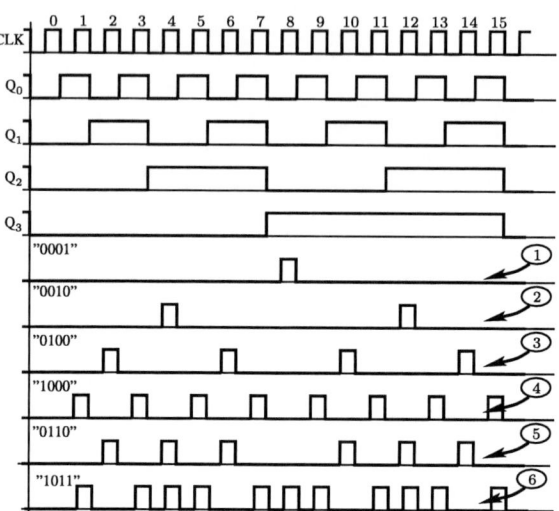

Fig. 5. Rate Multiplier Waveforms

$$O_A = RATE[3] \wedge Q_0$$
$$O_B = RATE[2] \wedge Q_1 \wedge \overline{Q_0}$$
$$O_C = RATE[1] \wedge Q_2 \wedge \overline{Q_1} \wedge \overline{Q_0}$$
$$O_D = RATE[0] \wedge Q_3 \wedge \overline{Q_2} \wedge \overline{Q_1} \wedge \overline{Q_0}$$
$$Rout = (O_A \vee O_B \vee O_C \vee O_D) \wedge CLK$$

If the input is $RATE[3..0] = "1000"$, then only the O_A term is active and it can be seen that the $(Q_0 \wedge CLK)$ term changes state 8 times in a frame of 16. This is shown as trace 4 in Fig. 5. On the other hand if $RATE[3..0] = "0100"$, then the O_B term is active and $(Q_1 \wedge \overline{Q_0} \wedge CLK)$ is active for 4 out of 16 clocks pulses. This is shown as trace 3 in Fig. 5. Similarly if $RATE[3..0] = "0010"$ then two pulses are output at R_{out} and only one pulse for $RATE[3..0] = "0001"$. If $RATE[3..0]$ is a combination of the above 4 possibilities then the output is the corresponding combination. Trace 5, for example, in Fig. 5 is a combination of "0100" and "0010" while trace 6 is a combination of "1000", "0011" and "0001".

The schematic in Fig. 3(a) has been modified to include the rate multiplier. With the DC link voltage at 100V, the $RATE$ input was adjusted to produce various sinusoidal outputs. Fig. 6 plots the measured spectrum for 50Hz and 60Hz at the inverter output. Based on the measured spectrum, the total harmonic distortion (THD) has also been calculated and presented in Table I. The worst-case THD is 4% and clearly very acceptable for power drives in industrial applications.

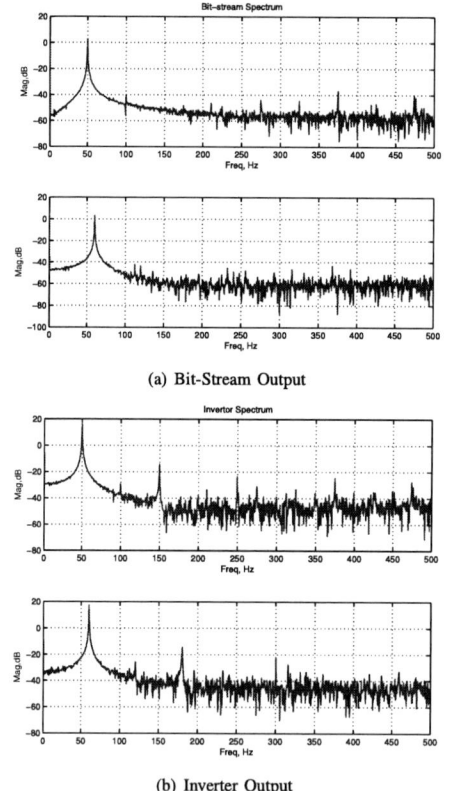

(a) Bit-Stream Output

(b) Inverter Output

Fig. 6. Spectral Response

142

TABLE I
THE TOTAL HARMONIC DISTORTIONS FOR BIT-STREAM AND INVERTER
OUTPUTS

Frequency	Bit-stream	Inverter
20	1%	1.4%
25	0.6%	1.7%
30	0.9%	2.2%
40	0.9%	2.6%
50	0.6%	3.2%
60	0.8%	4%

IV. CONCLUSIONS

A simple technique for the generation of high quality sinusoidal signals has been presented. The technique uses bit-streams and digital logic hardware implemented on an FPGAs. The technique has been applied to produce a variable frequency sinusoidal waveforms and demonstrated using a single phase inverter. The generator produces sinusoids with THD between 1.4% to 4%. The simplicity and ease of implementation on FPGAs makes this technique attractive in a wide variety of applications including induction and permanent magnet synchronous motor drives.

REFERENCES

[1] B.-T. Huang, K.-Y. Lee, and Y.-S. Lai, "Design of a two-stage ac/dc converter with standby power losses less than 1 w," in *Power Conversion Conference - Nagoya, 2007. PCC '07*, 2007, pp. 1630–1635.

[2] M. Begovic, A. Pregelj, A. Rohatgi, and C. Honsberg, "Green power: status and perspectives," *Proceedings of the IEEE*, vol. 89, no. 12, pp. 1734–1743, 2001.

[3] S. Rahman, "Green power: what is it and where can we find it?" *Power and Energy Magazine, IEEE*, vol. 1, no. 1, pp. 30–37, 2003.

[4] F. Blaabjerg, J. Pedersen, and P. Thoegersen, "Improved modulation techniques for pwm-vsi drives," *IEEE Transactions on Industrial Electronics*, vol. 44, no. 1, pp. 87–95, 1997.

[5] A. von Jouanne, P. Enjeti, and W. Gray, "Application issues for pwm adjustable speed ac motor drives," *IEEE Industry Applications Magazine*, vol. 2, no. 5, pp. 10–18, 1996.

[6] H.-J. Kim, H.-D. Lee, and S.-K. Sul, "A new pwm strategy for common-mode voltage reduction in neutral-point-clamped inverter-fed ac motor drives," *IEEE Transactions on Industry Applications*, vol. 37, no. 6, pp. 1840–1845, 2001.

[7] J. Carrasco, L. Franquelo, J. Bialasiewicz, E. Galvan, R. PortilloGuisado, M. Prats, J. Leon, and N. Moreno-Alfonso, "Power-electronic systems for the grid integration of renewable energy sources: A survey," *IEEE Transactions on Industrial Electronics*, vol. 53, no. 4, pp. 1002–1016, 2006.

[8] J. L. Rodriguez-Amenedo, S. Arnalte, and J. C. Burgos, "Automatic generation control of a wind farm with variable speed wind turbines," *Power Engineering Review, IEEE*, vol. 22, no. 5, pp. 65–65, 2002.

[9] D. Schulz, R. Fabis, and R. Hanitsch, "A three-phase power electronic converter for grid integration of distributed generation," in *Power Tech Conference Proceedings, 2003 IEEE Bologna*, vol. 1, 2003, pp. 8 pp. Vol.1–.

[10] A. Kwasinski, P. Krein, and P. Chapman, "Time domain comparison of pulse-width modulation schemes," *Power Electronics Letters, IEEE*, vol. 1, no. 3, pp. 64–68, 2003.

[11] J. Holtz, "Pulsewidth modulation for electronic power conversion," *Proceedings of the IEEE*, vol. 82, no. 8, pp. 1194–1214, 1994.

[12] P. C. Loh, F. Blaabjerg, and C. P. Wong, "Comparative evaluation of pulsewidth modulation strategies for z-source neutral-point-clamped inverter," *IEEE Transactions on Power Electronics*, vol. 22, no. 3, pp. 1005–1013, 2007.

[13] J. Espinoza, G. Joos, J. Guzman, L. Moran, and R. Burgos, "Selective harmonic elimination and current/voltage control in current/voltage-source topologies: a unified approach," *IEEE Transactions on Industrial Electronics*, vol. 48, no. 1, pp. 71–81, 2001.

[14] L. G. Franquelo, J. Napoles, R. C. P. Guisado, J. I. Leon, and M. A. Aguirre, "A flexible selective harmonic mitigation technique to meet grid codes in three-level pwm converters," *IEEE Transactions on Industrial Electronics*, vol. 54, no. 6, pp. 3022–3029, 2007.

[15] H. Lu, W. Qu, X. Cheng, Y. Fan, and X. Zhang, "A novel pwm technique with two-phase modulation," *IEEE Transactions on Power Electronics*, vol. 22, no. 6, pp. 2403–2409, 2007.

[16] C. Lascu, L. Asiminoaei, I. Boldea, and F. Blaabjerg, "High performance current controller for selective harmonic compensation in active power filters," *IEEE Transactions on Power Electronics*, vol. 22, no. 5, pp. 1826–1835, 2007.

[17] M. Saeedifard, H. Nikkhajoei, R. Iravani, and A. Bakhshai, "A space vector modulation approach for a multimodule hvdc converter system," *Power Delivery, IEEE Transactions on*, vol. 22, no. 3, pp. 1643–1654, 2007.

[18] A. K. Gupta and A. M. Khambadkone, "A space vector modulation scheme to reduce common mode voltage for cascaded multilevel inverters," *IEEE Transactions on Power Electronics*, vol. 22, no. 5, pp. 1672–1681, 2007.

[19] F. Gao and M. Iravani, "Dynamic model of a space vector modulated matrix converter," *Power Delivery, IEEE Transactions on*, vol. 22, no. 3, pp. 1696–1705, 2007.

[20] H. Karshenas, H. Kojori, and S. Dewan, "Generalized techniques of selective harmonic elimination and current control in current source inverters/converters," *IEEE Transactions on Power Electronics*, vol. 10, no. 5, pp. 566–573, 1995.

[21] N. D. Patel, S. K. Nguang, and G. G. Coghill, "Neural network implementation using bit streams," *IEEE Transactions on Neural Networks*, vol. 18, pp. 1488–1504, 2007.

Control Strategies of the Quasi-Resonant DC-Link Inverter

Sławomir Mandrek[*] and Piotr J. Chrzan[†]

[*] Det Norske Veritas, Sopot, Poland, e-mail: *slawomir.mandrek@ dnv.com*
[†] Gdansk University of Technology, Faculty of Electrical and Control Eng. Gdańsk, Poland, *pchrzan@ely.pg.gda.pl*

Abstract—Control strategies of a parallel quasi resonant dc link voltage inverter (PQRDCLI) for electrical drive applications are considered. The main objectives of the proposed strategies are feasibility and robust operation of the PQRDCLI vis-à-vis output voltage dU/dt limitation and stable duration of zero voltage intervals. The first strategy is based on controlling equal charge and discharge of the input series capacitor bank. By applying in the second strategy, external dc link voltage power supply converters one can precisely stabilize both the output voltage derivatives and zero voltage interval. Finally, in the third sensorless strategy – constant intervals are applied with the floating capacitor midpoint potential variations.

Keywords—Resonant converter, ZVS converters, Control of Drive

I. INTRODUCTION

The fundamental topology of voltage source inverters is usually based on a bridge scheme consisting of six bilateral semiconductor switches. Usefulness of this class of converters, particularly in electrical drive applications has been confirmed by numerous advantages, e.g.: relative simple topology, available modulation strategies, robust detection of failures or fast commutation rates. However, during commutation, undesirable high rate of change at the power transistors may provoke harmful environmental phenomena, such as: electromagnetic interference EMI emissions (conducted and radiated), increase of bearing currents, overvoltage spikes at motor terminals or motor winding insulation degradation. A solution diminishing these effects has been attained in a series of papers originated by Divan et al. [3,4], particularly through the development of active clamped resonant dc-link inverter topologies, where the bus line voltage decreases resonantly to zero, enabling soft zero-voltage switching (ZVS) of inverter states.

Applying pulse width modulation (PWM) techniques, further progress has been obtained by the quasi resonant dc-link inverter (QRDCLI) structures, where an initial instant of each resonant cycle is commanded externally from the microcontroller [1,2,5,6]. However, this requires very precise timing control of auxiliary active elements of resonant inverter. As each output voltage pulse initiation, conventionally controlled by the PWM modulator, must be preceded by the auxiliary resonant circuit operation. Many new QRDCLI topologies apply more that two additional transistors or require uncontrolled resonant oscillations in a whole zero voltage period. Moreover, if quasi-resonant operation is conditioned by triggering thresholds of inductor current or capacitor voltage, then ultra rapid acquisition and conversion of voltage/current signals is needed.

Therefore, in this paper for previously described [8] parallel QRDCLI – new gate timing control strategies are developed. They feature various complexity but also the attractive implementation issues for quasi-resonant DC link operation. The considered PQRDCLI topology as depicted in Fig. 1 consists of only two transistors with parallel diodes, additional diode D_3, a resonant inductor L and resonant capacitor C_f. The capacitor C_f is normally present in hard commuted inverters as a snubber capacitor, hence this element does not extend traditional topology of the inverter. Also series connection of input capacitors C_1 and C_2 is very often applied. This structure allows ZVS for inverter transistors and also gives ZVS/ZCS for auxiliary circuit transistors T_1 and T_2. The main advantage of considered quasi-resonant inverter is a reduced number of active elements in comparison with other topologies [7, 11], relatively simple and robust controllability with limitation of output inverter voltage dU/dt. The developed gate control strategies have been tested for motor and generator modes of operation.

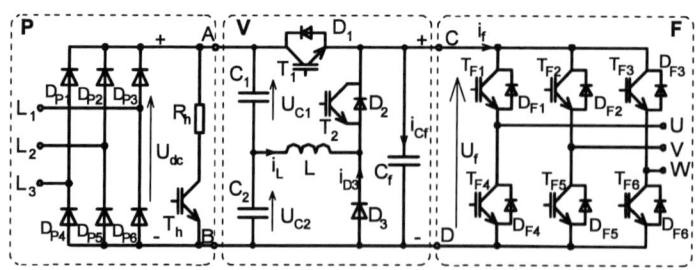

Fig. 1. Indirect frequency converter with a PQRDCLI;
(P- rectifier with breaking chopper R_h, T_h; V – parallel quasi resonant circuit, F – inverter).

978-1-4244-1741-4/08/$25.00 ©2008 IEEE 144

II. CONTROL STRATEGIES

Every operation of the quasi-resonant circuit is initiated by a command signal from the voltage modulator (Fig.2). First, the auxiliary transistor T_2 is turned-on in advance of Δt_f period, before disconnecting of main transistor T_1, which is turned-off with a constant rate Δt_{fmax}.

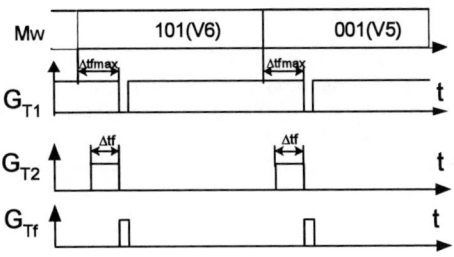

Fig. 2. Gate control strategy;
Mw – space vector index, G_{Tx} – transistor gate signals.

In the first two control strategies: forward period Δt_f corresponding to the mode M1 as in Fig. 4 is a linear function of load current I_o

$$\Delta t_f = \frac{L}{(1-k)\cdot U_{dc}}\left(I_p - I_o\right) \qquad (1)$$

Assuring constant discharge current I_p of capacitor C_f, (1) is valid for motor and generator modes taking into account change of load current sign: $\pm I_0$ ($I_p > I_0$ provides safe time margin for DSP operation).

After initiation of resonant cycle, a decrease of DC link voltage u_f to zero triggers voltage comparator U_{f_enb} as in Fig. 3. Its threshold level U_{f_ref} is a compromise between ZVS of inverter transistors and noise separation of power electronics circuits. Respectively, recovery of the DC link voltage supply is sensed by the comparator U_{dc_enb} enabling the transistor T_1 to be turn-on at ZVS conditions.

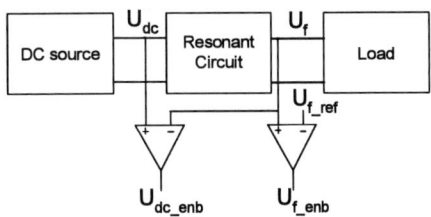

Fig. 3. Voltage comparators configuration.

A. Capacitor voltage control

In this strategy the C_1-C_2 capacitor divider midpoint voltage U_{C2} control is needed. At the *motor mode* ($I_o>0$), inverter transistors T_f are gated on during the zero voltage period t_{ZVM} to conduct increasing inductor current i_L up to the predetermined limit value. Following Fig. 4

$$i_L(t_5) - i_L(t_2) = 2I_p \qquad (2)$$

hence

$$t_{ZVM} = 2I_p L/(kU_{dc}) \qquad (3)$$

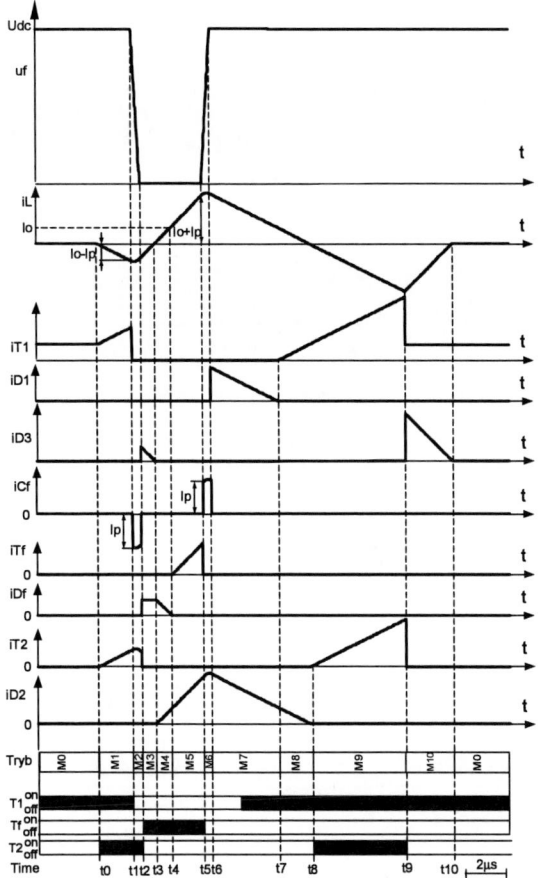

Fig. 4. The PQRDCLI transient waveforms at motor mode ($I_0>0$).

Initiation of the second turn-on period of T_2 (M9-M10) stabilizes U_{C2} voltage and zero mean inductor current i_L.

Similarly, in the *generator mode* ($I_o<0$), in order to assure an equal charge and discharge of capacitors C_1-C_2, the sequencing of M5-M7 (Fig.5) can stabilize the inductor current i_L mean value over the whole commutation cycle at zero value. From this condition, the zero-voltage period t_{ZVG} becomes depended on the load current I_o. That is,

$$t_{ZVG} = \frac{L}{kU_{dc}}2\left(I_p + |I_o|\right) \qquad (4)$$

B. Auxiliary voltage sources

In this case, constant voltage in C_1-C_2 divider is ensured by two external power supply converters PSU1-2 feeding the capacitors (Fig.6). Energy of the unbalance divider is transferred between capacitors. Zero voltage period is independent of load current value and direction (3). Respecting initiation procedure given by (1), the rate of change of the inverter output voltage is expressed as in

$$\frac{du_f}{dt} = \frac{I_p}{C_f} \qquad (5)$$

Auxiliary voltage source strategy provides stabilization of the output voltage rate and zero voltage periods (Fig.7).

145

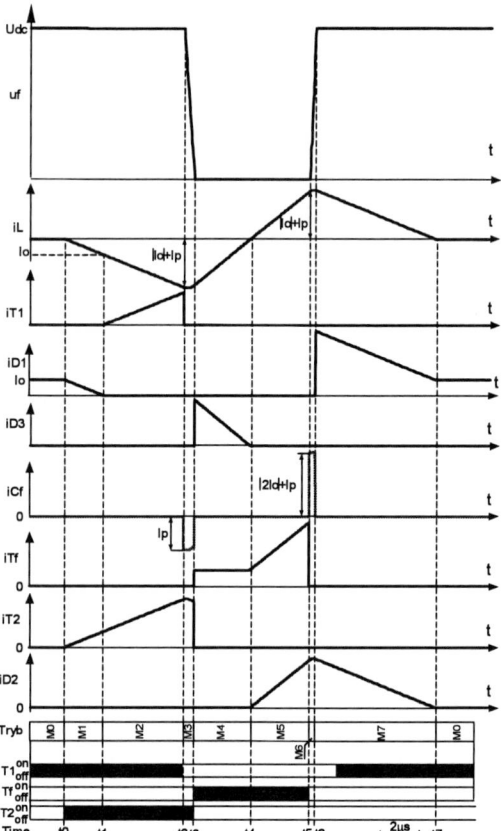

Fig. 5. The PQRDCLI transient waveforms at generator mode (I_0<0).

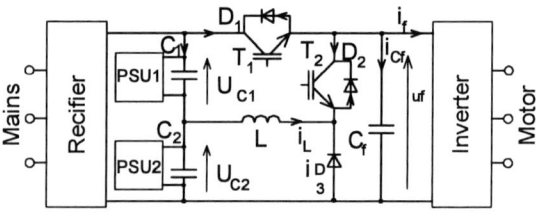

Fig. 6. Quasi-resonant DC-link inverter structure with auxiliary voltage source converters PSU1, PSU2.

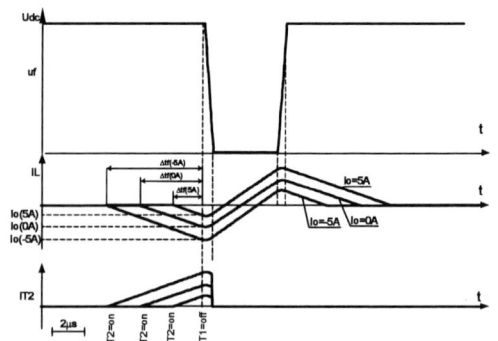

Fig. 7. Simulation results; independent of load current - rate of change of the inverter output voltage.

C. Constant subperiods

In this sensorless strategy, the Δt_f period, the auxiliary transistor T_2 turn-on and the inverter transistors T_f turn-on periods are constant, calculated for rated load conditions. The C_1-C_2 midpoint voltage is floating with its steady-state mean value depended on the transistor T_2/T_f turn-on ratio reference and load current. Subcircuits in Fig. 8 illustrate sequential energy exchange cycles between capacitor bank C_1-C_2 and inductor L. In the first subcircuit of Fig.8a, energy is transferred from the capacitor C_1 to the inductor L. Capacitor voltage slightly decreases. When the transistor T2 is turned-off (Fig.8b), the inductor energy recharges the capacitor C_2. Next, when inverter transistors T_f are turned-on (Fig.8c), the capacitor C_2 discharges, increasing inductor current i_L. Stored inductor energy is transferred back to the capacitor C_1 in subcircuit of Fig.8d.

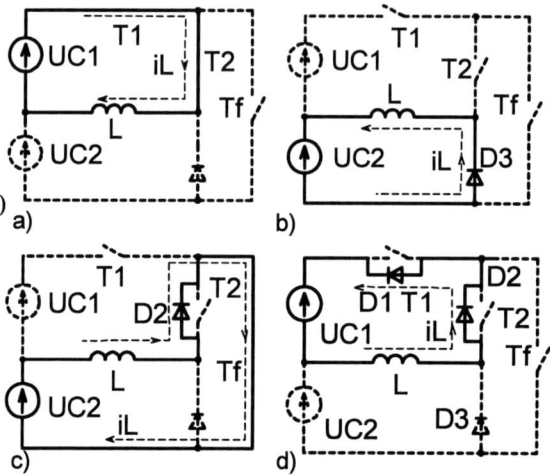

Fig. 8. Quasi-resonant DC-link operation subcircuits;
a), c) active and b) d) freewheel periods.

Applying constant t_{T2on}, t_{Tfon} turn-on periods, the steady state capacitor bank C_1-C_2 midpoint voltage U_{C2} reaches equilibrium potential for the above energy exchange cycles. It corresponds to the inductor current mean zero value (Fig.9). Assuming zero load current I_o, one can prove, the following expression

$$\frac{U_{C2}}{U_{C1}} = \frac{t_{T2on}}{t_{Tfon}} \qquad (6)$$

III. EXPERIMENTAL RESULTS

Experimental unit used for two first strategies consists of the PQRDCLI circuit fed from a double power supply unit $2xUdc$ to enable bi-directional load current tests. Inverter bridge has been replaced by pair of the T_f transistor with the flywheel diode D_f. Load conditions constitute the series inductor L_0 with a half bridge switching leg. Control strategy with constant subperiods has been effectively tested in a second experimental unit based on the three phase PQRDCLI with induction motor load. Control system was implemented by applying fixed-point TMS320F2407A signal processor with gate array logic (GAL) interface.

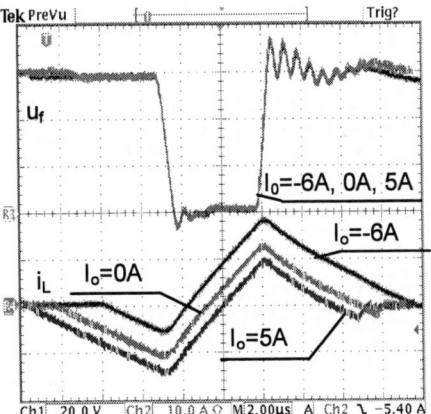

Fig. 9. Auxiliary voltage sources; u_f voltage and i_L current waveforms for subsequent load conditions: I_o=-6A, 0A and 5A

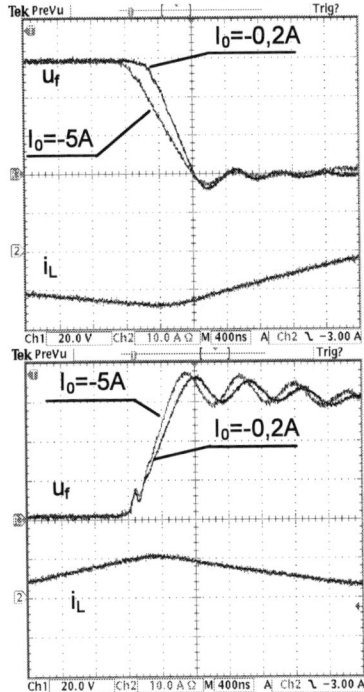

Fig. 10. Constant subperiods; I_o=-0,2A and I_o=-5A, a) fall, b) rise of u_f.

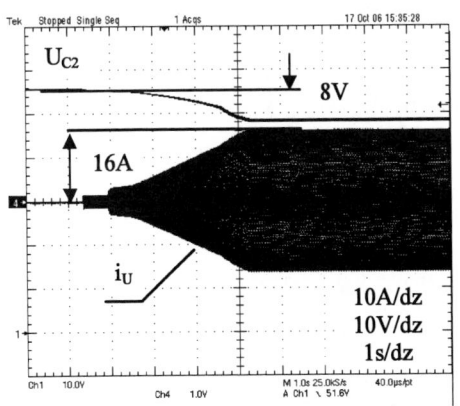

Fig. 11. Constant subperiods - midpoint potential U_{C2} at load condition.

IV. CONCLUSIONS

Three control strategies: A) capacitor voltage control, B) auxiliary voltage sources, C) constant subperiods, have been experimentally verified. Each of these methods ensures quasi-resonant operation providing ZVS/ZCS conditions. Due to resonant rate of change of the inverter voltage, dU/dt has been effectively reduced in all three strategies (~150V/μs). Strategy A guarantees the series capacitor voltage balance and can be advantageous for multilevel inverter topologies. Using strategy B, the inverter voltage is precisely controlled including very low modulation index, occurring at low motor speed. Finally, strategy C, providing attractive sensorless operation, gives floating capacitors C_1-C_2 midpoint potential due to actual load conditions.

REFERENCES

[1] P. Chlebiš, "New concept of vector-modulated quasi-resonant dc link inverter", in *Proc. Int. Conf. on Power Electr. Motion Control EPE-PEMC'00* Košice Slovak Rep. Sept. 2000, vol. 2, pp. 61-64.

[2] J. W. Choi, S. K. Sul, "Resonant link bidirectional power converter: Part I – Resonant circuit", *IEEE Trans. on Ind. Applications*, vol. 10, no. 4, July 1995, pp. 479-484.

[3] D.M. Divan, L. Malesani, P. Tenti, V. Toigo, "A Synchronized DC link converter for soft-switched PWM", *IEEE Trans. on Ind. Appl.*, vol. 29, no. 5 Sept.-Oct. 1993, pp. 940-948.

[4] D.M. Divan, G. Skibinski, "Zero-switching-loss inverters for high-power applications", *IEEE Trans. on Ind. Appl.*, vol. 25, no. 4, July/Aug. 1989, pp. 634-643.

[5] J. He, N. Mohan, B. Wold, "Zero-Voltage-Switching PWM Inverter for high-frequency DC-AC power conversion", *IEEE Trans. on Ind. Appl.* vol.29, no 5 Sept./October 1993, pp. 959-968.

[6] M.Krogemann, J.C.Clare, „A soft switching parallel Quasi Resonant DC-Link Inverter with Modified asynchronous space vector PWM", *Sixth International Conference on Power Electronics and Variable Speed Drives*, Nottingham, UK, IEE Conf. Publ. No. 429, 23-25 September 1996, pp. 208 – 213.

[7] M. Kurokawa, Y. Konishi, M. Nakaoka, "Evaluations of voltage-source soft-switching inverter with single auxiliary resonant snubber" *Electric Power Applications, IEE Proceedings*, Vol.148, No.2 March 2001. Volume 148, Issue 2, Mar 2001, pp. 207 – 213.

[8] S.Mandrek, P.J.Chrzan, "Quasi-resonant dc-link inverter with a reduced number of active elements," *IEEE Transactions on Industrial Electronics*, vol. 54, no. 4 August 2007, pp. 2088-2094.

[9] S.Mandrek, P.J.Chrzan, "Critical evaluation of resonant dc voltage link inverters for electrical drives," *Electr. Power Quality and Utilization*, t. 10 z. 1/2, Dec. 2004, pp. 5-12.

[10] S. Mandrek, P.J. Chrzan, "Indirect frequency converter with quasi-resonant dc link circuit", *(in polish) Patent Application A1 (21)*, P-372473, Polish Patent Office Bulletin 2006, no 16, p.25.

[11] J. Yoshitsugu, M. Ando, E. Hiraki, K. Inoue, M. Nakaoka, „A Consideration on Soft Switching Inverter Using A Single Active Resonant Snubber for AC Motor Drive", *9'th International Conference on Power Electronics and Motion Control*, EPE-PEMC 2000 Košice, Czech Republic, Volume 1, pp. 94-103.

Consideration for Input Current-Ripple of Pulse-link DC-AC Converter for Fuel Cells

Kentaro Fukushima*, Tamotsu Ninomiya†, Masahito Shoyama*,

Isami Norigoe**, Yosuke Harada** and Kenta Tsukakoshi**

* Kyushu University, Fukuoka, Japan, e-mail: *fukushima@ckt.ees.kyushu-u.ac.jp*
† Nagasaki University, Nagasaki, Japan
** EBARA DENSAN, Tokyo, Japan

Abstract—This paper mentions the static characteristics of pulse-link DC-AC converter for fuel cells, and considers the input current-ripple reduction method. Fuel cells have weakness about current-ripple because the chemical reaction time is much slower than commercial frequency. Therefore, the input current-ripple reduction is essential factor in the DC-AC converter for fuel cells applications. Input current-ripple from fuel cells gives damage the fuel consumption and life time. The conventional DC-AC converter has large smoothing capacitor between boost converter stage and PWM converter stage, in order to reduce input current-ripple. That capacitor prevents from reduction the size of unit. Authors have proposed a novel topology called as pulse-link DC-AC converter. The pulse-link DC-AC converter topology is no need to insert large capacitor. Furthermore, the series-connected LC circuit between two stages connected in parallel works as ripple canceling. This paper shows the mechanism of current-ripple reduction.

Keywords—DC-AC Converter, Pulse-link, Fuel Cells, Current-Ripple.

I. BACKGROUND

Environmental issues, such as global heating and energy shortage, have recently become a big international problem. In Japan, about 30% of amount of CO_2 emission is produced by energy conversion department [1]. Therefore, some new clean energy is strongly demanded. One of the new clean energy system using fuel cells has been collecting global interests. When fuel cells generate electricity, a large amount of thermal energy arises at the same time. So, the cogeneration system using both electricity and thermal energy is now researched actively around the world. Especially, a home-use cogeneration system with fuel cells is developed from the stream of what is called distributed power system or micro-grid power systems. Here, the voltage provided by fuel cells is DC, and the power distribution at home is now AC. Therefore, a DC-AC converter is needed for the home-use cogeneration system.

The specifications for DC-AC converter for fuel cells are 3 terms. Firstly, DC-AC converter boosts input voltage from fuel cells to the level of commercial voltage. The voltage from fuel cells is DC, and is generally lower than commercial voltage. Secondly, DC-AC converter has isolation structure between fuel cells stage and load stage from the point of safety. Thirdly, DC-AC converter should reduce input current-ripple. Third-term is special term for fuel cells application. Current-ripple in fuel cells is serious problem. The current-ripple in fuel cells gives damage to the fuel capacity and life span because chemical reaction time is much slower than commercial frequency [2, 3, 4].

In the conventional DC-AC converter for fuel cells, a large smoothing capacitor is inserted in parallel between the boost DC-DC converter and PWM inverter. This capacitor has some rolls. One is to smooth output voltage, another is to reduce input current-ripple. This capacitor absorbs the variation from AC. However, this capacitor disturbs the size reduction of this unit.

To overcome problem, authors has proposed a novel DC-AC converter topology called as Pulse-link DC-AC converter [5, 6]. In this topology, the first-stage boost converter provides a series boosted voltage pulses directly to the second-stage PWM inverter. Therefore, a large capacitor for the smoothed DC power source is not needed. This concept has known as High frequency link or pulse DC link [7, 8]. Furthermore, in order to reduce the current-ripple, a series connected LC circuit is inserted in parallel between two stages.

This paper focuses the mechanism of input current-ripple reduction used by series connected LC circuit. Moreover, the duty ratio of switch is controlled by detecting input current. As the result, input current-ripple is less than 1 Amp.

II. STEADY-STATE ANALYSIS

Fig. 1 shows the proposed circuit topology. As mentioned above, this topology has two stages, and this converter provides boosted pulsed voltage directly to PWM inverter. And between two stages, series LC circuit is connected in parallel in order to reduce current-ripple. The value of the capacitor using this LC circuit is less than the conventional one.

Fig. 2 shows the switching sequences model of this converter in commercial frequency. This converter has 5 switches. Switch Q_1 controls the boost pulse from input voltage. And, from S_1 to S_4 are PWM inverter switches. S_1 and S_4 are controlled to make output voltage sinusoidal waveform, while S_2 and S_3 are decided the plus/minus of output voltage. And control combination of S_1, S_3 and S_2, S_4 is a pair. Q_1 and S_1/S_4 are synchronous at rising time. As the result, there are three states in one switching period.

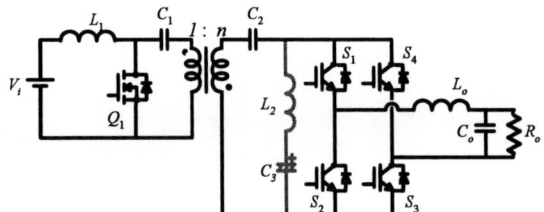

Fig. 1. Pulse-link DC-AC converter topology.

A. Operating State1 (Q_1:ON, S_1:ON, S_3:ON)

Fig. 3(a) shows the equivalent circuit of state 1. State equation on state 1 is written as below:

$$\begin{cases} v_{L_1} = V_i - r_{Q_1}\hat{i}_{L_1} - nr_{Q_1}\hat{i}_{L_2} - nr_{Q_1}\hat{i}_{L_o} \\ v_{L_2} = \hat{v}_{C'} - \hat{v}_{C_3} - nr_{Q_1}\hat{i}_{L_1} - n^2r_{Q_1}\hat{i}_{L_2} - n^2r_{Q_1}\hat{i}_{L_o} \\ v_{L_o} = \hat{v}_{C'} - \hat{v}_{C_o} - nr_{Q_1}\hat{i}_{L_1} + (r_{s1} + r_{s4} - n^2r_{Q_1})\hat{i}_{L_2} - (r_{s1} + r_{s4})\hat{i}_{L_o} \\ i_{C'} = -\hat{i}_{L_2} - \hat{i}_{L_o} \\ i_{C_3} = \hat{i}_{L_2} \\ i_{C_o} = \hat{i}_{L_o} - \dfrac{\hat{v}_{C_o}}{R_o} \end{cases}$$

(1)

Where, $C' = \dfrac{C_1 C_2}{C_1 + n^2 C_2}$.

B. Operating State2 (Q_1:ON, S_1:OFF, S_3:ON)

Fig. 3(b) shows the equivalent circuit of state 2. State equation on state 2 is written as below:

$$\begin{cases} v_{L_1} = V_i - r_{Q_1}\hat{i}_{L_1} - nr_{Q_1}\hat{i}_{L_2} \\ v_{L_2} = \hat{v}_{C'} - \hat{v}_{C_3} - nr_{Q_1}\hat{i}_{L_1} - n^2r_{Q_1}\hat{i}_{L_2} \\ v_{L_o} = -\hat{v}_{C_o} - nr_{Q1} - (r_{s4} + r_{D3})\hat{i}_{L_o} \\ i_{C'} = -\hat{i}_{L_2} \\ i_{C_3} = \hat{i}_{L_2} \\ i_{C_o} = \hat{i}_{L_o} - \dfrac{\hat{v}_{C_o}}{R_o} \end{cases}$$

(2)

C. Operating State3 (Q_1:OFF, S_1:OFF, S_3:ON)

Fig. 3(c) shows the equivalent circuit of state 3. In this state, input power is charged C_2 through the diodes of four PWM switches. State equation on state 3 is written as below:

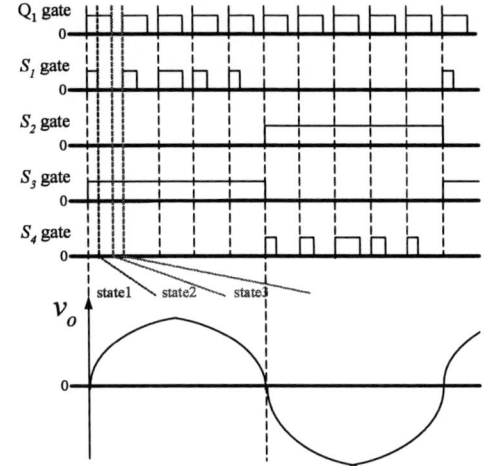

Fig. 2. Switching sequences of the converter.

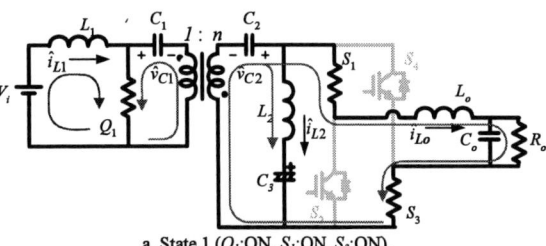

a. State 1 (Q_1:ON, S_1:ON, S_3:ON).

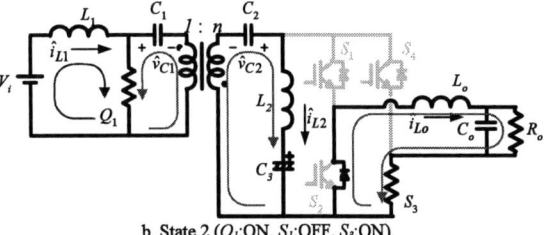

b. State 2 (Q_1:ON, S_1:OFF, S_3:ON).

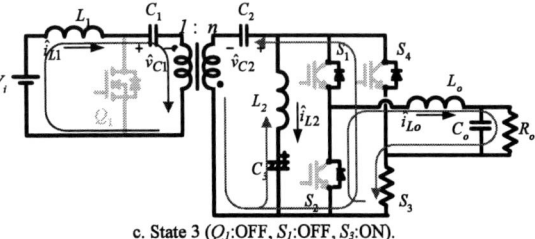

c. State 3 (Q_1:OFF, S_1:OFF, S_3:ON).

Fig. 3. Equivalent circuit of each states.

$$\begin{cases} v_{L_1} = V_i - \dfrac{1}{n}\hat{v}_{C'} - \dfrac{1}{n}\left(r_{D1'} + r_{D2'}\right)\hat{i}_{L1} - \left(r_{D1'} + r_{D2'}\right)\hat{i}_{L_2} \\[2mm] v_{L_2} = -\hat{v}_{C_3} - \dfrac{1}{n}\left(r_{D1'} + r_{D2'}\right)\hat{i}_{L_1} - \left(r_{D1'} + r_{D2'}\right)\hat{i}_{L_2} - r_{D2'}\hat{i}_{L_o} \\[2mm] v_{L_o} = -\hat{v}_{C_o} - \dfrac{1}{n}r_{D1'}\hat{i}_{L_1} - r_{D2'}\hat{i}_{L_2} - \left(r_{D2'} + r_{D3'}\right)\hat{i}_{L_o} \\[2mm] i_{C'} = \dfrac{1}{n}\hat{i}_{L_1} - \hat{i}_{L_2} \\[2mm] i_{C_3} = \hat{i}_{L_2} \\[2mm] i_{C_o} = \hat{i}_{L_o} - \dfrac{\hat{v}_{C_o}}{R_o} \end{cases}$$

$$(3)$$

D. Steady State

From above equations, the state-averaging vector is written below by using state space averaging method. Here, on-resistance and conduction losses are neglected.

$$\frac{dX}{dt} = \begin{bmatrix} 0 & 0 & 0 & -\dfrac{\frac{1}{n}\left(1 - D_{Q1}\right)}{L_1} & 0 & 0 \\[2mm] 0 & 0 & 0 & \dfrac{D_{Q1}}{L_2} & -\dfrac{1}{L_2} & 0 \\[2mm] 0 & 0 & 0 & \dfrac{ds}{L_o} & 0 & -\dfrac{1}{L_o} \\[2mm] \dfrac{\frac{1}{n}\left(1 - D_{Q1}\right)}{C'} & -\dfrac{1}{C'} & -\dfrac{d_s}{C'} & 0 & 0 & 0 \\[2mm] 0 & \dfrac{1}{C_3} & 0 & 0 & 0 & 0 \\[2mm] 0 & 0 & \dfrac{1}{C_o} & 0 & 0 & -\dfrac{1}{C_o R_o} \end{bmatrix} X + \begin{bmatrix} \dfrac{1}{L_1} \\ 0 \\ 0 \\ 0 \\ 0 \\ 0 \end{bmatrix} V_i$$

$$(4)$$

, where $X = \begin{bmatrix} \hat{i}_{L1} & \hat{i}_{L2} & \hat{i}_{Lo} & \hat{v}_{C'} & \hat{v}_{C3} & \hat{v}_{Co} \end{bmatrix}^T$.

From equation (4), the steady-state characteristics are shown below:

$$V_{c'} = \frac{n}{1 - D_{Q1}}V_i \qquad (5)$$

$$V_o = \frac{nd_s(t)}{1 - D_{Q1}}V_i \qquad (6)$$

$$I_{L1} = \frac{d_s^{\,2}}{R_o\left(1 - D_{Q1}\right)^2}V_i \qquad (7)$$

Furthermore, from Equation (5), the peak voltage pulse that is input to PWM inverter (v_{inv_in}) is written below

$$v_{inv\_in} = \frac{n}{1 - D_{Q1}}V_i \qquad (8)$$

Here, D_{Q1} is duty ratio of switch Q_1. And, $d_s(t)$ is duty ratio of PWM inverter switch of S_1/S_4. $d_s(t)$ is changed shown as equation (9) in order to make output voltage to be sinusoidal waveforms.

$$d_s(t) = d_{s1\_max} \cdot \left| \sin\left(2\pi \cdot 50t\right) \right| \qquad (9)$$

TABLE I.
CIRCUIT PAPAREMTER VALUES

Symbol	Description	value
Vi	Input voltage	20[V]
$L1$	Input inductance	400[uH]
$L2$	Middle inductance	1[mH]
LM	Magnetizing inductance	400[uH]
$C1$	primary-side capacitance	3[mF]
$C2$	Secondary-side capacitance	330[uF]
$C3$	Middle capacitance	300[uF]
n	Turn ratio	3
Lo	Output inductance	3[mH]
Co	Outout capacitance	9.4[uF]
fs	Switching frequency	30[kHz]

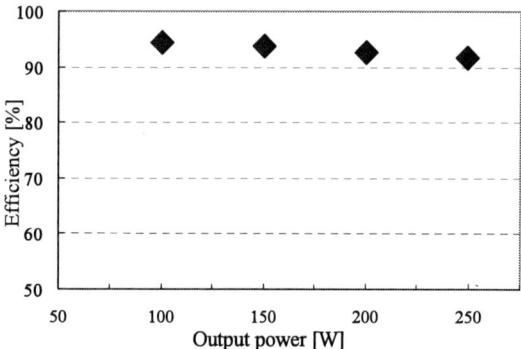

Fig. 4. Characteristics of Efficiency vs. Output power.

Moreover, the relationship of D_{Q1} and d_{s_max} is limited Equation (10), because PWM inverter is provided voltage only when Q_1 is ON.

$$D_{Q1} \geq d_{s1\_max} \qquad (10)$$

From this limit, pulse output voltage is regarded as constant voltage viewing from PWM inverter.

III. EXPERIMENTAL RESULTS

A. Circuit parameter values

To evaluate the performance of the circuit, the experimental circuit is implemented with the specifications and parameters in Table I. From table I, C_1 is 3[mF], and it is aluminum electrolytic capacitor. C_1 is decided from the allowable current. Primary-side is flown large current, so capacitance of C_1 becomes large value. However, primary-side is low voltage, so the size of aluminum electrolytic capacitor is not so large even if the value is large because withstand-voltage is low. Therefore, large value of aluminum electrolytic capacitor is used at C_1 in this experiment.

150

B. Characteristics of Efficiency

Fig. 4 shows the characteristics of efficiency vs. output power. Here, it is measured when output voltage is regulated $100 \pm 1 [V_{rms}]$. From fig. 4, it is considered that the efficiency is more than 90 [%]. This topology can convert DC to AC with high efficiency.

C. Waveforms of output voltage and input current

Fig. 5 shows the experimental waveforms of output voltage (v_o), input current (i_i), and inductor current of L_2 (i_{L2}) when output power is 100[W]. And table II shows the experimental measurement. From those results, it is considered that output voltage is achieved to output commercial voltage. Furthermore, it is considered that inductor current of i_{L2} oscillates low frequently with zero crossing.

IV. INPUT CURRENT-RIPPLE

Steady-state equation of inductor current of L_2 is equation (7), and shown below, again.

$$I_{L1} = \frac{d_s^2}{R_o(1-D_{Q1})^2} V_i$$

When substitutes equation (9) for above equation, I_{L2} is written below:

$$I_{L1} = \frac{d_{s\_max}^2 (1-\cos(2\pi \cdot 100t))}{2R_o(1-D_{Q1})^2} V_i \qquad (11)$$

Input current is equal to I_{L1}. From this equation, it is considered that input current oscillates low frequency as double commercial frequency.

For the comparison, the experimental waveforms when series L_2C_3 circuit is not inserted are shown in fig. 6. The output load condition is same. When L_2C_3 circuit is not inserted in the topology, input current-ripple is 6.11[A_{p_p}]. On the other hand, in the case when L_2C_3 circuit is inserted, input current-ripple is reduced at 4[A_{p_p}] from fig. 5. And it is sure that efficiency is not depended whether series L_2C_3 circuit or not. From the result, it is considered that series L_2C_3 circuit is effective about input current-ripple reduction. However, it is not enough to reduce input current ripple. Here is mentioned reduction methods of input current- ripple.

A. Optimization of the series L_2C_3 circuit parameter values

Input current-ripple is occurred by output commercial frequency of output voltage. If the resonant frequency of series L_2C_3 circuit connected the circuit in parallel is synchronized 100[Hz], input current-ripple will be reduced. Impedance of series L_2C_3 circuit is written below equation:

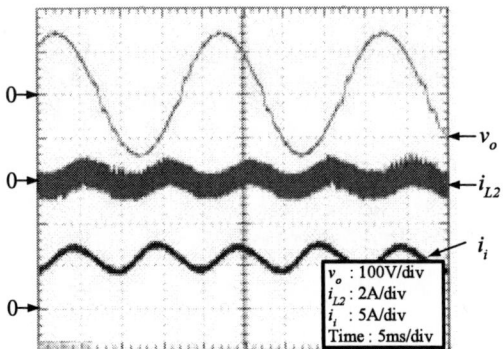

Fig. 5. Experimental waveforms of v_o, i_i, and i_{L2}.

TABLE II.
EXPERIMENTAL MEASUREMENT AT P_o=100[W]

Symbol	Description	value
Vi	Input voltage	20[V]
Ii	Input current	5.3[A]
Vo	Output voltage	100[V(rms)]
P_o	Output resistance	100[W]
η	Efficiency	94[%]

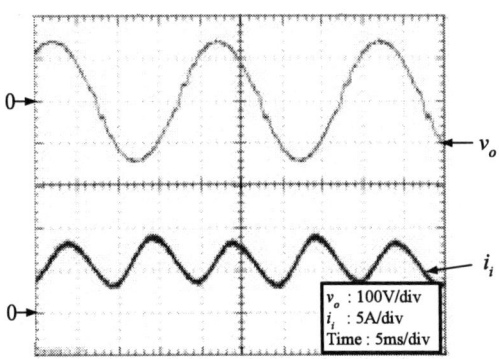

Fig. 6. Experimental waveforms of v_o, and i_i (without L_2C_3).

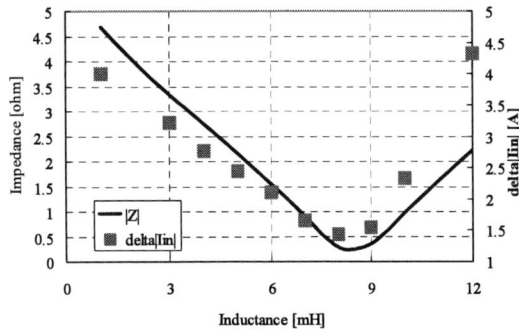

Fig. 7. Inductance values vs. Impedance and Current-ripple.

$$|Z| = \left| j\left(\omega L - \frac{1}{\omega C}\right)\right| \qquad (12)$$

Here, inductance is being changed because capacitor is not recommended large capacitance value.

Fig. 7 shows the characteristics of series L_2C_3 impedance and input current-ripple measurement by changing of L_2. Here, the value of C_3 is 300[uF], and $\omega = 2 \cdot \pi \cdot 100$ [rad/s]. From fig. 8, it is considered that impedance $|Z|$ curve and experimental measurement of input current-ripple is agreed well.

Furthermore, the experimental waveforms at L_2=8[mH] is shown in Fig. 8. From Fig. 8, inductor current of L_2 is oscillated with opposite phase of output semi-sinusoidal voltage. This thing means that when output load is light, extra energy from input power is stored by series L_2C_3, and when output power is heavy, series L_2C_3 provides with input power. From the result, this series L_2C_3 circuit is regarded as pulse energy tank, and works as ripple canceling circuit.

B. Sensed current and controlled duty ratio of switch

To reduce input-current-ripple further, D_{Q1} is controlled by detecting input current. In the experiment, FPGA is used and it controls with A/D converter shown Fig. 9. In the experiment, current sensor detects input current. 1[A] is converted to 0.125[V] used by current sensor. And the converted voltage is input to A/D converter. 0.01[V] is corresponding to 1 binary data at A/D converter. The duty ratio signal of switch Q_1 ($Q_{1signal}$) which is binary data is calculated by below equation:

$$Q_{1signal} = Q_{1signa\_ref} + k\left(I_{in} - I_{ref}\right) \qquad (13)$$

, where $Q_{1signal\_ref}$ is corresponding to the binary data that duty ratio of switch Q_1 (D_{Q1}) is 0.7, and I_{ref} is converted reference input current to binary data.

Fig. 10 shows the experimental waveforms of output v_o, i_i, and i_{L2} when D_{Q1} is controlled by detecting input current. From fig. 10, it is observed that input current i_i is almost canceled the low frequency ripple. Ripple is less than 1[A].

V. CONCLUSION

This paper analyzed Pulse-link DC-AC converter for fuel cells and considered input current-ripple. This converter provides the boosted voltage pulses directly to the PWM inverter, and L_2 and C_3 circuit is inserted to reduce input-current ripple. This topology converts DC to AC with high efficiency. Furthermore, the mechanism of input current-ripple is cleared and input current-ripple reduction methods are shown. Input current-ripple is caused by commercial frequency.

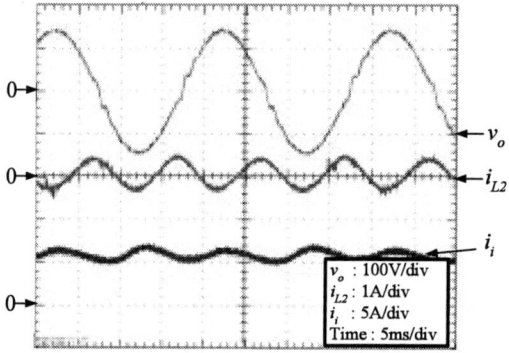

Fig. 8. Experimental waveforms of v_o, i_i, and i_{L2} (L_2=8mH).

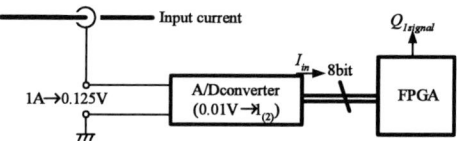

Fig. 9. Current sensing component block.

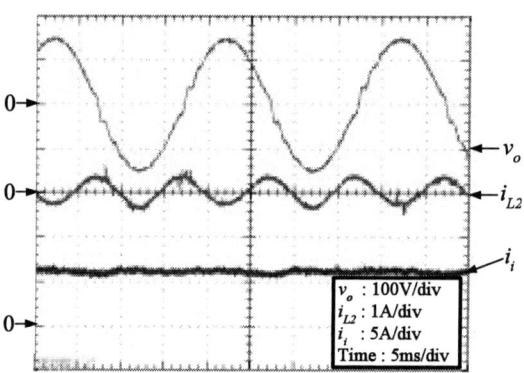

Fig. 10. Experimental waveforms of v_o, i_i, and i_{L2}.
(with input current sensed)

Series L_2 and C_3 circuit works as current ripple canceling if parameters are optimized as resonant condition. Moreover, the method which is sensed input current of this converter and controlled D_{Q1} is shown to cancel input-current ripple. When the control is worked, the input-current ripple is improved to be reduced less than 1[A].

REFERENCES

[1] "The GHGs Emissions Data of Japan", Greenhouse Gas Inventory Office of Japan, by way of Japan Center for Climate Change Actions, http://www.jccca.org/

[2] S. Moon, J. Lai, S. Park and C. Liu, "Impact of SOFC Fuel Cell Source Impedance on Low Frequency AC Ripple," Power Electronics Specialists Conference, Proc. of IEE PESC 2006, pp.2037-2042, Jun. 2006.

[3] W. Choi, P.N. Enjeti and J.W. Howze, "Development of an Equivalent Circuit Model of a Fuel Cell to Evaluate the Effects of Inverter Ripple Current," Proc. of IEEE APEC 2004, pp. 255-361, Feb. 2004.

[4] G. Fontes, C. Turpin, R. Saiset, T. Meynard, and S. Astier, "Interactions between fuel cells and power converters Influence of current harmonics on a fuel cell stack," Proc. of PESC 2004, pp. 4729-4735, 2004.

[5] K. Fukushima, T. Ninomiya, S. Abe, I. Norigoe, Y. Harada, K. Tsukakoshi, and Z. Dai, "Steady-State Characteristics of a novel DC-AC Converter for Fuel Cells," Proc. of IEEE INTELEC 2007, pp.904-908, 2007.

[6] K. Fukushima, T. Ninomiya, I. Norigoe, Y. Harada, K. Tsukakoshi, and Z. Dai, "Characteristics of a Pulse-link Inverter for Fuel Cells," Proc. of ICPE'07, pp.1066-1070, 2007.

[7] P. T. Krein, R. S. Balog, and X. Geng, "High-Frequency Link Inverter for Fuel Cells Based on Multiple-Carrier PWM," IEEE Transaction on PE, Vol. 19, No. 5, pp. 1279-1288, Sep. 2004.

[8] D. Chen and L. Li, "Novel Static Inverters With High Frequency Pulse DC Link," IEEE Transaction on PE, Vol. 19, No. 4, pp. 971-978, Jul. 2004.

New Practical Approach to Input Current Shaping in AC-DC Power Converters

Kuno Janson, Viktor Bolgov, Lauri Kütt, Ants Kallaste, Heigo Mõlder

Tallinn University of Technology/Department of Fundamentals of Electrical Engineering and Electrical machines,
Tallinn, Estonia, e-mail: *kunojanson@staff.ttu.ee*

Abstract— In principle, a distortion of a waveform shape of AC/DC converter current could be corrected by use of a fast-controlled impedance-matching transformer. Changes in the converter input impedance can be eliminated by means of the varying transformation ratio of this transformer. The paper considers a practical performance of the fast-controlled impedance-matching transformer on a basis of a non-conducting converter. It is found that the shape of current waveform can be made sinusoidal if the instant power passing through a converter is proportional to the squared instant value of the supply voltage and/or the output voltage and the output current of the converter are in inverse proportion. A non-conducting converter with alternating of parallel and series resonance meeting requirements is considered. The converter topology, an operating principle and output voltage control are described. The basics of the converter calculation and the converter current waveform modeled on computer are presented.

Keywords— AC-DC power resonant converter with parallel and series resonances, virtual impedance-matching transformer, power factor correction, power conversion harmonics, current waveform shaping.

I. Introduction

A lot of solutions were proposed to correct a power factor of AC/DC converters. The power factor equal to unity can be obtained by means of a controlled additional converter [1]. The additional converter can be used almost always, but it is relatively complicated, expensive and decreases power efficiency. The improving of the main converter or its control circuit is cheaper, although this way cannot provide the power factor equal to unity. This way is not also universal. In many cases, a partial correction of the power factor with simpler devices is the most expedient [2]. Passive correction circuits are enough simple. One of those is the converter with alternating of parallel and series resonances (PSA converter) [3]. The PSA converter topology contains an inductor and a capacitor, which are oversized and expensive at the utility frequency. The PSA converter can also operate at higher frequencies comparing to mains frequency. In that case, the inductor and the capacitor are relatively small and cheap. The output voltage and the output current of the PSA converter are approximately in inverse proportion at maximum power. That is why changes in load impedance do not noticeably change the power supplied to the load. The load power is approximately proportional to the squared input voltage of the PSA converter. These properties allow passive correction of the power factor [4]. It should be possible to create a PSA converter operating

at higher frequency that corrects the power factor passively.

II. High Frequency Converter with Alternating of Parallel and Series Resonance

A PSA converter on basis of diode rectifier operating at power line frequency is shown in Fig. 1. The converter includes a transformer, which has two secondary windings in series connection (W_2 and W_3). The winding W_2 and the inductor L are operating like an inductive phase-shifting circuit, which current lags an input voltage. The winding W_3 and the capacitor C are combined into another phase-shifting circuit with a current leading an input voltage. A coupling point m of the windings W_2 and W_3 is directly connected to the diode rectifier. The output current and power of the PSA converter against the output voltage are given in Fig. 2.

These curves are obtained at the sinusoidal input voltages $U_s=1.0$ and $U_s=0.5$. The reactances of the inductor L and the reactance of the capacitor C determine a maximal current of the PSA converter. If the voltage decreases twice than the currents are also reduced twice and the power is diminished four times. That is why the maximum of the power curve $P_{S0.5}$ (Fig. 2) obtained at twice reduced voltage is four times less comparing to the one of the curve P_S gained at rated voltage ($E_{0.5}$ and E points). Thus, a maximal power of a PSA converter is proportional to a squared input voltage and the PSA converter meets the first requirement for the current shaping by a passive converter. Voltage drops over rectifier diodes must be taken into account at voltages lower than 10 V. In that case, the squared relationship for the power is affected a little bit.

If the operational point of the converter moves from point A to point B on the curve I_d (Fig. 2) then the power of the converter behaves according to the *CED* curve. The

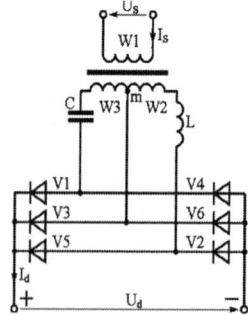

Figure 1. Passive converter with alternating of parallel and series resonance.

978-1-4244-1741-4/08/$25.00 ©2008 IEEE   154

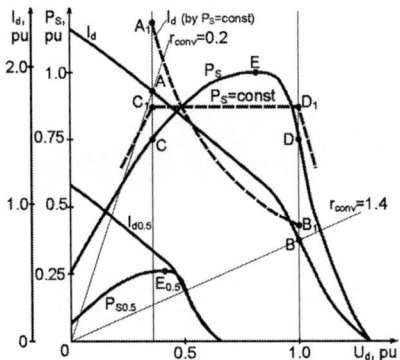

Figure 2. PSA converter: output current I_d and power P_S vs. output voltage U_S.

powers corresponding to the points C and D are by 30% less than the power at the E point. The average power over the span is shown as a horizontal dashed line between points C_1 and D_1. If we need a constant power between the points C_1 and D_1 then the output current has to vary according to an exponential curve between the points A_1 and B_1. If there is the constant power between the points C_1 and D_1 then a load resistance r_{conv} can vary from 0.2 to 1.4 and the input impedance of the converter is constant in relation to mains. The constant input impedance is a sufficient condition for obtaining the sinusoidal shape of the supplied current waveform. As the curve of the output current in the consider domain (the curve AB) differs from the exponential one A_1B_1 corresponding to the constant power, there is some distortion of the current waveform at abruptly and deeply variable loads.

Considering PSA converter behavior, one should take into account the fact that its operational point has to move between points A and B (Fig. 2) during ¼ of a supply voltage cycle. To meet this requirement, an operation frequency of a PSA converter has to be higher than mains frequency. The higher frequency is obtained by an inverter connected before the PSA converter (Fig. 3).

The increased frequency gives another useful benefit. The increased frequency allows to overcome one of the main disadvantages of a PSA converter operating at the utility frequency, namely the high weight, size and cost of the inductor L and the capacitor C_1 can be reduced. The frequency of the inverter must be nearly constant during operation, otherwise reactances of the L and C_1 elements are not equal anymore, but that condition is of importance for proper operation of the PSA converter. The supply voltage of the inverter is not a constant DC voltage as usually used; it pulsates between zero and a maximal value of the mains voltage. If the operational frequency of the inverter is quite high in comparison with the utility

frequency, then the output voltage of the inverter is a square-wave (a meander), which amplitude is modulated according to instant values of the mains voltage. At a maximum of the mains voltage, the inverter supplies a maximal square-wave voltage $U_{invpr90}$ to the transformer T. In the same time, the secondary voltage of the transformer is as follows

$$U_{invse90} = U_{invpr90}/K_T. \qquad (1)$$

where K_T is the transformation ratio of the transformer; $K_T = w_1/(w_2 + w_3)$. Under idealized conditions, the value $U_{invpr90}$ is equal to the amplitude of the mains voltage $u_S(t)$

$$U_{invpr90} = \sqrt{2}\ U_s. \qquad (2)$$

Behavior of a PSA converter differs at the square-wave supply voltage in relation to the sinusoidal one (Fig. 4).

Power characteristics are obtained at equal amplitudes of the square-wave voltage (curve P_{PSAme}) and the sinusoidal one (curve P_{PSAS}). The powers are given in p.u. as related to the reference power S_B used in the PSA converters. The reference power is defined as a sum of the reactive powers in the capacitive and inductive branches at a sinusoidal supply voltage and shorted input terminals of the rectifier bridge

$$S_B = U_{w2}\ I_{LK} + U_{w3}I_{CK}, \qquad (3)$$

where U_{w2} and U_{w3} are the voltages over the secondary windings of the transformer and I_{LK} and I_{CK} are the short-circuit currents of those windings. In the considered case, we can choose $U_{w2} = U_{w3}$ and $I_{LK} = I_{CK}$, i.e. the total reactance of the inductive branch $x_{2\Sigma}$ is equal to the total reactance of the capacitive branch $x_{3\Sigma}$ and we can use the same reactance $x = x_{2\Sigma} = x_{3\Sigma}$ for the both branches. If we take into account $I_{LK} = U_{w2}/x$ and $I_{CK} = U_{w3}/x$, then the reference power can be found as follows

$$S_B = U^2_{w2+w3}/2x, \qquad (4)$$

where U_{w2+w3} is the total secondary voltage of the transformer.

The total reactance of the inductive branch is

$$x_{2\Sigma} = x_S + x_L, \qquad (5)$$

where x_S is the leakage reactance of the transformer and x_L is the reactance of the inductor at a sinusoidal voltage of an inverter frequency.

The total reactance of the capacitive branch is

$$x_{3\Sigma} = x_C - x_S, \qquad (6)$$

where x_C is the reactance of the capacitor also obtained at a sinusoidal voltage of an inverter frequency.

A top point of the power curve $P_{PSAme} = f(U_d)$ of a PSA converter is usually called rated operational point. Fig. 4 shows the curve under the rated power $P^*_{PSAmen} = 1.38$ p.u. and the rated voltage $U^*_{dPSAn} = 0.75$ p.u. That rated power in p.u. is given in relation to the reference power S_B.

$$P^*_{PSAmen} = \frac{P_{PSAmen}}{S_B}, \qquad (7)$$

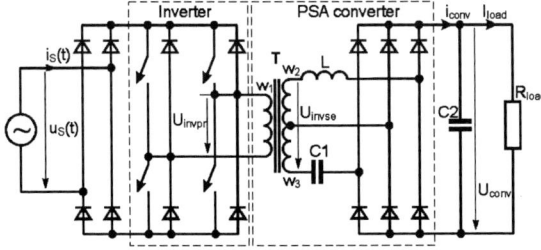

Figure 3. PSA converter: output current I_d and power P_S vs. output voltage U_S.

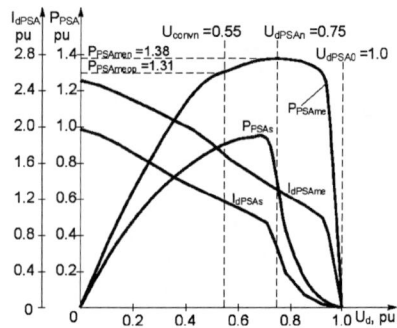

Figure 4. PSA converter behavior at sinusoidal supply voltage (P_{PSAs}, I_{dPSAs}) and square-wave one (P_{PSAme}, I_{dPSAme}) with equal amplitudes.

where P_{PSAmen} is the actual power of the converter obtained from the converter simulation (or measuring) results. Equations (4) and (7) give

$$P_{PSAmen} = U_{w2+w3}^2 \cdot P_{PSAmen}^* / 2x , \qquad (8)$$

where U_{w2+w3} is the effective value of a sinusoidal voltage, which amplitude is equal to the one of the square-wave voltage $U_{invse90}$ used to define P_{PSAmen}^*. It means $\sqrt{2}U_{w2+w3} = U_{invse90}$ and (8) can be rewritten as follows:

$$P_{PSAmen90} = U_{invse90}^2 \cdot P_{PSAmen}^* / 4x . \qquad (9)$$

The power $P_{PSAmen90}$ in (9) is defined by help of two voltages. Firstly, the supply voltage $U_{invse90}$ corresponds to amplitude of a mains voltage and, secondly, the output voltage is the rated voltage of the PSA converter U_{dPSAn} (Fig. 4). Since the supply voltage of the PSA converter of the considered topology varies sinusoidally according to the mains voltage $U_{invse}(t)=U_{invse90}\sin\omega t$ then the power of the PSA converter varies in time in the same way

$$P_{PSAmen}(t) = U_{invvse90}^2 \cdot \sin\omega t \cdot P_{PSAmen}^* / 4x . \qquad (10)$$

Besides the power, the open-circuit voltage of the PSA converter U_{dPSA0} also varies with time. The open-circuit voltage in Fig. 4 corresponds to a time instant when an instant value of the mains voltage is maximal. As a first approximation, the open-circuit voltage is equal to the supply voltage of the PSA converter

$$U_{dPSA0}(t)=U_{invse}(t). \qquad (11)$$

Since the voltage U_{invse} supplied to the considered circuit is sinusoidal in time then the open-circuit voltage varies also in time according to sinusoidal law

$$U_{dPSA0}(t) = U_{invse90} \cdot \sin\omega t . \qquad (12)$$

The power curves of the PSA converter for different time instants (from p_{conv90} to $p_{conv18.4}$) calculated by help of (10) and (12) are shown in Fig. 5b. Besides, curves of the output current of the converter are also given there. The curves of the output current are found taking into account that, at a short-circuit, the output current of the PSA converter varies in time together with the mains voltage according to sinusoidal law.

The characteristics of the PSA converter obtained above allow to analyze an operation of the power supply shown in Fig. 3. In the studied case, the curves of instant

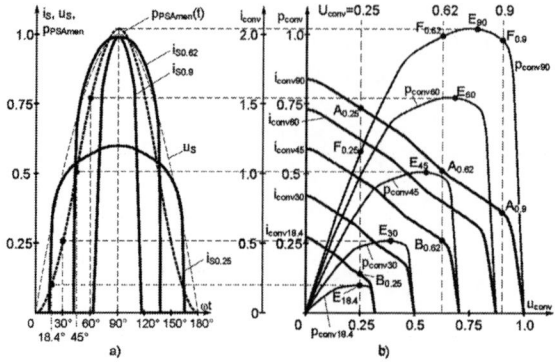

Figure 5. Curves of rated power, mains voltage and mains current of PSA converter (a) and changes of operational point (b) at output voltage equal to 0.25, 0.62 and 0.9 of open-circuit voltage.

power are not horizontal straight lines as it is required to get a sinusoidal waveform of the mains current, but they begins and ends at the zero power level. That is why the shape of the mains current waveform depends on a value of the output voltage U_{conv} at which the considered power supply operating (more exactly, on a voltage over the capacitor C_2). The output voltage can be chosen to provide a cross-point of the power curve p_{conv90} and a vertical line corresponding to the output voltage near the maximal power (for example, point $F_{0.62}$ under voltage $U_{conv}=0.62$). However the cross-point (operational point) can be chosen at a lower power (e.g. point $F_{0.25}$). As seen, the maximum value of the throughput power of the converter (at an instant of the maximal mains voltage) depends on the chosen output voltage. If the maximum value of the power decreases then the average power and the output power of the rectifier also go down.

If the output voltage of the converter varies then, besides the power, the shape of the mains current waveform does the same. The curves of the mains current $i_s(t)$ in Fig. 5a are calculated by dividing the instant power found in Fig. 5b with instant values of the mains voltage $u_s(t)$. The vertical straight line $U_{conv}=0.62$ crosses over the curves of the mains current i_{conv90} (point $A_{0.62}$), i_{conv60} and i_{conv45} (point $B_{0.62}$) but do not over the curves i_{conv30} and $i_{conv18.4}$. That means there is no current drawn from mains over a span 0°...30°. The mains current is consumed between nearly 30° and 45° and its waveform is distorted ($i_{s0.62}$ in Fig. 5a). If zero points of the curves p_{conv30} and $p_{conv18.4}$ were obtained at an output voltage higher than 0.62 (these power curves would be wider) then such distortion would not happen. At the output voltage $U_{conv}=0.9$, there is no mains current within the span 0°...60°, the current arises between nearly 60° and 90°. In this case, the distortion is even higher ($i_{s0.9}$). The distortion pattern of the current waveform is other at lower output voltages. At $U_{conv}=0.25$, the mains current arises already before 18.4°. However, the cross-point $F_{0.25}$ of the power curve p_{conv90} and the direct line corresponding to the output voltage are much lower than maximal power level (point E_{90}). That is why the instant power is lower and the waveform of the current $i_{s0.25}$ is flattened out from the top at the time instant when the mains voltage is maximal. The distortion of the current waveform substantially depends on a relationship between the rated voltage and the open-circuit voltage of the power supply, or the relative rated voltage U_{convn}^*. The open-circuit voltage of

156

the power supply is equal to the one of the PSA converter $U_{dPSA0/90}$ at the time instant when the mains voltage is maximal. Thus, the relative rated voltage is

$$U^*_{convn} = U_{convn} / U_{dPSA0/90}. \qquad (13)$$

One can conclude from Fig. 5 that, in the case of a usual non-controlled PSA converter, the power curve segment with a constant power is too narrow to provide a mains current waveform with very low distortion. That segment is several times larger for PSA converters with control of the capacitive branch by means of by-pass thyristors [5].

III. Voltage Stabilization for Power Supply of PSA Converter and General Characterization of PSA Converter

The idealized passive converter studied above and also the PSA converter are supposed to provide more or less constant output power. That means load changes unavoidably results in voltage variations. It is a useful feature in some cases, for example in the power supply of the electric arc. However, the constant output voltage of a rectifier is usually required and the voltage level must frequently be controlled. At the same time, the load resistance can be changed from infinity to some minimal value. If these requirements are satisfied then the output power of a rectifier varies.

If we need to keep a constant voltage of a rectifier by help of a PSA converter at a changing load then the power of the converter has to be varied. That means, inter alia, that the amplitude of the instant power p_{conv90} has to be changed at the maximum of the instant mains voltage (point E_{90} in Fig. 5b). According to (9), this power can be varied by changing the supply voltage of the PSA converter. The voltage can be changed by applying a pulse-width modulation to the inverter. Fig. 6 demonstrates the curves of the output current and power of a PSA converter when the duty ratio of the inverter square-waves is changed within a span D=1.0 ... 0.1.

As a first approximation, the current and power are proportional to the duty cycle. The open-circuit voltage of the converter does not substantially change at the variation of the duty cycle. If a feedback control circuit is used that decreases the duty cycle at an increase of the rectifier voltage then the rectifier voltage can be stabilized [6].

Rectifier designing requires the rated voltage U_{convn}, the rated current I_{loadn} and therewith the rated power P_{convn} and rated load resistance R_{loadn} to be pre-given. The mains voltage U_s is also pre-given. If the relative rated voltage of

the rectifier U^*_{convn} is chosen (it could be approximately 0.5 ... 0.65) then the transformation ratio of transformer can be found by help of (1), (2), (11) and (12)

$$K_T = \frac{\sqrt{2}U_S \cdot U^*_{convn}}{U_{convn}}. \qquad (14)$$

Besides the finding of the relative rated voltage, the power curve of a PSA converter (Fig. 4) allows to determine the operational point power $P_{PSAmeop}$ that is a bit lower than the rated power of the PSA converter P_{PSAmen}. This power decrease can be taken into account by the factor of the top power decrease

$$K_{ptop} = P_{PSAmeop} / P_{PSAmen}. \qquad (15)$$

At the time of the maximal mains voltage, the instant power of the idealized converter is equal to the doubled average power over a half-cycle of a sinusoid curve. If we assume that the average power of the converter is equal to the load power and the PSA converter is idealized one then

$$P_{PSAmeop90} = 2P_{convn}. \qquad (16)$$

At the time of the maximal mains voltage, (16) can be rewritten by help of (15)

$$P_{PSAmen90} = 2P_{convn} / K_{ptop}. \qquad (17)$$

The reactances of the PSA converter branches $x = x_{2\Sigma} = x_{3\Sigma}$ can be found by means of (9), (11), (13) and (17)

$$x = \frac{R_{loadn} \cdot P^*_{PSAmen} \cdot K_{ptop}}{8(U^*_{convn})^2}. \qquad (18)$$

If the reactances of leakage inductors are pre-set then the capacity of the capacitor C_1 and inductivity of the inductor L can be determined by use of (18), (5), (6) and the chosen inverter operational frequency.

When the parameters of the transformer T, inductor L and capacitor C_1 are determined then the rectifier shown in Fig. 3 can be simulated with a computer. Equations (14) and (18) are found without taking into account any losses in the circuit and the distinction of a PSA converter from an idealized converter. That is why the transformer, inductor and capacitor parameters have to be corrected according to the simulation results.

IV. Results of Power Supply Simulation

The next parameters of the power supply were pre-given for the simulation: the rated voltage U_{convn}=33 V, the rated current I_{loadn}=50 A and the mains voltage U_s=230 V. The relationship between the rated voltage and the open-circuit voltage (U^*_{convn}) of 0.55 was chosen giving P^*_{PSAmen}=1.38; K_{ptop}=0.94 according to Fig. 4. The total reactance of the PSA converter branches x=0.3539 Ω was calculated basing on that data and (18), as well as the transformation ratio K_T=5.421 was determined by help of (14). The leakage reactance values of the both transformer secondary windings were chosen as a half of the total reactance of a branch (x_S=x/2). Thus, the reactance of the capacitor is x_C=x+0.5x=0.5309 Ω and its capacity is

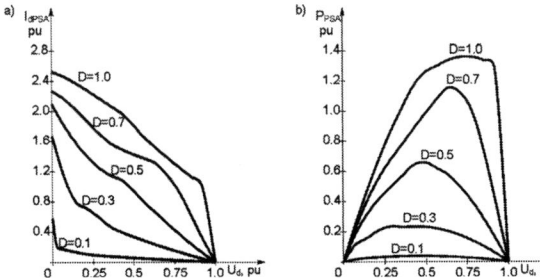

Figure 6. Curves of output current (a) and output power (b) at changes of duty cycle D of square-waves supplied to converter.

9.994 µF at an inverter frequency of 30 kHz. In the same way is found for the inductive branch that a sum of leakage and inductor inductances is $L_S + L_1 = 1.877$ µH.

The simulation of the power supply with the pre-given parameters by means of PSCAD program gave the load current $I_{load} = 41$ A and voltage $U_{conv} = 27$ V at the rated load $R_{loadn} = 0.66$ ☐ and the smoothing capacitor C_2 with the capacity of 50000 µF. The calculated values of the voltage and current are less than pre-given ones. That is because (14) and (18) are derived for lossless converter. The pre-given values can be obtained by use of iterative matching if we correct either the total reactance x or the transformation ratio K_T or both of them. In the considered case, if we boost the secondary voltage of the transformer in proportion with the ratio of the pre-given and calculated voltages then the new transformation ratio of the transformer

$$K_{T1} = K_T \cdot \frac{U_{conv}}{U_{convn}} = 5.421 \frac{27}{33} = 4.435 . \quad (19)$$

When the new transformation ratio is used then the voltage and the current of the power supply differ from the pre-given values by less than 1%.

The shape of the mains current waveform is shown in Fig. 7 while the voltage of the emf source modeling the load of the power supply is changing from zero to open-circuit value. If the voltage is zero, i.e. the load circuit is shorted, then the mains current is of the lowest distortion. The distortion rises as the voltage increases. That is fundamentally in agreement with the structure shown in Fig. 5. The distortion factor of the mains current (THD$_i$) is of 25...30% under rated operational conditions. That is 3 times less than in the case of an usual diode rectifier with a capacitive output [7]. The phase shift between a current and voltage is not seen in Fig. 7. The power factor is nearly 0.96 under rated operational conditions. The short-circuit mains current is less than the rated one because of PSA converter capability to restrict a current. That is why short-circuit proof of the power supply can generally be provided without use of current restricting feed-back loops or protection devices.

V. CONCLUSIONS AND SUMMARY

1) The current consumed by a diode rectifier can basically be made sinusoidal by means of a converter interconnected between a rectifier bridge and smoothing filter that keeps its throughput power in proportion to a squared instant power and provides inverse proportion between an output current and output voltage. This converter substituted for a virtual transformer could be a passive converter without control circuits. A non-controlled converter with alternating of parallel and series resonance (PSA converter) connected behind an inverter can approximately function as a passive converter.

2) A constant power segment of power curve of a usual non-controlled PSA converter is not wide enough to avoid the distortion of the mains current waveform at low instant values of a mains voltage. That distortion is maximal under open-circuit conditions. The distortion diminishes when the load current rises and the output voltage consequently decreases. Total harmonic distortion of the mains current of the single-phase rectifier is of 25...30%.

3) The output voltage of the power supply of the inverter and the PSA converter goes substantially down under impact of the output current. The short-circuit current of the power supply is 1.5 ... 1.8 times the rated one if no auxiliary tools are used. That is useful, e.g. for a power supply of electric arc. The output voltage of the power supply can also be stabilized if a pulse-width modulation of the inverter and proper control fed-back loops are used. At the same time, a possible rise of the output voltage of the power supply is avoided if the duty ratio of the output voltage square-waves is decreased.

4) As compared to a usual widespread pulsed rectifier, the rectifier topology with a PSA converter has an additional capacitor and inductor, as well as six diodes instead of two high frequency rectifying diodes. The use of the modest tools allows decreasing 3 times distortion of the mains current and automatically restricting currents in all circuit elements to rated levels at shorted output of the rectifier. The changes in a topology do not substantially affect a power factor.

REFERENCES

[1] S. Busquets-Monge, J.-C. Crebier, S. Ragon, E. Hertz, D. Boroyevich, Z. Gürdal, A. Arpilliere, and D. K. Linder, "Design of a boost power factor correction converter using optimization techniques," *IEEE Trans. Power Electron.*, vol. 19, pp. 1388-1396, Nov. 2004.

[2] H.-F. Liu and L.-K. Chang, "Flexible and low cost design for a flyback AC/DC converter with harmonic current correction," *IEEE Trans. Power Electron.*, vol. 20, pp. 17-24, Jan. 2005.

[3] K. Janson and J. Järvik, "AC-DC converter with parametric reactive power compensation," *IEEE Trans. Ind. Electron.*, vol. 46, pp. 554-562, June 1999.

[4] K. Janson, V. Bolgov, T. Vinnal, and J. Järvik, "New theoretical approach to input current shaping in ac-dc power converters," in *Proc. 5th Int. Conf. Compatibility in Power Electronics*, CD-ROM, May - June 2007.

[5] K. Janson, J. Järvik, E. Sepping and J. Shklovski, "DC current source mode in converter with alternating of parallel and series resonance," in *Proc. 10th European Conf. Power Electronics and Applications*, CD-ROM, Sept. 2003.

[6] K. Janson, J. Järvik and J. Shklovski, "Power factor correction method for AC/DC converters and corresponding converter," PCT patent application PCT/EE2005/000010, Dec. 2005.

[7] N. Mohan, T.M. Undeland, W.P. Robbins, *Power Electronics: Converters, Applications, and Design*. John Wiley & Sons, 1995.

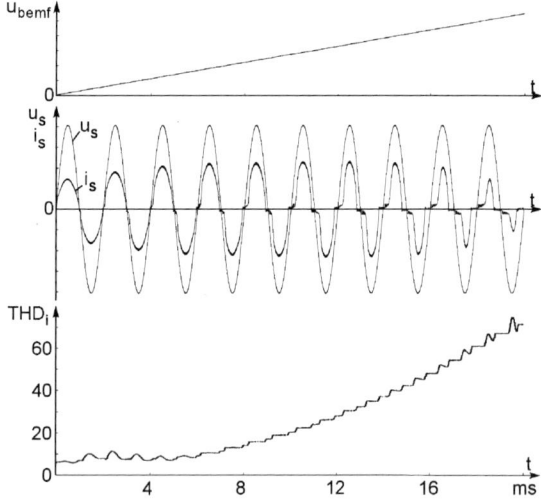

Figure 7. Waveforms of mains voltage and mains current, THD$_i$ of mains current at voltage over load model changing from zero to open-circuit value.

LLCC-PWM Inverter for Driving High-Power Piezoelectric Actuators

Rongyuan Li, Norbert Fröhleke, Joachim Böcker

Institute of Power Electronics and Electrical Drives, Paderborn University, Paderborn, Germany
Li@lea.upb.de, Froehleke@lea.upb.de, Boecker@lea.upb.de

Abstract— In this contribution a novel LLCC-PWM inverter is presented for driving ultrasonic high power piezoelectric actuators. The proposed system of a pulse-width modulated inverter and LLCC-type filter is designed in a way to reduce the total harmonic distortion of the motor voltage and to locally compensate for the reactive power of piezoelectric actuators. In order to limit the switching frequency, a pulse width modulation using elimination technique of selected harmonics is designed and implemented on a FPGA. Due to local compensation of reactive power and high dynamic behavior of LLCC PWM inverter, the whole power supply shows an optimal performance at minimized volume and weight compared to LC and LLCC resonant converters.

Keywords— Piezo actuators, sonotrode, piezoelectric converter, high-frequency power converter.

I. INTRODUCTION

The challenge of an appropriate power supply for a ultrasonic piezoelectric actuator arises from the following reason: A piezoelectric actuator is known to exhibit a distinct capacitive behavior. On a closer inspection, the electrical behavior is even more complicated. It even depends on the frequency-dependent interactions between actuator and load. Previous works on ultrasonic motors have shown that the quality factor as measure of the system damping has a strong influence on the converter topology to be chosen [1] [3] [4] [5].

Over the last decade electronic power supplies for piezoelectric systems are well studied and applied using different kinds of resonant converter concepts. A resonant inverter with LLCC-type output filter presented in [3] and [4] shows advanced characteristics and best suited properties in respect to efficiency, stationary and dynamic behavior, as well as to control and commissioning efforts. The drawbacks of these resonant inverters are the large volume, heavy and costly magnetic components of the resonant filter such as transformer and inductor [6], especially in case of driving piezoelectric actuators in the range of some kW.

Therefore, power converters which do not require heavy inductors are of great interest. In order to reduce the size and the weight of magnetic components, a carrier based PWM controlled inverter with LC filter was investigated [5]. It was shown that LC-PWM inverters are more suitable for weakly damped piezoelectric vibration systems such as bond sonotrodes, where none or small reactive power is delivered at the operating point by the inverter. Thanks to the high switching frequencies, the inductivity L_s can be decreased significantly compared to

those used in resonant filters. This results in smaller and lighter components. However, the high switching frequency of PWM inverters results consequently in high switching losses and might be in conflict with EMC issues.

To overcome these drawbacks, the proposed LLCC-PWM inverter shown in Fig. 1 is investigated to excite the high power piezoelectric actuator, in which a LLCC filter circuit is utilized and operated in PWM controlled mode. In order to eliminate selected harmonics, suitable switching angles of the PWM can be calculated. The novel solution offers significant advantages to improve the performance of the power supply as follows:

1. The reactive power of the piezoelectric actuator is compensated locally, by placing the inductor L_p close to the actuator. Hence, cables of considerable length between output transformer and actuators can be rated only with respect to the real power.

2. Due to the PWM method and reactive power compensation, the output filter shows optimized performance at minimized volume and weight, compared to the classical resonant inverters presented in [3] [4].

3. The total harmonic distortion (THD) of the piezoelectric actuator voltage is reduced without increasing the switching frequency compared with a LC-PWM inverter. The importance of a low THD result from the fact that life time of the piezo stack suffers from too high THD of the exciting voltage.

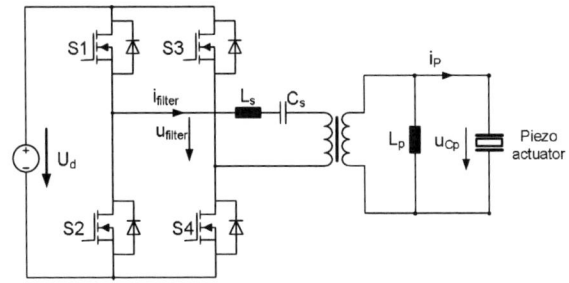

Fig. 1 Proposed LLCC-PWM controlled inverter

A power supply prototype capable of converting a 270 V DC input voltage to an AC output voltage with maximum amplitude of 270 V is implemented including frequency and magnitude control. Experimental results are presented for a standard multilayer piezoelectric actuator driven in a frequency range of 30 kHz to 40 kHz.

II. High Power Piezoelectric Actuator

High power piezoelectric actuators are used to build various kinds of piezoelectric system like ultrasonic motors or sonotrodes for ultrasonic machining. Due to their high power density, they are applied more and more in aircrafts and industry [1] [7].

These actuators are normally constructed using the solid-state piezoelectric ceramic actuator which converts electrical energy directly into mechanical energy through linear motion, and are able to withstand high pressures of up to 100 kN.

Moreover, in case of preloaded mechanical system, the actuators can be supplied with AC voltage sources solely omitting DC-biasing and could be operated in their mechanical resonance frequency. This ensures a more efficient operation.

Different from conventional electromagnetic actuators, piezoelectric actuator do not utilize magnetic fields and so they do not behave inductively. However, their distinct capacitive behavior has to be considered when designing the power supply and its control scheme.

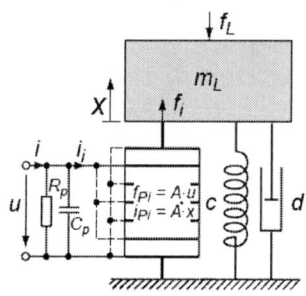

(a) Simple mechanical modeling of piezoelectric actuator

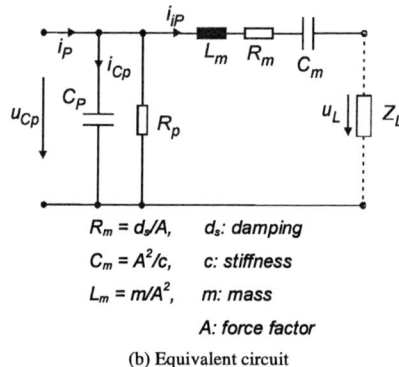

$R_m = d_s/A,$ d_s: damping

$C_m = A^2/c,$ c: stiffness

$L_m = m/A^2,$ m: mass

A: force factor

(b) Equivalent circuit

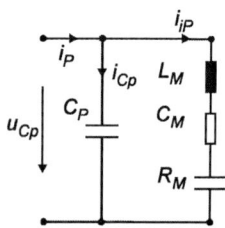

(c) Equivalent circuit including load

Fig. 2 Equivalent circuit of a piezoelectric actuator

For the analysis of piezoelectric motors the load is represented by the well known equivalent circuit, depicted in Fig. 2(a), where the capacitance of the piezoelectric material is represented by C_p, the dielectric losses within the ceramics are represented by R_p and can be neglected due to their minor contribution. The mechanical vibration system is described by a series resonant circuit L_m-C_m-R_m, which represent inertial mass, stiffness and mechanical damping shown in Fig. 2(b). Z_L is representing the mechanical load, which can be approximately modeled by an equivalent electrical impedance consisting of inductance L_L, capacitance C_L, and resistance R_L of the load [4] [7]. By combining these elements with the L_m, C_m, R_m of the actuator, the whole mechanical resonant circuit results as shown in Fig. 2(c) with following parameters:

$$L_M = L_m + L_L, \ C_M = \frac{C_m C_L}{C_m + C_L}, \ R_M = R_m + R_L$$

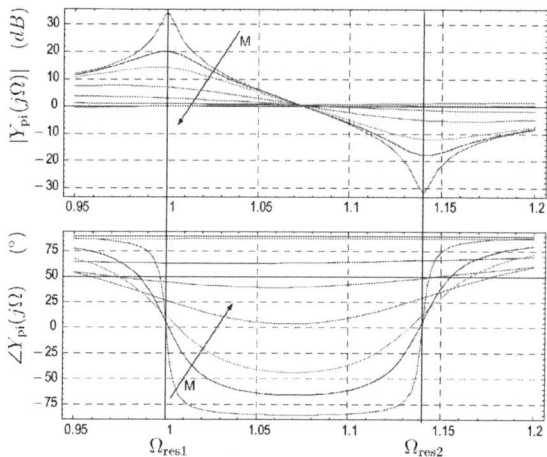

Fig. 3 Frequency characteristic of actuator system input admittance.
(Admittance ratio M, equals 0.001, 0.05, 0.01, 0.5, 1, 2, 5, 10.)

The input admittance $Y(j\Omega)$ frequency characteristic is shown in Fig. 3, where the normalized frequency $\Omega = \omega/\omega_{res1}$ is used. Effects of load changes are depicted in Fig. 3, with the admittance ratio $M = R_M C_P \omega_{res1}$.

Two resonance frequencies at

$$\Omega_{res1} = \omega_{res1} / \omega_{res1} = 1 \text{ and}$$

$$\Omega_{res2} = \omega_{res2} / \omega_{res1}$$

can be observed. The series resonance frequency $\omega_{res1} = 1/\sqrt{L_M C_M}$ is the resonance frequency of the mechanical equivalent circuit and is determined only by the mechanical parameters. In contrast to ω_{res1}, the

parallel resonance frequency $\omega_{res2} = \left(L_M \dfrac{C_M C_P}{C_M + C_P}\right)^{-1/2}$

is caused by the capacitance of the piezoelectric ceramics. Note, that $\omega_{res2} > \omega_{res1}$. Operation at that parallel resonance frequency requires high input voltage at low input current.

III. POWER SUPPLY DESIGN

The electrical characteristic of equivalent mechanic resonant circuit decides about some design consideration of the power inverter.

1. The fundamental frequency of the power inverter is usually chosen in the proximity of the resonant frequency of the mechanical vibration. In order to reduce the high supply voltage, the operation range of piezoelectric actuator is chosen near to the series resonance frequency $\omega_{res1} = 1/\sqrt{L_M C_M}$. Then, the switching frequency of the inverter is also determined.

2. The real power and reactive power required by the piezoelectric actuator have to be delivered by the inverter; the power ratio and the efficiency are determined by the operational point of the piezoelectric actuator.

3. The actuator capacitance, originating from the high number of piezo stacks, varies with temperature, resulting from differing operating conditions. This obviously complicates the power supply design.

Under power supply aspects the proposed inverter should thus provide a robust and highly dynamic operating behavior to guarantee a stable operation within a certain frequency and capacitance range. The above mentioned requirements are taken care of by designing control and filter.

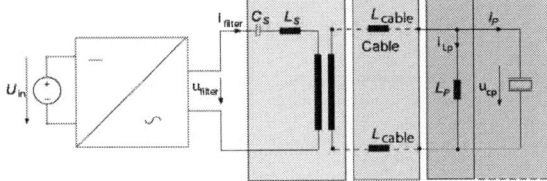

Fig. 4 LLCC type filter Topologies

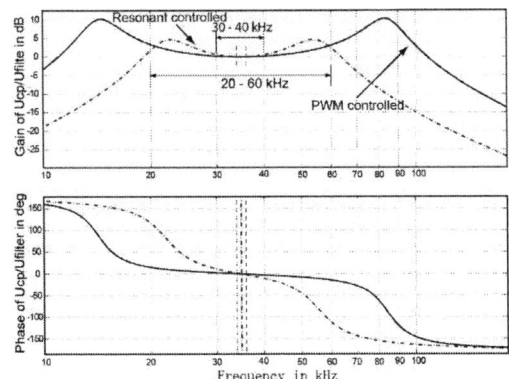

Fig. 5 Frequency characteristic u_{Cp}/u_{filter}

A. Filter topology

The inverter output filter plays an important role as coupling between power supply and actuator. A LLCC filter topology is shown in Fig. 4, with a transformer being integrated into the resonant circuit to provide galvanic isolation. Of course, the inverter resembles the LLCC

resonant inverter, but due to the different modulation scheme, these inverters behave differently, which can be seen from frequency characteristic shown in Fig. 5.

By closer inspection, the advantages of LLCC filters such as robustness to parameter fluctuations of e.g. the piezoelectric capacitance, as well as simple controllability compared with LC filters become obvious. The second electrical resonant point of the filter can be set at a higher frequency compared with resonantly modulated inverters, considering the smaller low order harmonics of the PWM output voltage. Hence, the LLCC filter results smaller and lighter compared with the LC type, and a better dynamic response is obtained.

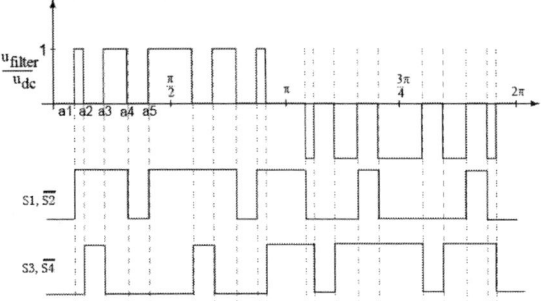

Fig. 6 Inverter output voltage with harmonic eliminated pulse width modulation

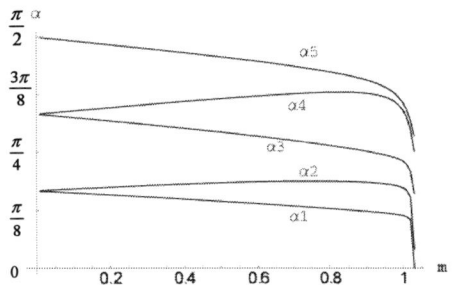

Fig. 7 Switching angles for eliminating 3rd, 5th, 7th and 9th harmonic vs. modulation index

B. PWM with elimination of specific harmonics

Harmonic elimination techniques are described in [8] [9]. The switching angles are pre-calculated, so that low order voltage harmonics applied to the load are eliminated or reduced. Since the remaining harmonics are located at higher frequencies, a LLCC filter with a small series inductor L_s is sufficient.

In order to eliminate the harmonics of orders 3, 5, 7, and 9, five switching angles are used in a quarter period as depicted in Fig. 6. In Fig. 7, the resulting angles versus modulation index are shown.

A FPGA device (Xilinx Vertex4-xc4vsx35) is used as pulse width modulator. The algorithms to calculate the switching angles are implemented by using Xilinx's System-Generator tool on Matlab-Simulink platform

The simulations with pre-calculated switching angles are performed, and the resulting spectrum of the inverter output voltage is shown in Fig. 8 with reference value of 1. As expected, 3rd, 5th, 7th, 9th harmonic components do not occur.

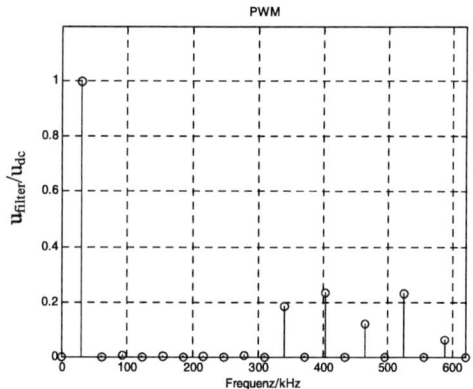

Fig. 8 Spectrum of PWM output voltage

$$\begin{bmatrix} \dot{x}_{el,s} \\ \dot{x}_{el,c} \end{bmatrix} = \begin{bmatrix} A_{LLCC\_Og} & \omega_B \\ -\omega_B & A_{LLCC\_Og} \end{bmatrix} \begin{bmatrix} x_{el,s} \\ x_{el,c} \end{bmatrix} + \begin{bmatrix} B_{LLCC\_Og} & 0 \\ 0 & B_{LLCC\_Og} \end{bmatrix} \cdot \begin{bmatrix} u_{in,s} \\ u_{in,c} \end{bmatrix}$$

$$\begin{bmatrix} u_{pi,s} \\ u_{pi,c} \end{bmatrix} = \begin{bmatrix} C_{LLCC\_Og} & 0 \\ 0 & C_{LLCC\_Og} \end{bmatrix} \begin{bmatrix} x_{el,s} \\ x_{el,c} \end{bmatrix}$$

where

$$x_{el,s} = \begin{bmatrix} i_{Ls,s} & u_{Cs,s} & i_{Lp,s} & u_{pi,s} \end{bmatrix}^T$$

$$x_{el,c} = \begin{bmatrix} i_{Ls,c} & u_{Cs,c} & i_{Lp,c} & u_{pi,c} \end{bmatrix}^T$$

$$A_{LLCC\_Og} = \begin{bmatrix} -\dfrac{R_s}{L_s} & -\dfrac{1}{L_s} & 0 & -\dfrac{1}{L_s} \\ \dfrac{1}{C_s} & 0 & 0 & 0 \\ 0 & 0 & 0 & \dfrac{1}{L_p} \\ \dfrac{1}{C_p} & 0 & -\dfrac{1}{C_p} & -\dfrac{1}{C_p R_p} \end{bmatrix}, \quad B_{LLCC\_Og} = \begin{bmatrix} \dfrac{1}{L_s} & \dfrac{1}{L_s} \\ 0 & 0 \\ 0 & -\dfrac{1}{L_p} \\ 0 & \dfrac{1}{C_p R_p} \end{bmatrix}$$

IV. CONTROL SCHEME

A. Averaging model of MM-USM for power supply control design

In order to reduce the simulation time and facilitate the control design, a model with idealized switching behavior of the power circuitry is assumed. The generalized averaging method [10] [11] is employed to approximate the original state variables by Fourier series representations of an adequate order. Fig. 9 illustrates the averaging model and the control diagram.

In view of the principle of piezoelectric energy conversion, the averaging model is divided into an electrical and a mechanical subsystem. Each subsystem can be analyzed using methods for linear systems.

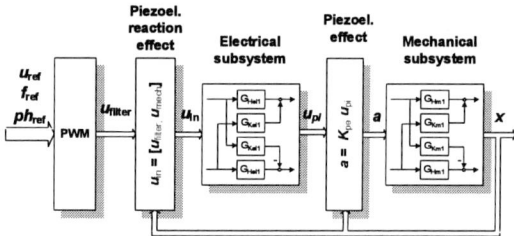

Fig. 9 Averaging model for power supply control design

The state variables of the electrical subsystem are $x_{el} = \begin{bmatrix} i_{Ls}, u_{Cs}, i_{Lp}, u_{Pi} \end{bmatrix}$, input variables are $[u_{filter}, u_{mech}]$, and output variable is the voltage of piezoelectric actuator u_{pi}. The electrical subsystem is described by a state space equation.

By a 1st-order Fourier series representation, every quantity $x_{el}(t)$ can be expressed as

$$x_{el}(t) = x_{el,s}(t) \sin(\omega_B t) + x_{el,c}(t) \cdot \cos(\omega_B t)$$

with slowly time-varying Fourier coefficients $x_{el,s}(t)$, $x_{el,c}(t)$. The state space equations of the electrical subsystem and piezoelectric mechanical subsystem represented by Fourier coefficients result as follows:

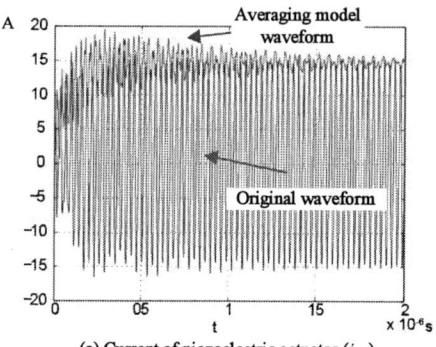

(a) Current of piezoelectric actuator (i_{Pi})

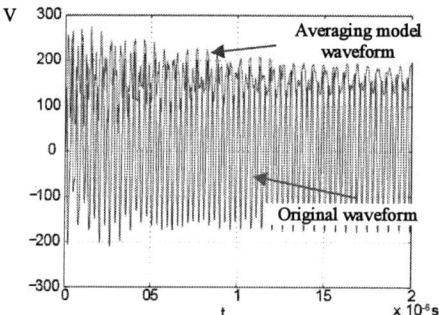

Fig. 10 Simulation waveforms of averaging model

These state space equations represent an averaging model of the tangential mode power supply and the piezoelectric actuator. Since the Fourier coefficients vary much slower in time than the original quantities, this model is more suitable for means of control design.

Simulation results of a step response of voltage u_{filter} from 0 to 180V are presented in Fig. 10. From performed simulation we notice some oscillatory effect in the averaging model and original system. The reason originates from the resonance of the mechanical subsystem, indicating the trend, the more damping exists in the mechanical subsystem, and the smaller the oscillation amplitude is excited.

B. Feed-forward voltage control

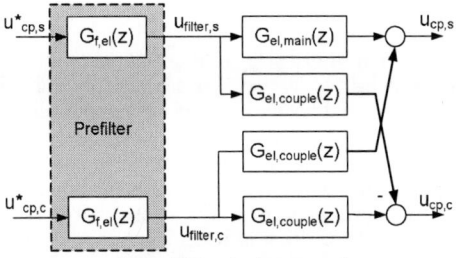

(a) Feed-forward voltage control

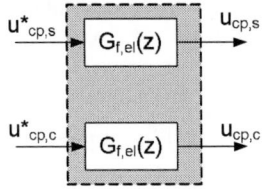

(b) Simplified electrical subsystem with feed-forward control

Fig. 11 Voltage control diagram

The proposed voltage control scheme is presented in Fig. 11. The electrical subsystem is replaced by a system consisting of main transfer function $G_{el,main}(z)$ and coupled transfer function $G_{el,couple}(z)$, while a second order Butterworth filter

$$G_{f,el}(z) = \frac{1 + 2z^{-1} + z^{-2}}{1 + b_1 z^{-1} + b_2 z^{-2}}$$

is used as pre-filter.

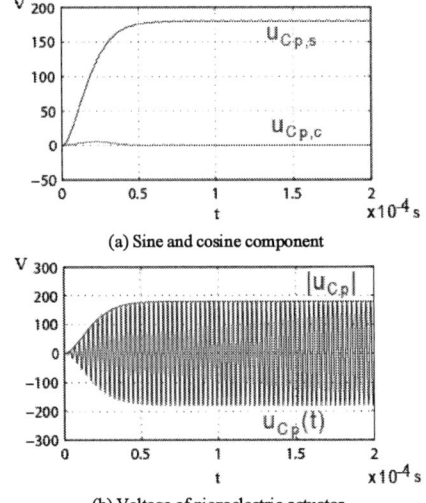

(a) Sine and cosine component

(b) Voltage of piezoelectric actuator

Fig. 12 Simulation results of voltage control

Due to the robust characteristics (see Fig. 5) of the LLCC filter circuit in the proximity of ω_B, the electrical subsystem can be simplified by the pre-filter characteristic. This provides a simple inner loop transfer function in order to design outer mechanical oscillation amplitude control.

In Fig. 12 step response simulation results of the voltage control are presented, with the amplitude calculated by $|u_{Cp}| = \sqrt{u_{Cp,s}^2 + u_{Cp,c}^2}$. Comparing simulation results in Fig. 12 with Fig. 10 we notice that the oscillatory effects are well damped by feed-forward voltage control.

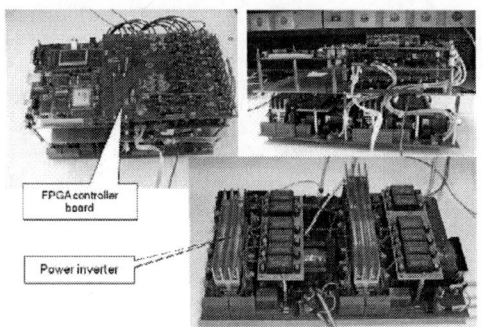

Fig. 13 Inverter prototype

V. IMPLEMENTATION

An experimental inverter prototype to deliver a power of 1.5 kW to the load was built to verify the operation principle, see Fig. 13. The rated output voltage is 270 V (amplitude) at a frequency of 33 kHz. The design of components and parameters are listed in Table 1 and Table 2.

Table 1 Component design of LLCC filter

Mechanical resonant frequency	f_m	33 kHz
Piezoelectric capacitor	C_p	137 nF
Parallel inductor	L_p	151 µH
Series inductor	L_s	45 µH
Series capacitor	C_s	440 nF

Table 2 Operating parameters of harmonic eliminated PWM inverter with LLCC filter

Real power	P		1480 W
Apparent power	S		1648 VA
Power factor			0.9
RMS current of L_s			7.5 A
current of switch S_1		RMS	4.8 A
		average	2.6 A

Considering that the first non-negligible harmonic components occur at a frequency of more than 300 kHz, then a light and small inductor L_s is sufficient, which is a strong argument for the selection of the circuitry when applied in aircrafts. For driving the piezoelectric elements in the tangential mode of the PIBRAC motor, the components of PWM controlled LLCC filter are given in Table 1.

163

Selected calculated variables of the harmonic eliminated PWM inverter with LLCC filter are listed in Table 2; the current and voltage of filter input and output are shown in Fig. 14. The power factor is larger than that of the LLCC resonant inverter, but the generated inverter losses are increased corresponding to using higher switching frequencies.

Due to the fact that the target motor is still under construction an equivalent load was used instead constituted by respective ohmic resistors and capacitors. Selected experimental waveforms are shown in Fig. 14, indicating good congruency with results gained by simulation.

Spikes appearing at the filter voltage u_{filter} will largely be reduced by exchanging utilized CoolMOS transistor against HF-MOSFET housing enhanced diodes in respect to reverse recovery.

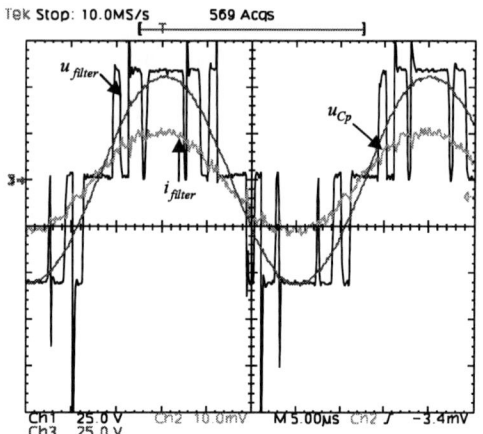

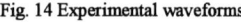

Voltage of piezo actuator u_{Cp}, Input voltage of filter u_{filter},

Input current of filter (i_{filter}, 2A/div)

Fig. 14 Experimental waveforms

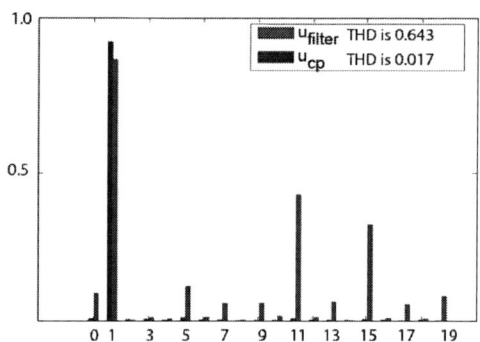

Fig. 15 Frequency spectrum of piezo actuator voltage u_{Cp} and filter input

voltage u_{filter},

The measured spectrum of the piezo actuator voltage u_{Cp} and filter input voltage u_{filter} is shown in Fig. 15. Note, that the 3rd, 5th, 7th, 9th harmonic components are larger than the simulation results in Fig. 8. This is caused by the delay time of MOSFET drivers, which reduce the switching angles slightly compared to calculation.

Even though the higher harmonic components are dominated in the harmonic distortion of u_{filter}, we notice that the higher harmonic components of u_{Cp} are well suppressed with increasing frequency due to the band-pass characteristic of the LLCC filter circuit.

VI. CONCLUSION

The LLCC PWM inverter is more complex, but yields less THD, smaller and lighter filter components and larger bandwidth compared with a resonant operated inverter. A thoroughly designed LLCC filter exhibits the required robustness in respect to parameter variations, and enables the reduction in size of filter components and the utilization of cable parasitics. The cable weight is also minimized by the reduction of RMS-current rating arising from local reactive power compensation. A FPGA is employed as controller by reason of its flexibility, fast and parallel processing characteristics. The operation of the power supply system is verified by simulation and measurement.

ACKNOWLEDGMENT

Thanks belong to the European Community for funding the PIBRAC project under AST4-CT-2005-516111 as well as our project partners.

REFERENCES

[1] T. Schulte, N. Fröhleke, "Development of power converter for high power piezoelectric motors", Aupec 2001.

[2] J. Audren, , D. Bezanere, "Vibration motors", United States Patent 6628044, Sep 2003.

[3] F.-J. Lin, R.-Y. Duan, H.-H. Lin, "An Ultrasonic Motor Drive Using LLCC Resonant Technique" Proc. of IEEE Power Electronics Specialists Conference (PESC) 1999, vol. 2, pp. 947-952.

[4] Th. Schulte, "Power Converters and Control Concepts for Traveling Wave Type Ultrasonic Motors", Doctorate thesis (in German), Fortschritt Berichte VDI, Reihe 21, Nr. 363, Paderborn, 2004.

[5] C. Kauczor, N. Fröhleke, "Inverter Topologies for Ultrasonic Piezoelectric Transducers with High Mechanical Q-Factor", IEEE PESC Conference, 2004.

[6] H.-D. Njiende, N. Fröhleke, "Optimization of Inductors in Power Converter Feeding High Power Piezoelectric Motors", Aupec 2001.

[7] R. Li, N. Fröhleke, H. Wetzel, J. Böcker, "Investigation of Power Supplies for a Piezoelectric Brake Actuator in Aircrafts", Int. Power Electronics and Motion Control Conference (IPEMC), Aug. 2006, Shanghai, China.

[8] J. Sun, S. Beineke, and H. Grotstollen, "Optimal PWM based on real-time solution of harmonic elimination equations," IEEE Trans. Power Electron., vol. 11, pp. 612–621, July 1996.

[9] D. Czarkowski, D.-V. Chudnovsky, G.-V. Chudnovsky, I. -W. Selesnick, "Solving the Optimal PWM Problem for Single-Phase Inverters". IEEE Transactions on Circuits and System - I: Fundamental Theory and Applications, Vol. 49, No. 4, April 2002.

[10] Maas, J.; Schulte, T.; Fröhleke, N.: Model-Based Control for Ultrasonic Motors. IEEE/ASME Transaction on Mechatronics, Vol. 5, Nr. 2, 2000, S. 165-180.

[11] Sanders, S. R.; Noworolski, J. M.; Lui, X. Z.; Verghese, G. C.: Generalized Averaging Method for Power Conversion Circuits. IEEE Transactions on Power Electronics, Vol. 6, Nr. 2, 1991, S. 251-258.

Modelling and Analysis of a Matrix-Reactance Frequency Converter Based on Buck-Boost Topology by DQ0 Transformation

Paweł Szcześniak[*], Zbigniew Fedyczak[*] and Marius Klytta[†]

[*] University of Zielona Góra / Institute of Electrical Engineering, Zielona Góra, Poland,
e-mail: _P.Szczesniak@iee.zu.zgora.pl_, _Z.Fedyczak@iee.zu.zgora.pl_.
[†] University of Applied Sciences / Faculty of Electrical Engineering, Giessen, Germany,
e-mail: _Marius.klytta@ei.fh-giessen.de_.

Abstract—This paper deals with a three-phase matrix-reactance frequency converter (MRFC). The analysed MRFC topology is based on buck-boost matrix-reactance chopper (MRC) one with source synchronous connected switches (LSCS) set arranged as in the step-up matrix converter (MC). The MRFC in question makes it possible to obtain a load output voltage much greater than the input voltage. Presented in this paper is a description a method for the analysis of the steady and transient state properties of presented MRFC. The static and dynamic characteristics of the presented converter under the control strategy proposed by Venturini are fully analysed on the basis of the circuit model development by the DQ0 transformation. Various static converter characteristics such as voltage and current gain, input power factor are completely analysed. Transition characteristics are also analysed by a small-signal model. The usefulness of the models is verified through computer simulations with good agreements.

Keywords—Matrix-reactance frequency converter, matrix converter, DQ0 transformation.

I. INTRODUCTION

One among the most desirable features of AC frequency converters is the generation of load voltage with arbitrary amplitude and frequency [1]. In recent years, matrix converters (MC) have received considerable attention as a competitor to the commonly used pulse width-modulated voltage-source inverter (PWM-VSI). One disadvantage of the MC is the voltage transfer ratio, which is limited to 0.5 of the input voltage [2] at linear voltage transformation, and to 0.866 or 1.053 at low-frequency load voltage deformations for space-vector or fictitious DC link control strategy concepts respectively [3], [4]. For application in the industrial drive field the maximum available magnitude of the output voltage should be even a little greater than the amplitude of the input voltage. In the case of the variable speed drive system for an induction motor, a reduction of the supply voltage by 10% means 20% loss of torque capability, which cannot be accepted in most applications. For application in FACTS a conventional auxiliary transformer employed to increase the MC output voltage is necessary. In reference [5] the concept of the step-up

This work was supported by Polish Ministry of Science and Higher Education, Project No. N510 036 32/3380.

MC is presented. In a circuit with this converter input inductors and output capacitors are used that function as input current sources and output voltage sources respectively. The voltage transfer ratio can be much greater than one, though voltage gain and input power factor cannot be controlled independently, they can be controlled by properly selected control parameters. Another concept of frequency converters based on buck-boost matrix-reactance chopper (MRC) is presented in references [6]-[9]. This concept is developed by authors and in reference [8] the generation concept of whole family of the matrix-reactance frequency converters based on unipolar PWM AC MRC is presented. The MRFC topologies are generated by use of the MRC one-cycle switched circuit models with suitable voltage and current sources introduced instead of the capacitors and inductors respectively. One of the synchronous connected switches (SCS) sets of the MRC is replaced by the step-down [2] or step-up [5] matrix-connected switches (MCS) set, offering the possibility of load voltage frequency change. A crucial fact is that all generated MRFCs have the possibility to obtain a load output voltage much greater than the input voltage. Actual problem deals with discussed MRFCs depend on researches of the properties these converters.

In this paper we obtained the analytic expressions for the voltage and current gain and input power factor of the MRFC based on buck-boost MRC, using DQ0 transformation technique [10]-[12]. By applying the DQ0 transformation, the three-phase-balanced MRFC is transformed into a simple single-phase circuit model that does not contain a switch element. The analysis results are obtained for a classical Venturini control strategy. The main aim of this paper is description of the basic static and dynamic properties in order to an evaluation of the presented converter usefulness as an AC/AC frequency changer.

II. DESCRIPTION OF ANALYSED CONVERTER

A. MRFC Based on Buck-Boost MRC

The schematic diagram of the MRC with buck-boost topology is shown in Fig. 1 [9], [13] whereas one-cycle switched circuit models of this converter for two switches states are shown in Fig. 2 [8]. In these models (Fig. 2) suitable voltage and current sources are taken into consideration instead of capacitors and inductors respectively. During the first switch state the source synchronous-connected switches (SSCS) are turn on and the load synchronous-connected switches (LSCS) are turned off. During the second switch-state the switches are in the inverse states.

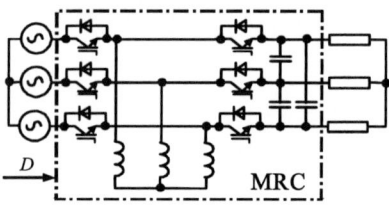

Fig. 1. Three-phase unipolar PWM AC MRC based on buck-boost topology

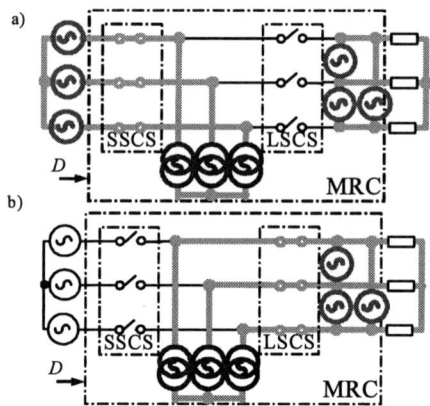

Fig. 2. One-cycle switched circuit models of three-phase unipolar PWM AC MRC based on buck-boost topology, a) for SSCS on and LSCS off, b) for SSCS off and LSCS on

As is visible from Fig. 2, in each of the switch states, the SCS sets take part in two types of electrical energy transfer between the inner voltage and current sources: the first one, as electrical energy is transferred from the voltage to the current sources; the second one as electrical energy is transferred from the current to the voltage sources. From an analysis of the electrical energy transfers it follows that one of SCS sets can be replaced by a MCS set, offering the possibility of the load voltage frequency change. The use of step-down or step-up of the MCS set is dependent on input and output voltage or current sources configurations [5].

During the first switches state electrical energy is transferred from the voltage to the current sources. In this case SSCS set can be replaced by a matrix connected switches (MCS) set with step-down configuration, offering the possibility of the load voltage frequency change. The schematic diagram of such MRFC (buck-boost I) is shown in Fig. 3 [7], [8]. The MCS output voltages u_a, u_b, and u_c are formed according to (1). Furthermore in both circuits the input low-pass filter L_F, C_F is used in order to reduce the source current deformation.

During the second switches state electrical energy is transferred from the current to the voltage sources. In this case LSCS set can be replaced by a matrix connected switches (MCS) set with step-up configuration [5]. The concept of such MRFC (buck-boost II) is presented in [8].

$$\begin{bmatrix} u_a \\ u_b \\ u_c \end{bmatrix} = \begin{bmatrix} s_{aA} & s_{aB} & s_{aC} \\ s_{bA} & s_{bB} & s_{bC} \\ s_{cA} & s_{cB} & s_{cC} \end{bmatrix} \begin{bmatrix} u_A \\ u_B \\ u_C \end{bmatrix} = \mathbf{T} \begin{bmatrix} u_A \\ u_B \\ u_C \end{bmatrix}, \qquad (1)$$

where: $s_{jK} = \begin{cases} 1, & \text{switch } S_{jK} \text{ closed} \\ 0, & \text{switch } S_{jK} \text{ open} \end{cases}$ -switch state function, $j = \{a, b, c\}$, $K = \{A, B, C\}$, $\mathbf{T}$ - instantaneous transfer matrix.

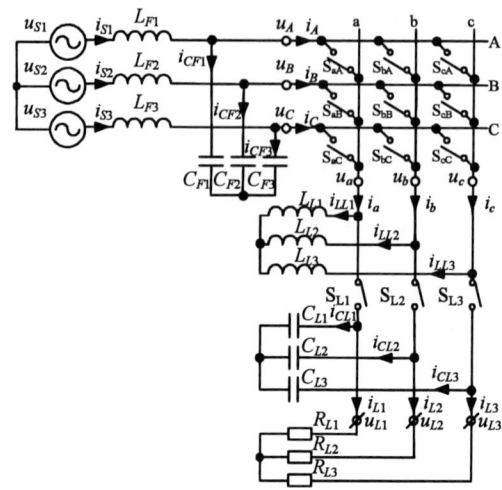

Fig. 3. MRFC based on buck-boost MRC

B. Control strategy

A description of control strategy in general form is shown in Fig. 4a. The classical Venturini control strategy is taking into consideration with low frequency transfer matrix described by (2) [2]. Exemplary time waveforms of the control signals, illustrating operation of the discussed MRFC is shown in Fig. 4b.

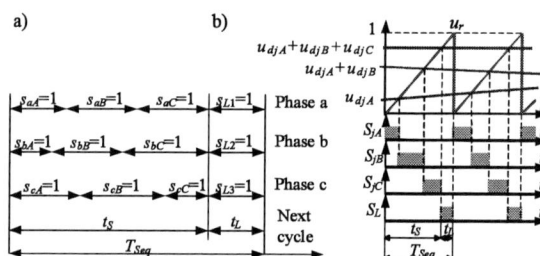

Fig. 4. Control strategy: a) general form of control strategy description, b) exemplary time waveforms of the control signals for switches in one phase

$$\mathbf{M} = \begin{bmatrix} d_{aA} & d_{aB} & d_{aC} \\ d_{bA} & d_{bB} & d_{bC} \\ d_{cA} & d_{cB} & d_{cC} \end{bmatrix}, \qquad (2)$$

where:

$d_{aA} = d_{bB} = d_{cC} = D_S(1 + 2q\cos(\omega_m t + \varphi))/3$,

$d_{aB} = d_{cA} = d_{bC} = D_S(1 + 2q\cos(\omega_m t + \varphi - 2\pi/3))/3$,

166

$d_{aC} = d_{bA} = d_{cB} = D_S(1 + 2q\cos(\omega_m t + \varphi - 4\pi/3))/3$,

d_{jK} – low frequency component of the MCS switch state function, $D_S = t_S/T_{Seq}$ – sequence pulse duty factor, $\omega_m = \omega_L - \omega$, ω, ω_L – pulsation of the supply and load voltages respectively, q – voltage gain.

III. MODELLING OF CONSIDERED TOPOLOGY

A. Averaged State Space Model

Assuming that all switches are ideal, inductors and capacitors are linear and a so called running average operator is used [14], then the mathematical models of the proposed MRFC, for low frequency transfer matrix of MCS according to classical Venturini control strategy expressed by (2), is described by the matrix differential equation (3).

$$\frac{d\overline{\mathbf{x}}}{dt} = \mathbf{A}(t)\overline{\mathbf{x}} + \mathbf{B}(t), \quad (3)$$

where:

$\overline{\mathbf{x}} = [\overline{i}_{S1}\ \overline{i}_{S2}\ \overline{i}_{S3}\ \overline{i}_{LL1}\ \overline{i}_{LL2}\ \overline{i}_{L3}\ \overline{u}_{CF1}\ \overline{u}_{CF2}\ \overline{u}_{CF3}\ \overline{u}_{L1}\ \overline{u}_{L2}\ \overline{u}_{L3}]^T$ - vector of the averaged state variables, $\mathbf{A}(t)$, $\mathbf{B}(t)$ – matrix and vector of MRFC parameters:

$\mathbf{A}(t)=$

$$\begin{bmatrix}
0 & 0 & 0 & 0 & 0 & 0 & \frac{1}{L_{F1}} & 0 & 0 & 0 & 0 & 0 \\
0 & 0 & 0 & 0 & 0 & 0 & 0 & \frac{1}{L_{F2}} & 0 & 0 & 0 & 0 \\
0 & 0 & 0 & 0 & 0 & 0 & 0 & 0 & \frac{1}{L_{F3}} & 0 & 0 & 0 \\
0 & 0 & 0 & 0 & 0 & 0 & \frac{d_{aA}}{L_{L1}} & \frac{d_{aB}}{L_{L1}} & \frac{d_{aC}}{L_{L1}} & \frac{(1-D_S)}{L_{L1}} & 0 & 0 \\
0 & 0 & 0 & 0 & 0 & 0 & \frac{d_{bA}}{L_{L2}} & \frac{d_{bB}}{L_{L2}} & \frac{d_{bC}}{L_{L2}} & 0 & \frac{(1-D_S)}{L_{L2}} & 0 \\
0 & 0 & 0 & 0 & 0 & 0 & \frac{d_{cA}}{L_{L3}} & \frac{d_{cB}}{L_{L3}} & \frac{d_{cC}}{L_{L3}} & 0 & 0 & \frac{(1-D_S)}{L_{L3}} \\
\frac{1}{C_{F1}} & 0 & 0 & \frac{d_{aA}}{C_{F1}} & \frac{d_{bA}}{C_{F1}} & \frac{d_{cA}}{C_{F1}} & 0 & 0 & 0 & 0 & 0 & 0 \\
0 & \frac{1}{C_{F2}} & 0 & \frac{d_{aB}}{C_{F2}} & \frac{d_{bB}}{C_{F2}} & \frac{d_{cB}}{C_{F2}} & 0 & 0 & 0 & 0 & 0 & 0 \\
0 & 0 & \frac{1}{C_{F3}} & \frac{d_{aC}}{C_{F3}} & \frac{d_{bC}}{C_{F3}} & \frac{d_{cC}}{C_{F3}} & 0 & 0 & 0 & 0 & 0 & 0 \\
0 & 0 & 0 & \frac{(1-D_S)}{C_{L1}} & 0 & 0 & 0 & 0 & 0 & \frac{1}{R_1 C_{L1}} & 0 & 0 \\
0 & 0 & 0 & 0 & \frac{(1-D_S)}{C_{L2}} & 0 & 0 & 0 & 0 & 0 & \frac{1}{R_2 C_{L2}} & 0 \\
0 & 0 & 0 & 0 & 0 & \frac{(1-D_S)}{C_{L3}} & 0 & 0 & 0 & 0 & 0 & \frac{1}{R_3 C_{L3}}
\end{bmatrix}$$

$\mathbf{B}(t) = [u_{S1}\ u_{S2}\ u_{S3}\ 0\ 0\ 0\ 0\ 0\ 0\ 0\ 0\ 0]^T$.

It should be noted that equation (3), as a result of averaging, is continuously non-stationary one.

From equations, (3) we can easily obtain the three-phase averaged circuit model (Fig. 5). The equivalent circuit of switches would comprise ideal transformers, whose turns ratios depend on the duty ratios.

B. The ABC-DQ0 Transformation

Sinusoidal time-varying systems can be changed to time-invariant system by the DQ0 transformation [10], [11]. The DQ0 transformation of the variables is given as follows:

$$\mathbf{x}_{dq0} = \mathbf{K}\mathbf{x}_{abc}, \quad \mathbf{x}_{abc} = \mathbf{K}^{-1}\mathbf{x}_{qd0}, \quad (4)$$

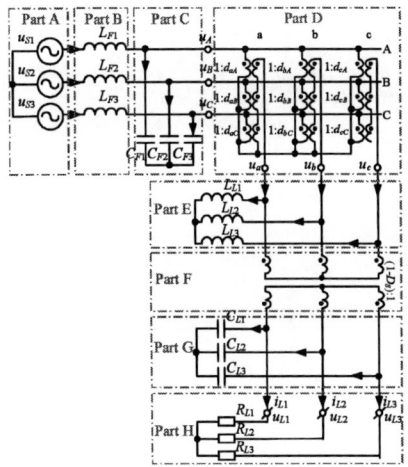

Fig. 5. Averaged circuit model of the considered MRFC

$$\mathbf{K} = \sqrt{\frac{2}{3}}\begin{bmatrix} \cos(\omega t) & \cos(\omega t - 2\pi/3) & \cos(\omega t + 2\pi/3) \\ \sin(\omega t) & \sin(\omega t - 2\pi/3) & \sin(\omega t + 2\pi/3) \\ 1/\sqrt{2} & 1/\sqrt{2} & 1/\sqrt{2} \end{bmatrix}, \quad (5)$$

where: $\mathbf{x}_{abc} = [x_a,\ x_b,\ x_c]^T$, $\mathbf{x}_{qd0} = [x_q,\ x_d,\ x_0]^T$, x_d - forward (rotating) phasor, x_q - backward (rotating) phasor, x_0 - zero-sequence component.

As in the presented topology, there are two input and output work frequencies, we also have two transform matrices $\mathbf{K}_S = \mathbf{K}$ and $\mathbf{K}_L$ expressed by (6) [5], [12].

$$\mathbf{K}_L = \sqrt{\frac{2}{3}}\begin{bmatrix} \cos(\omega_L t) & \cos(\omega_L t - 2\pi/3) & \cos(\omega_L t + 2\pi/3) \\ \sin(\omega_L t) & \sin(\omega_L t - 2\pi/3) & \sin(\omega_L t + 2\pi/3) \\ 1/\sqrt{2} & 1/\sqrt{2} & 1/\sqrt{2} \end{bmatrix}. \quad (6)$$

The circuit DQ0 transformation is obtained by the following procedures:

- Partition of the averaged circuit model into basic subcircuits.
- Transformation of each of the subcircuits into DQ0 equivalent circuits based on the DQ0 transformation equations.
- Reconstruction of the transformed subcircuits by connecting the nodes of adjacent subcircuits.

C. Partition of the circuit into basic subcircuits

We can divide the averaged circuit model of the presented MRFC into several fundamental subcircuits along the dotted lines indicated in Fig. 5. After partitioning, we obtain eight basic subcircuits.

D. Transformation of Basic Subcircuits into DQ0 Equivalent Circuits

1) Circuit DQ0 Transform of Three-Phase Voltage Sources Set (Part A)

For a three-phase balanced voltage source set, the procedure is as follows:

$$\mathbf{u}_{Sdq0} = \mathbf{K}_S\mathbf{u}_S = \mathbf{K}_S U_S\begin{bmatrix} \sin(\omega t + \varphi_1) \\ \sin(\omega t - 2\pi/3 + \varphi_1) \\ \sin(\omega t + 2\pi/3 + \varphi_1) \end{bmatrix} = U_S\begin{bmatrix} \sin\varphi_1 \\ \cos\varphi_1 \\ 0 \end{bmatrix}, (7)$$

where $\mathbf{u}_S$ is the vector of the voltage sources. Thus, the DQ0 transformed circuits of the voltage source set is shown in Fig. 6a.

2) Circuit DQ0 Transform of Three-Phase Source and Load Inductor Sets (Part B and Part E)

Using basic principles from circuit theory, the source inductors (Part B) are modelled by equation (8):

$$L_F \dot{\mathbf{i}}_{LFabc} = \mathbf{u}_{LFabc}, \qquad (8)$$

where $L_{F1} = L_{F2} = L_{F3} = L_F$. Application of (4) and (5) to (8) yields:

$$L_F \left(\mathbf{K}_S^{-1} \dot{\mathbf{i}}_{LFqd0} + \dot{\mathbf{K}}_S^{-1} \mathbf{i}_{LFqd0} \right) = \mathbf{u}_{LFabc}. \qquad (9)$$

Finally, the DQ0 transform of source inductors can be formulated as:

$$L_F \dot{\mathbf{i}}_{LFqd0} = -L_F \mathbf{K}_S \left(\dot{\mathbf{K}}_S^{-1} \right) \mathbf{i}_{LFqd0} + \mathbf{K}_S \mathbf{u}_{LFqd0} =$$
$$= -L_F \omega \begin{bmatrix} 0 & 1 & 0 \\ -1 & 0 & 0 \\ 0 & 0 & 0 \end{bmatrix} \mathbf{i}_{LFqd0} + \mathbf{u}_{LFqd0} \qquad , \quad (10)$$

and the circuit models are shown in Fig. 6b. The DQ0 "inductor" is represented by real dynamic inductor L_F in series with an imaginary static reactor $\pm j\omega L_F$. Since the voltage and current of the static reactor obeys Ohm's law, the reactor is replaced by a lossless resistor symbol [11].

Similar, equations and circuit models apply to the load inductor set (Part E):

$$L_L \dot{\mathbf{i}}_{LLabc} = \mathbf{u}_{LLabc}, \qquad (11)$$

where $L_{L1} = L_{L2} = L_{L3} = L_L$. From expression (4), (6), and (11) obtain:

$$L_L \dot{\mathbf{i}}_{LLqd0} = -L_L \mathbf{K}_L \left(\dot{\mathbf{K}}_L^{-1} \right) \mathbf{i}_{LLqd0} + \mathbf{K}_L \mathbf{u}_{LLqd0} =$$
$$= -L_L \omega_L \begin{bmatrix} 0 & 1 & 0 \\ -1 & 0 & 0 \\ 0 & 0 & 0 \end{bmatrix} \mathbf{i}_{LLqd0} + \mathbf{u}_{LLqd0} \qquad . \quad (12)$$

Figure 6c illustrates the DQ0 components of load inductors.

3) Circuit DQ0 Transform of Three-Phase Source and Load Capacitor Sets (Part C and Part G)

For the source capacitors circuit (Part C), the differential equations are in the following form:

$$C_F \dot{\mathbf{u}}_{CFabc} = \mathbf{i}_{CFabc}, \qquad (13)$$

where $C_{F1} = C_{F2} = C_{F3} = C_F$. Taking into account expressions (4), (5) and (13), the DQ0 transform of source capacitors is defined as follows:

$$C_F \left(\dot{\mathbf{K}}_S^{-1} \mathbf{u}_{CFqd0} + \mathbf{K}_S^{-1} \dot{\mathbf{u}}_{CFqd0} \right) = \mathbf{i}_{CFabc}, \qquad (14)$$

$$C_F \dot{\mathbf{u}}_{CFqd0} = -C_F \mathbf{K}_S \left(\dot{\mathbf{K}}_S^{-1} \right) \mathbf{u}_{CFqd0} + \mathbf{K}_S \mathbf{i}_{CFqd0} =$$
$$= -C_F \omega \begin{bmatrix} 0 & 1 & 0 \\ -1 & 0 & 0 \\ 0 & 0 & 0 \end{bmatrix} \mathbf{u}_{CFqd0} + \mathbf{i}_{CFqd0} \qquad . \quad (15)$$

For the load capacitors circuit (Part G), the DQ0 transform is defined as follows:

$$C_L \dot{\mathbf{u}}_{CLabc} = \mathbf{i}_{CLabc}, \qquad (16)$$

$$C_L \left(\mathbf{K}_L^{-1} \mathbf{u}_{CLqd0} + \mathbf{K}_L^{-1} \dot{\mathbf{u}}_{CLqd0} \right) = \mathbf{i}_{CLabc}, \qquad (17)$$

$$C_L \dot{\mathbf{u}}_{CLqd0} = -C_L \mathbf{K}_L \left(\dot{\mathbf{K}}_L^{-1} \right) \mathbf{u}_{CLqd0} + \mathbf{K}_L \mathbf{i}_{CLqd0} =$$
$$= -C_L \omega_L \begin{bmatrix} 0 & 1 & 0 \\ -1 & 0 & 0 \\ 0 & 0 & 0 \end{bmatrix} \mathbf{u}_{CLqd0} + \mathbf{i}_{CLqd0} \qquad . \quad (18)$$

The DQ0 transformed circuit of source and load capacitor sets are shown in Fig. 6d and Fig. 6e, respectively. Similar as with inductors, the DQ0 "capacitors" are represented by real dynamic capacitors C_F and C_L in parallel with imaginary static reactors $\pm 1/(j\omega C_F)$, and $\pm 1/(j\omega_L C_L)$.

4) Circuit DQ0 Transform of Matrix Switches Set (Part D)

If the switching function of the matrix switches is defined by (2) then the DQ0 transformation of the nine switch matrix is given as follows [5], [12]:

$$\mathbf{u}_{Lqd0} = \mathbf{K}_L \mathbf{u}_{Labc} = \mathbf{K}_L \mathbf{M} \mathbf{u}_{SABC} = \mathbf{K}_L \mathbf{M} \mathbf{K}_S^{-1} \mathbf{u}_{Sqd0} = \mathbf{M}_{dq0} \mathbf{u}_{Sqd0}, \quad (19)$$

$$\mathbf{M}_{qd0} = \mathbf{K}_L \mathbf{M} \mathbf{K}_S^{-1} = D_S \begin{bmatrix} q & 0 & 0 \\ 0 & q & 0 \\ 0 & 0 & 1 \end{bmatrix}. \qquad (20)$$

The DQ0 transformed circuit of matrix switches set is shown in Fig. 6f.

5) Circuit DQ0 Transform of Load Switches Set (Part F)

If the switching function of the load switches is defined as:

$$\mathbf{M}_L = \begin{bmatrix} (1-D_S) & 0 & 0 \\ 0 & (1-D_S) & 0 \\ 0 & 0 & (1-D_S) \end{bmatrix}, \qquad (21)$$

then the DQ0 transform is described as follows (Fig. 6g):

$$\begin{aligned} \mathbf{u}_{Sabc} &= \mathbf{M}_L \mathbf{u}_{Labc} \\ \mathbf{K}_L^{-1} \mathbf{u}_{Sdq0} &= \mathbf{M}_L \mathbf{K}_L^{-1} \mathbf{u}_{Ldq0} \\ \mathbf{u}_{Sdq0} &= \mathbf{M}_L \mathbf{u}_{Ldq0} \end{aligned} \qquad . \quad (22)$$

6) Circuit DQ0 Transform of Three-Phase Load Resistor Set (Part H)

Assuming that, $R_{L1}=R_{L2}=R_{L3}=R_L$, the procedure of DQ0 transform of the resistor set is as follows (Fig. 6h):

$$\mathbf{u}_{Lqd0} = \mathbf{K}_L \mathbf{u}_{Labc} = \mathbf{K}_L R_L \mathbf{i}_{Labc} = R_L \mathbf{i}_{Lqd0} \qquad (23)$$

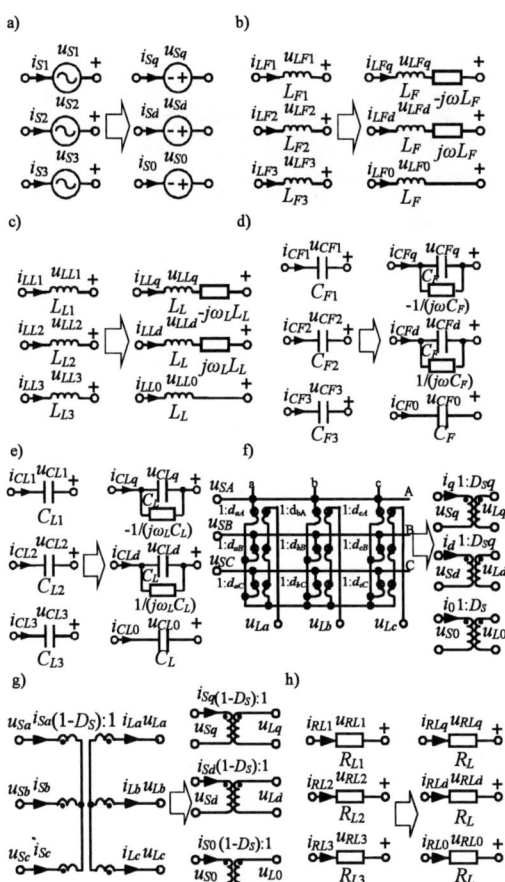

Fig. 6. DQ0 transformation of: a) voltage sources, b) source inductors, c) load inductors, d) source capacitors, e) load capacitors, f) matrix switches, g) load switches, h) load resistors

E. Circuit Reconstruction

The equivalent DQ0 circuit models of the presented MRFC based on buck-boost MRC (Fig. 3) are obtained as shown in Fig. 7 by rejoining of the DQ0 transformed subcircuits. Therefore, the three-phase circuit in Fig. 3 can be represented by three single-phase subcircuits for forward, backward and zero-sequence components.

Furthermore, assuming that the initial phase of input voltages equals zero $\varphi_1=0$, and that the circuit is symmetrical and balanced, we obtain [6], [12]:

$$\mathbf{u}_{Sdq0} = \mathbf{K}_S \mathbf{u}_S = U_S \begin{bmatrix} 0 \\ 1 \\ 0 \end{bmatrix}. \qquad (24)$$

The equivalent circuits have been simplified from three circuits to one circuit, which is shown in Fig. 8.

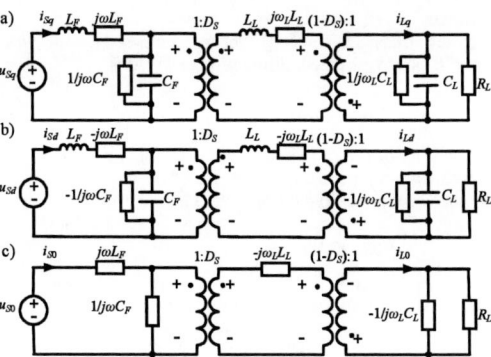

Fig. 7. DQ0 transformation of three phase MRFC based on buck-boost MRC (Fig. 1); a) forward sequence component, b) backward sequence component, c) zero-sequence component

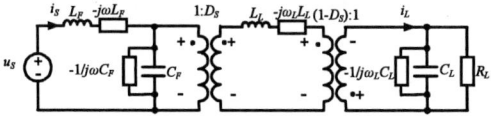

Fig. 8. DQ0 transformation of three phase MRFC based on buck-boost MRC (Fig. 1) for $\varphi_1=0$, and balanced-symmetrical circuit condition

IV. STEADY STATE ANALYSIS

The steady state model is obtained simply by eliminating the reactive elements. With reference to Fig. 9, the inductors seem to be short and capacitors open. The steady state characteristics can be obtained by considering the circuit model of the presented MRFC. For steady state analysis a single-phase circuit model is divided into four terminal networks (Fig. 9) [9], [15]. With reference to Fig. 9 four-terminal chain equations in complex form can be written as (25).

$$\mathbf{A_{LF}} = \begin{bmatrix} 1 & j\omega L_F \\ 0 & 1 \end{bmatrix}, \qquad \mathbf{A_{LS}} = \begin{bmatrix} 1 & j\omega_L L_L \\ 0 & 1 \end{bmatrix},$$

$$\mathbf{A_{CF}} = \begin{bmatrix} 1 & 0 \\ j\omega C_F & 1 \end{bmatrix}, \qquad \mathbf{A_{CL}} = \begin{bmatrix} 1 & 0 \\ j\omega_L C_L & 1 \end{bmatrix}, \qquad (25)$$

$$\mathbf{A_{TR1}} = \begin{bmatrix} \dfrac{1}{qD_S} & 0 \\ 0 & qD_S \end{bmatrix}, \qquad \mathbf{A_{TR2}} = \begin{bmatrix} (1-D_S) & 0 \\ 0 & \dfrac{1}{(1-D_S)} \end{bmatrix}.$$

Fig. 9. Steady state equivalent circuit for MRFC based on buck-boost MRC; $\mathbf{A_{LF}}$, $\mathbf{A_{LS}}$ chain matrix for the source and load inductors respectively, $\mathbf{A_{TR1}}$, $\mathbf{A_{TR2}}$ chain matrix for source and load transformer respectively, $\mathbf{A_{CF}}$, $\mathbf{A_{CL}}$ chain matrix for the source and load capacitors respectively

VI. Conclusions

The steady state and small signal mathematical and circuit models of MRFC with buck-boost topology have been elaborated. Furthermore, the steady state characteristics and transient responses of the analysed circuit have also been investigated. Simulation test results, obtained for MRFC with idealized switches, have confirmed that elaborated models can be useful with respect to steady state and transient responses of MRFC topology. The validity of the proposed models will be the subject of future investigations of the presented MRFC with active load and for a closed control system too.

Appendix A

Table I.
Theoretical and Simulation Tests Circuit Parameters

Parameter	Symbol	Value
Supply voltage/frequency	U_S/f	230 V/50 Hz
Switching frequency	f_S	5 kHz
Inductances	L_{F1} - L_{F3}, L_{L1} - L_{L3}	0.5 mH
Capacitances	C_{F1} - C_{F3}, C_{L1} - C_{L3}	50 μF
Load resistance	R_L	10 Ω

References

[1] P. W. Wheeler, J. Rodriguez, J. C. Clare, L. Empringham, and A. Weinstejn, "Matrix converters: A technology review," *IEEE Trans. on Ind. Electron.*, vol. 49, No. 2, pp. 276–288, April 2002.

[2] M. Venturini and A. Alesina, "The generalized transformer: a new bi-directional sinusoidal waveform frequency converter with continuously adjustable input power factor," *Conf. Record, PESC'80*, pp. 242 252.

[3] L. Huber and D. Borojevic, "Space vector modulated three-phase to three phase matrix converter with input power factor correction," *IEEE Trans. on Ind Appl.*, vol. 31, No. 6, pp. 1234–1246, Nov./Dec. 1995.

[4] D. Casadei, G. Serra, A. Tanti, and L. Zaroi, "Matrix converter modulation strategies: a New general approach based on space-vector representation of switch state," *IEEE Trans. on Ind. Electronics*, vol. 49, No. 2, pp. 370–381, April 2002.

[5] W. H. Kwon and G. H. Cho, "Analysis of static and dynamic characteristics of practical step-up nine-switch matrix converter," *Proc. Inst. Elect. Eng. B*, vol. 140, pp. 139–146, 1993.

[6] G. S. Zinoviev, A. Y. Obuchov, W. A. Otchenasch, and W. I. Popov, "Transformerless PWM AC boost and buck-boost converters," (In Russian), *Technicznaja Elektrodinamika*, T2, pp. 36–39. Nacjonalnaja Akademia Nauk Ukrainy, Kijev 2000.

[7] Z. Fedyczak and P. Szcześniak, "Study of matrix-reactance frequency converter with buck-boost topology," *PELINCEC 2005*, Warsaw 2005, CD-ROM.

[8] Z. Fedyczak, P. Szcześniak, and I. Korotyeyev, "Generation of matrix-reactance frequency converters based on unipolar matrix-reactance choppers," *Proc. of PESC'08*, Rhodes 2008.

[9] Z. Fedyczak, PWM AC voltage transforming circuits, (In Polish), Zielona Góra University Press, Zielona Góra, 2003.

[10] C. T. Rim, D. Y. Hu, and G. H. Cho: "Transformers as Equivalent Circuits for Switches: General Proofs and D-Q Transformation – Based Analyses," *IEEE Trans. on Ind. Apl.* vol. 26, No. 4, 1990.

[11] J. Chen and D. T. Ngo, "Graphical phasor analysis of three-phase PWM converters," *IEEE Trans. On Power Electron.*, vol. 16, No.5, pp. 659–666, 2001.

[12] W. H. Kwon and G. H. Cho, "Analysis of non-ideal step down matrix converter based on circuit DQ transformation," *IEEE Power Electronics Specialists Conference*, PESC'91, Cambridge, USA, pp. 825–829, 1991.

[13] Z. Fedyczak, R. Strzelecki, and K. Sozański, "Review of three-phase PWM AC transformer topologies and applications," *Conf. Proc. of SPEDAM 2002*, pp. B5–19–24, Ravello 2002.

[14] R. D. Middlebrook and S. Ćuk, "A general unified approach to modelling switching-converter power stages," PESC'76 Conf. Rec. pp.18–34, 1976.

[15] Z. Fedyczak, "Four-terminal chain parameters of averaged AC models of non-isolated matrix-reactance PWM AC line conditioners," *Archives of Electrical Engineering*, vol. 50, No. 4, pp. 395-409, 2001

Appendix B

Small Signal Transfer Functions of Analyzed MRFC

$$\mathbf{\underline{G}}_{qd0\,\hat{i},\hat{u}} = \frac{\begin{bmatrix} D_S^2 q^2\left(1+R_L C_L(s+j\omega_L)\right)+C_F\left(R_L D_1^2+L_L\left(1+R_L C_L(s+j\omega_L)\right)(s+j\omega_L)\right)(s+j\omega) \\ D_S q\left(1+R_L C_L(s+j\omega_L)\right) \\ R_L D_1^2 + L_L\left(1+R_L C_L(s-j\omega_L)\right)(s-j\omega_L) \\ D_1 D_S q R_L \end{bmatrix}}{R_L C_F C_L L_F L_L \det(s\mathbf{I}-\mathbf{A})} = \frac{\begin{bmatrix} G_1 \\ G_2 \\ G_3 \\ G_4 \end{bmatrix}}{G_3 + G_1 L_F(s+j\omega)},$$

$$\mathbf{\underline{G}}_{qd0\,\hat{i},\hat{d}} = \mathbf{\underline{G}}_{qd0\,\hat{i},\hat{u}} +$$

$$+\frac{\begin{bmatrix} -qR_L\left(G_2-D_S G_2\right)-qG_6\left(qD_S G_3-D_S G_4+j\omega_L L_L G_2\right) \\ q\left[D_1\left(1+D_S G_6\right)R_L-X_{LL}G_6+R_L C_L(s+j\omega_L)\left(D_1 R_L-X_{LL}G_6\right)\left(1+C_F L_F(s+j\omega)^2\right)+L_F\left(qD_S G_2 G_6+C_F\left(R_L D_1\left(1+D_S G_6\right)-X_{LL}G_6\right)(s+j\omega)\right)(s+j\omega)\right] \\ qL_F(s+j\omega)\left(D_S G_6\left(qG_3-G_4\right)+G_2\left(R_L-D_S R_L+X_{LL}G_6\right)\right) \\ -q\left[R_L\left(L_L G_6\left(sD_S+j\omega_L(2D_S-1)\right)-D_1^2 R_L\right)+L_F\left(qD_S G_6\left(qD_S^2 R_L-G_4\right)-R_L C_F\left[D_1^2 R_L-G_6 L_L\left(sD_S+j\omega_L(2D_S-1)\right)\right](s+j\omega)\right)(s+j\omega)\right] \end{bmatrix}}{G_5\left(G_3+G_1 L_F(s+j\omega)\right)},$$

where: $D_1=(D_S-1)$, $X_{LL}=j\omega_L L_L$, $G_5=\omega L_F\left(q^2 D_S^2\left(R_L\omega_L C_L-j\right)+D_1^2 R_L\omega C_F\right)+\omega_L L_L\left(R_L\omega_L C_L-j\right)\left(1-C_F L_F\omega^2\right)-D_1^2 R_L$, $G_6=1+j\omega_L C_L R_L$,

$$\det(s\mathbf{I}-\mathbf{A})=\frac{R_L D_1^2+L_L\left(1+R_L C_L(s+j\omega_L)\right)(s+j\omega_L)+L_F\left[D_S^2 q^2\left(1+R_L C_L(s+j\omega_L)\right)+C_F\left(R_L D_1^2+L_L\left(1+R_L C_L(s+j\omega_L)\right)(s+j\omega_L)\right)(s+j\omega)\right]}{R_L C_F C_L L_F L_L}.$$

Assuming that all variables have two components: a running constant component (the averaged value in the switching period T_{Seq}), which is marked by upper case letter, and a perturbation one marked by lower case letter, which is covered by the sign "^":

$$\mathbf{u} = \mathbf{U} + \hat{\mathbf{u}}, \quad \mathbf{x} = \mathbf{X} + \hat{\mathbf{x}}, \quad d = D_S + \hat{d} \qquad (33)$$

The small signal state space equations are expressed as follows [14]:

$$\frac{d}{dt}(\mathbf{X} + \hat{\mathbf{x}}) \approx \mathbf{A}\hat{\mathbf{x}} + \mathbf{B}\hat{\mathbf{u}} + \left[(\mathbf{A}_1 - \mathbf{A}_2)\mathbf{X} + (\mathbf{B})\mathbf{U}\right]\hat{d} \qquad (34)$$

where $\mathbf{A}_1 = \mathbf{A}(D_S = 1)$, $\mathbf{A}_2 = \mathbf{A}(D_S = 0)$.

According to (34) Laplace transform of a small signal state-space equation is expressed as (35).

$$s\hat{\mathbf{x}}(s) = \mathbf{A}\hat{\mathbf{x}}(s) + \mathbf{B}\hat{\mathbf{u}}(s) + \left[(\mathbf{A}_1 - \mathbf{A}_2)\mathbf{X} + (\mathbf{B})\mathbf{U}\right]\hat{d}(s) \quad (35)$$

After rearrangement there is:

$$\hat{\mathbf{x}}(s) = (s\mathbf{I} - \mathbf{A})^{-1}\left\{\mathbf{B}\hat{\mathbf{u}}(s) + \left[(\mathbf{A}_1 - \mathbf{A}_2)\mathbf{X} + (\mathbf{B})\mathbf{U}\right]\hat{d}(s)\right\} = $$
$$= \underline{\mathbf{G}}_{qd0_{\hat{\mathbf{x}}, \hat{\mathbf{u}}}}(s)\hat{\mathbf{u}}(s) + \underline{\mathbf{G}}_{qd0_{\hat{\mathbf{x}}, \hat{d}}}(s)\hat{d}(s) \qquad (36)$$

We obtain two following perturbation transform functions:

$$\underline{\mathbf{G}}_{qd0_{\hat{\mathbf{x}}, \hat{\mathbf{u}}}}(s) = \frac{\hat{\mathbf{x}}_{qd0}(s)}{\hat{\mathbf{u}}_{qd0}(s)}, \quad (37) \qquad \underline{\mathbf{G}}_{qd0_{\hat{\mathbf{x}}, \hat{d}}}(s) = \frac{\hat{\mathbf{x}}_{qd0}(s)}{\hat{d}(s)} \quad (38)$$

Detailed definitions of equations (37) and (38) are presented in Appendix B.

Presented in Fig. 13 are the transient responses of state variables at a step change of the load frequency from 25Hz to 50Hz, for summarized pulse duty factor equal $D_S = 0.75$.

variables at a step change of the sequence pulse duty factor D_S from 0.5 to 0.75, for $f_L = 25$Hz are presented.

Figures 13, 14 and 15 show good consistency of calculation and simulation test results. The obtained results confirm that small signal models can be useful for transient response analysis of the described MRFC.

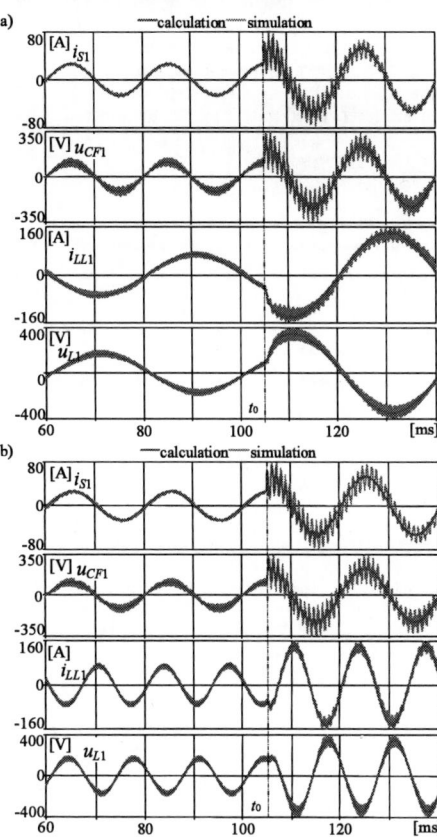

Fig. 14. Transient responses of states variables at step change of the supply voltage at $D_S = 0.75$ a) for output frequency 25Hz b) for output frequency 75Hz

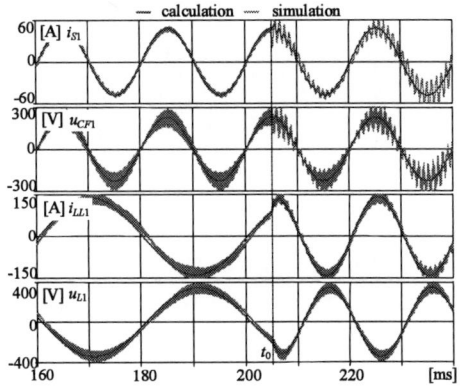

Fig. 13. Transient responses of states variables at step change of the output frequency f_L from 25Hz to 50Hz, for $D_S = 0.75$

Represented in Fig. 14 are the calculation and simulation test results of transient responses of state variables for two different output frequencies, 25 and 75Hz. Presented in both cases is the step change of the input voltages from 50% to 100% of their nominal values in time moment t_0 and pulse duty factor equal $D_S = 0.75$. Whereas, in Fig. 15 the transient responses of state

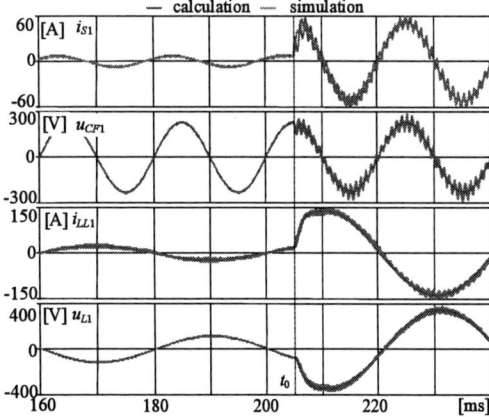

Fig. 15. Transient responses of states variables at step change of the sequence pulse duty factor D_S from 0.5 to 0.75, for $f_L = 25$Hz

VI. Conclusions

The steady state and small signal mathematical and circuit models of MRFC with buck-boost topology have been elaborated. Furthermore, the steady state characteristics and transient responses of the analysed circuit have also been investigated. Simulation test results, obtained for MRFC with idealized switches, have confirmed that elaborated models can be useful with respect to steady state and transient responses of MRFC topology. The validity of the proposed models will be the subject of future investigations of the presented MRFC with active load and for a closed control system too.

Appendix A

TABLE I.
Theoretical and Simulation Tests Circuit Parameters

Parameter	Symbol	Value
Supply voltage/frequency	U_S/f	230 V/50 Hz
Switching frequency	f_S	5 kHz
Inductances	L_{F1} - L_{F3}, L_{L1} - L_{L3}	0.5 mH
Capacitances	C_{F1} -C_{F3}, C_{L1} -C_{L3}	50 μF
Load resistance	R_L	10 Ω

References

[1] P. W. Wheeler, J. Rodriguez, J. C. Clare, L. Empringham, and A. Weinstejn, "Matrix converters: A technology review," *IEEE Trans. on Ind. Electron.*, vol. 49, No. 2, pp. 276–288, April 2002.

[2] M. Venturini and A. Alesina, "The generalized transformer: a new bi-directional sinusoidal waveform frequency converter with continuously adjustable input power factor," *Conf. Record, PESC'80*, pp. 242 252.

[3] L. Huber and D. Borojevic, "Space vector modulated three-phase to three phase matrix converter with input power factor correction," *IEEE Trans. on Ind Appl.*, vol. 31, No. 6, pp. 1234–1246, Nov./Dec. 1995.

[4] D. Casadei, G. Serra, A. Tanti, and L. Zaroi, "Matrix converter modulation strategies: a New general approach based on space-vector representation of switch state," *IEEE Trans. on Ind. Electronics*, vol. 49, No. 2, pp. 370–381, April 2002.

[5] W. H. Kwon and G. H. Cho, "Analysis of static and dynamic characteristics of practical step-up nine-switch matrix converter," *Proc. Inst. Elect. Eng. B*, vol. 140, pp. 139–146, 1993.

[6] G. S. Zinoviev, A. Y. Obuchov, W. A. Otchenasch, and W. I. Popov, "Transformerless PWM AC boost and buck-boost converters," (In Russian), *Technicznaja Elektrodinamika*, T2, pp. 36–39. Nacjonalnaja Akademia Nauk Ukrainy, Kijev 2000.

[7] Z. Fedyczak and P. Szcześniak, "Study of matrix-reactance frequency converter with buck-boost topology," *PELINCEC 2005*, Warsaw 2005, CD-ROM.

[8] Z. Fedyczak, P. Szcześniak, and I. Korotyeyev, "Generation of matrix-reactance frequency converters based on unipolar matrix-reactance choppers," *Proc. of PESC'08*, Rhodes 2008.

[9] Z. Fedyczak, PWM AC voltage transforming circuits, (In Polish), Zielona Góra University Press, Zielona Góra, 2003.

[10] C. T. Rim, D. Y. Hu, and G. H. Cho: "Transformers as Equivalent Circuits for Switches: General Proofs and D-Q Transformation – Based Analyses," *IEEE Trans. on Ind. Apl.* vol. 26, No. 4, 1990.

[11] J. Chen and D. T. Ngo, "Graphical phasor analysis of three-phase PWM converters," *IEEE Trans. On Power Electron.*, vol. 16, No.5, pp. 659–666, 2001.

[12] W. H. Kwon and G. H. Cho, "Analysis of non-ideal step down matrix converter based on circuit DQ transformation," *IEEE Power Electronics Specialists Conference*, PESC'91, Cambridge, USA, pp. 825–829, 1991.

[13] Z. Fedyczak, R. Strzelecki, and K. Sozański, "Review of three-phase PWM AC transformer topologies and applications," *Conf. Proc. of SPEDAM 2002*, pp. B5–19–24, Ravello 2002.

[14] R. D. Middlebrook and S. Ćuk, "A general unified approach to modelling switching-converter power stages," PESC'76 Conf. Rec. pp.18–34, 1976.

[15] Z. Fedyczak, "Four-terminal chain parameters of averaged AC models of non-isolated matrix-reactance PWM AC line conditioners," *Archives of Electrical Engineering*, vol. 50, No. 4, pp. 395-409, 2001

Appendix B

Small Signal Transfer Functions of Analyzed MRFC

$$
\mathbf{G}_{qd0\hat{\mathbf{x}},\hat{\mathbf{u}}} = \frac{\begin{bmatrix} D_S^2 q^2\left(1+R_L C_L\left(s+j\omega_L\right)\right)+C_F\left(R_L D_1^2+L_L\left(1+R_L C_L\left(s+j\omega_L\right)\right)\left(s+j\omega_L\right)\right)\left(s+j\omega\right) \\ D_S q\left(1+R_L C_L\left(s+j\omega_L\right)\right) \\ R_L D_1^2+L_L\left(1+R_L C_L\left(s-j\omega_L\right)\right)\left(s-j\omega_L\right) \\ D_1 D_S q R_L \end{bmatrix}}{R_L C_F C_L L_F L_L \det(s\mathbf{I}-\mathbf{A})} = \frac{\begin{bmatrix} G_1 \\ G_2 \\ G_3 \\ G_4 \end{bmatrix}}{G_3+G_1 L_F\left(s+j\omega\right)},
$$

$$
\mathbf{G}_{qd0\hat{\mathbf{x}},\hat{d}}=\mathbf{G}_{qd0\hat{\mathbf{x}},\hat{\mathbf{u}}}+
$$

$$
+\frac{\begin{bmatrix} -qR_L\left(G_2-D_S G_2\right)-qG_6\left(qD_S G_3-D_S G_4+j\omega_L L_L G_2\right) \\ q\left[D_1\left(1+D_S G_6\right)R_L-X_{LL}G_6+R_L C_L\left(s+j\omega_L\right)\left(D_1 R_L-X_{LL}G_6\right)\left(1+C_F L_F\left(s+j\omega\right)^2\right)+L_F\left(qD_S G_2 G_6+C_F\left(R_L D_1\left(1+D_S G_6\right)-X_{LL}G_6\right)\left(s+j\omega\right)\right)\left(s+j\omega\right)\right] \\ qL_F\left(s+j\omega\right)\left(D_S G_6\left(qG_3-G_4\right)+G_2\left(R_L-D_S R_L+X_{LL}G_6\right)\right) \\ -q\left[R_L\left(L_L G_6\left(sD_S+j\omega_L\left(2D_S-1\right)\right)-D_1^2 R_L\right)+L_F\left(qD_S G_6\left(qD_S^2 R_L-G_4\right)-R_L C_F\left[D_1^2 R_L-G_6 L_L\left(sD_S+j\omega_L\left(2D_S-1\right)\right)\right]\left(s+j\omega\right)\right)\left(s+j\omega\right)\right] \end{bmatrix}}{G_5\left(G_3+G_1 L_F\left(s+j\omega\right)\right)},
$$

where: $D_1=\left(D_S-1\right)$, $X_{LL}=j\omega_L L_L$, $G_5=\omega L_F\left(q^2 D_S^2\left(R_L\omega_L C_L-j\right)+D_1^2 R_L\omega C_F\right)+\omega_L L_L\left(R_L\omega_L C_L-j\right)\left(1-C_F L_F\omega^2\right)-D_1^2 R_L$, $G_6=1+j\omega_L C_L R_L$,

$$
\det(s\mathbf{I}-\mathbf{A})=\frac{R_L D_1^2+L_L\left(1+R_L C_L\left(s+j\omega_L\right)\right)\left(s+j\omega_L\right)+L_F\left[D_S^2 q^2\left(1+R_L C_L\left(s+j\omega_L\right)\right)+C_F\left(R_L D_1^2+L_L\left(1+R_L C_L\left(s+j\omega_L\right)\right)\left(s+j\omega_L\right)\right)\left(s+j\omega\right)\right]}{R_L C_F C_L L_F L_L}.
$$

A Modular AC/DC Rectifier Based on Cascaded H-bridge Rectifier

H. Iman-Eini[1,2], Sh. Farhangi[1], JL. Schanen[2],

1: School of Electrical and Computer Engineering, University of Tehran, Tehran, Iran
2: G2ELab, ENSIEG B.P.46, 38402 St Martin d'Heres cedex, Grenoble, France
Email: Imaneini@g2elab.inpg.fr

Abstract— In this paper, a modular AC/DC rectifier for converting the medium voltage (MV) to the low voltage (LV) levels is presented. The proposed rectifier includes two stages: the cascaded H-bridge (CHB) converter and the isolated DC/DC converters. The CHB converter is directly connected to the MV levels and it eliminates the necessity for heavy and bulky step-down transformers, corrects the input power factor, and maintains the voltage balance across the capacitor voltages. The second stage includes the high frequency parallel-output DC/DC converters, which are connected to the primary DC buses. The second stage provides the galvanic isolation, regulates the output voltage, and attenuates the low frequency voltage ripple ($2f_{line}$) generated by the first stage. The active load-current sharing technique is utilized to balance the load power among the parallel converters. The detailed analysis for modeling and control of the proposed structure is presented. The validity and performance of the proposed topology is verified by a laboratory scaled prototype.

Keywords— Output-parallel converters, modular topologies, cascaded H-bridge rectifier, active load-current sharing.

I. INTRODUCTION

Due to rapid decline in cost and the availability of highly reliable, low loss semiconductors with high frequency switching capability, it can be expected that power electronics products and solutions will play an important role in the future electrical networks. It is possible, e.g., to replace typical AC/DC rectifiers, including step down MV/LV transformers, with the multilevel converters to size down the rectifiers and to decrease the negative impact of the rectifiers on the supply network.

The multilevel converters achieve high-voltage switching by means of a series of voltage steps, each of which lies within the ratings of the individual power devices [1]. Among the multilevel converters, the cascaded H-bridge topology (CHB) is particularly attractive in high-voltage applications, because it requires the least number of components to synthesize the same number of voltage levels. Additionally, due to its modular structure, the hardware implementation is rather simple and the maintenance operation is easier than alternative multilevel converters.

Different applications for the CHB converter have been proposed in the literature [2-5]. In this paper, a modular step-down AC/DC converter based on CHB converter is presented. The converter structure is shown in Fig.1. The proposed design eliminated the necessity for standard step-down transformer, which is both heavy and bulky. The application of modular topology and cascaded H-bridge rectifier makes it possible to connect the modules in series directly at medium voltage level. The proposed structure can be utilized as an active rectifier in Uninterruptable Power Supplies (UPS) or in high-power Adjustable Speed Drives (ASDs). Another application is in the Power Electronic based Transformers with DC links (PETs) [6].

The main control challenges about the modular AC/DC converter are the voltage and power balancing among the series-input and parallel-output converters, respectively. Several references have studied the DC-bus voltage balance problem in the CHB converters, and proposed different control strategies using low-frequency modulation techniques, like in [7, 8]. Some other references have proposed control methods to achieve active load current sharing among the parallel-output converters [9, 10]. However, the application of the later methods for the AC/DC converters is not straightforward.

In this paper, a new control strategy is presented to ensure that the capacitor voltages converge to the reference value, even if the series H-bridges do not match perfectly together or have different power losses. The input current is programmed to be sinusoidal and in phase with the input voltage. However, it is possible to adjust the input power factor to control both the active and reactive powers. An active load-current sharing method is also presented to balance the load power among the parallel cells. The performance of the converter and validity of the new approach are verified by experimental results on a laboratory scaled prototype.

II. SYSTEM CONFIGURATION

The power structure shown in Fig. 1 includes two parts: the front-end CHB converter and the parallel-output DC/DC converters which are explained in the following paragraphs.

978-1-4244-1741-4/08/$25.00 ©2008 IEEE

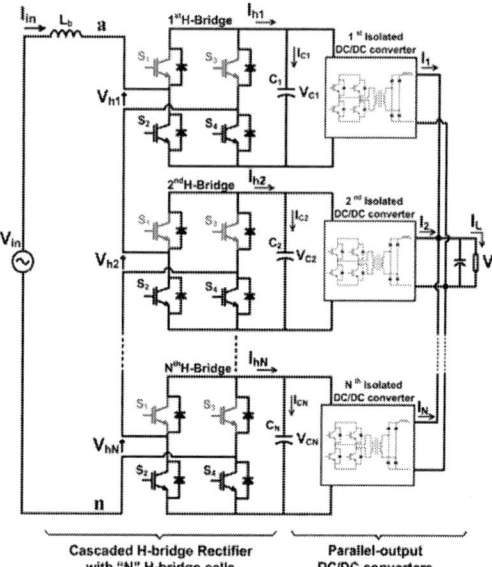

Fig.1 A modular AC/DC converter based on CHB converter (upper switches in the CHB converter can be replaced by the fast diodes, in unidirectional applications)

A. Cascaded H-Bridge Converter

As it can be seen from Fig.1, CHB converter has N H-bridge cells connected in series. Each H-bridge consists of four power switches (with anti-parallel diodes) and a DC bus capacitor. Each capacitor feeds a high-frequency DC/DC converter. It is worth noting that the unidirectional rectifier can be realized from bidirectional rectifier by turning off the upper switches of the H-bridge cells (or by replacing the upper switches with relatively fast diodes). In this work, the bidirectional CHB converter is analyzed, and the results can be used for the unidirectional converter as well.

In Fig. 1, the AC terminal voltage of the rectifier, V_{an}, can be written as follows:

$$V_{an} = V_{h1} + V_{h2} + \cdots + V_{hN} \qquad (1)$$

$$V_{hi} = h_i \cdot V_{Ci} \quad , i = 1,2,...,N \qquad (2)$$

where V_{hi}, V_{Ci}, and h_i are the AC terminal voltage, the capacitor voltage, and the switching function of the ith H-bridge (or cell), respectively. Assuming $V_{C1} = V_{C2} = \cdots = V_{CN} = V_C$, where V_C is the reference voltage of DC buses, each cell can generate three voltage levels: $+V_C$, $-V_C$, and zero on the AC side. So, using N H-bridge cells a maximum of 2N+1 different voltage levels are obtained to synthesize V_{an}. Applying Kirshhoff's Voltage Law (KVL) at the input voltage loop yields:

$$V_{in} = V_{an} + L_b \frac{dI_{in}}{dt} \qquad (3)$$

where V_{in} is the input voltage, I_{in} the input current, and L_b the input inductance which is used to shape the input current. Applying Kirshhoff's Current Law (KCL) for each cell leads to:

$$I_{hi} = h_i \cdot I_{in} \quad , i = 1,...,N \qquad (4)$$

where I_{hi} is the current of ith H-bridge and is a function of the input current. The equations (1)-(4) describe a Linear Time Varying system with one input (V_{in}) and N+1 states (V_{C1} to V_{CN} and I_{in}). The CHB controller should determine the switching functions, h_1 to h_N, to maintain voltage balancing.

B. Parallel-Output DC/DC Converters

The second part of the rectifier (Fig.1) contains the parallel-output DC/DC converters. These converters are connected to the distinct DC buses of the CHB rectifier, and are joined to each other at the output side. This extremely modular structure prepares a low-voltage and highly-stable DC output. Additionally, it provides the galvanic isolation between the input and output, regulates the output voltage, and reduces the low frequency voltage ripple generated by the first part, which is inherent to the power factor correction.

In the second stage, different type of DC/DC converters can be utilized; however, the full-bridge converter is the best one in the terms of efficiency and voltage stress [11]. We apply the interleaving technique among the parallel-output cells to achieve several advantages such as lower current ripple in the output capacitor, faster transient response to load changes [12], and improved power handling capabilities. In addition, the filter requirement is reduced leading to a higher power density, and possibly higher efficiency of the overall system [10].

Several attempts have been made to combine the functions of the PFC stage (first part) and the isolation stage (second part) to reduce the number of solid-state switches and overall cost. However, the advantage of the two-stage approach is that the CHB converter provides regulated intermediate bus voltage, which facilitates the design optimization of both converters with respect to efficiency. Since in a two stage front-end converter the design of both stages can be more easily optimized, the overall performance is improved as compared to a single-stage topology. Thus, we use a two stage design and by a proper design, the flexibility, reliability and performance of the whole converter are improved.

III. MODELLING AND CONTROL OF THE MODULAR AC/DC CONVERTER

The modular AC/DC converter consists of two parts and each part can be controlled independently from the other one because the DC bus capacitors (C_1 to C_N) are large enough to decouple the operation and the control of both stages, as illustrated in Fig.1.

A. Control of the Cascaded H-bridge Converter

The main challenges associated with the cascaded rectifier control are shaping the input current, controlling the input power factor, and holding the DC bus voltages at the desired reference value. The CHB converter, in the rectification mode, aims to achieve N equal DC voltages across the capacitors (C_1 to C_N). However, this can become difficult if the series H-bridges have slightly different characteristics or have different power losses. When implementing very high voltage converters, even the parasitic stray capacitances to earth can lead to unwanted scaling effects and leads to unequal voltage distribution among the series connected converters.

Fig.2 shows the basic block diagram of the proposed controller. It consists of the analog and digital parts. The analog controller generates the PWM signal, Q. This controller consists of two control loops: the inner current loop and the outer voltage loop. The voltage loop contains a PI controller to regulate the total voltage of DC buses to the reference value, i.e. $\sum V_{Ci} = N \cdot V_C$. The digital controller generates a square-wave synchronized signal, V'_{in}, from the input voltage. The sync signal has the same frequency as the input voltage and its phase is adjusted by the digital controller. The square-wave signal is filtered by a low pass Butterworth filter and its output is multiplied by the output of PI controller. Using this method, a pure sinusoidal reference is generated and the input power factor (or the reactive power) is also controlled. After generating the reference current, the inner current loop programs the input current to follow the reference current $I_{in}{}^*$.

The task of digital controller is to keep voltage balancing across the DC link capacitors. This controller determines the appropriate switching functions, h_1 to h_N, for the series-connected H-bridges. Each switching function h_i (i=1, ..., N) corresponds to four operating modes: "0", "+1", "-1", and PWM. The switching functions are determined by the digital controller and are applied to the H-bridge cells.

In Fig.2(b), the operating mode "0" corresponds to the conduction of bottom switches (S_2, S_4). In modes "+1" and "-1" the diagonal switches (S_1, S_4) and (S_2, S_3) are turned on, respectively. In mode "+1" the corresponding AC terminal voltage is "+V_C" and in mode "-1" it is "-V_C". In PWM mode, the gate signals, g_1 to g_4, drive the corresponding cell. These signals are obtained from Q, the output of analog controller, as follows:

$$g_1 = S\overline{Q}, \quad g_2 = \overline{V\,S} \tag{5}$$

$$g_3 = \overline{S}Q, \quad g_4 = \overline{\overline{S}Q} \tag{6}$$

where S is the sign of input voltage and it is one if the input voltage is positive; otherwise it is zero.

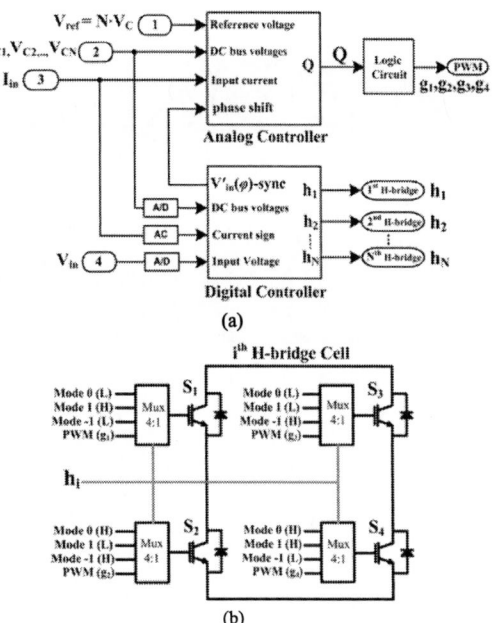

Fig 2 (a) Block diagram of the proposed controller for the CHB converter, (b) drive circuit for i^{th} H-bridge cell

To take advantages of both low frequency (stepped modulation) and high frequency (PWM) modulation techniques, the hybrid modulation method is employed which is shown in Fig.3. In this method, the input voltage V_{in} is divided into equal sections with the scale of V_C (V_C is the reference of primary DC links). The voltage region K is then defined as follows (see Fig.3):

$$(K-1) \cdot V_C < |V_{in}| < K \cdot V_C \,\,, K = 1,...,N \tag{7}$$

Region K is where the magnitude of input voltage, $|V_{in}|$, lies between $(K-1)V_C$ and KV_C. The minimum number of cells to synthesize the multilevel waveform, V_{an}, is equal to the closest integer greater than (V_m/V_C), where V_m is the peak input voltage. The following benefits can be achieved by utilizing the hybrid modulation technique:

- Considerable reduction in size and volume of the input inductance L_b; because the input inductance will not tolerate voltage more than V_C.
- Reduction in THD and EMI at the input side.
- Low switching loss; because at each time only one cell works in high frequency mode.

The digital controller performs the control algorithm to maintain the voltage balancing across the DC link capacitors, while the analog controller regulates the sum of DC bus voltages to $N \cdot V_C$. The proposed control rules, defined hereafter, aims to synthesize the waveform shown in Fig.3 and to maintain the voltage balancing across the capacitors.

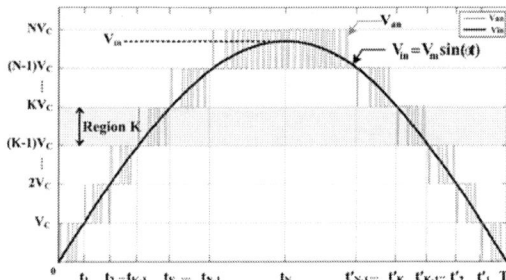

Fig.3 Definition of voltage regions for K=1 ,..., N

1. If $V_{in} > 0$, $I_{in} > 0$, and the voltage region is K, then (K-1) cells with the lowest DC bus voltage are chosen to be charged in mode "+1", the Kth cell in PWM mode, and the rest in mode "0".

2. If $V_{in} > 0$, $I_{in} < 0$, and the voltage region is K, then (K-1) cells with the highest DC bus voltage are chosen to be discharged in mode "+1", the Kth cell in PWM mode, and the rest in mode "0".

3. If $V_{in} < 0$, $I_{in} > 0$, and the voltage region is K, then (K-1) cells with the highest DC bus voltage are chosen to be discharged in mode "-1", the Kth cell in PWM mode, and the rest in mode "0".

4. If $V_{in} < 0$, $I_{in} < 0$, and the voltage region is K, then (K-1) cells with the lowest DC bus voltage are chosen to be charged in mode "-1", the Kth cell in PWM mode, and the rest in mode "0".

To perform the above rules, the digital controller takes the voltage and current samples with the sampling frequency f_o ($f_o < f_{PWM}$). Then, the region of input voltage, K, is updated according to (7), the control algorithm is performed, and the appropriate switching functions, h_1 to h_N, are determined. The switching functions are applied to the H-bridge cells and the corresponding operating modes are selected by the multiplexers (Fig 2(b)). This procedure is repeated in the next sampling periods. Thus, the voltage of DC buses is controlled by adjusting the average current fed to the H-bridge cells, over the mains half-cycle.

B. Modeling of the Parallel-Output DC/DC Converters

The isolated DC/DC converter in the isolation stage is modeled under the following assumptions:

1) Each power switch in the on state is modeled by a constant voltage source $V_{on,s}$ and a series resistance r_{ds} and in the off state by an infinite resistance.

2) Each diode in the on state is modeled by a constant voltage source $V_{on,d}$ and a series resistance r_d , and in the off state by an infinite resistance.

3) The parasitic capacitances are neglected.

4) Isolation transformer is assumed to be ideal.

5) The DC/DC converter works in the CCM mode.

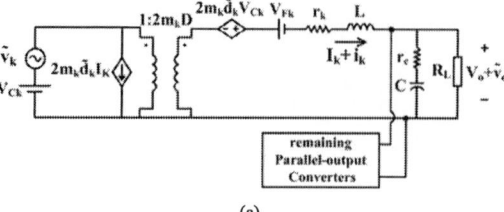

(a)

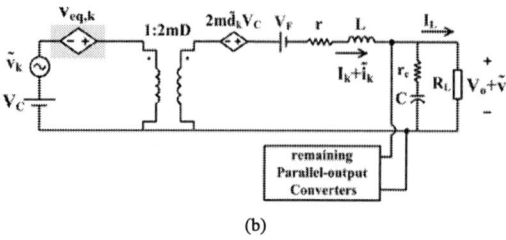

(b)

Fig.4 (a) Large signal averaged model of the kth parallel cell , where $m_k=m+m_{pk}$, $r_k=r+r_{pk}$ and $V_{Ck}=V_C+V_{pk}$).(b) Simplified mode of Fig.4(a) when the parasitic terms and input DC offset of the kth cell is modeled with the DC voltage source $V_{eq,k}$

To simplify the representation of equations, in the remainder of this manuscript we use the small letter 'x' as a large signal parameter, the capital letter 'X' as a DC value, and '$\tilde{x}$' symbol as a small signal variable, i.e. $x = X + \tilde{x}$.

According to the above assumptions and the modeling method discussed in [13, 14], the large signal averaged model of the PWM full bridge converter is realized and shown in Fig.4(a). The illustrated transformer in Fig 4 is an analytical model that is valid for the entire frequency domain.

In Fig.4, V_{CK}, V_O, and I_K are the DC values of the input voltage, output voltage and the load current, respectively. Also $\tilde{v}_k$, $\tilde{v}_o$ and $\tilde{i}_k$ are the corresponding small signal values. The variable D represents the duty cycle of the full-bridge converter and is defined as: $D = t_{on}/T_s \leq 0.5$, where t_{on} is the time duration in which the diagonal switches are on, and T_s is the switching period. Also $\hat{d}$ represents the corresponding small signal value of the duty cycle. The parameters m, r_c, V_F and r are the transformer ratio, equivalent series resistance of the capacitor, the equivalent averaged voltage drop and the equivalent averaged resistance, respectively. For V_F and r, it is derived:

$$V_F = 2(V_{on,d} + V_{on,s} \cdot 2mD) \qquad (8)$$

$$r = r_L + (0.5+D)(2r_d + r_{t2}) + 2Dm^2(2r_{ce} + r_{t1}) \qquad (9)$$

where r_L is the equivalent series resistance of the inductor, r_{t1} the resistance of the primary winding and r_{t2} the resistance of the secondary winding. Equations (8) and (9) have been derived using the principle of energy conservation [13].

In practice, when N ideal DC/DC converters are paralleled at the output side, the load current is divided equally among them. However, a very small mismatch among the parallel converters can lead to unequal load current sharing. The main sources of mismatch are from the transformers ratio (due to the, e.g., leakage flux), the primary voltage offsets, and the equivalent series resistances. Although this issue will not lead to instability and runaway condition in the modular structure because of the CHB rectifier control, but it will cause the unequal thermal stress on the components of parallel converters and low performance of the converter.

We assume that the transformer ratio (m_k), the cell input voltage (V_{Ck}), and the equivalent averaged resistance (r_K) of the kth cell can be written as:

$$m_k = m + m_{pk}, \quad m_{pk} \ll m \quad , k = 1, \ldots, N \quad (10)$$

$$V_{Ck} = V_C + V_{pk}, \quad V_{pk} \ll V_C \;, \quad k = 1, \ldots, N \quad (11)$$

$$r_k = r + r_{pk} \quad , \quad r_{pk} \ll r \;, \quad k = 1, \ldots, N \quad (12)$$

where m, V_C and r are the desired values of each H-bridge and m_{pk}, V_{pk} and r_{pk} are the parasitic terms or deviations from the nominal values. Considering m_k, V_{Ck} and r_k in the large-signal model shown in Fig.4(a) and analyzing the circuit, a simplified model is achieved, where the effect of parasitic terms is modeled with a controlled DC voltage source, $V_{eq,k}$, as

$$V_{eq,k} = \frac{\left(2m_{pk}(D + \tilde{d}_k)V_{Ck} + r_{pk}I_k\right)}{2mD} + V_{pk}$$

$$\approx \frac{m_{pk}}{m}V_C + \frac{r_{pk}I_k}{2mD} + V_{pk} \quad , k = 1, \ldots, M \quad (13)$$

where $V_{eq,k}$ is a controlled DC voltage source that represents the effects of mismatches and the input DC offset of the kth converter cell. Fig.4(b) shows the simplified model of the kth cell according to (13). In the simplified scheme, the controlled current source has been deleted; because, it is parallel with the input voltage and does not affect the output voltage, v_o, and the cell current, i_k.

The DC voltage, $V_{eq,k}$, acts such as a DC offset and causes the current offset among the parallel cells. The amount of current offset, $I_{off,k}$ ($k = 1, .., N$), is obtained as follows:

$$\begin{bmatrix} I_{off1} \\ I_{off2} \\ \vdots \\ I_{offN} \end{bmatrix} = \frac{2mD}{Nr} \begin{bmatrix} (N-1) & -1 & -1 & -1 \\ -1 & (N-1) & -1 & -1 \\ & & \vdots & \\ -1 & -1 & -1 & (N-1) \end{bmatrix} \cdot \begin{bmatrix} V_{eq1} \\ V_{eq2} \\ \vdots \\ V_{eqN} \end{bmatrix} \quad (14)$$

From (14), it is concluded that a little voltage offset in an H-bridge cell will cause large current offsets among the parallel cells. For example, if the number of parallel cells is N = 5, r = 0.6 Ω, and 2mD ≈ 1, then an offset of 3 V (0.5 %) in the first cell will cause the following offsets: I_{off1} = +4 A and I_{off2} to I_{off5} = -1 A.

C. Control of the parallel-output converters

The controller of the parallel-output converters has two main functions: regulation of the output voltage and the load current sharing among the parallel converters. The proposed controller is shown in Fig.5. This controller uses a voltage control loop and N similar current control loops.

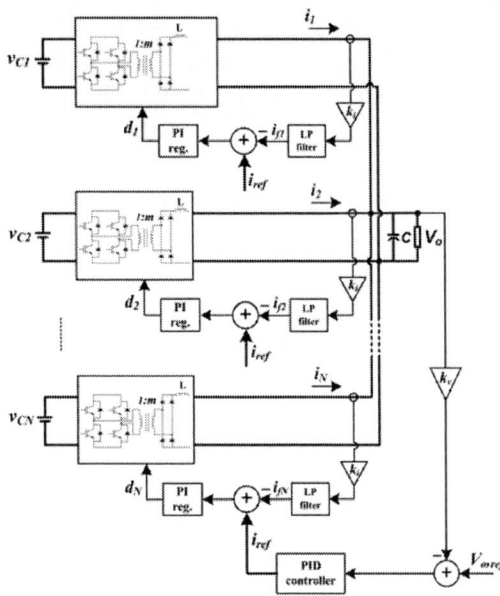

Fig 5 Proposed scheme for control of parallel-output cells

In Fig.5, the current controllers are designed to provide equal load current sharing among the parallel cells. As it can be seen, the inductor current of the kth cell is sensed and scaled to the proper magnitude. Then, it passes through a low pass filter. This filter is utilized to limit the loop bandwidth and to decrease the noise power. The filter output is compared with the reference current i_{ref} generated by the voltage controller. The error signal is entered to the PI regulator and its output determines the switching duty cycle d_k. The duty cycle expression can be written as follows:

$$d_k = D + D_{off,k} + \tilde{d} \quad , k = 1, \ldots, N \quad (15)$$

where d_k is the duty cycle of the kth cell, D is the DC term, and $D_{off,k}$ is a small DC term which is generated by the kth current controller to cancel the offset term defined in (14). According to Fig.4(b), $D_{off,k}$ is derived as follows:

$$2m(D + D_{off,k}) \cdot (V_C + V_{eq,k})$$

$$= 2m(DV_C + D_{off,k}V_C + DV_{eq,k} + D_{off,k}V_{eq,k}) \quad (16)$$

In (16), the last term is negligible in comparison with the other terms. As the desired voltage is $2mDV_C$, the remaining terms are set to zero. So, it is obtained:

$$2m(D_{off,k}V_C + DV_{eq,k}) = 0 \rightarrow D_{off,k} = -\frac{DV_{eq,k}}{V_C} \quad (17)$$

As a result, the effects of mismatches and the input voltage offsets (modeled as a part of $V_{eq,k}$) are canceled by the inner current loops and equal DC operating points are achieved for all cells, as follows

$$I_k = I_L / N \quad , \quad V_{Ck} = V_C \quad , \quad D \approx \frac{V_O}{V_C} \cdot \frac{1}{2m} \cdot \frac{R_L + r/N}{R_L} \quad (18)$$

It is worth noting that the variable $\tilde{d}$ in (15) is a small signal duty cycle which corresponds to the input voltage and load current variations. This term controls the dynamical behavior of the parallel cells. Also, the external voltage loop, shown in Fig. 5, is used to regulate the output voltage of parallel converters to the reference value $V_{o,ref}$. The output of PID controller is used as reference current for all parallel cells. This controller compensates the load side variations.

IV. VERIFICATION BY EXPERIMENTAL RESULTS

In this section, the validity of the proposed approach and the designed controllers are verified by experimental results on a laboratory scaled prototype. The hardware prototype is a 1800 W single-phase AC/DC converter based on a 7-level CHB converter which is illustrated in Fig.6. The prototype consists of three series converters at the input side and three parallel cells at the load side. The input AC voltage is 230 Vrms and it can vary between 70 and 260 Vrms. The reference voltage for intermediate DC buses is $V_C = 125$ V and the output voltage $V_o = 100$ V. The digital control unit is implemented based on a TMS 320F2812 DSP controller. Other principal parameters of the prototype are given in Table I.

It is worth noting that the low voltage power MOSFETs (MOSFET+Internal Body Diode with the break down voltage of 200 V) have been intentionally used in the prototype to demonstrate a scale-down of the real situation. However, in medium voltage levels, the IGBTs would be the best choice owing to better voltage and current ratings.

Table I. Principal parameters of the experimental prototype

Parameter or Component	Symbol	Value
Number of series H-bridges (cells)	N	3
Nominal power	P	1800 W
Input AC voltage	V_{in}	70-260 V rms
Intermediate DC Bus voltage	V_C	125 V
Output DC voltage	V_o	100 V
Input line inductance	L_b	2 mH
Switching frequency of parallel cells	f_s	15 kHz
Intermediate DC bus capacitor	C_i	1 mF, 250V
Output capacitor	C_o	470 uF, 160V
Output inductor	L_o	220 uH
Steady state Duty cycle	D	0.4
Isolation transformer turn ratio	M	1:1
Solid-state power switches	S_i	IRFP250, 200V
Secondary diodes	D_i	BYW81, 200V
Equivalent average resistance	r	0.26 Ω

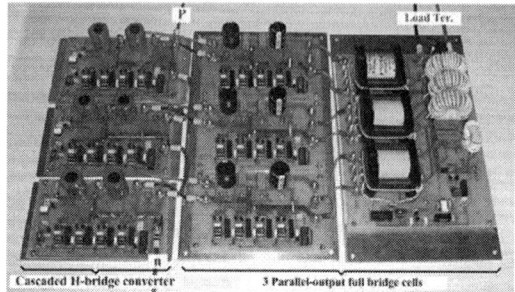

Fig.6 Modular step-down AC/DC rectifier based on the cascaded H-bridge converter (V_{in}=230 Vrms, V_C=125 V, V_O=100 V, N=3, P=1800 W)

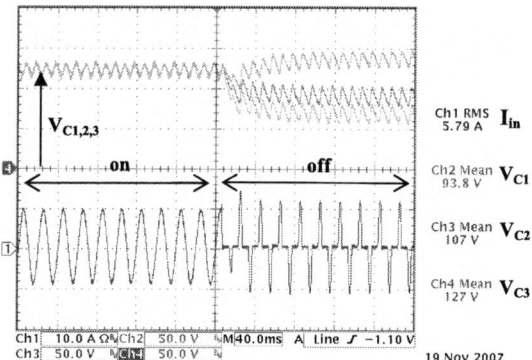

Fig.7 DC bus voltages and the input current waveforms when the CHB controller is on and off (Ch$_1$:I$_{in}$-10 A/div, Ch$_{2,3,4}$:V$_{C1}$ to V$_{C3}$ -50 V/div)

The first experiment investigates the voltage balancing and the current control mechanisms in two cases: with and without the CHB controller. In this experiment, the CHB converter is connected to the resistive loads and the loads are selected as P$_1$ = 625 W, P$_2$ = 488 W and P$_3$= 312 W. The voltage of DC buses and the waveform of input current are shown in Fig.7.

According to Fig.7, when the CHB converter control is turned off, the voltage of DC buses diverge from the reference value V$_C$ = 125 V. Without the controller, the current waveform is distorted, low order harmonics appear, and the power loss increases significantly. In addition, the power factor reduces considerably, leading to more generation of reactive power and reduction of real power capacity.

Fig.8 shows the operation principle of the CHB converter. Using 3 series connected H-bridges, a 7-level voltage waveform is obtained to synthesize the AC terminal voltage V$_{an}$. The red waveform shows the input current I$_{in}$. As it can be seen, the input current is in phase with the AC terminal voltage V$_{an}$. Hence, the power factor is close to one and the current waveform is sinusoidal.

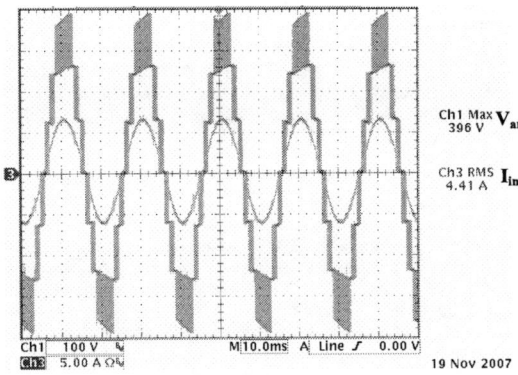

Fig 8 Operation principle of the CHB converter: AC terminal voltage V_{an}, and the input current I_{in}

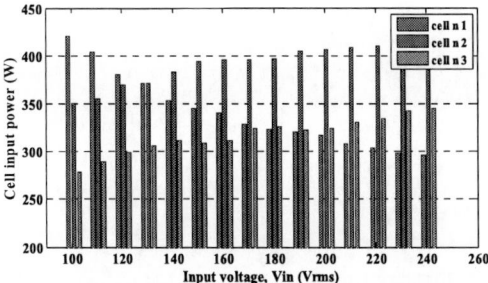

Fig.9(a) Parallel cells input power versus the AC input voltage, without the current controllers

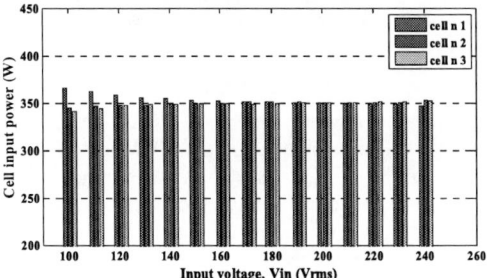

Fig.9(b) Parallel cells input power versus the AC input voltage, in presence of current controllers

The second experiment investigates the power balancing mechanism among the parallel-output cells. Here, the parallel-output cells are connected to the CHB converter and fed by the primary DC buses. In Fig.9, the average input power transferred to the parallel cells is demonstrated as a function of the source voltage. To investigate the controller behavior, the power and voltage points are measured using the power analyzer and are shown for two cases: with the current control loops in Fig.9(a) and without the current control loops in Fig.9(b).

As it can be seen from Fig.9(a), when the current control loops are not utilized, the load power is divided unequally among the parallel cells and is dependent on the converter operating point. On the other hand, without the current controllers, the load sharing among the parallel cells is different, even the cells have been made in the same manner. This is due to the slight mismatches of the converters and the DC offsets at the primary DC buses.

Fig.9(b) shows the cells powers when the current loops are utilized. It is observed that the load power is divided equally and symmetrically among the parallel cells. In this test, each current loop corrects the cell duty cycle according to (17) and the current sharing is well done.

Next experiment verifies the dynamical behavior of the CHB converter control and the current controllers under load steps. In this experiment, the input voltage is 230 V and the load power changes between 900 W and 1800 W, in a step-wise manner. The input DC bus voltage of a cell (e g , V_{CI}), the output voltage (V_o), and the load current are shown in Fig .10 .

As it can be seen from Fig.10, in spite of quick load variations, the CHB converter controls the voltage of primary DC buses and they follow the reference voltage (V_C =125 V) by a DC offset less than 1%. Also, the output voltage closely follows the voltage reference V_o = 100 V and the isolation stage attenuates significantly the voltage ripple at the primary DC buses.

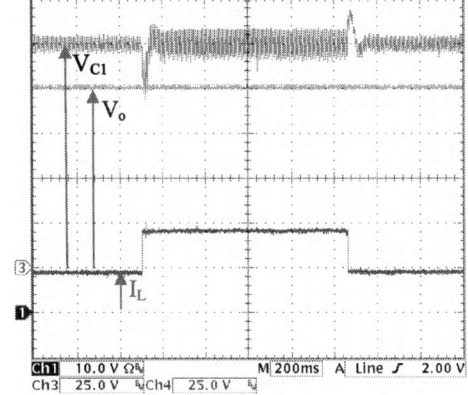

Fig .10 Dynamical behavior of the CHB converter control and the current controllers under load step: primary DC bus voltage (red), output DC bus voltage (green), and the total load current (blue)

In the last experiment, the interleaving technique is realized by out of phase driving of each paralleled converters. The result of interleaving is an effective increase in ripple frequency and also ripple reduction of the output capacitor (see Fig.11). An active clamp circuit is also applied to limit the voltage overshoots on the secondary diodes. The illustrated waveforms in Fig.11 confirm the usefulness of the proposed circuit. Without the active clamp circuit, the secondary diodes should tolerate voltage overshoots as high as 200 V.

179

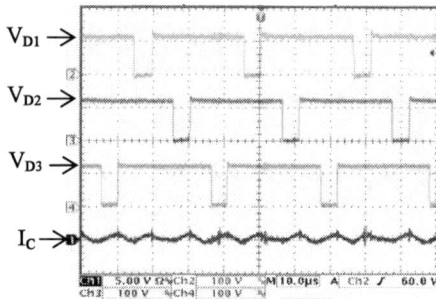

Fig.11 Applying the interleaving technique and the secondary diode clamp circuit. Ch_1: the output capacitor current (5 A/div), $Ch_{2,3,4}$: secondary diode voltage waveforms (100 V/div)

V.CONCLUSION

In this paper, a modular step-down AC/DC rectifier was presented. The modular structure offers important advantages during design, testing, manufacturing and service stages. The same modules can be used for different voltages (or currents) by stacking (or paralleling) the appropriate number of modules depending on the working voltage (or power). The proposed design eliminated the necessity for standard step-down transformer, which is both heavy and bulky. The application of modular topology and cascaded H-bridge rectifier makes it possible to connect the modules in series directly at medium voltage level.

Additionally, a new control strategy was presented to ensure that the primary capacitor voltages converge to the reference value, even if the series H-bridges do not match perfectly or have different power losses. The proposed controller programs the input current to be sinusoidal and in phase with the input voltage. Analytical formulas were derived to calculate the effects of mismatches on the equal load current sharing. The validity of the design and the rectifier performance were verified by the experimental results.

REFRENCES

[1] J. S. Lai and F. Z. Peng, "Mutilevel converters – A new breed of power converters," *IEEE Trans. Ind. Appl.*, vol. 32, no. 3, pp. 509–517, May/Jun. 1996.

[2] Han, B. Bae, S. Baek, and G. Jang, "New Configuration of UPQC for Medium-Voltage Application," *IEEE Trans. Power Delivery*, vol. 21, no. 3, pp. 1438-1444, July 2006.

[3] S. Bernet, "Recent developments of high power converters for industry and traction applications," in *IEEE Tran. Power Electronics*, vol. 15, no. 6, pp. 1102 - 1117, Nov. 2000.

[4] A. Dell'Aquila, M. Liserre, V. G. Monopoli, and P. Rotondo, "An Energy-Based Control for an n-H-Bridges Multilevel Active Rectifier," *IEEE Trans. Industrial Electronics*, vol. 52, no. 3, pp. 670 – 678, June 2005.

[5] A. Rufer, N. Schibli, C. Chabert, and C. Zimmermann, "Configurable front-end converters for multicurrent locomotives operated on 16 2/3 Hz ac and 3 kV dc systems," in *IEEE Trans. on Power Electronics*, vol. 18, no. 5, pp. 1186- 1193, September 2003.

[6] H. Iman-Eini, Sh. Farhangi, and J-L. Schanen, "Design of power electronic transformer based on cascaded H-bridge

multilevel converter, " in *Proc. IEEE Int. Symposium on Industrial Electronics (ISIE 2007)*, pp. 877-882 .

[7] A. J. Watson, P. W. Wheeler, and J. C. Clare, "A complete harmonic elimination approach to DC link voltage balancing for a cascaded multilevel rectifier, " *IEEE Trans. Ind. Electron.*, vol. 54, no. 6, pp. 2946 – 2953, December 2007.

[8] P. Zanchetta, D. Gerry, V. G. Monopoli, J. C. Clare, and P. W. Wheeler, "Predictive current control for multilevel active rectifiers with reduced switching frequency," *IEEE Trans. Ind. Electron.*, vol. 55, no. 1, pp. 163 – 172, January 2008.

[9] J.W. Kim, J.S. Yon, and B.H. Cho, "Modeling, control, and design of input series output parallel converter for high-speed trainpower system," *IEEE Trans. Industrial Electronics*, vol. 48, no. 3, pp. 536 - 544, June 2001

[10] R. Giri, V. Choudhary, R. Ayyanar, and N. Mohan, "Common-duty-ratio control of input-series connected modular DC-DC converters with active input voltage and load-current sharing," *IEEE Trans. Industry Application*, vol. 20, no. 6, pp. 1101 –1111, July 2006.

[11] M.T. Aydemir, A. Bendre, and G.Venkataramanan, "A critical evaluation of high power hard and soft switched isolated DC-DC converters," *in Industry Applications Conf.* IAS 2002, vol. 2, pp. 1338-1345.

[12] S. Ang and A. Oliva, Power Switching Converters. CRC Press, 2nd Edition, 2005, Chap. 6.

[13] D. Czarkowski, M.K. Kazimierczuk, "SPICE compatible averaged models of PWM full-bridge DC-DC converter, " in proc. *IEEE Power Electronics and Motion Control conf.*, vol. 1, pp. 488 – 493, Nov. 1992.

[14] V. Vorperian, "Simplified analysis of PWM converters using model of PWM switch. Continuous conduction mode," *IEEE Trans. Aerospace and Electronic systems*, vol. 26, No. 3, pp. 490 - 496, May 1990.

Low Loss Soft Switching Boost Converter

So-Ri Park[1], Sang-Hoon Park[2], Chung-Yuen Won[†] and Yong-Chae Jung[3]

[1,2,3] School of Information and Communication Engineering, Sungkyunkwan University
300 Cheoncheon-dong, Jangan-gu, Suwon, Gyeonggi-do, 440-746, Korea
[1] e-mail: efu27@skku.edu
[2] e-mail : marohachi@skku.edu
[†] e-mail : won@yurim.skku.ac.kr

[3] Department of Electronic Engineering, Namseoul University,
21 Maeju-ri, Seonghwan-eup, Cheonan, 330-707, Korea
[4] e-mail: ychjung@nsu.ac.kr

Abstract — A new soft switching boost converter is proposed in this paper. The conventional boost converter generates switching losses at turn on and off. Because of those, the whole system efficiency is reduced. The proposed converter utilizes soft switching method using an auxiliary switch and resonant circuit. Therefore, the converter reduces switching losses lower than the hard switching. The proposed soft switching boost converter can be applied to photovoltaic system, power factor correction and so forth.

Keywords — Photovoltaic, Power Factor Correction, Soft Switching, ZVS Converters

I. INTRODUCTION

Recently, switch mode power supply has been smaller and lighter because the switching frequency is higher. But as the switching frequency is higher, the periodic loss increases at turn on and turn off. As a result, this loss brings increasing loss of whole system.

Many converters using resonance to reduce switching loss have been presented in papers [1-6]. Among them, many researches using resonance are presented about ZVZCS (Zero Voltage Zero Current Switching) converter that ZVS and ZCS perform simultaneously [1-4].

However, the auxiliary circuit for resonance increases the complexity of circuit and the circuit cost. For some resonant converters with auxiliary switch, main switch achieves soft switching but auxiliary switch performs hard switching. Thus, these converters cannot improve the whole system efficiency owing to switching loss of auxiliary switch [1].

A new soft switching boost converter with an auxiliary switch and resonant circuit is proposed in this paper. The resonant circuit is consisted of a resonant inductor, two resonant capacitors, two diodes and an auxiliary switch. It makes partial resonant path for main switch perform soft switching at zero voltage. Moreover, the auxiliary switch also achieves soft switching by resonant circuit. Compared with other soft switching converters, the

proposed converter improves the whole system efficiency by reducing switching loss than other converters in the same frequency. In this paper, some simulation results are presented for 600[W], 30[kHz] prototype boost converter using IGBT. And then it shows the experimental results to verify the operational principle of the proposed circuit.

II. PROPOSED SOFT SWITCHING BOOST CONVERTER

A. Configuration of the Proposed Converter

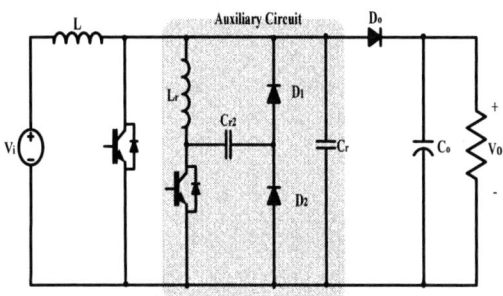

Fig. 1 Circuit diagram of the proposed converter.

Fig 1 shows the proposed low loss soft switching boost converter. The main switch and the auxiliary switch of the proposed circuit are capable of soft switching by an auxiliary switching block, consisting of an auxiliary switch, two resonant capacitors, a resonant inductor and two diodes.

B. Equivalent circuit analysis

Key waveforms of the proposed converter are shown in Fig. 2. Each of the main switch and auxiliary switch PWM waveforms are S_1 and S_2. Main switch and auxiliary switch current, voltage, the resonant inductor current are analyzed as Fig. 2.

The circuit operation in one switching cycle can be divided into nine stages, as shown in Fig. 3.

978-1-4244-1741-4/08/$25.00 ©2008 IEEE

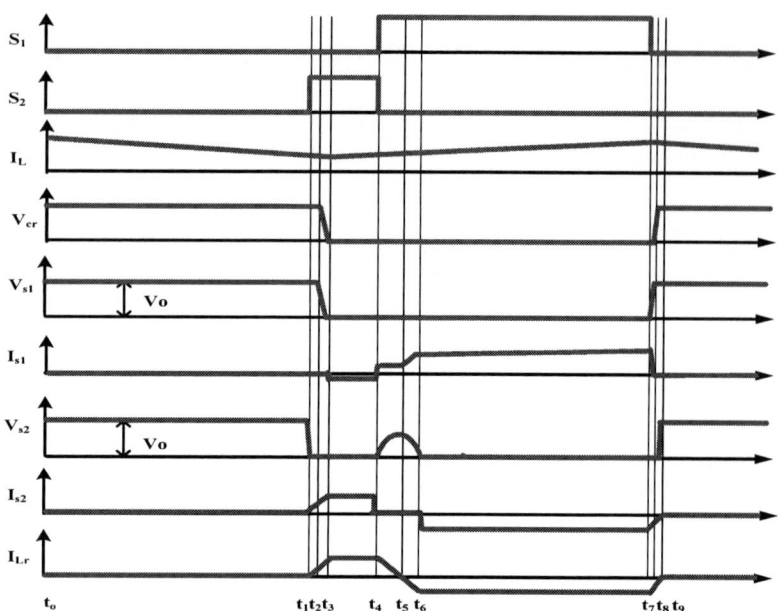

Fig. 2 Key waveforms of the proposed converter.

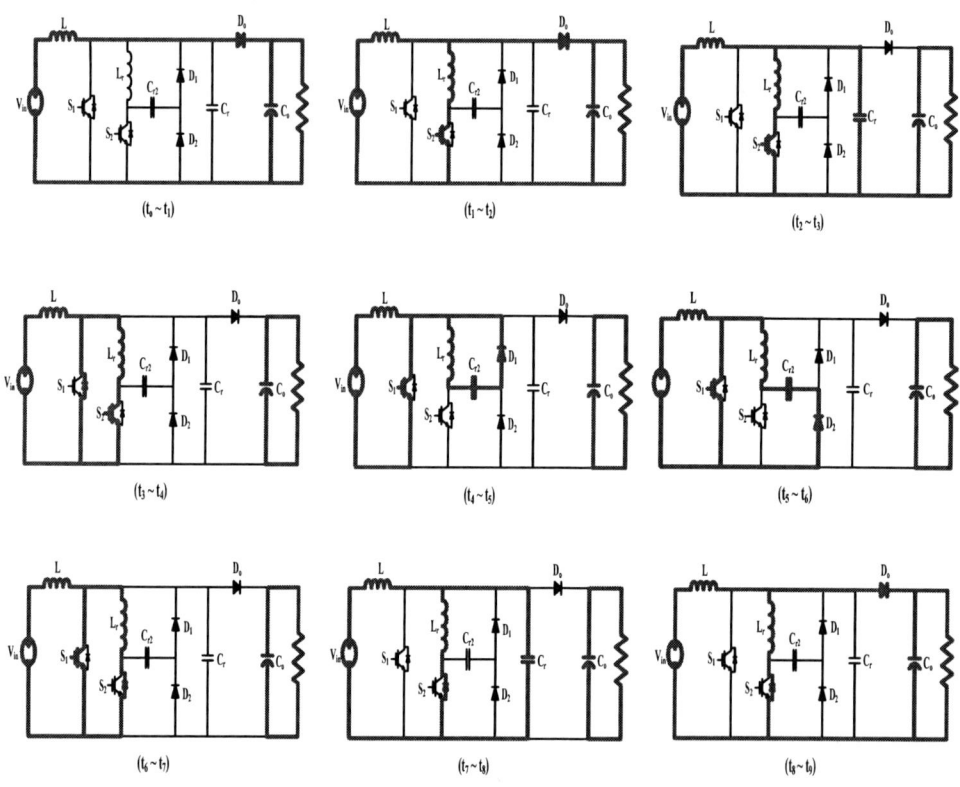

Fig. 3 Equivalent circuits during one switching period.

- **Stage 1 (t₀ - t₁)**

Main and auxiliary switches turn off. The energy of main inductor transfers to load through main diode. When the auxiliary switch turns on, the stage 2 begins. Main inductor current in this mode is expressed as the equation (1).

$$I_L = I_L(t_9) - \frac{V_o - V_{in}}{L}t \tag{1}$$

- **Stage 2 (t₁ - t₂)**

Turning on the auxiliary switch, the resonant inductor current begins to increase linearly from zero. The resonant inductor current I_{Lr} is equal to main inductor current at t_2, stage 2 is finished. At this interval, main and resonant inductor currents are given by equation (2) and (3).

$$I_{Lr}(t) = \frac{V_o}{L_r}t \tag{2}$$

$$I_L(t) \approx I_{min} \tag{3}$$

- **Stage 3 (t₂ - t₃)**

As soon as the resonant inductor current and the main inductor current are equal, main diode is turned off. Then the resonant capacitor C_r is discharged through resonant path C_r and L_r. Finishing the resonance, the resonant capacitor voltage equals to zero. And stage 3 is finished at t_3. At t_2, the resonant capacitor voltage equal to output voltage V_o. Thus the time interval for two currents becoming equal since t_1 is determined as equation (4).

$$t_{12} = \frac{I_{Lm}}{(V_o / L_r)} \tag{4}$$

The resonant period of resonant capacitor and resonant inductor, is

$$t_r = \frac{\pi}{2}\sqrt{L_r C_r} \tag{5}$$

The resonant impedance is $Z_r = \sqrt{\dfrac{L_r}{C_r}}$. And the resonant inductor current and resonant capacitor C_r voltage are given by,

$$I_{Lr}(t) = I_{min} + \frac{V_o}{Z_r}\sin\omega_r t \tag{6}$$

$$V_{cr}(t) = V_o \cos\omega_r t \tag{7}$$

- **Stage 4 (t₃ - t₄)**

As soon as resonant capacitor C_r voltage equals to zero, the body-diode of main switch is naturally turned on. When the body-diode is turned on, the main switch

voltage equals to zero. At that time the turn-on signal is given to the main switch with zero voltage condition. In this stage, the main inductor current is given by,

$$I_L(t) = I_{min} + \frac{V_{in}}{L}t \tag{8}$$

- **Stage 5 (t₄ - t₅)**

At stage 4, main switch turns on under zero voltage. At that time, auxiliary switch is turned-off on the same condition. This is the beginning of stage 5. In this stage resonant inductor L_r and resonant capacitor C_{r2} start a resonance. After half-resonance of L_r and C_{r2}, the current of L_r is zero. Then stage 5 is finished and C_{r2} has been fully charged by resonance.

$$I_L(t) = I(t_4) + \frac{V_{in}}{L}t, \; I_{Lr} = I_{Lr}(t_3)\cos\omega_a t \tag{9}$$

$$\omega_a = \frac{1}{\sqrt{L_r C_{r2}}}, \; Z_a = \sqrt{\frac{L_r}{C_{r2}}} \tag{10}$$

- **Stage 6 (t₅ - t₆)**

After stage 5 finishes, the current flow of the resonant inductor L_r changes backward and this stage starts. At stage 6, the reverse resonance of L_r and C_{r2} through main switch and D_2 occurs. In this time, C_{r2} voltage is fall to zero by resonance. Then the resonance of L_r and C_{r2} is finished and C_{r2} voltage is zero. During the stage 6, the resonant capacitor voltage is discharged as equation (12).

$$V_{Cr2}(t) = Z_a I_{Lr}(t_3)\sin\omega_a t \tag{11}$$

$$V_{Cr2}(t_5) = Z_a I_{Lr}, \; V_{Cr2}(t_6) = 0 \tag{12}$$

- **Stage 7 (t₆ - t₇)**

After C_{r2} voltage becomes zero, the body diode of auxiliary switch is turned on. And the current flows through body diode resonant inductor - main switch freewheeling path. By PWM algorithm, the main switch turned off, this stage is ended. In this interval the amount of resonant inductor current is equal at t_3. But the current flow is backward.

$$I_L(t) = I_L(t_6) + \frac{V_{in}}{L}t \tag{13}$$

$$I_{Lr}(t) = -I_{Lr}(t_3) \tag{14}$$

- **Stage 8 (t₇ - t₈)**

The sum of the two inductor currents charges resonant capacitor C_r in this stage. If resonant capacitor voltage equals to output voltage, this stage is finished.

$$I_{Lr} = I_L(t_7) - \{I_L(t_7) + I_{Lr}(t_3)\}\cos\omega_r t \tag{15}$$

$$Z_r \{I_L(t_7) + I_{Lr}(t_3)\} > V_o \tag{16}$$

The equation (16) is a zero voltage condition.

● Stage 9 (t₈ - t₉)

At t_8, the resonant capacitor C_r has been charged and the main diode voltage is zero. Therefore, the main diode turns on and the resonant inductor current decreases linearly toward zero. After the current equals zero, the stage 9 is ended and next switching cycle starts. In this stage, the main inductor current and the resonant inductor current are given by equation (17), (18).

$$I_L(t) = I_L(t_7) - \frac{V_o - V_{in}}{L} t \qquad (17)$$

$$I_{Lr} = -I_{Lr}(t_3) + \frac{V_o}{L_r} t \qquad (18)$$

C. Delay time and voltage conversion characteristics.

To achieve the zero voltage switching, a delay time (D_{elay}) of main switch PWM is required. The minimum delay time must be satisfied the following equation. The time is consisted of the resonant time between L_r and C_r, and the time that the resonant inductor current becomes input current.

$$T_{Delay} \geq \frac{I_{in} L_r}{V_o} + \frac{\pi}{2} \sqrt{L_r C_r} \qquad (19)$$

During the delay time, the auxiliary switch is turned on. And the PWM signals of main and auxiliary switch are as the Fig.4.

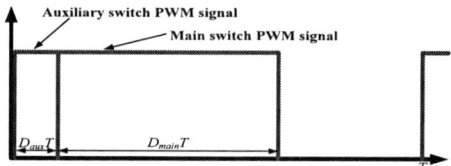

Fig. 4 The PWM signals.

$$V_L = V_{in}\left(D_{aux} - \frac{T_r}{T}\right) + V_{in}\left(D_{main} + \frac{T_r}{T}\right) + (V_{in} - V_o)(1 - D_{aux} - D_{main}) \qquad (20)$$

In steady state operation, the main inductor voltage V_L is given by the equation (20). After the delay time, the energy is accumulated the main inductor. It is as the turn on the switch at the conventional boost converter. Thus the auxiliary turned on is the effect of the total duty is longer.

At the equation (20), T_r is the resonant time between the resonant inductor L_r and resonant capacitor C_r. D_{main} is main switch duty ratio and D_{aux} is auxiliary switch duty ratio. From the main inductor voltage, the voltage conversion ratio is defined as follow.

$$\frac{V_o}{V_{in}} \approx \frac{1}{1 - (D_{aux} + D_{main})} \qquad (21)$$

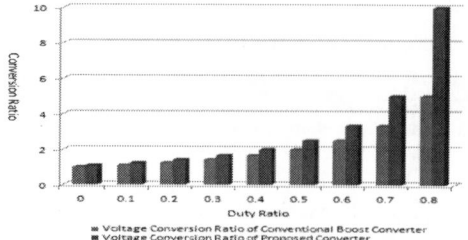

Fig. 5 Voltage conversion ratio.

Fig. 5 is the voltage conversion ratio to main switch duties of two converters. The green bar is to the conventional boost converter and the red one is to the proposed converter in this paper. The last one has duty margin because of the auxiliary switch turned on. Thus it can be established the higher voltage under the same duty ratio.

III. SIMULATION

Simulation parameters are presented in Table. 1. In this paper, the proposed circuit is simulated by PSIM 6.0 developed by Powersim Inc.

Fig. 6 shows the schematic diagram for simulation. The simulation performed under 30[kHz] switching frequency and 130~170[V] input voltage. To control the output voltage V_o, V_o is sensing. Resonant capacitor voltage is also sensing for zero voltage switching of main switch. Main switch is turned on and off when resonant capacitor C_r voltage is zero. Thus, PWM signal is generated by multiple of two sensing voltages.

Table I. Simulation parameters.

Input voltage	V_i	130~170[V]
Output voltage	V_o	400[V]
Switching frequency	f_s	30[kHz]
Resonant capacitor1	C_r	3.3[nF]
Resonant capacitor2	C_{r2}	30[nF]
Resonant inductor	L_r	20[μH]
Main inductor	L_m	560[μH]

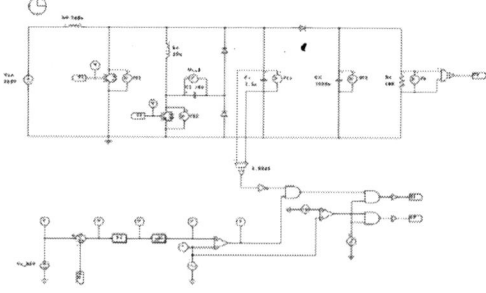

Fig. 6 Schematic diagram for simulation.

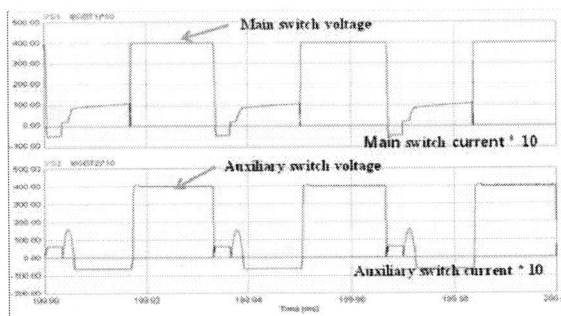

Fig. 7 Waveforms of main and auxiliary switch.
(Upper: main switch voltage and current, Lower: auxiliary switch voltage and current)

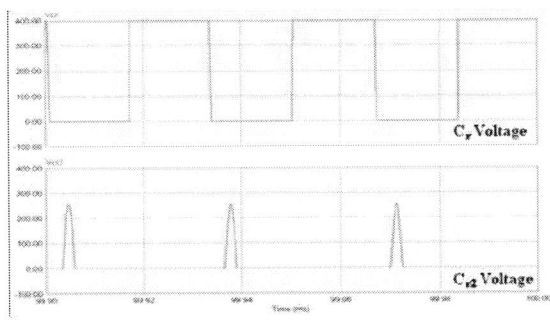

Fig. 8 Waveforms of resonant capacitor voltage.
(Upper: resonant capacitor C_r voltage, Lower: resonant capacitor C_{r2} voltage)

Fig. 7 illustrates two pairs of switch voltage and current waveforms. Before the main switch turns on, the body diode is turned on. As a result, the main switch is capable of zero voltage switching. Auxiliary switch also achieves soft switching. Fig. 8 shows two resonant capacitor voltage waveforms.

By resonance with resonant inductor, the resonant capacotor C_{r2} is charged and discharged like a sine waveform. At input voltage 130~170[V], output voltage is controlled to 400[V]. In this case, the output power is 600[W].

IV. EXPERIMEENAL RESULT

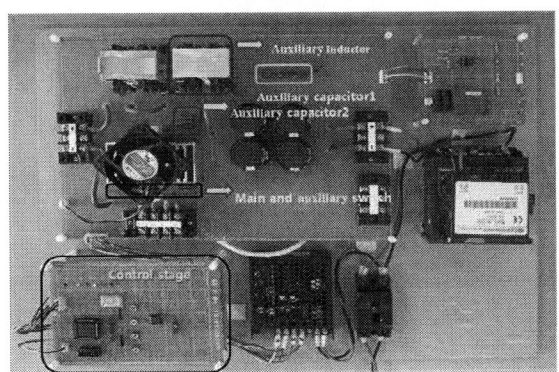

Fig. 9 Experimental setup.

Fig. 9 is a photograph of experimental setup. There are inductors, switches, diodes and capacitors. Control stage is consisted of the TL494 PWM controller and EPM7064 EPLD. EPLD used for delay of the main switch PWM which is controlled by TL494. Delay time is defined by the equation (19). At the equation, L_r is 20[μH], C_r is 3.3[nF]. Thus the minimum delay time is 0.5[μsec].

Figure 10, 11, 12 are experimental waveforms of two switch and main diode.

Fig. 10 is the voltage and current waveform of the main switch and Fig. 10(b) is zoomed in from the Fig. 10(a). Before the main switch is turned on, the body diode offers freewheeling path. Thus, it is able to confirm that the main switch is turned on under the zero voltage condition through the Fig. 10(b).

Fig. 11 is the auxiliary switch waveform. Auxiliary switch turns on and off under the zero voltage as well. Because of the resonance between C_{r2} and L_r, the resonant capacitor C_{r2} is discharged. Thus the voltage of auxiliary switch is equal to zero and the body diode is turned on. The waveform Fig. 11(b) is zoomed in from Fig. 11(a) at the zero voltage switching point. In the circle interval of Fig. 11(a) the resonance between the resonant L_r and capacitor C_{r2} occurs, and the part current charges the parasitic capacitor of the switch after the switch turns off. And before the body diode is turned on, this current charges the parasitic capacitor. But during this interval, switch is off state. Thus this current does not related to loss.

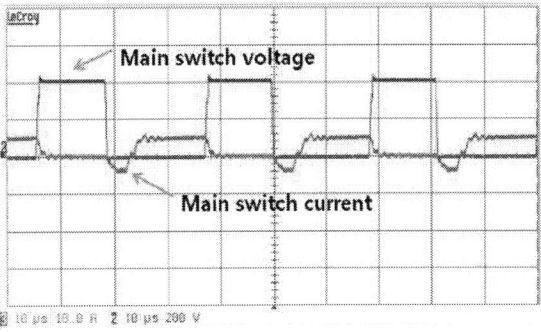

(a) (Y-axis : Voltage – 200[V], Current – 10[A], X-axis : 10[μs])

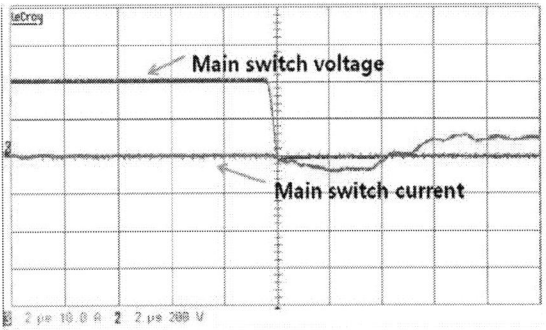

(b) Zoom in (Y-axis : Voltage – 200[V], Current – 10[A], X-axis : 10[μs])
Fig. 10 Main switch voltage and current waveforms.

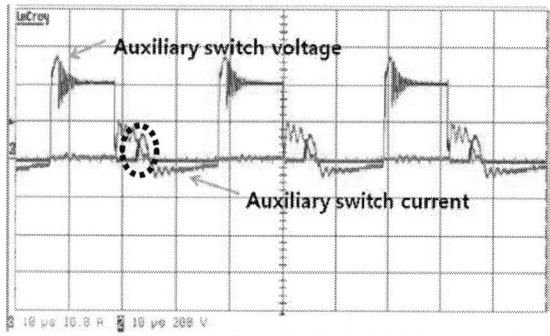

(a) (Y-axis: Voltage – 200[V], Current – 10[A], X-axis: 10[μs])

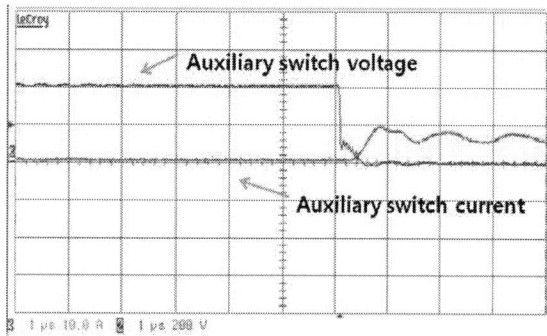

(b) Zoom in (Y-axis: Voltage – 200[V], Current – 10[A], X-axis: 1[μs])
Fig. 11 Auxiliary switch Voltage and current waveforms.

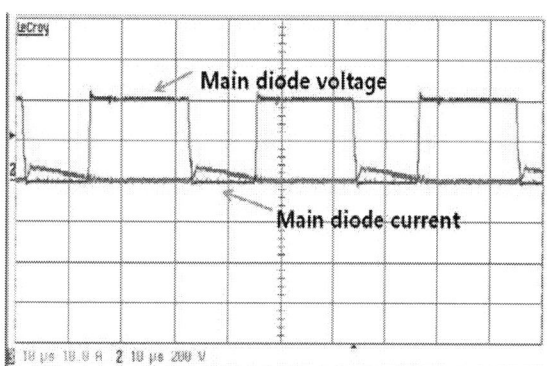

(a) (Y-axis: Voltage – 200[V], Current – 10[A], X-axis: 10[μs])

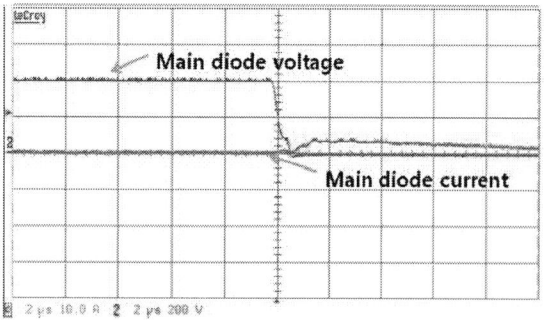

(b) Zoom in (Y-axis: Voltage – 200[V], Current – 10[A], X-axis: 2[μs])
Fig. 12 Main diode voltage and current waveforms.

Table II.
Total efficiencies of the circuits in the hard-switching and
the proposed soft-switching converter.

I_o	Load	Switching	Pi (W)	Po (W)	Effi. (η)
0.4[A]	26.67[%]	H	177.5	160.0	90.12
		S	168.3	160.0	95.06
0.8[A]	53.33[%]	H	353.4	320.0	90.55
		S	315.2	320.0	95.17
1.2[A]	80.00[%]	H	533.3	480.0	91.01
		S	501.3	480.0	95.58
1.5[A]	100.0[%]	H	657.1	600.0	91.31
		S	623.5	600.0	96.23

For the hard switching and the proposed soft switching boost converters, total efficiencies are measured for various load currents in Table II. As shown in Table II, the efficiency of the proposed converter is higher than hard switching converter.

V. CONCLUSION

In this paper, a new soft switching boost converter is proposed using auxiliary switch and resonant circuit. Main switch performs soft switching in zero voltage by resonant capacitor and inductor, and so is the auxiliary switch. The proposed converter has been analyzed in detail. The operation principles and theoretical analysis of the proposed converter have been confirmed by simulation and prototype of 600[W] and 30[kHz]. The proposed converter is suitable for high efficiency converter, photovoltaic DC/DC converter, power factor correction and so on.

ACKNOWLEDGMENT

This work is the outcome of a Manpower Development Program for Energy & Resources supported by the Ministry of Knowledge and Economy (MKE)

REFERENCES

[1] Jain, N., Jain, P.K., Joos, G., "A zero voltage transition boost converter employing a soft switching auxiliary circuit with reduced conduction losses", Power Electronics, IEEE Transactions on Volume 19, Issue 1, Jan. 2004.
[2] Erning D.W., Hefner A.R. Jr., "IGBT model validation for soft-switching applications", IEEE Transactions on Industry Applications, Volume 37, Issue 2, March-April 2001.
[3] Jain N., Jain P., Joos G., "Designing a zero voltage transition boost converter for power factor corrected modular telecom rectifiers", Twenty-Third International Telecommunications Energy Conference, INTELEC 2001. 14-18 Oct. 2001.
[4] Saha, S.S., Majumdar, B., Halder, T., Biswas, S.K., "New Fully Soft-Switched Boost-Converter with Reduced Conduction Losses", Power Electronics And Drive Systems 2005 international confer., On Volume 1, pp. 107 – 112, 16-18 Jan. 2006.
[5] Bodur H., Bakan A. F., "A new ZVT-ZCT-PWM DC-DC converter", IEEE Transactions on Power Electronics, Volume 19, Issue 3, pp. 676-684, May 2004.
[6] Xinke Wu, Junming Zhang, Xin Ye, Zhaoming Qian, "Analysis and Derivations for a Family ZVS Converter Based on a New Active Clamp ZVS Cell", IEEE Transactions on Industrial Electronics, Volume 55, Issue 2, pp. 773 – 781, Feb. 2008.

Methods for Experimental Assessment of Component Losses to Validate the Converter Loss Model

Yi Wang, Sjoerd de Haan and Jan Abraham Ferreira
Electrical Power Processing Group
Delft University of Technology, The Netherlands
Email: yi.wang@tudelft.nl, s.w.h.dehaan@ewi.tudelft.nl and j.a.ferreira@ewi.tudelft.nl

Abstract—This paper introduces a novel loss model concept for performance evaluation and design optimization of power electronics converters based on MathCAD sheet. A dual active bridge (DAB) converter (12 V/360 V, 1 kW and 25 kHz) is used as the test platform. The practical methods for extracting useful component parameters used in loss model and for experimentally assessing losses to verify the loss model are elaborated. With the verified loss model, design optimizations of the DAB converter are done theoretically.

Keywords—Loss model, loss measurement, system optimization, dual active bridge converter.

I. INTRODUCTION

Power loss is the issue that must be considered in the performance evaluation and cooling method design of power electronics systems. Many loss drivers exist in the power electronic system. The voltage, current, switching frequency and temperature all have significant impacts on component losses. These impacts are so nonlinear that it is quite obscure to analyze the combined influence of several varied loss drivers on the system losses. The other factors like the count of paralleled semiconductors, the count of interleaved converter units and different modulation methods make the system loss analysis even more complicated. Therefore, synthetic analysis of the system loss in multi-domain to obtain the optimum operating point is always not straightforward and time-consuming.

Many [1] [2] [3] presented their loss models or loss calculation platforms in order to pre-analyze and optimize the power electronics systems before putting them into real operation. [1] made a loss model only considering the conduction losses in the system to evaluate and compare the performance of three-level converter and full bridge converter, missing the switching loss and core loss which take a large part in the high frequency system losses. In [2], the circuit loss parameters such as the switching energy of used semiconductors and loss map of the core are first calibrated as functions of operating points, e.g. current and flux density magnitude ΔB. The stray inductance effect on the switching energy are pleasantly considered here. The losses are calculated incorporating the calibrated loss parameters and the circuit waveforms from the circuit simulation software. [3] utilizes Pspice to generate the interested waveforms, MathCAD to calculate the loss in semiconductors and Maxwell to calculate the

loss in magnetics. These loss models are verified to be able to get reliable results, however, they can only calculate the loss at one set of operating point (input/output voltage, load and frequency) in one simulation cycle and the varying effect of temperature on the on-state resistance of the semiconductors are not considered.

This paper first proposes a novel loss model completely based on MathCAD sheet (section II). It is able to synthetically and analytically calculate and analyze the component losses and system losses as functions of voltages, power, switching frequency, designed temperature of the heatsink on the semiconductors. Different modulations are also considered in the loss model. This loss model concept is implemented and validated on a DAB converter (12 V - 360 V, 1 kW and 25 kHz). The key part of this paper is that the practical methods for extracting useful component parameters used in loss model (section III) and for experimentally assessing losses to verify the loss model (section IV) are emphasized in this paper. The application of this loss model in design will be presented in section V.

II. A NOVEL LOSS MODEL

The basic idea of the proposed loss model is to program in the MathCAD in such a way that the component losses and system losses could be defined as a function of many variables: operating points (v_i, v_o and P), designed circuit parameters (e.g. switching frequency f_s, or leakage inductance of transformer L_{lk} and the count of paralleled switches N_{sw}) and heatsink temperature T_{sk}. If more than one modulation methods exist, an index of *modulation* is set in the loss model for switching among the modulation methods. The outputs of the loss model includes the component losses, system losses and even the junction temperature of semiconductors if needed. Basically, there two main steps in the loss modeling: waveforms modeling and loss calculation. All of the input variables to the loss model exert influences on both steps. This loss model concept could be seen in Fig. 1. In the following, two steps in the loss modeling are elucidated.

A. Waveforms modeling

Generally speaking, it is voltage and current in the circuit that directly results in the losses on electronic com-

978-1-4244-1741-4/08/$25.00 ©2008 IEEE

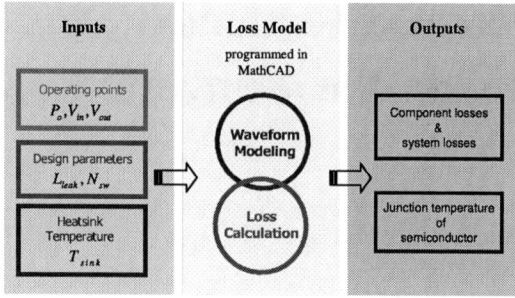

Fig. 1. The proposed loss model concept

ponents, so it is natural to necessitate accurate modeling of voltages and currents in component loss models.

Ideal voltage and current waveforms on components in power electronics converter can be always mathematically modeled according to given modulation principle as a function of variables like time, voltages (v_i, v_o), power, switching frequency and other known component parameters like capacitance, inductance, transformer turn ratio. For the piece-wise linear waveforms which are common in DC-DC converter, they should be modeled in all time-pieces in one cycle.

The current waveform function Eq. (5) of semiconductor switch in flyback converter is given here as an waveform modeling example. The switch on-time is defined as:

$$t_{on}(v_i, v_o, f_{sw}) = \frac{v_o}{(v_o + N \cdot v_i)}) \cdot f_s \qquad (1)$$

and the duty ratio of the semiconductor is:

$$D(v_i, v_o, f_s) = t_{on}(v_i, v_o, f_s) \cdot f_s \qquad (2)$$

where

- flyback transformer turns ratio from the primary to secondary: N. It is assumed known here.
- magnetizing inductance of flyback transformer: L_m. It is assumed known here.
- switching frequency: f_s.
- I_{peak} and I_0 are the highest and lowest current value in the current waveform, they are functions of v_i, v_o, f_s and power.

I_{peak} and I_0 functions are defined as follows:

$$I_{peak}(v_i, v_o, f_s, P) = \frac{N \cdot P}{[1 - D(v_i, v_o, f_s)]v_o} + \frac{v_o[1 - D(v_i, v_o, f_s)]}{2L_m \cdot N} \qquad (3)$$

$$I_0(v_i, v_o, f_s, P) = I_{peak}(v_i, v_o, f_s, P) - \frac{v_i \cdot t_{on}(v_i, v_o, f_s)}{L_m} \qquad (4)$$

The semiconductor current waveform (Eq. (5)) in one switching cycle can be drawn in MathCAD by fixing the input variables at desired values. Modeling the waveforms in this way, one can continue to calculate the RMS current or instantaneous current at any interested time point in one cycle. Voltage waveforms on components and flux density waveform on magnetic core could be modeled in similar way. Note that all the waveforms are defined in the steady state. The current and/or voltage waveforms on components which are main loss sources in the system should always be calculated to feed the following loss calculation with required current and voltage information.

To bring the modeling waveform more close to the real one, non-ideal factors have to be considered in the model. For example, the on-resistance of semiconductor loaded with the different heatsink temperatures and currents will generate varied voltage drop, therefore, the equivalent input or output voltages will be changed in circuit like flyback, full bridge converter, etc. To transfer the same load, the current waveform will be different from the ideal case for sure. This effect can be integrated in the waveform modeling by calculating the equivalent terminal voltage [4] and then the model has one more input variable: temperature of heatsink on semiconductor.

Different modulation methods generate different waveforms in the circuit, so an index for switching among different modulation methods can be also regarded as an input variable for converters with multiple modulation control in varying operating ranges.

Other design parameters of interests in the DAB converter [5] could be the input variables as well. For instance, the number of paralleled switches and the leakage inductance.

B. Loss calculation

Besides the voltage and current characteristics, the physical properties of these components which results in losses should also be well known before the loss calculation. The key component parameters related to the losses are the on-state resistance and switching energy of the semiconductor, DC resistance and the wire arrangement of transformer or inductor windings, magnetic core parameters, ESR of capacitors. Some of these parameters are temperature and/or current dependent. If design interests

$$i_{sw}(v_{in}, v_o, P, f_s, t) = \begin{cases} I_0(v_i, v_o, f_s, P) + \dfrac{v_i}{L_m} \cdot t & \text{if } 0 \leq t < t_{on}(v_i, v_o, f_s) \\ \dfrac{I_{peak}(v_i, v_o, f_s, P) - \dfrac{v_o}{L_m \cdot N} \cdot [t - t_{on}(v_i, v_o, P, f_s)]}{N} & \text{if } t_{on}(v_i, v_o, f_s) < t \leq 1/f_s \end{cases} \qquad (5)$$

don't focus on fixed voltage and power ranges, it is very useful to model the current and temperature dependent effect.

In this paper the verification platform of the loss model concept is the DAB converter with high input current at rated load (about 90 A), so the calibration of on-resistance of low voltage side semiconductor are extremely necessary, for both waveform modeling and loss calculation. Two independent variables having effects on on-resistance are heatsink temperature and current, which are also the input variables to the loss model. The calibration of on-resistance will be presented in next section.

The analytical methods to reliably calculate the losses in power electronic converters are well recorded in the literature and they are suitable to be programmed in MathCAD sheet. The details of implementing these calculation method in MathCAD will not be presented in this paper. The main outputs of the loss model are component losses and system losses. The component losses includes semiconductor losses, transformer and/or inductor losses and capacitor losses and system losses is the sum of all the losses above. Since all the loss functions involve waveform functions, they share the same input variables. The general form of loss functions, e.g. for the DAB converter, is:

$$P_{loss}(v_i, v_o, P, f_s, L_{lk}, T_{sk}, N_{sw}, modulation) \quad (5)$$

The quantities in the parenthesis includes all independent input variables. The combined effect of these variables on the losses could be simply and quickly calculated and analyzed by plotting the loss graphes. If the semiconductor heatsink thermal resistance from ambient to the semiconductor case is known, the junction temperature of chip can be calculated as one output of the loss model as well.

III. KEY COMPONENT PARAMETERS EXTRACTION

The component parameters that are important to the loss calculating accuracy should be calibrated before building the loss model.

A. DAB converter introduction

The proposed loss model concept will be validated on a high current DAB converter with

- the rated input(output) voltage = 12 V (360 V)
- the rated output power = 1 kW
- the switching frequency = 25 kHz
- the transformer turns ratio from secondary to primary = 30

The circuit diagram of DAB converter is shown in Fig. (2). The input bridge consists of 12 MOSFETs with 3 in parallel at each switch position and the output bridge has 4 IGBTs. MOSFETs in two legs of MOSFET bridge are enveloped in two MOSFET power module packages (SK80MD055) and IGBTs in one module (SK15GH067). To improve the current capacity of standard SK80MD055 module, the chips inside were changed to Vishay SUM110N06 by custom design. In between two

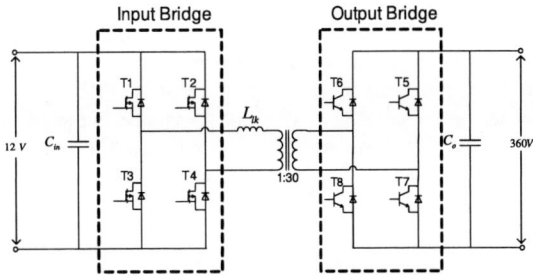

Fig. 2. Dual active bridge circuit diagram

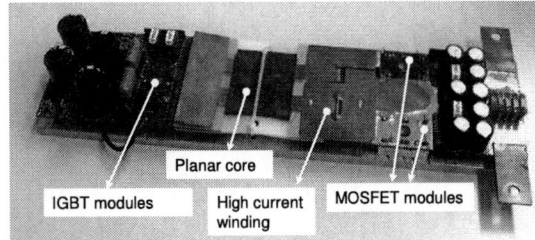

Fig. 3. The picture of DAB converter prototype

switch bridges is the high frequency transformer. DAB utilizes the leakage inductance L_{lk} of the transformer to transfer the energy [5] and it has three modulation methods which could be used in different operating ranges [6] [19] and only rectangular modulation is implemented at present. The picture of the DAB converter is shown in Fig. 3.

Since the input current are quite high (more than 90 A at rated load), the on-resistance of MOSFETs R_{DS-on} generates a large amount of conduction heat. By rough calculation, the conduction loss in MOSFETs make up about 40% of the system losses on this DAB converter. On the contrary, a switching frequency of 25 kHz for the selected fast MOSFETs is not a key loss driver. Therefore, calibrating the R_{DS-on} is the most important before loss calculation.

B. R_{DS-on} calibration

R_{DS-on} of MOSFET is a function of flowing current and junction temperature. Unfortunately, it is very difficult to directly measure the junction temperature. This part proposes a useful and practical method to correlate the R_{DS-on} to the MOSFET case temperature and current. In the experiment with the constant case temperature which can be also regarded as the heatsink temperature, the R_{DS-on} tends to bear an approximate second order dependency on the drain current I_d, then R_{DS-on} can be defined as:

$$R_{DS-on}(I_d) = A + B \cdot I_d + C \cdot I_d^2 \quad (6)$$

A set of A, B, C represents the curve coefficients at certain heatsink temperature. They can be obtained by a 2nd order polynomial fitting from the measured $R_{DS-on} - I_d$ curve.

To include the temperature dependency, A, B and C can be made heatsink temperature dependent (similar to the method in [7]). If only two heatsink temperatures (T_0, T_1) are sampled, the temperature dependency can be linearly modeled. More temperatures are sampled causes higher order of the dependency and more accurate results. Linear temperature dependency can be defined as:

$$A(T_{sk}) = a_0 + a_1 \cdot T_{sk}$$
$$B(T_{sk}) = b_0 + b_1 \cdot T_{sk} \qquad (7)$$
$$C(T_{sk}) = c_0 + c_1 \cdot T_{sk}$$

with

$$a_0 = \frac{A_{T_0} \cdot T_1 - A_{T_1} \cdot T_0}{T_0 - T_1}$$
$$a_1 = \frac{A_{T_0} - A_{T_1}}{T_0 - T_1}$$
$$b_0 = \frac{B_{T_0} \cdot T_1 - B_{T_1} \cdot T_0}{T_0 - T_1}$$
$$b_1 = \frac{B_{T_0} - B_{T_1}}{T_0 - T_1}$$
$$c_0 = \frac{C_{T_0} \cdot T_1 - C_{T_1} \cdot T_0}{T_0 - T_1}$$
$$c_1 = \frac{C_{T_0} - C_{T_1}}{T_0 - T_1} \qquad (8)$$

where A_{T_0}, B_{T_0}, C_{T_0} and A_{T_1}, B_{T_1}, C_{T_1} are two sets of coefficients in Eq. 6 at heatsink temperatures T_0 and T_1, respectively. Then the R_{DS-on} function would be

$$R_{DS-on}(I_d, T_{sk}) = A(T_{sk}) + B(T_{sk}) \cdot I_d + C(T_{sk}) \cdot I_d^2 \qquad (9)$$

In the experiment to calibrate R_{DS-on} of MOSFET, the substrate of one MOSFET module was tightly connected to the temperature plate of Julabo heating circulator (Model: FP50) (see Fig. 4a). The plate temperature is controlled by the heating circulator and so is the module substrate temperature. All the MOSFETs in the module are turned on with the gate voltage 12 V which is the same to that in DAB converter. Therefore, three MOSFETs in parallel at top position and at the bottom positions are all connected in series, causing the module resistance to be 2/3 on-resistance of a single MOSFET chip (see Fig. 4b). The current is injected into the drains of top MOSFETs from DC power supply. For R_{DS-on} calibration, the temperature plate was fixed at 40 °C and 55 °C, respectively, and current was increased at a step of 20 A. The current value is observed by measuring the voltage drop on a 0.1 mΩ shunt in series with the module. R_{DS-on} of a single MOSFET chip is then calculated as 1.5 times of the ratio from measured voltage to current. Two R_{DS-on}- I_d curves were obtained at at 40 °C and 55 °C. From these two curves, all the coefficients required in Eq. 9 are set by curve fitting and Eq. 8. To verify the calibrated R_{DS-on} function, R_{DS-on}- I_d curve is measured again with substrate at 60 °C and it matches well with the calibrated function (see Fig. 5)

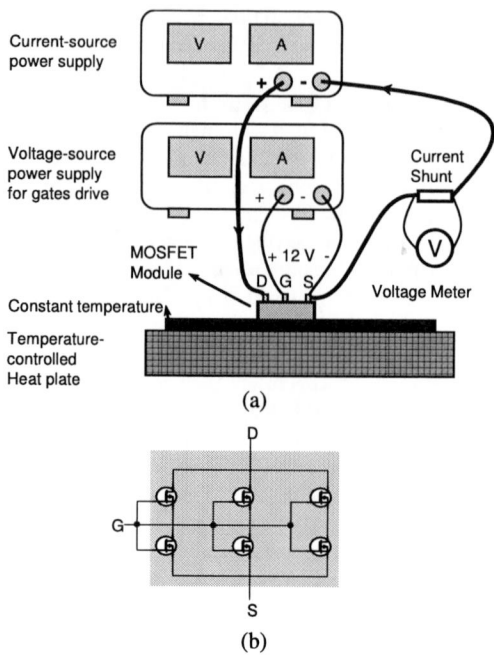

(a)

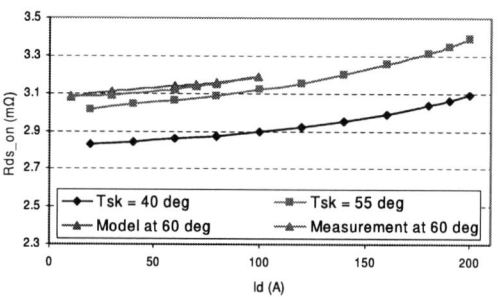

(b)

Fig. 4. (a). The setup of calibrating R_{DS-on} of MOSFET chip. (b). The connection of the MOSFET chips in the module during the calibration

Fig. 5. $R_{DS-on}(I_d, T_{sk})$ curve fitting at 40 °C and 55 °C and verification at 60 °C heatsink temperature

C. Other loss-related parameters

Since switching losses of the fast MOSFETs at 25 kHz will be much smaller than the conduction loss in this high current case, the switching times (t_r and t_f) could simply use the information in the MOSFET chip datasheet without sacrificing much accuracy. For the same reason, the information from the datasheet are used for IGBT loss calculation. For more accurate switching losses modeling under very high frequency, one could refer to [7] [10] [11]. [7] calibrated the switching energy as a function of the junction temperature and the drain current. [10] and [11] took the parasitic elements into account. The proposed loss model is able to incorporate these methods for more precise loss at high frequency operation.

The transformer is another main loss source due to the high current and high turns ratio. The DC resistances of

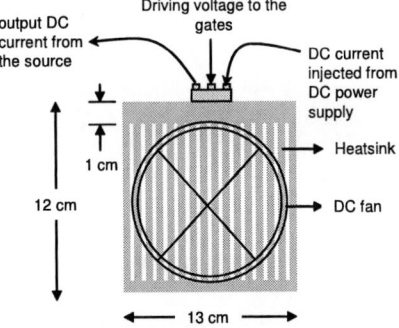

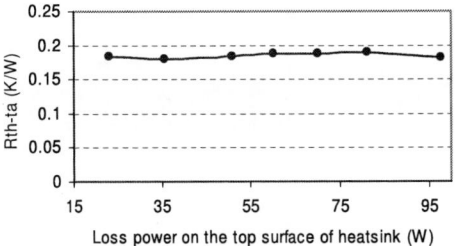

Fig. 7. R_{th-ta} is approximately constant (averaged at 0.185 K/W) at the constant air-blow speed and varying loss power on the top surface. The driving voltage of the heatsink fan is 45 Vdc.

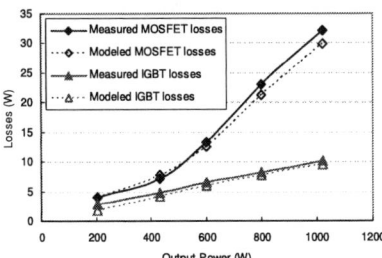

Fig. 8. Comparisons between measured and modeled losses of MOSFETs and IGBTs (with $T_{sk} = 31\,^{\circ}$C and rated input/output voltages)

Fig. 6. The setup to measure the thermal resistance of the heatsink R_{th-ta}. One MOSFET module is placed in the middle of the heatsink top surface and is driven in the resistive mode.

two windings are important for loss calculations and are measured in advance by feeding certain DC current to both windings and measuring the resultant voltage drops. The results are 1 Ω and 0.5 mΩ for the high and low voltage windings.

The capacitor losses cover a small part of the total losses, so the ESR of input and output capacitors are simply measured with impedance analyzer at the fundamental current ripple frequency 50 kHz.

IV. EXPERIMENTAL VERIFICATION

The loss model are validated on component level and system level with very practical experiments.

A. Semiconductor Losses

Many methods [12] [13] using digital oscilloscope to measure the loss in semiconductors were recorded. They are susceptible to the measuring error and time-consuming [18]. This paper addresses a practical and easy way to measure the loss in the MOSFET module.

The basic idea is that the thermal resistance of a plate-fin heatsink with constant air-blowing speed is constant from the inlet air to the surface on which the modules are mounted and the module losses can be calculated by measuring the temperature difference between inlet air and heatsink surface, with known heatsink thermal resistance.

Fig. 6 shows the setup to measure the thermal resistance of the used heatsink from top surface to the inlet air (R_{th-ta}). The dimensions of the heatsink are 13 cm x 12 cm x 10 cm, with 1 mm thick fin and 4 mm fin pitch and the driving voltage of the mounted DC air-blowing fan is 45 V. One MOSFET module is mounted in the middle of the heatsink top surface. The pins connection is the same to Fig. 4b. The gate voltage is lowered to a range that the MOSFETs work in the resistive mode like a resistor. If the fan-driving voltage is kept constantly at 45 V, the R_{th-ta} is found to be approximately the same at 0.185 K/W with varying loss power in the module (see fig.7).

This calibrated heatsink placed under the MOSFET and IGBT modules are used to measure the semiconductor

losses, one heatsink for two MOSFET modules and another one for the IGBT module. The loss is calculated as Eq. (10).

$$P_{sw} = \frac{T_{sk} - T_{inlet}}{0.185 K/W} \qquad (10)$$

where T_{sk} is the heatsink temperature below the module substrate and T_{inlet} is the temperature of the heatsink inlet air. Due to the large cooling power and small thermal resistance, the temperature rise of the heatsink under the module substrates at the rated output power 1 kW is less than 10 $^{\circ}$C. In the loss model, the module heatsink temperature T_{sk} is set at 31 $^{\circ}$C. One could found that the measured losses and modeled losses of the semiconductor modules shown in Fig. 8 matches quite well. The reason of the existing difference could be: (1) the over-ideal calculation of the MOSFET switching losses; (2) When calibrating R_{th-ta}, one module is placed on the heatsink. But, there are two modules on the heatsink when measuring the MOSFET losses and R_{th-ta} should be changed somewhat due to the varied heat spreading pattern; (3)the simple use of information from IGBT datasheet to calculate the losses in IGBT module and (4) the measurement errors. Considering relatively small calculation and parameter extraction effort, this loss model is accurate enough for practical design.

B. Transformer Losses

The transformer losses consist of winding losses and core losses. With the copper foils and PCB traces such planar conductors as the windings, 1D analytical calculation for modeling the winding losses [14] is precise

191

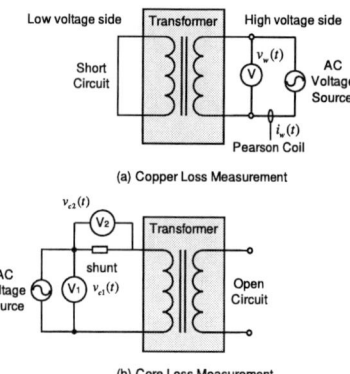

(a) Copper Loss Measurement

(b) Core Loss Measurement

Fig. 9. The circuit diagram for measuring the transformer losses

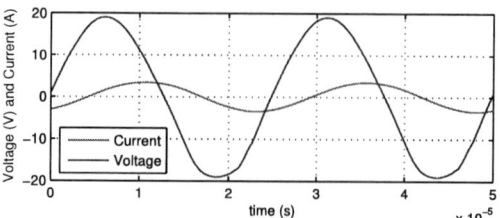

Fig. 10. Voltage $v_w(t)$ and current $i_w(t)$ waveforms at 40 kHz

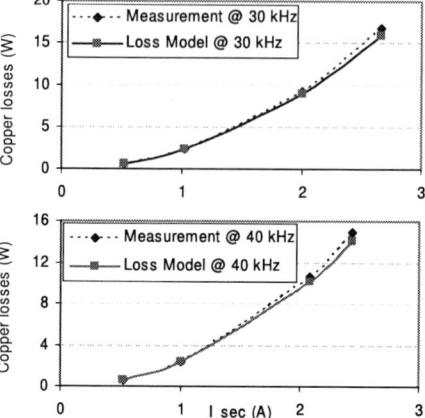

Fig. 11. Comparisons between the measured and modeled core loss at 30 kHz and 40 kHz

enough and suitable for being built in the winding loss model. In the winding loss model, the copper losses covered by the first 20 harmonics of the modeled transformer current waveform are calculated and summed up as the overall winding losses. The modified Steinmetz equation [15] which improves the Steinmetz equation [16] by takeing the magnetic flux changing rate into account is utilized for calculating the ferrite core loss with non-sinusoidal excitation.

During the normal DAB circuit operation, it is difficult to experimentally estimate the loss in the high frequency transformer. This following part introduces how to facilitate and speed up the verification of the transformer loss model in a practical way.

1) Winding Loss: Since the winding losses are calculated as the superposition of losses in the first 20 current harmonics, the calculating accuracy can be guaranteed if the loss model is accurate at certain harmonics. To measure the loss in the transformer winding under certain sinusoidal current excitation, additional tests were done and the circuti diagram of the setup is shown in Fig. 9.

The high current winding is shorted and the low current side is fed with sinusoidal voltage source under certain frequency (Fig. 9a). The current and terminal voltage waveforms are recorded by digital oscilloscope (sampling rate: 10020 sample/s) via Pearson coil (model: 2877) and probe. Since the core excitation voltage is very low due to the shorted winding, the loss in the core can be neglected and the instantaneous product of measured voltage and current can be integrated over one period to obtain the copper loss power (see Eq. 12). Measured voltage and current waveforms at 40 kHz are shown in Fig.10. The measured and modeled winding losses under the excitation currents at 30 kHz and 40 kHz shown in Fig. 11 illustrate the good match between the loss model and measurement.

$$P_w = \frac{1}{T} \cdot \int_0^T i_w(t) \cdot v_w(t)\, \mathrm{d}t \qquad (11)$$

2) Core Loss: Assuming a good accuracy of modified Steinmetz equation, only the parameters in original

Steinmetz equation for core material (3C90) given in the literature are validated for a range of excitation frequency. Similar method in winding loss measurement can be used here in principle, but owing to very small loss in the core the phase shift between terminal voltage and current are very close to 90 degree, which causes the power calculation (instantaneous v and i product integration) very sensitive to the sampling delay of the digital oscilloscope and therefore unacceptably high error [18].

To decrease the phase shift between the voltage and current and gain higher power calculation accuracy, a well-calibrated resistor ($R_{shunt} = 1.13\ \Omega$) is connected in series with the low voltage winding of the transformer in the setup shown in Fig. 9b. The high voltage side is open such that the current flowing through the low voltage winding is only resulted from the core loss. The voltage measurement crosses the resistor and transformer winding and the current is indicated by measuring the voltage drop on the resistor. The core loss is calculated as:

$$P_w = \frac{1}{T} \cdot \int_0^T \frac{v_{c1}(t) \cdot v_{c2}(t)}{R_{shunt}}\, \mathrm{d}t - \frac{1}{T} \cdot \int_0^T \frac{v_{c2}(t)^2}{R_{shunt}}\, \mathrm{d}t \qquad (12)$$

The measured and modeled core loss at 20 kHz, 35 kHz and 100 kHz are compared in Fig. 12. A satisfactory match can be observed except around 12 V input voltage at 20 kHz curve. The mismatch is because the used AC voltage source cannot provide enough reactive power

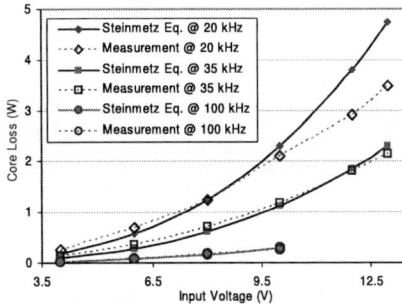

Fig. 12. Comparisons between the measured and modeled core losses at 30 kHz and 40 kHz

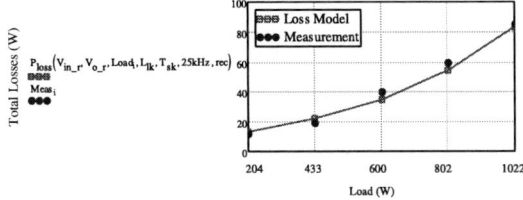

Fig. 13. Comparisons between the measured and modeled system loss related to the load power

and makes the voltage waveform distorted from sinusoid. Anyway, the use of given Steinmetz equation parameters for 3C90 can be justified.

C. System Losses

As the final phase of experimental verification, the system loss has to be validated for also including other losses, like capacitor losses. A power analyzer (Delta PZ4000) was used to measure the DAB system loss by subtracting the measured output power from the input power. Fig. 13 shows the measured and modeled system losses as a function of load in MathCAD plot. The appointed operating points and components variable are:

- Rated inputvoltage: V_{in_r} = 12 V.
- Rated outputvoltage: V_{o_r} = 360 V.
- the switching frequency : 25 kHz.
- The leakage inductance L_{lk} = 125 μH.
- The temperature of MOSFET heatsink T_{sk} = 31 °C.
- Rectangular modulation method (rec).

Considering the measurement error, the loss model could be considered accurate. Note that the losses in the driving circuit are not included in the loss model, this is due to the fact that the driving circuit is powered from a separate power supply. Due to the high current level, the parasitic resistances from input terminals to MOSFETs are measured and the related losses are incorporated in the loss model as well.

V. APPLICATIONS OF THE LOSS MODEL

The variables in the power loss function can be changed to the interested values such that the influences of the operating points or component parameters on the system

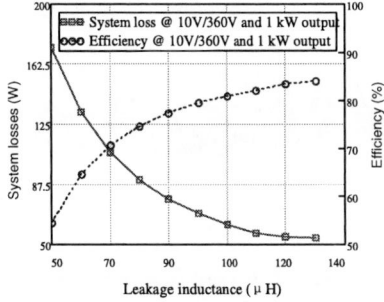

Fig. 14. Determination of the optimal transformer leakage based on the loss model

losses can be evaluated and compared. Several system evaluation and optimization works have been done based on the loss model.

A. Optimal transformer Leakage Inductance

In the DAB converter with rectangular modulation method, smaller transformer leakage inductance utilizes smaller RMS current in the circuit to transfer certain amount of power at the rated input/output voltages and therefore generates less conduction losses [5]. However, if the V_i/V_o deviates away from transformer ratio (1:30), smaller leakage inductance causes higher RMS current and losses in the circuit. The criteria of an optimal leakage inductance in our case is to make sure that the system efficiency at rated output power should be higher than 70% when $V_i : V_o$=10 V : 360 V. Fig. 14 shows the losses in MOSFETs P_{MOS} and the system efficiency η at the rated power, related to the transformer leakage inductance L_{lk} from 10 μH to 145 μH. Their function variables are given below:

$$P_{MOS}(10V, 360V, 1000W, L_{lk}, 60K, 25kHz, rec)$$
$$\eta(10V, 360V, 1000W, L_{lk}, 60K, 25kHz, rec)$$

The leakage inductance to achieve 70% system efficiency is about 70 μH and the losses in MOSFETs is about 110 W at $V_i: V_o$ = 10V: 360 V and rated output power. At rated conditions, DAB converter with 70 μH leakage transformer can achieve 93% efficiency and this is verified experimentally.

B. Maximum power rating of the custom-designed MOS-FET modules

Since the MOSFET modules were custom-designed with the MOSFET chips changed to achieve higher power rating than standard SK80MD055 module, the maximum continuous power rating of the new module should be predicted as the design reference. The maximum MOS-FET junction temperature is 125 °C and the given module thermal resistance from substrate to the junction is 0.33 K/W. If the designed heatsink temperature is 80 °C, the

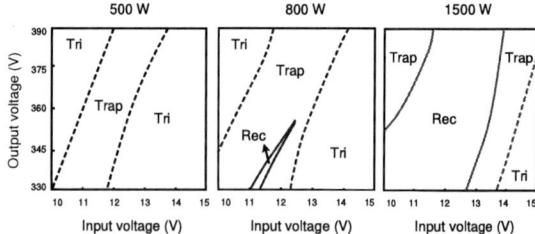

Fig. 15. Operating ranges of three modulation methods under 500 W, 800 W and 1.5 kW load ($L_{lk} = 125~\mu$H)

maximum continuous power loss of one module will be:

$$P_{mosfet} = \frac{125\,^{\circ}\text{C} - 80\,^{\circ}\text{C}}{0.33 K/W} = 136.4 W \qquad (13)$$

Predicted by the loss model, the output power, which generates 136.4 W in one MOSFET module is 2.2 kW(1.07 kW) when the V_i is 12 V(10 V) and V_o is 360 V, is the maximum continuous converter output power that the MOSFET modules can handle if the heatsink temperature stays at 80 °C.

C. Operating ranges of three modulation methods

Three modulation methods are available for DAB converter control [6] [19], classified as the rectangular, trapezoidal and triangular method which are named according to the modulated transformer current shapes. Due to the varied modulated current shapes and different soft-switching switch counts, one of them produces less losses than other two in certain operating ranges. Their waveforms are all modeled in the loss model and system losses under three modulations are calculated. Based on this loss model, the operating range boundaries among three modulations is estimated and shown in Fig. 15.

Other DAB performance optimization can be also done using the proposed loss model, such as the optimal number of MOSFETs in parallel, the optimal transformer turns ratio and the optimal switching frequency if the same MOSFET chips are used. All of the mentioned optimizations have not been verified experimentally and this will be the future work.

VI. CONCLUSIONS

This paper presents a novel loss model concept based on MathCAD sheet. A DAB converter (12V/360V, 1 kW and 25 kHz) is used as the test platform of this loss model. The experimental methods to extract parameters for the loss model and to validate the loss model are addressed in details. These methods are practical, fast and accurate. Based on the verified loss model, several circuit optimizations have been done.

ACKNOWLEDGMENT

The authors would like to thank A. van Zwam and B. Roodenburg for their supports and suggestions in this work.

REFERENCES

[1] J.Y. Bae, Y. Kim, D.H. Lee, S.D. Kwon, P.S. Kim and D.H. Han "A study on the loss model and characteristic comparison of three-level converter and full-bridge converter through the conduction loss analysis of power devices", *IECON 2004*, vol.3, pp. 2314-2320, Nov. 2004.

[2] Y. Hayashi, K. Takao, T. Shimizu, H. Ohashi, "Power loss design platform for high output power density converters", *Power Electronics and Applications, 2007 European Conference on*, pp.1-10, 2-5 Sept. 2007.

[3] S. Deng, H. Mao, T. Wu, S. Xiao, I. Batarseh, "Power losses estimation platform for power converters", *IEEE 19th Annual Applied Power Electronics Conference and Exposition*, vol.3, pp. 1784-1789, 2004.

[4] Y. Wang, S. W. H. de Haan and A. van Zwam,"Analysis of sensitivity of the performance of interleaved flyback converter to the principal design parameters", *7th International Conference on Power Electronics*, Daegu, Korea, pp. 85-89, Oct. 2007.

[5] M.H.Kheraluwala, R.W. Gascoigne, D.N. Divan, E. D. Baumann, "Performance characterization of a high-power dual active bridge dc-to-dc converter", *IEEE transactions on Industry Application*, vol. 28, no. 6, pp. 1294-1300, Nov./Dec. 1992.

[6] N. Schibli, "Symmetrical multilevel converters with two quadrant dc-dc feeding",Ph. D dissertation, Ecole Polytechnique Federale de Lausanne, Suisse, Switzerland, 2000

[7] U. Drofenik and J. Kolar, "A general scheme for calculating switching- and conduction-losses of power semiconductors in numerical circuit simulations of power electronic systems", *Proc. of the 5th Int. Power Electron. Conference*, Niigata, Japan. April, 2005.

[8] Y. Yorozu, M. Hirano, K. Oka, and Y. Tagawa, "Electron spectroscopy studies on magneto-optical media and plastic substrate interface", *IEEE Transl. J. Magn. Japan*, vol. 2, pp. 740–741, August 1987.

[9] M. Young, *The Technical Writer's Handbook*. Mill Valley, CA: University Science, 1989.

[10] K. Takao, H. Irokawa, Y. Hayashi and H. Ohashi, "Overall circuit loss design method for integrated power converter",*4th International Conference on Integrate Power Systems, CIPS 2006*, Naples, Italy, 7-9 Jun. 2006.

[11] Y. Bai, Y. Meng, A. Q. Huang, F.C.Lee, "A novel model for MOSFET switching loss calculation", *The 4th International Power Electronics and Motion Control Conference, 2004. IPEMC 2004*, vol.3, pp. 1669-1672, 14-16 Aug. 2004

[12] G. Cauffet and J. P. Keradec. "Digital oscilloscope measurements in high-frequency switching power electronics", *IEEE Transactions on Instrumentation and Measurement*, vol. 41, no. 6, pp.856-860, Dec 1992.

[13] N. Locci, F. Mocci and M. Tosi, "Measurement of instantaneous losses in switching power devices", *IEEE Transactions on Instrumentation and Measurement*, vol. 37, no. 4, pp.541-546, Dec 1988.

[14] J. T. Strydom and J. D. van Wyk, "Improved loss determination for planar integrated power passive modules",*17th Annual IEEE Applied Power Electronics Conference and Exposition, 2002. APEC 2002*, vol.1, pp.332-338, 2002.

[15] M. Albach, T. Durbaum and A. Brockmeyer "Calculating core losses in transformers for arbitrary magnetizing currents a comparison of different approaches", *27th Annual IEEE Power Electronics Specialists Conference, 1996. PESC '96*, vol.2, pp.1463-1468, 23-27 Jun 1996.

[16] C. P. Steinmetz, "On the law of hysteresis", *Proc. IEEE*, vol. 72, pp. 196C221 , Feb. 1984.

[17] Application Note, "Design of planar power transformers", *www.ferroxcube.com/appl/info/plandesi.pdf*, Ferroxcube company.

[18] C. Xiao, G. Chen and W. G. Odendaal, "Overview of power loss measurement techniques in power electronics systems," *Industry Applications Conference, 2002. 37th IAS Annual Meeting. Conference Record of the*, vol.2, pp. 1352-1359, 2002

[19] F. Krismer, S. Round and J. W. Kolar, "Performance Optimization of a High Current Dual Active Bridge with a Wide Operating Voltage Range", *Proceedings of the 37th Power Electronics Specialists Conference*, Jeju, Korea, June 18 - 22, CD ROM, ISBN: 1-4244-9717-7, 2006.

Modified multistage semiconductor-Fitch generator topology with magnetic compression

Stanisław Kalisiak*, Marcin Hołub†

* Szczecin University of Technology, Electrical Engineering Department, Szczecin, Poland, e-mail: *kal@ps.pl*
† Szczecin University of Technology, Electrical Engineering Department, Szczecin, Poland, e-mail: *mholub@ps.pl*

Abstract— For non-thermal plasma technology corona discharge devices high-voltage, high-current pulses are used with very high demands considering rising voltage slopes. Many solid state pulse power modulator (SSPPM) system topologies are known however most include a high power transformer compromising the overall system efficiency. A modified Fitch generator topology is introduced enlarging the output voltage to supply voltage ratio to theoretically the factor of three. Moreover the output voltage waveform enables the magnetic pulse compressor cross-section minimization with the factor of 0,67 due to a unique output voltage waveform. Test stand results are given for a 10-stage construction and a single stage magnetic compressor, power switch dynamic parameters influence on systems efficiency is discussed.

Keywords— Pulsed power converter, Fitch generator, plasma supply systems, resonant converter.

I. INTRODUCTION

Modern plasma technology systems often require high voltage, high current, short duration pulse power supplies [1,2,3,4]. Although fast development of modern power switches (most of all thyristors and LTTs) results in high blocking voltage ratings, costs and rise and fall times are often hard to adopt in pulsed power systems, most of all considering sources with short rising and falling edge demands. Many system topologies have been introduced in order to multiply the output voltage/blocking voltage ratio when compared to single switch. Most important ones include the Marx topology [5] depicted partly in Fig. 1 and the Fitch topology [6] (also called resonant charge transfer Marx topology).

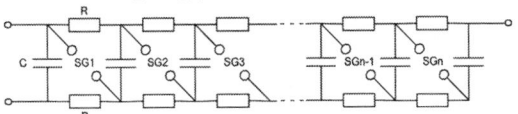

Fig. 1. A classical Marx [5] topology pulse power source. SG – spark gap switch, C – capacitor, R – charging resistance.

Modern designs include solid state switches instead of spark-gap or gaseous switches (as thyratrons) because of the well known advantages as lifetime, control apparatus simplification and output frequencies. However the overall tendency is that with rising blocking voltage ratings dynamic parameters of the switches are heavily decreased. In order to maintain high output voltage capabilities with fast switch responses modular construction was developed enabling output voltage multiplication without the use of a high voltage transformer.

II. MODIFIED, MULTILEVEL-FITCH CONVERTER TOPOLOGY

A modification of the Fitch topology is proposed incrementing the voltage multiplication factor of a single module to three. Solid state switch based construction is presented in Fig. 2, a single stage clarifying basics of operation is depicted in Fig. 3.

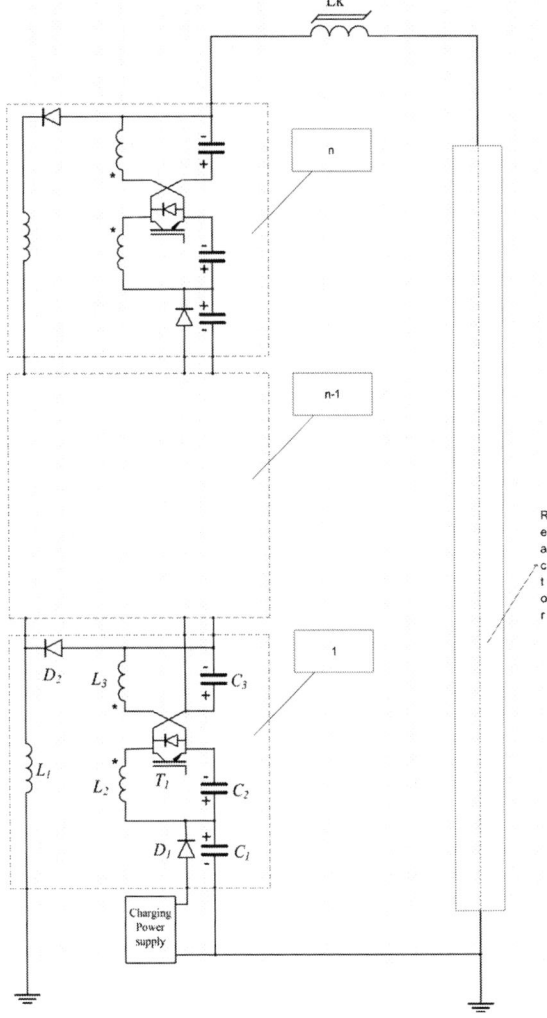

Fig. 2. Proposed, modified Fitch generator topology.

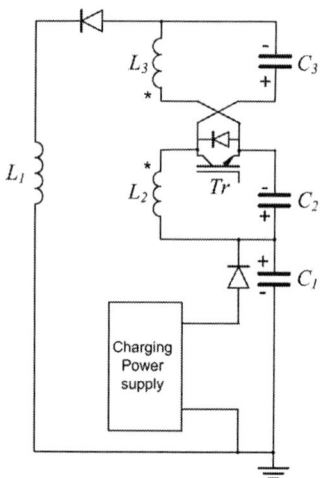

Fig. 3. Single power stage of the pulsed power system.

Principle of operation can be discussed using schematic waveforms from Fig. 4. While the power switch is not operated the output voltage U_{out} is equal or close to zero and the voltage amplitude across capacitors C_1, C_2, C_3 is matched to charging voltage and the polarization agrees with the one depicted in Fig. 3.

When the solid state switch (Figures include an IGBT transistor) is operated output voltage U_{out} is equal to the sum of capacitor voltages $U_{C1}+U_{C2}+U_{C3}$, that is the instant voltage value at the beginning of the recharge process equals $-U_C$.

The recharging process begins and the voltage changes according to (1):

$$U_{out} = U_C(1 - 2\cos\omega t), \qquad (1)$$

where $\omega = \dfrac{1}{\sqrt{LC}}$ when $L_2 = L_3 = L$ and $C_1 = C_2 = C_3 = C$.

[Fig. 4 plot: U_{out} [p.u.] vs t, showing charging curve rising from $-U_C$ [p.u.] to $3U_C$ [p.u.] at T/2, with Charging and Discharging capacitor diagrams]

Fig. 4. Principle of a single stage operation.

The condition to recharge capacitors simultaneously can be achieved by magnetic coupling of recharge inductances L_2

and L_3. For a serial connection of n number of stages (Fig. 2) following output voltage equation can be obtained:

$$U_{out(n)} = n \cdot U_C(1 - 2\cos\omega t) \qquad (2)$$

As can be derived from resonant capacitor recharge conditions with high quality factor after the half-period of:

$$\frac{T}{2} = \pi\sqrt{LC} \qquad (3)$$

overall output voltage reaches:

$$U_{out(max)} = 3 \cdot n \cdot U_C. \qquad (4)$$

Equation 4 summarizes the advantage of the topology introduced, a fact of tripling the supply voltage on a single voltage stage.

III. INFLUENCE OF DYNAMIC SWITCH PARAMETERS ON SYSTEM EFFICIENCY

In case of short pulse formation in high voltage circuits solid-state switch dynamic losses have a major efficiency impact. In case of the described topology an equivalent circuit can be used depicting the serial connection of resulting capacity and inductance with the power switch (Fig. 5a). In the converter prototype constructed Ixys IGBT IXDN 55N120 D1 transistors were used, with a nominal rise time of $t_r=70ns$. Considering voltage-current waveforms of the switching element an analytical approach can be used as depicted in Fig. 5b.

A. Simplified approach

Considering waveforms depicted in Fig. 5. a simplified energy loss analysis can be led taking linear collector current and collector-emitter voltage waveforms during switching. Because of the system equivalent circuit mainly turn-on losses will have a strong impact on system efficiency. As can be denoted from theoretical turn-on losses equation energy loss can be described as:

$$E_{loss} \cong \frac{t_r^2}{L} \cdot U_{C0} \cdot \left(\frac{U_{C0} + 2U_{CE(sat)}}{6} \right) \qquad (5)$$

and in respect to the energy stored in capacitor at the beginning of the recharge process E_{stored}:

$$\frac{E_{loss}}{E_{stored}} \cong \frac{t_r^2}{U_{C0} \cdot L \cdot C} \cdot \left(\frac{U_{C0} + 2U_{CE(sat)}}{3} \right) \qquad (6)$$

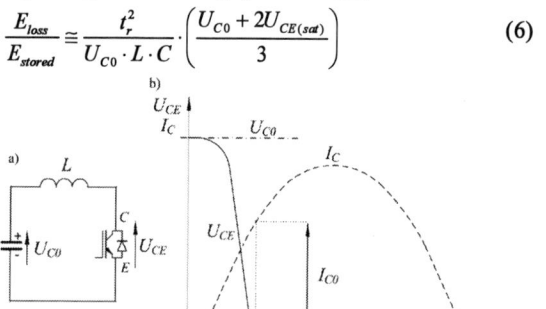

Fig. 5. a) - equivalent circuit, b) – power switch voltage-current analytical waveforms.

A graphical illustration of the dependency (6) is presented in Fig. 6 together with the calculated recharge half-period. Calculations were led using Ixys transistor parameters, $C = 44nF$, $U_{C0} = 600V$.

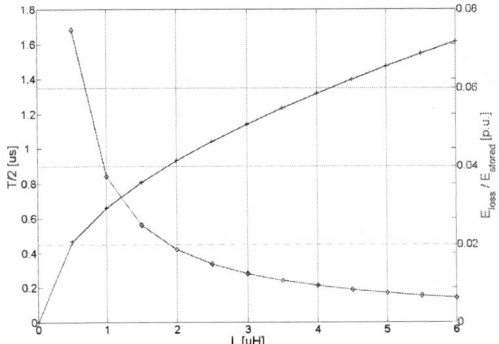

Fig. 6. Simplified, analytical energy loss ratio and recharge half-period.

As can be denoted for the inductances above 3µH the energy loss ratio should not exceed 1,5%, which also is in good agreement with results presented in [7]. Further output pulse shortening, and in consequence resonant inductance L minimization leads to dramatic energy loss enlargement, furthermore power switch critical ratings limit the current rising slope. In consequence the energy efficiency suddenly drops if the $T/2 / t_r$ factor is lower than approximately 10. For the inductance values above 3µH no significant efficiency improvement is achieved and the output pulse length is increased, which is undesirable in the analyzed case. Of course dynamical switching parameters of the switch are crucial as described in (6), the influence of switching-on time was investigated with the means of different gate resistance values R_G. Measured values are presented in Fig. 7.

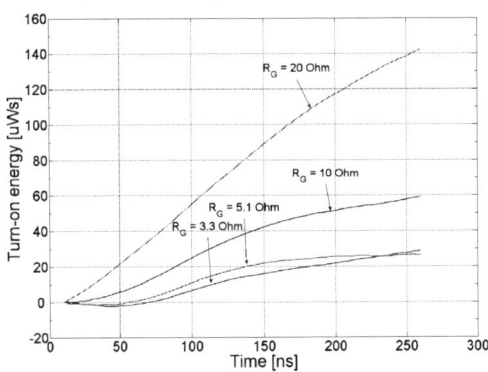

Fig. 7. Measured turn-on energy loss as a function of gate resistance.

B. Quality factor approach

Alternative analysis can be led using not energy balance equations or each of its components into account but a summarized, mean value based on the energy balance before and after the recharge half-period. The quality factor Q for the equivalent circuit presented in Fig. 5a) is given by:

$$Q = \frac{\omega_0 L}{R}, \tag{7}$$

where

$$\omega_0 = \frac{1}{\sqrt{LC}}. \tag{8}$$

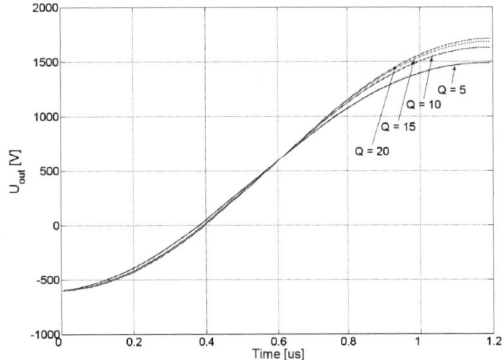

Fig 8. Quality factor – output voltage waveform influence

Considering the quantities given in (7) and (8) an equation can be derived for the output voltage U_{out} waveform as a function of supply voltage U_{C0} and the quality factor Q:

$$U_{out} = U_{C0}\left[1 - 2\cos(\omega t) \cdot e^{\frac{-t\omega_0}{2Q}}\right] \tag{9}$$

A graphical illustration of single module operation for the supply voltage $U_{C0}=600V$ is given in Fig. 8.

The dependency was also examined using the test-stand measurements. Circuit damping was adjusted using different transistor gate resistances therefore changing the switching-on behavior as seen in Fig. 7. Results of a single capacitor recharge voltage U_C waveform are depicted in Fig. 9, tests were led using supply voltage of 600V. Gate resistance was adjusted between $R_G=3.3\Omega$ and $R_G=100\Omega$.

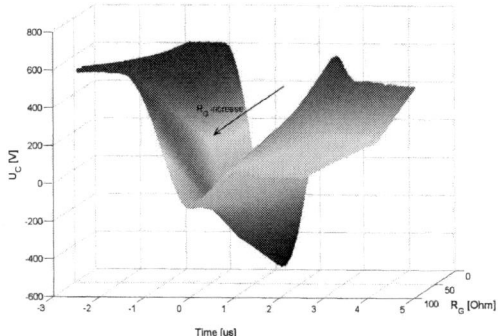

Fig. 9. Capacitor recharge voltage U_C as a function of time and gate resistance R_G.

From (9) voltage recharge ratio can be obtained describing the percentage of initial capacitor voltage U_{C0} transfer after the recharge half-period ($U_{C(t=T/2)}$). The analytical solution is given by:

$$\frac{U_{C(t=T/2)}}{U_{C0}} = \frac{1}{3}\left(1 + 2e^{\frac{-\pi}{\sqrt{4Q-1}}}\right) \tag{10}$$

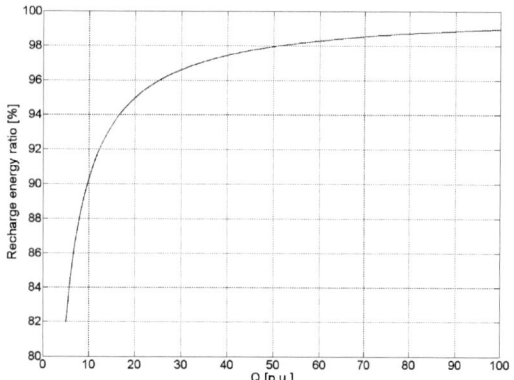

Fig. 10. Voltage recharge ratio as a function of quality factor Q.

A graph showing the recharge ratio described in (10) is shown in Fig. 10 for the quality factor Q range between 5 and 100. All the dependencies given are for a single module of the construction as depicted in Fig. 3.

As can be denoted from Fig. 10 module's quality factor should be possibly high, with the minimal value of approximately $Q \approx 25$ for high system efficiency. A series of measurements led, depicted in Fig. 9, leads to summarization of transistor t_r ratings influence on systems quality factor and efficiency given in Table 1 (measurements were led for a single module, $L_2 = L_3 = 3{,}5\mu H$, $C_1 = C_2 = C_3 = 44nF$ as depicted in Fig. 3.):

TABLE I.
IGBT TRANSISTOR DYNAMIC PARAMETERS INFLUENCE ON SYSTEMS QUALITY FACTOR AND RECHARGE EFFICIENCY

R_G [Ω]	U_{C0} [V]	$U_{C(t=T/2)}$ [V]	Q [p.u.]	E_{loss}/E_{stored} [%]
100	602	-120	1	96
51	598	-354	3	65
20	602	-480	6,94	36
10	600	-534	13,48	21
5,1	595	-568	33,82	9
3,3	600	-572	32,87	9

As can be denoted for nominal gate resistance of $5{,}1\Omega$ the practically verified quality factor reaches almost 34, which compared to Fig. 10 should result in energy efficiency of approximately 97%. Except for transistor turn-on (Fig. 6) losses also ohm (current displacement) losses and magnetic material properties are visible.

IV. MAGNETIC COMPRESSION

Compression of output pulses is often necessary in order to obtain required timing parameters of the output pulse, most of all considering $\frac{dU_{out}}{dt}$ factor. Plasma systems using pulsed corona discharge phenomena require very short output pulses of a few hundred nanoseconds [8]. Additional applications of voltage magnetic compressors include typical systems were pulsed power systems or solid state pulse power modulators (SSPPM) are used, that is laser supplies [9], Z-pinch supplies [10].

Introduced topology exhibits a unique property of magnetic system size reduction. In case of serially charged capacitors in the moment of magnetic compression following condition has to be fulfilled:

$$\int_0^{T/2} U_{out(n)} dt = N \cdot \Delta B \cdot A_S , \qquad (11)$$

where N is the number of turns, ΔB is the change of magnetic field and A_S is the magnetic core cross-section. Because, as can be denoted from Fig. 4, the integral voltage value is lowered through the initiating and ending negative voltage for the same ΔU_{out} smaller cross-sections of pulse compressors can be chosen. Fig. 11 depicts output voltage waveforms of a classical Fitch topology compared to the proposed, modified topology. Calculations were led for identical U_{out} voltage peak values, U_{C0} of 600V, inductance and capacitor values as described in paragraph III.B.

As can be noted from Fig. 11 for the same peak voltage values voltage integrals will vary, that is the ration of integral values of proposed topology to a classical Fitch converter can be described as:

$$\frac{\int_0^{T/2} U_{C0}(1 - 2\cos\omega_0 t)dt}{\int_0^{T/2} 1{,}5 U_{C0}(1 - \cos\omega_0 t)dt} = \frac{\pi\sqrt{LC} \cdot U_{C0}}{1{,}5\pi\sqrt{LC} \cdot U_{C0}} \approx 0{,}67 \qquad (12)$$

In consequence the magnetic compressor material's cross section can be reduced with the coefficient of approximately 0,67 when compared to classical Fitch generator output waveforms.

V. TEST STAND CONSTRUCTION AND TEST RESULTS

Proposed topology was designed and constructed in Laboratory of Power Electronic Converters, Szczecin University of Technology. A 10-stage converter was developed with a DSP control system. For the resonant recharge modules WIMA FKP capacitors were used, a parallel connection of two 22nF capacitors for a maximum of 6000V for a single LC branch. A "litz" wire was used for inductance windings in order to minimize the damping coefficients of the system. RM14 cores were used, measured inductance was in the range of 3,5 μH, stray resistance (Fluke PM6304 bridge, for 100kHz) was in the range of 10mΩ. In consequence the quality factor for a single stage reaches 34 .As previously mentioned Ixys

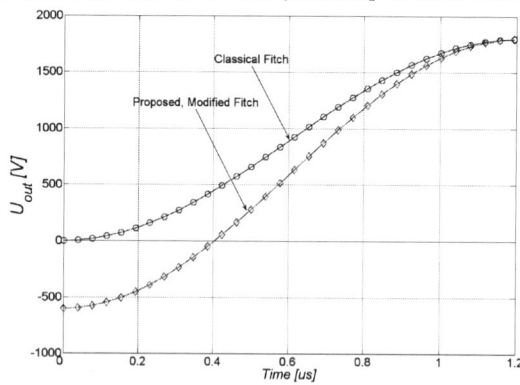

Fig. 11. Half-period output voltage waveforms of a classical Fitch converter and the proposed, modified topology.

Fig. 12. Converter construction of the modified Fitch topology pulse power source, 10-stage construction.

IXDN 55N120 D1 IGBTs were implemented. A SMPS charging power supply was used in order to supply driver stages of consecutive modules. Prototype construction is depicted in Fig. 12.

Measurements were led using LeCroy Wave Runner 6100A digital oscilloscope with PPE 20kV voltage probe and Fluke ISM 50/10 passive current shunt. Typical output voltage waveforms for different supply voltages are presented in Fig. 13.

As can be denoted maximal voltage reaches approximately 17,3kV, considering supply voltage of 600V the practical conversion factor is 2,88 per stage. Because of poor dynamical power diode behaviour RCD snubbers were used across IGBT transistors. Difference between theoretical and practical voltage multiplication factor is caused mainly by transistor losses, limited quality factor and supply voltage distribution among consecutive stages.

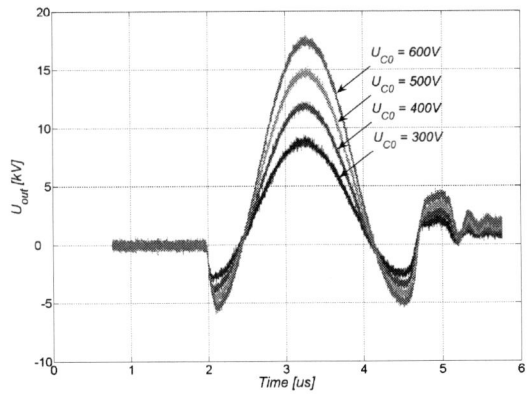

Fig. 13. Typical output voltage waveforms for different supply voltages, 10-stage construction.

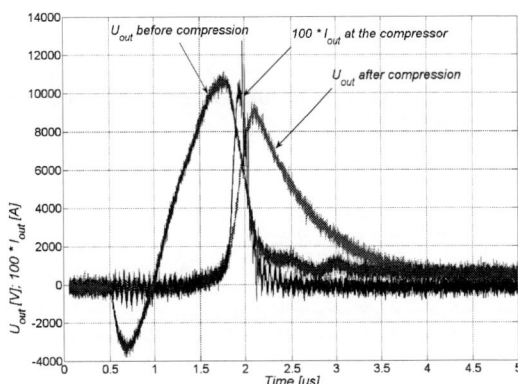

Fig. 14. Single stage magnetic compression output waveforms.

A single stage magnetic compressor circuit was implemented with following parameters: C_k = 1,466 nF, $L_{k(initial)}$ = 1,21 mH. Fig. 14 depicts measured output voltage and current waveforms.

Spark gap switches were used as plasma load, voltage rise time ratings were improved with the factor of approximately 2,3, which is comparable to results obtained in [11]. Digital oscilloscope registered rise time values (10% - 90% of the waveform amplitude) shown an improvement from 690 ns for the voltage rising slope before the magnetic compressor to 300ns after compression, better results are obtainable when dedicated nanocrystalline or amorphous materials can be used [12].

Considering the recharge current half-period the pulse compression coefficient is equal to approximately four.

VI. SUMMARY AND CONCLUSIONS

Paper presented introduces a modified, solid state Fitch-generator topology with increased voltage conversion factor and improved output voltage form in terms of magnetic pulse compression circuit dimensions. Test stand analysis proved the voltage conversion factor of 2,88 and magnetic pulse compression circuit material's cross section minimization by the factor of 0,67. Both analytical description and test results had proven the correctness of converter construction concept.

Power, solid state switch dynamical parameters influence on system's efficiency is discussed as both simplified energy loss approach and quality factor discussion. Both analytical discussion (Fig. 6) and measurements led had proven that dynamical transistor parameters influence strongly the overall energy efficiency of the converter, used IXYS transistors, with their nominal t_r=70ns belong to the group of fast switching devices, however much faster switches are reported in literature (IXYS DE475-102N21A, Infineon SPP20N65C3) but are relatively expensive.

Considering control circuitry special attention has to be taken because of high voltages and surge voltage output, light-fibre control signal transmission and carefully designed charging voltage supply are necessary.

Main drawbacks of the topology used include both the number and necessary quality of passive resonant construction elements. Moreover fast power switches have to be used with possibly large collector-emitter blocking voltage ratings. Poor dynamical diode behaviour implies the use of additional snubber circuitry therefore minimizing the voltage conversion factor. Great attention has to be used in order to achieve possibly high circuit quality factor ratings, careful design and precise construction are necessary.

REFERENCES:

[1] Stephan Roche: "Solid State Pulsed Power Systems", *Physique & industrie*, 2003 [www.physiqueindustrie.com]

[2] Pokryvailo, A.; Yankelevich, Y.; Wolf, M.; Abramzon, E.; Shviro, E.; Wald, S.; Welleman, A.: "A 1 KW pulsed corona system for pollution control applications", *Pulsed Power Conference, 2003. Digest of Technical Papers PPC-2003. 14th IEEE International Volume 1*, 15-18 June 2003 Page(s): 225 - 228 Vol.1

[3] G. Lombardi1, N. Blin-Simiand, F. Jorand, L. Magne, S. Pasquiers, C. Postel and J. -R. Vacher: "Effect Of Propene, n - Decane, and Toluene Plasma Kinetics on NO Conversion in Homogeneous Oxygen-Rich Dry Mixtures at Ambient Temperature", *Plasma Chemistry and Plasma Processing, publisher Springer Netherlands*, Volume 27, Number 4 August, 2007, Pages 414-445

[4] Lawless, P.A.; Yamamoto, T.; Poteat, S.; Boss, C.; Nunez, C.M.; Ramsey, G.H.; Engels, R.: "Characteristics of a fast rise time power supply for a pulsed plasma reactor", *Industry Applications Society Annual Meeting*, 1993., Conference Record of the 1993 IEEE, Volume , Issue , 2-8 Oct 1993 Page(s):1875 - 1881 vol.3

[5] Erwin Marx: "Verfahren zur Schlagpruefung von Isolatoren und anderen elektrischen Vorrichtungen", *Patentschrift nr. 455933*, 13 Feb. 1928

[6] Richard Anthony Fitch et al.: "Electrical Pulse Generators", *US Patent nr 3,366,799*, 30 Jan. 1968

[7] Semikron application manual, *"Features of switches"*, section 3.8.3.3, pp. 239 - 244, http://www.semikron.com/internet/index.jsp?sekId=229

[8] Sung-Duck Jang, Yoon-Gyu Son, Jong-Seok Oh and Moo-Hyun Cho, Dong-Jun Koh: "Pulsed Plasma Process for Flue Gas Removal from an Industrial Incinerator by Using a Peak 200-kV, 10-kA Pulse Modulator", *Journal of the Korean Physical Society*, Vol. 44, No. 5, May 2004, pp. 1157-1162

[9] H.M. von Bergmann, P.H.Swart: "Thyristor-driven pulsers for multikilowatt average power lasers", *IEE Proceedings –B*, Vol. 139, No.2, march 1992, pp. 123-130

[10] N. Kishi, M.Watanabe, N.Izuka, J.Fei, T.Kawamura, A.Okino, K.Horioka, E.Hotta: "Improvement of high power gas jet type Z pinch plasma light source for EUV lithography", *proceedings of 28th ICPIG*, July 15-20 2007, Prague, Czech Republic

[11] C.H.Smith, D.M. Nathasingh: "Magnetic Characteristics of amorphous metal saturable reactors in pulse power systems", *proceedings of European Particle Accelerator Conference*, Berlin, Germany, 24-28 march 1992, pp. 1603-1605.

[12] R. Burdt, R.D. Curry, K. McDonald, P. Melcher, R. Ness, Ch. Huang: "Evaluation of nanocrystalline materials, amorphous metal alloys and ferrites for magnetic pulse compression applications", *Journal of applied Physics,* No 99, 08D911, 2006, pp.99-101.

Modeling and Measuring Results of a Shunt Current Source Active Power Filter with Series Capacitor

P. Parkatti[*], M. Salo[**], and H. Tuusa[***]

[*]Department of Electrical Energy Engineering, Tampere, Finland, e-mail:perttu.parkatti@tut.fi
[**]Department of Electrical Energy Engineering, Tampere, Finland, e-mail:mika.salo@tut.fi
[***]Department of Electrical Energy Engineering, Tampere, Finland, e-mail:heikki.tuusa@tut.fi

Abstract—In this paper, a method for reducing the voltage stresses on switching components and power losses in a shunt current source active power filter (CSAPF) is presented. The method is based on the series capacitor structure which is used to block the fundamental supply voltage component. The computational capacitor voltage balancing is studied to compensate the effects of the phase asymmetry. The frequency domain behavior of the harmonic control and main circuit is examined. The results show that the voltage stresses and power losses are lower in the CSAPF with a series-connected capacitor than in the conventional topology.

Keywords—active filter, current source inverter, harmonics, efficiency.

I. INTRODUCTION

In recent years active power filters and harmonic compensators have been widely studied. When compensation is carried out with pulse width modulation, the switching components are exposed to high voltage stresses. High voltage stresses increase power losses and impose greater demands on switching components, which makes compensation equipment more expensive. The power losses reduce the harmonic filtering efficiency, which increases the filtering costs. Practically every switching device has its maximum voltage durability, thus there is always a limited voltage that can be active filtered without a transformer.

Several active power filter topologies with lower voltage stresses or power losses have been developed and studied; for example hybrid active filters [1], series active filters [2-3] and active filters with a series-connected capacitor [4-9]. This paper concentrates on the shunt CSAPF with a series-connected capacitor which was previously presented in [8] and [9].

The advantages of the CSAPF with series-connected capacitor over conventional topology are lower voltage stresses, the option to use switching devices with a lower voltage rating, and the decreased switching power losses which these provide.

The disadvantages of the topology are increased compensation currents, a more complex control system and main circuit, and the tendency of the capacitor voltage to become unbalanced when load currents are asymmetrical. Also, the reactive currents cannot be controlled as much as with the topology without a series capacitor (conventional CSAPF).

The main circuit, control principle and transfer functions of the CSAPF control are examined. A computational method for balancing the capacitor voltages of the CSAPF with series-connected capacitor topology is presented. After that, prototype measurements are performed with 400 V supply and 10 kVA load, and compensation results are presented.

II. CSAPF WITH SERIES-CONNECTED CAPACITOR

Fig. 1 presents the main circuit of the CSAPF connected with a series-connected capacitor. The main principle of the circuit is the use of the capacitors C_2 that make possible the reduction of the fundamental supply voltage over switching devices. Voltage balancing resistors may be needed in parallel with the C_1 capacitors and resonance damping resistors in parallel with the inductors L_f and C_2. However, the resistors are not presented in Fig. 1 and Fig. 2.

When computational balancing of capacitor voltage and resonance damping [10] are used, these resistors are not needed.

The equivalent circuit of the CSAPF with a series-connected capacitor is presented in Fig. 2. The topology makes it possible to control the voltages of the capacitor C_1. The capacitor C_2 is used as a voltage divider to reduce the fundamental component. L_f is needed to filter out a high frequency current ripple.

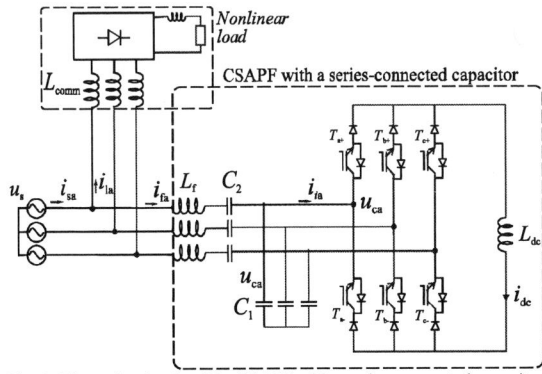

Fig. 1. The main circuit of the CSAPF with a series-connected capacitor.

978-1-4244-1741-4/08/$25.00 ©2008 IEEE

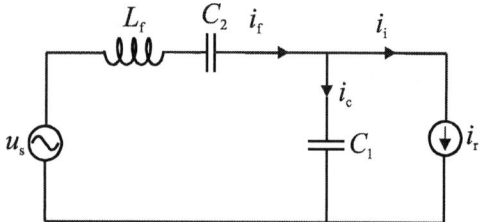

Fig. 2. Equivalent circuit of the CSAPF with a series-connected capacitor.

Without control, the circuit reduces switching voltages inversely to the ratio of capacitor impedances [8].

The fundamental voltage of the inverter bridge can be controlled by adjusting the reactive current level with an inverter as stated in [8].

III. CSAPF CONTROL SYSTEM

Fig. 3 shows the block diagram of the control system for the CSAPF with series-connected capacitor. The control system is mostly based on [11].

The control system is based on the feedforward of the load current harmonics. The active filter current control is implemented in the synchronous reference frame rotating with the supply voltage vector u_s. The supply voltage reference angle is determined with a phase-locked loop (PLL) from supply voltages.

In Fig. 3, underlined variables refer to space vectors and the superscript * to the reference values. Subscripts s, l, h, f, c, dc, and i refer respectively to supply, load, high-pass filtered, delay-compensated, capacitor, dc-link, and CSAPF switching bridge variables. d and q refer respectively to the direct and quadrature axis components of a synchronous reference frame.

The load currents i_l are measured and transformed into the two-axis components with the block "3→2". After this, the values are transformed into a reference frame rotating with the supply voltage vector with the block $e^{-j\theta}$. Now, the 50 Hz current component can be seen as a dc quantity in the reference frame. The load current harmonics are extracted with high-pass filters "HPF" (8) and multiplied by the value 1.2. The control delay compensation block "CDC" compensates the control delays caused by the digital control system [12]. The block output is the reference for the harmonics compensating filter currents $i^*_{f d,q}$.

The dc-link current control is implemented in the synchronous reference frame rotating with the voltage vector $\underline{u}_C$ of the capacitor set C_1. The $\underline{u}_C$ voltage vector is determined by measuring all phase voltages from the capacitors. The dc-link current controller output is calculated with a PI(e^2) controller, which gives the fundamental active current component i^*_{cd0} in the capacitor reference frame. The reference is converted to the variables $i^*_{dc,d0}$ and $i^*_{dc,q0}$ which rotate with the supply voltage vector reference frame.

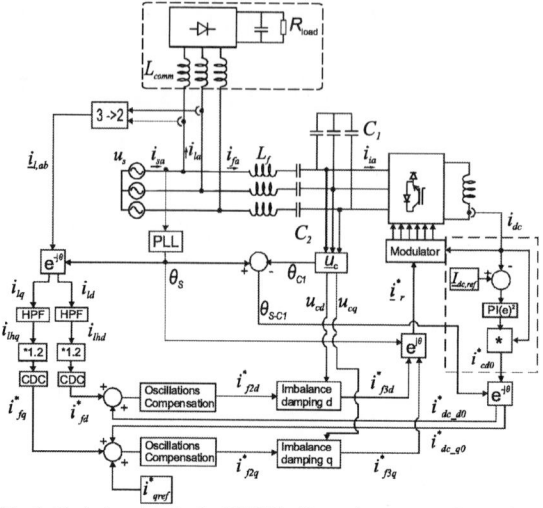

Fig. 3. Control system for the CSAPF with a series-connected capacitor.

To minimize the capacitor set C_1 voltages, the voltage u_{C1} at the fundamental frequency should be set at zero. This can be done by setting constant value i^*_{qref} to the required reactive current value. An ideal reactive current reference can be defined by calculating the current i_i, that the inverter has to produce to decrease the capacitor C_1 voltage to zero.

Because the open loop control of the active filter currents may cause resonance in the LC-filter, this can be computationally damped to bring current references to the "Oscillations Compensation" block (6). After that, the capacitor voltage imbalance damping values are calculated in the block "Imbalance damping"

The current-source PWM bridge modulates unipolar dc current. The modulation method is presented as a whole in [13]. There are six active vectors and three zero vectors. In the ideal case, the current flows through two switching devices at the same time. The modulation has to be carried out so that the PWM bridge does not break the current path. To ensure this, overlapping of the modulation signals is needed in practice.

The transfer function of the CSAPF with a series-connected capacitor is examined. Table 1 shows the modeled component values used. An additional resonance damping resistor is used in parallel with capacitors C_2. In parallel with capacitors C_1 only a leakage resistance is modeled.

The frequency behavior of the inductor L_f was modeled with the first-order series of the Foster method [14]. Equation (1) shows the transfer function of the inductor.

$$Z_{Lf}(s) = \frac{(sL_f + R_{series\_Lf}) * R_{parallel\_Lf}}{(sL_f + R_{series\_Lf}) + R_{parallel\_Lf}} \quad (1)$$

Capacitors 1 and 2 were modeled as capacitance, leakage resistance, and parasite resistance, as shown in Equations (2) and (3).

$$Z_{C1}(s) = \frac{(1/sC_1 + R_{series\_C1}) * R_{parallel\_C1}}{(1/sC_1 + R_{series\_C1}) + R_{parallel\_C1}} \quad (2)$$

$$Z_{C2}(s) = \frac{(1/sC_2 + R_{series\_C2}) * R_{parallel\_C2}}{(1/sC_2 + R_{series\_C2}) + R_{parallel\_C2}} \quad (3)$$

The impedance of the series-connected capacitor and inductor set can be calculated using (4).

$$Z_{LfC2}(s) = Z_{Lf}(s) + Z_{C2}(s) \quad (4)$$

The filter currents behave according to (5) when inverter currents are controlled.

$$F_s(s) = \frac{i_f}{i_i} = \frac{Z_{C1}(s)}{Z_{C1}(s) + Z_{LfC2}(s)} \quad (5)$$

It is shown in [6] that a transfer function G(s) on the stationary frames is changed to a transfer function G(s+jω) on the frames rotating at an angular frequency of ω. Therefore, the transfer function from the inverter currents to the supply currents on the rotating frames at the fundamental frequency can be obtained as follows:

$$F_{s,dq}(s+j\omega) = \frac{Z_{C1}(s+j\omega)}{Z_{C1}(s+j\omega) + Z_{LfC2}(s+j\omega)} \quad (6)$$

where ω is 2*π*50 Hz.

The solid line in Fig. 5 shows the bode diagram i_f / i_i from inverter phase currents i_i to filter currents i_f.

With the component values shown in Table 1, the control bandwidth is insufficient without computational damping methods. Current oscillation damping control has to be used to carry out the compensation correctly. The damping equations derived in [10] can be written in discrete form as

$$G_{DAMP}(z) = \frac{(0.2+0.4k)z^4 + (0.4-0.4k)z^3}{z^4}$$
$$+ \frac{(0.3-0.2k)z^2 + (0.1+0.2k)z}{z^4}, \quad (7)$$

where k is calculated from the resonant frequency of the passive filter

$$k = \frac{C2*C1*L_f}{(C2-C1)} * \frac{1}{t_s^2} \quad (8)$$

and t is a discrete sample time of the microcontroller (50 μs).

TABLE I

CSAPF COMPONENT VALUES.

Value	L_f	C_1	C_2	L_{dc}	L_{comm}	R_{load}	C_{load}
L (mH)	2.3	-	-	140	1.5	-	-
C (uF)	-	6	30	-	-	-	1500
R_{series} (Ω)	0.2	20 m	20 m	0.4	0.2	25	20 m
$R_{parallel}$ (Ω)	45	100 k	5 k	-	-	-	-

The harmonic compensation currents are divided inversely proportionally to the impedances Z_{C1} and Z_{LfC2}. Therefore the damping control value has to be multiplied approximately by the value 1.2.

A bode diagram of the damping control $G_{DAMP}(z)$ multiplied by the value 1.2 is presented in Fig. 5 as a dashed line. The control system values are also high-pass filtered to remove the 50 Hz component with discrete filter (9).

$$G_{HPF}(z) = \frac{z^{400} - z}{400(z^{400} - z^{399})} \quad (9)$$

If the inverter currents are controlled using load current feedforward with the discrete oscillation damping method, current flows through passive circuit $F_s(s)$. Fig. 4 shows the transfer function model from load currents to CSAPF currents. The control bandwidth from load currents to the active power filter currents in rotating reference frame can be calculated by $G_{filt}=i_f/i_l=1.2*G_{DAMP}*G_{HPF}*F_{s,dq}$, which is shown in Fig. 6 as a dashed line. The solid line in Fig. 6 shows the bode diagram from load currents $I_{i(d,q)}$ to supply currents $I_{s(d,q)}$ when feedforward control is running. It can be calculated by $G_{sup}=i_s/i_l=1-i_f/i_l$.

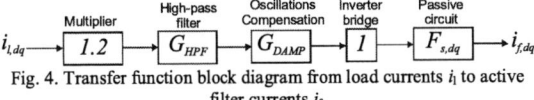

Fig. 4. Transfer function block diagram from load currents i_l to active filter currents i_f.

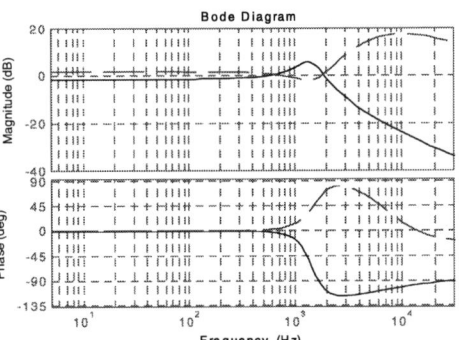

Fig. 5. Bode diagram from inverter phase currents i_i to filter currents i_f is shown with a solid line (-). Bode diagram from control system input to output is shown with a dashed line (--).

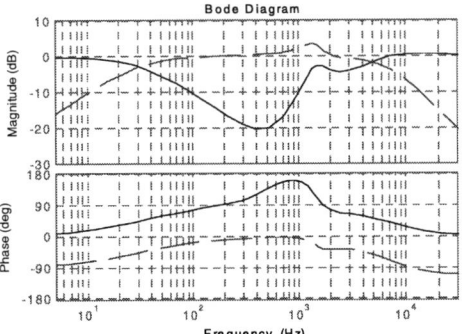

Fig. 6. Bode diagram from load currents i_l to supply currents i_s is shown as a solid line (-) and bode diagram from load currents i_l to filter currents i_f when feedforward is on, is shown as a dashed line (--).

The damping ratio for the 5th and 7th harmonics is approximately -20 dB. The 50 Hz frequencies can be seen as dc quantity in Figs 4 and 5. It can be seen that the control does not have effects on 50 Hz frequencies because of the high-pass filter.

IV. A COMPUTATIONAL METHOD FOR BALANCING CAPACITOR VOLTAGES

When an asymmetrical load is being compensated, the compensation currents are also asymmetrical. This causes imbalance in the capacitor set C_1. Also, quick transitions in active filter currents cause capacitor voltage imbalance. If there is too much voltage imbalance the dc-link current becomes very difficult to control and increased capacitor voltages may damage the switching devices.

The voltage imbalance can be reduced by using balancing resistors in parallel with capacitors C_1. The problem with this method is increased power losses. If we need to ensure that the C_1 voltage balance remains within tolerable limits, 1 kΩ balancing resistors are needed. According to the calculations and prototype measurements performed, the balancing resistors cause an approximate 100 W increase in CSAPF power losses, which seems to reduce the efficiency by approximately 1 percentage unit. This is one of the main reasons why computational voltage balancing method is very useful.

The control system can be equipped with an active damping which creates a virtual resistance in parallel with the capacitors. This way capacitor voltage imbalance can be reduced [15]. The idea in imbalance damping is to computationally calculate the currents that the resistors would cause if they were connected in parallel with the capacitors C_1. Virtual resistor currents can be calculated by $I_v = U_{c1}/R_v$ and are added to the harmonic compensation current references. The unbalance damping method increases the total harmonic distortion a little, but also helps to keep the system stable.

In Fig. 3, the computational imbalance damping reference is calculated in a block "Inbalance damping d,q". When a sudden situation in load transition state occurs, the capacitor voltages drift to imbalance. This may cause voltage rise beyond the capacitor's voltage durability limits. The computational capacitor voltage balancing method damps the voltage imbalance and the risk of overvoltage can be avoided.

V. EXPERIMENTAL MEASUREMENTS

The operation of the CSAPF was experimentally verified with the prototype measurements. The measurements were performed in 400 V mains. The prototype was designed to compensate 25 A peak phase currents in 400 and 690 V mains. The power class of the prototype is 20 kVA in 690 V mains.

Table 1 shows the component values used. The values in the table attempt to describe the real frequency behavior of the components. An additional resonance

damping resistor is used in parallel with capacitors C_2. In C_1 there is only leakage resistance of the capacitor. The virtual resistance of computational imbalance damping was 220 Ω. The frequency behavior of the inductor L_f is modeled using the first-order series Foster method [14]. Capacitors 1 and 2 were modeled as capacitance, leakage resistance, and parasite resistance. With these values, the simulation model in [8] and [9] matches these experimental results.

The reactive power value used was calculated by setting the capacitor C_1 voltage at zero. Now the capacitor current $I_{C1}=0$ V. Now it can be seen that i_i must be equal to i_f. When u_{C1} is 0 V, inverter currents can be calculated by $i_i=i_f=u_s/Z_{LfC2}$. When values from Table 1 are used, the i^*_{qref} would be 2.6 A in an ideal case. In following measurements, i^*_{qref} was 2.0 A.

Figs. 7 and 8 present the prototype measurement results for the case where a 10 kVA RL-type load was compensated in 400 V mains with the previously described control system and component values. THD$_{2kHz}$ of power system current during compensation was reduced from 27 % to 5.1 %. Load power was 10400 W and power taken from the supply was 10650 W. Total power values were measured using LEM NORMA D 6100 and Yokogawa WT 1030 digital power analyzers. The efficiency of the active power with this load was 97.7 %.

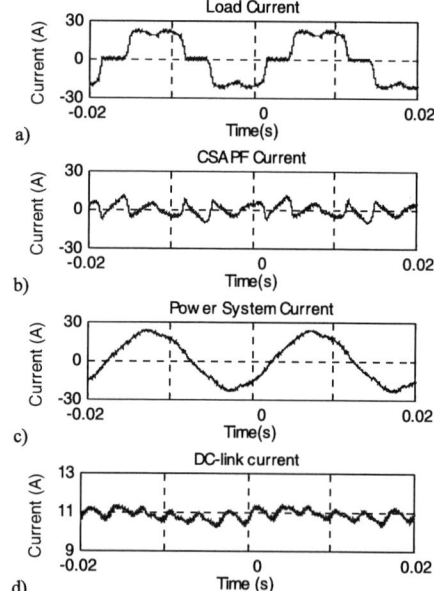

Fig. 7. Measured (a) load current i_l, (b) CSAPF current i_f, (c) power system current i_s, and (d) dc-link current i_{dc}.

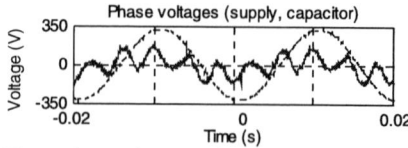

Fig. 8. Phase voltage u_s is presented as a dashed line. The solid line presents the IGBT bridge voltage u_c.

Figs. 9 and 10 present the prototype measurement results for when a 10 kVA RC-type load was compensated in 400 V mains. THD$_{2kHz}$ of power system current reduced from 33 % to 8.4 %. Load power was 10060 W and power taken from the supply was 10350 W. The efficiency of the active power with the 10 kVA RC-type load was 97.2 %. The efficiency values are equivalent to the simulated efficiencies stated in [9].

Figs. 11 and 12, 13, and 14 present the results for the case where a very distorted and asymmetric load with nominal power 7.1 kW is compensated. The computational capacitor voltage balancing method was used. Without any balancing, the dc-current control becomes very difficult to control and the voltage in some capacitors may also rise too high and damage the components.

The compensated phase power imbalance was approximately 13 % when the phase powers were P_a 2.4 kW, P_b=2.2 kW, and P_c=2.5 kW.

Fig. 11 shows the distorted load currents. The current asymmetry can be easily seen because the current waveforms and heights are unequal. In general terms, asymmetry is caused by unequal supply phase voltages, asymmetric commutation inductors, and unequal diode threshold voltages.

When these load current harmonics and phase asymmetry are compensated, the capacitor voltages also become asymmetric. Fig. 12 presents the C_1 currents during compensation. Computational capacitor voltage balancing reduces phase asymmetry and stabilizes the dc-link current control but also increases the THD$_{2kHz}$ of the supply currents. Fig. 13 shows the compensation results for the previously mentioned asymmetric load of 7.1 kW: a) load current, b) CSAPF current, c) supply current, and d) dc-link current. THD$_{2kHz}$ of supply currents decreased from 56 % to 10 % when compensation was running. Fig. 14 presents the phase A supply voltage and capacitor voltage.

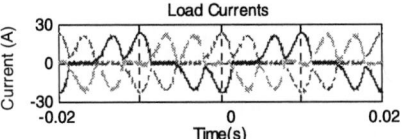

Fig. 11. Asymmetric and distorted load currents i_{la}, i_{lb}, and i_{lc}.

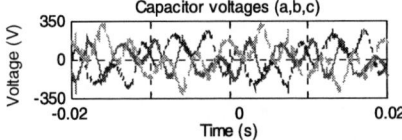

Fig. 12. Capacitor voltage unbalance caused by compensation of asymmetric load distorted load currents u_{ca}, u_{cb}, and u_{cc}.

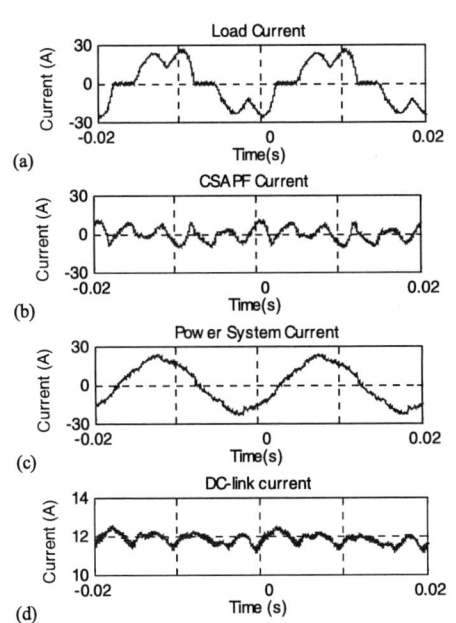

Fig. 9. Measured (a) load current i_l, (b) CSAPF current i_f, (c) power system current i_s, and (d) dc-link current i_{dc}.

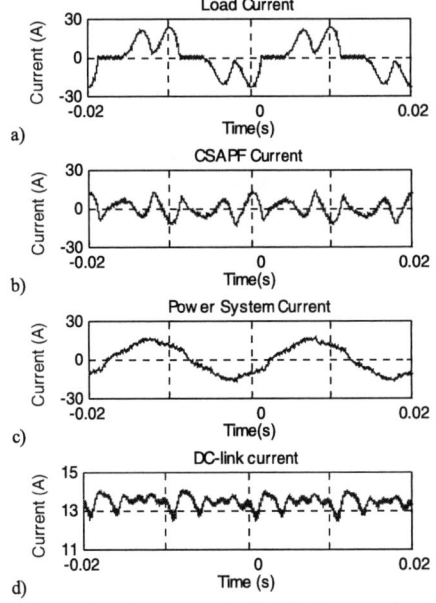

Fig. 13. Measured (a) load current i_l, (b) CSAPF current i_f, (c) power system current i_s, and (d) dc-link current i_{dc}.

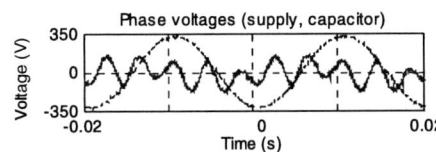

Fig. 10. Phase voltage u_s is presented as a dashed line. The solid line presents the IGBT bridge voltage u_c.

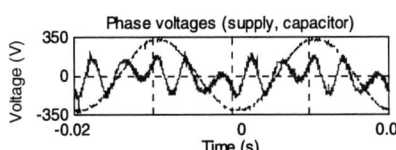

Fig. 14. Phase voltage u_s is presented as a dashed line. The solid line presents the IGBT bridge voltage u_c.

Compensation of asymmetry may decrease the fundamental voltage of one capacitor and increase the fundamental voltage of other capacitors. Because of this, the phase of one capacitor may turn to negative if the reactive current reference is too high. This can be avoided by decreasing the reactive current reference when asymmetry is compensated. The efficiency of the active power filter was decreased to 96 % when load current asymmetry was compensated, because i^*_{qref} can only be 1.5 A to keep the dc-link current control stable. The dc-link current also had to be greater.

VI. CONCLUSIONS

The main circuit of the CSAPF with a series-connected capacitor was presented. Control principle, transfer functions and computational capacitor voltage balancing method were studied. After that, prototype measurements were performed in 400 V supply with a nonlinear 10 kVA and 7 kVA unbalanced nonlinear diode bridge load. Compensation results were presented.

The results showed that the load current harmonics can be compensated with the series-capacitor-connected CSAPF. Also, phase asymmetry can be compensated within certain limits. The THD$_{2kHz}$ of the supply currents was decreased to below 10 %. It can be assumed that harmonic distortion can be further decreased by improving the control system. It was possible to achieve a measured prototype efficiency of over 97 % thanks to reduced phase voltages. Computational capacitor voltage balancing decreases imbalance and helps to keep the system stable.

It can be concluded that CSAPF with a series-connected capacitor is a useful topology for reducing voltage stresses and power losses in harmonic compensation if the load is almost symmetric. If there is a lot of asymmetry, the benefits achieved in the case of power losses and voltage stresses are decreased.

REFERENCES

[1] M. Salo; S. Pettersson.," Current-Source Active Power Filter with an Optimal DC Current Control", Power Electronics Specialists Conference, 2006. PESC '06. 37th IEEE 18-22 June 2006 Page(s):1 – 4

[2] J. Turunen; M. Salo, H. Tuusa, "Comparison of series hybrid active power filters based on experimental tests", Power Electronics and Applications, 2005 European Conference on 11-14 Sept. 2005 Page(s):10 pp.

[3] L. Moran, P. Werlinger, J. Dixon, and R. Wallace, "A series active power filter which compensates current harmonics and voltage unbalance simultaneously", Power Electronics Specialists Conference, 1995. PESC '95 Record, 6th Annual IEEE, vol. 1, 18-22 June 1995 pp.222 - 227 vol.1

[4] H. Akagi, S. Srianthumrong, and Y. Tamai, "Comparisons in circuit configuration and filtering performance between hybrid and pure shunt active filters", Industry Applications Conference, 2003. 38th IAS Annual Meeting. Conference Record of the vol. 2, 12-16 Oct. 2003 pp.1195 - 1202 vol. 2

[5] H.-L. Jou, J.-C. Wu, Y.-J. Chang, and Y.-T. Feng, "A novel active power filter for harmonic suppression", in Power Delivery, IEEE Transactions vol. 20, Issue 2, Part 2, April 2005 pp.1507 – 1513.

[6] S. Srianthumrong, H. Akagi, "A medium-voltage transformerless AC/DC power conversion system consisting of a diode rectifier and a shunt hybrid filter", Industry Applications, IEEE Transactions on Volume 39, Issue 3, May-June 2003 Page(s):874 - 882

[7] R. Inzunza and H. Akagi, "A 6.6-kV transformerless shunt hybrid active filter for installation on a power distribution system", Power Electronics, IEEE Transactions on Volume 20, Issue 4, July 2005 Page(s):893 – 900 Digital Object Identifier 10.1109/TPEL.2005.850951

[8] P. Parkatti, M. Salo and H. Tuusa. "A Novel Vector Controlled Current Source Shunt Active Power Filter with Reduced Component Voltage Stresses.", Power Electronics Specialists Conference, 2007. PESC 2007. IEEE 17-21 June 2007 Page(s):1121 - 1125

[9] P. Parkatti, M. Salo and H. Tuusa. " A novel vector controlled current source shunt active power filter and its comparison with a traditional topology", Power Electronics and Applications, 2007 European Conference on 2-5 Sept. 2007 Page(s):1 - 9

[10] M. Salo and H. Tuusa, "A vector controlled current-source PWM rectifier with a novel current damping method", in IEEE Trans. Power. Electron., pp. 464-471, Vol. 15, No. 3, May 2000.

[11] M. Salo and H. Tuusa, "A novel open-loop control method for a current-source active power filter", in IEEE Trans. Power. Electron., pp.:313-321, vol. 50, Issue 2, April 2003.

[12] M. Routimo, M. Salo and H. Tuusa, "A novel simple prediction based current reference generation method for an active power filter", Power Electronics Specialists Conference, 2004. PESC 04. 2004 IEEE 35th Annual Volume 4, 2004 Page(s):3215 – 3220.

[13] T. Halkosaari and H. Tuusa, "Optimal vector modulation of a PWM current source converter according to minimal switching losses", 31st Annual Power Electronics Specialists Conference. PESC'00, vol. 1, pp. 127-132, 2000.

[14] F. de Leon, A. Semlyen, "Time domain modeling of eddy current effects for transformer transients", Power Delivery, IEEE Transactions on Volume 8, Issue 1, Jan. 1993 Page(s):271 - 280

[15] P.A. Dahono, "A method to damp oscillations on the input LC filter of current-type AC-DC PWM converters by using a virtual resistor", Telecommunications Energy Conference, 2003. INTELEC '03. The 25th International 19-23 Oct. 2003 Page(s):757-761

A Multi-Drive System Based on a Two-stage Matrix Converter

Dinesh Kumar, Patrick W Wheeler, Jon C Clare, Lee Empringham

University of Nottingham, Nottingham, UK e-mail: *pat.wheeler@nottingham.ac.uk*

Abstract— **Multi-drive systems can be used in situations where several motors are mechanically connected. Such systems have applications in both industrial and aerospace applications, where they can be used to provide system redundancy. In conventional systems the power converter for the motor drive is based on a rectifier and inverter, leading to the requirement for large passive components such as DC-link capacitors. This paper proposes a multi-drive system based on a two-stage direct power converter. In this topology a direct converter uses two output stages with a single input bridge, eliminating the need for large passive components. Simulation results are included to demonstrate the feasibility of the proposed topology.**

Keywords—**Multi-drive, motor control unit, direct power converter, matrix converter, adjustable speed drive.**

I. INTRODUCTION

A multi-motor drive system can be created from industrial motor drives using a common DC bus. The common bus is used to supply the drive modules with DC power and each drive will use an inverter for the DC to AC power conversion. The output of each inverter is then connected to a motor. The DC power can be derived from a single supply unit (rectifier). Multi-drive systems have many benefits, for example reducing cables lengths, reduced supply currents, reduced components counts and increased availability. Examples of traditional multi-drive topologies using DC-link based converter topologies are shown in Fig. 1 and Fig. 2.

The limitation of these topologies is that they require large DC-link components [1,2]. These DC-link capacitors often have a short life time compared to other electronic devices. As a result, the overall converter lifetime is reduced..

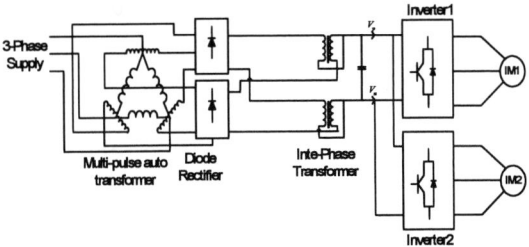

Fig. 1. Multi-pulse based multi-drive system

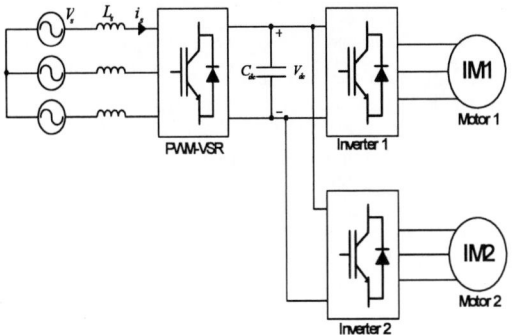

Fig. 2 Indirect AC-AC converter based multi-drive system

In application, such as aerospace, where the size and the weight are critical issues, these DC-link capacitor based converters have significant disadvantages due to the size of the capacitors..

In recent years Direct or Matrix Converters have attracted the attention of researchers. One variation on this topology is the Two-stage Matrix Converter [3]. It is possible to develop this Two-stage Matrix Converter to form a compact multi-drive power converter. As Matrix Converters have no DC-link energy storage the resulting topology has the potential for a significant improvement in power density.

II. DIRECT POWER CONVERTER

Cycloconverters were the first well known ac to ac direct power conversion topologies. However, naturally commutated cycloconverters have serious limitations with regards to the input power factor, output frequency and higher distortions in the input and output waveform [4].

The matrix converter has been a promising technique for direct ac to ac power conversion [5]. This topology overcomes the limitations present in the naturally commutated cycloconverters through the use of forced commutated switching devices. The matrix converter provides the sinusoidal supply currents, variable frequency variable amplitude load voltages, adjustable input power factor irrespective of the load and four quadrant operations. Most importantly, the elimination of the DC-link energy storage provides the potential for more compact ac to ac converter topology which may have advantages in applications such as aerospace. These promising benefits fuelled research in the last few decades and have demonstrated matrix converter technology for various motor drive applications. A matrix converter comprises of nine bidirectional switches as shown in Fig. 3.

978-1-4244-1741-4/08/$25.00 ©2008 IEEE

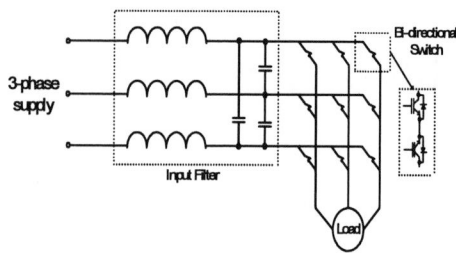

Fig. 3 Conventional matrix converter topology

Another approach for the direct ac to ac conversion is the two-stage matrix converter. The two-stage matrix converter is an indirect form of the matrix converter, providing similar input/output performances of the conventional matrix converter [6].

The topology of two-stage matrix converter is shown in Fig. 4. This topology consists of a rectification stage, which is essentially a 3/2-phase matrix converter. This rectification stage is directly connected to an inversion stage, which is a standard inverter bridge. Unidirectional polarity of the voltage is required in the link between the rectification and inversion stage. This link stage is often refer to as a 'variable DC-link' with the observation that there are no passive components and that this DC-link voltage is not stiff or constant.

The inversion stage uses six unidirectional switches with anti-parallel diodes. The function of this stage is similar to a traditional voltage source inverter. The inversion stage uses the variable DC-link voltage, to generate the variable amplitude and variable frequency voltage demand for the inductive load.

The function of the rectification stage is to switch between the three line-to-line voltages in order to provide sinusoidal input current whilst at the same time providing the required 'DC-link' voltage. Zero vectors are inherently produced by the inversion stage, because a zero output voltage vector is equivalent to connecting all outputs to the same DC-link polarity. Only one line-to-line voltage is selected at a time and applied with the correct polarity to the inversion stage. This operation is functionality analogous to a matrix converter controlled by an indirect modulation approach [7].

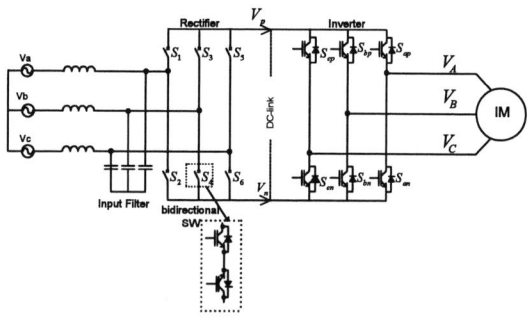

Fig. 4 Two-stage matrix converter topology

The two-stage direct power converter has some limitations, for example it is not possible to produce

rotating vectors. The topology has some advantages over conventional matrix converter such as the possibility of a reduced number of switches and a different loss profile compare to a matrix converter.

A. Space vector modulation for conventional matrix converter

An indirect space vector modulation (SVM) is often used for matrix converters [8], in which a combination of the two adjacent vectors and a zero-vector to synthesize a reference vector of variable amplitude and angle. The proportion between the two adjacent vectors gives the direction and the zero-vector duty-cycle determines the magnitude of the reference vector. The input current vector, I_{in}, that corresponds to the rectification stage and output voltage vector V_{out} that corresponds to the inversion stage are the reference vectors.

The input reference vector, I_{in}, is synthesized with two adjacent vectors I_γ and I_δ. The duty cycle of I_γ and I_δ, are given by (1), where the rectification's modulation index m_I is set to unity and θ^*_{in} is the angle of I_{in} within the respective sector.

$$d_\gamma = m_I . \sin\left(\frac{\pi}{3} - \theta^*_{in}\right) \text{ and } d_\delta = m_I . \sin\theta^*_{in} \qquad (1)$$

The output reference vector V_{out} is synthesized with two adjacent vectors V_α and V_β. The duty cycle of V_α and V_β, are given by (2), where the inversion's modulation index is m_U and θ^*_{out} is the angle of V_{out} within the respective sector.

$$d_\alpha = m_U . \sin\left(\frac{\pi}{3} - \theta^*_{out}\right) \text{ and } d_\beta = m_U . \sin\theta^*_{out} \qquad (2)$$

To obtain a correct balance of the input currents and the output voltages in the same switching period, the modulation pattern should combine the rectification (γ-δ-0) and inversion (α-β-0) vectors uniformly, producing the following switching pattern: α γ- α δ- β γ-0. The input current vector I_{in} is the reference of the rectification and output voltage vector V_{out} is the reference of the inversion stage shown in Fig. 5(a-b).

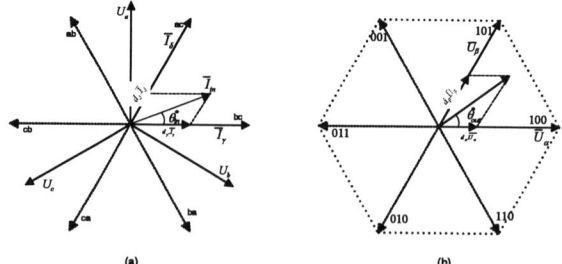

Fig. 5 Generation of reference vectors using space vector modulation: (a) rectification stage, (b) inversion stage

In order to combine the duty-cycles of the two virtual stages and to obtain the same behavior for the single stage matrix converter, it is necessary to obtain the duty cycles of the combined rectifier and inverter switching states as a cross product of their respective duty cycles as shown in (3), while the duration of the zero-vector is completing the switching sequence given in (4).

$$d_{\alpha\gamma} = d_{\alpha}.d_{\gamma}, \;\; d_{\alpha\delta} = d_{\alpha}.d_{\delta}, \;\; d_{\beta\delta} = d_{\beta}.d_{\delta}, \;\; d_{\beta\gamma} = d_{\beta}.d_{\gamma} \quad (3)$$

$$d_0 = 1\text{-}(d_{\alpha\gamma} + d_{\alpha\delta} + d_{\beta\delta} + d_{\beta\gamma}) \quad\quad\quad\quad (4)$$

The duration of each sequence is then found by multiplying the corresponding duty-cycle to the switching period.

B. Space vector modulation for two-stage matrix converter

The implementation of the SVM modulator for two-stage matrix converter can be easily translated from the conventional matrix converter with the following amendments: in the rectification stage, the zero vector is eliminated and the switching sequence consists only of the two adjacent active current vectors, therefore compared to (1), they should be resized (5) such that they will occupy the whole switching period:

$$d_{\gamma}^{R} = \frac{d_{\gamma}}{d_{\gamma}+d_{\delta}} \;\; \text{and} \;\;\;\; d_{\delta}^{R} = \frac{d_{\delta}}{d_{\gamma}+d_{\delta}} \quad (5)$$

These duty-cycles multiply with the switching period and the resulting ON-times directly drive the rectification stage switches. Since the average voltage in the DC-link is not constant anymore due to the cancellation of the zero-vector in the rectification stage, it is necessary to calculate its value to compensate the modulation index of the inversion stage;

$$V_{PN\text{-}avrg} = d_{\gamma}^{R}.V_{line\text{-}\gamma} + d_{\delta}^{R}.V_{line\text{-}\delta} \quad (6)$$

$$m_{U} = \sqrt{2}.\frac{V_{out}}{V_{PN\text{-}avrg}} \quad\quad\quad\quad (7)$$

The inverter stage may use a double-sided asymmetric PWM switching sequence $0_{\gamma}\text{-}\alpha_{\gamma}\text{-}\beta_{\gamma+\delta}\text{-}\alpha_{\delta}\text{-}0_{\delta}$, but with unequal sides because each side corresponds to a rectification switching sequence which has a different DC-link voltage. Therefore, the value of the modulation index m_{U} in (2) has to be corrected with the momentary average DC-link voltage $V_{PN\text{-}avrg}$ (6), which takes into account its variation. The inversion stage duty-cycles are given in (8) and (9);

$$d_{0\gamma} = \frac{d_{\gamma}.[1\text{-}(d_{\gamma}+d_{\delta}).(d_{\alpha}+d_{\beta})]}{d_{\gamma}+d_{\delta}} \;\; \text{and} \;\; d_{\alpha\gamma} = d_{\gamma}.d_{\alpha} \quad (8)$$

$$d_{\beta(\gamma+\delta)\alpha\delta} = (d_{\gamma}+d_{\delta})d_{\beta} \;\; \text{and} \;\; d_{\alpha\delta} = d_{\delta}.d_{\alpha} \quad (9)$$

The switching pattern for two-stage matrix converter is shown in Fig. 6.

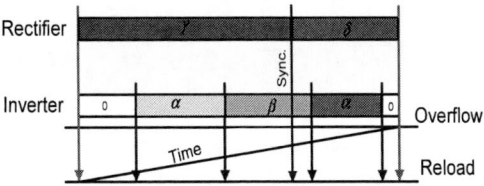

Fig. 6 switching pattern for two-stage matrix converter

III. PROPOSED MULTI DRIVE SYSTEM

This paper present a two-stage matrix converter based multi-drive system with two motors on single shaft. In this topology the rectification stage is shared by two induction motors. Each motor is controlled using an independent output bridge, each utilising an indirect vector field oriented control scheme [10]. The overall configuration of proposed system is shown in Fig. 7.

Fig. 9 shows the vector control scheme for the complete system. It should be noted that each motor has an independent current control loop. The reference for these two current control loops is derived from a single speed control loop, the required torque producing current being halved to give the reference for each motor.

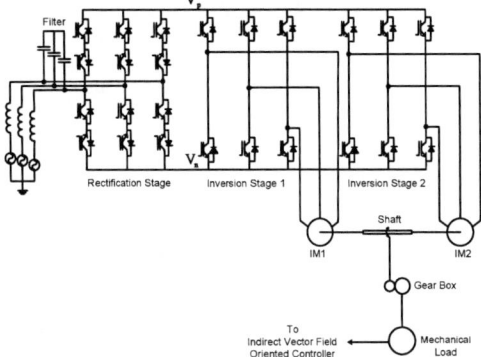

Fig. 7 Multi-drive system based on two-stage matrix converter

A. Space vector modulation for Multi-Drive System

In order to draw the sinusoidal input currents from the supply the average output power must be constant. This is fulfilled in the case of symmetrical, sinusoidal loads. Using the two stage matrix converter topology it is possible to connect several inversion stages to the same variable DC-link, assuming that the switching patterns of the inversion stages are synchronized with the rectification stage. The space vector modulation for the multi-motor drive topology can be derived from the modulation use for the two-stage matrix converter topology given in Section II.

In the rectification stage, the zero-vector is eliminated and the switching sequence consists only of the two adjacent current vectors (line-to-line voltages). The zero-vectors are applied by each inversion. A central control unit can then be used to control the motors in this multi-motor drive system. If both output bridges require a low modulation index, alternative modulation techniques can be used to provide lower switching losses by reducing the average DC-link voltage [9]. By using (1), the adjusted duty-cycles of the rectification stage are given by;

$$d_\delta^R = \frac{d_\delta}{d_\gamma + d_\delta} \quad \text{and} \quad d_\delta^R = \frac{d_\delta}{d_\gamma + d_\delta} \qquad (10)$$

where the modulation index of the rectification side is unity.

These duty-cycles directly drive the rectification stage. Due to cancellation of zero-vector in the rectification stage the average voltage in the DC-link is not constant anymore, so it is necessary to calculate its value to compensate the modulation indexes of the inversion stages;

$$V_{PN} = d_\gamma . V_{line-\gamma} + d_\delta . V_{line-\delta} \quad \text{and} \quad m_U^k = \sqrt{2} . \frac{V_{out}^k}{V_{PN}} \qquad (11)$$

The inverter stages use a doubly-sided switching sequence; 0-α-β-β-α-0 , which should be symmetrical, because each side should apply on each of the rectification sequence duty-cycles, where:

$$d_\alpha^k = m_U^k . \sin\left(\frac{\pi}{3} - \theta_{out}^{*k}\right) \text{ and } d_\beta^k = m_U^k . \sin(\theta_{out}^{*k}) \qquad (12)$$

Therefore, the duty-cycles for the switching sequences of any of the "k" inversion stage can be found;

$$d_0^k = \frac{d_\gamma . \left[1 - (d_\gamma + d_\delta).(d_\alpha^k + d_\beta^k)\right]}{d_\gamma + d_\delta} \qquad (13)$$

$$d_1^k = d_\gamma . d_\alpha^k, \qquad d_2^k = (d_\gamma + d_\delta) d_\beta^k \qquad (14)$$

$$d_3^k = d_\delta . d_\alpha^k, \qquad (15)$$

$$d_4^k = \frac{d_\delta . \left[1 - (d_\gamma + d_\delta).(d_\alpha + d_\beta)\right]}{d_\gamma + d_\delta} \qquad (16)$$

$$= 1 - d_0^k - d_1^k - d_2^k - d_3^k$$

Fig. 8 shows the switching pattern for the rectification stage and the inversion stages for proposed multi-drive system.

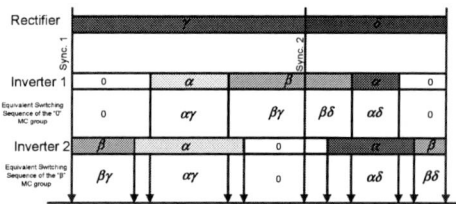

Fig.8 switching pattern for multi-drive system based on two-stage matrix converter

A. Indirect vector field orineted control scheme

Induction motors are most widely used motors in industrial motion control and domestic appliances due to their reliability, robustness and simplicity. The major limitation of Induction motors is poor dynamic behavior because of rotor flux is in relative with stator flux, which depends on slip speed, and hence both are not always orthogonal position. In order to overcome this indirect vector field oriented control can be used to provide control of the motor speed. The aim of the vector control scheme is to decouple the vectors of armature flux and field current independently, so that both field current and torque are controlled independently and good torque response can then be obtain.

Equations (17)-(19) are the fundamental equations for vector control and allows the induction motor to act like a separately excited DC machine with decoupled control of torque and flux, making it possible to operate the induction motor as a high-performance four-quadrant servo drive.

$$\omega_{sl} = \left(\frac{1}{\tau_r i_{mrd}}\right) i_{sq} \qquad (17)$$

$$\tau_r = \frac{R_r}{L_r} \qquad (18)$$

$$T_e = \frac{2}{3}\left(\frac{P}{2}\right)\frac{M^2}{L_r} i_{sd} i_{sq} \qquad (19)$$

Fig. 9 shows the block diagram for the vector control scheme for matrix converter based multi-drive system where three output currents for each machine and the rotor position are measured. The motor speed ω_r is measured by determining the rate of change of the position encoder output and compared to the demanded speed ω_r^*. The resulting speed error is then processed by a proportional-integral (PI) controller to produce the reference current, i_{sq}^*.

In the vector control the 3-phase stator line currents $(i_{sa1}(t), i_{sb1}(t), i_{sc1}(t))$ and $(i_{sa2}(t), i_{sb2}(t), i_{sc2}(t))$ are transformed to the stationary two-axis currents, $(i_{s\alpha1}(t), i_{s\beta1}(t))$ and $(i_{s\alpha2}(t), i_{s\beta2}(t))$. These are then transformed into the rotating d-q axis currents, $(i_{sd1}(t), i_{sq1}(t))$ and $(i_{sd2}(t), i_{sq2}(t))$. The equivalent complex operator $e^{-j\theta}$ is used for transformation. In order to get the rotating d-q axis currents, the value of flux angle is necessary, which is determined by summing the rotor position signal and the reference slip position obtained by integrating equation (17). The inverse transformation of d-q axis values to the instantaneous stator reference frame is represented by the complex operator $e^{j\theta}$ [10].

The two current controllers are used for each motor, which employ PI control process the errors of $(i_{sd1}(t), i_{sq1}(t))$ and $(i_{sd2}(t), i_{sq2}(t))$ to give $\left(v_{sd1}', v_{sq1}'\right)$ and $\left(v_{sd2}', v_{sq2}'\right)$. Voltage compensation terms are added to the output of each current controller to get the resulting

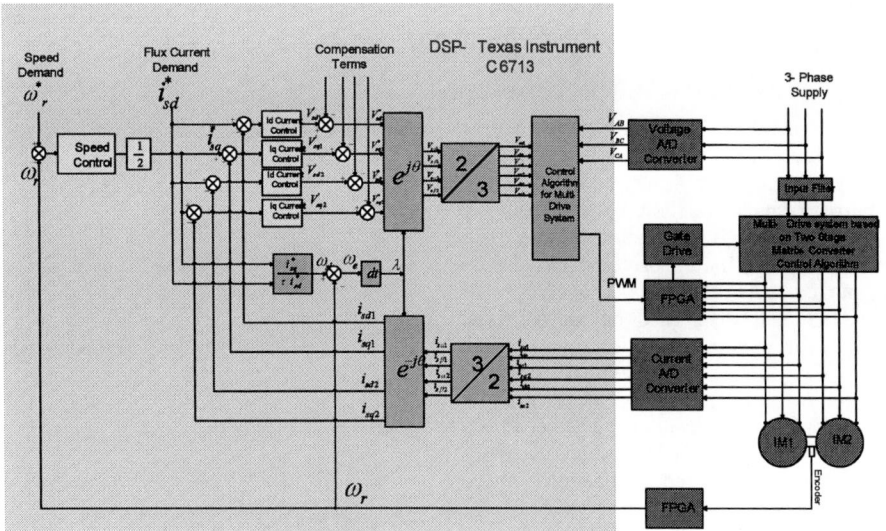

Fig. 9. System Configuration for Matrix converter based Multi-Drive system.

reference signals $\left(v_{sd1}^{*}, v_{sq1}^{*}\right)$ and $\left(v_{sd2}^{*}, v_{sq2}^{*}\right)$. These voltages are then converted to stationary two axis voltages $\left(v_{s\alpha1}^{*}, v_{s\beta1}^{*}\right)$ and $\left(v_{s\alpha2}^{*}, v_{s\beta2}^{*}\right)$ using the complex operator $e^{j\theta}$. The three phase voltages are obtained by using $\alpha\beta$ to 3-phase transformation. These three phase voltages $\left(V_{sa1}, V_{sb1}, V_{sc1}\right)$ and $\left(V_{sa2}, V_{sb2}, V_{sc2}\right)$ are used as the input signals for the space vector modulation algorithm for matrix converter based multi-drive system.

It should be noted that each motor has an independent current control loop. The reference for these two current control loops is derived from a single speed control loop, the required torque producing current being halved to give the reference for each motor.

The motor and a space vector modulation strategy have been implemented on a DSP (TI C6713). This is connected to the A/D converters via an FPGA. This FPGA also handles all the three-step output current direction based current commutation strategy for the converter system and generation of control signals for the IGBTs in the converter circuit.

IV. SIMULATION

The viability of the proposed motor drive system is confirmed through the simulation studies performed using SABER. A number of simulation tests have been performed to confirm the correct operation of the multi-drive system.

Fig. 10 illustrates the dynamic performance of the shaft speed in acceleration mode from standstill to 300rpm, a 50% of rated load applied at 1.0sec and removed at 2.0sec during positive reference constant speed.

Fig. 11 shows the waveforms of motor shaft speed, q-axis currents and phase currents, while the reference speed is reversely given from 300rpm to -300rpm and d-axis current is set at 2.5A. The 50% of rated load is applied to the motor at 1.0sec during positive reference constant

speed and removed at 3.0sec during negative reference constant speed.

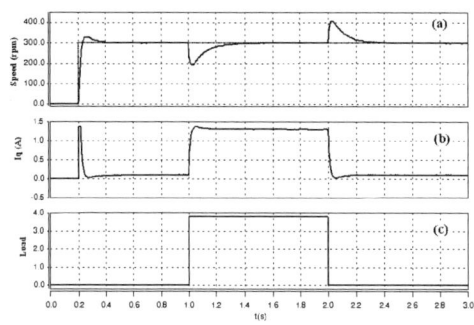

Fig.10. Simulation results in case of speed changes with step from standstill to 300rpm: (a) motor shaft speed, (b) q-axis current of motor 1 and (c) applied load

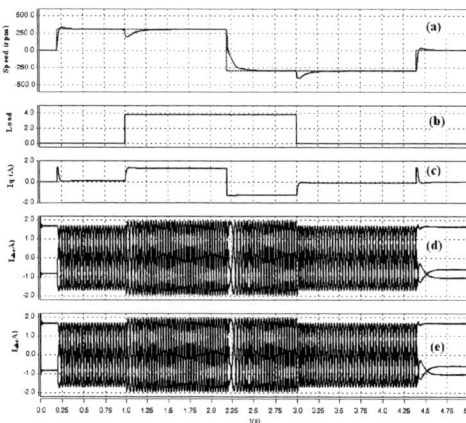

Fig.11 Simulation results in case of speed reversal from 300rpm to -300rpm: (a) motor shaft speed, (b) q-axis current, (c) applied load, (d) 3-phase currents of motor 1 and (e) 3-phase currents of motor 2.

The simulation results show that the actual speed follows the reference speed well and also q axis current.

V. CONCLUSION

The design and operation of a two-stage matrix converter topology which allow the implementation of a multi-motor drive system has been described in the paper. The feasibility and performance of the proposed topology has been demonstrated through simulation results. The matrix-converter based multi-motor drive system has a number of advantages over the conventional power converter based topologies when used on an AC bus, giving the potential for a substantial improvement in power density. As higher temperature power semiconductor switching devices such as silicon carbide devices become available, this technology will become a good candidate for power conversion system in many applications, especially in aerospace applications where volume and weight are at a premium.

REFERENCES

[1] A Baghramian, A J Forsyth, 2Averaged-Value Models of Twelve-Pulse Rectifiers for Aerospace Applications," *Second International Conference on Power Electronics, Machines and Drives*, vol 1, 31ᵗ Mar - 2ⁿᵈ Apr, 2004

[2] Z.Liu: Dynamic analysis of centre-driven web winder controls, Proc. IEEE Industrial Application Society, *Annual Meeting IAS*, Phoenix, USA, 1999, pp. 1388-1396.

[3] C. Klumpner and F. Blaabjerg, "A new cost-effective multi-drive solution based on a two-stage direct power electronic conversion topology," *in Proc. IAS'02*, vol. 1, 2002, pp. 444–452.

[4] C.L. Neft and C.D. Schauder, "Theory and design of a 30-hp matrix converter," *IEEE Transactions on Industry Applications*, vol. 28, no. 3, pp. 248 – 253, 1992.

[5] P.W. Wheeler, J. Rodriguez, J.C. Clare, L. Empringham, and A. Weinstein, "Matrix converters: a technology review," *IEEE Transactions on Industrial Electronics*, vol. 49, no. 2, pp. 276 – 288, 2002.

[6] L. Wei and T.A. Lipo, "A novel matrix converter topology with simple commutation," *Proc. of Industry Applications Society Annual Meeting*, vol. 3, pp. 1749 – 1754, 2001.

[7] B.K. Bose, "Power electronics-a technology review," *Proc. of IEEE*, vol. 80, no. 8, pp. 1303 – 1334, August 1992.

[8] D. Casadei, G. Serra, A. Tani, and L. Zarri, "Matrix converter modulation strategies: A new general approach based on space-vector representation of the switch state," IEEE Trans. Ind. Electron., vol. 49, pp. 370–381, Apr. 2002.

[9] L. Helle, S. Munk-Nielsen, "A novel loss reduced modulation strategy for matrix converters", Proc of PESC'01, vol. 2, pp 1102-1107, 2001.

[10] P Vas: Vector Control of AC Machines, Clarendon Press, Oxford 1990.

Characteristics of the Single Active Bridge Converter with Voltage Doubler

Andreas Averberg*, Axel Mertens[†]
Leibniz University Hannover,
Institute for Drive Systems and Power Electronics, Hannover, Germany
*e-mail: *averberg@ial.uni-hannover.de*
[†]e-mail: *mertens@ial.uni-hannover.de*

Abstract—This paper investigates the single active bridge dc-dc converter with voltage doubler. An analytical investigation is given, completely describing the operating behaviour of the converter. Herewith, the impact of the transformer's properties and a calculation of the current stress of the semiconductors and passive devices is shown. The results are compared to the single active bridge with full bridge rectifier. With regard to a small ripple current at the converter input, the results can be used as a tool for an optimised converter design. All results are compared to simulations. A 1.2 kW prototype was built, and measurement results are given.

Keywords—DC power supply, Fuel cell system, High frequency power converter.

I. INTRODUCTION

Due to environmental sustainability, nowadays renewable energy becomes more and more important. Fuel cells are only one option for the future. At medium power levels they have a relatively low output voltage, a high output current and large variations in output voltage under variable load conditions [1], [2], [3]. In order to make fuel cells usable for applications with a higher voltage demand, a dc-dc converter with high voltage gain can be used. Many dc-dc converters have been proposed and investigated during the last decade [4]. In [5], an overview of different single phase topologies for small distributed power generators is given. For fuel cell applications, a voltage-fed full bridge, operating with a high transmission ratio transformer, and full bridge rectifier are proposed for single stage converters [6], [7], [8] or two stage converters with a boost converter connected upstream [9]. In [10], the transformer's leakage inductance is used for voltage boosting by substituting two rectifier diodes with IGBTs.

However, the transformer's winding ratio and its leakage inductance affect the operating behaviour of the topology, hence the value of the output voltage, which is investigated in [8]. It is shown in [11], that the transformer's transmission ratio should be held as small as possible to receive a good efficiency of the converter. Thus, the single active bridge converter with voltage doubler rectifier might be a further alternative. In [12], an overview of circuit topologies with voltage doubler rectifier is given. For a fuel cell application, a converter with voltage doubler is used in [13].

This paper deals with the single active bridge converter with voltage doubler shown in Fig. 1, operating

with a phase-shifted pwm. A full analytical investigation of the converter is given. Herewith, the impact of the transformer's winding ratio and its leakage inductance on the operating behaviour of the topology is shown. The current stress of the passive components as well as of all semiconductor devices is emphasised and compared to the single active bridge with full bridge rectifier. All results are proved by simulations. A 1.2 kW prototype was built and experimental results are given in this paper. In the

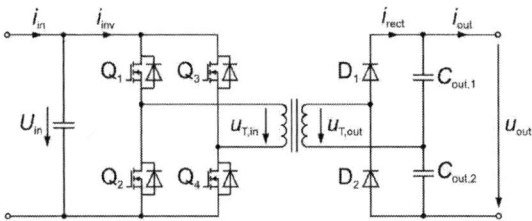

Fig. 1. Single active bridge converter with voltage doubler SAB_VD

following, the single active bridge converter with voltage doubler rectifier is denoted as SAB_VD. The analytical and experimental results are compared to the single active bridge converter with full bridge rectifier. This converter has two diodes instead of the capacitors $C_{\text{out},1}$ and $C_{\text{out},2}$ in its rectifier circuit. It is denoted as SAB_FB in this paper and shown in Fig. 2.

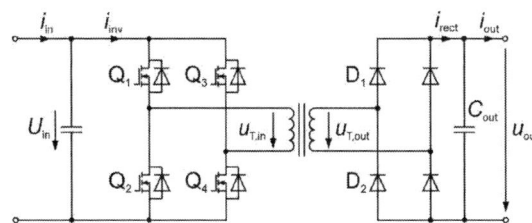

Fig. 2. Single active bridge converter with full bridge rectifier SAB_FB

II. ANALYSIS OF THE CONVERTER

When modeling the high frequency transformer by its leakage inductance only, the equivalent circuit diagram shown in Fig. 3 is obtained. All secondary values have to be referred to the primary side. Two different operating modes can be distinguished. In the discontinuous conduction mode (DCM), the rectifiers output current

978-1-4244-1741-4/08/$25.00 ©2008 IEEE

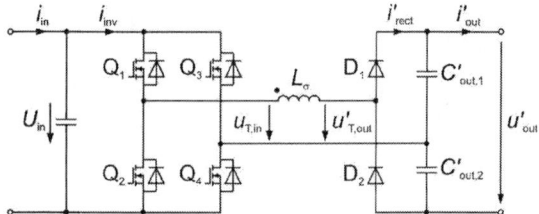

Fig. 3. Equivalent circuit diagram of the SAB_VD

i'_{rect} is zero for more than one moment during the first half period. During $0 < t < \frac{T}{2}$, i'_{rect} is equal to the current through the leakage inductance $i_{L,\sigma}$, which is shown in Fig. 4 (a). During $\frac{T}{2} < t < T$, i'_{rect} is zero. In Fig. 4 (a), additionally the voltage waveforms of the transformer's input and output voltage $u_{\text{T,in}}$ and $u'_{\text{T,out}}$ as well as the output voltage u'_{out} are given. The second operating mode is at the border between continuous and discontinuous conduction mode and denoted as border mode (BM). Fig. 4 (b) shows the waveforms. During the time $t_1 < t < \left(\frac{T}{2} + t_1\right)$, the rectifier current is equal to $i_{L,\sigma}$. For the remaining time, i'_{rect} is zero. The continuous conduction mode (CCM) does not exist for this converter due to the absent inductor at the converter output. By

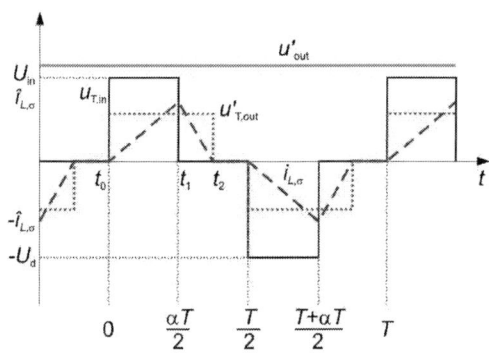

(a) Discontinuous conduction mode (DCM)

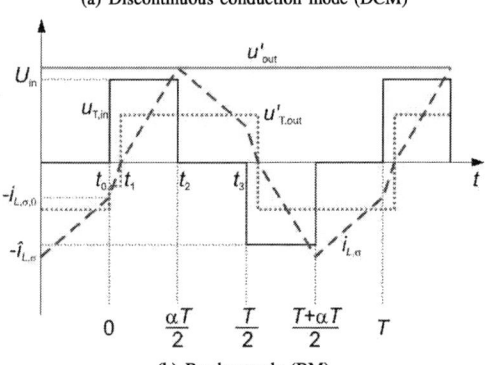

(b) Border mode (BM)

Fig. 4. Voltage and current waveforms

using the waveforms given in Fig. 4 and considering the conducting devices for every single section in both operating modes, the output voltages of the topology can be calculated.

In BM, the first half period can be divided into three different states. At $t = 0$, switches Q_1 and Q_4 are closed, switches Q_2 and Q_3 are open. The current through the leakage inductance $i_{L,\sigma}$ is negative, hence the body diodes $D_{Q,1}$ and $D_{Q,4}$ are conducting. Fig. 5 is obtained.

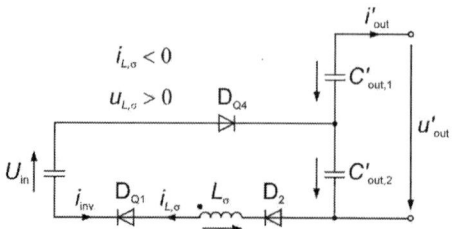

Fig. 5. Border mode, state 1

The current $i_{L,\sigma}(t)$ can be calculated as:

$$i_{L,\sigma}(t) = i_{L,\sigma}(0) + \frac{U_{\text{in}} + u'_{C,\text{out},2}}{L_\sigma} \cdot t$$

$$0 = i_{L,\sigma}(0) + \frac{U_{\text{in}} + 0.5 \cdot u'_{\text{out}}}{L_\sigma} \cdot t_1 \quad (1)$$

At $t = t_1$, the current $i_{L,\sigma}$ crosses zero. For $t_1 < t < \frac{\alpha T}{2}$, Q_1 and Q_4 are conducting and state 2 (Fig. 6) is active.

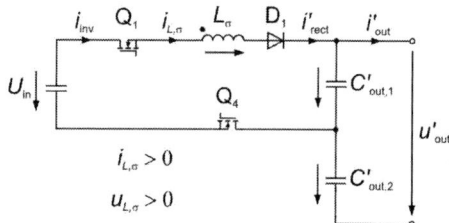

Fig. 6. Border mode, state 2

It is obtained:

$$i_{L,\sigma}(t) = \frac{U_{\text{in}} - u'_{C,\text{out},1}}{L_\sigma} \cdot (t - t_1)$$

$$i_{L,\sigma}\left(\frac{\alpha T}{2}\right) = \frac{U_{\text{in}} - 0.5 \cdot u'_{\text{out}}}{L_\sigma}\left(\frac{\alpha \cdot T}{2} - t_1\right) \quad (2)$$

When α is the converter's duty cycle, at $t = \frac{\alpha \cdot T}{2}$, Q_4 will be turned off and Q_3 will be turned on. The current commutates to the body diode $D_{Q,3}$ and $u_{\text{T,in}}$ turns to zero. State 3 is shown in Fig. 7.

It is received:

$$i_{L,\sigma}(t) = i_{L\sigma}\left(\frac{\alpha T}{2}\right) - \frac{u'_{C,\text{out},1}}{L_\sigma} \cdot \left(t - \frac{\alpha T}{2}\right)$$

$$i_{L,\sigma}\left(\frac{T}{2}\right) = i_{L\sigma}\left(\frac{\alpha T}{2}\right) - \frac{0.5 \cdot u'_{\text{out}}}{L_\sigma} \cdot (1 - \alpha) \cdot \frac{T}{2} \quad (3)$$

In the stationary case, it is obtained:

$$-i_{L,\sigma}\left(\frac{T}{2}\right) = i_{L,\sigma}(0) \quad (4)$$

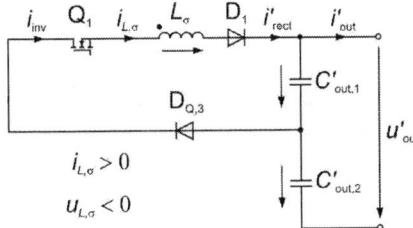

Fig. 7. Border mode, state 3

Q_1 will be turned off and Q_4 will be turned on at $t = t_3 = \frac{T}{2}$. The same considerations can be done for the second half period. By solving the equations (1) to (4), expressions for the three unknown variables $i_{L,\sigma}(0)$, $i_{L,\sigma}\left(\frac{\alpha T}{2}\right)$ and t_1 can be found. Herewith, the converter output current can be calculated as:

$$i'_{out} = \frac{1}{T} \cdot \int_{t_1}^{\frac{T}{2}+t_1} i_{L,\sigma} dt$$

$$i'_{out} = \frac{U_{in}^2 \cdot (2 - \alpha) \cdot \alpha - (0.5 \cdot u'_{out})^2}{16 \cdot f \cdot L_\sigma \cdot U_{in}} \qquad (5)$$

Rearranging this equation and converting the referred values back to the secondary side leads to the output voltage

$$u_{out,BM} = \frac{2}{N} \sqrt{(2 - \alpha) \cdot \alpha \cdot U_{in}^2 - 16 \cdot f \cdot \frac{i_{out}}{N} \cdot U_{in} \cdot L_\sigma} \qquad (6)$$

where N is the winding ratio $\frac{w_1}{w_2}$. The same calculations can be done for the DCM. The result is

$$u_{out,DCM} = \frac{2}{N} \cdot \frac{\alpha^2 \cdot U_{in}^2}{\alpha^2 \cdot U_{in} + 8 \cdot f \cdot \frac{i_{out}}{N} \cdot L_\sigma} \qquad (7)$$

It becomes obvious that the leakage inductance causes a voltage drop at the converter's output which rises with the switching frequency and the load current.

In [8], the SAB_FB was investigated. The equations for the output current in BM and DCM are as follows:

$$u_{out,BM} = \frac{1}{N} \sqrt{(2 - \alpha) \cdot \alpha \cdot U_{in}^2 - 8 \cdot f \cdot \frac{i_{out}}{N} \cdot U_{in} \cdot L_\sigma} \qquad (8)$$

$$u_{out,DCM} = \frac{1}{N} \cdot \frac{\alpha^2 \cdot U_{in}^2}{\alpha^2 \cdot U_{in} + 4 \cdot f \cdot \frac{i_{out}}{N} \cdot L_\sigma} \qquad (9)$$

A comparison brings out that the output voltage of SAB_VD is twice the output voltage of SAB_FB for no load conditions only. With rising load current, the voltage drop in the SAB_VD topology is larger than the voltage drop in the SAB_FB topology.

Fig. 8 (a) shows the dependency of the output voltage over the inverted winding ratio of the transformer for three different duty cycles. Further parameters are: input voltage $U_{in} = 28 \, \text{V}$, output current $i_{out} = 1.5 \, \text{A}$,

switching frequency $f = 60 \, \text{kHz}$ and leakage inductance $L_\sigma = 350 \, \text{nH}$. As can be seen, a maximum occurs at a special winding ratio, and increasing the number of secondary turns further leads even more to an increased voltage gain. The crosses mark results taken from a simulation. The simulated and calculated results fit exactly. For $\alpha = 0.7$, the curve u_{out} over $\frac{w_2}{w_1}$ is additionally given for the SAB_FB in Fig. 8 (b). Although no voltage doubler is used, the maximum voltage gain is exactly the same, in this case 1040 V. The SAB_FB offers the possibility to insert an output inductor L_{out}. For the sake of completeness, in Fig. 8 (b) u_{out} over $\frac{w_2}{w_1}$ is also given for $L_{out} = 0.3 \, \text{mH}$. In this case, the continuous conduction mode CCM may occur.

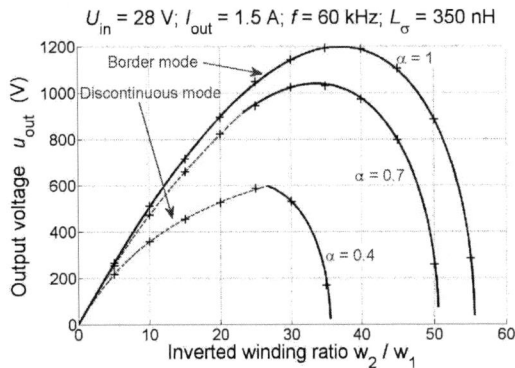

(a) SAB_VD

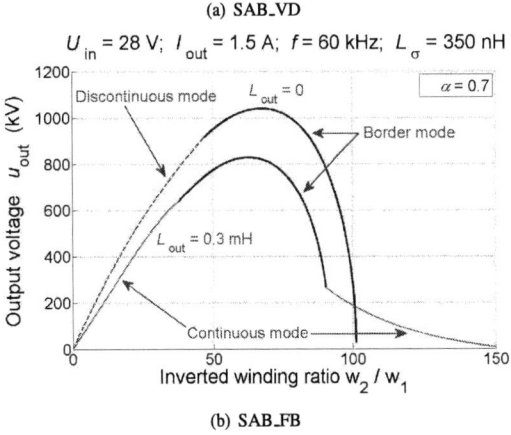

(b) SAB_FB

Fig. 8. Output voltage over inverted winding ratio

As can be shown, the maximum voltage gain is obtained always in BM or at the border between BM and DCM. It is possible to find the winding ratio which gives the best voltage gain by building the derivation of (6), and set it to zero. Equation (10) is obtained.

$$\frac{1}{N_{opt}} = \frac{1}{12} \cdot \frac{U_{in} \cdot \alpha \cdot (2 - \alpha)}{i_{out} \cdot f \cdot L_\sigma} \qquad (10)$$

N_{opt} is the winding ratio, which leads to the highest overall voltage gain. By inserting equation (10) into (6), the maximum possible output voltage for a given output current can be calculated.

$$U_{\text{out,max}} = \frac{\sqrt{3}}{36} \cdot \frac{U_{\text{in}}^2 \cdot \alpha \cdot (2-\alpha) \cdot \sqrt{(\alpha \cdot (2-\alpha))}}{i_{\text{out}} \cdot f \cdot L_\sigma} \quad (11)$$

Fig. 9 shows the curve progressions of the maximum reachable output voltage over the output current for four different leakage inductances. Fig. 9 is valid for both, SAB_VD and SAB_FB. When using SAB_FB with an additional output inductor, the results can be slightly different due to CCM operation of the converter. This is shown in [8]. Multiplying u_{out} with i_{out} gives the output power, which is also shown in Fig. 9. Since in (11) i_{out}

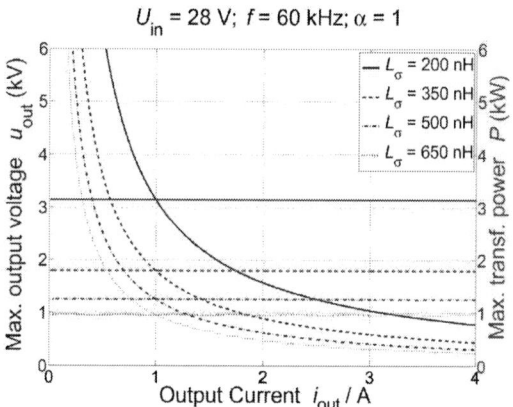

Fig. 9. Maximum values of output voltage and transferred power

only appears in the denominator, the maximum output power over the whole current range is constant. This means, the voltage drop across the leakage inductance due to high frequency operation limits the possible output power of the converter for a given output current.

III. EFFICIENCY OF THE CONVERTER

A. Dispartment of the loss mechanisms

As described in section II, by solving the equations (1) to (4), expressions for the three unknown variables $i_{L,\sigma}(0)$, $i_{L,\sigma}\left(\frac{\alpha T}{2}\right)$ and t_1 can be found. By inserting equation (6), the duty cycle α can be eliminated. The results can be used for pre-estimating mean values and rms values of the semiconductor currents, the semiconductors' turn off currents and finally the converter's efficiency in dependency of the output power in the form of i_{out} and u_{out}. This is presented here for the BM.

Fig. 10 shows the voltage and current waveforms of the switches Q_1 to Q_4 for BM. If the MOSFET of switch Q_x is conducting, it is marked by an index M: $i_{\text{M,x}}$, if the anti-parallel body diode is conducting, it is marked by an index B: $i_{\text{B,x}}$. Due to the phase-shifted pwm, all mean and rms values of the switches Q_1 and Q_2 are equal and all mean and rms values of the switches Q_3 and Q_4 are equal, too.

Typically, a MOSFET has a better on-state resistance $R_{\text{DS,on}}$ than its intrinsic body diode. For this reason, it is expedient to distinguish between the MOSFET's current

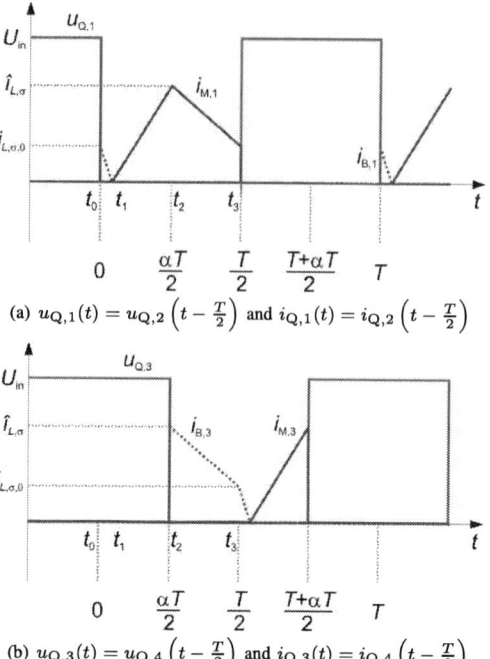

(a) $u_{\text{Q,1}}(t) = u_{\text{Q,2}}\left(t-\frac{T}{2}\right)$ and $i_{\text{Q,1}}(t) = i_{\text{Q,2}}\left(t-\frac{T}{2}\right)$

(b) $u_{\text{Q,3}}(t) = u_{\text{Q,4}}\left(t-\frac{T}{2}\right)$ and $i_{\text{Q,3}}(t) = i_{\text{Q,4}}\left(t-\frac{T}{2}\right)$

Fig. 10. Voltage and current waveforms of the switches Q_1 to Q_4 in BM

i_{M} and the current through its body diode i_{B}. With the knowledge of the current waveforms of Fig. 10 and the calculated characteristic values of section II, it is possible to calculate the current mean values individually. The results for the switch Q_1 are given in equations (12) and (13). The variable X stands for the triple $f \cdot L_\sigma \cdot i_{\text{out}}$. The same calculations can be accomplished for the rms values. Since the calculation for $I_{\text{M,1}}$ results in a long term, only $I_{\text{B,1}}$ is shown in (14). Equation (15) arises for the turn off current of switch Q_1.

The currents for the other switches as well as the currents through the rectifier diodes can be calculated analogously. With the described method, expressions for the currents are obtained, which depend only on the pre-known parameters $I\left(U_{\text{in}}, u_{\text{out}}, i_{\text{out}}, f, L_\sigma, \text{N}\right)$.

To pre-estimate the transformer's copper losses, the current $i_{L,\sigma}$ has to be calculated, which can be done as before. Furthermore, iron losses appear in the transformer, which are a function of the magnetic flux density B_μ and the switching frequency f. When B_μ and f are known, they can be taken from the core material's data sheet.

$$P_{\text{Fe}} = f\left(B_\mu, f\right) \cdot V_{\text{Fe}} \quad (16)$$

Here V_{Fe} is the transformer's volume, A_{Fe} is its cross section. The magnetic flux density can be substituted by the magnetomotive force, which is the integral of the transformer input voltage $u_{\text{T,in}}$ over one half period.

$$B_\mu = \frac{\psi_\mu}{\text{w}_1 \cdot A_{\text{Fe}}} \quad (17)$$

$$\bar{i}_{M,1} = \frac{\left(U_{in}^2 - \frac{u_{out} \cdot N}{2}\right)\sqrt{U_{in}^2 - 16 \cdot U_{in} \cdot \frac{X}{N} - \left(\frac{u_{out} \cdot N}{2}\right)^2} - U_{in}^3 + 8U_{in}\left(3 \cdot U_{in}\frac{X}{N} + \frac{u_{out}N}{2}\right) + U_{in}\left(\frac{u_{out} \cdot N}{2}\right)^2}{16 \cdot U_{in}^2 \cdot f \cdot L_\sigma} \tag{12}$$

$$\bar{i}_{B,1} = \frac{\left(\sqrt{U_{in}^2 - 16 \cdot U_{in} \cdot \frac{X}{N} - \left(\frac{u_{out} \cdot N}{2}\right)^2} - U_{in} + \frac{u_{out} \cdot N}{2}\right)^2 \cdot \left(U_{in} + \frac{u_{out} \cdot N}{2}\right)}{32 \cdot U_{in}^2 \cdot f \cdot L_\sigma} \tag{13}$$

$$I_{B,1} = \frac{\sqrt{3}}{24} \cdot \sqrt{\frac{\left(U_{in} - \sqrt{U_{in}^2 - 16U_{in}\frac{X}{N} - \left(\frac{u_{out} \cdot N}{2}\right)^2} - \frac{u_{out} \cdot N}{2}\right)^3 \cdot \left(U_{in} + \frac{u_{out} \cdot N}{2}\right)^2}{U_{in}^3 \cdot f^2 \cdot L_\sigma^2}} \tag{14}$$

$$i_{Q,1,off} = i_{L,\sigma}(0) = \frac{\left(\sqrt{U_{in}^2 - 16U_{in}\frac{X}{N} - \left(\frac{u_{out} \cdot N}{2}\right)^2} - U_{in} + \frac{u_{out} \cdot N}{2}\right) \cdot \left(U_{in} + \frac{u_{out} \cdot N}{2}\right)}{4 \cdot U_{in} \cdot f \cdot L_\sigma} \tag{15}$$

$$\psi_\mu = \frac{1}{2} \cdot \int_0^{\frac{T}{2}} u_{T,in}(t)\, dt = \frac{\alpha}{4 \cdot f} \cdot U_{in} \tag{18}$$

This leads to

$$P_{Fe} = f\left(\frac{\alpha \cdot U_{in}}{4 \cdot f \cdot w_1 \cdot A_{Fe}}\right) \cdot V_{Fe} \tag{19}$$

As before, equations (6) and (7), respectively can be used to eliminate the variable α, which is not known automatically. An expression

$$P_{Fe} = f\left(U_{in}, u_{out}, i_{out}, f, L_\sigma, N, w_1, A_{Fe}, V_{Fe}\right) \tag{20}$$

is obtained.

With these equations, all main losses of the converter can be pre-estimated.

B. The converter in a fuel cell application

In fuel cell systems, the converter input voltage, which is equal to the fuel cell output voltage, depends on the instant power. Loading the fuel cell leads to a reduced output voltage. The crosses in Fig. 11 mark measured values of the output voltage of a 1.2 kW Nexa fuel cell stack [14]. The ideal open circuit potential of one

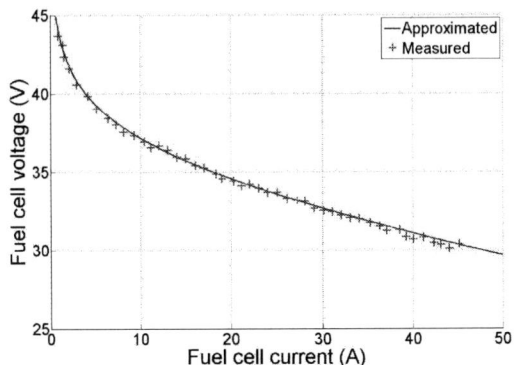

Fig. 11.   Characteristic curve of the Nexa fuel cell system

single cell is given by the Nernst potential E. During

operation, activation related losses η_{act}, ohmic losses η_{ohm} and mass transport related losses η_{conc} occur. These loss mechanisms depend on the fuel cell's current [1].

$$u_{FC}(i_{FC}) = \text{Number of Cells} \cdot$$
$$\cdot \left(E - \eta_{act}(i_{FC}) - \eta_{ohm}(i_{FC}) - \eta_{conc}(i_{FC})\right) \tag{21}$$

Based on (21), a mathematical model for the fuel cell output voltage was built. The solid line in Fig. 11 shows the approximated characteristic.

By interlinking the mathematical model of the fuel cell with the mathematical model of the converter, the characteristics of the complete system can be investigated. The upper graphs in Fig. 12 show the calculated development of the converter losses over the output power when it is connected to the fuel cell system. Due to the pure analytic investigation, the calculation is extremely fast. It takes only a few seconds to get the results for the predicted losses over the full range of power.

In this case, the switching frequency is $f = 60\,\text{kHz}$ and the leakage inductance $L_\sigma = 308\,\text{nH}$. The output voltage is held on a constant level of $u_{out} = 600\,\text{V}$. The losses are divided into rectifier losses, transformer losses, inverter switching losses and inverter ohmic losses. Four different cases are given. The number of primary windings w_1 is 2, the number of secondary windings w_2 varies from 23 to 26, hence $\frac{1}{N}$ varies from 11.5 to 13. As can be seen, over the whole power range, the topology with the smallest number of secondary turns brings out the best efficiency. This is due to a higher duty cycle α which is given in the lower graph of Fig. 12. Having equal current mean values, a higher duty cycle results in lower values of the effective currents and lower turn off currents. Accordingly, the inverter losses decrease. Coinstantaneously, a higher duty cycle results in a higher transformer voltage stress, leading to higher transformer losses. But in this case, the over all losses decrease. As can be seen in the course of α, the reserve for $\frac{1}{N} = 11.5$ is little (for nominal power $P = 1200\,\text{W}$, α is nearly 1). For this reason, the inverted winding ratio of the prototype was chosen to 12.

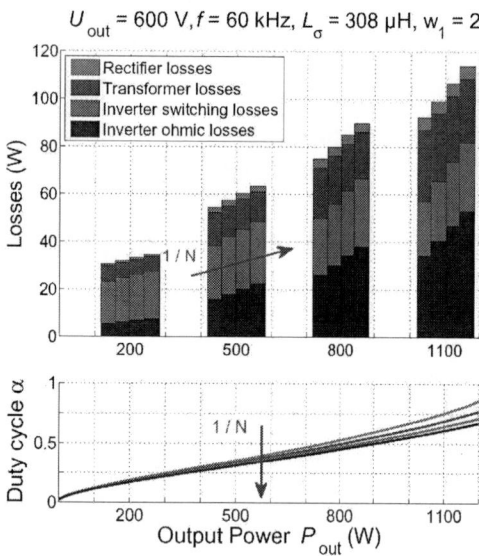

Fig. 12. Efficiency over the input current

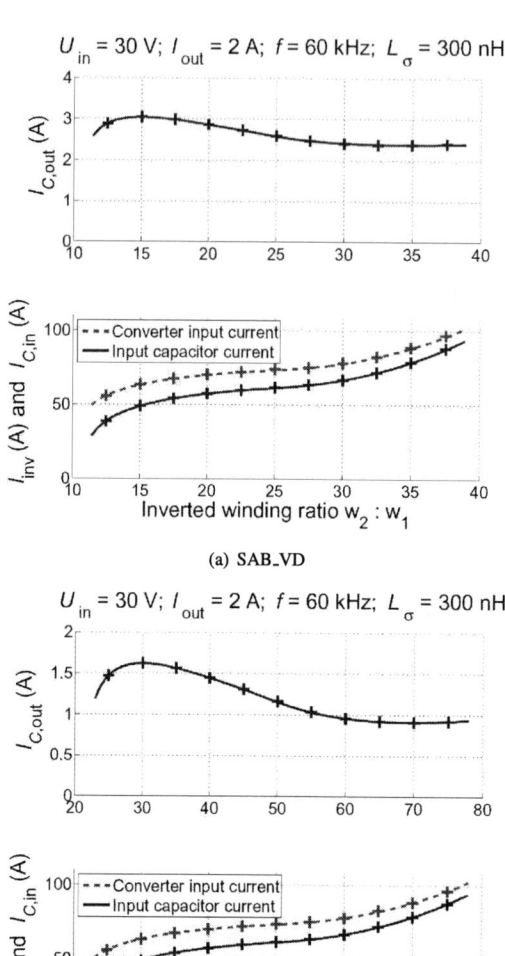

(a) SAB_VD

(b) SAB_FB

Fig. 13. rms currents of input and output capacitor and inverter rms current

IV. RIPPLE CURRENT

The effect of inverter ripple current on fuel cell stack performance and stack lifetime remains uncertain. In [15], it is mentioned that at least converter frequencies under 120 Hz or ripple factors above 4 % have a negative impact on the fuel cell. The double layer capacitor of the fuel cell can be used to smooth the current only up to a certain rms value as described in [16]. However, it remains to be usefull to be able to calculate this current. In the following, the behaviour of the inverter input current i_{inv} is described under different topology parameters. This current is equal to the fuel cell's current when using no input capacitor. Otherwise, the input capacitor smoothes the input current and provides a current ripple itself. Due to the fact, that a high rms current results in a larger operating temperature which leads - especially for electrolytic capacitors - to shorter capacitor lifetimes, the input and output capacitors' rms currents are calculated, too.

For a given output power, the mean values of the specific converter currents, e.g. i_{inv} or $i_{L,\sigma}$ are the same, independent of the other topology parameters. The rms values of these currents vary due to the different current ripples. Hence, the rms value is a quantity for the current ripple. These values can be achieved as described in section III-A.

The calculated rms current at the upper output capacitor $C_{\text{out},1}$ is shown in the upper graph of Fig. 13 (a). The rms current of $C_{\text{out},2}$ is the same. The lower graphs show the rms currents at the converter input and at the input capacitor. Simulations are done to prove the calculations. Their results are marked by the crosses. The assumed parameters are an input voltage of $U_{\text{in}} = 30$ V, an output voltage of $u_{\text{out}} = 600$ V, an output current of $i_{\text{out}} = 2$ A and a leakage inductance of $L_\sigma = 300$ nH. The switching

frequency is $f = 60$ kHz. As can be seen, for small winding ratios, $I_{C,\text{out}}$ is about 3 A. It becomes lower with increasing winding ratio. The currents at the converter input evolve contrarily. Due to the fact that fuel cell converters have a large input current and small output current, besides the findings in [8], [11] and section III-B, this is another reason to implement a winding ratio as small as possible.

To give a comparison to the SAB_FB topology, in Fig. 13 (b) the same curves are shown for the SAB_FB again. As before, the crosses mark simulation results to approve the calculations. Besides the fact that the winding ratio is doubled, the input currents are exactly the same. The current $I_{C,\text{out}}$ is smaller but not exactly halved. The curve shows the same trend but in a wider range.

V. TRANSFORMER DESIGN

For the transformer an ETD/59/31/22 core was used. To identify a reasonable winding ratio, ten transformers with different winding ratios are built. The development of L_σ with rising numbers of secondary turns w_2 is given in Fig. 14 for ten transformers. As can be seen, this is a slightly linear rising curve. In section II, it is presented

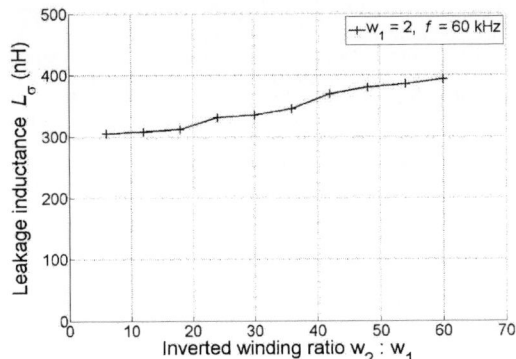

Fig. 14. Measured leakage inductance for a transformer with various secondary turns

how the leakage inductance influences the operating behaviour of the converter. Especially for applications with high voltage gain, it is important to reduce the leakage inductance L_σ as far as possible. In section III-B, it is stated out, that a small inverted winding ratio brings out a good efficiency. This fits to the course of L_σ.

Compared to the SAB_FB, the inverted winding ratio of the SAB_VD can be halved, hence a slightly smaller leakage inductance occurs. On the other hand, the voltage drop due to the high frequency operation is higher for the SAB_VD (cp. equations (6), (8) and (7), (9) respectively). However, the transformer's volume and its losses for both topologies are the same. The iron losses are a function of the magnetic flux density B_μ which can be calculated by equation (18). When the secondary turns are halved due to the voltage doubler, the duty cycle α has to be the same as in the SAB_FB leading to the same magnetic flux density, hence to the same iron losses. Since the current in the secondary turns of the SAB_VD will be twice the current in the SAB_FB, the copper area has to be doubled, which leads to a similar transformer volume.

VI. EXPERIMENTAL RESULTS

For the single active bridge with voltage doubler, a 1.2 kW prototype was built. Parameters of the important electrical devices can be taken from I. Based on the findings of section III-B and V respectively, the transformer's winding ratio is chosen to $\frac{w_1}{w_2} = \frac{2}{24}$. This corresponds to a leakage inductance $L_\sigma = 308$ nH. For the two output capacitors, the transformer's magnetizing current achieves a balancing effect depending on its negative feedback [17].

Measured voltage and current waveforms for the discontinuous operating modes are shown in Fig. 15 (a).

TABLE I
PARAMETERS OF THE PROTOTYPE

Inverter		Rectifier		Transformer	
$R_{DS,on}$	4.5 mΩ	$R_{F,D}$	50 mΩ	A_{Fe}	368 mm^2
$R_{F,B}$	5 mΩ	$U_{F,D}$	1 V	V_{Fe}	51200 mm^3
$U_{F,B}$	0.5 V				
Q_{GS}	46 nC				
Q_{GD}	65 nC				
U_{Gate}	15 V				
R_{Gate}	5 Ω				

These are the transformer input voltage $u_{T,in}$ and input current $i_{T,in}$, which is equal to $i_{L,\sigma}$, when neglecting the transformer's magnetizing current. Furthermore, the output voltage is given. Fig. 15 (b) shows the waveforms for the border mode. For the transformer input current,

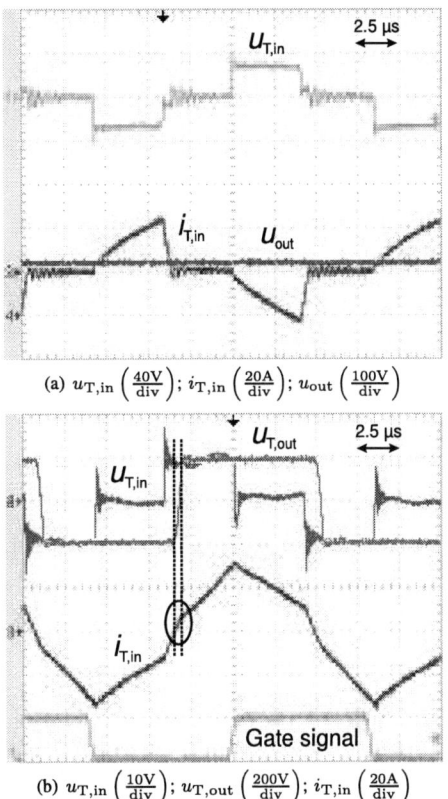

(a) $u_{T,in} \left(\frac{40V}{div}\right)$; $i_{T,in} \left(\frac{20A}{div}\right)$; $u_{out} \left(\frac{100V}{div}\right)$

(b) $u_{T,in} \left(\frac{10V}{div}\right)$; $u_{T,out} \left(\frac{200V}{div}\right)$; $i_{T,in} \left(\frac{20A}{div}\right)$

Fig. 15. Current and voltage waveforms in the two operating modes

there is one difference to the ideal waveform marked by the circle in Fig. 15 (b). An offset in $i_{T,in}$ occurs. The reason might be the charging time at the rectifier diodes as can be seen in the waveform of $u_{T,out}$.

The converter was connected to a NEXA fuel cell system [14]. As described in section III-B, this fuel cell system has an output voltage in a range of 44 V at no load down to 31 V and 40 A at full load (cp. Fig. 11). In the upper graph of Fig. 16, the measured efficiency of the converter is given when it is supplied by the fuel cell voltage. The output voltage is held on a constant value of $u_{out} = 600$ V. Under full load, the voltage gain is

$\frac{600\,\text{V}}{31\,\text{V}} = 19.4$. The efficiency reaches 92.7 %. The dashed line shows the calculated efficiency, using the analytic results from section II. As can be seen, calculated and experimental results fit reasonably well.

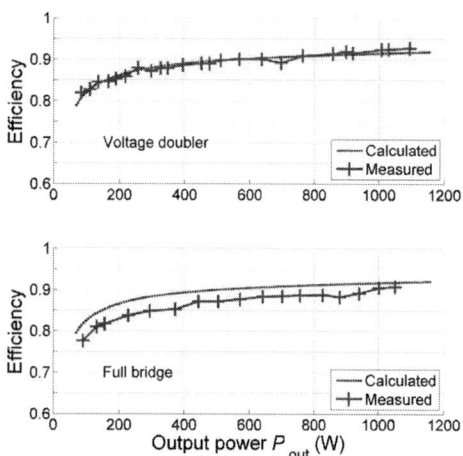

Fig. 16. Efficiency over the output power

In [11], an analytic investigation of the single active bridge with full bridge rectifier is given. Based on these results, the efficiency of the SAB_FB was calculated when using the same devices as for the SAB_VD. The only difference is the number of secondary transformer turns $w_2 = 48$. The result of the calculation is presented in the lower graph of Fig. 16. Both calculated efficiencies of SAB_VD and SAB_FB respectively, are nearly the same. Additionally, measured results taken from a prototype of the SAB_FB are shown, which was built with the same devices as the prototype of the SAB_VD. The experimental results are marked by the crosses. The trend fits to the calculation, but the measured curve shows a slightly lower efficiency in the range of approximately two percent.

VII. CONCLUSION

In this paper, a complete analytic investigation of the single active bridge with voltage doubler is presented. Both possible operating modes DCM and BM are involved into the analysis. Based on the results, the maximum voltage gain and the maximum power which can be transferred by this topology are emphasised. These values depend on the leakage inductance and the converter's switching frequency. Furthermore, a tool for pre-estimating the converter losses divided in the several loss mechanisms is presented. Due to the pure analytic investigation, this tool is extremely fast. The benefits of this tool are shown for the converter working in a fuel cell system. All calculations are approved by simulation results. The impact of the transformer's winding ratio on the converter's current ripple as well as on the input and output capacitors is stated out. A prototype is built and experimental results for the converter working in a

fuel cell system are given. At a voltage gain of 19.4, an efficiency of 92.7 % is reached. Furthermore, all results are compared to the single active bridge with full bridge rectifier.

REFERENCES

[1] W. Vielstich, H. A. Gasteiger, and A. Lamm, *Handbook of Fuel Cells - Fundamentals, Technology and Applications.* John Wiley and Sons, Ltd., 2003, vol. 1: Fundamentals and Survey of Systems.

[2] R. Ramakumar, "Fuel cells-an introduction," in *Power Engineering Society Summer Meeting, 2001. IEEE*, vol. 1, July 2001.

[3] M. Ellis, M. Von Spakovsky, and D. Nelson, "Fuel cell systems: efficient, flexible energy conversion for the 21st century," *Proceedings of the IEEE*, vol. 89, no. 12, pp. 1808–1818, Dec. 2001.

[4] M. Aydemir, A. Bendre, and G. Venkataramanan, "A critical evaluation of high power hard and soft switched isolated dc-dc converters," in *Industry Applications Conference, 2002. 37th IAS Annual Meeting. Conference Record of the*, vol. 2, 13-18 Oct. 2002, pp. 1338–1345vol.2.

[5] Y. Xue, L. Chang, S. B. Kjr, J. Bordonau, , and T. Shimizu, "Topologies of single-phase inverters for small distributed power generators: An overview," in *IEEE Transactions on Power Electronics, Vol. 19, No. 5, September 2004*, 2004, pp. 1305–1314.

[6] J. Wang, M. Reinhard, F. Peng, and Z. Qian, "Design guideline of the isolated dc-dc converter in green power applications," in *Power Electronics and Motion Control Conference, 2004. IPEMC 2004. The 4th International*, vol. 3, 14-16 Aug. 2004, pp. 1756–1761Vol.3.

[7] H. Xu, L. Kong, and X. Wen, "Fuel cell power system and high power dc-dc converter," *Power Electronics, IEEE Transactions on*, vol. 19, no. 5, pp. 1250–1255, Sept. 2004.

[8] A. Averberg and A. Mertens, "Analysis of a voltage-fed full bridge dc-dc converter in fuel cell systems," in *Power Electronics Specialists Conference, 2007. PESC '07. IEEE 38th*, 2007, pp. 286–292.

[9] J. Lee, J. Jo, S. Choi, and S.-B. Han, "A 10-kw sofc low-voltage battery hybrid power conditioning system for residential use," *Energy Conversion, IEEE Transactions on*, vol. 21, no. 2, pp. 575–585, June 2006.

[10] R. Sharma and H. Gao, "Low cost high efficiency dc-dc converter for fuel cell powered auxiliary power unit of a heavy vehicle," *Power Electronics, IEEE Transactions on*, vol. 21, no. 3, pp. 587–591, May 2006.

[11] A. Averberg and A. Mertens, "Design considerations of a voltage-fed full bridge dc-dc converter with high voltage gain for fuel cell applications," in *Power Electronics and Applications, 2007 European Conference on*, 2007.

[12] J. Salmon, "Circuit topologies for single-phase voltage-doubler boost rectifiers," in *Applied Power Electronics Conference and Exposition, 1992. APEC '92. Conference Proceedings 1992., Seventh Annual*, 23-27 Feb. 1992, pp. 549–556.

[13] J. Wang, F. Peng, J. Anderson, A. Joseph, and R. Buffenbarger, "Low cost fuel cell converter system for residential power generation," *Power Electronics, IEEE Transactions on*, vol. 19, no. 5, pp. 1315–1322, Sept. 2004.

[14] *Nexa Power Module Users Manual*, Ballard Power Systems Inc., 2003.

[15] R. S. Gemmen, "Analysis for the effect of inverter ripple current on fuel cell operating condition," in *Transactions of the ASME, Vol. 125, May 2003*, 2003.

[16] G. Fontes, C. Turpin, R. Saisset, T. Meynard, and S. Astier, "Interactions between fuel cells and power converters influence of current harmonics on a fuel cell stack," in *Power Electronics Specialists Conference, 2004. PESC 04. 2004 IEEE 35th Annual*, vol. 6, 20-25 June 2004, pp. 4729–4735Vol.6.

[17] Y. Gu, L. Hang, Z. Lu, Z. Qian, and D. Xu, "Voltage doubler application in isolated resonant converters," in *Industrial Electronics Society, 2005. IECON 2005. 32nd Annual Conference of IEEE*, 6-10 Nov. 2005, p. 5pp.

Analysis of Capacitor Dividers for Multilevel Inverter

Oleg Sivkov , Jiri Pavelka

Czech Technical University, Faculty of Electrical Engineering/Dept. of Electric Drive and Traction, Prague, Czech Republic, ollerrg@seznam.cz, sivkoo1@feld.cvut.cz, pavelka@fel.cvut.cz

Abstract— This paper analyses a possibility for realization of multilevel inverters. After general description of all famous solutions, the focus is paid on Diode Clamped Multilevel Inverters (DCMI) and Flying Capacitor Multilevel Inverters (FCMI). The comparison of topological structure differences and control strategies is presented. The special focus is paid to balancing of voltages on capacitors. Switching states and their transitions of three-level inverter allow to balance capacitor voltages in both types of inverters, DCMI and FCMI, where DCMI requires less capacitor power size than FCMI to achieve the same capacitor voltage swinging. However in higher-level DCMI (more than 3) capacitor voltage balancing generally is not able. FCMI control strategy is possible for all level. Special control strategy for five-level DCMI capacitor voltage stabilization is briefly described. Advantages and disadvantages of DCMI and FCMI are compared for the same output power.

Keywords—**High Voltage power converters, Active filter, Adjustable speed drive, Multilevel converters, Converter circuit, IGBT, IGCT, Converter control, Design, Unified Power Flow controller.**

I. Introduction

At present time voltage multilevel inverters become more and more popular and find a great range of applications. They are found in new areas of medium and high voltage applications such as frequency inverter for high voltage adjustable speed drives, inverter for high voltage compensations, for high voltage Unified Power Flow Controllers (UPFC), for high voltage active power filters etc.

Four solutions for multilevel inverters are used. They are:

multilevel inverters with magnetic coupling (special multi-winding transformer);

independent sources for each level;

level voltage stabilization using other auxiliary power circuits;

level voltage stabilization using switching control strategy for own multilevel inverter devices.

Multilevel inverters with magnetic coupling requires special multi-winding transformer, but the control strategy for PWM is very simple. Each level uses the separated two-level inverter with rectangular shape output voltage. This output voltages are shifted and summarized in multi-winding transformer. All two-level inverters are supplied from common dc-source. This type of multilevel inverter

is used in large compensator with the power in-order hundreds of MVAr.

Independent sources for each level require N-1 mutually isolated sources and each source must be capable for 2-quadrant operation. It mostly requires a compound transformer supply and control rectifier is needed to return the reactive energy back to the network.

Level voltage stabilization using other auxiliary power circuits is acceptable in higher power converters. It demands additional power devises and concrete connection depends on the number of levels and a type of load.

Switching control strategy for own multilevel inverter devices doesn't demand any special sources and allows to stabilize voltage on capacitor dividers. To ensure the inverter's proper function it is enough a simple two-level source [2]. The other three mentioned solutions are not considered here.

The purpose of this paper is to analyse an output multilevel voltage and a capacitor voltage in two of the most popular types of multilevel inverters; they are Diode Clamped Multilevel Inverter (DCMI) and Flying Capacitor Multilevel Inverter (FCMI).

II. Three-Level Diode Clamped Inverter Topology And Divider Capacitor Voltage Balance

"Fig. 1" shows the scheme of the simplest diode-clamped three-phase three-level inverter. This topology has an advantage over two-level one in case of serious connected devices. Couples of clamping diodes Da1, Da2; Db1, Db2; Dc1, Dc2 warrant the distribution of DC circuit supply voltage Udc among the couples of serious connected switches.

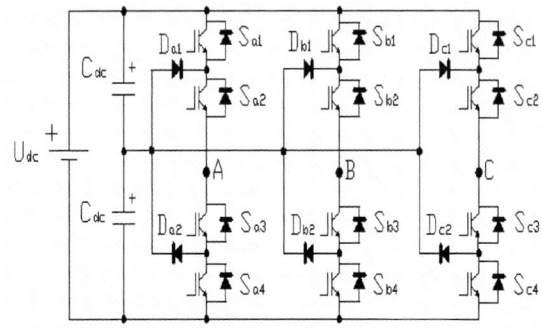

Fig. 1. Scheme of three-phase three-level diode-clamped inverter.

978-1-4244-1741-4/08/$25.00 ©2008 IEEE

The power switching devices represent an Insulate Gate Bipolar Transistors (IGBTs). This topology can ensure three voltage levels. TABLE I shows all possible switching states of switches S_1, S_2, S_3, S_4 and inverter output voltage U_{out} in each phase. The last row represents the inverter output voltage in relative values (p. u.) that produces three levels for different states of switches. The switching state of inverter is –1 when two lowest switches S_3 and S_4 are ON (switching state 1 in the table); +1 when two highest switches S_1 and S_2 are ON (state 3). The inverter's output voltage is zero when two switches S_2 and S_3 are ON (state 2); in this state the current flows through one of that switches depending on its direction. According to the "Fig. 2" in state 2 if the current direction is positive, it flows through the diode D_1 and switch S_2, if the current is negative it flows through the switch S_3 and diode D_2.

TABLE I
SWITCHING STATES OF THREE-LEVEL DIODE-CLAMPED
INVERTER

Switch	Switching state		
	1	2	3
S_1	OFF	OFF	ON
S_2	OFF	ON	ON
S_3	ON	ON	OFF
S_4	ON	OFF	OFF
U_{out}	-1	0	+1

"Fig. 2" shows the possible switching states of three-phase three-level DCMI. Triplet number defines the switching state of three phases for demanded voltage vector. We can see that some voltage vectors can be obtained by two or three triplets.

The number of different vectors is 19 and number of triplets is 27. That allows choosing triplets according to another condition, for example, stabilization of divider capacitor voltages.

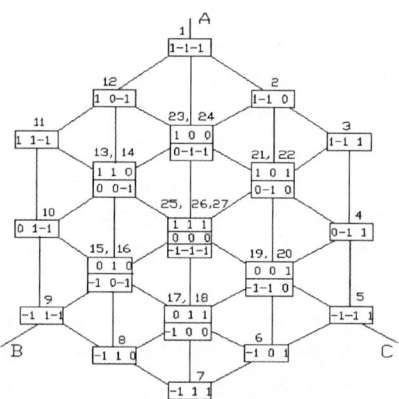

Fig. 2. Space vectors of three-phase three-level DCMI.

The behavior of above mentioned inverter circuit was studied using MATLAB Simulink. The MATLAB Simulink model of the equivalent circuit "Fig. 1" and the equivalent circuit of the load "Fig.3" were designed and tested.

The load is represented by resistance R_L, reactance X_L and voltage source U_i. The voltages in each phase are shifted with an angle of $-\dfrac{2}{3}*\pi$ of each other and correspond the "equations (1) – (3)"

$$U\_iA = Uimax*sin(\omega t). \tag{1}$$

$$U\_iB = Uimax*sin(\omega t - \frac{2}{3}*\pi). \tag{2}$$

$$U\_iC = Uimax*sin(\omega t - \frac{4}{3}*\pi). \tag{3}$$

where Uimax is the load voltage amplitude.

The voltage equation according to the second Kirhgoff's law is defined according to the "equation (4)":

$$Uout = U\_i + Iout*R + Iout*X. \tag{4}$$

from where the output current Iout can be defined.

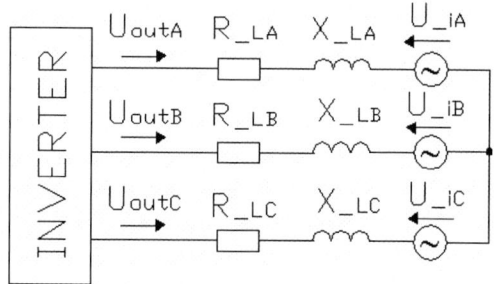

Fig.3. Scheme of three-phase equivalent inverter load.

The considered circuit is for the active load, i. e. the load with the voltage source.

TABLE II shows input parameter values for program Simulink in three-level DCMI. All the parameters except frequency f are considered in a relative values and capacitor voltage swinging ΔU_C is in percent. Output current consists of active Ireal and reactive Iimag components.

Output voltage amplitude Umax and output current components Ireal and Iimag are the fixed parameters in our case. Load voltage and the angle between output and load voltages are the variable parameters and they are calculated from the equations at the TABLE III for program Simulink as well. The vector of load voltage also consists of active and reactive components Ureal, Uimag. If the active load current component is equal 1 and the reactive 0 than the type of load is pure resistive. When I_{real} = 0 and I_{imag} = -1 the type of load is pure inductive.

Resistances R_{LX} and reactances X_{LX} represent an impedances separation of load source and inverter source. They are chosen R=X in all cases.

"Fig. 4" shows the simulation results in Simulink of output voltage Uout, load voltage U_i, output current Iout and voltage on upper capacitor Uc1 and lower capacitor

222

Uc2. After some while of inverter start up the clamping capacitor voltage fixes to a steady state value 0.95 that is equal to the output voltage as it must be. The input dc voltage is distributed on the clamping capacitors (the half 0.95 on the upper capacitor C, the other half 0.95 – on the on the lower capacitor C) before the inverter is on, because the capacitor dividers are connected at the input of the inverter. Thus the clamping capacitors are charged before the inverter is put into operation.

TABLE II
INPUT PARAMETER VALUES FOR MATLAB SIMULINK PROGRAM IN THREE-LEVEL DCMI

f=50	network frequency, [Hz]
N=14	switching triangular carrier frequency per period, [number/period]
X=0.5	reactance, [p.u.]
R=0.5	resistance, [p.u.]
C=0.12	clamping capacitor, [p.u.]
Rc=0.001	resistance of capacitor circuit, [p.u.]
U_dc=1.9	direct voltage source, [p.u.]
Umax=0.95	amplitude of output voltage, [p.u.]
Uc0=0.95	initial capacitor voltage, [p.u.]
Ireal=1	active component of output current, [p.u.]
Iimag=0	reactive component of output current, [p.u.]

TABLE III
USED EQUATIONS OF VARIABLE INPUT PARAMETERS FOR MATLAB SIMULINK PROGRAM IN THREE-LEVEL DCMI

omega=2*pi*f	angular frequency
L=x/omega	inductance
Uireal=Umax-R*Ireal+x*Iimag	active component of load voltage
Uiimag=-(x*Ireal+R*Iimag)	reactive component of load voltage
Uimax=sqrt(Uireal*Uireal+Uiimag*Uiimag)	full load voltage amplitude
z=sqrt(R*R+x*x)	impedance
r=Iimag/Ireal	load ratio
fi=atan2(Uiimag,Uireal)	angle between output load voltage and load voltage
Iout = sqrt(Ireal*Ireal + Iimag*Iimag)	output current

"Fig. 5" gives a close look into the results in the "Fig. 4". It is well visibly all three levels of output load voltage and shape of output load current that is almost phase-to-phase with output load voltage because of forced resistive load and its shape doesn't repeat voltage shape because of inductance in impedance. Clamping capacitor voltage is stabilized with the swinging over ± 0.3%. It is also visibly in "Fig. 5" that the voltage is distributed symmetrically in one to another capacitors.

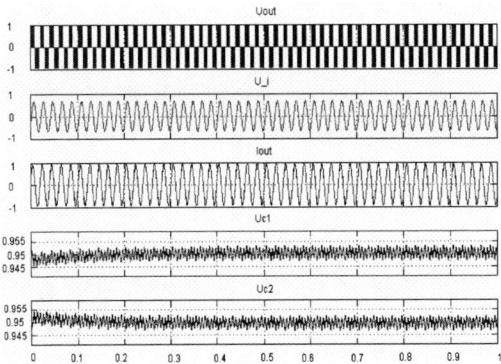

Fig. 4.   Output voltage Uout, load voltage U_i, output current Iout and clamping capacitor voltages Uc1 and Uc2 of three-level DCMI with resistive load I$_{real}$=1, I$_{imag}$=0, N=14, Z=0.707 and C=0.12.

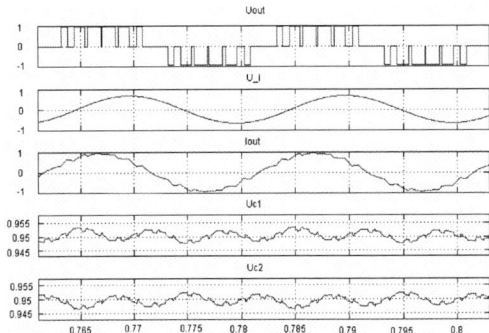

Fig. 5.   A zoom-in of output voltage, load voltage, output current and clamping capacitor voltages shown in "Fig. 4".

In case of inductive load Ireal=0, Iimag=-1 the output current Iout is 90^0 delays from output voltage Uout, and in capacitive load Ireal=0, Iimag=1 the output current is 90^0 in front of output voltage.

III.   THREE-LEVEL FLYING CAPACITOR INVERTER TOPOLOGY AND DIVIDER CAPACITOR VOLTAGE BALANCE

Scheme of single phase of three-level FCM is in the "Fig. 6". In comparison to three-level DCMI the flying capacitor C is connected between average switches in each phase. That's allows to stabilize capacitor voltage in each phase independently. The zero level on output is the difference of supply voltage and capacitor voltage.

TABLE IV shows all possible switching states of switches S$_1$, S$_2$, S$_3$, S$_4$ in three-level FCMI and inverter output voltage U$_{out}$. The switching states of switches are ON or OFF and of three-level inverter output voltage are –1, 0 and 1.

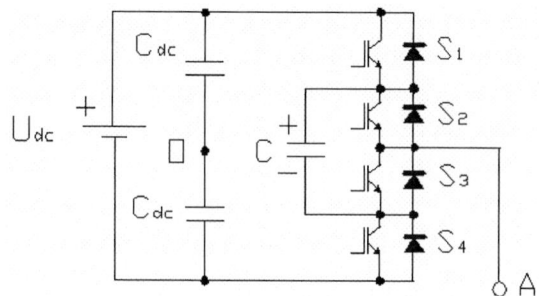

Fig. 6. Scheme of single-phase of Three-level Flying Capacitor inverter.

TABLE IV
SWITCHING STATES OF THREE-LEVEL FLYING CAPACITOR INVERTER

Switch	Switching state			
	1	2	3	4
S_1	OFF	OFF	ON	ON
S_2	OFF	ON	OFF	ON
S_3	ON	OFF	ON	OFF
S_4	ON	ON	OFF	OFF
Uout	−1	0	0	+1

Switching states of one phase of three-level FCMI are in "Fig. 7". The upper parts show the switching state number (from 1 to 4) the lower part informs about increasing (+), decreasing (-), or without changing (0) of the capacitor voltage for positive current polarity. According to that figure and the TABLE IV we can see that if two upper switches are ON the output load voltage is +1, if two lower switches are ON the output load voltage is −1, in these two states capacitor voltage is without changing. There are two zero states: state 2 if S_2 and S_4 are ON – the capacitor is charged, state 3 if S_1 and S_3 are ON – the capacitor voltage is discharged. The present switching state can be only changed to a state that is connected to it by a bidirectional bond corresponding to a level changing in the output voltage waveform. It is seen that by each change to zero level is possible to choose the capacitor voltage increasing or decreasing.

Three-phase load for FCMI is the same as for DCMI ("Fig.3") for sure. Parameters of three-level FCMI same as in an example with three-level DCMI are considered in relative values except frequency f. The table of input parameters for three-level FCMI is analogical to the TABLE II the input parameters of three-level DCM. The equations of variable input parameters, the choosing of load type for three-level FCMI are also the same as for three-level DCMI (TABLE III).

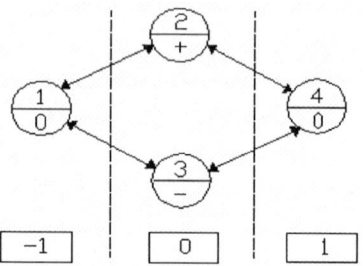

Fig. 7. Switching states in Three-level FCMI.

"Fig. 8" shows the simulation results in Simulink with an output voltage U_out, load voltage U_i, output current I_out and capacitor voltage Uc. Flying capacitor voltage fluctuates over the value 0.95 that is equal to the value of the output voltage amplitude. The initial capacitor voltage is taken 0.95 therefore it starts from that value. If the initial capacitor voltage is taken 0, the capacitor voltage starts from zero and only after some time it reaches the value 0.95 because here the capacitor is charged after the inverter starts functioning. That's inaccessible from the overvoltage on the devices. That's why the capacitor must be charged before the inverter starts functioning. It is not presented here. "Fig. 9" gives a close look into the results in the "Fig. 8". Output load voltage ensures all three levels, the current has its corresponding shape and flying capacitor voltage is stabilized.

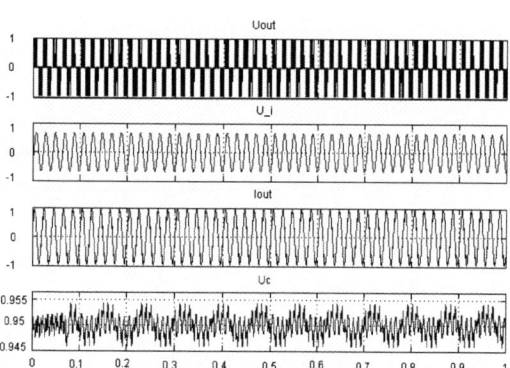

Fig. 8. Output voltage U_out, load voltage U_i, output current I_out and clamping capacitor voltages of three-level FCMI with resistive load I_{real}=1, I_{imag}=0, N=14 and Z=0.16.

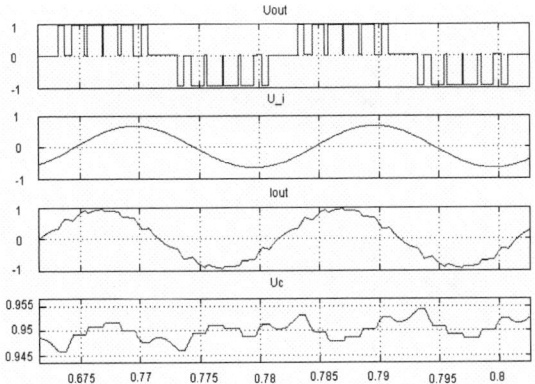

Fig. 9. A zoom-in of output voltage, load voltage, output current and capacitor voltages shown in "Fig. 8".

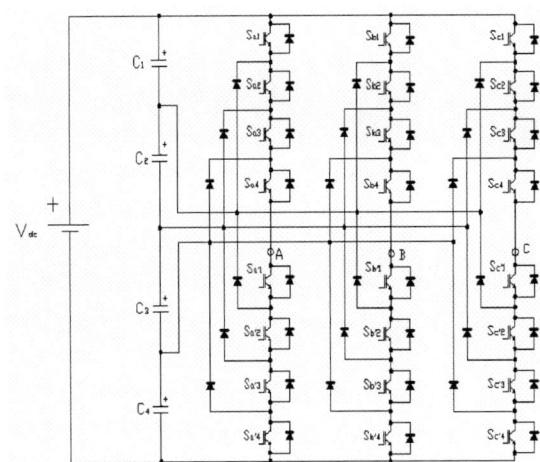

Fig. 10 Scheme of three-phase Five-level Diode Clamped Inverter.

IV. FIVE-LEVEL DIODE CLAMPED INVERTER TOPOLOGY AND CAPACITOR VOLTAGE BALANCE

Generally in N-level DCMI topology a series connection of N-1 capacitors can be used as the sources that create N-2 inner points with different potentials. The capacitors form an ideal voltage divider, a point between two capacitors is a DC potential, and may be connected to the output point by switching an appropriate switches. This characteristic permits the generation of N voltage levels.

"Fig. 10" shows the scheme of three-phase Five-level Diode Clamped Inverter. In comparison to three-level DCMI topology it has two times more the capacitor dividers on the input and each phase has twice greater the number of IGBT with reverse diodes and three times more the number of feed back diodes.

The speciality of this topology is that the capacitor voltage balance in higher than three levels of the inverter is generally not possible. The reason is the unequal load sharing among switches and capacitors. Separated sources or additional circuitry must be used. The capacitor voltage stabilization in five-level DCMI is proved in papers [3] using the additional power hardware and new PWM strategy.

To work with five-level DCMI and its additional circuits to achieve capacitor voltage balance is not our purpose.

The number of vectors is 3*(N-1)*N+1, and number of triplets is N^3 in DCMI.

The analogical scheme to "Fig. 2" (for three-level diode inverter) can be drawn for higher-level inverter with more vectors and more triplets. In five-level DCMI the number of vectors is 61 and number of triplets is 125.

A voltage balance of this type of inverter using the advanced PWM with further power circuits is described in papers [3] and describing the problem of higher than three level inverter is not our goal because it requires further power circuits. The problem of capacitor voltage balance using further power circuits in higher than five-level DCMI becomes more complicated.

V. FIVE-LEVEL FLYING CAPACITOR INVERTER TOPOLOGY AND CAPACITOR VOLTAGE BALANCE

Flying capacitor multilevel topology with N levels needs N-2 flying capacitors that are connected between the switches in each phase. The lower levels on the outputs are achieved as the difference of supply voltage and capacitor voltage.

The scheme of three-phase Five-level Flying Capacitor Inverter is in the "Fig. 11". In comparison to Three-level Flying Capacitor Inverter each phase has twice-greater number of IGBTs with reverse diodes and three times more flying capacitors. This topology as three-level one allows independent voltage control in each phase. The output levels like +0.5, 0, -0.5 is the difference of supply and capacitor voltage.

Switching states of one phase in five-level inverter are in "Fig.12". Same as in case with 3-level DCMI upper parts show the switching state number and the lower part – increasing (+), decreasing (-), or without changing (0) capacitor voltages. In 5-level inverter there are three flying capacitors and therefore at the lower part there are three signs that correspond to every flying capacitor polarity. If all four upper IGBTs are ON the output voltage is +1, if all four lower IGBTs are ON the output voltage is –1, the capacitors don't charge or discharge, they are constant. In the states 2-15 the current flows through any of that three flying capacitors effecting them to charge or discharge. Thus the lower voltage levels –0.5, 0, 0.5 appear at the output. The present switching state can be only changed to a state that is connected to it by a bidirectional bond corresponding to a level changing in the output voltage waveform. This occurs because only one couple of switches can switch in one moment.

All the flying capacitor values in each phase are accepted the same C1 = C2 = C3 = C. And dividing capacitor voltages are defined in equations: "(5)" defines the voltage on the capacitor C1, "(6)" defines the voltage on the capacitor C2 and "(7)" defines the voltage on the capacitor C3.

$$Uc1 = \frac{3}{4} * Udc. \qquad (5)$$

$$Uc2 = \frac{1}{2} * Udc. \qquad (6)$$

$$Uc3 = \frac{1}{4} * Udc. \qquad (7)$$

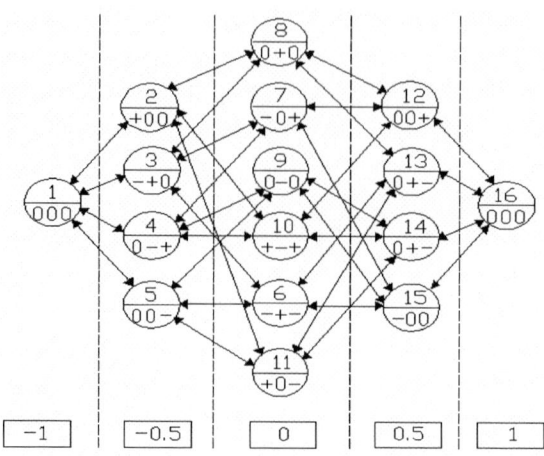

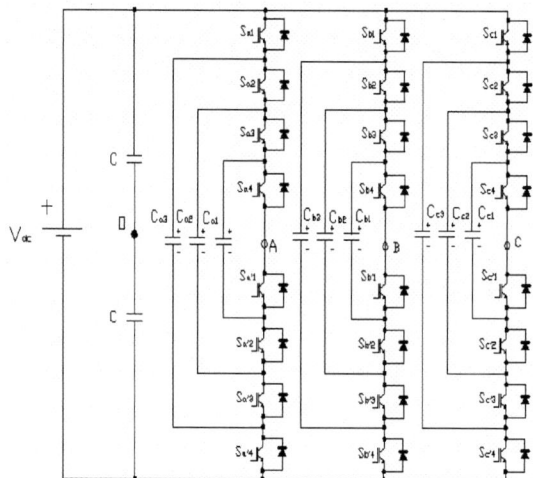

Fig. 11. Scheme of three-phase Five-level Flying Capacitor Inverter.

Fig.12. Switching states in five-level FCMI

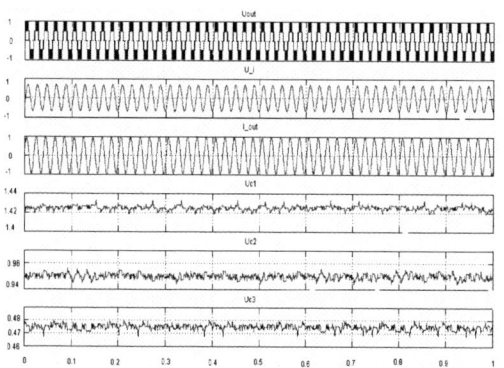

Parameters of five-level FCMI same as in examples mentioned above are considered in relative values except frequency f.

The three-phase load of this inverter topology is the same as in three-level ones (Fig.3). Consequently the load equations in five-level FCMI are the same (1, 2, 3) and the load type can be chosen in the same way as in three-level inverters.

"Fig. 13" shows the simulation results in Simulink with an output voltage U_out, load voltage U_i, output current I_out and capacitor voltages Uc1, Uc2 and Uc3 that are stabilized.

Flying capacitor voltage Uc1 fluctuates over the value 1.425 that is according to the "equation 5", Uc2 fluctuates over the value 0.95 ("equation 6"), Uc3 – over the value 0.475 ("equation 7"). As in three-level inverter all initial capacitor voltages are taken their final values, and the case when initial capacitor voltage is zero will be in thesis. "Fig. 14" gives a close look into the results in the "Fig. 13".

Fig. 13.   Output voltage Uout, load voltage U_i, output current Iout and flying capacitor voltages Uc1, Uc2 and Uc3 of five-level FCMI with resistive load I$_{real}$=1, I$_{imag}$=0, N=24 and Z=0.707

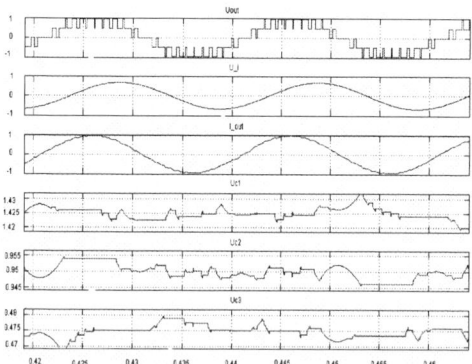

Fig. 14.   A zoom-in of output voltage Uout, load voltage U_i, output current Iout and capacitor voltages Uc1, Uc2 and Uc3 shown in "Fig. 13"

226

VI. CONTROL STRATEGY

The switching states of three-level diode clamped inverter are in "Fig. 2". PWM strategy of output voltage is supposed. This strategy determines the required vector position. The basic idea of capacitor voltage balancing strategy is to choose better triplet from possible triplets from the criteria of capacitor voltage stability. Only one couple of transistors of one phase can switch (one of them ON, the other one OFF depending on needed switching state of phase) during one switching. The transition from 13 to 14 triplets is explained as an example. In switching state 13 the current flows through the lower capacitor and it charges and hence the voltage moves upstairs; in state 14 the upper capacitor is charged, the voltage moves downstairs and thus capacitor voltage is balanced. The difference between triplets 13, 14 of the same vector is that to get from 13 to 14 the vector can get only through the path: 13 – 11 – 12 – 14. From triplet 14 to 13 the vector can move in opposite order of triplets 14 – 12 – 11 – 13. Vector can't go from the triplet 13 to the triplet 14 or 12 same as from the triplet 14 it can't go to the triplet 13 or 11 because only one phase and only two IGBTs can switch in one moment.

Control strategy in FCMI differs from DCMI that capacitor voltage is balanced in each phase of inverter independently and therefore it can be considered in one phase. One phase of three-phase three-level FCMI is in the "Fig. 6". It works according to the switching of IGBTs $S_1 ... S_4$ in the table 4, when the corresponding switch is ON the current flows through this switch in direct way or in back way through the reverse diode depending on the current direction. Thus three levels on the output of inverter are ensured. Zero level is ensured in the states 2 and 3 when the current flows through the flying capacitor. It is the result of difference between supply and capacitor voltages. When the output load voltage is equal to zero the capacitor can be charged or discharged depending on the current direction.

The strategy of five-level DCMI is no possible for voltage balance in capacitors because in comparison to three-level DCMI, the load currents in different phases have different values and flow in different parts of capacitor divider. The papers [3] prove the possibility of capacitor voltage balance in five-level DCMI with auxiliary power circuits and new PWM strategy. However in higher than five-level DCMI the capacitor voltage equalising becomes even more problematic.

On the contrary, in five-level FCMI the switching states transition is realized according to the "Fig.12" in each phase independently. "Fig.12" shows the output voltage states and the states of all three capacitors if they charge, discharge or constant. The levels +0,5, 0 and –0.5 are ensured when the current flows through the any of the capacitors C_1, C_2 or C_3 the switching state number of which are shown at this figure. The capacitor voltage balance is possible here without any auxiliary power circuits. The "Fig. 13" and "Fig. 14" show the simulation results where capacitor voltage balance is ensured. It is possible in all the level numbers of FCMI.

VII. CAPACITOR VOLTAGE SWINGING COMPARISON

The capacitor voltage swinging ΔU_C can be defined as the "ratio (8)".

$$\Delta U_C = \frac{(U_{max} - U_{min})/2}{(U_{max} + U_{min})/2}. \tag{8}$$

The final dependences of capacitor voltage swinging ΔU_C on the size of capacitors are on the "Fig.15" and "Fig.16". They show what size of capacitor C can be used for the determination the same value of capacitor voltage swinging ΔU_C in both types of inverters.

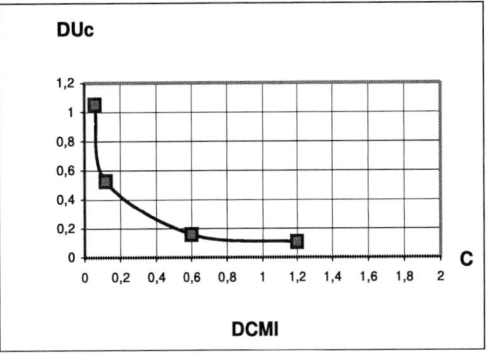

Fig.15. Dependence of capacitor voltage swinging ΔU_C on capacitor size C in Three-Level DCMI.

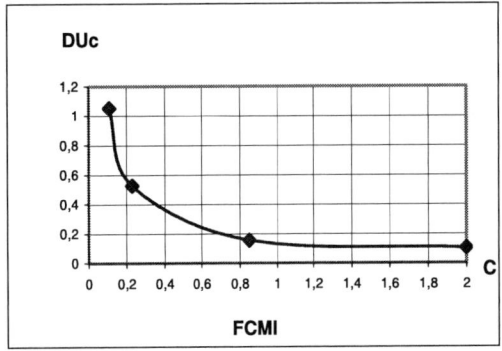

Fig.16. Dependence of capacitor voltage swinging ΔU_C on capacitor size C in Three-Level FCMI.

The results are in the TABLE V. The left columns of the TABLE V of both DCMI and FCMI show the corresponding capacitor sizes is needed to get required voltage swinging. The right columns show the total capacitance power. In DCMI there are two capacitors, in FCMI there are three capacitors.

The relation of the capacitor powers of FCMI to DCMI is in the last right column. It shows that three-level FCMI requires 2 – 2.6 times larger size of capacitors than three-level DCMI to get the same capacitor voltage swinging. It says that three-level DCMI requires about one half of capacitors as three-level FCMI.

TABLE V

DEPENDENCE OF CAPACITOR VOLTAGE SWINGING ΔU_C ON CAPACITOR SIZE IN BOTH TYPES OF THREE-LEVEL INVERTERS.

ΔU_c	DCMI		FCMI		$C_{\Sigma\,FCMI}\,/$ $C_{\Sigma DCMI}$
	C	$C_\Sigma=2*C$	C	$C_\Sigma=3*C$	
0.2	0.42	0.84	0.58	1.74	2.07
0.4	0.27	0.54	0.37	1.11	2.06
0.6	0.16	0.32	0.25	0.75	2.34
0.8	0.11	0.22	0.18	0.54	2.54

The calculations were done in relative units for both types of inverters. It is considered that such comparison is correct because for the same input parameters (output voltage amplitude Umax, network frequency f, switching carrier frequency N) the output parameters (load voltage U_i, output current I_out) have the same amplitude and shape (comparing "Fig. 5" and "Fig. 9").

A capacitor voltage swinging comparison for DCMI and FCMI with the levels higher than 3 is not possible because the control strategies are not possible for such DCMI.

VIII. CONCLUSION

A concise comparison of two types of multi-level inverters is presented.

In three-level inverters output voltage is ensured and the capacitor voltage is stabilized in both types of these inverters. The simulation results of both types of these inverters are realized and compared. The final dependences of capacitor voltage swinging on the size of capacitors ("Fig.15" and "Fig.16") show that three-level inverters FCMI requires higher capacitance to get the same capacitor voltage swinging. And the table 5 shows that three-level FCMI needs 2-2.6 times higher capacitor power size than DCMI to achieve the same results.

However in higher-level DCMI general capacitor voltage balance strategy is no possible, and only the special methods are applied for capacitor voltage balance. In the contrary in five-level FCMI the capacitor voltage balance is possible without any auxiliary circuits, the simulation results are done with it ("Fig. 13" and "Fig. 14"). The capacitor voltage stabilization is possible in all the levels of FCMI.

Now the experimental equipment of five-level FCMI is in the laboratory of the department "Electric drive and traction" in CTU and it is supposed that results of experiment measurement will be published next year.

ACKNOWLEDGMENT

Research described in the paper was supervised by Prof. J. Pavelka, FEE CTU in Prague and supported by the Czech Ministry of Education under MŠMT grant No.MSM 6840770017.

REFERENCES

[1] V. T. Tran: "Multilevel Inverters for high voltage drive", PhD thesis, pp. 10–14, 83–88, Czech Technical University, Prague, September 2003.

[2] Timothy L. Skvarenina: "The Power Electronics", Handbook, industrial electronic series, ISBN 0-8493-7336-0, pp. 6-7–6-15, 2002.

[3] G. Borghetti, M. Carpanto, M. Marchesoni, P. Tenca and L. Vaccaro: "A new balancing technique with power losses minimization in Diode-Clamped Multilevel Converters", Universita degli Studi di Genova – Dipartimento di Ingegneria Elettrica, pp. 2,3,7–10, Genova, Italy, September 2007.

[4] Nguyen Van Nho, Hong-Hee Lee: "Carrier PWM algorithm for Multileg Multilevel Inverter"Department of Electrical and Electronic Engineering, HCMUT, Hochiminh, pp.3–7 Vietnam, September 2007.

[5] V. T. Tran, "Control strategy for Flying Capacitor Multilevel Inverter", in POSTER 2003 – Book of Entended Abstracts, (Prague), p. PE30, CTU, Faculty of Electrical Engineering, May 2003.

[6] J. Pavelka, "High Voltage Multilevel Flying Capacitor Type" in EPE-EMC proceedings [CD-ROM], ISBN 953-184-047-4, (Cavtat and Dubrovnik, Croatia), September 9-11 2002.

[7] X. Yuan and I. Barbi, "Fundamentals of a New Diode Clamping Multilevel Inverter", IEEE Transactions on Power electronics, vol. 15, pp. 711–718, July 2000.

[8] J. H. Seo, C. H. Choi, and D. S. Hyin, "A New Simplified Space-vector PWM Method for Three-level Inverters"IEEE Transactions on Power electronics, vol. 16, pp. 545–550, July 2001.

Space Vector Modulation for a Capacitor Clamped Multi-level Matrix Converter

Xu Lie, Jon C. Clare, Patrick W. Wheeler, Lee Empringham

PEMC Group, School of Electrical & Electronic Engineering, the University of Nottingham, UK

Abstract—as an array of controlled bi-directional semiconductor switches, the Matrix Converter allows direct AC-AC conversion without an intermediate DC link. The Matrix Converter has several attractive advantages that have been investigated in the last twenty years. Multi level topologies have become increasingly popular in recent years due to high power quality, high-voltage capability, low harmonics and low EMI issues. This paper is concerned with applying multi-level topology and Space Vector Control algorithms to direct AC-AC power converters. Results from a small scale experimental prototype are given to validate the theoretical findings.

Key words:
Matrix converter: MC;
Multi-level Matrix Converter: MMC;
Space Vector Modulation: SVM;
Flying capacitor(s): FC(s);

I. INTRODUCTION

Matrix Converters are forced commutated AC-AC converters without an intermediate DC stage which can output sinusoidal waveforms with arbitrary amplitude and frequency, while maintaining sinusoidal input currents. A basic 3×3 MC consists of an array of nine bi-directional switches arranged so that the output lines of the converter can be connected to any of the inputs. With controllable phase, the converter can operate with unity displacement factor and is inherently capable of regeneration. The converter has the potential to be considerably smaller than conventional technologies since there are no large energy storage elements [1]-[3].

The features mentioned above, together with continued development have started to see the MC being exploited in practical applications. Although relatively large MCs have been built and operated successfully [10], the application of the traditional MC is limited to low or medium voltage level with the voltage rating restriction of power semiconductor devices in practical. However, it is well known that multi-level technology is a good solution for high voltage, high power conversion. Multi level converters can reach higher voltage level with low cost and low voltage devices, while reducing the size and cost of the filters and increasing the performance of the converter because of the staircase shaped outputs [4]. It is therefore interesting to look at the potential for using multi-level techniques in direct AC-AC converters either for reasons of increased power handling capability, or for improved waveform quality. There are generally 3 kinds of multi level technologies: diode clamped, H-bridge cascade and capacitor clamped. Previous papers considering multi-level direct converters [7] [11] have generally considered that circuits using capacitor clamping are favorable. This paper concentrates on the development of a SVM method that can be applied to such a converter. The theoretical development is discussed in detail and the experimental results from a low power prototype are provided to validate the method.

II. SPACE VECTOR MODULATION IN MULTI LEVEL MATRIX CONVERTER

A. Multi-level Matrix Converter (MMC) topology

The MMC topology considered is given in Fig.1 [7] [11] [12]. Compared with a conventional MC, the Δ connected capacitors create 3 extra middle voltage levels $(V_A+V_B)/2$, $(V_B+V_C)/2$ and $(V_C+V_A)/2$ at output phases u, v, and w. The voltage across the capacitors C_1, C_2 and C_3 are controlled to be $(V_A-V_B)/2$, $(V_B-V_C)/2$ and $(V_A-V_C)/2$ respectively by the related switching pairs. For example, when S_{Aa1}, S_{Ba2} are conducting and i_u is in the direction shown in Fig 1, C1 is charged by i_u or alternatively discharged by i_u when S_{Ba1} and S_{Aa2} are switched on. The output voltage V_u remains at the same value regardless of which switch pair is on since (1):

$$V_u = (V_A + V_B)/2 = V_A - V_{AB}/2 = V_B + V_{AB}/2 \qquad (1)$$

In the MMC case, there are 6 different voltage states at every output leg [7] [11] as shown in Table I.

Similarly to a conventional MC [5] [6], the MMC must obey the same constraints on the switching pattern to avoid short circuit at the input side or open circuits at the output side. Using switching functions (S=1 for ON, S=0 for OFF), the mathematical expression is:

$$\begin{cases} S_{Ai1} + S_{Bi1} + S_{Ci1} = 1 \\ S_{Ai2} + S_{Bi2} + S_{Ci2} = 1 \end{cases}, \ i \in (a,b,c) \qquad (2)$$

In practice, charging/discharging of three capacitors must be controlled so as to keep a sinusoidal shaped voltage across the capacitor when the mid-voltage is applied [4] [11] [12]. To simplify the analysis of modulation strategies, the capacitors are considered to be ideal voltage sources of the appropriate values.

Obviously, the total available switching pairs are 9^3=729; all possible voltage states on an output phase is 6^3=216; and all possible current states is 27 (the same as conventional MC).

978-1-4244-1741-4/08/$25.00 ©2008 IEEE

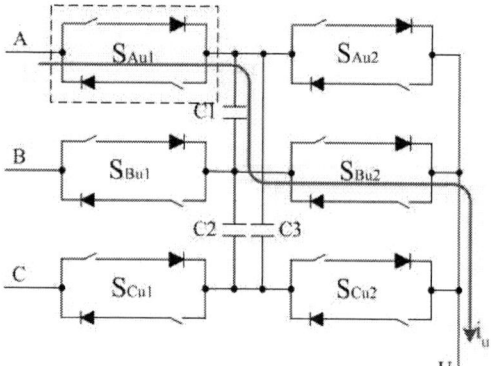

Fig.1 MMC topology, output phase u (the same for output phase v and w)

Table I: The possible output voltage & input current state

Switches conducting on	Output Voltage State
S_{Ai1} & S_{Ai2}	V_A
S_{Bi1} & S_{Bi2}	V_B
S_{Ci1} & S_{Ci2}	V_C
S_{Ai1} & S_{Bi2} or S_{Bi1} & S_{Ai2}	$(V_A+V_B)/2$
S_{Bi1} & S_{Ci2} or S_{Ci1} & S_{Bi2}	$(V_B+V_C)/2$
S_{Ci1} & S_{Ai2} or S_{Ai1} & S_{Ci2}	$(V_C+V_A)/2$
Switches conducting on	**Input Current State**
S_{Ai1}	i_u injected into A
S_{Bi1}	i_u injected into B
S_{Ci1}	i_u injected into C

B. Vector grouping in Space Vector Modulation

From Table I, it is obvious that each output leg can output V_A, V_B, V_C (named as full amplitude inputs) or $(V_A+V_B)/2$, $(V_B+V_C)/2$, $(V_C+V_A)/2$ (named as half amplitude outputs). In a space vector frame representation, from (3), we can deduce that the amplitude of the 216 voltage states could be either $2V_{in}/\sqrt{3}$ (named as full amplitude vector), V_{in} (named as middle amplitude vector) or $V_{in}/\sqrt{3}$ (named as half amplitude vector) with different switching pairs.

$$V_o \square \left| \frac{1}{3} \left[(2V_1 - V_2 - V_3) \square j\sqrt{3}(V_2 - V_3) \right] \right| \quad (3)$$

After tedious calculation, all of the states can be split into 10 groups as follows.

Group I: Each output line is connected to a different full amplitude input line. The space vector output voltage angle is time varying. Hence they are not been used in SVM.

Group II: Each output line is connected to a different half amplitude input line. Again the space vector output voltage angle is time varying and this Group is not used.

Group III: All of the three output legs are connected to a common full amplitude input leg. They are the so-called zero vectors of the MC and MMC.

Group IV: All of the three output phases are connected to a common half amplitude input phase. These are the extra zero vectors in MMC. Every state consists of 8 switching pairs. However, not all of these vectors produce zero current at the input and only the ones which do are used.

Group V: Two output lines are connected to a common full amplitude input line, and the remaining output line is connected to one of the other full amplitude input lines. These 18 non-zero vectors are the same full amplitude vectors as

produced by the conventional MC SVM. Their voltage angles are distributed at 0°, 60°, 120°, 180°, 240° and 300° and every state has only one related switching pair.

Group VI: Two output lines are connected to a common half amplitude input, and the remaining one is connected to one of the other half amplitude input lines. These 18 non-zero vectors have the same voltage vector angle distribution as *Group V* but have only half the amplitude ($V_{in}/\sqrt{3}$) of the conventional MC space vectors. Furthermore, every state consists of 8 possible switching pairs with different input current angles as listed in Table II.

As mentioned before, the voltage across C_1, C_2 and C_3 needs to be controlled to be $V_{AB}/2$, $V_{BC}/2$ and $V_{AC}/2$ to maintain the mid-voltages at the required value. From Fig 1, it is obvious that the capacitors are charged or discharged by the load current and the capacitor control can be fulfilled by a proper control of switching pairs in SVM. However, the current state in the space vector frame and the current vector angle is unavoidably changed by capacitor control from Table II. In order to maintain the input current space vector as desired, the current vector angle needs to be kept constant or at zero state during state changes for capacitor control.

Group VII: two output lines are connected to a common full amplitude input line, and the remaining output line is connected to a half amplitude input line. The voltage angles of the 27 vectors are distributed at 0°, 60°, 120°, 180°, 240° and 300°, every state has 2 switching pairs. The amplitude of 9 vectors are V_{in} and $V_{in}/\sqrt{3}$ for the other 18 (Table III).

Group VIII: two output lines are connected to different full amplitude inputs, and the remaining one is connected to a half amplitude input. Among all the vectors, only 18 of them produce constant current vector angles which distribute at 30°, 90°, 190°, 210°, 270° and 330°. Every state has 2 switching pairs with the amplitude of V_{in} (Table IV).

Group IX: two output lines are connected to a common half amplitude input line, and the remaining output line is connected to a full amplitude input line. All states provide constant space vector voltage angles which distribute at 0°, 60°, 120°, 180°, 240° and 300°. Every state has 4 switching pairs. The voltage vector amplitude can be V_{in} or $V_{in}/\sqrt{3}$ with the related switching pairs. With the current vector restriction, the vectors of $V_o \square V_{in}$ can not be used in MMC SVM since their current angles are not fixed for capacitor control. In this case, *Group IX* will reduce to 18 vectors with the amplitude of $V_o \square V_{in}/\sqrt{3}$ (Table V).

Group X: two output lines are connected to a different half amplitude input line, and the remaining output line is connected to a full amplitude input line. All the output space vector angles of above states are variable and this Group is not used.

230

Table II: The state in Group VI

Voltage State	Switching Pairs	Voltage angle	Current angle
$V_u, V_v, V_w =$ $(V_C+V_A)/2$; $(V_B+V_C)/2$; $(V_B+V_C)/2$ $V_o = V_{in}/\sqrt{3}$	$S_{Cu1} S_{Au2} S_{Bv1} S_{Cv2} S_{Bw1} S_{Cw}$	0 or π	π/2 or 3π/2
	$S_{Cu1} S_{Au2} S_{Bv1} S_{Cv2} S_{Cw1} S_{Bw2}$		π/2 or 3π/2
	$S_{Cu1} S_{Au2} S_{Cv1} S_{Bv2} S_{Bw1} S_{Cw2}$		π/2 or 3π/2
	$S_{Cu1} S_{Au2} S_{Cv1} S_{Bv2} S_{Cw1} S_{Bw2}$		*zero*
	$S_{Au1} S_{Cu2} S_{Cv1} S_{Bv2} S_{Bw1} S_{Cw2}$		*5π/6 or 11π/6*
	$S_{Au1} S_{Cu2} S_{Bv1} S_{Cv2} S_{Cw1} S_{Bw2}$		Not fixed
	$S_{Au1} S_{Cu2} S_{Cv1} S_{Bv2} S_{Bw1} S_{Cw2}$		Not fixed
	$S_{Au1} S_{Cu2} S_{Cv1} S_{Bv2} S_{Cw1} S_{Bw2}$		π/6 or 7π/6

Table III: The state in Group VII

Voltage State	Switching Pairs	Voltage angle	Current angle
$V_u, V_v, V_w =$ $V_A; V_A; (V_A+V_B)/2$ $V_o = V_{in}/\sqrt{3}$	$S_{Au1} S_{Au2} S_{Av1} S_{Av2} S_{Aw1} S_{Bw2}$	π/3 or 4π/3	zero
	$S_{Au1} S_{Au2} S_{Av1} S_{Av2} S_{Bw1} S_{Aw2}$		5π/6 or 11π/6
$V_u, V_v, V_w =$ $V_A; V_A; (V_B+V_C)/2$ $V_o = V_{in}$	$S_{Au1} S_{Au2} S_{Av1} S_{Av2} S_{Bw1} S_{Cw2}$	π/3 or 4π/3	5π/6 or 11π/6
	$S_{Au1} S_{Au2} S_{Av1} S_{Av2} S_{Cw1} S_{Bw2}$		π/6 or 7π/6

Table IV: The state in Group VIII

Voltage State	Switching Pairs	Voltage angle	Current angle
$V_u, V_v, V_w =$ $V_A; V_B; (V_A+V_B)/2$ $V_o = V_{in}$	$S_{Au1} S_{Au2} S_{Bv1} S_{Bv2} S_{Aw1} S_{Bw2}$	7/6 or 11π/6	7/6 or 11π/6
	$S_{Au1} S_{Au2} S_{Bv1} S_{Bv2} S_{Bw1} S_{Aw2}$		7/6 or 11π/6

Table V: The state in Group VIX

Voltage State	Switching Pairs	Voltage angle	Current angle
$V_u, V_v, V_w =$ $V_A;$ $(V_A+V_B)/2;$ $(V_A+V_B)/2$ $V_o = V_{in}/\sqrt{3}$	$S_{Au1} S_{Au2} S_{Av1} S_{Bv2} S_{Aw1} S_{Bw2}$	0 or π	zero
	$S_{Au1} S_{Au2} S_{Av1} S_{Bv2} S_{Bw1} S_{Aw2}$		5π/6 or 11π/6
	$S_{Au1} S_{Au2} S_{Bv1} S_{Av2} S_{Aw1} S_{Bw2}$		5π/6 or 11π/6
	$S_{Au1} S_{Au2} S_{Bv1} S_{Av2} S_{Bw1} S_{Aw2}$		5π/6 or 11π/6

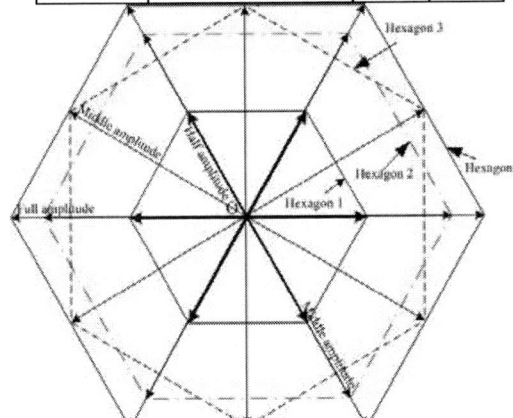

Fig.2 MMC output voltage space vector (Hexagons)

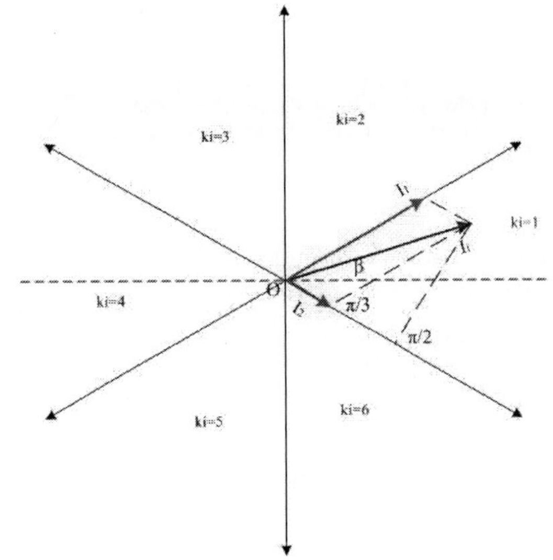

Fig. 3 Input current space vector

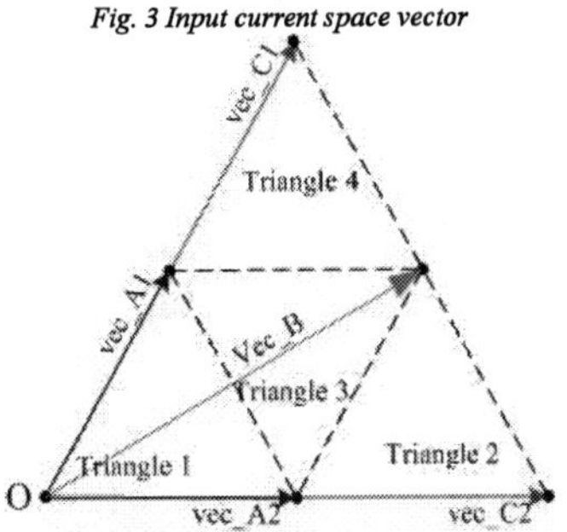

Fig. 4 MMC output voltage space vector sector

C. Space vector diagram

Based on the previous analysis, the MMC output voltage space vectors can be presented as shown in Fig 2. The whole cycle is divided into 4 hexagons, 12 sectors (30° each) [12].

• Hexagon 1 (include *Group VI, VII and IX*):

$Amplitude_{Hexagon\_1} = V_{in}/\sqrt{3}$, the angle of output space vectors distribute at 0°, 60°, 120°, 180°, 240° and 300°.

• Hexagon 2 (include *Group VII and IX*):

$Amplitude_{Hexagon\_2} = V_{in}$, the angle of output space vectors distribute at 0°, 60°, 120°, 180°, 240° and 300°.

• Hexagon 3 (*Group VIII*):

$Amplitude_{Hexagon\_3} = V_{in}$, the angle of output space vectors distribute at 30°, 90°, 190°, 210°, 270° and 330°.

• Hexagon 4 (*Group III*):

$Amplitude_{Hexagon\_4} = 2V_{in}/\sqrt{3}$, the angle of output space vectors distribute at 30°, 90°, 190°, 210°, 270° and 330°.

At the input current space vector side, the whole space vector cycle can be split into 6 equal sectors (named as ki, shown in Fig 3) as the conventional MC does [1] [5].

Compared with MC, the MMC allows composition of the target vector in hexagons 1, 2, 3 and 4 while only the vectors in hexagon 4 exists in the MC. Furthermore, the SVM could be carried out in every 30° sector. In the proposed MMC SVM, the modulation is implemented in 4 triangles per 60°. We name the selected vectors as vec_A (include vec_A1, vec_A2), vec_B and vec_C (include vec_C1 and vec_C2) as indicated in Fig 4 (The subscript sign _A, _B and _C indicate the vectors with half, middle or full amplitude respectively). We have:

• If the target vector is rotated into *triangle 1*, vec_A1 and vec_A2 will be used to synthesize the target vector.

• If the target vector is in *triangle 2*, vec_A2, vec_C2 and vec_B will be used.

• If the target vector is in *triangle 3*, vec_A1, vec_A2, vec_C1 and vec_C2 will be used.

• If the target vector is in *triangle 4*, vec_A1, vec_C1 and vec_B will be used.

D. Duty cycles

All of the space vectors provide fixed angles but variable amplitudes in MMC SVM. Their amplitudes can be expressed as in (4):

$$amp = \left| \frac{1}{3}\left[(2V_1 - V_2 - V_3) + j\sqrt{3}(V_2 - V_3) \right] \right| = \sqrt{\text{Re}^2 + \text{Im}^2} \quad (4)$$

V_1, V_2, V_3 are the voltage of the output leg at any instant. With some simplification, the amplitudes of these vectors are listed in table VI.

As discussed in [5] [6], MC and MMC duty cycles are under the restriction that:

$$\begin{cases} \left(I_i^1 \cdot amp1 \cdot d1 + I_i^2 \cdot amp2 \cdot d2\right) \bullet j \cdot e^{j\beta} \cdot e^{j(k_i-1)\pi/3} = 0 \\ \left(I_i^3 \cdot amp3 \cdot d3 + I_i^4 \cdot amp4 \cdot d4\right) \bullet j \cdot e^{j\beta} \cdot e^{j(k_i-1)\pi/3} = 0 \end{cases} \quad (5)$$

From (5), we can easily derive the following formula as shown in (6):

$$\frac{amp1 \cdot d1}{amp2 \cdot d2} = \frac{amp3 \cdot d3}{amp4 \cdot d4} = \frac{\cos(\beta - \pi/3)}{\cos(\beta + \pi/3)} \quad (6)$$

In (5) and (6), d1, d2, d3 and d4 are 4 duty cycles and β is the current vector angle calculated from the bisecting line of every sector.

If we assume the voltage vector angle α is calculated from the beginning line of every sector, the following duty cycle formulas can be deduced in triangle 1, 2, 3, and 4.

Table VI: Space vector amplitude

Vector Number	Current sector	Amplitude Equations
amp1_A & amp3_A	1 or 4	$V_{in}\sin(\pi/3 + \beta_{in})/\sqrt{3}$
	2 or 5	$V_{in}\sin(\beta_{in})/\sqrt{3}$
	3 or 6	$V_{in}\sin(\pi/3 - \beta_{in})/\sqrt{3}$
amp2_A & amp4_A	1 or 4	$V_{in}\sin(\pi/3 - \beta_{in})/\sqrt{3}$
	2 or 5	$V_{in}\sin(\pi/3 + \beta_{in})/\sqrt{3}$
	3 or 6	$V_{in}\sin(\beta_{in})/\sqrt{3}$
amp1_B & amp3_B	1 or 4	$V_{in}\sin(\pi/3 + \beta_{in})$
	2 or 5	$V_{in}\sin(\beta_{in})$
	3 or 6	$V_{in}\sin(\pi/3 - \beta_{in})$

amp2_B & amp4_B	1 or 4	$V_{in}\sin(\pi/3 - \beta_{in})$
	2 or 5	$V_{in}\sin(\pi/3 + \beta_{in})$
	3 or 6	$V_{in}\sin(\beta_{in})$
amp1_C & amp3_C	1 or 4	$2V_{in}\sin(\pi/3 + \beta_{in})/\sqrt{3}$
	2 or 5	$2V_{in}\sin(\beta_{in})/\sqrt{3}$
	3 or 6	$2V_{in}\sin(\pi/3 - \beta_{in})/\sqrt{3}$
amp2_C & amp4_C	1 or 4	$2V_{in}\sin(\pi/3 - \beta_{in})/\sqrt{3}$
	2 or 5	$2V_{in}\sin(\pi/3 + \beta_{in})/\sqrt{3}$
	3 or 6	$2V_{in}\sin(\beta_{in})/\sqrt{3}$

The amp1, amp2, amp3 and amp4 denote the sequence in application which is the same as the definition in [5].
The subscript sign _A, _B and _C are of the same meaning as mentioned before.

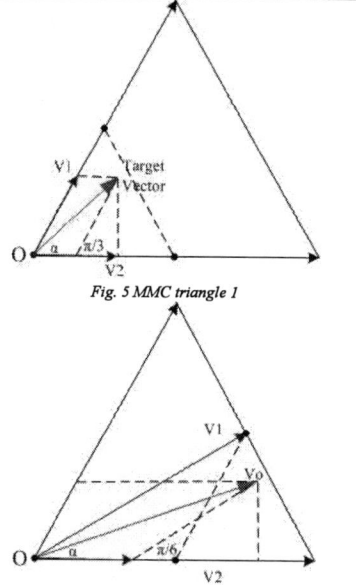

Fig. 5 MMC triangle 1

Fig. 6 MMC triangle 2

• In triangle 1 (Fig 5), we have:

$$\begin{cases} V_o \cos(\alpha) = V_1 \cos(\pi/3) + V_2 \\ V_o \sin(\alpha) = V_1 \sin(\pi/3) \end{cases} \&$$

$$\begin{cases} V_1 = amp1 \cdot d1 + amp2 \cdot d2 \\ V_2 = amp3 \cdot d3 + amp4 \cdot d4 \end{cases} \quad (7)$$

From (7), the duty cycles in triangle 1 are as follows:

$$d1 = 2V_o \sin(\alpha)\cos(\beta - \pi/3)/\left[\sqrt{3}amp1\_A\cos(\beta)\right]$$

$$d2 = 2V_o \sin(\alpha)\cos(\beta + \pi/3)/\left[\sqrt{3}amp2\_A\cos(\beta)\right]$$

$$d3 = 2V_o \sin(\pi/3 - \alpha)\cos(\beta - \pi/3)/\left[\sqrt{3}amp1\_A\cos(\beta)\right] \quad (8)$$

$$d4 = 2V_o \sin(\pi/3 - \alpha)\cos(\beta + \pi/3)/\left[\sqrt{3}amp2\_A\cos(\beta)\right]$$

$$d0 = 1 - d1 - d2 - d3 - d4$$

• In triangle 2 (Fig 6), we have:

$$\begin{cases} V_o \cos(\alpha) = V_1 \cos(\pi/6) + V_2 \\ V_o \sin(\alpha) = V_1 \sin(\pi/6) \end{cases} \&$$

$$\begin{cases} V_1 = amp1\_B \cdot d1 + V_{in} \cdot amp2\_B \cdot d2 \\ V_2 = amp1\_A \cdot d3 + amp2\_A \cdot d4 \\ \quad + amp1\_C \cdot d5 + amp2\_C \cdot d6 \\ d1 + d2 + d3 + d4 + d5 + d6 = 1 \end{cases} \quad (9)$$

Define the parameter a, b, q, k as follows:

$$a = 2V_o \cdot \sin(\alpha) \cdot \cos(\pi/3 - \beta)/[\cos(\beta) \cdot amp1\_B]$$
$$b = 2V_o \cdot \sin(\pi/6 - \alpha) \cdot \cos(\pi/3 - \beta)/[\cos(\beta) \cdot amp1\_A]$$
$$q = [amp1\_A \cos(\pi/3 + \beta)]/[amp2\_A \cos(\pi/3 - \beta)]$$
$$k = 1/(1 + q) \qquad (10)$$

Then the duty cycles are:

$$\begin{cases} d1 = a; & d2 = q \cdot d1; \\ d3 = 2k - 2a - b; & d4 = q \cdot d3; \\ d5 = a + b - k; & d6 = q \cdot d5; \end{cases} \qquad (11)$$

- In triangle 4 (Fig 7), we have:

$$\begin{cases} V_o \cos(\alpha) = V_1 \cos(\pi/6) + V_2 \\ V_o \sin(\alpha) = V_1 \sin(\pi/6) \end{cases} \&$$

$$\begin{cases} V_1 = amp1\_A \cdot d1 + amp2\_A \cdot d2 \\ \quad + amp1\_C \cdot d3 + amp2\_C \cdot d4 \\ V_2 = amp1\_B \cdot d5 + amp2\_B \cdot d6 \\ d1 + d2 + d3 + d4 + d5 + d6 = 1 \end{cases} \quad (12)$$

Define the parameter a, b, q, k as follows:

$$a = 2V_o \sin(\pi/6 - \alpha) \cdot \cos(\pi/3 - \beta)/[\cos(\beta) \cdot amp1\_B]$$
$$b = 2V_o \sin(\alpha) \cdot \cos(\pi/3 - \beta)/[\cos(\beta) \cdot amp1\_A]$$
$$q = [amp1\_A \cdot \cos(\pi/3 + \beta)]/[amp2\_A \cdot \cos(\pi/3 - \beta)]$$
$$k = 1/(1 + q) \qquad (13)$$

Following a similar approach as triangle 2, the duty cycles are expressed as:

$$\begin{cases} d1 = 2k - 2a - b; & d2 = q \cdot d1; \\ d3 = a + b - k; & d4 = q \cdot d3; \\ d5 = a; & d6 = q \cdot d5; \end{cases} \qquad (14)$$

- In triangle 3 (Fig 8), we have:

$$\begin{cases} V_o \cos(\alpha) = V_1 \cos(\pi/3) + V_2 \\ V_o \sin(\alpha) = V_1 \sin(\pi/3) \end{cases} \&$$

$$\begin{cases} V_1 = amp1\_A \cdot d1 + amp2\_A \cdot d2 \\ \quad + amp1\_C \cdot d3 + amp2\_C \cdot d4 \\ V_2 = amp1\_A \cdot d5 + amp2\_A \cdot d6 \\ \quad + amp1\_C \cdot d7 + amp2\_C \cdot d8 \\ d1 + d2 + d3 + d4 + d5 + d6 + d7 + d8 = 1 \end{cases} \quad (15)$$

Define the parameters a, b, q, k as follows:

$$a = 2V_o \sin(\alpha) \cos(\pi/3 - \beta)/\left[\sqrt{3}\cos(\beta)amp1\_A\right]$$
$$b = 2V_o \sin(\pi/3 - \alpha)\cos(\pi/3 - \beta)/\left[\sqrt{3}\cos(\beta)amp1\_A\right]$$
$$q = [amp1\_A \cos(\pi/3 + \beta)]/[amp2\_A \cos(\pi/3 - \beta)] \qquad (16)$$
$$k = \frac{a + b + a \cdot q + b \cdot q - 1}{2 - a - b - a \cdot q - b \cdot q}$$

Then the duty cycles are:

$$\begin{cases} d1 = \dfrac{a}{1 + 2k}; & d2 = q \cdot d1; \\ d3 = k \cdot d1; & d4 = q \cdot d3; \\ d5 = \dfrac{b}{1 + 2k}; & d6 = q \cdot d5; \\ d7 = k \cdot d5; & d8 = q \cdot d7; \end{cases} \qquad (17)$$

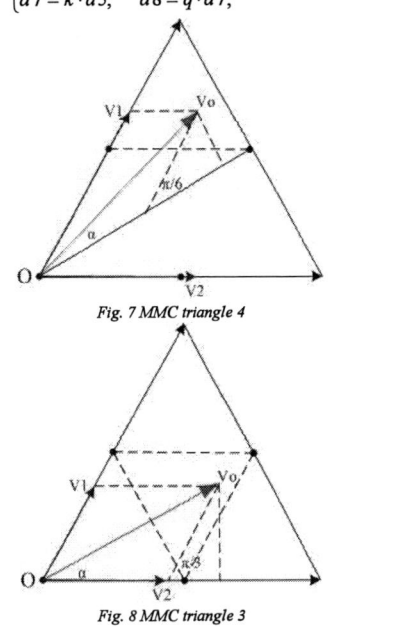

Fig. 7 MMC triangle 4

Fig. 8 MMC triangle 3

E. 4-step commutation:

The MMC 4-step commutation can be regarded as two synchronous MC 4-step commutations. With the same clock pulse, the two MC 4-step commutations can be controlled independently. The MC 4-step commutation is discussed in detail in [1], [8] and [9].

F. Flying capacitor voltage control:

In MMC flying capacitor (FC) control, the voltage vector angle and current vector angle must be kept unchanged during the charging and discharging process. In this case, only the switching pairs under this restriction can be selected. In a SABER simulation, the voltages across the FCs are monitored by voltage sensors and compared with the relative reference voltage. With the feed back information, the switching pairs can be switched at opposite state to change the current direction flow through the FCs until the FCs voltages reach the required values (Fig. 9).

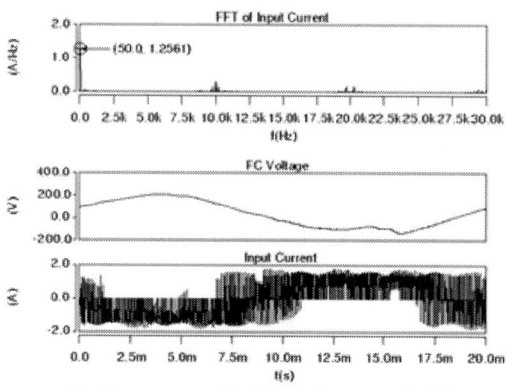

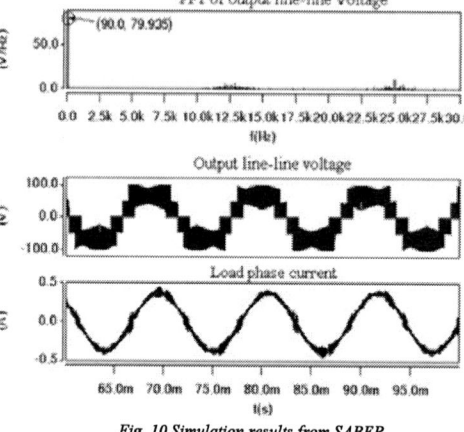

Fig. 9 Input current and its FFT & voltage across the FC

The charging and discharging process obeys:

$$V_C = (1/C) \int i_C dt \qquad (18)$$

In (18), the current injected into the FC is the (normally sinusoidal) load current i_u, i_v or i_w. Clearly the FC will unavoidably lose control when the related current has a very small value. Furthermore, the input current direction is inevitably changed instantaneously since the charging and discharging process is achieved by using different switching pairs, which result in reversed current direction at the input (Fig. 9). However, the FFT shows that the harmonics at the input side are limited at an acceptable range.

III. SIMULATION AND EXPRIMENTAL RESULTS

In SABER simulation, the input is a 100V (peak of line-line voltage) 50Hz 3-phase voltage source; output is 3-phase 4mH+124Ω load; the sampling frequency is 12.5 kHz; the voltage transfer ratio is 80%.

In the experimental work, small 2:1 ratio transformers are used to instead of the flying capacitors to supply a stable voltage. This is due to the lack of sufficient voltage transducers in the experimental rig. The experimental environment is as listed below:

Input: 50 Hz, 100V (peak of line-line voltage);
Target output: 90 Hz, 80V (peak of line-line voltage);
Voltage transfer ratio: 80%;
Switching frequency: 12.5 KHz;
Load: 3 phase LC load, 4 mH+124 Ω per phase.

Figure 9 shows simulation results for the input side illustrating the flying capacitor voltage and the input current spectrum. Note that the distortion of the FC voltage is caused when the related phase current is very small. It is hoped to improve on this issue in future work. Figure 10 shows simulation results for the output side, illustrating clearly the multilevel nature of the output voltage. Figures 11-13 show similar results for the low-power experimental prototype, again illustrating the multi-level behavior.

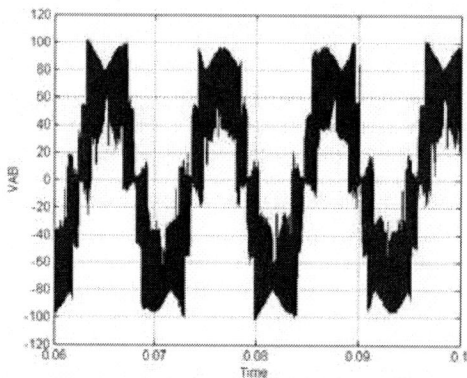

Fig. 10 Simulation results from SABER.

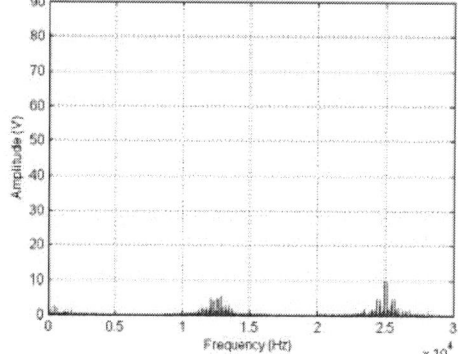

Fig. 11 Experimental result: line-line voltage (unit V: S)

Fig. 12 Experimental result: FFT of line-line voltage

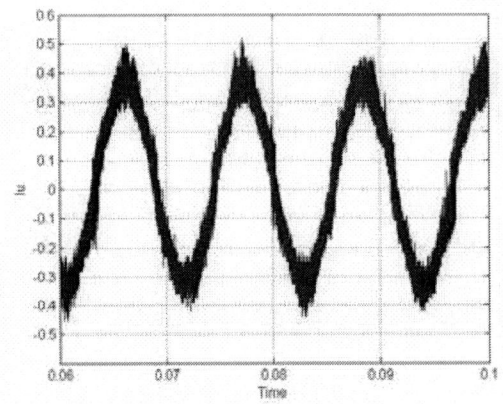

Fig. 13 Experimental result: load current (unit A: S)

IV. CONCLUSION

Multi-level converters have well established advantages in high power applications. Matrix (or direct AC-AC) converters have potential advantages due to the lack of bulk energy storage. It is therefore interesting to look at the potential for using multi-level techniques in direct AC-AC converters either for reasons of increased power handling capability, or for improved waveform quality. This paper has concentrated on the development of a space-vector modulation method for a capacitor clamped multi-level matrix converter. The space vector representation and modulation of the converter has been described in detail and results of simulations run in SABER are presented. These results are validated by experimental results from a low power prototype. Future papers will concentrate on practical implementation of capacitor balancing and input power quality.

REFERENCE

[1] Patrick W. Wheeler, José Rodríguez, Jon C. Clare, Lee Empringham and Alejandro Weinstein, "Matrix Converters: A Technology Review", IEEE Transactions on Industrial Electronics, VOL. 49, NO. 2, April 2002.

[2] A. Alesina and M. G. B. Venturini, "Analysis and design of optimum amplitude nine-switch direct AC–AC converters", IEEE Trans.v Power Electron., vol. 4, pp. 101–112, Jan. 1989.

[3] L. Huber and D. Borojevic, "Space vector modulation with unity input power factor for forced commutated cycloconverters", in Conf. Rec. IEEE-IAS Annul. Meeting, 1991, pp. 1032–1041.

[4] Y. H. Liu, J. Arrillaga, N. R. Watson, "Capacitor Voltage Balancing in Multi-Level Voltage Reinjection (MLVR) Converters", IEEE Transaction on Power Delivery, VOL. 20, NO. 2, APRIL 2005.

[5] Domenico Casadei, Giovanni Serra, Angelo Tani and Luca Zarri, "Matrix Converter Modulation Strategies: A New General Approach Based on Space-Vector Representation of the Switch State", IEEE Transactions on Industrial Electronics, VOL. 49, NO. 2, April 2002.

[6] D. Casadei, G. Grandi, G. Serra, and A. Tani, "Space vector control of matrix converters with unity input power factor and sinusoidal input/output waveforms", in Proc. EPE Conf., vol. 7, Brighton, U.K., Sept. 13–16, 1993, pp. 170–175.

[7] Yong SHI, Xu YANG, Qun HE, and Zhaoan WANG. , "Research on a Novel Multi-level Matrix Converter", 2004 35th Annual IEEE Power Electronics Specialists Conference.

[8] Patrick W. Wheeler, Jon C. Clare, Lee Empringham and Michael Bland, "Gate Drive Level Intelligence and Current Sensing for Matrix Converter Current Commutation", IEEE Transactions on Industrial Electronics, VOL. 49, NO. 2, April 2002

[9] L. Empringham, P. Wheeler, and J. Clare, "Bi-directional switch current commutation for matrix converter applications", in Proc.PEMC Prague, Sept. 1998, pp. 42–47

[10] Thomas F. Podlesak, Dimosthenis C. Katsis, Patrick W. Wheeler, Jon C. Clare, Lee Empringham, and Michael Bland, "A 150-kVA Vector-Controlled Matrix Converter Induction Motor Drive" , IEEE TRANSACTIONS ON INDUSTRY APPLICATIONS, VOL. 41, NO. 3, MAY/JUNE 2005

[11] Yong Shi, Xu Yang, Qun He, and Zhaoan Wang, "Research on a Novel Capacitor Clamped Multilevel Matrix Converter", IEEE TRANSACTIONS ON POWER ELECTRONICS, VOL. 20, NO. 5, SEPTEMBER 2005

[12] Patrick W. Wheeler, Xu Lie, Meng Yeong Lee, Lee Empringham, Christian Klumpner, Jon Clare, "A Review of Multi-level Matrix Converter Topologies" , Power Electronics, Machines and Drives PEMD 2008

New Family of Matrix-Reactance Frequency Converters Based on Unipolar PWM AC Matrix-Reactance Choppers

Zbigniew Fedyczak, Paweł Szcześniak and Igor Korotyeyev

University of Zielona Góra, Institute of Electrical Engineering, Zielona Góra, Poland,
e-mail: Z.Fedyczak@iee.uz.zgora.pl; P.Szczesniak@iee.uz.zgora.pl; I.Korotyeyev@iee.uz.zgora.pl.

Abstract—This paper deals with three-phase direct matrix-reactance frequency converters (MRFC) based on unipolar PWM AC matrix-reactance choppers (MRC). The topologies of the proposed MRFC are based on a three-phase unipolar MRC structure. Each MRC with conventional topology has two synchronous-connected switches (SCS) sets. In the MRFC, unlike the MRC topology, one of SCS sets is replaced by a matrix-connected switches (MCS) set in order to make possible of the load voltage frequency change. Six new topologies of the MRFC based on MRC boost, buck-boost, Ćuk, Zeta or SEPIC structures are presented. Through the generation concept of the proposed converters both the description of above-mentioned converter topologies and general description of the control strategies are presented. The structure of the proposed MRFC contains a three-phase matrix converter (MC), which is introduced instead of the source or load SCS used in unipolar MRC. The step-down or step-up of the MC set is dependent on the input and output voltage or current source configurations. Analysis determining the location where the MC should be introduced is realized by means of the one-cycle switched models with suitable voltage and current sources introduced instead of the capacitors and inductors respectively. Furthermore, exemplary results of the simplified theoretical analysis, based on the averaged state space method, as well as simulation test results obtained for a classical Venturini control strategy of MC, are also presented as an initial verification of the properties of the proposed converters.

Keywords—Matrix converter, matrix-reactance chopper, matrix-reactance frequency converter.

I. INTRODUCTION

One among the most desirable features of AC frequency converters is the generation of load voltage with arbitrary amplitude and frequency [1]. In recent years, matrix converters (MC) have received considerable attention as a competitor to the commonly used pulse width-modulated voltage-source inverter (PWM-VSI). The real development of MC starts with the work of Venturini and Alesina [2]. As is well known, the MC provides sinusoidal input and output waveforms, bidirectional power flow, controllable input power factor, and more compact design [1]-[7]. One disadvantage of the MC is the voltage transfer ratio, which is limited to 0.5 of

the input voltage [2] at linear voltage transformation, and to 0.866 or 1.053 at low-frequency load voltage deformations for space-vector or fictitious DC link control strategy concepts respectively [1], [3]–[7]. For application in the industrial drive field the maximum available magnitude of the output voltage should be even a little greater than the amplitude of the input voltage. In the case of the variable speed drive system for an induction motor, a reduction of the supply voltage by 10% means 20% loss of torque capability, which cannot be accepted in most applications. For application in FACTS a conventional auxiliary transformer employed to increase the MC output voltage is necessary. In reference [15] the concept of the step-up MC is presented. In a circuit with this converter input inductors and output capacitors are used that function as input current sources and output voltage sources respectively. The voltage transfer ratio can be much greater than one, though voltage gain and input power factor cannot be controlled independently, they can be controlled by properly selected control parameters. Another concept of frequency converters with buck-boost voltage transformation is presented in references [9]–[12]. A general description of the matrix-reactance frequency converter (MRFC) based on PWM AC matrix-reactance chopper (MRC) with buck-boost topology is proposed. In this MRFC the source switches are arranged as in the MC. Such an approach gives the possibility to obtain the load output voltage much greater than the input voltage. The conception of the MRFC proposed in references [9]–[12] is continuously developed by authors.

In this paper the generation concept of whole family MRFC based on unipolar PWM AC MRC is presented. The MRFC topologies are generated by use of the MRC one-cycle switched circuit models with suitable voltage and current sources introduced instead of the capacitors and inductors respectively. One of the synchronous-connected switches (SCS) sets of the MRC is replaced by the step-down [2] or step-up [15] matrix-connected switches (MCS) set, offering the possibility of load voltage frequency change. Systematic application of the proposed method has generated six new topologies of the MRFC. A crucial fact is that all generated MRFCs have the possibility to obtain a load output voltage much greater than the input voltage.

The main aim of this paper is to present both the concept of topologies generation and the general description of such converters. A selected theoretical analysis and test results simulation is presented as an initial verification of the basic features of the proposed converters.

This work was supported by Polish Ministry of Science and Higher Education, Project No. N510 036 32/3380.

II. CONCEPT OF PROPOSED CONVERTERS

A. Basic Converter Structure

The topologies of the presented MRFC are based on a three-phase unipolar MRC structures. Each unipolar MRC with conventional topology has two synchronous-connected switches (SCS) sets [13], [14]. In general, only two switch states of the MRC are used. In the MRFC, unlike with the MRC topology, one of SCS sets is replaced by a matrix-connected switches (MCS) set and 28 (27 + 1) switch states can be used. It allows the possibility of obtaining load voltage frequency change. A general structure of a basic MRFC is shown in Fig.1.

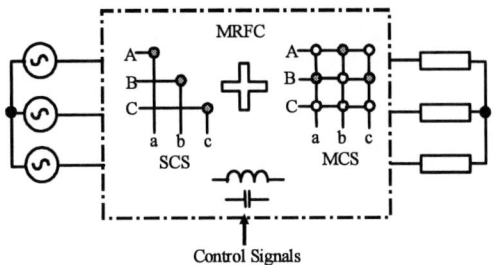

Fig. 1. General structure of basic matrix-reactance frequency converters, SCS - synchronous-connected switches, MCS – matrix-connected switches

B. Topologies Generation

No formal synthesis procedure is given for the derivation of the proposed MRFC. As mentioned above, the proposed MRFC topologies are generated by use of unipolar MRC one-cycle switched circuit models. These models are constructed on the basis of the switched circuit models of unipolar MRC shown in Fig. 2 [13], [14].

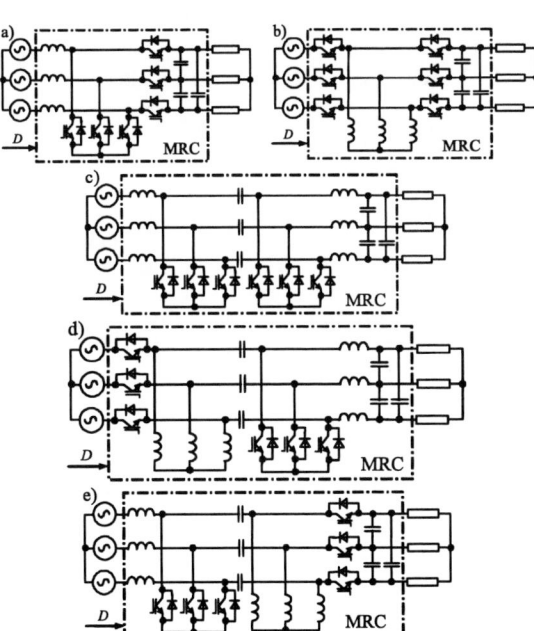

Fig. 2. Three-phase unipolar PWM AC MRC based on, a) boost, b) buck-boost, c) Ćuk, d) Zeta, e) SEPIC topology

The one-cycle switched circuit models of the discussed MRC (Fig. 2) for two switch states are shown in Figs. 3-7. In these models suitable voltage and current sources are taken into consideration instead of capacitors and inductors respectively. During the first switch state the source synchronous-connected switches (SSCS) are turn on and the load synchronous-connected switches (LSCS) are turned off. During the second switch-state the switches are in the inverse states.

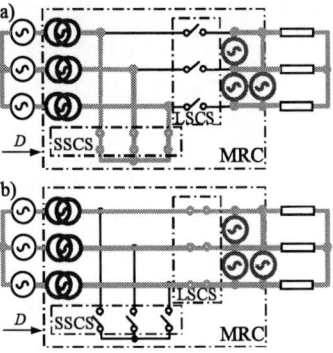

Fig. 3. One-cycle switched circuit models of three-phase unipolar PWM AC MRC based on boost topology, a) for SSCS on and LSCS off, b) for SSCS off and LSCS on

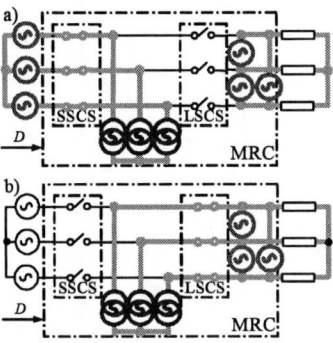

Fig. 4. One-cycle switched circuit models of three-phase unipolar PWM AC MRC based on buck-boost topology, a) for SSCS on and LSCS off, b) for SSCS off and LSCS on

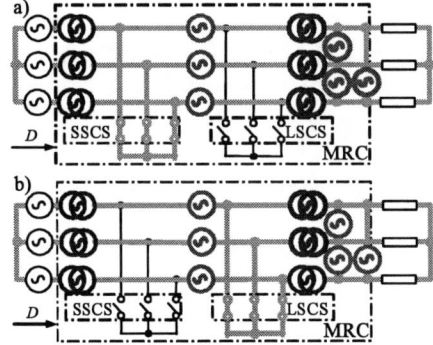

Fig. 5. One-cycle switched circuit models of three-phase unipolar PWM AC MRC based on Ćuk topology, a) for SSCS on and LSCS off, b) for SSCS off and LSCS on

As is visible from Figs. 3-7, in each of the switch states, the SCS sets take part in two types of electrical energy transfer between the inner voltage and current sources: the first one, as electrical energy is transferred from the voltage to the current sources; the second one as electrical energy is transferred from the current to the voltage sources. From an analysis of the electrical energy transfers it follows that one of SCS sets can be replaced by a MCS set, offering the possibility of the load voltage frequency change. The use of step-down or step-up of the MCS set is dependent on input and output voltage or current sources configurations, which are shown in Fig. 8 [15].

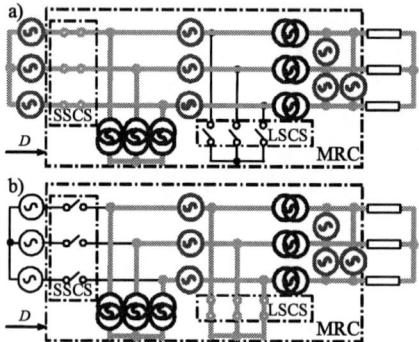

Fig. 6. One-cycle switched circuit models of three-phase unipolar PWM AC MRC based on Zeta topology, a) for SSCS on and LSCS off, b) for SSCS off and LSCS on

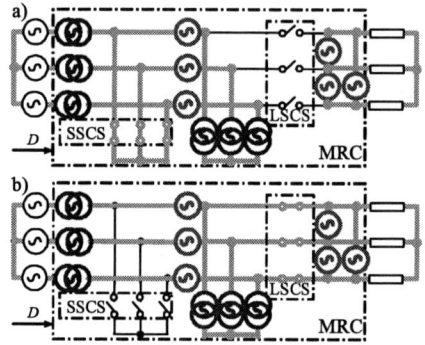

Fig. 7. One-cycle switched circuit models of three-phase unipolar PWM AC MRC based on SEPIC topology, a) for SSCS on and LSCS off, b) for SSCS off and LSCS on

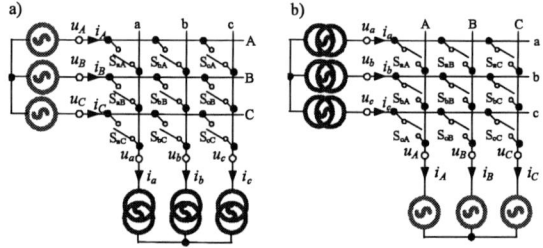

Fig. 8. Matrix-connected switches sets with, a) step-down configuration, b) step-up configuration

C. General Description of Control Strategies

A description of the control strategies of the proposed MRFC, in general form, is illustrated in Fig. 9. Topologies of the proposed MRFC, which are indicated in the capture of the Fig. 9, are presented in detail in the next section. In each switching sequence T_{Seq} there are two time periods, t_S and t_L. In the period t_S the source switches (SSCS or MCS if used) are switching whereas in the period t_L the load switches (LSCS or MCS if used) are switching.

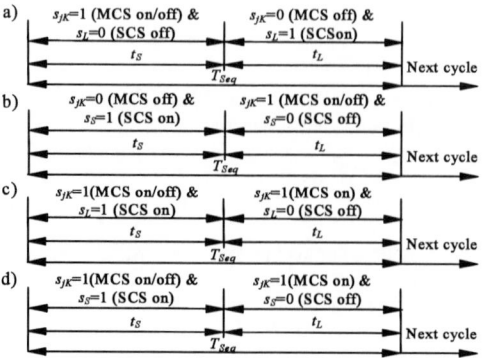

Fig. 9. General form of the control strategy, a) type 1 (buck-boost I, Zeta I), b) type 2 (boost, buck-boost II, SEPIC II), c) type 3 (Ćuk II, SEPIC I), c) type 4 (Ćuk I, Zeta II)

It is both essential and obvious that the switching transfer matrix, for output-input voltages and currents relationships has the form as in (1) for step-down MCS set configuration (Fig. 8a) or has the form as in (2) for step-up MCS set configuration (Fig. 8b) [1], [2].

$$
\begin{bmatrix} u_a \\ u_b \\ u_c \end{bmatrix} = \begin{bmatrix} S_{aA} & S_{aB} & S_{aC} \\ S_{bA} & S_{bB} & S_{bC} \\ S_{cA} & S_{cB} & S_{cC} \end{bmatrix} \begin{bmatrix} u_A \\ u_B \\ u_C \end{bmatrix} = \mathbf{T} \begin{bmatrix} u_A \\ u_B \\ u_C \end{bmatrix}, \quad (1)
$$

$$
\begin{bmatrix} i_A \\ i_B \\ i_C \end{bmatrix} = \begin{bmatrix} S_{aa} & S_{bA} & S_{cA} \\ S_{aB} & S_{bB} & S_{cB} \\ S_{aC} & S_{bC} & S_{cC} \end{bmatrix} \begin{bmatrix} i_a \\ i_b \\ i_c \end{bmatrix} = \mathbf{T}^T \begin{bmatrix} i_a \\ i_b \\ i_c \end{bmatrix}, \quad (2)
$$

where: $s_{jK} = \begin{cases} 1, & \text{switch } S_{jK} \text{ closed} \\ 0, & \text{switch } S_{jK} \text{ open} \end{cases}$, $j = \{a, b, c\}$, $K = \{A, B, C\}$.

III. DESCRIPTION OF DERIVED TOPOLOGIES

A. MRFC Based on Boost MRC Topology

A new topology of the MRFC based on boost MRC is shown in Fig. 10a. In this circuit the LSCS set is replaced by the MCS set with step-up configuration. The MCS output currents i_A, i_B, and i_C are formed according to (2) and give the possibility of load voltage frequency change. A description of control strategy, in general form, and exemplary time waveforms of the control signals, illustrating operation of the discussed circuit are shown in Figs. 10b and 10c respectively.

B. MRFC Based on Buck-Boost MRC Topology

Two topologies of the MRFC based on buck-boost MRC (new buck-boost II) are shown in Figs. 11a and 12a. In the buck-boost I topology (Fig. 11a) the SSCS set is replaced by the MCS set with step-down configuration. The MCS output voltages u_a, u_b, and u_c are formed according to (1). In the buck-boost II topology (Fig. 12a) the LSCS set is replaced by the MCS set with step-up configuration. The MCS output currents i_A, i_B, and i_C are formed according to (2). Both topologies also give the possibility of the load voltage frequency change. Furthermore in both circuits the input low-pass filter L_F, C_F is used in order to reduce the source current deformation. A description of control strategy, in general form, and exemplary time waveforms of the control signals, illustrating operation of the discussed circuits are shown in Figs. 11b, 12c and 12b, 12c respectively.

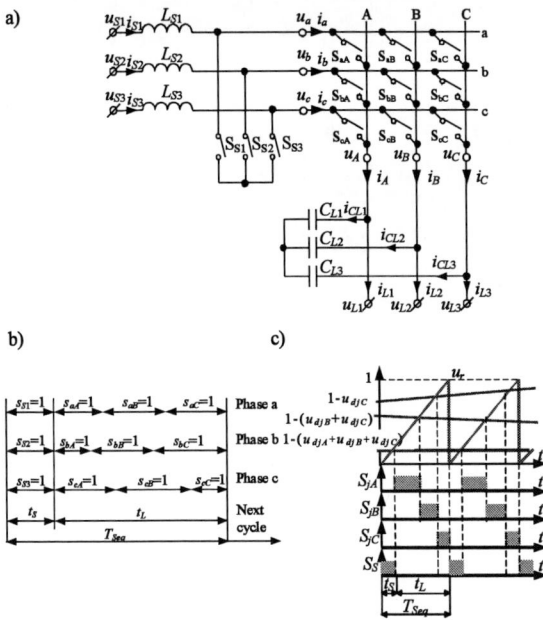

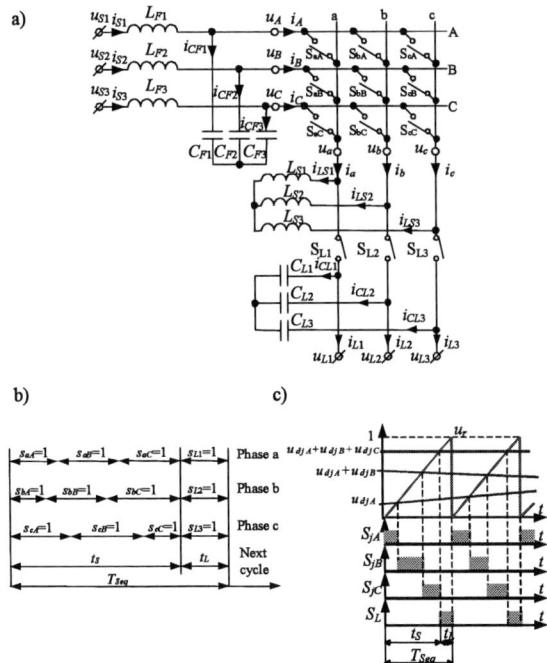

Fig. 10. MRFC based on boost MRC, a) schematic diagram, b) general form of control strategy description, c) exemplary time waveforms of the control signals

Fig. 11. MRFC buck-boost I topology based on buck-boost MRC, a) schematic diagram, b) general form of control strategy description, c) exemplary time waveforms of the control signals

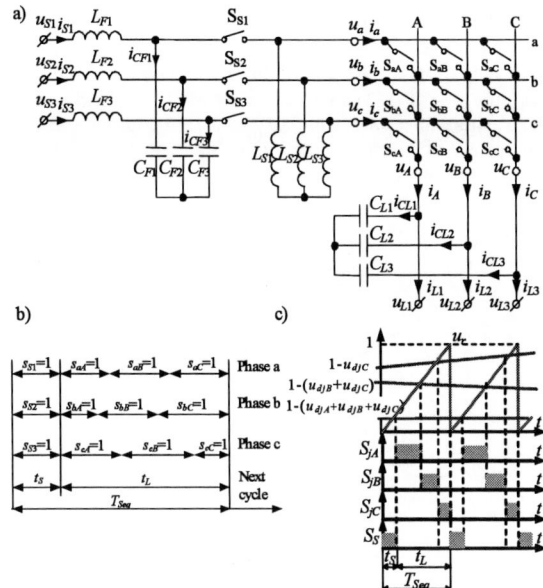

Fig. 12. MRFC buck-boost II topology based on buck-boost MRC, a) schematic diagram, b) general form of control strategy description, c) exemplary time waveforms of the control signals

C.   MRFC Based on Ćuk MRC Topology

Two topologies of the MRFC based on □uk MRC (new □uk II) are shown in Figs. 13a and 14a.

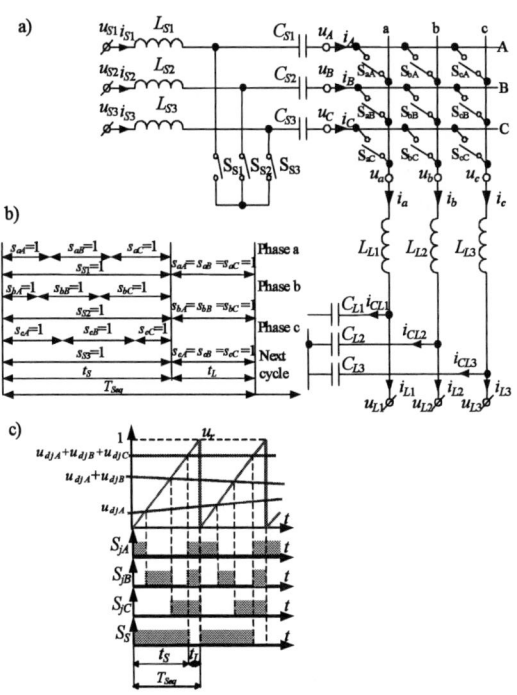

Fig. 13. MRFC □uk I topology based on □uk MRC, a) schematic diagram, b) general form of control strategy description, c) exemplary time waveforms of the control signals

In the □uk I topology (Fig. 13a) the LSCS set is replaced by the MCS set with step-down configuration. The MCS output voltages u_a, u_b and u_c are formed according to (1). In the □uk II topology (Fig. 14a) the

SSCS set is replaced by the MCS set with step-up configuration. The MCS output currents i_A, i_B and i_C are formed according to (2). It should be noted that in both circuits in time period t_S switches of the MCS and SCS are switched simultaneously. Similarly to previous topologies a description of control strategy, in general form, and exemplary time waveforms of the control signals, illustrating operation of the discussed circuits are shown in Figs. 13b, 13c and 14b, 14c respectively.

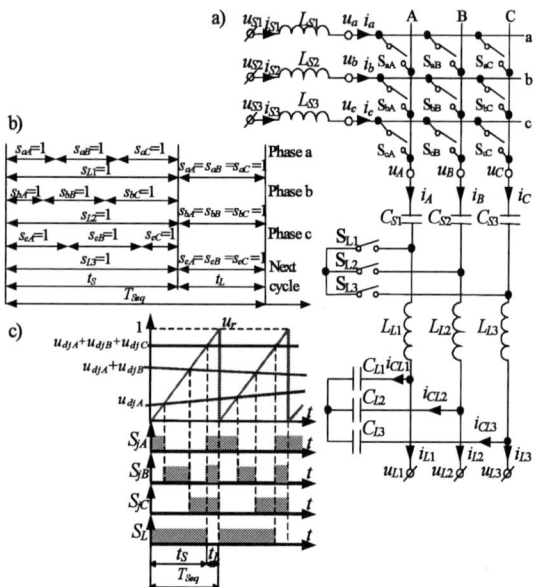

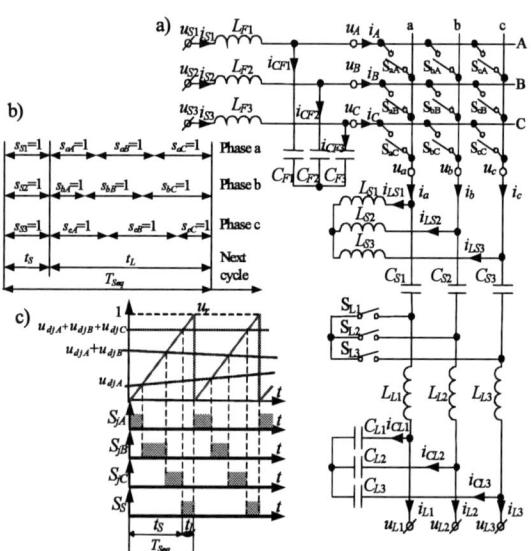

Fig. 14. MRFC Ćuk II topology based on Ćuk MRC, a) schematic diagram, b) general form of control strategy description, c) exemplary time waveforms of the control signals

D. MRFC Based on Zeta MRC Topology

Two topologies of the MRFC based on Zeta MRC (new Zeta II) are shown in Figs. 15a and 16a.

Fig. 15. MRFC Zeta I topology based on Zeta MRC, a) schematic diagram, b) general form of control strategy description, c) exemplary time waveforms of the control signals

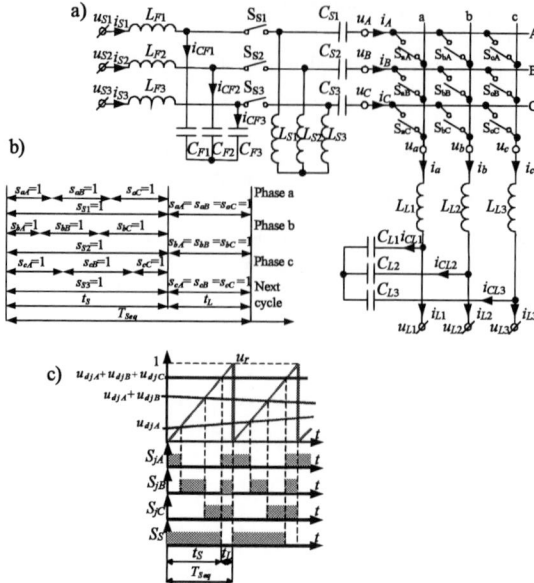

Fig. 16. MRFC Zeta II topology based on Zeta MRC, a) schematic diagram, b) general form of control strategy description, c) exemplary time waveforms of the control signals

In both Zeta I and Zeta II topologies (Figs. 15a and 16a) the SSCS (Zeta I) or LSCS (Zeta II) set is replaced by the MCS set with step-down configuration. Furthermore in both circuits the input low-pass filter L_F, C_F is used. The MCS output voltages u_a, u_b and u_c are formed according to (1). It should be noted that in Zeta II circuit, similarly as in Ćuk topology, in time period t_S switches of the MCS and SCS are switched simultaneously. A description of control strategy, in general form, and exemplary time waveforms of the control signals are shown in Figs. 15b, 15c and 16b, 16c respectively.

E. MRFC Based on SEPIC MRC Topology

Two new topologies of the MRFC based on SEPIC MRC are shown in Figs. 17a and 18a.

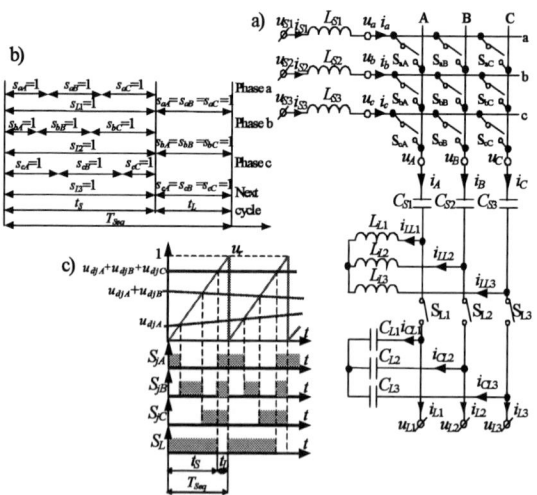

Fig. 17. MRFC SEPIC I topology based on SEPIC MRC, a) schematic diagram, b) general form of control strategy description, c) exemplary time waveforms of the control signals

240

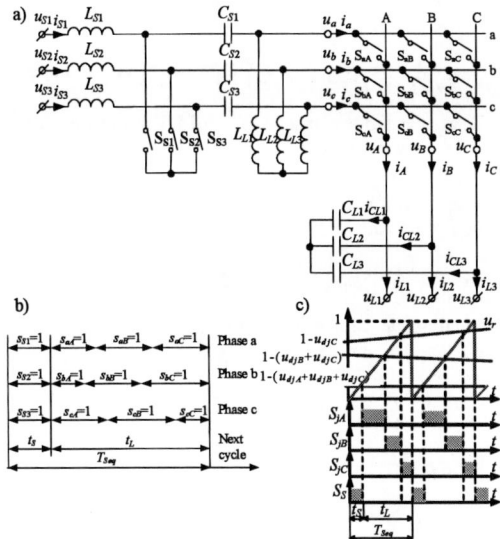

Fig. 18. MRFC SEPIC II topology based on SEPIC MRC, a) schematic diagram, b) general form of control strategy description, c) exemplary time waveforms of the control signals

In both SEPIC I and SEPIC II topologies (Figs. 17a and 18a) the SSCS (Zeta I) or LSCS (Zeta II) set is replaced by the MCS set with step-up configuration. The MCS output currents i_A, i_B and i_C are formed according to (2). In SEPIC I circuit in time period t_S switches of the MCS and SCS are also switched simultaneously. A description of control strategy, in general form, and exemplary time waveforms of the control signals, which illustrates operation of the discussed circuits are shown in Figs. 17b, 17c and 18b, 18c respectively.

F. Basic Elements Specification

In the Table I a specification of the basic elements of the proposed MRFC is presented. In general, in all topologies the number of transistors and diodes is the same. Furthermore, according to expectations the most cost-effective seems to be topologies without input filters (boost family topologies).

TABLE I.
BASIC ELEMENTS OF PROPOSED MRFC

Topology	Inductances	Capacitors	Transistors RB-IGBT/IGBT	Diodes
Boost	3	3	18+3	3
Buck-Boost I	6	6	18+3	3
Buck-Boost II	6	6	18+3	3
Ćuk I	6	6	18+3	3
Ćuk II	6	6	18+3	3
Zeta I	9	9	18+3	3
Zeta II	9	9	18+3	3
SEPIC I	6	6	18+3	3
SEPIC II	6	6	18+3	3

IV. MATHEMATICAL MODELS

A. Averaged State Space Equations

Assuming that all switches are ideal, inductors and capacitors are linear and a so called running averaging operator is used [16], [17], then the mathematical models of the proposed MRFCs, are described by the matrix differential equation (3).

$$\frac{d\overline{\mathbf{x}}}{dt} = \mathbf{A}(t)\overline{\mathbf{x}} + \mathbf{B}(t), \qquad (3)$$

where: $\overline{\mathbf{x}}$ - vector of the averaged state variables, $\mathbf{A}(t)$, $\mathbf{B}(t)$ - matrix and vector of MRFC parameters.

B. Modeling and Analysis of the Select Topology

The analysed MRFC topology is based on boost matrix-reactance chopper (Fig. 10). The MCS output currents i_A, i_B, and i_C are formed according to (2). Then, the low frequency transfer matrix of MCS according to classical Venturini control strategy is expressed by (4).

$$\mathbf{M} = \begin{bmatrix} d_{aA} & d_{bA} & d_{cA} \\ d_{aB} & d_{bB} & d_{cB} \\ d_{aC} & d_{bC} & d_{cC} \end{bmatrix}, \qquad (4)$$

where: $d_{aA} = d_{bB} = d_{cC} = (1 + 2q\cos(\omega_m t))/3$,
$d_{aB} = d_{cA} = d_{bC} = (1 + 2q\cos(\omega_m t - 4\pi/3))/3$,
$d_{aC} = d_{bA} = d_{cB} = (1 + 2q\cos(\omega_m t - 2\pi/3))/3$,

d_{jK} – low frequency component of the MCS switch state function, $D_S = t_S / T_{Seq}$ - sequence pulse duty factor, $\omega_m = \omega_L - \omega$, ω, ω_L - pulsation of the supply and load voltages respectively, q – voltage gain.

The mathematical model of the analysed MRFC, described by the matrix differential equation (3) is defined as follows:

$$\overline{\mathbf{x}} = [\overline{i}_{S1} \quad \overline{i}_{S2} \quad \overline{i}_{S3} \quad \overline{u}_{L1} \quad \overline{u}_{L2} \quad \overline{u}_{L3}]^T,$$

$$\mathbf{A}(t) = \begin{bmatrix} -\dfrac{R_{LS1}}{L_{S1}} & 0 & 0 & \dfrac{d_{aA}}{L_{S1}} & \dfrac{d_{aB}}{L_{S1}} & \dfrac{d_{aC}}{L_{S1}} \\ 0 & -\dfrac{R_{LS2}}{L_{S2}} & 0 & \dfrac{d_{bA}}{L_{S2}} & \dfrac{d_{bB}}{L_{S2}} & \dfrac{d_{bC}}{L_{S2}} \\ 0 & 0 & -\dfrac{R_{LS3}}{L_{S3}} & \dfrac{d_{cA}}{L_{S3}} & \dfrac{d_{cB}}{L_{S3}} & \dfrac{d_{cC}}{L_{S3}} \\ -\dfrac{d_{aA}}{C_{L1}} & -\dfrac{d_{bA}}{C_{L1}} & -\dfrac{d_{cA}}{C_{L1}} & \dfrac{-1}{R_{L1}C_{L1}} & 0 & 0 \\ -\dfrac{d_{aB}}{C_{L2}} & -\dfrac{d_{bB}}{C_{L2}} & -\dfrac{d_{cB}}{C_{L2}} & 0 & \dfrac{-1}{R_{L2}C_{L2}} & 0 \\ -\dfrac{d_{aC}}{C_{L3}} & -\dfrac{d_{bC}}{C_{L3}} & -\dfrac{d_{cC}}{C_{L3}} & 0 & 0 & \dfrac{-1}{R_{L3}C_{L3}} \end{bmatrix},$$

$$\mathbf{B}(t) = \begin{bmatrix} (U_m/L_{S1})\cos(\omega t) \\ (U_m/L_{S2})\cos(\omega t + 2\pi/3) \\ (U_m/L_{S3})\cos(\omega t + 4\pi/3) \\ 0 \\ 0 \\ 0 \end{bmatrix}.$$

In order to obtain stationary averaged state space model constant-power d-q transformation [18], [19] in two frequencies form (5) – (7) is used.

$$\mathbf{K} = \begin{bmatrix} \mathbf{K}_S^{-1} & 0 \\ 0 & \mathbf{K}_L^{-1} \end{bmatrix}, \qquad (5)$$

$$\mathbf{K}_S = \sqrt{\frac{2}{3}} \begin{bmatrix} \cos(\omega t) & \cos(\omega t - 2\pi/3) & \cos(\omega t + 2\pi/3) \\ \sin(\omega t) & \sin(\omega t - 2\pi/3) & \sin(\omega t + 2\pi/3) \\ 1/\sqrt{2} & 1/\sqrt{2} & 1/\sqrt{2} \end{bmatrix}, (6)$$

$$\mathbf{K}_L = \sqrt{\frac{2}{3}} \begin{bmatrix} \cos(\omega_L t) & \cos(\omega_L t - 2\pi/3) & \cos(\omega_L t + 2\pi/3) \\ \sin(\omega_L t) & \sin(\omega_L t - 2\pi/3) & \sin(\omega_L t + 2\pi/3) \\ 1/\sqrt{2} & 1/\sqrt{2} & 1/\sqrt{2} \end{bmatrix}, (7)$$

where: $\mathbf{K}_S$, $\mathbf{K}_L$ - are the d-q transformation matrices defined for pulsation of the supply and load voltages, ω and ω_L respectively.

Assuming that $L_{S1}=L_{S2}=L_{S3}=L_S$, $R_{LS1}=R_{LS2}=R_{LS3}=R_{LS}$, $C_{L1}=C_{L2}=C_{L3}=C_L$, $R_{L1}=R_{L2}=R_{L3}=R_L$ and taking into consideration substitution (8) to (3) we obtain equation (8) with new state variables.

$$\bar{\mathbf{x}} = \mathbf{KY}, \qquad (8)$$

$$\frac{d\mathbf{K}}{dt}\mathbf{Y} + \mathbf{K}\frac{d\mathbf{Y}}{dt} = \mathbf{A}(t)\mathbf{KY} + \mathbf{B}(t). \qquad (9)$$

Multiply equation (9) by inverse matrix $\mathbf{K}^{-1}$ and take into account (10) we obtain mathematical model of the discussed MRFC in the d-q frame expressed by (11).

$$\Omega = \mathbf{K}^{-1}\frac{d\mathbf{K}}{dt} = \begin{bmatrix} \Omega_S & 0 \\ 0 & \Omega_L \end{bmatrix}. \qquad (10)$$

$$\frac{d\mathbf{Y}}{dt} = \left(\mathbf{K}^{-1}\mathbf{A}(t)\mathbf{K} - \Omega\right)\mathbf{Y} + \mathbf{K}^{-1}\mathbf{B}(t), \qquad (11)$$

where:

$$\Omega_S = \begin{bmatrix} 0 & \omega & 0 \\ -\omega & 0 & 0 \\ 0 & 0 & 0 \end{bmatrix}, \quad \Omega_L = \begin{bmatrix} 0 & \omega_L & 0 \\ -\omega_L & 0 & 0 \\ 0 & 0 & 0 \end{bmatrix}. (12)$$

Defining new matrix and vector of the discussed MRFC parameters as in (13) finally we obtain stationary averaged state space model expressed by (14).

$$\mathbf{K}^{-1}\mathbf{A}(t)\mathbf{K} = \mathbf{A} \quad \text{and} \quad \mathbf{K}^{-1}\mathbf{B}(t) = \mathbf{B}. \qquad (13)$$

$$\frac{d\mathbf{Y}}{dt} = (\mathbf{A} - \Omega)\mathbf{Y} + \mathbf{B}, \qquad (14)$$

where:

$$\mathbf{A} = \begin{bmatrix} \mathbf{A}_{11} & \mathbf{A}_{12} \\ \mathbf{A}_{21} & \mathbf{A}_{22} \end{bmatrix}, \quad \mathbf{B} = \begin{bmatrix} \sqrt{3/2}(U_m/L_S) & 0 & 0 & 0 & 0 & 0 \end{bmatrix}^T,$$

$$\mathbf{A}_{11} = -\frac{R_{LS}}{L_S} \begin{bmatrix} 1 & 0 & 0 \\ 0 & 1 & 0 \\ 0 & 0 & 1 \end{bmatrix}, \quad \mathbf{A}_{22} = -\frac{1}{R_L C_L} \begin{bmatrix} 1 & 0 & 0 \\ 0 & 1 & 0 \\ 0 & 0 & 1 \end{bmatrix},$$

$$\mathbf{A}_{12} = -\frac{1-D_S}{L_S} \begin{bmatrix} q\cos(\varphi) & q\sin(\varphi) & 0 \\ q\sin(\varphi) & q\cos(\varphi) & 0 \\ 0 & 0 & 1 \end{bmatrix},$$

$$\mathbf{A}_{21} = \frac{1-D_S}{C_L} \begin{bmatrix} q\cos(\varphi) & q\sin(\varphi) & 0 \\ q\sin(\varphi) & q\cos(\varphi) & 0 \\ 0 & 0 & 1 \end{bmatrix}.$$

The solution of the equation (14) as the value of vector $\mathbf{Y}$ is described by (15).

$$\mathbf{Y} = e^{(\mathbf{A}-\Omega)t}\mathbf{Y}_0 + (\mathbf{A}-\Omega)^{-1}\left(e^{(\mathbf{A}-\Omega)t} - \mathbf{I}\right)\mathbf{B}, \qquad (15)$$

where: $\mathbf{I}$ is the unit matrix, $\mathbf{Y}_0$ – initial values of the transformed state variables.

After rearranging (15), according to (8), we obtain a final description of the state variables in discussed MRFC in the *abc* frame, which is expressed by (16).

$$\bar{\mathbf{x}} = \mathbf{K}e^{(\mathbf{A}-\Omega)t}\mathbf{Y}_0 + \mathbf{K}(\mathbf{A}-\Omega)^{-1}\left(e^{(\mathbf{A}-\Omega)t} - \mathbf{I}\right)\mathbf{B} \qquad (16)$$

The steady-state values of the averaged state variables obtained from (16) are described by (17) [18].

$$\bar{\mathbf{x}} = -\mathbf{K}(\mathbf{A}-\Omega)^{-1}\mathbf{B}. \qquad (17)$$

V. Selected Simulation Test Results

A. Circuit Parameters

Simulation studies have been carried out with the help of the program PSpice and relevant circuit parameters, which are presented in Table II.

TABLE II.
CALCULATION AND SIMULATION TESTS CIRCUIT PARAMETERS

Parameter	Symbol	Value
Supply voltage	U_S	230 V
Supply frequency	f	50 Hz
Switching frequency	f_S	5 kHz
Inductances	L_{F1} - L_{F3}, L_{S1} - L_{S3}, L_{L1} - L_{L3}	0.5 mH
Resistance of inductances	R_{LS1}, R_{LS2}, R_{LS3}	0.01Ω
Capacitances	C_{F1} - C_{F3}, C_{S1} - C_{S3}, C_{L1} - C_{L3}	50 μF
Load resistance	R_L	10 Ω

B. Voltage Relationships and Input Power Factor

In Fig. 19 there are shown exemplary time waveforms of the load voltages in MRFC based on buck-boost MRC (Fig. 11) for different settings of these voltages for $q = 0.5$ and for different values of the sequence pulse duty factor D_S. Following in Figs. 20-21 are shown a voltage gain and input power factor simulation characteristics as a function of sequence pulse duty factor D_S at f_L=25 Hz and q =0.5. for all proposed MRFCs. As is visible from Fig. 20, for all proposed MRFC's the voltage gain greater than one can be obtained. We can see from Fig. 21 that modification of the control strategy is needed for input power factor improvement.

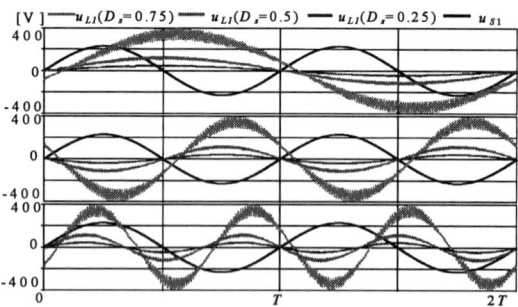

Fig. 19. Phase load voltage time waveforms for MRFC based on buck-boost MRC (for f_L = 25, 50, 75 Hz from top to bottom, at $q = 0.5$) for different D_S

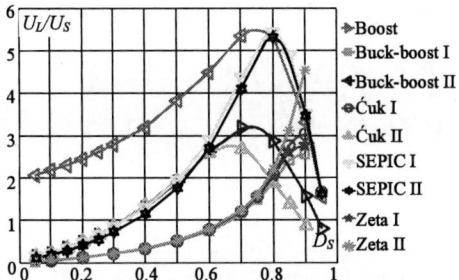

Fig. 20. Voltage gain characteristic for proposed MRFCs as a function D_S for $f_L = 25$ Hz at $q = 0.5$

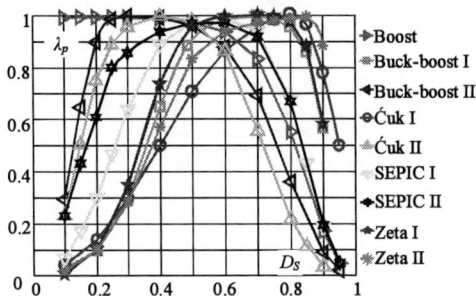

Fig. 21. Input power factor characteristic for proposed MRFCs as a function D_S for $f_L = 25$ Hz at $q = 0.5$

The characteristics of voltage gain and input power factor as functions of load voltage setting frequency and pulse duty factor D_S, obtained by means of (17) for MRFC based on boost MRC, are shown in Fig. 22. For the purpose of comparison these characteristics are presented together with ones obtained by means of simulation investigations of the presented circuit with idealized switches. As is visible from Fig. 22 the calculation and simulation test results demonstrate good coincidence what confirm usefulness of used analytical method.

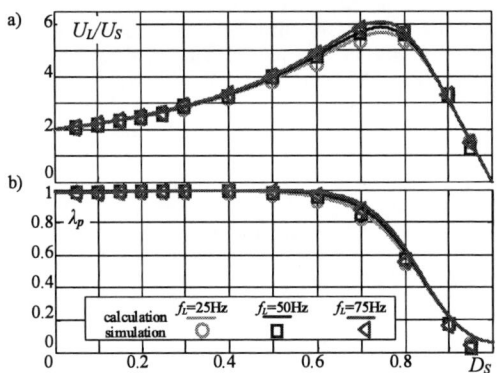

Fig. 22. The 2D simulation and calculation characteristics of the voltage gain a) and input power factor b) for MRFC based on boost MRC as a function of the pulse duty factor D_S for $q = 0.5$

VI. CONCLUSIONS

To obtain AC/AC frequency conversion possibility with buck-boost voltage transformation the concept of the new family of the MRFCs based on unipolar MRCs has been presented. The proposed MRFC topologies are generated by respect introduction of the MCS instead of one of the SCS used in MRC. The hitherto prevailing results of the

theoretical analysis and simulation tests confirm the presented conception. Further research will be focused on a detailed theoretical analysis of the proposed MRFCs. Furthermore, the formulation of an improved control strategy and experimental verification of results obtained are also planned.

REFERENCES

[1] P. W. Wheeler, J. Rodriguez, J. C. Clare, L. Empringham, and A. Weinstejn, "Matrix converters: A technology review," *IEEE Trans. on Ind. Electron.*, vol. 49, No. 2, pp. 276–288, April 2002.

[2] M. Venturini and A. Alesina, "The generalized transformer: a new bi-directional sinusoidal waveform frequency converter with continuously adjustable input power factor," *Conf. Record, PESC'80*, pp. 242-252.

[3] L. Huber and D. Borojevic, "Space vector modulated three-phase to three phase matrix converter with input power factor correction," *IEEE Trans. on Ind Appl.*, vol. 31, No. 6, pp. 1234–1246, Nov./Dec. 1995.

[4] D. Casadei, G. Serra, A. Tanti, and L. Zaroi, "Matrix converter modulation strategies: a New general approach based on space-vector representation of switch state," *IEEE Trans. on Ind. Electronics*, vol. 49, No. 2, pp. 370–381, April 2002.

[5] M. Apap, J. C. Clare, P. W. Wheeler, and K. J. Bradley, "Analysis and comparison of AC-AC Matrix converter control strategies," *Proc. of PESC' 2003*, pp. 1287–1292, Cairns 2003.

[6] L. Helle, K. B. Larsen, A. H. Jorgensen, S. Munk-Nielsen, and F. Blaabjerg, "Evaluation of modulation schemes for three-phase to three-phase matrix converters," *IEEE Trans. on Ind. Electronics*, vol. 51, No. 1, pp. 158–170, Febr. 2004.

[7] J. W. Kolar, F. Schafmeister, S. D. Round, and H. Ertl, "Novel three-phase AC-AC sparse matrix converter," *IEEE Trans. on Power Electronics*, vol. 22, No. 5, pp. 1649–1661, Sept. 2007.

[8] Y-D. Yoon and S-K. Sul. "Carier-based modulation technique for matrix converter," *IEEE Trans. on Power Electronics*, vol. 21, No. 6, pp. 1691–1703, Nov. 2006.

[9] G. S. Zinoviev, A. Y. Obuchov, W. A. Otchenasch, and W. I. Popov, "Transformerless PWM AC boost and buck-boost converters," (In Russian), *Technicznaja Elektrodinamika*, T2, pp. 36–39. Nacjonalnaja Akademia Nauk Ukrainy, Kijev 2000.

[10] Z. Fedyczak and P. Szcześniak, "Study of matrix-reactance frequency converter with buck-boost topology," *PELINCEC 2005*, Warsaw 2005, CD-ROM.

[11] Z. Fedyczak, P. Szcześniak, and M. Klytta, "Matrix-reactance frequency converter based on buck-boost topology," 12th *Conf. EPE-PEMC*, Portoroż 2006, CD-ROM, pp. 763–768.

[12] Z. Fedyczak, I. Korotyeyev, P. Szcześniak, "Generation of matrix-reactance frequency converters based on unipolar matrix-reactance choppers," *Proc. of PESC'08*, Rhodes 2008, (in progress).

[13] Z. Fedyczak, PWM AC voltage transforming circuits, (In Polish), Zielona Góra University Press, Zielona Góra, 2003.

[14] Z. Fedyczak, R. Strzelecki, and K. Sozański, "Review of three-phase PWM AC transformer topologies and applications," *Conf. Proc. of SPEDAM 2002*, pp. B5–19–24, Ravello 2002.

[15] W. H. Kwon and G. H. Cho, "Analyses of static and dynamic characteristics of practical step-up nine-switch convertor," *IEE. Proc.-B*, vol. 140, No. 2, March 1993.

[16] R. D. Middlebrook and S. Ćuk, "A general unified approach to modelling switching-converter power stages," *PESC'76 Conf. Rec.* pp. 18–34, 1976.

[17] I. Y. Korotyeyev and Z. Fedyczak, "Steady-state modelling of basic unipolar PWM AC line matrix-reactance choppers," *COMPEL*, vol. 24, No. 1, 2005, pp. 55-68.

[18] I. Korotyeyev, Z. Fedyczak, and P. Szcześniak, "Steady and Transient States Analysis of Matrix-Reactance Frequency Converter Based on Boost PWM AC Matrix-Reactance Chopper," ISNCC, Łagów, Poland, 2008, (in progress).

[19] I. Y. Korotyeyev and Z. Fedyczak, "Analysis of transient and steady state processes in three-phase symmetric matrix-reactance converter system," *Technicznaja Elektrodinamika*, Nacjonalnaja Akademia Nauk Ukrainy, Kijev 2008.

Consideration of Conduction Losses for the Series Resonant Converter by Means of a Simple Extension to the SPA Approach

Alexander Bucher, Thomas Duerbaum, Daniel Kuebrich and Markus Schmid

Chair of Electromagnetic Fields, Friedrich-Alexander-University Erlangen-Nuremberg, Erlangen, Germany

e-mail: A.Bucher@emf.eei.uni-erlangen.de

Abstract—The cumbersome derivation of the steady-state characteristics of resonant converters can be simplified by means of the state-plane analysis. Under the assumption of ideal components, a closed-form solution can be derived in case of the series resonant converter above the resonant frequency. However, losses due to parasitic resistances cannot be easily included within this approach. Nevertheless, a more precise prediction of the converter's output characteristics taking the conduction losses into account is desirable. Therefore this paper describes a simple extension to the regular approach that leads to better agreement with measurements. This approach is based on the results derived under ideal assumptions, thus avoiding a more complicated and tedious analysis including the conduction losses.

Keywords—Resonant Converter, Modelling, ZVS Converter

I. INTRODUCTION

A. Motivation

Offering the realisation of soft-switching without additional efforts, resonant converters are especially suited for higher power levels and high switching frequencies. As a consequence of the latter, the inductive components which often take up a great share of volume can be reduced in size. This satisfies the requirements for a high degree of miniaturisation for many of today's applications [1]. The schematic of an ideal full-bridge series resonant converter (SRC) is shown in Fig. 1. The exact solutions for the SRC were derived in the time domain in [2] and [3]. A methodology to simplify the derivation of steady-state characteristics is the State-Plane Analysis (SPA) which was applied to the SRC in [4], [5] and [6]. For the frequency domain, solutions can be found in [7] and [8]. An accuracy comparison of the exact solution compared to the results of the well-known First Harmonic-Approximation (FHA) published in [9] can be found in [10]. As done by previous authors, the solution of the SPA is briefly derived under ideal assumptions in the following section. Based upon these results, a simple extension to the obtained equations is presented which improves the agreement of the calculated voltage conversion ratio to measured results. The results of the analysis of the ideal converter are extended in order to account for conduction losses without the need to reanalyze the converter. Measurements showing the significant improvement in the prediction of the steady-state characteristics of a prototype are discussed in the final section.

B. Assumptions

For the results presented in section 2 all active as well as all passive components are assumed as ideal. The full-bridge generates a square-wave voltage of 50 % duty cycle which is applied to the resonant tank and to the primary of the transformer. The value of C_o is assumed to be large and therefore V_o is virtually ripple-free, which also makes I_o a dc quantity. Under the assumptions of ideal rectifier diodes, the voltage across the rectifier bridge $v_B(t)$ is a square-wave voltage which switches in time with the zero-crossings of the transformed tank current $i_L(t)$. As shown by previous authors, ZVS is achieved if the switching frequency of the full-bridge is higher than the tank's resonant frequency. Only one mode of operation occurs within this frequency range which is a continuous conduction mode (see [4] for instance). In order to preserve clarity and to ensure the desirable feature of ZVS, the frequency range is limited to frequencies above the resonant frequency f_0.

C. Normalization

A key element making the state-plane analysis a useful tool for calculating resonant LC-converters is the necessary normalization. It simplifies the analysis and the results can be easily applied to any converter design. Following the nomenclature of previous authors, normalized quantities are represented by replacing their letter while keeping the same subscript. The letter V denoting voltages is replaced by the letter M, I denoting currents is replaced by J. Table 1 shows the resulting normalized terms. The normalization voltage can be chosen freely, but if V_i is selected as normalization voltage, the dc conversion ratio

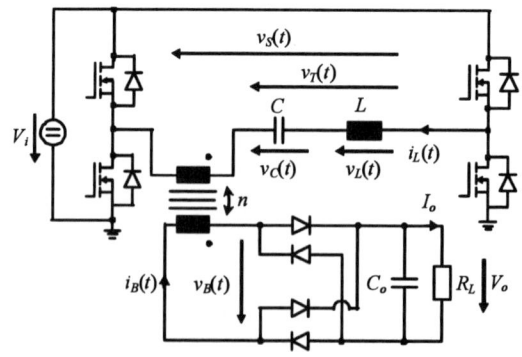

Fig. 1 Ideal full-bridge series resonant converter.

978-1-4244-1741-4/08/$25.00 ©2008 IEEE

$M_o = V_o/V_i$ is automatically included in the derived results. Another important term is the angle γ, which is defined to be

$$\gamma = \pi/F . \tag{1}$$

II. STATE-PLANE-ANALYSIS

The steady-state waveforms of the series resonant converter above the resonant frequency can be divided in four subintervals as shown in [2]-[6]. Depending on the flow direction of $i_L(t)$ and the state of the switches, $v_B(t)$ and $v_S(t)$ change sign, resulting in four possible values for the voltage $v_T(t)$ across the resonant tank, one for each subinterval. For the duration of each subinterval, the converter can be represented by a dc circuit as indicated in Fig. 2. The solutions of the occurring differential equations for the state variables in normalized form are given by

$$m_C(t) = R\cos(\omega_0 t - \varphi_0) + M_T$$
$$j_L(t) = -R\sin(\omega_0 t - \varphi_0) \quad . \tag{2}$$

It can be shown that the resulting trajectory is circulating clockwise within the m_C-j_L-plane. Thus the complete trajectory shown in Fig. 3 consists of four segments of circles representing the corresponding subinterval, see [4]-[6]. The centres $(M_T, 0)$ of the segments of circles representing the four different subintervals are determined by the states of the input square-wave voltage $v_S(t)$ and the state of the output rectifier bridge. The radius R and starting angle φ_0 of each subinterval depend on the power that is to be transferred by the converter.

Since only frequencies above the tank's resonant frequency are taken into account for this paper, the tank displays inductive behaviour. Therefore the inductor current $i_L(t)$ lags the square wave voltage $v_S(t)$. At $t = 0$, $v_S(t)$ changes sign and is then positive until $t_2 = T_S/2$. With the inductor current $i_L(t)$ lagging $v_S(t)$, the trajectory starts in the third quadrant. At $t = t_1$, $i_L(t)$ becomes zero and thus $v_B(t)$ has to change sign too, resulting in a new centre and radius for the next segment of the trajectory. The following segment has to be added at $t = t_2$, when $v_S(t)$ switches again to become negative. The state-plane trajectory has to be point symmetric in steady-state. Otherwise the transferred power would increase or decrease from one switching cycle to the next.

With these preliminary considerations, the complete state-plane trajectory of Fig. 3 can be constructed. The triangle $\triangle DEF$ is characteristic for a certain combination of M_o and J_o with corresponding values for the occurring angles and lengths. Thus the steady-state solution of the series resonant converter can be derived by analysing the geometric relationships between the quantities describing this triangle. The radii R_1 and R_2 are given by the corresponding center and the peak capacitor voltage

TABLE I.
CIRCUIT AND NORMALIZED TERMS

Circuit variable	Symbol	Normalized variable
Resonant frequency	$\omega_0 = 1/\sqrt{LC}$	
Characteristic impedance	$R_0 = \sqrt{L/C}$	
Output voltage	V_o	$M_o = V_o/V_i$
Output current	I_o	$J_o = I_o R_0/V_i$
Capacitor voltage	$v_C(t)$	$m_C(t) = v_C(t)/V_i$
Inductor current	$i_L(t)$	$j_L(t) = i_L(t) \cdot R_0/V_i$
Switching frequency	$f_s = 1/T_s$	$F = f_s/f_0$
Load resistor	R_L	$Q = R_0/R_L$

$m_C(t_1) = M_{C,p}$ with

$$R_1 = 1 + nM_o + M_{C,p} \quad \text{and} \quad R_2 = 1 - nM_o + M_{C,p}. \tag{3}$$

The length of the third side of the triangle $\triangle DEF$ is found to be

$$1 - nM_o + |-1 - nM_o| = 2. \tag{4}$$

By applying the law of cosines to $\triangle DEF$ a relationship between the radii of the triangle and the angle δ is derived with

$$4 = R_1^2 + R_2^2 - 2R_1 R_2 \cos\delta, \tag{5}$$

where δ is given by

$$\delta = \pi - \alpha - \beta = \pi - \gamma. \tag{6}$$

In order to derive the converter's output characteristic a

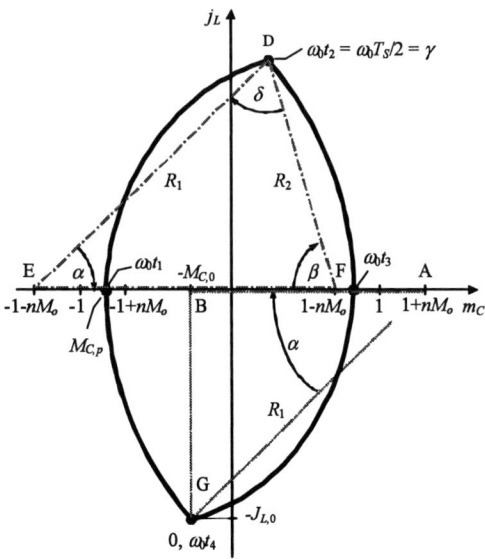

Fig. 3 State-plane trajectory of the ideal series resonant converter for $F \geq 1$.

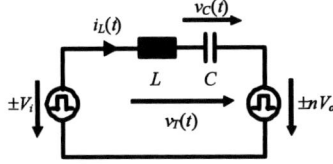

Fig. 2 Circuit representing one subinterval.

245

further expression is necessary in order to relate $M_{C,p}$ to the dc output current. The charge transferred by the converter during one half of the switching cycle is given by

$$I_o = n \cdot \overline{|i_L(t)|} = \frac{n}{T_S/2} \int_{t_1}^{t_3 = t_1 + \frac{T_S}{2}} i_L(t)\mathrm{d}t = \frac{2n}{T_S}\Delta Q \quad (7)$$

During t_1 and $t_1 + T_S/2$ the tank capacitor is charged by the inductor current which leads to

$$i_L(t) = C\frac{\mathrm{d}u_C(t)}{\mathrm{d}t} \quad\Rightarrow\quad \Delta Q = C \cdot 2V_{C,p} . \quad (8)$$

With (8) substituted in (7) one obtains for the normalized terms

$$M_{C,p} = \frac{J_o \gamma}{2n} . \quad (9)$$

With (3) to (6) and (9) the output characteristic describing the relationship between the dc output voltage M_o and the corresponding dc output current J_o can be derived as

$$M_o^2 n^2 \sin^2\left(\frac{\gamma}{2}\right) + \left(\frac{J_o\gamma}{2n}+1\right)^2 \cos^2\left(\frac{\gamma}{2}\right) = 1 . \quad (10)$$

For a given load characteristic as a function of M_o vs J_o, the resulting voltage conversion ratio can then be calculated from Eq. (10). In case of ohmic loads (with a normalized load characteristic of $J_o = M_o \cdot Q$) the resulting value for M_o can be calculated analytically for a given switching frequency f_S (included in γ) and for a given load R_L (included in Q).

$$M_o = \frac{-\dfrac{Q\gamma}{n} + \sqrt{\left(\dfrac{Q\gamma}{n}\right)^2 + 4z\tan^2\left(\dfrac{\gamma}{2}\right)}}{2z} \quad (11)$$

$$\text{with } z = n^2\tan^2\left(\frac{\gamma}{2}\right) + \left(\frac{Q\gamma}{2n}\right)^2 .$$

With the calculated value for M_o the waveform of the tank current $j_L(t)$ for the first half of the switching cycle is given by Eq. (12). The radii of the corresponding segments of circles can be easily extracted by analyzing the triangle $\triangle ABG$ and the trajectory at $t = t_1$ respectively. The positive starting values for the capacitor voltage $M_{C,0}$ (with $m_C(t_0) = -M_{C,0}$) and for the inductor current $J_{L,0}$ (with $j_L(t_0) = -J_{L,0}$) are given by

$$J_{L,0} = (1 - nM_o + M_{C,p})\sin(\gamma - \alpha)$$
$$M_{C,0} = M_o J_o \gamma / 2 . \quad (13)$$

III. APPROXIMATE EXTENSIONS TO ACCOUNT FOR CONDUCTION LOSSES

The exact solution obtained by means of the SPA provides good insight into the steady-state characteristics of the SRC and it can therefore be considered as a very valuable tool for its analysis. Nevertheless, noticeable deviations between the predicted values of the SPA and measurements on a real prototype can occur because of the assumptions of ideal components. However, problems arise if loss mechanisms shall be included within this approach, because the resulting trajectories become more complicated. With resistances taken into account, the segments of the trajectory are not longer circular. Thus the steady-state state-plane trajectory cannot be as easily constructed as in case of ideal assumptions. In order to avoid this inconvenience, the results of the ideal SPA provide a basis for an extension which obtains a better agreement of the theoretical results with measurements. These approximate extensions presented in this section account for two of the main parasitics of the SRC, which can be related to the input and to the output of the converter. Thus their impact on the steady-state characteristics of the converter can be examined separately.

A. Output

The parasitic elements which are considered by the proposed approach are shown in Fig. 4 for the converter's output. The rectifier diodes are modelled by their unavoidable forward voltage drop together with an incremental resistance. The load voltage V_L is considered to be ripple-free, thus I_o is also a dc quantity. For every switching half cycle two diodes are conducting. Therefore the output voltage of the idealized rectifier bridge V_o is $2 \cdot V_{th}$ higher than the load voltage with an additional voltage drop across the incremental resistances $2 \cdot R_{dio}$ of the conducting diodes. If the resulting voltage ripple across these resistances is neglected, the output voltages can be averaged with

$$-M_o + 2M_{th} + \frac{J_o}{Q_{dio}} + M_L = 0 \quad \text{with} \quad Q_{dio} = \frac{R_0}{2R_{dio}} . \quad (14)$$

The normalized output voltage across the load M_L is given by $M_L = J_o/Q$ and thus (14) can be rewritten as

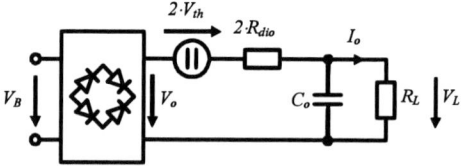

Fig. 4 Parasitics taken into account for the output of the converter.

$$j_L(t) = \begin{cases} \sqrt{(1 + M_{C,0} + nM_o)^2 + J_{L,0}^2}\, \sin(\omega_0 t - \varphi_1) & \text{for} \quad 0 \le t < t_1 \quad \text{with} \quad \varphi_1 = \pi + \arctan\left(\dfrac{J_{L,0}}{1 + M_{C,0} + nM_o}\right) \\[2em] (1 - nM_o + M_{C,p})\sin(\omega_0 t - \alpha) & \text{for} \quad t_1 \le t < t_2 \quad \text{with} \quad \alpha = \arccos\left(\dfrac{1 + M_o + M_o M_{C,p}}{1 + M_o + M_{C,p}}\right) \end{cases} \quad (12)$$

$$J_o = Q'(M_o - M'_{th}) \quad \text{with} \quad M'_{th} = 2M_{th}$$

$$\text{and} \quad \frac{1}{Q'} = \frac{2R_{dio}}{R_0} + \frac{1}{Q}. \tag{15}$$

With Eq. (15) substituted in Eq. (10), an expression for the voltage conversion ratio M_o can be obtained which approximatively accounts for the nonlinearity of the rectifier diodes. The result is given in (16). Thus the parasitics of the rectifier diode bridge are taken into account by separating their nonlinear characteristic, resulting in a nonlinear load characteristic with an idealized diode bridge. This solution for the corrected steady-state output voltage M'_o of the rectifier bridge can still be solved analytically without the need for numerical methods if only parasitics for the output rectifier diodes are considered. In this context the correction is denoted by an additional apostrophe with respect to V_o in Eq. (16). In the last step, the corrected output voltage of interest across the load can then be calculated from

$$M_L = \frac{1}{Q} \cdot \frac{M'_o - M'_{th}}{\frac{1}{Q_{dio}} + \frac{1}{Q}} = \frac{M'_o - M'_{th}}{\frac{Q}{Q_{dio}} + 1}. \tag{17}$$

B. Input

In order to account for conductions losses at the input of the converter, a relationship between the actual value of the amplitude V_i of the square-wave voltage and the corresponding dc value V_g of the source has to be found. In reality, the tank is driven with a square-wave voltage with an amplitude different from the voltage V_g due to conduction losses in the power switches and the resistance R of the resonant tank. This situation is depicted in Fig. 5. The resulting voltage drop across these resistances increases or decreases the actual amplitude V_i of the square-wave voltage applied to the tank, depending on the flow direction of $i_L(t)$. The angles α and β represent these subintervals in the state-plane. For switching frequencies far off the resonant frequency f_0, these angles approach nearly the same value as can be seen in Fig. 3. Thus the increase and decrease of the effective amplitude V_i of the resulting square-wave voltage is averaged out. But this situation changes for switching frequencies near the resonant frequency. An exemplary state-plane trajectory is shown in Fig. 6. During the first switching half-cycle the angle β is significantly larger in value than the angle α Thus V_i is reduced according to Fig. 5 during the major part of the switching period.

Based on these considerations an approximate relationship between V_g and V_i can be derived if the tank inductor current is averaged during one switching half-cycle with

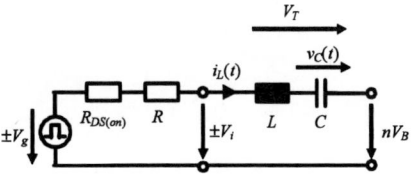

Fig. 5 Parasitics taken into account for the input of the converter.

$$V_i \approx \frac{R_0 V_g}{R_0 + (R_{DS(on)} + R)J_{L,av}}. \tag{18}$$

In case of ideal components with $R_{DS(on)} = 0$ and $R = 0$, the amplitude of the square-wave voltage V_i equals the dc input voltage V_g. For high average input currents V_i is reduced according to Eq. (18) resulting in a lower voltage conversion ratio. Since the average input current $J_{L,av}$ is depending on the actual load of the converter, the corrected value of V_i cannot be calculated directly. In order to calculate the corrected output voltage by means of (16) and (17) it is therefore necessary to derive a solution for the corrected value of V_i in a first step.

The average input current $J_{L,av}$ during one half-cycle can be derived by means of the corresponding waveform given in (12) with

$$J_{L,av} = \frac{R_1[\cos(\alpha - \varphi_1) - \cos(\varphi_1)]}{\gamma} + \frac{R_2[\cos(\gamma - \alpha) - 1]}{\gamma}. \tag{19}$$

In a next step, the equations obtained from the ideal SPA have to be denormalized with respect to V_i. From (12) one obtains for the angles α and φ_1

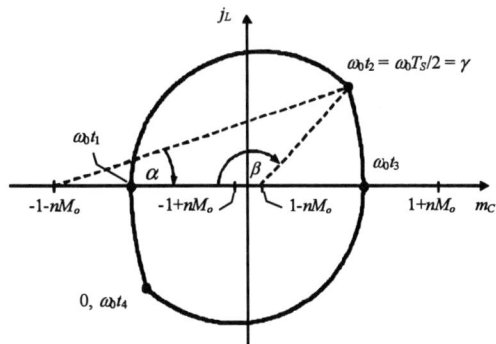

Fig. 6 Exemplary state-plane trajectory of the ideal series resonant converter for $F = 1.2$, $M_o = 0.87$ and $Q = 1$.

$$M'_o = \frac{-AQ'\gamma \cos^2\left(\frac{\gamma}{2}\right) + \sqrt{A^2 Q'^2 \gamma^2 \cos^4\left(\frac{\gamma}{2}\right) - 4\left[\sin^2\left(\frac{\gamma}{2}\right) + Q'^2 \frac{\gamma^2}{4}\cos^2\left(\frac{\gamma}{2}\right)\right] \cdot \left[A^2 \cos^2\left(\frac{\gamma}{2}\right) - 1\right]}}{2\left[\sin^2\left(\frac{\gamma}{2}\right) + Q'^2 \frac{\gamma^2}{4}\cos^2\left(\frac{\gamma}{2}\right)\right]} \tag{16}$$

$$\text{with} \quad A = 1 - M'_{th} Q' \frac{\gamma}{2}$$

$$\alpha = \arccos\left(\frac{V_i + nV_o + V_o J_o \dfrac{\gamma}{2}}{V_i + nV_o + V_i J_o \dfrac{\gamma}{2n}}\right) \qquad (20)$$

and

$$\varphi_1 = \pi + \arctan\left[\frac{\left(V_i - nV_o + V_i M_{C,p}\right)\sin(\gamma - \alpha)}{V_i + V_i M_{C,0} + nV_o}\right]. \qquad (21)$$

The nonlinear load characteristic of (15) has to be considered for the terms $V_i M_{C,p}$ and $V_i M_{C,0}$ in Eq. (21) together with (13) which leads to

$$V_i M_{C,p} = Q'\left(V_o - V_{th}'\right)\frac{\gamma}{2n} \qquad (22)$$

and

$$V_i M_{C,0} = V_i M J_o \frac{\gamma}{2} = V_o Q'\left(\frac{V_o}{V_i} - \frac{V_{th}'}{V_i}\right)\frac{\gamma}{2}. \qquad (23)$$

If Eq. (22) and Eq. (23) are substituted in Eq. (21) and Eq. (20), φ_1 as well as α become functions in terms of V_o and V_i. With Eq. (19) in mind, similar expressions for the radii R_1 and R_2 have to be found. The radius R_1 is given by

$$R_1 = \frac{-J_{L,0}}{\sin\varphi_1} =$$
$$= -\left[1 - \frac{nV_o}{V_i} + Q'\left(\frac{V_o}{V_i} - \frac{V_{th}'}{V_i}\right)\frac{\gamma}{2n}\right]\frac{\sin(\gamma - \alpha)}{\sin\varphi_1}, \qquad (24)$$

which is the desired representation together with α and φ_1. The radius of the second subinterval can be extracted directly from the x-axis of Fig. 3 and is given by

$$R_2 = n\frac{V_o}{V_i} - 1 - Q'\left(\frac{V_o}{V_i} - \frac{V_{th}'}{V_i}\right)\frac{\gamma}{2n}. \qquad (25)$$

With (19) to (25) substituted in (18), a relationship between V_i and V_o has been derived, which approximatively takes into account the conduction losses at the input. The resulting nonlinear equation is given in (26). The correction of the input resistances can then be combined with the correction of the output losses presented in the previous section. By means of Eq. (16), V_o can be eliminated in (26), resulting in a nonlinear equation for the corrected amplitude V_i of the square-wave voltage. The solution of this nonlinear equation has to be found numerically. Thus the mathematical complexity is increased if the conduction losses of the input shall be considered.

IV. MEASUREMENTS

A verification of the derived results was carried out on a prototype with a dc input voltage of $V_g = 30$ V for different values of the tank's quality factor. Exemplary measured waveforms are shown in Fig. 7 with a measured voltage conversion ratio 10% below the predicted value.

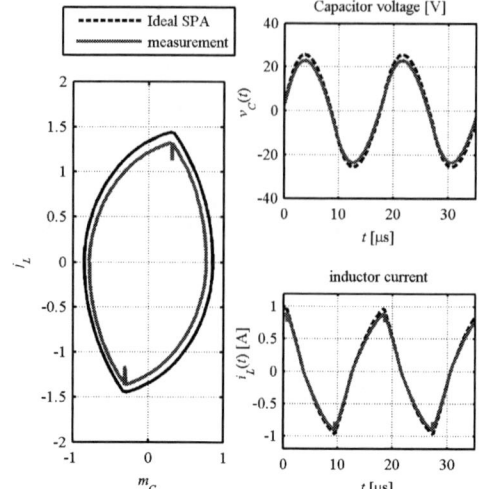

Fig. 7 Exemplary measured waveforms.

The waveforms and the corresponding peak values are predicted by means of the SPA with a high degree of accuracy, as can be seen by the right half of the figure. Nevertheless, a deviation between the SPA prediction and the measurement becomes visible with respect to the state-plane trajectory. The losses neglected by the ideal SPA result in a measured trajectory with smaller area for a given switching frequency.

For $Q = 2.2$, the measured voltage conversion ratio against the normalized switching frequency is shown in Fig. 8. As already mentioned, significant deviations from the predicted values for M_L are observed in case of the ideal SPA. Especially for frequencies near the resonant frequency with high currents at the input and the output the differences become noticeable. For the proposed correction of the output by means of a nonlinear load characteristic, a significant improvement is obtained. This improvement is further increased if the input correction is added. The relative deviation of the complete correction compared to the ideal SPA is shown in Fig. 9 and Fig. 10. This relative error E_{rel} is defined to be

$$E_{rel} = \frac{M_{calc} - M_{meas}}{M_{meas}}, \qquad (1)$$

with M_{meas} as the measured voltage conversion ratio for the load and M_{calc} as the predicted value for M_L. In comparison to the uncorrected results of the ideal SPA, a significant improvement is achieved by means of the correction. Especially for frequencies below $F = 1.2$ the relative error is drastically minimized, increasing the accuracy of the prediction of the voltage conversion ratio. The error curves for the corrected value of M_L are more evenly dis-

$$\frac{V_i}{V_g} - \frac{R_{DS(on)} + R}{\gamma R_0 V_g}\left[V_i - nV_o + Q'\left(V_o - V_{th}'\right)\frac{\gamma}{2n}\right]\cdot\sin(\gamma - \alpha)\cdot\left[\frac{\cos\alpha}{\tan\varphi_1} + \sin\alpha - \tan\varphi_1\right] +$$
$$+ \frac{R_{DS(on)} + R}{\gamma R_0 V_g}\left[nV_o - V_i - Q'\left(V_o - V_{th}'\right)\frac{\gamma}{2n}\right]\cdot\left[\cos(\gamma - \alpha) - 1\right] \approx 1 \qquad (26)$$

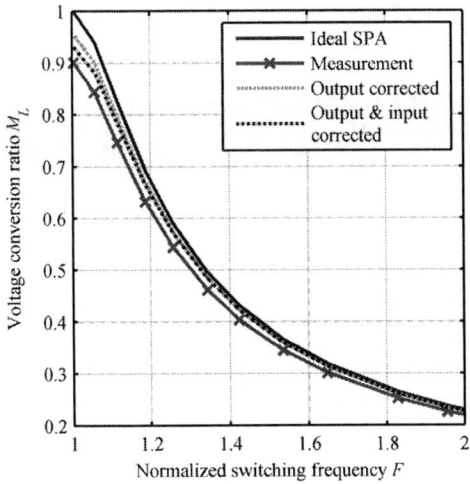

Fig. 8   Measured voltage conversion ratios for a quality factor of $Q = 2.2$.

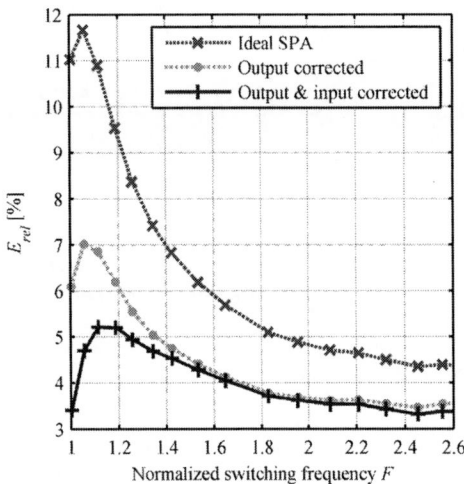

Fig. 9   Relative deviation of measured voltage conversion ratios for a quality factor of $Q = 2.2$.

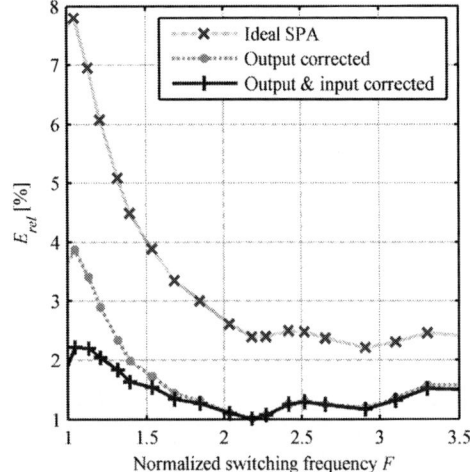

Fig. 10 Relative deviation of measured voltage conversion ratios for a quality factor of $Q = 1.45$.

tributed, without the distinct peak around the resonance observed for the ideal SPA. The remaining error values according to Fig. 9 and Fig. 10 are below several percent. For high switching frequencies both error curves approach the same value, indicating that other parasitics than the addressed resistances are responsible for these deviations. Especially the capacitances of the semiconductors have to be mentioned in this context since the charging intervals of these parasitics show an increasing influence on the occurring waveforms for short switching intervals.

V. CONCLUSION

Instead of including resistances in the analysis of the SRC and thus making the calculation of steady-state characteristics unnecessarily complicated, the proposed approach integrates loss mechanisms into the results based on ideal assumptions. Concerning the nonlinearity of the output rectifier diodes, the mathematical complexity is not increased compared to the ideal analysis. However, a slight increase in case of input resistances has to be accepted in order to benefit from the increase in accuracy the proposed approach offers.

REFERENCES

[1] J. A. Sabate, D. Kustera and S. Sridhar, "Cell-Phone Battery Charger Miniaturization", *Industry Applications Conference Record*, pp. 3036 – 3043, 2000.

[2] V. Vorpérian, "Analysis of Resonant Converters", Ph. D. Thesis, California Institute of Technology, Pasadena, May 1984.

[3] A. F. Witulski and R. W. Erickson, "Steady-State Analysis of the Series Resonant Converter", IEEE Transactions on Aerospace and Electronic Systems, vol. 21, pp. 791 – 799, November 1985.

[4] R. Oruganti, "State-Plane Analysis of Resonant Converters", Ph. D. Thesis, Virginia Polytechnic Institute and State University, Blacksburg, March 1987.

[5] S. G. Trabert and R. W. Erickson, "Steady State Analysis of the Duty Cycle Controlled Series Resonant Converter", Power Electronics Specialists Conference, pp. 545-556, 1987.

[6] I. E. Batarseh, "Analysis and Design of High-Order Parallel Resonant Converters", Ph. D. Thesis, University of Illinois, Chicago, April 1990.

[7] J.-C. Li and Y.-P. Wu, "Closed-Form Expressions for the Frequency-Domain Model of the Series Resonant Converter", IEEE Transactions on Power Electronics, vol. 5, pp. 337 – 335, July 1990.

[8] A. K. S. Bhat, "A Generalized Steady-State-Analysis of Resonant Converters Using Two-Port Model and Fourier-Series Approach", IEEE Transactions on Power Electronics, vol. 13, January 1998.

[9] R. L. Steigerwald, "A Comparison of Half-Bridge Resonant Converter Topologies", IEEE Transactions on Power Electronics, vol. 3, pp. 174 – 182, April 1988.

[10] A. Bucher, T. Duerbaum and D. Kuebrich, "Comparison of First Harmonic Approximation with exact Solution in case of a Series Resonant Converter", Proceedings of Power Conversion Intelligent Motion Conference, pp. 751 – 756, 2006.

Validation and Comparison of different PWM Converter Small Signal Models

Alexander Bucher[*], Markus Schmid[*], Lukas Bendkowski[†] and Thomas Duerbaum[*]

[*] Friedrich-Alexander University of Erlangen-Nuremberg, Erlangen, Germany,
e-mail: *a.bucher@emf.eei.uni-erlangen.de*
[†] Semikron Elektronik GmbH & Co. KG, Nuremberg, Germany

Abstract—**In order to determine which small signal model most accurately describes PWM converters, a comparison with a switched model is necessary. Due to parasitic effects the comparison with hardware measurements does not clarify this question. Therefore published small signal models based on ideal components are compared with a time domain simulation of a switched converter using the same idealization in this paper. The transfer function is determined by emulating the quadrature detector used in practical measurement set-ups. The paper shows the theoretical background, the implementation in SPICE and the comparison with traditionally found transfer functions.**

Keywords— **Modeling, Simulation, Converter control.**

I. INTRODUCTION

The selection of an appropriate controller of a switch-mode power supply (SMPS) in consideration of stability and dynamic requirements requires good knowledge about the characteristics of the system to be controlled. In order to describe the dynamic behavior of the system, several analysis and modeling methods such as state-space averaging [1,2,3] or circuit averaging methods [4,5,6,7] are available. However, the small-signal models obtained by these different approaches show slightly different behavior. For the purpose of assessing the accuracy of the transfer functions derived from these models a comparison with a switched circuit is necessary. Nevertheless, measurements of transfer functions are in general compromised by other problems, often caused by parasitic effects of the used components. Thus, this paper presents a method to determine the theoretical transfer function of a SMPS without using averaging steps by means of simulation. This goal is achieved by considering the converter's switching action. On the one hand, this method allows the desired comparison with the theoretically derived transfer functions by means of ideal components. On the other hand, the influence of non-ideal components can be investigated. A quadrature detector to determine the transfer function is implemented in SPICE to resemble practical measurement set-ups. Based upon the theoretical background, the implementation in SPICE as well as the evaluation and comparison of theoretical transfer functions is carried out. Above all, advice is given with regard to some eventually occurring difficulties.

II. THEORY OF OPERATION

The intention of the paper is to determine the transfer function of a SMPS taking the switching action into account by means of simulation. Thus, the switching action of the power converter is simulated by a transient simulation tool, e.g. SPICE. In order to determine the small-signal behaviour a small disturbance is superimposed on the relevant steady-state control parameter. Fig. 1 depicts the procedure to determine the control-to-output transfer function. The small signal disturbance

$$d(t) = \hat{d} \cdot \sin(\omega_x t) , \qquad (1)$$

with the small amplitude $\hat{d}$ and a frequency of $\omega_x = 2\pi f_x$ is added to the steady state duty-cycle D. Simulation with SPICE results in the steady-state output voltage V_o with an additional small signal disturbance of

$$v_o(t) = \hat{v}_o \cdot \sin(\omega_x t + \varphi_o) \qquad (2)$$

with the amplitude $\hat{v}_o$ and the phase shift φ_o.

With regard to the small-signal analysis only the disturbance $v_o(t)$ is important. Thus the output signal has to be high-pass filtered. Additionally, the disturbances V_i and I_{load} are depicted in Fig. 1, which are kept constant if the control-to-output transfer function is calculated.

The knowledge of both values $\hat{v}_o$ and φ_o in relation to the amplitude and phase angle of the input disturbance $d(t)$ evaluated at several frequencies f_x gives the desired small signal transfer function. In order to extract both values $\hat{v}_o$ and φ_o from the transient simulation the frequency domain representation of the output disturbance

$$\underline{\hat{V}}_o = v_{o,\mathrm{Re}} + \mathrm{j} \cdot v_{o,\mathrm{Im}} = \left| \underline{\hat{V}}_o \right| \cdot \mathrm{e}^{\mathrm{j}\varphi_o} \qquad (3)$$

with the real part $v_{o,\mathrm{Re}}$ and the imaginary part $v_{o,\mathrm{Im}}$ is helpful to determine the amplitude

$$\left| \underline{\hat{V}}_o \right| = \sqrt{v_{o,\mathrm{Re}}{}^2 + v_{o,\mathrm{Im}}{}^2} = \hat{v}_o \qquad (4)$$

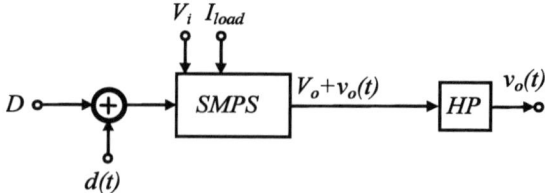

Fig. 1 Simulation procedure to determine the control-to-output transfer function

978-1-4244-1741-4/08/$25.00 ©2008 IEEE

and the phase

$$\varphi_o = \arctan\frac{v_{o,\text{Im}}}{v_{o,\text{Re}}} \qquad (5)$$

Furthermore, the correlation between the time and the frequency domain representation is required. The time domain output disturbance is related to the frequency domain by

$$\begin{aligned}
v_o(t) &= \hat{v}_o \cdot \sin(\omega_x t + \varphi_o) \\
&= \text{Im}\left\{\hat{v}_o \cdot e^{j(\omega_x t + \varphi_o)}\right\} = \text{Im}\left\{\underline{\hat{V}}_o \cdot e^{j\omega_x t}\right\}.
\end{aligned} \qquad (6)$$

From (3) the time domain representation can be described using a sine and cosine wave of the discrete frequency f_x with

$$\begin{aligned}
v_o(t) &= \text{Im}\left\{\hat{v}_o \cdot e^{j(\omega_x t + \varphi_o)}\right\} = \\
&= \text{Im}\left\{(v_{o,\text{Re}} + j \cdot v_{o,\text{Im}}) \cdot e^{j\omega_x t}\right\} = \\
&= v_{o,\text{Re}} \cdot \sin(\omega_x t) + v_{o,\text{Im}} \cdot \cos(\omega_x t).
\end{aligned} \qquad (7)$$

Finally, the quadrature detector of Fig. 2 delivers the real and the imaginary part of the simulated output disturbance by multiplying the output disturbance by a sine or cosine wave of the discrete frequency f_x and low-pass filtering the result.

Equation (8) and (9) illustrate the approach for one frequency point. The real part of the small signal solution can be calculated by

$$\begin{aligned}
&\left[v_o(t) \cdot 2\sin(\omega_x t)\right] * h_{LP}(t) = \\
&= \left[(v_{o,\text{Re}} \sin(\omega_x t) + v_{o,\text{Im}} \cos(\omega_x t)) \cdot 2\sin(\omega_x t)\right] * h_{LP}(t) = \\
&= \left[v_{o,\text{Re}} 2\sin^2(\omega_x t) + v_{o,\text{Im}} \cos(\omega_{f_x} t) 2\sin(\omega_x t)\right] * h_{LP}(t) = \\
&= \left[v_{o,\text{Re}}(1 - \cos(2\omega_x t)) + v_{o,\text{Im}} \cdot \sin(2\omega_x t)\right] * h_{LP}(t) = \\
&\approx v_{o,\text{Re}}.
\end{aligned} \qquad (8)$$

Accordingly, the imaginary part is given by

$$\begin{aligned}
&\left[v_o(t) \cdot 2\cos(\omega_x t)\right] * h_{LP}(t) = \\
&= \left[v_{o,\text{Re}} \sin(2\omega_x t) + v_{o,\text{Im}}(1 + \cos(2\omega_x t))\right] * h_{LP}(t) = \\
&\approx v_{o,\text{Im}}.
\end{aligned} \qquad (9)$$

Since both parts of the complex representation are known, it is now possible to determine both the gain and the phase at the discrete frequency f_x according to

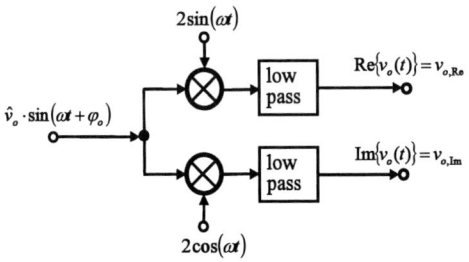

Fig. 2 Quadrature detector

$$\underline{G}_{vd} = \frac{\left|\hat{V}_o\right| \cdot e^{j\varphi_o}}{\left|\underline{d}\right| \cdot e^{j0°}} \qquad (10)$$

Hence, the simulation of the SMPS with superimposed small signal disturbance at the frequency f_x and the appliance of the quadrature detector allows to determine one particular point of the small signal transfer function. Conducting this simulation for multiple frequencies results in the desired bode-plot characteristic.

III. IMPLEMENTATION IN SPICE

In order to determine the small signal behaviour of the SMPS, the SMPS itself has to be simulated in the first step. A simple dc-dc boost converter serves as an example for this purpose. The SPICE realization of the idealized boost converter (without depicting the values and the models of the particular components) and the high-pass filter according to Fig. 1 is shown in Fig 3.

In case that the control-to-output transfer function is simulated the switching pattern of the switch is determined by the voltage controlled voltage source d. The duty-cycle itself is composed by the DC voltage V_D (steady state duty-cycle) with a superimposed small sine wave V_d of the discrete frequency f_x. The steady-state duty-cycle is realized with a voltage level between 0 V and 1 V, the sine wave has an amplitude of 1 mV. The sum of both voltages is compared to a saw tooth voltage pulsating with the switching frequency. As long as the sum v(ton) is larger than the actual level of the saw tooth voltage v(saw), the switch will be in the on-state. Thereby, the transient SPICE simulation of the SMPS delivers an output voltage V_o which exhibits a disturbance superimposed on the steady-state output voltage. Afterwards, the output voltage is filtered by the high-pass so that only the small signal disturbance is processed by the quadrature detector according to Fig. 2.

Fig. 4 depicts all relevant parts of the corresponding simulation model. The high-pass filtered output voltage of

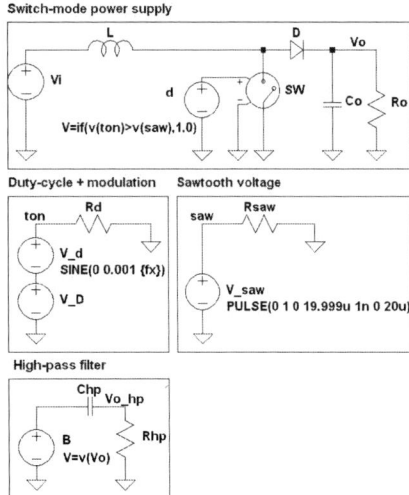

Fig. 3 Simulation model of an ideal boost converter in SPICE

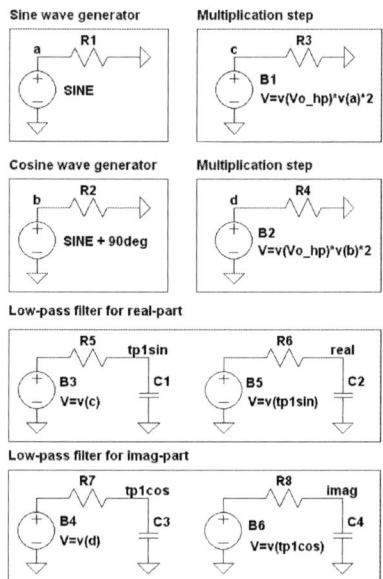

Fig. 4 Model of the quadrature detector

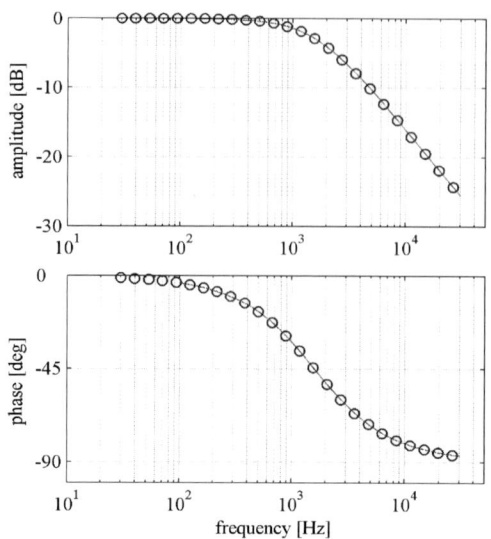

Fig. 5 Simulated and analytical transfer characteristic of the low-pass to be investigated
(dotted: simulation, red line: analytical solution)

the SMPS is multiplied by a sine and cosine signal respectively, with an additional factor of two. The resulting voltages are filtered by a simple 2nd order low-pass filter in order to obtain both the real and the imaginary part of the output voltage disturbance in the frequency domain. A post-processing of both values according to (10) delivers the relevant parameters of the transfer characteristics.

An easy way to verify the SPICE implementation of the quadrature detector is for instance the simulation of the transfer characteristic of a well-known circuit like a simple 1st order low-pass filter. The cut-off frequency of the low-pass to be simulated ($R_o = 100\,\Omega$ and $C_o = 1\,\mu F$) can be calculated with the given values from

$$f_{low-pass} = \frac{1}{2\pi \cdot R_o \cdot C_o} = 1.59\,\text{kHz} \ . \tag{11}$$

In order to emulate the dc voltage typical for SMPSs the input signal of the low-pass filter is realized by a voltage source together with a dc offset and a superimposed sinusoidal voltage. Conducting this simulation multiple times for frequency points in the relevant range and comparing these values to the analytically derived transfer function

$$G_{LP}(j\omega) = \frac{1}{1 + j\omega C_0 R_0} \tag{12}$$

results in Fig. 5. It can be seen that the simulation results obtained by the investigated approach correspond to the analytical solution.

IV. SIMULATION OF TRANSFER FUNCTIONS AND COMMON PROBLEMS

In this chapter simulated transfer functions and the comparison to analytically found ones will be presented. Furthermore, some common problems will be discussed.

As already mentioned above, an idealized dc-dc boost converter serves as a demonstrator for the proposed method according to Fig. 3. Thus, no parasitic elements are considered at the beginning. Two exemplary operating points for continuous conduction mode (CCM) and discontinuous conduction mode (DCM) are given in Table 1.

For the results of this section only CCM operation is considered.

The switch SW and the diode D are both idealized with a forward threshold voltage of 10 mV, a on-resistance of 10 mΩ and an off-resistance of 1 MΩ. Since the control-to-output transfer characteristic in CCM is simulated in the first instance, a sine wave with an amplitude of $\hat{d} = 0.001$ is superimposed on the steady state duty-cycle of D=0.515. Thus, the small signal requirement is fulfilled. To obtain feasible results the simulation is carried out with a maximum step size of 10 ns for a duration of at least 1 s. Above all the settling time of the used filter elements has to be considered. Hence, the simulation data is only evaluated from 0.95 s to 1 s. Here, especially the high-pass after the SMPS to be simulated (see Fig. 3) is responsible for the long simulation time. Compared to this duration the subsequent evaluation with help of the quadrature detector is done quickly.

Fig. 6 shows the results of this simulation and the comparison to the traditionally found analytical solution from [1].

TABLE I.
SIMULATION PARAMETERS OF IDEAL BOOST CONVERTER

	CCM	DCM
V_i	15 V	15 V
D	0.515	0.31
L	150 μH	150 μH
C_0	47 μF	47 μF
R_0	45.9 Ω	365 Ω
f_s	50 kHz	50 kHz

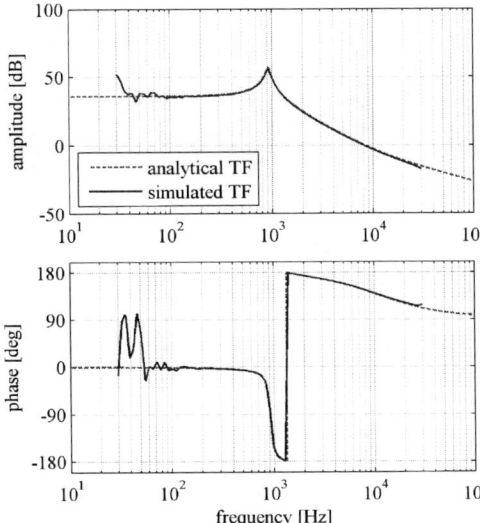

Fig. 6 Simulated control-to-output transfer function
(first attempt)

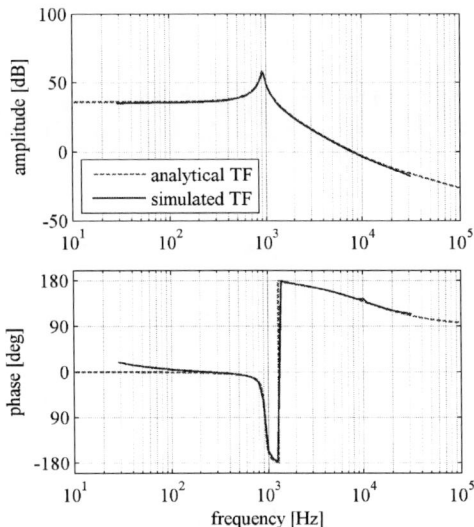

Fig. 7 Simulated control-to-output transfer function with adapted
evaluation time

It can be seen, that both curves match quite closely except for frequencies below 100 Hz and above 25 kHz. Because of the nyquist-shannon sampling theorem the simulation can only deliver valid results up to half the switching frequency, in this case up to $f_{max} = f_s/2 = 25$ kHz. In the lower frequency range another aspect plays a major role: Due to the mentioned fixed evaluation time from 0.95 s to 1 s the simulated data will be analyzed for 50 ms. If e.g. a disturbance of 50 Hz with a cycle duration of 20 ms is simulated, the simulation data will be analyzed for 2.5 times the cycle duration. This fractional consideration of the simulation data results in the observed variations at lower frequencies. For higher frequencies this aspect is less significant.

Due to this effect, the evaluation time is adapted to the actual frequency of the disturbance in the next step. Thus, it is guaranteed that only integer multiples of the modulation period are analyzed. Fig. 7 shows the result of this simulation.

It is obvious that the problem at lower frequencies is reduced dramatically, but the curves still do not match completely. At a frequency of 50 Hz for instance the simulated transfer function has a phase value of approximately 11°. It can be figured out, that the used high-pass is responsible for that characteristic: The used values of $R_{hp} = 50\ \Omega$ and $C_{hp} = 320\ \mu$F implicate a phase displacement of 11.25°. Thus, an adequate dimensioned high-pass filter will improve the situation. Fig. 8 shows the final simulation with an adapted high-pass characteristic.

V. COMPARISON OF TRANSFER FUNCTION

The final part of this paper is dedicated to the major benefit of the proposed method: The ability to compare and even to rank the quality of small-signal transfer functions derived from traditional modeling methods with the likewise theoretical switch model of the same idealizations. It was found, that in CCM all traditional models predict the same behavior. Furthermore it can be seen

from Fig. 8 that the small signal behavior derived by simulating the switching action matches exactly. Hence, there is no need to investigate this operational mode and the attention can be turned to DCM.

With regard to DCM operation several theoretical models to determine the small signal behavior exist. In [8] the terms reduced-order model, full-order model and full-order-corrected model (new full order model) are used to determine different levels of abstraction which are a result of different derivation methods of the diode's freewheeling period $d_2 \cdot T_s$ with the help of the time course of the inductor current $i_L(t)$ and the inductors volts-second balance. Fig. 9 depicts the inductor current $i_L(t)$ and the relevant interval $d_2 \cdot T_s$.

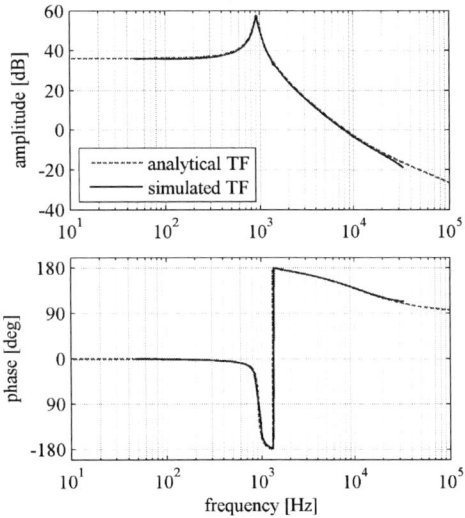

Fig. 8 Simulated control-to-output transfer function of idealized dc-dc
boost converter in CCM

The reduced-order model according to [2] (traditional state-space averaging approach) is based on the assumptions that since the inductor current is zero at the beginning and at the end of one switching cycle, that is

$$i_L(t=0) = i_L(t=T_S) = 0, \qquad (13)$$

the discrete derivation of the inductor current $\mathrm{d}\overline{i_L}/\mathrm{d}t$ from switching cycle to switching cycle is zero according to

$$\frac{\mathrm{d}\overline{i_L}}{\mathrm{d}t} = \frac{i_L(t=T_S) - i_L(t=0)}{T_S} = 0 \ . \qquad (14)$$

As the derivation is zero it is assumed that no volts-seconds are applied on the inductor during the particular switching period, so

$$d_1(t) \cdot U_{on} + d_2(t) \cdot U_{off} + d_3(t) \cdot 0 = 0, \qquad (15)$$

whereas the voltage U_{on} and U_{off} are responsible for the increasing and decreasing inductor current respectively. The volts-second balance is then used to calculate the duration of the period of a conducting diode $d_2 \cdot T_s$. Thus, the reduced-order model does not consider accurately the high frequency inductor dynamics and the behavior at higher frequencies cannot be described exactly. From (14) it can be seen that the inductor current is no longer a state-variable and has lost its dynamic character.

The full-order model according to [5] (average switch model) also makes use of the volt-second balance of (15) to calculate the missing parameter $d_2(t)$. Thus an equivalent circuit with controlled voltage and current sources is developed to substitute the power switch and the diode. Afterwards the resulting circuit of the power converter, thus the average circuit model of the power switch and the diode in combination with the remaining components including the inductor is used to determine a small signal model by perturbating the relevant signals. Hence, the full-order model includes the inductor but is still based on the error-prone calculation of $d_2 \cdot T_s$.

The full-order-corrected model of [8] is by contrast based on an alternative method to derive the missing value of $d_2 \cdot T_s$. Instead of the volts-second balance it uses the average current of the inductor. With reference to Fig. 9 the average inductor current of the particular switching cycles is given by

$$\overline{i_L} = \frac{i_{pk}}{2} \big(d_1(t) + d_2(t)\big) = \frac{U_{on}}{2L} d_1(t)\big(d_1(t) + d_2(t)\big)T_S, \quad (16)$$

with a maximum inductor current of

$$i_{pk} = \frac{U_{on}}{L} d_1(t) T_S . \qquad (17)$$

The missing value $d_2(t)$ can be figured out finally to

$$d_2(t) = \frac{2L\overline{i_L}}{U_{on} d_1(t) T_S} - d_1(t). \qquad (18)$$

If e.g. the standard modeling technique of [5] is combined with the expression in (18) an improved small signal model results which allows for an accurate description of the high-frequent dynamics of the SMPS.

In addition to the three just mentioned models a fourth model, the average switch inductor model (ASIM) of [7] is proposed for DCM. The idea is to replace the switching parts of the converter, thus the switch and the diode, as well as the inductor by an equivalent circuit which emulates their behavior. The development of this model directly incorporates the above found expression of (16) to describe the average current of the inductor. As a consequence it can be assumed, that the ASIM will deliver the same result as the just mentioned full-order corrected model. Fig 10 shows a comparison of both models in case of the idealized boost converter in DCM (see Table 1).

Nevertheless, all the models and resulting small signal transfer functions are based on averaging steps. The purpose of the paper is now to clarify the quality of these theoretical models by means of the proposed switched small-signal model with the example of a dc-dc boost converter in order to predict the exact behavior at higher frequencies. This idealized converter is simulated for the DCM operation according to Table 1. The results are depicted in Fig. 11.

First of all it can be seen that for lower frequencies all models match quite well. But as expected, for higher frequencies the transfer functions, especially the predicted phase, differs noticeably. In Fig. 12 the curves are zoomed for higher frequencies. With respect to the phase angle it is obvious that the reduced order model does not predict the behavior at higher frequencies correctly. Therefore, it

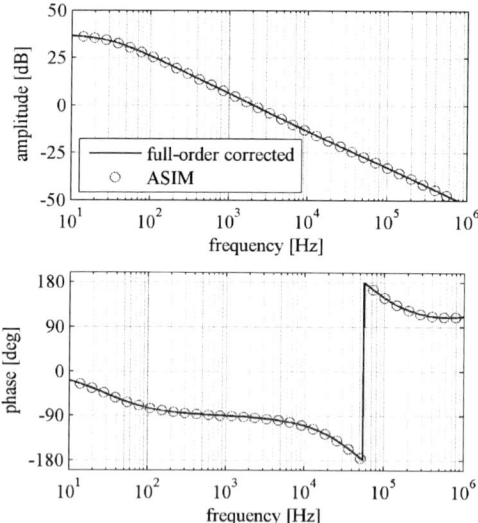

Fig. 10 Comparison of full order corrected model and average switch inductor model (ASIM)

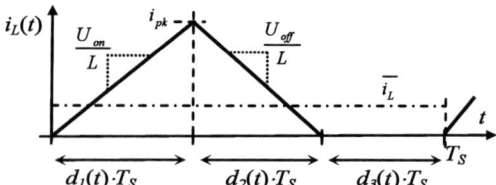

Fig. 9 Time course of the inductor current of one switching cycle

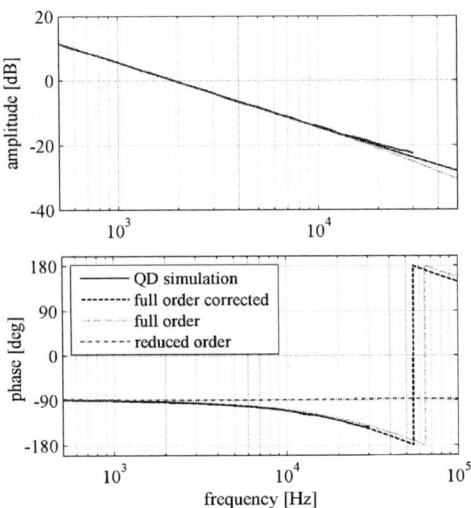

Fig. 11 Comparison of control-to-output transfer function based on different small-signal models in DCM with ideal components

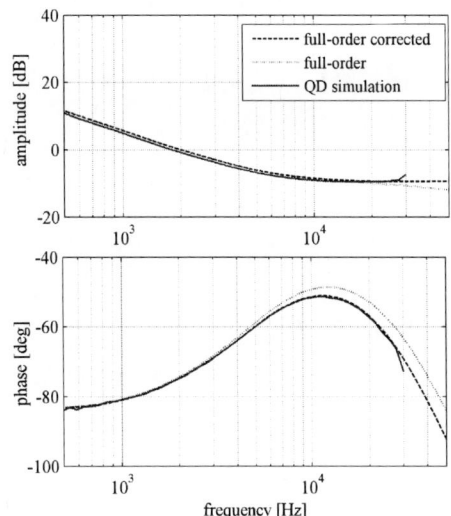

Fig. 13 Comparison of control-to-output transfer function based on different small-signal models in DCM with parasitic elements

will not be further considered. With respect to the full-order and the full-order-corrected model no significant differences can be recognized. Remark: The results of the quadrature detector simulation are only valid up to 25 kHz (half the switching frequency).

The next step is the consideration of the inductor's and the capacitor's parasitic resistance. The ESR of the capacitor was modeled with $R_C = 0.58 \, \Omega$, the series resistor of the choke was taken into account with $R_L = 0.5 \, \Omega$. The results of the simulations are depicted in Fig. 13. The curves indicate that if the behavior at high frequencies is important, the transfer function should, as expected, be preferably modeled by means of the full-order corrected model.

VI. CONCLUSION

The paper presents the use of SPICE simulation with a high degree of idealization. Some common pitfalls of the transfer function calculation have been demonstrated and solved. The proposed method allows a comparison of analytically derived transfer functions. Especially, in DCM where the traditional models differ significantly at higher frequencies the proposed switched simulation could confirm the validity of the full-order-corrected small-signal model.

REFERENCES

[1] R.D. Middlebrook, S. Cuk, "A General Unified Approach to Modelling Switching-Converter Power Stages," *IEEE Power Electronics Specialists Conference*, 1976, pp.18–34.

[2] R.D. Middlebrook, S. Cuk, "A General Unified Approach to Modelling Switching DC-to-DC Converters in Discontinuous Conduction Mode," *IEEE Power Electronics Specialists Conference*, Record, 1977, pp.36–57.

[3] R.W. Erickson, D. Maksimovic, *Fundamentals of Power Electronics*, 2nd edition, Springer Science+Business Media, Inc. 2001.

[4] G.W. Wester, R.D. Middlebrook, "Low-Frequency Characterization of Switched dc-dc Converters," *IEEE Transactions on Aerospace and Electronic Systems*, Vol. 9, Issue 3, 1973, pp. 376-385

[5] V. Vorperian: "Simplified Analysis of PWM Converters Using the Model of the PWM Switch: Parts I and II," *IEEE Transactions on Aerospace and Electronic Systems*, Vol. AES-26, pp. 490-505, 1990.

[6] S. Ben Yaakov, "SPICE simulation of PWM DC-DC convertor systems: voltage feedback, continuous inductor conduction mode," *Electronics Letters*, Vol. 25, Issue 16, 1989

[7] S. BenYaakov, Y. Amran and F. Huliehel, "A Unified Spice Compatible Average Model of PWM Converters," *IEEE Transactions on Power Electronics*, Vol. 6, pp. 585-594, 1991.

[8] J. Sun, D. M. Mitchell, M. F. Greuel, P. T. Krein and R. M. Bass: "Averaged Modeling of PWM Converters Operating in Discontinuous Conduction Mode," *IEEE Transactions on Power Electronics*, Vol. 16, No. 4, pp. 482-492, 2001.

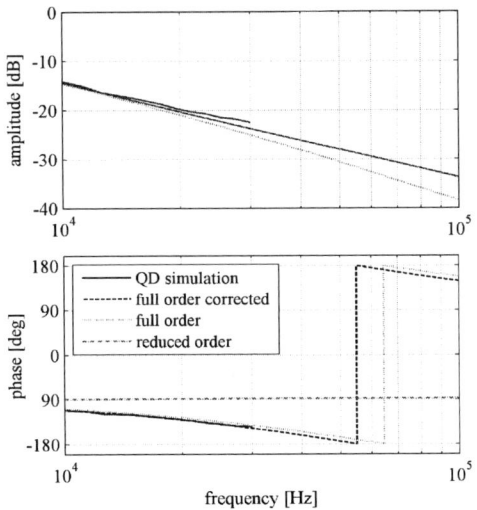

Fig. 12 Differences of alternatively derived control-to-output transfer functions at higher frequencies

255

Dynamic Behaviour of a Series – Connected Multilevel Converter with Interleaved Switching

C. Fahrni[*], A. Rufer[*]

[*] Laboratoire d'électronique industrielle, STI-LEI-EPFL, Lausanne, Switzerland
e-mail: claude.fahrni@epfl.ch, alfred.rufer@epfl.ch,

Abstract— This paper presents an analysis and design method for the current control of a series connected multilevel converter with interleaved switching. Based on a conventional design method for the parameters of a PI controller, the variable response of a current control circuit with one channel is presented, in dependency of the intervention time of the step on the set value. Then, a model for the equivalent transfer function of the control circuits of the interleaved solution is presented. A new approach for the design of the parameters of the current controller is proposed. The resultant step response in closed loop operation of the system will show the no more fluctuating response regarding overshoot.

Keywords— Converter control, multilevel converters, interleaved converter, modeling.

I. INTRODUCTION

Design methods for the parameters of a current controller of a classical switching converter have been described for long time, as the examples of the optimal gain or phase margin criteria, as described in [1], [2]. In these application examples, the controller design strategy consists in compensation of so called dominant system time constant, and to consider an approximation of the transfer functions for the PWM modulators as a first order polynomial using a mean value of the variable delay introduced by the modulator.

For multilevel converters using series connected H-bridges as described in [3], [4], the modulation strategy uses normally interleaved carriers resulting in a higher definition of the output voltage. Fig. 1 shows a 3 channel series connected multilevel converter with a common current controller. In this paper, a theoretical analysis of the control design of such a converter is presented. The study shows that the interleaved modulation can be modelled in a similar way as for classical modulators, leading to a better correspondence of the model and the real behaviour of the converter. This property allows to chose a slightly different strategy for the design of the controller, or to chose a different type of controller as usually.

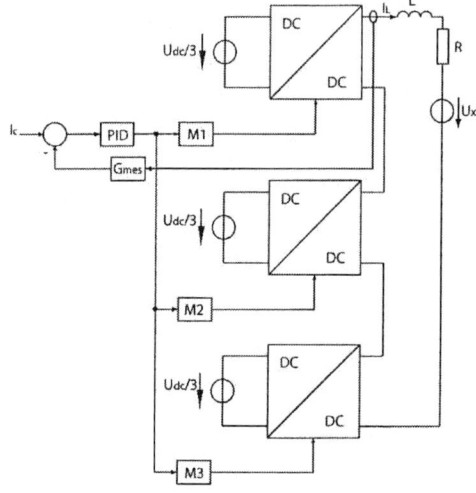

Fig. 1: Series connected multilevel converter with interleaved switching

II. PROPERTIES OF THE CLOSED LOOP CONTROL OF A CLASSICAL PWM CONVERTER

A classical current control system using a single H bridge converter is shown in Fig. 2, where the converter load is given by a passive RL circuit connected in series with an equivalent voltage source. As indicated in the figure, the converter is modulated with a PWM unit (M) and the load current is controlled with the help of a PI controller. For the design of the controller, a usual method based on an open loop analysis of the control circuit is done. The corresponding model with transfer functions is shown in Fig. 3.

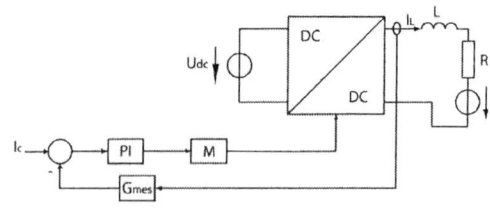

Fig. 2: Controlled system of a converter feeding a L-R load

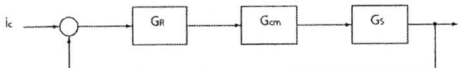

Fig. 3: Model of the system with transfer functions

In Fig. 3, G_s represents the system transfer function, G_{cm} an equivalent transfer function representing the delay introduced by the PWM modulation circuit, and G_R the PI controller. The open loop transfer function is given by (1):

$$G_0 = G_R \cdot G_{cm} \cdot G_s = \frac{1 + sT_n}{sT_i} \cdot \frac{k_{cm}}{1 + T_{cm}} \cdot \frac{k_s}{1 + sT_a} \quad (1)$$

The corresponding frequency responses are represented in Fig. 4.

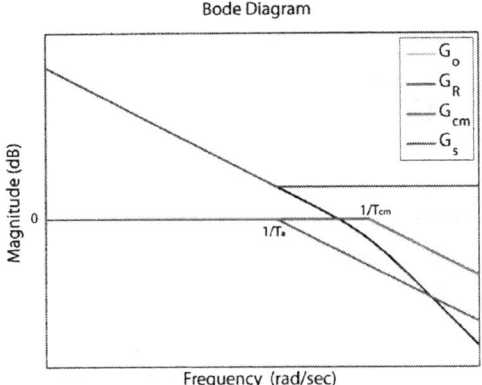

Fig. 4: Frequency responses

When the dominant time constant of the system T_a is compensated by the correlation time constant T_n of the PI controller, the remaining open-loop transfer function becomes:

$$G_0 = \frac{k_{cm} \cdot k_s}{sT_i \cdot (1 + sT_{cm})} \quad (2)$$

The value of the second parameter of the PI controller which is the integration time constant T_i is chosen in order to get a zero-crossing pulsation equal to the half of the breakpoint pulsation of $1/T_{cm}$.

In the previous forms, the equivalent transfer function G_{cm} of the PWM modulation was chosen as a first order approximation of the variable switching delay as:

$$G_{cm} = e^{-sT_{cm}} \cong \frac{1}{1 + sT_{cm}} \quad (3)$$

where T_{cm} is calculated as the mean value between 0 and the maximum delay T_p. T_p is the period of the modulation. As shown in Fig. 5, the delay can vary between 0 and T_p, dependent on the time of occurrence of the input transient.

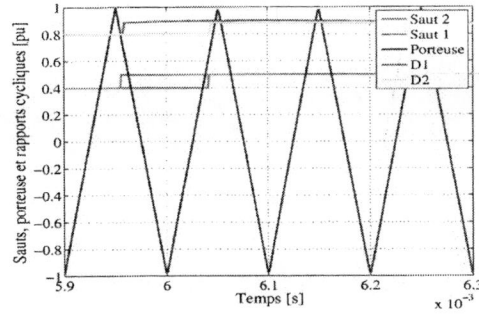

Fig. 5: Different occurrences for the step control of the PWM modular

Because of the design of the phase margin of the system is done with a mean value for T_{cm}, the result is a fluctuating response in the closed loop behaviour of the controlled system. Fig. 6 shows the different responses in closed loop control when the occurrence of the step on the current set value varies according Fig. 5.

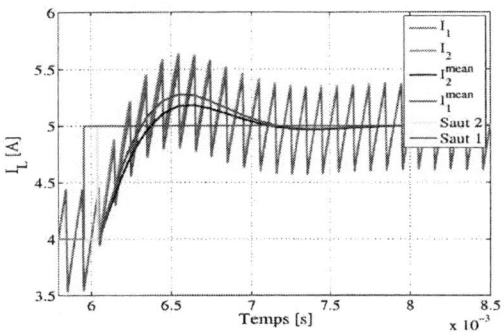

Fig. 6: Time response in dependency of the occurrence time of the step function of the current set-value

III. EQUIVALENT TRANSFER FUNCTION OF THE MULTILEVEL INTERLEAVED CONVERTER

Equation (3) has given the approximated form of the delay introduced by a single channel PWM modulator. The variable delay can be represented as shown in Fig. 7a, where this delay varies between 0 and T_p. By considering the series connected multichannel interleaved converter, the step response of the voltage control through the multi-carrier PWM modulator can be represented as shown in Fig. 7b. A variable time delay is always present, but concerns only a part of the output voltage U_{tot}/n, where n is the number of series connected cells. After that first small step, the other channels deliver each their contribution with an additional delay of each T_p/n. The resulting output voltage shows so the typical ramp behaviour of Fig. 7b.

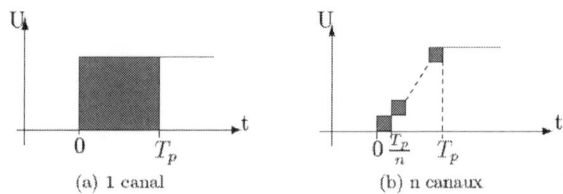

Fig. 7: Step response of the voltage control by PWM modulation

a) single channel converter
b) series connected multilevel converter

A resultant transfer function of the multilevel interleaved modulation can be calculated with the help of the curves given in Fig. 8. First a fluctuating term is considered and modelled as a mean value of the appearing delay, as was considered in (3) but with a reduced value given by the switching period of the converters divided by the number of channels

$$G'_{cm} = e^{-sT_{cm}/n} \cong \frac{1}{1+sT_{cm}/n} \qquad (4)$$

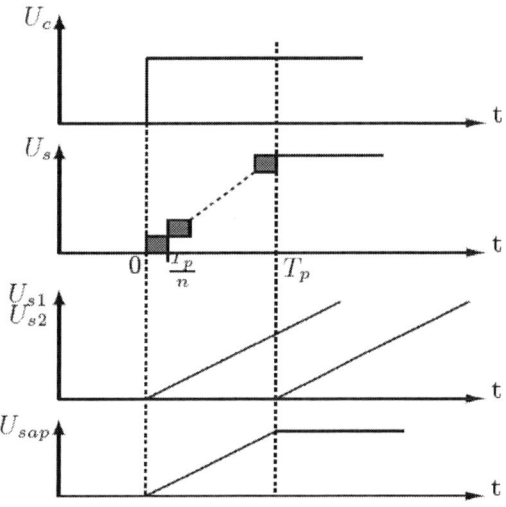

Fig. 8: Approximation of the interleaved PWM modulation of a series connected multilevel converter

Then, the superposition of the following steps appearing at the output of the converter is approximated by a ramp function (U_{sap} in Fig. 8) with a ramp time equal to T_p as:

$$\frac{1}{s} \cdot G_b = \frac{1}{s^2 T_p} - \frac{1}{s^2 T_p} \cdot e^{-sT_p} = \frac{1}{s^2 T_p} - \frac{1}{s^2 T_p} \cdot \frac{1}{1+sT_p} \qquad (5)$$

leading to: $G_b = \dfrac{1}{1+sT_p}$ (6)

The transfer functions of (4) and (6) can then be used as a new model for the series connected multichannel converter with interleaved switching. This new model of the PWM control is represented in Fig. 9.

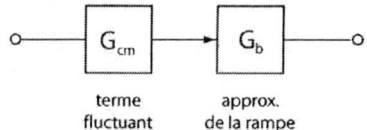

terme   approx.
fluctuant   de la rampe

Fig. 9: Block diagram of the model for the interleaved modulation with series connected converters

For the control of this type of converter, a PID controller can now be used, where T_n is used to compensate the dominant time constant T_a of the system as was already done in the classical method (1), and where T_v is used for the compensation of T_p.

$$G_0 = G_R \cdot G_{cm} \cdot G_s =$$
$$= \frac{(1+sT_n)\cdot(1+sT_v)}{sT_i} \cdot \frac{k_{cm}}{1+T_{cm}/n} \cdot \frac{1}{1+sT_p} \cdot \frac{k_s}{1+sT_a} \qquad (7)$$

$$G_0 = \frac{k_{cm}\cdot k_s}{sT_i \cdot (1+sT_{cm}/n)} \qquad (8)$$

With this strategy, the resulting open-loop transfer function becomes of the same form as in eq. (2), but with a strongly reduced small time constant. The calculation of the remaining integration time constant T_i is then done in dependency of that small time constant T_{cm}/n.

IV. PROPERTIES OF THE CLOSED LOOP CONTROL OF THE INTERLEAVED MULTILEVEL CONVERTER

As it was done for the single channel converter and described in Fig. 5, the step function is also triggered with different positions inside of the switching period T_p, as indicated in Fig. 10. The result on the behaviour of the closed current control is shown in Fig. 11, where the fluctuation in the step response is now eliminated. According to the open–loop transfer function of equation (8), the rise time in Fig. 11 should be theoretically faster. However, the real implementation of the PID controller in an application with such a high ripple on the control value leads to restrictions on the final open-loop gain.

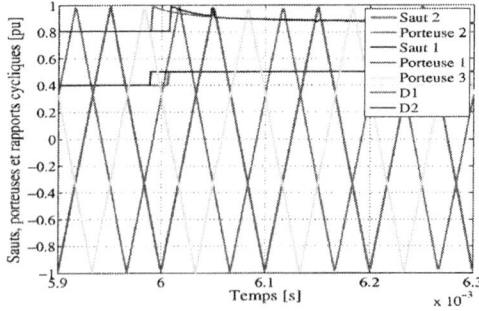

Fig. 10: Different occurrences for the step control of the interleaved multilevel modulator

258

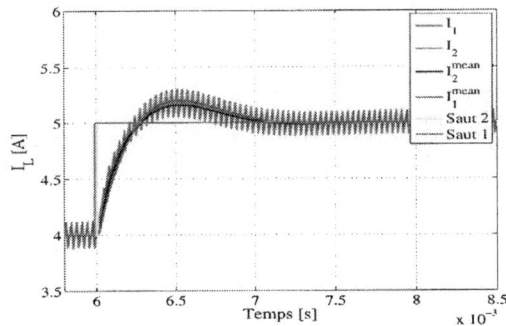

Fig. 11: Step response of current control for an interleaved multilevel converter

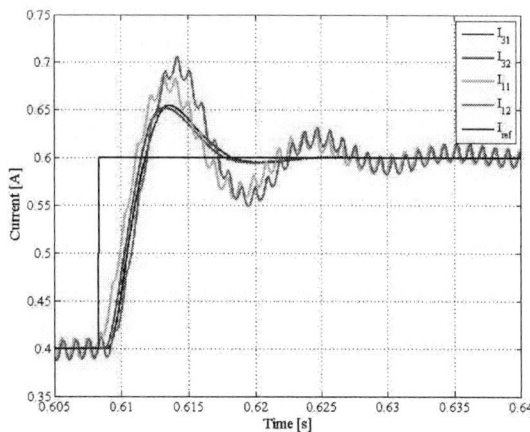

Fig. 13: Experimental step responses

V. EXPERIMENTAL RESULTS

Experiments have been done on a system composed by one, respectively three H-bridge converters. The electronic boards (control and power) are represented in Fig. 12. The values of the RL-load are L=4.8mH and R=5Ω. The modulator is implemented in an FPGA and the control algorithms are implemented in a DSP. The schemes of the experimental set-up correspond to the schemes represented in Fig. 1 and Fig. 2.

Fig. 12: Experimental set-up

In the first case, one converter is feeding an RL-load. The PI-controller is designed with the method presented in paragraph II. Two current steps are produced on the current reference, one with a small delay between the variation of the reference and the variation of the output and the second with a longer delay. Then, two steps with different delays are generated on the system with three interleaved converters controlled by a PID-controller. The parameters are almost identical as those for the PI-controller, but a derivative term is added. That is the reason why the response is less oscillating. In Fig. 13, the step responses are represented in pu. I_{ref} is the current reference, I_{11} is the response to the step with a short delay and I_{12} is the response to a long delay, both for the system with only one converter. I_{31} and I_{32} are the response to a step with a short, respectively long, delay for the system with three interleaved converters. The differences between the step responses are smaller for a system with interleaved converter. Furthermore, the system with interleaved converters is faster.

VI. CONCLUSION

A finer method, compared to the classical one, to design current controllers for multilevel interleaved converters is presented in this paper. It allows reach a better step response and even a better dynamic. This method is based on a new model elaborated for interleaved converters, starting from the property of a smaller fluctuating delay, and from the fact that the voltage contributions from each series connected cell is applied to the output with a given sequence. Simulation results and then experimental records show the validity of the theory. This new tool can be very helpful to design controllers for performing and very accurate and fast systems. A deeper study with much more details of the context of the present development is given by [5] and [7].

REFERENCES

[1] Bühler, H., Réglage des systèmes d'électronique de puissance, Presses Polytechniques et Romandes, ISBN 2-88074 341-9, CH1015 Lausanne

[2] Basilio, J.C.; Matos, S.R., Design of PI and PID controllers with transient performance specification Education, IEEE Transactions on, Volume 45, Issue 4, Nov 2002 Page(s): 364 – 370

[3] N. Schibli. Symmetrical multilevel converters with two quadrant DC-DC feeding. PhD thesis, Lausanne, 2000, http://biblion.epfl.ch/EPFL/theses/2000/intranet/EPFL_TH2220.pdf

[4] M. Marchesoni; M. Mazzucchelli, Multilevel converters for high power AC drives: a review, Industrial Electronics, 1993. Conference Proceedings, ISIE'93 - Budapest., IEEE International Symposium on 1993 Page(s):38 – 43.

[5] C. Fahrni, A. Rufer, F. Bordry, and J. P. Burnet. A novel 60 MW Pulsed Power System based on Capacitive Energy Storage for Particle Accelerators. In EPE 2007 : 12th European Conference on Power Electronics and Applications, 2007.

[6] Destraz B, Louvrier Y., Rufer A., High efficiency Interleaved Multi-Channel DC-DC Converters Dedicated to Mobile Applications

[7] C. Fahrni "Principe d'alimentation par convertisseurs multiniveaux à stockage integé – Application aux accélérateurs de particules" Ph D Thesis, Nr. 4034, EPFL, Ecole Polytechnique Fédérale de Lausanne, Lausanne Switzerland. http://biblion.epfl.ch/EPFL/theses/2008/4034/EPFL_TH4034.pdf

Simple Analysis of a Flying Capacitor Converter Voltage Balance Dynamics for DC Modulation

A. Ruderman [1], B. Reznikov [2], and M. Margaliot [3]

[1] Elmo Motion Control Ltd., [2] General Satellite Corporation, [3] Tel Aviv University
aruderman@elmomc.com, reznikovb@spb.gs.ru, michaelm@eng.tau.ac.il

Abstract–Flying capacitor multilevel PWM converter with a natural voltage balance is an attractive multilevel converter choice because it requires no voltage balance control effort. Flying capacitor converter practically does not suffer from voltage balance imposed performance limitations as opposed to multiple point clamped converter. Voltage balance dynamics analytical research methods reported to date deal mostly with an AC modulation case and are essentially based on a frequency domain analysis using double Fourier transform. Therefore, these methods require high mathematical skills, are not truly analytical and rather difficult to use in an everyday practice by electrical engineer. In this paper, we consider a DC modulation case to demonstrate that a straightforward time domain approach based on switching intervals piece-wise analytical solutions makes it easy to obtain time-averaged discrete and continuous models for voltage balance dynamics simulation. A primitive single-phase single-leg three-level converter analytical investigation yields a surprisingly simple accurate expression for capacitor charge / discharge related time constant revealing its dependence on inductive load parameters, carrier frequency, and duty ratio.

I. INTRODUCTION

Multilevel converters are being progressively used for medium and high voltage / power applications. Multilevel converter topologies, modulation strategies, and performances have been extensively studied over the past two decades [1, 2]. Flying Capacitor (FC) multilevel PWM converter is an attractive choice due to the natural voltage balance property.

In Multiple Point Clamped (MPC) converter used for front end applications, it is possible to achieve capacitors voltage balance only for a limited operation envelope in terms of modulation index – load displacement angle. In such a MPC converter, maximal possible modulation index M=1 is achieved for pure reactive (inductive) load only (zero power factor). With load power factor approaching unity, maximal modulation index is theoretically compromised to about M=0.55 for a three-phase and M=0.63 for a single-phase converter because of the performance limitations that originate from capacitor voltage balance [3].

Suppose ideally smooth, ripple free load current, relatively high switching frequency, and appropriate phase shifted voltage modulation strategy. Under the above assumptions, a flying capacitor is charged and discharged by the same amount of load current on time intervals of equal durations. It is clear that the capacitor voltage is thus oscillating around some average value that is defined by the capacitor initial voltage and no any voltage balance conclusion may be drawn from this simplistic model. Therefore, it is recognized that FC converter

voltage balance process is actually driven by the load current high order harmonics.

Reported FC converter analytical voltage balance research methods mostly deal with an AC modulation [4-8]. However, from a methodology perspective it would be correct to get started with a more simple DC modulation case.

The reported FC converter voltage balance analysis methods are essentially based on frequency domain transformations. This selection of the analysis tools probably comes from the recognition of the important role of load current high order harmonics in the whole voltage balance process. However, intuitively it seems somewhat artificial to use "intermediate" frequency domain methods for derivation of linear time invariant models to be analyzed in time domain.

As a result, the reported FC converter voltage balance methods are not "truly analytical" and not easy to understand. The usage of the frequency domain methods for building linear time invariant voltage balance models makes it difficult to gain a thorough insight into FC converter capacitor voltage balance physical mechanisms and apply this approach in an everyday engineering practice.

In this paper, we apply a straightforward time domain approach based on "sewing" analytical transient solutions of consecutive PWM period switching subintervals to derive DC modulated FC converter voltage balance dynamics models - both discrete and continuous. This approach seems most adequate for linear switched (variable structure) systems that are naturally formed by idealized FC converters along with their linear active-inductive loads.

By means of this technique, we study a primitive single capacitor single leg three-level FC converter. We obtain surprisingly simple accurate expression for a capacitor charge / discharge related time constant by applying small parameter technique. The small parameter naturally arises due to the fact that converter switching is performed at a frequency that is much higher than equivalent RLC-circuit natural one. In other words, this means that current and voltage ripples are relatively low that is always true for a good practical converter. This approach allows revealing the capacitor charge rate dependence on inductive load parameters, carrier frequency, and duty ratio.

The suggested analytical approach may be easily adopted for a DC modulated FC converter with an arbitrary number of voltage levels (cells). We have to admit that a generalization to an AC modulation case is not trivial and requires an adequate technique to perform time averaging on a fundamental period.

978-1-4244-1741-4/08/$25.00 ©2008 IEEE

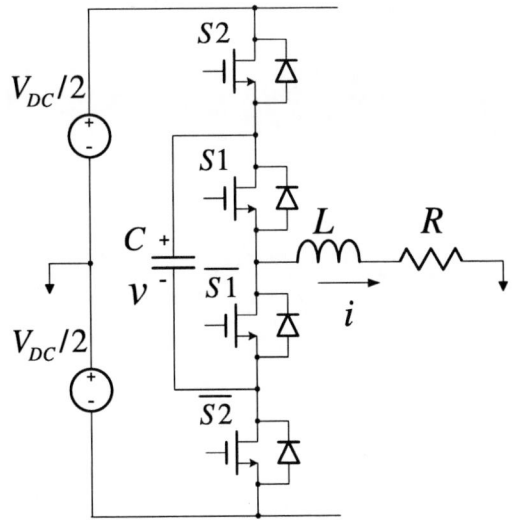

Fig. 1. Single-phase three-level FC converter with RL-load

However, an insight into voltage balance mechanism gained for DC modulation may be useful for AC modulation as well and we solidify this statement with corresponding examples.

II. SINGLE-PHASE SINGLE-LEG THREE-LEVEL FC CONVERTER TOPOLOGY AND MODULATION STRATEGY

A single-phase single-leg three-level FC converter with active-inductive load is given in Fig. 1.

Fig.1 converter voltage modulation strategy is demonstrated in Fig. 2,a. Instantaneous voltage command V_{COM} is scanned by two opposite phase triangular wave carrier signals to define converter switching instants. Carrier wave $s1$ is responsible for switching a complementary switch pair $S1 - \overline{S1}$; $s2$ - for $S2 - \overline{S2}$.

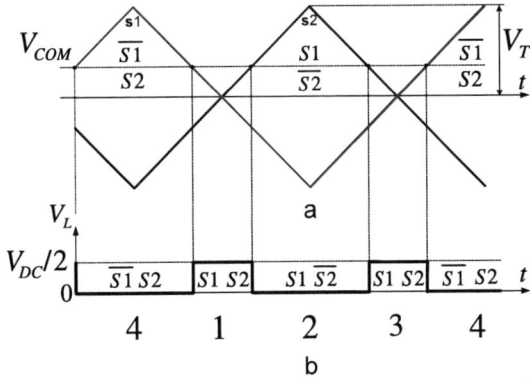

Fig. 2. Voltage modulation process (a) and output voltage waveform with switching states (b)

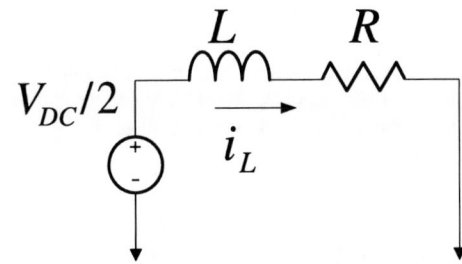

Fig. 3. FC converter topology on intervals 1 and 3 ($S1S2$) (the capacitor is disconnected)

Fig. 2,b shows converter switching states and output voltage waveform assuming equal voltage sources, ideal switches, and the capacitor voltage $v \quad V_{DC}/2$.

As readily seen from Fig.2, a switching period is comprised of four intervals. Assuming ideal switches, intervals 1 and 3 generate the same converter topology shown on Fig. 3 – note that the capacitor is disconnected and keeps its initial voltage unchanged. Both intervals have the same duration

$$\Delta t_1 \quad \Delta t_3 \quad \frac{D}{2} T_{PWM} \tag{1}$$

where T_{PWM} - PWM period; $D \quad V_{COM}/V_T$ - PWM duty ratio (normalized DC voltage command).

FC converter topologies on the intervals 2 and 4 are shown in Fig. 4,a, b. The difference between the topologies is the opposite polarity of both voltage source and capacitor.

The duration of both intervals is

$$\Delta t_2 \quad \Delta t_4 \quad \frac{(1-D)}{2} T_{PWM}, \tag{2}$$

$$\Delta t_1 \quad \Delta t_2 \quad \Delta t_3 \quad \Delta t_4 \quad T_{PWM}.$$

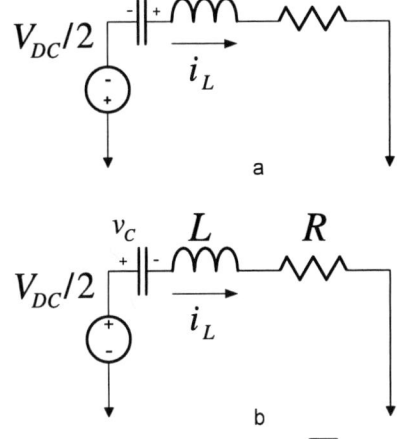

Fig. 4. FC converter topologies on intervals 2 (a - $S1\overline{S2}$) and 4 (b - $\overline{S1}S2$)

III. FC Converter Dynamics Modeling

On each switching interval, FC converter may be modeled as a linear time-invariant system. The FC converter under consideration is a second order switched linear system with load inductor current and capacitor voltage being state variables. Therefore, assuming constant DC voltages, on each interval the converter behavior is described by the following state equation

$$X(t) \quad A_j(t)X(0) \quad B_j(t)\frac{V_{DC}}{2}; X(t) \quad \begin{bmatrix} i(t) \\ v(t) \end{bmatrix}, \quad (3)$$

$$X(0) \quad \begin{bmatrix} i(0) \\ v(0) \end{bmatrix} \text{ - the interval initial conditions; } j \quad 1,2,3,4.$$

The state space equations' matrices in (3) are obtained by analytically solving ordinary linear time-invariant differential equations on different switching intervals.

On interval 1 (Fig.3),

$$A_1(t) \quad \begin{bmatrix} \exp(-t/T_L) & 0 \\ 0 & 1 \end{bmatrix}; \quad (4)$$

$$B_1(t) \quad \begin{bmatrix} 1-\exp(-t/T_L) \ /R \\ 0 \end{bmatrix}, \quad (5)$$

where $T_L \quad L/R$ - load time constant.

On interval 1, the capacitor is disconnected and keeps its initial voltage. The converter dynamic behavior on interval 3 is identical to that of interval 1 (Fig.3):

$$A_3(t) \quad A_1(t); \\ B_3(t) \quad B_1(t). \quad (6)$$

Now, suppose oscillating step response of the equivalent LCR-circuit Fig.4 ("small" resistance). Then on interval 2

$$A_2(t) \quad \exp(-\alpha t)A_2'(t);$$

$$A_2'(t) \quad \begin{bmatrix} \cos(\omega t)-\dfrac{\alpha}{\omega}\sin(\omega t) & \dfrac{1}{\omega L}\sin(\omega t) \\ -\dfrac{1}{\omega C}\sin(\omega t) & \cos(\omega t) \quad \dfrac{\alpha}{\omega}\sin(\omega t) \end{bmatrix}; \quad (7)$$

$$B_2(t) \quad \begin{bmatrix} -\dfrac{1}{\omega L}\exp(-\alpha t)\sin(\omega t) \\ -\exp(-\alpha t)\left(\dfrac{\alpha}{\omega}\sin(\omega t) \quad \cos(\omega t)\right) \quad 1 \end{bmatrix}, \quad (8)$$

where
$$\alpha \quad 0.5R/L \quad 1/(2T_L);$$
$$\omega \quad \sqrt{\omega_0^2-\alpha^2};$$

$$\omega_0^2 \quad 1/(LC); R \quad 2\sqrt{L/C}.$$

On interval 4,

$$A_4(t) \quad \exp(-\alpha t)A_4'(t);$$

$$A_4'(t) \quad \begin{bmatrix} \cos(\omega t)-\dfrac{\alpha}{\omega}\sin(\omega t) & -\dfrac{1}{\omega L}\sin(\omega t) \\ \dfrac{1}{\omega C}\sin(\omega t) & \cos(\omega t) \quad \dfrac{\alpha}{\omega}\sin(\omega t) \end{bmatrix}; \quad (9)$$

$$B_4(t) \quad \begin{bmatrix} \dfrac{1}{\omega L}\exp(-\alpha t)\sin(\omega t) \\ -\exp(-\alpha t)\left(\dfrac{\alpha}{\omega}\sin(\omega t) \quad \cos(\omega t)\right) \quad 1 \end{bmatrix}, \quad (10)$$

Recalling (1), (2), we now have a discrete FC converter model for calculating state variables at the switching instants:

$$X(t_1) \quad A_1(\Delta t_1)X(0) \quad B_1(\Delta t_1)\frac{V_{DC}}{2};$$

$$X(t_2) \quad A_2(\Delta t_2)X(t_1) \quad B_2(\Delta t_2)\frac{V_{DC}}{2};$$

$$X(t_3) \quad A_3(\Delta t_3)X(t_2) \quad B_3(\Delta t_3)\frac{V_{DC}}{2}; \quad (11)$$

$$X(t_4) \quad A_4(\Delta t_4)X(t_3) \quad B_4(\Delta t_4)\frac{V_{DC}}{2};$$

$$X(t_5) \quad A_1(\Delta t_1)X(t_4) \quad B_1(\Delta t_1)\frac{V_{DC}}{2};$$

....

where

$$t_1 \quad \Delta t_1;$$
$$t_2 \quad \Delta t_1 \quad \Delta t_2;$$
$$t_3 \quad \Delta t_1 \quad \Delta t_2 \quad \Delta t_3;$$
$$t_4 \quad \Delta t_1 \quad \Delta t_2 \quad \Delta t_3 \quad \Delta t_4 \quad T_{PWM}; \quad (12)$$
$$t_5 \quad T_{PWM} \quad \Delta t_1$$

....

Example 1. Consider single-phase single capacitor three-level FC converter with the following converter and load parameters:

$$V_{DC} \quad 100V;$$
$$R \quad 1Ohm; L \quad 0.25mH; C \quad 100uF; \quad (13)$$
$$T_{PWM} \quad 300us \ (f_{PWM} \quad 3.33kHz).$$

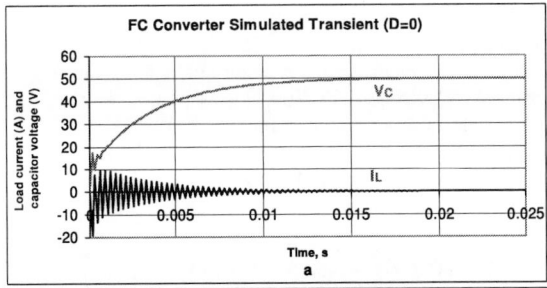

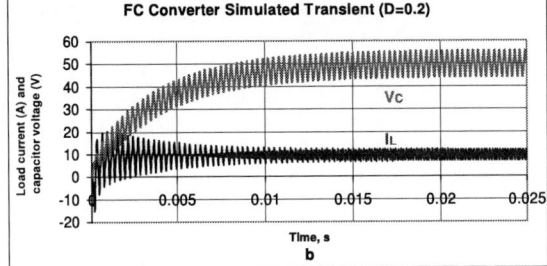

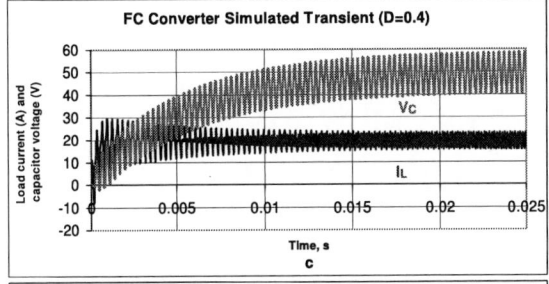

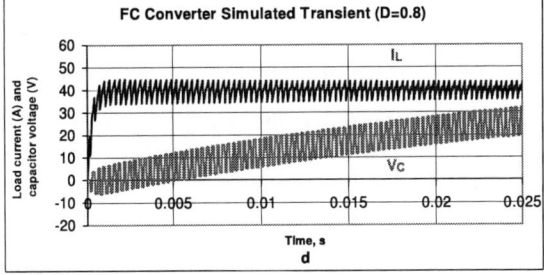

Fig. 5. FC converter load current and capacitor voltage simulation for zero initial conditions and different duty ratios: a – D=0; b – D=0.2; c – D=0.4; d – D=0.8

Converter dynamics simulation results obtained using Excel by programming formulas (11), (12) are presented in Fig. 5. Though demonstrated are results for zero initial conditions, average load current and capacitor voltage always converge to

$$i(\infty) \quad \frac{V_{DC}D}{2R};$$

$$v(\infty) \quad \frac{V_{DC}}{2} \tag{14}$$

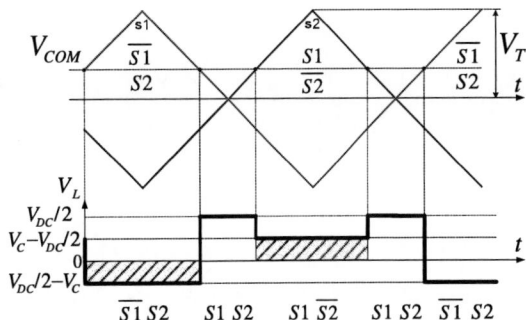

Fig. 6. Average load voltage does not depend on capacitor voltage (equal shaded areas shown)

for any set of initial conditions.

Here are more observations from FC converter dynamics modeling experience along with their physical interpretation. Intuitively, we expect the FC converter time averaged model to behave like a 2nd order linear time invariant system. However, time averaged load current transient behavior is practically characterized by a single load time constant

$$T_L \quad L/R \tag{15}$$

without any dependence on operating conditions (duty ratio). This is because the average load voltage (instantaneous load voltage example for ideally balanced capacitor voltage $V_{DC}/2$ is shown in Fig. 2) actually does not depend on capacitor voltage as illustrated by Fig. 6.

An average capacitor voltage curve contains two exponential terms – a small fast exponent with the load time constant and a large slow exponent. The large dominating capacitor charge time constant T_C increases with a duty ratio increase – slowly for duty ratios $0 \quad D \quad 0.5$ and dramatically for $D \quad 0.6$. The explanation is that, for the duty ratio approaching unity, capacitor time constant strives to infinity, because for $D \quad 1$ the capacitor is totally disconnected.

The FC convertor modeling approach based on analytical solutions of corresponding linear time-invariant differential equations on different switching intervals is not limited to a single-phase single-leg three-level (single capacitor) converter with DC modulation.

If the FC converter level count is more than three (two and more flying capacitors), still there is no any problem to obtain simple analytical solutions for individual switching interval differential equations. The reason is that, in multiple flying capacitors case, as different capacitors are connected in series, the overall system of equations still has the 2nd order because the capacitor voltages are linearly dependent.

For AC modulation, switching intervals durations (1), (2) are not constant - they vary with sinusoidal voltage command. For negative commanded voltages there is additional switching topology (Fig.7) instead of that of Fig.3.

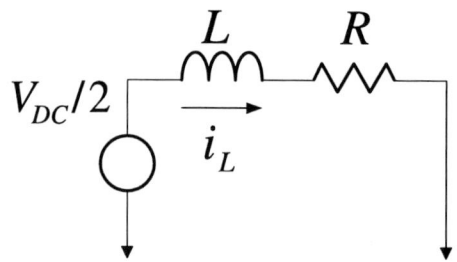

Fig. 7. FC converter topology for negative output voltage ($\overline{S1S2}$) (the capacitor is disconnected)

The FC converter simulation results for combined DC-AC unipolar ($D \geq 0$) excitation

$$D(t) \quad D_0 - D_0 \cos(\omega_f t) \quad (16)$$

are presented in Fig. 8.

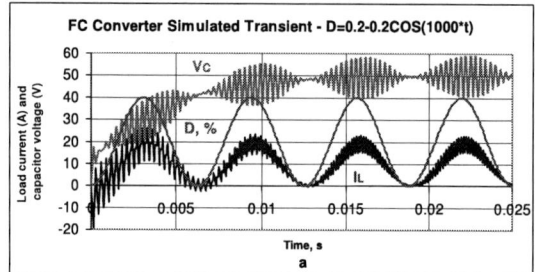

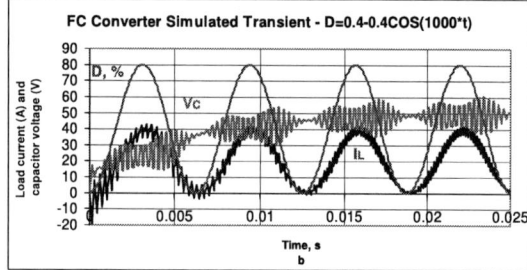

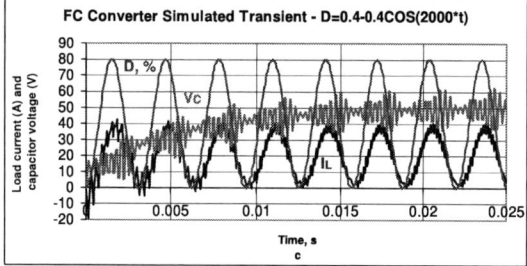

Fig. 8. FC converter load current and capacitor voltage simulation for combined DC-AC excitation: a - $D(t)$ $0.2-0.2COS(1000 \cdot t)$; b - $D(t)$ $0.4-0.4COS(1000 \cdot t)$; c - $D(t)$ $0.4-0.4COS(2000 \cdot t)$

As expected for inductive load, voltage (duty ratio) AC component leads the current one. Capacitor voltage balance dynamics still shows aperiodic behavior with a dominant slow exponent. With average duty ratio increase, capacitor charge rate slightly slows down (Fig. 8, a, b).

For AC frequency increase, the capacitor charge rate remains unchanged (Fig. 8, b, c). This likely holds unless the capacitor charge time constant is larger than an AC period.

IV. FC CONVERTER DYNAMICS ANALYSIS

This section is devoted to analytical solutions for DC PWM modulation. Suppose we take initial conditions of the interval 1 and substitute them, along with the interval 1 duration, into interval 1 dynamic solution (3)-(5). This will give us initial conditions for the interval 2. By substituting them with the interval 2 duration into interval 2 dynamic solution (3), (7), (8), we obtain initial conditions for the interval 3. Carrying out the same procedure for the intervals 3, 4, we complete a PWM period and obtain a linear difference vector equation in the form

$$X(t \quad T_{PWM}) \quad A(D)X(t) \quad B(D)\frac{V_{DC}}{2}, \quad (17)$$

where

$$A(D) \quad A_4 A_3 A_2 A_1; \\ B(D) \quad A_4 \quad A_3 \quad A_2 B_1 \quad B_2 \quad B_3 \quad B_4. \quad (18)$$

This procedure may be considered as a kind of averaging on a PWM period (in "fast" time) to obtain averaged equations in "slow" time.

While calculating the matrices A and B in (17), we started with the switching interval 1. However, there are three more possibilities to get started with the intervals 2, 3, and 4 that will, generally speaking, generate different system matrices.

Assuming system (17), (18) stability, a steady state solution may be found as

$$X(\infty) \quad I - A(D)^{-1} B(D)\frac{V_{DC}}{2}, \quad (19)$$

I - unity matrix.

Not only the solution (19) is a function of the duty ratio, it is also the intervals order (1234, 2341, 3412, 4123) dependent. For good practical converter, current and voltage ripples should be low and the solution (19) is supposed to be close to (14) for any intervals order.

Using "on average" derivatives approximations in "slow" time

$$\frac{di}{dt} \approx \frac{i(t \quad T_{PWM}) - i(t)}{T_{PWM}}; \\ \frac{dv}{dt} \approx \frac{v(t \quad T_{PWM}) - v(t)}{T_{PWM}}, \quad (20)$$

we obtain the FC converter "averaged" differential equations in the form

$$\frac{dX}{dt} = \frac{1}{T_{PWM}}[A(D) - I]X + \frac{1}{T_{PWM}}B\frac{V_{DC}}{2}. \quad (21)$$

Note that in (21) we use an actual PWM period and there is no need to perform a limit transition $T_{PWM} \to 0$.

An accurate solution and switched simulation will show twice transistor switching frequency ripples in both load current and capacitor voltage. These ripples are filtered out in the averaged model (21).

Again, four discrete (and matching continuous) models are possible dependent on the initial switching interval selection. However, it may be shown that the averaged switched system characteristic polynomial and eigenvalues (natural frequencies) don't depend on specific initial interval selection.

A. Time Constants for Zero Duty Ratio DC PWM

Consider first zero duty ratio $D = 0$. Then the PWM period is comprised of the two intervals - 2 and 4 - with durations $\Delta t_2 = \Delta t_4 = T_{PWM}/2$ (unity matrices A_1, A_3 in (18)). The resulting matrix A characteristic polynomial does not depend on the intervals 2 and 4 order -

$$P(\lambda) = \lambda^2 - 2\exp(-\alpha T_{PWM})\left(1 - 2\frac{\alpha^2}{\omega^2}\sin^2\left(\frac{\omega T_{PWM}}{2}\right)\right)\lambda \quad (22)$$

$$+\exp(-2\alpha T_{PWM}).$$

The roots of (22) may be presented as

$$\begin{aligned}\lambda_1 &= \exp(-\alpha T_{PWM})\lambda_1'; \\ \lambda_2 &= \exp(-\alpha T_{PWM})\lambda_2',\end{aligned} \quad (23)$$

$0 < \lambda_1 < \lambda_2 < 1$, with $\lambda_{1,2}'$ being the roots of

$$P'(\lambda) = \lambda^2 - 2\left(1 - 2\frac{\alpha^2}{\omega^2}\sin^2\left(\frac{\omega T_{PWM}}{2}\right)\right)\lambda + 1, \quad (24)$$

$0 < \lambda_1' < 1 < \lambda_2'$.

Once we get the discrete system (17) eigenvalues (23), we calculate equivalent continuous system time constants as

$$\begin{aligned}T_1 &= -T_{PWM}/\ln(\lambda_1); \\ T_2 &= -T_{PWM}/\ln(\lambda_2).\end{aligned} \quad (25)$$

A small parameter

$$\delta = (\alpha/\omega)\sin 0.5\omega T_{PWM} \quad (26)$$

naturally arises in (22), (24) because the switching frequency is much higher than the natural frequency of the equivalent RLC-circuit $\omega T_{PWM} \ll 1$.

Using square root

$$\sqrt{1+x^2} = 1 + \frac{1}{2}x^2 + o(x^3)$$

and logarithm

$$\ln(1+x) = x - \frac{1}{2}x^2 + \frac{1}{3}x^3 + o(x^3)$$

series expansions, we approximate the logarithms of (23) as

$$\begin{aligned}\ln\lambda_1 &= -\alpha T_{PWM} - 2\delta - (1/3)\delta^3 + o(\delta^4); \\ \ln\lambda_2 &= -\alpha T_{PWM} + 2\delta + (1/3)\delta^3 + o(\delta^4).\end{aligned} \quad (27)$$

Expanding sine function for the small parameter (26)

$$\sin x = x - \frac{1}{6}x^3 + o(x^4),$$

we eventually obtain the approximate expressions for the two time constants (25) for $D = 0$:

$$T_1(0) = \frac{L}{R}, \quad (28)$$

$$T_2(0) = 48\frac{L}{R}\frac{LC}{T_{PWM}^2}. \quad (29)$$

The "small" time constant (28), as we expected based on simulation experience and instantaneous load voltage analysis, is actually the load associated one and does not depend on the capacitance.

The "large" capacitor charge time constant (29) may be interpreted as

$$T_2(0) = 48\frac{LC}{T_{PWM}^2}T_1(0)$$

or capacitor equivalent charge resistance my be viewed as

$$R_C = 48R\frac{T_1^2}{T_{PWM}^2} = \frac{48}{R}\frac{L^2}{T_{PWM}^2}.$$

According to our numerical calculations, for reasonable converter parameters that provide low ripples of load current and capacitor voltage, large time constant expression (29) holds with an excellent accuracy.

Example 2. Consider a FC converter with the parameters (12). Its transient for $D = 0$, zero DC bus voltage and zero initial load current is shown in Fig. 9.

Capacitor charge / discharge time constant according to (29) amounts to

$$T_2(0) = 3.33ms.$$

This is what can be observed from capacitor transient graphs - charge (Fig. 5, a) and discharge (Fig. 9).

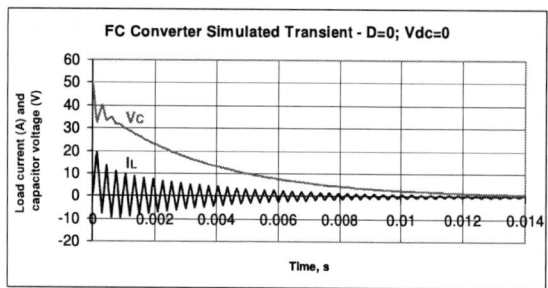

Fig. 9. Capacitor discharge for $V_{DC} \quad 0V$; $D \quad 0$; $i(0) \quad 0A$

Starting from section III, we assumed throughout this section oscillating step response of the equivalent LCR-circuit (Fig.4). For aperiodic step response ("large" resistance $R \quad 2\sqrt{L/C}$),

$$\alpha \quad 0.5R/L; \; \alpha_1 \quad \sqrt{\alpha^2 - \omega_0^2}; \; \omega_0^2 \quad 1/(LC)$$

and characteristic polynomial is different

$$P(\lambda) \quad \lambda^2 - 2\exp - \alpha T_{PWM}\left(1 \quad 2\frac{\alpha^2}{\alpha_1^2}sh^2\left(\frac{\alpha_1 T_{PWM}}{2}\right)\right)\lambda \quad (30)$$

$$\exp - 2\alpha T_{PWM} \; .$$

However, the small parameter analysis of (30), (23) shows that approximate time constants expressions (28), (29) hold.

B. Time Constants for an Arbitrary Duty Ratio DC PWM

Now consider a general case of an arbitrary DC PWM duty ratio $0 \le D \le 1$. First, it can be shown that the small load associated time constant does not practically depend on duty ratio meaning that (28) holds for any duty ratio

$$T_1(D) \quad T_1(0) \quad \frac{L}{R}, \; 0 \le D \le 1.$$

Second, we found a small parameter approximation of the slow exponent decay factor (inverse time constant) in the following form:

$$\alpha_2(D) \quad \alpha_2(0)(1 - 3D^2 \quad 2D^3);$$

$$\alpha_2(0) \quad 1/T_2(0) \quad \frac{RT_{PWM}^2}{48L^2C} \qquad (31)$$

(observe $\alpha(1) \quad 0$ and $\alpha'(1) \quad 0$).

Therefore, the equivalent capacitor charge time constant for an arbitrary duty ratio is given by

$$T_2(D) \quad \frac{T_2(0)}{1 - 3D^2 \quad 2D^3}. \qquad (32)$$

The capacitor charge time constant (32) increases with duty ratio and, for $D \to 1$, $T_2(D) \to \infty$ because for unity duty ratio the capacitor is totally disconnected.

C. Time Constants Estimation for AC PWM

For DC PWM, FC converter models (17), (22) obtained by averaging on PWM period are linear time invariant. For AC PWM, FC converter models (17), (22) are linear time-variable and, unfortunately, we can't directly apply the power of linear time-invariant systems analysis.

However, if the load fundamental frequency is relatively high, then, assuming a quasi-static behavior, the FC converter linear time-invariant model for AC modulation may hopefully be obtained by appropriate fast dynamics time averaging on a fundamental period. We have to acknowledge that, in general, it may be not trivial because an adequate dynamics averaging mathematical tool is required.

For the single-capacitor FC converter (Fig.1), we obtained analytical time constant (decay factor) expressions (30)-(32) for DC modulation. For AC PWM, it is reasonable to assume that the expression (30) for the small duty ratio independent time constant holds.

For an exponent with a fast periodical decay factor variation

$$\exp - \alpha(t)t \; , \qquad (33)$$

where $\alpha(t)$ period equals T, an averaged decay factor amounts to

$$\alpha_0 \quad \frac{1}{T}\int_0^T \alpha(t)dt \qquad (34)$$

and equivalent time constant is $T_0 \approx 1/\alpha_0$ given that $T_0 \quad T$.

Now consider sinusoidal PWM with modulation index M, $0 \le M \le 1$, and relatively high fundamental frequency (that is still essentially lower than a switching one)

$$D(t) \quad M \sin(\omega_f t) \qquad (35)$$

for the sine arguments

$$0 \le \omega_f t \le \pi \qquad (36)$$

(not to enter into negative duty ratio problem).

Then, to obtain the large time constant for AC PWM, we must average the DC PWM decay factor (31) on the half period (36) (the quarter period will also do). This way, for AC PWM anticipated slow exponent decay factor is

$$\alpha_2(M) \quad \frac{RT_{PWM}^2}{48L^2C}\left(1 - \frac{3}{2}M^2 \quad \frac{8}{3\pi}M^3\right) \qquad (37)$$

and the equivalent capacitor charge time constant

$$T_2(M) \quad \frac{48L^2C}{RT_{PWM}^2\left(1 - \frac{3}{2}M^2 \quad \frac{8}{3\pi}M^3\right)}. \qquad (38)$$

The comparison of accurate DC PWM and anticipated AC PWM capacitor charge decay factors and time constants is given in Fig.10 and Fig.11 respectively.

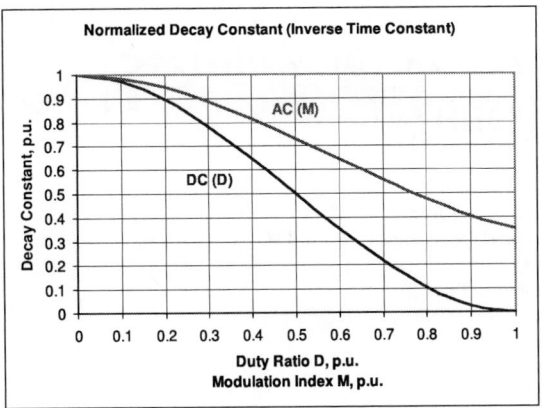

Fig. 10. Normalized capacitor charge exponential decay constant for DC and AC modulation

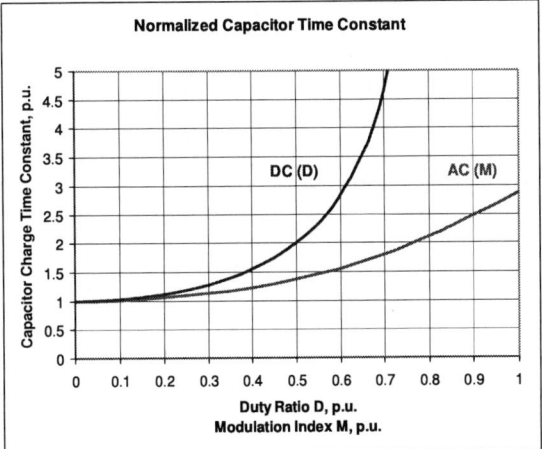

Fig. 11. Normalized capacitor charge time constant for DC and AC modulation

Note twice DC PWM time constant increase for 50% duty ratio and about 3 times AC PWM time constant increase for 100% modulation index (overmodulation is not considered).

One can also easily calculate the equivalent capacitor charge time constant for combined DC-AC modulation (16) by averaging on AC period. The comparison of Fig. 8, b and c confirms the assumption that the fundamental frequency has no impact on the capacitor charge rate unless the capacitor charge time constant is larger than an AC period.

V. CONCLUSION

While the reported FC converter voltage balance dynamics research mostly deals with an AC modulation and is based on heavy frequency domain transformations, DC modulation analysis seems to be a missing intermediate link in physical understanding of FC converter transient behavior.

A general time domain approach to construct a family of FC converter models – both discrete and continuous – by sewing analytical solutions for consecutive switching intervals is applicable to modeling a multilevel multiphase converter and is not limited to DC PWM. However, for DC PWM the model is linear time-invariant that makes voltage balance dynamics analysis straightforward.

For a primitive single capacitor three-level FC converter, we obtained extremely simple and accurate expression for a capacitor related time constant by applying small parameter technique. The small parameter naturally arises for practical converters with low current and voltage ripples. The capacitor time constant formula reveals the capacitor charge rate dependence on inductive load parameters, carrier frequency, and duty ratio and does not depend on the equivalent LCR-circuit transient behavior. This time constant increases with DC PWM duty ratio and strives to infinity with the duty ratio approaching unity (disconnected capacitor).

We have to admit that a generalization to an AC modulation case is not trivial and an adequate mathematical technique to perform FC converter dynamics averaging across the AC trajectories is required. However, an insight into the voltage balance mechanism gained from a DC PWM consideration may definitely be useful for an AC PWM as well. For a single capacitor single leg three-level FC converter, we estimated AC PWM associated capacitor charge time constant dependence on modulation index by averaging that obtained for DC PWM on the AC fundamental period.

ACKNOWLEDGMENT

The first and second authors gratefully acknowledge Elmo Motion Control and General Satellite Corporation management respectively for on-going support to advanced applied power electronics research and development.

REFERENCES

[1] J. S. Lai and F. Z. Peng, "Multilevel Converters - New Breed of Power Converters," *Proc. IEEE Ind. Appl. Society Annual Meeting*, 1995, pp. 2348–2356.

[2] D.G. Holmes, T.A. Lipo, Pulse Width Modulation for Power Converters: Principles and Practice. Hoboken, NJ: John Wiley, 2003.

[3] M. Marchesoni and P. Tenca, "Theoretical and Practical Limits in Multilevel MPC Inverters with Passive Front Ends," *Proc. European Conf. on Power Electronics and Applications (EPE)*, Graz, Austria, Aug. 27-29, 2001.

[4] T. Meynard, M. Fadel, and N. Aouda, "Modeling of Multilevel Converters," *IEEE Trans. Ind. Elec.*, vol. 44, no. 3, pp.356-364, June 1997.

[5] X. Yuang, H. Stemmler, and I. Barbi, "Self-Balancing of the Clamping-Capacitor-Voltages in the Multilevel Capacitor-Clamping-Inverter under Sub-Harmonic PWM Modulation," *IEEE Trans. Power Electron.*, vol. 16, no. 2, pp. 256-263, March 2001.

[6] R. Wilkinson, H. de Mouton, and T. Meynard, "Natural Balance of Multicell Converters: the Two-Cell Case," *IEEE Trans. Power Electron.*, vol. 21, no. 6, pp. 1649-1657, November 2006.

[7] R. Wilkinson, H. de Mouton, and T. Meynard, "Natural Balance of Multicell Converters: the General Case," *IEEE Trans. Power Electron.*, vol. 21, no. 6, pp. 1658-1666, November 2006.

[8] B.P. McGrath and D.G. Holmes, "Analytical Modeling of Voltage Balance Dynamics for a Flying Capacitor Multilevel Converter," *Proc. IEEE Power Electronics Specialists Conference (PESC)*, Orlando, FL, June 17-21 2007, pp. 1810-1816.

Simulation of Simplified Seven Level Multilevel Converter Circuit.

[1]Gerardo Ceglia ; [2]Víctor Guzmán; [3]Carlos Sánchez; [4]Fernando Ibáñez; [5]Julio Walter; [6]María Giménez

[1,2,5,6] Universidad Simón Bolívar University (Venezuela),
Grupo de Electrónica de Potencia, Valle de Sartenejas, Vía Baruta, Apto. 1080A
Caracas, Venezuela
[3,4] Universidad Politécnica de Valencia (Spain)
Grupo de Sistemas Avanzados en Ingeniería Energética (SAVIE),
Instituto de Ingeniería Energética.
Valencia, España
Email [1,2,5,6]: [1]gceglia@usb.ve, [2]vguzman@usb.ve, [5]jwalter@usb.ve, [6]mgimenez@usb.ve
Email [3,4]: [3]csanched@eln.upv.es, [4]fibanez@eln.upv.es
Tel. [1,2,5]: +58 212 9063630, Fax. [1,2,5]: +58 212 9063631
Tel. [3,4]: +34 963 877007, Ext.- 76082

Abstract: At present multilevel converters are technically interesting due to their high power handling capabilities, low output harmonics level and reduced requirements in blocking voltages in the switching devices ratings and lower commutation stresses and losses. The multilevel converter configurations now in use have as their main disadvantage their circuit complexity, requiring a great number of power devices and passive components in their implementation, and increasing control circuit complexity. System costs is rather high, and therefore the multilevel inverters are considered cost effective only in very high power applications. In this work a new seven level inverter having a reduced component count is presented, based upon the H bridge with auxiliary switch 5 level architecture. This new configuration may be of interest for applications working at lower and medium power levels. Also a new seven level inverter controller is introduced. The combination of the new power converter topology and the new controller circuit reduces both system cost and complexity

Keywords: Multilevel, Converter, IGBTs, Pspice.

I. INTRODUCTION.

Multilevel inverters configurations were first proposed as compromise solutions for applications requiring blocking voltages higher than those provided by the switching devices offered in the market. The two basic multilevel topologies with fixed and floating capacitors, presented in [1], were developed into diverse design having three [2-3], four [4-5], five, [6-16], seven or more levels [17]. In all these variants, the main disadvantage found when the number of levels is increased is the great number of power switches required (IGBTs, MOSFET, etc); especially when three phase inverters are concerned [18-28]; due to this the actual power circuits implemented are complex and expensive.

An additional disadvantage is the complex control circuitry required to drive the many switches involved in multilevel inverter operation. Not only the number of gate drive circuits is high, but since it is necessary to ensure that the DC levels in all the capacitors are balanced, their coordination is a complex task that must be performed by a powerful high performance (and therefore expensive) processor.

Overall these characteristics (complex power circuitry, high number of gate drivers and high computational load) combine to drive both system complexity and cost to very high levels, making the multilevel inverter a solution that could be applied only in very high power applications such as marine motor drives, massive chemical industry drives and high power transmission systems [19-28] where the overall capital invested was so high that the power converter cost was not the main consideration.

The continuing development of high power, high switching frequency devices such as IGBTs (Insulated-Gate Bipolar Transistors) working at 3.3kV, 4.5kV and 6.5kV and IGCT (Insulated-Gate Commutated Thyristors) working at 4.5 kV or 6 kV [29-30] has improved overall converter performance, and renew the interest in multilevel topologies, that may be able to compete in the market with the standard two level PWM converters at lower power ranges.

Even taking into account this tendency to lower the prize level at which multilevel converters can compete with standard configurations, the prize difference will remain a strong disadvantage, unless the complexity issue is solved, both at the power circuit and the modulator circuit levels.

This work presents the initial PSPICE simulated results obtained when the new multilevel converter topology presented in [32] is applied to the design of a seven levels converter. As demonstrated below, this new configuration produces a significant reduction in the number of power devices and gate drivers required. To further reduce overall system complexity, a new multilevel control structure that can be implemented using low cost FPGA circuits is proposed and simulated to drive the power converter topology.

The initial simulation results prove that the design ideas work as expected, at least at the circuit simulation level, and further tests using a scaled down circuit prototype are under way.

II. THE NEW MULTILEVEL TOPOLOGY

Figure 1 shows the complete power circuit used in the seven level inverter.

The H bridge is formed by the four main power devices, DISP1 to DISP4. A capacitor voltage divider, formed by C1, C2 and C3, provides all the voltages required by the multilevel converter operation. The two auxiliary switches, formed respectively by the controlled switch DISP5 and the four diodes, D5 to D8, and the controlled switch DISP6 and the four diodes, D9 to D12 connects as required the center point of the left hand half bridge to points B or C in the capacitive voltage divider.

Table 1 shows the switching combinations that generate the seven output voltage levels, which are: VS, 2VS/3, VS/3, 0, -VS/3, -2VS/3, -VS.

Figure 2 shows in heavy lines the current paths defined by the switching combinations listed in Table 1. The required voltage output levels are generated as follows:

1. VS level (fig. 2a): DISP1 is ON, connecting the positive load terminal to the positive DC supply terminal (Vs), and DISP4 is ON, connecting the negative load terminal to ground; the load voltage is Vs. If load current is negative it will flow through the auxiliary diodes D1 and D4.
2. 2VS/3 level (fig. 2b): Auxiliary switch DISP5 is ON, connecting the positive load terminal to point A either through diodes D5 and D8 if load current is positive, as shown in the figure, or through diodes D6 and D7 if load current is negative. DISP4 is ON, connecting the negative load terminal to ground; the load voltage is 2Vs/3.
3. VS/3 level (fig. 2c): Auxiliary switch DISP6 is ON, connecting the positive load terminal to point A,

either through diodes D9 and D12 if load current is positive, as shown in the figure, or through diodes D10 and D11 if load current is negative. DISP4 is ON, connecting the negative load terminal to ground; the load voltage is Vs/2.

4. Zero level: This level can be produced by two switching combinations. In fig. 2g the two main switches DISP3 and DISP4 are on, short circuiting the load. The same output will result if the other two main switches, DISP1 and DISP2 are ON.
5. -VS/3 level (fig. 2f): Auxiliary switch DISP5 is ON, connecting the positive load terminal to point A, either through diodes D6 and D7 if load current is negative, as shown in the figure, or through diodes D5 and D8 if load current is positive. DISP2 is ON, connecting the negative load terminal to Vs (fig. 2f); the load voltage is -Vs/2.
6. -2VS/3 level (fig. 2e): Auxiliary switch DISP6 is ON, connecting the positive load terminal to point A, through diodes D10 and D11 if load current is negative, as shown in the figure, or through diodes D9 and D12, in load current is positive. DISP4 is ON, connecting the negative load terminal to ground; the load voltage is – 2Vs/3.
7. -VS level (fig. 2d): DISP2 is On, connecting the load negative terminal to Vs, and DISP3 is ON, connecting the load positive terminal to ground; the load voltage is -Vs. If load current is negative it will flow through the auxiliary diodes D2 and D3.

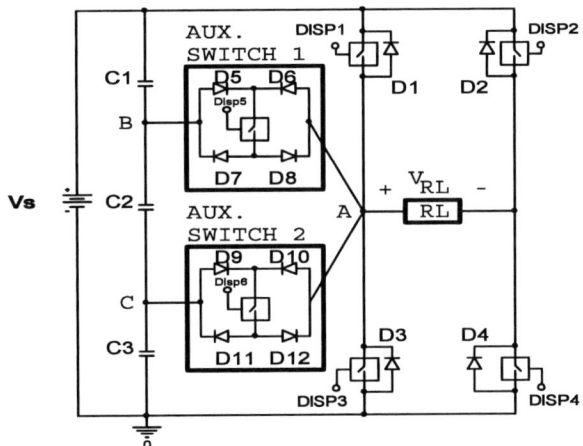

Fig. 1. Proposed 7 level H bridge inverter.

TABLE I.
SWITCHING COMBINATIONS REQUIRED TO GENERATE THE SEVEN VOLTAGE LEVELS IN THE OUTPUT WAVEFORM

DISP1	DISP2	DISP3	DISP4	DISP5	DISP6	D5	D6	D7	D8	D9	D10	D11	D12	V_{RL}
on	off	off	on	off	off	off	off	off	off	off	off	off	off	VS
off	off	off	on	on	off	on	off	off	on	off	off	off	off	2VS/3
off	off	off	on	off	on	off	off	off	off	on	off	off	on	VS/3
off	off	on	on	off	off	off	off	off	off	off	off	off	off	0
off	on	off	off	on	off	off	on	on	off	off	off	off	off	-VS/3
off	on	off	off	off	on	off	off	off	off	on	on	off	-2VS/3
off	on	on	off	off	off	off	off	off	off	off	off	off	off	-VS

269

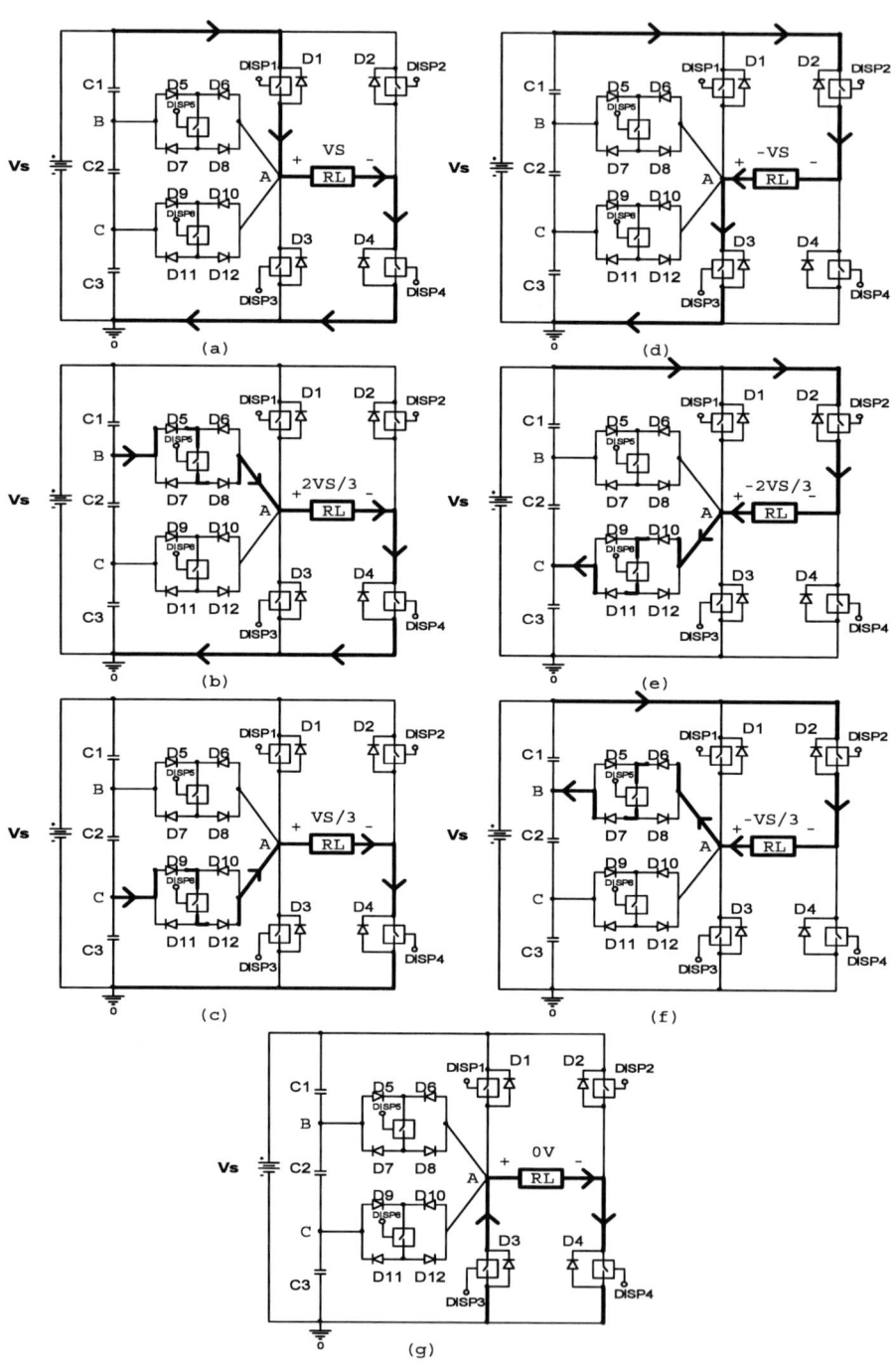

Fig. 2. Switching combinations required to generate the output voltage, VRL, delivered to the load in each of the seven output intervals. The heavy line shows the current path through the converter switches and the load. a) VRL = VS, b) VRL = 2 VS/3, c) VRL = VS/3, d) VRL = 0 , e) VRL = -VS/3, f) VRL =-2 VS/3, g) VRL = -VS.

Table 1, calculated according to [33], presents the number of components required to implement a 7-levels inverter using the new topology and three previously defined ones: the two that can be considered as the standard multilevel stages, the Diode Clamped and the Capacitor Clamped configurations and a new and highly improved multilevel stage, the Asymmetric Cascade configuration.

TABLE II.
COMPARISON BETWEEN FOUR DIFFERENT 7-LEVEL INVERTERS TOPOLOGIES

Multilevel Inverter type	H bridge, Auxiliary switch	Diode Clamped	Capacitor Clamped	Asymmetric Cascade
Main switches	4	36	36	36
Required blocking voltage	Vs/2	Vs/7	Vs/7	Vs/7
Antiparallel diodes	8	36	36	36
Auxiliary switches	2	36	-	-
Required blocking voltage	Vs/3	-	-	-
Auxiliary diodes	4	-	-	-
Switches, total	6	36	36	36
Diodes, total	12	72	36	36
Capacitors	3	7	17	9

The proposed configuration has a significant advantage over the other three in the number of all required component types. Compared with the best of the other configurations, it requires only six controlled switches instead of the thirty six required by the other configurations (83% reduction), only twelve diodes instead of the thirty six required by the capacitor clamped or the asymmetric cascade configurations (66% reduction), and only three capacitors instead of the nine required by the asymmetric cascade configuration (66% reduction). In the other hand, the proposed configuration is at a disadvantage when the required voltage ratings are compared: in the new configuration the main power switches are required to block one half of the main supply voltage and the auxiliary switches one third the main supply voltage, as opposed to 1/7 of the main supply voltage in the other configurations. The new topology has a very significant advantage in those applications where circuit complexity is the main problem and will be at a disadvantage where maximum reduction in device blocking voltage is the main design aim, hence the new configuration is adequate for medium power level applications.

III. CONVERTER MODULATOR

Since the new power stage topology with multiple auxiliary switches is based on the simplified five level topology with auxiliary switch presented in [31], it was decided to perform the initial validation tests in SPICE driving it with the PSPICE model of the Sigma-Delta Adaptive modulator presented in [32], already developed and tested for the five level topology. To do this the sigma delta modulator's two level digital output is combined with the information provided by the twelve activation windows defined in figure 3 in relation to the required ideal voltage output sinusoidal waveform.

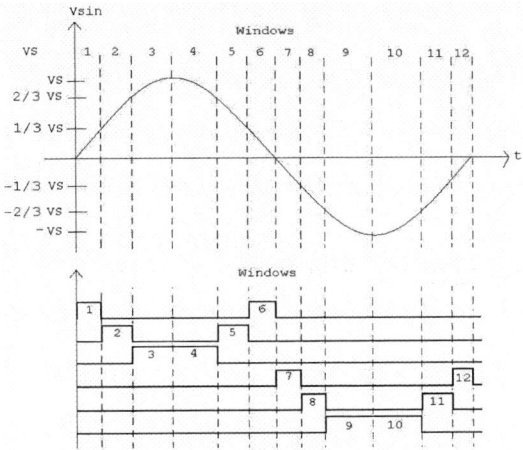

Fig. 3. Seven level converter activation windows.

The circuit required to generate the activation windows and combine this information with the sigma-delta modulator output, called the "Multilevel Drive Signal Generator" is presented in block diagram in figure 4. This block uses as inputs the ideal sine wave reference (a 50 Hz one in this example) and the digital output provided by the sigma-delta generator. The timing references extracted by the "windows" blocks are combined with the modulation information provided by the sigma-delta block, and the twelve signals defining the sequence of twelve PWM voltage bands that forms the seven level converter output are combined in the Decoder Switch Block to finally generate the six drive signals for the six power switches: the four "main switches" in the H bridge (DISP 1 to 4) and the two "auxiliary switches" DISP 5 and 6).

IV. SIMULATED RESULTS.

As the first step in the testing of the new seven level converter, the operation of the new topology seven level inverter power stage, the sigma-delta modulator and the multilevel drive signal generator were simulated in ORCAD-PSPICE.
System operation was simulated at two different switching frequencies: 10 KHz and 200 KHz. These frequencies were selected since they are representative of the ones in use in the two extreme application segments: high power inverters, where switching losses are a main concern and the switching frequency must therefore be low, and low to medium power inverters where switching losses are less important and low output distortion is

paramount, driving switching frequencies to the higher available values.

The seven voltage levels used in the simulations are VS = 150V, 2/3 VS = 100V, 1/3 VS = 50V, 0V, -1/3 VS = -50V, -2/3 VS = -100V, -VS = -150V. These values were selected to ensure compatibility with the inverter test rig available in the lab, thereby simplifying the performance of the initial prototype tests.

Figures 8 and 10 show the output voltage waveform taken from the load terminals (V_{RL} in fig. 1) when the converter fundamental output frequency is set at 50 Hz and the converter commutation frequency is set at 10 KHZ (fig. 8) and at 200 kHz (fig. 10). In both cases it is clearly visible that both the power stage topology and the "Multilevel Drive Signal Generator" are working as designed, since the simulated output is identical to the ideal output defined for a seven-level converter.

Figures 9 and 11 show the simulated seven level converter output voltage waveform harmonic spectrum as calculated by the ORCAD-PSPICE program using a FFT algorithm when the converter fundamental output frequency is set at 50 Hz and the converter commutation frequency is set at 10 KHZ (fig. 9) and at 200 kHz (fig. 11). The graphs show that the output waveform in both cases is very clean, with only the fundamental frequency (50Hz) present as a significant component in the low and medium frequency range, up to 10 KHz.

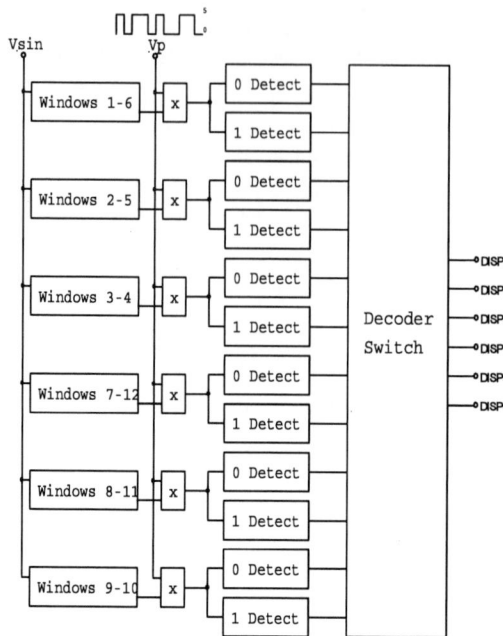

Fig. 4. Multilevel Drive Signal generator

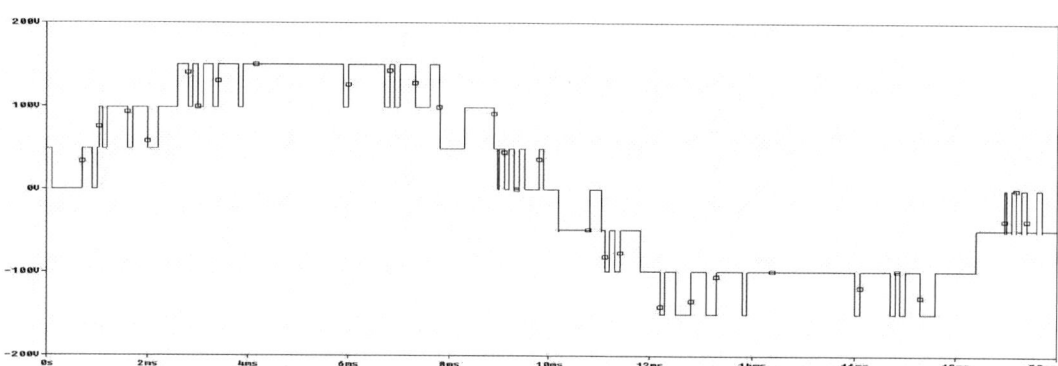

Fig. 5. Simulated output voltage waveform. Seven level converter operating at 10khz switching frequency.

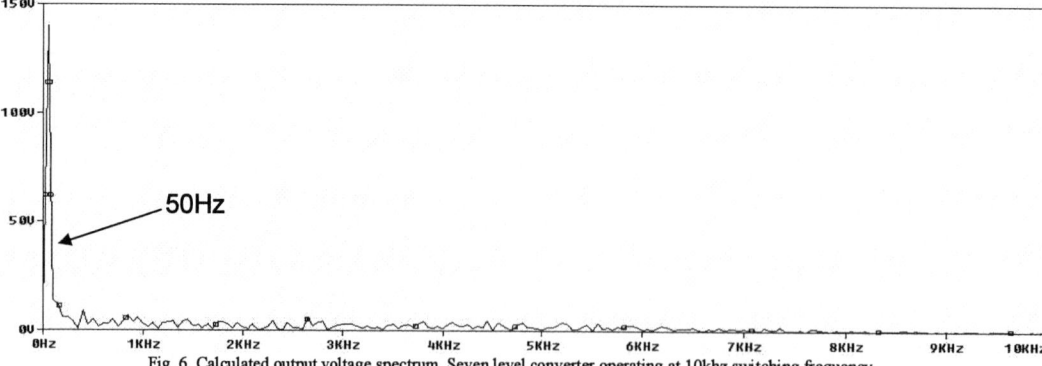

Fig. 6. Calculated output voltage spectrum. Seven level converter operating at 10khz switching frequency.

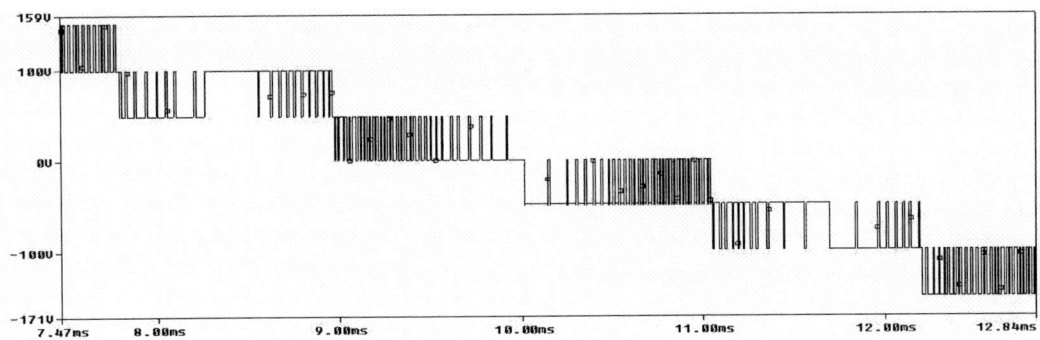

Fig. 7. Simulated output voltage waveform close-up. Seven level converter operating at 200khz switching frequency.

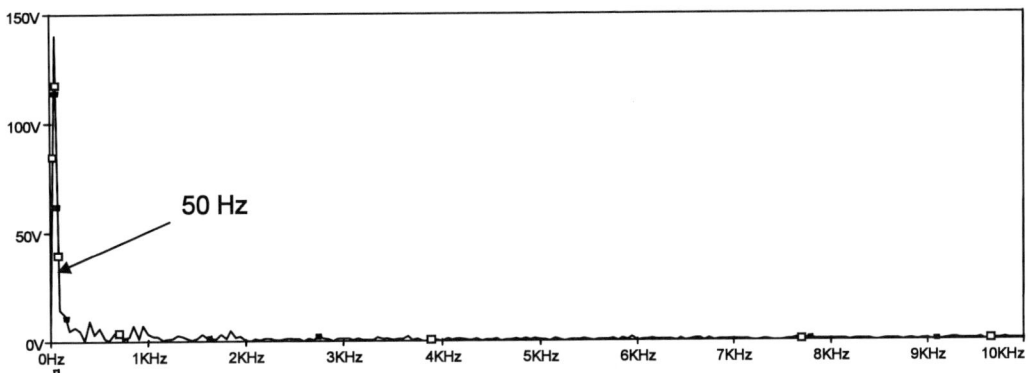

Fig. 8. Calculated output voltage spectrum. Seven level converter operating at 200khz switching frequency.

CONCLUSIONS

1.- Simulated results prove that the proposed seven level converter based upon the H bridge plus auxiliary switch five level inverter operates as proposed.

2.- Simulated results prove that the proposed Multilevel Drive Signal Generator can be used in combination with a standard sigma-delta generator to produce an operational seven level controller.

3.- The combination of the new seven level power stage and the new multilevel controller produces a multilevel converter configuration requiring less components than the configurations already presented in the literature.

4.- In the proposed new seven level power stage the main power switches are required to block one half of the main supply voltage and the auxiliary switches one third the main supply voltage, as opposed to 1/7 of the main supply voltage in the other configurations, hence this configuration is not the best for very high power applications where maximum reduction in device blocking voltage is the main design aim.

4.- Since the simulated results present no unexpected problems, and confirm the predicted advantages, the new configuration development will proceed to the next stage: actual circuit tests in a reduced scale laboratory prototype, aimed at studying the long term stability in the capacitor's voltages when the inverter is connected to a variable load.

REFERENCES.

[1] Jose Rodriguez, Jih-Sheng Lai, Fang Zheng Peng. "Multi-nivel Inverter: A Survey of Topologies, Controls, and applications". *IEEE Trans. On Industrial Electronics*, Vol. 49, No.4, August 2002.

[2] Emilio J. Bueno, Roberto Garcia, Marta Marrón, Felipe Espinosa. "Modulation Techniques Comparación for Three Levels VSI Converters". 0-7803-7474-06/02/$17.00 2002 IEEE.

[3] H. du Toit Mouton. "Natural Balancing of Three-Level Neutral-Point-Clamped PWM Inverters". *IEEE Trans. on Industrial Electronics*, Vol. 49, No.5, October 2002.

[4] Keith Corzine, Xiaomin Kou, James R. Baker. "Dynamic Average-Value Modeling of Four-Level Drive System". *IEEE Trans. on Industrial Electronics*, Vol. 18, No.2, October 2003.

[5] Gautam Sinha, Thomas A. Lipo. "A Four-Level Inverter Based Drive with a Passive Front End". *IEEE Trans. on Power Electronics*, Vol. 15, No.2, March 2000.

[6] C.K.Lee, Joseph S.K.Leung, S.Y. Ronb Hui, Henry Shu-Hung Chung, "Circuit-Level Compararison of STATCOM Technologies". *IEEE Trans. On Power Electronics*, Vol. 18, No.4, July 2003.

[7] Ying Cheng, Chang Qian, Mariesa L. Crow, Steve Pekarek, Stan Atcitty, "A Comparison of Diode-Clamped and Cascaded Multilevel Converters for a STATCOM With Energy Storage". *IEEE Trans. On Industrial Electronics*, Vol. 53, No.5, October 2006.

[8] Cassiano Rech, José Renes Pinheiro, "Hybrid Multilevel Converters: Unified Analysis and Design Considerations". *IEEE Trans. On Industrial Electronics*, Vol. 54, No.2, August 2007.

[9] Amit Kumar Gupta, Ashwin M. Khambadkone, "A Space Vector Modulation Scheme to Reduce Common Mode Voltage for ascaded Multilevel Inverters". *IEEE Trans. On Power Electronics*, Vol. 22, No.5, July 2007.

[10] Mingyao Ma, Lei Hu, Alian Chen, Xiangning He, "Reconfiguration

of Carrier-Based Modulation Strategy for Fault Tolerant Multilevel Inverters". *IEEE Trans. On Power Electronics*, Vol. 22, No.5, September 2007.

[11] Diego E Soto-Sanchez and Tim C, Green. "Voltage Balance and Control in a Multi-Level Unified Power Flow Controller". *IEEE Trans. on Power Delivery*, Vol. 16, No.4, October 2001.

[12] Takashi Ishida, Kouki Matsuse, Kyoaki Sasagawa, Lipei Huang. "Fundamental Characteristics of a Five- Level Double Converter for Induction Motor Drive". 0-7803-6401-5/00/$10.00 2000 IEEE.

[13] Jochen von Bloh, Rik W. De Doncker. "Control Strategies for Multilevel Voltage Source Converter for Medium-Voltage DC Transmission Ssytems". 0-7803-6456-2/00/$10.00 2000 IEEE.

[14] Ying Cheng, Mariesa L. Crow. "A diode-Clamped Multi-level Inverter for the StatCom/BESS". 0-7803-7322-7/02/$17.00 2002 IEEE.

[15] F. Tourkhani, P. Viarouge, T.A. Meynard. "A Simulation-Optimization System for the Optical Design of a Multilevel Inverter". *IEEE Trans. on Power Electronics*, Vol. 14, No.6, November 1999.

[16] Keith A. Corzine, Xioamin Kou. "Capacitor Voltage Balancing in full Binary Combination Schema Flying Capacitor Multilevel Inverters". *IEEE Power Electronic Letters*, Vol. 1, No.1, March 2003.

[17] Feel-Soon Kang, Sung-Jun Park, Man Hyung Lee, Cheul-U Kim, "An Efficient Multilevel-Synthesis Approach and Its Application to a 27-Level Inverter". *IEEE Trans. On Industrial Electronics*, Vol. 52, No.6, December 2005.

[18] Yiqiang Chen, Bakari Mwinyiwiwa, Zbigniew Wolanski, Boon-Teck Ooi, "Unified Power Flow Controller (UPFC) Based on Chopper Stabilized Diode-Clamped Multilevel Converters". *IEEE Trans. On Power Electronics*, Vol. 15, No.2, March 2000.

[19] Leon M. Tolbert, Fang Zheng Peng, Thomas G. Habetler, "Multilevel PWM Methods at Low Modulation Indices". *IEEE Trans. On Power Electronics*, Vol. 15, No.4, July 2000.

[20] Yiqiao Liang, C. O. Nwankpa, "A Power-Line Conditioner Based on Flying-Capacitor Multilevel Voltage-Source Con verter with Phase-Shift SPWM". *IEEE Trans. On Industrial Applications*, Vol. 36, No.4, July/August 2000.

[21] Keith A. Corzine, James R. Baker, "Reduced-Parts-Count Multilevel Rectifiers". *IEEE Trans. On Industrial Electronics*, Vol. 49, No.4, August 2002.

[22] Mansour Hashad, Jan Iwaszkiewicz, "A Novel Orthogonal-Vectors-Based Topology of Multilevel Inverters". *IEEE Trans. On Industrial Electronics*, Vol. 49, No.4, August 2002.

[23] Keith Corzine, *Member, IEEE*, Xiaomin Kou, *Student Member, IEEE*, and James R. Baker, "Dynamic Average-Value Modeling of a Four-Level Drive System". *IEEE Trans. On Power Electronics*, Vol. 18, No.2, July 2003.

[24] Poh Chiang Loh, Donald Grahame Holmes, Yusuke Fukuta, Thomas A. Lipo, "Reduced Common-Mode Modulation Strategies for Cascaded Multilevel Inverters". *IEEE Trans. On Industrial Applications*, Vol. 39, No.5, September/October 2003.

[25] Brendan Peter McGrath, Donald Grahame Holmes, Thomas Lipo, "Optimized Space Vector Switching Sequences for Multilevel Inverters". *IEEE Trans. On Power Electronics*, Vol. 18, No.6, November 2003.

[26] Poh Chiang Loh, Donald Grahame Holmes, Yusuke Fukuta, Thomas A. Lipo, "A Reduced Common Mode Hysteresis Current Regulation Strategy for Multilevel Inverters". *IEEE Trans. On Power Electronics*, Vol. 19, No.1, January 2004.

[27] Xiaomin Kou, Keith A. Corzine, Yakov L. Familiant, "A Unique Fault-Tolerant Design for Flying Capacitor Multilevel Inverter". *IEEE Trans. On Power Electronics*, Vol. 19, No. 4, July 2004.

[28] V. T. Somasekhar, K. Gopakumar, M. R. Baiju,, "A Multilevel Inverter System for an Induction Motor With Open-End Windings". *IEEE Trans. On Industrial Electronics*, Vol. 52, No.3, June 2005.

[29] S. Bernet, "Recent Developments for High Power Converter for Industry and Traction Applications". *IEEE Trans. on Power Electronics* . Vol. 15 No. 6, pp. 1102-1117, Nov 2000.

[30] A. Nagel, S. Bernet y P.K. Steimer. "A 24 MVA Inverter using IGCT Series Connection for Medium Voltage Applications". *Proceeding of the IEEE IAS Annual Meeting*. Chicago, Oct. 2001.

[31] Gerardo Ceglia, Víctor Guzmán, *Member, IEEE*, Carlos Sánchez, Fernando, "A New Simplified Multilevel Inverter Topology for DC–AC Conversion". *IEEE Trans. On Power Electronics*, Vol. 21, No. 5, July 2006.

[32] C. Sánchez, F. Ibáñez, M. Alcañiz, J. Polo, and R. Masot, "Analysis of Sigma–Delta modulation techniques in low frequency DC–AC converters," in *Proc. IEEE 34th PESC'03*, Jun. 2003, pp. 507–512.

[33] R. Pindaro Rico, J. pou Felix, "Convertidores multinivel CC/CA Topologias básicas," in *mundo Electronico*, Jun. 2002, pp. 28–35.

SEPP High-Frequency Inverter Incorporating an Auxiliary Switch and Its Performance Evaluation

H.Ogiwara*, Y.Fujita*, R.Urabe*, M.Itoi*, T.Sugai*, M.kuwata * and M.Nakaoka**
* Deparment of Electrical and Electronic Engineering, Ashikaga Institute of Technology,
268-1 Omae-cho, Ashikaga, Tochigi, 326-8558 Japan
E-mail: ogiwara@ashitech.ac.jp
** Kyungnam University, Korea

Abstract : This paper presents a single ended push-pull (SEPP) high frequency inverter incorporating a reverse blocking active auxiliary quasi-resonant circuit using bipolar mode static induction transistors (BSITs) as active switches. This inverter can realize soft switching of the active switches over wider output power regulation range compared with the conventional SEPP inverter. It is performed by incorporating an auxiliary active switch to the conventional SEPP inverter to feed a required current for the soft switching operation of the main switches. The current fed to the main switches during their short switching period assists their soft switching operation over wide output power regulation range. The detailed evaluation of its operational principle and characteristics is carried out with the aid of computer aided simulation and the experimental result obtained by a bread board.

Keywords :High frequency power converter, Induction heating, Soft switching, Power supply

I. Introduction

Recently, soft switching PWM power conversion systems have been studied extensively. They are operated under zero voltage switching (ZVS) or zero current switching (ZCS) of their switching devices by using an active auxiliary quasi-resonant snubber circuit. Especially, the soft switching power conversion systems expected as the next generation power conversion techniques are classified by the connecting methods of the active auxiliary sub-resonant snubber circuit to the power conversion equipments.[1]

Our proposed high frequency inverter is the conventional PWM controlled single ended push-pull (SEPP) high frequency inverter connected by a reverse blocking active auxiliary sub-switch. By using this sub-switch, a soft switching operation over wide output range of the SEPP inverter can be attained under PWM output power regulation. This paper describes the operational principle and its characteristic behavior of our proposed inverter with the aid of computer aided simulation, and detailed evaluation of the actual characteristics of the inverter. Next, the observed waveforms by using BSITs as the active switches are shown revealing the effectiveness of the inverter.

II. Fundamental circuit and its operational principle

The circuit constitution of high frequency inverters are classified into various types, according to their low costs, output power capacities, soft switching ranges of their output powers and output regulation systems.[1] Figure 1 illustrates the conventional SEPP

inverter which contains two active power semiconductor switching devices (SW1/D1 and SW2/D2), two loss-less capacitances (C1 and C2), a series tuned resonance compensating capacitor (Co) and a induction heating equivalent load (Lo and Ro) [2]. Figure 2 explains the asymmetrical waveforms of the PWM gate pulse driving signals fed to both gates of the active switches to regulate output power of the inverter. As can be seen in the figure, the signals are different at high output power regulation and low output power regulation. Here, we define the duty ratio D of the signal as follows:

$$D = ton1/T \qquad (1)$$

where ton1 and T are the on time of SW1 and one period time of the output power frequency ,respectively.

Next, we determine each signal and positive direction of the currents and voltages as directed in Fig.1 and list the symbols and rating values of circuit parameters in Table 1. According to the above items, the computer aided simulation is carried out to analyze

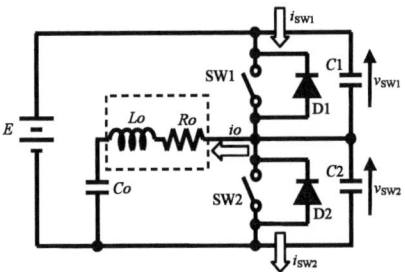

Fig. 1.   Conventional SEPP high frequency inverter.

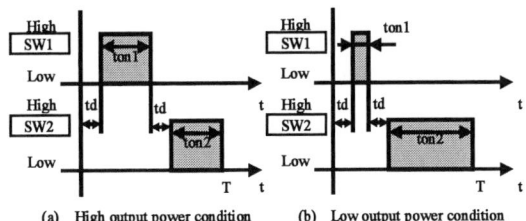

(a)   High output power condition   (b)   Low output power condition

Fig. 2.   Asymmetrical PWM driving signals.

978-1-4244-1741-4/08/$25.00 ©2008 IEEE   275

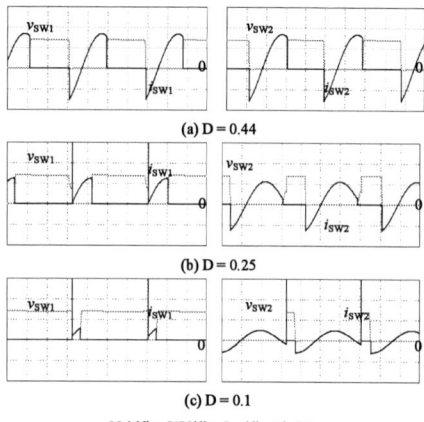

(a) D = 0.44

(b) D = 0.25

(c) D = 0.1

20A/div, 50V/div, 5µs/div, td=1.5µs

Fig. 3. * Simulation results for various D.

Table 1. * Circuit parameters and ratings.

Items	Symbol	Value
Source voltage	E	70V
Loss-less snubber capacitance	$C1,C2$	33nF
Series resonant capacitance	Co	0.35µF
Equivalent resistance for induction heating	Ro	0.597Ω
Equivalent inductance for induction heating	Lo	28.7µH
Operating frequency	fo	52.3kHz
Rated output power	Po	650W

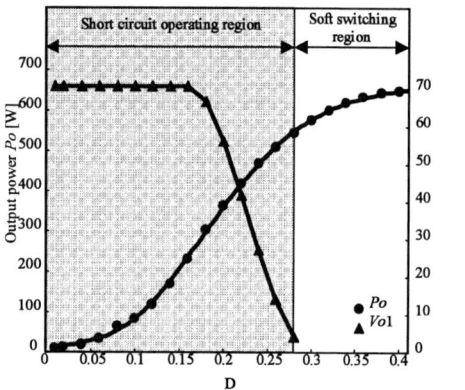

Fig. 4. Output power and residual voltage characteristics.

the steady state operation of the inverter based on the ideal device action. As detected in the table, note that the equivalent induction heating load values are adopted as the load values for the simulation. Figure 3 shows the results of the simulation analysis at D=0.1, 0.25 and 0.44 respectively, taking into account of the ratio of observed waveforms of the currents at switching times. Hence, the maximum value of D was determined as 0.44.

As seen in the figure, the calculated wave forms of the currents and voltages at various value of D mentioned above. For D=0.44, the soft switching operation is attained at turn on and off times of both switches.

In case of D=0.25 and 0.1, however, the residual voltage of SW1 at its turn on time. As a result, short circuit phenomena are observed at both conditions. In addition, the rising of Vsw2 can not smoothly performed for D=0.25 and a short circuited current is generated for D=0.1.

Figure 4 shows the relation between the regulated output power and the residual voltage (Vo1) at the turn on time of SW1. The residual voltage is the voltage which exists in the loss-less capacitor (C1) at the turn on time of SW1. When this voltage vanishes at D=0.44 as seen in Fig.3, it indicates that the natural conversion of the current is attained. Namely, the soft switching of the circuit is smoothly carried out. On the other hand, when this residual voltage has a finite value, it shows that the charge accumulated during the dead time can not be completely eliminated. As a result, when SW1 is turned on during this time, a short circuit is formed between C1 and SW1. Therefore, all the charge accumulated in C1 flows into the active switch SW1, resulting in destruction of the switch.

In addition, the turn-off loss occurs in SW2. We call this phenomenon as a short circuited mode. The occurrence of this mode depends mainly on the magnitude of the current which generates the period of the turn on time of SW1 and the turn off time of SW2. The magnitude of this current determines that the accumulated charges in C1 and C2 can be completely eliminated or not. The output power regulation system of this circuit can not attain over wide range of D. As seen in Fug.4, the output power decreases from 650 W continuously with decreasing D, however, below D<0.28 the residual voltage remains. Accordingly, the soft switching region for output power regulation is limited a narrow range of 0.28<D<0.44. This range correspond to the output power from 550W to 650 W. This fact means that wide range output power regulation can not be attained in this inverter. To improve this situation, the soft switching region can be widened by elongating the dead time of the switches. This method, however, results in decrease of output power of the inverter. Therefore, a novel high frequency inverter is required which can operate under soft switching operation with wider output power regulation range. It can be obtained by feeding a sufficient current required for soft switching operation of the switches during the period between the turn on time of SW1 and the turn off time of SW2.

III. Constitution and operational principle of the proposed circuit

3.1 Circuit constitution

Figure 5 illustrate the circuit constitution of the proposed circuit of the high frequency inverter. As seen in the figure, an electrically neutral point is introduced in the circuit by installing two dividing capacitor (Cs1 and Cs2) between the source voltage. A resonant current converting circuit to promote soft switching operation of the switches with an auxiliary reverse blocking active sub switch (SW3) is connected between the above neutral point and that of the main switches. This circuit is shown in the enclosed dotted lines in the figure. [3][4]. Figure 6 shows the PWM driving signals fed to the three switches. The times td1 and td2 shown in the figure are the dead times of the switches, respectively. Moreover, the gate signal is fed to SW3 is supplied before the time tb and trun off before the trun off gate signal fed to SW1.

As already mentioned, the residual voltage remains at low value of D for the conventional SEPP inverter. So, the soft switching operation can not conducted at the turn on time of SW1 and the turn off time of SW2. However, soft switching is attained at the turn off time of SW1 and the turn on time of SW2. Accordingly, sufficient currents are fed to both main active switches only at a short period between the turn on time of SW1 and turn off time of SW2 by operating the auxiliary active switch SW3 in our proposed circuit.

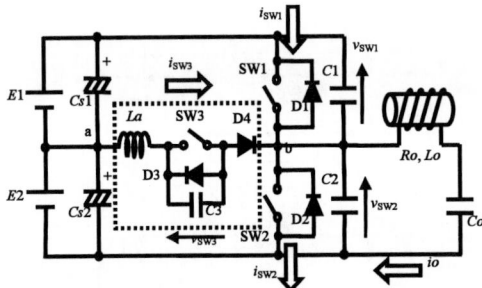

Fig. 5. Proposed SEPP high frequency inverter with auxiliary switch.

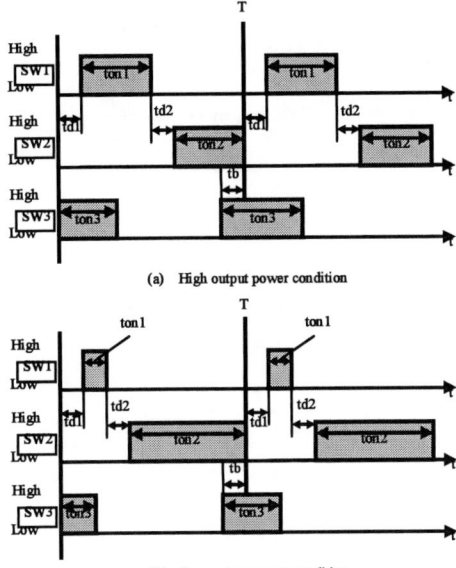

(a) High output power condition

(b) Low output power condition

Fig. 6. PWM driving signals for proposed circuit.

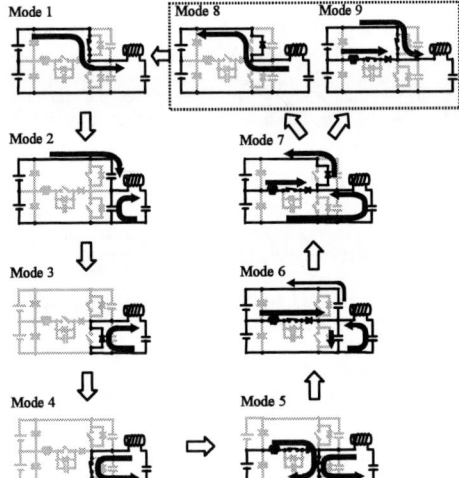

Fig. 7. Circuit operating mode classification in steady state.

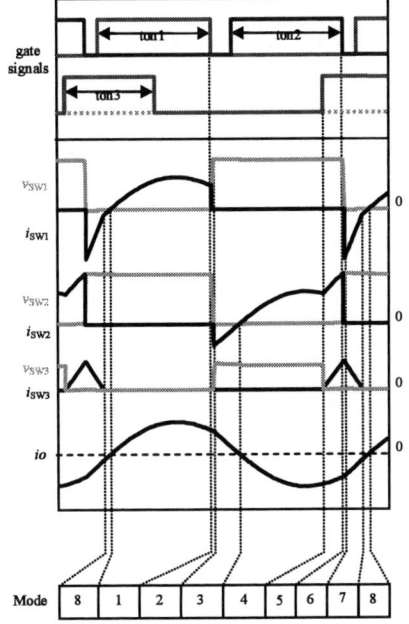

Fig. 8. Operating mode and operating classification under D=0.44.

3.2 Operational principle

We determine each signal and positive direction of the currents and voltages of our proposed circuit as directed in Fig.5. Next, Figure 7 illustrates schematically the steady state transition circuit diagram. In our circuit, its operation is repeated from mode 1 to mode 8 periodically at a high value of D, however, at a low value of D , the mode 7 transfers into mode 9 and returns to the initial mode 1. Each steady state operational mode is explained with the aids of Fig.7 and Fig.8 at D=0.44.

Mode 1: When SW1 turns on, a loop from E1, SW1, Ro, Lo, Co, E2 to E1 is formed supplying an electrical input power from the source voltage into the load. Namely, the mode 1 is the power supplying mode.

Mode2: When SW1 turns off at a time determined by the value of D, the charge accumulated in C2 connected in parallel to SW2 begins to discharge. The voltage across SW2(Vsw2) decreases to zero at a resonant gradient determined by the loop from Ro,Lo,Co,C2 to Ro. At the same time, the voltage across SW1 begins to increase. Finally, SW1 turns off at ZVS turn off.

Mode 3: When Vsw2 reaches zero, D2 turns on. A quasi resonant

277

mode occurs. By turning on SW2 during the period of turn on time of D2, SW2 turn on at ZVS&ZCS turn on mode.

Mode 4: When the current flowing through D2 vanishes, the current is converted into SW2. Here, the quasi resonant mode occurs as same as that of the mode 3.

Mode 5: By turn on SW3 during the on period of SW2, newly, an auxiliary resonant circuit constituting from the source voltage dividing capacitor and the resonant current converting inductance La begins to operate. This circuit begins to feed the current ($i$sw3).

Mode 6: After the inverter period, SW2 turns off. Then, the current flowing through the load and the inductance La begins to charge C2. Thus, the voltage (Vsw2) begins to increase at a gradient and reaches E1+E2. At the same time, Vsw1 decreases to zero. In this mode, the resonant current is insufficient in the conventional SEPP inverter at a low value of D. However, ZVS&ZCS turn on of SW1 becomes possible since a new current is supplied into SW2 by operating the auxiliary resonant circuit from the mode 5.

Mode 7: When the charging of C2 is completed, Vsw1 vanishes and D1 turns on. Then, $i$sw3 flowing through the auxiliary resonant circuit is fed into D1. At the same time, a power regeneration loop occurs and $i$sw3 begins to decrease.

Mode 8: The current $i$sw3 decreases from zero to negative. However, it is blocked by the reverse blocking diode (D4) connected in series to SW3 so that it vanishes completely. The power regeneration loop continues to operate at this time. Moreover, if SW1 is turned on during the period of the mode 7 and the mode 8, ZVS&ZCS turn on of SW1 becomes possible. When the value of D is sufficiently large, the above mode is repeated periodically and the high frequency power is supplied into the load.

Mode 9: This mode may occur when the value of D is low. Namely, it occurs when $i$sw3 flowing through La is larger than that flowing through D1. However, soft switching operation can be attained without any problem.

Figure 9 shows the mode transition including its steady state waveforms at D=0.07. Also, Figure 10 shows schematically the enlarged steady state operating waveforms, in which $i$sw3 is shown as $-i$sw3 to compare their sizes. As seen in the figure, when $i$sw3 flows through the auxiliary resonant circuit, $i$sw1 changes from negative value to positive value. Therefore, all of he current flowing through the auxiliary resonant circuit is supplied to the load, indicating that this mode is the power supplying mode. In this case, the waveform of $i$sw1 is slightly deformed, however, no effect does not occur on the soft switching operation of the switches or the inverter operation.

The steady state waveforms of the voltages and currents at D=0.44 shown in Fig.8 reveal that $i$sw3 reaches zero at first and then $i$sw1 reaches zero afterward. However, at low values of D, this behavior is reversed. As described above, we can find that ZVS&ZCS operation of SW1 at its turn on time and ZVS operation of SW2 at its turn off time are realized for the low value of D in our proposed inverter. This fact indicates that our proposed inverter can attain to widen the soft switching operation range compared with that of the conventional SEPP high frequency inverter. Especially, this inverter can maintain the current required for the soft switching of SW2 just before its turn off time so that the conduction loss and the peak current of the switch can be suppressed considerably.

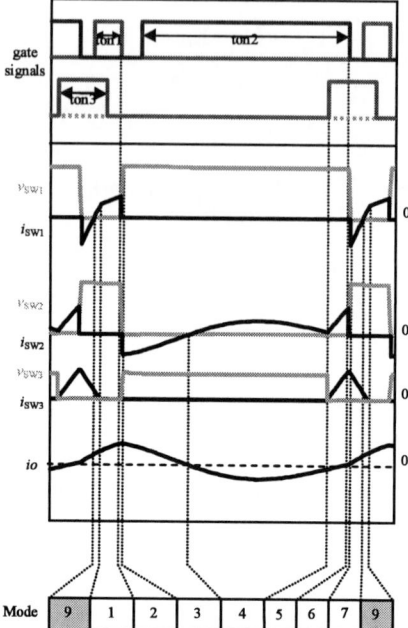

Fig. 9.   Operating mode and operating classification under D=0.07.

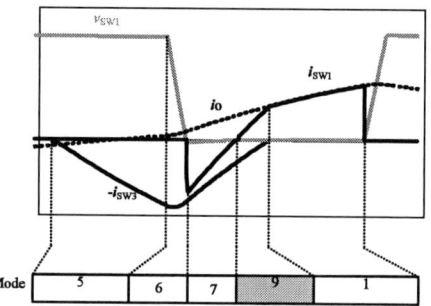

Fig. 10.   Steady state operating waveforms after magnification.

3.3 Steady state operation

We conduct the steady state operation analysis of our proposed inverter by computer aided simulation. Table 2 lists the circuit parameters and their ratings utilized for the analysis. We adopt the values of the resistance of the switches as 10 mΩ since on-resistance of BSITs used for the switches are 10 mΩ as mentioned afterward. In addition, the stray impedances (resistance and inductance) due to the mounting circuits are all neglected. The dead times of the active main switches should be determined as follows: Namely, td1 and td2 are short and long as possible, respectively. In this circuit, the short circuited mode does not occur at the turn on time of SW1 and the discharging mode can be completed at the turn off time of SW2 when td2 is sufficiently long. In conclusion, we determine the times td1=1μs,td2=1.5μs, and tb=1.8μs,respectively, as stated afterward.

Figure 11 shows the steady state waveforms of the currents and the voltages obtained by simulation under the condition that D=0.44. As seen in the figure, each main active switch operates under soft switching condition at all turn on and off times. Figure 12 shows the

278

steady state waveforms of the currents and the voltages obtained by simulation under the condition that D=0.07. As can be seen, both main switch SW1 and SW2 operate under ZVS&ZCS turn on and ZVS turn off even at the low value of D, which is different from the results shown in Fig.3 (b) and (c) for the conventional SEPP inverter. In addition, the auxiliary switch attains ZCS turn off and ZVS&ZCS turn off operation.

3.4 Output power characteristics

Figure 13 shows the output power characteristics obtained by simulation at the various value of D. In this figure shows the rated output power of 650 W. As seen in Fig.13, output power regulation range under soft switching condition is extremely widened ,compared with that of the conventional SEPP high frequency inverter described already, namely, 0.006<D<0.44 corresponding to output power from 30 W to 650 W.

Table 2.　Circuit parameters and ratings for proposed circuit.

Item	Symbol	Value
Source voltage	$E1, E2$	35V
Loss-less snubber capacitance	$C1, C2, C3$	33nF
Series resonant capacitance	Co	0.35μF
Equivalent resistance for induction heating	Ro	0.597Ω
Equivalent inductance for induction heating	Lo	28.7μH
Inductance for auxiliary inductor	La	3.0μH
On resistance for switches	r_{on}	10mΩ
Operating frequency	Fo	52.3kHz
Rated output power	Po	650W

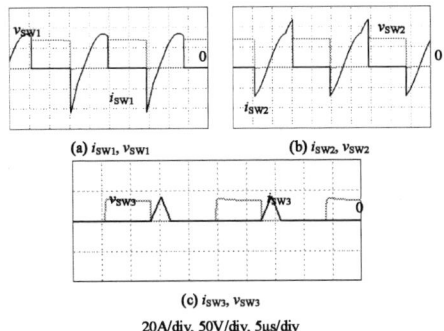

(a) i_{SW1}, v_{SW1}　　　　(b) i_{SW2}, v_{SW2}

(c) i_{SW3}, v_{SW3}

20A/div, 50V/div, 5μs/div

Fig. 11.　Simulation results under D=0.44.

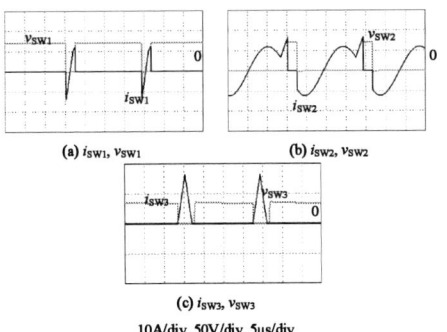

(a) i_{SW1}, v_{SW1}　　　　(b) i_{SW2}, v_{SW2}

(c) i_{SW3}, v_{SW3}

10A/div, 50V/div, 5μs/div

Fig. 12　Simulation results under D=0.07.

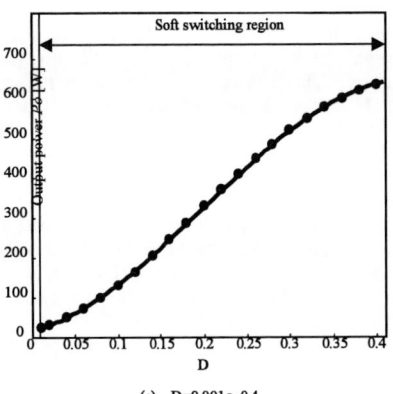

(a)　D=0.001～0.4

Fig. 13.　Output power characteristics for proposed circuit.

Table 3.　Absolute maximum ratings of B-SIT(SBM155).

Item	Symbol	Value
Drain to gate voltage	V_{DGO}	150V
Drain current	I_D	50A
Gate current	I_G	5A
Total power dissipation	P_T	200W
Drain to source on-resistance	$R_{DS(on)}I_G=0.5A,I_D=50A$	10mΩ

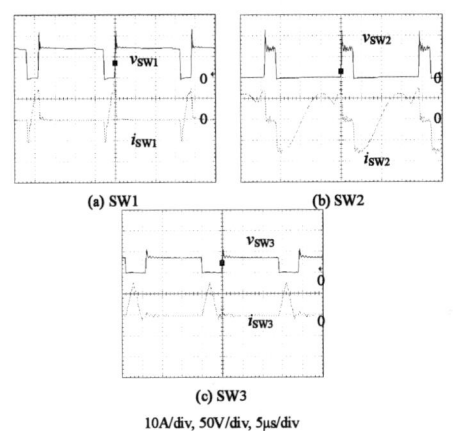

(a) SW1　　　　(b) SW2

(c) SW3

10A/div, 50V/div, 5μs/div

Fig. 14. Observed waveforms under D=0.1.

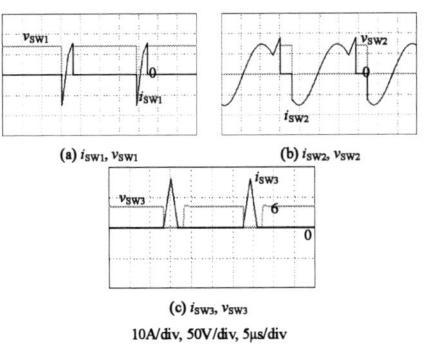

(a) i_{SW1}, v_{SW1}　　　　(b) i_{SW2}, v_{SW2}

(c) i_{SW3}, v_{SW3}

10A/div, 50V/div, 5μs/div

Fig. 15　Simulation results under D=0.1.

IV. Experimental results

We adopted BSITs with extremely low on resistances as the switching devices and utilized the circuit parameters listed in table2 in our experiment. Table 3 show the absolute maximum ratings of BSIT. Figure 14 shows the observed wave forms of the voltages across the switches Vsw1, Vsw2 and Vsw3 and the currents flowing through the above switches (isw1, isw2 and isw3), respectively. Our experiment was carried out under D=0.1, taking into consideration the turn on time of BSIT.

The output power efficiency (the ratio of output power to input power) reached 95 percent. We compare the waveforms across SW1 and that of the current flowing through SW1 shown in Fig.14(a) with those of the conventional inverter shown in Fig.3(c). A short circuited phenomenon is observed at the turn on time of SW1 in the conventional inverter, however, ZVS&ZCS turn on of the switch is realized in our inverter without occurring any short circuited phenomena.

In addition, rising of the voltage across SW2 of the conventional SEPP inverter at its turn off time shown in Fig.3(b) is not smoothly carried out, and a short circuited current occurs as shown in Fig3(c). On the other hand, the above behavior are improved in our inverter as seen in Fig.14(b). In conclusion, the problems of the conventional SEPP inverter shown in Fig.3(b) and (c) are found improved by incorporating an auxiliary sub switch and a resonant current injecting to circuit for promote soft switching of the main switches. Figure 15 shows the result of simulation analysis under the same condition of our experiment at D=0.1. As seen in the figure, we can find that both results agree with sufficiently. It confirms also the validity of the simulation analysis.

V. Conclusions

The content of this paper and the obtained results are described as follows:

(1) The detailed characteristic evaluation, the operational principle and its characteristic behavior of our proposed inverter were analyzed by using simulation analysis. As a result, the soft switching operation of the conventional SEPP high frequency inverter over wider output power regulation range, though, the number of its parts increased.

(2) Its operational principle of our inverter was based on the method to maintain the required current for soft switching operation of the main switches. Therefore, we could suppress the peak current values of the switches considerably. As a result, conduction losses of the switches could be decreased so that a high power conversion efficiency could be expected. The steady state operational transition and detailed waveforms were calculated with the aid of simulation analysis for D=0.44 and D=0.07. The result indicates that each switching device operates under soft switching mode over wider duty ratio.

(3) The experimental result was shown at D=0.1, taking into account the turn on time of BSITs utilized for switching devices of our inverter. The obtained result could explain sufficiently the simulation result. Accordingly, the validity of the operational principle of our inverter and simulation result could be revealed. In addition, the problems of the conventional SEPP inverter were found improved by the experimental result.

We described the operational principle and its characteristics on the basis of the analyzed results. In future, we will improve this inverter which can supply a higher output power and operate at a higher frequency.

References

[1] A.H. Weinberg and L. Ghislanzoni:" A New Zero Voltage and Zero Current Power-Switching Techniques", IEEE Trans. on Power Electronics, Vol.7,No.4, pp.655-665 (1992).

[2] R.L.Steigerwald:" A Comparison of Half-Bridge Resonant Converter Topologies", IEEE Trans. On Power Electronics,Vol.3, No.2,pp.174-182 (1988).

[3] R.W.De Doncker and J.P.Lyons : " The Auxiliary Resonant Commutated Pole Converter", Proc.of IEEE IAS Conf.Vol.2,pp.1228-1235(1996-10).

[4] H.Ogiwara, M.Itoi and M.Nakaoka:" PWM-controlled soft-switching SEPP high-frequency inverter for induction heating applications", IEE Proc.,Electr. Power Appl. 151,No.4, pp.404-413 (2004).

Multiphase coupled converter models dedicated to transient response and output voltage regulation studies

Nadia Bouhalli[*], Marc Cousineau[*], Emmanuel Sarraute[*] and Thierry Meynard[*]

[*]Université de Toulouse, LAPLACE, CNRS, INPT, ENSEEIHT,
2 Rue Charles Camichel B.P. N° 7122 31071 Toulouse Cedex 7, France
e-mail : nadia.bouhalli@laplace.univ-tlse.fr

Abstract— In order to study transient response and output voltage regulation in multiphase coupled buck converter, it is proposed two models of interleaved coupled buck converter. These two models provide accurate current and voltage waveforms for any value of duty cycle. In the first part, the two proposed models are described in details. In the second part, it is shown the interest of this approach to study dynamic behaviour and determine compensation filters for voltage regulation in a multiphase coupled buck converter.

Keywords— Interleaved converters, Multiphase coupled buck converter, Converter control, Regulation.

I. INTRODUCTION

High current low voltage power converters with fast response are needed for powering digital systems such as microprocessors which require more than 100 A at under than 1V [1-2]. As special power supply for the microprocessor is the voltage regulator module (VRM). The industry standard VRM topology used is the multiphase buck converter [3], [4] and [5]. Increasing demand for higher current, lower voltage and faster dynamic response imposes a challenge for multi-phase buck converter designs. Coupling phases of interleaved buck has the potential to reduce steady-state power losses while maintaining or even improving dynamic performances [6], [7], [8], [9] and [10]. Coupling phases gives different equivalent inductances for transient response 'L$_{tr}$' and steady-state operation 'L$_{ss}$' [11] and [12] and it is unclear which one should be used to design filters compensation.

In the literature [6] and [13], only when the duty cycle, D$_1$ is lower than the inverse of the number of phases (D$_1$ < 1/q, q: the number of parallel phases), the multi-phase buck (Fig. 1) is equivalent to a single phase buck with the effective duty cycle is equal to 'q' times the actual duty cycle ($D_{eq} = q.D_1$), the switching frequency seen from the output is equal to 'q' times the actual switching frequency ($F_{sweq} = q.F_{sw}$) and the input voltage is the actual input voltage divided by the number of phases ($Vin_{eq} = Vin/q$).

In this paper, we propose two accurate models of interleaved coupled buck converter, for any value of duty cycle, in order to show which equivalent inductance must be used to design compensation filters for voltage regulation. In the first part of this paper, the two proposed models are described in details. In the second part, it's shown the interest of these models to study dynamic behaviour and determine compensation filters for voltage closed loop regulation in a multiphase coupled buck converter.

II. MULTIPHASE COUPLED BUCK CONVERTER DESCRIPTION

Interleaved converters are widely employed in various applications such as automotive 42/14V systems and Voltage Regulator Module (VRM). The well known advantage of this system is an increase of the apparent frequency of the current ripple applied across the input and output filters; associating 'q' identical commutation cells fed by interleaved control signals (equal duty cycles, phase-shifts 2π/q), the apparent frequency for the input filters and the output capacitor is 'q' times the switching frequency. In addition, an improvement of dynamic behaviour can be obtained in interleaved converters [6]. Two main solutions can be applied to interconnect interleaved commutation cells: by means of either uncoupled/independent inductors or intercell transformers (Fig. 1). By the first mean of interleaving, the increase of frequency concerns only the input filters

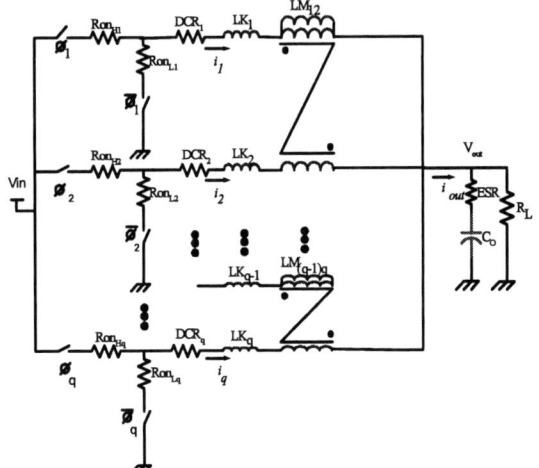

Fig. 1. q-phase coupled buck converter

and the output capacitor but the phase current is at the switching frequency 'F_{sw}'. By the second mean of interleaving, also the apparent frequency of the current of phases is multiplied by 'q' [6], [7], [8], [9] and [10]. Fig.2 shows the output current waveform of the multiphase coupled converter: the apparent frequency of the current is q times the switching frequency. From (Fig.2) it can be seen that for each $(1/q)T_{sw}$, the same waveform of the output current occurs for different value of duty cycle (D_1, D_2, D_3,…). It can be noted that any duty cycle can be written as a function of the duty cycle $D1 < 1/q$: $D = D_1 + (k-1)/q$ (where $k = 1, 2,..., q$). The main idea in this paper, is to develop two simple models of interleaved coupled buck converter using the fact that D is equal to D_1 modulo 1/q: the first model is an equivalent uncoupled interleaved buck converter using only equivalent leakage inductances present in each phase. It is dedicated to study transient response. The second model is equivalent to a single buck converter used to simplify filters design for output voltage regulation.

III. EQUIVALENT MODELS FORMULATION

In this section, it will be demonstrated that for any value of duty cycle, the multiphase coupled buck converter (Fig. 1) can be reduced to a simple uncoupled parallel buck converters or to a simple single buck converter. To simplify explanation a two-phase coupled converter will be analyzed, but the same method can be applied to a greater number of phases.

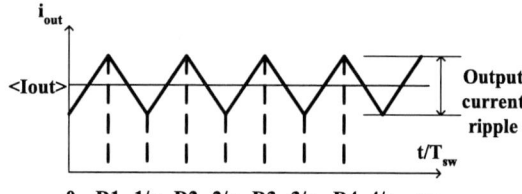

Fig. 2. q-phase coupled buck converter output current ripple

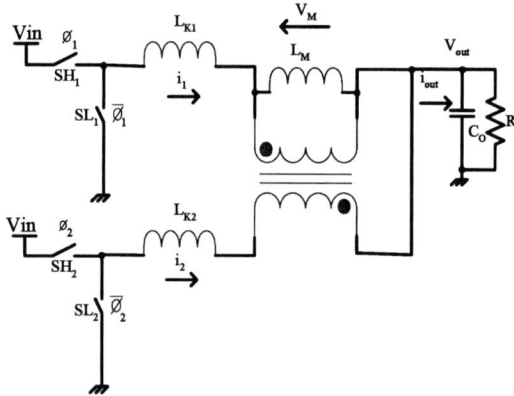

Fig. 3 . A simplified schematic of two- phase coupled buck

A. Case 1: Duty cycle equal to $D_1 < 1/2$

Fig. 3 is a simplified schematic of two-phase coupled converter topology in which the intercell transformer is represented by two leakage inductances L_{K1} et L_{K2}, the ideal transformer and magnetizing inductance L_M. For the purpose of this analysis, it is assumed that $L_{K1} = L_{K2} = L_K$ and the following relationship should be noted:

$$i_{out} = i_1 + i_2 \tag{1}$$

$$V_{LK1} = L_K (di_1 / dt) \tag{2}$$

$$V_{LK2} = L_K (di_2 / dt) \tag{3}$$

$$V_M = L_M (di_M / dt) \tag{4}$$

$$i_M = i_1 - i_2 \tag{5}$$

Equation (5) can be written as the following:

$$V_M = L_M (di_1 / dt - di_2 / dt) \tag{6}$$

From Fig.4, there are four distinct states of the phase currents and the output current when the duty cycle is equal to $D_1 < 1/2$.

1) State one: from t = 0 to t = $D_1 T_{sw}$

State one of the two-phase coupled converter covers the phase-one on time from t = 0 to t = $D_1.T_{sw}$. During state one, phase one is connected to the input voltage (S_{H1} closed and S_{L1} open) and phase two is connected to ground (S_{H2} open and S_{L2} closed) and writing the voltage equations around the two current loops, we have:

$$V_{LK1} = V_{in} - V_{out} - V_M \tag{7}$$

$$V_{LK2} = V_M - V_{out} \tag{8}$$

Substituting (7) and (8) into (6), we have:

$$V_M = V_{in} L_M / (L_K + 2L_M) \tag{9}$$

To simplify the proceeding analysis a ratio, $p = L_M / L_K$ [8] will be used and (9) reduces to:

$$V_M = V_{in} (p / (1 + 2p)) \tag{10}$$

Substituting (10) into (7) and (8), we have:

$$\frac{di_1}{dt} = (V_{in} / L_K)(1 - p/(1+2p) - D_1) \tag{11}$$

$$\frac{di_2}{dt} = (V_{in} / L_K)(p/(1+2p) - D_1) \tag{12}$$

282

$$\frac{d_{iout}}{dt} = 2(Vin)\frac{(1-2D_1)}{(L_K)} \qquad (13)$$

From (11), (12) and (13), it can be noted that although the coupling ratio (p) does not affect the output current ripple, it has a significant effect on the phase current ripple slope. using a perfect coupling, the coupling ratio is increased toward infinity:

$$V_M = V_{in}/2 \qquad (14)$$

Equations (11) and (12) become equal to:

$$\frac{di_1}{dt} = \frac{di_2}{dt} = (\frac{V_{in}}{2})(\frac{1-2D_1}{L_K}) \qquad (15)$$

$$\frac{di_1}{dt} = \frac{di_2}{dt} = (Vin_{eq})(\frac{1-D_{eq}}{L_K}) \qquad (16)$$

Furthermore, using (1) we have an output current of:

$$\frac{d_{iout}}{dt} = 2(Vin_{eq})\frac{(1-D_{eq})}{(L_K)} \qquad (17)$$

With: $Vin_{eq} = V_{in}/2$ and $D_{eq} = 2D_1$

From (16), the strong coupling makes the current wave forms in the two phases indistinguishable and each phase is equivalent to a single buck converter with the equivalent inductance is the leakage inductance present in each phase. This formulation does no effect on studying the transient response, as it is shown in [6]: the transient response is determined exclusively by the leakage inductances.

2)State Two: from $t = D_1T_{sw}$ to $t = T_{sw}/2$

State two covers the off time from $t = D_1T_{sw}$ to $t = T_{sw}/2$. During state two, the input is disconnected entirely (S_{H1} and S_{H2} open) and both windings are tied to ground (S_{L1} and S_{L2} closed). Looking in Fig. 4 and writing the voltage equations around the two current loops:

$$V_{LK1} = -V_{out} - V_M \qquad (18)$$

$$V_{LK2} = V_M - V_{out} \qquad (19)$$

In state two $V_M = 0$ and we have:

$$\frac{di_1}{dt} = \frac{di_2}{dt} = (-V_{in}/L_K)D_1 \qquad (20)$$

Equation (20) can be written as the following:

$$\frac{di_1}{dt} = \frac{di_2}{dt} = (-\frac{V_{in}}{2})(\frac{2D_1}{L_K}) \qquad (21)$$

$$\frac{di_1}{dt} = \frac{di_2}{dt} = -(Vin_{eq})(\frac{D_{eq}}{L_K}) \qquad (22)$$

Using (1) the output current is equal to:

$$\frac{d_{iout}}{dt} = -2(Vin_{eq})\frac{D_{eq}}{(L_K)} \qquad (23)$$

From (22) the current waveforms in the two phases are identical and each phase is equivalent to a single buck converter

3)State three: from $t = T_{sw}/2$ to $t = T_{sw}/2 + D_1T_{sw}$

In this state, L_{K2} is now connected to the input and L_{K1} is connected to the ground. By symmetry, we obtain the same waveforms in (15), (16) and (17) as in state one, with just inverting the two currents.

4)State four: from $t = T_{sw}/2 + D_1T_{sw}$ to T_{sw}

This state is identical to state two and the same waveforms (22) and (23) are obtained.

It can be noted that when the duty cycle $D_1 < 1/2$ (2: the number of phases) the two phase and the output currents are increasing in state one and decreasing in state two and that this cycle repeats again in states three and four. The 2-phase coupled buck (Fig. 3) is equivalent to a 2 single parallel buck converters where the phase-shift between two phases is equal to zero, with equivalent duty cycle '$D_{eq} = 2.D_1$', equivalent switching frequency '$F_{sweq} = 2.F_{sw}$', equivalent input voltage '$Vin_{eq} = Vin/2$'.

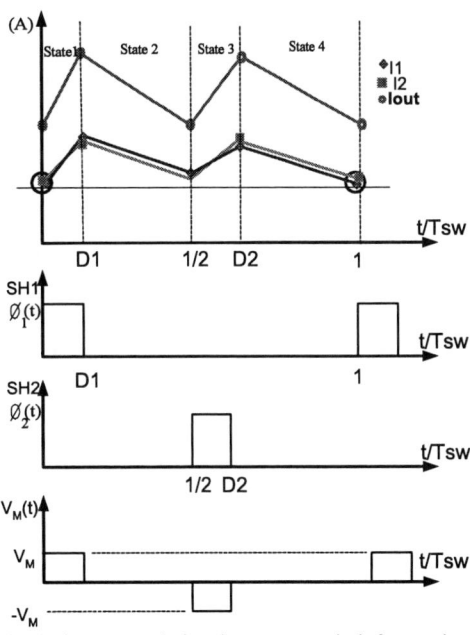

Fig. 4. Phase current ripple and output current ripple for two-phase coupled buck converter for a duty cycle equal to $D_1 < 1/2$

B. Case 2: Duty cycle equal to $D_2 > 1/2$

Now, the same study will be done with a duty cycle $D_2 = 1/2 + D_1$

1) State one: from $t = 0$ to $t = D_1 T_{sw}$

During state one, phase one and phase two are connected to the input voltage (S_{H1} and S_{H2} closed) and (S_{L1} and S_{L1} open): in this case $V_M = 0$ and writing the voltage equations around the two current loops:

$$V_{LK1} = V_{in} - V_{out} \qquad (24)$$

$$V_{LK2} = V_{in} - V_{out} \qquad (25)$$

With $V_{out} = D_2 V_{in}$:

$$\frac{di_1}{dt} = \frac{di_2}{dt} = ((V_{in} - D_2 V_{in})/L_K) \qquad (26)$$

As $D_2 = D_1 + 1/2$, (26) becomes:

$$\frac{di_1}{dt} = \frac{di_2}{dt} = (V_{in}/L_K)(0.5 - D_1) \qquad (27)$$

$$\frac{di_1}{dt} = \frac{di_2}{dt} = (Vin_{eq})(\frac{1 - D_{eq}}{L_K}) \qquad (28)$$

$$\frac{d_{iout}}{dt} = 2(Vin_{eq})\frac{(1 - D_{eq})}{(L_K)} \qquad (29)$$

With: $Vin_{eq} = V_{in}/2$ and $D_{eq} = 2D_1$

It can be noted in this state that the obtained expressions of phase and output currents are the same as in state one with duty cycle $D_1 < 1/2$.

2) State Two: from $t = D_1 T_{sw}$ to $t = T_{sw}/2$

$$V_{LK1} = V_{in} - V_{out} - V_M \qquad (30)$$
$$V_{LK2} = V_M - V_{out} \qquad (31)$$

$V_M = Vin(p/(1+2p))$, $V_{out} = D_2 V_{in}$ and $D_2 = D_1 + 1/2$:

$$\frac{di_1}{dt} = (V_{in}/L_K)(1 - D_2 - p/(1+2p)) \qquad (32)$$

$$\frac{di_2}{dt} = (V_{in}/L_K)(p/(1+2p) - D_2) \qquad (33)$$

$$\frac{d_{iout}}{dt} = 2(Vin)\frac{(1 - 2D_1)}{(L_K)} \qquad (34)$$

From (32), (33) and (34), it can be noted that although the coupling ratio (p) does not affect the output current ripple, it has a significant effect on the phase current ripple.

If we suppose that we have a perfect coupling: the coupling ratio is increased toward infinity: $V_M = V_{in}/2$, (32) and (33) become equal to:

$$\frac{di_1}{dt} = \frac{di_2}{dt} = -(Vin_{eq})(\frac{D_{eq}}{L_K}) \qquad (35)$$

$$\frac{d_{iout}}{dt} = -2(Vin_{eq})\frac{D_{eq}}{(L_K)} \qquad (36)$$

It can be noted in this state that the obtained expressions of phase and output currents, are the same as in state two with duty cycle $D_1 < 1/2$.

3) State three: from $t = T_{sw}/2$ to $t = T_{sw}/2 + D_1 T_{sw}$

This state is identical to state one and the same waveforms (28) and (29) are obtained.

4) State four: from $t = T_{sw}/2 + D_1 T_{sw}$ to T_{sw}

By symmetry, we obtain the same relations (32), (33) and (34) as in state two, with just inverting the two currents.

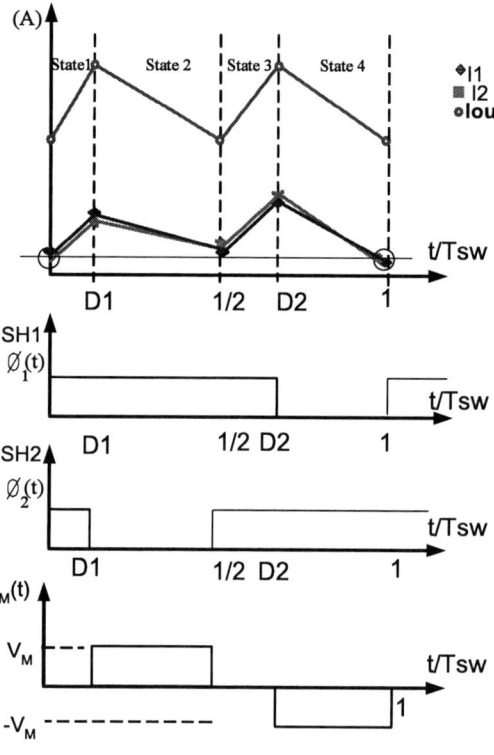

Fig. 5. Phase current ripple and output current ripple for a duty cycle equal to $D_2 > 1/2$

It can be noted that when the duty cycle $D_2 > 1/2$, the two phase and the output currents are increasing in state one and decreasing in state two and that this cycle repeats again in states three and four. The 2-phase coupled buck is equivalent to 2 single parallel buck converters where the phase-shift between two phases is equal to zero, with equivalent duty cycle '$D_{eq} = 2.D_1$', equivalent switching frequency '$F_{sweq} = 2.F_{sw}$', equivalent input voltage '$Vin_{eq} = V_{in}/2$'.

TABLE I.
CURRENT RIPPLE FOR TWO DIFFERENT VALUE S OF DUTY CYCLE

Duty cycle	$\dfrac{di_1}{dt} = \dfrac{di_2}{dt}$ in State 1	$\dfrac{di_1}{dt} = \dfrac{di_2}{dt}$ in State 2
$D_1 < 1/2$	$(Vin_{Eq})(\dfrac{1-D_{eq}}{L_K})$	$-(Vin_{Eq})(\dfrac{D_{eq}}{L_K})$
$D_2 = D_1 + 1/2$ $D_2 > 1/2$	$(Vin_{Eq})(\dfrac{1-D_{eq}}{L_K})$	$-(Vin_{Eq})(\dfrac{D_{eq}}{L_K})$

In table I (it was considered only state 1 and 2 since there is the same behavior in states 3 and 4) are summarized the different phase current ripple in two-phase coupled buck converter for two different duty cycle: $D_1 < 1/2$ and $D2 > 1/2$ ($D_2 = 1/2 + D_1$). It was shown that for the two different duty cycles the same model can be used

To simplify explanation a two-phase coupled converter was analyzed in this paper, but the same method can be applied to a greater number of phases. For any number of phases 'q' and for any value of duty cycle $D = D_1 + (k-1)/q$ (where $k = 1,2,..., q$ and D1 < 1/q), the multi-phase buck (Fig. 1) is equivalent to a q-single parallel buck converters where the phase-shift between two successive phases is equal to zero, with equivalent duty cycle 'Deq = q.D1', equivalent switching frequency 'Fsweq = q.Fsw' and equivalent input voltage 'Vineq = Vin/q' . The inductance per phase is the sum of leakage inductances present in each phase. The on-time is unchanged, but the apparent duty cycle is 'q' time D_1. It can be noted that $V_{out} = D.V_{in} = (D_1 + (\dfrac{k-1}{q})V_{in})$ with $k = 1,2,..., q$, so to have the same output voltage an additional voltage source V_k has to be added in the proposed model with $V_k = (D - D_1).V_{in} = (\dfrac{k-1}{q})V_{in}$ as it is shown in Fig. 6. This first model is very helpful to study transient response for any duty cycle $D = D_1 + (k-1)/q$ but in order to study output voltage regulation with usual tools, a second equivalent model is proposed (Fig.7). With equal leakage inductances 'L_k', the equivalent multiphase uncoupled buck converter (Fig. 6) becomes equivalent to a single buck converter with only one inductance equal to L_k/q .

If unmatched phases provide different mean current values in each branch, model of Fig. 6. can be used with additional branch resistors providing effects of mismatch. Fig. 8. shows this model with expression of resistors.

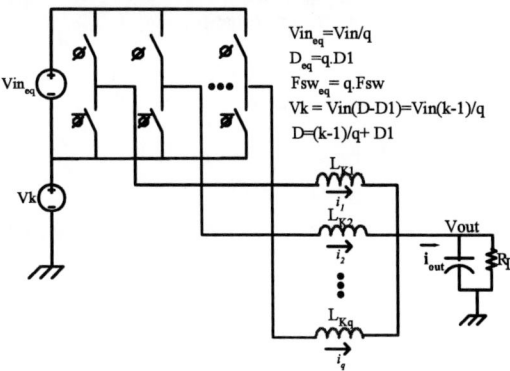

Fig. 6. Equivalent multiphase model of multiphase buck converter using Vin_{eq}, D_{eq}, Fsw_{eq} and V_k

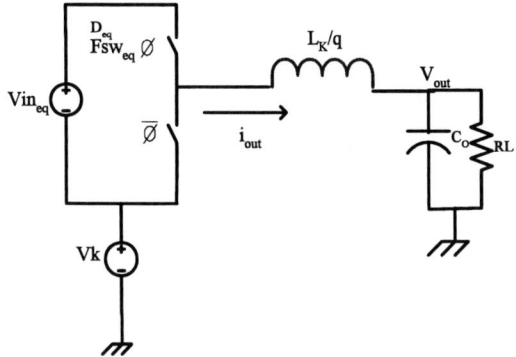

Fig. 7. Equivalent single phase buck converter of multiphase buck converter using Vin_{eq}, D_{eq}, Fsw_{eq} and V_k

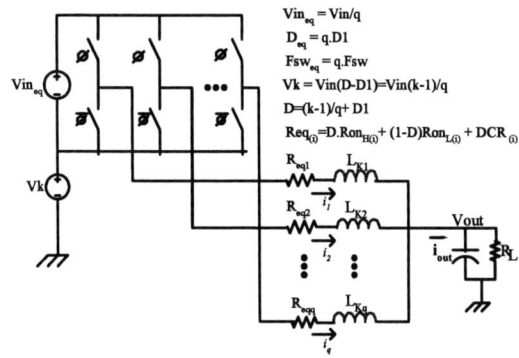

Fig. 8. Equivalent single phase buck converter

IV. VALIDATION OF THE TWO EQUIVALENT MODELS

In order to validate the two models presented in the previous section (multiphase and simple buck), comparisons between real coupled system and models have to be done by simulation. A five coupled-phase buck converter is studied (q = 5). In Fig. 9 is shown the electrical scheme of this converter in cyclic cascade configuration [14]. In the simulation, V_{in}=6V, V_{out} =3.3V, I_{out}=15A, R_{load}= 0.22, L_K =1uH, L_M=150uH and Fsw =50KHz. In this case, it can be noted that duty-cycle D > 1/5.

Using general expression of duty-cycle presented in section II , D can be formulated in this example with the following expression:

$$D = \frac{k-1}{q} + D1 = 0.55 \qquad (37)$$

with D=0.55, k=3, D1=0.15

Fig 10 and Fig 11 show respectively responses of the two models for a fast load transient. It can be noted that voltage and current signal waves are very similar. Current slopes are accurate and transient time responses are the same. Model switching frequency is five time higher than real system frequency (see signals V(HS_model) and V(HS)).

For all cases with odd number for 'k', comparing the real system and the multiphase model, phase current slopes are the same. With even number for 'k', slopes are inverted. This behaviour has no impact on accuracy of mean values and time responses. For all 'k' value cases, output current waves are exactly the same.

Now, simple buck model can be used to perform closed loop analysis in voltage mode in order to provide a accurate and stable output voltage value. Such a simple model allows writing expression of electrical system to compensate. Overall open-loop gain expression is the following:

$$G(s).H(s) = \frac{Vin.K(s)}{Vp} \cdot \frac{1 + esr.Co.s}{1 + \left(\dfrac{Leq}{Rload} + esr.Co\right)s + Leq.Co.s^2} \qquad (38)$$

with Leq=2.L_K/q, K(s) : compensation gain stage, Rload : output resistive charge, Co : output filter capacitor and esr : equivalent serial resistor of capacitor Co.

Note that term '2.L_K' in expression of Leq represent the sum of leakage inductances present in each phase of converter (Fig. 9).

Expression (38) is a well-known voltage-mode buck open-loop gain equation. A second order behavior providing a resonant frequency at f_n has to be compensate to guaranty stability.

$$fn = \frac{1}{2.\pi.\sqrt{Leq.Co}} \qquad (39)$$

Fig 12 describes a possible filter K(s) providing 2 zeros and 1 pole used to get a good phase margin, large bandwidth, good attenuation at f_n and very high gain at low frequencies. With L_K=6uH, Lm=1.4mH, C=5x100uF and R_{load}=0.2Ω, following values are computed:

R1=1KΩ, C1=10nF, R2=10KΩ, C2=3uF and C3=100pF.

Fig 13 shows output signal behavior for the closed-loop system. Error voltage 'Verror' reach the good value of 0.75 corresponding to expecting '5.D1' value.

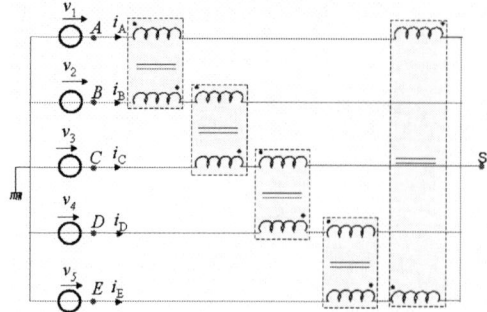

Fig. 9. Electrical scheme oh 5-coupled buck converter

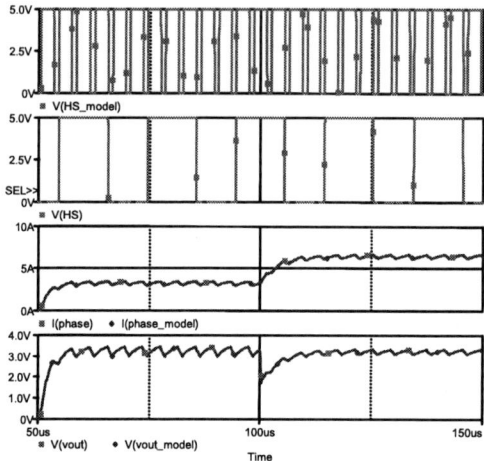

Fig. 10. Load transient response for both real five coupled-phase system and multiphase model.

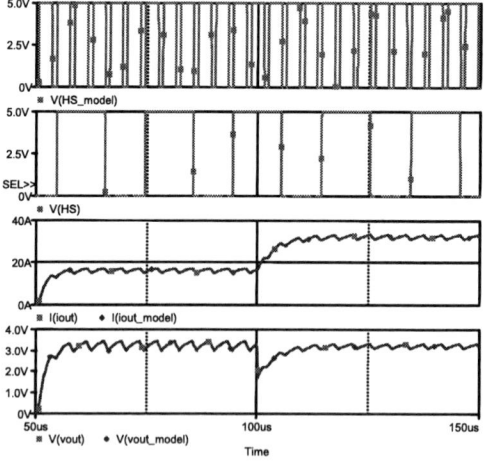

Fig. 11. Load transient response for both real five coupled-phase system and simple buck model

286

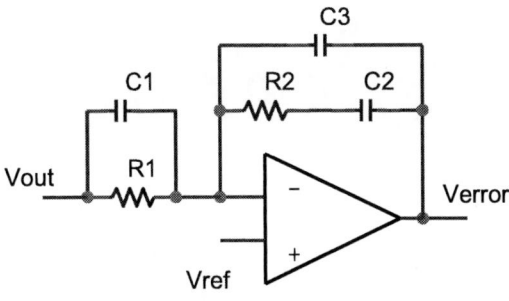

Fig. 12. K(s) filter used for compensation purpose

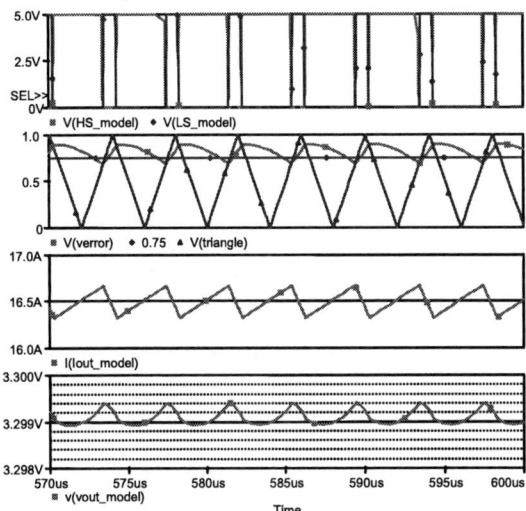

Fig. 13. Steady-state signal waveforms of closed-loop system using simple buck model

Obtained signals are very similar to those provided by the real system of Fig. 9.

V. CONCLUSION

In this article, two accurate models of complex multiphase coupled buck, for any value of duty cycle, have been described. Method used to develop those models is an analytical approach providing accurate slopes and mean values for all system signals. Simulation results validate them in a steady and dynamic state for a five-coupled buck. Validity range of multiphase model is given by factor '$p=L_K/L_M$' (higher than possible). Simple buck model is accurate for any value of 'p'. It can be noted that using the single buck converter model, it is possible to design compensation filters for any value of duty cycle. Further works have to be done in order to get a dynamic 'V_K' source, following duty-cycle variation for accurate closed-loop transient load analysis.

REFERENCES

[1] B. Rose, "Voltage regulator technology requirements", in 4th Annual Intel Technology Symposium, Sept. 1990.

[2] E. Stanford, "Device requirements for low voltage, fast transient response regulators used to power future microprocessors and other low voltage logic chips", Proceedings of the Sixteenth International High Frequency Power Conversion Conference, Rosemont, IL, September, 2001, pp. 1-10.

[3] G. Schuellein, "Current sharing of redundant synchronous buck regulators powering high performance microprocessors using the V^2 control method", in Thirteenth Annual Applied Power Electronics Conf. and Exposition, 1998, vol. 2, pp. 853–9.

[4] A.M.Wu, Jinwen Xiao, D. Markovic, and S.R. Sanders, "Digital PWM control: application in voltage regulation modules", in *30th Annual IEEE Power Electronics Specialists Conf.*, 1999, pp. 77–83 vol.1.

[5] Jinwen Xiao, A.V. Peterchev, and S.R. Sanders, "Architecture and IC implementation of a digital VRM controller", in *32nd Annual IEEE Power Electronics Specialists Conf.*, 2001.

[6] J. Li, C. R. Sullivan, and A. Schultz, "Coupled inductors design optimization for fast- response low voltage DC-DC converters", IEEE Applied Power Electronics Conference, Dallas, March 2002.

[7] J. Czogalla, J. Li, C. R. Sullivan, " Automotive application of multiphase coupled-inductor DC-DC converter", Proc. IAS"03, vol. 3, pp. 1524, 2003.

[8] J. Li, A. Stratakos, A. Schultz, and C. R. Sullivan, "Using coupled inductors to enhance transient performance of multi-phase buck converters", APEC'04, vol. 2, pp. 1289-1293, 2004.

[9] P.-L Wong, P Xu, B. Yang, and F. C. Lee, "Performance improvements of interleaving VRMs with coupling inductors", APEC'00, pp. 973-978, 2000.M. Young, *The Technical Writer's Handbook*. Mill Valley, CA: University Science, 1989.

[10] Pit-Leong Wong, Q. Wu, Peng Xu, Bo Yang, and F.C. Lee, "Investigating coupling inductors in the interleaving QSW VRM", in Proceedings of APEC 2000 - Applied Power Electronics Conf., pp. 973–8 vol.2.

[11] Y. Dong, M. Xu, and F. C. Lee, "DCR Current Sensing Method for Achieving Adaptive Voltage Positioning (AVP) in Voltage Regulators with Coupled Inductors", PESC '06, 2006.

[12] P.-L Wong, P Xu, B. Yang, and F. C. Lee, "Performance improvements of interleaving   VRMs with coupling inductors", APEC'00, pp. 973-978, 2000.M. Young, *The Technical Writer's Handbook*. Mill Valley, CA: University Science, 1989.

[13] P. Zumel, O. Garcia, J.A. Cobos, and J. Uceda," Tight magnetic coupling in multiphase interleaved converters based on simple transformers", Proc. APEC"05, vol. 1, pp. 385, 2005.

[14] I. G. Park and S. I. Kim, "Modeling and analysis of multi-intercell transformers for Connecting power converters in parallel", PESC'97, vol. 2, pp. 1164-1170, 1997.

A 13.56 MHz Current-output-type Inverter Utilizing An Immittance Conversion Element

Yosei Sakamoto*, Keiji Wada* Toshihisa Shimizu*
*Tokyo Metropolitan University / Department of Electrical and Electronics Engineering
1-1 Minami Osawa, Hachioji, Tokyo, Japan
e-mail: *shimizut@tmu.ac.jp*

Abstract—High frequency plasma processing techniques are utilized in the field of industrial manufacturing. High frequency plasma generators are traditionally composed of analog amplifiers, although their conversion efficiencies are limited up to 50 %. A current-output-type inverter, composed of an immittance conversion element, that enables constant current injections to the loads and megahertz plasma discharging with high conversion efficiency. A novel resonant gate driver is also proposed to achieve not only 10 MHz range operation, but a low driving power. Design and analysis of the proposed inverter was carried out, resulting in an increasing of efficiency up to 72.7 % at 13 MHz.

Key Words– High Frequency Converter, Resonant Converter, ZCS Converters, High speed drive

I. INTRODUCTION

High frequency plasma processing techniques are utilized in the field of industrial manufacturing, such as separators for fuel cells, color filters for liquid crystal displays, and semiconductors. In order to supply high-frequency powers to plasma generators, analog amplifiers with output frequencies in the megahertz range, are traditionally used, even though the conversion efficiency is limited up to 50 %. Therefore, a lot of work has been reported to increase a conversion efficiency [1] ∼ [9].

In order to realize the high conversion efficiency of megahertz range inverters, there are several problems that require solving. One major problem is that the conventional gate drive circuits which generate square-wave voltages can not drive MOSFETs in the megahertz range, because the peak current is limited by the increasing inductive reactance component of the gate drive circuit due to the high frequency. Therefore resonant gate drive techniques were adopted, as reported in references [1] ∼ [3].

Generally, a resonant gate drive circuit is composed of a resonant inductor and an input capacitor of the MOSFET switch, and consumes less power than conventional gate drivers do. However, the gate driving power on the resonant gate drive circuit still reduces the conversion efficiency of the system when the switching frequency of the inverter is increased to the megahertz range. Therefore, a revised resonant gate drive circuit suitable for megahertz operation is proposed.

Another problem is that conventional voltage source inverters are not suitable for plasma discharging applications, because stable current injection capability within the megahertz range is required. In order to improve the current injection characteristics, the authors have presented a current-output-type high frequency inverter based on an immittance conversion element [3] ∼ [6]. The current-output-type high frequency inverter can produce constant currents, so that high voltage can be generated by adopting an LC parallel resonant load of which the resonant frequency is adjusted to an operation frequency of the inverter. The feature of the inverter in [3] is the push-pull type structure, where the immittance conversion element is utilized instead of a center-tapped transformer. Therefore the source terminals of each MOSFET are directly connected to the ground line, which enables the MOSFETs to be driven by a gate drive circuit without galvanic isolation between the control circuit and the main circuit. However, the operation frequency of this inverter is limited to 1 MHz because a conventional resonant gate drive circuit is utilized.

In this paper, a current-output-type inverter is proposed that utilizes a novel resonant gate drive circuit. Design and analysis of both the inverter and the gate driver are carried out, and the experimental results show that the conversion efficiency is increased up to 72.7 % at 13 MHz.

II. A CURRENT-OUTPUT-TYPE INVERTER UTILIZING AN IMMITTANCE CONVERSION ELEMENT

A. Main circuit configuration

Fig. 1 and Table. I show the proposed inverter circuit configuration and its parameters, respectively. The proposed inverter circuit consists of two MOSFETs, Q_1, Q_2, an immittance conversion element, T, load resistor, R, output capacitor, C_o, output inductor, L_o, and a gate driver, G. The immittance conversion element is composed of two coaxial cables, T_1, T_2, whose lengths are adjusted to one quarter of the propagation wavelength of the inverter operation frequency. In order to construct a center-tapped configuration, the sending end of the immittance conversion element is configured such that the two external conductors of the coaxial cables, T_1, T_2, are connected to the high side of the DC power source and each internal conductor is connected to each drain of Q_1 or Q_2. At the receiving end of the immittance conversion element, the internal conductor and external conductor of each coaxial cable are connected in inverse parallel and these two terminals form the output terminals. In order to measure the output power of the system, an RF power meter (type 4421, Bird) is used. The power

978-1-4244-1741-4/08/$25.00 ©2008 IEEE

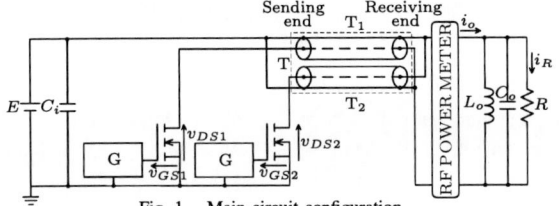

Fig. 1.   Main circuit configuration.

TABLE I

CIRCUIT PARAMETERS.

MOSFETs Q_1, Q_2	BLF245 (65V/6A:PHILIPS)
Immittance conversion element T	3D-2V
Length of coaxial cables T_1, T_2	3.70m
The characteristic impedance Z_ω	50 Ω
Input DC Voltage E	10 ~ 15V
Input Capacitance C_i	136 μF
Output Inductance L_o	6.92 μH
Output Capasitance C_o	5 ~ 110 pF
Output Resistance R	22 ~ 44 Ω
Switching and Output Frequency f_s	12.34 MHz

meter is connected between the output terminals of the immittance conversion element and the load. An LCR parallel resonant circuit is connected as a load, of which the resonant frequency is adjusted to the output frequency of the inverter, and this enables a sinusoidal current to be supplied to the load resistor, R.

B. Operation principle of the proposed inverter

Fig. 2 shows the equivalent circuit configuration of the proposed inverter for four different modes. In this equivalent circuit the input DC voltage source, E, is separately replaced by two input DC voltages, E_1, E_2, on each side of the circuit. The MOSFETs, Q_1 and Q_2, are replaced by ideal switches, S_1, S_2. The LC parallel resonant load is removed and only Resistance, R, is connected to the output terminal of the inverter.

The switches, S_1 and S_2, are alternately turned on and off with a duty ratio of 0.5.

To make the understanding of the operation principle easy, the value of the input DC voltage source, E_2, is assumed to be zero.

It is also assumed that at initial operation there is no transient current or voltage in the transmission lines, T_1 or T_2, or load resistor, R. The terminals of T_1 and T_2 are connected to the load resistor, R, at the receiving end, and to the set of DC voltage source and switch at the sending end. To explain the operation principle, the one switching cycle T_{sw} is divided into four modes. The details of each mode are explained as follows.

⟨Mode1 : $0 \leq$ t $\leq \frac{T_{sw}}{4}$⟩

When the switch, S_1, is turned on at time $t = 0$, a propagating voltage wave, with a value of E, starts propagating from the sending end of the transmission line, T_1. The propagating wave arrives at the receiving end at $t = \frac{T_{sw}}{4}$. Reflection and transmission phenomenon then occurs at the load resistor, R, based on the reflection coefficient, Γ, due to the impedance difference between

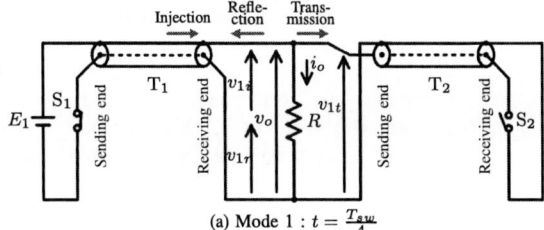

(a) Mode 1 : $t = \frac{T_{sw}}{4}$

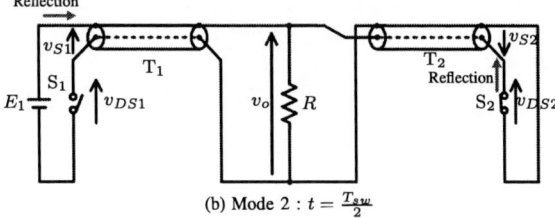

(b) Mode 2 : $t = \frac{T_{sw}}{2}$

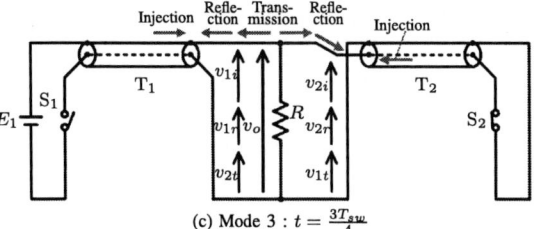

(c) Mode 3 : $t = \frac{3T_{sw}}{4}$

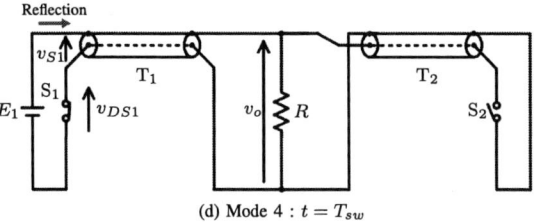

(d) Mode 4 : $t = T_{sw}$

Fig. 2.   The equivalent circuit cofiguration of the proposed inverter.

Z_ω and resistor, R, where Z_ω represents the characterisitc impedance of T_1, T_2. The reflection coefficient, Γ is given by the following equation.

$$\Gamma = -\frac{Z_\omega}{Z_\omega + 2R} \qquad (1)$$

The injection voltage, v_{1i}, at the receiving end of T_1, the reflection voltage, v_{1r}, and the transmission voltage, v_{1t}, to the trasmission line, T_2, and the resultant output voltage v_o at load point are expressed as:

$$v_{1i}\left(\tfrac{T_{sw}}{4}\right) = E \qquad (2)$$
$$v_{1r}\left(\tfrac{T_{sw}}{4}\right) = \Gamma E \qquad (3)$$
$$v_{1t}\left(\tfrac{T_{sw}}{4}\right) = (1+\Gamma)E \qquad (4)$$
$$v_o\left(\tfrac{T_{sw}}{4}\right) = v_{1i}\left(\tfrac{T_{sw}}{4}\right) + v_{1r}\left(\tfrac{T_{sw}}{4}\right) = (1+\Gamma)E$$
$$= (1+\Gamma) \cdot v_{1i}\left(\tfrac{T_{sw}}{4}\right). \qquad (5)$$

The output voltage, v_o, has a positive value, because the absolute value of the reflection coefficient is smaller than 1.

⟨Mode2 : $\frac{T_{sw}}{4} \leq$ t $\leq \frac{T_{sw}}{2}$⟩

Each reflection or transmission voltages, v_{1r}, or v_{1t}, arrives at each the sending end of T_1 or T_2, respectively. Since switch, S_1, is turned off at $t = \frac{T_{sw}}{2}$, the reflection voltage, v_{1r}, then reflects again at the sending end of T_1 with a reflection coefficient of $\Gamma = 1$. This reflection voltage, v_{S1r}, propagates again towards the receiving end of T_1 and is expressed as:

$$v_{S1r}\left(\tfrac{T_{sw}}{2}\right) = \Gamma E. \qquad (6)$$

Therefore, the voltage, v_{S1}, appearing on the sending end of T_1 and the voltage, v_{DS1}, appearing across S_1 are expressed as:

$$v_{S1}\left(\tfrac{T_{sw}}{2}\right) = v_{1r}\left(\tfrac{T_{sw}}{4}\right) + v_{S1r}\left(\tfrac{T_{sw}}{2}\right) = 2\Gamma E \qquad (7)$$

$$v_{DS1}\left(\tfrac{T_{sw}}{2}\right) = E_1 - v_{S1}\left(\tfrac{T_{sw}}{2}\right) = (1 - 2\Gamma)E. \qquad (8)$$

The transmission voltage, v_{1t}, arriving at the sending end of T_2 at $t = \frac{T_{sw}}{2}$, reflects with a reflection coefficient of $\Gamma = -1$, because S_2 is then turned on. The reflected voltage, named as v_{S2r}, propagates towards the receiving end of T_2 and is expressed as:

$$v_{S2r}\left(\tfrac{T_{sw}}{2}\right) = -(1 + \Gamma)E. \qquad (9)$$

Hence, the voltage, v_{S2}, appearing on the sending end of T_2 and the voltage, v_{DS2}, appearing across S_2 are as follows.

$$v_{S2}\left(\tfrac{T_{sw}}{2}\right) = v_{1t}\left(\tfrac{T_{sw}}{4}\right) + v_{S2r}\left(\tfrac{T_{sw}}{2}\right) = 0 \qquad (10)$$

$$v_{DS2}\left(\tfrac{T_{sw}}{2}\right) = 0 \qquad (11)$$

⟨Mode3 : $\frac{T_{sw}}{2} \leq t \leq \frac{3T_{sw}}{4}$⟩

The reflection voltages, v_{S1r}, v_{S2r}, arrive at each receiving end at $t = \frac{3T_{sw}}{4}$. The arrived voltages, v_{1i}, v_{2i} on T_1 and T_2 are expressed as follows.

$$v_{1i}\left(\tfrac{3T_{sw}}{4}\right) = v_{S1r}\left(\tfrac{T_{sw}}{2}\right) = \Gamma E \qquad (12)$$

$$v_{2i}\left(\tfrac{3T_{sw}}{4}\right) = v_{S2r}\left(\tfrac{T_{sw}}{2}\right) = -(1 + \Gamma)E \qquad (13)$$

Reflection and transmission phenomenon occurs again at the load point based on the reflection coefficient, Γ, given by Eq. (1). The transmission voltage, v_{1t}, from T_1 to T_2, and the reflection voltage, v_{2r}, from T_2 to T_2 are expressed as follows.

$$v_{1t}\left(\tfrac{3T_{sw}}{4}\right) = (1 + \Gamma) \cdot v_{1i}\left(\tfrac{3T_{sw}}{4}\right)$$
$$= (1 + \Gamma)\Gamma E \qquad (14)$$

$$v_{2r}\left(\tfrac{3T_{sw}}{4}\right) = \Gamma \cdot v_{2i}\left(\tfrac{3T_{sw}}{4}\right)$$
$$= -(1 + \Gamma)\Gamma E \qquad (15)$$

The two voltages, v_{1t} and v_{2r}, that propagate to T_2 have same amplitude and opposite polar character, respectively. Thus, the propagating wave towards T_2 is negated.

On the other hand, the transmission voltage, v_{2t}, from T_2 to T_1, and the reflection voltage, v_{1r}, from T_1 to T_1 are expressed as follows.

$$v_{2t}\left(\tfrac{3T_{sw}}{4}\right) = (1 + \Gamma) \cdot v_{2i}\left(\tfrac{3T_{sw}}{4}\right)$$
$$= -(1 + \Gamma)^2 E \qquad (16)$$

$$v_{1r}\left(\tfrac{3T_{sw}}{4}\right) = \Gamma \cdot v_{1i}\left(\tfrac{3T_{sw}}{4}\right)$$
$$= \Gamma^2 E \qquad (17)$$

Therefore, the voltage wave propagating towards the sending end of T_1 is expressed as:

$$v_{1r}\left(\tfrac{3T_{sw}}{4}\right) + v_{2t}\left(\tfrac{3T_{sw}}{4}\right) = -(1 + 2\Gamma)E. \qquad (18)$$

The resultant output voltage, v_o, at the load point is expressed as:

$$v_o\left(\tfrac{3T_{sw}}{4}\right) = v_{1i}\left(\tfrac{3T_{sw}}{4}\right) + v_{1r}\left(\tfrac{3T_{sw}}{4}\right) + v_{2t}\left(\tfrac{3T_{sw}}{4}\right)$$
$$= -(1 + \Gamma)E$$
$$= -(1 + \Gamma)v_{1i}\left(\tfrac{T_{sw}}{4}\right). \qquad (19)$$

It should be noted that the output voltage, v_o, given by Eq. (19), has an opposite polar character of the one given by Eq. (5), and both amplitudes are the same.

⟨Mode4 : $\frac{3T_{sw}}{4} \leq t \leq T_{sw}$⟩

The wave expressed by Eq. (18) propagates and arrives at the sending end of T_1 as S_1 is turned on at $t = T_{sw}$. The propagating wave reflects with a reflection coefficient of $\Gamma = -1$ because the impedance of the voltage source, E_1, is zero and S_1 is in the on state. The reflected voltage, v_{S1r}, is then expressed as:

$$v_{S1r}(T_{sw}) = -\left\{v_{2t}\left(\tfrac{3T_{sw}}{4}\right) + v_{1r}\left(\tfrac{3T_{sw}}{4}\right)\right\}$$
$$= (1 + 2\Gamma)E. \qquad (20)$$

Therefore, the voltage, v_{S1}, appearing on the sending end of T_1 and the voltage, v_{DS1}, appearing across S_1 are expressed as follows.

$$v_{S1}(T_{sw}) = E + v_{S1r}(T_{sw}) + v_{2t}\left(\tfrac{3T_{sw}}{4}\right) + v_{1r}\left(\tfrac{3T_{sw}}{4}\right)$$
$$= E - (1 + 2\Gamma)E + (1 + 2\Gamma) = E \qquad (21)$$

$$v_{DS1}(T_{sw}) = 0 \qquad (22)$$

At the second cycle of T_{sw} both the voltage, E, supplied from the voltage source, E_1, and the reflection voltage, v_{S1r}, expressed by Eq. (20) starts propagating from the sending end of T_1. The input voltage, v_{1i}, is then expressed as:

$$v_{1i}\left(\tfrac{T_{sw}}{4} + T_{sw}\right) = E + v_{S1r}(T_{sw})$$
$$= E + (1 + 2\Gamma)E. \qquad (23)$$

In the same way, v_{1i} at the third cycle of T_{sw} is expressed as:

$$v_{1i}\left(\tfrac{T_{sw}}{4} + 2T_{sw}\right) = E + (1 + 2\Gamma)\left\{E + (1 + 2\Gamma)E\right\}$$
$$= E + (1 + 2\Gamma)E + (1 + 2\Gamma)^2 E. \qquad (24)$$

Hence v_{1i} at the n-th cycle of T_{sw} is given by the following equation.

$$v_{1i}(n) = \sum_{k=1}^{n} E(1 + 2\Gamma)^{k-1} \qquad (25)$$

As for the output voltages shown in Eqs. (5) and (19) of the first cycle, the output voltages appearing for the first and fourth modes of the second cycle are expressed as:

$$v_o\left(\tfrac{T_{sw}}{4} + T_{sw}\right) = (1 + \Gamma) \cdot v_{1i}\left(\tfrac{T_{sw}}{4} + T_{sw}\right)$$
$$= (1 + \Gamma)\left\{E + (1 + 2\Gamma)E\right\}$$
$$= E(1 + \Gamma)\left\{1 + (1 + 2\Gamma)\right\} \qquad (26)$$

$$v_o\left(\tfrac{3T_{sw}}{4} + T_{sw}\right) = -(1+\Gamma)\cdot v_{1i}\left(\tfrac{T_{sw}}{4} + T_{sw}\right)$$
$$= -(1+\Gamma)\{E + (1+2\Gamma)E\}$$
$$= -E(1+\Gamma)\{1 + (1+2\Gamma)\} \qquad (27)$$

In the same way, v_o at the third cycle of T_{sw} is expressed as follows.

$$v_o\left(\tfrac{T_{sw}}{4} + 2T_{sw}\right) = (1+\Gamma)\cdot v_{1i}\left(\tfrac{T_{sw}}{4} + 2T_{sw}\right)$$
$$= (1+\Gamma)\{E + (1+2\Gamma)E + (1+2\Gamma)^2 E\}$$
$$= E(1+\Gamma)\{1 + (1+2\Gamma) + (1+2\Gamma)^2\} \qquad (28)$$

$$v_o\left(\tfrac{3T_{sw}}{4} + 2T_{sw}\right) = -(1+\Gamma)\cdot v_{1i}\left(\tfrac{T_{sw}}{4} + 2T_{sw}\right)$$
$$= -(1+\Gamma)\{E + (1+2\Gamma)E + (1+2\Gamma)^2 E\}$$
$$= -E(1+\Gamma)\{1 + (1+2\Gamma) + (1+2\Gamma)^2\} \qquad (29)$$

Thus the output, v_o, at the n-th cycle of T_{sw} is expressed as:

$$v_o(n) = \sum_{k=1}^{n} E(1+\Gamma)(1+2\Gamma)^{k-1}. \qquad (30)$$

Considering Eqs. (25) and (30), the amplitudes of injection voltage, V_{1i}, at the receiving end of T_1 and output voltage, V_o, and output current, I_o, on the steady state are expressed as follows.

$$V_{1i} = \lim_{n\to\infty} v_{1i}(n) = -\frac{E}{2\Gamma} = \frac{Z_\omega + 2R}{2Z_\omega}E \qquad (31)$$

$$V_o = \lim_{n\to\infty} v_o(n) = -\frac{1+\Gamma}{2\Gamma}E = \frac{R}{Z_\omega}E \qquad (32)$$

$$I_o = \frac{V_o}{R} = \frac{E}{Z_\omega} \qquad (33)$$

The same results are obtained in the case when E_1 and E_2 are assumed to be 0 and E, respectively. Therefore the resultant amplitudes of the output voltage and output current increase to twice of that expressed by Eqs. (32) and (33), respectively.

It should be noted that the amplitude of the output current, I_o, shown in Eq. (33) only depends on the the characteristic impedance, Z_ω, of the transmission lines, T_1, T_2, and the input DC voltage, E, of the inverter and it does not depend on the load resistance of R. In other words, the inverter is supplied the input DC voltage and outputs a fixed amplitude current with a square waveform like the current source. This is the reason why the proposed inverter is referred to as "a current-output-type inverter." It should also be noted that the output voltage changes according to the load impedance, and a high voltage output suitable for plasma discharging can be generated.

III. PROPOSED RESONANT GATE DRIVER

A. Gate drive circuit configuration

Fig. 3 shows the proposed resonant gate drive circuit configuration. The MOSFETs, Q_1, Q_2, can be driven directly by the gate drive circuit without galvanic isolation, because the source terminals of each MOSFET switch is connected to a common ground.

In order to realize a series resonant configuration, the gate terminals of MOSFETs, Q_1 and Q_2, a resonant inductor, L_r, and the secondary winding of the transformer

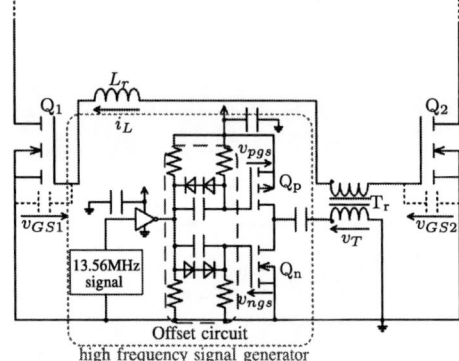

Fig. 3. Gate drive circuit configuration.

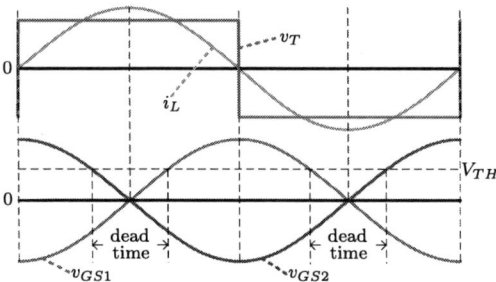

Fig. 4. Waveforms of the gate drive circuit.

for drive signal transmission are connected in series. The output terminal of the high frequency signal generator, surrounded by the dotted line in Fig. 3, is connected to the primary winding of the transformer.

Generally, when MOSFETs, Q_1, Q_2, are driven alternately, two control signals, one of which is periodically inverted, are required. Then logic ICs are often utilized, although the turn on or off delay times of each logic IC are not completely the same. Hence, a dead time between each switching is required in order to compensate imbalance of the delay times. The ratio of the dead times to a switching cycle becomes large by increasing switching frequency, which makes the realilzation of 10 MHz range operation difficult. However, the proposed gate driver can drive two MOSFETs with a common signal, because the series resonant circuit is composed of two MOSFETs and other resonant components, making 10 MHz range operation possible.

In addition, equivalent dead times are obtained as shown in Fig. 4, where $v_{GS1}, v_{GS2}, i_L, v_T,$ and V_{TH} represent the gate-source voltages of the MOSFETs, Q_1 and Q_2, the current through the resonant inductor, L_r, the primary side voltage of the transformer, T_r, and the threshold voltage of MOSFETs, Q_1, Q_2, respectively.

Since the discharge energy that comes from the gate-source capacitor on one MOSFET is transferred to the gate-source capacitor on the other MOSFET and charges it, the loss on the above series resonant circuit is very small, and the resultant gate driving power supplied through the transformer is minimized.

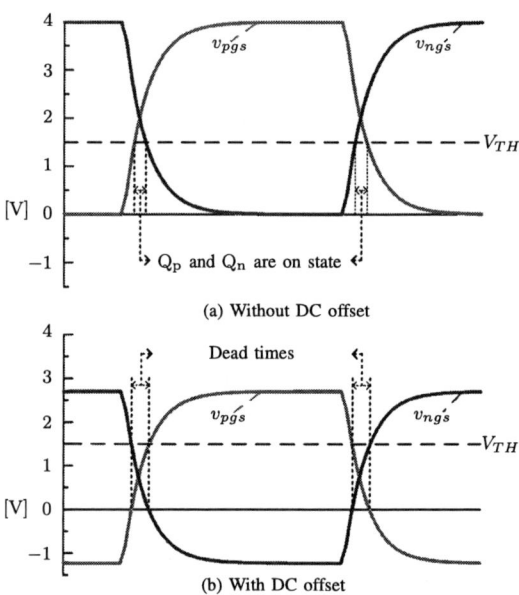

(a) Without DC offset

(b) With DC offset

Fig. 5. Simulated waveforms of v_{pgs}, v_{ngs}

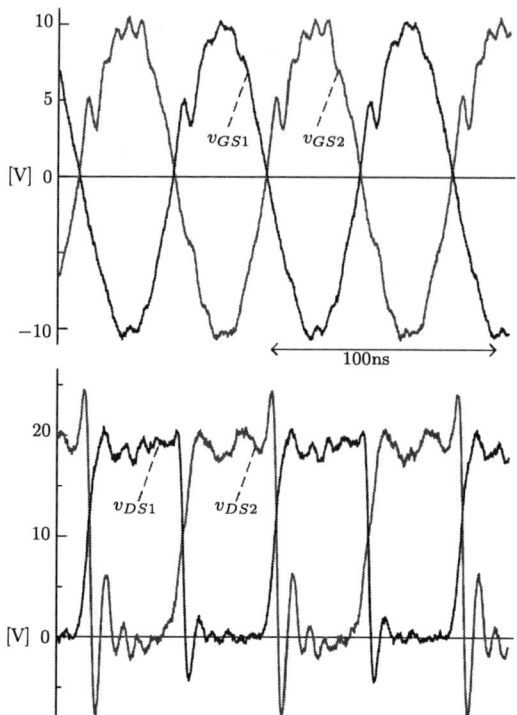

Fig. 6. Experimental waveforms of MOSFETs

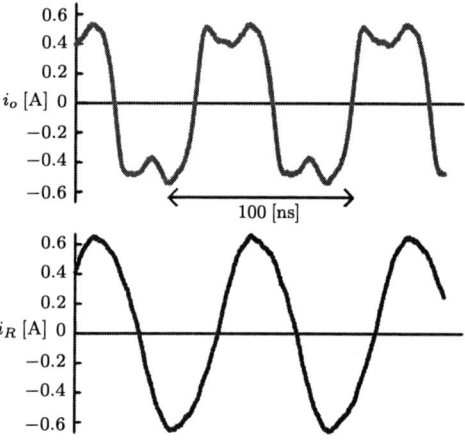

Fig. 7. Experimental waveforms of the output currents.

In Fig.3, the area surrounded by the dashed line shows the offset circuit, where Q_p and Q_n represent the MOSFETs used to construct the high frequency signal generator, and v_{pgs} and v_{ngs} represent their gate-source voltages, respectively. The offset circuit injects the DC offsets into v_{pgs} and v_{ngs} in order to prevent the state where both MOSFETs Q_p and Q_n are on. As shown in Fig. 5 (a), where V_{TH} represents the threshold voltages of MOSFETs Q_p and Q_n, there are times when both Q_p and Q_n are simultaneously in the on state. However, in Fig. 5 (b), with the DC offset, it is found that only either Q_p or Q_n is in the on state.

IV. EXPERIMENTAL RESULT

A prototype of the inverter with the proposed resonant gate drive circuit was developped based on the configurations shown in Figs. 1 and 3, and the circuit parameters given in Table. I. The experimental results for the prototype are given as follows.

Fig. 6 shows the experimental waveforms of v_{GS1}, v_{GS2}, v_{DS1} and v_{DS2}, that is the gate-source voltages and drain-source voltages of the main switches, Q_1, Q_2, respectively. Adequate dead times are inserted in each gate-source voltage, hence stable operation with high frequency switching around 13 MHz is realized, as shown in Fig. 6. However, v_{GS1} and v_{GS2} are distorted at around 5 V during the increase of the voltages due to the Miller effect. Moreover, v_{DS2} has excessive surge voltages at the instant of each turn on of Q_2. This is because the lengthening the connecting wire from the drain terminal of Q_2, which enables the drain current of Q_2 to be measured, increases the resultant parasitic inductance. The Miller effects and the surge voltages increase the switching losses, therefore further improvement is required for increase of the conversion efficiency.

Fig. 7 shows the experimental waveforms of the output current, i_o, and the current, i_R, flowing through the load resistor, R, under the condition that the input DC voltage, E, is 12.7 V and the characteristic impedance, Z_ω, of the transmission lines, T_1 and T_2 are 50 Ω and the load resistance is 44 Ω. It is expected that the output current, i_o, and the current, i_R, flowing through the load resistor, R, are square and sinusoidal waves, respectively. As shown in Fig. 7, the amplitude of the output current, i_o is about 0.5 A. This coincides well with the theoretical value 0.51 A that is calculated with Eq. (33). However, the output current, i_o, has a squarish waveform which includes some excess harmonic components. This is because that the

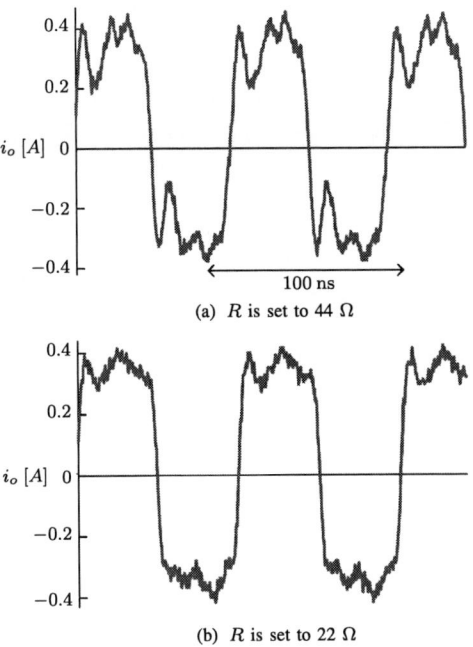

(a) R is set to 44 Ω

(b) R is set to 22 Ω

Fig. 8. Experimental waveforms of the output current when the output resistance is changed.

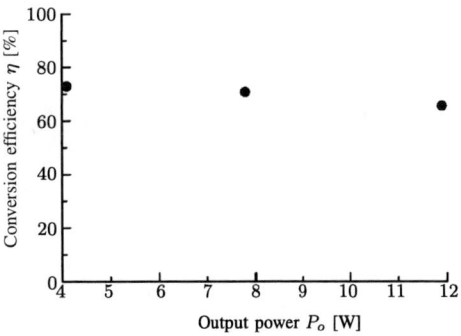

Fig. 9. Conversion efficiency when the output power is changed.

induced voltage at the sending end of the immittance conversion element is not a pure square waveform. The load current, i_R, flowing through the load resistor has a sinusoidal waveform with a little harmonic component. Usually, measured waveforms contain some excess harmonic distortion in such a high frequency region. Then it can be said that the measured waveforms verify the basic operation priciple.

Unfortunately, output voltage on the prototype is not high enough for required plasma discharge. This is because that relatively low resistance value on R is used, and the quality factor of the parallel resonant circuit is reduced. If much higher resistance on R is connected and makes the quality factor high, higher resonant voltage can be generated. In this case, higher current must be supplied from the sending end of the immittance conversion element, and the current rating of the MOSFETs utilized in the main circuit should be increased. This will be executed in our next experimental setup.

Fig. 8 (a) and (b) show experimental waveforms of the output current, i_o, under the condition that the input DC voltage, E, is set to 10.0 V and, R is set to 22 and 44 Ω, respectively.

The amplitudes of both waveforms are approximately 0.4 A, and are almost the same as the theoretical value calculated with Eq. (33). Since the experimental values of the currents shown in Fig. 8 are constant, regardless of how large the load resistance is, the feature of the inverter, referred to as "the current-output characteristic", is verified.

Fig. 9 shows the conversion efficiency, η, under the condition that the output power, P_o, is changed. The

conversion efficiency, η, is defined by the following equation, where P_G and P_i represent the gate driving power and the DC input power, respectively.

$$\eta = \frac{P_o}{P_i + P_G} \times 100 \quad [\%] \tag{34}$$

Since the gate driving power, P_G, on the proposed gate drive circuit is under 0.4 W, it is confirmed that the gate driving power has little infuence on the conversion efficiency. Regarding the measurement of the output power, the forward and reflected output powers are measured with an RF power meter. The resultant output power is calculated by subtracting the reflected power from the forward power in this case. The maximum conversion efficiency value is 72.7 %, so that considerable improvement of the conversion efficiency is realized in copmarison with analog amplifiers. The LCR parallel resonant circuit is then connected as a load, and the output power and the operating frequency are 4.1 W and 12.35 MHz, respectively.

V. CONCLUSION

A high conversion efficiency inverter for megahertz plasma discharging was presented. The proposed current-output-type inverter utilizes an immittance conversion element and a novel resonant gate drive circuit. The features of the proposed inverter clarified that the inverter was suitable for megahertz plasma discharging with a high conversion efficiency. The design and analysis of both the inverter and gate driver were carried out, and the experimental results revealed an increase of the conversion efficiency up to 72.7 % at approximately 13 MHz.

ACKNOWLEDGMENT

This research project was suppported by KAKENHI (Grant-in-Aid for Exploratory Research) (No. 20656050).

REFERENCES

[1] M. P. Theodoridis, S. V. Mollov: "Robust MOSFET Driver for RF, Class-D Inverters," IEEE TRANSACTIONS ON INDUSTRIAL ELECTRONICS, VOL. 55, NO. 2, FEBRUARY 2008.

[2] Y. Chen, F. C. Lee, L. Amoroso, and H. -P. Wu: "Resonant MOSFET Gate Driver With Efficient Energy Recovery," IEEE TRANSACTION ON POWER ELECTRONICS, VOL. 19, NO. 2, MARCH 2004.

[3] T. Shimizu, H. Kinjyo, K. Wada: "A Novel High-frequency Current Output Inverter based on an Immittance Conversion Element and a Hybrid MOSFET-SiC Diode Switch," IEEE PESC, 2003, pp. 2003-2008.

[4] T. Shimizu, M. Shioya: "Characteristics of Electric Power Transmission on High-Frequency Inverter Having Distributed Constant Line," IEEE TRANSACTIONS. ON INDUSTRIAL ELECTRONICS, Vol. 38, No. 2, pp.119-131, 1992.

[5] H. Ohguchi, T. Shimizu, et. al. : "A High Frequency Electric Ballast for HID Lamps based on a 114 Long Distributed Constant Line," TRANSACTION ON POWER ELECTRONICS, Vol. 13, No. 6, pp.1023-1029, 1998.

[6] H. Ohguchi, T. Shimizu, M. Tamate, R. Shimotaya, H. Takagi, and M. Ita: "13.56MHz Current Source Generator based on Third Harmonic Power Transmission using Immittance Conversion Topology and Investigation on Novel Immittance Conversion Element," IEEE Proc. of ISIE'2000, pp. 477-481,2000.

[7] J. W. Phinney, D. J. Perreault, and J. H. Lang: "Radio-Frequency Inverters With Transmission-Line Input Networks," IEEE TRANSACTIONS ON POWER ELECTRONICS, VOL. 22, NO. 4, JULY 2007.

[8] H. Fujita, H. Akagi: "Control and Performance of a Pulse-Density-Modulated Series-Resonant Inverter for Corona Discharge Processes," IEEE TRANSACTIONS ON INDUSTRY APPLICATIONS, VOL. 35, NO. 3, MAY/JUNE 1999.

[9] W. Saito, T. Domon, I. Omura, M. Kuraguchi, Y. Takada, K. Tsuda, and M. Yamaguchi: "Demonstration of 13.56-MHz Class-E Amplifier Using a High-Voltage GaN Powere-HEMT," IEEE ELECTRON DEVICE LETTERS, VOL. 27, NO. 5, MAY 2006.

Voltage Fed Zero-Voltage Zero-Current Switching PWM DC-DC Converter

Jaroslav Dudrik, Vladimír Ruščin
Department of Electrical, Mechatronic and Industrial Engineering,
Technical University of Košice, Letná 9, 04200 Košice,
Slovak Republic,
E-mail: jaroslav.dudrik@tuke.sk, phone: +421 55 6022276, fax: +421 55 6330115
E-mail: vladimir.ruscin@tuke.sk, phone: +421 55 6022296, fax: +421 55 6330115

Abstract – A new zero-voltage zero-current switching full-bridge phase-shifted PWM converter with controlled output rectifier is presented in this paper. IGBT switches are used in the high-frequency inverter of the DC-DC converter. Zero-voltage turn-on and zero-current turn-off for all power switches of the inverter is achieved for full load range from no-load to short circuit by using new secondary energy recovery clamp and modified PWM control strategy. Moreover by adding energy recovery clamp the zero-current turn-on and zero-voltage turn-off for rectifier switch is ensured. The principle of operation is explained and analysed and simulation results are presented.

Keywords - Soft switching, ZVZCS converters, switched-mode power supply

I. INTRODUCTION

The soft switching PWM converters are very suitable for high voltage, high power applications where IGBTs are predominantly used as power switches.

The conventional phase shifted PWM converters are often used in many applications because their topology permits all switching devices to operate under zero-voltage switching by using circuit parasitics such as power transformer leakage inductance and devices junction capacitance.

However, because of phase-shifted PWM control, the converter has a disadvantage that circulating current flows through the power transformer and switching devices during freewheeling intervals.

The circulating current is a sum of the reflected output current and transformer primary magnetizing current. Due to circulating current, RMS current stresses of the transformer and switching devices are still high compared with those of the conventional hard-switching PWM full-bridge converter. To decrease the circulating current to zero and thus to achieve zero-current switching, various snubbers, auxiliary circuits and/or clamps connected mostly at the secondary side of power transformer are applied [1] – [10].

The snubbers and/or clamps are necessary to secure disconnection of the secondary side of the power transformer, as it is shown in a very simplified version in Fig.1. and Fig.3.

The disconnection of the secondary windings is usually achieved by application of the reverse bias for the output rectifier (Fig.1.) or using controlled rectifier (Fig.3.).

Consequently, both primary and secondary currents of the transformer become zero. Only a low magnetizing current circulates during freewheeling interval as shown in Fig.2. Thus, the RMS current of the transformer and switches are considerably reduced.

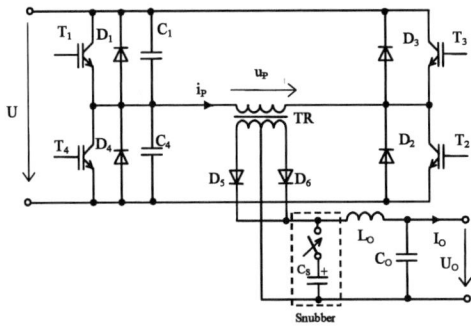

Fig.1. Principle of the ZVZCS converter operation

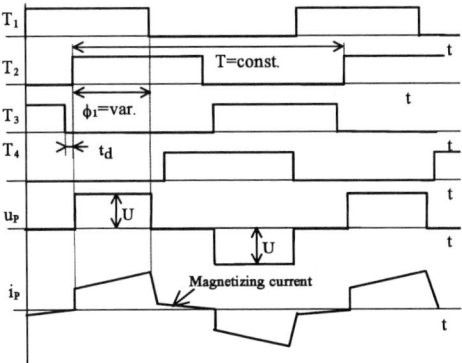

Fig.2. Operation waveforms of ZVZCS PWM converter

The inverter switches operate under zero-voltage switching either in one leg (converter in Fig.1) or in both legs of the converter (converter in Fig.3).

However, the optimal switching for IGBTs is zero-voltage turn-on and mainly zero-current turn-off due to elimination of the current tail influence, which has considerable high involvement in creation of the IGBT turn-off losses.

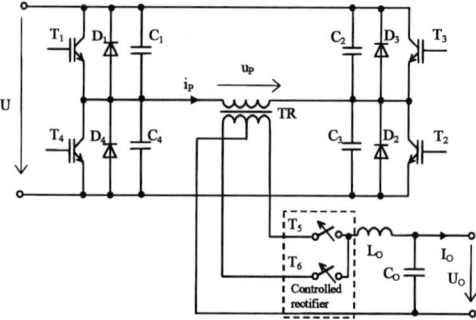

Fig.3. Principle of the ZVS converter operation

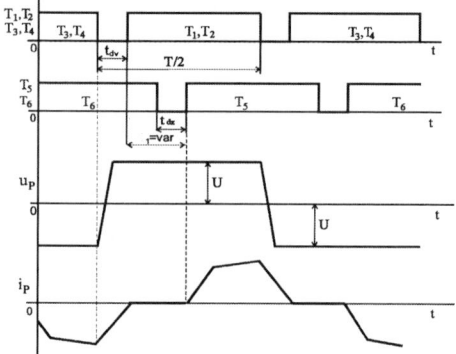

Fig.4. Operation waveforms of ZVS PWM converter

II. POWER CIRCUITS OF THE PROPOSED CONVERTER

To avoid the problems mentioned above, the topology of the following ZVZCS converter was proposed.
The DC-DC converter shown in Fig.5 consists of high-frequency inverter, power transformer, output rectifier, output secondary switch and output filter.

The main part of the converter includes high frequency full-bridge inverter consisting of four ultrafast IGBT's T_1-T_4 and freewheeling diodes D_1-D_4. The secondary winding of the high-frequency step-down power transformer TR is connected through a fast recovery rectifier D_5, D_6 and secondary switch T_S to output filter consisting of smoothing choke L_0 and capacitor C_0.
The converter is controlled by modified pulse-width modulation (Fig.6), and consequently the zero-voltage turn-on and zero-current turn-off all of the transistors T_1-T_4 in the inverter are reached.

The semiconductor switch T_S in the secondary side is used to reset secondary and consequently also primary current. The transistor T_S operates with double switching frequency. At turn-off of the switch T_S the energy stored in leakage inductance is clamped by D_C and C_C and then transferred trough D_S and L_S to the load. By using non-

dissipative turn-off snubber to reduce turn-off losses of the transistor T_S, the overall efficiency is increased.
The additional energy recovery clamp is very simple, consisting of only few components and so the additional cost is not high.

III. OPERATION PRINCIPLE

The basic operation of the proposed soft switching converter has nine operating modes (intervals) within each half cycle. The switching diagram and operation waveforms are shown in Fig. 6.

It is assumed that all components and devices are ideal.
The turn-off snubber used for decreasing turn-off losses of the secondary switch was not included into circuit analysis.

Interval (t_0-t_1): The transistors T_1, T_2 and T_S are turned on at t_0. The primary current (only magnetizing current) flows through diodes D_1, D_2 and consequently the transistors T_1 and T_2 are turned on with ZVS.
The collector current of the transistor T_S starts to flow in the loop T_S-C_C-D_S-L_S-L_O-C_O and capacitor C_C is discharged. So, the rise of the collector current is in resonant way with the resonant frequency ω_{R1} different at no-load and short circuit in a range:

$$\sqrt{(L_O + L_{CS}) \cdot \frac{C_O \cdot C_C}{C_O + C_C}} \le \omega_{R1} \le \sqrt{(L_O + L_{CS}) \cdot C_C}$$

(1)

Interval (t_1-t_2): The transformer leakage inductance L_{LP} reflected to the primary side causes that primary current i_P is linearly increased with the slope U/L_{LP} while the secondary voltage u_S is zero as a result of commutation between output freewheeling diode D_O and rectifier diode D_5.
The discharging of the clamp capacitor C_C causes the current overshoot at turn-on of the transistor T_S, which maximum is limited by the value of the smoothing inductance current i_{LO}.

Interval (t_2-t_3): At t_2 the commutation between diode D_5 and output freewheeling diode D_O is finished. At t_3 the clamp capacitor current commutates to clamp diode D_C.
Interval (t_3-t_4): Transistors T_1 and T_2 are conducting and the energy is delivered from the source to the load via power transformer TR, diode D_5 and smoothing choke L_O and from inductance L_S in the loop L_S-L_O-C_O-D_C-D_S. So, the smoothing inductance current is a sum of the secondary current and inductance L_S current:

$$i_O = i_S + i_{LS}$$

(2)

Interval (t_4-t_5): The primary current increases with the slope:

296

$$\frac{di_p}{dt} = \frac{U - n \cdot U_O}{L_{LP} + n^2 \cdot L_O} + \frac{U}{L_m} \qquad (3)$$

Where $n = \dfrac{N_P}{N_S}$ is power transformer turns ratio and L_m magnetizing inductance of the power transformer TR.

__Interval (t_5-t_6):__ At t_5 the secondary transistor T_S turns off. At that time the commutation between transistor T_S and clamp diode D_C occurs and charging of the clamp capacitor C_C starts. This commutation time can be neglected, because only neglected parasitic inductance of wires is in the commutation loop T_S-D_C-C_C. Afterwards the commutation between D_C, D_5 and output freewheeling diode D_O starts. Because in the commutation path a relatively large leakage inductance of the transformer is found, the commutation is slow.

In the mentioned commutation path the resonance occurs and rise of the current depends on the resonant frequency ω_{R2}:

$$\omega_{R2} = \sqrt{(L_O + L_{LS}) \cdot \frac{C_O \cdot C_C}{C_O + C_C}} \quad \text{for } R_O = \infty \qquad (4)$$

$$\omega_{R2} = \sqrt{(L_O + L_{LS}) \cdot C_C} \qquad \text{for } R_O = 0 \qquad (5)$$

During the commutation the energy stored in the leakage inductance is transferred to the clamp capacitor C_C and consequently an over-voltage ΔU_S appears on secondary voltage.

Its value can be calculated from equation (the output current ripple is neglected):

$$\frac{1}{2} L_{LS} \cdot I_O^2 = \frac{1}{2} C_C \cdot U_{CC}^2 \qquad (6)$$

where L_{LS} is the transformer leakage inductance reflected to the secondary side and U_{CC} is maximum clamp capacitor voltage.

Then

$$\Delta U_S = U_{CC} - U \frac{N_S}{N_P} \qquad (7)$$

__Interval (t_6-t_7):__ Only small magnetizing current i_m flows through primary winding of TR. The output current flows trough output freewheeling diode D_O.

__Interval (t_7-t_8):__ In this interval the transistors T_1 and T_2 are turned off with ZCS. Only small magnetizing current i_m is switched off by transistors T_1 and T_2. The magnetizing current charges or discharges the internal output capacitances $C_{OSS1} - C_{OSS4}$ of the IBGT transistors $T_1 - T_4$ respectively.

The minimum dead time t_d for the transistors in the leg is given by:

$$t_{d,\min} \geq t_{recom} \qquad (8)$$

where t_{recom} is minority carrier recombination time of IGBTs due to stored charges that could not be removed at turn-off process.

When we take into account also charging and discharging of the capacitances $C_{OSS1} - C_{OSS4}$ by magnetizing current, then minimum dead t_d for achieving of zero voltage turn-on must be:

$$t_{d,\min} \geq \frac{4 C_{OSS} \cdot U}{I_{m,\max}} \qquad (9)$$

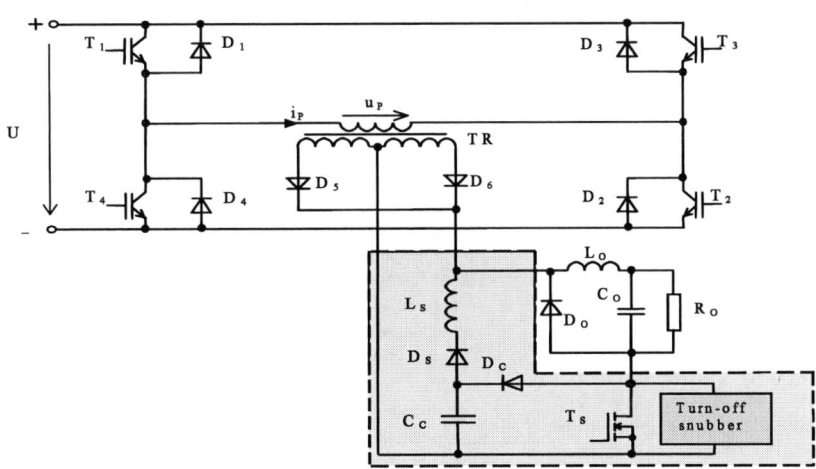

Fig.5. Scheme of the proposed ZCS PWM DC-DC converter

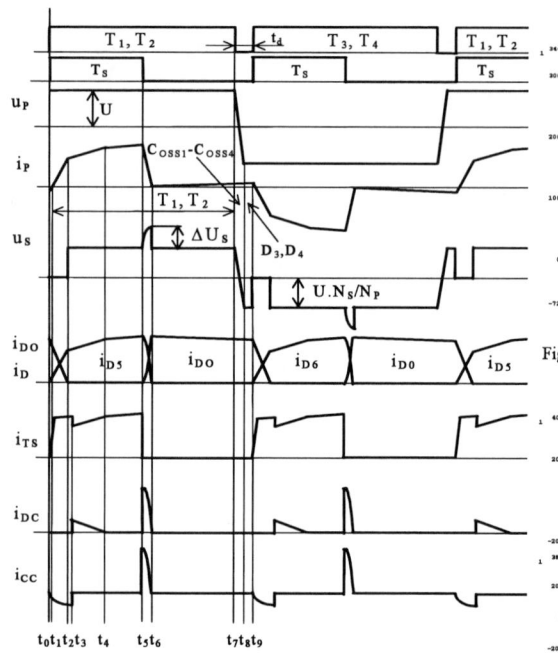

Fig. 6 Operation waveforms of the converter

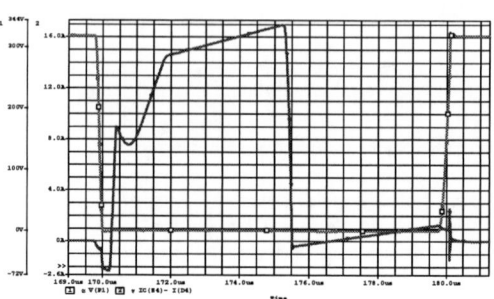

Fig. 7. Switch (transistor T_4 + diode D_4) voltage u_{CE4} and switch current $i_{C4}+i_{D4}$.

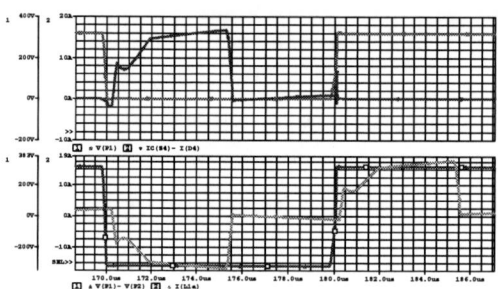

Fig. 8. Switch voltage u_{CE4} and switch current $i_{C4}+i_{D4}$ (upper waveforms) Power transformer TR primary voltage u_p and primary current i_p. (bottom waveforms)

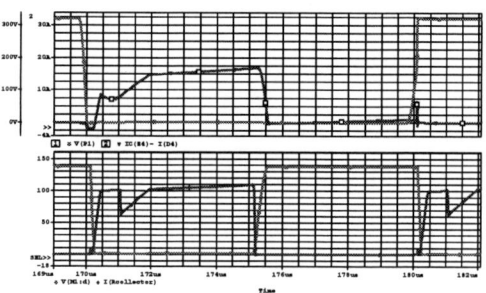

Fig. 9. Switch voltage u_{CE4} and switch current $i_{C4}+i_{D4}$ (upper waveforms) Collector voltage u_{DS} and collector current i_D of the transistor T_S (bottom waveforms)

Interval (t_8-t_9): At t_8 the freewheeling diodes D_3, D_4 starts to lead primary current and thus conditions for the zero-voltage turn-on for the transistors T_3 and T_4 are set up.

So, in the end of the interval the situation from first interval is repeated for the transistors T_3 and T_4, which turn-on at zero-voltage during conduction of the freewheeling diodes D_3, D_4 (at t_9).

IV. SIMULATION RESULTS

A simulation model in programme Orcad was created to verify the properties of the proposed converter. The simulations were made at input voltage U = 320V.
Parameters:
Transformer TR parameters:
Turns ratio n = 6.5,
Magnetizing inductance L_m = 800 μH,
Leakage inductance L_{LP} = 5 μH.
Clamp circuit parameters:
Clamp capacitor C_C = 220 nF,
Clamp inductance L_S = 1 μH.

The following waveforms were obtained at resistive load.

Fig.7 shows switch voltage u_{CE4} and switch current $i_{C4}+i_{D4}$ during turn-on and turn-off of the transistor T_4 in the converter. The switch (transistor T_4 including diode D_4) is turned-on under zero-voltage because at turn-on of the transistor T_4 its freewheeling diode D_4 is in on-state. Moreover the rate of rise of the collector current is limited by the leakage inductance L_{LP} of the transformer.

The transistor turn-off losses are negligible because transistor T_4 turns-off only small magnetizing current (about 1 Amp in this case) as can be seen in Fig.7.

Fig. 8 shows primary voltage u_P and current i_P of the power transformer TR at output load current above I_0 = 100A (bottom waveforms) in comparison with switch voltage u_{CE4} and switch current i_{C4} of the transistor T_4 (upper waveforms).

After turn-off of the transistor T_4 only a small magnetizing current flows through primary winding of transformer. Maximum magnetizing current $I_{m,max}$ is approximately 1 Amp. Depending on the dead time it is high enough for charging or discharging output capacitances $C_{OSS1} - C_{OSS4}$ of the IGBT switches and thus to achieve zero-voltage turn-on.

Collector voltage u_{DS} and collector current i_D of the secondary transistor T_S (bottom waveforms) is shown in

298

Fig. 9. The secondary switch (transistor T_S) is turned-on under zero-current due to influence of the leakage inductance of the transformer L_{LS} reflected to the secondary side and clamp inductance L_S.

The turn-off loss is reduced by clamp capacitor C_C acting as the non-dissipative snubber as it is evident in Fig. 9.

The clamp diode current is displayed in Fig. 10 together with secondary voltage. Sum of the collector current and diode clamp current equals the value of the smoothing inductance current.

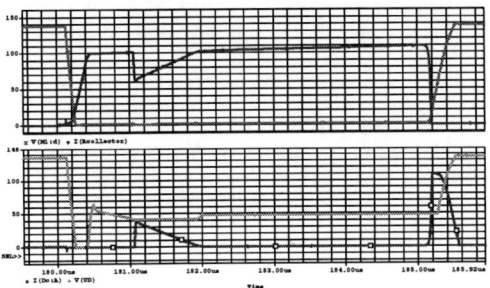

Fig. 10. Collector voltage u_{DS} and collector current i_D of the secondary transistor T_S (upper waveforms)
Rectified secondary voltage u_d of the power transformer TR and clamp diode current i_{DC} (bottom waveforms)

During commutation between secondary diode, output freewheeling diode, and secondary switch the secondary voltage and accordingly rectified secondary voltage is zero. At turn-off of the secondary switch the secondary and also rectified voltage rises as a result of energy stored in leakage inductance. The over-voltage can be decreased to acceptable value by proper design of the clamp capacitor and clamp inductance.

For completeness Fig. 11 shows also the clamp capacitor current and clamp inductance current (bottom waveforms).

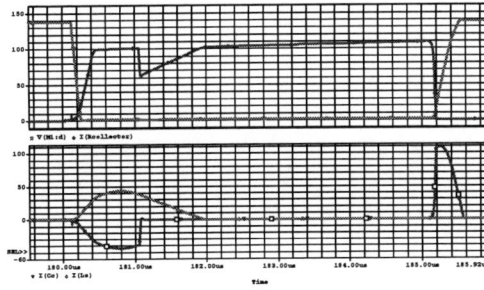

Fig. 11. Collector voltage u_{DS} and collector current i_D of the secondary transistor T_S (upper waveforms)
Clamp capacitor current i_{CC} and snubber inductance current i_{LS} (bottom waveforms)

V. CONCLUSION

Promising results were obtained for IGBT transistors in the full-bridge inverter using the secondary side energy recovery clamp in combination with modified PWM control. Soft switching and reduction of circulating currents in the proposed converter are achieved for full load range.

At proper design it is possible to utilize the magnetizing current of power transformer for charging or discharging output capacitances of the IGBT switches and thus zero-voltage turn-on of the IGBTs to achieve.

If the magnetizing current is not high enough for charging or discharging output capacitances of the IGBT switches, during chosen dead time, then at least zero-current turn-on is reached as a result of leakage inductance of the power transformer.

The IGBT transistors are turned-off almost under zero current. Only small magnetizing current of the power transformer is turned-off by IGBT transistors.

The main task of the proposed secondary energy recovery clamp is transfer of the leakage inductance energy to the load at turn-off of the secondary switch.

Moreover it ensures zero current turn-on and zero voltage turn-off of the secondary switch.

Because this function of clamp is not fully effective when clamp inductance current is continuous, an additional turn-off snubber is employed to improve turn-off process of the secondary switch.

Finally, it is possible to say, that IGBTs in the full bridge inverter operate at almost ideal switching conditions – ZV turn-on and ZC turn-off.

Soft switching of the secondary switch and leakage energy transfer to the load is ensured by energy recovery clamp containing only non-dissipative components.

ACKNOWLEDGMENT

This work was supported by Slovak Research and Development Agency under projects APVV-0095-07 and APVV-0287-07.

REFERENCES

[1] E. S. Kim, K. Y. Joe, M. H. Kye, Y. H. Kim, and B. D. Yoon, " An Improved Soft Switching PWM FB DC-DC Converter for Reducing Conduction Losses," *in Record, IEEE PESC'96*, Vol. I., pp. 651-656.

[2] J. G. Cho, G. H. Rim, and F. C. Lee, "Zero Voltage and Zero Current Switching Full Bridge PWM Converter Using Secondary Active Clamp, " *in Record, IEEE PESC' 96*, Vol. I., pp. 657-663.

[3] K. H. Rinne, K. Theml, and O. McCarthy, " An Improved Zero-Voltage and Zero-Current Switching Full Bridge Converter," *in Record, EPE' 95*, Vol. 2., pp. 725-730.

[4] M. Horváth, and J. Borka ," Welding Technology and Up-to-date Energy Converters," *in Record EDPE 2005*, Dubrovnik, Croatia, September 24-26, 2005. CD-Proc. E05-06.

[5] J. Leuchter, and P. Bauer, "Analysis of Losses in the Power Indirect Converters", *in Record, Circuits Theory Symposium*, ISBN 80-7231-011-9, Brno, 2005, pp. 117-120, (in Czech).

[6] A. Tereň, I. Feňo, and P. Špánik, "DC/DC Converters with Soft (ZVS) Switching", *in Record, ELEKTRO 2001*, section - Electrical Engineering. Žilina 2001, Slovakia, pp. 82 – 90.

[7] P. Chlebiš, *Soft Switching Converters*, Monograph, VŠB-TU Ostrava, Ostrava, Czech republic, 2004, (in Czech).

[8] J. Hamar, and I. Nagy,"Bi-directional Resonant Buck & Boost Converter", *in Record, ELECTROMOTION*, Romania, Oct.-Dec. 2001, Vol.8, No. 4, pp.189-195.

[9] N. D. Trip, "A New Active Snubber for DC-DC Boost Converters", *in Record, 8th International Conference on Engineering of Modern Electric System*, Section Electronics, Oradea, Romania, May 2005, pp.124-127.

[10] S. Petrov, "Expectations of Resonant Converters Utilization as Welding Power Sources", *Schematics* No. 7, July 2006, pp.30-33 (in Russian).

[11] D. Milly and V. Maxim, "Simulation and Analysis of Power Converter Input Currents. International Computer Science Conference", *in Record MicroCAD'98*, February 25-26, Miskolc, Hungary, 1998, pp. 63 – 68.

[12] J. G. Cho, J. W. Baek, D. W. Yoo, H. S. Lee, and G. H. Rim, "Novel Zero-Voltage and Zero-Current Switching (ZVZCS) Full Bridge PWM Converter Using Transformer Auxiliary Winding," *in Record, IEEE PESC'97*, Vol. I., pp. 227-232.

[13] J. G. Cho, J. W. Baek, Ch. Y. Jeong, and G. H. Rim, "Novel Zero-Voltage and Zero-Current Switching Full Bridge PWM Converter Using a Simple Auxiliary Circuit," *IEEE Trans. on Industry Applications*, Vol. 35, pp. 15-20, 1999.

[14] R. Liu, "Comparative Study of Snubber Circuits for DC-DC Converters Utilized in High Power Off-line Power Supply Applications," *in Record, IEEE APEC'99*, pp.821-826.

[15] J. Dudrik and P. Dzurko, "An Improved Soft-Switching Phase-Shifted PWM Full-Bridge DC-DC Converter" *in Record, EPE-PEMC'2000*, Vol. 2, 2000, Košice, pp. 65-69.

[16] J. Dudrik, J., P. Špánik, and N.-D. Trip, "Zero Voltage and Zero Current Switching Full-Bridge DC-DC Converter with Auxiliary Transformer", *IEEE Trans. on Power Electronics*, Vol.21, No.5, 2006, pp. 1328 – 1335.

[17] J. Dudrik, *High Frequency Soft Switching DC-DC Power Converters*, Monograph, Elfa, Košice, Slovakia, 2007(in Slovak).

PWM Spectrum Evaluation and Over-Modulation Phenomena in a Three-Phase Inverters - Analytical Approach

Miro Milanovič*

*University of Maribor/ FERI, Maribor, Slovenia, e-mail: *milanovic@uni-mb.si*

Abstract—This paper provides a comprehensive spectrum analysis of three-phase inverters' output voltages. The output voltages were generated by triangular Pulse Width Modulation algorithm (PWM). This approach is presented in order to improve engineering education based on the classic analytical approach where the advantages of Bessel function series and calculation of Fourier coefficients have been used. The over-modulation phenomena were also considered. This analysis offers a complete quantitative and qualitative knowledge of those PWM signals necessary for high harmonics influence study when different passive and active loads are connected to inverters' outputs.

I. INTRODUCTION

An inverter takes power from a dc source, and supplies the ac load such as a utility grid. The pulse width modulation process is very promising and has historically been used for more than 50 years. It is described in many text books [1]-[6]. The important role in adjustable speed AC drives play the different kind of dissipations dependent on current harmonic spectra and, which is actually dependent, on voltage harmonic spectra. Due to the changeable frequency the filter function is hidden in the PWM algorithm. By using the over-modulation phenomena the magnitude range of the first voltage harmonic can be enlarged.

Teaching undergraduate students PWM algorithms is a difficult task. In many text books, authors describe a modern approach based on Fast Fourier Transformation (FFT). Only in [4] it is possible to learn the basics of the analytical harmonic spectra analysis but it requires the full students attention. This paper explores the analytical spectral analysis and over-modulation phenomena of a PWM signal obtained by sinusoidal triangle modulation, in detail.

II. DESCRIPTION OF MODULATION PRINCIPLES

A. Duty cycle function

The three-phase inverter circuit is shown in Fig. 1. In order to generate three-phase voltages u_{A0}, u_{B0} and u_{C0} at the converter outputs it is necessary to introduce the switching (existing) function described in [1]. It was used for a mathematical description of the modulation function.

$$H_{ij} = \begin{cases} 1, & S_{ij} = \text{ON} \\ 0, & S_{ij} = \text{OFF} \end{cases} \quad (1)$$

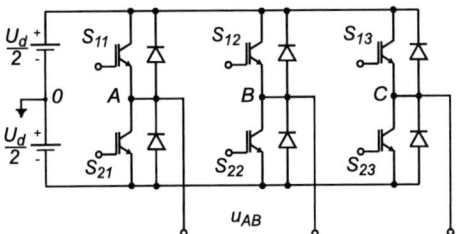

Fig. 1. Three-phase inverter

where index $i = 1, 2$ denotes the upper or lower transistor in half-bridge and index $j = 1, 2, 3$ denotes the first, second or third half-bridge. The converter phase-output voltages can be described as combination of the switching events at the converter half-bridges output points, as follows:

$$\begin{aligned} u_{A0} &= H_{11}U_d/2 + H_{21}\left(-U_d/2\right) \\ u_{B0} &= H_{12}U_d/2 + H_{22}\left(-U_d/2\right) \\ u_{C0} &= H_{13}U_d/2 + H_{23}\left(-U_d/2\right). \end{aligned} \quad (2)$$

According to converter scheme the half-bridge switches should not be switched-on at the same time. This requirements are expressed as:

$$\begin{aligned} H_{11} + H_{21} &= 1 \\ H_{12} + H_{22} &= 1 \\ H_{13} + H_{23} &= 1. \end{aligned} \quad (3)$$

On the other hand the expected "phase" voltages u_{A0}, u_{B0} and u_{C0} are:

$$\begin{aligned} u_{A0} &= \hat{U} \cos(\omega_o t) \\ u_{B0} &= \hat{U} \cos(\omega_o t + 2\pi/3) \\ u_{C0} &= \hat{U} \cos(\omega_o t + 4\pi/3), \end{aligned} \quad (4)$$

where ω_o, $\hat{U}$ are angular frequency and the magnitude of the desired voltages respectively. The switching function (1) and it's properties (3) simplify the notation of the phase-voltages u_{A0}, u_{B0} and u_{C0} from (2) into:

$$\begin{aligned} u_{A0} &= (2H_{11} - 1)\,U_d/2 \\ u_{B0} &= (2H_{12} - 1)\,U_d/2 \\ u_{C0} &= (2H_{13} - 1)\,U_d/2. \end{aligned} \quad (5)$$

In order to calculate the duty cycle functions (Fig. 2) from (4) and (5) the average value of H_{ij} needs to be considered at the time interval T_s. The time interval T_s

978-1-4244-1741-4/08/$25.00 ©2008 IEEE

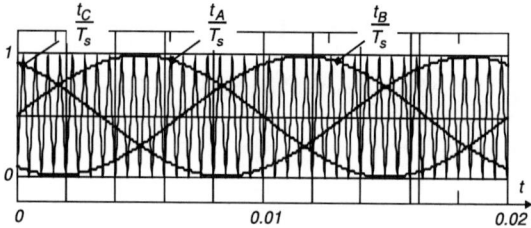

Fig. 2. The duty cycle functions and three-angle modulation signal.

must be chosen to satisfy the condition: $2\pi/T_s \gg \omega_o$. So it follows:

$$
\begin{aligned}
\langle H_{11} \rangle = & \ D_A(t) & = \frac{t_A}{T_s} & = \frac{1}{2} + m_I \cos(\omega_o t) \\
\langle H_{12} \rangle = & \ D_B(t) & = \frac{t_B}{T_s} & = \frac{1}{2} + m_I \cos\left(\omega_o t + \frac{2\pi}{3}\right) \\
\langle H_{13} \rangle = & \ D_C(t) & = \frac{t_C}{T_s} & = \frac{1}{2} + m_I \cos\left(\omega_o t + \frac{4\pi}{3}\right)
\end{aligned}
\tag{6}
$$

where $m_I = \hat{U}/U_d$ represents the modulation index ($m_I \in (0, 1/2)$). The switching intervals t_A, t_B in t_C indicate time interval where appropriate switches are switched-on. **Duty cycle functions** $D_A(t)$, $D_B(t)$ and $D_C(t)$ must be always positive. In order to change $D_x(t)$ (x denotes the phases A, B, and C) into switching sequence, an auxiliary triangle function u_{TA} needs to be introduced. Fig. 2 shows the triangle signal, and duty ratio signal, respectively. The switching signal (sequences) for the all switches were obtained by a comparison of duty cycle function signal $D_x(t)$ and triangle signal $u_{TA}(t)$ as it is shown in Fig. 3. When the switching signal is provided for the converter switches, the two-level voltages at the converter output appear.

$$
u_{x0}(t) = \begin{cases} +U_d/2, & D_x(t) \leq u_{TA}, \Rightarrow \delta_x(t) = 1 \\ -U_d/2, & D_x(t) \geq u_{TA}, \Rightarrow \delta_x(t) = 0 \end{cases}
\tag{7}
$$

The signal $\delta_x(t)$ is implicitly described in (7).

$$
\delta_x(t) = \begin{cases} 1, & S_{11}, S_{22} = \text{ON} \\ 0, & S_{11}, S_{22} = \text{OFF} \end{cases}
\tag{8}
$$

By the help of (5) the phase-output voltage can be described as:

$$
u_{x0} = (2\delta_x(t) - 1) U_d/2.
\tag{9}
$$

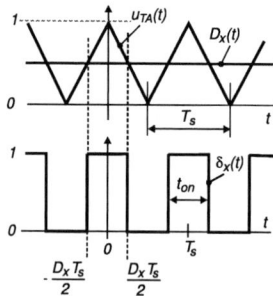

Fig. 3. Triangle signal and triggering pulse train generation.

It is also evident from Fig. 3, that the signal $\delta_x(t)$ in (9) is non-smooth and in order to evaluate the u_{x0} spectrum the Fourier analysis of the pulse train function $\delta_x(t)$ should be considered. It is well known that any periodic signal of period T_s can be expanded into Fourier trigonometric series form:

$$
\delta_x(t) = \frac{a_0}{2} + \sum_{n=1}^{\infty} a_n \cos(n\omega_c t) + b_n \sin(n\omega_c t)
\tag{10}
$$

where ω_c is a triangle signal frequency $\omega_c = 2\pi/T_s$, and the coefficients a_0, a_n and b_n form a set of real numbers uniquely associated with the function $\delta_x(t)$. After analysis it follows:

$$
\delta_x(t) = D_x(t) + \frac{2}{\pi} \sum_{n=1}^{\infty} \frac{\sin(n\pi D_x(t))}{n} \cos(n\omega_c t).
\tag{11}
$$

By introducing A, B, and C instead of x (11) expand into three Fourier series.

$$
\delta_A(t) = D_A(t) + \frac{2}{\pi} \sum_{n=1}^{\infty} \frac{\sin(n\pi D_A(t))}{n} \cos(n\omega_c t)
\tag{12}
$$

$$
\delta_B(t) = D_B(t) + \frac{2}{\pi} \sum_{n=1}^{\infty} \frac{\sin(n\pi D_B(t))}{n} \cos(n\omega_c t)
\tag{13}
$$

$$
\delta_C(t) = D_C(t) + \frac{2}{\pi} \sum_{n=1}^{\infty} \frac{\sin(n\pi D_C(t))}{n} \cos(n\omega_c t)
\tag{14}
$$

The coefficients of (10) calculation:

According to well known Fourier formulas:

$$
a_n = \frac{2}{T_s} \int_0^{T_s/2} (f(t) + f(-t)) \cos(n\omega t) dt
\tag{15}
$$

$$
b_n = \frac{2}{T_s} \int_0^{T_s/2} (f(t) - f(-t)) \sin(n\omega t) dt,
\tag{16}
$$

where $\omega = 2\pi/T_s$. The coefficients a_0, a_n and b_n can be evaluation. According to Fig. 3, $f(t)$ is even function and is defined as:

$$
f(t) = \begin{cases} 1, & t \in \left(-\frac{D_x T_s}{2}, \frac{D_x T_s}{2}\right) \\ 0, & \text{everywhere else} \end{cases}
$$

Due to even properties of the function $f(t)$ the coefficients $b_n = 0$. Coefficient a_0 is:

$$
a_0 = \frac{2}{T_s} \int_0^{D_x T_s/2} 2f(t) dt = \frac{2}{T_s} \int_0^{D_x T_s/2} 2 dt = 2D_x
$$

and coefficient a_n is:

$$
a_n = \frac{2}{\pi} \int_0^{D_x T_s/2} f(t) \cos(n\omega t) d\omega t = \frac{2}{\pi} \frac{\sin(n\pi D_x)}{n}
$$

302

B. Output voltage spectral analysis for Three-phase inverter

So in order to analyze all three voltages u_{A0}, u_{B0} and u_{C0}, it follows from (9) and by using (12), (13) and (14) that the phase-voltages are:

$$u_{A0}(t) = (2D_A - 1)\frac{U_d}{2}$$
$$+\frac{4}{\pi}\frac{U_d}{2}\sum_{n=1}^{\infty}\frac{1}{n}\sin(n\pi D_A)\cos(n\omega_c t) \quad (17)$$

$$u_{B0}(t) = (2D_B - 1)\frac{U_d}{2}$$
$$+\frac{4}{\pi}\frac{U_d}{2}\sum_{n=1}^{\infty}\frac{1}{n}\sin(n\pi D_B)\cos(n\omega_c t) \quad (18)$$

$$u_{C0}(t) = (2D_C - 1)\frac{U_d}{2}$$
$$+\frac{4}{\pi}\frac{U_d}{2}\sum_{n=1}^{\infty}\frac{1}{n}\sin(n\pi D_C)\cos(n\omega_c t) \quad (19)$$

Two of them are enough for calculation of line to line voltage (u_{AB}) and it's spectrum. From (17) and (18) ($u_{AB} = u_{A0} - u_{B0}$) yields:

$$u_{AB} = U_d(D_A - D_B) +$$
$$+\frac{4U_d}{2\pi}\sum_{n=1}^{\infty}\frac{1}{n}\Big(\sin(n\pi D_A) - \sin(n\pi D_B)\Big)\cos(n\omega_c t)$$
$$(20)$$

Calc. of $(D_A - D_B)$ and $(\sin(n\pi D_A) - \sin(n\pi D_B))$:

In order to evaluate the spectrum lines from (20), the terms $(D_A - D_B)$ and $(\sin(n\pi D_A) - \sin(n\pi D_B))$ are considered. This requires the trigonometric function manipulation knowledge, so it follows:

$$D_A - D_B = \left(\frac{1}{2} + m_I\cos(\omega_R t)\right)$$
$$-\left(\frac{1}{2} + m_I\cos\left(\omega_o t + \frac{2\pi}{3}\right)\right)$$
$$= m_I\left(\cos(\omega_o t) - \cos\left(\omega_o t + \frac{2\pi}{3}\right)\right)$$
$$= m_I\left(\frac{\sqrt{3}}{2}\sin\omega_o t + \frac{3}{2}\cos\omega_o t\right)$$
$$= m_I\sqrt{3}\sin\left(\omega_o t + \frac{\pi}{3}\right)$$

The evaluation of $(\sin(n\pi D_A) - \sin(n\pi D_B))$ requires also the trigonometric function manipulation and some knowledge of the Bessel function properties. It follows:

$$\sin(n\pi D_A) - \sin(n\pi D_B)$$
$$= \sin\left(\underbrace{\frac{n\pi}{2}}_{\alpha} + \underbrace{n\pi m_I\cos(\omega_o t)}_{\beta}\right)$$
$$- \sin\left(\underbrace{\frac{n\pi}{2}}_{\alpha} + \underbrace{n\pi m_I\cos\left(\omega_o t + \frac{2\pi}{3}\right)}_{\gamma}\right)$$

By using:

$$\sin(\alpha + \beta) - \sin(\alpha + \gamma) =$$
$$= \sin\alpha\cos\beta + \cos\alpha\sin\beta - \sin\alpha\cos\gamma - \cos\alpha\sin\gamma$$
$$= \sin\alpha(\cos\beta - \cos\gamma) + \cos\alpha(\sin\beta - \sin\gamma)$$

it yields:

$$\sin(n\pi D_A) - \sin(n\pi D_B) =$$
$$= \sin\left(\frac{n\pi}{2}\right)\Big[\cos\Big(\underbrace{n\pi m_I}_{x}\underbrace{\cos(\omega_o t)}_{A}\Big)$$
$$- \cos\Big(\underbrace{n\pi m_I}_{x}\underbrace{\cos\left(\omega_o t + \frac{2\pi}{3}\right)}_{B}\Big)\Big]$$
$$+ \cos\left(\frac{n\pi}{2}\right)\Big[\sin\Big(\underbrace{n\pi m_I}_{x}\underbrace{\cos(\omega_o t)}_{A}\Big)$$
$$- \sin\Big(\underbrace{n\pi m_I}_{x}\underbrace{\cos\left(\omega_o t + \frac{2\pi}{3}\right)}_{B}\Big)\Big]$$

The expressions in square bracket can be replaced by Bessel series of the first kind. It can be written:

$$\cos(x\cos A) = J_0(x) - 2J_2(x)\cos(2A)$$
$$+2J_4(x)\cos(4A) - \dots$$
$$\sin(x\cos A) = 2J_1(x)\cos(A) - 2J_3(x)\cos(3A)$$
$$+2J_5(x)\cos(5A) - \dots$$

$$\cos(x\cos B) = J_0(x) - 2J_2(x)\cos(2B)$$
$$+2J_4(x)\cos(4B) - \dots$$
$$\sin(x\cos B) = 2J_1(x)\cos(B) - 2J_3(x)\cos(3B)$$
$$+2J_5(x)\cos(5B) - \dots$$

where: $x = n\pi m_I$, $A = \omega_o t$ and $B = \omega_o t + \frac{2\pi}{3}$. So after replacing:

$$\sin(n\pi D_A) - \sin(n\pi D_B) =$$
$$= \sin\left(\frac{n\pi}{2}\right)\Big[J_0(x) - 2J_2(x)\cos(2A)$$
$$+2J_4(x)\cos(4A) - \dots$$
$$-J_0(x) + 2J_2(x)\cos(2B) - 2J_4(x)\cos(4B) + ..\Big]$$
$$+ \cos\left(\frac{n\pi}{2}\right)\Big[2J_1(x)\cos(A) - 2J_3(x)\cos(3A)$$
$$+2J_5(x)\cos(5A) - \dots$$
$$-2J_1(x)\cos(B) + 2J_3(x)\cos(3B)$$
$$-2J_5(x)\cos(5B) + \dots\Big] \quad (21)$$

After rearranging (21) it follows:

$$\sin(n\pi D_A) - \sin(n\pi D_B) =$$
$$= \cos\left(\frac{n\pi}{2}\right)2J_1(x)[\cos(A) - \cos(B)]$$
$$- \sin\left(\frac{n\pi}{2}\right)2J_2(x)[\cos(2A) - \cos(2B)]$$
$$- \cos\left(\frac{n\pi}{2}\right)2J_3(x)[\cos(3A) - \cos(3B)]$$
$$+ \sin\left(\frac{n\pi}{2}\right)2J_4(x)[\cos(4A) - \cos(4B)]$$
$$+ \cos\left(\frac{n\pi}{2}\right)2J_5(x)[\cos(5A) - \cos(5B)] + . $$
$$(22)$$

In (22) the term $\cos(2A) - \cos(2B)$ appears ($A = \omega_o t$ and $B = \omega_o t + 2\pi/3$). It can be rearranged by well known formula:

$$\cos(n\omega_o t) - \cos\left(n\left(\omega_o t + \frac{2\pi}{3}\right)\right) = Q\sin(n\omega_o t + \varphi)$$
$$(23)$$

where Q and φ are:

$$Q = \sqrt{\left(\sin\left(n\frac{2\pi}{3}\right)\right)^2 + \left(1 - \cos\left(n\frac{2\pi}{3}\right)\right)^2}$$

$$\varphi = \operatorname{atan}\left(\frac{1 - \cos\left(n\frac{2\pi}{3}\right)}{\sin\left(n\frac{2\pi}{3}\right)}\right)$$

According to (23) it follows from (22):

$$\begin{aligned}
\sin\left(n\pi D_A\right) &- \sin\left(n\pi D_B\right) = \\
&= \cos\left(\tfrac{n\pi}{2}\right) 2J_1(x)\sqrt{3}\left[\sin\left(\omega_o t + \tfrac{\pi}{3}\right)\right] \\
&- \sin\left(\tfrac{n\pi}{2}\right) 2J_2(x)\sqrt{3}\left[\sin\left(2\omega_o t + \tfrac{\pi}{3}\right)\right] \\
&+ \sin\left(\tfrac{n\pi}{2}\right) 2J_4(x)\sqrt{3}\left[\sin\left(4\omega_o t + \tfrac{\pi}{3}\right)\right] \\
&+ \cos\left(\tfrac{n\pi}{2}\right) 2J_5(x)\sqrt{3}\left[\sin\left(5\omega_o t + \tfrac{\pi}{3}\right)\right] + ..
\end{aligned}$$

(24)

In (20) appears $\left[\sin\left(n\pi D_A\right) - \sin\left(n\pi D_B\right)\right]\cos(n\omega_c t)$, *so it can be written:*

$$\begin{aligned}
\left[\sin\left(n\pi D_A\right)\right. &\left.- \sin\left(n\pi D_B\right)\right]\cos(n\omega_c t) = \\
&= \cos\left(\tfrac{n\pi}{2}\right) 2J_1(x)\sqrt{3}\left[\sin\left(\omega_o t + \tfrac{\pi}{3}\right)\right]\cos(n\omega_c t) \\
&- \sin\left(\tfrac{n\pi}{2}\right) 2J_2(x)\sqrt{3}\left[\sin\left(2\omega_o t + \tfrac{\pi}{3}\right)\right]\cos(n\omega_c t) \\
&+ \sin\left(\tfrac{n\pi}{2}\right) 2J_4(x)\sqrt{3}\left[\sin\left(4\omega_o t + \tfrac{\pi}{3}\right)\right]\cos(n\omega_c t) \\
&+ \cos\left(\tfrac{n\pi}{2}\right) 2J_5(x)\sqrt{3}\left[\sin\left(5\omega_o t + \tfrac{\pi}{3}\right)\right]\cos(n\omega_c t) \\
&+ ..
\end{aligned}$$

(25)

In next step the formula

$$\sin\alpha\cos\beta = \frac{1}{2}\left[\sin\left(\alpha - \beta\right) + \sin\left(\alpha + \beta\right)\right]$$

is used and (25) can be written as follows:

$$\begin{aligned}
\left[\sin\left(n\pi D_A\right)\right. &\left.- \sin\left(n\pi D_B\right)\right]\cos(n\omega_c t) = \\
&= \cos\left(\tfrac{n\pi}{2}\right) J_1(x)\sqrt{3}\left[\sin\left(\left(n\omega_c - \omega_o\right)t + \tfrac{\pi}{3}\right)\right. \\
&\left. + \sin\left(\left(n\omega_c + \omega_o\right)t + \tfrac{\pi}{3}\right)\right] \\
&- \sin\left(\tfrac{n\pi}{2}\right) 2J_2(x)\sqrt{3}\left[\sin\left(\left(n\omega_c - 2\omega_o\right)t + \tfrac{\pi}{3}\right)\right. \\
&\left. + \sin\left(\left(n\omega_c + 2\omega_o\right)t + \tfrac{\pi}{3}\right)\right] \\
&+ \sin\left(\tfrac{n\pi}{2}\right) 2J_4(x)\sqrt{3}\left[\sin\left(\left(n\omega_c - 4\omega_o\right)t + \tfrac{\pi}{3}\right)\right. \\
&\left. + \sin\left(\left(n\omega_c + 4\omega_o\right)t + \tfrac{\pi}{3}\right)\right] \\
&+ \cos\left(\tfrac{n\pi}{2}\right) 2J_5(x)\sqrt{3}\left[\sin\left(\left(n\omega_c - 5\omega_o\right)t + \tfrac{\pi}{3}\right)\right. \\
&\left. + \sin\left(\left(n\omega_c + 5\omega_o\right)t + \tfrac{\pi}{3}\right)\right] \\
&+ ..
\end{aligned}$$

The above expression is suitable for substitution in (20).

By using of trigonometric functions manipulation and advantages of the Bessel series it is possible to calculate from (20) the inverter output voltages spectra, modulated by three-angle pulse width modulation algorithm. So (20) can be written in another form:

$$\begin{aligned}
u_{AB} = &\ U_d m_I \sqrt{3}\sin\left(\omega_o t + \tfrac{\pi}{3}\right) \\
&+ \frac{2U_d\sqrt{3}}{\pi}\sum_{n=1}^{\infty}\frac{1}{n}\Bigg\{\cos\left(\tfrac{n\pi}{2}\right)J_1(x) \\
&\cdot\left[\sin\left(\left(n\omega_c - \omega_o\right)t - \tfrac{\pi}{3}\right)\right. \\
&\left. - \sin\left(\left(n\omega_c + \omega_o\right)t + \tfrac{\pi}{3}\right)\right] \\
&- \sin\left(\tfrac{n\pi}{2}\right)J_2(x)\left[\sin\left(\left(n\omega_c - 2\omega_o\right)t - \tfrac{\pi}{3}\right)\right.
\end{aligned}$$

$$\begin{aligned}
&\left. - \sin\left(\left(n\omega_c + 2\omega_o\right)t + \tfrac{\pi}{3}\right)\right] \\
&+ \sin\left(n\tfrac{\pi}{2}\right)J_4(x)\left[\sin\left(\left(n\omega_c - 4\omega_o\right)t - \tfrac{\pi}{3}\right)\right. \\
&\left. - \sin\left(\left(n\omega_c + 4\omega_o\right)t + \tfrac{\pi}{3}\right)\right] \\
&+ \cos\left(n\tfrac{\pi}{2}\right)J_5(x)\left[\sin\left(\left(n\omega_c - 5\omega_o\right)t - \tfrac{\pi}{3}\right)\right. \\
&\left. - \sin\left(\left(n\omega_c - 5\omega_o\right)t + \tfrac{\pi}{3}\right)\right]...\Bigg\}
\end{aligned}$$

(26)

The procedure was verified by analyzing of the PWM signal, where voltage $u_{AB}(t)$ was synthesized, by next set of data: $U_d = 513\ V$, frequency of synthesized signal $f_o = 50\ Hz \Rightarrow \omega_o = 2\pi f_o$, frequency of triangular signal $f_c = 2\ kHz \Rightarrow \omega_c = 2\pi f_c$, and modulation index $m_I = 0.5$. The spectrum lines were calculated from (26) and indicated in Table I where a_f represents the magnitude of the fundamental component and the other coefficients are indicated by a_{ij}. Fig 4 shows the voltage u_{AB} and it's spectrum lines.

TABLE I
CALCULATED SPECTRAL LINES

$a_f(f_o)$	$m_I U_d \sqrt{3}$	444 V
$a_{12}(f_c \pm 2f_o)$	$\frac{2U_d\sqrt{3}}{\pi}\sin\left(\frac{\pi}{2}\right)J_2\left(\pi m_I\right)$	141 V
$a_{14}(f_c \pm 4f_o)$	$\frac{2U_d\sqrt{3}}{\pi}\sin\left(\frac{\pi}{2}\right)J_4\left(\pi m_I\right)$	8 V
$a_{21}(2f_c \pm f_o)$	$\frac{U_d\sqrt{3}}{\pi}\cos\left(\pi\right)J_1\left(2\pi m_I\right)$	80 V
$a_{25}(2f_c \pm 3f_o)$	$\frac{U_d\sqrt{3}}{\pi}\cos\left(\pi\right)J_5\left(2\pi m_I\right)$	15 V

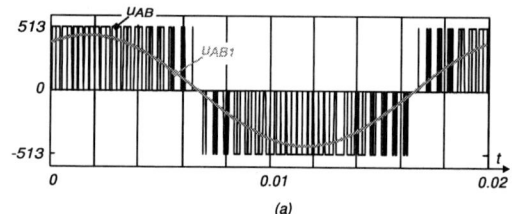

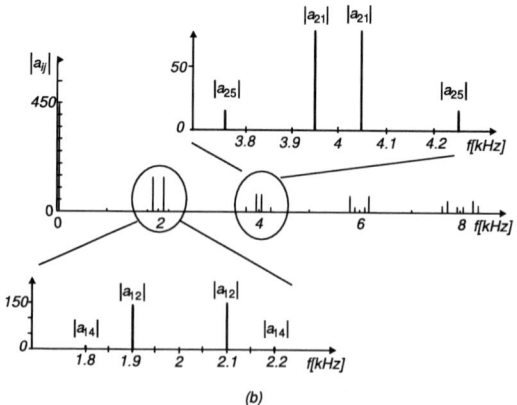

Fig. 4. Line voltage u_{AB} and it's spectrum lines

C. Over-modulation phenomena

Some applications require the largest magnitude of power supply voltage as is possible to generate by changing of the modulation index in range $m_I \in (0, 1/2)$. The over-modulation appears when the switching function excess the magnitude of the high-frequency triangle signal, so when $m_I > 1/2$ (Fig. 5) the over-modulation phenomena appears. In order to define the new duty cycle function the expected phase-voltage can be expressed:

$$u_{A0(sin)} = \begin{cases} \hat{U}\sin\omega t; & 0 \le \omega t \le \alpha \\ U_d/2; & \alpha \le \omega t \le (\pi - \alpha) \\ \hat{U}\sin\omega t; & (\pi - \alpha) \le \omega t \le (\pi + \alpha) \\ -U_d/2; & (\pi + \alpha) \le \omega t \le (2\pi - \alpha) \\ \hat{U}\sin\omega t; & (2\pi - \alpha) \le \omega t \le 2\pi \end{cases}$$

$$(27)$$

where α is evaluated from Fig. 5 as:

$$\alpha = \arcsin\left(\frac{U_d}{2\hat{U}}\right) = \arcsin\left(\frac{1}{2m_I}\right).$$

The voltage indicated in (27), and also the other phase-voltage u_{B0} and u_{C0} are non-smooth and can be expressed by Fourier series:

$$
\begin{aligned}
u_{A0(sin)} &= b_1\cos\omega t + b_3\cos 3\omega t + b_5\cos 5\omega t ... \\
u_{B0(sin)} &= b_1\cos\left(\omega t + \frac{2\pi}{3}\right) + b_3\cos 3\left(\omega t + \frac{2\pi}{3}\right) \\
&\quad + b_5\cos 5\left(\omega t + \frac{2\pi}{3}\right) + ... \\
u_{C0(sin)} &= b_1\cos\left(\omega t + \frac{4\pi}{3}\right) + b_3\cos 3\left(\omega t + \frac{4\pi}{3}\right) \\
&\quad + b_5\cos 5\left(\omega t + \frac{4\pi}{3}\right) + ...
\end{aligned}
$$

$$(28)$$

The coefficients b_n are evaluated by well known formulas for Fourier analysis (15) and (16).

Calculation of coefficients b_n :

Due to odd function properties of (27) ($f(-\omega t) = -f(\omega t)$) the coefficient $a_n = 0$, and

$$
\begin{aligned}
b_n &= \frac{2}{\omega T}\int_0^{T/2} 2f(\omega t)\sin(n\omega t)d\omega t \\
&= \frac{2}{\pi}\Big[\int_0^\alpha \hat{U}\sin\omega t\sin(n\omega t)d\omega t + \int_\alpha^{\pi-\alpha}\frac{U_d}{2}\sin(n\omega t)d\omega t \\
&\quad + \int_{\pi-\alpha}^\pi \hat{U}\sin\omega t\sin(n\omega t)d\omega t\Big]
\end{aligned}
$$

after short calculation it follows:

$$
\begin{aligned}
b_n &= \frac{2U_d}{\pi}\Big[m_I\Big(\frac{\sin(1-n)\alpha}{2(1-n)} - \frac{\sin(1+n)\alpha}{2(1+n)} \\
&\quad + \frac{\pi}{2}\frac{\sin(1-n)\pi}{(1-n)\pi} - \frac{\sin(1-n)(\pi-\alpha)}{2(1-n)} \\
&\quad + \frac{\sin(1+n)(\pi-\alpha)}{2(1+n)}\Big) + \frac{1}{n}\cos n\alpha\Big].
\end{aligned}
$$

$$(29)$$

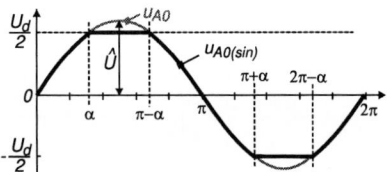

Fig. 5. Over-modulated phase-voltage $u_{A0(sin)}$

From (29) can be extracted the terms for b_1, b_3, b_5, and ..., as follows:

$$
\begin{aligned}
b_1 &= \frac{2U_d}{\pi}\left[m_I\left(\alpha - \frac{1}{2}\sin 2\alpha\right) + \cos\alpha\right] \\
b_3 &= \frac{2U_d}{\pi}\left[m_I\left(\frac{\sin 2\alpha}{2} - \frac{\sin 4\alpha}{4}\right) + \frac{1}{3}\cos 3\alpha\right] \\
b_5 &= \frac{2U_d}{\pi}\left[m_I\left(\frac{\sin 4\alpha}{4} - \frac{\sin 6\alpha}{6}\right) + \frac{1}{5}\cos 5\alpha\right] \\
&\vdots
\end{aligned}
$$

$$(30)$$

Due to the over-modulation phenomena the expected converter output voltages defined in (5) should be modified by considered (28):

$$
\begin{aligned}
u_{A0(sin)} &= (2H_{11} - 1)U_d/2 \\
u_{B0(sin)} &= (2H_{12} - 1)U_d/2 \\
u_{C0(sin)} &= (2H_{13} - 1)U_d/2.
\end{aligned}
$$

$$(31)$$

According to above modification, two duty cycle functions are:

$$D_A(t) = \frac{t_A}{T_s} = \frac{1}{2} + \frac{1}{U_d}u_{A0(sin)} \tag{32}$$

$$D_B(t) = \frac{t_A}{T_s} = \frac{1}{2} + \frac{1}{U_d}u_{B0(sin)} \tag{33}$$

From (32) and (33) the line to line voltage u_{AB} spectrum can be evaluated. As follows from (17), (18), and (20) the line to line voltage u_{AB} can be expressed:

$$
\begin{aligned}
u_{AB}(\omega t) &= b_1\sqrt{3}\sin(\omega t + \pi/6) \\
&\quad + b_5\sqrt{3}\sin(5(\omega t + 5\pi/6)) - \cdots \\
&\quad + \text{Bessel term}
\end{aligned}
$$

$$(34)$$

From (34) is evident that the voltage u_{AB} is described as a sum of the low frequency components (spectrum line coefficients b_1, b_5 and others caused by over-modulation) and the high frequency components (Bessel term). The over-modulation has also some influence to the high frequency spectrum lines (spectrum line next to the multiple of the frequency f_c) but this phenomena will not be considered here. In (34) is also indicated that the third harmonic is disappearing from spectrum. This is due to the three-phase voltage system symmetry. In the single-phase inverter circuit the knowledge of the third voltage harmonic is welcom due to it's possible compensation. For certain modulation index m_I the 3rd, 5th and other voltage u_{AB} spectrum harmonics line is possible to calculate. From (30), (34), and (26) the peak

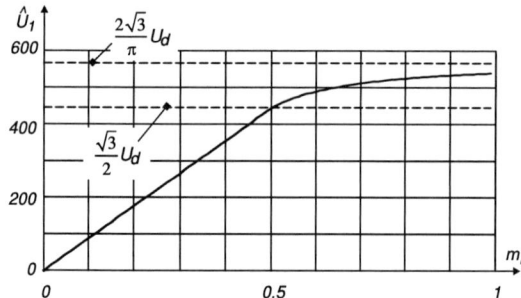

Fig. 6. First harmonic versus modulation index

value of the first harmonic versus modulation index can be described by:

$$\hat{U}_1 = \begin{cases} m_i \sqrt{3} U_d; \ m_I \in (0, 1/2) \\ \frac{2\sqrt{3}U_d}{\pi} \left[m_I \left(\alpha - \frac{\sin 2\alpha}{2} \right) + \cos \alpha \right]; \ m_I \geq 1/2 \end{cases}$$

(35)

From (35) it is possible to evaluate the magnitude of the line to line voltage first harmonic. Fig. 6 shows the first harmonic components versus modulation index. After m_I exceeds $1/2$ the magnitude increase as follows from (35). The procedure was verified by analyzing of the voltage $u_{AB}(t)$, by next set of data: $U_d = 513 \ V$, frequency of synthesized signal $f_o = 50 \ Hz \Rightarrow \omega_o = 2\pi f_o$, frequency of triangular signal $f_c = 2 \ kHz \Rightarrow \omega_c = 2\pi f_c$, and modulation index $m_I = 1.0$. The voltage u_{AB} first harmonic was calculated $\hat{U}_1 = 541 \ V$, as was confirmed by FFT-analysis. Fig. 7 shows the over-modulated line to line voltage and it's spectrum.

III. CONCLUSION

Step by step PWM spectrum analysis for undergraduate and graduate students who studied power-electronics and electric-machine inverters have been described. The three-phase inverter modulation algorithm and over-modulation phenomena have been explored.

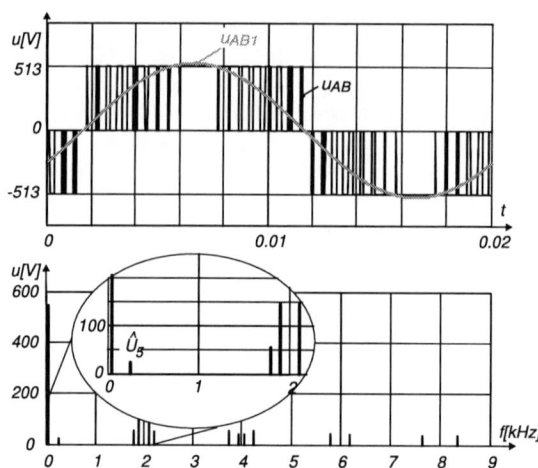

Fig. 7. Over-modulated line to line voltage u_{AB}, and it's spectrum

The students deepest understanding of the PWM modulation processes can be reached by such approach. The results of this analysis is appropriate for further investigation of the PWM processes as are harmonic influences to the losses of the different types of ac motor drives. The analysis can be easily spread to the single-phase inverters and its' filter design.

REFERENCES

[1] P. Wood, *Switching Power Converters*, Van Nostrand, New York, 1981.
[2] W. Leonhard, *Control of Electrical Drives*, Springer-Verlag, Berlin. 1985.
[3] N. Mohan, T. Undeland, W. Robbins, *Power Electronics, Devices, Converter, Application and Design*, John Wiley and Sons. New York, Singapore, Toronto, Brisbane, 1989.
[4] P. T. Krein, *Elements of Power Electronics*, Oxford Univ. Press, 1998
[5] R.W. Ericson and D. Maksimovic, *Fundamentals of Power electronics*, Kluwer Academic Publisher, 2001
[6] H.W. Van der Broeck, Analysis and realization of pulse width modulator based on voltage space vektors, *IEEE-IAS Annual Meeting*, pp. 244-251, October 1986.

Experimental Study of a Matrix Converter Excited Doubly-Fed Induction Machine in Generation and Motoring

Ivan Shapoval[*], Jon Clare[†] and Eduard Chekhet[*]

[*] Institute of Electrodynamics of the Ukrainian National Academy of Sciences, Kyiv, Ukraine, e-mail: chk@ied.org.ua

[†] The University of Nottingham, Nottingham, United Kingdom, e-mail: Jon.Clare@nottingham.ac.uk

Abstract—**Full-scale experimental testing of the 7.5 kW doubly-fed induction machine controlled by matrix converter is reported. A number of doubly-fed induction machine and matrix converter control algorithms have been implemented in real time using a DSP-controller. The experimental rig used to control the doubly-fed induction machine is described. Experimental results demonstrate that the doubly-fed induction machine control algorithms guarantee perfect torque tracking of positive and negative trajectories of torque reference under the condition of unity stator side power factor.**

I. INTRODUCTION

The vector controlled doubly-fed induction machine (DFIM) is an attractive solution for high performance, restricted speed range drives and energy generation applications [1]. The typical connection scheme of DFIM is shown in Fig. 1. For limited speed variations around the synchronous speed of the induction machine, the power handled by the converter at the rotor side is a small fraction (depending on slip) of the overall converted power.

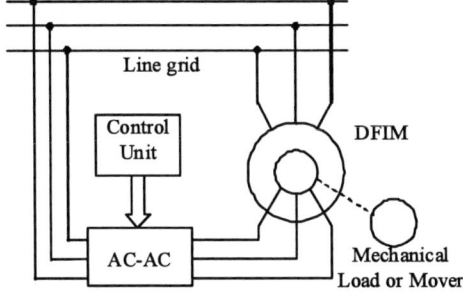

Fig. 1. The typical connection scheme of DFIM.

The fundamentals of DFIM vector control are presented in [1] and widely used in different developments [2]-[7]. In both motor and generator applications the DFIM is able to provide torque production together with stator side power factor control. If a suitably controlled AC/AC converter is used to supply the rotor side of the DFIM, the overall system can be controlled with low harmonic distortion in the stator and rotor sides. Moreover, when the DFIM is used as a variable-speed drive in the dynamic braking mode, the slip power is regenerated by the converter to the supply grid, resulting in highly efficient energy conversion.

Two approaches are possible to supply the DFIM rotor circuit: a standard AC-DC-AC power converter having a vector controlled input rectifier and a direct AC-AC matrix converter (MC) solution. Some simulation results of MC application for DFIM control have already been reported in literature [8]-[10].

The aim of this paper is to present results from experimental testing of a MC excited DFIM as a generator and motor. The concept of indirect stator flux orientation has been implemented in a similar way to that used for indirect rotor flux orientation in a squirrel cage induction machine, in order to solve the full order DFIM control problem.

An intensive experimental study shows that high performance torque tracking is achievable keeping stator side power factor at unity level during energy generation and drive regimes. Soft connection (almost transient-less) of the DFIM stator to line grid is achieved using proposed excitation-synchronization control algorithm during initialization stage of DFIM operation.

The paper is organized as follows. Section II presents general configuration of torque tracking control algorithm for DFIM. In Section III the short description of MC control algorithm is given. Results of experimental testing of the DFIM with MC are given in Section IV.

II. DFIM CONTROL ALGORITHM

The equivalent two-phase model of the symmetrical DFIM with connected to line stator, represented in stator voltage-vector oriented frame (*d-q*) is

$$\dot{\varepsilon} = \omega,$$
$$\dot{\omega} = \left(\mu p_n \left(\psi_{1q} i_{2d} - \psi_{1d} i_{2q} \right) - T_L \right) / J,$$
$$\dot{\psi}_{1d} = -\alpha_1 \psi_{1d} + \omega_1 \psi_{1q} + \alpha_1 L_m i_{2d} + U,$$
$$\dot{\psi}_{1q} = -\alpha_1 \psi_{1q} - \omega_1 \psi_{1d} + \alpha_1 L_m i_{2q}, \qquad (1)$$
$$\dot{i}_{2d} = -\gamma_2 i_{2d} + \omega_2 i_{2q} + \alpha_1 \beta \psi_{1d} - $$
$$- \beta p_n \omega \psi_{1q} - \beta U + u_{2d} / \sigma_2,$$
$$\dot{i}_{2q} = -\gamma_2 i_{2q} - \omega_2 i_{2d} + \alpha_1 \beta \psi_{1q} + \beta p_n \omega \psi_{1d} + u_{2q} / \sigma_2,$$

where $\left(u_{2d}, u_{2q} \right), \left(i_{2d}, i_{2q} \right), \left(\psi_{1d}, \psi_{1q} \right)$ are rotor voltages, rotor currents and stator fluxes, T_L is a driving torque, generated by the prime mover, U and ω_1 are stator (line) voltage amplitude and angular frequency, ε and ω are angular position and rotor speed, $\omega_2 = \omega_1 - \omega$ is slip angular frequency, p_n is number of pole pairs. Positive constants related to DFIM electrical parameters are defined as:

978-1-4244-1741-4/08/$25.00 ©2008 IEEE

$$\alpha_1 = R_1/L_1 \,, \ \sigma_2 = L_2\left(1 - L_m^2/L_1 L_2\right), \ \beta = L_m/L_1\sigma_2 \,,$$
$$\gamma_2 = R_2/\sigma_2 + \alpha_1\beta L_m \,, \ \mu = 3L_m/2L_1 \,,$$

where R_1, R_2, L_1, L_2 - resistance and inductance of stator and rotor respectively, L_m - mutual inductance.

When the DFIM is used as generator, the torque T_L in the first equation of (1) is a driving torque, generated by the prime mover, and obeying the mechanical system dynamics, whose general representation is

$$\dot{\omega} = \left(T - T_L\right)/J \,,$$
$$T_L = k_{\omega m}\left(\omega - \omega_m^*\right),$$
(2)

where $k_{\omega m} > 0$ is the speed controller gain of the prime mover and $\omega_m^* > 0$ is the prime mover speed reference.

Electromagnetic torque T of the DFIM is the load torque for the mechanical system of the primary energy converter. The main control objective of the DFIM operating as a generator is to produce the desired generated torque $T^*(t)$ independently of ω.

Assuming the rotor current-fed condition, the following torque-flux control algorithm is constructed

Torque control algorithm

$$i_{2d} = T^*/\mu\psi^*$$
(3)

Flux level control algorithm

$$i_{2q} = \left(\alpha_1\psi^* + \dot{\psi}^*\right)/\alpha_1 L_m$$
(4)

with the flux reference calculated from

$$\omega_1\psi^* + \alpha_1 L_m i_{2d} + U = 0$$
(5)

The flux reference ψ^* computed from (3) and (5) is equal to

$$\psi^* = \left(-U - \sqrt{U^2 - 4\left(\frac{2}{3}\right)\omega_1 R_1 T^*}\right)\Big/2\omega_1$$
(6)

In [6] it is shown that torque-flux control algorithm (3) – (6) guaranties global asymptotic exponential torque tracking together with asymptotic stator flux orientation, given by condition

$$\lim_{t\to\infty}\psi_{1d} = 0, \ \lim_{t\to\infty}\left(\psi_{1q} - \psi^*\right) = 0$$
(7)

From (4), (6) and (7) it can be concluded that, during the steady state (with $\dot{T}^* = 0$), $\lim_{t\to\infty}\psi_{1q} = \psi^* = L_m i_{2q}$, which implies that $\lim_{t\to\infty}i_{1q} = 0$, and operation with zero stator side reactive power is achieved.

In actual DFIM rotor currents are not available as control inputs and the torque-flux controller outputs (i_{2d}, i_{2q}) in (3) and (4) can only represent desired trajectories (i_{2d}^*, i_{2q}^*) for the real currents i_{2d}, i_{2q}. The rotor voltage vector $u_2 = \left(u_{2d}, u_{2q}\right)^T$ is the only physically available control input of DFIM. The current loop control algorithm should be designed to guarantee that current tracking errors

$$\tilde{i}_{2d} = i_{2d} - i_{2d}^* \,,$$
$$\tilde{i}_{2q} = i_{2q} - i_{2q}^* \,,$$
(8)

asymptotically decay to zero.

Following [6] the current controller control algorithm is defined as

$$u_{2d} = \sigma\left(\gamma i_{2d}^* - \omega_2 i_{2q}^* + \beta\omega\psi^* + \beta U + \dot{i}_{2d}^* - k_i\tilde{i}_{2d} - x_d\right),$$
$$\dot{x}_d = k_{ii}\tilde{i}_{2d},$$
$$u_{2q} = \sigma\left(\gamma i_{2q}^* + \omega_2 i_{2d}^* - \alpha\beta\psi^* + \dot{i}_{2q}^* - k_i\tilde{i}_{2q} - x_q\right),$$
$$\dot{x}_q = k_{ii}\tilde{i}_{2q},$$
(9)

where i_{2d}^*, i_{2q}^* are rotor currents represented in a $(d\text{-}q)$ reference frame; k_i and k_{ii} are positive proportional and integral gains of current controllers; ψ^* is stator flux reference; x_d, x_q are integral components of current controllers.

The block diagram of the proposed controller is shown in Fig. 2.

In contrast to squirrel cage induction machines, the DFIM is supplied from both the stator and rotor sides. A special initialization procedure is required in order not to violate sensible limits of machine operation and to ensure that the rotor/stator currents as well as the required rotor voltages are inside the given limits. The adopted starting sequence for the DFIM generator is as follows.

The primary mover is started first, with the DFIM disabled. When the mechanical speed is sufficiently close to the synchronous speed the control unit, acting on the rotor voltages, imposes currents in the rotor in order to produce on the open stator windings an induced voltage vector which is opposite to that of the line-voltage (machine "excitation"). During this stage the excitation control algorithm acts to synchronize the stator EMF vector to the line voltage vector (both amplitude and phase). When synchronization is achieved the stator circuit is connected to the line grid, ensuring a soft transient. The control unit starts to perform the proposed control algorithm with a zero torque reference (machine "connection"). At this point a torque reference can be applied.

The rotor current control algorithm is constructed as

$$u_{2d} = L_2\left(i_{2d}^* R_2/L_2 - \left(\omega_1 - \omega\right)i_{2q}^* - k_i\tilde{i}_{2d} + v_d\right),$$
$$u_{2q} = L_2\left(i_{2q}^* R_2/L_2 + \left(\omega_1 - \omega\right)i_{2d}^* - k_i\tilde{i}_{2q} + v_q\right).$$
(10)

For constant current references the EMF equations and the error dynamics of rotor currents during excitation become:

$$E_d = L_m\left[-\left(R_2/L_2 + k_i\right)\tilde{i}_{2d} - \omega\tilde{i}_{2q} - \omega_1 i_{2q}^* + v_d\right],$$
$$E_q = L_m\left[-\left(R_2/L_2 + k_i\right)\tilde{i}_{2q} + \omega\tilde{i}_{2d} + \omega_1 i_{2d}^* + v_q\right],$$
(11)

$$\dot{\tilde{i}}_{2d} = -\left(R_2/L_2 + k_i\right)\tilde{i}_{2d} + \omega\tilde{i}_{2q} + v_d, \quad \dot{v}_d = -k_{ii}\tilde{i}_{2d},$$
$$\dot{\tilde{i}}_{2q} = -\left(R_2/L_2 + k_i\right)\tilde{i}_{2q} - \omega\tilde{i}_{2d} + v_q, \quad \dot{v}_q = -k_{ii}\tilde{i}_{2q}.$$
(12)

To define the EMF reference it can be noted that the voltage line vector is aligned with the d-axis, therefore EMF references are

$$E_d^* = U,$$
$$E_q^* = 0.$$
(13)

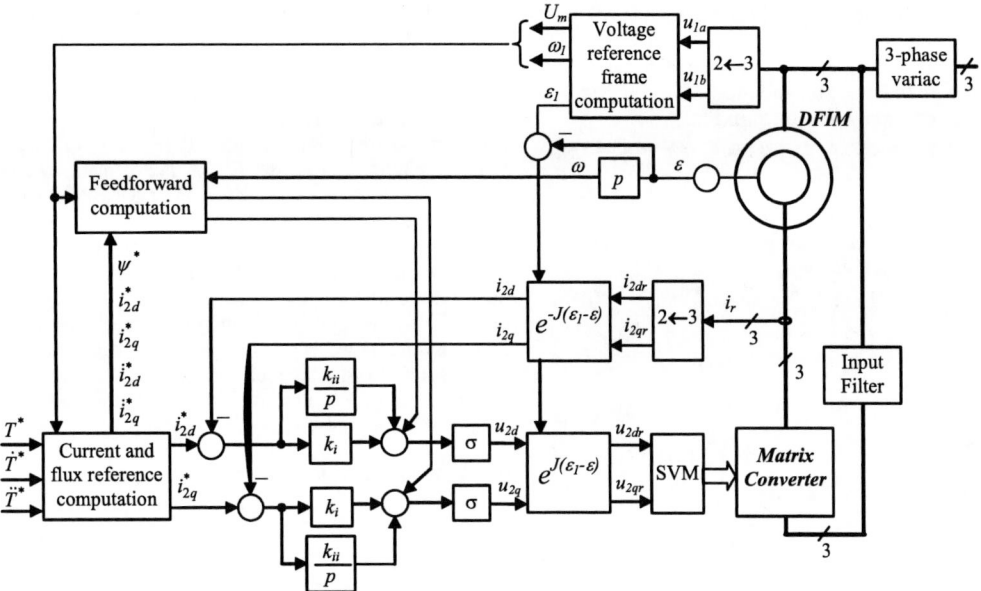

Fig. 2. Block diagram of the torque tracking stator side power factor stabilizing controller.

Consequently, the references for rotor current are

$$i_{2d}^* = 0,$$
$$i_{2q}^* = -U/L_m\omega_1. \qquad (14)$$

From (11) and (12) it can be concluded that synchronization is achieved with transient performance defined by the dynamics of the rotor current subsystem (12). Note that: a) the current references given by (14) are the same as in (3), (4), (6) with $T^* = 0$; and b) the structure of the current controller (10) is a part of the general current controller (9), with additional integral actions.

III. MATRIX CONVERTER CONTROL ALGORITHM

Space vector modulation (SVM) of the MC is based on the instantaneous space-vector representation of output voltage and input current [11], [12]. Through SVM, the matrix converter generates appropriate voltage waveforms for exciting the DFIM rotor. The averaged values of the voltage reference vector are obtained as the result of synthesis from five adjacent stationary vectors (four non-zero and one zero) [11], [12]. As a result of alternate operation on each SVM period the line voltages form an "averaged" voltage to create the output voltage vector.

The SVM algorithm has the following steps:
- first, on the basis of information about the instantaneous input voltage during each SVM cycle the moment of switching from one combination of voltage to another is determined;
- after that on the basis of output voltage vector the required sector is determined;
- duty-cycles and the corresponding time intervals are computed;
- finally, the reference space output voltage vector is formed at the beginning of the next SVM cycle.

Commutation strategies for a MC can be based on two approaches. The first, based on the current direction information and the second, based on measured AC input phase voltages relationship [11], [12]. In this work a commutation strategy based on the current direction information is used.

IV. EXPERIMENTAL RESULTS

A. The Experimental Rig

Torque tracking control algorithms in generation and motoring have been experimentally tested using a slip-ring induction motor with ratings: power 7.5 kW; current 17.5 A; voltage 380 V; speed 1460 rpm; stator resistance $R_1 = 0.45\ \Omega$; rotor resistance $R_2 = 0.2\ \Omega$; stator inductance $L_1 = 0.161$ H; rotor inductance $L_2 = 0.095$ H; mutual inductance $L_m = 0.088$ H; number of pole pairs $p_n = 2$.

The experimental tests were carried out using an experimental rig, whose overall layout is shown in Fig. 3. The experimental rig includes:

1. A 7.5 kW slip-ring induction motor supplied by a matrix converter, operating at 12.5 kHz switching frequency.

2. A current (speed) controlled DC motor, used to provide the load torque to the DFIM, during drive operation, or to stabilize the speed of the rotor shaft, when the DFIM is used as a generator.

3. A DSP-based, real-time controller implemented using FPGA MC control board with TMS320C6711 DSK connected to PC.

4. LEM current and voltage sensors for measuring the analogue signals.

5. An incremental encoder with resolution 2500ppr, used to measure rotor position and speed.

6. A personal computer, acting as operator interface for programming, debugging, program downloading, virtual oscilloscope and automation function during the experiments.

The power circuit of the matrix converter has been developed and built in the Power Electronics, Machines and Control (PEMC) Group of the University of Nottingham around the EUPEC FM35R12KE3 Matrix Converter module [11], [13]. The 18 IGBTs and 18 diodes in this module are rated at 1200V and 35Amps. The Matrix Converter requires an input filter uses three 2uF capacitors and three 1mH inductors. This filter has not been optimized for the matrix converter operating conditions and, hence,

309

the input current waveform quality is lower than expected. In order to protect the matrix converter power devices during experimental tests, stator voltages and hence rotor voltages were limited to 120V line-to-line through a 3-phase variac on the supply, as shown in Fig. 3.

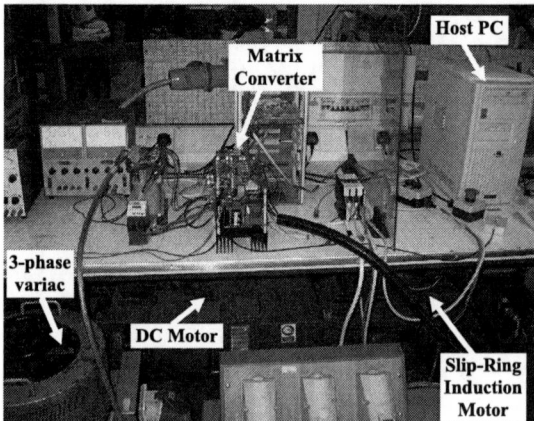

Fig. 3. Photograph of the experimental rig.

B. Control Platform

The control of the MC is implemented with the help of interaction between a digital signal processor (TMS320C6711 DSK board) and a field programmable gate array (FPGA board). For fast data processing Texas Instruments TMS320C6711 DSP board with Actel ProASIC A500K050 FPGA is used in the control board. The C6711 DSK features a 150 MHz clock and is capable of executing 900 million floating-point operations per second. It has a parallel port controller which is able to interface to standard parallel port on a host PC. The host PC provides the user interface to the DSP. The DSP and FPGA based control platform used in the MC is developed by the PEMC group. The FPGA on this board is operated with 10 MHz clock frequency. The FPGA board connected to the DSP board via an expansion port connector.

All the calculations related to the space vector modulation, data manipulations and host interfacing were performed in the DSP. The PWM pulse generation, the commutation control, the watchdog and other software protection items were implemented in the FPGA. Data acquisition and pulse generation are coordinated by the FPGA. On the control platform the analog measurement signals are encoded to digital form. The FPGA is operated with 10 MHz clock frequency and is used to retrieve data from the nine analog-to-digital channels and communicate with the DSP. These digital data are read by the DSP. The output signals resulting from the calculation performed by the DSP are the switching control signals. The switching signals are stored in the FPGA register in the format of the switching state vector and time. The major function of the FPGA is to output the switching state vector and time when the next interrupt occurs (the interrupt occurs every 80μs). Then the PWM pulses are generated and transmitted to the gate driver board.

The control system also includes hardware protection circuits in case of overload. The hardware-based instantaneous overcurrent protection circuit is built in the FPGA board. This protection circuit is based on the use of comparators in which the reference voltage can be adjusted to the maximum peak current allowed in the system to protect the IGBTs under the short circuit or loss of controls. When the measured current is higher than the maximum peak current, the comparator will provide an instantaneous trip signal to the FPGA board and stop the switching pulses. Also the FPGA board has a watchdog timer circuit to protect the MC if the DSP-FPGA network experiences deadlock.

The input data required by the control system is supplied from the measurement boards. This data includes the two line-to-line input MC voltages, two line-to-line mains/stator voltages, three output MC currents. The current measurements use LEM LA55-P current transducers to measure all instantaneous currents: the three output MC currents. In order to measure the line-to-line voltages, the voltage transducer LEM LV25-P is used.

C. Software and Host Interface

The software for all control algorithms of the DFIM and MC is written in C programming language using Code Composer Studio [14]. Code Composer Studio (CCStudio) software is a fully integrated development environment (IDE) supporting Texas Instruments DSP platforms.

The host PC provides the user interface to the network with a link to the DSP. While the DSP is performing a routine calculation the control reference can be set and also the instantaneous control variable can be monitored via the host PC. In addition, it is used to capture and transfer the data variables passed back to the computer for monitoring purposes. The host program in C programming language is developed by the PEMC group.

D. Experimental Tests

Experimental results, reported in Figs. 4 and 5, was performed to investigate system behaviour during torque tracking in generator and motor modes. The sequence of operation during this test is shown in Fig. 4. The DFIM, already connected to the line grid, is required to track a trapezoidal torque reference, which starts at t = 0.2s from zero initial value reaches the value of -3 Nm at t = 0.3s, and then, at t = 0.8s increases up to 3 Nm. Note that flux value, required to track torque trajectory with unity power factor at stator side is not a constant, as it is usually assumed neglecting stator resistance in field oriented solutions. From Fig. 4, it can be noted that the primary mover speed varies within 10-20rpm (no integral action is adopted in the primary mover speed controller). Transient of DFIM variables during torque tracking trajectory that is changed from generator mode to motor mode under synchronous speed is shown in Fig.5. Rotor current errors are controlled at zero level.

Stator current reactive component is stabilized at zero level during all transients indicating for high decoupling properties of the proposed controller. As result, the stator phase current has a phase angle opposite to the line voltage one and shows a low content of high order harmonics. Matrix converter input/output voltages and currents during undersynchronous speed of DFIM are shown in Fig. 6. The controller gains during all of the tests are set at $k_i = 500$; $k_{ii} = 80000$.

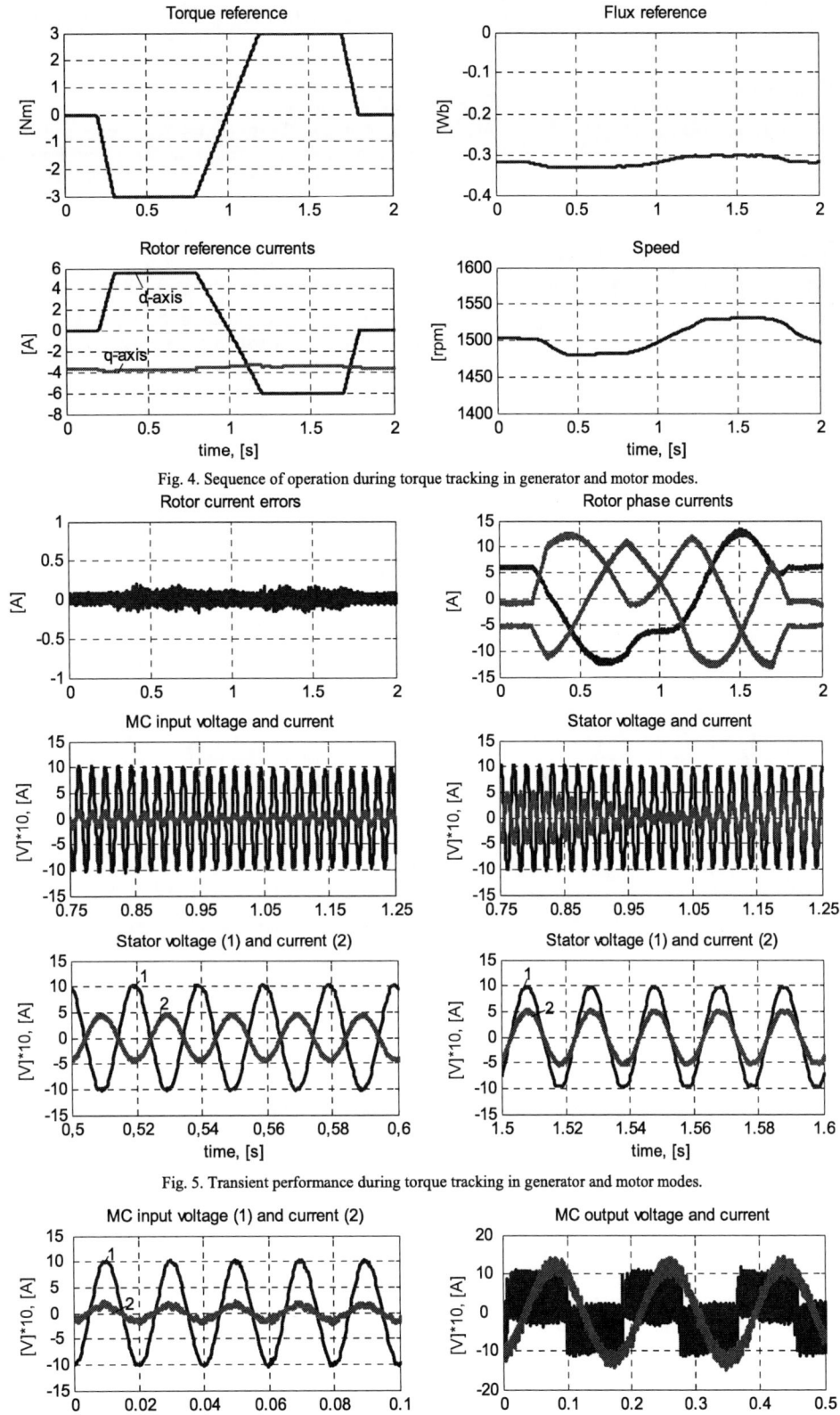

Fig. 4. Sequence of operation during torque tracking in generator and motor modes.

Fig. 5. Transient performance during torque tracking in generator and motor modes.

Fig. 6. Matrix converter input/output waveforms during undersynchronous speed of DFIM.

Satisfactory waveforms of input DFIM stator side currents and MC input currents are obtained which are considered in Figs. 7 and 8. The spectrum of the current waveforms of the previous pictures is shown in the same Figures.

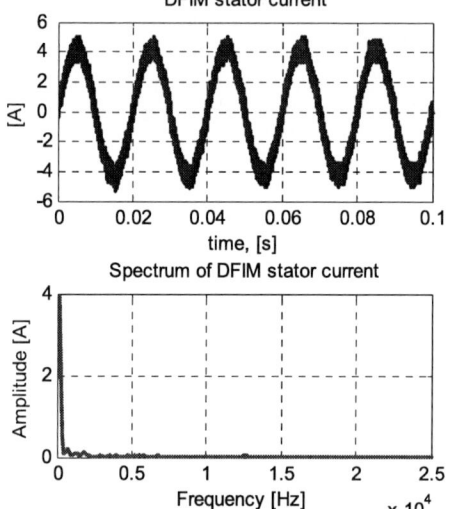

Fig. 7. DFIM stator phase current and its spectrum.

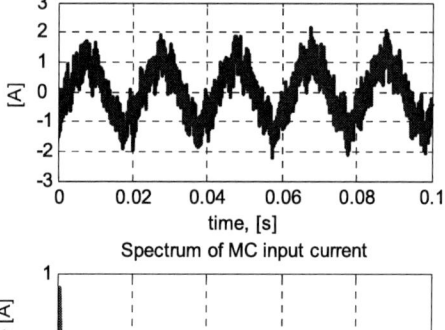

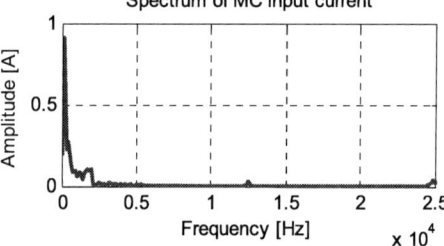

Fig. 8. MC input phase current and its spectrum.

V. CONCLUSIONS

Results of experimental testing of the MC-fed DFIM are presented. All control algorithms (MC and DFIM) have been implemented in real-time using DSP-controller. The experimental rig used to control DFIM has been presented. The rig was used to confirm the control methods proposed for the DFIM. The structure of the rig including all hardware and circuits were described.

The control of the experimental rig is implemented with the help of interaction between a digital signal processor (TMS320C611 DSL board) and field programmable gate array (FPGA board). Data acquisition and pulse generation are coordinated by an FPGA. The software controlling the processor is written in the C programming language using Code Composer Studio as it is specifi-

cally designed for programming and supporting Texas Instruments DSP platforms.

It is demonstrated by experiments that: high performance torque tracking is guaranteed under condition of unity stator side power factor and that satisfactory waveforms of input DFIM stator side currents and MC input currents are obtained. The main conclusion from the performed experimental study is that control proposed technical solutions for MC and DSP controller development are suitable for practical application in high performance DFIM based electromechanical systems.

ACKNOWLEDGMENT

The authors acknowledge gratefully the financial support of the INTAS within the framework of Fellowship Grant for Young Scientists Nr. 05-109-4411.

REFERENCES

[1] W. Leonhard, *Control of Electric Drives*, Springer-Verlag, Berlin, 1997.

[2] R. Pena, J.C. Clare, G.M. Asher, "Doubly Fed Induction Generator using Back-to-Back PWM Converters and Its Applications to Variable-Speed Wind-Energy Generation", *IEE Proceedings of Electric Power Applications*, vol.143, no.3, May 1996, pp.231-241.

[3] R. Pena, R. Cardenas, E. Escobar, J. Clare, P. Wheeler, "Control System for Unbalanced Operation of Stand-Alone Doubly Fed Induction Generators", *IEEE Trans. on Energy Conversion*, vol. 22, no.2, June 2007, pp.544-545.

[4] R. Cardenas, R. Pena, J. Clare, G. Asher, J. Proboste "MRAS Observers for Sensorless Control of Doubly-Fed Induction Generators", *IEEE Trans. on Power Electronics*, vol. 23, no. 3, May. 2008, pp.1075-1084.

[5] R. Pena, R. Cardenas, J. Proboste, J. Clare, G. Asher, "Wind–Diesel Generation Using Doubly Fed Induction Machines", *IEEE Trans. on Energy Conversion*, vol. 23, no.1, March 2008, pp.202-214.

[6] S. Peresada, A. Tilli, A. Tonielli, "Robust Active-Reactive Control of a Doubly-Fed Induction Machine", *Proc. of IEEE - IECON'98*, Aachen, Germany, Sept. 1998, pp.1621-1625.

[7] S. Peresada, A. Tonielli "High Performance Robust Speed-Flux Tracking Controller for Induction Motor", *International Journal of Adaptive Control and Signal Processing*, vol.14, 2000, pp.177-200..

[8] L. Zhang, C.Watthanasarn, "A matrix converter excited doubly-fed induction machine as a wind power generator", *Seventh International Conference on Power Electronics and Variable Speed Drives* (Conf. Publ. No. 456), 1998, pp. 532 -537.

[9] K.Ghedamsi, D.Aouzellag, E.M. Berkouk, "Application of matrix converter for variable speed wind turbine driving a doubly fed induction generator", *Proc. of International Symposium on Power Electronics, Electrical Drives, Automation and Motion*, SPEEDAM2006, May 2006, pp.1201–1205.

[10] Qi Wang, Xiaohu Chen, Yanchao Ji, "Control for Maximal Wind Energy Tracing in Matrix Converter AC Excited Brushless Doubly-Fed Wind Power Generation System", *Proc. of IEEE Industrial Electronics Conference*, IECON 2006 - 32nd Annual, Nov. 2006, pp. 718-723.

[11] P.W. Wheeler, J. Rodriguez, J.C. Clare, L. Empringham, A. Weinstein, "Matrix converters: a technology review", *IEEE Trans. on Industrial Electronics*, vol.49, no.2, April 2002, pp.276-288.

[12] E. Chekhet, V. Mikhalsky, V. Sobolev, I. Shapoval, "Control and commutation technique for matrix converters", *Technical electrodynamics. Special issue "Problems of modern electrical engineering", Ukraine*, 2006, Vol. 1, pp.56-67.

[13] M. Hornkamp, M. Loddenkoetter, M. Muenzer, O. Simon, and M. Bruckmann, "EconoMAC the first all-in-one IGBT module for matrix converters", *in Proc. PCIM*, 2001, pp. 417-422.

[14] *Code Composer Studio User's Guide*, Texas Instruments - Literature Number: SPRU328b, 2000.

Effect of Type and Interconnection of DG Units in the Fault Current Level of Distribution Networks

H.R. Baghaee [*], M. Mirsalim [*][+] (*IEEE Senior Member*), M. J. Sanjari [*], and G.B. Gharehpetian [*]

[*] Center of Excellence in Power Engineering, Amirkabir University of Technology, Tehran, Iran
[+] Also, with the Department of Engineering, St. Mary's University, San Antonio, TX, USA
, e-mails: hrbaghaee@aut.ac.ir, mmirsalim@aut.ac.ir, m_j_sanjari@aut.ac.ir, and grptian@aut.ac.ir

Abstract— Fundamental requirements for the connection of distributed generation resources to the network are not only power quality constraints, but also voltage regulation and the total fault level, which should remain below the network desired value. This constraint is often the main limiting factor for the interconnection of these resources units to the grids. In the presented paper, the impact of installation of distributed resources in the distribution systems from the perspective of increase in the fault contribution will be discussed and comparative study will be performed to analyze the effect of type and interconnection of distributed generation unit on the fault current contribution of the distribution systems. Simulation results indicate that the increase in fault currents is often greater in the synchronous machine implementation versus a comparable inverter based design.

Keywords— Distributed Generation, Distribution System, Fault Current, Power Electronic.

NUMENCLATURE

Symbol	Definition
V	Bus RMS Voltage
θ	Bus Voltage angle
f_n	Network Frequency
P_g	Generator real power
Q_g	Generator reactive power
v_g	Instantaneous generator internal Voltage
θ_g	Generator Voltage angle
i_g	Instantaneous generator current
f_g	Generator Output Frequency
R_g	Generator internal resistance
X_g	Generator internal reactance
v_i	Instantaneous converter input Voltage
v_o	Instantaneous converter output Voltage
i	Instantaneous converter output current
v_{dc}	Converter DC link Voltage
A_m	Converter magnitude modulation index
α_m	Converter angle modulation index
R_f	Filter resistance
X_f	Filter reactance

I. INTRODUCTION

The ever-increasing energy consumption has created increased interest in green power generation systems. Moreover, due to steady progress in power deregulation and utility restructuring and because tight constraints are imposed on the construction of new transmission lines for long-distance power transmission, interest in distributed generation (DG) systems installed near load centers is increasing. The benefits provided by DGs are not only improved power quality and system reliability, but also loss reduction. However, if certain minimum standards for control, installation and protection are not maintained, power system operations may be adversely impacted by the use of DGs. Thus, DGs should meet various operating requirements of the utilities or the power system operators [1-5].

A fundamental requirement for the connection of DG resources to the network, besides voltage regulation and power quality constraints [6], is the total fault level, which is mainly obtained by the short-circuit contribution of both upstream grid and the DG, should remain below the network design value. This constraint is often the main limiting factor for the interconnection of DG units to the grids [7-10]. To facilitate the interconnection of DGs to a distribution system, standards are being developed [11, 12]. An engineering analysis is usually needed to assess the impact of the DG on the operation of the system [13-16].

Distribution systems are mainly characterized by power losses and a design short circuit capacity (SCC), i.e. a maximum acceptable fault current, related to the switchgear used and to the thermal and mechanical withstands capability of the equipment and utilities. In medium voltage (MV) and low voltage (LV) radial networks, the fault current contribution of the upstream grid is practically determined by the short-circuit impedance of the HV/MV or MV/LV transformers, which is selected as low as possible to improve voltage regulation and the overall power quality performance of the network. So, the short-circuit capacity of existing distribution networks, especially at the MV level, is close to the design value, leaving little margin for the connection of even moderate amounts of DG. Short-circuit calculations for switchgear selection and protection coordination are performed according to established national and international practices, most important and widely accepted being the ANSI/IEEE and IEC Standards [20-22].

The requirement of not exceeding the design short-circuit capacity should be satisfied at every point of the distribution system under maximum fault current conditions. In typical radial networks, fed by a HV/MV (or MV/LV) substation, this condition normally needs to be checked at the MV (or LV) bus bars of the substation. The contribution of individual DG sources, on the other

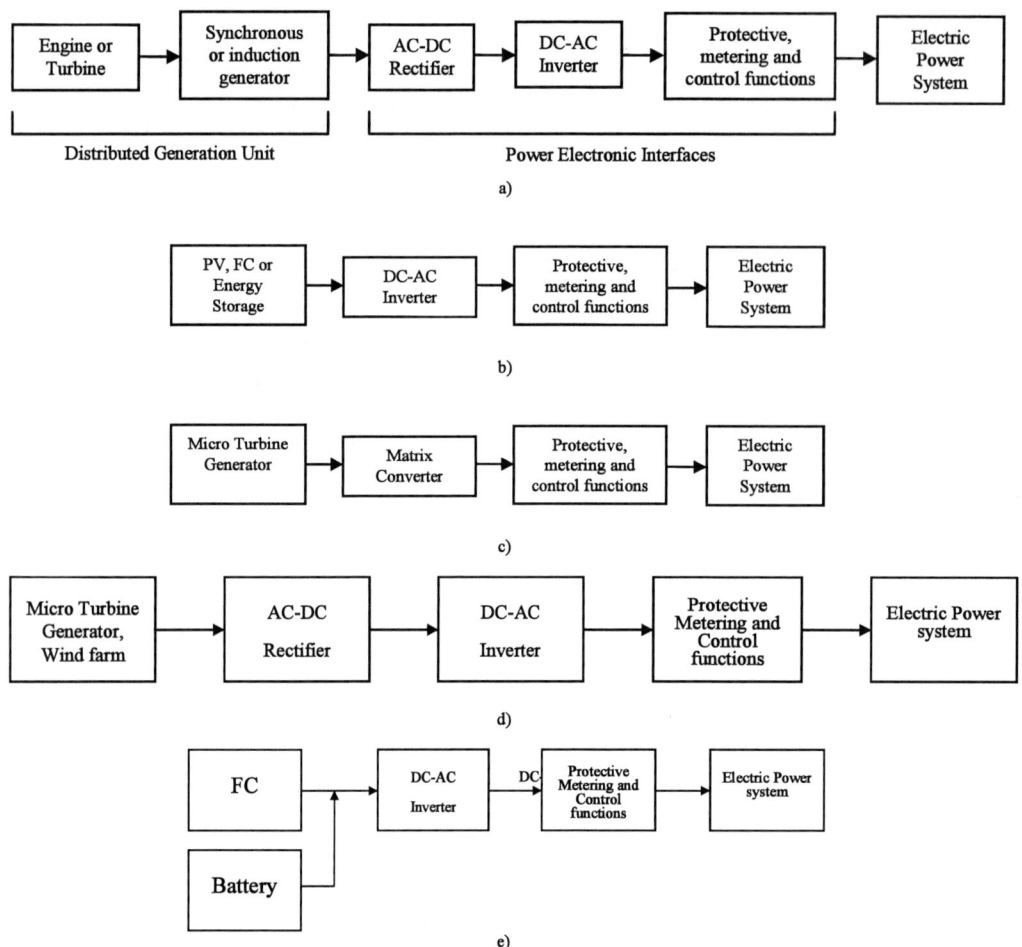

Fig. 1 Different types of DG interconnection

hand, reduces to a much smaller degree at remote network nodes, because their internal impedance is relatively high compared to the impedance of the network lines. The resulting fault level of distribution system is the pharos sum of the maximum fault currents from the upstream grid, through the step-down transformer, and the various generators (and possibly motors) connected to the network.

On account of the fact that the procedure of fault current limitation requires such a flexible and authentic device which can act rapidly. Due to their speed of response and flexibility, power-electronic (PE) converter systems of electronically-coupled DG units are the prime candidates to perform the require control and/or protection functions to meet the micro-grid objectives [6].

The goal of this paper is to attempts to compare effect of different DG sources and their interconnection to the grid in the contribution to increase the fault level of distribution systems based on a general dynamic model for DG units and their interconnections. The models have been developed in MATLAB/Simulink environment based on reference frame theory [28]. Then, effect of type and interconnection of DG units has been studied and results have been presented.

II. DG INTERCONNECTION IANTERFACES

The electric output of DG units can be connected to the electrical power system via three basic interconnection interfaces. The block diagram of different interconnections of DG units has been shown in Fig. 1.

A. Synchronous Generator

Synchronous generators are used with most reciprocating engines and most high power turbines (gas, steam, and hydro). In a synchronous machine, the electrical frequency of induced voltage depends on the speed of rotation of the generator.

B. Induction Generator

Induction generators are typically only used in wind turbines and some low-head hydro applications. There are two types of rotor designs available: cage-rotor and

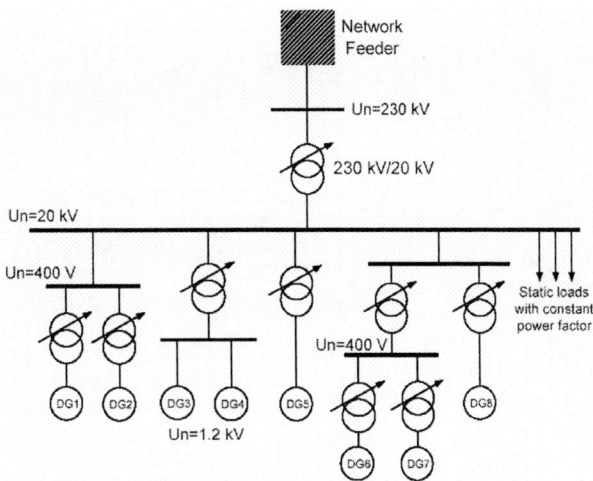

Fig. 2. Test System

wound-rotor. The advantage of the cage-rotor induction generator is the lower cost compared to a synchronous generator, but induction generators require a supply of VARs either from capacitors, from the electric power system, or from power electronic-based reactive compensator to operate [29]. The doubly fed induction generator (DFIG) has added advantages, however, is more expensive.

C. Power Electronic Interfaced DG Units

With the fast development of solid-state-based packages, power electronic (PE) devices can now convert almost any form of electrical energy to a more desirable and usable form. Another benefit of PE coverers is their extremely fast response times. PE interfaces can respond to power quality events or fault conditions within in the sub-cycle range. PE-based inverters are widely used in micro turbines generators (MTG), fuel cells (FC), photovoltaic (PV) and fuel cell combined with an energy storage system like battery-, some wind turbines, and energy storage systems. This high-speed response can enable advanced applications such as the operation of intentional islands (micro grids) for high-reliability applications and reducing fault level currents of distributed generation [8].

The PE interface can also contain protective functions for both the distributed energy system and the local electric power system that allow paralleling and disconnection from the electric power system. These functions would typically meet the IEEE Std. 1547 interconnection requirements [11], but can be set more sensitive depending on the situation and utility interconnection requirements. Fig. 1 shows block diagram of the DG system and PE.

III. MODELLING ON DG UNITS

Fig. 2 shows the system under study which involves different type of DGs connected to the grid via different methods. The type and interconnection methods of DGs are presented in Table. 1.

A. Modeling of DG1

TABLE I.
TYPE AND INTERCONNECTION METHODS OF DGS

DG No.	Type of DG	Interconnection
1	Diesel generator	Directly
2	Micro turbine	AC-DC-AC
3	Micro turbine	Matrix converter
4	PV	DC-AC inverter
5	Wind farm	AC-DC-AC
6	Fuel cell	DC-AC inverter
7	Wind farm	Directly
8	Fuel cell + battery	DC-AC inverter

DG1 which is connected to the grid shown in Fig. 2 is a diesel generator and the transient model mentioned in [28] has been used to model its transient behavior during the fault.

B. Modeling of DG2

Fig. 1 d demonstrates a block diagram of DG2. The three-phase voltage equation of the generator-side is

$$v_{pg2} = R_{G2}i_{pg2} + L_{G2}d(i_{pg2}) + v_{pi2} \qquad (1)$$

where subscript p denotes a, b and c phase components of variables, d is the d/dt operator, $R_{G2}=diag\{R_{g2},\ R_{g2},\ R_{g2}\}$ and $L_{G2}=diag\{L_{g2},\ L_{g2},\ L_{g2}\}$. The voltage vectors of the generator and the converter input side are denoted by v_{pg2} and v_{pi2}, respectively.

To transfer the generator-side instantaneous variables to a rotating reference frame (RRF), transformation matrix is chosen such that d and q the and components of the generator-side current are proportional to the converter instantaneous real and reactive power components [30,31]. Thus, the generator-side variables are transferred to a $dq0$ frame by

$$f_{tg2} = K_{g2}f_{pg2} \qquad (2)$$

where subscript t denotes q, d and o components of variables and the transformation matrix is

$$K_{g2} = \begin{bmatrix} \cos(\theta_{g2}) & \cos\left(\theta_{g2} - \dfrac{2\pi}{3}\right) & \cos\left(\theta_{g2} + \dfrac{2\pi}{3}\right) \\ \sin(\theta_{g2}) & \sin\left(\theta_{g2} - \dfrac{2\pi}{3}\right) & \sin\left(\theta_{g2} + \dfrac{2\pi}{3}\right) \\ \dfrac{1}{2} & \dfrac{1}{2} & \dfrac{1}{2} \end{bmatrix} \quad (3)$$

$$\theta_{g2}(t) = \int_0^t \omega_{g2}(\tau)d\tau + \theta_{g2} \quad (4)$$

where θ_{g2} is is phase-angle of the generator phase-a voltage and ω_{g2} is the generator angular frequency. Substituting for v_{pg2}, i_{pg2} and v_{pi2} from (2) in (1), yields

$$K_{g2}^{-1}v_{tg2} = R_{G2}K_{g2}^{-1}i_{tg2} + L_{G2}d\left(K_{g2}^{-1}i_{tg2}\right) + K_{g2}^{-1}v_{ti2} \quad (5)$$

Multiplying both sides of (5) by K_{g2} we have

$$v_{tg2} = R_{G2}i_{tg2} + L_{G2}d\left(i_{tg2}\right) + K_{g2}d\left(K_{g2}^{-1}\right)L_{g2}i_{tg2} + v_{ti2} \quad (6)$$

where

$$v_{tg2} = V_{mg2}M \quad (7)$$

$$K_{g2}d\left(K_{g2}^{-1}\right) = \omega_{g2}N \quad (8)$$

$$M = \begin{bmatrix} 0 & 1 & 0 \end{bmatrix}^T \quad (9)$$

$$N = \begin{bmatrix} 0 & 1 & 0 \\ -1 & 0 & 0 \\ 0 & 0 & 0 \end{bmatrix} \quad (10)$$

V_{mg2} in (7) is the amplitude of the internal voltage of DG2. To express the rectifier ac-side voltages v_{ti2}, (6), in terms of A_{mi2} and α_{mi2}, the switching functions of the rectifier, S_i, can be defined as:

$$S_1(t) = A_{mi2}\sin(\theta_{mi2}(t)) \quad (11)$$

$$S_2(t) = A_{mi2}\sin\left(\theta_{mi2}(t) - \frac{2\pi}{3}\right) \quad (12)$$

$$S_3(t) = A_{mi2}\sin\left(\theta_{mi2}(t) + \frac{2\pi}{3}\right) \quad (13)$$

where $\theta_{mi2}(t) = \omega_{g2}t + \alpha_{mi2}$ and

$$A_{mi2} = \frac{2}{V_{dc}}((L_{g2}\omega_{g2}I_{dg2} + R_{g2}I_{qg2})^2 + \qquad (14)$$
$$(V_{mg2} + L_{g2}\omega_{g2}I_{qg2} - R_{g2}I_{dg2})^2)^{1/2}$$

$$\alpha_{mi2} = tg^{-1}\left(-\frac{L_{g2}\omega_{g2}I_{dg2} + R_{g2}I_{qg2}}{V_{mg2} + L_{g2}\omega_{g2}I_{qg2} - R_{g2}I_{dg2}}\right) + \theta_{g2} \quad (15)$$

Based on (11) to (13), the three phase voltages at the rectifier ac-side can be written as the following.

$$v_{ai2}(t) = \frac{1}{2}A_{mi2}V_{dc}\sin(\theta_{mi2}(t)) \quad (16)$$

$$v_{bi2}(t) = \frac{1}{2}A_{mi2}V_{dc}\sin\left(\theta_{mi2}(t) - \frac{2\pi}{3}\right) \quad (17)$$

$$v_{ci2}(t) = \frac{1}{2}A_{mi2}V_{dc}\sin\left(\theta_{mi2}(t) + \frac{2\pi}{3}\right) \quad (18)$$

Substituting for v_{ai2}, v_{bi2} and v_{ci2} from (16) to (18) in (2), yields

$$v_{ti2} = \frac{1}{2}A_{mi2}V_{dc}\begin{bmatrix} \sin(\alpha_{mi2} - \theta_{g2}) & \cos(\alpha_{mi2} - \theta_{g2}) & 0 \end{bmatrix}^T \quad (19)$$

Substituting for v_{tg2}, $Kg2d(K_{g2}^{-1})$ and v_{ti2} from (7), (8) and (19) in (6), we obtain

$$V_{mg2}M = R_{G2}i_{tg2} + L_{G2}d(i_{tg2}) + L_{g2}\omega_{g2}Ni_{tg2}$$
$$+\frac{1}{2}A_{mi2}V_{dc}\begin{bmatrix} \sin(\alpha_{mi2} - \theta_{g2}) & \cos(\alpha_{mi2} - \theta_{g2}) & 0 \end{bmatrix}^T \quad (20)$$

The equation set representing the circuit between the inverter and the micro-grid system of Fig. 2, in RRF, is

$$v_{p2} = R_{F2}i_{p2} + L_{F2}d(i_{p2}) - v_{po2} \quad (21)$$

where subscript p denotes a, b and c phase components of variables, $R_{F2} = diag\{R_{f2}, R_{f2}, R_{f2}\}$ and $L_{F2} = diag\{L_{f2}, L_{f2}, L_{f2}\}$. Similar to the discussions presented for the generator-side circuit equations of Fig. 2, i.e., (1) to (19) that finally concluded (20), we transfer (21) to a frame and deduce

$$-V_{mg2}M = R_{F2}i_{t2} + L_{F2}d(i_{t2}) + L_{f2}\omega_n Ni_{t2}$$
$$-\frac{1}{2}A_{mo2}V_{dc}\begin{bmatrix} \sin(\alpha_{mo2} - \theta_2) & \cos(\alpha_{mo2} - \theta_2) & 0 \end{bmatrix}^T \quad (22)$$

where subscript t denotes q, d and 0 components of variables, ω_n is the grid angular frequency, V_{m2} and θ_2 are the amplitude and phase angle of DG2 output terminal voltage respectively and

$$A_{mo2} = \frac{2}{V_{dc}}((L_{f2}\omega_n I_{d2} + R_{f2}I_{q2})^2 + \qquad (23)$$
$$(V_{m2} - L_{f2}\omega_n I_{q2} + R_{f2}I_{d2})^2)^{1/2}$$

$$\alpha_{mo2} = tg^{-1}\left(-\frac{L_{f2}\omega_n I_{d2} + R_{f2}I_{q2}}{V_{m2} - L_{f2}\omega_n I_{q2} + R_{f2}I_{d2}}\right) + \theta_2 \quad (24)$$

The dq-based, fundamental-frequency, model of DG2 unit is provided by (20) and (22), and can be used for the steady-state and dynamic analysis of the unit.

C. Modellig of DG3

Fig. 1 c demonstrates a block diagram of DG3 which is connected to the micro grid shown in Fig. 2. The three-phase voltage equation of the grid-side of DG3 is

$$v_{p3} = R_{F3}i_{p3} + L_{F3}d(i_{p3}) + v_{po3} \quad (25)$$

where subscript p denotes a, b and c phase components of variables, d is the d/dt operator, $R_{F2} = diag\{R_{f2}, R_{f2}, R_{f2}\}$, $L_{F2} = diag\{L_{f2}, L_{f2}, L_{f2}\}$ and v_{po3} and v_{p3} are three phase voltages at the converter output and output terminals, respectively.

The abc variables in (25) are transformed to a rotating reference frame by

$$v_{t3} = K_3 f_{p3} \quad (26)$$

where subscript t denotes , and components of variables and the transformation matrix is.

$$K_3 = \begin{bmatrix} \cos(\theta_3) & \cos\left(\theta_3 - \dfrac{2\pi}{3}\right) & \cos\left(\theta_3 + \dfrac{2\pi}{3}\right) \\ \sin(\theta_3) & \sin\left(\theta_3 - \dfrac{2\pi}{3}\right) & \sin\left(\theta_3 + \dfrac{2\pi}{3}\right) \\ \dfrac{1}{2} & \dfrac{1}{2} & \dfrac{1}{2} \end{bmatrix} \quad (27)$$

$$\theta_3(t) = \int_0^t \omega_n(\tau)d\tau + \theta_3 \quad (28)$$

θ_3 is is phase-angle of the generator phase-a voltage at DG3 terminal. Substituting for abc variables based on (26) in (25), yields

$$-v_{t3} = R_{F3}i_{t3} + L_{F3}d(i_{t3}) + K_3d\left(K_3^{-1}\right)L_{f3}i_{t3} + v_{to3} \quad (29)$$

where

$$v_{t2} = V_{m2}M \quad (30)$$

$$K3d\left(K_3^{-1}\right) = \omega_n N \quad (31)$$

Matrices M and N are defined by (9) and (10), respectively and V_{m3} is the voltage amplitude at DG3 output terminal. The fundamental-frequency switching functions of the matrix converter are [32]:

$$\begin{bmatrix} S_1(t) \\ S_2(t) \\ S_3(t) \end{bmatrix} = -\frac{2}{3}A_{m3}\begin{bmatrix} \cos(\theta_{m2}(t)) \\ \cos\left(\theta_{m2}(t)-\frac{2\pi}{3}\right) \\ \cos\left(\theta_{m2}(t)+\frac{2\pi}{3}\right) \end{bmatrix} \quad (32)$$

where $\theta_{m3}(t)=\omega_{m3}t+\alpha_{m3}$. Assuming that the input terminal voltages of the converter are

$$\begin{bmatrix} v_{ai3}(t) \\ v_{bi3}(t) \\ v_{ci3}(t) \end{bmatrix} = V_{mi3}\begin{bmatrix} \sin(\theta_{i3}(t)) \\ \sin\left(\theta_{i3}(t)-\frac{2\pi}{3}\right) \\ \sin\left(\theta_{i3}(t)+\frac{2\pi}{3}\right) \end{bmatrix} \quad (33)$$

where $\theta_{i3}(t)=\omega_{g3}t+\alpha_{i3}$, then the output phase-a voltage of the matrix converter is [30]:

$$v_{ao3} = \begin{bmatrix} S_1(t) & S_2(t) & S_3(t) \end{bmatrix}\begin{bmatrix} v_{ai3}(t) \\ v_{bi3}(t) \\ v_{ci3}(t) \end{bmatrix} \quad (34)$$

Substituting for S_1, S_2 and S_3 from (32) and for v_{ai3}, v_{bi3} and v_{ci3} from (33) in (34), yields

$$v_{ao3} = V_{mo3}\sin(\omega_n t + \alpha_{o3}) \quad (35)$$

where $V_{mo3}=A_{m3}V_{mi3}$ and $\alpha_{o3}=\alpha_{m3}-\alpha_{i3}$. Substituting for v_{ao3}, v_{bo3} and v_{co3} based on (35) in (26), where and v_{bo3} and v_{co3} are 120 and 120 out of phase with respect to v_{ao3}, we have

$$v_{to2} = \frac{1}{2}V_{mo3}\begin{bmatrix} \sin(\alpha_{o3}-\theta_3) & \cos(\alpha_{o3}-\theta_3) & 0 \end{bmatrix}^T \quad (36)$$

Substituting for v_{t3}, $K_3d(K_3^{-1})$ and v_{to3} from (30), (31) and (36) in (29), yields

$$-V_{m3}M = R_{F3}i_{t3} + L_{F3}d(i_{t3}) + L_{f3}\omega_n N i_{t3}$$
$$-V_{mo3}\begin{bmatrix} \sin(\alpha_{o3}-\theta_3) & \cos(\alpha_{o3}-\theta_3) & 0 \end{bmatrix}^T \quad (37)$$

The abc variables of G3, are transferred to a frame by

$$v_{tg3} = K_{g3}f_{pg3} \quad (38)$$

where

$$K_{g3} = \begin{bmatrix} \cos(\theta_{g3}) & \cos\left(\theta_{g3}-\frac{2\pi}{3}\right) & \cos\left(\theta_{g3}+\frac{2\pi}{3}\right) \\ \sin(\theta_{g3}) & \sin\left(\theta_{g3}-\frac{2\pi}{3}\right) & \sin\left(\theta_{g3}+\frac{2\pi}{3}\right) \\ \frac{1}{2} & \frac{1}{2} & \frac{1}{2} \end{bmatrix} \quad (39)$$

$$\theta_{g3}(t) = \int_0^t \omega_{g3}(\tau)d\tau + \theta_{g3} \quad (40)$$

θ_{g3} is the phase-angle of the generator phase-a voltage. Similar to the discussion presented for (37), voltage equations of the generator-side circuit, in the $dq0$ frame, are

$$V_{mg3}M = R_{G3}i_{tg3} + L_{G3}d(i_{tg3}) + L_{g3}\omega_{g3}N i_{tg3}$$
$$+V_{mi3}\begin{bmatrix} \sin(\alpha_{mi3}-\theta_{g3}) & \cos(\alpha_{mi3}-\theta_{g3}) & 0 \end{bmatrix}^T \quad (41)$$

where V_{mg3} and ω_{g3} are the amplitude of internal voltage and the angular frequency of DG3, respectively. Usually, a capacitor is used at the input of the matrix converter to maintain the voltage. The capacitor current equation, is

$$i_{pC3} = C_{g3}d(v_{pi3}) \quad (42)$$

Substituting for abc variables from (38) in (42), yields

$$i_{tC3} = C_{g3}d(v_{ti3}) + C_{g3}\omega_{g3}N v_{ti3} \quad (43)$$

In (42) and (43), subscripts p and t denote $\{a, b, c\}$ and $\{q, d, 0\}$ variables, respectively. Equations (37), (41), and (43) represent a fundamental-frequency $dq0$-based model of the DG3 unit and can be used for the dynamic analysis of the DG unit.

D. Modellig of DG4

Fig. 1 b demonstrates a block diagram of DG4 which is connected to the micro grid shown in Fig. 2. The PV system can be represented by a series combination of a DC source and internal resistance and inductance. The DC voltage at the inverter DC-side is

$$v_{i4}(t) = V_{dc4} - (R_{g4} + dL_{g4})i_{g4} \quad (44)$$

the three phase voltages at the inverter ac-side are

$$v_{ao4}(t) = \frac{1}{2}A_{mo4}v_{i4}\sin(\theta_{mo4}(t)) \quad (45)$$

$$v_{bo4}(t) = \frac{1}{2}A_{mo4}v_{i4}\sin\left(\theta_{mo4}(t)-\frac{2\pi}{3}\right) \quad (46)$$

$$v_{co4}(t) = \frac{1}{2}A_{mo4}v_{i4}\sin\left(\theta_{mo4}(t)+\frac{2\pi}{3}\right) \quad (47)$$

where $\theta_{mo4}(t)=\omega_{g4}t+\alpha_{mo4}$. Substituting for v_{ai2}, v_{bi2} and v_{ci2} from (45) to (47) in (2), yields

$$v_{to4} = \frac{1}{2}A_{mo4}V_{dc}\begin{bmatrix} \sin(\alpha_{mo4}-\theta_{g4}) & \cos(\alpha_{mo4}-\theta_{g4}) & 0 \end{bmatrix}^T \quad (48)$$

The equations representing the circuit between the inverter and terminal of the micro-grid system of Fig. 2, in the RRF, are

$$v_{p4} = R_{F4}i_{p4} + L_{F4}d(i_{p4}) - v_{po4} \quad (49)$$

where subscript p denotes a, b and c phase components of variables, $R_{F4}=diag\{R_{f4}, R_{f4}, R_{f4}\}$ and $L_{F4}=diag\{L_{f4}, L_{f4}, L_{f4}\}$. Like the discussions presented for the generator-side circuit equations, i.e., (40) to (45), we transfer (45) to a RRF and so we have

$$-V_{mg4}M = R_{F4}i_{t4} + L_{F4}d(i_{t4}) + L_{f4}\omega_n N i_{t4}$$
$$-\frac{1}{2}A_{mo4}V_{dc}\begin{bmatrix} \sin(\alpha_{mo4}-\theta_2) & \cos(\alpha_{mo4}-\theta_4) & 0 \end{bmatrix}^T \quad (50)$$

where subscript t denotes q, d and 0 components of variables and V_{m4} and ω_n are the amplitude of terminal voltage and angular frequency, respectively. θ_4 is the voltage angle of terminal voltage of DG4.

The dq-based, fundamental-frequency, model of DG2 unit is provided by (50), and can be used for dynamic analysis of the unit.

E. Modellig of DG5

Modeling of DG5 has been performed using the induction generator's model mentioned in [28] and also the modeling of AC-DC-AC conversion system mentioned in part B.

F. Modellig of DG6

Dynamic modeling of Fuel cell and the PE interface is like the PV model mentioned in part D with different parameter values.

317

G. Modellig of DG7

Modeling of DG7 has been performed using the induction generator's model mentioned in [28].

H. Modellig of DG8

Dynamic modeling of Fuel cell and battery system and related the PE interface is like the fuel cell model mentioned in part F with different parameter values.

IV. SIMULATION RESULTS

The transient models of DG units discussed in section III have been implemented in MATLAB/Simulink environment and tested on a test system shown in Fig. 2. The goal of this simulation is to analyze the contribution of each DG unit in the fault and then compare them form view point of fault contribution. The DG units have been exposed to various fault situations. To analyse the effect of DGs in fault contribution in the grid, two different indices is introduced as:

$$Index1 = \int_{T_{start}}^{T_{end}} e^{\alpha t} \left(\frac{I_{fault}}{I_{Load}}\right) dt \qquad (51)$$

$$Index2 = \int_{T_{start}}^{T_{end}} e^{-\beta t} \left(\frac{I_{fault}}{I_{Load}}\right) dt \qquad (52)$$

Where T_{end}=1sec and the fault starts at t=0.5 sec. Greater the alpha can lead to have more important effect of steady state (alpha=3.31). Also, greater beta can result in more important effect of transient (beta=2.64). This simulation has been performed for the three phase fault at different locations of the network. The results of the simulation have been presented in Table. 2 this results show the effect of type and interconnection of the DG units in fault contribution in the grid. Results show that PE interface can be effective in fault current reduction in the grid having distributed resources.

V. CONCLUSION

In this paper, the impact of installation of distributed resources in the distribution systems from the perspective of increase in the fault contribution was discussed and comparative study was performed based on two indices to show the transient and steady state effect of fault current caused by DG units and also analyze the effect of type and interconnection of distributed generation unit on the fault current contribution of the distribution systems. Simulation results indicate that the increase in fault currents is often greater in the synchronous and induction machine implementation versus a comparable inverter based design.

REFERENCES

[1] S. S. Venkata, A. Pahwa, R. E. Brown, and R. D. Christie, "What future distribution engineers need to learn," *IEEE Trans. Power Syst.*, vol. 19, no. 1, pp. 17–23, Feb. 2004.
[2] T. Ackermann, G. Andersson and L. Soder, "Distributed generation: a definition", Elsevier Electric Power System Research, 2001, Vol. 57, pp. 894-895.
[3] C. Wang and M. H. Nehrir, "Analytical approaches for optimal placement of distributed generation sources in power systems," *IEEE Trans.Power Syst.*, vol. 19, no. 4, pp. 2068–2076, Nov. 2004.
[4] F. V. Edwards, G. J. W. Dudgeon, J. R. McDonald and W. E. Leithead, "Dynamics of distribution network with Distributed

TABLE II.
DG UNITS CONTRIBUTIONS IN FAULT CURRENT OF THE SYSTEM

No. of DG	Index 1	Index 2
1	20.02	28.21
2	19.07	26.54
3	18.02	24.32
4	15.2	22.35
5	13.2	20.18
6	13.02	18.56
7	12.58	17.32
8	11.52	16.56

Generation", IEEE Power Engineering Society Summer Meeting, 2000, Vol. 2, pp. 1032-1037
[5] R. C. Dugan and T. E. McDermott, "Operating Conflicts for Distributed Generation Interconnected with Utility Distribution Systems", IEEE Industry Applications Magazine, 2002, Vol. 8, No. 2, pp. 19–25.
[6] S.A. Papathanassiou, "A Technical Evaluation Framework for the Connection of DG to the Distribution Network", Elsevier Electric Power System Research, Vol 77, January 1 2007, pp. 24–34.
[7] R. A. Walling, R. Saint, R. C. Dugan, J. Burke, L. A. Kojovic, "Summary of Distributed Resources Impact on Power Delivery Systems", , IEEE Transactions on Power Delivery, Accepted for future publication Volume PP, Issue 99, 2007 Page(s):1 – 10
[8] B. Kroposki, C. Pink, R. DeBlasio, H. Thomas, M. Simoes and P.K. Sen, "Benefits of power electronic interfaces for distributed energy systems", IEEE Power Engineering Society General Meeting, June 2006, pp. 18-22.
[9] R. C. Dugan and T. E. McDermott, "Operating conflicts for distributed generation on distribution systems," in Proc. Rural Electric Power Conf., 2001, pp. A3/1–A3/6.
[10] N. Nimpitiwan and G. T. Heydt, "Fault current issues for market driven power systems with distributed generation," in Proc. North Amer. Power Symp., Moscow, Idaho, Aug. 2004, pp. 400–406.
[11] IEEE Standard for Interconnecting Distributed Resources with Electric Power Systems, IEEE Std 1547-2003.
[12] IEEE Recommended Practice for Utility Interface of Photovoltaic (PV) Systems, IEEE std. 929-2000, 2000.
[13] J. C. Gomez and M. M. Morcos, "Coordinating overcurrent protection and voltage sags in distributed generation systems," IEEE Power Eng Rev., vol. 22, no. 2, pp. 16–19, Feb. 2002.
[14] R. C. Dugan and T. E. McDermott, "Distributed generation," IEEE Ind. Appl. Mag., vol. 18, no. 2, pp. 19–25, Apr./May 2002.
[15] T. Ackermann and V. Knyazkin, "Interaction between distributed generation and the distribution network: Operation aspects," in Proc. IEEE T&D Conf., 2002, pp. 1357–1362.
[16] P. Barker and R.W. DeMello, "Determining the impact of DG on power systems, radial distribution," in Proc. IEEE Power Eng. Soc. Summer Meeting, 2000, pp. 1645–1656.
[17] M. T. Doyle, "Reviewing the impact of distributed generation on distribution system protection," in Proc. IEEE Power Eng. Soc. Summer Meeting, 2002, pp. 103–105.
[18] A. Girgis and S. Brahama, "Effect of distributed generation on protective device coordination in distribution system," in Proc. Large Engineering Systems Conf. Power Engineering, Jul. 2001, pp. 115–119.
[19] S. K. Salman and I. M. Rida, "Investigating the impact of embedded generation on relay setting of utilities' electrical feeders," IEEE Trans. Power Del., vol. 16, no. 2, Apr. 2001.
[20] A.J. Rodolakis, "A comparison of North American (ANSI) and European (IEC) fault calculation guidelines", IEEE Transactions on Industry Applications, Vol 29, May/June 1993, pp. 515–521.
[21] G. Knight and H. Sieling, "Comparison of ANSI and IEC 909 short-circuit current calculation procedures", IEEE Transactions on Industry Applications, Vol 29, May/June 1993, pp. 625–630.
[22] A. Berizzi, S. Massucco, A. Silvestri and D. Zaninelli, "Short-circuit current calculation: a comparison between methods of IEC and ANSI standards using dynamic simulation as reference", IEEE Transactions on Industry Applications, Vol 30, July/August 1994, pp. 1099–1106.

[23] IEC 60909-0, "Short-circuit currents in three-phase a.c. systems—Part 0: calculation of short-circuit currents", 2001.

[24] IEC 60909-1, "Short-circuit currents in three-phase a.c. systems—Part 1: factors for the calculation of short-circuit currents according to IEC 60909-0", 2002.

[25] IEC 60909-2, "Electrical equipment—Data for short-circuit current calculations in accordance with IEC 909 (1988)", 1992.

[26] IEC 60909-3, "Short-circuit currents in three-phase a.c. systems—Part 3: currents during two separate simultaneous line-to-earth short circuits and partial short-circuit currents flowing through earth", 2003.

[27] IEC 60909-4, "Short-circuit currents in three-phase a.c. systems—Part 4: examples for the calculation of short-circuit currents", 2000.

[28] P. C. Krause, "Analysis of electric machinery", McGraw-Hill Publications, 1986.

[29] M. Mirzaei, M. Mirsalim, and E. Abdollahi, "Analytical Modeling of Axial Air-Gap Solid Rotor Induction Machines Using a Quasi-Three-Dimensional Method", IEEE Transactions on Magnetics, vol.43, no.7, July 2007.

[30] H. Nikkhajoei, "Matrix converter and its application in a micro-turbine based generation system," Ph.D. dissertation, Univ. Toronto, Toronto, ON, Canada, 2004.

[31] H. Nikkhajoei and R. Iravani, "Steady-State Model and Power Flow Analysis of Electronically-Coupled Distributed Resource Units", IEEE Transactions on Power Delivery, Jan. 2007, Vol. 22, pp. 721-728.

[32] H. Nikkhajoei and R. Iravani, "A matrix converter based micro-turbine distributed generation system," IEEE Trans. Power Del., vol. 20, no. 3, pp. 2182–2192, Jul. 2005.

G.B. Gharehpetian *(IEEE Member)* was born in Tehran, in 1962. He received his BS and MS degrees in electrical engineering in 1987 and 1989 from Tabriz University, Tabriz, Iran and Amirkabir University of Technology (AUT), Tehran, Iran, respectively, graduating with First Class Honors. In 1989 he joined the Electrical Engineering Department of AUT as a lecturer. He received the Ph.D. degree in electrical engineering from Tehran University, Tehran, Iran, in 1996. As a Ph.D. student he has received scholarship from DAAD (German Academic Exchange Service) from 1993 to 1996 and he was with High Voltage Institute of RWTH Aachen, Aachen, Germany. He held the position of Assistant Professor in AUT from 1997 to 2003, and has been Associate Professor since 2004. Dr. Gharehpetian is a Senior Member of Iranian Association of Electrical and Electronics Engineers (IAEEE), member of IEEE and member of central board of IAEEE. Since 2004 he is the Editor-in-Chief of the Journal of IAEEE. The power engineering group of AUT has been selected as a Center of Excellence on Power Systems in Iran since 2001. He is a member of this center and since 2004 the Research Deputy of this center. Since November 2005 he is the director of the industrial relation office of AUT. He is the author of more than 200 journal and conference papers. His teaching and research interest include power system and transformers transients, FACTS devices and HVDC transmission.

BIOGRAPHIES

Hamid Reza Baghaee *(IEEE Student Member' 2008)* received the BSc degree in Electrical Engineering from Kashan University in 2006. Currently he is graduate student of Power Engineering in Amirkabir University of Technology. His research interests are power system dynamic and control, HVDC & FACTS devices, Distributed Generation (DG) and application of Artificial Intelligence in power systems.

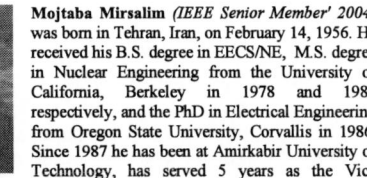

Mojtaba Mirsalim *(IEEE Senior Member' 2004)* was born in Tehran, Iran, on February 14, 1956. He received his B.S. degree in EECS/NE, M.S. degree in Nuclear Engineering from the University of California, Berkeley in 1978 and 1980 respectively, and the PhD in Electrical Engineering from Oregon State University, Corvallis in 1986. Since 1987 he has been at Amirkabir University of Technology, has served 5 years as the Vice Chairman and more than 7 years as the General Director in Charge of Academic Assessments, and currently is a Full Professor in the department of Electrical Engineering where he teaches courses and conducts research in energy conversion and CAD, among others.

His special fields of interest include the design, analysis, prototyping, and optimization of electric machines, renewable energy, FEM, and hybrid vehicles. Mirsalim is the author of more than 100 international journal and conference papers and three books on electric machinery and FEM. He is the founder and at present, the director of the Electrical Machines & Transformers Research Laboratory at http://ele.aut.ac.ir/EMTRL/Homepage.htm

Mohammad Javad Sanjari received the BSc degree in Electrical Engineering from Amirkabir University of Technology in 2006. Currently he is graduate student of Power Engineering in Amirkabir University of Technology. His research interests are power system dynamic and control, power system security assessment, HVDC & FACTS devices, Distributed Generation (DG) and application of Artificial Intelligence in power systems.

An Isolated Full-Bridge DC/DC Converter with Bidirectional Communication Capability

Lon-Kou Chang[*] and Ru-Shiuan Yang[†]

National Chiao Tung University/Electrical and Control Engineering, Hsinchu, Taiwan
[*] e-mail: *lkchang@cc.nctu.edu.tw*
[†] e-mail: *rushiuan.ece92g@nctu.edu.tw*

Abstract—**This work presents a novel isolated full-bridge DC/DC converter with bidirectional communication capability. The transformer in the proposed converter is utilized as an isolation interface for transferring energy and data. Power delivery and forward data transfer are conducted simultaneously by modifying the full-bridge switching phase. Backward data transfer is realized by manipulating the amplitude of the resonant signal through modulating the impedance of the resonant tank at the secondary side of the transformer. Finally, the operation principle of the proposed converter was verified on a 280mW prototype operating at 400kHz from a 12V DC input.**

Keywords—**converter circuit, data transmission, industrial communications, resonant converter, switched-mode power supply, transformer.**

I. INTRODUCTION

In numerous applications, such as in medical instruments and telecommunications, isolated interfaces are required for safety. In conventional approaches, power conversion and communication function are realized with independent interface circuits. For isolated DC/DC power conversion, a transformer is applied as power transfer interface. For data communication, extra pulse transformers, optical-couplers, or capacitors are applied as data transfer interfaces [1]–[4]. This work presents a novel isolated full-bridge DC/DC converter with bidirectional communication capability. A safe design with reduced cost and device counts is achieved by using a common isolated transformer that facilitates power and data transfer.

Operations of the full-bridge converter in providing bidirectional communication are as follows. (a) Power delivery and forward data transfer can be made simultaneous by altering the full-bridge switching phase. A positive or negative voltage phase across the primary winding of the transformer can deliver a 0 or 1 signal of forward datum, respectively. (b) When in discontinuous conduction mode (DCM) operation, backward data transfer can be achieved through the resonant operation provided by transformer inductance and parasitic capacitors in the bridge stage when the transformer is demagnetized completely. Via the proposed design, no additional supplied power is needed to transfer backward data.

Generally, an L-C resonant tank is employed in the technologies, such as zero-voltage switching converters [5]–[7], series L-L-C resonant converters [8], [9], and

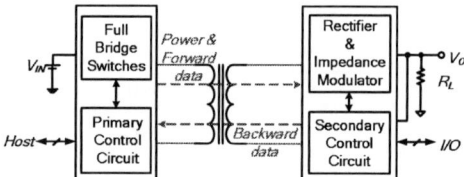

Fig. 1. Block diagram of the proposed power converter

parallel resonant converters [10], [11], to achieve zero-voltage soft switching to increase power conversion efficiency. In the proposed design, the resonant technology is utilized for different purpose, data communication. In the proposed design, backward data transfer is attained by manipulating the amplitude of a resonant signal through modulating the impedance of the resonant tank on transformer secondary side. Therefore, backward datum can be retrieved by detecting the amplitude of the resonant signal across the primary winding. Via these two data transfer and retrieve technologies, isolated bidirectional data communication is accomplished.

II. CIRCUIT AND OPERATION PRINCIPLE

Fig. 1 presents the block diagram of the proposed isolated DC/DC converter including the converter and data communication stages. The data communication stage includes a primary control circuit and secondary control circuit to achieve bidirectional communication. The primary control circuit is employed to transfer forward data from the *Host* side to the *I/O* side through the transformer and simultaneously transfer power to the load. The secondary control circuit is utilized to transfer backward data from the *I/O* side to the *Host* side through the transformer. Additionally, the primary and secondary control circuits are also in charge of receiving data transmitted from opposite sides.

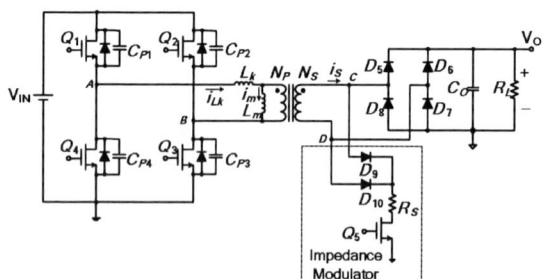

Fig. 2. Proposed DC/DC converter stage including the impedance modulator

978-1-4244-1741-4/08/$25.00 ©2008 IEEE

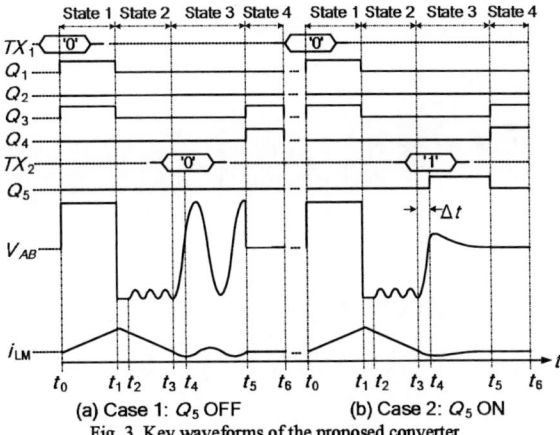

(a) Case 1: Q_5 OFF   (b) Case 2: Q_5 ON

Fig. 3. Key waveforms of the proposed converter

TABLE I.
CONTROL TABLE OF THE ON-TRANSISTORS RELATED TO THE
TRANSMITTED DATUM

State Datum	State 1	State 2	State 3	State 4
$TX_1 = 0$	Q_1, Q_3			Q_3, Q_4
$TX_1 = 1$	Q_2, Q_4			Q_3, Q_4
$TX_2 = 0$				
$TX_2 = 1$			Q_5	

Fig. 2 presents the converter stage of the proposed power converter with the following 4 primary circuit blocks: (1) an isolated transformer that transfers power and data; (2) a full-bridge switching stage, including transistors Q_1, Q_2, Q_3, and Q_4, with body diodes and parasitic capacitors that generates a switching signal according to the forward datum TX_1 sent from the *Host* side; (3) a voltage rectifier, including diodes D_5, D_6, D_7, and D_8, and a output capacitor C_O that provides rectified supply voltage to the load and associated secondary control circuit; and, (4) an impedance modulator, including diodes D_9 and D_{10}, and transistor Q_5, that manipulates the impedance of the transformer according to the backward datum TX_2 given by the *I/O* side.

There are four operational states in a switching cycle for the proposed converter during bidirectional communication (Fig. 3). We assume that all transistors have zero on-resistances, and load current is constant. Table 1 shows the control table of the on-transistors related to the transmitted data, TX_1 and TX_2, in each operation state. The duty ratios of the four operational state are denoted by $D_{1, 2, 3, \text{or } 4}$.

1) State 1 [$t_0 - t_1$]: *Duration of power and forward data transfer*

In this state, power is transferred from the input DC bus (V_{IN}) to the load through the full-bridge converter. According to the transistor switching control rules as shown in Table 1, voltage polarity of the full-bridge converter output V_{AB}, is manipulated by the logic state of forward datum TX_1. Under the voltage polarity arrangement, power and forward datum TX_1 can be transferred simultaneously to the secondary side through the isolated transformer. After rectifying the

coupled voltage V_{CD} across the transformer secondary winding, a DC output voltage (V_O) is acquired that provides power for the circuits on the transformer secondary side. Meanwhile, the transferred forward datum is retrieved simply by a level-detect circuit on transformer secondary side.

When transmitted forward datum TX_1 is 0, transistors Q_1 and Q_3 are turned on, resulting in $V_{AB} = V_{IN}$; otherwise, $V_{AB} = -V_{IN}$ for the case of $TX_1 = 1$. When $V_{AB} = V_{IN}$, diodes D_5 and D_7 are turned on, and the increasing rate of the magnetizing inductance current of the transformer is given by

$$\frac{i_m(t)}{dt} = \frac{n(V_O + 2V_D)}{L_m} \tag{1}$$

where V_O is average output voltage, and V_D is the forward conduction voltage of each rectifier diode.

Since voltage across leakage inductance of the transformer is very small, such that $V_{Lk} \ll V_{Lm}$, the voltage on the magnetizing inductance is approximate to input voltage; i.e., $V_{IN} \approx n(V_O + 2V_D)$, where n is the turns ratio of the primary winding N_P to the secondary winding N_S. Moreover, leakage inductance can be utilized to limit the maximum current increasing rate of switching components. Once the current slop limit, $(di_{Lk}/dt)_{max}$, is given, minimum leakage inductance can be determined from the following equation:

$$\left. \frac{di_{Lk}}{dt} \right|_{max} = \frac{V_{in} - n[(V_O - \Delta V_O / 2) + 2V_D]}{L_k} \approx \frac{n\Delta V_O}{2L_k} \tag{2}$$

where ΔV_O is output voltage ripple.

2) State 2 [$t_1 - t_3$]: *Duration of transformer demagnetization*

The case of $TX_1 = 0$ is assumed for illustrating the circuit operation. When $t = t_1$, all of the full-bridge transistors are turned off, the leakage inductance of the transformer is demagnetized through the body diodes of transistor Q_2 and Q_4. Thus, the voltage V_{AB} is equal to $-V_{IN}$ and results in $V_{Lk} \approx -2V_{IN}$. This high V_{Lk} causes i_{Lk} reducing fast. After i_{Lk} is less than i_{Lm}, L_m starts to demagnetize through secondary winding of the transformer and shortly i_{Lk} reduces to zero. Thus the rectifier diodes at secondary side, D_6 and D_8 are turned on and result in the voltage V_{Lm} being $-n(V_O + 2V_D)$. In contrast, V_{Lm} will be $n(V_O + 2V_D)$ if $TX_1 = 1$.

We assume all parasitic capacitors of the bridge stage have the same value $C_P (C_P = C_{P1} = C_{P2} = C_{P3} = C_{P4})$. At $t = t_2$, leakage inductance L_k has demagnetized completely and turns to resonate with the equivalent parasitic capacitor C_P across terminals A and B. Since voltage $V_{AB} = -V_{IN}$ at $t = t_2$, and $V_{Lm} = -n(V_O + 2V_D) \approx -V_{IN}$, oscillation amplitude of V_{AB} is extremely small.

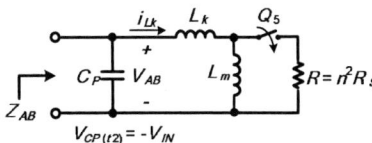

Fig. 4. Equivalent circuit of the proposed converter in State 3

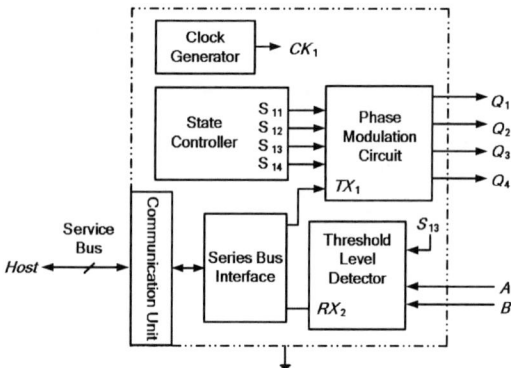

Fig. 5. Primary control circuit

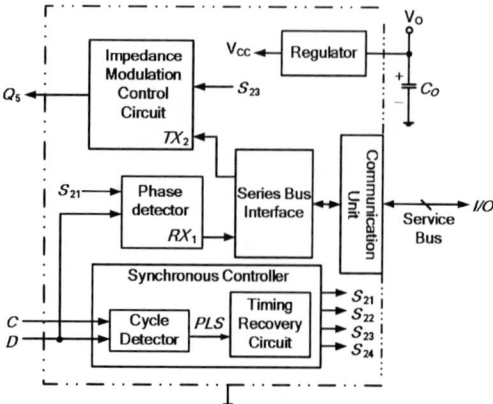

Fig. 6. Secondary control circuit

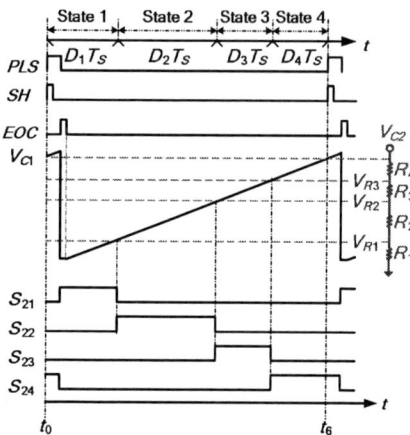

Fig. 7. Associated waveforms of the synchronous controller

3) State 3 [$t_3 - t_5$]: *Duration of L-L-C resonance and backward data transfer*

Fig. 4 shows the equivalent circuit of the proposed converter in State 3. In this state, the resonant quality factor (Q factor) is determined by the conduction state of transistor Q_5.

At $t = t_3$, L_m demagnetizes completely and starts resonating with L_k and C_P. Thus L_k, L_m and C_P form an L-L-C resonant tank with resonant frequency of

$$\omega_{r1} = \frac{1}{\sqrt{C_p(L_m + L_k)}}. \tag{3}$$

Since i_{Lk} and i_{Lm} are 0 at t_3 and transistor Q_5 is off, the amplitude of the resonant signal V_{AB} will be equal to $V_{AB}(t_2)$.

At $t = t_4 = t_3 + \Delta t$, transistor Q_5 turns on according to the logic state of transmitted backward datum TX_2, where Δt is a short delay caused by the synchronization between primary and secondary control circuits. When $TX_2 = 1$, transistor Q_5 is turned on and resistor R_S is connected to terminals C and D. This connection changes the equivalent impedance across terminals A and B, and results in a low quality factor and high oscillation damping (Fig. 3). Conversely, when $TX_2 = 0$, transistor Q_5 is turned off at t_4; thus, infinite Q factor will be obtained and V_{AB} will have a large oscillation amplitude that is approximate to V_{IN}. By detecting the oscillation amplitude of V_{AB}, backward datum can be retrieved. Moreover, no supplied power is required to transfer backward data.

Based on the operation theories described in States 1 and 2, it can be concluded that the times required for magnetizing and demagnetizing L_m are approximately equal. In order to reset the transformer completely, duty ratios must satisfy criterion $D_2 \geq D_1$.

4) State 4 [$t_5 - t_6$]: *Duration of control circuit synchronization*

In this state, Q_3 and Q_4 are both on, and voltages across the primary and secondary windings of the transformer will remain 0 until time t_6. Thus, V_{AB} will have an abrupt edge at the end of this state and the operation will return to State 1. This abrupt edge can be used as a synchronization signal on the transformer secondary side.

Fig. 5 presents a feasible primary control circuit of the proposed power converter. The primary control circuit comprises the following 5 primary circuit blocks: (1) a clock generator that provides a constant clock signal CK_1 for the primary control circuit; (2) a communication unit and a series bus interface for receiving/transferring data from/to the *Host* side; (3) a state controller that generates signals $S_{11,12,13, \text{ and } 14}$ to control operation timing of the proposed converter; (4) a phase-modulation circuit that controls the full-bridge switching phase and transmits the forward datum TX_1; and, (5) a threshold-level detector that retrieves the backward datum into RX_2 by detecting the voltage level of the primary winding voltage $|V_{AB}|$ measured in State 3.

Fig. 6 presents a feasible secondary control circuit for the proposed power converter. The secondary control circuit comprises the following 5 primary circuit blocks: (1) a communication unit and series bus interface that receives/transfers data from/to the *I/O* side; (2) a regulator that supplies power to the circuit on the transformer secondary side; (3) a synchronous controller that generates synchronous state signals $S_{21,22,23, \text{ and } 24}$ to synchronize the operation timing between the primary and secondary control circuits; (4) an impedance modulator that transfers the backward datum TX_2; and, (5) a phase detector that retrieves the forward datum into RX_1 by detecting the voltage of terminal D on the transformer secondary winding in State 1. When V_D exceeds the reference voltage V_{REF2}, then RX_1 is set to 1; otherwise, RX_1 is set to 0.

Synchronized timing control is important to successfully retrieving forward data sent from the primary control circuit. Fig. 7 shows waveforms produced by the synchronous controller of the secondary control circuit. When a new switching cycle starts and causes an edge-change to V_{CD}, a synchronization pulse signal (PLS) is generated. By sensing the positive and negative edges of the PLS, a sample-and-hold (SH) signal and an end-of-cycle (EOC) signal are generated. The SH signal is used for storing the sampled voltage of V_{C1} at time t_0. The EOC signal discharges capacitance C_1 and initiates the charging cycle of C_1. By using a constant charging current, V_{C1} will be linearly charged up. The duty ratios of the four operational states $D_{1,2,3,\ and\ 4}$ are designed as constants. Duty ratios regeneration is based on the relationship of $R_1 : R_2 : R_3 : R_4 = D_1 : D_2 : D_3 : D_4$. Therefore, synchronous operational timing control signals $S_{21,\ 22,23,\ and\ 24}$ of the secondary control circuit can be generated easily.

III. DESIGN CONSIDERATIONS

Component parameters L_k, L_m, C_P and R must be selected carefully to maximize the performance and reliability of backward data transfer. Once the equivalent parasitic capacitor C_P and leakage inductance of transformer L_k have been measured, magnetizing inductance L_m and resistance of the impedance modulator R_S can be derived. All components are assumed to have zero parasitic resistances.

The two cases of the proposed converter operated in State 3 (Fig. 4), according to the logic state of transmitted backward datum TX_2, are discussed as follows.

1) Case 1 (Q_5 OFF)

The Q factor of a resonant circuit is defined as the ratio of maximum stored energy to energy loss per cycle, and is further equivalent to

$$Q = \omega_r \left(\frac{\text{maximum energy stored}}{\text{average power dissipated}} \right). \quad (4)$$

In Case 1, C_P, L_k, and L_m form the resonant tank with resonant frequency ω_{r1}. As no resistive element exists in the resonant circuit, the Q factor in Case 1 will be approximately infinite. Restated, amplitude of resonant signal V_{AB} will not decay over time. To ensure that backward data can be correctly retrieved by the threshold-level detector of the primary control circuit, at least 1/2 resonant cycle should exist in the duration of State 3. Accordingly, minimum magnetizing inductance L_m can be obtained from

$$\sqrt{C_p(L_m + L_k)} < \frac{D_3 T_S}{\pi}. \quad (5)$$

2) Case 2 (Q_5 ON)

When transistor Q_5 turns on, a resistance R is parallel to magnetizing inductance L_m, where R is equivalent to $n^2 R_S$. Thus, the resonant frequency and the Q factor of the resonant circuit changes. Since the imaginary part of resonant circuit impedance, $\text{Im}(Z_{AB})$, is zero at resonance,

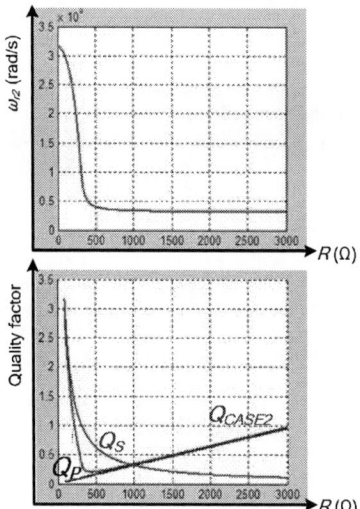

Fig. 8 (a) ω_{r2} vs. R. (b) Q factor vs. R.

resonant frequency in Case 2 can be obtained by

$$\omega_{r2} = \sqrt{\frac{-m + \sqrt{m^2 - 4n}}{2}} \quad (6)$$

where

$$m = \frac{C_p R^2 (L_m + L_k)^2 - L_m^2 L_k}{C_p L_m^2 L_k^2}, \quad n = -\frac{R^2 (L_m + L_k)}{C_p L_m^2 L_k^2}.$$

Thus, the quality factor of the resonant circuit of Case 2, Q_{CASE2}, can be obtained from the Q factor definition and (6), given as

$$Q_{CASE2} = \omega_{r2} C_p \, \text{Re}\left(Z_{AB}\right)$$
$$= \frac{\omega_{r2}^3 L_m^2 R C_p}{\left[\omega_{r2}^2 C_p R(L_m + L_k) - R\right]^2 + \omega_{r2}^2 L_m^2 \left(\omega_{r2}^2 C_p L_k - 1\right)^2}. \quad (7)$$

As seen in (7), Q_{CASE2} will be infinite when resistance R is infinite. This mathematical result is the same as that in Case 1 since the resonant circuit is the same. Moreover, when setting R as 0, the resonant tank contains only C_P and L_k, and Q_{CASE2} will be infinite too. Thus, Q_{CASE2} has a valley point (Fig. 8). Therefore, choosing an appropriate resistance R is extremely important to achieving a reliable backward data transfer process.

To determine the appropriate resistance R, the following two approximation methods can be employed: (1) when R is a small value satisfying $X_{Lm} \gg R$, then L_m can be neglected and the resonant circuit in Case 2 can be approximated to a second-order series R-L_k-C_P resonant circuit with quality factor $Q_S = (\sqrt{(L_k/C_P)})/R$; and, (2) when R is a large value satisfying $X_{Lk} \ll |R//j\,X_{Lm}|$, then L_k can be neglected and the resonant circuit in Case 2 can be approximated to a second-order parallel R-L_m-C_P resonant circuit with quality factor $Q_P = (\sqrt{(C_P/L_m)})*R$.

The resonant signal of a second-order resonant circuit has a decay function $e^{-\alpha t}$, where $\alpha = \omega_r/(2Q)$. When the second-order resonant circuit is operated at a critical damped condition such that $Q = 0.5$, then $e^{-\alpha t}$ will be as small as 0.04 at time $t = \pi/\omega_r$. Therefore, the design satisfying $Q < 0.5$, the overdamped condition, ensures that backward data can be retrieved by the threshold level detector of the primary control circuit correctly. Similarly, one can also prove that $Q < 0.5$ generates a great data retrieval capability in Case 2.

An example is given below to illustrate the steps to obtain an optimum R value. We assume $C_P = 10$pF, $L_k = 1$uH, and $L_m = 100$uH. The relation curves of $\omega_{r2}(R)$ and $Q(R)$ with respect to resistance R can then be obtained from (6) and (7) (Fig. 8). For design optimization, a minimal Q value is the best. By finding the minimum value of Q_{CASE2}, $R = R_{min} = 440\Omega$ can be obtained. Furthermore, the Q_{CASE2} curve closely matches the Q_P curve of the parallel R-L_m-C_P resonant tank when R is grater than 500Ω (Fig. 8). Therefore, finding the maximum resistance R_{max} that satisfies $Q_{CASE2} < 0.5$ can be replaced by finding the maximum resistance that satisfies $Q_P < 0.5$. This fact provides a convenient design consideration for finding an appropriate R value through neglecting the leakage inductance L_k. A computation that yields the maximum resistance such that $R_{max} = 1582\Omega$ satisfies the condition of $Q_P = 0.5$. Finally, an appropriate range of R is acquired that provides a reliable backward data retrieval process.

IV. EXPERIMANETAL RESULATS

An experimental prototype has been built to verify the operation principles of proposed design. The specifications of the prototype are as follows:

V_{in} : 12V

I_O : 0 ~ 70mA

V_O : 4 ~ 12V

f_S : 400kHz

$D_1 : D_2 : D_3 : D_4 = 4 : 5 : 4 : 3$

The converter stage shown in Fig. 2 consists of the following components:

Q_1, Q_2, Q_3, Q_4, Q_5 : FQU2N60C N-channel MOSFETs

$D_5, D_6, D_7, D_8, D_9, D_{10}$: SB160 diodes

$C_O = 10\mu$F

Transformer : ferrite ring core TN9.6/6.3-3F3,

$$N_P : N_S = 2:1,\ L_m = 98.7\mu H,$$
$$\text{and } L_k = 0.97\mu H$$

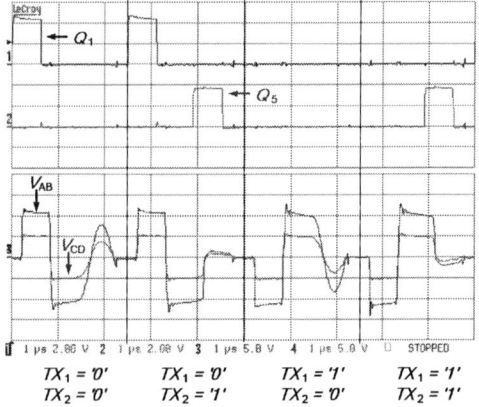

| $TX_1 = '0'$ | $TX_1 = '0'$ | $TX_1 = '1'$ | $TX_1 = '1'$ |
| $TX_2 = '0'$ | $TX_2 = '1'$ | $TX_2 = '0'$ | $TX_2 = '1'$ |

Fig. 9. Experimental waveforms for the proposed design

At first, $L_m = 98.7$uH was chosen. The measurement result shows that the resonant period of V_{AB} is 1us in State 3 at the condition $f_S = 100$kHz. Therefore, the equivalent parasitic capacitor of the switches across the terminals A and B of the full-bridge stage should be $C_P = 254$pF, and the resonant frequency satisfies the requirement given by (5). Finally, The resistance of impedance modulator $Rs = R/n^2 = 22\Omega$ was chosen according to (6) and (7) to achieve a reliable backward data transfer.

Since the transformer operates at DCM, magnetic saturation problem can be solved. Average output voltage V_O is 4.3V when load-current is 70mA, and maximum output voltage is 12V at zero-load condition.

Fig. 9 presents the experimental results of the proposed converter in four possible data-transmission cases. In Fig. 9, probe 1 shows the turned-on signal of Q$_1$; probe 2 shows the turned-on signal of Q$_5$; probe 3 shows the voltage across the transformer primary winding (V_{AB}); and, probe 4 shows the voltage across the transformer secondary winding (V_{CD}). For forward data transmission, the voltage phase of V_{AB} has already been successfully presented according to logic states of the forward datum TX_1 transmitted in State 1. For backward data transmission, the amplitudes of oscillation signals as mentioned in State 3 have also been successfully modulated by the transmitted backward datum TX_2 and can be clearly identified by a threshold level as expected. Finally, all the waveforms have well matched the theoretical ones.

V. CONCLUSION

The isolated full-bridge DC/DC converter with bi-directional communication capability has been presented with illustrations of its operations and analyses. A novel backward data transfer circuit which is achieved by manipulating the amplitude of the resonant signal is presented. An efficient approach for finding an appropriate range of modulation resistance to achieve reliable backward data communication is also presented. Experimental results have demonstrated that the proposed converter provide isolated power conversion and bi-directional communication via the same switching cycle. In conclusion, the proposed power converter provides high isolation capability by using an isolated transformer and has a small device footprint.

ACKNOWLEDGMENT

The authors would like to thank the National Science Council of the Republic of China, Taiwan, for financially supporting this research under Contract No. NSC 96-2221-E-009-239-MY3.

REFERENCES

[1] Jeffrey D. Wilkinson, "Input isolation circuit for computer-controlled medical device," *US Patent*, no.5267150, Nov. 30, 1993.

[2] Chengwu Chen and Michael A. Gley, "Scheme for isolating a computer system from a data transmission network," *US Patent*, no.5473552, Dec. 5, 1995.

[3] Michael J. Gambuzza, "Data access arrangement for a digital subscriber line," *US Patent*, no.6226331, May 1, 2001.

[4] Raphael Rahamim and Grey Beutler, "Modem a digital high voltage isolation barrier," *US Patent*, no.6351530, Feb. 26, 2002.

[5] G. Hua and F. C. Lee, "Soft-switching techniques in PWM converters", *IEEE Trans. Industrial Electronics*, vol. 42, pp. 595–603, 1995.

[6] Quan Li and P. Wolfs, "A leakage-inductance-based ZVS two-inductor boost converter with integrated magnetics," *IEEE Power Electronics Letters*, vol. 3, pp. 67–71, 2005.

[7] Tiecheng Sun, Xueqin Zhu, Hongpeng Liu, Lian Liang, and Peng Gao, "A Novel ZVS PWM FB DC/DC Converter Using Auxiliary Resonant Net," in *Proc. 12th International Power Electronics and Motion Control Conference*, pp. 728–732, 2006.

[8] K. Siri and C. Q. Lee, "Constant switching frequency LLC-type series resonant converter," in *Proc. 32nd Midwest Symposium on Circuits and Systems*, vol. 1, pp. 513–516, 1989.

[9] Yilei Gu, Zhengyu Lu and Zhaoming Qian, "A Novel LLC Resonant Converter Topology: Voltage Stresses of All Components in Secondary Side Being Half of Output Voltage," in *Proc. of CES/IEEE 5th International Power Electronics and Motion Control Conference*, vol. 1, pp. 1–5, 2006.

[10] C.Q. Lee, R. Liu, and I. Batarseh, "Parallel resonant converter with LLC-type commutation," *IEEE Trans. Aerospace and Electronic Systems*, vol. 25, pp. 844–847, 1989.

[11] A. Bucher, T. Durbaum, D. Kubrich, and A. Stadler, "Comparison of Different Design Methods for the Parallel Resonant Converter," in *Proc. 12th International Power Electronics and Motion Control Conference* , pp. 810–815, 2006.

Efficiency and Power Losses in PM BLDC Motor with Variable Bridge/half-bridge Structure Electronic Commutator

K. Krykowski [*], A. Bodora [†]

Silesian University of Technology/Department of Power Electronics, Electrical Drives and Robotics,
Gliwice, Poland
[*] e-mail: *krzysztof.krykowski@polsl.pl*
[†] e-mail: *aleksander.bodora@polsl.pl*

Abstract— In practice we often meet with electrical drives, where it is necessary to operate with higher speed, while load torque is less than rated torque. This type of operation is called "constant power operation" and may be defined as second range of natural motor torque-speed characteristic. In case of motors excited with permanent magnets magnetic flux is constant, and while supply voltage remains constant, motor speed is also (approximately) kept constant.

In practice, different solutions are attempted in order to achieve increased rotating speed of brushless motors at decreased torque. One of the methods, which make possible PM BLDC motor operation at increased speed and decreased torque, is use of electronic commutator with switchable bridge-half-bridge structure. The investigation presented in the paper has been aimed at estimation of losses and efficiency in the drive consisting of brushless motor operating with variable structure commutator. Results of theoretical analysis, computer simulations and laboratory tests have shown that application of electronic variable structure commutator, which makes possible operation in the second speed range without increasing motor power and supply source power, causes an insignificant decrease in overall drive efficiency.

Keywords— brushless dc motor, electronic commutator, drive system, efficiency.

I. INTRODUCTION

In practice we often meet with electrical drives, where it is necessary to operate with higher speed, while load torque is less than rated torque. This type of operation is called "constant power operation" and may be defined as second range of natural motor torque-speed characteristic. Change of speed range in electric motor is related to change of idle run speed. In the second operation range, as idle run speed increases, electromagnetic torque decreases. This is usually achieved by decreasing excitation flux. In case of motors excited with permanent magnets magnetic flux is constant, so there is no simple way of decreasing flux in the air gap.

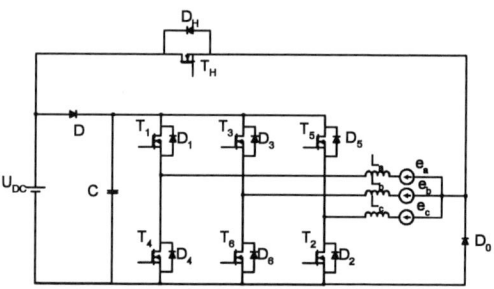

Fig.1. Diagram of variable structure electronic commutator

In practice, different solutions are attempted in order to achieve increased rotating speed of brushless motors at decreased torque [1],[2],[3],[4],[5],[6].

One of the methods which make possible PM BLDC motor operation at increased speed and decreased torque is use of electronic commutator with switchable bridge-half-bridge structure [7],[8].

The circuit is shown in Fig.1, and analytically determined torque-speed curves of bridge and half-bridge commutator are shown in Fig.2.

If we look at Fig.1, then apart from transistor bridge (T1-T6) we may also observe a fully-controlled valve TH, which connects motor's neutral point with positive terminal of supply source, D01 diode, which connects negative terminal of supply source to motor's neutral point and link (buffer) circuit consisting of diode (D) and capacitor (C).

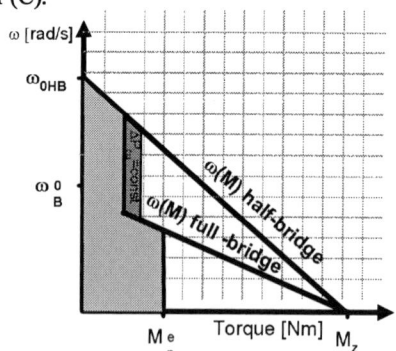

Fig.2. Torque-speed curves for PMBLDC motor drive with variable structure commutator

The properties of PM BLDC motor with electronic commutator and torque-speed curves have been discussed in [7], [8].

The acting principle and torque-speed characteristics of PM BLDC motor with variable structure commutator have been presented in papers [7], [8]. A significant factor determining suitability of using this type of electronic commutator is its efficiency. Since available references do not provide any information on this point, we have tried ourselves to determine power losses and efficiency of PM BLDC motor with switchable structure commutator. Computer simulations and lab tests with a 1 kW motor have supplemented analytical investigation.

II. DRIVE LOSSES

Efficiency of the converter-motor assembly is mostly influenced by motor and converter losses and it may be formulated as:

$$\eta_{ECM} = \eta_{EC}\eta_M \qquad (1)$$

Since PM BLDC motor cannot operate without commutator present, the losses for converter and motor have not been treated separately. Converter losses are mostly due to voltage drop across power electronics valves. In case of PM BLDC motor the most significant losses are copper losses (P_{losCu}) and iron losses (P_{losFe}). Other losses such as e.g. mechanical losses are also present.

Another approach is to classify losses into constant losses (speed and frequency dependent) and variable losses (load-dependent). Variable losses cover copper losses, valve losses, wiring losses and losses in diode D of buffer capacitor circuit. In case of commutator with bridge structure these losses may be expressed as:

$$P_{losB\,\mathrm{var}} = \sum_{k=1}^{3} 2(R_{on} + R_s)I_{skRMS}^2 \qquad (2)$$

and for electronic commutator connected directly to the source or

$$P_{losBv\,\mathrm{var}} = \sum_{k=1}^{3} 2(R_{on} + R_s)I_{skRMS}^2 + U_{DF}I_{DC} \qquad (3)$$

for variable structure commutator connected into bridge. Often it is more convenient to introduce average converter output current. This current is related to the source current by approximate formula:

$$I_d \approx \frac{I_{DC}}{D} \qquad (4)$$

and is equal to the average valve current for valve conducting time.
Therefore:

$$I_{T(AV)} = I_d \qquad (5)$$

and valve average current is defined as:

$$I_{T(AV)} = \frac{3}{T} \int_{t_s}^{t_s + T/3} |i_T| dt \qquad (6)$$

where t_s is valve's switch-on time instant.
The RMS valve current value for valve conducting time may be defined as:

$$I_{TRMS} = \sqrt{\frac{3}{T} \int_{t_s}^{t_s + T/3} i_T^2 dt} \qquad (7)$$

Mixing formulas (6) and (7) together we arrive at current form factor defined for valve conducting period as:

$$k_{iT} = \frac{I_{TRMS}}{I_{T(AV)}} \qquad (8)$$

Introducing this factor and taking formula (5) into account, variable losses may be determined as:

$$P_{losB\,\mathrm{var}} = \sum_{k=1}^{3} 2(R_{on} + R_s)k_{iT}^2 I_d^2 \qquad (9)$$

for commutator operating in bridge structure supplied directly from the source or

$$P_{losBv\,\mathrm{var}} = \sum_{k=1}^{3} 2(R_{on} + R_s)k_{iT}^2 I_d^2 + U_{DF}I_{DC} \qquad (10)$$

for variable structure commutator connected into bridge structure. Relation for losses in both these configurations may be given as:

$$P_{losBv\,\mathrm{var}} = P_{losB\,\mathrm{var}} + U_{DF}I_{DC} \qquad (11)$$

Constant losses may be determined if frequency-dependent losses (iron losses) and speed-dependent losses are known. In case of typical PM BLDC motors we may assume with satisfactory accuracy that these losses are proportional to speed. Then, in case of bridge:

$$P_{loscon} = \frac{\omega}{\omega_n} P_{loscon} \qquad (12)$$

Torque corresponding to these losses (loss torque) may be defined as:

$$T_{los} = \frac{P_{loscon}}{\omega_n} P_{loscon} \qquad (13)$$

Total electromagnetic torque is the sum of load torque and loss torque in accordance with formula:

$$T_e = T_m + T_{los} \qquad (14)$$

and is related to motor's average current in the following way:

$$T_e = 2k_f I_d \qquad (15)$$

while average current is related in turn to source current by approximate formula (4).
Total losses are equal to sum of constant and variable losses and are expressed as:

$$P_{losB} = P_{losB\,\mathrm{var}} + P_{losBcon} \qquad (16)$$

In case of half-bridge commutator variable losses may be expressed as:

$$P_{losHB\,\mathrm{var}} = \sum_{k=1}^{3} (R_{on} + R_s)I_{skRMS}^2 \qquad (17)$$

commutator operating in bridge structure supplied directly from the source or

$$P_{losHBv\,\mathrm{var}} = \sum_{k=1}^{3} (2R_{on} + R_s)I_{skRMS}^2 \qquad (18)$$

for variable structure commutator connected into bridge structure.

Introducing switch current form factor for conducting period the following formula is obtained:

$$P_{losHB\,var} = \prod_{k=1}^{3} (R_{on} + R_s)k_{iT}^2 I_d^2 \qquad (19)$$

and

$$P_{losHBv\,var} = \prod_{k=1}^{3} (2R_{on} + R_s)k_{iT}^2 I_d^2 \qquad (20)$$

While discussing (9), (10), (19) and (20) it must be noted that distortion coefficient for valve current depends on motor operational conditions and may be expressed roughly as:

$$k_{iTB} = 1 \div 1.05 \qquad (21)$$

for converter connected into bridge structure and

$$k_{iTHB} = 1 \div 1.35 \qquad (22)$$

for converter connected into half-bridge structure.

III. IMPACT OF VARIABLE STRUCTURE ELECTRONIC COMMUTATOR ON DRIVE LOSSES AND EFFICIENCY

Drive user is always interested in getting the answer to the question, how choice of structure influences efficiency of electronic commutator-motor assembly. If different losses are known, overall efficiency may be defined as:

$$\eta_{ECMB} = \frac{P_{Mout}}{P_{Mout} + P_{los}} \qquad (23)$$

Mixing formulas (11), (16) and (23) together the following formulas are obtained:

$$\eta_{ECMB} = \frac{P_{Mout}}{P_{Mout} + P_{losB}} \qquad (24)$$

and

$$\eta_{ECMBv} = \frac{P_{Mout}}{P_{Mout} + P_{losB} + U_{DF}I} \qquad (25)$$

They apply to electronic commutator-PM BLDC motor assembly efficiency in case of electronic commutator connected directly to the dc voltage supply and variable structure electronic commutator connected into bridge structure. If we compare equations (24) and (25) and transform them a little, the following formula is obtained:

$$\frac{\eta_{ECMBv}}{\eta_{ECMB}} = 1 - \frac{U_{DF}}{U_{DCF}} \qquad (26)$$

and it makes possible estimation of variable structure impact of drive efficiency, if commutator is connected into bridge. Mixing formulas (16), (19), (20) and (23) together the following formulas are obtained:

$$\eta_{ECMHB} = \frac{P_{Mout}}{P_{Mout} + P_{losB}} \qquad (27)$$

and

$$\eta_{ECMHBr} = \frac{P_{Mout}}{P_{Mout} + P_{losB} + R_{on}k_{iT}^2 I_d^2} \qquad (28)$$

They apply to electronic commutator-PM BLDC motor assembly efficiency in case of electronic commutator connected directly to the dc voltage supply and variable structure electronic commutator connected into half-

bridge structure. If we compare equations (27) and (28) and transform them a little, the following formula is obtained:

$$\frac{\eta_{ECMBv}}{\eta_{ECMB}} = 1 - \frac{R_{on}k_{iT}^2}{D} \frac{I_{DC}}{U_{DC}} \qquad (29)$$

and it makes possible estimation of variable structure impact of drive efficiency, if commutator is connected into half-bridge.

Basing on these relationships, efficiency characteristics of PM BLDC motor rated at 1 kW and supplied from 24 V dc source have been calculated. These characteristics are shown in Fig.3.

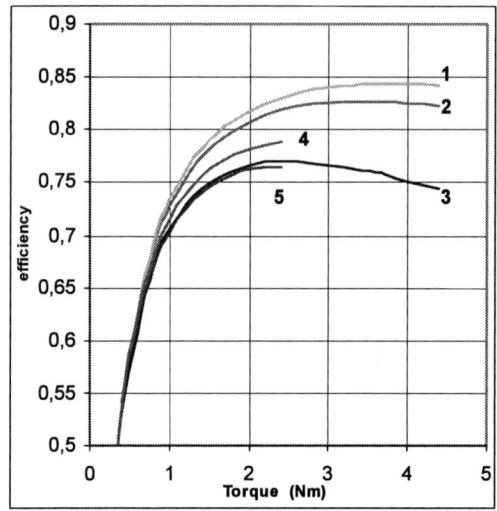

Fig. 3. Efficiency of tested PM BLDC motor with variable structure commutator vs. load torque for bridge structure (n = 1000 rpm (1), n = 2000 rpm (2), n = 2600 rpm (3)) and for half-bridge structure (n = 3500 rpm (4), n = 4500 rpm (5))

IV. LABORATORY TEST STAND

In order to verify the theoretical investigation, lab and simulation tests have been run. In lab tests, prototype PM BLDC motor has been used, rated at UDC = 24V, nn = 2600 rpm, Mn = 3.7 Nm, Pn = 1 kW, fn = 130Hz. Motor construction allows for second range operation (half-bridge) at the maximum (limit) parameters: nn2 = 4500 rpm, Mn2 = 2 Nm, Pn2 = 950 W, . Phase resistance and inductance are equal to and, respectively. Motor has been supplied from battery with 120 Ah capacitance and 24 V voltage and loaded with dc generator. No-load losses have been determined from the measurements. These losses change proportionately to speed and they are equal to Pbj = 102W at n = 3000rpm.

a)

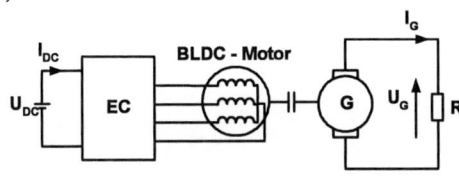

Fig.4. Laboratory test stand with PM BLDC motor operating in two
speed ranges and loaded with dc generator
a) schema of laboratory stand
b) view of laboratory stand

During lab tests PM BLDC motor has been loaded with dc machine operating as generator and rated at 1.2 kW. Circuit diagram is shown in Fig.4.

Shaft power has been determined as sum of generator output power and generator losses, in accordance with

$$P_{Mout} = U_G I_G + P_{Glos} \qquad (30)$$

and generator losses have been determined basing on the manufacturer's data.

Three SUP/SUB75N08-09L transistors connected in parallel have constituted one branch of the analysed converter. Valve conducting resistance for one branch has been equal to $R_{on} = 2,9 - 5,8m\Omega$, while voltage drop across diode has been equal to $U_{DF} = 0,2 - 0,4V$.

V. COMPUTER MODEL

The electrical circuit shown in Fig.1 has been transformed into a three-phase computer circuital model of PM BLDC motor with variable structure commutator (Fig.5).

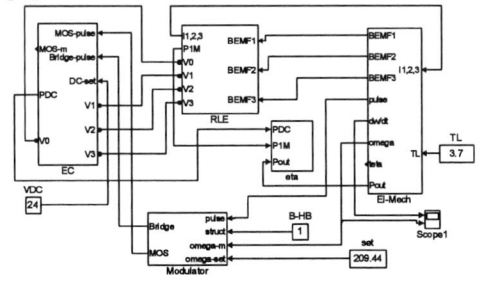

Fig. 5. The 3-phase circuital model of PM BLDC motor in basic version
- block diagram

This model is based on circuital model presented in [9] and contains several blocks.

"RLE" block contains motor's main circuits, "El-Mech" block contains relationships describing drive's coefficients linking mechanical and electrical properties as well as mechanical characteristics, "Set" block is used to set angular speed ω, "EC" block contains variable structure commutator, "B-HB" block sets commutator structure, "Modulator" block contains PWM modulator and speed controller, "eta" is efficiency block and "VDC" block is used to set supply voltage.

Output terminals of "Modulator" block are used to control appropriate transistors of electronic commutator, "Bridge" output is responsible for bridge transistors and "MOS" output controls TH transistor responsible for changing commutator structure.

VI. SIMULATION AND LABORATORY VERIFICATION

To verify efficiency analysis series of simulations and lab tests have been run. Difference between simulation results and theoretical results shown in Fig.6 has practically amounted to nil.

a)

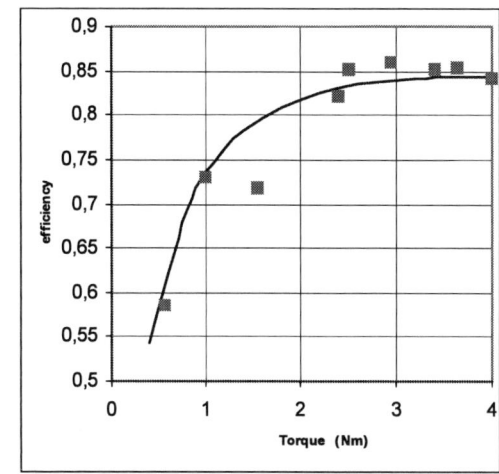

b)

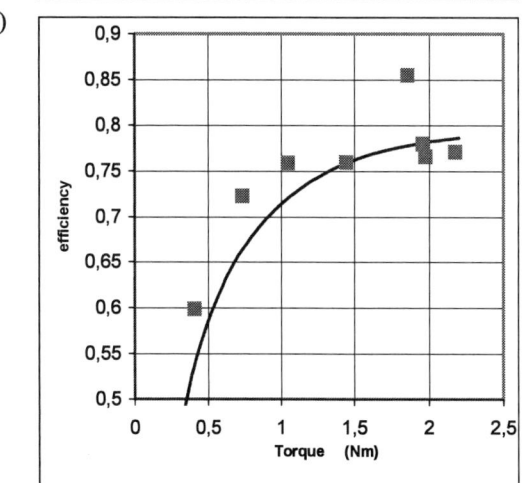

Fig. 6. PM BLDC motor efficiency vs. load torque:
a) for variable structure electronic commutator operating in bridge
structure and at speed equal to rated speed n = nn = 2600 rpm
b) for variable structure electronic commutator operating in half-bridge
structure and at speed equal to n = 1,73 nn = 4500 rpm

In case of lab tests difference in results has been greater, attaining several per cent or sometimes more. Some of measurement results of efficiency vs. load torque for typical operational conditions, i.e. for bridge structure and at rated speed $n_n = 2600$ rpm, load changing from 10 to 110 per cent of the rated load, and for half-bridge structure and at speed $n = 4500$ rpm, load changing from 10 to 60 per cent of the rated load have been shown in Fig.6.

VII. FINAL CONCLUSIONS AND REMARKS

Simulations and lab tests have shown that benefits arising from application of electronic variable structure commutator, which makes possible operation in the second speed range are attained at the insignificant cost of overall drive efficiency decrease of the order of c. 1 per cent.

The investigation has shown that if source power is limited and it is necessary to enlarge drive speed range, drive system consisting of PM BLDC motor and electronic commutator is an attractive and valuable design alternative.

REFERENCES

[1] Cros J., Paynot C., Figueroa J., Viarouge P., Multi-Star PM brushless DC motor for traction applications 10th European Conference on Power Electronics and Applications, 2-4 September 2003,

[2] Krishnan R., Electric Motor Drives, Modeling, Analysis and Control, Prentice Hall, New Jersey 2001.

[3] Nipp, E.; Alternative to field-weakening of surface-mounted permanent-magnet motors for variable-speeddrives, Industry Applications Conference, 1995. Thirtieth IAS Annual Meeting, IAS '95. Conference Record of the 1995 IEEE , Volume: 1 , 8-12 Oct 1995.

[4] Sakurai T, Sakurai M, Arimo M., Driving Method for Delta-Connected Brushless- Sensorless Motor - Power Conversion Conference - Nagaoka 1997.

[5] J. S. Lawler, J. M. Bailey, J. W. McKeever, and J. Pinto, "Limitations of the Conventional Phase Advance Method for Constant Power Operation of the Brushless DC Motor," submitted to IEEE Journal of Industry Applications, 2001.

[6] J. S. Lawler, J. M. Bailey, J. W. McKeever, and J. Pinto, "Extending the Constant Power Speed Range of the Brushless DC Motor Through Dual Mode Inverter Control – Part II: Laboratory Proofof- Principle," submitted to the IEEE Journal of Industry Applications, 2001.

[7] Krykowski K., Bodora A.: Variable structure bridge/half-bridge electronic commutator for PM BLDC Motor supply. ISIE2005. Dubrovnik

[8] Krykowski K., Bodora A.: Properties of the electronic commutator designed for two zone operation of PM BLDC motor drive. ISIE2005. Dubrovnik

[9] Krykowski K., Hetmańczyk J.: The circuital model of PM BLDC. Prace Naukowe Politechniki Śląskiej. Elektryka 2007/4 (204), Gliwice 2007.

Analysis of a device for converting a unipolar input voltage into two symmetric bidirectional output voltages with a magnetically coupled coil

Felix. A. Himmelstoss[*], Member, IEEE, Wilhelm Kraeftner[†]

[*]Technikum Wien, Vienna, Austria, e-mail: felix.himmelstoss@technikum-wien.at
[†]VDO Automotive AG, Vienna, Austria, e-mail: wilhelm.kraeftner@siemens.com

Abstract—**A new principle of a DC-DC converter with a magnetic coupled coil is analyzed. The speciality of this converter is that it transforms a unipolar input voltage into two symmetrical bidirectional DC voltages with only one power switch. The value of the output voltage can be varied by the duty cycle of the switch. The generation of two symmetrical voltages, one positive and one negative related to ground, makes it possible to invert a DC voltage to a single phase AC network with only three switches. For the operation of a class - D amplifier, for multi quadrant operation of DC motors or actuactors a symmetrical bidirectional supply voltage can also be very useful. After basic analysis in continuous inductor current mode, dimensioning of the converter components, a state space model and linearized transfer functions for the control of the converter are derived.**

Index Terms— **DC-DC power conversion, DC-AC power conversion, energy conservation, power conversion**

I. INTRODUCTION

THE basic structure of the here treated DC-DC converter [1] with a magnetically coupled coil is shown in Fig. 1.

The unipolar input voltage U_{IN} is converted with only one power switch S_1 into two symmetric bidirectional output voltages U_{C1} and U_{C2}. The value of the voltages can be adjusted by the duty cycle of the active switch S_1. The output voltage can be both higher and lower than the input voltage. The converter is treated and analyzed as a PWM structure with hard switching. Low input impedance can be achieved by the capacitor C_{IN} in parallel to the DC-supply. As storage element for the energy transfer a core with two magnetically coupled coils L_1 and L_2 is used. During on time T_{ON} of the switch S_1 the energy taken from the DC supply U_{IN} is stored via the coil L_1 as magnetic energy in the core, while both diodes D_1 and D_2 are blocked. When the switch S_1 opens both voltages across coil L_1 and L_2 change the sign because of the coupling and the core is demagnetized via D_1 and D_2. The magnetic energy is now transferred to the capacitors C_1 and C_2 and to the parallel connected loads R_{L1} and R_{L2}. The same capacitors values, same numbers of turns for the coils and R_{L1} equals R_{L2} means symmetric bi-directional voltages at the output. The

dimensioning of the components, a non-linear and a linearized model, the transfer functions for DC-DC converter mode for symmetric loads will be derived.

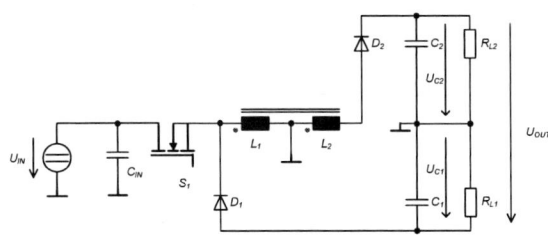

Fig. 1. Converter structure

II. BASIC ANALYZES

The basic analysis is done with idealized components (no parasitic resistors, no leakage inductances, ideal switches), symmetric load (R_{L1} equals R_{L2}) and for the continuous mode in steady state condition. In addition, the buffer capacitors C_1 and C_2 are assumed to be large enough so that the voltage across them stays constant during a switching period. The switch of the converter is PWM controlled, therefore two equivalent circuits for the on-time T_{ON} (Fig. 2) and the off-time T_{OFF} (Fig. 3) of the switch S_1 can be drawn.

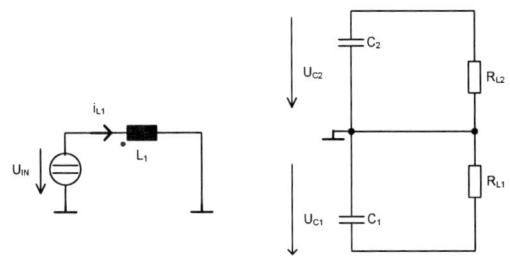

Fig. 2. Equivalent circuit during T_{ON}

978-1-4244-1741-4/08/$25.00 ©2008 IEEE

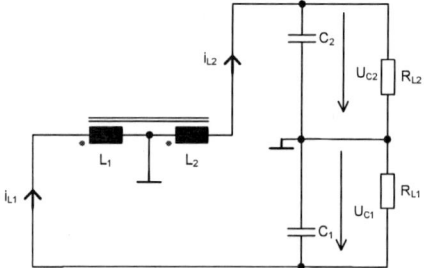

Fig. 3. Equivalent circuit during T_{OFF}

Now it is possible to see that during the T_{ON} state the energy from the supply is converted by the coil L_1 into magnetic energy in the core and the output is supplied only by the capacitors (Fig. 2). During T_{OFF} the core is demagnetized and energy is transferred into the output capacitors and the load, because of the conductive state of the diodes (Fig. 3).

A good idea to start is to consider the voltages across the coils. On the assumption that steady state occurs the area below the voltage curves must be equal for the two switch stages T_{ON} and T_{OFF}.

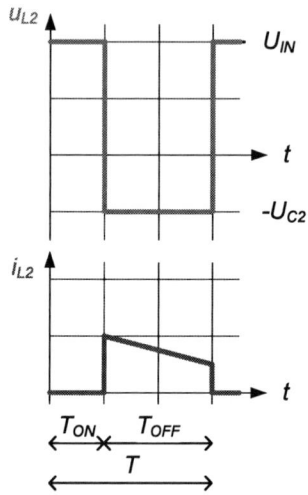

Fig. 4. Voltage across and current through coil L_1

Starting with coil L_1, during T_{ON} the input voltage U_{IN} and during T_{OFF} the negative voltage from the capacitor C_1 is applied across L_1 (Fig. 4). Therefore, the following equations can be derived

$$T_{ON} \cdot U_{IN} = T_{OFF} \cdot U_{C1} \qquad (1)$$

$$U_{C1} = \frac{d}{1-d} \cdot U_{IN} . \qquad (2)$$

The average of the current stays constant during a switching period. The difference between coils L_1 and L_2 is that through the first coil a current flows in both states T_{ON} and T_{OFF} (Fig. 4), through the second one a current only flows during the demagnetizing condition (Fig. 5). Due to the stray inductance of the tapped inductor, the current through the coil

differs from the ideally constructed curves. In appendix 3 oscillograms of the currents through the windings of a small 50 W converter are shown. A bifilar winding with little stray inductance was used.

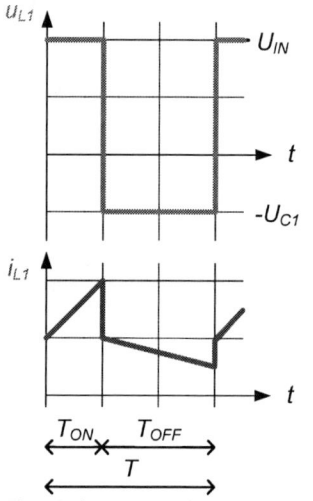

Fig. 5. Voltage across and current through coil L_2

As the coil ratio equals one the input voltage stays also across L_2 during T_{ON}. During the off time of the switch (T_{OFF}) the negative voltage from the capacitor C_2 applies across the coil L_2. Therefore, the following equation can be specified

$$U_{C2} = \frac{d}{1-d} \cdot U_{IN} . \qquad (3)$$

For the complete output voltage U_{out} following equation is valid

$$U_{OUT} = U_{C1} + U_{C2} . \qquad (4)$$

So the voltage transformation value M between U_{OUT} and U_{IN} can be written as

$$M = \frac{U_{OUT}}{U_{IN}} = 2 \cdot \frac{d}{1-d} . \qquad (5)$$

Depending on the duty cycle d the voltage transformation value M has the following gradient (Fig.6).

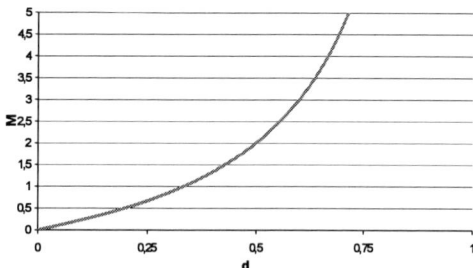

Fig. 6. Voltage transformation value M as function of the duty cycle d

The next step is the dimensioning of the main components of the converter, the coils and the capacitors.

To get a good magnetic coupling between the coils, a bifilar winding technique should be used. For the inductor value of the coils the current ripple and the minimum load current are decisive.

In accordance with Fig. 3 and Fig. 4, the current through the diodes D_1 and D_2 can be drawn (Fig. 7).

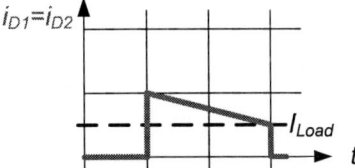

Fig. 7. Current through diode D_1 and D_2

As steady state is assumed, the voltage across the output capacitor stays constant during the switching period T. So during the T_{OFF} state the full output power has to be provided by the capacitors. That means that the average current through the diodes of a switching periode equals the load current I_{Load} (Fig. 7). To assure continuous conduction mode (CCM), the sum of the current through the diodes D_1 and D_2 should never become zero.

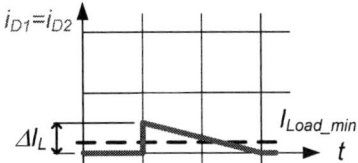

Fig. 8. Ripple current ΔI_L and minimum load current I_{Load_min}

The lowest allowed load current I_{Load_min} is therefore equivalent to the mean value of the diode current at the border between the continuous and discontinuous mode. The ripple of the inductor current during T_{OFF} is therefore (Fig.8)

$$\Delta I_L = \frac{2 \cdot I_{Load\_min}}{1-d} \cdot \qquad (6)$$

Fig. 9. Ripple on the inductor current i_{L1}

The minimum inductor value L_1 can be calculated to

$$L_1 \geq \left(\frac{U_{IN}}{U_{IN} + U_{C1}} \right)^2 \cdot \frac{U_{C1}}{4 \cdot f_S \cdot I_{Load\_min}} \cdot \qquad (7)$$

The capacitors are discharged during the on time of the switch by the load current. Another important factor for dimensioning is the maximum acceptable variation of the capacitor voltage ΔU_{C1}. Therefore, the capacitor value is

$$C_1 \geq \frac{I_{Load\_max}}{\Delta U_{C1} \cdot f_s} \cdot \frac{U_{C1}}{U_{IN} + U_{C1}} \qquad (8)$$

Due to the symmetry of the converter, L_2 and C_2 have the same value as coil L_1 and capacitor C_1, respectively.

III. DYNAMIC MODEL REPRESENTATION

In the model the resistive and magnetic loses are neglected. That means no equivalent series resistance for the coils and capacitors have been taken into consideration. Also losses during the on time of the active S_1 (due to the on resistance r_{DSON}) and of the passive switches D_1 and D_2 (due to the dynamic resistor r_{BE} and the threshold voltage) are not implemented and we also assume ideal magnetic coupling. In the appendix A2 a complete model of the converter with losses is given.

For the drawn equivalent circuits Fig. 1 for T_{ON} and Fig. 2 for T_{OFF}, two differential equation systems can be derived. The state variables of the system are the magnetic flux Φ in the coil core and the voltages across the capacitors u_{C1} and u_{C2}. For the on time of the switch the following system is valid

$$\frac{d\Phi}{dt} = \frac{U_{IN}}{N_1}$$

$$\frac{du_{C1}}{dt} = -\frac{u_{C1}}{R_{L1} \cdot C_1}$$

$$\frac{du_{C2}}{dt} = -\frac{u_{C2}}{R_{L2} \cdot C_2} \cdot \qquad (8)$$

For the T_{OFF} interval the following state variable equations can be given

$$\frac{d\Phi}{dt} = -\frac{u_{OUT}}{N_1 + N_2} = -\frac{u_{C1}}{N_1 + N_2} - \frac{u_{C2}}{N_1 + N_2}$$

$$\frac{du_{C1}}{dt} = \frac{N_1^2}{(N_1 + N_2) \cdot L_1 \cdot C_1} \cdot \Phi - \frac{u_{C1}}{R_{L1} \cdot C_1}$$

$$\frac{du_{C2}}{dt} = \frac{N_2^2}{(N_1 + N_2) \cdot L_2 \cdot C_2} \cdot \Phi - \frac{u_{C2}}{R_{L2} \cdot C_2} \cdot \qquad (9)$$

The system equations for the two states can be combined when the system time constant is long compared to the switching period T=1/ f_S.
The first equation has to be weighted with,

$$d = \frac{T_{ON}}{T} \qquad (10)$$

and the second one with,

$$1 - d = \frac{T_{OFF}}{T} \cdot \qquad (11)$$

The result of the weighting process is the following nonlinear equation, which describes the full dynamic behavior of the idealized converter

$$\frac{d}{dt}\begin{pmatrix}\Phi\\u_{C1}\\u_{C2}\end{pmatrix}=\begin{bmatrix}0 & \dfrac{d-1}{N_1+N_2} & \dfrac{d-1}{N_1+N_2}\\[2mm] \dfrac{(1-d)\cdot N_1^2}{(N_1+N_2)\cdot L_1\cdot C_1} & -\dfrac{1}{R_{L1}\cdot C_1} & 0\\[2mm] \dfrac{(1-d)\cdot N_2^2}{(N_1+N_2)\cdot L_2\cdot C_2} & 0 & -\dfrac{1}{R_{L2}\cdot C_2}\end{bmatrix}\cdot\begin{pmatrix}\Phi\\u_{C1}\\u_{C2}\end{pmatrix}+\begin{pmatrix}\dfrac{d}{N_1}\\0\\0\end{pmatrix}\cdot u_{IN} \tag{12}$$

To use the linear controller techniques, a linearization around the working point of the system is necessary. Therefore the following ansatz is used:

$$\Phi=\Phi_0+\hat{\Phi}$$
$$u_{C1}=U_{C10}+\hat{u}_{C1}$$
$$u_{C2}=U_{C20}+\hat{u}_{C2}$$
$$d=D_0+\hat{d}\,. \tag{13}$$

The index zero denotes the working point and the $\wedge$ labels a disturbance around the working point. The result is the linearisation of equation (12):

$$\frac{d}{dt}\begin{pmatrix}\hat{\Phi}\\\hat{u}_{C1}\\\hat{u}_{C2}\end{pmatrix}=\begin{bmatrix}0 & \dfrac{D_0-1}{N_1+N_2} & \dfrac{D_0-1}{N_1+N_2}\\[2mm] \dfrac{N_1^2\cdot(1-D_0)}{(N_1+N_2)\cdot L_1\cdot C_1} & -\dfrac{1}{R_{L1}\cdot C_1} & 0\\[2mm] \dfrac{N_2^2\cdot(1-D_0)}{(N_1+N_2)\cdot L_2\cdot C_2} & 0 & -\dfrac{1}{R_{L2}\cdot C_2}\end{bmatrix}\cdot\begin{pmatrix}\hat{\Phi}\\\hat{u}_{C1}\\\hat{u}_{C2}\end{pmatrix}$$
$$+\begin{bmatrix}\dfrac{D_0}{N_1} & \dfrac{U_{C10}+U_{C20}}{N_1+N_2}+\dfrac{U_{IN0}}{N_1}\\[2mm] 0 & -\dfrac{N_1^2\cdot\Phi_0}{(N_1+N_2)\cdot L_1\cdot C_1}\\[2mm] 0 & -\dfrac{N_2^2\cdot\Phi_0}{(N_1+N_2)\cdot L_2\cdot C_2}\end{bmatrix}\cdot\begin{pmatrix}\hat{u}_{IN}\\\hat{d}\end{pmatrix} \tag{14}$$

Depending on which output voltage should be controlled (U_{C1} or U_{C2} or U_{OUT}), the related small signal transfer function can be derived. Regarding the following voltage controller design, we work with the transfer function $G_s(s)=U_{OUT}(s)/D(s)$. Therefore the linearized state space system from equation (14) has to be Laplace transformed

$$\begin{bmatrix}s & \dfrac{1-D_0}{N_1+N_2} & \dfrac{1-D_0}{N_1+N_2}\\[2mm] \dfrac{(D_0-1)\cdot N_1^2}{(N_1+N_2)\cdot L_1\cdot C_1} & s+\dfrac{1}{R_{L1}\cdot C_1} & 0\\[2mm] \dfrac{(D_0-1)\cdot N_2^2}{(N_1+N_2)\cdot L_2\cdot C_2} & 0 & s+\dfrac{1}{R_{L2}\cdot C_2}\end{bmatrix}\cdot\begin{pmatrix}\hat{\Phi}(s)\\\hat{U}_{C1}(s)\\\hat{U}_{C2}(s)\end{pmatrix}=$$
$$=\begin{bmatrix}\dfrac{D_0}{N_1} & \dfrac{U_{C10}+U_{C20}}{N_1+N_2}+\dfrac{U_{IN0}}{N_1}\\[2mm] 0 & -\dfrac{N_1^2\cdot\Phi_0}{(N_1+N_2)\cdot L_1\cdot C_1}\\[2mm] 0 & -\dfrac{N_2^2\cdot\Phi_0}{(N_1+N_2)\cdot L_2\cdot C_2}\end{bmatrix}\cdot\begin{pmatrix}\hat{U}_{IN}(s)\\\hat{D}(s)\end{pmatrix}. \tag{15}$$

The system (15) has to be enlarged with the output equation in the complex variable domain

$$\hat{U}_{OUT}(s)=\begin{bmatrix}0 & 1 & 1\end{bmatrix}\cdot\begin{pmatrix}\hat{\Phi}(s)\\\hat{U}_{C1}(s)\\\hat{U}_{C2}(s)\end{pmatrix}. \tag{16}$$

The matrix of all possible transfer functions can be calculated according to

$$\mathbf{G}(s)=\mathbf{c}^{T}\left[s\cdot I-\mathbf{A}\right]^{-1}\cdot\mathbf{b}\,. \tag{17}$$

The transfer function (the coefficients are given in the appendix A1) between output voltage and duty cycle is

$$G_S(s)=\frac{\hat{U}_{OUT}(s)}{\hat{D}(s)}=\frac{a_2\cdot s^2+a_1\cdot s+a_0}{s^3+b_2\cdot s^2+b_1\cdot s+b_0} \tag{18}$$

and with the following symmetric converter setup

$$L_1=L_2=58\ \mu H$$
$$C_1=C_2=1000\ \mu F$$
$$D_0=0{,}67$$
$$N_1=N_2=17$$
$$U_{IN}=12\ V$$
$$U_{C1}=-24\ V$$
$$U_{C2}=24\ V$$
$$R_{L1}=R_{L2}=12{,}5\Omega \tag{19}$$

the Bode plot of the small signal behaviour can be drawn (Fig. 10)

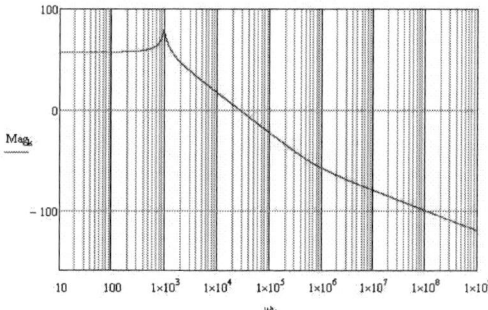

Fig. 10. Frequency characteristics $\mathrm{Mag}_k(\omega_k)=20\log\left|G_S(\omega_k)\right|$

and also the phase characteristics (Fig. 11)

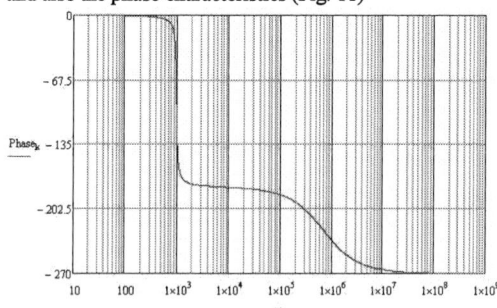

Fig. 11. Phase characteristics $\mathrm{Phase}_k(\omega_k)=\arg(G_S(\omega_k))$

The system has a non-minimum phase behavior, which is typical for converter structures with the possibility to step-up the input voltage.

IV. CLOSED LOOP SYSTEM

In Fig. 12 a Simplorer block diagram of the closed loop system is presented.

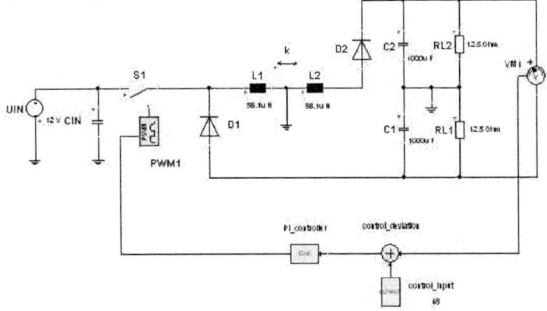

Fig. 12. Block diagram of the controlled converter

Starting from the idealized converter description, a simple PI voltage controller (Fig. 13) can be designed, which is sufficient to stabilize the converter. The step response to a reference value step (applied to the converter model with included parasitic resistances) is shown in Fig. 14. A comprehensive study about the control of this special converter will be published in the future.

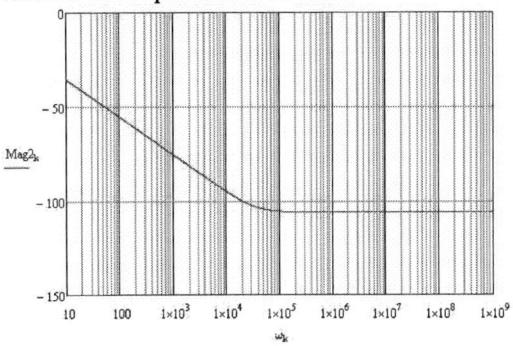

Fig. 13. Frequency characteristics of the controller

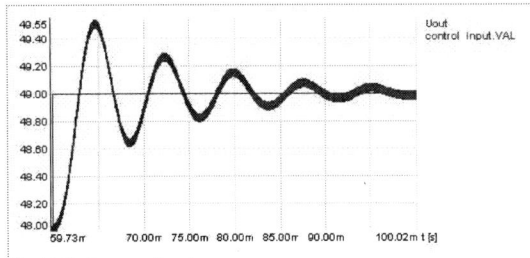

Fig. 14. Reference value step

V. CONCLUSIONS

A new DC-DC converter with a magnetically coupled coil that transforms a unipolar input voltage with only one power switch into two symmetrical DC voltages was presented. The value of the output voltage can be varied by the duty cycle of the switch. The existence of two symmetrical voltages, one positive and one negative related to ground, makes it possible to invert a DC voltage to a single phase AC network by only three switches (Fig. 15).

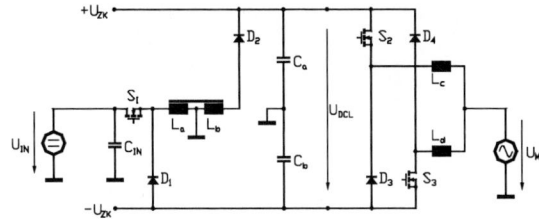

Fig. 15. DC-AC converter

The possibility of transforming the voltage up or down in the first converter stage, enables the control of an AC inverter with an additional degree of freedom, which reduces the switching losses of the following inverter stage. So the combination of the discussed new DC-DC converter principle with common AC inverter stages provides additional benefit [1]. For the operation of a class - D amplifier, for multi quadrant operation of DC motors or actuators a symmetrical bidirectional supply voltage can also be very useful.

REFERENCES

[1] F. Himmelstoss and K. Edelmoser: *"Conversion circuit changing input voltage into two symmetrical unipolar voltages* (in German)", Austrian Patent, filed 2004.01.29, AT 413.912 B, granted 2005.10.15.

[2] N. Mohan, T. Undeland, W.P. Robbins: *Power Electronics*, New York: John Wiley & Sons, 2003.

ACKNOWLEDGEMENT

This work is dedicated to Prof. Dr. Franz C. Zach in commemoration of his 65th birthday.

VI. APPENDIX

A 1: Coefficients of (18), the matrix elements are from (14)

$$b_2 = -\left(A_{11} + A_{22} + A_{33}\right)$$

$$b_1 = A_{11}A_{22} + A_{11}A_{33} + A_{22}A_{33} - A_{21}A_{21} - A_{13}A_{31} - A_{23}A_{32}$$

$$b_0 = -A_{11}A_{22}A_{33} - A_{12}A_{23}A_{31} - A_{13}A_{21}A_{32} + A_{12}A_{21}A_{33} + \\ + A_{13}A_{22}A_{31} + A_{11}A_{23}A_{32}$$

$$a_2 = B_{22} + B_{32}$$

$$a_1 = -\left(A_{11} + A_{33}\right)B_{22} + A_{21}B_{12} + A_{23}B_{32} - \\ - \left(A_{11} + A_{22}\right)B_{32} + A_{31}B_{12} + A_{32}B_{22}$$

$$a_0 = A_{11}A_{33}B_{22} + A_{23}A_{31}B_{12} + A_{13}A_{21}B_{32} - A_{21}A_{33}B_{12} -$$
$$- A_{13}A_{31}B_{22} - A_{11}A_{23}B_{32} + A_{11}A_{22}B_{32} + A_{12}A_{31}B_{22} +$$
$$+ A_{21}A_{32}B_{12} - A_{12}A_{21}B_{32} - A_{22}A_{31}B_{12} - A_{11}A_{32}B_{22}$$

A 2: Model of the converter with parasitic resistances
System matrix A

$$A_{11} = \left[\frac{r_C^2}{(R_L + r_C)} - (r_L + r_D + r_C) \right] \cdot \frac{(1-d)}{2 \cdot L} - \frac{r_s + r_L}{L} \cdot d$$

$$A_{12} = \frac{R_L \cdot (d-1)}{(R_L + r_C) \cdot 2 \cdot N}$$

$$A_{13} = \frac{R_L \cdot (d-1)}{(R_L + r_C) \cdot 2 \cdot N}$$

$$A_{21} = \frac{R_L}{(R_L + r_C) \cdot C_1} \cdot \frac{N}{2 \cdot L} \cdot (1-d)$$

$$A_{22} = -\frac{1}{(R_L + r_C) \cdot C_1}$$

$$A_{31} = \frac{R_L}{(R_L + r_C) \cdot C_2} \cdot \frac{N}{2 \cdot L} \cdot (1-d)$$

$$A_{33} = -\frac{1}{(R_L + r_C) \cdot C_2}$$

Input matrix B

$$B_1 = \frac{d}{N}$$

System matrix for the linearized model

$$A_{11} = \left[\frac{r_C^2}{(R_L + r_C)} - (r_L + r_D + r_C) \right] \cdot \frac{(1-D_0)}{2 \cdot L} - \frac{r_s + r_L}{L} \cdot D_0$$

$$A_{12} = \frac{R_L \cdot (D_0 - 1)}{(R_L + r_C) \cdot 2 \cdot N}$$

$$A_{13} = \frac{R_L \cdot (D_0 - 1)}{(R_L + r_C) \cdot 2 \cdot N}$$

$$A_{21} = \frac{R_L}{(R_L + r_C) \cdot C_1} \cdot \frac{N}{2 \cdot L} \cdot (1-D_0)$$

$$A_{22} = -\frac{1}{(R_L + r_C) \cdot C_1}$$

$$A_{31} = \frac{R_L}{(R_L + r_C) \cdot C_2} \cdot \frac{N}{2 \cdot L} \cdot (1-D_0)$$

$$A_{33} = -\frac{1}{(R_L + r_C) \cdot C_2}$$

Input matrix for the linearized model

$$B_{11} = \frac{D_0}{N}$$

$$B_{12} = \frac{1}{N} \cdot U_{IN0} + \frac{R_L \cdot (U_{C10} + U_{C20})}{2 \cdot (R_L + r_C) \cdot N} +$$
$$+ \left[\frac{r_L + r_D + r_C}{2 \cdot L} - \frac{r_C^2}{(R_L + r_C) \cdot 2 \cdot L} - \frac{r_s + r_L}{L} \right] \cdot \Phi_0$$

$$B_{22} = -\frac{R_L}{(R_L + r_C) \cdot C_1} \cdot \frac{N}{2 \cdot L} \cdot \Phi_0$$

$$B_{32} = -\frac{R_L}{(R_L + r_C) \cdot C_2} \cdot \frac{N}{2 \cdot L} \cdot \Phi_0$$

A 3: Oscillograms through the windings

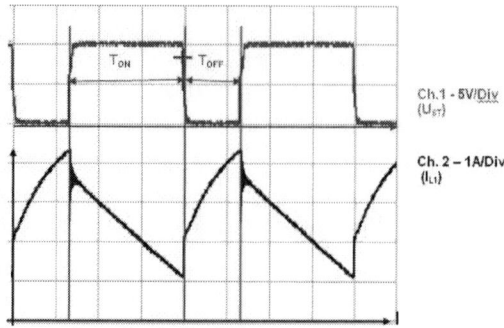

Current through winding N1

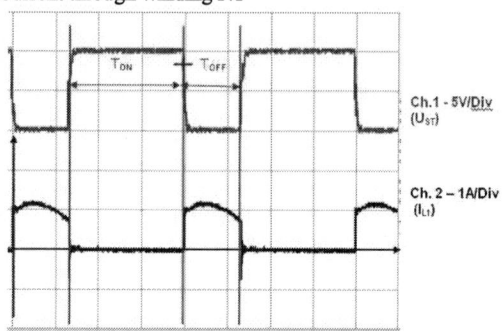

Current through winding N2, mind the influence of the stray inductance

Invariant Modulation Strategy for Two-stage Direct Power Converter

Radiy Bekbudov

IEEE member, Canada, e-mail: *radbek@ieee.org*

Abstract—The modulation strategy with the current invariance to input voltage disturbances and novel modulation algorithms for Two-stage Direct Power Converter are presented. Based on theoretical approach, the conditions for getting the current invariance are obtained. The novel modulation algorithms with minimum input current distortions are developed according to the proposed strategy. The switching sequence of the converter with power losses optimization procedure is specified. The performance efficiency of the proposed strategy and modulation algorithms in comparison with a conventional one are shown by simulation of the conversion procedure under balanced and unbalanced mains.

Keywords—AC/AC converter, modulation strategy, robustness.

I. INTRODUCTION

A number of recent publications indicate that the efforts of developers are concentrated on the so-called Sparse Matrix Converter, or Alternative Converter, or Two-Stage Direct Converter [1-5]. Among the existing names for this type of converter, the Two-stage Direct Converter (TDC) mostly fully reflects its functional properties, which is the main criterion by the author's opinion.

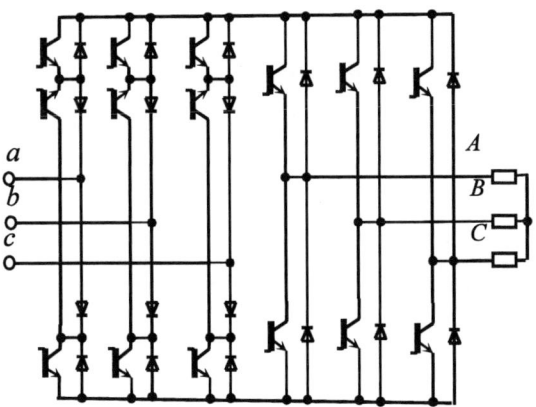

Fig.1 Two-stage Direct Converter with 15-IGBTs

In comparison with the conventional matrix converter, TDC, undoubtedly, has the advantage of lowered volume and cost, but on the other hand, TDC has the disadvantage of static and dynamic losses [2-4]. The very important property such as the current invariance of the converter to input voltage disturbances still remains obscure, although it negatively affects the quality of the input and output currents.

The conventional modulation strategy is fully presented in [1-4]. The disadvantage of this strategy is the impossibility to achieve the simultaneous non-distorted input and output currents even though there are no voltage disturbances in input mains. In this paper the novel modulation strategy and novel algorithms are proposed to provide the current invariance of converter under input voltage unbalance and get a minimum distortions into input and output currents.

II. INVARIANT MODULATION STRATEGY

The harmonic content of output current mainly depends on the spectrum of the synthesized output voltage space vector if the load is linear and balanced. For the input current, the approach is different due to the fact that its harmonic content depends on the spectra of input voltage and input power. In general, a power may be expressed by

$$ P = \mathrm{Re} \left\{ \vec{U} \cdot \vec{I}^{*} \right\} $$

where * is the sign of conjugation.

The power spectrum can be presented as a convolution of the spectra of its components [6]

$$ S_P(\omega) = \frac{1}{2\pi} \int_{-\infty}^{\infty} S_V(u) S_I(\omega - u) du \quad (1) $$

where $S_P(\omega)$ is a power spectrum; $S_V(u)$, $S_I(\omega-u)$ are spectra of voltage and current.

If one of the components of the spectrum is distorted, then additional harmonics will occur in power spectrum, i.e. the power is non-stationary. There are no insuperable restrictions to get simultaneously non-distorted input and output currents in spite of non-stationarity of input power and stationarity of output power.

When transferring energy from the main to the load the power gradient can be presented by the common equation

$$ \nabla P = \vec{U} \cdot \nabla \left| \vec{I}^{*} \right| + \vec{I}^{*} \cdot \nabla \left| \vec{U} \right| \le 0 \quad (2) $$

In order to get commonness of analysis and taking into account the direction of power transfer, the formal transit from vector's form (2) to the scalar mixed derivatives can be performed

$$ \frac{\partial^2 P}{\partial x \partial t} = \frac{\partial^2 \left| \vec{U} \right|}{\partial x \partial t} \cdot \left| \vec{I} \right| + \frac{\partial^2 \left| \vec{I} \right|}{\partial x \partial t} \cdot \left| \vec{U} \right| \le 0 \quad (3) $$

The equation (3) allows getting the expression for the providing of simultaneous non-distortions into input and output currents provided that the power losses are absent

978-1-4244-1741-4/08/$25.00 ©2008 IEEE

$$\frac{1}{|I|}\frac{\partial^2 |I|}{\partial x \partial t} = \frac{1}{P}\frac{\partial^2 P}{\partial x \partial t} - \frac{1}{|U|}\frac{\partial^2 |U|}{\partial x \partial t} \qquad (4)$$

Provided the output current is undistorted the condition of simultaneous non-distortions of currents may be presented as

$$\frac{1}{|I|}\frac{\partial^2 |I|}{\partial x \partial t} = const \qquad (5)$$

Expression (5) is the stationarity condition of current vector amplitude and it can be applied to both the whole conversion and any stage of conversion. In addition, for any stage of conversion the following ratio can be applied

$$\Delta P = P_1 - P_2$$

where P_1, P_2 are input and output power.

If $\Delta P = const$, then input and output power are either stationary or non-stationary and their spectra are the same excluding the value of zero harmonic

$$S_1^P(\omega) = S_2^P(\omega)\big|_{\omega>0} \qquad (6)$$

The analysis of (4) is shown to provide the condition (5) the definite ratios of parameters are required. The main variants of possible parameters ratios are presented in Tabl.1

TABLE I.
MAIN VARIANTS OF POSSIBLE PARAMETERS RATIOS

Parameters	Variants			
	I	II	III	IV
$|U_1|$	≠const	≠const	≠const	const
$\frac{1}{P}\frac{\partial^2 P}{\partial x \partial t}$	0	≠const	≠const	0
$\frac{1}{|U|}\frac{\partial^2 |U|}{\partial x \partial t}$	const	$\frac{1}{P}\frac{\partial^2 P}{\partial x \partial t}$	≠const	≠const
$|U_2|$	≠const	const	const	const
$|I_2|$	const	const	const	const
$\frac{1}{|I|}\frac{\partial^2 |I|}{\partial x \partial t}$	const	0	≠const	≠const
$|I_1|$	const	const	≠const	≠const

' U_1, I_1 are input parameters of stage;U_2, I_2 are output parameters of stage

The variant **III** corresponds to the performance of conventional matrix converter under unbalanced main. The variant **IV** corresponds to the rectification stage with conventional algorithm, when the input voltage is balanced. The analysis of Tabl.1 is shown to provide the condition (5) the only two stages of conversion are needed: the combination of variants **I** and **II**. The variants **I** and **IV** correspond to the rectification stage, and the difference between them is a conversion algorithm. For the variant **I** the conversion is occurred without distortion of input vector amplitude

$$\frac{1}{|U|}\frac{\partial^2 |U|}{\partial x \partial t} = const$$

or, in other words, saving spectrum excluding the value of zero harmonic

$$S_1^{|U|}(\omega) = S_2^{|U|}(\omega)\big|_{\omega>0} \qquad (7)$$

The difference between variants **II** and **III** is condition

$$\frac{1}{P}\frac{\partial^2 P}{\partial x \partial t} - \frac{1}{|U|}\frac{\partial^2 |U|}{\partial x \partial t} = 0$$

This condition corresponds to the performance of the inverter, when there is a physical conduction between input and output currents

$$\frac{\partial^2 |I|}{\partial x \partial t} = 0$$

or, in other words, their spectra are identical

$$S_1^{|I|}(\omega) \equiv S_2^{|I|}(\omega) \qquad (8)$$

Thus, only two-stage conversion allows providing the current invariance. The invariant strategy is based on strict simultaneous providing the conditions of (6) and (7) during the first stage and condition of (8) during the second stage. If one of the conditions (6)-(8) is achieved up to ε-level (ε≥0), the ε-invariance can be only provided during the performance. To get the required quality of input and output currents value of ε should be reduced up to the necessary level.

III. INVARIANT MODULATION ALGORITHM

The conventional modulation algorithm for a rectification stage is presented in [1,3]. According to this algorithm, the active duty-cycles γ_i are expressed by

$$\gamma_1' = \gamma_1/(\gamma_1 + \gamma_2),$$
$$\gamma_2' = \gamma_2/(\gamma_1 + \gamma_2), \gamma_0' = 0 \qquad (9)$$

where

$$\gamma_1 = Sin(\pi/3 - \theta), \gamma_2 = Sin(\theta),$$
$$\gamma_0 = 1 - \gamma_1 - \gamma_2, \theta = 2\pi f_0 t$$

Unfortunately, this algorithm distorts the dc-link voltage, and as a result, the spectrum mainly contains the sixth harmonic of input voltage frequency f_0. Thus, this modulation algorithm does not meet the proposed strategy because of unsatisfactory ε-level of condition (7), although the condition (6) is provided ($\gamma_0' = 0$). To eliminate this deficiency, the equation of ideal conversion algorithm should be taken into account

$$\gamma_1 Cos\theta + \gamma_2 Cos(\pi/3 - \theta) - C = 0 \big|_{\gamma_1 + \gamma_2 \le 1} \qquad (10)$$

where $0 \le \theta \le \pi/3$; $C = \sqrt{3}/2$; γ_1 and γ_2 belong to (9).

The equation (10) responds to the condition (7), but does not respond to the strict requirement for rectification stage, because of $\gamma_1 + \gamma_2 \le 1$, and, as a result, the condition (6) is breached. So, the equation (10) has to be changed to the common one with a strict restriction responding to the requirement (6)

$$\gamma_1' \cos\theta + \gamma_2' \cos(\pi/3 - \theta) - C = \Delta C \Big|_{\gamma_1' + \gamma_2' = 1} \quad (11)$$

where $\Delta C = \varphi(\theta) > 0$.

The equation (11) contains restriction for rectification stage, but the additional term ΔC leads to breach of (7), and it looks like the ideal solution of (11) for simultaneous providing of (6) and (7) does not exist. The term ΔC defines a level of dc-voltage distortions, so the limited value of ΔC can be considered as a ε-level of (7)

$$\varepsilon = Sup(\Delta C/C) \to 0 \quad (12)$$

The common equation (11) has a family of solutions depending on a value of (12), and only solutions with less calculations are preferred for application. For example, the one of solutions can be presented by

$$\gamma_1' = \begin{cases} \gamma_1 - \gamma_0, & (\gamma_1 \geq \gamma_2) \\ \gamma_1 + 2\gamma_0, & (\gamma_1 < \gamma_2) \end{cases} \quad (13)$$

$$\gamma_2' = 1 - \gamma_1'$$

where $\gamma_1, \gamma_2, \gamma_0$ belong to (9).

Algorithm (13) provides level of $\varepsilon = 0.04$. Another solution that provides level of $\varepsilon = 0.0008$ is closed enough to desired value of (12), but much more calculation expenses are required. Based on the local average of dc-link voltage and its spectrum the performance comparison of proposed algorithms (with $\varepsilon = 0.04$ and $\varepsilon = 0.0008$) and conventional one are presented in Fig.2.

The analysis of the presented results shows that based on common equation (11) the proposed algorithms bring less distortions into the spectrum of dc-link voltage, save the power efficiency. As a result, they meet the requirements of the proposed modulation strategy better than the conventional algorithm (9). By the way, algorithm (9) can be one of the solution family (11) with $\varepsilon = 0.155$. The common equation (11) contains both restrictions (6) and (7), but the last one is only achieved up to level $\varepsilon > 0$.

Thus, an ε-invariance of voltage conversion can be provided only for the rectification stage of TDC.

IV. OPTIMIZATION OF POWER LOSSES

Zero current pauses at the inversion stage are needed to provide a safety commutation between phases at the rectification stage. In case of a full-scale conversion, the conventional approach is based on the changes of modulation index **m** of the inverter up to the required limit [1]. The limitation of modulation index up to m=1-δ reduces the power efficiency of TDC. In addition, the applicable commutation sequences [1,4,5] utilize a different number of switching during the duty cycle, that results in different switching losses. To optimize the power losses the dynamic of zero current pause for the commutation sequences [4] with only six switchings should be taken into consideration. According to this approach, the modulation index **m** can be reduced in a short time during the period of output frequency. The main aspects of this approach will be considered below.

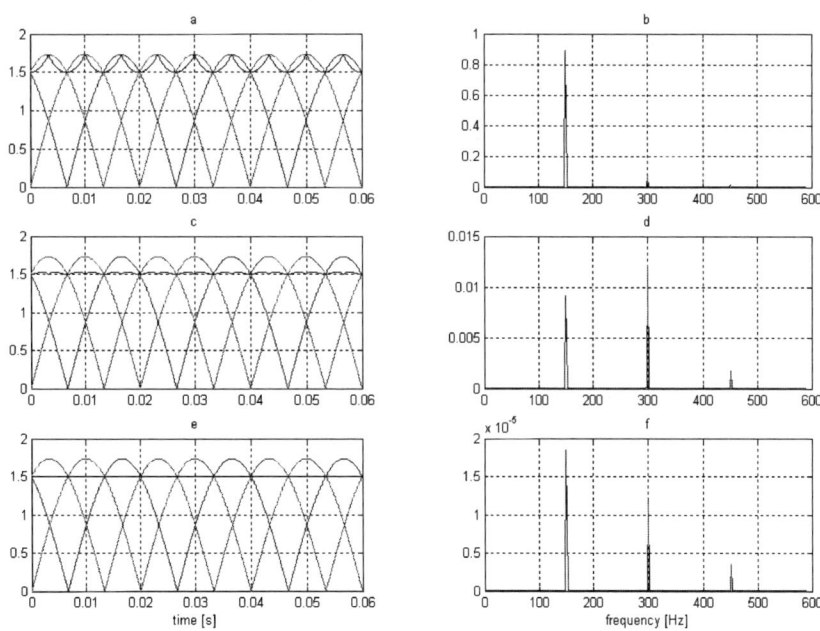

Fig.2. Performance comparison of conventional algorithm with $\varepsilon = 0.155$ **(a, b)** and proposed algorithms with $\varepsilon = 0.04$ **(c, d)** and with $\varepsilon = 0.0008$ **(e, f)**: **(a, c, e)** local average values of dc-link voltage with line-to-line voltage; **(b, d, f)** spectra of local average values of dc-link voltage without zero harmonics [the amplitudes of voltages and spectra are normalized, the spectra **(d, f)** are normalized to spectrum of **(b)** for better comparison].

For the inversion stage, the active duty-cycles d_i are usually calculated according to expression

$$d_1 = m \cdot Sin(\pi/3 - \alpha), d_2 = m \cdot Sin(\alpha),$$
$$d_0 = 1 - d_1 - d_2, \alpha = \omega t, 0 < m \le 1. \tag{14}$$

The zero pause of (14) may be expressed by another way

$$d_0 = d_0' + d_0'',$$
$$d_0' = 1 - d_{10} - d_{20}, d_0' = (1-m) \cdot (d_{10} + d_{20}) \tag{15}$$

where $d_{10} = Sin(\pi/3 - \alpha), d_{20} = Sin(\alpha)$.

The component d_0' does not depend on the modulation index **m**, but it takes close to zero values around the middle of any synthesized sector of output voltage vector. This component may be used for the zero current pauses except that area. To cover that area the other component d_0'' can be applied. In this case the index modulation should be smoothly changed up to $m = 1 - \delta_0$. The value δ_0 defines a dead zone for the safety commutation of input phase and it is calculated based on (15)

$$\delta_0 = \frac{T_\delta}{T_S} = \min(d_0'') \bigg|_{m = 1 - \delta_0} \tag{16}$$

where T_δ, T_S are minimum time and period of switching. Provided that ½ of d_0' is applied the zero current pause d_{00} can be calculated

$$d_{00} = \begin{cases} \dfrac{d_0}{2}, (d_0 \ge 2\delta_0) \\ \dfrac{d_0'}{2} + d_0'' \bigg|_{m = 1 - \delta}, (d_0 < 2\delta_0) \end{cases} \tag{17}$$

To keep d_{00} =const for lower term of (17) the index modulation should be calculated by smoothly changing of δ ($0 \le \delta \le \delta_0$) while d_{00} is in dead zone δ_0. The TDC switching sequences of duty-cycles is presented in Fig.3.

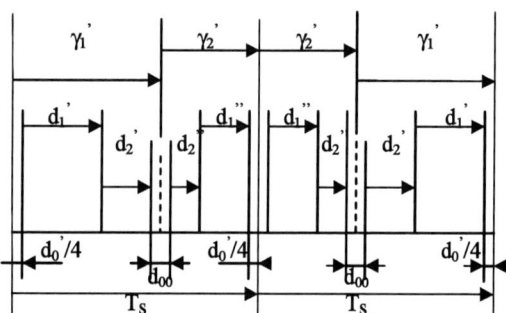

Fig.3. Switching sequences of duty-cycles

The presented in Fig.3 duty-cycles are calculated by

$$d_1' = \gamma_1' \cdot d_1, d_1'' = \gamma_2' \cdot d_1$$
$$d_2' = \gamma_1' \cdot d_2, d_2'' = \gamma_2' \cdot d_2 \tag{18}$$

A smoothly changing of the modulation index up to $m=1-\delta_0$ leads to the optimization of power losses of TDC, on one side, and, to appearance of additional distortions in output current, on the other side. As it is shown in below, based on simulation the anxiety concerning extra distortions is groundless. In case the required value $m \le 1$

- δ_0, the duty-cycle d_{00} of switching sequences can be simply replaced by $d_0/2$ of (14).

V. SIMULATION RESULTS

The two-stage conversion is realized in Matlab based on imitation like a mathematical procedure operating with vectors. All the simulation parameters concerning amplitudes of voltage, current, power and their spectra are normalized and presented in relative quantities. The amplitude A_i of spectrum harmonic after normalization is described by

$$A_i^N = A_i / \sum_{i=1,n} A_i$$

The parameters corresponding to the input voltage main and frequencies are following:

$$U_{ph}^i = 1; f_{in} = 25Hz; f_S = 2.5kHz; f_{out} = 60Hz.$$

The L-R circuit (R=1 Ω) is considered as a load with $\cos \varphi = 0.8$ at the frequency f=500 Hz. This value of frequency is selected for the better comparison of main distortions and their influence into input and output currents. To make the viewing process easier, only the low frequency parts of spectra are presented. For objective comparison of algorithms performances the non-distorted dc link current are used. This mode is performed by the synthesis of output voltage vector with frequency f_{out}= 0 Hz to eliminate any influence of output current distortions into input current. It can be considered as zero-invariance mode (ε=0) at the inversion stage. The compensation algorithm is applied for the unbalanced main mode to obtain stationary output power.

A. Balanced Main

Simulations results for conventional and proposed algorithms with regular and optimized switching procedure under balanced main are presented in Fig.4 - Fig.6. Analysis of presented results shows that:

1) The conventional modulation algorithms distorts dc link voltage more than proposed algorithms (Fig.4 (**b, h**)- Fig.6 (**b, h**)). The algorithm with ε=0.0008 brings distortions that are close to noise level.

2) The spectra of dc link power and input current amplitudes are similar that confirms the condition (6) (Fig.4 (**k, n**) - Fig.6 (**k, n**)). The contributions of 6^{ths} harmonics of input and output frequencies into input current are more significant. The returning of the input frequency distortions into input current approximately is 7 times less for algorithm with ε=0.0008 than for the conventional one (Fig.4 (**n**), Fig.6 (**n**)). The passing of output frequency distortions into input current is the same for all algorithms.

3) The spectrum of output current amplitude contains approximately 10 times less input frequency distortions for the proposed algorithm with ε=0.0008 than for conventional one (Fig.4 (**q**), Fig.6 (**q**)).

4) The optimized switching procedure unexpectedly improves the spectra of currents amplitudes by reducing contribution of 6th harmonic of output frequency. This side effect is more noticeable into output current (Fig.4 (**q, r**) - Fig.6 (**q, r**)), especially for the proposed modulation algorithms. For example, the spectrum of output current amplitude for optimized procedure is approximately becomes 10 times less and close to the noise level as it is shown in Fig.6 (**r**).

340

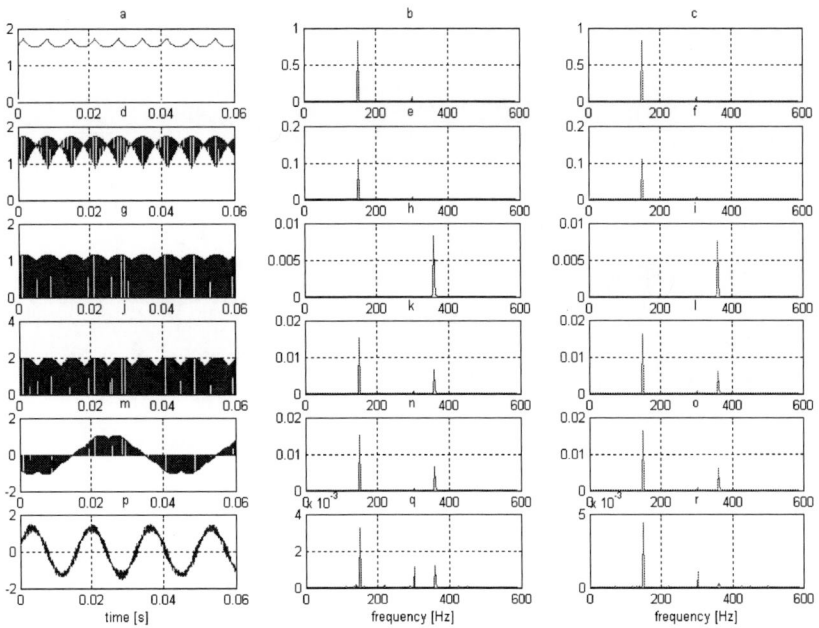

Fig.4 Simulation results for conventional algorithm with ε =0.155 under balanced main; (**a, d, j, m, p**) input and output normalized values and normalized spectra of their amplitudes without zero harmonics: (**b, e, h, k, n, q**) with regular and (**c, f, i, l, o, r**) with optimized switching procedure accordingly; (**a**) local average value of dc link voltage; (**d**) dc link voltage; (**g**) dc link current; (**j**) dc link power; (**m**) input current; (**p**) output current.

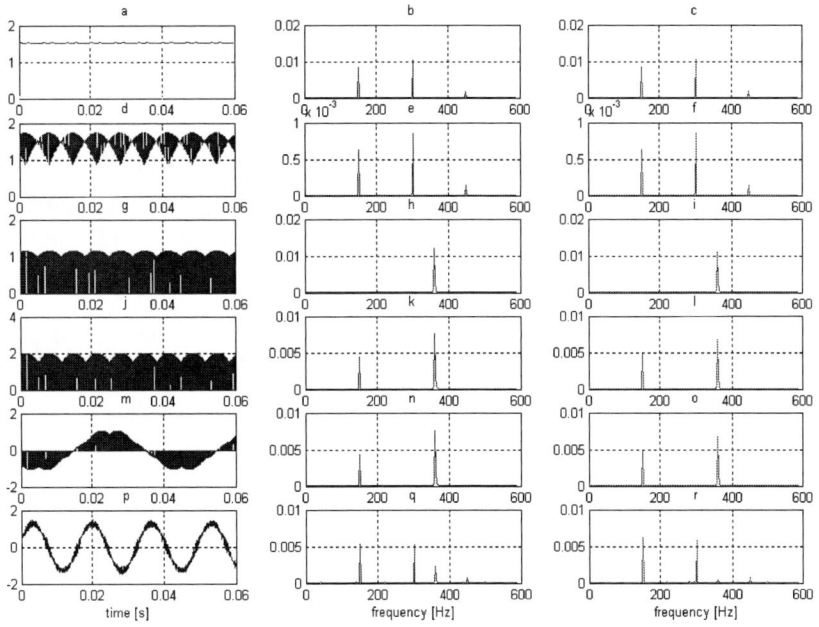

Fig.5 Simulation results for proposed algorithm with ε =0.04 under balanced main; (**a, d, j, m, p**) input and output normalized values and normalized spectra of their amplitudes without zero harmonics: (**b, e, h, k, n, q**) with regular and (**c, f, i, l, o, r**) with optimized switching procedure accordingly; (**a**) local average value of dc link voltage; (**d**) dc link voltage; (**g**) dc link current; (**j**) dc link power; (**m**) input current; (**p**) output current.

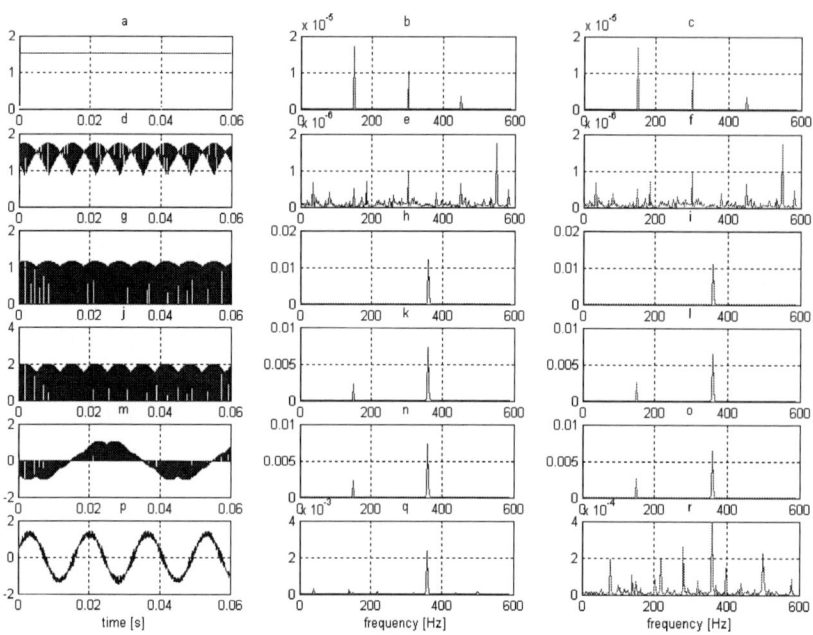

Fig.6 Simulation results for proposed algorithm with ε =0.0008 under balanced main; **(a, d, j, m, p)** input and output normalized values and normalized spectra of their amplitudes without zero harmonics: **(b, e, h, k, n, q)** with regular and **(c, f, i, l, o, r)** with optimized switching procedure accordingly; **(a)** local average value of dc voltage; **(d)** dc link voltage; **(g)** dc link current; **(j)** dc link power; **(m)** input current; **(p)** output current.

The efficiency of optimized procedure is evaluated based on averaging of index modulation m=1-δ₀ (with δ₀=0.01) during synthesis of output voltage and for applying parameters the average value is **m~** =0.9975, that is equal m =1-δ₀/4 for the regular switching procedure.

5) The influence of 12th harmonic of input frequency into output current is significant just for algorithms with ε=0.155 and ε=0.04 (Fig.4 **(q)**, Fig5 **(q)**).

B. Unbalanced Main

The unbalanced main is simulated based on the approach [7]. According to that approach the unbalanced input main is presented by symmetrical components

$$V_i^{'} = V_i + \Delta V_i^{*} = V_i \cdot \{e^{-j\omega_0 t} + \Delta e^{j\omega_0 t}\} \qquad (19)$$

where Δ is the normalized amplitude of conjugate component. The value Δ=0.06 is applied for simulation of unbalanced main. Simulations results for conventional and proposed algorithms under unbalanced main with output frequencies f_out=0 Hz and f_out=60 Hz are presented in Fig.7 - Fig.9.

Analysis of presented results shows:

1) The spectra of dc link power and input current amplitudes are not similar because of the appearing of the 2nd harmonic of input frequency into dc link. However, the condition (6) is provided because of input voltage and power both contain this harmonic too. The relative contributions of the 6th harmonics of input and output frequencies into spectrum of input current remain the same as with a balanced main (Fig.4 **(n)** –Fig.9 **(n)**).

2) The condition (7) is provided only for the proposed algorithms unlike for the conventional one as shown in Fig.7 **(e)** - Fig.9 **(e)**

3) The dc current mode at the TDC output provides the zero-invariance (ε =0) for the inversion stage. Due to this, mode all distortions caused by performance of modulation algorithm at the rectification stage is presented into output current (Fig.7(r), Fig.8(r)). The zero-invariance at the inversion stage in combination with ε-invariance (ε =0.0008) at the rectification stage allows achieve TDC current invariance as it is shown in Fig.9 **(o, r)**. The spectra of dc link and output currents amplitude are identical except the harmonics amplitudes due to the filtering in the load. This is the confirmation of providing the condition (8).

4) There is no 2nd harmonic of input frequency into output current due to the operating of algorithm compensation. In addition, operating of this algorithm causes the reducing of relative contributions of the 12th harmonic of input frequency into output currents. The performance of compensation algorithm can be evaluated by comparison of Fig.4-6 **(h)** and Fig.7-9 **(h)**.

5) The presented waveforms of input current are not filtered. The filtering at the input of TDC makes the spectrum of input current better.

VI. CONCLUSION

The invariant strategy in combination with the proposed algorithms provides the currents ε-invariance of TDC and minimizes the distortions into input and output currents.

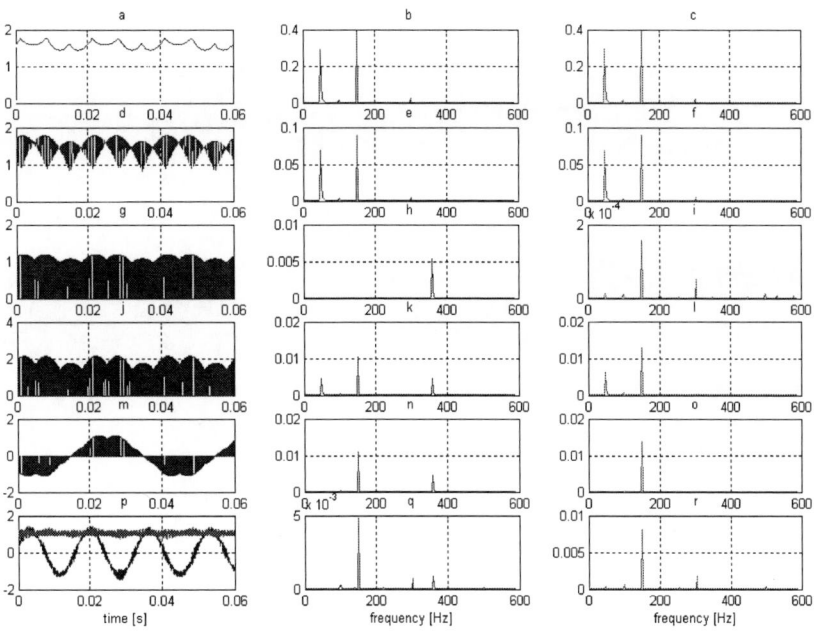

Fig.7 Simulation results for conventional algorithm with ε =0.155 under unbalanced main; (a, d, j, m, p) input and output normalized values and normalized spectra of their amplitudes without zero harmonics: (b, e, h, k, n, q) with output frequency f=60Hz and (c, f, i, l, o, r) with output frequency f=0 Hz accordingly; (a) local average value of dc link voltage; (d) dc link voltage; (g) dc link current; (j) dc link power; (m) input current; (p) output currents.

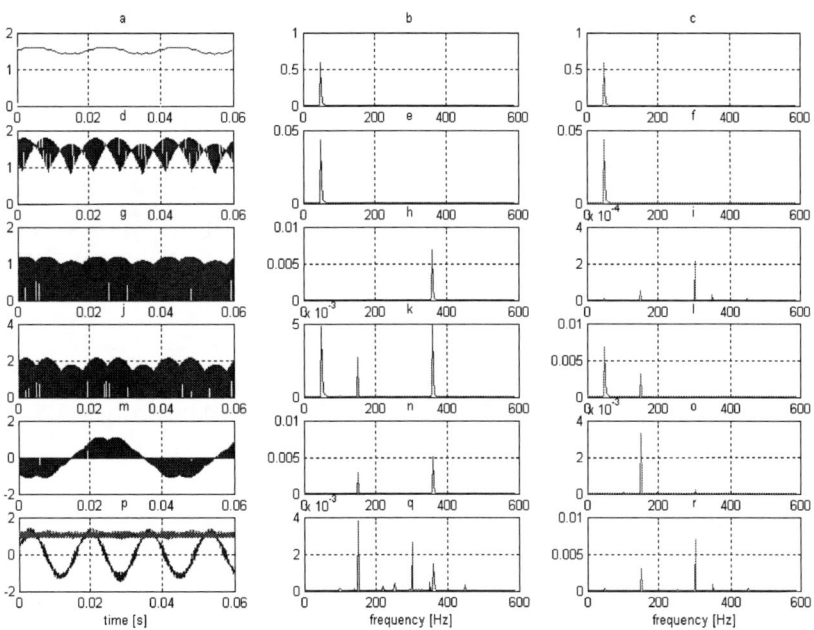

Fig.8 Simulation results for conventional algorithm with ε =0.04 under unbalanced main; (a, d, j, m, p) input and output normalized values and normalized spectra of their amplitudes without zero harmonics: (b, e, h, k, n, q) with output frequency f=60Hz and (c, f, i, l, o, r) with output frequency f=0 Hz accordingly; (a) local average value of dc link voltage; (d) dc link voltage; (g) dc link current; (j) dc link power; (m) input current; (p) output currents.

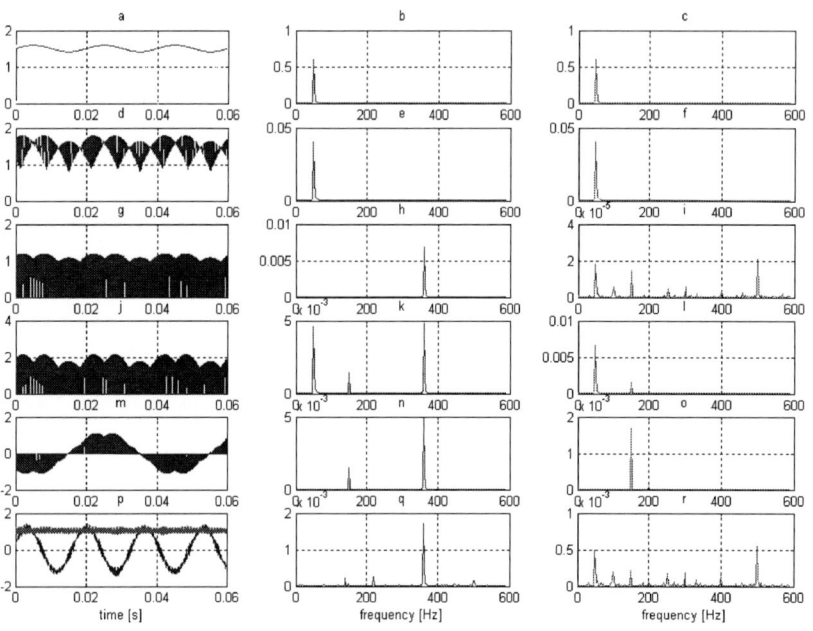

Fig.9 Simulation results for conventional algorithm with ε =0.0008 under unbalanced main: **(a, d, j, m, p)** input and output normalized values and normalized spectra of their amplitudes without zero harmonics: **(b, e, h, k, n, q)** with output frequency f=60Hz and **(c, f, i, l, o, r)** with output frequency f=0 Hz accordingly; **(a)** local average value of dc link voltage; **(d)** dc link voltage; **(g)** dc link current; **(j)** dc link power; **(m)** input current; **(p)** output currents

The ε-invariance is provided by two stages: rectification and inversion. The input current is more sensitive to distortions of rectification algorithm. The output current is more sensitive to distortions of inversion stage caused by performance of conventional space vector modulation (SVM). The output current is more sensitive to distortions of inversion stage caused by performance of conventional space vector modulation (SVM). This disadvantage of SVM algorithm can be partly improved by applying of optimized switching procedure mainly aimed to optimization of power losses.

The simulation has shown that the utilization of invariant modulation strategy with proposed algorithms allows getting a better performance of TDC under balanced and unbalanced input voltage mains than with conventional algorithm at the rectification stage.

The performance of the proposed modulation strategy is verified by simulation in ideal operating conditions such as the switching distortions are absent, the load is balanced and linear and the algorithm compensation is perfect. Obviously, that obtained results will be changed in real performance of TDC, but the main tendency will remain: the proposed invariant modulation strategy can be only achieved by employing the two-stage conversion. Thus, TDC has an undoubtedly advantage versa the conventional matrix converter.

ACKNOWLEDGMENT

The author does not have opportunity to perform experimental researches in laboratory environment, but he hopes that the others researchers will obtain experimental confirmation of the proposed modulation strategy and, thanks them in advance.

REFERENCES

[1] J.W.Kolar, M. Baumann, F. Schafmeister, H.Ertl "Novel Three-Phase AC-DC-AC Sparse Matrix Converter - Part I and Part II", *proc. APEC*, Vol .2, pp.777-791 (2002)

[2] J.W.Kolar, S.Rimond, F. Schafmeister,M. Heldwein, E.Pereira, L.Serpa " Comparison of Performance and Realization Effort of a Very Sparse Matrix Converter to a Voltage DC Link PWM Inverter with Active Front End", IEEJ Trans. IA, Vol.126, No.5, 2006

[3] C.Klumpner , F.Blaaberg "Two Stage Direct Power Converter: an Alternative to the Matrix Converter", IEE Seminar on (Digest No. 2003/10100), Volume, Issue, April 2003 -P: 7/1 - 7/9

[4] Chekhet E., Sobolev V., Mikhalsky V., Shapoval I., Polischuk S. Development Tendencies of Matrix Converters for the Induction Motor Drive // Bulletin of NTU "KhPI", Problems of the Automated Electrodrive. Theory and Application. – 2005, No.45. – P. 32-37.

[5] L.Wei, T.A.Lipo, H.Chan "Matrix Converter Topologies with Reduced Number of Switches", *IEEE Trans. Ind. Applicat.*, Vol.36, No.1, January/February 2000

[6] L. Franks, Signals Theory, Prentice-Hall, Inc.: Englewood Cliffs, NJ, 1969

[7] F.Blaaberg, D.Casadei, C.Klumpner, M.Matteini "Comparison of Two Current Modulation Strategies for Matrix Converter under Unbalanced Input Voltage Conditions", *IEEE Trans.Ind.Electron.* Vol.49, No.2, April 2002

 **Radiy Bekbudov**. He received Diploma of Electrical Engineer from St. Petersburg Electro-technical University "LETI", Russia, in 1977 and PhD in Electrical Engineering from Institute of Electrodynamics, Kiev, Ukraine, in 1997. From 1977 to 1988 he was a Research Engineer at the Central Design Office "Arsenal", Kiev. Since 1988 to 2001 he has been with the Department of Transformation and Stabilization of Electromagnetic Processes of Institute of Electrodynamics. His research interest is a robust control of direct power converters and electrical drives. Dr. Bekbudov is a member of IEEE, Canada.

Experimental Study of A Multicell ac/ac Converter Balancing Circuit

Robert Stala*, Andrzej Mondzik[†]

AGH University of Science and Technology, Department of Electrical Drive and Industrial Equipment,
Krakow, Poland, * e-mail: stala@agh.edu.pl, [†] e-mail: mondzik@agh.edu.pl

Abstract— The paper presents the experimental study of the balancing process forced by passive RLC circuit in a multicell ac/ac converter. To ensure its proper operation the converter employs voltages over the capacitors within its internal topology. Appropriate charging and reversing charge of cells' capacitors is achieved by means of the passive balancing circuit - a series RLC circuit connected at the converter output. The main issue is the correct choice of the balancing circuit wave impedance and the converter parameters (switching frequency, capacitances of cells capacitors) that allows maintaining a correct shape and proportion of voltages across cells capacitors at minimum balancing current.

Keywords—AC/AC converter, converter circuit, multilevel converters, converter control.

I. INTRODUCTION

A multicell ac/ac converter is a step-down, multilevel ac voltage controller (Fig. 1, Fig. 2). A conception of operation of the multicell ac/ac converter bases on the flying capacitors topology and is different than ac chopper [24] or cycloconverters. The ac/ac multicell converter can be applied as voltage regulator in general purposes of ac supply or for line voltage stabilization (Fig. 2).

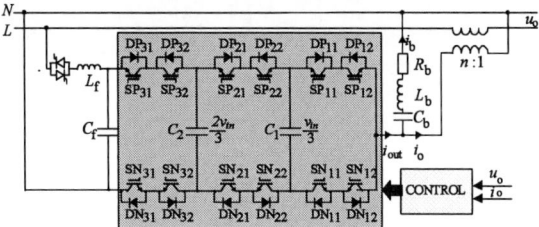

Fig. 2. Line voltage stabilization with use of multicell ac/ac converter

Fig. 1 shows the diagram of a three-cell ac/ac converter circuit and examples of the input voltage (v_{in}), output voltage (u_{out}) and voltages across cell capacitors (u_{C2}, u_{C1}) waveforms and control signals of the cell 3.

The multicell ac/ac converter operation is similar to that of a dc/dc converter [1], [2], [3], [5] but differs in closing the switches groups in half-cycle intervals, depending on the input voltage sign. The converter topology is configured according to the dc/dc circuits' topology, both for the positive and negative input voltage. The rms output voltage in a multicell ac/ac converter is controlled by varying the duty factor D of the active switches control pulses [1], [2], [3].

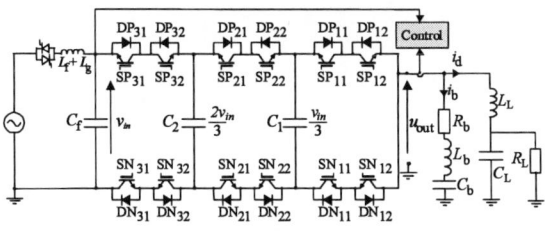

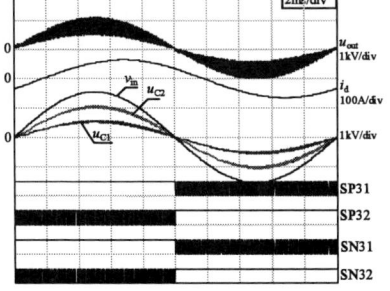

Fig. 1. A single-phase three-cell ac/ac converter circuit; waveforms of the input voltage (v_{in}), output voltage (u_{out}), voltages across cells' capacitors (u_{C2}, u_{C1}), load current (i_d) and control signals of the cell 3 (SP31, SP32, SN31, SN32)

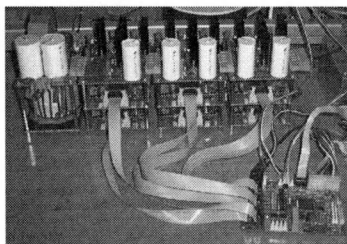

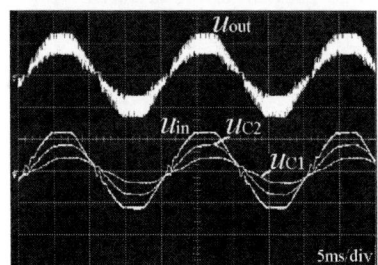

Fig. 3. The laboratory setup of a three-cell ac/ac converter with FPGA controller; the measurements results: output voltage (u_{out}), input voltage (v_{in}), voltages across cells' capacitors (u_{C2}, u_{C1}), f_0=7052Hz, D=0.82

978-1-4244-1741-4/08/$25.00 ©2008 IEEE 345

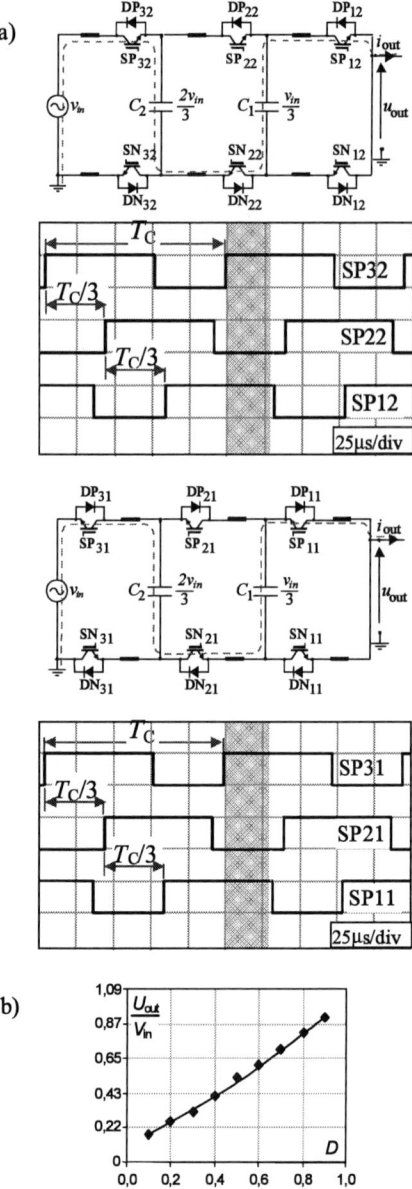

a)

b)

Fig. 4. (a) the modulation conception in the ac/ac three-cell converter – active switches and example of the current path for positive and negative input voltage, (b) relative rms value of the output voltage vs. duty cycle ($D=t_{on}/T_C$)

The control functions for a given group of switches (P or N) are shifted by 1/3 of a cycle. In multicell converter topologies, both dc/dc and ac/ac ones, following cell source voltages ratio should be maintained:

$$V_n=(nV_{in})/N \qquad (1)$$

where: n – the cell index, N – the number of cells in the converter.

The balancing process allows maintaining a constant ratio of cells capacitors voltages with respect to the input voltage in an ac/ac multicell converter in entire period of the input voltage [1], [2], [3].

In case of improper voltage sharing on the cells' capacitors: $u_{Cn}\neq (nV_{in})/N$, the maximum voltages across switches increase above their rated values. The modulation of the output voltage can be also improper. Stabilization of the cells' capacitors voltages (u_{Cn}) at the appropriate level can be reached by means of a series RLC circuit connected to the converter output terminals, i.e. in parallel with load ($R_bL_bC_b$ – the balancing circuit) [1], [2], [3], [4], [5], [7]. This method does not involve any additional measurements or specific control.

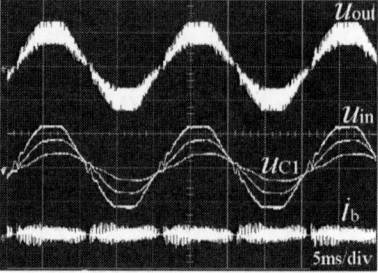

Fig. 5. The three-cell ac/ac converter— the measurement results of the output voltage, the cells' capacitors voltages and balancing current waveforms under no-load conditions

The resonant frequency of the balancing circuit should be equal to the switching frequency of a single cell (f_C):

$$f_0=f_C \qquad (2)$$

where: f_C – frequency of a single cell control,
f_0 – resonant frequency of the balancing circuit.

Output voltage (u_{out}) has ac component of $3f_C$ frequency (Fig. 6, Fig. 7).

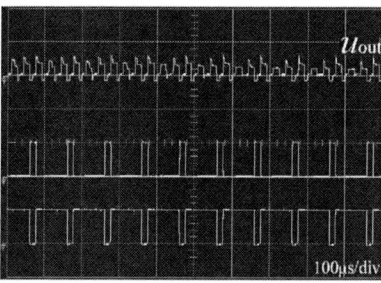

Fig. 6. The three-cell ac/ac converter— the measurement results of the output voltage the example of one active cell control signals SP and SN for $f_C=f_0$, under no-load conditions

Fig. 7 presents an example of output voltage (u_d) in case of proper and improper cells' voltages. The difference of those states determines the ac component of f_C frequency ($u_{d(b)}$). When $u_{Cn}\neq(mV_d)/n$, an additional component ($u_{d(b)}$) of f_C frequency arises in the output voltage. The $u_{d(b)}$ component makes current flow trough the balancing circuit (i_b) and the cells' capacitors.

In the ac/ac converter the input voltage varies causing unbalance state but the f_C component of the balancing current disappears near amplitude of the signal (Fig. 8, Fig. 9).

346

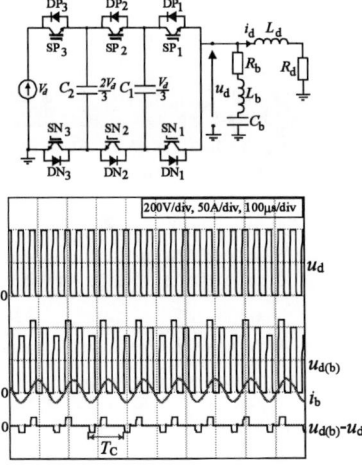

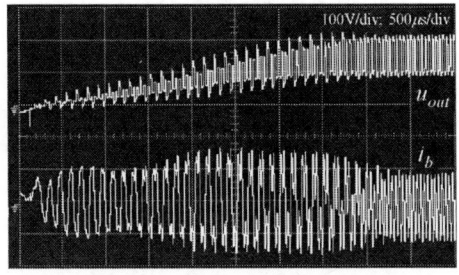

Fig. 7. (a) three-cell dc/dc converter with cells' capacitors and balancing circuit, (b) IsSpice waveforms output voltage under balance conditions (u_{C1}=400V, u_{C2}=800V) (u_d), unbalance conditions (u_{C1}=350V, u_{C2}=750V) ($u_{d(b)}$), balancing current (i_b) and the difference between output voltage in these two states (u_d-$u_{d(b)}$)

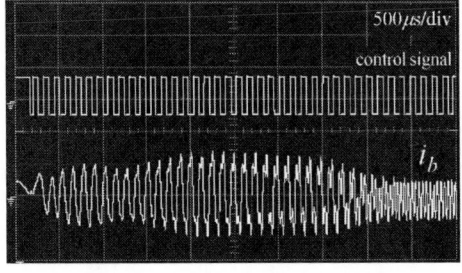

Fig. 8. The three-cell ac/ac converter— the measurement results of the output voltage and the balancing current at first 5ms of the input voltage period in 230V, 50Hz system.

Fig. 9. The three-cell ac/ac converter— the measurement results of one active control signal and the balancing current at first 5ms of the input voltage period in 230V, 50Hz system.

In [1], [3] it is proven that voltage variation on the cells' capacitors C_n is as follows:

$$\frac{du_1}{dt} = \frac{1}{C_1}\left(i_b + i_d\right)\left(SP2 - SP1\right) \qquad (3)$$

$$\frac{du_2}{dt} = \frac{1}{C_2}\left(i_b + i_d\right)\left(SP3 - SP2\right) \qquad (4)$$

Fig. 10 presents waveforms in case of active SP1 and SP2 signals and their difference. Fig. 11 presents spectrum of the (SP2-SP1) signal and spectrum of (i_b+i_d) current under unbalance conditions. From results presented in Fig. 11 it follows that f_C component exists in the spectrum of the (SP2-SP1) signal and the output current (under unbalance conditions). Thus, according to (3) and (4) under unbalance state the average value of the flying capacitors voltages changes. The f_C component in the output current is forced by the f_C component in the output voltage. In [1] and [3] it is proven analytically that under unbalance state i.e. when the voltage sharing on the flying capacitors is different than in relation (1), the f_C component arise in the output voltage of the 3-cell converter.

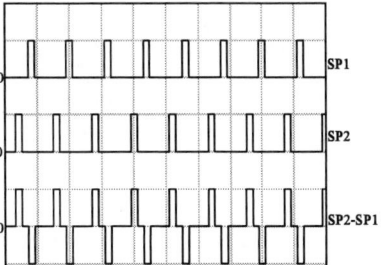

Fig. 10. Control signals SP2, SP1 and their difference (SP2-SP1);

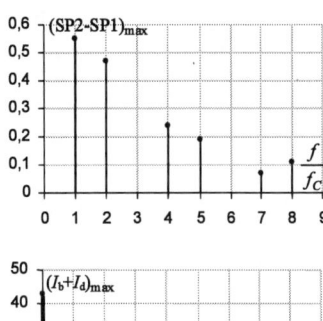

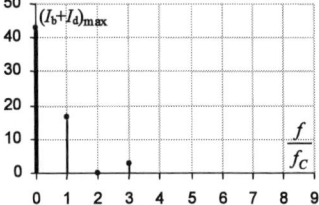

Fig. 11. Spectrum of (i_b+i_d) for exemplary unbalance conditions.

II. EFFECTIVENESS OF THE BALANCING PROCESS FORCED BY PASSIVE RLC CIRCUIT

The effectiveness of the balancing with passive RLC circuit in a multicell ac/ac converter depends on the balancing circuit parameters, the converter parameters (cell capacitors capacitances) and the converter operating conditions. Some tests of those problems were presented in [3]. Tab. I and Tab. II present results of measurements of the balancing process in 230V/50Hz system in cases of different wave impedances of the balancing circuit. The tests were made at nearly the same switching frequency, the same duty cycle (D=0.48) and flying capacitors capacitances (C_C=8µF). In all cases the switching frequency was adjusted to the resonance frequency of the balancing circuit (f_C=f_0).

Once the capacitances of cell capacitors have been determined for a given range of load and the output voltage control (duty cycle D range), the next step is the choice of the balancing circuit parameters, i.e. its resonance frequency and wave impedance. The resonance frequency determines the switching frequency, whereas the wave impedance impacts the rate of energy transfer from the balancing circuit to cells' capacitors [3].

TABLE I.
INFLUENCE OF THE BALANCING CIRCUIT WAVE IMPEDANCE ON STABILIZATION OF THE CELL CAPACITORS VOLTAGES. NO-LOAD CONDITIONS, SWITCHING FREQUENCY: f_C=f_0

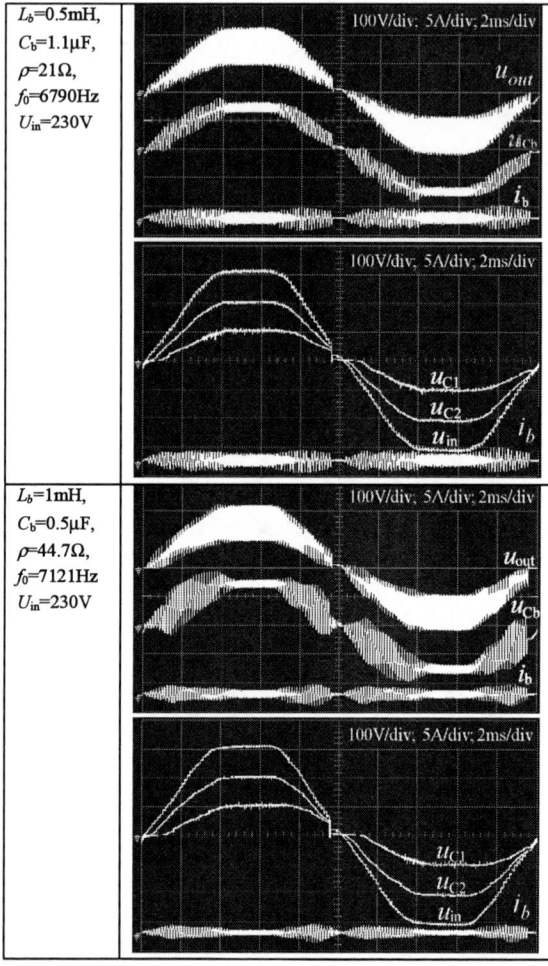

The proper choice of the balancing circuit wave impedance is important for the correct operation of the balancing circuit. From the results of measurements carried out for selected values of wave impedance of the balancing circuit, the following conclusions can be derived:

- for resonance frequency range $f_0 \approx$ 7kHz and within the $\rho \in$ (20Ω ÷ 40Ω) the voltages across the cell capacitors maintain an approximate shape of the input voltage (v_{in}) (Tab. I),

- increasing the balancing circuit wave impedance (increasing L_b and decreasing C_b) causes delay of the voltages on the flying capacitors in relation to the input voltage. The delay is well apparent in the waveforms presented in Tab. II (ρ=81.6Ω),

- increasing the balancing circuit wave impedance reduces the balancing current but causes important increasing amplitude of the f_C component in voltage on the balancing circuit capacitor (u_{Cb}).

TABLE II.
THE BALANCING PROCESS AT WAVE IMPEDANCE OF THE BALANCING CIRCUIT ρ= 81.6Ω. NO-LOAD CONDITIONS, SWITCHING FREQUENCY: f_C=f_0

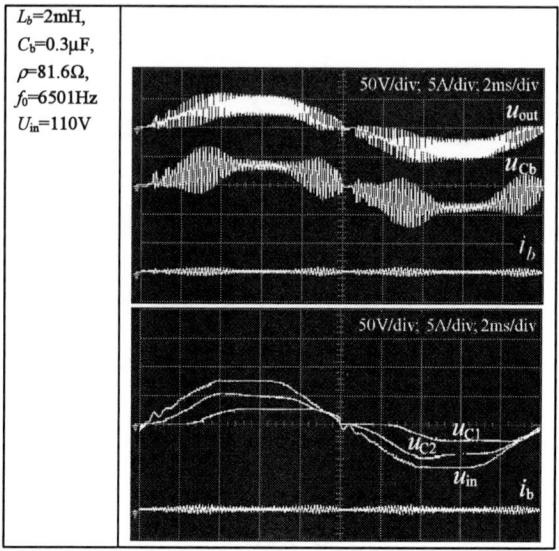

III. CONCLUSIONS

The balancing circuit in a multicell ac/ac converter should ensure a constant ratio of cells capacitors voltages to the ac input voltage. It is a series RLC circuit with resonance frequency equal to the switching frequency of a single cell (f_C), connected at the converter output. The balancing current flow is a consequence of the input voltage variations, since they result in the switching frequency component in the output voltage.

An improperly chosen switching frequency, i.e. different from the balancing circuit resonant frequency, reduces the effectiveness of the balancing process. This results in an improper voltage sharing between cells' capacitors, what causes increase of maximum values of across the switches and also change the shape of modulated output voltage.

Choosing the low wave impedance of the balancing circuit has important advantages. In such case the flying capacitors voltages have proper shape. Besides, the size and cost of the balancing circuit choke can be reduced. Amplitude of the f_C component in voltage on the balancing circuit capacitor also decreases when wave impedance of the balancing circuit decreases. Increasing the balancing circuit wave impedance reduces the balancing current value.

REFERENCES

[1] T. A. Meynard, H. Foch, P. Thomas, J. Courault, R. Jakob, M. Nahrstaedt, "Multicell Converters: Basics Concepts and Industry Applications". *IEEE Trans. Ind. Electron.* Vol. 49, pp. 955-964, Oct. 2002.

[2] T. A. Meynard, H. Foch, F. Forest, C. Turpin, F. Richardeau, L. Delmas, G. Gateau, T.A. Lefeuvre, "Multicell Converters: Derived Topologies". *IEEE Trans. Ind. Electron.* Vol. 49, pp. 978-987, Oct. 2002.

[3] S. Pirog S., R. Stala, "Selection of Parameters for a Balancing Circuit of DC-DC and AC-AC Multicell Converters". *11th European Conference on Power Electronics and Applications*, 11-14 September 2005 – Dresden, Germany, CD Proceedings.

[4] R. H. Wilkinson, T. A. Meynard, H. du Toit Mouton, "Natural Balance of Multicell Converters: The Two-Cell Case", IEEE Trans. Power Electron. Vol. 21, pp. 1649 - 1657, Nov. 2006.

[5] S. Pirog, M. Baszynski, J. Czekonski, S. Gasiorek, A. Mondzik, A. Penczek, R. Stala, "Multicell DC/DC Converter with DSP/CPLD Control. Practical Results", 12th International Power Electronics and Motion Control Conference (EPE-PEMC 2006), Aug. 2006, pp: 677 - 682

[6] B. P. McGrath, T. Meynard, G. Gateau, D. G. Holmes, "Optimal Modulation of Flying Capacitor and Stacked Multicell Converters Using a State Machine Decoder", IEEE Trans. Power Electron. Vol. 22, pp. 508-516, March 2007.

[7] H. du Toit Mouton, "Natural Balancing of Three-Level Neutral-Point-Clamped PWM Inverters", *IEEE Trans. Ind. Electron.* Vol. 49, pp. 1017-1025, Oct. 2002.

[8] G. Gateau, M. Fadel, P. Maussion, R. Bensaid, T. A. Meynard, "Multicell Converters: Active Control and Observation of Flying-Capacitor Voltages". *IEEE Trans. Ind. Electron.* Vol. 49, pp. 998-1008, Oct. 2002.

[9] C. Turpin, P. Baudesson, F. Richardeau, F. Forest, T. A. Meynard, "Fault Management of Multicell Converters". *IEEE Trans. Ind. Electron.* Vol. 49, pp. 988-997, Oct. 2002.

[10] Xiaoming Yuan, H. Stemmler, and I. Barbi, "Self-Balancing of the Clamping-Capacitor-Voltages in the Multilevel Capacitor-Clamping-Inverter under Sub-Harmonic PWM Modulation", *IEEE Trans. Power Electron*, Vol. 16, pp. 256-263, March 2001.

[11] Dae-Wook Kang, Byoung-Kuk Lee, Jae-Hyun Jeon, Tae-Jin Kim, and Dong-Seok Hyun, "A Symmetric Carrier Technique of CRPWM for Voltage Balance Method of Flying-Capacitor Multilevel Inverter, *IEEE Trans. Ind. Electron.*, Vol. 52, pp. 879-888, June 2005.

[12] L. Zhang and S.J. Watkins, "Capacitor voltage balancing in multilevel flying capacitor inverters by rule-based switching pattern selection", IET Electr. Power Appl., 2007, 1, (3), pp. 339-347

[13] F. Forest, T. A. Meynard, S. Faucher, F. Richardeau, Jean-Jacques Huselstein, and C. Joubert, "Using the Multilevel Imbricated Cells Topologies in the Design of Low-Power Power-Factor-Corrector Converters", *IEEE Trans. Ind. Electron.*, Vol. 52, pp. 151-161, Feb. 2005.

[14] Miguel F. Escalante, Jean-Claude Vannier, and Amir Arzandém "Flying Capacitor Multilevel Inverters and DTC Motor Drive Applications", IEEE TRANSACTIONS ON INDUSTRIAL ELECTRONICS, VOL. 49, NO. 4, AUGUST 2002.

[15] Xiaomin Kou, Keith A. Corzine, and Yakov L. Familiant, "A Unique Fault-Tolerant Design for Flying Capacitor Multilevel Inverter", IEEE TRANSACTIONS ON POWER ELECTRONICS, VOL. 19, NO. 4, JULY 2004.

[16] Chen Zhao Junming Zhang Xinke Wu Zhaoming Qian, "The Analysis of the Charge Unbalance in Flying Capacitors of a Novel Three-level ZVS Converter", EEEE PEDS 2005.

[17] Bor-Ren Lin and Chun-Hao Huang, "Implementation of a Three-Phase Capacitor-Clamped Active Power Filter Under Unbalanced Condition", IEEE TRANSACTIONS ON INDUSTRIAL ELECTRONICS, VOL. 53, NO. 5, OCTOBER 2006.

[18] Dietmar Krug, Steffen Bernet, Seyed Saeed Fazel, Kamran Jalili, and Mariusz Malinowski, "Comparison of 2.3-kV Medium-Voltage Multilevel Converters for Industrial Medium-Voltage Drives", IEEE TRANSACTIONS ON INDUSTRIAL ELECTRONICS, VOL. 54, NO. 6, DECEMBER 2007.

[19] R. Ruelland, G. Gateau, T. A. Meynard, J.-C. Hapiot, "Design of FPGA-based emulator for series multicell converters using co-simulation tools", *IEEE Trans. Power Electron.* Vol. 18, pp. 455 - 463, Jan. 2003.

[20] T. A. Meynard, M. Fadel, and N. Aouda, "Modeling of Multilevel Converters", IEEE TRANSACTIONS ON INDUSTRIAL ELECTRONICS, VOL. 44, NO. 3, JUNE 1997.

[21] Ó. López, J. Álvarez, J. Doval-Gandoy, F. D. Freijedo, A. Nogueiras, A. Lago, and C. M. Peñalver, "Comparison of the FPGA Implementation of Two Multilevel Space Vector PWM Algorithms", IEEE TRANSACTIONS ON INDUSTRIAL ELECTRONICS, VOL. 55, NO. 4, APRIL 2008.

[22] M. Aime, G. Gateau, and T. Meynard. "Implementation of a peak current control algorithm within a FPGA". IEEE Transactions on Industrial Electronics, Vol. 53, NO. 6, NOV. 2006.

[23] D. H. Jang, G. H. Choe, "Improvement of Input Power Factor in AC Choppers Using Asymmetrical PWM Technique", IEEE TRANSACTIONS ON INDUSTRIAL ELECTRONICS, VOL. 42, NO. 2, APRIL 1995,

[24] J. H. Kim, B. D. Min, B. H. Kwon and S. C. Won, "A PWM Buck–Boost AC Chopper Solving the Commutation Problem", IEEE TRANSACTIONS ON INDUSTRIAL ELECTRONICS, VOL. 45, NO. 5, OCTOBER 1998.

[25] J. H. Youm, and B. H. Kwon, "Switching Technique for Current-Controlled AC-to-AC Converters", IEEE TRANSACTIONS ON INDUSTRIAL ELECTRONICS, VOL. 46, NO. 2, APRIL 1999.

[26] P. W. Wheeler, J. C. Clare, M. Apap, and K. J. Bradley, "Harmonic Loss Due to Operation of Induction Machines From Matrix Converters", IEEE TRANSACTIONS ON INDUSTRIAL ELECTRONICS, VOL. 55, NO. 2, FEBRUARY 2008.

[27] D. Chen, "Novel Current-Mode AC/AC Converters With High-Frequency AC Link", IEEE TRANSACTIONS ON INDUSTRIAL ELECTRONICS, VOL. 55, NO. 1, JANUARY 2008.

A Comparison and Optimum Design of Reluctance-Controlled Classical Load-Resonant Converters

Stefan V. Mollov[*] and Michael P. Theodoridis[†]

[*] ADENEO, ADETEL Group, 69136 Ecully, France, e-mail: *S.V.Mollov@gmail.com*
[†] Technological Educational Institute of Athens, Agiou Spiridonos & Milou, Greece, e-mail: *M.P.Theodoridis@gmail.com*

Abstract— **Fundamental frequency analysis is used to obtain the steady-state operation of the three most popular load resonant converters, when operated with limited frequency range due to a variable inductance. The power throughput is regulated by electronically varying the reluctance of the resonant inductor. The developed design equations indicate that the converters operate efficiently in the region of maximum control gain, with a phase around 35 degrees. The validity of the proposed design equations and the identified optimum operating conditions are confirmed through time-domain simulations and measurements from a 200W, 500kHz prototype of a series-parallel resonant converter.**

Keywords— **Resonant converter, Passive component integration, Converter control.**

I. INTRODUCTION

Load-resonant converters [1] have distinct advantages when operation at high frequency is required. The current of the resonant tank naturally reverses polarity, enabling soft commutation, Zero Voltage Switching (ZVS) when commutation is above the load-resonance and Zero Current Switching when commutation is below the load-resonance.

There are two principal options for controlling the power throughput of these converters. One popular way is to modulate the amplitude of the harmonics introduced to the resonant tank. This is achieved either by modulating the duty ratio of the commutating pole (usually implemented in half-bridge configurations) [2] or by introducing a phase-shift between the commutating poles in a full-bridge converter [3]. The advantage of this control scheme is the constant frequency operation – the generated harmonic content is predictable and easy to suppress. In addition, the device gate drivers are easier to design. In both cases however, soft commutation and efficient operation are difficult to maintain if the load or the supply voltage changes are to be responded to.

We therefore turn our attention to the more conventional type of power control, whereby the switching frequency is modulated. Deviation of this frequency from the load-resonant frequency results in diminishing voltage and power transfer [4], without the loss of favourable soft commutation. Unfortunately, wide frequency variation is required to keep the output voltage constant, ultimately impeding the design of efficiently operating resonant topologies.

In this work we propose to modify the load-resonant frequency in order to perform the power control. Although the switching frequency is kept constant, the effect of modulating the value of any of the resonant components will be identical to that of the conventional frequency control. Either reactive component can be potentially controlled, however, in order to obtain a fast-acting closed-loop control, the value of the respective resonant component needs to be controlled electronically, rather than mechanically. The capacitance variation is thus seen unfeasible for control purposes and for the time being. We therefore turn our attention to controlled inductive components.

A recent development is the advent of thin-film inductive components for power electronic applications [5,6]. The component core is built from micro-machined thin-film laminations. The use of thin-films is prompted by their favorable magnetic properties, particularly the lack of hysteresis loss [7], and results in high-density efficient inductors and transformers. A property of some frequently used thin-films is the ability to control the effective permeability, by introducing a DC field in the direction of the easy axis of magnetization [8], which can be provided by a bias winding wrapped around the thin-film component. Due to the high power density of these devices, the inductance of the bias winding is rather small, making it relatively easy to modify the inductivity of the magnetic component, with fast response times.

The present work does not develop the design of this type of components. Rather, we intent to identify the design considerations for the most popular load-resonant converters employing controllable inductors. The aim is to identify the values of the resonant tank components, for which changing the inductor value results in the largest voltage ratio (output to input voltage) deviation and to evaluate the resulting steady-state performance.

II. SERIES RESONANT CONVERTER

The equivalent fundamental frequency circuit is shown in Figure 1, the arrow indicating the fact that the resonant inductor is variable. The half-bridge voltage is substituted with an equivalent sinusoidal voltage source with amplitude $2V_{DC}/\pi$ and the rectifier-filter-load combination – with an effective resistance $R_{EFF} = 8R/\pi^2$.

978-1-4244-1741-4/08/$25.00 ©2008 IEEE

$$v_{in} = \frac{2}{\pi} V_{DC} \sin(\omega t)$$

Figure 1. Series Resonant converter fundamental frequency circuit with variable inductor.

The voltage conversion ratio comes from the conventional expression (for full-bridge this expression would be multiplied by 2):

$$M = \frac{4R\omega C}{\sqrt{64R^2\omega^2 C^2 + \pi^4 (\omega^2 LC - 1)^2}} \quad (1)$$

The voltage ratio $M = V_o / V_{DC}$ is plotted against L in Figure 2. Here the switching frequency ω and the resonant capacitor C have assumed values, kept constant. The x-axis is normalized to such value of resonant inductor, L_o, for which the switching frequency equals the resonant frequency $\omega_0 = \frac{1}{\sqrt{L_0 C}}$. The quality factor is defined at this inductor value as $Q = \frac{\omega_0 L_0}{R}$.

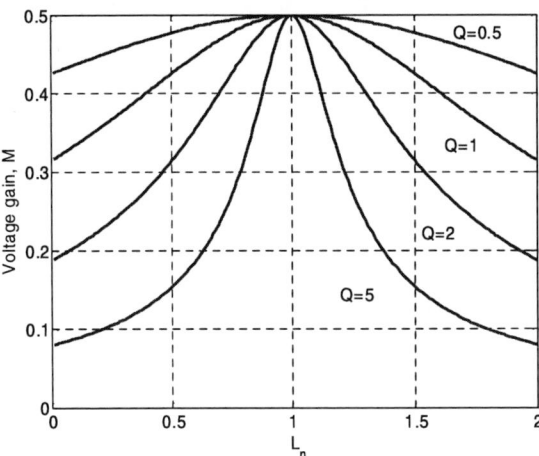

Figure 2. Variation of the Voltage conversion ratio M with resonant inductor L.

The curves look very much the same as these for the Series Resonant Converter (SRC) with frequency control. High Q-factor defines steeper decrease of voltage ratio, as operation is forced away from resonance. Very far from resonance, the converter becomes uncontrollable, as large variation of L does not result in conceivable variation in M. This is more pronounced for lighter loading, low Q.

The first derivative of (1) with respect to L will give the low frequency control gain:

$$\frac{\partial M}{\partial L} = -\frac{4\omega^3 RC^2 \pi^4 (LC\omega^2 - 1)}{\left(64\omega^2 R^2 C^2 + \pi^4 (LC\omega^2 - 1)^2\right)^{3/2}} \quad (2)$$

The second derivative, solved for L, will indicate the conditions for which the smallest change of L results in

the largest deviation of M, this is the largest low-frequency control gain:

$$\frac{\partial^2 M}{\partial L^2} = \frac{8\pi^4 \omega^5 C^3 \left(\pi^4 (\omega^2 LC - 1)^2 - 32\omega^2 R^2 C^2\right)}{\left(\pi^4 (\omega^2 LC - 1)^2 + 64\omega^2 R^2 C^2\right)} = 0 \quad (3)$$

The solutions to (3) are:

$$L_1 = \frac{\pi^2 + 4\sqrt{2}\omega RC}{\pi^2 \omega^2 C} \quad (4a)$$

$$L_2 = \frac{\pi^2 - 4\sqrt{2}\omega RC}{\pi^2 \omega^2 C} \quad (4b)$$

The solutions in (4) are then substituted into (1) to identify the corresponding values for the resonant capacitor. For this converter the result is $M = 1/\sqrt{6}$, indicating that any value of the resonant capacitor will satisfy (1) and (4). Close inspection of the curves in Figure 2 shows that they exhibit an inflection point (the curve changes from convex to concave) at this value of $M = 1/\sqrt{6}$. There are two solutions to (3) due to the fact that maximum gain occurs above resonance (eq.(4a)) and below resonance (eq.(4b)) at this same level of voltage ratio.

The low-frequency gain can be computed by substituting (4) into (2):

$$LFG_{SRC} = -\frac{\sqrt{3}\pi^2}{72} \frac{\omega}{R} \quad (5)$$

The expression in (5) has a positive sign for the solution in (4b), indicating operation below resonance.

The gain is inversely proportional to the load resistance. This is explained with the fact that heavier loads result in steeper voltage ratio curves (Figure 2). The gain is also proportional to the switching frequency. This is to be expected, as the same deviation of L will detune the resonant tank more if the operating frequency is higher. The third characteristic of (5) is constant (for given R and ω) and this is due to the fact that it always occurs for constant $M = 1/\sqrt{6}$.

The impedance, loading the equivalent voltage source in Figure 1 is :

$$Z = j\omega L + \frac{1}{j\omega C} + R_{eff} \quad (6)$$

The solutions from (4) are substituted into (6) to identify the power factor of the resonant tank. It appears to be constant and independent of the values of the resonant components. As long as the relationships in (4) are met, the angle is $\varphi = \arctan(1/\sqrt{2}) \approx 35°$.

The above angle is reasonably large, as it would normally allow adequate energy for the recharge of snubber capacitors in Zero Voltage Switching mode. Simultaneously, its value is sufficiently low, so that only small reactive energy is circulated in the resonant tank, therefore we could expect reasonably high converter efficiency and benign ratings for the resonant components.

III. PARALLEL RESONANT CONVERTER

The fundamental frequency equivalent circuit of the inductor-controlled Parallel Resonant Converter (PRC) is identical to that of the conventional PRC. The half-bridge voltage is denoted by a source $v_{in} = V_{DC}\dfrac{2}{\pi}\sin\omega t$ and an equivalent resistor $R_{eff} = R\dfrac{\pi^2}{8}$ substitutes the rectifier-load combination.

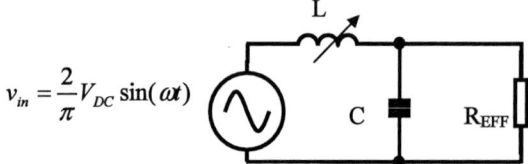

$$v_{in} = \frac{2}{\pi} V_{DC}\sin(\omega t)$$

Figure 3. Equivalent fundamental frequency circuit of inductor-controlled PRC.

The voltage conversion ratio is derived in (7):

$$M = \frac{4}{\sqrt{\pi^4 - \pi^4 2LC\omega^2 + \pi^4 L^2 C^2\omega^4 + \dfrac{64\omega^2 L^2}{R^2}}} \tag{7}$$

The above expression is plotted against the resonant inductor in Figure 4. The plot is generated in the same way as for Figure 2, however the quality factor now is

$$Q = \frac{R}{\omega_0 L_0}$$

Figure 4. Voltage conversion ratios of the inductor-controlled PRC.

The largest voltage ratios are no longer closely coupled to the resonant frequency, as it is for the SRC. The load-resonant frequency for this converter is modified by the load resistor. Low Q-factors (heavy loading) reduce the load-resonant frequency and the selectivity of the resonant tank, whereby the maximum voltage ratio occurs for normalized inductances smaller than unity.

Moreover, most of the voltage ratio range is above 0.5, indicating the tendency of the PRC to operate as a step-up converter. Higher Q-factors result in steeper slopes.

The first derivative of the voltage conversion ratio with respect to the resonant inductor is:

$$\frac{\partial M}{\partial L} = -\frac{4R\omega^2\left(64L + \pi^4 R^2 C\left(\omega^2 LC - 1\right)\right)}{\left(\pi^4\omega^4 L^2 C^2 R^2 - 2\pi^4\omega^2 LCR^2 + 64\omega^2 L^2 + \pi^2 R^2\right)^{3/2}} \tag{8}$$

The procedure for identifying the values of the resonant components for maximum control gain is as follows. The second derivative of the expression for the voltage ratio is equated to zero and solved for the resonant inductor L. This solution is then substituted into (7) and solved for C to calculate the values of the resonant capacitor. At this stage the expression for C is only a function of ω, R and M, see (9). This solution is then substituted into the expression for the resonant inductor obtained from the second derivative.

This results in the following real positive values for the resonant components:

$$C = \frac{\sqrt{2}\sqrt{3M^2\pi^4 - 32}}{\pi^2 R\omega} \tag{9}$$

$$L_1 = \frac{\sqrt{3M^2\pi^4 - 32} + 4}{3\sqrt{2}\pi^2 M^2}\frac{R}{\omega} \tag{10a}$$

$$L_2 = \frac{\sqrt{3M^2\pi^4 - 32} - 4}{3\sqrt{2}\pi^2 M^2}\frac{R}{\omega} \tag{10b}$$

The above resonant components are real and positive for voltage ratios bigger than $M > 4/\pi^2 \approx 0.41$. This is attributed to the tendency of the PRC to operate as a step-up converter, which is practically always capable of providing voltage ratios of at least $M = 0.5$.

Equation (10) indicates that to maintain a predetermined voltage conversion ratio, the resonant inductor should follow the load resistance. This opens an interesting possibility for control. Provided the inductor is so designed, that its inductance linearly depends on the bias current, the converter will automatically respond to load changes, when the bias winding is placed in series with the load.

Figure 5 demonstrates the steady-state performance of the PRC using the values for the resonant components as calculated in (9) and (10), which are the operating points for which maximum control gain is attained. The normalized frequency $\omega_n = \omega\sqrt{LC}$ becomes

$$\omega_{n1} = \frac{\sqrt{4 + \sqrt{3M^2\pi^4 - 32}}\left(3M^2\pi^4 - 32\right)^{1/4}}{\sqrt{3}M\pi^2} \tag{11a}$$

$$\omega_{n2} = \frac{\sqrt{-4 + \sqrt{3M^2\pi^4 - 32}}\left(3M^2\pi^4 - 32\right)^{1/4}}{\sqrt{3}M\pi^2} \tag{11b}$$

and is plotted in the top trace in Figure 5 (eq.(11b) is for the resonant inductor L_2).

The bottom trace plots the Q-factor $Q = R/\sqrt{\dfrac{L}{C}}$ (eq.(12b) is for the resonant inductor L_2):

$$Q = \frac{\sqrt{6}M\left(3M^2\pi^4 - 32\right)^{1/4}}{\sqrt{4 + \sqrt{3M^2\pi^4 - 32}}} \tag{12a}$$

$$Q = \frac{\sqrt{6}M\left(3M^2\pi^4 - 32\right)^{1/4}}{\sqrt{-4 + \sqrt{3M^2\pi^4 - 32}}} \tag{12b}$$

The broken lines are using L_2 for the solution for the resonant inductor.

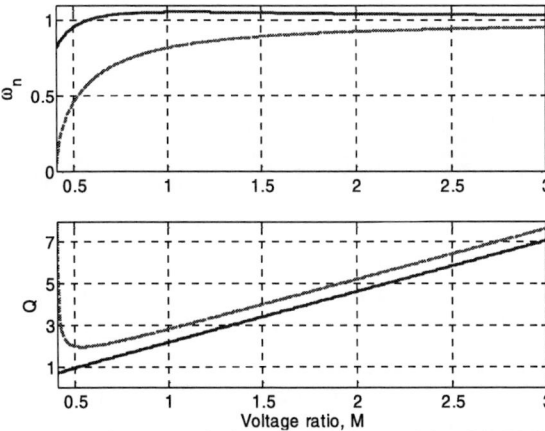

Figure 5. Maximum control gain steady-state characteristics of the PRC. Solid lines denote operation above resonance, dashed lines – below resonance.

The normalized frequency ω_n is very close to unity for most of the voltage ratio range considered. However, below $M = 1$, this frequency is far below unity. The operation is still above resonance for L_1 however, as this occurs for relatively small Q-factors (bottom trace). It is below resonance for L_2 for the entire range of M. The discontinuity observed in the bottom trace for L_2 is due to denominator in (12b). It occurs for very low Q-factors, heavy loading, where the selectivity of the resonant tank has diminished. For voltage ratios above 0.5, the Q-factor increases linearly.

The low-frequency control gain is obtained by substituting the solutions given by (9) and (10) into the first derivative (8):

$$LFG_{PRC} = -M^3 \frac{\pi^2}{2\sqrt{2}} \frac{\omega}{R} \tag{13}$$

and is negative, for operation above resonance. For operation below resonance, (13) is positive (using the value for L_2). The low-frequency gain is this time dependent on the cube of the operational voltage ratio. The gain therefore favors operation at high voltage ratios, where high Q-factors are required and the control curves (Figure 4) are steeper.

The resonant tank impedance is:

$$Z = j\omega L + \frac{R_{eff}}{1 + j\omega C R_{eff}} \tag{14}$$

Substituting the resonant components values from (9) and (10) into (14) results in the operational phase of $\varphi = \arctan\left(1/\sqrt{2}\right) \approx 35°$ (negative for the L_2). We could therefore expect efficient operation above resonance with ZVS maintained.

IV. SERIES-PARALLEL RESONANT CONVERTER

The equivalent fundamental equivalent circuit of the inductor-controlled Series-Parallel Resonant Converter (SPRC) is identical to that of the conventional SPRC, Figure 6. The equivalent load resistor R_{eff} has the same

value as that in the PRC, as both converters employ a voltage-fed rectifier.

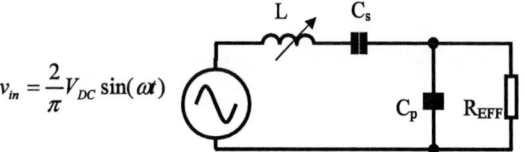

Figure 6. Fundamental equivalent circuit of the inductor-controlled SPRC.

In the following discussion the parallel resonant capacitor is assumed to be equal to the series resonant capacitor. It has been generally recognized that this design offers a compromise between the necessary control frequency range and good part-load efficiency.

The voltage conversion ratio for the half bridge SPRC is given by the expression in (16):

$$M = \frac{4R\omega C}{\sqrt{\left(\pi^2 R\omega C\right)^2 \left(\omega^2 LC - 2\right)^2 + 64\left(1 - \omega^2 LC\right)^2}} \tag{16}$$

This expression is plotted in Figure 7 in a manner identical to that for the SRC.

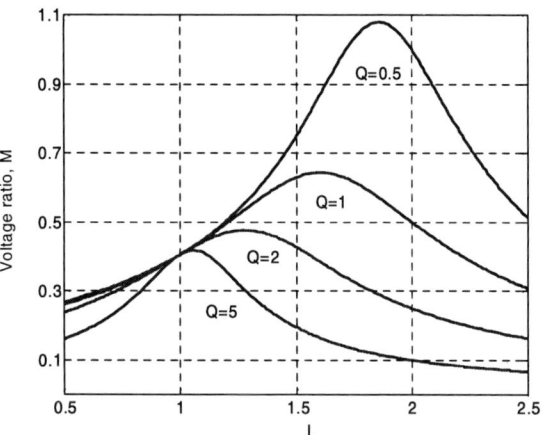

Figure 7. Voltage ratios for the inductor-controlled SPRC.

The SPRC is a hybrid structure, and inherits some of the properties from both the SRC and PRC. For example, the load-resonant frequency is increased as the load gets lighter. For no load condition, the resonance occurs at $\omega_0\sqrt{2}$ (provided series and parallel capacitors are equal). This can be observed also in Figure 7, for low Q-factors ($Q = Z_o/R$) the resonance tends towards $L_n = 2$. This translates in resonant frequency $\omega = 1/\sqrt{2L_0C}$, which is offset by $\sqrt{2}$ times form the resonant frequency at heavy load.

Heavy loading tends to shunt the parallel resonant capacitor, therefore the SPRC behaves in fashion similar to that of the SRC. Light load, low Q-factor, forces the converter to behave more like PRC, and the voltage ratios can exceed $M = 0.5$ by far. The transition between the two occurs for intermediate Q-factors (between approximately 1 and 3), and is characterized by shallow curves, as it could be clearly seen in Figure 7.

353

The procedure for identifying the values of the resonant components is identical to that for the PRC and results in one positive real value of the resonant capacitor:

$$C = \frac{\sqrt{2}\sqrt{3M^2\pi^4 - 32}}{\pi^2 \omega R} \qquad (17)$$

There are two solutions for the resonant inductors:

$$L_1 = \frac{R}{\omega}\frac{\sqrt{2}}{3\pi^2}\frac{-16 + 3M^2\pi^4 + 2\sqrt{3M^2\pi^4 - 32}}{M^2\sqrt{3M^2\pi^4 - 32}} \qquad (18a)$$

$$L_2 = \frac{R}{\omega}\frac{\sqrt{2}}{3\pi^2}\frac{-16 + 3M^2\pi^4 - 2\sqrt{3M^2\pi^4 - 32}}{M^2\sqrt{3M^2\pi^4 - 32}} \qquad (18b)$$

All resonant components require that $M > 1/\sqrt{6}$. This reflects the hybrid nature of the converter, specifically the fact that the voltage ratio has a minimum value (as for the PRC). Note that the solutions for the resonant inductor have the same structure as for the PRC, carrying the same implications for the control law opportunity regarding the load variations.

The solution in (18b) is smaller than the one in (18a) and, since they will resonate with the same capacitor, we could reasonably expect that it is for operation below resonance.

The normalized frequency and the Q-factor for the above solutions and for operation above resonance are:

$$\omega_n = \omega\sqrt{LC} = \frac{\sqrt{6}}{3\pi^2}\frac{\sqrt{-16 + 3M^2\pi^4 + 2\sqrt{3M^2\pi^4 - 32}}}{M} \qquad (19)$$

$$Q = \sqrt{\frac{L}{C}}/R = \frac{1}{\sqrt{3}}\frac{\sqrt{-16 + 3M^2\pi^4 + 2\sqrt{3M^2\pi^4 - 32}}}{M\sqrt{3M^2\pi^4 - 32}} \qquad (20)$$

The solutions for operation below resonance have the same form as (19) and (20), but differ by having a negative sign in front of the inner square root in the numerators. These expressions are plotted in Figure 8, solid lines denoting operation above resonance.

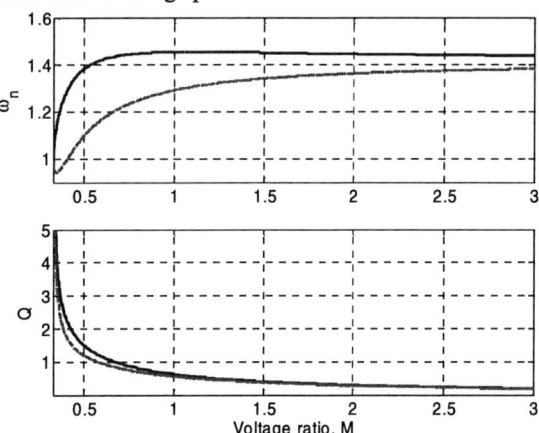

Figure 8. Steady-state characteristics of the inductor-controlled SPRC. Broken lines indicate operation below resonance.

For maximum control gain, the converter needs to be operated with high Q-factor and relatively close to the resonant frequency, when operation close to or below $M = 0.5$ is required. Higher voltage ratios require operation with fast diminishing Q-factors, and the

operating frequency is close to the load-resonant frequency typical for lightly loaded resonant tank.

The low-frequency control gain is obtained by substituting equations 17 and 18 into the expression for the first derivative:

$$LFG_{SPRC} = -M^3 \frac{\omega}{R}\frac{\pi^2\sqrt{2}}{2\sqrt{2}} \qquad (21)$$

It is positive for operation below resonance and is identical to that of the PRC.

The impedance loading the equivalent voltage source in Figure 6 is:

$$Z = j\omega L + \frac{1}{j\omega C} + \frac{R_{eff}}{1 + j\omega C R_{eff}} \qquad (22)$$

Substituting (17) and (18) into (22), allows us to compute the impedance phase at $\varphi = \arctan(1/\sqrt{2}) \approx 35°$, negative for operation below resonance.

V. Experimental Results

A 200W prototype of a SPRC was built, with resonant capacitor ratio equaling unity and operating at 500kHz. Using equations (17) and (18a) the resonant components were calculated as $C_S = C_P = 23.57\,nF$ and $L = 8.22\,\mu H$, the input voltage being 100 V and the output 50 V.

The (controlled) resonant inductor was constructed from three cores with grade 3F3 ferrite. Each core has 10 common turns to form the resonant inductor and 21 turns on each core for the biasing coil. The three core are then combined to form the controlled inductor, by connecting the resonant windings in parallel and the biasing windings in series [9]. The control characteristics of this inductor were measured with an impedance analyzer and the results are shown in Figure 9. Each core contains two ring cores R58/40/12.

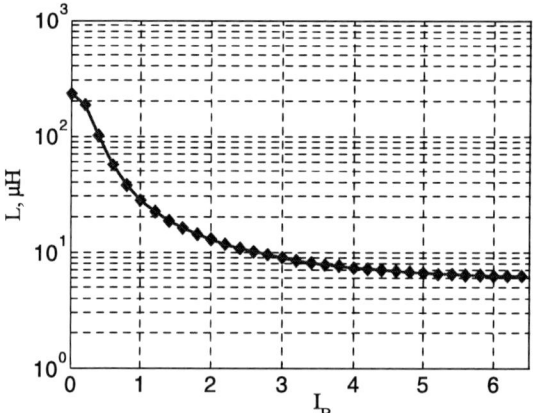

Figure 9. Variable inductor control characteristics.

The primary objective for the design of the controlled inductor was to obtain the widest control range for reasonable control (bias) current I_B. No other attempt was made to optimize the inductor performance. The control curve in Figure 9 indicates that the inductor value

354

can be varied by more than an order of magnitude for a bias current of less than 2 A.

The voltage ratio M was then measured against the inductance variation. The corresponding values of M and the bias current were recorded. The data in Figure 9 was used to produce the x-axis in Figure 10. The solid line plots the theoretical prediction for M from equation (16) and the crosses indicate the measured data.

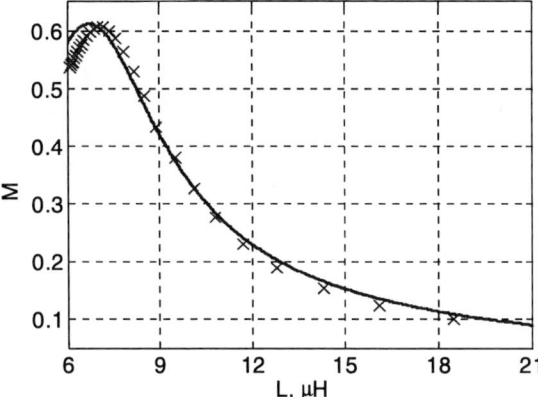

Figure 10. Predicted and experimental voltage ratios. Crosses indicate prototype measurement data.

Both sets of data align rather well, the divergences being most obvious around the resonance. The measured data suggests somewhat higher Q-factor, with steeper slopes just below resonance.

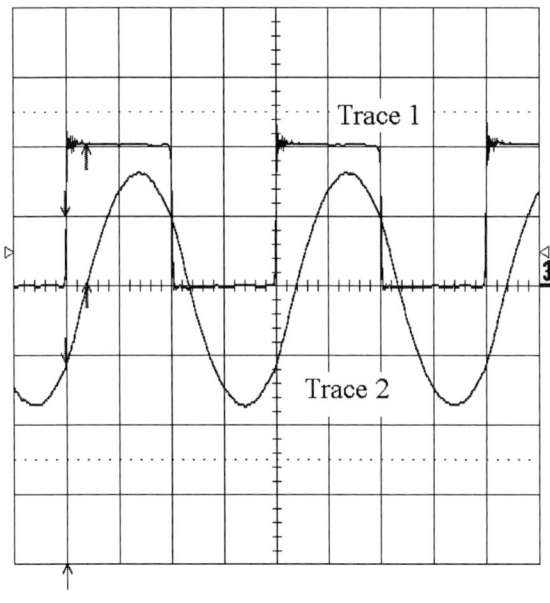

Figure 11. Measured waveforms. Trace 1: 50V/div; Trace 2: 5A/div, Time: 500ns/div.

The dominant effect explaining these discrepancies is associated with losses. As resonance is approached, zero voltage switching is lost. As the resonant inductor gets smaller, forcing operation below resonance, the associated switching losses remain relatively constant. However, the power throughput is being reduced, therefore the predicted and the experimental data appear to diverge as operation is further away and below resonance.

The measured half-bridge voltage (Trace 1) and the resonant current (Trace 2) from the prototype operating at the point of maximum control gain are shown in Figure 11. The value of the resonant inductor was adjusted so that the voltage ratio is M=0.5. The measured angle (Figure 11) is 34.2 degrees, which is very close to the predicted 35 degrees. A time-domain simulation using the calculated resonant component values resulted in a voltage ratio just above 0.499 at 35 degrees, confirming the viability of the developed analysis.

VI. CONCLUSIONS

Advanced manufacturing technologies, such as thin-film deposition techniques, become more accessible and enable new approaches to power transfer control. The present work is a design-oriented analysis of the optimum operating conditions for inductor-controlled load-resonant converters. This type of converters narrows the control frequency range and in certain instances (with targeted resonant inductor design) could negate the need for an electronic feedback loop.

The developed analysis indicates that the inductor-controlled converters operate efficiently in the region of maximum control gain, with an angle around 35 degrees. The agreement between time-domain simulations and the prediction for the operating conditions is excellent. The analysis results also align well with results from a 200W prototype of a series-parallel resonant converter.

The developed design procedures would be similar for the case where the resonant capacitor is the controlled component. Due to the topological and behavioral symmetry of the resonant circuits, the operating angle, for the operating point of maximum control gain, is likely to be the same, around 35 degrees.

ACKNOWLEDGMENT

The authors would like to acknowledge the funding for this work provided by EECE, University of Birmingham, UK.

REFERENCES

[1] Steigerwald, R.L.; "A comparison of half-bridge resonant converter topologies" *Power Electronics, IEEE Transactions on,* Volume: 3, Issue: 2, April 1988 Pages: 174 – 182.

[2] Forsyth, J.; Mollov, V.; "Resonant converter for high-power-factor rectification" *Electronics Letters,* Volume: 36, Issue: 18, 31 Aug. 2000, Pages: 1516 – 1518.

[3] Kazmierczuk, M.K Czarkowski, D.; Thirunarayan, N.; "A new phase-controlled parallel resonant converter" *Industrial Electronics, IEEE Transactions on,* Volume: 40, Issue: 6, Dec. 1993, Pages: 542 – 552.

[4] Bhat, A.K.S.; "A generalized steady-state analysis of resonant converters using two-port model and Fourier-series approach" *Power Electronics, IEEE Transactions on,* Volume: 13, Issue: 1, Jan. 1998, Pages: 142 – 151.

[5] Prabhakaran, S.; Sullivan, C.R.; Venkatachalam, K.; "Measured electrical performance of V-groove inductors for microprocessor

power delivery" *Magnetics, IEEE Transactions on*, Volume: 39, Issue: 5, Sept. 2003, Pages:3190 – 3192.

[6] Sato, F.; Ono, T.; Wako, N.; Arai, S.; Ichinose, T.; Oba, Y.; Kanno, S.; Sugawara, E.; Yamaguchi, M.; Matsuki, H.; "All-in-one package ultracompact micropower module using thin-film inductor" *Magnetics, IEEE Transactions on*, Volume: 40, Issue: 4, July 2004, Pages: 2029 – 2031.

[7] Sullivan, C.R.; Sanders, S.R.; "Design of microfabricated transformers and inductors for high-frequency power conversion" *Power Electronics, IEEE Transactions on*, Volume: 11, Issue: 2 , March 1996, Pages:228 – 238.

[8] Saleh N., "Variable Microelectronic Inductors", *IEEE Transactions on Components, Hybrids and Manufacturing Technology*, Vol. CHMT-2, No. 1, March 1978.

[9] Medini, D.; Ben-Yaakov, S.; "A current-controlled variable-inductor for high frequency resonant power circuits", *Applied Power Electronics Conference and Exposition, 1994*, APEC '94 Conference Proceedings 1994, Ninth Annual, 13-17 Feb. 1994, Pages: 219 - 225 vol.1.

Capacitor Clamped Multilevel Matrix Converter Controlled with Venturini Method

Janina Rząsa

*Faculty of Electrical and Computer Engineering, Rzeszow University of Technology,
2 W. Pola Street, 35-959 Rzeszow, Poland, e-mail: *jrzasa@prz.rzeszow.pl*

Abstract— The article is discussion with Authors of [21], who have presented proposal of multilevel matrix converter scheme. Because the Authors have shown results of research involved only with one phase output multilevel matrix converter and suggested application one of known modulation methods to control of three phase scheme, this paper undertakes this problem. New elements of power scheme are added. Venturini modulation method is examined. Improvement of waveforms synthesized at output and reduction of reverse voltages on semiconductor elements is achieved only for the case when multilevel matrix converter operates at output frequency equal the supply voltage frequency.

Keywords— matrix converter, multilevel converters, Pulse Width Modulation (PWM)

I. INTRODUCTION

Matrix converter is a direct frequency changer with forced commutation, controlled by means of pulse width modulation method. Since the 80s of the last century, the three-phase to three-phase circuit enjoys significant and still growing popularity. Many authors focus in their work on description of control methods of the matrix. Two types of approach are used in elaboration of control method: the indirect approach [7, 11, 17], in which the direct frequency converter is seen as a combination of a rectifier circuit and a voltage inverter; and the direct approach [1-3, 17-20, 22-24]. Among control methods one can distinguish the so-called Venturini methods [1-2, 6, 17-20, 22-24], vector methods [3, 11-13, 17] and scalar methods [14, 17]. Matrix converter operates at sinusoidal output and input waveforms with possibility to control the input phase angle. Filtration is necessary only with respect to output voltage and input current distortion components, frequencies of which are equal and higher than the switching frequency. An advantage of the matrix converter consists in possibility to realize the matrix of switches as integrated semiconductor power modules thanks to lack of link circuits with passive elements. Certain limitation in application of matrix converter consists in low value of the circuit's supply voltage utilization factor. In the classical matrix converter, RMS value of the fundamental component of the synthesized phase output voltage equals 0.866 of the phase supply voltage RMS value at the most, while the maximum reverse voltage in bidirectional switches reaches a value equal to the amplitude of the line supply voltage.

In last few years, by analogy to multilevel inverters, several solutions concerning multilevel matrix converters were published [4, 21]. Application of multilevel energy conversion in matrix converters is aimed at reduction of voltage rating of the switches with respect to supply voltages and further improvement of synthesized current and voltage waveforms. Authors of [21] analyse a multilevel capacitor clamped matrix converter, presenting general assumptions related to operation of a circuit with three-phase output as well as simulation and laboratory measurement results for a circuit with single-phase output. At the same time, they suggest possibility to use control methods known for classical matrix converter in systems with three-phase output.

The present paper is an attempt to join in the discussion on possibility and consequences of use of such control methods. In the first place, we discuss the most classical control method with use of Venturini modulation. The multilevel capacitor clamped matrix converter circuit considered in [21] is extended by the present author with additional clamp capacitors and balancing circuits, connected in parallel to the load. Additional capacitors ensure symmetry of voltages and symmetry of current loop. Balancing circuits provide automatic maintenance of required voltage levels on those capacitors.

II. MULTILEVEL ENERGY CONVERSION IN POWER ELECTRONICS CIRCUITS

Multilevel structure of converters should ensure: more precise reproduction of required output waveforms; synthesis of high-value output voltages with use of multiple intermediate stages; and reduction of maximum voltage values and voltage steepness values to which semiconductor power elements are exposed.

Multilevel matrix converter considered in this paper may be counted among the so-called Capacitor-Clamped Multilevel Converters, known also as Flying Capacitor Multilevel Converters or Multicell Converters. Proper operation of capacitor-clamped converters requires maintenance of appropriate voltage levels on those capacitors. The capacitors charging process depends on switching algorithm of semiconductor switches as well as the character of the source supplying the circuit and character of the load. Based on the literature [8–9, 15-16], one can establish the following conditions of correct operation of multicell converters:

– a converter is supplied from a voltage source and loaded with a receiver of inductive character, treated in theoretical models as a current source;

– current source frequency is much less than the switching frequency which allows to assume that during the switching cycle, output current does not change;

– average value of currents of a clamp (cell) capacitor within the cycle corresponding to the switching frequency equals zero;

– voltage on capacitors of individual cells is given by

978-1-4244-1741-4/08/$25.00 ©2008 IEEE   357

$$U_{Cn} = \frac{U_i}{p} n \qquad (1)$$

where p is number of cells, and U_i is the RMS value of voltage supplying the multicell converter;

— capacitance value of the capacitor should be determined based on admissible value of voltage ripple on that capacitor ΔU_C (2)

$$C = \frac{I_o}{\Delta U_C p f_s} \qquad (2)$$

where f_s is the switching frequency of the converter's switches;

— signals controlling switches in individual cells of the converter should be shifted with respect to each other by angle $2\pi/p$.

Reasoning given in [8] shows at the same time that when a multicell converter circuit operates with fulfillment of the above-listed conditions concerning the character of power circuit and control method, then the process of automatic maintenance of voltages on capacitors at required levels is observed and, in transient conditions, the state of balance of those voltages occurs as a result of flow of current component with switching frequency. That leads to the conclusion that in order to accelerate the process of charging capacitors up to required voltage level, one should use an additional resonance circuit connected in parallel to the load. Parameters of that circuit should be selected so that resonance frequency was equal to the switching frequency. Use of such circuit, and thus acceleration of the process of setting of voltage on capacitors, is of particular importance in case where one deals with synthesis of AC voltage at output as in multicell inverters.

A special challenge is application of multilevel energy conversion of AC voltage into AC voltage with other parameters, as in multicell AC voltage regulators or multilevel matrix converters. It is important in such circuits that instantaneous value of voltage on capacitors in individual cells be always an appropriate portion of the instantaneous value of input voltage. In case where the output voltage, and therefore also the output current, is variable, successive current pulses recharging a capacitor in a multicell circuit differ in their instantaneous values. In turn, application of the pulse modulation method is the reason for which the pulses differ more or less in their width, depending on applied switching frequency. In such conditions it is more difficult to ensure the appropriate division of voltages on cell capacitors and the fundamental task in the circuit design process consists in proper selection of the circuit providing automatic maintenance of voltages on clamp capacitors and application of appropriate control methods (switching algorithm) to semiconductor elements.

III. PRINCIPLE OF OPERATION OF THE MULTILEVEL MATRIX CONVERTER

Conventional matrix converter, composed of nine fully controllable bidirectional switches, as a circuit realizing conversion of electric energy by means of direct connection of terminals of a three-phase load with three phases of a supply source, must fulfil the assumption that at inductive character of the load, the supply must have a character of voltage sources. The multilevel matrix converter circuit (Fig. 1) considered in this paper was designed on the grounds of conclusions derived from analysis of operation of classical matrix converter, multilevel converters, in particular from analysis of operation of multicell converters taking into account specificity of operation of a direct frequency changer circuit.

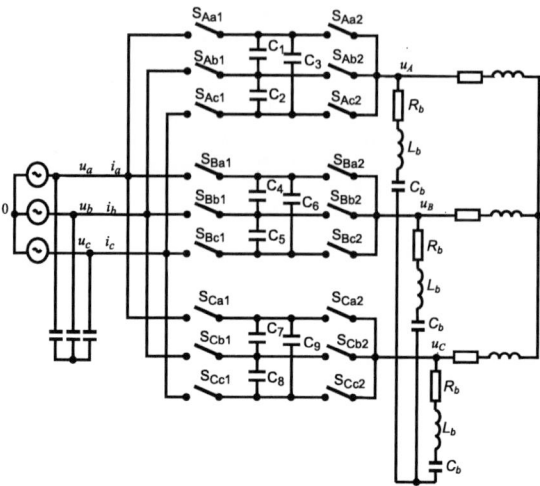

Fig. 1. Topology of a multilevel matrix converter

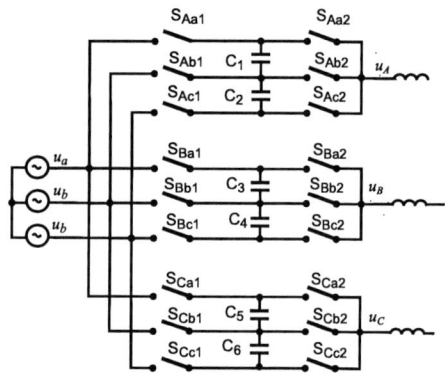

Fig. 2. Topology of a multilevel matrix converter proposed in [21]

In comparison with the circuit proposed in [21] (Fig. 2), three clamp capacitors have been added as well as circuits providing automatic maintain of voltages on clamp capacitors, consisting of R_b resistance, L_b inductance and C_b capacitance connected in series. Parameters of the circuits providing automatic balance of voltages on capacitors were selected so that resonance frequency of those circuits is equal to the switching frequency of the switches. Capacitance of clamp capacitors, according to formula (2), depends on the value of load current, switching frequency, number of cells in the multilevel converter and assumed value of voltage ripple on clamp capacitors. Selection of admissible values of voltage ripple ΔU_C was made taking into account the fact that it must not be too small, as the capacitance of the clamp capacitor would be too large and the capacitor would not be able to recharge up to required voltage value.

Establishment of a switch controlling algorithm in a multilevel matrix converter supplied from a three-phase voltage source and with a resistance-inductance load, requires fulfillment of the same conditions as in the classical converter, which means that one and only one of supply phases may be connected with each of output phases. In ideal conditions of charging, voltages on clamp capacitors should be equal to:

$$
\begin{aligned}
u_{C1} &= u_{C4} = u_{C7} = (u_a - u_b)/2 \\
u_{C2} &= u_{C5} = u_{C8} = (u_b - u_c)/2 \\
u_{C3} &= u_{C6} = u_{C9} = (u_c - u_a)/2
\end{aligned}
\tag{3}
$$

As a result of use of clamp capacitors, there exist additional current flow paths between input and output phases and additional intermediate voltage levels in waveforms of output voltage can be generated (TABLE I). Taking into account only admissible states of switches in a multilevel matrix converter, each of output phases may be connected with three supply phases in nine different ways corresponding to converter configurations. These configurations arise from 'on' or 'off' states of bidirectional switches, existence functions of which for output phase A are defined in TABLE I.

TABLE I.
INSTANTANEOUS VALUES OF A OUTPUT PHASE VOLTAGES

Switch state ('on': 1, 'off': 0)						Instantaneous value of voltage in output phase A
S_{Aa1}	S_{Aa2}	S_{Ab1}	S_{Ab2}	S_{Ac1}	S_{Ac2}	
a) 1	1	0	0	0	0	u_a
b) 0	0	1	1	0	0	u_b
c) 0	0	0	0	1	1	u_c
d) 1	0	0	1	0	0	$(u_a+u_b)/2$
e) 0	1	1	0	0	0	$(u_a+u_b)/2$
f) 0	0	0	1	1	0	$(u_b+u_c)/2$
g) 0	0	1	0	0	1	$(u_b+u_c)/2$
h) 0	1	0	0	1	0	$(u_a+u_c)/2$
i) 1	0	0	0	0	1	$(u_a+u_c)/2$

Considering a three-phase to three-phase matrix converter, one has to take into account $9^3 = 729$ possible circuit configurations which can be used practically in the process of synthesis of output voltages and synthesis of voltages of clamp capacitors determining intermediate levels of supply voltages.

IV. CONTROL BY MEANS OF VENTURINI METHOD

In the classical matrix converter controlled with Venturini modulation method [1, 2, 17], output voltages u_A, u_B, u_C are determined by means of modulation functions m_{ij} (4), representing conduction cycle width variations of individual switches in time. Slow-varying modulation functions m_{ij} in the formula (4) correspond to discrete switch state functions. In practice these are non-negative continuous functions, instantaneous values of which do not exceed unity.

$$
\begin{bmatrix} u_{A0} \\ u_{B0} \\ u_{C0} \end{bmatrix} =
\begin{bmatrix} m_{Aa} & m_{Ab} & m_{Ac} \\ m_{Ba} & m_{Bb} & m_{Bc} \\ m_{Ca} & m_{Cb} & m_{Cc} \end{bmatrix}
\begin{bmatrix} u_a \\ u_b \\ u_c \end{bmatrix}
\tag{4}
$$

To ensure that neither branches with inductances may be interrupted nor circuits containing voltage sources short-circuited, the sum of modulation functions of three switches from single output phase has to be equal to unity at any instant of time (5).

$$
\begin{aligned}
m_{Aa} + m_{Ab} + m_{Ac} &= 1 \\
m_{Ba} + m_{Bb} + m_{Bc} &= 1 \\
m_{Ca} + m_{Cb} + m_{Cc} &= 1
\end{aligned}
\tag{5}
$$

Taking into account the fundamental requirement concerning direct energy conversion, consisting in equality of instantaneous powers on input and output clamps of the switching matrix, authors of [1] have determined modulation functions for individual switches of the classical matrix converter by means of a heuristic method. Assuming that required output voltages are sinusoidal with given angular frequency ω_o, modulation function for one of the switches is determined by formula (6), in which φ_i — input displacement angle; φ_o — output phase angle; $k_U = U_i/U_o$; $\alpha_1 = 0.5\,(1-\theta)$; $\alpha_2 = 0.5\,(1+\theta)$; $\theta = \tan(\varphi_i)/\tan(\varphi_o)$; U_i and U_o — RMS values of input and output voltage, respectively.

$$
\begin{aligned}
m_{Aa}(t) = {} & \frac{\alpha_1}{3}\left[1 + 2k_U\cos\!\big((\omega_o + \omega_i)t\big)\right] \\
& + \frac{\alpha_2}{3}\left[1 + 2k_U\cos\!\big((\omega_o - \omega_i)t\big)\right]
\end{aligned}
\tag{6}
$$

Use of modulation function according to formula (6) allows to regulate input displacement angle within range $-\varphi_o < \varphi_i < \varphi_o$ and the RMS value of the required output voltage within range $0 < U_o < 0.5\,U_i$. Increase of the range of output voltage variations up to value of $0.866\,U_i$ is connected with assumption that in required waveforms of output phase voltages, presence of components with frequencies equaling tripled input and output frequency is admissible [1,6].

Replacement of continuous modulation function with discrete existence functions representing states of switches for the classical matrix converter can be in practice realized in the Venturini modulation method by means of comparison of the analog signal corresponding to function $m_{ij}(t)$ with the carrier signal in the natural modulation process.

By analogy to switch control method used in multilevel inverters, application of natural modulation with displacement of carrier signals (phase shifted PWM method, [16]) is proposed in the examined multilevel matrix converter. Displacement of carrier signals involved in control of switches S_{ij1} and switches S_{ij2} is $T_s/2$. Duty cycles in which switches S_{ij1} are switched-on, arise from comparison of corresponding modulation functions with one of carrier signals (Fig. 3). A carrier signal shifted by half of the cycle T_s determines duty cycles of S_{ij2} switches.

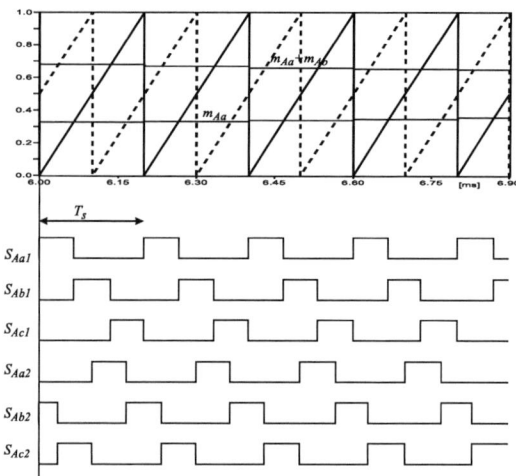

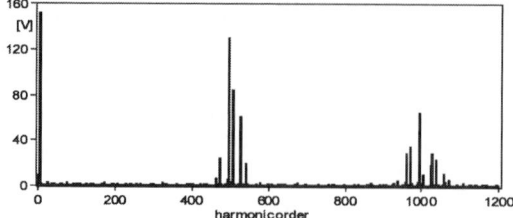

MC's PlotXY - Fourier chart(s). Copying date: 2008-05-21
File MM.pl4   Variable v:X   -VXP [peak]
Initial Time: 0,05501   Final Time: 0,155

Fig. 5. Harmonic spectrum of load phase voltage in a classical matrix converter at $\varphi_i = -\varphi_o, f_o = 60$ Hz, $f_{carr} = 5$ kHz

Currents of output phases are sinusoidal (Fig. 6), while voltage waveforms on switches take instantaneous values equal to corresponding line voltages, thus maximum value of that voltage is equal the amplitude of the line supply voltage (Fig. 7).

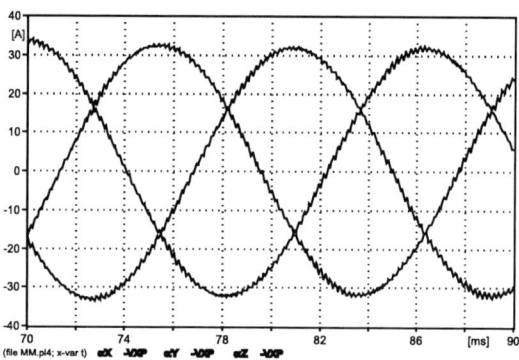

Fig. 6. Currents of three load phases in a classical matrix converter at $\varphi_i = -\varphi_o, f_o = 60$ Hz

Fig. 3. Switch conduction cycles of output phase A in a multilevel matrix converter

V. RESULTS OF SIMULATION TESTS

A. Classical matrix converter

Simulation tests were realized in ATP-EMTP program. The matrix of bidirectional switches was modeled as a matrix of ideal switches controlled with signals generated in TACS subroutine. The supply grid is represented by sinusoidal voltage sources with RMS value of 220 V and frequency of 50 Hz, while the load consists in star-connected resistance and inductance elements with values of 2 Ω and 10 mH. Carrier frequency is $f_{carr} = 5$ kHz.

To demonstrate effects obtained in matrix converter through application of multilevel energy conversion, Figs. 4–7 present waveforms of voltages and currents in a classical circuit. Fig. 4 represents phase voltage waveforms of a three-phase load supplied from a classical matrix converter. Harmonic spectrum (Fig. 5) of this waveform, besides fundamental component with assumed output frequency, contains also distortion components gathered as side bands around multiples of the carrier frequency.

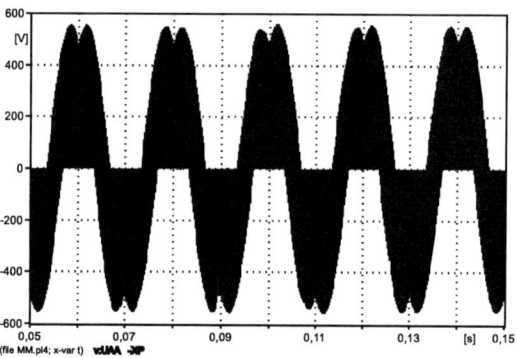

Fig. 7. Voltage on a switch in a classical matrix converter at $\varphi_i = -\varphi_o$, $f_o = 60$ Hz

B. Multilevel matrix converter – analyse of output waveforms

Next stage of research consisted in simulations performed with use of a model according to diagram presented in Fig. 1. To compare results obtained in proposed circuit some of waveforms received from simulation carried out for scheme from [21] (Fig.2) are presented. Parameters of supply and load are the same as in classical matrix converter. Capacity of clamp capacitors

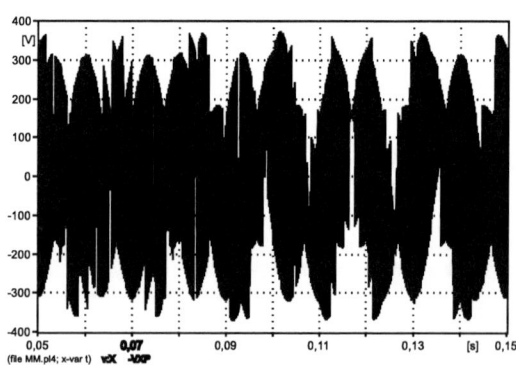

Fig. 4. Waveform of load phase voltage in a classical matrix converter at $\varphi_i = -\varphi_o, f_o = 60$ Hz, $f_{carr} = 5$ kHz

360

in the examined model was calculated according to formula (2), and than corrected in subsequent simulations in such a way that with the examined method and control strategy and at frequency of synthesized output waveforms equaling 50 Hz, approximately equal distribution of voltages to two switches connected in series between the supply phase and the output phase was obtained. As a result, for the examined circuit capacitance of capacitors C_1–C_9 was selected as amounting to 10 µF. The circuit providing automatic maintenance of voltages on clamp capacitors was designed under assumption that carrier signal frequency in all simulation models amounted to 5 kHz, thus the elements of the circuits were given the following values: $R_b = 2\,\Omega$, $L_b = 1$ mH, $C_b = 1$ µF.

In case of application of the proposed by present author, scheme of matrix converter (Fig. 1), the obtained simulation results show that the control with use of pulse width modulation with phase-shifted carrier signals modulated by means of slow-varying Venturini functions gives positive results. An effect of such control consists in improvement of the form of synthesized output voltage (Fig. 8,) in comparison with the classical circuit (Fig. 4), consisting in fact that distortion components connected with carrier frequency (Fig. 9) are considerably decreased. Currents of output phases (Fig.10) are sinusoidal, like in the classical converter. Current of balance circuit in branch with R_b, L_b, and C_b consists of only high order component connected with currier frequency (Fig. 11). Waveform of output voltage (Fig. 12) and its harmonic spectrum (Fig. 13) obtained for circuit shown in Fig. 2 do not present similar features.

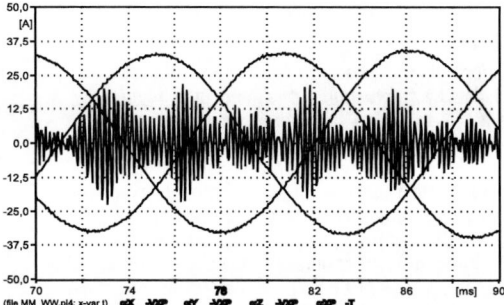

Fig. 10. Waveform of 3 phases load currents and balance circuit current in a multilevel matrix converter (Fig. 1) at $\varphi_i = -\varphi_o$, $f_o = 60$ Hz

MC's PlotXY - Fourier chart(s). Copying date: 2 008-05-21
File MM_WW.pl4   Variable c:XP   -T [peak]
Initial Time: 0,05201   Final Time: 0,152

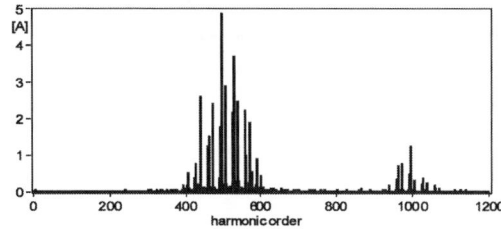

Fig. 11. Harmonic spectrum of balance circuit current in a multilevel matrix converter (Fig. 1) at $\varphi_i = -\varphi_o$, $f_o = 60$ Hz

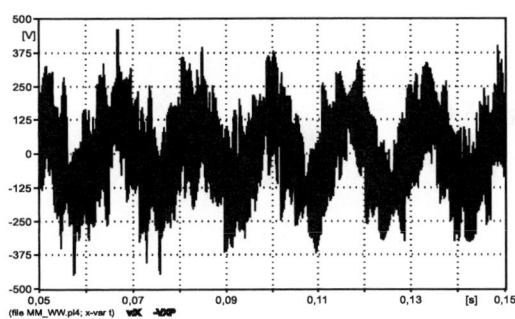

Fig.8. Waveform phase load voltage in a multilevel matrix converter (Fig.1) at $\varphi_i = -\varphi_o$, $f_o = 60$ Hz, $f_{carr} = 5$ kHz

MC's PlotXY - Fourier chart(s). Copying date: 2 008-05-21
File MM_WW.pl4   Variable v:X   -VXP [peak]
Initial Time: 0,052   Final Time: 0,152

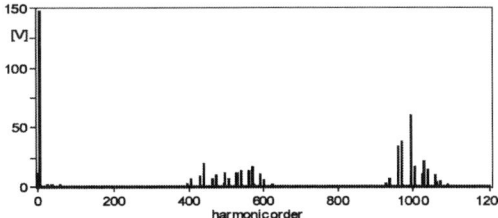

Fig. 9. Harmonic spectrum of phase load voltage in a multilevel matrix converter (Fig.1) at $\varphi_i = -\varphi_o$, $f_o = 60$ Hz, $f_{carr} = 5$ kHz

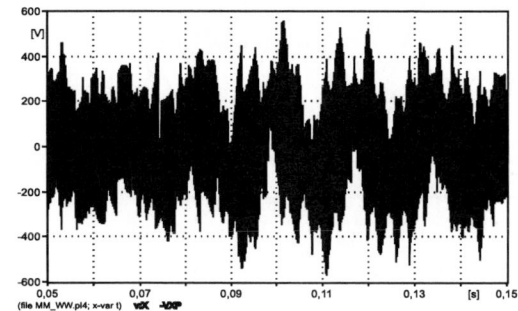

Fig. 12. Waveform of phase load voltage in a multilevel matrix converter (Fig.2), at $\varphi_i = -\varphi_o$, $f_o = 60$ Hz

MC's PlotXY - Fourier chart(s). Copying date: 2 008-05-21
File MM_WW.pl4   Variable v:X   -VXP [peak]
Initial Time: 0,05201   Final Time: 0,152

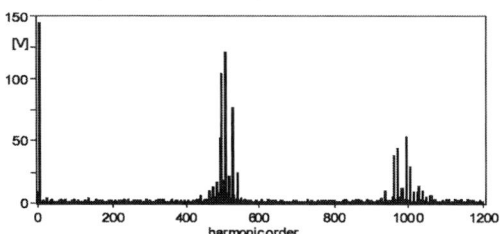

Fig. 13. Harmonic spectrum of phase load voltage in a multilevel matrix converter (Fig.2), at $\varphi_i = -\varphi_o$, $f_o = 60$ Hz

C. *Multilevel matrix converter – analyse of voltage on the switchs*

Realization of the second requirement imposed on multilevel energy conversion, consisting in reduction by half of voltages to which semiconductor elements of the circuit are exposed, requires more detailed explanation. On the grounds of the simulation, a positive result in the form of significant reduction of voltages on converter's switches to values only insignificantly exceeding the half of amplitude of line supply voltages was obtained only for the circuit shown in Fig 1 at output frequency equaling the supply frequency ($f_o = f_i = 50$ Hz). Voltages on switches depend on voltages on clamp capacitors. Waveforms of voltages on the capacitors, at constant parameters R_b, L_b, C_b, depend on input displacement angle assumed in control method. The best results were achieved with $\varphi_i = \varphi_o$ (Fig. 14). Voltages on clamp capacitors were virtually sinusoidal, distorted only with components of frequencies connected to the carrier frequency. However, amplitudes of fundamental components in voltages of three capacitors C_1, C_2, and C_3, connected with one output phase were not equal to each other, as result of which voltages on switches S_{Aa1} (Fig.15) and S_{Aa2} were higher than half amplitude of line supply voltage. Asymmetric set of three voltages u_{c1}, u_{c2}, u_{c3} could be interpreted as two symmetric sets of voltages: positive and negative phase sequence.

In case where $\varphi_i \neq \varphi_o$ and $f_o = 50$ Hz waveforms of voltages on capacitors (Fig. 16, Fig.18), contain: the fundamental component, which in three phase set of u_{c1}, u_{c2}, u_{c3} form positive and negative phase sequence voltages; the components connected with the switching frequency; and low order odd harmonics of 50 Hz frequency which, especially in case where $\varphi_i = -\varphi_o$, increase peak values of voltages on switches (Fig. 17, Fig.19) above the value equal to half of the line supply voltage. The maximum value of these voltages are however more less than corresponding to them peak values of switch voltage (Fig.20) in converter discussed in [21].

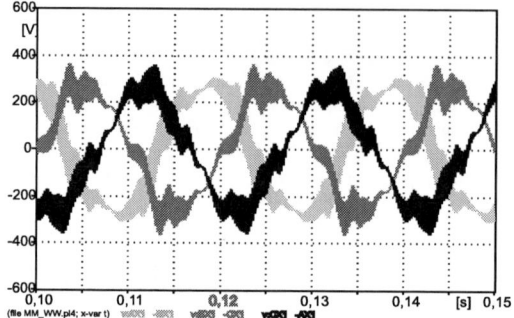

Fig. 16. Waveforms of voltages on capacitor C_1 C_2 and C_3 in a multilevel matrix converter (Fig.1) at $\varphi_i = 0$, $f_o = 50$ Hz

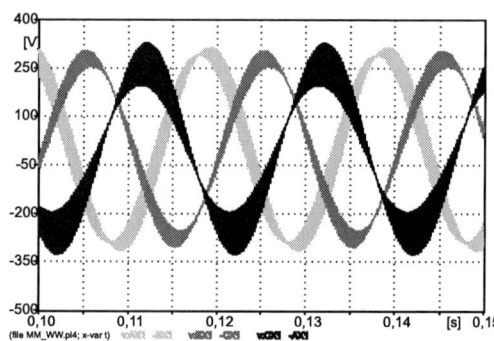

Fig. 14. Waveforms of voltages on capacitors C_1, C_2 and C_3 in a multilevel matrix converter (Fig.1) at $\varphi_i = \varphi_o$, $f_o = 50$ Hz

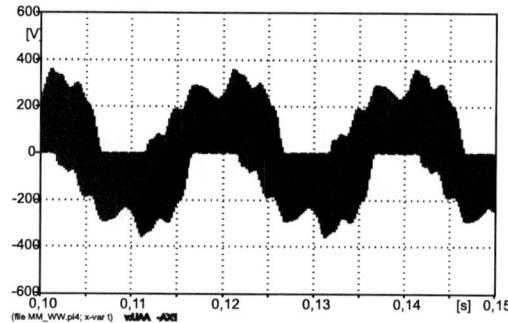

Fig. 17. Waveforms of voltages on switch S_{Aa1} in a multilevel matrix converter (Fig.1) at $\varphi_i = 0$, $f_o = 50$ Hz

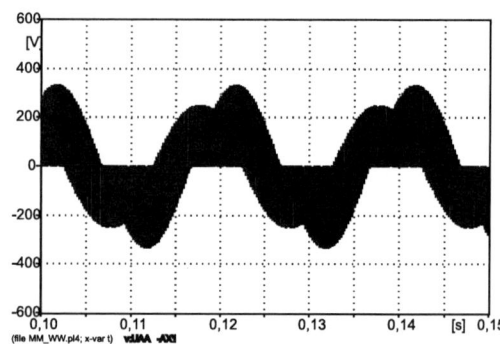

Fig. 15. Waveforms of voltage on switch S_{Aa1} in a multilevel matrix converter (Fig.1) at $\varphi_i = \varphi_o$, $f_o = 50$ Hz

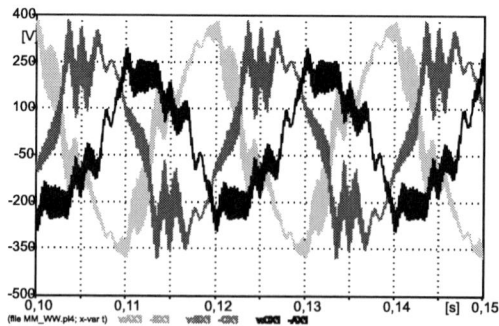

Fig. 18. Waveforms of voltages on capacitor C_1 C_2 and C_3 in a multilevel matrix converter (Fig.1) at $\varphi_i = -\varphi_o$, $f_o = 50$ Hz

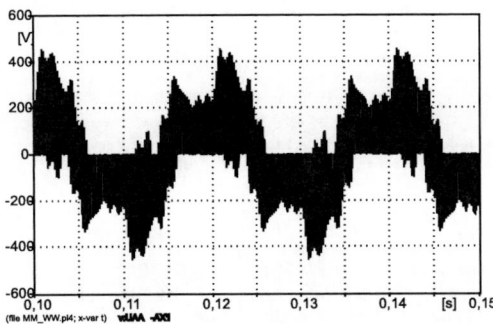

Fig. 19. Waveforms of voltages on switch S_{Aa1} in a multilevel matrix converter (Fig.1) at $\varphi_i = -\varphi_o, f_o = 50$ Hz

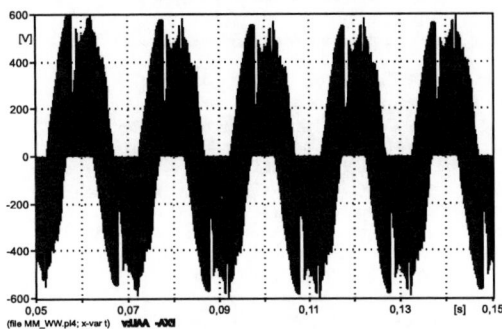

Fig. 20. Waveforms of voltages on switch S_{Aa1} in a multilevel matrix converter (Fig.2) at $\varphi_i = -\varphi_o, f_o = 50$ Hz

In case where a matrix converter, controlled by means of Venturini modulation method, operates at $f_o \neq f_i$, voltages on clamp capacitors, apart from fundamental component with supply frequency, f_i, and high-frequency distortion components, contain also components with frequencies f_o and $f_i + 2f_o$ and $f_o + 2f_i$ when $\varphi_i = -\varphi_o$ (Fig. 21); f_o and $f_i - 2f_o$ and $f_o - 2f_i$ when $\varphi_i = \varphi_o$; and f_o and $f_i \pm 2f_o$ and $f_o \pm 2f_i$ when $\varphi_i = 0$. These are the same components which occur in switch currents (Fig. 22).

MC's PlotXY - Fourier chart(s). Copying date: 2008-05-14
File MM_WW.pl4   Variable v:AX1 -BX1 [peak]
Initial Time: 0,05201   Final Time: 0,152

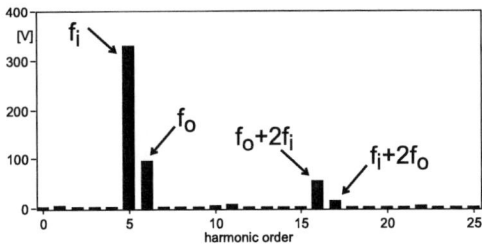

Fig. 21. Low-frequency components of voltage on capacitor C_1 in a multilevel matrix converter (Fig.1) at $\varphi_i = -\varphi_o, f_o = 60$ Hz

MC's PlotXY - Fourier chart(s). Copying date: 2008-05-14
File MM_WW.pl4   Variable c:UAA -AX1 [peak]
Initial Time: 0,05201   Final Time: 0,152

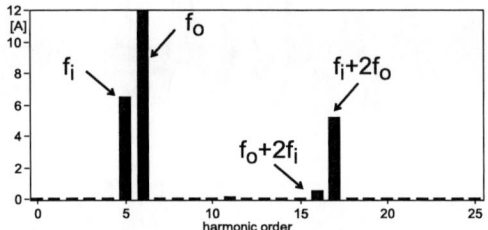

Fig. 22. Low-frequency components of current of switch S_{Aa1}; in a multilevel matrix converter (Fig.1) at $\varphi_i = -\varphi_o, f_o = 60$ Hz

D. *Justification of obtained simulation results*

Presence of the above-mentioned distortion components in switch currents can be explained by means of the mechanism of direct energy conversion occurring in the matrix converter and necessity to balance instantaneous power at input and output terminals. Each of three phase output branches is supplied from three input phases through corresponding three branches of bidirectional switches. Balance of instantaneous powers between three supply phases with sinusoidal voltages and each of output phases individually, requires occurrence of distortion components with frequencies of $f_i \pm 2f_o$ in these three phases (precisely, in branches containing the switches) [5, 19]. Similarly, balance of instantaneous power measured at terminal of one of input phases from which three output phases are supplied, requires occurrence of distortion components with frequencies of $f_o \pm 2f_i$ in each of switch branches connected to the given input terminal. As it follows from analysis carried out in [19], in case of the supply-receiver symmetry, current distortion components with frequencies $f_i \pm 2f_o$ inflowing to one of output phases, e.g. phase 'A', from three different feeding phases a,b,c through switches S_{Aa2}, S_{Ab2} and S_{Ac2}, form a symmetric three-phases system of the same or opposite order, which means that they do not occur in phase currents of the receiver. Similar situation exists at input terminals with current distortion components of frequencies $f_o \pm 2f_i$ flowing through three switch branches, e.g. S_{Aa1}, S_{Ba1} and S_{Ca1}, connected to the same feeding phase 'a', which causes that the supply current is sinusoidal. Clamp capacitor currents, representing differences of currents of switches connected to form one branch, contain the same distortion components, which dependently on value of output frequency and input displacement factor occur as low order harmonics or negative phase sequence voltage causing asymmetry in fundamental component. These components flow through capacitors and result in occurrence of distortion components of the same frequencies in capacitor voltages.

As a result of clamp capacitors voltage distortions caused by low frequency components, peak values of voltages on two switches between one of input phases and one of output phases are not equal to each other and in some cases differ significantly from the value equaling half the amplitude of the line voltage supplying the converter (Fig. 22), while output waveforms are synthesized in accordance with adopted assumptions, i.e. output currents are sinusoids with assumed frequency.

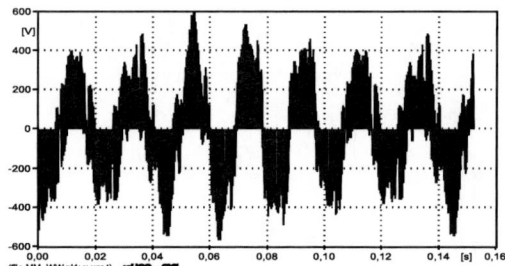

Fig. 22. Voltage on switch S_{Aa1} in a multilevel matrix converter (Fig.1) at $\varphi_i = -\varphi_o$, $f_o = 60$ Hz

The performed tests entitle us to conclude that employment of a multilevel energy conversion in the here proposed matrix converter circuit with use of Venturini control method fulfills imposed requirements only to some extend. That is, the method allows to synthesize output waveforms with much less distortions, than in the classical converter and multilevel matrix converter offered by authors of [21]. Reduction of voltage rating of power transistors used as bidirectional switches is possible only for a special case of operation of the circuit, when it is assumed that frequency of synthesized waveforms is equal to the supply grid frequency. Imposed input displacement angle may be changed in principle, however it must be remembered that when $\varphi_i \neq \varphi_o$, then instantaneous voltages of the switches may exceed the value equal to half of the line supply voltage amplitude.

VI. SUMMARY

Application of multilevel conversion in capacitor clamped matrix converter is possible under assumption that the circuit proposed in [21] will be extended with three capacitors and additional circuits providing automatic maintenance of voltages on clamp capacitors.

It is possible to use a method employed in control of multilevel voltage inverters and consisting in modulation with carrier signal shift. Application of slow-varying Venturini modulation functions gives positive results consisting in improvement of waveforms of the synthesized output voltages and currents regardless of φ_i and f_o values.

Reduction of voltage rating of power transistors used as bidirectional switches by half is possible only when $f_o = f_i$.

The results obtained at $f_o = f_i$ entitle us to conclude that such converter can find its application in for instance compensation of reactive power where the output frequency is equal to the supply frequency.

REFERENCES

[1] Alesina A., Venturini M.: Analysis and Design of Optimum–Amplitude Nine–Switch Direct AC–AC Converters. *IEEE Trans. on Power Electronics, Vol. 4, No. 1, Jan.* 1989.

[2] Alesina A., Venturini M.: Solid–State power conversion: a Fourier analysis approach to generalized transformer synthesis. *IEEE Trans. Circuits Syst., Vol. CAS*-28, 319–330, *Apr.* 1981.

[3] Casadei D.: Seminar on "Matrix converter" *Power Electronics and Intelligent Control for Energy Conservation PELINCEC 2005*, Warsaw, October 16–19, (2005).

[4] Erickson R. W., Al-Naseem O.A.: A New Family of Matrix Converters. http://ece-www.colorado.edu/~rwe/papers/IECON01.pdf

[5] Gyugyi L., Pelly B.R., *Static Power Frequency Changers*, New York: John Wiley & Sons, 1976.

[6] Holmes D.G., Lipo T.A.: Implementation of a controlled rectifier using ac–ac matrix converter theory. *IEEE Trans. Power Electronics, Vol. 7, No.1, Jan.* 1992, 240–250.

[7] Huber L., Borojevic D.: Space Vector Modulated Three–Phase to Three–Phase Matrix Converter with Input Power Factor Correction. *IEEE Trans. on Ind. Applicat., Vol. 31, No. 6, Nov./Dec.* 1995, 1234–1246.

[8] Meynard, H. Foch T.A., Thomas P., Courault J., Jakob R., Nahrstaedt M.: Multicell Converters: Basic Concepts and Industry Applications, *IEEE Trans. On Ind. Electronics.* Vol. 49, No. 5, pp. 978 – 987, October (2002).

[9] Meynard T.A., Fadel M., Aouda N.: Modeling of Multilevel Converters, *IEEE Transactions on Power Electronics,* Vol.44, No.3, pp. 356–364, June (1997).

[10] P.Nielsen: *The Matrix Converter for an Induction Motor Drive. PhD thesis*, Aalborg: University, Danish Academy of Technical Sciences, Danfoss A/S Transmission Division.

[11] Nielsen P., Blaabjerg F., Pedersen J.K.: New protection issues of a matrix converter: design considerations for adjustable–speed drives. *IEEE Trans. on Ind. Applicat., Vol. 35, No. 5, Sept.–Oct.* 1999, 1150–1161.

[12] Nielsen P., Blaabjerg F., Pedersen J.K.: Space Vector Modulated Matrix Converter with Minimized Number of Swithings and a Feedforward Compensation of Input Voltage Unbalance. *Proc. of PEDES '96, Vol.* II, 833–839.

[13] Nielsen P., Casadei D., Serra G., Tani A.: Evaluation of the Input Current Quality by Three Different Modulation Strategies for SVM Controlled Matrix Converters Under Input Voltage Unbalance. *Proc. of PEDES '96, Vol.* II, 794–800.

[14] Oyama J., Xia X., Higuchi T., Yamada E.: Displacement Angle Control of Matrix Converter, *PESC'97. IEEE Power Electronics Specialists Conference.* St. Louis, June 22–27, (1997), *pp.* 1033–1039.

[15] Piróg S.: Multicells converters (in Polish), *Przegląd Elektrotechniczny,* pp. 537–543, LXXIX 9/2003.

[16] Rodriguez J.: Tutorial on Multilevel Converters, *Power Electronics and Intelligent Control for Energy Conservation PELINCEC 2005*, Warsaw, October 16–19, (2005).

[17] Rząsa J.: *Selected Methods of Input Current and output Voltage Waveforms in Matrix Converters (in Polish). PhD thesis,* Warsaw: Warsaw University of Technology, (2001).

[18] Rząsa J.: Effect of control strategy in classical Venturini modulation method on waveform of common mode voltage in matrix converter (in Polish), *Przegląd Elektrotechniczny,* pp. 12–20, (2006), Nr 1.

[19] Rząsa J.: Relationships between input and output current in the matrix converter (MC), *3rd International Workshop Compatibility in Power Electronics "CPE 2003"* Gdańsk-Sobieszewo, 28–30 May (2003), 121–123.

[20] Rząsa J.: Multilevel Matrix Converter Controlled with Venturini Method, (in Polish), *Przegląd Elektrotechniczny,* pp. 57 – 64, (2007), Nr 2.

[21] Shi Y., Yang X., He Q., Wang Z.: Research on Novel Capacitor Clamped Multilevel Matrix Converter, *IEEE Trans. On Power Electronics,* Vol. 20, No. 5, pp. 1055–1066, September (2005).

[22] Venturini M., Alesina A., The Generalised Transformer: A New Bidirectional Sinusoidal Waveform Frequency Converter With Continuously Adjustable Input Power Factor, *IEEE Power Electronics Specialists Conf. PESC'80,New York,* 1980, *pp.* 242–252.

[23] Wheeler P.W., Clare J.C., Empringham L. and Bland M., Matrix Converters: The Technology and Potential for Exploitation. *The Drives and Controls Power Electronics Conference, London, Section 5, March* 2001.

[24] Wheeler P. W., Rodrigez J., Clare J.C., Empringham L., Wenstein A., Matrix Converter: A Technology Review. *IEEE Transactions on Industrial Electronics.* Vol.49, No.2, April 2002, pp. 276-287.

Reliability Consideration for a High Power Zero-Voltage-Switching Flyback Power Supply

Arash Rahnamaee [†*], Jafar Milimonfared [*], Kaveh Malekian [*], Mohammad Abroushan [*]

[*] Electrical Engineering Department, Amirkabir University of Technology
(Tehran Polytechnic University), Tehran, Iran

[†] Corresponding author: *rahnamaee@ieee.org*

Abstract—In this study, a high power zero-voltage-switching flyback power supply is presented. An appropriate efficiency is achieved by designing an auxiliary circuit for reduction of dynamic losses in the power switch. Due to the importance of reliability in the switch mode power supplies, reliability assessment is discussed for this zero-voltage-switching flyback converter in details. This paper illustrates that flyback topology has suitable reliability in the high power applications. Although by implementing auxiliary circuit in this topology, softer switching for the main switch is achieved, simplicity of the flyback converter is reduced. As a result, the control and power stages are more complicated. Also, reliability calculations demonstrate that due to soft switching and, consequently, reduction of power switch stresses, zero-voltage-switching flyback has a proper failure rate.

Keywords—Flyback, reliability, switch mode power supply, zero-voltage-switching.

I. INTRODUCTION

FLYBACK switch mode power supply derivatives have been widely used in the industrial applications [1]-[3]. The flyback topology is rather simple and low cost. Excessive voltage and current stresses of switching components are a drawback to the use of the flyback. The transformer, which is used for energy storage and provides isolation, is the particular problem of this topology. Energy stored in the leakage inductance of main transformer leads to excessive turn-off voltage at the power switch. The conventional passive RCD snubber is used to provide softer switching conditions. This passive clamp dissipates leakage energy of transformer in the clamp resistor, resulting in reduction of the overall efficiency, and makes a trade off between voltage clamping and dissipation of its resistor. An active clamp topology is a proper remedy to overcome the RCD clamp. Resonant flyback switch mode power supplies were implemented to reduce switching losses and improve total efficiency of system [4]-[5]; however voltage stress that appears on main switch is too high and makes these structures less attractive. Various active clamp techniques have been discussed in details for forward [6]-[7] and flyback [8]-[10] converters. Active clamp circuits absorb leakage energy stored in the flyback transformer and minimize voltage stress of main switch. In addition, active clamps can provide zero-voltage-switching (ZVS) for the power switch and allow that flyback converter is used in higher power up to 1kW with proper efficiency. Flyback transformer current ripple in bidirectional

magnetizing current ZVS converters [11]-[12] are permitted to be negative in portion of the switching period. The negative magnetizing current discharges main switch capacitance, and ZVS operation of the primary switch can be achieved. Increasing the transformer current ripple in the high output power can reduce the reliability of the system. The ZVS flyback converter, which is explored mathematically in [13]-[14], is constructed at 1kW output power.

Reliability prediction, one of the most important forms of reliability analysis, is usually employed in order to evaluate inherent reliability of product design. A great number of deferent methods for reliability predictions of the electrical systems and equipments are proposed in [15]. The most common empirical reliability prediction approach is based on MIL-HDBK-217 [16]. Reliability prediction models of MIL-HDBK-217 are greatly accepted for many years. As power supplies are the main part of the electronic equipment, special attention must be considered to their reliability. The overall reliability of a system is strongly dependent on the reliability of its power supply. Usually, a system can incorporate some degree of redundancy to enhance reliability, but it is more difficult to incorporate redundant power supply. It is a necessary attribute for a produce to be fit for purpose for which it is designed and every bit as important as the actual functionality of the product.

According to the use of ZVS circuit, the flyback power supplies can be used in the high power application with suitable efficiency and failure rate. In the present paper, a 1kW power supply is constructed using ZVS flyback topology (see Fig.1). Also, failure rate computations are done on all components of the constructed flyback power supply and failure rate of whole system are achieved.

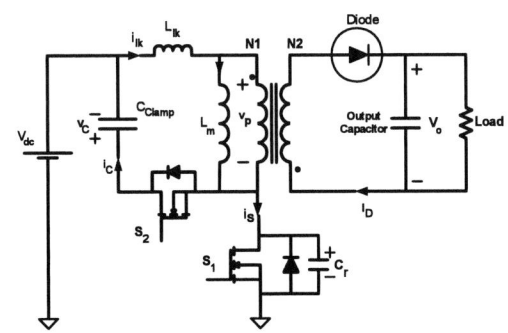

Fig. 1. Simplified schematic of ZVS flyback converter.

978-1-4244-1741-4/08/$25.00 ©2008 IEEE

II. ZVS FLYBACK POWER SUPPLY

Simplified schematic of active clamp incorporated to the flyback topology is shown in Fig. 1. The flyback transformer has been replaced by equivalent circuit including magnetizing inductance L_m, leakage inductance L_r, which represents the sum of primary referred leakage inductance of transformer and external resonant inductor. C_r is the resonant capacitor which is the parallel combination of the output capacitance of main switch and auxiliary switch. Switches S_1 and S_2 are shown with their body diodes. The clamp capacitor and auxiliary switch absorb the leakage energy stored in transformer and reduce the main switch voltage spike. The resonant between resonant capacitance C_r and inductance L_r makes the ZVS operation for both switches possible, however control and power stages of converter become more complicated. Advantages of ZVS operation are achievable with expense of additional switch and other components.

Fig. 2 illustrates key waveforms for the active clamp ZVS flyback converter. As mathematically discussed in [13], the power stage has six operation states shown in Fig. 2. Purpose of present paper is the reliability assessment of ZVS converter in high power output so that detailed operation of ZVS flyback converter is not considered thoroughly. In other words, after main switch is turned off during the time between t_3 and t_4, resonant inductance L_r and clamp capacitor C_{Clamp} resonant, so that energy stored in L_r becomes negative and voltage spike of the main switch is reduced. Auxiliary switch must be turned on, before L_r current becomes negative for ZVS operation of auxiliary switch, at t_4 sufficient energy must be stored in L_r for discharge of resonant capacitor C_r in the next interval (between t_4 and t_5). When switch voltage becomes zero, the body diode of main switch conducts the negative current of L_r. For assurance of ZVS of main switch, S_1 must be turned-on in the time between t_5 and t_6. Otherwise, the current of L_r reverses positive again and at least ZVS operation is partially lost. The energy stored in the leakage inductance L_r is recycled, so that output power can be increased with proper efficiency. Unlike RCD snubber, voltage spike suppression of main switch is done without any power dissipation. Characteristics of the constructed ZVS flyback switch mode power supply are given in Table I.

IGBTs have less conduction loss compared to MOSFETs at the same rating voltage. Excessive turn-off losses at high frequency operation are the main drawback of IGBTs. An important advantage of ZVS is that the severity of this problem can be reduced by adding external capacitor C_r of the IGBTs to slow down the increase rate of IGBT collector-emitter voltage when switch is turned off. As a result, the current and voltage turn-off overlap time of IGBT is decreased. The energy stored in C_r is discharged by resonant inductance before IGBT is turned on. So that energy stored in additional main switch capacitor is not dissipated when IGBT is turned on. The failure rate of IGBT decreases because working temperature is fallen down due to less dissipation. The constructed 1kW ZVS active clamp

flyback converter is shown in Fig. 3. Efficiency manner of constructed ZVS flyback converter (Fig. 4) depicts that flyback converter can achieve efficiency up to 90% in 1kW output power.

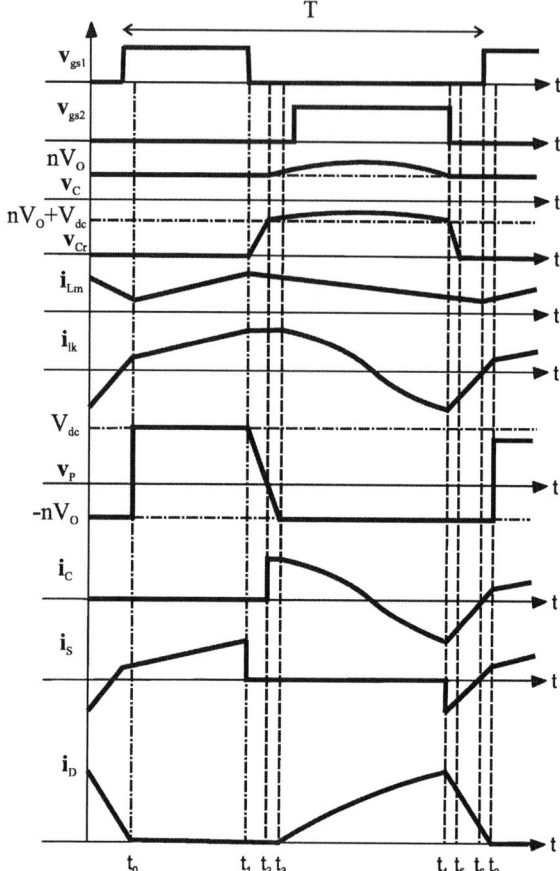

Fig. 2. Key waveforms of the ZVS active clamp converter.

Fig. 3. Constructed 1kW ZVS active clamp flyback converter.

TABLE I
CHARACTERISTICS OF THE CONSTRUCTED FLYBACK POWER SUPPLY

Input voltage	180-250 V_{ac}
Output voltage	48 V_{dc}
Output Current	21A
Switching Frequency	50kHz

366

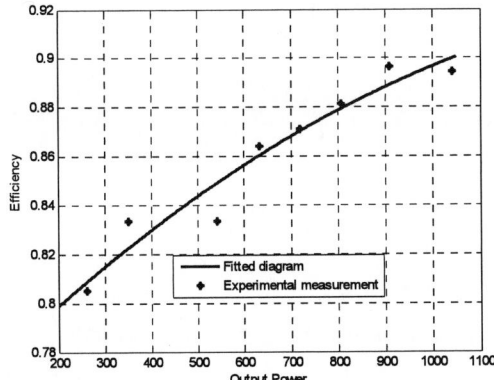

Fig. 4. Efficiency curve with respect to the output power.

III. RELIABILITY COMPUTATIONS FOR THE CONSTRUCTED ZVS FLYBACK

A. Concept of Reliability

A failure occurs when a component does not perform required function. Reliability is commonly characterized by the failure rate $\lambda(t)$. The failure rate is define as following relation;

$$Failure\ Rate = \frac{Number\ of\ failures}{Operating\ Time} \qquad (1)$$

Various statistical distributions are used in reliability prediction as discussed in [15]. Most of reliability prediction methods assume that reliability of electronic equipments can be described as follows;

$$R(t) = e^{(-\lambda t)} \qquad (2)$$

With this assumption, the failure the failure rate is constant and random [17]. Common approaches consider that the total failure rate of system is the sum of comprising components failure rate. For a system consisted of n components, the total failure rate can be calculated as;

$$\lambda_{total} = \lambda_1 + \lambda_2 + \ldots + \lambda_n \qquad (3)$$

The MMTF (mean time to failure), is used to indicate the reliability, it is defined as;

$$MTTF = \frac{1}{\lambda} \qquad (4)$$

$$R_{total}(t) = e^{-(\lambda_{total} t)} \qquad (5)$$

Failure rate for a high reliability part are typically around 10^{-8} failures per hour. Another indicator often used is the mean time between failures, MTBF. Strictly, this is the mean time between failures as equipment goes through successive cycles of failure and repair. MTBF is related to the MTTF as follows,

$$MTBF = MTTF + repair\ time \qquad (6)$$

Since repair is not normally possible for semiconductor parts, the MTTF and MTBF often have same meaning. As follows, failure rates of all components of constructed converter calculated and total failure rate of system are derived.

B. Reliability Prediction

Nowadays, the usage of the MIL-HDBK-217 is common in many areas and it is the most used reliability prediction method of the electronic components. Values included in the MIL-HDBK-217 have been built up from experience of component failures and, also, from the results of accelerated testing at the elevated stress. MIL-HDBK-217 contains two prediction methods: the "parts count" method, and the "parts stress" method. The part count prediction approach is applicable in the early stage of design when little information about the system is known. The parts stress prediction approach requires a greater amount of detailed information and is applicable during the later design progress when stresses and other environmental factors are known for each component [18]. As a design progress, more detailed design information is available and it is possible to estimate overall reliability by calculating the individual component failure rates. This method is known as part stress analysis as it is based on the operating stress of each component.

A variety of different models are used depending on the component type. For example, the part failure rate of semi-conductors in the parts stress method is defined as follows,

$$\lambda_p = \lambda_b \times \pi_E \times \pi_A \times \pi_Q \times \pi_R \times \pi_S \times \pi_C \qquad (7)$$

where, λ_p is the part failure rate; λ_b is the base failure rate; π_E is the environment factor; π_A is the application factor (linear or switched); π_Q is the quality factor; π_R is the power rating; π_S is the voltage stress factor; and π_C is the complexity factor.

1) Main switch: dynamic and static losses measurement of semiconductor components is essential for reliability prediction based on parts stress method in MIL-HDBK-217. The voltage and current waveforms of the main switch are illustrated in Fig. 5(a). Fig. 5(b), properly, shows the voltage and current overlap during turn-off and shows that because of adding external C_r at collector-emitter of IGBT, increase rate of v_{Cr} is decreased and dynamic dissipation of switch reduced. Fig 5(c) illustrates ZVS operation of the main switch. A BUP314D IGBT is used as a main switch in the constructed ZVS flyback power supply. Considering Fig. 7 and characteristic of this IGBT, the main switch losses can be calculated as follows,

$$P_{Loss} = P_{Static} + P_{Dynamic} \qquad (8)$$

$$v_{CE(sat)} = 3V \Rightarrow P_{Static} = 12.9W$$

$$P_{Dynamic} = \frac{1}{T_s} \int_{DT_s}^{DT_s + t_{overlap}} v_{switch} i_{switch} dt \qquad (9)$$

$$P_{Dynamic} = 34W \Rightarrow P_{Loss} = 46.9W$$

where D is duty cycle; T_s is switching period; and $t_{overlap}$ is current and voltage overlap duration for the main switch.

The base failure rate of IGBT is considered same as MOSFET [19], and other factors used in IGBT reliability computation are considered like ones related to BJT [20].

367

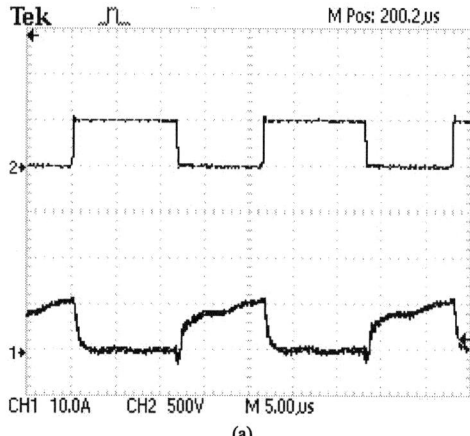

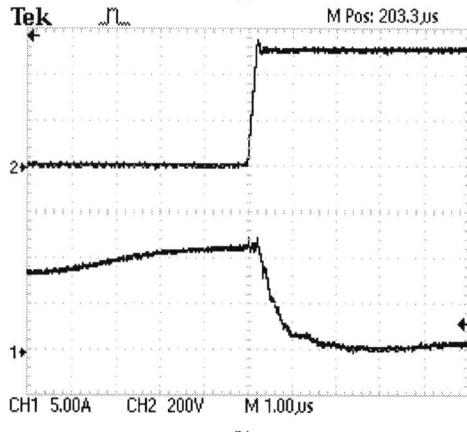

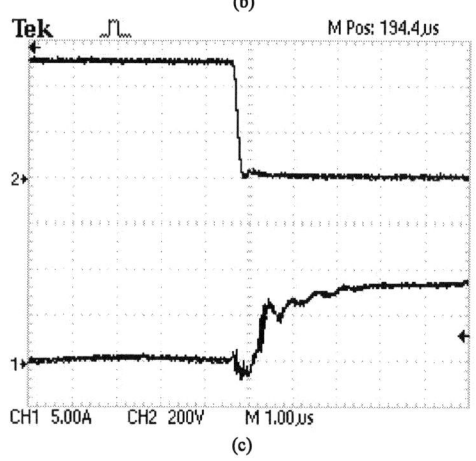

Fig. 5. Voltage and current waveforms of main switch (a) in the whole switching period (b) while switch is turning off (c) while switch is turning on.

Reliability factors are derived using measured values and MIL-HDBK-217F as follows,

$$\lambda_P = \lambda_b \times \pi_T \times \pi_A \times \pi_R \times \pi_S \times \pi_Q \times \pi_E \quad (10)$$

$$\lambda_b = 0.012$$

$$T_J = T_C + \theta_{JC} P_{Loss}$$

$$\theta_{JC} = 0.42^\circ C /W$$

$$T_J = 33^\circ C + 46.9 \times 0.42 = 52.70^\circ C$$

$$\pi_T = e^{\left(-2114\left(\frac{1}{T_J+273} - \frac{1}{298}\right)\right)} = 1.82$$

$$\pi_A = 0.7$$

$$\pi_R = 8$$

$$V_s = \frac{Applied\ V_{CE}}{Rated\ V_{CE}} = \frac{620}{1200} = 0.516$$

$$\pi_S = 0.045 e^{(3.1 \times V_s)} \approx 0.22$$

where V_{CE} is voltage across collector-emitter of IGBT; T_J is junction temperature; T_C is case temperature; and θ_{JC} is junction-to-case thermal resistance.

$$\pi_Q = 5.5$$

$$\pi_E = 6$$

$$\lambda_p = 0.012 \times 1.82 \times 0.7 \times 8 \times 0.22 \times 5.5 \times 6 = 0.88$$

$$failures\ /10^6\ Hours$$

By the use of active clamp circuit, the stress of IGBT, the main switch, is reduced and, as a result, a proper failure rate is obtained.

2) Auxiliary Switch: Fig. 6 illustrates voltage and current of the auxiliary switch. This switch must be turned on before current of the resonant inductance L_r becomes negative. As shown in Fig. 6, the MOSFET used as auxiliary switch operates at ZVS condition. The part failure rate for MOSFET can be predicted as follows;

$$\lambda_P = \lambda_b \times \pi_T \times \pi_A \times \pi_Q \times \pi_E \quad (11)$$

So the failure rate for auxiliary switch can be calculated as;

$$\lambda_b = 0.012$$

$$T_J = T_C + \theta_{JC} P_{Loss}$$

$$P_{loss} = 11W$$

$$\theta_{JC} = 0.57^\circ C /W$$

$$T_J = 28^\circ C + 11 \times 0.57 = 34.2^\circ C$$

$$\pi_T = e^{\left(-1925\left(\frac{1}{T_J+273} - \frac{1}{298}\right)\right)} = 1.21$$

$$\pi_A = 4$$

$$\pi_E = 6$$

$$\lambda_p = 0.012 \times 1.21 \times 4 \times 6 = 0.34\ failures\ /10^6\ Hours$$

3) Output diode: In order to calculate output diode failure rate, the same procedure like what applied for IGBT has been done. Output diodes current waveform is shown in Fig. 7. As shown in this figure, external resonant inductance L_r reduces decrease rate of the output diodes. Therefore, dynamic losses of the output diodes are reduced. For calculating diode static losses, the relations given in datasheet of applied diode, BYW99P-200 are used [21].

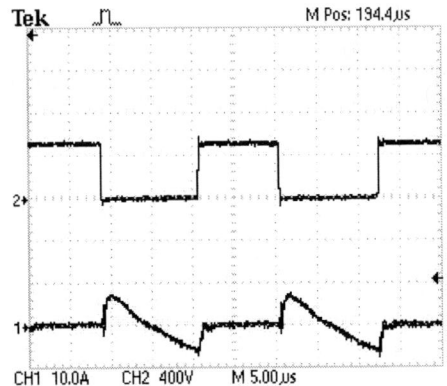

Fig. 6. Voltage and current waveforms of the auxiliary switch.

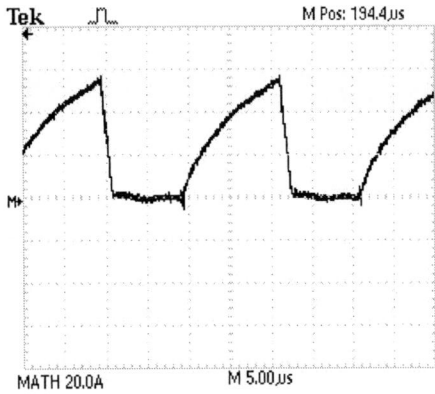

Fig. 7. Current waveform of output diodes.

$I_{AV}\left(Output\ Diode\right)=21A$

$I_{RMS}\left(Output\ Diode\right)=18A$

$P_{Static}=0.65\times I_{F(AV)}+0.016\times I_{F(RMS)}^{2}=18.83\,W$

$P_{Dynamic}=19\,W$

$P=19+18.83=37.83\,W$

$T_{JC}\left(Diode1\right)=P\left(Diode1\right)$
$\qquad\qquad\times\theta_{jc}\left(Per\ diode\right)+P\left(Diode\,2\right)\times\theta_{c}$

$\theta_{jc}\left(Per\ diode\right)=1.8°C\ /W$

$\theta_{c}=0.2°C\ /W$

$T_{J}=T_{C}+T_{JC}$

$T_{J}=38°C+\left(9.45\times1.8+9.45\times0.2\right)=56.91°C$

Failure rate of the output diode can be predicted using the above calculations.

$$\lambda_{P}=\lambda_{b}\times\pi_{T}\times\pi_{S}\times\pi_{C}\times\pi_{Q}\times\pi_{E}\qquad(12)$$

$\lambda_{b}=0.025\ Failures\ /10^{6}\ Hours$

$\pi_{T}=e^{\left(-3091\left(\frac{1}{T_{J}+273}-\frac{1}{298}\right)\right)}=2.72$

$V_{S}=\dfrac{Applied\ Voltage}{Rated\ Voltage}=\dfrac{120}{200}=0.6$

$\pi_{S}=V_{S}^{2.43}=0.3$

$\pi_{C}=1$

$\pi_{Q}=8\ Plastic\ Case$

$\pi_{E}=6.0$

For each diode, it can be written,

$\lambda_{P}=0.025\times2.72\times0.3\times1\times8\times6=0.97$

$$Failures\ /10^{6}\ Hours$$

4) Bridge diode: In the same way, the bridge diode reliability can be predicted as follows,

$\lambda_{b}=0.038\ Failures\ /10^{6}\ Hours$

$I_{AV\ (input\ stage)}=4.2A$

$P_{Loss}=2.73\,W$

$\theta_{JC}=2.2°C\ /W$

$T_{J}=T_{C}+\theta_{JC}P_{loss}=35+2.2\times2.73=41°C$

$\pi_{T}=1.8$

$V_{S}=\dfrac{310}{1000}=0.31$

$\pi_{S}=0.06$

$\pi_{C}=1$

$\pi_{Q}=5.5$

$\pi_{E}=6$

$\lambda_{p}=0.038\times1.8\times0.06\times1\times5.5\times6=0.14\ Failures\ /10^{6}\ Hours$

5) Power Transformer: The failure rate for a transformer is given as,

$$\lambda_{p}=\lambda_{b}\times\pi_{T}\times\pi_{Q}\times\pi_{E}\qquad(13)$$

The hot spot temperature is required to calculate π_{T}. So, reliability prediction procedure can done as follows,

$\lambda_{b}=0.049\ Failures\ /10^{6}\ Hours$

$T_{HS}=T_{A}+1.1(\Delta T)=\left(36-27\right)\times1.1+27=36.9°C$

$\pi_{T}=e^{\left(\frac{-.11}{8.617\times10^{-5}}\left(\frac{1}{T_{HS}+273}-\frac{1}{298}\right)\right)}=1.17$

$\pi_{Q}=3$

$\pi_{E}=6$

$\lambda_{p}=0.049\times1.17\times3\times6=1.03\ Failures\ /10^{6}\ Hours$

Same procedure can be done for the failure rate prediction of the resonant inductor.

$\lambda_{p}=0.022\times1.2\times3\times6=0.47\ Failures\ /10^{6}\ Hours$

6) Input and output capacitors: To predict the reliability of capacitors, the factors mentioned in the following relation should be derived,

$$\lambda_{p}=\lambda_{b}\times\pi_{T}\times\pi_{SR}\times\pi_{Q}\times\pi_{E}\times\pi_{cap}\times\pi_{V}\qquad(14)$$

where π_{cap} is capacitance factor; π_V is voltage stress factor; and π_{SR} is series resistance factor which is considered only for Tantalum capacitors.

$\lambda_b = 0.00012$

$T = 27°C$

$\pi_T = e^{\left(\frac{-0.35}{8.617\times10^{-5}}\left(\frac{1}{T+273}-\frac{1}{298}\right)\right)}$

$\pi_T = 1.1$

$\pi_{SR} = 1$

$\pi_Q = 3$

$\pi_E = 10$

For the output capacitors, it can be written,

$\pi_{cap} = C^{0.23} = 1000^{0.23} = 4.9$

$S = \dfrac{Actual\ Power\ Dissipation}{Rated\ Power} = \dfrac{48}{60} = 0.8$

$\pi_V = \left(\dfrac{S}{0.6}\right)^5 + 1 = 5.21$

$\lambda_p = 0.00012\times1.1\times1\times3\times10\times4.9\times5.2 = 0.1$

Failures $/10^6$ *Hours*

And in the same way for the input capacitors, the failure rate can be predicted as follows,

$\pi_{cap} = 560^{0.23} = 4.3$

$S = \dfrac{310}{400} = 0.775$

$\pi_V = 4.66$

$\lambda_p = 0.00012\times1.1\times1\times3\times10\times4.3\times4.66 = 0.08$

Failures $/10^6$ *Hours*

7) Clamp capacitor: In this work, MKT capacitors are applied as snubber capacitors. This kind of capacitor has a different base failure from electrolytic capacitors.

$\lambda_b = 0.00051\ Failures\ /10^6\ Hours$

$\pi_T = 1$

$\pi_{SR} = 1$

$\pi_Q = 3$

$\pi_E = 10$

$\pi_{cap} = C^{0.09} = (8.2)^{0.09} = 1.2$

$S = \dfrac{210}{250} = 0.84$

$\pi_V = \left(\dfrac{S}{0.6}\right)^5 + 1 = 6.37$

$\lambda_p = 0.00051\times1\times1\times3\times10\times0.93\times1.01 = 0.11$

Failures $/10^6$ *Hours*

9) Control section: Failure rate of the control section is negligible with respect to failure rate of power stage components. Failure rate of control section will be discussed in following,

Fuse:

$\lambda_p = \lambda_b \times \pi_E$

$\lambda_p = 0.01\times2 = 0.02\ Failures\ /10^6\ Hours$

IC:

$\lambda_p = \lambda_b \times \pi_T \times \pi_Q \times \pi_E$

$\lambda_p = 0.0048\times1\times2\times2 = 0.02\ Failures\ /10^6\ Hours$

Resistors:

$\lambda_p = \lambda_b \times \pi_P \times \pi_Q \times \pi_E$

$\lambda_p = 0.00045\times0.58\times10\times4 = 0.01\ Failures\ /10^6\ Hours$

Ceramic and other types of capacitor:

$\lambda_p = \lambda_b \times \pi_{CV} \times \pi_Q \times \pi_E$

$\lambda_p = 0.00075\times1\times10\times2 = 0.015\ Failures\ /10^6\ Hours$

Table II provides summarized results of reliability calculations as well as total failure rate of constructed active clamp ZVS flyback. As shown in this table, with regard to the ZVS operation of main switch, adding of external capacitor Cr and suppression of voltage spike on IGBT, system has found a proper failure rate.

TABLE II
FAILURE RATE CALCULATIONS RESULTS FOR BOTH POWER AND CONTROL STAGES

	Power Stage									Control Stage				
	Main Switch	Auxiliary Switch	Output Diode	Bridge Diodes	Transformer	Input Capacitors	Output Capacitors	Active Clamp Capacitor	External Inductance	Resistors	Capacitors	Fuse	IC	Others
Number	1	1	4	4	1	4	4	2	1	23	18	1	4	...
Failure rate per part (failures/10^6Hours)	0.88	0.34	0.97	0.14	1.03	0.08	0.1	0.11	0.47	0.01	0.015	0.02	0.01	0.04
Total failure rate (failures/10^6Hours)	0.88	0.34	3.88	0.56	1.03	0.32	0.4	0.22	0.47	0.23	0.27	0.02	0.04	0.04
Power stage: 8.1 (Failures/10^6Hours)										Control stage: 0.6 (Failures/10^6Hours)				
Total failure rate of system: 8.7 (Failures/10^6Hours)														

IV. CONCLUSION

In this paper, a 1kW active clamp ZVS flyback power supply has been designed and constructed. The reliability calculations have been done distinctly for all components of the constructed power supply. According these calculations, the active clamp ZVS flyback power supply has an appropriate reliability which is a result of power switch reduction. Using an active clamp circuit, the voltage spikes have been suppressed. Therefore, the efficiency of power supply has been suitable for a wide range of the output power. Also, this paper has illustrated the practical aspects for constructing a high power active clamp ZVS flyback converter. Finally, it has been proved that this topology can be used in applications with output power up to 1kW having proper efficiency and reliability.

REFERENCES

[1] Y. Gu, X. Gu, L. Hang, Z. Lu, and Z. Qian, "Improved Wide Range Dual Switch Flyback DC/DC Converters," *Applied Power Electronics Conference and Exposition*, vol. 1, pp. 654-660, 2004.

[2] H. Chen, W. Dong, Y. He, and Z. Qian, "Secondary Side Post Regulation Application in Multiple Outputs Flyback Converter," *Power Electronics and Drives Systems*, vol. 2, pp. 1273-1277, 2005.

[3] C.C. Wen and C.L. Chen, "Magamp application and limitation for multiwinding flyback converter," *Electric Power Applications, IEE*, vol. 152, pp. 517-525, 2005.

[4] C.T. Choii, C.K. Li, and S.K. Kok, "Control of an active clamp discontinuous conduction mode flyback converter," *Proc. IEEE Power Electronics and Drive Systems Conf.*, vol. 2, pp. 1120-1123, 1999.

[5] I.D. Jitaru and S. Birca-Galateanu, "Small-signal characterization of the forward-flyback converters with active clamp," *Proc. IEEE Applied Power Electronics Conf.*, vol. 2, pp. 626–632, 1998.

[6] B. Carsten, "Design techniques for transformer active reset circuits at high frequency and power levels," *in High Freq. Power Conversion Conf. Proc.*, pp. 235-245, 1990.

[7] C. Leu, G. Hua, and F. Lee, "Comparison of forward topologies with various reset schemes," *in Proc. 9th Ann. VPEC Power Electron. Sem.*, pp. 101-109, 1991.

[8] I. Jitaru, "Zero voltage PWM, double ended converter," *High Frequency Power Conversion Proceedings*, pp. 394-404, 1992.

[9] Y.S. Lee and B.T. Lin, "Adding active clamping and soft switching to boost-flyback single-stage isolated power-factor-corrected power supplies," *IEEE Trans. Power Electron.*, pp. 1017–1027, 1997.

[10] Y. Xi, P. K. Jain, and G. Joos, "A zero voltage switching flyback converter topology," *Power Electronics Specialists Conf.*, pp. 951-957.

11] H. Martin, "Topology for miniature power supply with low voltage and low ripple requirements," *U.S. Patent 4618 919*.

[12] K. Yoshida, T. Ishii, and N. Nagagata, "Zero voltage switching approach for flyback converter," *in Proc. 14th Int. Telecomm. Energy Conf.*, pp. 324-329, 1992.

[13] R. Watson, F.C. Lee, and G.C. Hua "Utilization of an Active-Clamp Circuit to Achieve Soft Switching in Flyback Converters," *IEEE Trans. On Power Electronics*, vol. 11, no. 1, Jan. 1996.

[14] B.R. Lin, C.-E. Huang, K. Huang, and D. Wang, "Design and implementation of zero-voltage switching flyback converter with synchronous rectifier," *IEE Proc.-Electr. Power Appl.*, vol. 153, no. 3, May 2006.

[15] IEEE Std 1413. 1™, *"IEEE Guide for Selecting and Using Reliability Predictions Based on IEEE 1413,"* IEEE Std, 2002.

[16] Department of Defense, *"Reliability Prediction Of Electronic Equipment,"* MIL-HDBK-217F, Washington DC, Notice 2, February 1995.

[17] A.Meyna, and B. Pauli, *"Taschenbuch Der Zuverlässigkeits- Und Sicherheitstechnik,"* München, Germany: Carl Hanser Verlag, 2003.

[18] M. Vintr, "Reliability Assessment for Components of Complex Mechanisms and Machines," *12th IFToMM World Congress*, Besançon (France), June 2007.

[19] D. Hirschmann, D. Tissen, S. Schröder, and R.W. De Doncker, "Reliability Prediction for Inverters in Hybrid Electrical Vehicles," *IEEE Trans. on Power Electronics*, vol. 22, no. 6, Nov. 2007.

[20] G. Chen, R. Burgos, Z. Liang, F. Lacaux, F. Wang, J.D. Wyk, W.G. Odendaal, and D. Boroyevich, "Reliability-Oriented Design Conside-rations for High-Power Converter Modules," *35th Annual IEEE Power Electronics Specialists Conference*, 2004.

[21] Datasheet of BYW99P/PI/W, *"High Efficiency Fast Recovery Rectifier Diodes,"* ST Corporation, Oct. 1999.

The Traction Drive Topology Using the Matrix Converter with Middle-Frequency Transformer

Martin Pittermann, Pavel Drábek, Marek Cédl

Department of Electromechanics and Power Electronics, University of West Bohemia, Plzen, Czech Republic
pitterma@kev.zcu.cz, drabek@kev.zcu.cz, alvist@kev.zcu.cz

Abstract— This paper presents research motivated by industrial demand for special traction drive topology devoted to minimization of traction transformer weight against topology with classical 50Hz traction transformer. The main attention has been given to the special traction drive topology for AC power systems: input high voltage trolley converter (single phase matrix converter) – middle frequency transformer – output converter (single phase voltage-source active rectifier + three phase voltage-source inverter) - traction motor. The control algorithm (Inserting of NULL vector of matrix converter, Two-value control of secondary active rectifier, PWM and etc.) of innovative traction topology with middle-frequency transformer has been described.

Keywords— Power converters for EV, Traction application, Matrix converter, High Voltage power converters, Transformer.

I. INTRODUCTION

At the beginning we can start describing configuration of the classical traction drive topology with the normal 50Hz (and especially 16,7 Hz for Germany and Austria) transformer situated at the input part of the AC traction vehicle as shown in Fig.1. Classical traction transformer is situated in the input part of the train vehicles and transformer outputs are connected to the traction converters for regulation the traction drives and auxiliary drives. Adjusting of the high level input trolley voltage to the applicable level and isolating character as well are the main aims of the traction transformer.

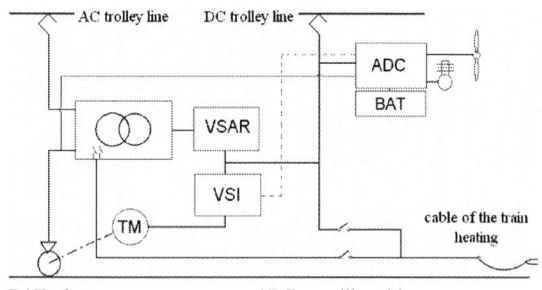

BAT – battery   ADC – auxiliary drives converter

VSAR – voltage-source active rectifier   VSI – voltage-source inverter

TM – traction motor

Fig. 1: The topology of the multi-system train vehicle with 50Hz
traction transformer at the input part.

The Czech railways dispose of two supply systems: DC supply system 3kV and AC supply system 25kV/50Hz. Therefore the Fig.1 shows the topology of multi-system loco train. The configuration is very similar to the classical AC loco train. Traction transformer is situated in the input part of the train vehicles and transformer outputs are connected to the traction converters for regulation the traction drives and auxiliary drives. In contrast with classical AC loco train there is additional connection to DC supply system which is directly connected to the DC link of TDC – main traction drive converter and supply VSI – voltage-source inverter.

II. TOPOLOGY WITH MIDDLE-FREQUENCY TRANSFORMER (MFT)

Today we can see several low power applications using converter topology with MFT (e.g. UPS, switching supply sources, welding machines, etc.). According to the growing of the power electronic area (mainly the price decreasing of semiconductor components) we can look forward to the future using this idea for high power devices (e.g. traction application) as well. Basic topology of the electric drive with MFT is in Fig.2:

1. Input converter (VTM in Fig.2) regulates input line voltage to the appropriate waveform for the MFT (e.g. AC course with high frequency)

2. Middle frequency transformer (MFT) galvanic insulates input and output and adjusts output voltage level

3. Output traction converter (SMM) modifies middle frequency course from transformer to suitable waveform for traction drive supply

Basically the configuration of the input (VTM) and output (SMM) converter can be realised arbitrarily. In case of input and output AC voltage converters VTM and SMM can be design as direct frequency converter (matrix converter [1],[5],[6],[7] or similar topology [2],[3]) or indirect frequency converter (rectifier + voltage source inverter [4]).

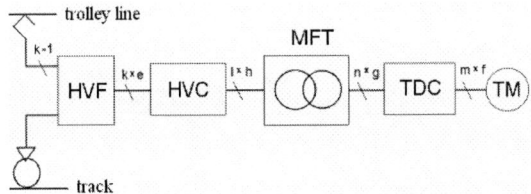

HVF – high voltage filter   HVC – high voltage converter

MFT – middle frequency transformer   TM – traction motor

TDC – main traction drive converter

Fig. 2. The block scheme of the electric drive topology with middle-frequency transformer

III. INPUT HIGH VOLTAGE CONVERTER REALIZATION

For designing high voltage converter it is necessary use high voltage semiconductor components (this should be idea in the future – e.g. SiC based material) or special converter topology ([2],[3],[4],[9] when input high voltage is spread on several active switches in serial connection. There are several variants to realize high voltage converter (it means that input voltage is higher than operating voltage of each semiconductor component):

a) Series connection of semiconductor components – series connection is very problematic and for realization it is necessary to guarantee voltage uniform spread – in steady state and transient state as well.

b) Special converter topology called "Modular Multi Level Converter" [2].

c) Converter topology based on current-source converters with input high voltage filter (consist of reactor and capacitor from current-source topology point of view).

d) Converter topology coming out of indirect frequency converters (it means 1f voltage-source active rectifier + 1f voltage-source inverter) [4].

In the following chapter we will discuss the train drive topology consist of middle-frequency transformer fed by single phase matrix converter as the input high voltage converter.

IV. SINGLE PHASE MATRIX CONVERTER

The single phase matrix converter is the one possibility of input high voltage converter based on current-source converter. Matrix converter can be understood as an alternative today standard indirect frequency converter (Active voltage rectifier – DC voltage bus – Voltage source inverter) for AC drives supplying. The absence of DC BUS (then any energy accumulator) of matrix converter is the main difference of these converters. Under matrix converter definition we can generally imagine system of bidirectional switches ordered to the matrix of size n * k (to connect n-phase supply source to the k-phase load (Fig.3).

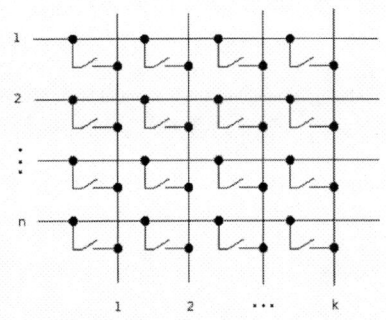

Fig. 3. Basic disposition of matrix converter n/k

Matrix of switches in the scheme consists of ideal bidirectional switches. In present time there is not any device to replace bidirectional switch but exist several variants how to substitute bidirectional switch by the common semiconductor components (Fig.4).

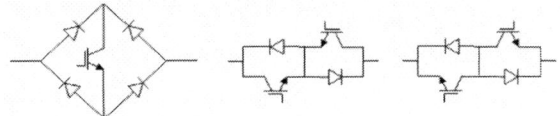

Fig.4. Variants of bidirectional switches

V. CONTROL OF THE TRACTION TOPOLOGY WITH SINGLE PHASE MATRIX CONVERTER

Fig. 5 shows scheme of the traction drive with single phase matrix converter as an input traction converter which supplies MFT. The circuit consists of input high voltage filter connected to the input of the matrix converter which supplies middle-frequency transformer. The output of the transformer connects single phase active voltage rectifier, three phase voltage source inverter supplying AC motor is added through the DC bus line.

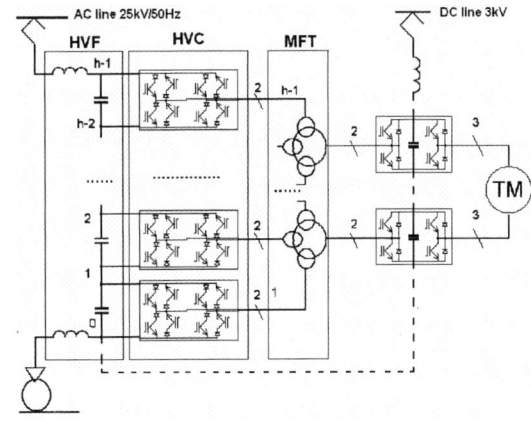

HVF – high voltage filter   HVC – high voltage converter

MFT – middle frequency transformer   TM – traction motor

Fig. 5. Traction topology with single phase matrix converter

Fig.6 schematically demonstrates basic ideas of operation principle (upper-hand part of the figure) of the traction topology with middle-frequency transformer. Input 50Hz (16,7Hz) sine-wave trolley voltage is cut to the middle-frequency voltage waveform (up to kHz) which is set on the input of the MFT. The secondary single phase voltage-source active rectifier processes this cut middle - frequency voltage waveform and control the DC bus line voltage used for supplying the three phase voltage-source inverter.

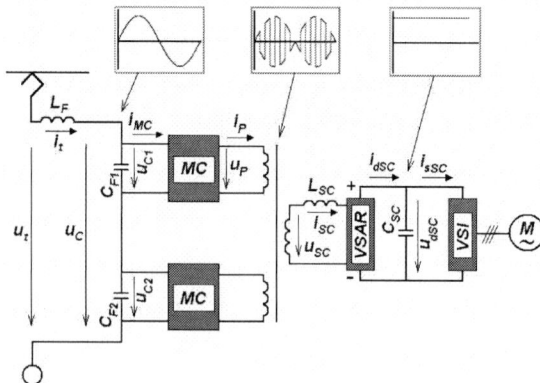

MC - 1f matrix converter (primary converter)

VSAR - secondary voltage-source active rectifier

VSI - secondary voltage-source inverter

L_F - reactor of input filter C_F - capacitor of input filter

L_{SC} - clamping choke of the secondary voltage-source active rectifier

Fig.6. Topology with single phase matrix converter as a primary traction converter.

In the input of the matrix converter we can see filter with capacitor and the inductive load (winding of the transformer) is connected at its output. These facts have to be taken into account at the control strategy of matrix converter – we cannot short circuit input terminals and disconnect output terminals at the same time.

Detailed scheme of 1f matrix converter is shown in Fig.7. Switching individual branches proceed by sequential crossing within switching states. More details about converter commutation you can find in [1]. To control matrix converter we use following stabile switching states:

0167 – Input of matrix converter is directly connected to the output (**state "1"**)

2345 – Input of matrix converter is reversely connected to the output (**state "-1"**)

0123 (or 4567) – input of the matrix converter is disconnect and the output is short circuit (so-called NULL vector) (**state "0"**)

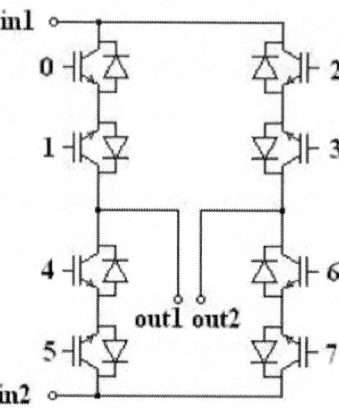

Fig.7. Detailed scheme of 1f matrix converter with individual transistors

Described traction topology consists of primary and secondary converter and then we have to consider appropriate control switching of both converters for regulation algorithms of traction topology with MFT. For example primary matrix converter we can control by Inserting NULL vectors, square-wave control, Two-value control, PWM etc. Secondary voltage-source active rectifier can be controlled by PWM, Two-value control etc. From several mentioned control algorithms of whole traction drive we introduce following variants:

1. Control by matrix converter by means of inserting NULL vectors [9],[10].
2. Control by secondary voltage-source active rectifier.
3. Control by matrix converter by means of two-value control.
4. Control by matrix converter by means of PWM.

In regard of finite page number in this paper we will discuss variants 2.

VI. CONTROL BY SECONDARY VOLTAGE-SOURCE ACTIVE RECTIFIER

Power circuit of traction topology with illustration of appropriate control diagram is shown in Fig. 8. The basic idea of this control algorithm is following: secondary voltage-source active rectifier takes sine-wave phase current from middle-frequency transformer (two-value control) and square-wave control of the matrix converter which put together the phase current of output active rectifier to the 50Hz sine wave taken from the trolley line. Amplitude and phase shift of the trolley line current is controlled only by output active rectifier.

Detailed block scheme of control algorithm of the matrix converter by secondary active rectifier with synoptical charts of mentioned control method is presented in Fig. 9.

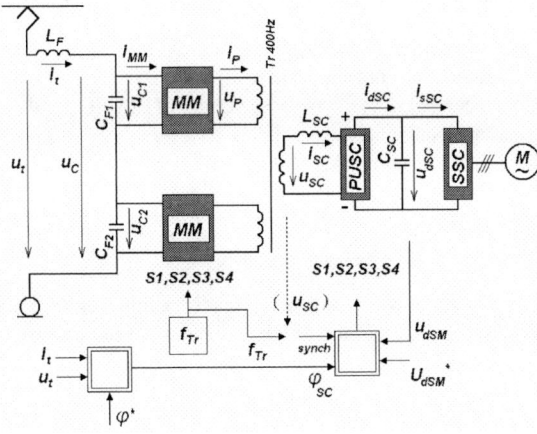

Fig.8. Power circuit of the drive with control circuit for control of the traction drive by output active rectifier

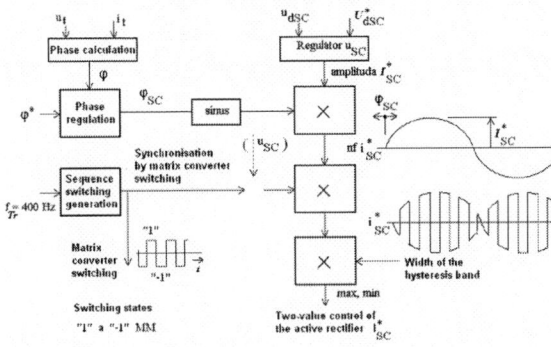

Fig.9. Block scheme of control algorithm of control of the traction drive by output active rectifier and square-wave control of matrix converter

Principle of the control algorithm is following. Measured φ and required φ* phase shift of trolley values are set to the regulator and its output φsc is used for shifting of the phase current of output active rectifier. Amplitude of the phase current of output active rectifier is set by regulator of the output DC bus voltage Udsc (upper and right hand part of the Fig.9). The matrix converter is simply controlled by square-wave control which put together the phase current of output active rectifier (400Hz square wave modulated by 50Hz sine wave) to the 50Hz sine wave taken from the trolley line.

Fig.11 and Fig.12 show appropriate simulation results of control algorithm of two-value control by output active rectifier. In the Fig. 11 you can see variables of traction trolley line – voltage and current of the trolley line (ut and it). In the figure the change from rectifier mode to the inverter mode and return back is presented. The digital filter is used in the control algorithm to detect phase shifting of between input values (input trolley voltage and current - Fig.10).

Fig.12 presents variables of secondary active rectifier – voltage usc (upper-hand part of the figure ... this voltage waveform is the same as output variables of the middle-

frequency transformer and outputs of the primary matrix converter) and input current isc (lower-hand part of the figure) with hysteretic band of the input current max, min and output voltage in the DC bus uc used for supplying secondary three phase voltage-source inverter for traction motor.

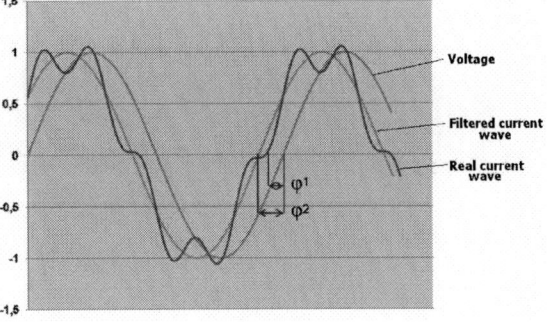

Fig.10. Reasons why it is important to filter waveform of the input trolley line current

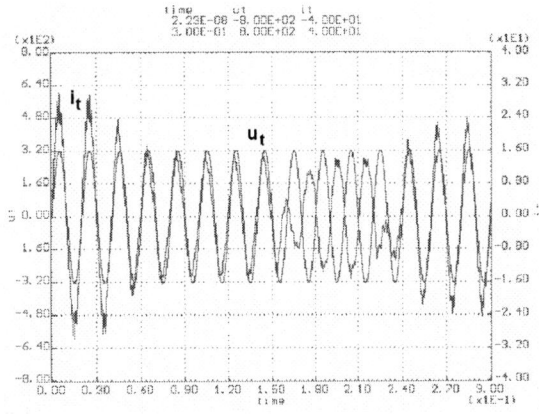

Fig.11. Simulation results of primary matrix converter (input and output variables) for following parameters: Ut=230V, P=3kW, LF=40mH, CF=50uF, LSC=1mH, CSC=4mF

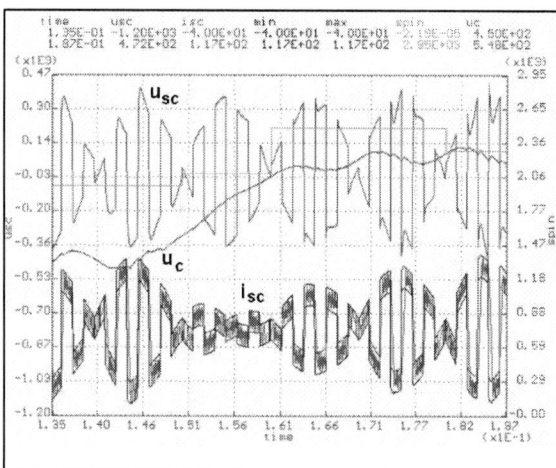

Fig.12. Simulation results of secondary active rectifier (input and output variables) for following parameters: Ut=230V, P=3kW, LF=40mH, CF=50uF, LSC=1mH, CSC=4mF

VII. CONCLUSION

Upon the industry demand variant of innovative topology of the traction vehicles fulfil the extensive weight reduce requirements have been discussed. The main attention has been given to the traction drive topology: input high voltage trolley converter (matrix converter) – middle frequency transformer – motor converter (single phase voltage-source active rectifier and three phase voltage-source inverter). Control methods of matrix converter using two-value control of secondary active rectifier and rectangular control of matrix converter have been proposed. These control methods ensure sinusoidal waveform and zero phase shift of taken current from trolley line.

From analysed control algorithms of whole traction drive the following variant has been proposed:

- Control of the traction topology by secondary voltage-source active rectifier and square-wave control of input matrix converter.

It means that secondary voltage-source active rectifier takes sine-wave phase current from middle-frequency transformer (two-value control) and square-wave control of the matrix converter which put together the phase current of output active rectifier to the 50Hz sine wave taken from the trolley line. Amplitude and phase shift of the trolley line current is controlled only by output active rectifier.

By simulation following advantages and disadvantages of control by matrix converter topology by control algorithm by secondary voltage-source active rectifier (square-wave control of input matrix converter) shows following advantages and disadvantages:

- Very simple control of matrix converter
- Good control of input filter
- Necessary the synchronization of PUSC and matrix converter
- Difficult control of PUSC

ACKNOWLEDGMENT

This research work has been made within research project of Ministry of Industry and Business of the Czech Republic No. MPO FT-TA2/035.

REFERENCES

[1] D. Q. Vinh, J. Fort, "Matrix converter – structure, switching and controlling," *In: XXIX. Electric Drive Symposium. Plzen 2005.*

[2] M. Glinka, R. Marquardt, "A New Single Phase AC/AC – Multilevel Converter for Traction Vehicles Operating on AC Line Voltage," *In: Proc. EPE'03, Toulouse 2003.*

[3] G. Kalvelage, P. Dubin, T. Lequeu, "Reduction of Mass and Volume of On-Board Multi-Input Voltage Converters Using SPARC Topology," *In: Proc. EPE'03, Toulouse 2003.*

[4] M. Victor, „Energieumwandlung auf AC – Triebfahrzeugen mit mittelfrequentransformator," *Fahrzeugtechnik eb 103 (2005) Heft 11, p. 505-510.*

[5] P. Wheeler, J. Rodriguez, J. Clare, L. Empringham, "Matrix Converters: A Technology Review," *IEEE Transactions on Industrial Electrinics, Vol. 49, No. 2, April, 2002.*

[6] Domenico Casadei, Jon Clare, Lee Empringham, Giovanni Serra, Angelo Tani, Andrew Trentin, Patrick Wheeler, Luca Zarri, "Large-Signal Model for the Stability Analysis of Matrix Converters," IEEE Trans. on Industrial Electronics, vol. 54, no. 2, pp. 939-950, Apr. 2007.

[7] P. Drabek, J. Fort, L. Piskac, M. Pittermann, F. Vondrasek, "Actual converter structeres : study of possibility application in traction,". *Research report ZČU Plzeň : 15.12.2006. 57 pgs.*

[8] M. Cedl, P. Drabek, J. Fort, M. Pittermann, F. Vondrasek, "The traction topology of direct converter with middle-frequency transformer (analyses of power electronic circuit of 4kW)". *Research report ZČU Plzeň : 2007. 17 pgs.*

[9] M. Pittermann, P. Drabek, M. Cedl, J. Fort, "The study of the traction drive with middle-frequency transformer". *In Transcom 2007. Žilina : University of Žilina, 2007. s. 159-162. ISBN 978-80-8070-694-4.*

[10] M. Pitterman, P. Drabek, M. Cedl, J. Fort, *The Study of the Traction Topology with the Middle-Frequency Transformer.* In: Proc. ISIE'08, Cambridge 2008.

Analysis of Multipulse Rectifiers with Modulation in DC Circuit in Vector Space Approach

Andrzej KAPŁON and Jarosław ROLEK

Kielce University of Technology/Chair of Power Electronics, Kielce, Poland, e-mail: a.kaplon@tu.kielce.pl

Abstract—Rectifiers consisting of parallel connected 6-pulse bridges with modulation in DC current circuit show properties of 24-, 36-pulse or multipulse rectifiers. The analysis of such rectifiers was conducted by means of the space vector method. A load unbalance of the component bridges of 12-pulse rectifier generates the intermediate position of space vector on a complex plane. An increase in the number of the intermediate positions approximates the space vector to the case of rotating one. In this paper topologies of multipulse rectifiers with modulation in DC current circuit are presented. The simulation results of investigations were obtained by means of the space vector for control of modulators.

Keywords—multi-pulse rectifiers, modulation, simulation, spectrum analysis.

I. INTRODUCTION

AC·DC q-pulse converters produce distortions and, in the case of converter supplied from an ideal voltage source and loaded by an ideal current source (smooth plots of DC current), generate into the power system high current harmonics of number h:

$$I_h = \frac{I_1}{h} = \frac{I_1}{kq \pm 1} \qquad (1)$$

where: k - any positive integer number, I_1 - current fundamental harmonic.

As results from (1) an increase in pulse number q is the basic method to eliminate high harmonic currents. One obtains an increase in q in classical solutions by series or parallel connection of p number of 3-phase bridges supplied by transformers with required phase shifting $\phi = \frac{2\pi}{q\,p}$ [1, 2, 3]. Another method consists in the use of rectifiers with modulators in DC current circuits. Modulators enable to obtain a multi-stair shape of supply phase currents which causes 12-pulse rectifiers to show properties of 24-, 36-pulse or multipulse ones.

An analysis of such systems can be performed by means of space vector [7].

A. Space Vector in Analysis of 6-pulse Rectifiers

The space vector referred to supply currents of m-phase power system without the neutral conductor is described by the following formula:

$$\underline{I} = I_\alpha + jI_\beta =$$
$$= \sqrt{\frac{2}{m}} \left(i_1 + \underline{a}\, i_2 + \underline{a}^2\, i_3 + \ldots + \underline{a}^{m-1}\, i_m \right) e^{-j\vartheta_1} \qquad (2)$$

where: $\underline{a} = e^{j\frac{2\pi}{m}}$ - unitary rotation vector, ϑ_1 - angle between current phase phasor and α-axis of complex plane.

Assuming that the q-pulse converter is supplied through a transformer from a symmetrical 3-phase voltage source, that the supply line is ideal and that the converter does not accumulate the energy, then in case of the current-source type load the supply current time plots are the functions of DC current I_d and have a q-stair shape described by the equation [2]:

$$i_k = A \cdot I_d \cdot cos\left(n\frac{2\pi}{q} + \xi + (k-1)\frac{2\pi}{m} \right) \qquad (3)$$

where: A – transformation ratio (ratio of peak values of input and output voltages of the supply system), $n = 0, 1, 2, \ldots, q-1$ – sector number, ξ - angle between adjacent peak values of the supply phase voltage and of the DC output voltage, m – number of phases of supply system, $k = 1, 2, \ldots, m$ – phase number.

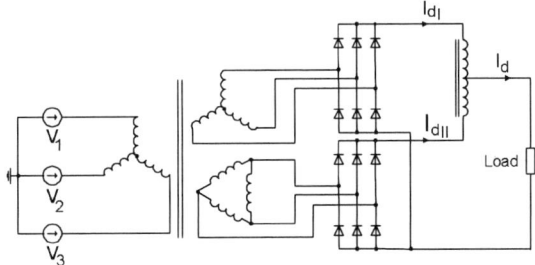

Fig. 1. Schema of 12-pulse rectifier.

In a period of supply voltage for the assumption of a simple commutation 6 configurations and adequately sectors of 3-phase bridge operation occur. In this case because of the shape of phase currents (3) the space current vector of the bridge remains immovable in every sector. For 12-pulse rectifier (Fig. 1) consisting of parallel connected two 6-pulse bridges supplied through transformers connected in Y/Y or Y/Δ, the space current vector is the vector sum of space current vectors of

978-1-4244-1741-4/08/$25.00 ©2008 IEEE   377

component bridges (Fig. 2). In the case shown in Fig. 1 load of bridges is balanced.

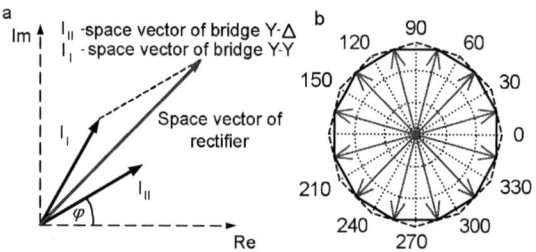

Fig. 2. Space vector (a) and 12-arm star (b) of rectifier in Fig. 1.

The constant total current load of the 12-pulse rectifier causes the arrow-head of space vector changing sectors to move on the sides of regular polygon, whose vertexes are determined by a symmetrical 12-arm star (solid line in Fig. 2b).

II. LOAD UNBALANCE OF COMPONENT BRIDGES OF 12-PULSE RECTIFIER

Load unbalance of component bridges of 12-pulse rectifier in every sector shifts the space vector (Fig. 3) in the direction of 6-arm star of a more loaded bridge. In this case space vector creates an unsymmetrical star and consequently the irregular polygon (solid line in Fig. 3). The constant total current load of the 12-pulse rectifier additionally causes the arrow-head of space vector changing sectors to move on the sides of this polygon.

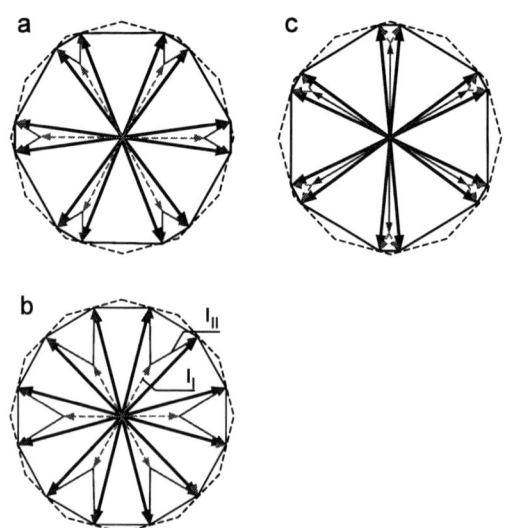

Fig. 3. Space vector by unbalanced (a, c) and balanced (b) load of component bridges of 12-pulse rectifier.

Thus, changing the load unbalance of component bridges one gets intermediate positions of the current space vector in each of 12 sectors on a complex plane. Putting e.g. two additional intermediate positions in each

of 12 sectors we get 36-pulse system and consequently 36-stair shape of line supply current (Fig. 5).

Figure 4 shows systems consisting of 6-pulse rectifiers they behave approximately as 24-pulse (Fig. 4a) and 36-pulse (Fig. 4b) rectifiers in consequence of alternate conducting of diodes and thyritors connected to terminals of the interphase transformer [4]. The depth of unbalance in load of each of component bridges depends on the number of turns of reactor segments. The number of turns of reactor segments results from the equality of ampere-turns for reactor configuration determined by a currently conducting valve [4]. Assuming a symmetrical voltage source and a current-source type load and neglecting the influence of commutation one gets the following phase supply currents of component bridges:

$$i_{1_{II}} = \frac{2 \cdot \sqrt{3}}{3} \cdot (0.5 \pm a) \cdot I_d \cdot cos(n\frac{2\pi}{q} + \frac{\pi}{12}) \qquad (4)$$

$$i_{1_I} = \frac{2 \cdot \sqrt{3}}{3} \cdot (0.5 \mp a) \cdot I_d \cdot cos(n\frac{2\pi}{q}) \qquad (5)$$

where: a – takes values: a_x ,$-a_x$ for converter in Fig. 4a and a_x ,$-a_x$, 0 for converter in Fig. 4b, taking into consideration that $a_x = \dfrac{z_y}{z_x}$, $n = 0, 1, 2, ..., q - 1$ – sector number.

a

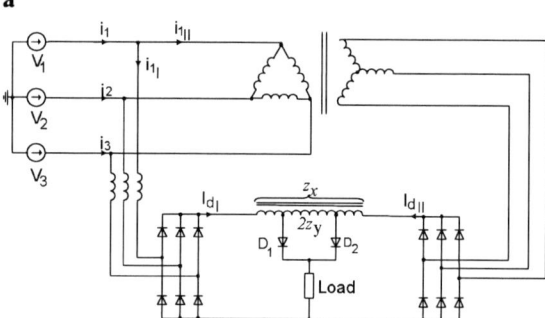

b

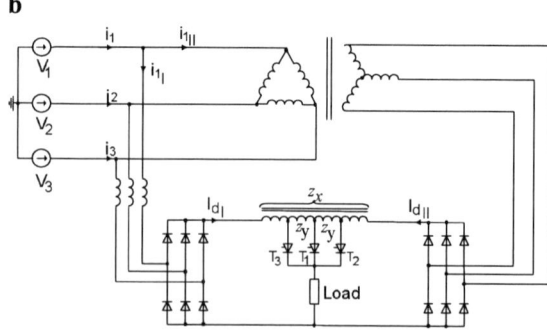

Fig. 4. Diode rectifier with modulator in the DC circuit:
a) 24-pulse, b) 36-pulse [4].

In comparison with rectifier in Fig. 1 having 12-arm star (heavy line in Fig. 5) the systems shown in Fig. 4a and 4b have adequately one and two additional positions of the current space vector in each of 12 sectors on a complex plane (light line in Fig. 5). Consequently, the space vector creates a 24- and 36-arm star respectively.

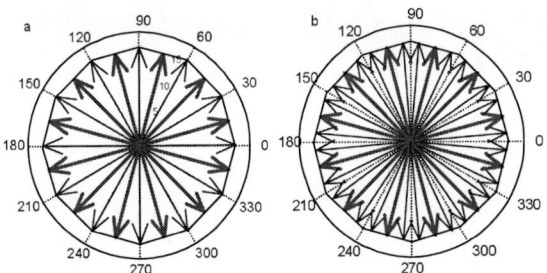

Fig. 5. Space vector star for rectifier: (a) in Fig. 4a, (b) in Fig. 4b.

The corresponding time plots of load DC current of each of component bridges showing their load unbalance are presented in Fig. 6.

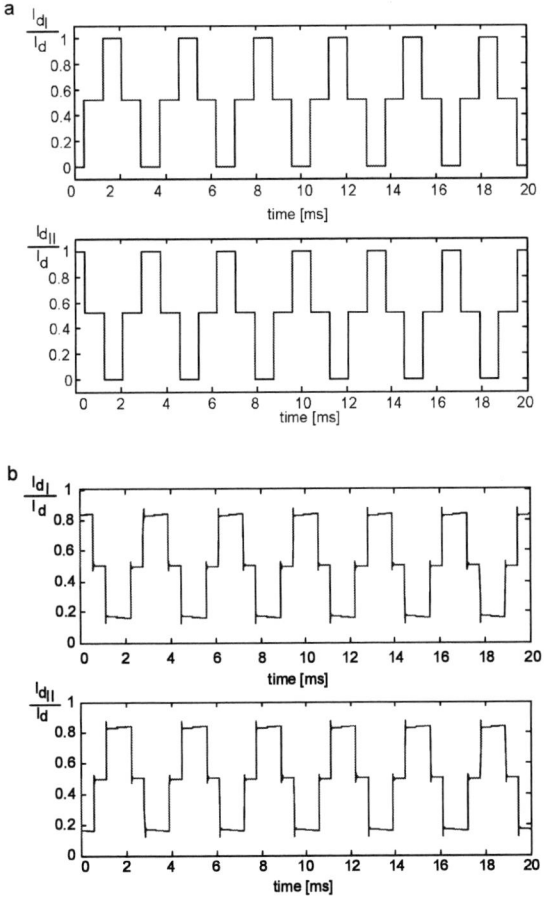

Fig. 6. Time plots of load currents of component bridges for rectifier: (a) in Fig. 4a, (b) in Fig. 4b.

Time plots and amplitude spectrums of phase supply currents of converter and of component bridges for 24-pulse and 36-pulse rectifiers from Fig. 4 are presented in Fig. 7. Amplitude spectrum of phase supply current of

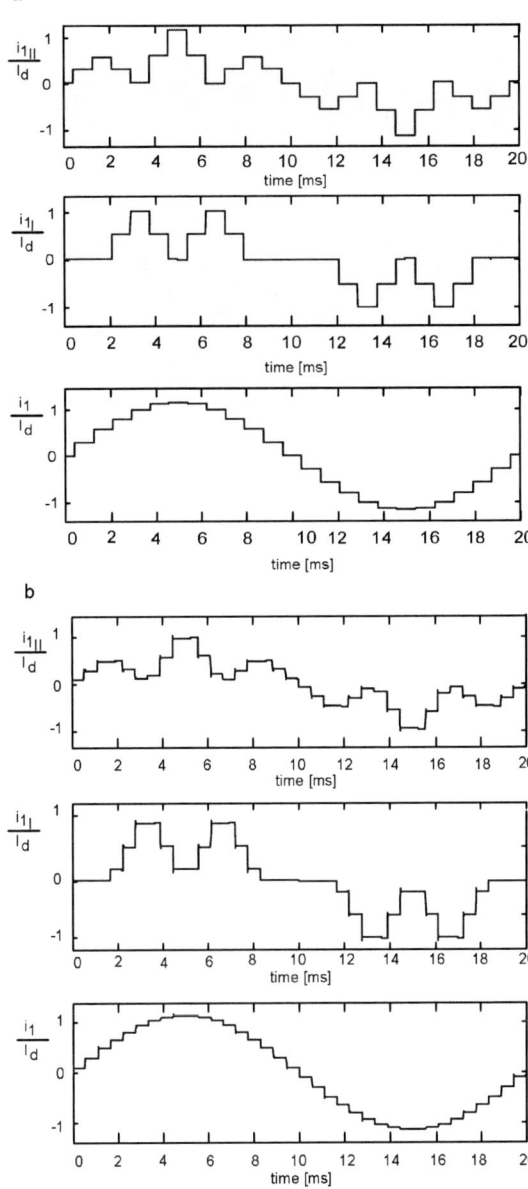

Fig. 7. Time plots of phase supply currents of component bridges and converter: (a) 24-pulse, (b) 36-pulse.

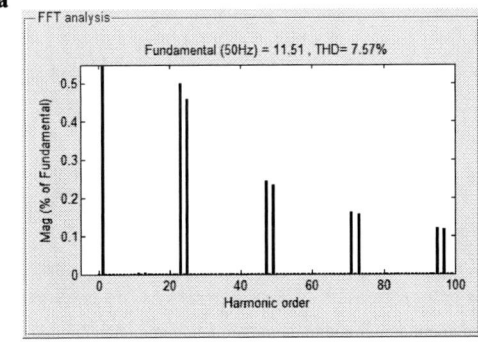

b

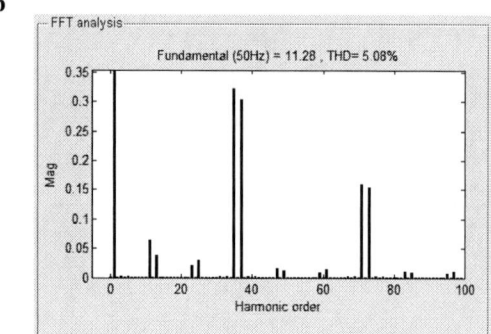

Fig. 8. Amplitude spectrums of phase supply current of converter:
(a) 24-pulse, (b) 36-pulse.

converter for 24-pulse and 36-pulse rectifiers from Fig. 4 are shown in Fig. 8. The total harmonic current distortion THD_i in these cases is equal to 7.57 % and 5.08 % respectively.

A further increase in the number of the intermediate positions approximates the space vector to the case of rotating one.

III. MULTI-PULSE RECTIFIERS WITH MODULATION IN DC CURRENT CIRCUIT

Any number of intermediate positions of the current space vector enables to receive a system shown in Fig. 9 [2, 5]. There, another type of modulation in DC current circuit having a form of interphase transformer is presented. Primary winding of the transformer is supplied from AC current modulator. The shape and magnitude of AC current transformed from primary to secondary winding determines the depth of load unbalance of rectifier component bridges. As a result any number of intermediate positions of the current space vector and than in the boundary case rotating space vector can be obtained.

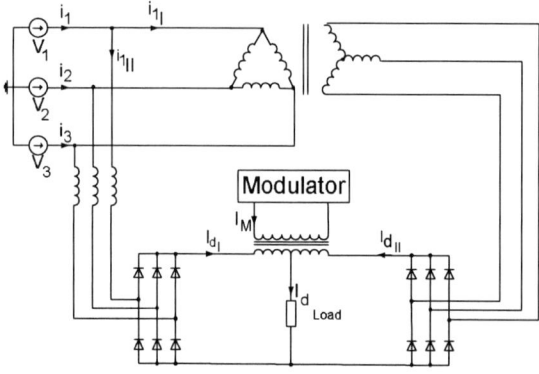

Fig. 9. Scheme of 12-pulse rectifier with interphase transformer in multi-pulse work mode.

Assuming a triangular shape of AC modulation current I_M one gets time plots of line currents of the converter and its component bridges presented in Fig. 10a.

a

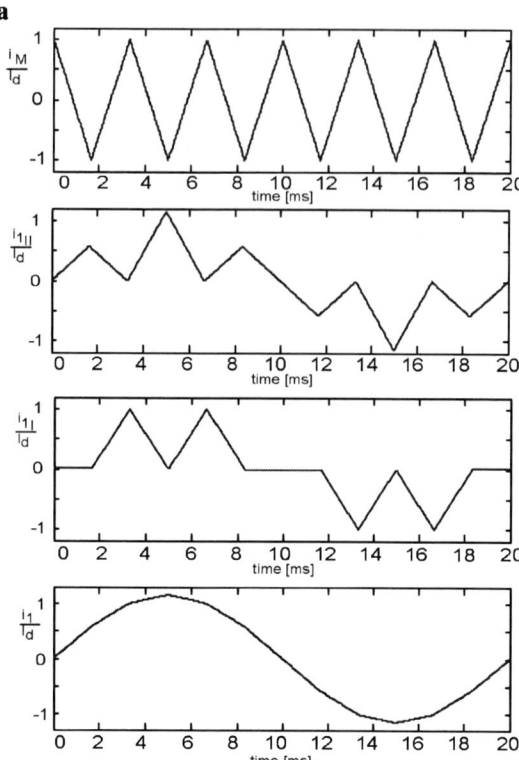

b

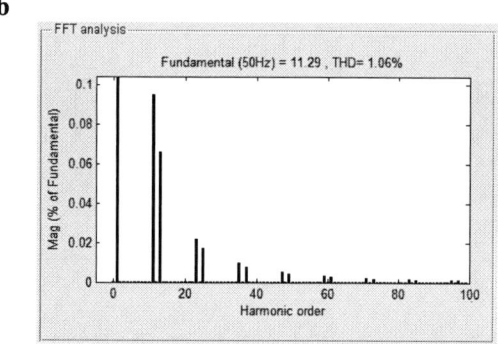

Fig. 10. Time plots of phase supply currents of converter and component bridges (a) and amplitude spectrum of converter phase supply current (b) for rectifier working in multi-pulse mode.

The shape of line current of the converter is close to sinusoidal one. In this case the current space vector is a rotating one but its arrow-head moves on the sides of regular polygon (Fig. 11a) due to constant total current load of the rectifier.

The total harmonic current distortion THD_i in this case is equal to 1.06 % (Fig. 10b). The equivalent VA rating (an equivalent double-wound transformer power rating) understood as an arithmetical average of AC loads of primary and secondary windings, does not exceed 2.35 % of DC load.

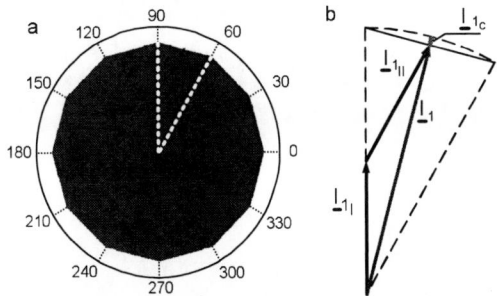

Fig. 11. Space vector of line current of rectifier from Fig. 8: (a) trajectory in multi-pulse mode, (b) in mode with additional load.

The circular shape of rotating current space vector trajectory requires keeping phasor module constant. This requirement can be satisfied by adequate adding of a complementary phasor (Fig. 11b). The complementary phasor and corresponding complementary line current are shown in Fig. 12. In practice complementary current and in consequence phasor one can be obtained by additional loading of the rectifier.

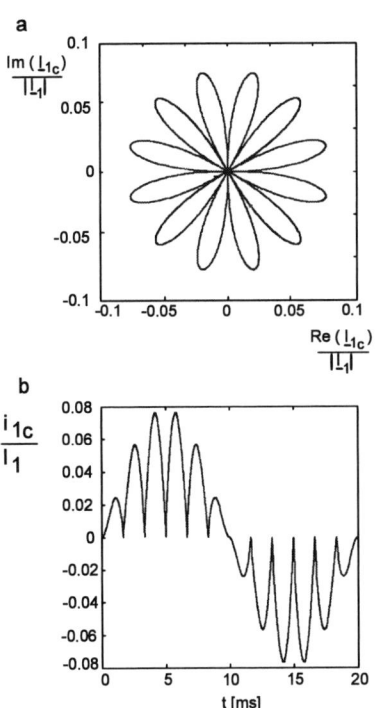

Fig. 12. Complementary: (a) phasor, (b) line current.

The nonlinear characteristic of the complementary phasor and corresponding complementary modulation current I_{Mc} could be approximated by triangular one (Fig. 13d). For this case simulation results of the rectifier working in multi-pulse mode (Fig. 9) are presented in Fig. 13a-c.

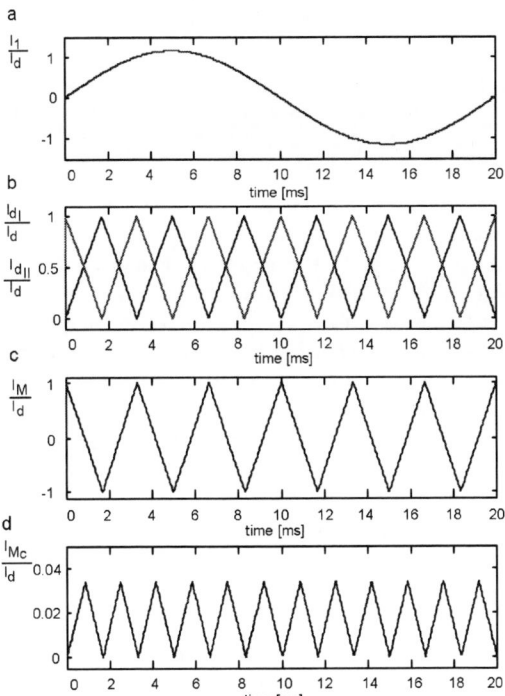

Fig. 13. Time plots of: (a) line current, (b) load currents of component bridges, (c) modulation current and (d) complementary modulation current for rectifier working in multi-pulse mode with additional load.

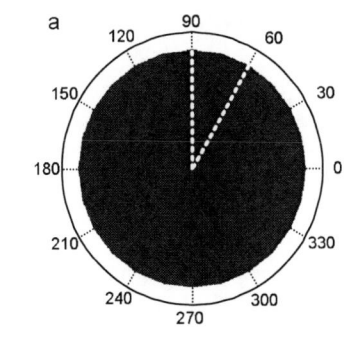

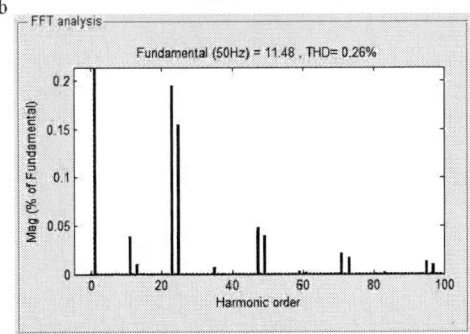

Fig. 14. Space vector trajectory (a) and amplitude spectrums (b) of line currents of rectifier working in multi-pulse mode with additional load.

381

The total harmonic current distortion THD_i is equal to 0.26 % (Fig. 14b), and the trajectory of space vector is close to circular (Fig. 14a).

Conclusions

The paper deals with an analysis of multipulse rectifiers with the use of the space vector method. An unbalance of loads of the component bridges of 12-pulse rectifier generates the intermediate positions of current space vector on a complex plane. An increase in the number of the intermediate positions approximates the space vector to the case of rotating one. Additional loading of the rectifier allows to receive a current space vector whose trajectory is close to circular one. Consequently, supply phase currents plots are close to sinusoidal ones. The described method can be used for the control of the modulator current in DC circuit of 12-pulse rectifier.

REFERENCES

[1] H. Tunia, A. Kaplon: "Electromagnetic Systems of Multipulse Converters", *Power Electronics and Electrical Drives selected problems*. PAN, PTETiS, Oficyna Wydawnicza Pol. Wrocławskiej, Wrocław, 2007, pp.: 11-30.

[2] H. Tunia: "Układy elektromagnetyczne prostowników wielopulsowych" VII Szkoła – Konferencja „*Elektrotechnika – Prądy Niesinusoidalne*", Lubiatów, 2004.

[3] A. Dereck Paice: "Power Electronic Converter Harmonics. Multipulse Methods For Clean Power", *IEEE PRESS*, New York, 1996.

[4] S. Miyairi, et. al.: "New Method for Reducing Harmonics Involved in Input and Output of Rectifier with Interphase Transformer", *IEEE Trans. on Industry Applications*, vol. IA-22, no. 5, 1986..

[5] R Strzelecki, H. Supronowicz: *Współczynnik mocy w systemach zasilania prądu przemiennego i metody jego poprawy*, Oficyna Wydawnicza Politechniki Warszawskiej, Warszawa, 2000.

[6] J. Rolek: „Układy prostowników wielopulsowych z modulacją w obwodzie prądu stałego". *Przegląd Elektrotechniczny*, Nr 05, Warszawa, 2008, pp.: 53-57.

[7] W Paszek: *Dynamika maszyn elektrycznych prądu przemiennego*, Helion, Gliwice, 1998.

High Efficiency Soft Switching Boost Converter for Photovoltaic System

Gil-Ro Cha[1], Sang-Hoon Park[2], Chung-Yuen Won[3], Yong-Chae Jung[4] and Sang-Hoon Song[5]

[1,2,3] School of Information and Communication Engineering, Sungkyunkwan University
300 Cheoncheon-dong, Jangan-gu, Suwon, Gyeonggi-do, 440-746, Korea
[1] e-mail : chagilro@skku.edu
[2] e-mail : marohachi@skku.edu
[3] e-mail : won@yurim.skku.ac.kr
[4] Department of Electronic Engineering, Namseoul University, 21 Maeju-ri
Seunghwan-eup, Cheonan, 330-707, Korea, ychjung@nsu.ac.kr
[5] Korea Testing Laboratory, 222-13 Guro3-dong Guro-gu, Seoul, 152-718, Korea, shsong@ktl.re.kr

Abstract— In this paper, a high efficiency soft switching boost converter is proposed for photovoltaic system. Using some resonant components, the circuit can be achieved the soft switching capability. Each of the switches in the proposed circuit performs ZV (Zero Voltage) or ZC (Zero Current) switching. Thus, the high efficiency characteristic can also be obtained, and then the size of the total system can be reduced. The operational modes of the proposed converter are explained in detail. And then some simulation results and some experimental results are presented for verifying the effectiveness of the proposed circuit.

Keywords—Photovoltaic, Soft Switching, Boost Converter, Resonant Converter

I. INTRODUCTION

PV (Photovoltaic) generation has constraints on the amount of sunlight, temperature, and other environmental conditions. Moreover, the energy conversion efficiency of PV generation is fairly low. In order to supplement for these weaknesses, there are many researches going on to increase the efficiencies of the solar cell and the electric power conversion device.

On the other hand, in cases of converters doing hard switching at a high frequency, the switching loss increases in proportion to the switching frequency. Thus, in order to reduce switching losses, the soft switching technology, which uses resonance by inductor and capacitor, has been actively researched [1-5].

This paper proposes a soft switching boost converter to improve the efficiency of the boost converter, which is an energy conversion device, to raise the low output voltage of solar array. Under the condition of zero-voltage and zero-current by inductor and capacitor resonance, soft switching can cut down the stress and loss produced at the switch.

In this paper, the detailed explanation of the operational modes is offered. After that, some simulations are performed to confirm the aforementioned operational explanation under the condition of 30 kHz switching frequency and 200V input voltage using a PSIM simulation tool. And then we adduce some experimental results to ascertain the validity of the proposed circuit.

II. PROPOSED SOFT SWITCHING BOOST CONVERTER

A. Proposed converter circuit

Fig 1 shows the proposed soft switching boost converter capable of minimizing the switching losses. A switch, two diodes, and the inductor and capacitor are added in the proposed circuit compared with the conventional boost converter. The two switches are controlled on and off simultaneously. Also, the switching loss is reduced by soft switching utilizing the resonance between inductor and capacitor.

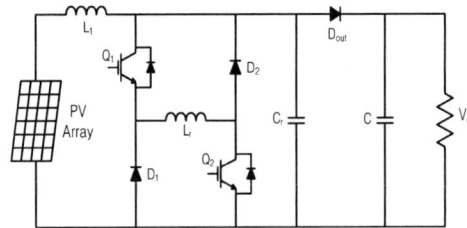

Fig. 1 The proposed soft switching boost converter

B. Operation mode analysis

To analyze the operational modes of the proposed circuit, it is explained in divided six modes according to the current paths, as shown in Fig. 2.

Mode 1 ($t_0 \leq t < t_1$)

Switch Q_1 and Q_2 are all off state, current cannot flow through switch Q_1 and Q_2. The main inductor current i_{L1} flows through the output diode D_{out}. The resonant capacitor voltage is equal to the output voltage.

$$i_{s1}(t) = i_{s2}(t) = i_{Lr}(t) = 0 \qquad (1)$$

$$i_{Dout}(t) = i_{L1}(t) \qquad (2)$$

978-1-4244-1741-4/08/$25.00 ©2008 IEEE

$$v_{Cr}(t) = V_o \qquad (3)$$

Mode 2 ($t_1 \leq t < t_2$)

Switch Q_1 and Q_2 are turned on with zero current condition simultaneously: current starts to flow in the resonant inductor L_r. The current flowing into the load through the output diode D_{out} gradually decreases, whereas the current flowing into the resonant inductor L_r increases. At the time period t_2, the current flowing in inductor L_1 and the current in resonant inductor L_r become equal. At this point, the current flowing to the load through output diode D_{out} becomes zero.

$$i_L(t) = i_L(t_1) - \frac{V_o - V_i}{L} t \qquad (4)$$

$$i_{Lr}(t) = \frac{V_o}{L_r} t \qquad (5)$$

$$i_L(t) = i_{Lr}(t) + I_{Dout}(t) \qquad (6)$$

$$i_L(t_2) = i_{Lr}(t_2) \qquad (7)$$

$$i_{Dout}(t_2) = 0 \qquad (8)$$

Mode 3 ($t_2 \leq t < t_3$)

The current flowing to the load through output diode D_{out} does not flow any longer since t_2 and the resonant capacitor C_r resonates with resonant inductor L_r. Through this resonance, the resonant capacitor voltage drops to zero from output voltage. This mode continues until the C_r voltage becomes zero.

$$i_L(t) \approx I_{min} \qquad (9)$$

$$i_{Lr}(t) = I_{min} + \frac{V_o}{Z_r} \sin \omega_r t \qquad (10)$$

$$v_{Cr}(t) = V_o \cos \omega_r t \qquad (11)$$

$$v_{Cr}(t_2) = V_o \qquad (12)$$

$$v_{Cr}(t_3) = 0 \qquad (13)$$

$$\omega_r = \frac{1}{\sqrt{L_r C_r}} \qquad (14)$$

$$Z_r = \sqrt{\frac{L_r}{C_r}} \qquad (15)$$

Mode 4 ($t_3 \leq t < t_4$)

After the abovementioned resonance is over, the current flowing through the resonant inductor L_r flows through diode D_1 and D_2. The current flowing through L_1 increases to accumulate energy in L_1, whereas the

resonant inductor current freewheels through two paths, Q_1-L_r-D_2 and D_1-L_r-Q_2.

$$v_{Cr}(t) = 0 \qquad (16)$$

$$i_L(t) = I_{min} + \frac{V_i}{L} \qquad (17)$$

$$i_{Lr}(t) = I_{Lr\,max} \qquad (18)$$

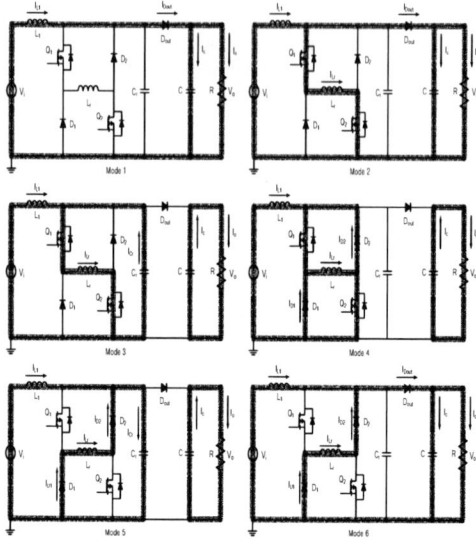

Fig. 2 The operation modes of the proposed circuit

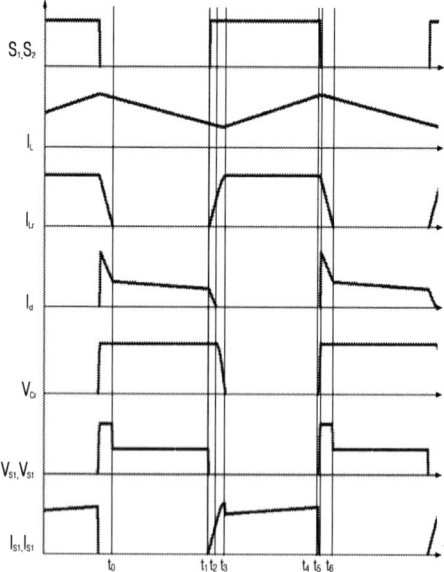

Fig. 3 The key waveforms

Mode 5 ($t_4 \leq t < t_5$)

In this mode, Q_1 and Q_2 are turned off with zero voltage condition simultaneously. The resonant capacitor C_r is charged by two inductor currents. This mode is maintained until the resonant capacitor v_{Cr} becomes equal to the output voltage. The output diode D_{out} stays turned off until the end of this mode.

$$i_L \approx i_{\max} \qquad (19)$$

$$i_{Lr}(t) = I_{\max} - (I_{\max} + I_{Lr\max})\cos\omega_r t \qquad (20)$$

$$v_{Cr}(t) = Z_r(I_{\max} + I_{Lr\max})\sin\omega_r t \qquad (21)$$

$$\omega_r = \frac{1}{\sqrt{L_r C_r}} \qquad (22)$$

$$Z_r = \sqrt{\frac{L_r}{C_r}} \qquad (23)$$

Mode 6 ($t_5 \leq t < t_6$)

The mode begins as C_r voltage equals to output voltage V_o. The main inductor current i_{L1} and the resonant inductor current i_{Lr} flow to the output through the output diode D_{out}. This mode is completed when the energy saved in resonant inductor L_r is transmitted to the output. As this process is the end of this mode, mode 1 of the next period starts from t_6.

$$i_{Dout} = i_{L1} + i_{Lr} \qquad (24)$$

$$i_{Lr}(t_6) = 0 \qquad (25)$$

$$v_{Cr}(t) = V_o \qquad (26)$$

Fig 3 shows key waveforms in a normal condition of the proposed soft switching boost converter.

III. SIMULATION RESULTS

The simulation is performed with 2kW and switching frequency 30 kHz. Fig. 4 shows a schematic diagram of the proposed soft switching boost converter.

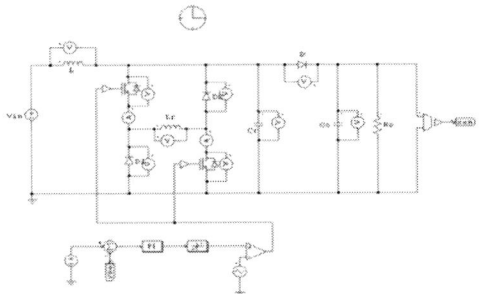

Fig. 4 The simulation schematic

Table. 1 The simulation parameters

Input voltage	V_i	200[V]
Output voltage	V_o	400[V]
Switching frequency	f_s	30[kHz]
Main inductor	L_m	560[μH]
Resonant inductor	L_r	40[μH]
Resonant capacitor	C_r	20[nF]

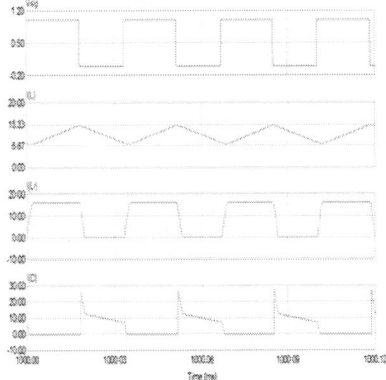

Fig. 5 The current waveforms of the main inductor, the resonant inductor and the output diode according to the gate pulse.

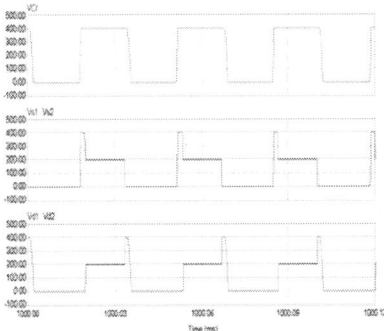

Fig. 6 The voltage waveforms of the resonant capacitor, the switches and the diodes

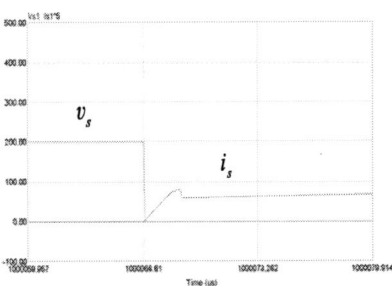

Fig. 7 The voltage and current waveforms of the switch at turn-on

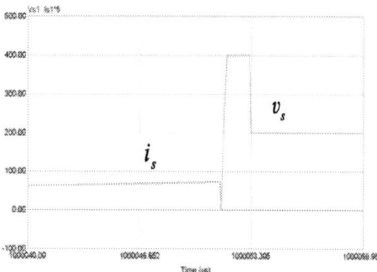

Fig. 8 The voltage and current waveforms of the switch at turn-off

Fig. 5 shows the waveforms of the main inductor current, resonant inductor current and output diode current according to switching pulses. Fig. 6 shows the voltage waveforms of switches, diodes and resonant capacitor under the steady state. In this figure, the voltages of switches and diodes are clamped to the output voltage by output capacitor.

Therefore voltage stresses of switching devices are decreased. When the switch is turned on, zero current switching takes place due to the resonant inductor. When it is turned off, zero voltage switching occurs owing to the resonant capacitor, as shown in Fig. 7 and Fig. 8. Zero voltage switching is done when the two freewheeling diodes are turned on after the discharge of the resonant capacitor. Also, zero current switching is done when the two freewheeling diodes are turned off after the resonant inductor current becomes zero.

IV. EXPERIMENTAL RESULTS

Fig. 9 shows a photograph of the experimental set of the proposed soft switching boost converter. We use IGBT FGA25N120A which has an output capacitance of about 180 pF. The drive circuit of the proposed soft switching boost converter is implemented by pulse width modulation control IC, TL494. Fig. 10 shows the experimental current waveforms of the main inductor and the resonant inductor, and the voltage of the drive signal.

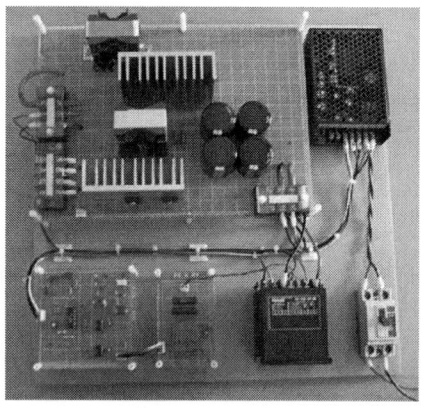

Fig. 9 The experimental soft switching boost converter

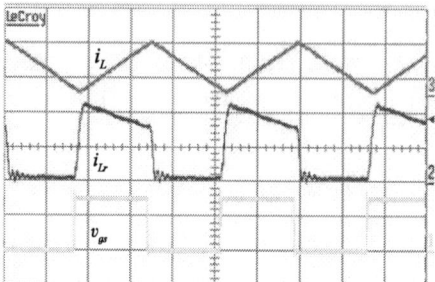

Fig. 10 The experimental current waveforms of the main inductor and the resonant inductor, and the experimental voltage waveform of the drive signal (main inductor current : 4A/div, resonant inductor current : 4A/div, drive signal : 10V/div, time : 10 μs /div)

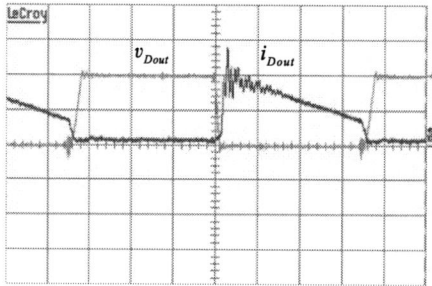

Fig. 11 The experimental voltage and cureent waveforms of the output diode (output diode voltage : 200V/div, output diode current : 4A/div, time : 5 μs/div)

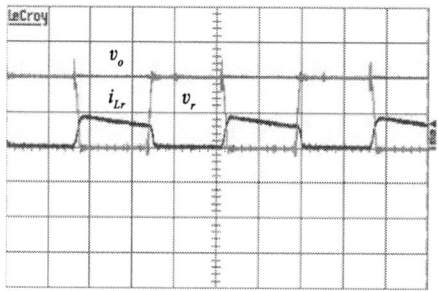

Fig. 12 The experimental voltage waveforms of the resonant capacitor and the output, and the experimental current waveform of the resonant inductor (output voltage : 200V/div, resonant capacitor voltage: 200V/div, resonant inductor current : 10A/div, time : 10 μs/div)

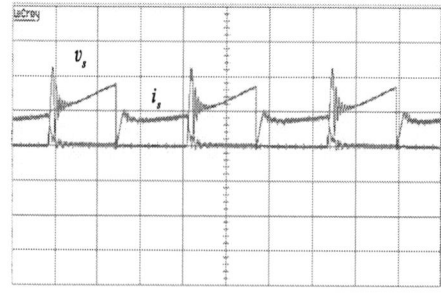

Fig. 13 The experimental voltage and cureent waveforms of the switch (switch voltage : 200V/div, switch current : 10A/div, time : 2 μs/div)

386

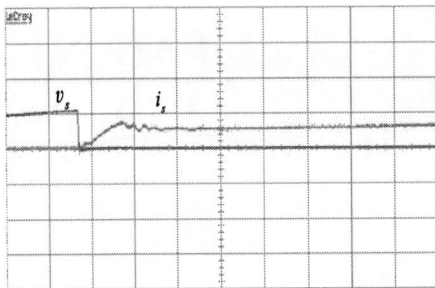

Fig.14 The experimental voltage and cureent waveforms of the switch (switch voltage : 200V/div, switch current : 10A/div, time : 2 μs/div)

Fig. 11 shows the experimental waveforms of the voltage and current of output diode. The output diode is turned on and off under the zero current condition and zero voltage condition. Fig. 12 shows the experimental waveforms of the resonant capacitor voltage, the output voltage and the resonant inductor current. Fig 13 shows the experimental waveforms of the switch voltage and current. Fig 14 shows the extension waveforms of the switch voltage and current when the switch is turned on. Therefore, it has been verified that all of the semiconductor devices of the proposed circuit have the soft switching characteristics.

V. CONCLUSION

This paper suggests a circuit constituted by adding an auxiliary switch, diode, inductor, and capacitor to the conventional boost converter circuit. The proposed soft switching boost converter achieves zero current switching by the resonant inductor when the switch is turned on. On the other hand, when the switch is turned off, it also achieves zero voltage switching by the resonant capacitor. Thus, the proposed circuit can be reduced the switching losses. In the paper, the proposed soft switching boost converter is analyzed, and its validity is proven through some simulation and experimental results.

ACKNOWLEDGMENT

This work is outcome of a Manpower Development Program for Energy & Resources supported by the Ministry of Knowledge and Economy (MKE)

REFERENCES

[1] Jung, Y.C.; Cho, J.G.; Cho, G.H.;, A new zero voltage switching resonant DC-link inverter with low voltage, Industrial Electronics, Control and Instrumentation, 1991. Proceedings. IECON'91., 1991 International Conference on 28 Oct.-1 Nov. 1991 Page(s): 308-313 vol. 1

[2] Yu-Ming Chang.; Jia-You Lee.; Wen-Inne Tsai and York-Yih Sun.;, Design and analysis of H-soft-switched converters, Electric Power Applications, IEE Proceedings, Volume 142, Issue 4, July 1995 Page(s): 255–261

[3] Jia-You Lee.; Yu-Ming Chang.; Wen-Inne Tsai and York-Yih Sun.;, A new soft switching transition PWM boost converter for power factor correction using parallel resonant tank, Industrial Electronics, Control, and Instrumentation, 1993. Proceedings of the IECON'93., International Conference on 15-19 Nov. 1993 Page(s): 942-947 vol. 2

[4] Yu-Ming Chang.; Jia-you Lee.; Wen-Inne Tsai and York-Yih Sun.;, An H-soft-switched cell for single-switch nonisolated DC-to-DC converters, Industrial Electronics, Control, and Instrumentation, 1993. Proceedings of the IECON'93., International Conference on 15-19 Nov. 1993 Page(s): 1077-1082 vol. 2

[5] Jain, N.; Jain, P.K.; Joos, G.;, A zero voltage transition boost converter employing a soft switching auxiliary circuit with reduced conduction losses, Power Electronics, IEEE Transactions on Volume 19, Issue 1, Jan. 2004 Page(s): 130-139

A POWER CONVERTER FOR FAULT TOLERANT MACHINE DEVELOPMENT IN AEROSPACE APPLICATIONS

Liliana de Lillo [1], Patrick Wheeler [2], Lee Empringham[3] Chris Gerada [4], Xiaoyan Huang [5]

[1,2,3]UNIVERSITY OF NOTTINGHAM,
Power Electronics, Machines and Control Group,
School of Electrical and Electronic Engineering,
Nottingham
URL: http://www.eee.nottingham.ac.uk/pemc

[1]*liliana.delillo@nottingham.ac.uk*, Tel: +44 115 8468840, Fax: +44 115 9515616;
[2]*pat.wheeler@nottingham.c.uk*, Tel: +44 115 9515591, Fax: +44 115 9515616;
[3]*lee.empringham@nottingham.ac.uk,* Tel: +44 115 9515541, Fax: +44 115 9515616;
[4]*chris.gerada@nottingham.ac.uk,* Tel: +44 115 9515541, Fax: +44 115 9515616;
[5]*eexxh@nottingham.ac.uk,* Tel: +44 115 9515541, Fax: +44 115 9515616;

Abstract – **This paper describes an experimental tool to evaluate and support the development of fault tolerant machines designed for aerospace motor drives. Aerospace applications involve essentially safety critical systems which should be able to overcome hardware or software faults and therefore need to be fault tolerant. A way of achieving this is to introduce variable degrees of redundancy into the system by duplicating one or all of the operations within the system itself. Looking at motor drives, multiphase machines such as multiphase brushless dc machines are considered to be good candidates in the design of fault tolerant aerospace motor drives. The paper introduces a multi-phase two level inverter using a flexible and reliable FPGA/DSP controller for data acquisition, motor control and fault monitoring to study the fault tolerance of such systems.**

Key words – **Multiphase Drive, DSP, Control of drive, Converter control, reliability**

I. INTRODUCTION

The last thirty years have witnessed an increasing amount of studies and research into the "all-electric aircraft" and the "more-electric aircraft" concepts [1,2] which aim to introduce more energy-efficient ways of converting and employing the power generated by aircraft engines. This would have, and has today where already implemented, had an impact on the design of the overall aircraft at every system level. More importantly at the development stage, each aircraft system design will have to respond to a series of safety analyses which will asses the reliability of each system. Certainly changes in aerospace industry will not occur without a long process of research work, which will certify their suitability in an environment where safety and therefore reliability are the major driving factors without forgetting that the most cost effective solutions would still be the most favourable ones.

Among the main areas which are addressed by the "more-electric aircraft" concept, the Flight Control System is a good example of the implementation of this concept, translated in the implementation of "electro-hydrostatic actuators" in parallel with the traditional hydraulically powered ones on the new generation Airbus A380 aircraft. The idea behind the use of an electro-hydrostatic or electro-mechanical actuator which basically replaces the actuator hydraulic power source with electric power simplifying the structure of the power delivery, needs to include a high degree of reliability to challenge the modern and well established aircraft structure.

Attention has been given to the power converter side in previous work [3] and also to the motor side in [4]. More knowledge needs to be gained on the electric motor drive system requirements with regards to the needs of safety critical applications such as aerospace applications. Also, more knowledge needs to be gained on the performance

978-1-4244-1741-4/08/$25.00 ©2008 IEEE

of the drive system and its control hardware, estimation and diagnostic algorithms.

The work which is described in its first stage in this paper aims to contribute with an experimental tool which will support the development of fault tolerant machines to extend the research into electric motor drives for aerospace applications. This paper introduces a research platform in the form of a six phase, two level inverter and a flexible FPGA/DSP controller for data acquisition, motor control and fault monitoring to study the fault tolerance of such systems.

Initial results of the open loop control of the inverter driving a six phase passive load are presented in the following sections together with results of the power converter separately driving each phase of a three phase permanent magnet machine.

II. FAULT TOLERANT DRIVES AND REDUNDANCY IN AEROSPACE APPLICATIONS

Extensive work has been conducted into the subject of fault tolerant three phase AC motor drives for industrial application and it is well documented in [5].

A fault tolerant system should have the ability to respond to any hardware or software failure keeping the minimum functionality which allows the system to continue to operate. This is crucial in safety critical aerospace applications.

One way of achieving fault tolerance already implemented in the flight control system [6] is the use of redundancy. Looking at the electric motor drive level which is generally constituted of a power converter, an electric motor and all the hardware and software which is necessary to implement the drive control, redundancy can be applied at a system level (for both the power converter and the motor) and at a control level.

The motor test development platform proposed in this paper will allow the validation of simulated results and the experimental analysis of the validity of machine designs which will be developed to respond to a high standard of reliability.

III. REQUIREMENTS AND DESIGN OF THE MOTOR TEST DEVELOPMENT PLATFORM

The motor test platform will be required to be versatile enough to drive different types of machines and therefore, different control methodologies and algorithms will need to be implemented. Below is a list of requirements for the development platform.

- Multi phase output
- Configurable to drive different motor types with different numbers of phases
- High speed controller
- High precision PWM generation
- Rapid code development design cycle

Fig. 1 below shows a schematic diagram of the six phase motor drive system. The schematic block includes the power circuit and its control and will be described individually in the following sections.

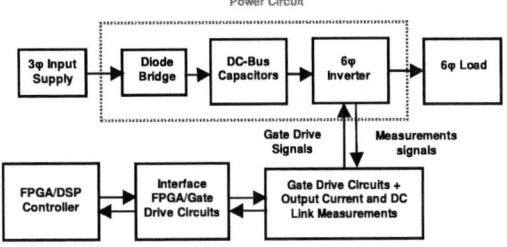

Fig. 1 Six phase Motor Drive System

A. Power Converter

The converter topology chosen to implement the development platform is a three phase diode bridge - Dc link - six phase two level voltage source inverter and can be seen in Fig. 2. The converter has been constructed using 100A, 1200V inverter leg modules. The power plane also integrates the gate drive circuits, output current measurement and DC-link measurement transducers.

Fig. 2 Power Converter: three phase diode bridge – DC link – six phase inverter

The connection between the controller and the gate drivers is provided by a fibre optic link which will increase noise immunity. Although employed for aerospace applications, the motor test development platform has been constructed for laboratory testing and therefore, low cost electrolytic capacitors are used to implement the DC link providing the 540V DC bus.

389

B. Controller

The controller was implemented using a high speed digital signal processor (DSP) and a high speed field programmable gate array (FPGA). The Texas instruments C6713 floating point processor and Actel, ProAsic3 FPGAs were used.

Fig. 3 shows the stack of boards starting from the bottom with the TI C6713 DSK. The FPGA card which also includes ten analogue to digital channels. Finally the top card is an interface card which will provide the fibre optic links for the gate drive signals, both a resolver and an encoder interface circuit for higher flexibility and finally, digital temperature sensor interface circuitry.

Fig. 3 High performance, DSP – FPGA based digital control platform

The control tasks are divided between the two main parts of the controller. The FPGA implements the high speed tasks such as the A2D interface, several high speed PWM generators, watchdog timer, trip/fault monitor and deadtime / commutation control whilst the high speed floating point DSP performs all of the modulation and motor control functions. A floating point processor is used to reduce development times since problems related to resolution and rounding when implementing controllers using integer processors are eliminated.

Measurement signal information from the current and voltage sensors, the temperature sensors and the rotary position resolver are employed not only to perform the control of the drive but also to provide the information necessary to generate, whenever necessary, the fault signals which can disable the converter.

C. Rapid code development

The DSP is programmed in C using the Code Composer integrated design environment. It is a fully integrated development and debugging suite.

The FPGA design has been carefully refined to aid code development. Different PWM generators have been designed and can be used depending on the application.

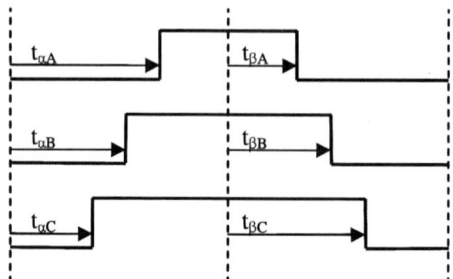

Fig. 4a Three phase PWM waveform, Timer representation

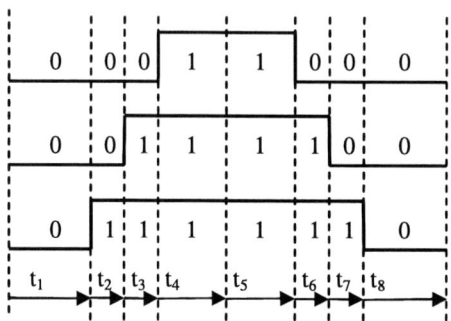

Fig. 4b Three phase PWM waveform, Space vector representation

Figures 4a and 4b show the same arbitrary three phase PWM waveform. The first is represented using on-times ($t_{\alpha X}$) and off-times ($t_{\beta X}$) for each PWM output phase. The waveform in Figure 4b is made up of the PWM vectors and the times (t_{1-8}) for each vector.

A timer / compare type of PWM generator can be used when the modulation algorithm is based on the natural sampling (fig.4a) method and a space vector PWM generator is used when the modulation strategy is based on the space vector representation (fig.4b).

This solution offers increased control flexibility and reduced code development time. For example, if using a space vector modulation algorithm with a typical timer based PWM unit (as is the case in many commercially available micro-processors) the PWM space vectors would not need to be converted into on and off times at the end of the PWM interrupt. The vectors and times can be sent directly to the vector PWM generator. This saves on code space and time and hence increases the potential interrupt resolution.

390

IV. Testing

A. Power Converter Initial testing - RL load

The converter has been connected, for initial testing, to a six phase passive RL load (R = 2 Ω, L = 4 mH). Fig. 5 below presents results of the initial testing of the power circuit.

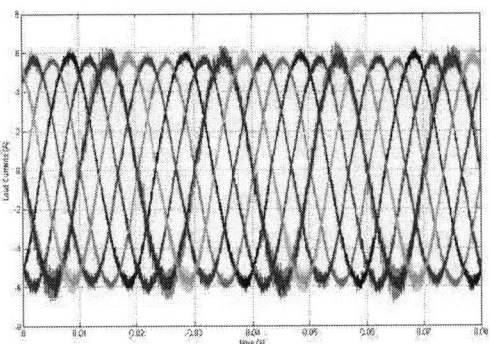

Fig. 5 Inverter initial tests: six output phase currents

The results in Fig.5 show the six output currents from the open loop test performed on the converter to verify the functionality of the power circuit and of the controller. A carrier-based sinusoidal PWM scheme has been implemented by considering six sinusoidal modulation waves, 60 degrees phase shifted from each other.

B. Preliminary tests on an individually driven, three phase permanent magnet synchronous machine

The next test with the multi-phase converter was performed on a three phase Permanent Magnet Synchronous Machine (PMSM) where each machine phase is driven by two of the converter output phases as shown in Fig.6. This configuration exhibits an increased degree of redundancy at a power converter level and increases potential reliability during either motor or converter failures.

The PMSM has been designed and manufactured by the PEMC group of the University of Nottingham, giving particular attention to fault tolerant features which are desired especially in aerospace applications.

The PMSM is a low speed, 20 pole machine, designed with high inductance per phase in order to be able to sustain the unbalance in the phase currents introduced by the eventual open or short circuit in one of the three motor phases. The machine is rated at 115V rms per phase and a maximum current of 24.4A.

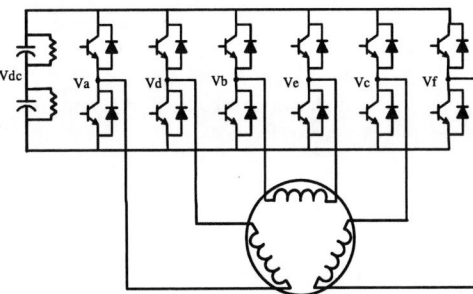

Fig. 6 Six phase inverter driving individually each motor phase of the PMSM

The machine has been partially commissioned with initial tests whose results are shown later in this section. More testing is required to have a complete knowledge of its behaviour under different speed and load conditions. For this reason the test of the open or short circuit of one of the three phases cannot be discussed at the time of writing of this paper.

The preliminary test of the multiphase converter driving the three phase machine described above has been implemented using d-q vector current control and speed control of the machine.

Fig. 7 shows results from a speed reversal test on the PMSM connected to the six phase inverter as described in Fig. 6 using a 540V DC link. The speed step change from -40 rad/s (mechanical) to 40 rad/s shows a fast response of the unloaded.

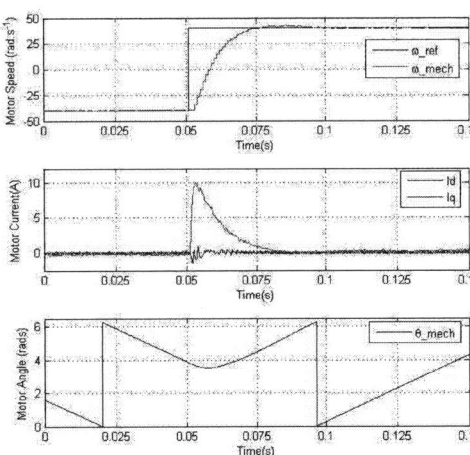

Fig. 7 Initial test results: speed reversal of the PMSM drive described in Fig. 6 – (from the top) Motor speed, phase currents and rotor mechanical angle

The d-q motor currents are also shown in Fig. 8 together with the motor phase currents. The last ones show little change due to no load applied to the machine while the available iq limit of 10A, is demanded during the speed reversal to generate the torque necessary. The demanded motor phase voltages, also shown in Fig. 8, are the demand voltages for three of the six inverter output phases V_{a_ref}, V_{b_ref}, and V_{c_ref}, the remaining three, V_{d_ref}, V_{e_ref} and V_{f_ref} (as shown in the circuit diagram in Fig. 6) are given a zero reference.

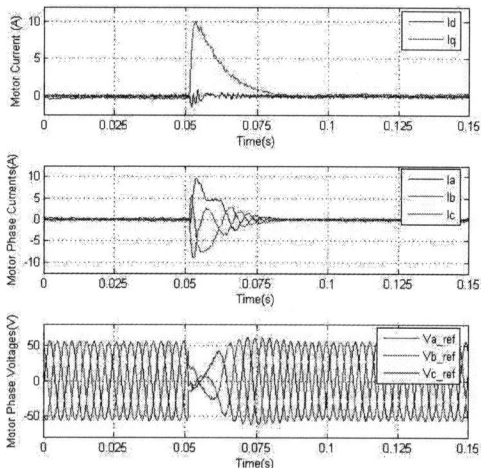

Fig. 8 Initial test results: speed reversal of the PMSM drive described in Fig. 6 – (from the top) id and iq currents, motor phase currents and phase voltage demands

Once the reference voltages are determined a carried-based modulation strategy has been implemented to generate the on and off times for the IGBTs of the six output phase inverter at a PWM frequency of 5kHz.

V. CONCLUSIONS

This paper introduces a motor test and development platform which will contribute to integrate the knowledge on the requirements of the design of electric motor drives which could be implemented in aerospace applications in regards to the "more electric aircraft" concept.

The test platform will therefore support the development of fault tolerant machines through experimental testing as well as to validate simulation results of the power converter side. Its FPGA/DSP based controller offers a great deal of flexibility and increased reliability.

Initial results of a six phase, two level inverter connected to a six phase RL load are presented to confirm the functionality of the converter and its control. The platform has also been used to drive and perform the vector control of an experimental Permanent Magnet Synchronous Machine, also wholly designed and built at the University of Nottingham. Initial results of the PMSM drive have been presented.

ACKNOWLEDGEMENTS

The authors would like to acknowledge the support provided by GE Aviation and EPSRC for this work as part of the SMARTPACT University Technology Strategic Partnership.

REFERENCES

[1] *Cronin M.J.J* , "**The All-Electric Aircraft** ", IEE Review, Vol.36(8), pp. 309-311, 1990

[2] *Jones R.I.* , "**The More-Electric Aircraft: the past and the future ?** ", Colloquium, Electrical Machines and Systems for the More Electric Aircraft, IEE Power Division, Vol. 99/180, pp 1/1-1/4, 1999

[3] Liliana de Lillo, "**A Matrix Converter Drive System for an Aircraft Rudder Electro-Mechanical Actuator**", PhD Thesis, March 2006

[4] Xiaoyan Huang; Bradley, K.; Goodman, A.; Gerada, C.; Wheeler, P.; Clare, J.; Whitley, C "**Fault-Tolerant Brushless DC Motor Drive For Electro-Hydrostatic Actuation System In Aerospace Application**" in: IAS 2006, 41st IAS Annual Meeting, Vol. 1, Oct. 2006, pp 473-480

[5] Brian A. Welchko, T. Lipo, T. M. Jahns, S.E. Schulz, "**Fault Tolerant Three-Phase AC Motor Drive Topologies: A Comparison of Features, Cost, and Limitations**" , IEEE Transactions on Power Electronics, Vol. 19, N. 4, July 2004, pp 1108 – 1116

[6] Ian Moir and Allan Seabridge, "**Aircraft Systems – Mechanical, Electrical, and Avionics Subsystems Integration**", Professional Engineering Publshing, UK, 2006

Optimal Bus Capacitance Design for System Stability in On-Board Distributed Power Architecture

Seiya Abe[*], Masahiko Hirokawa[**], Masahito Shoyama[*] and Tamotsu Ninomiya[***]

[*] Kyushu University, 744, Motooka, Nishi-ku, Fukuoka, 819-0395, Japan
[**]TDK Corporation, 2-15-7, Higashi-Ohwada, Ichikawa, Chiba, 272-8558, Japan
[***]Nagasaki University, 1-14, Bunkyo-Machi, Nagasaki, 852-8521, Japan

Abstract— Recently, the distributed power system is mainly used for the power supply system which requires the low-voltage / high-current output. The distributed power system consists of bus converter and POL. The most important factor is the system stability in bus architecture design. The overlap between the output impedance of bus converter and input impedance of POL causes system instability, and it has been an actual problem. Increasing the bus capacitor, system stability can be reduced easily. However, due to the limited space on the system board, increasing of bus capacitors is impractical. The urgent solution of the issue is desired strongly. This paper presents the output impedance design for on-board distributed power system by means of three control schemes of bus converter. The output impedance peak of the bus converter and the input impedance of the POL are analyzed, and it is conformed by experimentally for stability criterion. Furthermore, the optimal intermediate bus capacitance design for system stability is proposed.

Keywords— Distributed power system, bus converter design, output impedance, intermediate bus capacitance.

I. INTRODUCTION

Various LSI is used in the telecommunication application equipments and the driving voltage is various. On the other hand, increase of load current is also remarkable by advanced function of LSI. Since the present LSI is designed in accordance with semiconductor manufacture technology, the tolerance level of operation voltage is very narrow. Consequently, the voltage drop by the wiring impedance of power line causes malfunction of LSI. In order to reduce the malfunction of LSI by the voltage drop, it is proposed that the converter is arranging very close to the LSI. This converter is called POL. Thus, the power supply system which requires the low-voltage / high-current output has been changing from conventional centralized power system to distributed power system. The distributed power system consists of first-stage isolated DC-DC converter as a bus converter and second-stage non-isolated DC-DC converter as a POL. However, the instability phenomenon in a distributed power system is posing a problem recently. This is instability phenomenon resulting from overlapping between the output impedance of bus converter and the input impedance of POL. Increasing the bus capacitor, system stability can be reduced easily. However, due to the limited space on the system board, increasing of bus capacitors is impractical. The urgent solution of the issue is desired strongly, and the various discussion of system

stability has been reported[1-8]. Then, we also have reported the detailed discussion of system stability by control schemes of bus converter (Un-regulated, Semi-regulated and Full-regulated)[9-14]. However, so far, the detailed discussion of practical design and the optimal intermediate bus capacitance design of bus converter about on-board distributed power system has not been reported. This paper presents the optimal design of bus converter for on-board distributed power system by means of three control schemes of bus converter.

II. DISCRIMINATION OF STABILITY

Figure 1 shows the distributed power system consisting of bus converter and POL. Even if each converter has stable operation, the instability phenomenon may occur by connecting two converters in series. Bus converter and POL have input-to-output voltage transfer function Gvb(s) and Gvp(s), respectively. The overall input-to-output voltage transfer function Gvv(s) is given following equation;

$$G_{vv}(s) = \frac{G_{vb}(s)G_{vp}(s)}{1 + Z_o(s)/Z_{in}(s)} \quad (1)$$

where Zo (s) is the output impedance of bus converter, and Zin(s) is the input impedance of POL. From Eq. (1), the input and output impedance is greatly concerned with the system stability. The stability of closed-loop system is decided with the characteristics equation $1+Z_o(s)/Z_{in}(s)$. This means relation between Zo(s) and Zin(s) decides the system stability. This system may become unstable when both impedances are overlapped, as shown in Fig. 2 (a). It is necessary to eliminate this impedance overlap for system stability. However, eliminating this impedance overlap for all frequency range is very difficult.

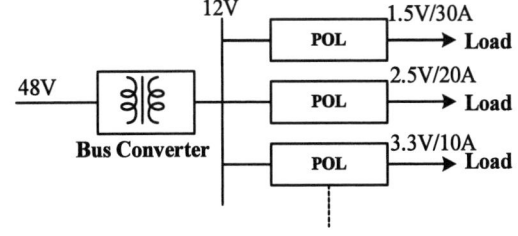

Fig. 1. On-board distributed power system.

On the other hand, it may have stable operation even if both impedances are overlapped. This is because the phase margin becomes large under the influence of the input impedance. When the bandwidth of POL is enough wider than the bandwidth of bus converter, if the peak value of output impedance becomes almost equal to the steady-state value of input impedance |Zin(0)| as shown in Fig. 2 (b), then this system becomes stability limit as shown in Fig. 2 (b). Moreover, this system becomes unstable if the peak value of output impedance exceeds |Zin(0)|. From mentioned above consideration, the new stability criterion can be defined as follows[15].

$$\begin{cases} \left| Z_{in}(0) \right| \ge Z_{o\_peak} & : \quad Stable \\ \left| Z_{in}(0) \right| < Z_{o\_peak} & : \quad Unstable \end{cases}$$

III. IMPEDANCE ANALYSIS

The half-bridge converter with the most popular circuit of the power-stage is used as a bus converter, and the synchronous buck converter with the most popular circuit is used as POL. Figure 3 and 4 show the circuit diagrams, respectively. The output impedance of bus converter and the input impedance of POL can be derived by applying the stage space averaging method[16,17].

A. Input Impedance

At first, the low-frequency value |Zin(0)| of input impedance is estimated. The input impedance of POL can be derived as following equation[18].

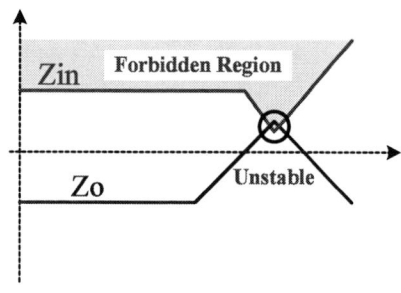

(a) Conventional discrimination.

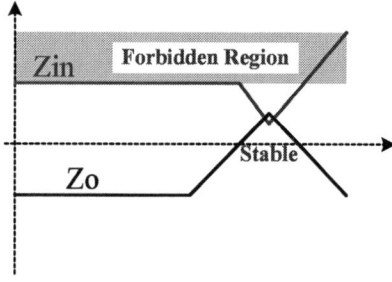

(b) New discrimination.

Fig. 2. Stability discrimination.

$$\frac{1}{Z_{in}(s)} = \frac{1}{Z_N(s)} \cdot \frac{T_p(s)}{1+T_p(s)} + \frac{1}{Z_D(s)} \cdot \frac{1}{1+T_p(s)} \quad (2)$$

From Eq. (2), the low-frequency value of input impedance |Zin(0)| is given by following equation.

$$\left| Z_{in}(0) \right|_{(dB\Omega)} \approx 20\log\left(\frac{R+r_L}{D^2} \right) \quad (dB\Omega) \quad (3)$$

|Zin(0)| has minimum value at rated load, so the estimation of |Zin(0)| must be at rated load. Next, the output impedance is examined.

B. Output Impedance

The output impedance of bus converter can be derived as following equations.

Open loop

$$Z_o(s) = \frac{s^2 L_b C_b r_{c_b} + s\left(L_b + C_b r_{L_b} r_{c_b} \right) + r_{L_b}}{s^2 L_b C_b + s C_b \left(r_{L_b} + r_{c_b} \right) + 1} \quad (4)$$

Closed loop

$$Z_{oc}(s) = \frac{Z_o(s)}{1+T_b(s)} \quad (5)$$

where,

$$T_b(s) = k \cdot PWM \cdot G_{dv_b}(s) \quad (6)$$

$$G_{dv_b}(s) = \frac{V_s}{P_b(s)}\left(sC_b r_{c_b} + 1 \right) \quad (Vs=Vin/2n) \quad (7)$$

$$P_b(s) = s^2 L_b C_b + s C_b \left(r_{L_b} + r_{c_b} \right) + 1 \quad (8)$$

k : sense gain products error amp. gain,
PWM : gain of the comparator.

In open loop case, the peak frequency is the same resonant frequency fp of the loop gain T(s) as shown in Fig. 5, and the peak value of the output impedance can be derived from Eq. (4).

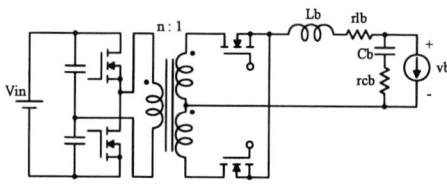

Fig. 3. Bus converter.

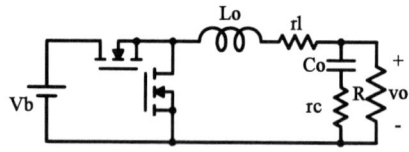

Fig. 4. Point of Load.

394

$$Z_{o\_peak} = \frac{L_b}{C_b \left(r_{c_b} + r_{L_b} \right)} \quad (9)$$

In closed loop case, the output impedance peak moves to crossover frequency fc as shown in Fig. 5. In this instant the peak value of the closed loop output impedance can be derive following equation.

$$Z_{oc\_peak} = \frac{L_b}{C_b \left\{ (1+\alpha) r_{c_b} + r_{L_b} \right\}} \quad (10)$$

where,

$$\alpha = |T(0)| = k \cdot PWM \cdot V_s \quad (11)$$

Moreover, from transfer function of loop gain, the crossover frequency fc is expressed as follows by means of peak frequency fp of loop gain.

$$f_c = \sqrt{1+\alpha}\, f_p \quad (12)$$

From Eq. (10) (12), the peak value of closed loop output impedance is expressed as follows.

$$Z_{oc\_peak} = \frac{L_b}{C_b \left\{ \left(\dfrac{f_c}{f_p} \right)^2 r_{c_b} + r_{L_b} \right\}} \quad (13)$$

As shown in Eq. (13), if fc is equal to fp, it becomes the same as Eq(9). Therefore, the peak value of output impedance is calculable by means of Eq. (13).

IV. OUTPUT IMPEDANCE SPECIFICATION

The output impedance characteristic of each control shames is different, and each bus converter has different operation. Therefore, the output impedance design suitable for the feature of each control method is required. From now, the output impedance design for each control shames is considered.

A. Un-regulated

In un-regulated case, the output impedance is the same as open-loop output impedance because of this control method has no control loop. In order to reduce the peak value of output impedance, it is effective to make inductance small or to enlarge capacitance.

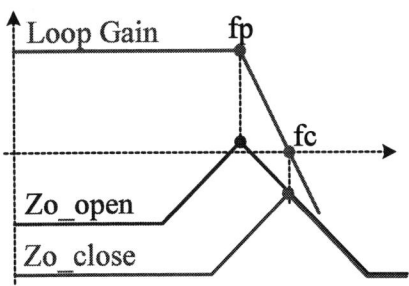

Fig. 5. Output impedance peak.

Generally, un-regulated bus converter is operated at maximum duty ratio. Therefore, the inductor of the bus converter can be reduced as small as possible to reduce the system instability. The peak value of output impedance is reducing with small inductor. Figure 6 shows the experimental result of the relation between the output impedance and inductance. Moreover, Fig. 7 shows the analytical and experimental results of the relation between the peak value of output impedance and inductance. Both results agreed well. As mentioned above, the peak of impedance is easily obtained from Eq. (13). However, this method depends on converter topology that has a double-ended circuit at secondary side such as half-bridge or full-bridge. Moreover, there are some limits such that high accurate input voltage or a POL with wide input range.

B. Semi-regulated

Semi-regulated bus converter has a control loop. However, regulation is related to variation of input voltage, therefore the output impedance is same as un-regulated case. In this case, the duty ratio is changed, and the inductor of the bus converter cannot be reduced. Therefore, very large bus capacitor is needed to reduce the peak value of output impedance.

Figure 8 shows the experimental result of the relation between the output impedance and capacitance. Moreover, Fig. 9 shows the analytical and experimental results of the relation between the peak value of output impedance and capacitance. Both results agreed well.

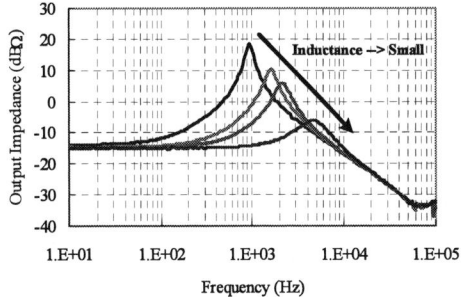

Fig. 6. Inductance and output impedance.

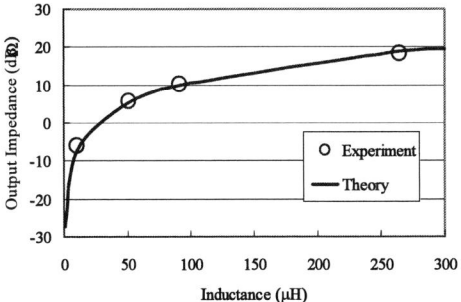

Fig. 7. Inductance and peak value of Zo.

395

In semi-regulated case, essentially it becomes very unstable and we have found that the demerit is very large capacitors are needed at the intermediate bus in order to be stable. However, it can be used at limited conditions such as wide input range (36-75V) and POL with low power (in other words, POL with very high input impedance).

C. Full-regulated

Full-regulated bus converter has a feedback loop, so the output impedance characteristic is changed. Therefore, output impedance can be made small with wide bandwidth. Figure 10 shows the experimental result of the relation between the output impedance and bandwidth. Moreover, Fig. 11 shows the analytical and experimental results of the relation between the peak value of output impedance and bandwidth. Both results agreed well.

Next, the relation between capacitance and output impedance peak is examined in closed loop case.

In closed loop case, if capacitance Cb changes to Cb+Cadd, peak frequency fp of loop gain is changed as follows.

$$f_p' = \frac{1}{2\pi\sqrt{L_b\left(C_b + C_{add}\right)}} \quad (14)$$

Therefore, the crossover frequency fc is changed as follows.

$$f_c' = \sqrt{1+\alpha}\, f_p' \quad (15)$$

Moreover, frequency ratio fc'/fp' is given as following equation.

$$\frac{f_c'}{f_p'} = \frac{f_c}{f_p} \quad (16)$$

From these results, output impedance peak can be expressed as following equation.

$$Z_{oc\_peak}' = \frac{k_{esr}C_b}{\left(C_b + C_{add}\right)} Z_{oc\_peak} \quad (17)$$

where, k

$$k_{esr} = \frac{(1+\alpha)r_{cb} + r_{Lb}}{(1+\alpha)r_{cb}' + r_{Lb}} \quad (18)$$

Figure 12 shows the experimental result of the relation between the output impedance and capacitance in closed loop case. Moreover, Fig. 13 shows the analytical and experimental results of the relation between the peak value of output impedance and capacitance in closed loop case. Both results agreed well. In this case, the total ESR is greatly changed by the additional capacitor. Moreover, in the case of closed loop, ESR has a great influence to the output impedance peak. Therefore, estimation of ESR is very important.

V. OPTIMAL DESIGN OF BUS CONVERTER

In order to evaluate the performance of this system, the experiment circuits are implemented using the specifications and parameters in Table 1.

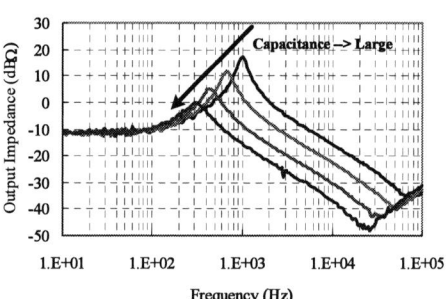

Fig. 8. Capacitance and output impedance.

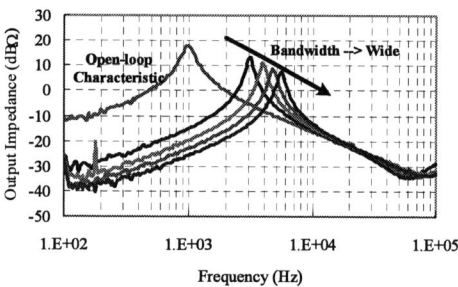

Fig. 10. Bandwidth and output impedance.

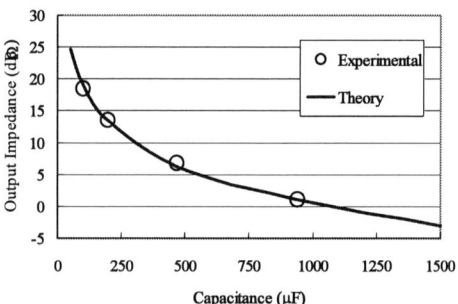

Fig. 9. Capacitance and peak value of Zo.

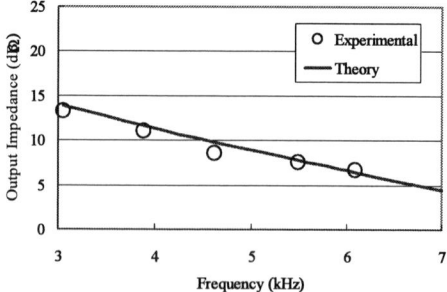

Fig. 11. Bandwidth and peak value of Zo

Here, the case with two POLs is discussed for actual example. The practical design process is shown below.

A. Input impedance estimation

The low-frequency value |Zin(0)| of input impedance is given by Eq. (3). The duty ratio is D=0.275 and output resistance is R=0.66(Ω) from the relation input and output. In this case, the |Zin(0)| is 20.6(dBΩ).

When two POLs of same condition are connecting in parallel, |Zin(0)| is 14.6(dBΩ). Figure 14 shows the experimental result of input impedance. The low-frequency value |Zin(0)| is around 15(dBΩ) as shown in Fig. 14.

The experimental results and analytical results are agreed well. If the stability margin is set to 6(dBΩ), then the peak value of output impedance must be set to 8.6(dBΩ).

B. Output impedance design

Figure 15 shows the output impedance characteristic of the basic case using the parameter of Table 1. As shown in Fig. 15, the peak value of the output impedance is around 18(dBΩ). From mentioned above calculation, the peak value of the output impedance needs to set around 8.6(dBΩ) for sufficient system stability.

TABLE I. CIRCUIT PARAMETERS

	Symbol	Description	Value
Bus Converter	Vin	Input Volotage	48V
	Vb	Bus Volotage	12V
	Lb	Output Inductor of Bus Converter	270µH
	Cb	Output Capacitor of Bus Converter	100µF
	rlb	Registance of Lb	300mΩ
	rcb	ESR of Cb	25mΩ
	kb	Feedback gain (with sence gain)	0.9
POL	Vo/Io	Output Condition	3.3V/5A
	Lo	Output inductor	2.8µH
	Co	Output capacitor	820µF
	rl	Registance of Lo	25mΩ
	rc	ESR of Co	10mΩ

In un-regulated case, the optimal inductance value is considered because the stability is improved by small inductance. From Eq. (13), the optimal inductance value can be derived as following equation.

$$L_{b\_optimal} = C_b \left(r_{C_b} + r_{L_b} \right) Z_{o\_peak} \quad (19)$$

where, the unit of |Zo_peak| is Ω.

Since the output impedance must be set to 8.6(dBΩ), the inductance value is set to around 87(µH) from Eq. (19). Figure 16 shows the experimental result of the output impedance with small inductance.

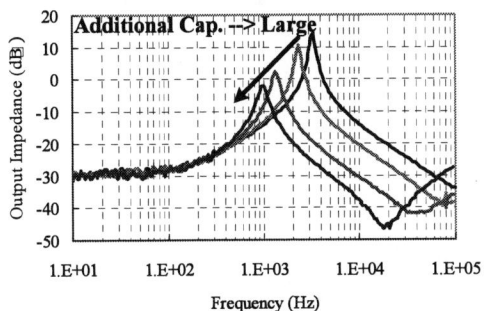

Fig. 12. Additional capacitance and output impedance

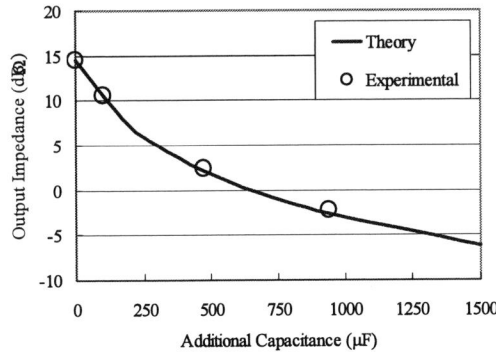

Fig. 13. Additional capacitance and peak value of Zo

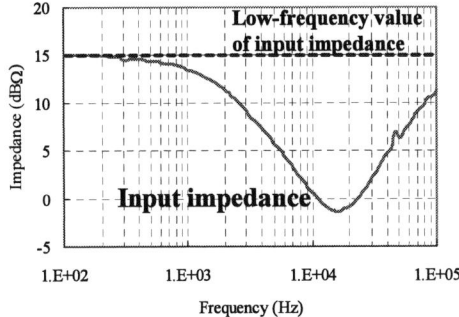

Fig. 14. Input impedance characteristic.

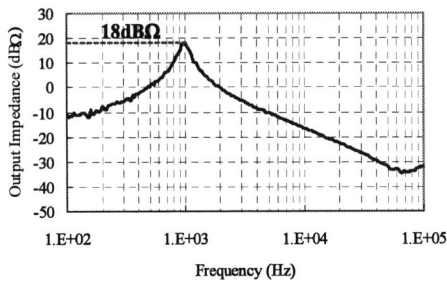

Fig. 15. Output impedance (Basic parameters).

The inductance value is around 90 (μH), and the peak value of output impedance is around 8.5 (dBΩ). The experimental results and analytical results are agreed well. Moreover, in open loop case, since the rL is generally larger than rc, the output impedance does not become smaller than rL as shown in Fig. 6. Therefore, the inductance value has minimum value. From Eq. (19), the minimum value of the inductance is given by following equation.

$$L_{b\_min} = C_b \left(r_{C_b} + r_{L_b} \right) r_{L_b} \quad (20)$$

In this case, the minimum value of inductance is around 10 (μH).

In semi-regulated case, the optimal capacitance value is considered because the stability is improved by large capacitance. From Eq. (13), the optimal capacitance value can be derived as following equation.

$$C_{b\_optimal} = \frac{L_b}{\left(r_{C_b} + r_{L_b} \right) Z_{o\_peak}} \quad (21)$$

where, the unit of |Zo_peak| is Ω.
Since the output impedance must be set to 8.6(dBΩ), the capacitance value is set to around 300(μF) from Eq. (16). In this case, the influence of ESR is considered. Because the ESR becomes small when the capacitor is connected in parallel.

Figure 17 shows the experimental result of output impedance with large capacitance. The capacitance value is 300 (μF), and the peak value of output impedance is around 8 (dBΩ). The experimental results and analytical results are agreed well.

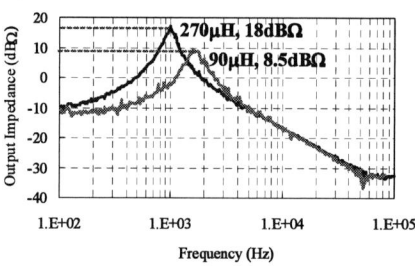

Fig. 16. Output impedance with small inductor.

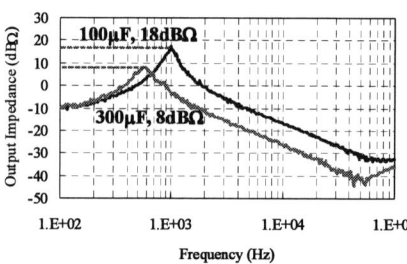

Fig. 17. Output impedance with large capacitor.

Moreover, the output impedance does not become smaller than rL as shown in Fig. 8. Therefore, the capacitance value has maximum value. From Eq. (21), the maximum value of the capacitance is given by following equation.

$$C_{b\_max} = \frac{L_b}{\left(r_{C_b} + r_{L_b} \right) r_{L_b}} \quad (22)$$

In this case, the maximum value of capacitance is around 2.8 (mF).

In full-regulated case, the optimal bandwidth is considered because the stability is improved by wide bandwidth. From Eq. (13), the optimal bandwidth can be derived as following equation.

$$f_{c\_optimal} = f_p \sqrt{\frac{\dfrac{L_b}{C_b Z_{oc\_peak}} - r_{L_b}}{r_{C_b}}} \quad (23)$$

where, the unit of |Zo_peak| is Ω.
Since the output impedance must be set to 8.6(dBΩ), the bandwidth is set to around 5.1kHz from Eq. (23). Figure 18 shows the experimental result of the output impedance with wide bandwidth. The bandwidth is around4.7kHz, and the peak value of output impedance is around 8.5 (dBΩ). The experimental results and analytical results are agreed well.

Next, the optimal capacitance is considered in closed loop case. From Eq. (17), the optimal additional capacitance can be derived as following equation.

$$C_{add\_optimal} = \left(\frac{k_{esr} Z_{oc\_peak}}{Z_{oc\_peak}'} - 1 \right) C_b \quad (24)$$

The basic parameters case, the closed loop output impedance peak is around 14.5(dBΩ). Since the output impedance must be set to 8.6(dBΩ), the additional capacitance is set to around 150μF from Eq. (24). Figure 19 shows the experimental result of the output impedance with additional capacitance. The capacitance is around 150μF, and the peak value of output impedance is around 9 (dBΩ). The experimental results and analytical results are agreed well.

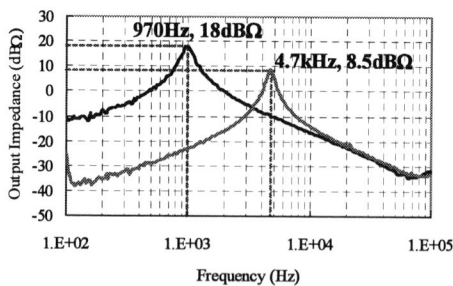

Fig. 18. Output impedance with wide bandwidth.

VI. Conclusions

This paper presents the output impedance design for on-board distributed power system by means of three control methods of bus converter. The output impedance peak of the bus converter and the input impedance of the POL were analyzed, and it was conformed by experimentally for stability criterion. As a result, the standard of the discrimination of stability on a frequency response of input and output impedance was clarified. Furthermore, the design process of each control method for system stability was proposed.

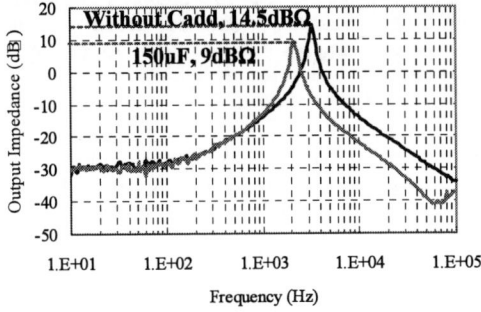

Fig. 19. Output impedance with additional capacitance

References

[1] C. M. Wildrick, F. C. Lee, B. H. Cho, B. Choi, "A Method of Defining the Load Impedance Specification for A Stable Distributed Power System", IEEE Transactions on Power Electronics Vol. 10. No. 3. May 1995, pp. 280-284.

[2] X. Feng, Z. Ye, K. Xing, F. C. Lee, D. Borojevic, "Individual Load Impedance Specification for a Stable DC Distributed Power System", IEEE Applied Power Electronics Conference (APEC) 1999, pp. 923-929.

[3] X. Feng, F. C. Lee, "On-line Measurement on Stability Margin of DC Distributed Power System", IEEE Applied Power Electronics Conference (APEC) 2000, pp. 1190-1196.

[4] M. P. Sayani, J. Wanes, " Analyzing and Determining Optimum On-Board Power Architectures for 48V-input Systems", IEEE Applied Power Electronics Conference (APEC) 2003.

[5] K. Hisanaga, K. Harada, "Stability Analysis of the Distributed Power System with Intermediate Bus Converter", IEICE Technical Report, Vol.103, No.199, pp.19-24, Jul. 2003 (in Japanese).

[6] K. Hisanaga, K. Harada, "Stability Analysis of the Distributed Power System with Intermediate Bus Converter (2nd Report)", IEICE Technical Report, Vol.103, No.652, pp.7-12, Feb. 2004 (in Japanese).

[7] Y. Ren, M. Xu, K. Yao, Y. Meng, F. C. Lee, J. Guo, "Two-Stage Approach for 12V VR", IEEE Applied Power Electronics Conference (APEC) 2004.

[8] J. Wei, F. C. Lee, "An Output Impedance-Based Design of Voltage Regulator Output Capacitors for High Slew-Rate Load Current Transients", IEEE Applied Power Electronics Conference (APEC) 2004.

[9] B. Choi, D. Kim, D. Lee, S. Choi, J. Sun, "Analysis of Input Filter Interactions in Switching Power Converters", IEEE Transactions on Power Electronics, Vol. 22, No. 2, March 2007, pp. 452-460.

[10] S. Abe, M. Hirokawa, T. Zaitsu, T. Ninomiya, "Stability Design of Bus Converter Following by POLs in Distributed Power System", IASTED Circuits, Signals,and Systems (CSS), 2005, pp552-557.

[11] S. Abe, H. Nakagawa, M. Hirokawa, T. Zaitsu , T. Ninomiya, "Comparison of System Stability in Distributed Power System Based on Control Method of Bus Converter", IASTED Energy and Power Systems (EPS) 2005, pp109-114.

[12] S. Abe, H. Nakagawa, M. Hirokawa, T. Zaitsu , T. Ninomiya, "System Stability of Full-Regulated Bus Converter in Distributed Power System", International Telecommunications Energy Conference (INTELEC) 2005, pp563-568.

[13] S. Abe, H. Nakagawa, M. Hirokawa, T. Zaitsu , T. Ninomiya, " Stability Improvement of Distributed Power System by Using Full-Regulated Bus Converter ", Annual Conference of the IEEE Industrial Electronics Society (IECON) 2005, pp2549-2553.

[14] S. Abe, T. Ninomiya, M. Hirokawa, T. Zaitsu , "Stability Comparison of Three Control Schemes for Bus Converter in Distributed Power System ", International Conference on Power Electronics and Drive Systems (PEDS) 2005, pp1244-1249.

[15] S. Abe, M. Hirokawa, T. Zaitsu , T. Ninomiya, " Stability Design Consideration for On-Board Distributed Power System Consisting of Full-Regulated Bus Converter and POLs ", IEEE Power Electronics Specialists Conference (PESC) 2006, pp2669-2673.

[16] R.D. Middlebrook, S. Cuk, "A General Unified Approach to Modeling Switching-Converter Power Stages," IEEE Power Electronics Specialists Conference (PESC) 1976, pp. 18-34.

[17] T. Ninomiya, M. Nakahara, T. Higashi, K. Harada, "A Unified Analysis of Resonant Converters," IEEE Transactions on Power Electronics Vol. 6. No. 2. April 1991, pp. 260-270.

[18] R. D. Middlebrook, "Input Filter Considerations in Design and Application of Switching Regulators", IAS'76, 1976, pp. 91-107.

Steady State Analysis of Hysteretic Control Buck Converters

L.K. Wong[*] and T.K. Man[†]

[*] National Semiconductor Corporation, Power Management Design Center, Hong Kong, e-mail: *LK.Wong@nsc.com*
[†] National Semiconductor Corporation, Power Management Design Center, Hong Kong, e-mail: *TK.Man@nsc.com*

Abstract— **This paper presents the analysis of a hysteretic control buck converter by means of variable structure system theory because a hysteretic control buck converter is inherently a variable structure system owing to the presence of switching actions. Analysis results show the relationship between the steady state performance and a number of parameters, in particular the output capacitor's ESR. If the ESR is too small, the output voltage ripple will increase significantly and a phase shift is resulted. Although these phenomena are commonly known in the field, there is no analytical result to articulate them.**

Keywords— **Sliding mode control, DC power supply, power management.**

I. INTRODUCTION

Hysteretic control power converters are inherently fast response and robust with simple design and implementation. They response to disturbances and load change right after the transient take place [5], so they give excellent transient performance. Also, they do not require components for the closed loop compensation network. This reduces the component count and solution size in implementation, and eliminates the design effort in adjusting component values for the network upon parameters (like input voltage, inductor, bulk capacitors) change. The above advantages make hysteretic control power converters a good solution for power supply.

In spite of the advantages, one major concern of using hysteretic control power converters is the stability issue. Although the design and analysis of fixed frequency PWM power converters have been well developed using the averaging and linearization approach [1], analytical analysis of hysteretic control power converters is rare, and focus on the large signal dynamic response [2-5]. Such method cannot reflect the steady state high frequency oscillations. Also, the commonly used bode plot in the frequency domain will be inaccurate above half of the operating frequency. Unlike PWM converters with fixed operating frequency, the operating frequency can be very low during some loading condition or transient. For a same converter, the operating frequency can be a few hundred kHz for full load, but lower than around 10 kHz at very light load. The use of linear system tools may not be appropriate for variable operating frequency converters.

There are some common phenomena for unstable hysteretic control power converters. First, the output voltage ripple is not in phase with the switching. Also, the output voltage swing is larger than the hysteretic band and the inductor current swing will be large. A large output voltage swing generates more noise, and an unexpected large swinging in the inductor current can saturate the inductor and damage the switches. Common averaging analysis method smoothes out the output voltage and inductor current waveforms so that the above unstable phenomena are averaged out and cannot be seen. Therefore non-linear control theories should be applied for analyzing hysteretic converters.

This paper presents the use of variable structure system (VSS) theory for hysteretic control buck converters to analyse the steady state performance. It is because the presence of switching actions in a switching mode power converter already forms a VSS. The theoretical background of the VSS will be addressed in section II. The analysis of hysteretic control buck converter will be detailed in section III. Section IV shows simulation results. It will be shown how the output capacitor's ESR affects the stability. Finally, a conclusion will be drawn in section V.

II. THEORETICAL BACKGROUND

A variable structure system (VSS) switches the system structure according to a hyper-plane, which is a function of states. Let a hyper-plane s of an n-th order system is defined as follows:

$$s = k^T x \tag{1}$$

where k is a gain vector and x is the state vector of the system, both of order n. Let there are two structures represented by two system dynamics $f_+(x)$ and $f_-(x)$, the overall system for the VSS is

$$\dot{x} = \begin{cases} f_+(x) & \text{if } s > 0 \\ f_-(x) & \text{if } s < 0 \end{cases} \tag{2}$$

To achieve a sliding mode on the hyper-plane, the stability, existence, and reachability [6, 7] should be considered. The stability of the hyper-plane can be consider by solving (1) at $s = 0$. To ensure the existence and reachability, the following condition should be satisfied:

$$s\dot{s} \leq -\eta|s| \tag{3}$$

where η is a positive constant. To satisfy (3), a switch with infinity speed is required so that the structure can be switched over at the instance when the states cross the hyper-plane. In such case, the states will stay on the plane and sliding mode occurs. If the stability condition is satisfied, the states will slide to the origin.

However, there is no switch that can achieve an infinite switching speed. If there is a delay in the switch, the switching does not occur at s equals 0. The structure in (2) will effectively be changed to

978-1-4244-1741-4/08/$25.00 ©2008 IEEE   400

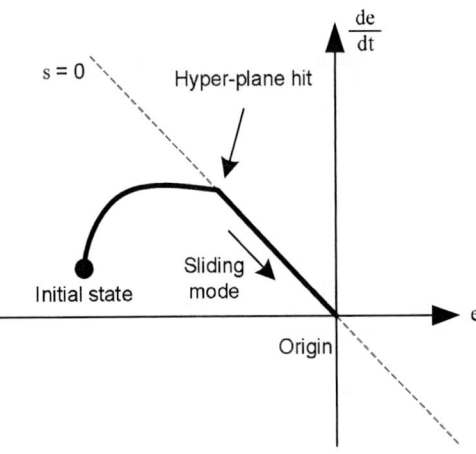

Fig. 1a. Infinity speed switching.

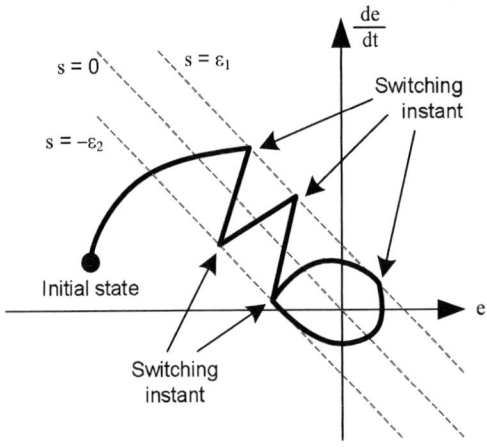

Fig. 1b. Switching with delay.

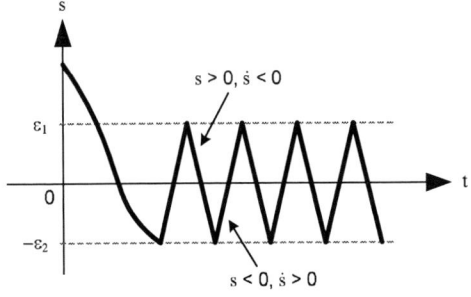

Fig. 2. Trajectory of s with delay switching.

$$\dot{x} = \begin{cases} f_+(x) \text{ if } s > \varepsilon_1 \\ f_-(x) \text{ if } s < -\varepsilon_2 \\ \text{unchanged if } -\varepsilon_2 < s < \varepsilon_1 \end{cases} \qquad (4)$$

where ε_1, ε_2 are positive, and unchanged here means that no switching occurs so $\dot{x}$ follows the current structure, which is either $f_+(x)$ or $f_-(x)$. Fig. 1a and 1b show the state trajectories of a VSS with a hyper-plane s on applying the control signal of (2) and (4) respectively. On applying the control signal of (2) with infinity switching speed, the state slides to the origin once the hyper-plane is hit. If there is a delay in the switch, i.e. effectively (4) is applied, chattering exists and a limit cycle appears around the origin at the steady state.

If a VSS satisfied the existence condition of (3), although the delay in the switch deviates the effective switching instant from the hyper-plane ($s = 0$), (3) still holds valid after the switching instant and before the states cross the plane. This can be shown from the waveform of Fig. 2 that the rate of change of s is always opposite to s (that s is decreasing when s is positive, and increasing when s is negative). This observation can be applied to analyse the steady state performance of a hysteretic control buck converter, the operation of which is similar to a variable structure system such that the structures switch according to a hyper-plane which is in terms of the output voltage, as shown in the following section.

III. HYSTERETIC CONTROL BUCK CONVERTER

The operation of a hysteretic control buck converter (Fig. 3) can be simply understood as follows. The switch S_1 turns on (S_2 works complementarily) when the output voltage V_{OUT} falls below a threshold $V_{REF} - \delta$ and turns off when V_{OUT} is higher than $V_{REF} + \delta$, where δ is corresponding to a hysteretic band. This operation contributes two structures as follows:

Structure 1: when S_1 is on and S_2 is off,

$$V_{IN} = L\frac{di_L}{dt} + V_C + R_C C\frac{dV_C}{dt} \qquad (5)$$

Structure 2: when S_1 is off and S_2 is on,

$$0 = L\frac{di_L}{dt} + V_C + R_C C\frac{dV_C}{dt} \qquad (6)$$

Hence, we have

$$QV_{IN} = L\frac{di_L}{dt} + V_C + R_C C\frac{dV_C}{dt} \qquad (7)$$

where Q is 1 and 0 for S_1 is on and off respectively. Furthermore, for both structures,

$$i_L = i_C + \frac{V_{OUT}}{R_{OUT}}$$

401

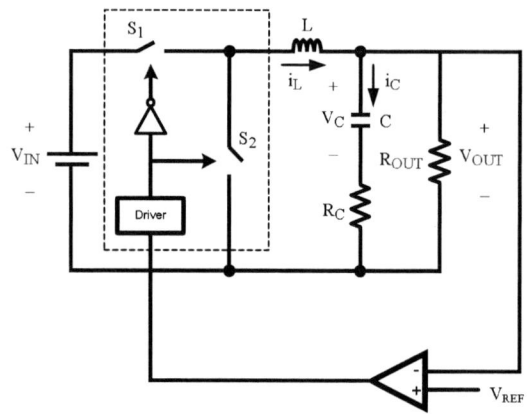

Fig. 3. Hysteretic control buck converter

Since

$$V_{OUT} = V_C + R_C i_C$$

$$i_C = C\frac{dV_C}{dt}$$

we have

$$i_L = C\frac{dV_C}{dt} + \frac{1}{R_{OUT}}\left(V_C + R_C C\frac{dV_C}{dt}\right)$$

$$\frac{di_L}{dt} = C\frac{d^2V_C}{dt^2} + \frac{1}{R_{OUT}}\left(\frac{dV_C}{dt} + R_C C\frac{d^2V_C}{dt^2}\right)$$

Substitute into (7),

$$\frac{d^2V_C}{dt^2} = \frac{R_{OUT}QV_{IN}}{LC(R_{OUT} + R_C)} - \frac{R_{OUT}V_C}{LC(R_{OUT} + R_C)}$$
$$- \frac{R_{OUT}R_C C + L}{LC(R_{OUT} + R_C)}\frac{dV_C}{dt}$$

(8)

Define a hyper-plane,

$$s = V_{REF} - V_{OUT}$$

$$= V_{REF} - V_C - R_C C\frac{dV_C}{dt}$$

Define a state e such that

$$e = V_{REF} - V_C.$$

Then

$$\frac{de}{dt} = -\frac{dV_C}{dt},$$

and

$$\frac{d^2e}{dt^2} = -\frac{d^2V_C}{dt^2},$$

$$s = e + R_C C\frac{de}{dt} \qquad (9)$$

Referring to the operation of a hysteretic control buck converter,

$Q = 1$ when S_1 is on, when $V_{OUT} < V_{REF} - \delta$, i.e. $s > \delta$,

$Q = 0$ when S_1 is off, when $V_{OUT} > V_{REF} + \delta$, i.e. $s < \delta$.

From (8), we have

$$\frac{d^2e}{dt^2} = -\frac{R_{OUT}QV_{IN}}{LC(R_{OUT} + R_C)} + \frac{R_{OUT}V_C}{LC(R_{OUT} + R_C)}$$

$$+ \frac{R_{OUT}R_C C + L}{LC(R_{OUT} + R_C)}\frac{dV_C}{dt}$$

$$= -\frac{R_{OUT}QV_{IN}}{LC(R_{OUT} + R_C)} + \frac{R_{OUT}(V_{REF} - e)}{LC(R_{OUT} + R_C)}$$

$$- \frac{R_{OUT}R_C C + L}{LC(R_{OUT} + R_C)}\frac{de}{dt}$$

Consider

$$s = \frac{de}{dt} + R_C C\frac{d^2e}{dt^2}$$

$$= \frac{de}{dt} - R_C\frac{R_{OUT}QV_{IN}}{L(R_{OUT} + R_C)}$$

$$+ R_C\frac{R_{OUT}(V_{REF} - e)}{L(R_{OUT} + R_C)} - R_C\frac{R_{OUT}R_C C + L}{L(R_{OUT} + R_C)}\frac{de}{dt}$$

$$= \frac{de}{dt} + \frac{R_C R_{OUT}(V_{REF} - QV_{IN})}{L(R_{OUT} + R_C)}$$

$$- \frac{R_C R_{OUT}}{L(R_{OUT} + R_C)}\left(s - R_C C\frac{de}{dt}\right)$$

$$- R_C\frac{R_{OUT}R_C C + L}{L(R_{OUT} + R_C)}\frac{de}{dt}$$

$$= \frac{R_{OUT}}{R_{OUT} + R_C}\frac{de}{dt} + \frac{R_C R_{OUT}(V_{REF} - QV_{IN})}{L(R_{OUT} + R_C)}$$

$$- \frac{R_C R_{OUT}}{L(R_{OUT} + R_C)}s$$

To satisfy (3), it is required that

$$s < 0 \quad \text{when } s > 0,$$

and

$$s > 0 \quad \text{when } s < 0.$$

402

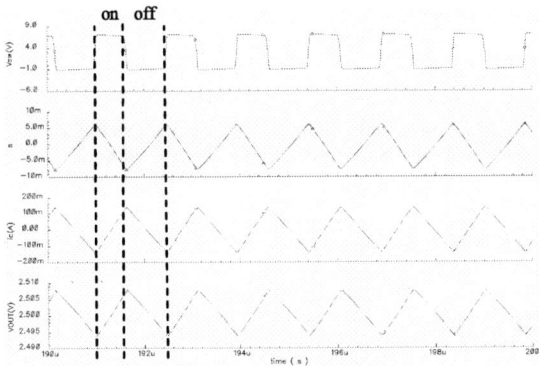

Fig. 4. Waveforms for a buck converter with $R_C = 50\text{m}\Omega$

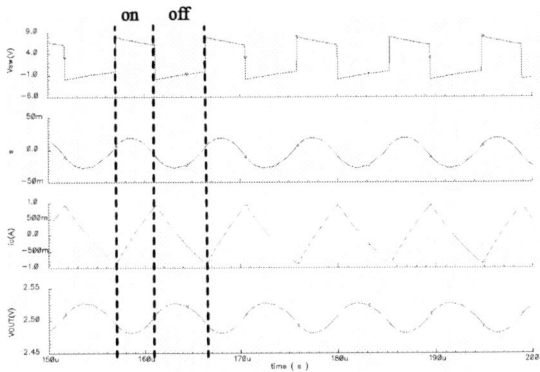

Fig. 5. Waveforms for a buck converter with $R_C = 5\text{m}\Omega$

For $s > 0$, to obtain the condition that $s < 0$,

$$\frac{R_{OUT}}{R_{OUT} + R_C}\frac{de}{dt} + \frac{R_C R_{OUT}(V_{REF} - QV_{IN})}{L(R_{OUT} + R_C)}$$

$$-\frac{R_C R_{OUT}}{L(R_{OUT} + R_C)}s < 0$$

Since R_C, R_{OUT} and L are all positive,

$$LR_{OUT}\frac{de}{dt} + R_C R_{OUT}(V_{REF} - V_{IN}) - R_C R_{OUT}s < 0$$

$$R_C > \frac{L\dfrac{de}{dt}}{V_{IN} - V_{REF} + s} = \frac{-L\dfrac{i_C}{C}}{V_{IN} - V_{REF} + s}$$

where i_C is the current of the output capacitor, which is rippled around the steady state point of 0A. Let $2I_{CMAX}$ be the maximum peak to peak value of the rippled current i_C, a sufficient condition for obtaining $s < 0$ for $s > 0$ is

$$R_{C+} = \frac{LI_{CMAX}}{C(V_{IN} - V_{REF})} \tag{10}$$

because

$$R_{C+} = \frac{LI_{CMAX}}{C(V_{IN} - V_{REF})}$$

$$> \frac{Li_C}{C(V_{IN} - V_{REF})}$$

$$> \frac{Li_C}{C(V_{IN} - V_{REF} + s)}$$

Similarly, for $s < 0$,

$$R_{C-} = \frac{LI_{CMAX}}{CV_{REF}} \tag{11}$$

As a result,

$$R_C > \max\{R_{C+}, R_{C-}\} \tag{12}$$

The above analysis shows a condition for satisfying (3), which is important for the steady state performance of a hysteretic control buck converter. If (3) is not satisfied, the output voltage ripple will increase significantly as shown in the results below.

IV. SIMULATION RESULTS

To illustrate the proposed analysis method, a simulation is carried out with a hysteretic control buck converter with the following parameters: $V_{IN} = 8\text{V}$, $V_{REF} = 2.5\text{V}$, $L = 10\mu\text{H}$, $C = 47\mu\text{F}$, $R_{OUT} = 2.5\Omega$. The parameter I_{CMAX} is set to 0.25A, and from (10) to (12), the minimum R_C is $21.3\text{m}\Omega$. Figs. 4 and 5 show waveforms of the hysteretic control buck converter with output capacitor's ESR equals to $50\text{m}\Omega$ and $5\text{m}\Omega$ respectively. The curves from the top to bottom of both figures show the waveforms of V_{SW}, s, i_C, and V_{OUT}. From Fig. 4, since the ESR is larger than the minimum R_C calculated above, a stable waveform is seen. The parameter s decreases immediately when S_1 turns on (when V_{SW} is at a high voltage level), and increases immediately when S_1 turns off. The ESR corresponding to Fig. 5 is smaller than the minimum R_C, an unstable waveform is obtained. It can be seen from Fig. 5 that s only decreases and increases respectively some time (but not immediately) after turning on and off S_1. Consequently, the output voltage ripple will increase significantly and a phase shift with reference to V_{SW} is resulted, which are common unstable phenomena observed for hysteretic control buck converter if the output capacitor's ESR is too small.

V. CONCLUSION

A variable structure system approach for analyzing the steady state stability of a hysteretic control buck converter has been proposed in this paper. Hysteretic control converters are simple in design and implementation, and give fast response. However, the lack of stability analysis is a major concern. In this paper, unstable in steady state has been articulated as the occurrence of large swinging in the output voltage and the inductor current. By design, the output voltage ripple should be limited to the hysteretic band. A large output voltage swing generates more noise than expect, and possibly damages circuits of the following stage. Also, a large swinging in the inductor current can be over the design limit in order to saturate and damage the inductor and the switches.

Owing to the large variation in operating frequency, the use of linearization technique and tools for linear systems may not provide an accurate analysis. Also, the unstable large swinging at steady state will be averaged out by the averaging method. Consequently, non-linear system techniques will be more appropriate for analyzing hysteretic control buck converters. Switching mode power converters inherently consist of two structures when the switch is turned on and off respectively. Therefore the variable structure system technique is used in the stability analysis. It has been shown that an adequate amount of the output capacitor's ESR is required for stability, which is known in industry but not analytically shown before.

REFERENCES

[1] R.D. Middlebrook and S. Cuk, "A general unified approach to modeling switching converter power stages," *IEEE PESC 1976 Record*, pp. 18-34, 1976.

[2] M. Castilla, L. Garcia de Vicuna, J.M. Guerrero, J. Matas, and J. Miret, "Design of voltage-mode hysteretic controllers for synchronous buck converters supplying microprocessor loads", *IEE Proc.-Electr. Power Appl.*, vol. 152, no. 5, pp. 1171-1178, September 2005.

[3] Jinbin Zhao, Terukazu Sato, Takashi Nabeshirna, Tadao Nakano, "Steady-state and dynamic analysis of a buck converter using a hysteretic PWM control", in *Proc. IEEE PESC '04 record.*, Aachen, Germany, 2004, pp. 3654-3658.

[4] T. Nabeshima, T. Sato, S. Yoshida, S Chiba, and K. Onda, "Analysis and design considerations of a buck converter with a hysteretic PWM controller," in *Proc. IEEE PESC '04 record.*, Aachen, Germany, 2004, pp. 1711-1716.

[5] Miguel Castilla, Luis Garcia de Vicuna, Josep M. Guerrero, Jose Matas, and Jaume Miret, "Designing VRM hysteretic controllers for optimal transient response", *IEEE Trans. Ind. Electr.*, vol. 54, no. 3, pp. 1726-1738, June 2007.

[6] R.A. DeCarlo, S.H. Zak, and G.P. Matthews, "Variable structure control of nonlinear multivariable systems: a tutorial," *IEEE Proc.*, vol. 76, iss. 3, pp. 212-232, Mar. 1988.

[7] J.J.E. Slotine and W. Li, *Applied Nonlinear Control*, Prentice-Hall, Inc. Englewood Cliffs, N.J. 1991

A Novel Control Method for IGBT Current Source Rectifier

Longcheng Tan[1,2], Yaohua Li[1], Ping Wang[1], Congwei Liu[1], Zixin Li[1,2], Yonggang Chen[1,2], Wei Xu[1,2]

1. Institute of Electrical Engineering Chinese Academy of Sciences
2. Graduate University of Chinese Academy of Sciences
P.O. Box 2703, 100190, Beijing, P.R.China
E-mail: lchtan@mail.iee.ac.cn

Abstract—A novel power factor control and power flow control method for IGBT current source rectifier is proposed in this paper. Unlike the conventional control method in the *a-b-c* three-phase stationary frame, the proposed control is based on the *d-q* rotating frame. The proposed power factor compensation scheme can provide maximum achievable power factor under all operating conditions, including the regenerative mode. What's more, the scheme is parameter insensitive, which can improve the system's stability. Due to the IGBT's switching character, the Space Vector Modulation (SVM) is adopted, so the fast dynamic response and the little harmonic content can be achieved. Simulation results are provided to verify the theoretical analysis.

Index Terms—current source rectifier, power factor control , space vector modulation, regenerative operation

I. INTRODUCTION

Current source rectifiers (CSR) are widely used as the front-end in power electronic systems due to its inherent short circuit protection capability and ruggedness. Because of direct interfacing with ac mains, the rectifiers must meet stringent specifications such as: low input current harmonics and high input power factor. Meanwhile, good current regulation and fast dynamic response are also required[1].

In current source rectifier just as in Fig. 1, the three-phase filter capacitor C_r is required to assist the commutation of switching devices and filter out the line current harmonics. However, the use of the filter capacitor makes the input power factor leading.

The prior power factor control method is proposed in [2], which enforces the current I_w delay the converter input voltage V_c to compensate the leading capacitor current so that the input power factor can be unity. This method has two main limitations. One is that the scheme is valid only when the unity power factor is achievable. Under light loads, the rectifier can not get the unity power factor. Another is that the method is parameter sensitive.

A better approach proposed by Xiao et al [3] allows for the full range of load currents while maintaining maximum achievable power factor by detecting the displacement angle between the phase voltage and line current. In a feedforward loop, the displacement angle is controlled to as small as possible by the modulation index through a proportional integral (PI) regulator, while in the feedback loop the dc current is controlled by the delay angle through another PI regulator. But the system also has several shortcomings. One

is that the dynamic response of the system is rather slow, and the dc current control is coupled with the power factor control. Another is that the design of the PI controller is rather difficult, especially in the regenerative mode, because of the coupling of the two loops[4].

In this paper, a novel power factor control technique for IGBT based current source rectifier is proposed. Using vector based power factor schemes, this approach can provide unity power factor when it is achievable and provide the highest possible power factor when it is unachievable whatever the operating conditions are. The technique is parameter insensitive and the system dynamic response is much faster.

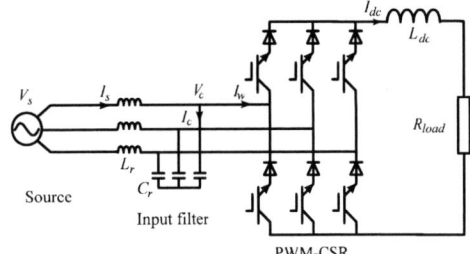

Fig. 1 Three phase current source rectifier topology

II. THE POWER COMPENSATION PRINCIPLE IN THE D-Q ROTATING FRAME

The synchronous reference frame is used to convert three phase sinusoidal quantities *a-b-c* into a static vector (*d-q*) through the following transform:

$$\begin{pmatrix} d \\ q \end{pmatrix} = \frac{2}{3} \begin{bmatrix} \cos(\theta) & \sin(\theta) \\ -\sin(\theta) & \cos(\theta) \end{bmatrix} \begin{bmatrix} 1 & -\frac{1}{2} & -\frac{1}{2} \\ 0 & \frac{\sqrt{3}}{2} & -\frac{\sqrt{3}}{2} \end{bmatrix} \begin{pmatrix} a \\ b \\ c \end{pmatrix} \quad (1)$$

So, the real power and the imaginary power can be defined in the synchronous reference frame as follows:

$$P = \frac{3}{2}\left(V_q \cdot I_q + V_d \cdot I_d\right) = |\mathbf{Vs}| \times Is_p$$

$$Q = \frac{3}{2}\left(V_q \cdot I_d + V_d \cdot I_q\right) = |\mathbf{Vs}| \times Is_r \qquad (2)$$

While the V_d, I_d and V_q, I_q are the d-axis component and the q-axis component of source voltage vector $\mathbf{V_s}$ and source current vector $\mathbf{I_s}$ respectively. The Is_p and Is_r are the power producing current component and the reactive current component of the source current $\mathbf{I_s}$ respectively.

The source voltage $\mathbf{V_s}$, the source current $\mathbf{I_s}$, the capacitor voltage $\mathbf{V_c}$, the capacitor current $\mathbf{I_c}$, the rectifier current $\mathbf{I_w}$, and the filter reactor voltage drop $\mathbf{V_{Lf}}$ are transformed to the source voltage synchronous reference frame and shown in Fig. 2. The impact of the filter capacitor on the displacement power factor can be shown graphically in the synchronous reference frame. Without power factor compensation, the resulting source current $\mathbf{I_s}$ leads the source phase voltage $\mathbf{V_s}$ as shown in Fig.2a. This results in a leading power factor, which is undesirable. With power factor compensation, quadrature current I_{wq} is forced into the rectifier as shown in Fig.2b, to compensate I_{cq}, which is the q-axis component of $\mathbf{I_c}$, so the resultant source current $\mathbf{I_s}$ can keep in phase with $\mathbf{V_s}$.

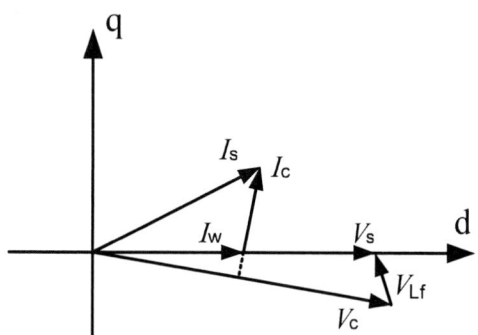

a) without power factor compensation

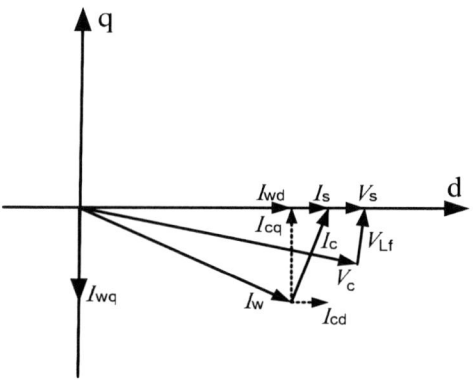

b) with power factor compensation

Fig. 2 The concise interpretation of power factor compensation

III. THE PROPOSED POWER FACTOR CONTROL METHOD

Based on the power factor compensation scheme discussed above, a novel power factor technique is proposed, shown in Fig.3. The detected source voltage V_s and line current I_s are transferred to the synchronous frame of the source voltage V_s. Therefore the power producing and reactive components of the line current can be calculated as follows:

$$Is_p = \frac{3}{2}\left(V_q \cdot I_q + V_d \cdot I_d\right)\bigg/|\mathbf{Vs}| = \frac{3}{2}I_d$$

$$Is_r = \frac{3}{2}\left(V_q \cdot I_d + V_d \cdot I_q\right)\bigg/|\mathbf{Vs}| = \frac{3}{2}I_q \qquad (3)$$

In the feedforward power factor control loop, the calculated reactive current Is_r is compared with the reactive current reference Is_r^* which is usually set at zero to get the unity power factor. The resultant error signal is used to control M_q through a PI regulator, which is the q-axis component of modulation index vector $\mathbf{M}$. In the feedback I_{dc} control loop, the dc side current is compared with the current reference I_{dc}^*, the error signal is used to control M_d through another PI regulator, which is the d-axis component of the modulation index vector $\mathbf{M}$.

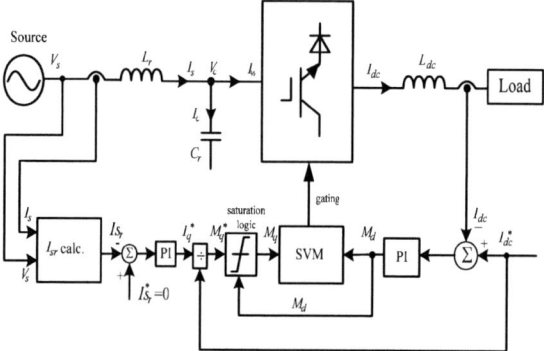

Fig. 3 Proposed power factor control scheme

Unity power factor is achieved when the reactive current Is_r is regulated to zero. Assuming that the source current $\mathbf{I_s}$ leads the source voltage $\mathbf{V_s}$, then through the comparison and the PI regulator, a negative M_q is got to compensate the leading current. Conversely, when there is a lagging source current, a positive M_q is got to compensate the lagging current. These processes continues until the unity power factor is reached when it's achievable.

Under some operating conditions such as a light load, the unity power factor is not achievable, that is the reactive current cannot be regulated to zero. In such case, the output of the PI regulator for M_q will keep increasing until the maximum value the system can offer. The maximum value of M_q is determined by the value of the M_d. Because the primary task of the rectifier is to control the dc side current, which is also the basis of the realization of the power factor control, so the M_d is allowed to any value up to saturation, while the allowed M_q's maximum value M_{qmax} can only be calculated as follows:

406

$$M_{q\max} = \sqrt{M_{\max}^2 - |M_d|^2} \qquad (4)$$

While the M_{max} is the allowed maximum modulation index, which is usually one.

IV. SIMULATION RESULTS AND DISCUSSIONS

The performance of the proposed method was evaluated by matlab simulation. The inverter is rated at 220 V, 50 Hz, and 20 kVA with L_r=0.04, C_f= 0.34, L_{dc}=1.04, R=0.6, all in per unit. The modulation technique is SVM[5], and the switching frequency of the rectifier is 6 kHz. As shown in Fig.4, at t=0.2s the rectifier reaches a steady-state operating point, where I_{dc}= I_{dc}^*=1.0, V_{dc}=0.6, and R=0.6, all in per unit. As shown in Fig.4(d) the source current I_s is in phase with the source voltage V_s with M_d=0.64 and M_q=-0.3, When t=0.3s, the current reference I_{dc}^* decreases to 0.6pu, in order to keep the dc voltage at 0.6pu, the load resistance is increased to 1.0pu. At 0.5s when the system reaches another steady state, just as shown in Fig.4(e) the input power factor is still unity. From Fig.4(g), we find that the reactive component of the source current can be controlled to zero, so the input power factor is still unity.

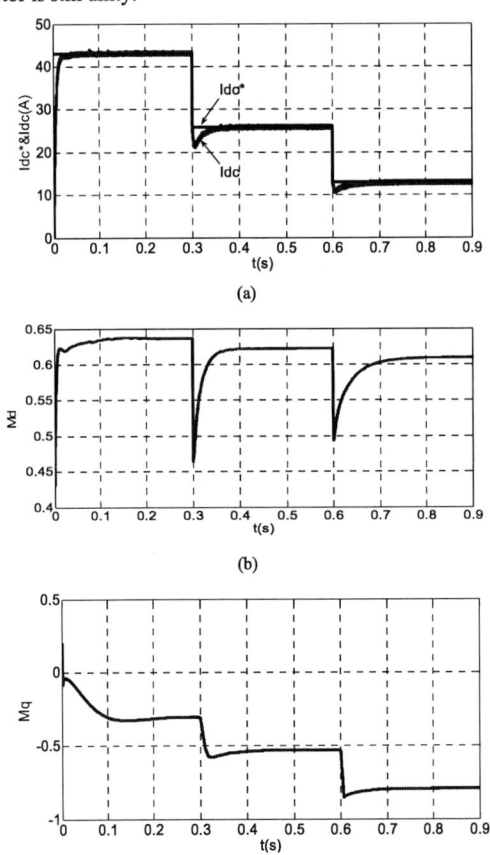

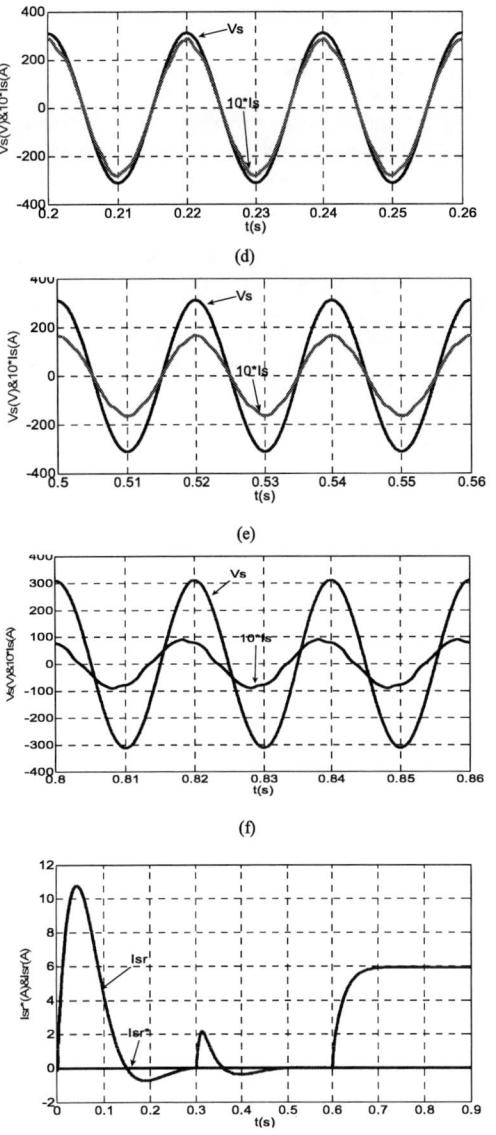

Fig. 4 Simulation results. During 0s to 0.3s, I_{dc}^*=1pu, 0.3s to 0.6s, I_{dc}^*=0.6pu, 0.6s to 0.9s, I_{dc}^*=0.3pu. (a) I_{dc}^* and I_{dc}, (b) M_d, (c) M_q, (d) source voltage (V_s) and source current (10*I_s) during 0.2s to 0.26s, (e) source voltage (V_s) and source current (10*I_s) during 0.5s to 0.56s, (f) source voltage (V_s) and source current (10*I_s) during 0.8s to 0.86s, (h) Is_r^* and Is_r

At t=0.6 s the current reference I_{dc}^* decreases to 0.3pu and at the same time the load resistance is increased to 2.0pu to keep the dc voltage at 0.6pu. When the rectifier reaches a new steady state, just as shown in Fig.4(g), the reactive current of source current cannot be regulated to zero, so the source current I_s leads the source voltage V_s, just as shown in Fig.4(f), because the dc side current I_{dc} is too small to compensate the leading current provided by the capacitor C_r. At this steady state, M_d=0.6 and M_q=-0.8, the modulation index has reached the maximum value, so the rectifier gets the maximum achievable power factor.

When the dc voltage changes its polarity, the rectifier will operate in the regenerative mode. The proposed technique also can produce a unity power factor when the converter operates in the regenerative mode. Fig.5 illustrates such a case. At t=0.3s, the rectifier reaches a steady state where $I_{dc}=I_{dc}^*$=1.0, V_d=0.6, and R=0.6, all in per unit. In order to simulate the regenerative operation, the load voltage starts to decrease at t=0.3s and then reaches -0.8pu at 0.4s. When the rectifier reaches another steady state at 0.5s with M_d=-0.19 and Mq=-0.35 just as shown in Fig.5(a), Fig.5(b) and Fig.5(c). From Fig.5(e), we can see that the reactive current Is_r is regulated to zero, so the input power factor is unity, which is verified in Fig.5(d), the source current and the source voltage are out-of-phase.

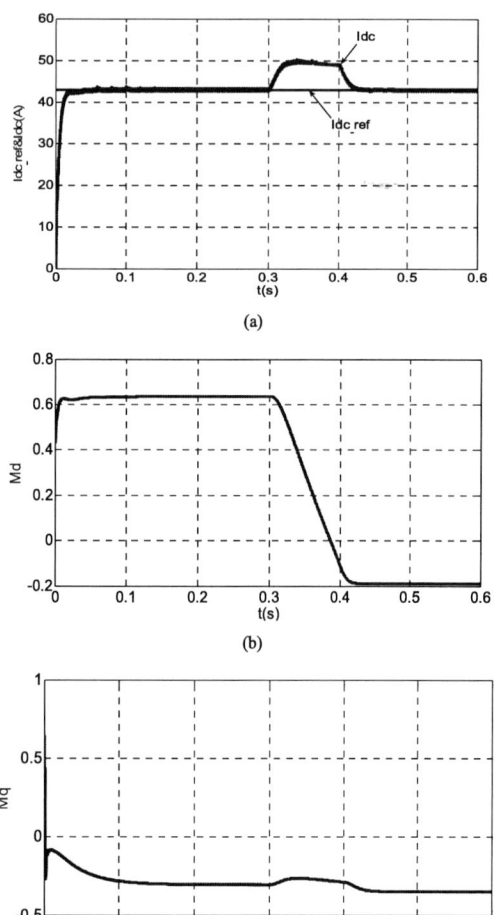

(a)

(b)

(c)

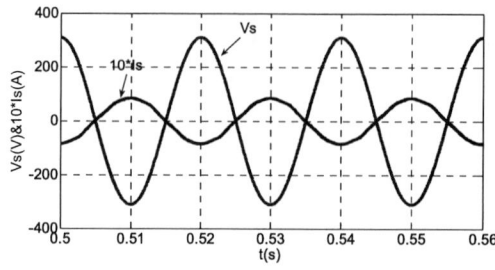

(d)

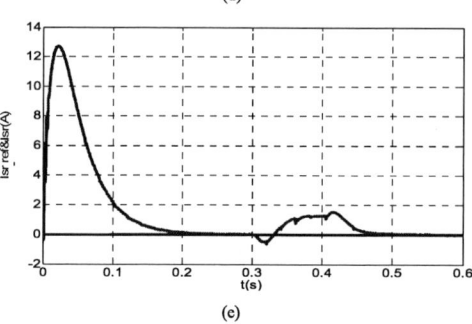

(e)

Fig. 5   Simulation results for regenerating operation. (a) I_{dc}^* and I_{dc}, (b) M_d, (c) M_q, (d) source voltage (V_s) and source current ($10*I_s$) during 0.5s to 0.56s, (e) Is_r^* and Is_r

V.   CONCLUSIONS

The proposed power factor control scheme for IGBT current source rectifier can provide the maximum achievable power factor under all operating conditions, including the regenerative mode, and this method does not require any system parameters. Unlike the conventional control technique, this control method is based on d-q rotating frame. In the feedforward loop, the reactive current is regulated through the q-axis component of modulation index vector by a PI regulator; meanwhile, in the feedback loop, the dc side current is controlled through d-axis component of the modulation index vector by another PI regulator. Additionally, due to the switching character of IGBT, the SVM is adopted in the system, so the current source rectifier has good current regulation and fast dynamic response.

REFERENCES

[1]   B.Wu, High-Power Converters and AC Drives. *IEEE Press – Wiley*, March 2006.
[2]   J.H.Choi, H. A. Kojori, and S. B. Dwan, " High power GTO-CSC based power supply utilizing SHE-PWM and operating at unity power factor," in *Proc, CCECE*, 1993.
[3]   Yuan Xiao, Bin Wu, Steven C. Rizzo, Reza Sotudeh, "A Novel Power Factor Control Scheme for High-Power GTO Current-Source Converter" *IEEE Trans. Ind. Applicat.*, vol. 34, no. 6, pp. 1278-1283, 1998.
[4]   Jason C. Wiseman, Bin Wu, "Active Damping Control of a High-Power PWM Current-Source Rectifier for Line-Current THD Reduction" *IEEE Trans. Ind. Electron.*, vol. 52, no.3, pp. 758-764, 2005.
[5]   Longcheng Tan, Yaohua Li, Ping Wang, "An Overmodulation Method for Space Vector PWM Current Source Inverters" *Conf. Rec. IEEE- ICIEA'07*, 2007, pp. 2431-2434.

A procedure to optimize the inductor design in boost PFC applications

Florent Liffran

Martek Power, Brétigny sur Orge, France, e-mail: *florent.liffran@martekpower.fr*

Abstract—This paper will explore the constraints for designing the inductor in a Boost Power Factor Corrected (PFC) converter working in continuous current mode (CCM). An original method will be proposed to optimize the inductor design by focusing on the possibilities offered by a given core in function of its winding parameters. Optimal current ripple and inductance value will be determined for each application. The influence of the switching frequency will also be quantified, as well as a pertinent comparison between the different powder core materials.

Keywords—Magnetic device, Passive component, Power factor correction, Switched mode power supply, High frequency power converter.

I. INTRODUCTION

Active PFC converters have become more and more popular due to international standards in line harmonics [1], and are used as front-end stage almost everywhere in AC line based applications. Thus, a lot of semiconductor manufacturers propose a line of PFC controllers. The more popular topologies include the well-known Flyback, classically dedicated to the lower power due to the transformer volume, and the Boost, typically used for power above some hundreds of watts.

In the Boost topology, the inductor is responsible for an important part of the volume and the losses of the converter, and it cost is not negligible. It appears consequently as a major component to work on to both improve the performances and reduce the volume of the boost PFC converter. Its design is classically realized either with gapped ferrite or powder core, those last being far more popular due to their higher saturation level (bringing more inductance in the same volume) and distributed airgap properties (less radiated noise and no eddy current in the close conductors) [2]. The design procedure described in fig. 1 is classically applied, and can be found among most of the cores suppliers [3] [4].

Classically, the switching frequency and the current ripple level are chosen from the operating conditions (input range, output power, etc.). This leads for the inductor to a Li^2 (stored energy) and then to a core with the help of selection charts. The choice between different materials is made by repeating this procedure, and comparing the obtained performances with their respective cost. Unfortunately, this approach can lead to a non-optimal design of the inductor, due to the following evident facts:

- The current ripple choice is often made without quantified arguments. The idea that there is of compromise between inductor size, switching losses, and differential mode (DM) noise generation is the only indicator. One can read that 20% is good, another that 50% is better… and finally the choice is somewhat arbitrary [5] [6].

- The relative permeabilities given by the selection chart do only depend on the Li^2 of the inductor (the stored energy being inversely proportional to the permeability), and do not consider the balance between the L and the i^2.

- The winding is chosen for a given saturation level which is also chosen arbitrarily.

- All these parameters are extracted from the worst case conditions, usually minimum input voltages to have the maximum current, but most of the PFC have to work for a much wider voltage range (85 to $265V_{AC}$). This leads to designs where the efficiency in nominal condition is not optimized.

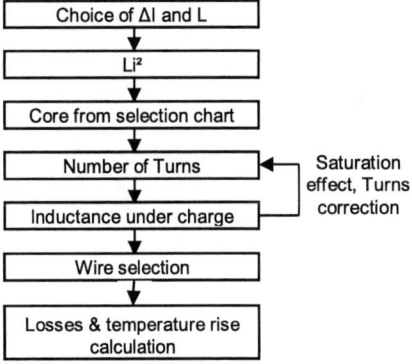

Fig. 1. Flowchart of inductor calculation

Therefore, this paper will propose to reverse this design cycle and to optimize the inductor characteristics offered by a given core. This is done through a proper parameterization of the PFC system and the inductor constraints. Then, the exercise will be repeated in several different conditions (core size, material, permeability, and switching frequency) to have quantified arguments to help the designs choices.

II. INDUCTOR PARAMETRISATION

First of all, the operating conditions must be defined. To illustrate the method, the following example is considered:

978-1-4244-1741-4/08/$25.00 ©2008 IEEE

	min	nominal	max
Input RMS voltage	85 V	115 V	265 V
Output DC voltage		390 V	
Output power		500 W	
Expected efficiency		92%	
Switching frequency		64 kHz	
Input RMS current (A)	6.4	4.7	2.1
Peak inductor current (A)	9.0	6.7	2.9

The proposed idea is to first choose a core, and explore the different design possibilities that can be realized with it. Several cores can then be evaluated through spreadsheet software, and the choice optimized. For that, we start from the number of turns N, from which are derived the other parameters. Afterwards, this operation is repeated for several different values of N to give a "core performance chart" for the considered application.

The steps are:

- H bias field calculation through the formula:

$$H = \frac{Ni}{L_e} \qquad (1)$$

Where:

i: peak inductor current

L_e: effective magnetic path length

- Saturation factor computation, to obtain the effective permeability under the peak bias current. For the need of repeated calculations, the $\mu_e = f(H)$ curves have to be approximated with a formula. An hyperbolic tangent (tanh) regression on the logarithm of the H gives very good results under the following form:

$$\mu_e = \frac{\mu_{initial}}{2}\left(1 + \tanh\left(-\alpha \log\frac{H}{\beta}\right)\right) \qquad (2)$$

Where:

β: value of H at $\mu_e = 0.5\,\mu_{initial}$

α: inflexion parameter to be tuned to fit the manufacturer's curves.

These parameters are different for each material and permeability. This regression offers the advantage to be valid for all H values, contrary to polynomial ones that are often proposed.

- Inductance under load calculation:

$$L = N^2 A_L \frac{\mu_e}{\mu_{initial}} \qquad (3)$$

(Tolerances on the A_L should be considered)

- Wire size selection, assuming a filling factor. Usually 20 to 40% are considered to be a good compromise. The single layer is also an option which can be interesting for cost and/or EMI purposes [7], but results in slightly higher copper losses when N is huge.

$$d_{max} = \sqrt{\frac{4 W_a F_F}{N\pi}} \qquad (4)$$

Where:

d_{max}: minimum diameter of the wire

W_a: window area of the core

F_F: filling factor

The final wire has of course to be chosen with a standard gauge, but (4) will give a good idea of the resultant performances. Other arrangements such as several wires in parallel are also possible.

- Winding electrical resistance R computation:

$$R = \rho_{CU}\frac{4 L_{turn}}{\pi d^2} \qquad (5)$$

Where:

ρ_{CU}: resistivity of copper at operating temperature

L_{turn}: mean length of a turn

d: diameter of the wire

Multiplied by the squared RMS input current, this gives the copper losses of the inductor. The effect of current ripple at switching frequency (modification of the RMS value of the current and skin effect) are neglected, since the ripple proportion is supposed to be small in CCM. If a high ripple level is chosen, a corrective factor can be necessary for the resistance calculation.

- Core losses calculation, based on the value at peak line voltage. The H field is taken at the maximum line current. Formulas are extracted from the magnetization curves with tanh functions, as for the $\mu_e = f(H)$, to obtain the B_{AC}. Then the iron losses are approximated with a classical Steinmetz formula under the form:

$$P_{vol} = P_0 B^\alpha f^\beta \qquad (6)$$

Where:

B: AC induction in the core

f: Induction frequency (switching frequency).

P_0, α and β: coefficients extracted from experimental logarithmic losses curves. They are usually provided by the core manufacturer for each material.

This method is not the best fitting over the whole range of B and f, but remains a good estimation for our considered range [8].

As β is usually close to 2, the iron losses can be expressed as an equivalent series resistor with small error. Thus the iron losses are proportional to the RMS value of the inductor current, and the value over a line frequency period is the one calculated at peak current divided by $\sqrt{2}$. This is simpler than other methods that consider the real duty cycle variations over the line frequency [9], but it is accurate enough for our needs of comparison.

The results can be plotted as losses versus inductance charts as shown in Figure 2, where *N* starts from 20 to 260 with steps of 10 turns between each data point.

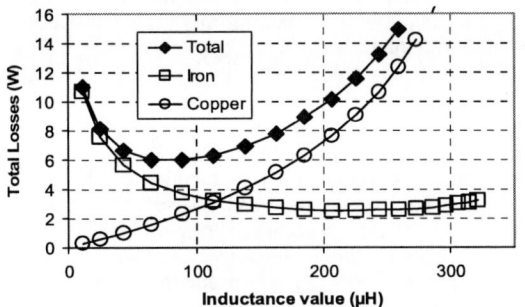

Fig. 2. Inductor losses versus inductance value for a MPP 26μ core (55932 from Magnetics®).

III. RESULTS ANALYSIS

At first, Figure 2 shows that copper and iron losses vary in opposite direction with the number of turns. Unlike the intuitive reasoning that more turns will give more *H*, more *B* and thus more iron losses, they decrease with *N*. This is due to the greater inductance value which reduces the current ripple, and remains true for moderate level of saturation; above 50% of permeability reduction (> 250μH on the graph) the additional turns provide small inductance and the iron losses tend to rise. On the contrary, the more turn, the thinner and the longer is the wire, which brings more copper losses. The consequence is the existence of a minimum in the total losses expression.

We can notice that the minimum does not necessarily correspond to copper losses equals to iron losses, as it can be suggested in some design guides. The ratio between iron and copper losses depends on a lot of parameters, such as filling factor and used material. That is why a fixed constraint on the repartition between copper and iron losses has low chances to lead to a minimum for the overall losses. This should not be taken as an input data for the design. It is, as for the number of turn and inductance value, an output of the design process, which shall be chosen at the end.

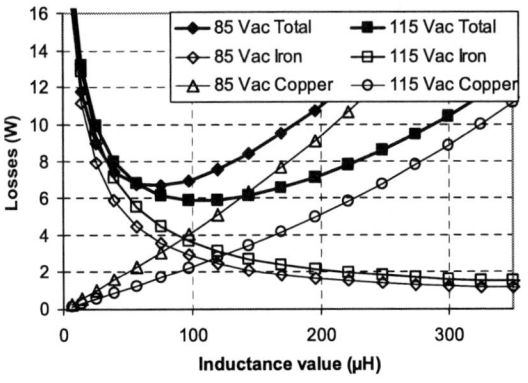

Fig. 3. Losses comparison on a 55326 MPP 26μ core, 10 turns per data point.

The inductance value can be chosen a bit above the minimum, because it will lower the current ripple. Copper losses are also easier to evacuate than iron losses, as the windings surrounds the core for toroids. Thus, a higher inductance value will facilitate the cooling, without penalizing the losses too much (the curve is quite flat around the minimum). On the other side, more turns will complicate the design a little bit.

It is interesting to see that this minimum in the losses does depend on the operating point. Figure 3 shows the same inductor, at two different input voltages: 85 V_{AC}, which is the worst case, and 115 V_{AC}, the nominal condition. Looking only at the 85 V_{AC}, the value of 100μH under load current (80 turns with Ø1mm wire) would have been a good choice, the losses being minimal. But with the 115 V_{AC} curve, we can see that the minimum has moved to higher values. 200μH (120 turns of Ø0.8mm wire) seems to be a better choice, the losses in nominal conditions being promoted face to the worst case. This help to save some energy over the life of the converter. In fact, even higher values can be preferred to lower the MOSFET losses and DM noise generation. The balance between copper and iron losses will also be better, giving an easier cooling. The side effect will be a more complicated winding due to the higher number of turns.

Limitations of the method

A constant fill factor of the core (40% in the presented curves) is assumed, which is not always realistic, especially when the wire diameter is too big and/or when a high number of turn is required. In such a case the calculations should be revised after a first pass, depending on the practical parameters. Similar formulas can be expressed for several cases, for example paralleled or flat wires. If an automated winding is preferred, a reserve in the window area for the shuttle must be considered.

The current ripple calculation does also not take into account its own effect on the saturation, because it would have required iterative steps to find the right value. The model can be considered as acceptable since the ripple is small in proportion of the main current and the core is not too much saturated. The iron losses should be experimentally verified once on the chosen design to check the validity of the model for the considered saturation level.

A last obvious limitation is the discontinuous conduction mode (DCM), when the ripple becomes higher than the average current. This can occurs for the low inductance values, that is why the ripple has to be checked to suppress the smaller values of *N*. DCM goes above the purpose of this paper, the equations needing to be changed, but can also be investigated with similar methods.

IV. CORE SELECTION PROCEDURE

A. Size selection

This method supports efficiently the choice of a core. Regarding the size, figure 4 shows without surprises that for the same conditions, the bigger is the core the lower are the losses, and the higher is the inductance.

Core	Outside diameter	Magnetic volume	Minimum losses	Corresponding Inductance
55256	40.8 mm	10.5 cm^3	4 W	180 µH
55326	36.7 mm	6.1 cm^3	4.9 W	120 µH
ratio		+ 58%	− 23%	+ 67%

The choice of the suppliers in the selection charts is usually focused on both saturation and thermal limits, and does not for example take into account transient situations of some seconds, where the losses are less important. In such cases a more suitable choice can be made.

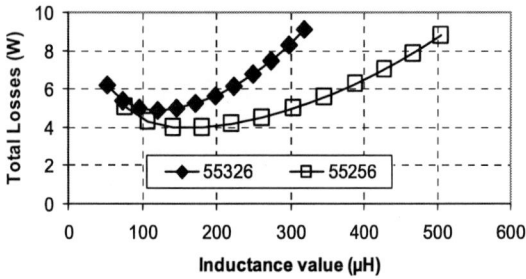

Fig. 4. Losses comparison for a 500W application at 85 V$_{AC}$ input, for two sizes of MPP 26µ cores. Turns vary from 50 to 160.

B. Temperature rise computation

From the losses and the core size, the temperature rise can be estimated. It does depend on the losses per unit of surface of the inductor. A precise prediction is difficult to obtain, several elements such as airflow or surrounding components affecting the power dissipation. Anyway, the following formula is commonly adopted to have estimation:

$$\Delta T = \left[\frac{P_{tot}}{S_{ext}} \right]^{0.833} \quad (7)$$

Where:

P_{tot}: total dissipated power in mW.

S_{ext}: Surface area of the winded core.

The resulting temperature elevation is in °C.

By computing this formula in the losses calculation spreadsheet, we can obtain the temperature curves as displayed in fig. 5.

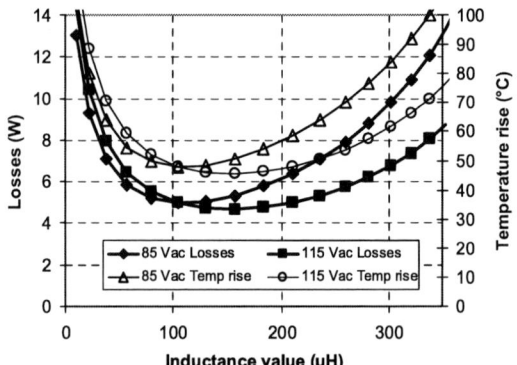

Fig. 5 Losses and temperature rise for a 55550 MPP 26µ core (33.8x19.3x11.4mm).

Both worst case and nominal conditions have been displayed. As the 85V$_{AC}$ input is supposed to be transient,

the temperature elevation will only be considered for 115V$_{AC}$.

If the maximum allowable temperature rise is 60°C, we can see that 300µH is the maximum inductance that can be obtained with this core.

C. Permeability selection

The choice of the permeability is also important. The classical selection methods usually give at the same time both core size and permeability in function of the Li^2, without taking into account the balance between the L and the i^2. The Boost topology bringing additional relationships between these parameters, the choice is not obvious. Figure 6 shows the curves for three Ø36.7mm MPP cores, with different permeabilities, in the 500W application at two operating conditions: 85V and 115 V$_{AC}$.

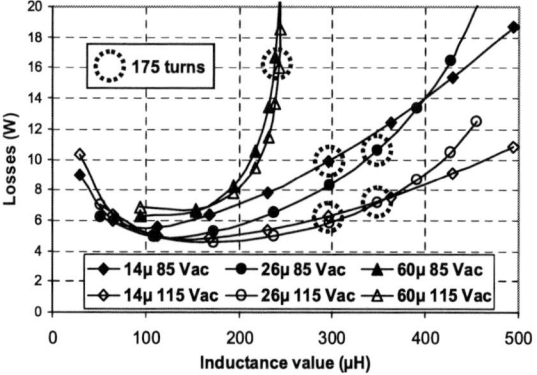

Fig. 6. Losses versus inductance for 3 different permeabilities of the same core, 500W output.

The Li^2 is comprised between 8mH.A^2 for 100µH to 40mH.A^2 for 500µH (9A peak current). Selection charts do propose a 60µ for such stored energy, but it appears clearly as not optimal and should be avoided. Between the 14µ and the 26µ the difference is slighter. At 85 V$_{AC}$, the 26µ has the advantage up to 400µH, providing lower losses and more inductance for the same number of turns. Above, the 14µ takes the advantage due to less saturation. Looking at the 115 V$_{AC}$ will confirm this statement, but the crossing point is lowered to 360 µH.

So surprisingly, if an especially high value of inductance is required with this core, the lower permeability will suit the best.

V. SWITCHING FREQUENCY STUDY

Curves have been plotted in figure 7 for a 55256 MPP core at different switching frequencies.

We can see that a high frequency is always favorable for the inductor, the different curves having no crossing point. This confirms the well-known idea that "the higher switching frequency, the smaller reactive components", but looks surprising with the frequency contribution to the iron losses. The explanation is in the current ripple reduction with the frequency due to the boost topology. In (6), α parameter (induction exponent) is bigger than β (frequency exponent), which makes the B more influent than the f. This effect is reduced for the highest inductance values, because of the saturation.

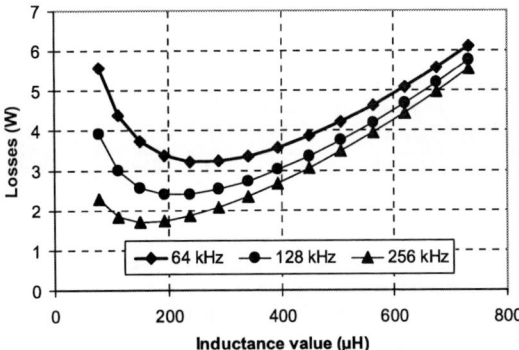

Fig. 7. Losses versus inductance for a 55256 MPP 26µ core, at 115 V$_{AC}$ input and 500W output.

Another statement is that the minimum of the losses is occurring for different values of L, and thus for different current ripple proportions. This is a major advantage of the proposed method: The choice of the current ripple value can be made with quantified arguments. Otherwise, as the optimal value is changing with the frequency, an arbitrary choice (the classical 20% for example) has small chances to be the best one.

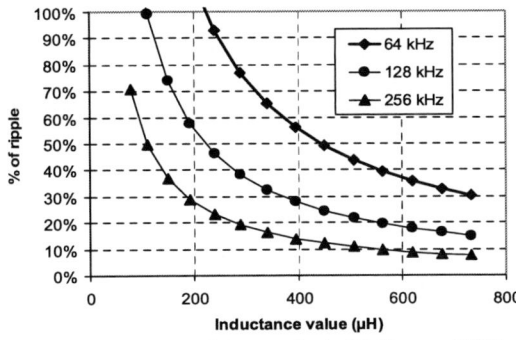

Fig. 8. Current ripple versus inductance for the 55256 core, at 115 V$_{AC}$ input and 500W output. 10 turns per data points.

Figure 8 shows the different percentages of current ripple in the same conditions as previously (ripple above 100% corresponds to DCM and shall be disregarded). The values for optimal losses are:

Switching frequency	N	L (µH)	Ripple
64 kHz	92	260	80%
128 kHz	85	215	50%
256 kHz	70	150	36%

Care should be taken when considering the configuration with a high amount of current ripple (as mentioned in the limitations of the method), because switching losses will be impacted, and copper losses can suffer from skin effect.

The choice of the switching frequency does of course not depend only on the inductor. It should consider a lot of other aspects, such as switching losses and DM noise generation. The quantification of the influence of this frequency over the inductor can help to make a better choice.

VI. MATERIALS COMPARISON

A. Available materials

Several mixes do exist to build powder core. The main differences between them are saturation induction, iron losses, and of course price. The basis is usually a mix of iron, characterized by a high saturation capability (pure iron saturates at 2T) but an also high losses level, and other metals to lower these losses. It is then mixed with non-magnetic alloys to obtain the distributed air gap properties, in various proportions to get the different permeability levels.

The main mixes are:

- Molybdenum Permalloy Powder (MPP): Mix of Nickel (~80%), Iron (17%) and Molybdenum (~3%), it exhibits the lowest losses and the highest price, due to a high nickel content.
- High Flux: Optimized for a higher saturation flux, it contains 50% of nickel.
- Sendust, also called Super-MSS, or KoolMµ (trademark from Magnetics®): Low cost version of the High Flux, without Nickel. The losses are a bit higher and saturation lower.

Orders of magnitude of the main characteristics are summarized in the tab bellow:

Material	Saturation flux (T)	Relative losses	Relative cost
MPP	0.75	1	~3
High Flux	1.5	~3	~2
Sendust	1.05	~2.5	1

Relative cost depends also on the size: for the smaller, the process takes an important place and the raw materials make a small difference, implying close prices. For bigger cores, as Nickel is much more expensive than the other metals the difference is more important. Other industrial considerations, such as ordered quantities, can also have a lot of influence on the price, so the final cost comparison should be conducted on real quotations.

These materials present also some differences, such as magnetostriction, temperature or frequency behavior. But as they are not of first importance in PFC applications, they will not be discussed here.

Beyond these considerations, the choice between the different materials is not obvious. A practical comparison will be conducted with the method previously developed, to analyze which material fits the best in each situation.

B. Basic comparison

In the application described in chapter II, the three different materials are experienced with the same permeability and core size:

Core	Outside diameter	Inside diameter	Height	Permeability	Material
55550	33.8 mm	19.3 mm	10.7 mm	26µ	MPP
58550	33.8 mm	19.3 mm	10.7 mm	26µ	High Flux
77550	33.8 mm	19.3 mm	10.7 mm	26µ	Sendust

The corresponding performances are plotted in Figure 9. We can see without surprise that MPP exhibits the lowest losses, then High Flux and finally Sendust. It is already interesting to see that the optimal inductance value is different for all material. The MPP offers the best advantage at low inductance value (200µH, 100 turns). This is due to the lower iron losses, which accommodates a higher ripple value. For higher inductance, the copper losses are growing faster than the iron losses lowers.

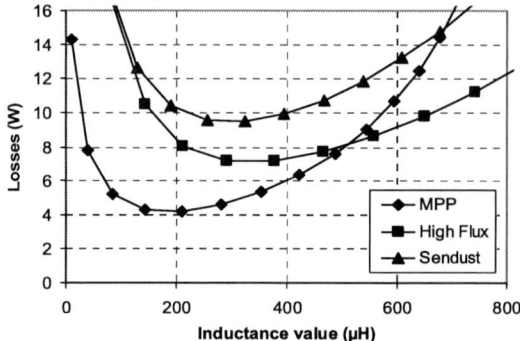

Fig. 9. Losses versus inductance for the 55550, 58550 and 77550 cores, at 115 V$_{AC}$ input and 500W output. 20 turns per data point.

The higher saturation of the High Flux contributes to make it interesting at high inductance. Sendust is similar with a losses curve vertically translated.

The exact contribution of each losses and saturation characteristics will now be analyzed in details.

C. Core losses influence

To study the influence of core losses, we will compare MPP with a fictive mix, having same saturation properties, but the losses formulas of High Flux and Sendust. The results are shown in Figure 10.

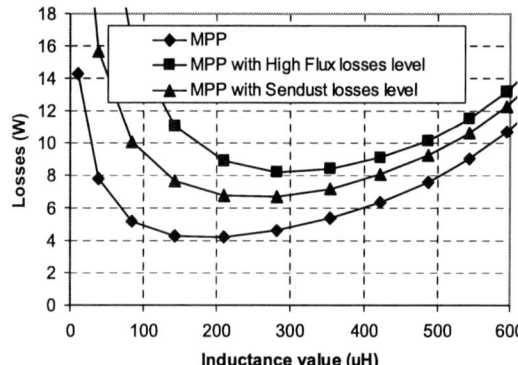

Fig. 10. Losses versus inductance for 55550 core, and fictive cores having same saturation characteristics, but different losses levels. 115 V$_{AC}$ input and 500W output, 20 turns per data point.

The losses level does of course translate vertically the curves, but the optimal inductance value changes also. As copper losses remains constant with the number of turns for a given core size, the lower core losses gives an advantage to lower turns (copper losses minimization).

In absolute values, we obtain 4.2W with MPP losses, 6.7W with Sendust-like, and 8.2W with High Flux-like. It makes a factor of 2 between MPP and High Flux-like, as the losses formula gives a value of 3.

Inductance	Losses formula	Copper losses	Iron Losses	Total
150µH	MPP	1.3 W	2.9 W	4.2 W
	High Flux	1.3 W	9.4 W	10.7 W
300µH	MPP	3 W	1.8 W	4.8 W
	High Flux	3 W	5.2 W	8.2 W

An appropriate choice of the inductance value to balance copper and iron reduces the overall losses. As this design remains dominated by iron losses, even with MPP formula, the difference remains important.

D. Saturation level influence

The saturation behavior of the material is much more difficult to analyze, because it is not only related to the DC saturation flux. Powder cores exhibit soft saturation characteristics, but not all in the same way: the permeability drop can be more or less progressive.

Figure 11 plots on the same graph the permeability versus magnetizing force for all three MPP, High Flux, and Sendust 60µ materials.

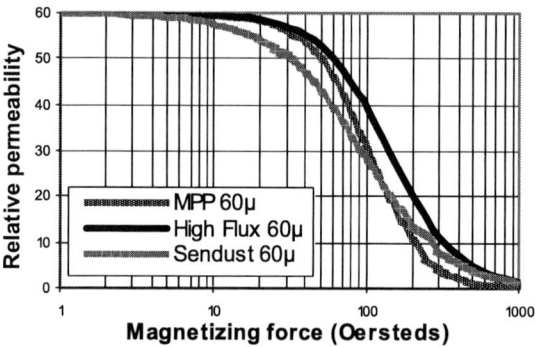

Fig. 11. Relative permeability versus magnetizing force for MPP 60µ, High Flux 60µ, and Sendust 60µ

First of all, we can notice that the three materials are not ordered in respect with their saturation flux. Without saturation, the permeability is defined as the ratio between Flux and Magnetization. Following that, the H level of saturation is given by:

$$H_{sat} = \frac{B_{sat}}{\mu_0 \mu_r} \quad (8)$$

We can crosscheck these values with the B_{sat} obtainable on the magnetization curves, and the H value at which 50% of the permeability is lost (that can be a definition of the H_{sat} in case of soft saturation).

Material	B$_{sat}$ (T)	H$_{sat}$ (Oersted)	H$_{\mu 50\%}$ (Oersted)
MPP	0.75	125	100
High Flux	1.5	250	140
Sendust	1.05	175	90

The B_{sat}, defined as the maximum attainable flux in the material, appears to be not closely related to the lost of permeability, represented by $H_{\mu.50\%}$.

In particular, Sendust exhibits a soft saturation behavior beginning at very small magnetization. As the mostly used part of the graph is for a lost of permeability comprised between 0 and 30%, this is a huge handicap. We can see on Figure 11 that above 150 Oe, the B_{sat} hierarchy is reestablished. The B_{sat} appears consequently of small importance compared with the soft saturation behavior.

Back to the losses curves, the diminution of μ makes directly higher losses, because less inductance is obtained and thus more current ripple. This can be observed on Figure 12, where MPP is compared with fictive materials, having same losses formulas but High Flux and Sendust saturation characteristics.

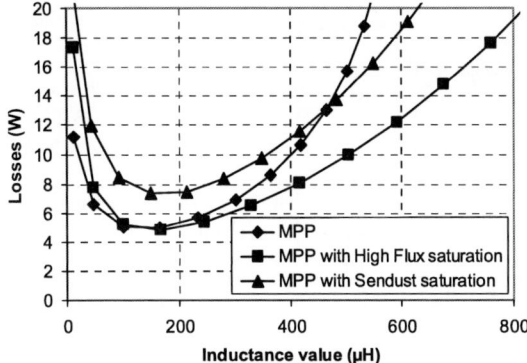

Fig. 12. Losses versus inductance for 55932 core, and fictive cores having same losses levels, but different magnetization characteristics. 115 V_{AC} input and 500W output, 20 turns per data point.

The hierarchy present on the $\mu = f(H)$ graph is preserved: High Flux characteristics are the best, and Sendust is the worst at the beginning, with a cross point with MPP after a certain inductance level.

To summarize, the soft saturation behavior (lost of permeability with magnetization) is a much more important characteristic than the saturation flux. Sendust disadvantage in the low saturation levels makes it much worst than MPP.

The overall losses appear to be more impacted by losses formula than by saturation behavior. Consequently, we can expect far better performances from MPP, followed by High Flux (advantaged in the higher inductances), and then Sendust, despite lower intrinsic core loss than High Flux.

VII. PERFORMANCE SIZE AND COST TRADEOFF

According to chapter IV A., core size is of first importance in inductor performance. Thus, poorer materials can be arranged in a way to have comparable performances as the better one. We will search what core size difference is necessary to obtain the same performances with the three materials.

Ripple level shall be considered carefully, because low switching frequency and/or low loss materials (such as MPP) can favor low levels of inductance. Such current ripple will influence the switching losses, and make skin effect not negligible any more. Continuous conduction occurs when peak-to-peak current ripple is less than DC current. In term of inductance, the condition is:

$$L > \frac{U_{in}^2}{2P_{in}f_{sw}} \quad (9)$$

In chapter's II example (500W, 64kHz, 85V_{AC}), inductances bellow 100μH shall be disregarded. To keep a current ripple under 50%, 200μH is the minimum inductance to be considered.

A. Design with MPP

When looking at the different permeabilities, 26μ is quickly chosen, being better than 14μ and 60μ in almost all sizes and conditions. Core size is more difficult: all give approximately the same minimum, around 5W, but at different inductance values, as shown on Figure 13.

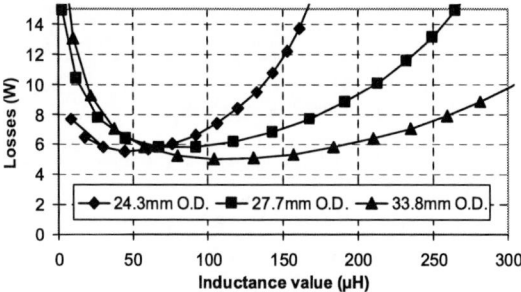

Fig. 13. Losses versus inductance for 55352, 55932, and 55550 cores. 85 V_{AC} input and 500W output, 10 turns per data point.

If we exclude inductances bellow 100μH, the 27.7mm diameter seems to be a good compromise. Forcing an inductance of 200μH to have a 50% ripple will give around 9.5W. This is not the optimum, and the MPP is not used at the best of its possibilities. A better configuration would have been a higher switching frequency or a lower system voltage.

9.5W represents 2% of efficiency lost in the inductor, which is a good compromise. If higher efficiencies are necessary, the 33.8mm has only 6W for 200μH.

B. Design with High Flux

The three same core sizes are plotted on Figure 14. The only size to allow less than 10W of losses as with MPP is 33.8mm diameter. As the minimum is quite flat, a value of 300μH can be obtained with these 10W.

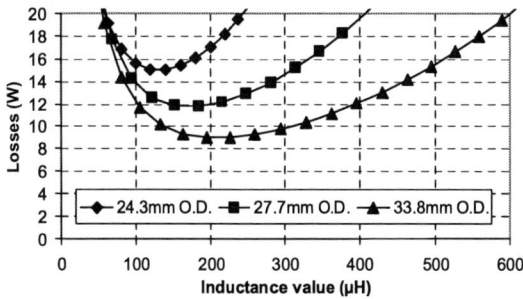

Fig. 14. Losses versus inductance for 58352, 58932, and 58550 cores. 85 V_{AC} input and 500W output, 10 turns per data point.

The choice of permeability is trickier, as losses levels are almost equivalent between the 14μ and 26μ. The advantage comes to the 26μ, because less turns are necessary, thus the construction is simplified. The 14μ would have an interest for the higher inductances, because of less saturation.

C. Design with Sendust

As seen previously, the Sendust has the same behavior than High Flux, with more losses. The permeability is kept at 26μ, as for MPP and High Flux. To have similar losses, larger cores are considered.

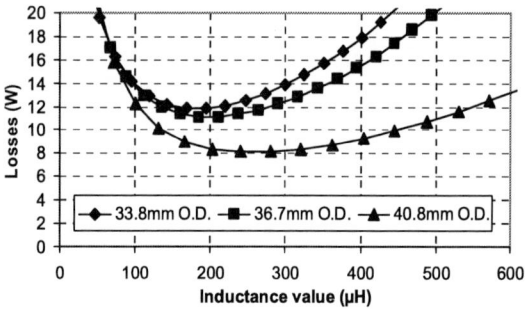

Fig. 15. Losses versus inductance for 77550, 77326, and 77256 cores. 85 V_{AC} input and 500W output, 10 turns per data point.

The 10W can be obtained only with a much larger core, of 40.8mm diameter. With it, we can take 450μH. Otherwise, 36.7mm will provide 250μH with 12W.

D. Overall comparison

The price of the different solutions is estimated through the volume of magnetic material and the relative cost. This is only a first approximation; a more complete study would include quotations for the cores from different suppliers.

The resulting performances are summarized bellow:

Core	Diameter	Inductance	Losses	Relative cost x Volume
MPP	27.7 mm	200μH	9.5W	12
High Flux	33.8 mm	300μH	10W	11
Sendust	36.7 mm	250μH	12W	6
Sendust	40.8 mm	450μH	8W	10.5

We can see that MPP offers without surprise the best performances in the smaller volume, but with a lower inductance value. The application example is not very favorable to MPP, because inductance had to be raised above the optimal value. Increasing the switching frequency would have lowered the current ripple, and thus provide an even smaller losses level. MPP has the more advantages over the other materials at higher frequencies.

High Flux appears as a good compromise, a relatively high inductance being required in this design. For almost the same cost and losses, inductance is 50% higher for a 50% bigger surface area. Decreasing the switching frequency would give even more advantages over MPP.

Sendust is an interesting alternative if cost is an issue, or if volume does not matter so much. The 40.4mm diameter option offers for the same price as MPP, with the double surface area, 125% more inductance and 20% fewer losses. The 36.7mm option offers a 50% cheaper inductor, for 20% more losses, with 75% bigger surface.

To conclude this comparison, we can say that MPP offers the best performances in the smaller volume for the high switching frequencies, where a small inductance is sufficient. If they are used at their optimal losses level, MPP cores offer low losses that can not be achieved with the other materials.

High Flux is advantaged when a high inductance is required, especially at low frequency or for applications with low voltage and high current. The comparison with MPP must be conducted case by case.

Sendust offers the lowest cost alternative, but with bigger cores. It can offer the same performances than High Flux, with usually 50% to 100% more surface area. Adapting the size will provide more or less compact, economic and efficient designs. For the highest power application, the price difference with the other materials becomes huge, that is why the biggest cores are often made from Sendust.

ACKNOWLEDGEMENT

A design method for CCM boost PFC inductors has been presented to help optimizing the choice of the magnetic material, permeability, size, and switching frequency. In particular, it helps current ripple and inductance selection.

The proposed method demonstrates that classical intuitive reasoning does often lead to non-optimal solutions. The obtained results also show that there are no absolute rules to choose between the different materials. The overall tradeoff is between performances, size, and cost, but all different configurations for voltage, current and switching frequency will give different preferences.

Future works can be focused on the relationship between inductance and switching losses in semi-conductors. The parameterization of the DM filter size with the inductance and switching frequency would also be very interesting. Thus, an optimization of the frequency will be possible regarding the volume and cost of the overall converter: input filter, power semi-conductors, inductor, and heatsinks.

REFERENCES

[1] O. Gracia, J.A. Cobos, "Single Phase Power Factor Correction : A Survey", IEEE Transactions on Power Electronics, Vol 18, N°3, May 2003, pp 749-755

[2] M.A. Swihart, "Inductor Cores – Material and Shape Choices", Magnetics®, www.mag-inc.com

[3] "Powder Core Selection Guide", Magnetics®, www.mag-inc.com

[4] "Powder Core Product Catalog", Arnold Magnetic Technologies, www.arnoldmagnetics.com

[5] P.C. Todd, "UC3854 Controlled Power Factor Correction Circuit Design", Unitrode Application Note U-134, www.ti.com

[6] R. Brown, M.Soldano, "PFC Converter Design with IR1150 One Cycle Control IC", International Rectifier Application Note AN-1077, www.irf.com

[7] S. Wang, F.C. Lee, W.G. Odendaal, "Single Layer Iron Powder Core Inductor Model and Its Effect on Boost PFC EMI Noise", PESC'03, pp. 847-852 vol2, 2003.

[8] C. Oliver, "A New Core Loss Model For Iron Powder Material", Switching Power Magazine, Vol3 Issue 2, 2002.

[9] L. Jinjun, T.G. Wilson, R.C. Wong, R. Wunderlich, F.C. Lee, "A Method for Inductor Core Loss Estimation in Power Factor Correction Application", APEC'02, pp 439-445 vol.1, 2002

Electric Vehicle Drive Inverters Simulation Considering Parasitic Parameters

Wen Huiqing[*†], Liu Jun[*], Zhang Xuhui[*†], Wen Xuhui[*]

[*]Institute of Electrical Engineering Chinese Academy of Sciences, Beijing, China

[†]Graduate University of Chinese Academy of Sciences, Beijing1, China

E-mail: whq@mail.iee.ac.cn

Abstract--**Due to always increasing commutation speed, parasitic parameters such as interconnection inductances and stray capacitors are directly linked to voltage surge, resonance and electromagnetic interference. To guide the selection of power components, assess low-inductive film capacitors and laminated bus bar, and in the end realize the high power density and high reliability for electrical vehicle inverters, a simulation platform based on Matlab/Simulink has been developed. Parameters of the simulation platform strictly accord with the practical working conditions, even parasitic existing in components and connections are also considered. This paper introduces the switching mechanism and the distribution of circuit parasitic. The system modeling and parameters confirmation of the simulation platform are presented in detail. Experimental results are shown to verify the effectiveness of the simulation platform.**

Keywords--**Electrical vehicle, Simulation, Voltage Source Inverters (VSI), Modeling, bus bar**

I. INTRODUCTION

Switches dynamic features have been greatly increased. IGBT presents a high voltage behavior (up to 3300V), di/dt (up to 5000 A/us) and dv/dt (up to 5000V/us). Thus parasitic elements such as the commutation loop stray inductance have become one of the most important parameters limiting the converter performance. In the hard-switched converters, the stray inductance is responsible for the dc bus overshoots and overvoltage spikes across IGBT during turn off [1]. In high power soft switching converters, the stray inductance may lead to non-zero voltage switching of the devices across the bus at device turn-off [2]. The high di/dt values produced by IGBT and diodes are indeed the main source for conducted and radiated noise emissions [3].

As one of the most important passive components in electric vehicle inverter, the dc bus capacitor need to deal with high-frequency pulsating current, voltage transient and over voltage due to regeneration [4, 5]. Among all these problems, the current handling capability is the major concern. A low-inductance high-frequency film capacitor can be used to replace the conventional electrolytic bulk capacitors to reduce size and enhance the reliability. This simulation platform can be used to analyze the current and voltage ripple. Laminated bus bar is usually adopted in the electric vehicle inverter to

minimize stray inductance. Besides, the parasitic inductance of bus bar L_{bus} is the main parts of the commutation loop stray inductance as shown in fig.1. So it is important to determine the stray inductor for laminated bus bar.

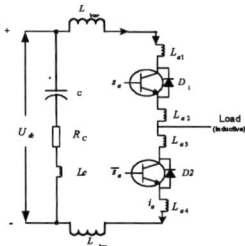

Fig.1. Distribution of the commutation loop stray inductance

Simulation plays an important role in the understanding the operating principle of the whole system and the function of some key components because it is more instructive and can save time and cost. Previous simulation platform only concerns the realization of control algorithm. Parasitic elements are usually omitted to save the simulation time [6, 7]. But in order to simulate accurately the actual conditions and guide the selection of some key components, a precise platform including main parasitic components will be built up. This paper will introduce a simulation platform including main parasitic and provide some useful experimental data for a better dc bus capacitor selection and laminated bus bar design.

In this paper, the complete circuit diagram of a traction motor drive inverter will be described in detail. The distribution of main parasitic will be introduced in the circuit diagram. Then parasitic extraction method through experimental measurements is presented. The accurate simulation platform based on Matlab/Simulink software has been developed. Finally, the simulated and experimental results are showed and compared. Through analysis the current ripple and voltage spike, the superiority with the used of the high-current low-inductance film capacitor is justified.

II. INVERTER CIRCUIT INCLUDING PARASITIC

Fig.2 shows the basic diagram of inverter for electric vehicles. The output currents of inverter should be as

978-1-4244-1741-4/08/$25.00 ©2008 IEEE   417

sinusoidal as possible to reduce the ripple content. The ripple contents of i_a, i_b and i_c depend on dc voltage, motor inductance, motor back electromagnetic force and the switching period [8]. The ripple ranges from 10% to 90% of full-load current based on different motor winding inductance and the modulation scheme. The battery current should be a smooth dc. Otherwise, the life of the battery will be greatly reduced.

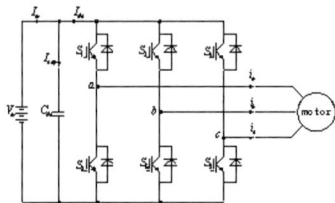

Fig.2. Basic diagram of inverter for electric vehicles

It is well known that there are parasitic parameters in power devices and connections. The complete circuit diagram of a traction motor drive inverter including main parasitic elements is shown in fig.3. R_{in1}, R_{in2} and L_{in1}, L_{in2} are the internal resistor and connecting inductance of battery. ESR_c and ESL_c are the equivalent series resistor and the equivalent series inductor of dc bus capacitors. The parasitic inductances exist between phase legs, L_{lk}, and all the device connections, as shown in the fig.3. The magnitude of the lumped parasitic inductance and the rate of current change are directly related with the magnitude of the voltage spike. Dc snubber capacitor (C_{sn1}-C_{sn3}), is used to suppress the voltage spike. The bus voltage magnitude and surge impedance of the cable parameters around the loop, such as cable resistance (R_a, R_b, R_c), self inductance (L_a, L_b, L_c), line-to-line capacitance (C_{ab}, C_{bc}, C_{ca}) will determine transient line-to-line cable charge current. $C_{hs-IGBT}$, $C_{hs-cable}$ represents separately the parasitic capacitor of IGBT module heat sink-to-ground and cable-to-ground, which will determine transient line-to-ground current. In the electric vehicle, body of the vehicle usually is viewed as "ground". It is obviously from this figure that these parasitic parameters can't be omitted in the analysis of IGBT transient characteristic and EMI performance of the vehicle.

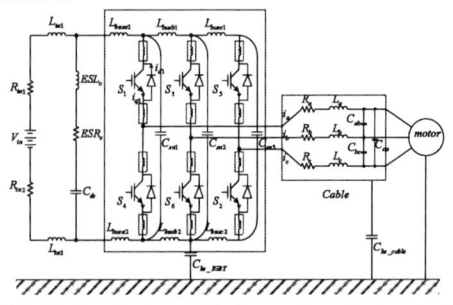

Fig.3. Distribution of main parasitic elements in traction inverter circuit

III. DETERMINATION OF PARASITIC PARAMETERS

The accuracy of parasitic parameters is the key of the accurate simulation platform, as pointing from the previous section. LCR meter HP4263B is used to derive parasitic of capacitors and connecting cables. Table 1 and Table 2 present parasitic of 16,500uF electrolytic bulk capacitor bank and 3,300uF film capacitor. It can be seen that the value of ESL has more change than the value of ESR with frequency. The ESL and ESR of 16,500uF electrolytic capacitor are determined as 266nH and 20.9 $m\Omega$ in the simulation platform considering the switching frequency is 10 kHz.

Table 1

Parasitic of 16,500uF electrolytic bulk capacitor bank

f(kHz)	ESL(nH)	ESR($m\Omega$)
0.1	3.01e5	19.51
0.12	2.12e5	21.11
1	3.24e3	21.39
10	266	21.59
100	208	24.87

Table 2.

Parasitic of 3,300uF film capacitor

f(kHz)	ESL(nH)	ESR($m\Omega$)
0.1	1532	2.8
0.12	1176	3
1	200	4.4
10	99	4.3
100	71.7	6.7

Compared with 16,500uF electrolytic bulk capacitor bank, the selected film capacitor (fig.4) has the same current ripple capability: 170 Arms at 10 kHz. But the film capacitor has low profile, larger heat sinking plate and especially low inductance. The rectangular block of film capacitor has a form factor similar to IGBT module and will greatly simplify the layout of dc bus bar. The bottom aluminum plate allows the heat sink to carry the heat away and maintain the operating temperature below the maximum heat sink temperature. The most important is that the ESR and ESL of the proposed film capacitor are greatly reduced, as shown in table 1 and table 2, which will greatly improve the performance of the whole system.

Fig.4 Photograph of the proposed low-inductance high-current handling film capacitor

Analytical method, finite element simulation and experiment test can all be used to derive the parasitic of laminated bus bar. Table 3 shows the measurement results of parasitic inductance of laminated bus bar.

Considering the switching frequency is 10 kHz, the parasitic inductance is determined as 180nH.Other parasitic such as the ESR of battery, the parasitic inductor of IGBT module may all be determined through experimental measurement.

TABLE 3
PARASITIC OF LAMINATED BUS BAR

f(kHz)	Ls(nH)
0.1	450
0.12	350
1	220
10	180
100	164

IV. SIMULINK MODELING OF ADJUSTABLE SPEED DRIVE SYSTEM

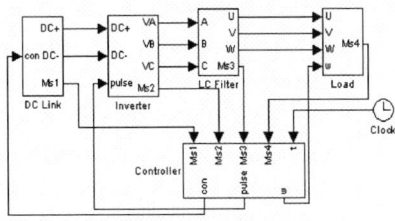

Fig.5.Simulink model for the adjustable speed drive system

Fig.5 shows the simulink model for the adjustable speed drive system. Because parasitic elements such as

equivalent series resistance (ESR), equivalent series inductance (ESL) of capacitor are frequency dependent [9, 10], look-up table or curving fitting methods are used to describe these parameters of other working frequency. The target system model of simulation platform based on Matlab/Simulink software includes five sub-models. DC link sub-model represents battery model and connected to inverter sub-model through positive and negative dc bus. Universal bridge model in Power system block is not adopted because parasitic should not be omitted based on the analysis above. The proposed inverter model constructed by discrete components makes the parameters setting convenient and can improve the accuracy of simulation. The internal construction of inverter sub-model is shown in Fig.6. The filter circuit constructed by inductor and capacitor is optional and the load is three-phase three phase squirrel cage induction motor. The measurement variables include the rotor current, speed and torque. The control module (controller) is an application module. The input variable of this module is the measurement output variable and simulation clock signal. The output control signals are sent separately to DC link sub-module and inverter sub-module. Several control methods are included in control sub-module such as carrier modulation, space vector modulation and selected harmonic PWM control method etc.

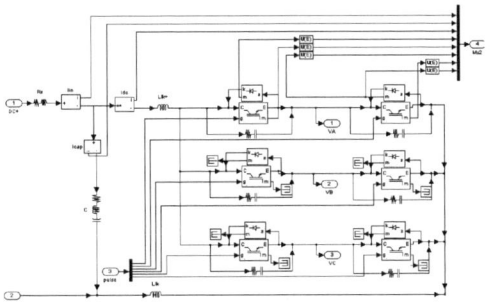

Fig.6.Diagram of internal construction of inverter sub-model

V. SIMULATED AND EXPERIMENTAL RESULTS

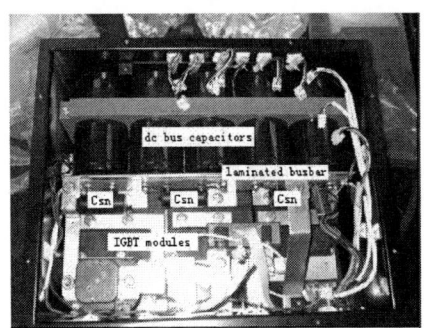

Fig.7. Photograph of the inverter assembly

The experimental set up used for the verification of the simulation results and the accuracy of the system model consists of a 100kW PWM IGBT inverter, 16,500uF electrolytic bulk capacitor bank (or 3,300uF film capacitor) and a maximum power 150kW induction motor. Fig.7 shows the photograph of the inverter assembly.

Fig.8 (a) shows the simulation waveform of the motor stator current and voltage. Compared with the experimental waveform as shown in Fig. 8 (b), the waveform of the motor line-to-line voltage is very similar. The voltage fluctuate caused by finite dc link capacitance is also reflected. The simulated waveform of output current has the similar characteristics with the experimental waveform except for high frequency common mode interference current.

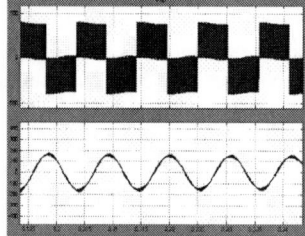

(a) Simulated waveforms

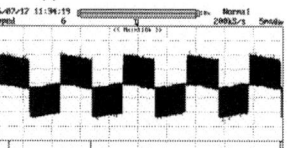

(b) Experimental waveforms

Fig.8. Comparison of output line-to-line voltage and phase current

The capacitor current ripple can also be obtained through the computer simulation and experimental test. As analyzed above, current ripple handling capability is an important factor in selection dc link capacitors. Fig.9 (a) shows the simulated capacitor current I_{cap}, battery current I_{in} and dc link current I_{dc}. Without the dc link capacitor, the battery current equals the dc link current, which is pulsating between 0 and the maximum load current in every switching period. The current ripple of battery will have a bad effect on the life of capacitors. The figure indicates that the dc link capacitors draw a current of I_{cap}, which absorbs substantial high frequency ripple current from I_{dc} and smoothes out the battery current. The detailed experimental current waveforms for several switching cycles are shown in Fig.9 (b). Under rated power output, the capacitor rms current is 128A and the battery input current is 297A. The ratio of the dc bus capacitor current I_{cap} and the dc input current I_{in} is 43%. The ratio will be higher under the peak power output. Compared with the conventional electrolytic bulk capacitor bank, the introduced low-inductance high frequency film capacitors have merits of longer life, can withstand negative voltage and have high ripple current. So it will have a bright future in the application of electric vehicles.

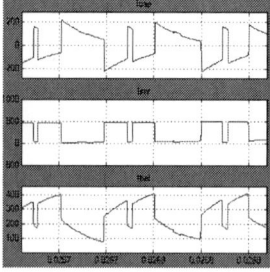

(a) Simulated waveform of input dc bus currents (from top to bottom: I_{cap}, I_{dc}, I_{in})

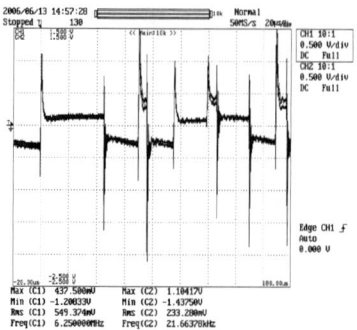

(b) Simulated current waveforms i_{cap} with expanded time scales

Fig.9. Comparison of Simulated and experimental waveforms with expanded time scales

VI. CONCLUSIONS

With switches dynamic features having been greatly increased, parasitic parameters such as interconnection inductances and stray capacitors are directly linked to voltage surge, resonance and electromagnetic interference. A precise platform including main parasitic components has been built up. Parasitic extraction method through experimental measurements is also presented. Experimental results are shown to verify the effectiveness of the simulation platform.

Based on this simulation platform, IGBT switching characteristic and the whole system performance can be assessed. The selection principle for dc bus capacitors will also be modified through this simulation. DC link capacitors account for a major fraction of the volume, weight and also cost of an inverter. Advantages of advanced film capacitor technology over conventional bulk capacitors for DC link application are review in this paper. The most important advantage is that the film capacitor has low ESR and high current handling capacity. Both simulated and experimental current waveforms indicate that dc bus capacitors need to handle high-frequency current with an rms value approximately 48% of the average dc bus current. Thus conventional design using multiple bulky electrolytic capacitors can be replaced with a single high-current film capacitor.

ACKNOWLEDGEMENT

This work was supported by the National Nature Science Foundation of China under Contract 50777060.

REFERENCES

[1] M.C.Caponet, F.Profumo, R.W.De Doncker, A.Tenconi, "Low stray inductance bus bar design and construction for good EMC performance in power electronic circuits," IEEE Transactions on Power Electronics, pages: 225~231, 2002.

[2] G.L.Skibinski, D.M.Divan, "Design methodology and modeling of low inductance planar bus structures," Fifth European Conference

on Power Electronics and Applications, pages: 98~105, 1993.

[3] Ji-hoon Jun, Chul-kyu Lee, Byung-il Kwon, "The analysis of bearing current using common mode equivalent circuit parameters by FEM," Proceedings of the Eighth International Conference on Electrical Machines and Systems, pages: 49~51, 2005.

[4] Huang Jimmy Huat Since, S.Sijher Taninder, Beh Jium Kai, "The impact of capacitors selection and placement to the ESL and ESR," International Symposium on Electronics Materials and Packaging, pages: 258~261, 2005.

[5] R. Grinberg, P.R.Palmer, "Advanced DC link capacitor technology application for a stiff voltage-source inverter," IEEE Conference on Vehicle Power and Propulsion, pages: 6~12, 2005.

[6] Jih-Sheng Lai, Xudong Huang, Shaotang Chen, "EMI characterization and simulation with parasitic models for a low-voltage high-current AC motor drive," IEEE Transaction on Industry Applications, 2004, pp.178-185.

[7] Weijun Huang, Haihong Qin, Huizhen Wang, Yangguang, Yan, "Optimum design of snubber capacitors in 9 kW three-phase inverter for doubly salient permanent magnet motor," Electrical Machines and Systems, 2005, ICEMS 2005, pp.296-299.

[8] Jih-Sheng Lai, H.Kouns, J.Bond, "A low-inductance DC bus capacitor for high power traction motor drive inverters," 37th IAS Annual Meeting on Industry Applications Conference, pages: 955~962, 2002.

[9] John D. Prymak, "Spice Modeling of Capacitors," 15th Capacitor and Resistor Technology Symposium, pages: 39~46, 1995.

[10] Jianbing Li, Yujie Shi, Zhongxia Liu, Dongfang Zhou, "Modeling, simulation and optimization design of PCB planar transformer," Proceedings of the Eighth International Conference on Electrical Machines and Systems, pages: 1736~1739, 2005.

DC-DC Converters with FPGA Control for Photovoltaic System

Jan Leuchter*, Pavel Bauer†, Vladimir Řerucha**and Petr Bojda***

*University of Defence / Electrical Engineering, Brno, Czech Republic, e-mail: *jan.leuchter@unob.cz*
† Delft University of Technology / Electrical Power Processing, Delft, The netherlands, e-mail: *p.bauer@ewi.tudelft.nl*
**University of Defence / Department of AD Systems, Brno, Czech Republic, e-mail: *vladimir.rerucha@unob.cz*
***University of Defence / Aerospace Electrical Systems, Brno, Czech Republic, e-mail: *petr.bojda@unob.cz*

Abstract— **The paper introduces on operating principle of power management of photovoltaic system with maximum power tracker control to achieve the maximum efficiency of such systems. The paper presents different topologies of basic photovoltaic concepts with Dc-Dc converters including control design of maximum power tracker by Matlab-Simulink. The paper will be present different topologies of converter concepts that improve the total efficiency value. PWM controlling of such sophisticated system can be achieved by Field-Programmable-Gate-Array (FPGA) that more details of set up PWM signals due to FPGA technologies is shown in this paper.**

Keywords—**Maximum Power Tracker, Photovoltaic System, FPGA, SEPIC.**

I. INTRODUCTION

The photovoltaic power system is very attractive source of electricity. The civilization of human beings is solely dependent on energy: Starting from the simplest energy consumption in the form of fire from wood to different types of energy sources these days. As the population of the world grew larger and larger and resources getting depleted, alternative energy sources that are environment friendly and at the same time that can fully satisfy the need of the population in general become inevitable. Besides to other renewable energy sources, the solar energy is one to be relied upon. The solar power can be utilized in the form of heat or electrical energy. Considering the case of harnessing the energy of the sun in the form of solar electric energy, the high cost and low efficiency of the photovoltaic cells, that convert the sun light to direct current electricity, stands as a major constraint for its full utilization. However, as the solar energy is believed as the real source of sustainable and clean energy; a lot is being done to improve the efficiency of the solar cells to use it in large scale.

By making use of the photovoltaic cells, arranging them in modules and arrays, we get dc voltage. Individual PV cells with cell areas ranging from small to approximately 225 cm², generally produce only a few watts or less at about half a volt. Thus, to produce large amounts of power, cells must be connected in series–parallel configurations. Such a configuration is called a module. The modules can be further connected to form larger units called arrays. When cells are connected together, normally they are incorporated into PV modules, which often combine as many as 40 cells in series to produce voltages in the range of 20 V and currents of several amperes.

When voltages and currents beyond the capability of an individual module are desired, the modules can be connected into arrays that will produce higher voltages and higher currents. Although most cells produce only a few watts, and most modules produce 10 to 300 W, most arrays produce a few thousand watts. The simplest configuration of photovoltaic system is shown in Fig. 1. There are shown a photovoltaic array and DC load.

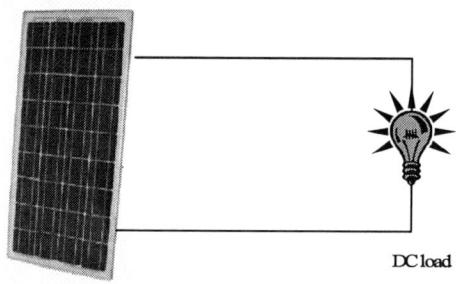

Fig. 1. Simplest configuration of photovoltaic system

The output VA characteristic of such 50 W array as a function of sun intensity is shown in Fig. 2. It is possible see dependences of input and output photovoltaic array parameters. The photovoltaic experimental module (M-S63-53) which is described by as follows parameters:

max. power	53/W/;
no-load voltage	21,5 /V/;
short current	3,42/A/;
optimum voltage	17,2 /V/.

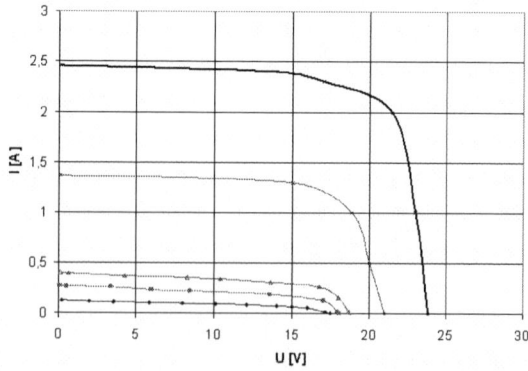

Fig. 2. AV characteristics of photovoltaic array M-S63-53 (measured)

Stand-alone photovoltaic systems may be as simple as a module connected to an electrical load mentioned above in Fig. 1. The next level of complexity involves the use of storage batteries to allow for the use of the photovoltaic-generated electricity when the sun is not shining. These systems also generally include an electronic controller to prevent battery overcharge and another controller to prevent over discharge of the batteries. Many charge controllers perform both functions. The configuration of photovoltaic system with battery is shown in Fig. 3. There are shown a output voltage and current of photovoltaic array. They are variable and controller set up constant value of DC voltage on the load. Charge and discharge controller can be solved as DC-DC bi-directional converters. The Bi-directional converter interfaces the battery voltage to the DC link of photovoltaic array and DC output of photovoltaic source. When the power flowing from the photovoltaic array is higher than required by the load, the bi-directional converter operates in the charge mode. If the delivered power is lower, converter operates in discharge mode and energy is pushed from battery. The energy flow from and to the buttery must be controlled. The converters topologies of Bi-directional converters are shown in part IV.

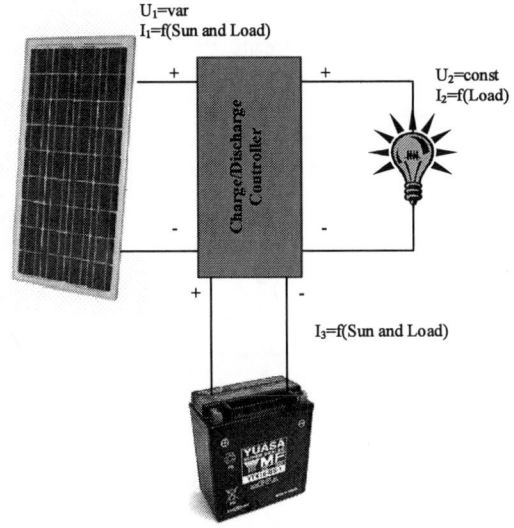

Fig. 3. System with battery backup

Based on the intended application, the voltage can be used as DC for some types of stand alone applications, or used to charge a battery that can be used as energy store. But in most of the cases, it is converted to AC, amplified into a desired voltage and frequency level and then connected to the grid or alternatively used to drive a machine that uses an AC source. If loads are AC, an inverter can be incorporated into the system to convert the DC from the photovoltaic array to AC. A wide range of inverter designs is available with a wide range of output waveforms ranging from square waves to relatively well-approximated sine waves with minimal harmonic distortion. Most good inverters are capable of operating at conversion efficiencies greater than 90% over most of their output power range. Fig 4. show a system with inverter and AC loads.

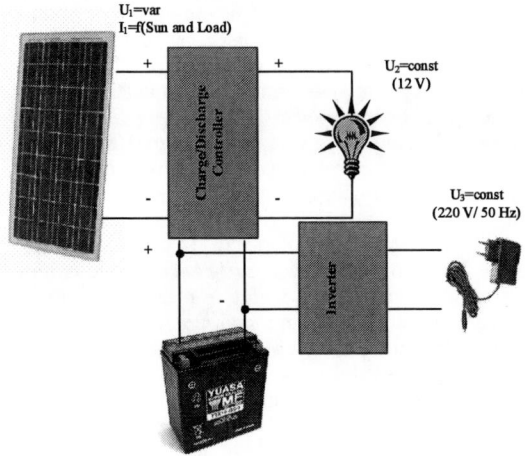

Fig. 4. System with DC and AC loads

The next level of complexity is to use a maximum power tracker or a linear current booster between module and load as a power matching device to ensure that the module delivers maximum power to the load at all times.

II. PHOTOVOLTAIC SYSTEM WITH MAXIMUM POWER TRACKER

The main idea of photovoltaic system with maximum power tracker is to achieve the maximum available power every time. A typical AV characteristic of a solar cell is shown in Fig. 2. The open-circuit voltage decreases only slightly of the load which is between 0 and 70% of maximum current of every curve. Under constant temperature, the locus of the maximum power points is an almost vertical line, sloping slightly toward the origin. Fig. 5 shows red curve in detail and next curve in this figure show the results of power dependence. It is possible to see that e.g. power of operating point of "A" is 10 W and if we set up the system by controlled current to the "B", we are able to archive 17 W which is 70% more. This increase is not negligible.

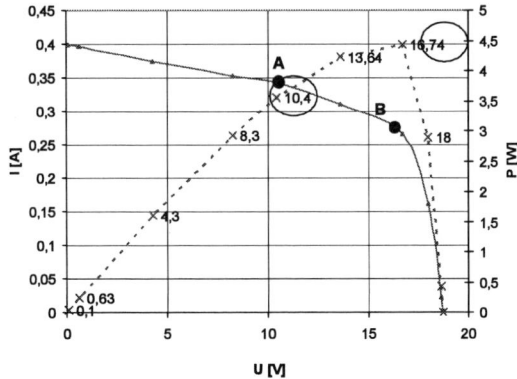

Fig. 5. AV and power output characteristics of photovoltaic array

Several techniques for tracking maximum power point tracker (MPPT) have been proposed, as described in below (part III). One method is commonly used to track the MPPT. The control system moves the operating point toward the maximum power point by periodically increasing or decreasing the voltage of photovoltaic array. Usually, by increasing or decreasing the duty ratio of on-state of switching device, the maximum power point is tracked. The variation of duty ratio is determined by considering its circuit parameters. So, Maximum power can be archived by feed back current controlled with criterion of maximum power where is measured current and voltage at the same time and the output power obtained by multiplication of current and voltage is compare with power value at moment before. Fig. 6 shows photovoltaic system using of maximum power tracker. Photovoltaic power system with maximum power tracker is consisting of the two semiconductor power converters:

1. DC-DC converter;
2. Bi-directional converter.

Both converters systems must be coordinated. The photovoltaic array must be loaded by current which required maximum power. Controllers compare current I_{opt}, calculated by maximum power tracker, with load current I^{2*}. If the load current I^{2*} is higher than I_{opt}, than battery must the rest value of current delivers to the load. If the difference is lower than surplus current can be saved to the battery. Both current ways provide a bi-directional converter which is controlled by duty D2 or D3. DC converter set up a constant required value of voltage on the load by D1.

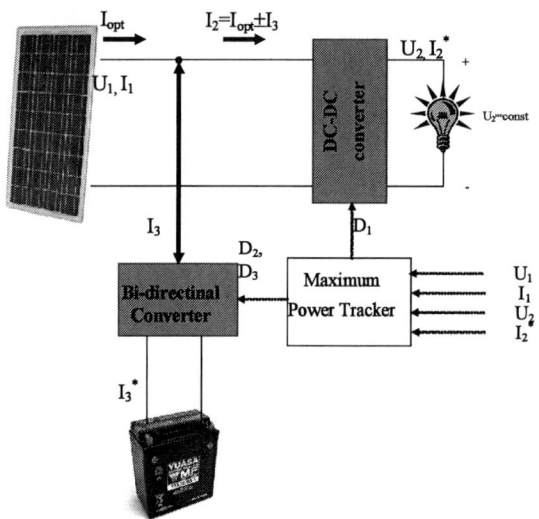

Fig. 6. System with Maximum Power Tracker

III.   TYPES OF PHOTOVOLTAIC SYSTEM WITH MAXIMUM POWER TRACKER

A variety of techniques has been developed in recent years about determine a operating point of maximum power tracker (MPPT).

Studies [6] indicate that the optimal operating voltage of a photovoltaic module is always very close to a fixed percentage of the open-circuit voltage. This implies that MPPT could simply use the open-circuit voltage to predict the optimal operating condition. This is called voltage-based MPPT, see Fig.7. Such method is simple and the results of the estimation of the optimum point are not really accurate.

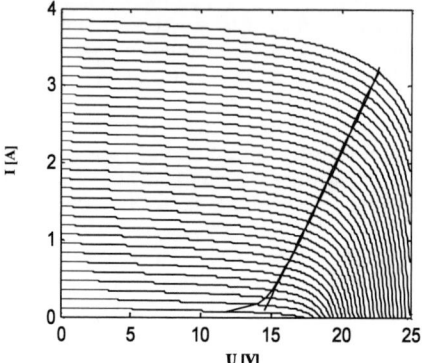

Fig. 7. MPPT is located as open circuit voltage

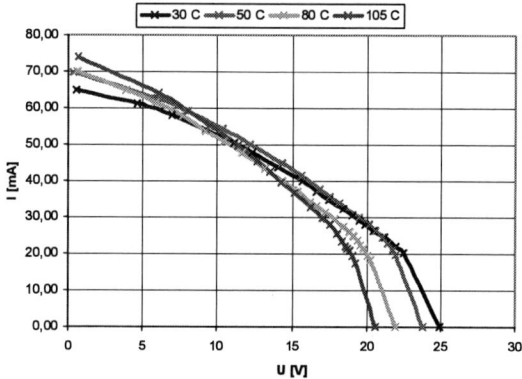

Fig. 8. AV characteristics as a temperature function

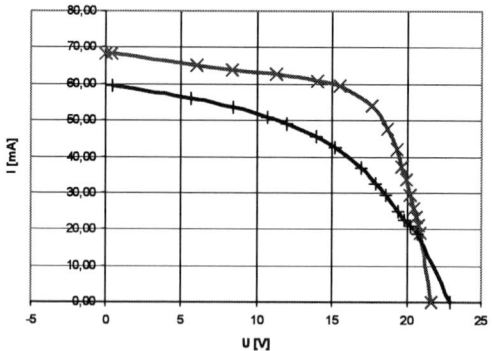

Fig. 9. AV characteristics as an ageing function (the photovoltaic modules weren't the identical and the same)

Another technique was developed according to a linear approximation between the maximum output power and the optimal operating current [6]. Such techniques is not really useful because the changes of AV characteristics of photovoltaic modules to change as well, as illustrated in Fig. 8 and 9, where is possible to see the important factor of temperatures variation effect and unpleasant effect of ageing, of photovoltaic modules. Next disadvantages of this method are that every photovoltaic panel which is produced is not the same quality and the system will be setting separately for every module.

- Similarly according method 1, MPPT method, based on the approximation of a linear relationship between the maximum power point (MPP) and the short-circuit current. The method estimates the operating point with better reliability in comparison with method previous.

- A fourth algorithm [6] measures and compares the output of two modules to track the MPP. Such topology of MPPT system is shown in Fig. 10. Both converters systems must be coordinated and control scheme of such system is complicated. - directional converter set it the maximum power tracker point by means of current controlled to achieved maximal efficiency of photovoltaic panel.

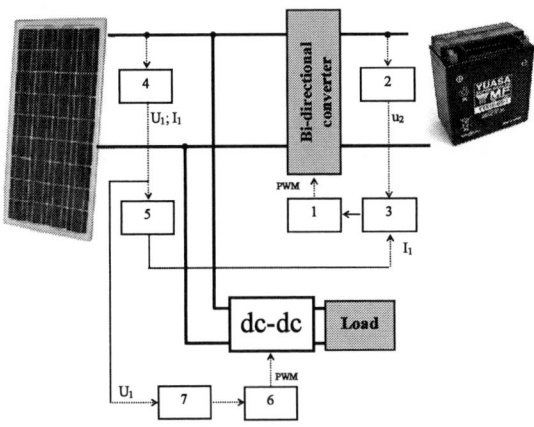

Fig. 10 Control part of photovoltaic system

(1. PWM generator I; 2: voltage probe II; 3: voltage control I; 4: voltage and current probe I; 5: current control I; 6: PWM generator II; 7: Voltage control I)

- Various other techniques [1] are also available to estimate the characteristics of photovoltaic output and track the MPP by way of some specific mechanisms such as a sliding-mode observer and neural network. [1].

IV. DC-DC CONVERTERS OF PHOTOVOLTAIC SYSTEM WITH MAXIMUM POWER TRACKER

The dc-dc converters are used in many regulated switch mode dc power sources and dc motor drive application. In dc-dc converters with a given input voltage, the average output voltage is controlled by setting the switch ON and OFF duration. As the name implies, a dc-dc converters produces an average output voltage higher or lower than dc input voltage U_{in}. The basic circuits of dc converters with inductor are shown in next figures (Fig. 11). [2]

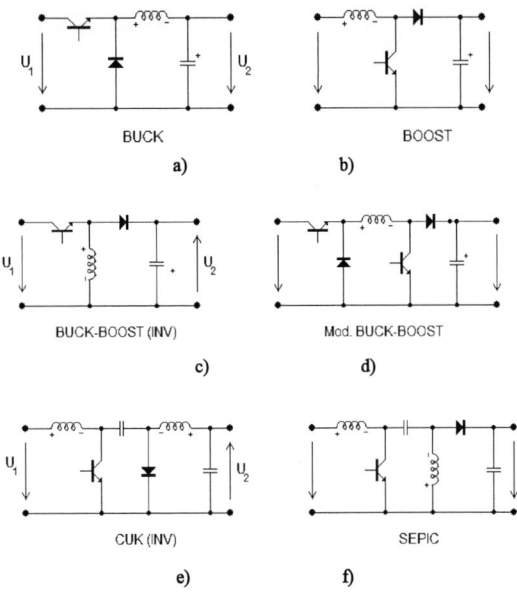

Fig. 11: Dc-dc converters

The first topology of dc-dc (Buck) produces a lower average output voltage than the dc input voltage. The second topology produces the output voltage higher (Boost). Both topologies are not interesting for photovoltaic application because there are required usually both way of voltage adjustment. Topologies Buck-Boost, Cuk and Sepic enable to regulate step-up or step-down according actual voltage and respects to common terminal of the input voltage. As a first step of the system realization, Sepic topology has been prepared with supercaps as shown in Fig. 11. An advantage of this converter is that SEPIC tolology are lower power losses than in the CUK converter. A photo and the main design data of the experimental converter is shown in Fig. 12.

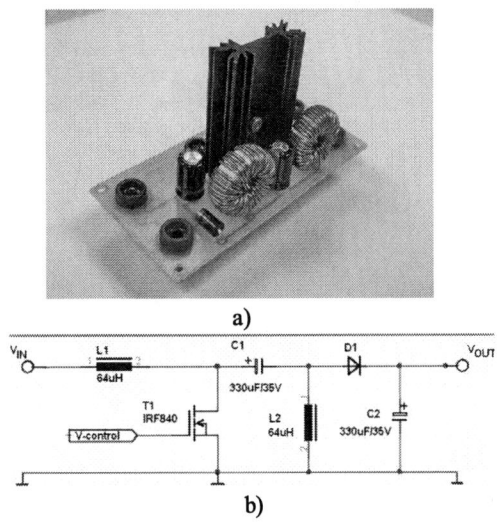

Fig. 12 A photo of the experimental SEPIC converter

425

V. BI-DIRECTIONAL DC/DC CONVERTERS FOR ENERGY BUFFER

As mentioned above (fig. 6 and 10) the energy buffer with bi-directional converter provide energy saving for next energy using during peak power delivery require when the sun intensity is not enough. The output characteristics of photovoltaic module were shown in Fig. 5. Output variable voltage of photovoltaic module corresponds to the variable intensity of sun and MPPT. Output voltage must be converted to the constant voltage or current to the battery charging by means of dc-dc converter. In the Fig. 13, 14 are shown two main topologies of bi-directional converters.

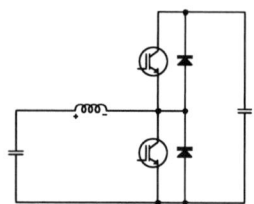

Fig. 13 Bi-directional converter with inductor (Buck and Boost)

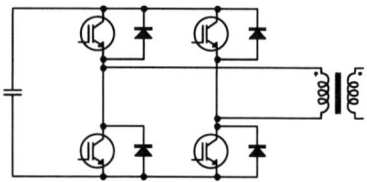

Fig. 14 Bi-directional converter with transformer (full-bridge)

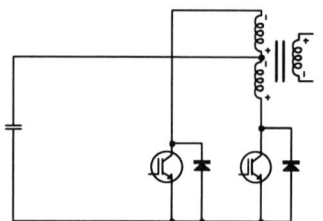

Fig. 15 Bi-directional converter with transformer (push-pull)

The first topology is widely known non-isolated bi-directional converter topology that is good at price. The main disadvantage of this topology is that the voltage gain is limited. Second topology of bi-directional converter with transformer is used as a topology full-bridge. Full-bridge topology makes it possible high voltage gain between input and output. Disadvantage of full-bridge topology can be high number of transistors and thereby the cost of converter is increasing. Cheaper solution with transformer is topology push-pull (Fig. 15). Disadvantage of push-pull is his transformer that is complicated.

A bi-directional converter topology which was shown in Fig. 13 it was used for our experimental workplace. This kind of converters is analyzed, tested and designed for application in the photovoltaic structure. The flow of energy from or to buffer must be controlled according requirements of power management. The DC/DC converter operates in BOOST mode during energy delivering from buffer to the dc line or in BUCK during buffer charging.

VI. ACCUMULATION OF ELECTRICAL ENERGY

With the development of new technologies on the area of accumulation are lot possibilities of to set the optimum battery buffer. For this purposes various types of electrical energy buffering methods and converters including buffers are used. Solution of energy buffers by means of accumulator (battery) can bring a lot of storage energy. Super-capacitors are the further solution of buffer problem. They found their place in many applications and are opening new area of energy accumulation. There are combined advantageous features in comparison to previous battery solution. Present supercapacitors are readily available with capacity 3500 F. Usually; the accumulators have limitation in number of life cycles. Supercapacitors, on the other hand, can be charged and discharged almost unlimited number of times. Energy of electrical field of capacitors can be expressed by universal equation (1) including the fact that supercapacitors are not discharged fully but only to half of nominal voltage U_{nom}.

$$W = \frac{1}{2} \cdot C \cdot \Delta U^2 = \frac{1}{2} \cdot C \cdot \left(U_{nom}^2 - \left(\frac{U_{nom}}{2} \right)^2 \right) \quad (1)$$

In Fig. 16 is shown the results of comparison of different energy storage battery (Lead-Acid, Li-Ion, Ni-Mh, Ni-Cd and Supercaps units (Maxwell) in relation to the price, weight and value of stored energy. Accumulators can bring very good relationship between stored energy and price. Generally it can be stated that supercapacitors can be charged and discharged almost an unlimited number of times, can be charged and discharged in terms of miliseconds, seconds or minutes, have very high power density, do not release any thermal heat during discharge due to negligible internal resistence, cannot be overcharged, are not affected by deep discharges as are chemical batteries, have a long lifetime (80% of capacity after 10 years, lifetime up to 20 years), the DC/DC round trip efficiency 80% to 95% in most applications, operating temperature range between -50C and 85C, are environment friendly.

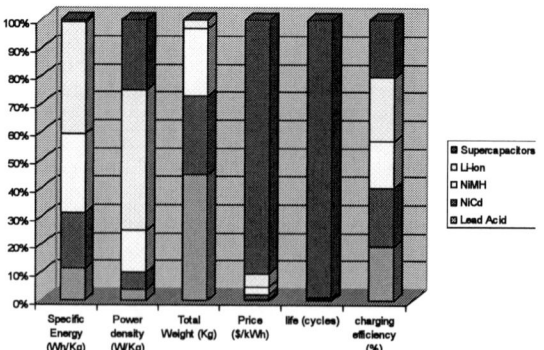

Fig. 16 Comparison of energy buffers [4]

VI. PHOTOVOLTAIC SYSTEM WITH SUPERCAPS

If will be used supercapacitor as energy buffer than the photovoltaic system with maximum power tracker can be change from fig. 10 to fig. 17. Supercapacitor can be injected to the dc-dc line as UPC source (in line) with comparison with battery from fig. 10 (system UPC - off line). Such system can bring easer solution with two identical dc-dc converter (f.g. SEPIC) which first one will be operated in current mode to set the optimum current during charging supercapacitor according MPPT. Second dc-dc converter will operate in voltage mode to achieve the constant output voltage 12 V. The photo of our working place is shown in fig. 18.

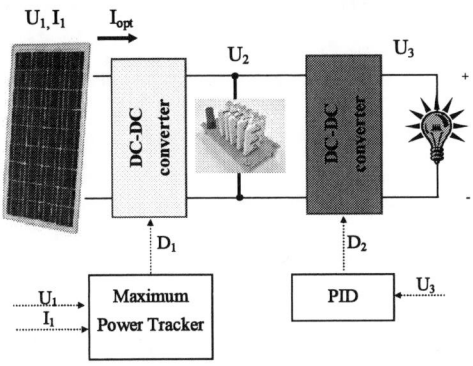

Fig. 17. System with Maximum Power Tracker and Supercaps

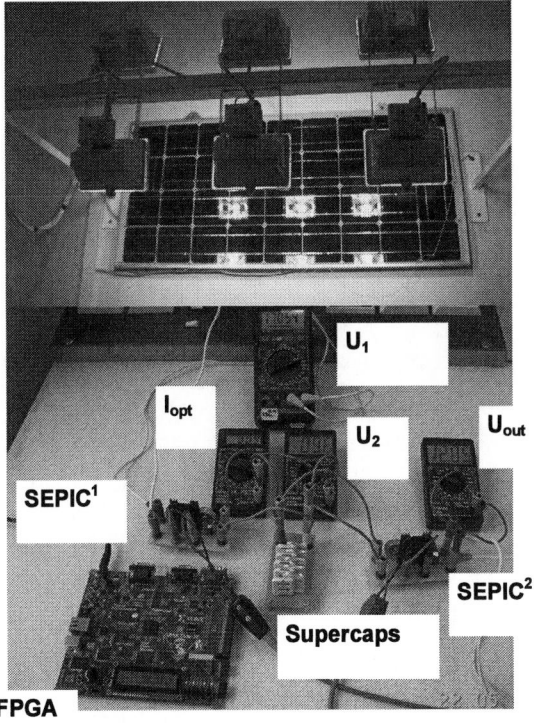

Fig. 18. The photo of photovoltaic workplace with Maximum Power Tracker and Supercaps

Recent technology of a control is of course based on Field-Programmable-Gate-Array (FPGA). Modern FPGA circuits allow incorporation of whole embedded system including a core of for instance a 32-bit microprocessor and peripherals such as an USB interface, etc. Clock frequencies of those circuits are approximately 100 MHz or above. These features create a lot of room for developing a reasonable control block with desired parameters. The final element of a DC-DC converter control block is, as it is mentioned above, of course PWM modulator. A flexibility of FPGA circuits paves the way for a design of the PWM modulator with a variable PWM frequency as well as a variable duty cycle. These possibilities require an adequate complex feedback control system which drives the parameters of PWM modulator. The control system also incorporates current-estimation units which can replace current sensors. Such a complex controller described above - usually called digital controller - contribute to significant improvement of final efficiency of the transformation of dc energy provided by the DC-DC converter.

VII. CONCLUSIONS

The paper presented our experimental photovoltaic system with maximum power Tracker. Model and control design of a photovoltaic system with SEPIC and bi-directional converters was shown. The control scheme which was described has been shown and the main goal of control system is achieved the maximum efficiency during supply, accumulation energy and delivery energy from battery to the load through buck converter. A PWM modulator has been successfully developed with FPGA technology.

REFERENCES

[1] Benjamin, V.P., Chong, Li Zhang, Abbas, D. Modeling and Control of a Bidirectional Converter for a Standalone Photovoltaick Power Plant, In EPE Conference 2007, Aalborg : EPE press, 2007, 8 p, ISBN: 9789075815108.

[2] Mohan, R., Power Electronics, New York : John Wiley and Sons, 1995, 800 p, ISBN: 0-471-58408-8.

[3] Diver, R., Andraka, C., Rawlinson, K., Moss, T., Goldberg, V., and Thomas, G. 2003. Status of the Advanced Dish Development System Project, paper no. ISEC2003-44237, Proceedings of the ASME International Solar Energy Conference, Kohala Coast, Hawaii, March 15–18.

[4] Marco Piemontesi, GE Digital Energy Riazzino, "Alternative Energy Storage Systems for UPS," PCIM 2006, Nuremberg-Germany.

[5] Lotker, M. 1991. Barriers to Commercialization of Large-Scale Solar Electricity: Lessons Learned form the LUZ Experience. Sandia National Laboratories, Albuquerque, NM, SAND91-7014.

[6] Chimento, F., Musumeci, S., Raciti, A., Sapuppo, C. A Control Algorithm for Power Converter in the Field of Photovoltaic Application, In EPE Conference 2007, Aalborg : EPE press, 2007, 8 p, ISBN: 9789075815108.

[7] Junior, P.V., Palheta, P.I.G, Silva, D.M., Nascimento, M.P., Costa, A.R., Filho, P.S.F., Siqueira, J.G. Applied Digital Control For Localiyation of the Maximum Power of Photovoltaic Generators, 1-4244-0431-2/06, IEEE.

ACKNOWLEDGMENT

The research work is supported by the Czech Ministry of Education, Youth and Sports (project no. OC169/COST542) and Research Program No VZ-0000403.

Control of a Converter with Superconductive Energy Storage Inductor

Rozanov Yurie Konstantinovich[*], Lepanov Michail Gennadevich[**], Kiselev Michail Gennadevich[***]

[*]Moscow Power Engineering Institute (Technical University), Department of Electric and Electronic Apparatuses, Moscow, Russian Federation, e-mail: *y.rozanov@mtu-net.ru*
[**]Moscow Power Engineering Institute (Technical University), Department of Electric and Electronic Apparatuses, Moscow, Russian Federation, e-mail: *lepanovm@mail.ru*
[***]Moscow Power Engineering Institute (Technical University), Department of Electric and Electronic Apparatuses, Moscow, Russian Federation, e-mail: *kiseliovmg@mpei.ru*

Abstract—**This paper presents a converter with superconductive inductor based on full controlled semiconductor devices with digital PWM control. The feature is continuous commutating current by space vector modulation method. A power flow can charge or discharge inductive storage and is changed according to the control low. It is given the computer model that demonstrates the main performance and operating modes**.

Keywords—**Current source inverter (CSI), converter control, space-vector modulation, Superconducting Magnetic Energy Storage (SMES).**

I. INTRODUCTION

There are many aims to use superconducting magnetic energy storages (SMES) in power energy systems: reserving of power supplies, damping oscillations of power generators, and providing static and dynamic stability [1]. Traditionally thyristor converters connect inductive storage with energy system. These converters control power flow from network to superconducting inductor (charging it) or vice-verse (discharging it). Also, it has many other functions. In these cases, the converter has many drawbacks which are decreased the power quality. The main of them are distortion of current on the network side and lower of power factor. Using of modern full controlled devices as GTO, IGCT, IGBT and others allow eliminating these disadvantages. Besides, we can provide operating of the converter on high voltage and with low switching frequency by connecting some modules in series and using multilevel topology. In this paper it is described the base topology – three-phase bridge converter realized on GTO. Converter's efficiency essentially depends on control system operating principle. Most of the modern control systems are microprocessor-based systems. Therefore, used control algorithms and methods determine converter performance. In this paper these questions are described in more detail and the results of computer simulation of current source converter are given.

II. PRINCIPLES OF THE CONTROL SYSTEM'S OPERATION

A. Algorithm of modulation control system

Specific feature of the converter with superconductive inductor is continuity of the current on the DC side all the time. Therefore, some switches have to be in conducting states simultaneously. It is possible using space vector modulating techniques [2]. Number of the combinations of switch's states for current source converter is more than the same in three-phase voltage source converter. It is effect of adding vectors providing the continuous current during commutation between active and zero vectors. The space-vector diagram for three-phase current converter is shown in Fig.1. There are six active vectors and three zero vectors. Positive and negative values of phase currents are given in the Table 1, where symbol "V" means that this switch has the state "ON".

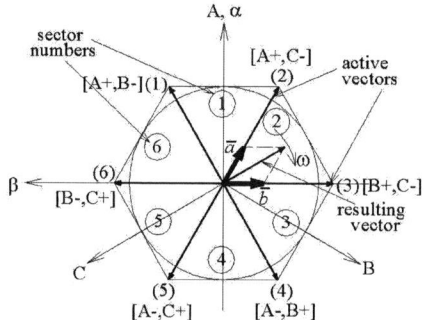

Fig. 1. The space-vector diagram for three-phase current converter

TABLE I.
COMBINATIONS OF SWITCH'S STATES FOR CURRENT SOURCE CONVERTER

State (vector) number	Switch's states						Current values		
	A+	A−	B+	B−	C+	C−	i_A	i_B	i_C
1	V	—	—	V	—	—	+	−	0
2	V	—	—	—	—	V	+	0	−
3	—	—	V	—	—	V	0	+	−
4	—	V	V	—	—	—	−	+	0
5	—	V	—	—	V	—	−	0	+
6	—	—	—	V	V	—	0	−	+
7	V	V	—	—	—	—	0	0	0
8	—	—	V	V	—	—	0	0	0
9	—	—	—	—	V	V	0	0	0

978-1-4244-1741-4/08/$25.00 ©2008 IEEE

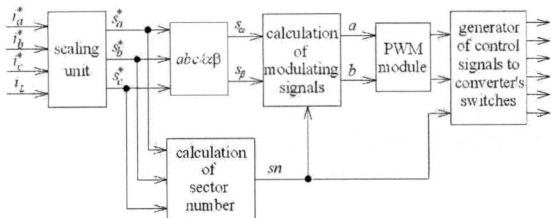

Fig. 2. The block diagram of modulation control system

In Fig. 1 and Table 1 A+, B+, C+ and A−, B−, C− are the switches of upper and lower groups of the bridge converter, respectively. States from 1 to 6 are active space vectors but other three are zero (see Table 1). The resulting current vector in every sector is produced by pulse width modulation (PWM) of two active vectors and one of the zero vectors (see Fig. 1).

The converter control system can be divided in two subsystems: the first subsystem calculates modulating currents, and the second subsystem generates the control signals to converter's switches. In Fig. 2 the block diagram of modulation control system is represented. The input signals of this subsystem are reference current signals and the inductive storage current. After scaling the reference signals have the amplitude less than unity. Then it is calculated the sector number in which there is reference vector in the current time. Control signals to converter's switches are produced by means of two PWM modulators. Modulating signals are calculated using signals a and b, which are proportional to relative operating times of active vectors during PWM period (see Fig. 1). In every sector these signals are determined by using the transformation from three-phase system to two-phase (a,b,c to α,β) as

$$\begin{cases} i_\alpha = i_a + i_b \cdot \cos\left(\frac{2\pi}{3}\right) + i_c \cdot \cos\left(\frac{4\pi}{3}\right), \\ i_\beta = i_b \cdot \sin\left(\frac{2\pi}{3}\right) + i_c \cdot \sin\left(\frac{4\pi}{3}\right) \end{cases} \quad (1)$$

The main feature of the modulation control system is that modulators control the switching of vectors, but not the switching of single switches as in voltage-source converters. The generator of control signals (see Fig. 2) is the programmable logic scheme that converts the output signals of PWM module to the switch control signals.

This simplifies the converter control by space-vector technique. The commutation between different vectors is provided through switching vectors. This ensures the continuity of inductive storage current.

Developed principles and algorithm of control signals producing are realized using a microcontroller. In Fig. 3, it is shown diagrams that explain operating principle of the modulation control system when the reference currents are sinusoidal signals with period T. Width of vector impulses that are generated by pulse-width modulators, determine the operation time of active vectors on every carrier period. In every sector the operation intervals of one of the active vectors during the PWM period is decreased, but operating times of another active vector is increased. Zero vectors operate when there are no active vector impulses.

B. Calculation of reference current signals

The calculation of active and reactive components of modulating currents is produced by the method of instantaneous power known as the "Akagi-Nabae theory" [3]. This theory also known as pq-theory based on Clark transformation, i.e. abc/αβ. According to it forward abc/αβ transformation is as follows:

$$\begin{bmatrix} u_\alpha(t) \\ u_\beta(t) \end{bmatrix} = \sqrt{\frac{2}{3}} \cdot \begin{bmatrix} 1 & -1/2 & -1/2 \\ 0 & \sqrt{3}/2 & -\sqrt{3}/2 \end{bmatrix} \cdot \begin{bmatrix} u_a(t) \\ u_b(t) \\ u_c(t) \end{bmatrix}. \quad (2)$$

Matrix (2) coefficient can be chosen ($\sqrt{2/3}$ here). From (2) the resulting vector is:

$$\bar{u} = u_\alpha + j \cdot u_\beta. \quad (3)$$

Consequently, reverse Clark transformation is as follows:

$$\begin{bmatrix} u_a(t) \\ u_b(t) \\ u_c(t) \end{bmatrix} = \sqrt{\frac{2}{3}} \cdot \begin{bmatrix} 1 & 0 \\ -1/2 & \sqrt{3}/2 \\ -1/2 & -\sqrt{3}/2 \end{bmatrix} \cdot \begin{bmatrix} u_\alpha(t) \\ u_\beta(t) \end{bmatrix}. \quad (4)$$

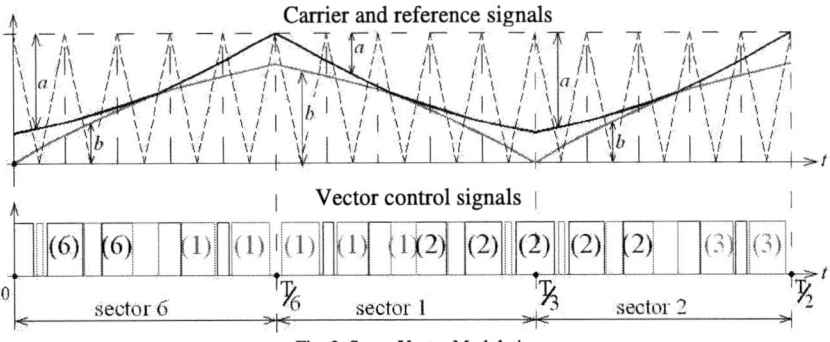

Fig. 3. Space Vector Modulation

429

According pq-theory one can show real and imaginary power in αβ frame as follows:

$$\begin{bmatrix} p(t) \\ q(t) \end{bmatrix} = \begin{bmatrix} u_\alpha(t) & u_\beta(t) \\ u_\beta(t) & -u_\alpha(t) \end{bmatrix} \cdot \begin{bmatrix} i_\alpha(t) \\ i_\beta(t) \end{bmatrix}. \tag{5}$$

Real and imaginary power can be represented as:

$$\begin{aligned} p &= \overline{p} + \tilde{p}, \\ q &= \overline{q} + \tilde{q}, \end{aligned} \tag{6}$$

where $\overline{p}$ and $\overline{q}$ are average components that correspond to fundamental active and reactive power, consequently; $\tilde{p}$ and $\tilde{q}$ are alternative components, caused by harmonics.

Thus, reference current signals i_α^*, i_β^* from this can be received:

$$\begin{bmatrix} i_\alpha^* \\ i_\beta^* \end{bmatrix} = \frac{1}{\left(u_\alpha^2 + u_\beta^2\right)} \cdot \begin{bmatrix} u_\alpha & u_\beta \\ u_\beta & -u_\alpha \end{bmatrix} \cdot \begin{bmatrix} p^* \\ q^* \end{bmatrix}, \tag{7}$$

where p^* and q^* are reference active and reactive power signals. Reference current signals in three-phase system are calculated by reverse Clark transformation.

One of the main advantages of pq-theory is that there is no need for phase synchronization with system voltages. Also, average components p and q can be easily detected (filtered) by means of low-order low-pass filters. The latter makes positive impact on system dynamic and stability.

In Fig. 4 the converter control system block diagram is shown. After voltages and currents abc/αβ transformation in stationary frame instantaneous real and imaginary power is calculated. Then the reference signals of active and reactive power are determined. The average values of instantaneous power are detected by means of low-pass filters (LPF). Then reference current components calculation in αβ-frame takes place. Reference current signals are transformed to modulating current signals that are received by modulation control system. Also, the average value of inductive storage current i_L is the input

signal of modulation system. This subsystem uses space-vector technique to generate control signals for converter switches.

III. MODELING OF FUNCTIONING IN TYPICAL OPERATING MODES

The model of the power converter circuit with inductive energy storage was done in the program MatLab/Simulink. The carrier frequency of PWM was 900 Hz. The simulation results are given for two operating modes: reactive power generation and damping oscillations of active power in the system. The attention was focused on the determination of the control system performance.

The model has following main elements: the grid as ideal three-phase source with voltages equal to 10 kV, frequency equal to 50 Hz and equivalent source inductance (1 mH), the superconductive inductor as reactor with inductance equal to 1 H and initial value of the storage current 500 A.

It is shown the results of simulation for mode of changing of active load in Fig. 5. On these diagrams, one can see that during the first period the consumed active power is the nominal equal to 6,7 MW. Then the step-like increasing of the load occurs. Converter provides deficit of power owing to energy stored in inductor. After it, the load decreases to 4,8 MW. In this period converter operates in the rectifier mode charging the reactor on the DC side. At the moment $t=0,14$ s the load current becomes nominal and converter active power equals to zero. Results of simulation confirm well converter's performance in dynamic modes. It is effect of digital control and principles of calculation reference currents in αβ-frame with pq-theory during transient processes.

The converter has good performance in mode of reactive power compensation. In Fig. 6 it is given changing of the system reactive power and reactive power of the converter, connected to the system, after connection of inductive load (at the moment $t=0,02$ s). The converter response in the mode of full inductive power compensation is no more than two periods of system frequency.

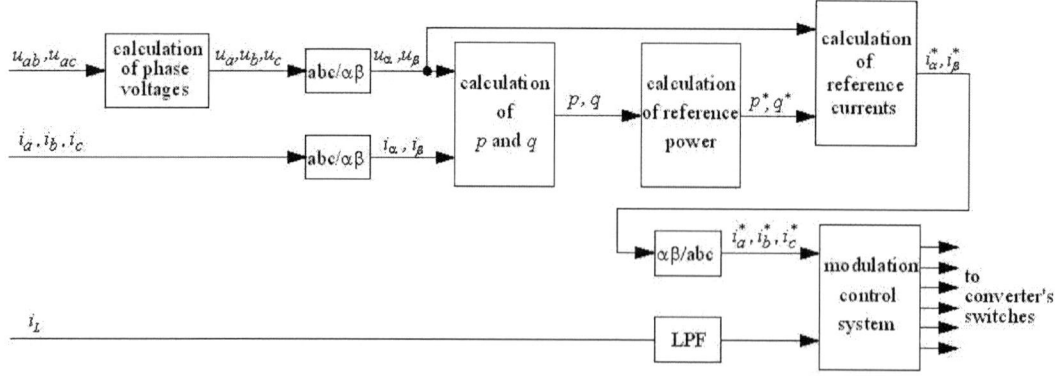

Fig. 4. The control system block diagram

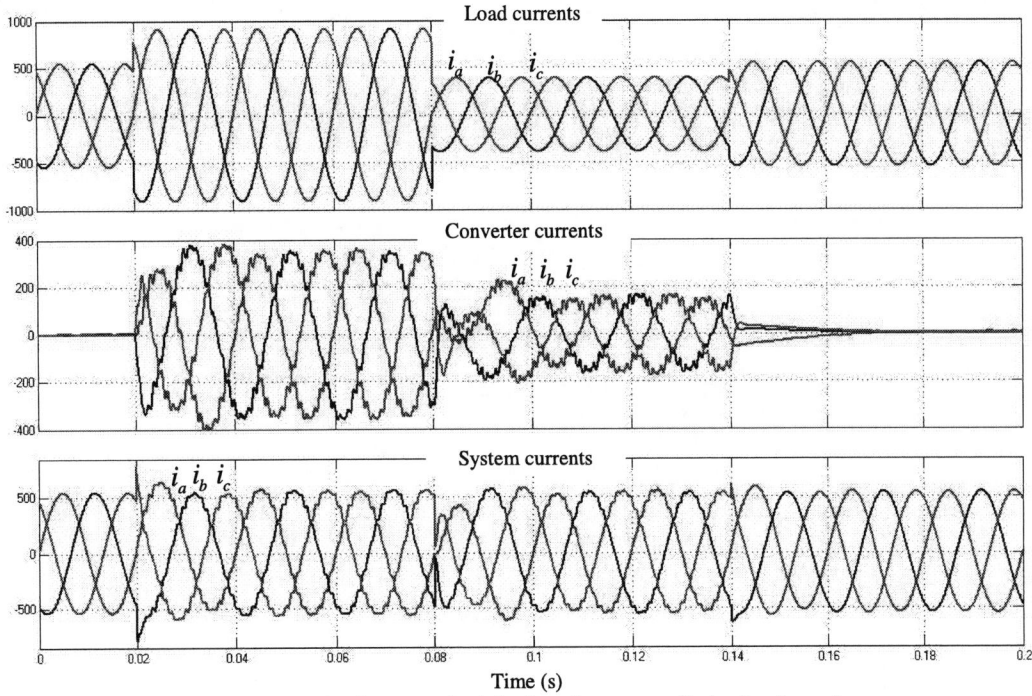

Fig. 5. The results of converter simulation for active power oscillation damping mode

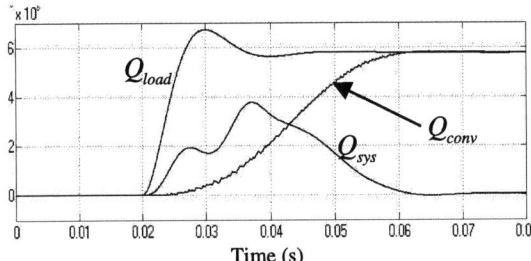

Fig. 6. Load reactive power (Q_{load}), converter reactive power (Q_{conv}) and reactive power of system (Q_{sys}).

IV. CONCLUSION

Recently it is growing the interest to superconducting magnetic energy storage. It stores the power for using it to provide adding power during overloads and other necessary cases. The power converter is used for connection between energy storage and power system. Developing of full controlled devices helps to decide the problem of power interchange successfully.

To ensure such performance as high operating speed and accuracy it is need to use new control methods. In this case the using of digital control and space-vector PWM give more advantages. But number of the combinations of switches states in three-phase current-source converter is more than the same in three-phase voltage converter. It was taken into account in the development of control algorithm. The control system based on space-vector modulation provides the utilization of inductive energy storage with maximum efficiency, accurate and reliable control. The results of modeling justified it. The control system is realized with PWM microcontroller that allows simplifying the vector control.

The calculation of active and reactive components of modulating currents is produced by the method of instantaneous power known as "Akagi theory". This ensures the fast response of the SMES converter in changing of active and reactive system power.

Some parts of the current-source converter were created on experimental model. Modeling the main processes in the converter was shown that the using of IGBT transistors as power switches with 16-bit microcontroller could give satisfactory converter performance in low and middle range of power.

REFERENCES

[1] C. Hsu and W. Lee, "Superconducting magnetic energy storage for power system applications," *IEEE Trans. Industry Applications,* vol. 29, no. 5, pp. 990-996, 1992.

[2] K. Zhou and D. Wang. "Relationship between space-vector modulation and three-phase carrier-based PWM: a comprehensive analysis", *IEEE Trans. Industrial Electronics*, vol. 49, no. 1, 2002.

[3] H. Akagi, Y. Kanazava, A. Nabae. "Generalized theory of the instantaneous reactive power in three-phase circuits". – Proceedings of the 1983 International Power Electronic Conference, Tokyo, Japan, 1983.

FPGA-based Controllers for Switching Converters

Karel Jezernik

University of Maribor / FERI, Maribor, Slovenia, *karel.jezernik@uni-mb.si*

Abstract—Nowadays, most digital control for motion and power electronics systems are based on DSPs. This paper presents a FPGA-based digital control for a DC-DC converter. The main difference from DSP-based solution is that FPGA allows simultaneous execution of all control procedures, enabling high performance and novel control methods. The control algorithm has been developed using a VHDL language or CAD state graphical approach based on the use of logical state diagram. The FPGA switching controller has been designed as simple as possible while maintaining good accuracy and dynamic response. Simulations and experimental results show the feasibility of the proposed method, opening interesting possibilities in motion and power electronics converter control.

Keywords: Converter control, Power supply

I. INTRODUCTION

Traditionally, power electronics circuits and systems have been controlled in industry using linear controllers combined with non-linear procedures like pulse width modulation (PWM). The models used for controller design are results of simplifications that include averaging the behavior of the system over time (to avoid modeling the switching) and linearizing around a specific operating point disregarding all constraints. To make the system operate in a reliable way for the whole operating range, the control circuits is subsequently augmented by a number of heuristic patterns. This procedure requires large development times and lacks theoretically backed guarantees for the operation of the system [1].

Switch mode dc-dc converters are switched circuits that transfer power from a dc input to a board. They are used in a large variety of applications due to their light weight, compact size, high efficiency and reliability. Since the dc voltage at the input is unregulated and the output power demand changes significantly over time (resulting in time – varying load), the control objective is to achieve output voltage regulation in the presence of input voltage and output load variation [2],[3],[4].

Hysteresis control is a fast and robust control approach but it is mostly applied only to simple converter systems. Considering such controllers as discrete-event systems opens a more systematic view and enables the controller design even for nonlinear systems. The goal of this paper is to continue to a more systematic design of hysteresis controllers from the view point of discrete-event systems so that it will be possible to benefit from their high robustness and dynamic response. A very interesting item is the controller hardware: event-driven controllers can

advantageously be realized on Field Programmable Gate Array (FPGA) with low effort.

Many authors report the use of FPGA in the field of switching converters recently [2],[5]. Shorter development cycles, lower costs and higher density are reported. Most authors experienced difficulties when migrant from sequential executed software of sampled control algorithm to parallel executed hardware of FPGA. Most of today control software is developed in Matlab, which uses floating point type for calculation purposes. Contrary to Matlab, an HDL modeling system used in design entry and simulation process by FPGA programming is based on using the variables that represent logic values. A power mapping between FPGA hardware with HDL software and mathematical model of experimental switching converter and nonlinear switching control of it, is extremely important.

However, with FPGA implementation, designer has the difficult task to characterize and describe the hardware architecture corresponding to the chosen control algorithm. FPGA designers must follow an efficient design methodology in order to benefit from the advantages of the FPGA and their powerful CAD tools. From software point of view, HDL modeling system is based on using the variables that request logic values.

II. CONTROL OF DC-DC BUCK CONVERTER

A. A Problem Statement

Fig. 1 shows the reference example used along the paper. The power stage is a buck converter. The buck converter parameters are the following: input voltage $U_d =$ 12 V, output voltage $u_o = 5$ V, $L = 400$ μH, $C = 100$ μF, load resistance is between $R_{min} = 5$ Ω and $R_{max} = 10$ Ω, and switching frequency $f_s = 1$ MHz.

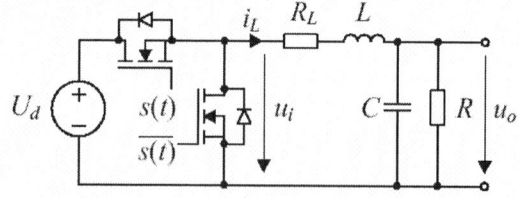

Fig. 1. Buck converter.

The control objective for dc-dc converters is to regulate the dc component of the output voltage to its reference. This objective has to be achieved subject to the constraints that are present, resulting from the converter topology. In particular, the manipulated variable (duty cycle) is

978-1-4244-1741-4/08/$25.00 ©2008 IEEE

bounded between zero one, and in the discontinuous current mode a state (inductor current) is constrained to be non-negative. Additional constraints are imposed as safety measures, such as current limiting or soft-starting, where the latter constitutes a constraint on the maximal derivative of the current during start-up. Moreover, the regulation has to be maintained despite gross changes in the load and the input voltage. We aim at developing novel control schemes that account for the complex properties of these systems and address the aforementioned control objectives [6].

B. Discrete-Event Control

A static power converter is a network compared of active electromagnetic components such as diodes, thyristors, transistors and by passive elements (R, L, C). Its function is to realize an energy transfer by switching the nonlinear elements thereby modifying the interconnection constraints at its work. There exist in two main control approaches:

- linear average control method with mainly open loop PWM modulation of energy transfer

- nonlinear switching control method like use of hysteresis controller [7],[8].

The average linear control approach is mainly used in today's solutions. The nonlinear switching approach deals explicitly with the variable interconnectivity of the converter by considering the nonlinear components being ideal switches. In this case the switching device is considered as a power-continuous element which improves a connections between energy source and corresponding loads.

In this paper we propose an event-driven models and analysis for the systematical enumerator of the possible configurations in the static power converters. This analysis is carried out by considering the switches as ideal (two binary states: On and OFF) and taking into account the commutations constraints. The dynamical discrete-event behavior of the converter are described by Petri nets. The states of the Petri nets are defined locally for the ideal switches [2],[9].

Switching converters present different network configurations. Applying Kirchoff's laws to each network configuration, a continuous time state-space equation is obtained for each configuration. The conditions for the transition from one configuration to another are included in the model which requires a prior knowledge of the converter operation. Thus, the power converter is described by a set of discrete states (configurations) with associated continuous dynamics. Assuming that the buck can operate in the continuous (CCM) and discontinuous (DCM) conduction modes, it has three configurations: *P1*, *P2*, and *P3*.

In configuration (Fig. 2) *P2*, the ideal switch $s(t)$ is closed and the $\overline{s}(t)$ is off. The dynamics of the inductor current $i_L(t)$ and the capacitor voltage $u_C(t)$ are

$$L\frac{di_L(t)}{dt} = U_d - u_0, \quad C\frac{du_C(t)}{dt} = i_L - \frac{u_0}{R} \quad (1)$$

Taking into account that capacitor voltage $u_C(t)$ and output voltage $u_0(t)$ are related via the following equation:

$$u_0 = u_C + R_C\,C\frac{du_C(t)}{dt}. \quad (2)$$

The following state-space model describes the idealized buck converter in configuration *P2*.

$$\frac{d}{dt}\begin{pmatrix} i_L \\ u_C \end{pmatrix} = \begin{pmatrix} \dfrac{-R_C\,R}{L(R+R_C)} & \dfrac{-R}{L(R+R_C)} \\ \dfrac{R}{C(R+R_C)} & \dfrac{-1}{C(R+R_C)} \end{pmatrix}\begin{pmatrix} i_L \\ u_C \end{pmatrix} + \begin{pmatrix} \dfrac{U_d}{L} \\ 0 \end{pmatrix}. \quad (3)$$

In configuration (Fig. 1) *P3*, $s(t)$ is open and $\overline{s}(t)$ is on. The following state-space model describes the idealized buck converter in configuration *P3*.

$$\frac{d}{dt}\begin{pmatrix} i_L \\ u_C \end{pmatrix} = \begin{pmatrix} \dfrac{-R_C\,R}{L(R+R_C)} & \dfrac{-R}{L(R+R_C)} \\ \dfrac{R}{C(R+R_C)} & \dfrac{-1}{C(R+R_C)} \end{pmatrix}\begin{pmatrix} i_L \\ u_C \end{pmatrix}. \quad (4)$$

In configuration (Fig. 1) *P3*, $s(t)$ is open and $\overline{s}(t)$ is off, the dynamics of the inductor current $i_L(t)$ and the capacitor voltage $u_C(t)$ are

$$i_L(t) = 0, \quad C\frac{du_C(t)}{dt} = -\frac{u_0}{R} \quad (5)$$

Fig. 2 shows the state diagram with the conditions that govern the transitions from one configuration to other. It is assumed that the $s(t)$ is closed when $sw = '1'$ and the $s(t)$ is open when $sw = '0'$.

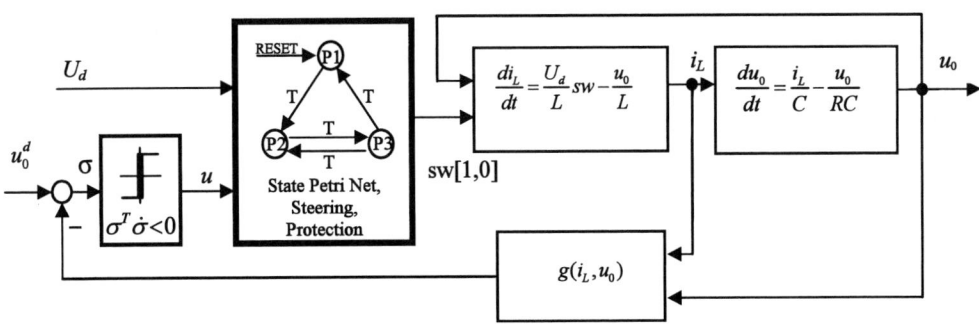

Fig. 2. Configuration state diagram for buck converter.

433

III. DES STABILITY ISSUE AND SWITCHING

A. Discrete-Event Controller

The common methods of modeling for switching converter are differential equations and difference equations. For Discrete-Event System (DES) the main method are state diagram and Petri net models. Coupling of discrete logical variable, represented with the transistor switch position and continuous state variable of control plant in hybrid systems is difficult to model. This problem can be solved by graphical state Petri net which represents both continuous and discrete parts of a hybrid system, in proposed paper DC-DC buck converter, via voltage vectors of state space modulator shown in Fig. 2.

The hierarchical model on Fig.1 consists of a DES controller, a continuous control process system and an interface which includes transistor switch (sw) and event recognizer. Controller determines discrete states (P1, P2, P3) according to specific rules and controls the process of continuous variable system. Control switch sw - transistor, is the executive body of DES controller, which chooses different continuous process according to DES controller orders. The event recognizer via Lyapunov stability issue ($\sigma \dot{\sigma} < 0$) reflects the real-time state of controlled system to DES controller in the form of logic value, and drive the change of DES state.

B. Calculating of Switching Frequency

Ideally, the converter will switch at infinite frequency. In presence of switching implementations (switching time constant and delay), this is not possible. The discontinuity in the feedback control produces a particular dynamic behavior in the vicinity of the surface trajectory known as chattering. If chattering is uncontrolled, the converter system will be self-oscillating at a very high switching frequency corresponding to the chattering dynamics.

This phenomena is main obstacle in using ideal sliding mode control in practical applications. This is undesirable as high switching frequency results in excessive switching losses, inductor and transformer core losses and EMI noise issue. Furthermore, with switching frequency being unpredictable, the design of the converter and the selection of components will be difficult.

In proposed discrete-event converter control, the control law is redefined with introduction in event controller a hysteresis band with the boundary condition K and $-K$ to create a dead region $-K \leq sw \leq K$ where no chattering of switching state can occur. The maximum switching frequency of discrete-event controller can therefore be controlled by varying K.

To control the switching frequency of the converter, the relationship between the hysteresis band K and switching frequency f_s must be known. From [9]:

$$\Delta t_{OFF} = 2 K L / u_o, \tag{6}$$

$$\Delta t_{ON} = 2 K L / (U_d - u_o), \tag{7}$$

where Δt_{OFF} and Δt_{ON} are respectively the turn-off and turn-on time duration of the switch. Therefore, the total period for one switching cycle is

$$T = \Delta t_{OFF} + \Delta t_{ON} = \frac{2 K L}{u_o (1 - u_o / U_d)}. \tag{8}$$

Since the cycle is repeated (cyclic) throughout the steady-state operation, the frequency of the converter can be expressed as

$$f_s = \frac{1}{T} = \frac{u_o (1 - u_o / U_d)}{2 K L}. \tag{9}$$

Considering that U_d and u_o are non-constant parameter consisting of respectively DC signals of $\overline{U}_d$ and $\overline{u}_o$ and time varying perturbations of $\tilde{U}_d$ and $\tilde{u}_o$, we can insert (9) into $f_s = \overline{f}_s + \tilde{f}_s$ using small signal approximation, where

$$\overline{f}_s = \frac{1}{T} = \frac{\overline{u}_o (1 - \overline{u}_o / \overline{U}_d)}{2 K L} \tag{10}$$

$$\tilde{f}_s = \frac{1}{T} = \frac{\tilde{u}_o (1 - 2\tilde{u}_o / (\overline{U}_d + \tilde{U}_d))}{2 K L} \tag{11}$$

with $\overline{f}_s$ representing the steady-state (nominal cyclic) switching frequency and $\tilde{f}_s$ representing the AC varying frequency of converter.

C. Switching Control

Switching control is a fast and robust control approach but it is mostly applied only to simple converter systems. Considering such controllers as discrete-event systems opens a more systematic view and enables the controller design. There are very valuable sliding mode approaches for the design of the sliding surface in order to ensure convergence or stability [3]. However, the approach does not pay much attention to a particular switching action and does not solve the condition problem of converter.

Considering a hysteresis controller as discrete-event dynamical system allow to forms in much more details on the switching actions and will enable a better understanding of the controller design. A discrete-event system reacts only if an event is recognized.

The task of the controller is to switch the driving voltage up or down, depending on the encountered event, in order to force the inductor current or output voltage u_o back into the tolerance band. With a simple hysteresis, the controller output is immediately the converter firing command so that controller state is identical with the controller output. This is not the case valid for discrete-event systems.

The set of possible controller states may be larger from the possible controller outputs. Distinguishing between controller output values and internal controller states is main contribution of discrete-event control approach. In proposed DC-DC converter controller there are in use three (P1, P2, P3) controller internal states and only one controller output (sw), Fig. 2.

The events are determined with voltage error as difference between reference and actual output voltage

$$\Delta u_o = u_o^d - u_o, \tag{12}$$

or inductor current control error as difference between allowed maximum and actual inductor current

$$\Delta i_L = i_{L\max} - i_L. \tag{13}$$

For stability issue both voltage and inductor current error form a Lyapunov function

$$V = \left[\Delta i_L, \Delta u_o\right]^T. \tag{14}$$

The derivative of control errors are

$$\frac{d}{dt}\left[\Delta i_L, \Delta u_o\right]^T = \left[-\frac{d\,i_L}{dt}, -\frac{d\,u_o}{dt}\right]^T, \tag{15}$$

which can be expressed with controller action $u = U_d \cdot sw$
for $sw = [1, 0]$.

The stability of buck converter will be satisfied if the derivative of Lyapunov function will have for any switch position negative value

$$\dot{V} = \left[\Delta i_L, \Delta u_o\right]^T \begin{bmatrix} (U_d\,sw - u_o)/L \\ -i_c\left(u_o^d - u_o\right)/C \end{bmatrix} < 0. \tag{16}$$

The derivative Lyapunov function will be determined with alternative switching between event determined with voltage and current error. The current controller take a role of over current protection and voltage controller execute regulation task of DC-DC converter. Fig. 3 shows the circuit diagram of voltage and current controller.

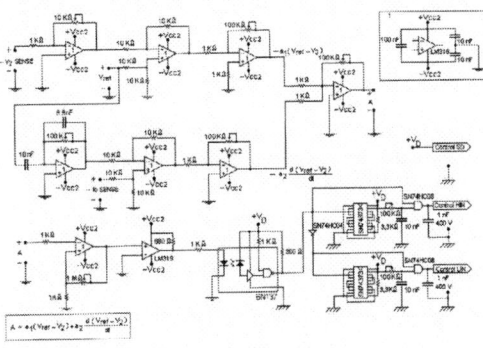

Fig. 3. Voltage and current control circuit.

IV. IMPLEMENTATION AND PROTECTIONS ISSUE

Substituting the common DSP solutions by FPGA-based ones means a trade-off between the DSP capacity for arithmetic operations and the FPGA concurrency. In order to explicit the FPGA concurrency new control algorithms must be developed, because adapting the DSP-oriented to FPGA would mean no special advantage. These new algorithms can be quite simple, like the switching control proposed, but they must be designed from the concurrency point of view.

An FPGA-based solution changes the design point of view. Arithmetic operations should be kept to the minimum to optimize the required logic resources. However, conditional execution should be exploited because of its hardware oriented nature. Conditional execution is also especially suitable for implementing protections, monitoring, steering, etc, because they only accurate under specific conditions and express logical event driven behavior.

An important advantage of using an FPGA for the controller is that some additional functions like protections, steering, monitoring etc., can be added with no additional resources and almost no drawback in performance. The FPGA concurrency allows to execute the logic dedicated to the main control and any additional logic denoted to protections simultaneously. In this way, there is almost no drawback in the controller performance because the control logic is executed as if there were no protections. Even more, the protections are executed continuously, instead of the periodic execution in a DSP. DSP solutions must keep the protections resources at a minimum because the main control is stopped while a protection is verified. In fact, any algorithm can be added to control one while there are available resources.

In the proposed FPGA controller following logic protections have been implemented:

- over- and under voltage, a limit to the output voltage. Whenever this limit is reached the switch is kept off in order to avoid possible damages.

- Over current, a limit to the input inductor current. This is a limit to the instant input current. Again, the switch is kept off when this limit is reached.

- Over temperature.

- Steering, enable, reset.

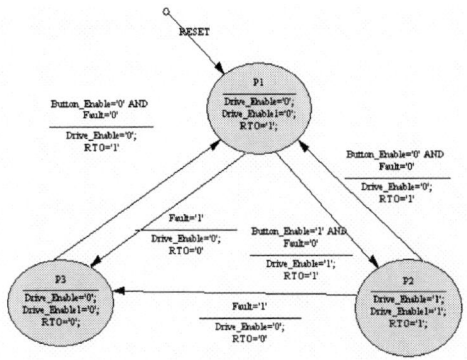

Fig.4. State CAD diagram.

V. RESULTS

A. Simulation Results

The DC-DC buck converter with discrete-event control was simulated in Matlab/Simulink environment. Some transients of inductor current i_L output dc voltage u_0, transistor-switch current i_{tr} and diode current i_d are presented in Fig. 5. The open loop output voltage transient exhibits high oscillation due resonance effect of LC elements in circuits by sudden output load change from high to low value, but closed loop behavior of it is asymptotic stable.

B. Experimental Results

The proposed approach is based on a fast parallel processing and is suitable for a FPGA implementation [7]. In such implementation it would be possible to reproduce an ideal sliding mode process.

The hardware target used for implementation is Xilinx Spartan 3 FPGA (Fig.6). The implemented architecture in form of state CAD diagram in the Integrated Software Environment (ISE^{TM}) that allows to take design from design entry through Xilinx device programming. State CAD on Fig. 4 is developed by mapping of PN-graph of DES controller on Fig. 2. After validating a diagram, State

CAD automatically generates simulatable and synthesizable HDL code directly from the state CAD diagram. Fig. 7 shows transients of change impact of output load on output voltage.

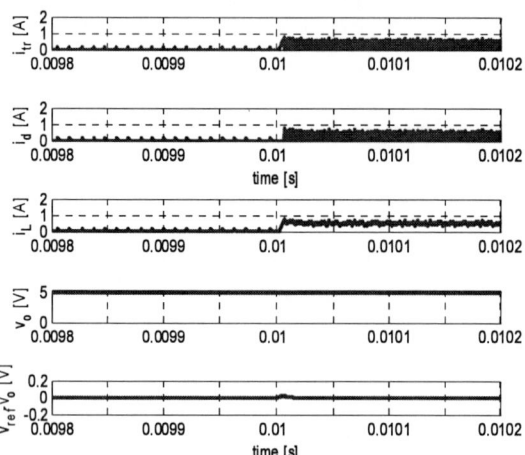

Fig. 5. Simulation results of discrete-event control.

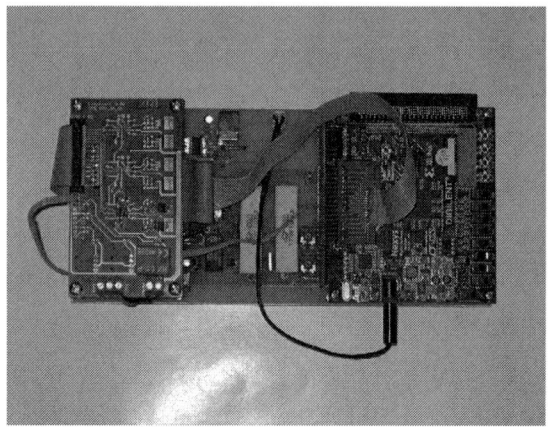

Fig. 6. Experimental system.

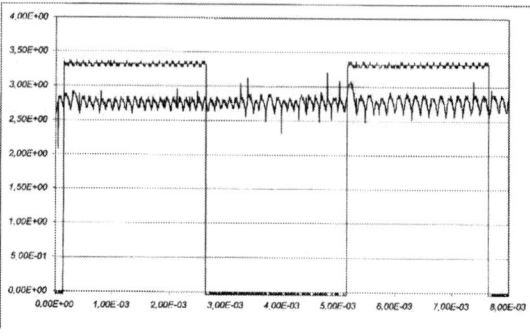

a)

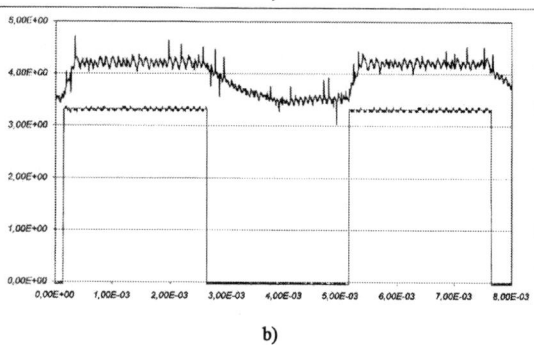

b)

Fig. 7. Experimental results by load change: a)voltage control, b)voltage and current control.

VI. CONCLUSION

A full digital controller for switching converters has been proposed. The most important difference from previously proposed digital controllers is that it is based on specific FPGA hardware instead of the common DSP solutions. The main advantage of this method is that the all the logic is executed continuously and simultaneously (concurrent operation) and new high speed algorithms can be used in this way.

However, complex arithmetic operations should be avoided whenever possible, because they need many resources. The other main characteristic of the switching controller is: simplicity.

Implementing the control algorithm in a hardware description language (HDL) allows high flexibility and technology independence. The same controller can be directly synthesized into any other FPGA or even in an ASSIC, or it can also be added to other logic blocks forming a more complex multi-task system in a single chip.

Solutions based on specific hardware, that allows high concurrency, are suitable to be used in power electronics and motion control applications.

REFERENCES

[1] J.G. Kassakian, M.F. Schlecht, and G.C. Verghese, *Principles of Power Electronics*. New York: Addison-Wesley, 1991.

[2] A. Polič and K. Jezernik, "Event-driven Current Control Structure for a Three Phase Inverter", *International Review of Electrical Engineering*, vol. 2, no. 1, pp. 28-35, jan.-Feb. 2007.

[3] V.I. Utkin, "Sliding Mode Control Design Principles and Applications to Electric Drives", *IEEE Transactions on Industrial Electronics*, vol. 40, no. 1, pp. 23-36, February 1993.

[4] M. Milanovič and D. Gleich, "Buck Converter Digitally Controlled by a Fuzzy State-space Controlled", *Journal of Science and Engineering. Series B, Applied Sciences and Engineering*, vol. 2, iss. 5-6, pp. 638-654, 2005.

[5] S. Berto, A. Paccagnella, M. Ceschia, S. Bolognani, and M. Zigliotto, "Potential and pitfalls of FPGA application in inverter drives - a case study". IEEE Int. Conf. on Industrial Technology, ICIT'03, vol.1, pp. 500-505, Dec. 2003.

[6] R. Venkataramanan, A. Sabanovic, and S. Cuk, "Sliding mode control of DC-to-DC converters", Proc. IEEE IECON, pp. 251-258, 1985.

[7] J.J.E. Slotine and W. Li, *Applied Nonlinear Control*. Englewood Cliffs, New Jersey; Prentice Hall, 1991.

[8] H. Sira-ramirez, "Sliding motion in bilinear switched networks", *IEEE Trans. Circuits Syst. I, Fundam. Theory Appl.*, vol. 34, no. 8, pp. 919-933, Aug. 1997.

[9] A. Polič and K. Jezernik, "Closed-loop Matrix based Model of discrete Event Systems for Machine Logic Control Design", *IEEE Transactions on Industrial Informatics*, vol. 1, no. 1, pp. 39-46, February 2005.

Gamesa DAC converter: the way for REE grid code certification

Itziar Martinez[*], Daniel Navarro[**]

[*] Gamesa Innovation & Technology, Zamudio, Spain e-mail: *itmartinez@gamesacorp.com*
[**]Gamesa Innovation & Technology, Pamplona, Spain, e-mail: *dnavarro@gamesacorp.com*

Abstract— On October 2006 REE (Red Electrica de España), one of Gamesa's major markets, publishes a new code version with new requirements regarding voltage ride through capabilities. All new windturbines installed from January 2008 on must fulfill this last version of the grid code. This issue becomes during 2007 a clear target for Gamesa: to get certification for this grid code. One of the determining active elements in this certification is the converter. DAC (Dip Active Converter) is the first self-made converter in Gamesa. For this converter the development process is presented: all the way from the first steps of the design to the final certification. Special emphasize is done in the development of the simulation model which follows a parallel process to the converter.

Keywords— Converter Control, Doubly Fed Induction Motor, Fault handling strategy, Wind Energy, Windgenerator systems.

Fig. 1.   Gamesa wind turbine during certification process

I. INTRODUCTION

As wind power generation is getting important percentages of the energy production, grid integration becomes a major issue. This results in a redefinition of the grid codes in many countries to cover this new scope. These new grid code versions require not only to keep connected during the fault, also to generate active or reactive power or even both.

In order to fulfill one of these new requirements, wind power manufacturers have to focus on new designs that are voltage ride-through capable. This is the case for the Gamesa DAC (Dip Active Converter) for the turbine with an electrical configuration DFIM (Double Fed Induction Machine).

As Gamesa first main market is the Spanish one and its grid code is mandatory for wind turbines installed from January 2008 on, the first aim of the DAC converter is to fulfill REE (Red Electrica de España) requirements.

Gamesa DAC converter is the first own design converter that goes through the certification process. In order to assure the certification success, an internal validation of the design is held before certifying. This internal validation is not only focused on the converter itself, but in its simulation model also.

Fig. 1. shows a Gamesa wind turbine during the certification process. For the certification, a dip generation unit is installed in the wind turbine. This unit performs the dip profiles stated in the certification process.

II. REE FULFILLMENT

There are two main documents related to the compliance of REE requirements. The first one is the grid code itself [1]. The second one is published by the AEE (Asociacion Empresarial Eolica) that describes the procedure for verification, validation and certification according to the grid code [2].

The document from AEE is very important as it states clearly the test that will be done to obtain the certification, translating the requirements in the Point of Common Coupling (PCC) to requirements in windturbine or FACT.

The procedure for verification, validation and certification describes two possible processes to get the certification. One is named "general process" and the other one is the "particular process". Both processes include voltage dip tests in windturbine.

The general process has the same requirements as specified in the grid code and requires a simulation model of the wind turbine. The particular process does not require a simulation model, but has some additional requirements and more restrictive limits.

REE grid code classifies the voltage dips into two groups: symmetrical and asymmetrical. Each of these groups has to comply with different requirements. Fig. 1. shows the limit curve for the phase to neutral voltage value for symmetrical dips. The generating unit has to ride-through voltage dips in the shadowed area. The same limit curve applies for asymmetrical dips changing the 0.2 pu limit to 0.6 pu. The value pu is related to the nominal value of the voltage measured as phase neutral value.

978-1-4244-1741-4/08/$25.00 ©2008 IEEE

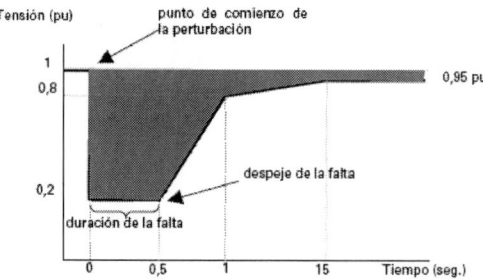

Fig. 2. Time-voltage curve that defines the voltage dip where the voltage ride through capability has to be assured.

The document from AEE defines three time zones for the voltage dip, as it is shown in Fig. 3. Zone A begins when the dip starts and lasts 150ms. Zone C begins in the fault clearance and lasts the shortest until the voltage recovers normal working condition or in 150ms. Zone B is the time between zone A and Zone C.

For the requirements evaluation, positive sequence fundamental values for active and reactive power and current have to be calculated. They are calculated according to the IEC 61400-4-20 [3].

A. Requirements for symmetrical dips

In the case of symmetrical dips, reactive current is demanded during zone B according to Fig. 4. . This reactive current in pu. is related to the total current during the dip. This is equivalent to a "cos phi" criteria during the dip.

The requirements for general and particular processes are shown in TABLE I. Apart from the reactive current in zone B, there are requirements regarding active and reactive power consumptions every 20ms, reactive energy consumption (Er) every 20ms and reactive current (Ir) every 20ms.

B. Requirements for asymmetrical dips

In the case of asymmetrical dips, reactive current is not demanded during zone B.

For asymmetrical dips, both general and particular processes have the same requirements. These are shown in TABLE II. There are requirements regarding active and reactive power consumptions every 20ms, active energy consumption (Ea) every 20ms and reactive energy consumption (Er) every 20ms.

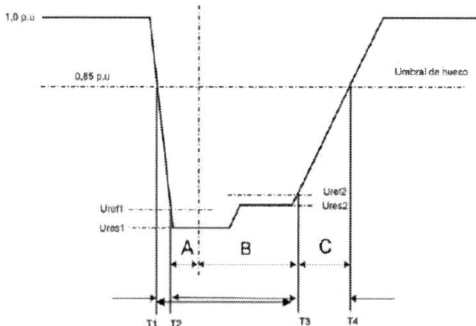

Fig. 3. Time zones in the voltage dip for requirements evaluation

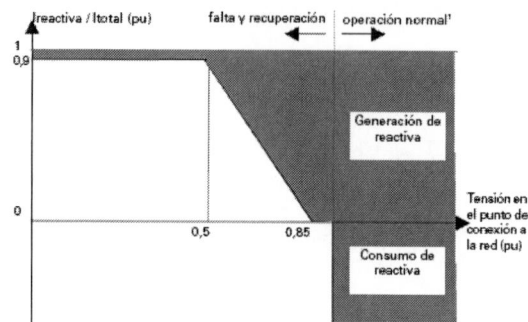

Fig. 4. Reactive current demand for symmetrical dips

TABLE I. SYMMETRICAL DIPS REQUIREMENTS ACCORDING TO GENERAL AND PARTICULAR PROCESS

SYMMETRICAL FAULTS		General process requirements	Particular process requirements
ZONE A	Net consumption Q (20ms)	< 0.6 pu	< 0.15 pu
ZONE B	Net consumption P (20ms)	< 0.1 pu	< 0.1 pu
	Net consumption Q (20ms)		< 0.05 pu
	Average Ir/Itot	Shadowed area fig.3	Shadowed area fig.3
ZONE C	Net consumption Er (20ms)	< 0.09 pu	< 0.09 pu
	Net consumption Ir (20ms)	< 1.5 pu	< 1.5 pu

TABLE II. ASYMMETRICAL DIPS REQUIREMENTS ACCORDING TO GENERAL AND PARTICULAR PROCESS

ASYMMETRICAL FAULTS		General process requirements	Particular process requirements
ZONE B	Net consumption P (20ms)	< 0.3 pu	< 0.3 pu
	Net consumption Q (20ms)	< 0.4 pu	< 0.4 pu
	Net consumption Ea (20ms)	< 0.045 pu	< 0.045 pu
	Net consumption Er (20ms)	< 0.04 pu	< 0.04 pu

III. GAMESA DAC CONVERTER

Gamesa DAC converter is designed to fulfill the grid code requirements regarding voltage dips. The same hardware is used for the compliance of the main grid codes, and just a firmware update is needed to cope with new grid code requirements.

The firmware of Gamesa DAC converter includes the necessary control strategies to maintain the stator of the double fed inductor generator connected to the grid and to comply with the different grid code requirements.

The development process for the converter begins with the development of a simulation platform. This is used for the design of the HW components and also to develop a control strategy to deal with the voltage dips.

IV. VALIDATION PROCESS

The aim of developing a solution for grid faults, validating it and certifying for REE in just one year is a very challenging issue. This aim is achieved following a very tight and optimized validation process that minimizes risks and costs while shortening the whole test duration.

The validation plan is based on two statements. The first one is that risks and change affection increase exponentially as the process goes on. Meanwhile the test condition controllability and unforeseen reaction capability reduces significantly.

The DAC validation process philosophy is shown in Fig. 5. This validation process has a detailed test plan with several stages:

1. Development of the HW and control design through simulation tools
2. Validation in Small Scale Testbench
3. Validation in Real Scale Testbench
4. Validation in Wind Turbine

This four step validation evaluates the technical solution in a progressive and deterministic way. It mitigates risks in the initial phases and allows a gradual approximation to the real context. The aim is to minimize the field works where meteorological events and logistics are adverse and the fundamental parameter, the wind, is not controllable. This validation process has allowed developing, validating and certifying the design in less than one year with only one month of work in wind farm.

This validation process is also necessary to verify the hardware and firmware design against the internal specification. Furthermore, it is necessary to tune and adjust the wind turbine model in different stages.

This wind turbine model is mandatory in order to apply for the general process in the certification. It is also a good description of the system to be shared with the grid utilities.

A. Development of the HW and control design through simulation tools

At this first stage the base of the development is defined (sensors, control strategy, etc) leading to the technical solution configuration.

The important parameters of the electrical model should be defined (saturation of inductances, high frequency effects…).

B. Validation in Small Scale Testbench

In order to validate the system in a small scale testbench, several conditions have to be taken into account:

1. There has to be a scaled down version of the converter hardware, in this case 80 times smaller in power. In the design of the scaled version it is taken into account the voltage level of the grid connection, the power and current limits of the converter and the dynamics of the system. The firmware is adapted to this change of power level.

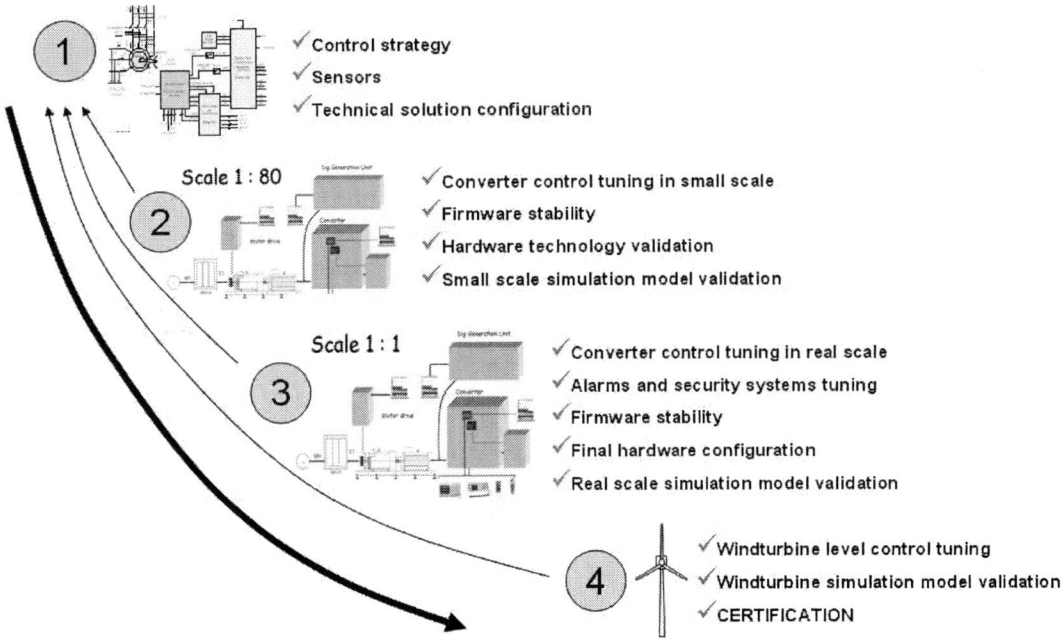

Fig. 5. DAC validation process philosophy

439

2. There has to be a scaled version of the generator that matches the current and power levels of the scaled converter in both stationary and transient functionality.

3. A motor with its control is needed in order to simulate the wind turbine driven by the wind.

4. It is necessary to have an equipment to generate voltage dips. It should be a scaled version of the certification equipment adapted to the voltage level of the grid connection and short-circuit impedance.

This Small Scale validation gives a good overview of the state of the converter and the firmware. It is quicker to test strategies, alarms, state-machine, etc in a scaled down version of the converter. It allows also an initial tuning and validation of the simulation model, as far as the converter is concerned.

This stage of the validation is the fastest one, several tens of dips maybe done in one hour. This gives the control tuning a very quick feedback of the influence of the different parameters. Components appearing in the currents and voltages characteristic of the transient behavior are correlated with the simulation outputs. A systematic way of tuning is gained in small scale test bench.

This is also the opportunity to check several measurements of critical elements during the voltage dip, which is a very noisy situation: rotor currents, DC-voltages, line currents etc. Also the real delays of some safety elements like the brake-chopper, contactors etc...are registered

Energy and power calculation to validate the certification requirements are calculated on-line to have a quick view of the compliance of the system.

C. Validation in Real Scale Testbench

The full scale converter can be tested in a controlled environment using a Real Scale Testbench. This allows a good validation level of the converter against the specification before setting the converter in field. It also permits the validation of the electrical part of the simulation model. In this stage is shown that the relevant parameters are accurately simulated for transient behavior.

In the Real Scale Testbench, the same generator and dip generator equipment as in the wind turbine can be used. In this way, the same simulation models can be also validated before testing them in the wind turbine.

During this stage, the converter control loops are tuned, the parameters for the alarms and security systems are fixed, the firmware is debugged to be error free and the converter hardware is checked.

A very critical point in all the converter strategy is the synchronization system. A deep work has been done in this way to have a stable and robust synchronization to the grid voltage. This is a key factor for the control under very deep voltage dips.

The real scale test bench allows to test easily all the different working points in a controlled environment.

D. Validation in Wind Turbine

Once the validation test results in Real Scale Test Bench are satisfactory, this means that the REE-requirements are proven to be fulfilled, the validation process in Wind Turbine starts.

The main difference with the previous stage is that the converter interacts with the other parts of the wind turbine, mainly with the mechanical systems. This interaction has to be very well defined to minimize generated loads. All the measurement noise is checked during the grid fault, as this stage is one of the ones having higher electrical noise in the system.

In this step of the process the wind turbine simulation model is fully checked in order to apply for the general process in certification.

V. CERTIFICATION PROCESS

The certification process for REE is described in the AEE document [2]. A test point (PE) is defined inside the dip generator equipment where voltage and current measurements are taken in every test and grid code compliment is checked. Fig. 6. shows the schematics of the dip generator unit located between the collector system (SC) and the wind turbine (AEG) or FACTS.

TABLE III shows the voltage dips and wind turbine conditions that are evaluated in the certification process. Each of the categories is tested thrice and the requirements have to be fulfilled in every test.

VI. RESULTS

During the validation process, the same voltage dips as in the certification process are performed with the same requirement evaluation. Results from these dips in the different validation scenarios are presented. In these results, the measurement point for the line voltage and current is located inside the converter. This point does not match with the certification measurement point. Between both points, a star-delta transformer is located.

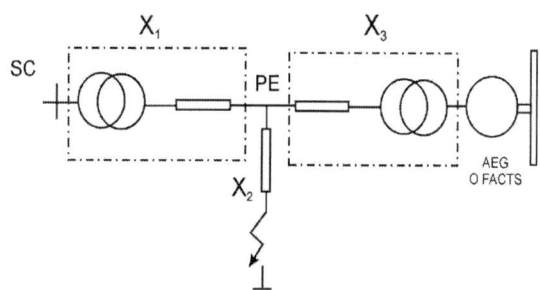

Fig. 6. Test point (PE) inside the dip generator unit

TABLE III. TEST CATEGORIES FOR REE CERTIFICATION PROCESS

CATEGORY	WIND TURBINE CONDITION	DIP TYPE
1	Partial load (10% - 30 % rated power)	Three phase dip with 20% residual voltage
2	Full load (> 80% rated power)	
3	Partial load (10% - 30 % rated power)	Two phase isolated dip with 60% residual voltage
4	Full load (> 80% rated power)	

While testing in Real Scale Test Bench and in Wind Turbine, the same measurement point as in the certification process is accessible. The measurements in this point are the input for an internal software tool that evaluates the fulfillment of the requirements in the grid code. The evaluation results are used to know the fulfillment status before the certification process takes place.

A. Results in Small Scale Test Bench

Fig. 7. and Fig. 11. show the line voltage and current for category 4 dip.

Fig. 15. and Fig. 19. show the line voltage and current for category 2 dip.

B. Results in Real Scale Test Bench

Fig. 8. and Fig. 12. show the line voltage and current for category 4 dip.

Fig. 16. and Fig. 20. show the line voltage and current for category 2 dip.

C. Results in Wind Turbine

Fig. 9. and Fig. 13. show the line voltage and current for category 4 dip.

Fig. 17. and Fig. 21. show the line voltage and current for category 2 dip.

D. Results in Wind Turbine simulation

Fig. 10. and Fig. 14. show the line voltage and current for category 4 dip.

Fig. 18. and Fig. 22. show the line voltage and current for category 2 dip.

E. Comments on the results

Although the differences in the set up of the three scenarios (small scale test bench, real scale testbench and wind turbine) are quite important, the results obtained for each of the voltage dips are very similar. Once debugged and tuned accordingly to the special characteristics of each scenario, the converter behaves in a very similar way in any type of voltage dips.

As it can be seen in both dips, the converter stays connected during the dips and injects reactive current as demanded by the grid code.

It is also shown the results of the windturbine simulation. These plots show that the results from the simulation are very similar to the real windturbine results. This data is used for validation of the model so that it can be used in further studies without repeating all the process again.

Simulation model validation is also mandatory in order to apply for the general process in REE certification.

VII. CONCLUSIONS

Gamesa DAC converter has been designed with the aim of fulfilling REE grid code requirements regarding voltage dips. This converter has been validated and certified following an intensive and detailed test plan in order to assure the success of the process. This test plan has several stages (small scale test bench, real scale test bench, wind turbine). In every stage, all possible tests are done, so that uncertainties are solved as soon as possible, and total testing time is optimized. The development of

the simulation platform reveals very useful by analyzing the control strategy and transient phenomena. The aim of developing this tool is also to have a clear picture of which grid codes can be fulfilled without having to follow the whole validation process again.

REFERENCES

[1] P.O 12.3: Requisitos de respuesta frente a huecos de tensión de las instalaciones eólicas.

[2] Procedure for verification, validation and certification of the requirements of the PO 12.3 on the response of wind farms in the event of voltage dips. AEE. Available: http://www.aeeolica.es/aee_actua_verificacion.php

[3] IEC 61400-4-30: Electromagnetic compatibility (EMC) Part 4-30: Test and measurement techniques. Methods for measuring the quality of supply.

(c) 2008 GAMESA INNOVATION AND TECHNOLOGY, S.L. Unipersonal (GAMESA). This article or presentation contains private and confidential information and is directed exclusively to its addressee. No part of this article or presentation may be reproduced, stored in a retrieval system, used in a spreadsheet or revealed or transmitted in any form by any means-electronic, mechanical, photocopying, recording or otherwise-without the previous written permission by GAMESA.

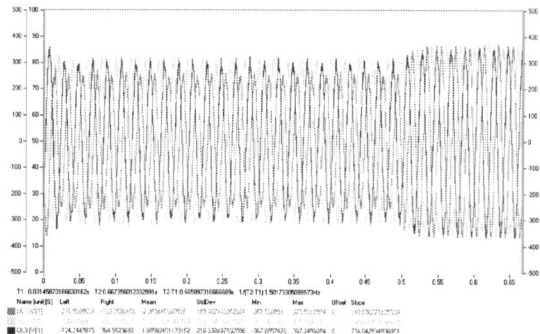

Fig. 7. Line voltage (phase to neutral) during a category 4 dip in the Small Scale Test Bench.

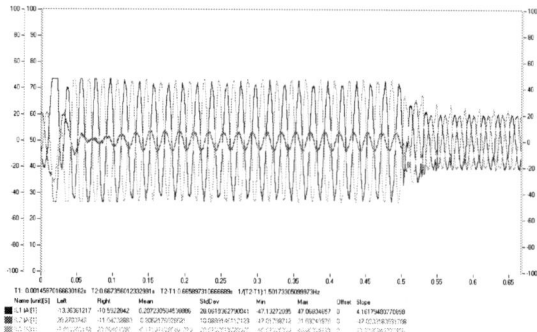

Fig. 11. Line current during a category 4 dip in the Small Scale Test Bench.

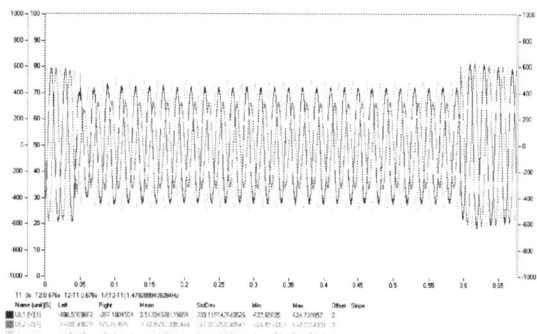

Fig. 8. Line voltage (phase to neutral) a category 4 dip in the Real Scale Test Bench

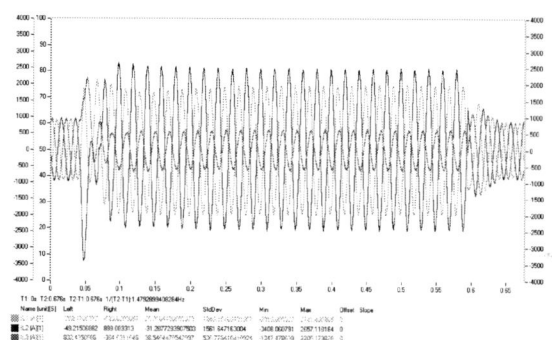

Fig. 12. Line current during a category 4 dip in the Real Scale Test Bench

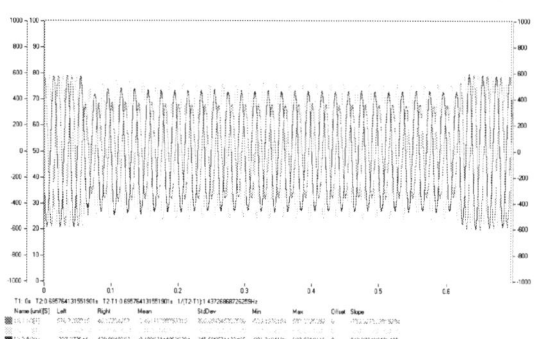

Fig. 9. Line voltage (phase to earth) during a category 4 dip in the wind turbine

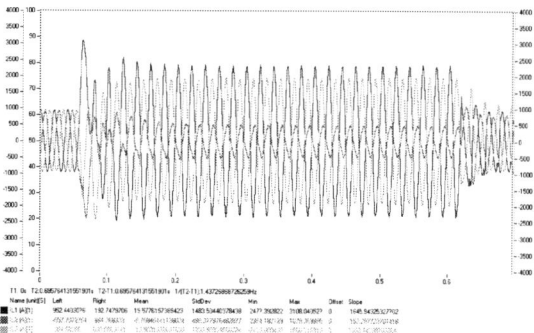

Fig. 13. Line current during a category 4 dip in the wind turbine

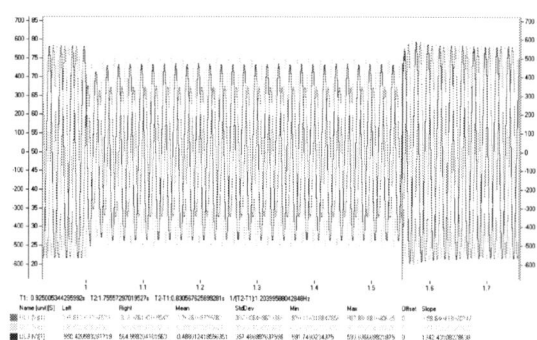

Fig. 10. Line voltage (phase to earth) during a category 4 dip in wind turbine simulation.

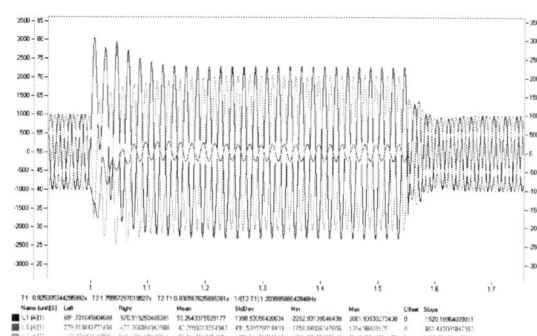

Fig. 14. Line current during a category 4 dip in wind turbine simulation

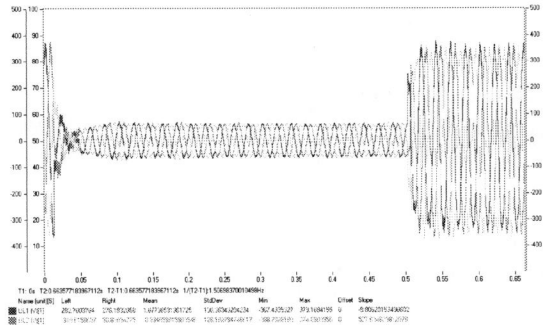

Fig. 15. Line voltage (phase to earth) during a category 2 dip in the Small Scale Testbench.

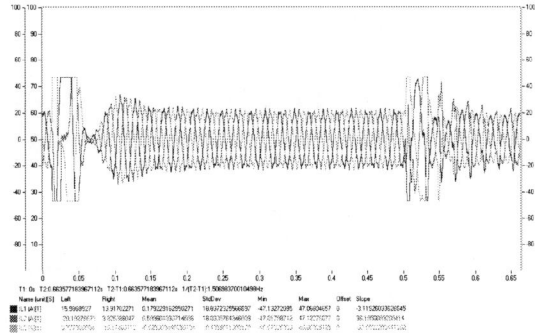

Fig. 19. Line current during a category 2 dip in the Small Scale Test Bench.

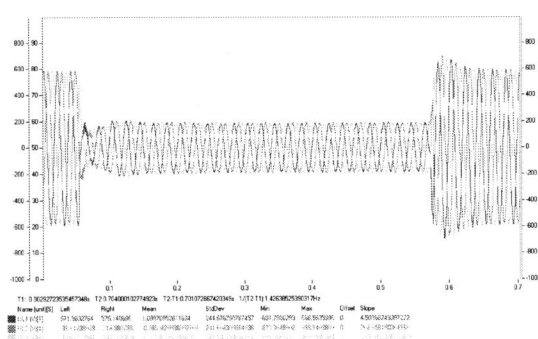

Fig. 16. Line voltage (phase to earth) during a category 2 dip in the Real Scale Testbench.

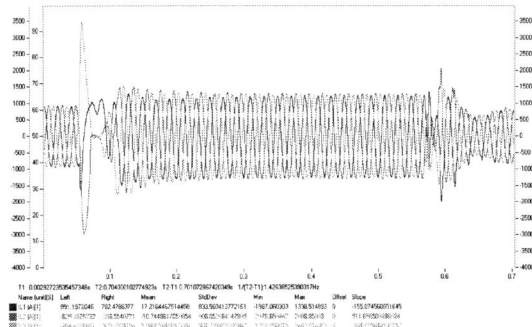

Fig. 20. Line current during a category 2 dip in the Real Scale Test Bench

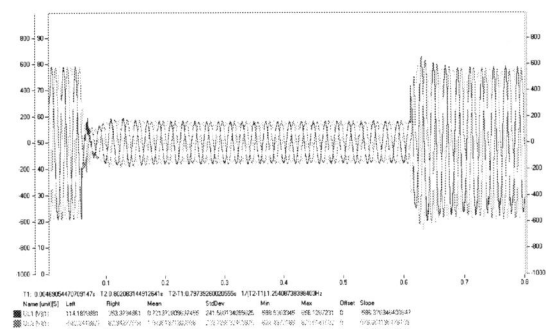

Fig. 17. Line voltage (phase to earth) during a category 2 dip in the wind turbine

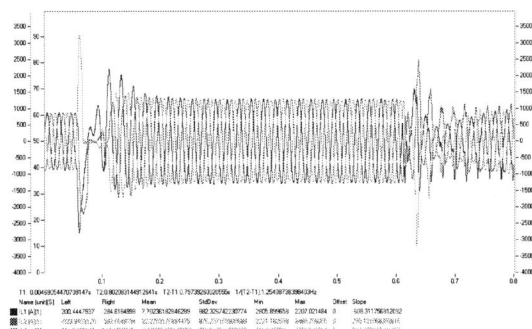

Fig. 21. Line current during a category 2 dip in the wind turbine

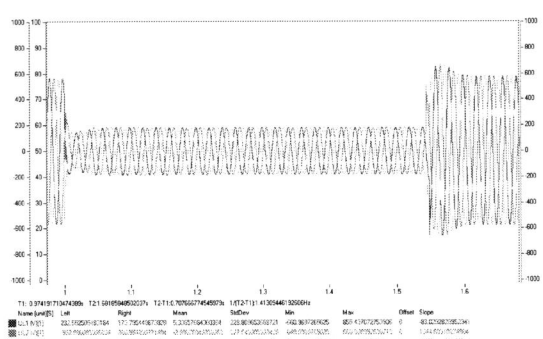

Fig. 18. Line voltage (phase to earth) during a category 2 dip in wind turbine simulation.

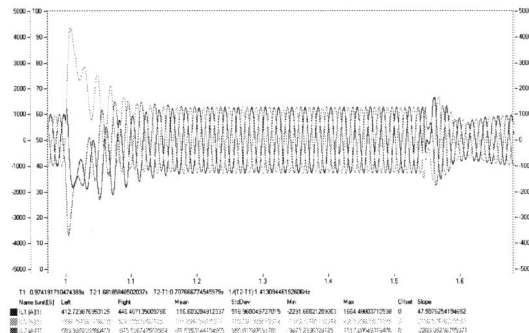

Fig. 22. Line current during a category 2 dip in wind turbine simulation.

Flatness-Based Voltage-Oriented Control of Three-Phase PWM Rectifiers

J. Dannehl, F.W. Fuchs

Institute of Power Electronics and Electrical Drives,
Christian-Albrechts-University of Kiel, D-24143 Kiel, Germany,
Phone: +49 (0) 431-880-6107, Email: jda@tf.uni-kiel.de

Abstract—**Flatness-based control is applied to the three-phase PWM-rectifier in synchronous reference frame. The DC-link voltage and reactive current are shown to be flat outputs of the full-order system. Two different approaches are presented. At first the DC-link voltage is controlled directly. The second employs inner current loops. Feed forward design based on system flatness is shown and discussed. In many applications the DC-link voltage and reactive current are controlled to constant values. In this case the direct flatness-based approach offers no advantages compared to conventional voltage-oriented PI-control whereas the second approach outperforms it with respect to the obtained control dynamic.**

Keywords— **Converter control, Non-linear control.**

I. INTRODUCTION

Three-phase grid-connected PWM rectifiers are often applied in regenerative energy systems and in adjustable speed drives when regenerative braking is required. Besides power regeneration they offer the control of the power factor as well as the DC-link voltage while emitting less current harmonics to the grid compared to passive diode rectifier bridges. A cascaded control structure with an outer DC-link voltage control and inner current control loops are commonly used. For L-filter grid connections the current control is mostly done with PI controllers in line voltage-oriented coordinates [1].

For many applications this so-called voltage-oriented control (VOC) is suitable and well working but for special applications research still goes on. Line voltage distortions like harmonics and unsymmetries for example are challenging the control. As PI controller can not reject sinusoidal disturbances additional control concepts are often necessary in order to meet the standards. Stability problems due to interactions with the fundamental VOC can occur, especially in weak grid conditions [2]. Another issue which often requires additional concepts is the resonance damping if LCL-filters are used as grid connection. The different control subsystems are mostly designed separately and interactions are often neglected. As the conventional VOC is a cascaded control it requires different time constants of the different loops. Therefore the DC link control bandwidth is limited. Furthermore the outer loop is tuned assuming the DC link voltage near to its constant reference. The inherited nonlinearity can lead to instability if the voltage variations are too high. An approache for minimizing the DC capacitance

is presented in [3]. When PWM rectifiers with reduced DC capacitance are used, a load step will cause a DC-link voltage dip. The smaller the capacitance the faster the controller has to react. In this case the time constants are getting closer to each other and the DC link voltage variations get higher. Stability problems may arise.

The application of nonlinear control strategies does not require different time constants of the DC link and current dynamics and the control design can be done for the full-order system without linearization around the constant DC voltage reference. Because of these reasons a faster control can be achieved which can be used for reducing the DC link capacitances. In [4], [5] and [6] the application of feedback linearization [7] for the PWM rectifier control yields faster control or smaller DC capacitors, respectively. Other nonlinear methods like Sliding Mode Control [8], passivity-based control [9] or the direct Lyapunov method [10] are also applied to the PWM rectifier control in order to improve the performance [11] [12] [10]. Another nonlinear method is the flatness-based control (FBC) [13] [14] which is successfully applied to motor control applications [15] [16] [17]. Recently, FBC is applied to the three-phase PWM rectifier [18] and [19] in order to achieve a higher control dynamic.

In [18] the PWM rectifier was shown to be flat with the reactive current and the stored system energy as flat outputs. In [19] the experimental validation is shown. The stored energy depends on the active and reactive currents as well as the DC link voltage. For the flatness-based control the reference trajectories of the flat outputs have to be derived. As the basic control objective is the control of the DC link voltage and the reactive power in terms of the reactive current the above choice requires calculations in order to formulate the reference trajectories of the flat outputs. In particular, the trajectory of the active current component has to be derived. Because the application in [19] is a D-STATCOM without DC load the active current is zero is steady state. Therefore the stored system energy is mostly related to the DC link voltage and the reactive current. But for transients the trajectory of the active current has to be derived which remains unclear from [19]. For other applications like adjustable speed drives with regenerative braking or distributed energy generation the active current is mostly nonzero and time varying.

In this paper the the DC link voltage and the reactive current are shown to be flat outputs itself. Its references

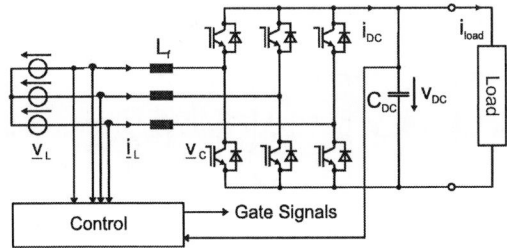

Fig. 1. PWM rectifier with L-Filter connected to the grid

TABLE I

SYSTEM PARAMETERS

Symbol	Quantity	value
V_L	Line voltage (phase-to-phase, rms)	400 V
ω	Line angular frequency	$2\,\pi\,50$ Hz
L_f	Filter inductance	6 mH
R_f	Resistance of filter inductor	100 mΩ
f_c	Switching/ control frequency	2 kHz
C_{DC}	DC link capacitance	2200 μF

can be used directly for the feed forward without pre-calculating other trajectories. Here two different flatness-based approaches are shown. First a direct control of the DC link voltage and reactive current without an inner active current loop is designed and discussed. The second approach is a cascade structure with an inner active current loop.

In section II the system description and modeling is shown which is followed by a discussion of the system flatness in section III. For the purpose of comparison the conventional VOC is shown in section IV. Both flatness-based control approaches are shown and analyzed in section V. Simulation results are presented and analyzed in section VI. Finally, a conclusion is given.

II. SYSTEM MODEL

The analyzed system is shown in Fig. 1. A three-phase IGBT voltage source converter is connected to the grid through a line-side filter. Here, the grid is modeled as an ideal, sinusoidal three-phase voltage source without line impedances and distortions like harmonics and unbalances. The DC link voltage and the line currents are measured for control purpose. The line voltages are measured for synchronizing with the grid. The system parameters can be found in Tab. I. Thereby the copper losses of the inductors are taken into account and modeled by R_f whereas its iron losses are neglected.

Here, the space vector notation is used [20]. The three-phase values are transformed into a two-phase stationary $\alpha\beta$-reference frame. Applying Kirchhoffs laws gives:

$$L_f \frac{\mathrm{d}}{\mathrm{dt}} \underline{i}_L^{\alpha\beta} = \underline{v}_L^{\alpha\beta} - \underline{v}_C^{\alpha\beta} - R_f \cdot \underline{i}_L^{\alpha\beta} \tag{1}$$

Transformation of (1) into the dq-reference frame rotating with the line voltage vector yields:

$$L_f \frac{\mathrm{d}}{\mathrm{dt}} \underline{i}_L^{dq} = \underline{v}_L^{dq} - \underline{v}_C^{dq} - R_f \cdot \underline{i}_L^{dq} - j\omega L_f \cdot \underline{i}_L^{dq} \tag{2}$$

The DC link voltage dynamic can be written as

$$C_{DC} \cdot \frac{\mathrm{d}v_{DC}}{\mathrm{dt}} = i_{DC} - i_{load} \tag{3}$$

In order to get a direct relation between (3) and (2) a power relation of the line side and the DC side can be used. The active line power of a three-phase system is given by (ϕ_I: phase angle between voltage and current):

$$P_{AC} = \cdot \frac{3\hat{v}_L \hat{i}_L}{\sqrt{2}\sqrt{2}} \cdot cos(\phi_I) = \frac{3}{2} \cdot \mathrm{Re}\left\{ \underline{v}_L^{\alpha\beta} \cdot \underline{i}_L^{\alpha\beta*} \right\}$$

$$= \frac{3}{2} \cdot \mathrm{Re}\left\{ \underline{v}_L^{dq} \cdot \underline{i}_L^{dq*} \right\} = \frac{3}{2} \cdot v_{Ld} i_{Ld} \tag{4}$$

Neglecting the filter and converter losses the active power balance of the line side (P_{AC}) and DC side ($P_{DC} = v_{DC} i_{DC}$) gives:

$$i_{DC} = \frac{3}{2} \cdot \frac{v_{Ld} i_{Ld}}{v_{DC}} \tag{5}$$

Finally, the DC link voltage dynamic can be written as

$$C_{DC} \cdot \frac{\mathrm{d}v_{DC}}{\mathrm{dt}} = \frac{3}{2} \cdot \frac{v_{Ld} i_{Ld}}{v_{DC}} - i_{load} \tag{6}$$

A similar derivation gives the reactive line side power:

$$Q_{AC} = -\frac{3}{2} \cdot v_{Ld} i_{Lq} \tag{7}$$

From (6) and (7) it becomes clear that the DC link voltage can be controlled by the d component of the line current and the reactive power by the q component. Note that i_{Lq} is commonly directly regulated to zero.

Defining the system state $\underline{x} = [i_{Ld}, i_{Lq}, v_{DC}]^T$ and the control input vector $\underline{u} = [v_{Cd}, v_{Cq}]^T$ the following state space representation can be derived out of (2) and (6):

$$\frac{\mathrm{d}}{\mathrm{dt}} \underline{x} = \begin{bmatrix} \dfrac{-R_f}{L_f} i_{Ld} + \omega i_{Lq} - \dfrac{v_{Cd}}{L_f} \\[2mm] \dfrac{-R_f}{L_f} i_{Lq} - \omega i_{Ld} - \dfrac{v_{Cq}}{L_f} \\[2mm] \dfrac{3}{2} \cdot \dfrac{v_{Ld} i_{Ld}}{C_{DC} v_{DC}} \end{bmatrix} + \begin{bmatrix} \dfrac{v_{Ld}}{L_f} \\[2mm] 0 \\[2mm] -\dfrac{i_{load}}{C_{DC}} \end{bmatrix} \tag{8}$$

III. FLATNESS OF THE SYSTEM

Consider the following system with the system state vector $\underline{x} = [x_1, ..., x_n]^T$ of order n and the system input vector $\underline{u} = [u_1, ..., u_m]^T$ of the order m:

$$\dot{\underline{x}} = f(\underline{x}, \underline{u}), \quad \underline{x}(0) = x_0 \tag{9}$$

A system is said to be flat if there exists a so-called flat output vector $\underline{y}$ of the same order as the input vector $\underline{u}$ (dim($\underline{y}$) = dim($\underline{u}$)) that can be expressed as functions of the system states ($\underline{x}$), the system inputs ($\underline{u}$) and a finite number of time derivatives (which are denoted by superscripts in brackets)

$$\underline{y} = \underline{\phi}\left(\underline{x}, u_1, ..., u_1^{(\alpha_1)}, ..., u_m, ..., u_m^{(\alpha_m)} \right) \tag{10}$$

and which fulfills the following two conditions [21]:

$$\underline{x} = \underline{\psi}_1\left(y_1, ..., y_1^{(\beta_1)}, ..., y_m, ..., y_m^{(\beta_m)} \right) \tag{11}$$

$$\underline{u} = \underline{\psi}_2\left(y_1, ..., y_1^{(\beta_1+1)}, ..., y_m, ..., y_m^{(\beta_m+1)} \right) \tag{12}$$

Note that the above is not the most general definition of flatness. For a detailed mathematical definition the reader is referred to [21] or [13]. For a flat system there exist many different outputs that can be chosen as flat outputs. As the DC link voltage and the reactive current are the quantities to control, here, it is shown that the these are flat outputs itself. Its references can be used directly for the feed forward without precalculating other trajectories. For the proof the disturbances (second term of the right hand side of (8) are neglected and the line voltage amplitude is assumed constant. For sake of simplicity in the following derivatives are denoted by dots ($\dot{x}$ instead of dx/dt for example). Consider as output vector

$$\underline{y} = [y_1, y_2]^T = [i_{Lq}, v_{DC}]^T = \underline{\phi}(x_1, x_3) \quad (13)$$

First the system state vector $\underline{x} = [i_{Ld}, i_{Lq}, v_{DC}]^T$ will be expressed as function of the flat outputs. Obviously two states are flat outputs themselves and as

$$\frac{dv_{DC}}{dt} = \dot{v}_{DC} = \dot{y}_2 = \frac{3}{2} \cdot \frac{v_{Ld} i_{Ld}}{C_{DC} y_2} \quad (14)$$

the d current component can be written as:

$$i_{Ld} = \frac{2}{3} \cdot \frac{C_{DC} y_2 \dot{y}_2}{v_{Ld}} \quad (15)$$

By using (13) and (15) the system state vector can be expressed as

$$\dot{x} = \begin{bmatrix} \dfrac{2}{3} \cdot \dfrac{C_{DC} y_2 \dot{y}_2}{v_{Ld}} \\ y_1 \\ y_2 \end{bmatrix} = \underline{\psi}_1(y_1, y_2, \dot{y}_2) \quad (16)$$

The second step is proving (12) whereas $\underline{u} = [v_{Cd}, v_{Cq}]^T$. Rearranging the first line of (8) gives

$$v_{Cd} = -L_f \cdot \left(\frac{di_{Ld}}{dt} + \frac{R_f}{L_f} i_{Ld} - \omega i_{Lq} \right) \quad (17)$$

Substituting i_{Ld} by (15) and rearranging gives

$$\begin{aligned} v_{Cd} &= \frac{2}{3} \cdot \frac{C_{DC} L_f}{v_{Ld}} (\dot{y}_2 \cdot \dot{y}_2 + y_2 \cdot \ddot{y}_2) \\ &\quad - \frac{2}{3} \cdot \frac{R_f C_{DC}}{v_{Ld}} \cdot y_2 \dot{y}_2 + \omega L_f y_1 \\ &= \psi_{2,1}(y_1, y_2, \dot{y}_2, \ddot{y}_2) \end{aligned} \quad (18)$$

Similarly, the second line of (8) combined with (15) gives

$$v_{Cq} = -L_f \left(\dot{y}_1 + \frac{2}{3} \cdot \frac{\omega C_{DC}}{v_{Ld}} y_2 \dot{y}_2 + \frac{R_f}{L_f} y_1 \right) \quad (19)$$

$$= \psi_{2,2}(y_1, \dot{y}_1, y_2, \dot{y}_2) \quad (20)$$

Combining (18) and (19) gives

$$\underline{u} = \underline{\psi}_2 = [\psi_{2,1}, \psi_{2,2}]^T = \underline{\psi}_2(y_1, \dot{y}_1, y_2, \dot{y}_2, \ddot{y}_2) \quad (21)$$

It can be seen from (16) and (21) that $\underline{y} = [i_{Lq}, v_{DC}]^T$ fulfils the conditions (11) and (12) and consequently is a flat output vector (with $\beta_1 = 0$ and $\beta_2 = 1$).

If a cascade control structure with inner current loops is used only the current dynamics of (8) are of interest for the current control. Obviously, in this case both current components are directly flat outputs and therefore $\underline{y} = [i_{Ld}, i_{Lq}]^T$ is a flat output vector.

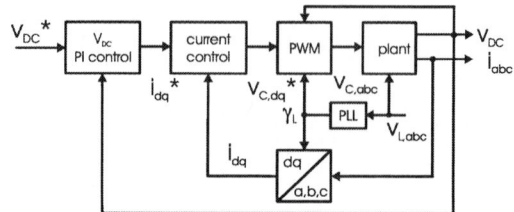

Fig. 2. Cascaded control structure of conventional voltage-oriented control

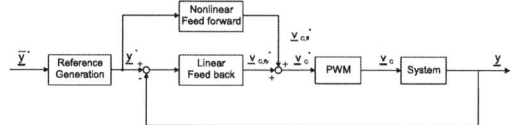

Fig. 3. Control structure of direct flatness-based DC link voltage and the reactive current control

IV. CONVENTIONAL VOLTAGE-ORIENTED CONTROL

The conventional voltage-oriented control consists of an outer DC link voltage control loop and inner current control loops [1]. The cascaded control structure is shown in Fig. 2. The outer loop regulates the DC link voltage to its constant reference V_{DC}^* by applying PI control. In order to prevent wind up problems in case of limitation of the current references an anti-wind up mechanism is used here [22]. For the design of the PI controller parameters (k_{DC}, T_{DC}) the inner active current loop is modeled as PT_1-lag element with the delay time of $T_{inner} = 4T_c$. Assuming the DC-link voltage near to its constant reference the PI controller can be tuned with the symmetrical optimum [23]:

$$k_{DC} = \frac{2}{3} \cdot \frac{C_{DC} V_{DC}^*}{2 T_{inner} v_{Ld}}; \quad T_{DC} = 4 T_{inner} \quad (22)$$

For the voltage-oriented current control PI controllers are used. In order to decouple the d and q current dynamics a decoupling by feedback is inserted [1]:

$$\begin{aligned} v_{Cd}^* &= \omega L_f i_{Lq} + k_I (i_{Ld} - i_{Ld}^*) \\ &\quad + \frac{k_I}{T_I} \int (i_{Ld} - i_{Ld}^*) \, dt \end{aligned} \quad (23)$$

$$\begin{aligned} v_{Cq}^* &= -\omega L_f i_{Ld} + k_I (i_{Lq} - i_{Lq}^*) \\ &\quad + \frac{k_I}{T_I} \int (i_{Lq} - i_{Lq}^*) \, dt \end{aligned} \quad (24)$$

Note that the line voltage can be compensated as well but this is omitted here. For the controller parameter design the converter is modeled as a delay of one switching period ($T_c = 1/f_c$). Tuning with symmetrical optimum gives

$$k_I = \frac{-L_f}{a_I T_c}; \quad T_I = a_I^2 T_c; \quad a_I = 3 \quad (25)$$

As the available converter output voltage is limited by the DC link voltage an anti-wind up mechanism is also used [22]. Here, the determination of the line voltage phase angle is done by a PLL algorithm. A survey of different synchronization solutions can be found in [24].

446

V. FLATNESS-BASED CONTROL

The flatness property can effectively be used for designing control algorithms. Basically, the control structure consists of a feed forward and a feedback part [14]. The key idea of the flatness-based control is to drive the system towards the reference trajectory by feed forward. The feedback part is inserted only in order to eliminate the deviations caused by disturbances and other nonidealities like model uncertainties, actuator limitations and other perturbations. Flatness allows to formulate the control input vector as function of only the flat output vector even for nonlinear systems, see (12). This expression is used for designing the feed forward. Under ideal conditions the feed forward can track the (time varying) reference if it is smooth enough. Otherwise due to derivatives in the references and limitations in the control inputs tracking errors would appear. Even if the references are smooth enough deviations from perfect tracking will appear due to disturbances, model uncertainties and other perturbations. Therefore feedback is inserted. As the system is near to the reference trajectory via feed forward, thus linearizable around it, the feedback can be designed with linear methods even for nonlinear systems [14]. As already mentioned the references should be smooth enough in order to make effectively use of the feed forward [25]. Therefore the references are smoothed in the block called reference trajectory generation.

In the following two different approaches are shown and analyzed. At first the DC link voltage and the reactive current are controlled directly and afterwards a cascaded control structure with an inner active current control loop will be used.

A. Direct Approach

In this section the direct control of the DC link voltage and the reactive current is designed and analyzed ($y = [y_1, y_2]^T = [i_{Lq}, v_{DC}]^T$). The control structure is shown in Fig. 3. The feed forward can be designed directly by using (18) and (19):

$$v_{Cd,ff}^* = \frac{2}{3} \cdot \frac{C_{DC}L_f}{v_{Ld}} \left(\dot{y}_2^* \cdot \ddot{y}_2^* + y_2^* \cdot \dddot{y}_2^* \right) \\ - \frac{2}{3} \cdot \frac{R_f C_{DC}}{v_{Ld}} \cdot y_2^* \ddot{y}_2^* + \omega L_f y_1^* \quad (26)$$

$$v_{Cq,ff}^* = -L_f \left(\dot{y}_1^* + \frac{2}{3} \frac{\omega C_{DC}}{v_{Ld}} y_2^* \ddot{y}_2^* + \frac{R_f}{L_f} y_1^* \right) \quad (27)$$

Note that the load current i_{load} is not included in the feed forward as it is assumed unknown. If it was measured or estimated it could be included for improving the DC link voltage control.

It is common practice to control the DC link voltage to a constant value and the reactive current constantly to zero. Therefore no real tracking is necessary with respect to the control variables. In this case the feed forward from (26) and (27) reduces to the following simple expressions

$$v_{Cd,ff}^* = \omega L_f y_1^* \quad (28)$$
$$v_{Cq,ff}^* = R_f y_1^* \quad (29)$$

It becomes clear from (28) that the coupling from the q component into the d path is decoupled. The voltage drop of the q path across the parasitic resistance of the inductor is compensated by (29). In case of constant references the core of the flatness-based structure, that is the feed forward, does not contribute remarkably to the system behavior. Most of the control has to be done of the feedback controller. As no tracking performance is required the direct flatness-based control is not suitable if the DC link voltage and reactive current is kept constant. Actually, the quantity which has to be tracked is the active current. Depending on the load current it has to be controlled to a certain value in order to stabilize the DC link voltage. Therefore a cascaded structure with an inner flatness-based active current control loop will be designed and analyzed in the next section.

As the direct approach is not well suited the design of the feedback and the reference smoothing is omitted here. The reader is referred to [19] for approaches. The direct approach is not further treated in this paper.

B. Cascaded Control

The cascaded control structure is shown in Fig. 4. From the first line of (16) the outer current feed forward can be directly derived:

$$i_{Ld,ff}^* = \frac{2}{3} \cdot \frac{C_{DC} v_{DC}^* \dot{v}_{DC}^*}{v_{Ld}} \quad (30)$$

Note again that the load current is not included in the feed forward because it is assumed unknown. In case of constant DC link voltage reference the feed forward does not contribute to the active current reference. As this is common practice there is no difference to the conventional voltage-oriented control since feedback is performing the tracking as well. For the same reason the PI controller already used for VOC can be used, see (22) for its parameters. The DC link voltage reference smoothing is not treated in this paper since it is constant. For the design of the inner voltage feed forward the first two lines of (8) are used:

$$v_{Cd,ff}^* = -L_f \frac{\mathrm{d}}{\mathrm{d}t} i_{Ld}^* - R_f i_{Ld}^* + \omega L_f i_{Lq}^* \quad (31)$$
$$v_{Cq,ff}^* = -L_f \frac{\mathrm{d}}{\mathrm{d}t} i_{Lq}^* - R_f i_{Lq}^* - \omega L_f i_{Ld}^* \quad (32)$$

The line voltage compensation is omitted here again. The active current reference depends on the load current which is time varying. For constant reactive current reference (32) reduces to:

$$v_{Cq,ff}^* = -\frac{R_f}{L_f} i_{Lq}^* - \omega L_f i_{Ld}^* \quad (33)$$

It can be seen that the flatness-based feed forward design yields a decoupling by feed forward whereas for the conventional VOC this is done by feedback, see (23) and (24). Comparing (31) and (33) with the control law of the conventional voltage-oriented control, see (23) and (24), shows that the FBC is additionally compensating the voltage drop across the parasitic filter resistances. As the FBC is using reference signals instead of measured

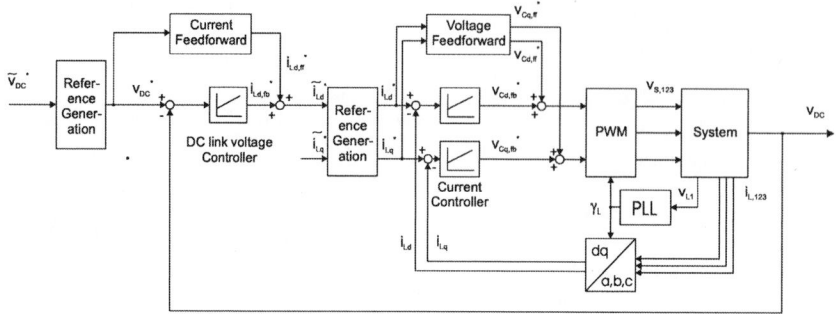

Fig. 4. Cascaded control structure of flatness-based control

ones the sensitivity against measurement noise and errors is lower compared to VOC. Another improvement of the FBC is the derivative part in (31) which increases the control bandwidth as will be shown later by simulations.

Combining (31) with the first line of (8) and (32) with the second line of (8) and neglecting the computation and PWM delay yields the dynamics of the current tracking errors $\Delta i_{Ld} = (i_{Ld}^* - i_{Ld})$ and $\Delta i_{Lq} = (i_{Lq}^* - i_{Lq})$:

$$\frac{\mathrm{d}\Delta i_{Ld}}{\mathrm{dt}} = -\frac{R_f}{L_f} \Delta i_{Ld} + \omega \, \Delta i_{Lq} + \frac{v_{Cd,fb}}{L_f} - \frac{v_{Ld}}{L_f} \quad (34)$$

$$\frac{\mathrm{d}\Delta i_{Lq}}{\mathrm{dt}} = -\frac{R_f}{L_f} \Delta i_{Lq} - \omega \, \Delta i_{Ld} + \frac{v_{Cq,fb}}{L_f} \quad (35)$$

For the feedback control design the couplings are neglected as the currents are assumed near to its reference due to the feed forward. Additionally, the line voltage is treated as disturbance. Finally, two decoupled first order PT_1-dynamics are achieved which can be controlled with PI controller as already used for the conventional voltage-oriented control. The difference to VOC is that (34) and (35) are the error dynamics, thus its references are zero. This makes clear again that the tracking is not done by the feedback but by the feed forward. Even if optimization for disturbance rejection instead of tracking behavior was possible, here the same PI controller as for VOC are used, see (25) for its parameters. Note that the computation and PWM delay is taken into account for the feedback design. So, the feed forward of (31) and (32) can be used as plug-in solution in order to improve the VOC. The anti-windup is included as well.

As already mentioned smooth references are important for the effective use of the feed forward. Here simple first order lag filters are used. See [19] [25] for improved methods and [17] for different simple ones. At best, the voltage limitations are directly included into the reference trajectory generation.

VI. SIMULATION RESULTS

In this section simulation results are shown and analyzed which are obtained with Matlab/Simulink. As emphasis is put on cascaded control structures and the DC link voltage control is the same for all approaches only the

current control behavior is analyzed here. Instead of a DC link capacitor and its voltage control here a constant DC voltage source is used. By that the influence of the outer DC link voltage control is canceled and the current control behavior can be pointed out more clearly. The three-phase IGBT converter, filter inductors and line voltage sources as well as the DC voltage source are modeled in PLECS, see Tab. I for the system parameters. The signals are sampled in the middle of each switching period and afterwards the control algorithm is executed. The PLL algorithm is executed with 20 kHz. The reference currents are smoothed with first-order lag filters which are discretized by the Tustin approximation with prewarping [22].

In Fig. 5 and 6 the response of the current control is shown for the different approaches. The reactive current reference is zero whereas a step from zero to 30 A in the active current reference is applied to the control system. Fig. 5(a) illustrates good tracking behavior of VOC even if only PI controllers are used and the references are not filtered. After 5 T_c the reference is reached the first time and an overshoot of 9 A can be seen. Couplings between the d and q components are clearly visible in Fig. 6(a). Filtering the reference yields only slightly less overshoot and a longer response time but the couplings are reduced. As can be seen in (23) and (24), in the conventional VOC an additional decoupling by feedback is used. Its effect can be seen in Fig. (5)(b) and (6)(b). Instead of improving the response a more oscillatory behavior in both current components is obtained with decreasing the response time. One reason can be the quite low control frequency. A compensation of the delay could improve the response. As the FBC in (31) and (32) shows a decoupling by feed forward another approach is shown here. Instead of using the measured current values the reference currents are used for the decoupling of VOC in (23) and (24). This approach is referred as VOC with feed forward decoupling whereas the other is called VOC with feedback decoupling. As can be seen in Fig. 5(c) and 6(c) the feed forward decoupling considerably improves the step response. The couplings are effectively reduced and a faster response is obtained as well. With smoother references the coupling are decreased but the response time is increased as well.

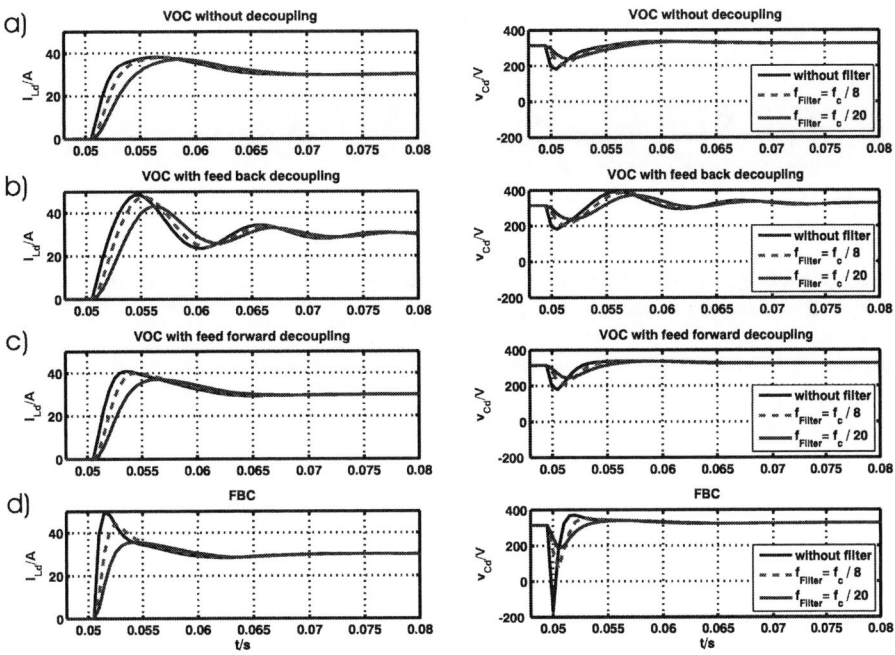

Fig. 5. Simulated responses to an active current reference step (applied at $t = 0.05s$) with different reference filters ($i_{Lq}^* = 0A$, $V_{DC} = const. = 700V$). Left: active current component i_{Ld}, right: d component of the converter output reference voltage u_{Ld}. a) Voltage-oriented control without decoupling, b) Voltage-oriented control with feedback decoupling, c) Voltage-oriented control with feed forward decoupling, d) Flatness-based control

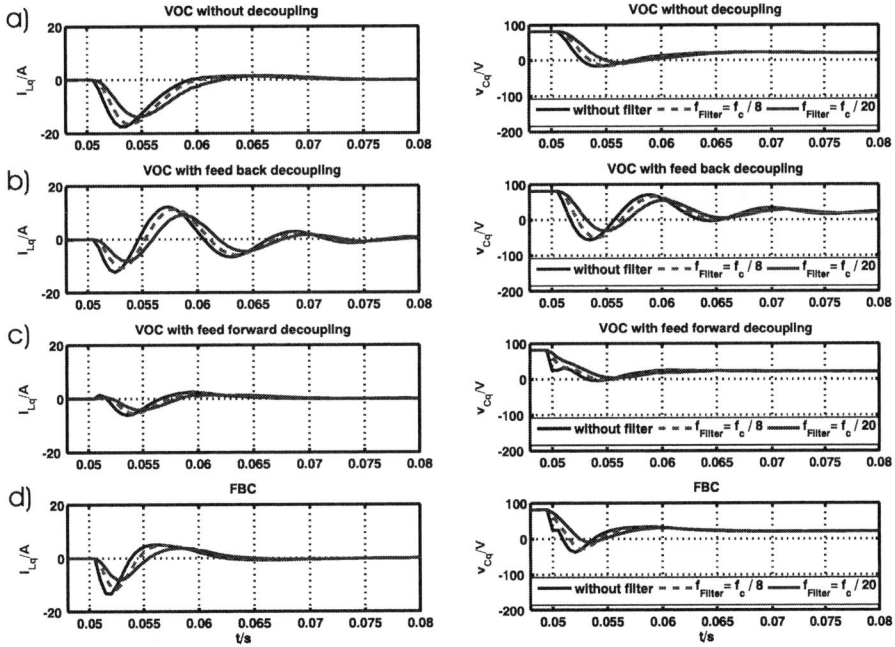

Fig. 6. Simulated responses to an active current reference step (applied at $t = 0.05s$) with different reference filters ($i_{Lq}^* = 0A$, $V_{DC} = const. = 700V$). Left: reactive current component i_{Lq}, right: q component of the converter output reference voltage u_{Lq}. a) Voltage-oriented control without decoupling, b) Voltage-oriented control with feedback decoupling, c) Voltage-oriented control with feed forward decoupling, d) Flatness-based control

449

Finally, the results achieved with FBC are shown in Fig. 5(d) and 6(d). In addition to the feed forward decoupling the derivative of the active current reference is included and the voltage drop across the parasitic resistances are compensated, see (31) and (32). Mostly the derivative part further decreases the response time. Without filtering the reference the d-current already exceeds the reference to 40 A. Thus, the fastest response is achieved but the overshoot is around 20 A which is more than 60 % overshoot. Applying a reference filter with a cut-off frequency of $f_c/8$ yields a lower overshoot but compared to the other approaches still the fastest response time is obtained. Compared to VOC with feed forward decoupling higher couplings are obtained but compared to the other approaches the couplings are still reduced.

VII. CONCLUSION

Flatness-based control of three-phase PWM rectifiers in synchronous reference frame is designed and analyzed. The DC link voltage and the reactive current are shown to be be flat outputs of the PWM rectifier system. Two different flatness-based control approaches are presented and analyzed. First the DC link voltage and the reactive current are controlled directly. The second approach employs an inner active current loop. The feed forward design based on the system flatness is shown and discussed. For the cascade structure simulations are carried out as well. For purpose of comparison the conventional voltage-oriented control is also shown and simulation results are presented.

In many applications the DC link voltage and the reactive current are controlled to constant values. In this case the direct flatness-based approach offers no advantages compared to the conventional voltage-oriented control. The second approach outperforms it as a faster step response is achieved with less coupling between both current components. In this case the good tracking behavior can be effectively used for the active current control. Filtering the reference currents yields a less oscillatory but still faster performance compared to the conventional control. The difference between decoupling by feed forward and feedback is shown by simulations.

VIII. ACKNOWLEDGMENT

This work has been funded by German Research Foundation (DFG).

REFERENCES

[1] Kazmierkowski, M.P., Krishnan, R., and Blaabjerg, F., *Control in Power Electronics: Selected Problems.* Oxford: Academic Press, 2002.

[2] Liserre, M., Teodorescu, R., and Blaabjerg, F., "Stability of photovoltaic and wind turbine grid-connected inverters for a large set of grid impedance values," *IEEE Transactions on Power Electronics*, vol. 21, no. 1, pp. 263– 272, January 2006.

[3] Winkelnkemper, M. and Bernet, S., "Impact of control model deviations on the dc link capacitor minimization in ac-dc-ac converters," in *IEEE Industrial Electronics, IECON 2005 - 31st Annual Conference on*, 2005, CD-ROM paper.

[4] Burgos, R.P., Wiechmann, E.P., and Holtz, J., "Complex statespace modeling and nonlinear control of active front-end converters," *IEEE Transactions on Industrial Electronics*, vol. 52, no. 2, pp. 363– 377, April 2005.

[5] Jinhwan Jung, Sun Kyoung Lim, and Kwanghee Nam, "A feedback linearizing control scheme for a pwm converter-inverter having a very small dc-link capacitor," in *Industry Applications Conference, 1998. Thirty-Third IAS Annual Meeting. The 1998 IEEE*, vol. 2, 1998, pp. 1497–503.

[6] Lee, T.S., "Input-output linearization and zero-dynamics control of three-phase ac/dc voltage-source converters," *IEEE Transactions on Power Electronics*, vol. 18, no. 1, pp. 11–22, January 2003.

[7] Isidori, A. , *Nonlinear Control Systems*, 2nd ed. Berlin: Springer-Verlag, 1989.

[8] Utkin, V.I., "Sliding mode control design principles and applications to electric drives," *IEEE Transactions on Industrial Electronics*, vol. 40, no. 1, pp. 23–36, February 1993.

[9] Ortega, R., Loria, A., Nicklasson, P.J., and Sira-Ramirez, H., *Passivity-based Control of Euler-Lagrange Systems.* London: Springer, 1998.

[10] Kömürcügil, H. and Kükrer, O., "Lyapunov-based control for three-phase pwm ac/dc voltage-source converters," *IEEE Transactions on Power Electronics*, vol. 13, no. 5, pp. 801–13, September 1998.

[11] K. Jezernik, "Vss control of unity power factor," *IEEE Transactions on Industrial Electronics*, vol. 46, no. 2, pp. 325–31, April 1999.

[12] Carrasco, J.M., Galvan, E., Escobar, G., Stankovic, A.M., and Ortega, R., "Passivity-based controller for a three phase synchronous rectifier," in *IEEE International Conference on Industrial Electronics, Control and Instrumentation*, vol. 4, 2000, pp. 2629–34.

[13] Fliess, M., Levine, J., Martin, P., and Rouchon, P., "Flatness and defect of nonlinear systems: introductory theory and examples," *International Journal of Control*, vol. 61, no. 6, pp. 1327–61, June 1995.

[14] Hagenmeyer, V., *Robust nonlinear tracking control based on differential flatness*, ser. Fortschritt-Berichte VDI: Reihe 8, Meß-, Steuerungs- und Regelungstechnik. Düsseldorf: VDI-Verlag, 2003.

[15] Hagenmeyer, V., Ranftl, A., and Delaleau, E., "Flatness-based control of the induction drive minimising energy dissipation," in *Nonlinear and Adaptive Control. Lecture Notes in Control and Information Sciences*, vol. 281, 2003, pp. 149–60.

[16] Martin, P. and Rouchon, P., "Two remarks on induction motors," in *CESA'96 IMACS Multiconference*, vol. 1, 1996, pp. 76–9.

[17] Dannehl, J. and Fuchs, F.W., "Flatness-based control of an induction machine fed via voltage source inverter - concept, control design and performance analysis," *IEEE Industrial Electronics, IECON 2006 - 32nd Annual Conference on*, pp. 5125–5130, Nov. 2006.

[18] Gensior, A., Rudolph, J. , and Güldner, H., "Flatness based control of three-phase boost rectifiers," in *Power Electronics and Applications, 2005 European Conference on*, 2005, CD-ROM paper.

[19] Song, E, Lynch, A.F., and Dinavahi, V., "Experimental validation of a flatness-based control for a voltage source converter," *American Control Conference, 2007. ACC '07*, pp. 6049–6054, 9-13 July 2007.

[20] Kazmierkowski, M.P. and Tunia, H., *Automatic Control of Converter-Fed Drives.* Elsevier, 1994.

[21] Rothfuß, R., *Anwendung der flachheitsbasierten Analyse und Regelung nichtlinearer Mehrgrößensysteme*, ser. Fortschritt-Berichte VDI : Reihe 8, Meß-, Steuerungs- und Regelungstechnik. Düsseldorf: VDI-Verlag, 1997.

[22] Aaström, K.J. and Hägglund, T., *PID Controllers: Theory, Design and Tuning*, 2nd ed. Instrument Society of America, 1995.

[23] Schröder, D., *Elektrische Antriebe 2, Regelung von Antriebssystemen*, 2nd ed. Berlin: Springer, 2001.

[24] Timbus, A., Liserre, M., Teodorescu, R., and Blaabjerg, F., "Synchronization methods for three phase distributed power generation systems. an overview and evaluation," *IEEE 36th Power Electronics Specialists Conference*, pp. 2474–2481, 2005.

[25] Zeitz, M., Graichen, K., and Meurer, T., "Vorsteuerung und Trajektorienplanung als Basis von linearen und nichtlinearen Folgeregelungen," in *GMA-Kongress "Automation als interdisziplinäre Herausforderung"*, Baden-Baden, 7./8. Juni 2005.

Control of a single phase H-Bridge multilevel inverter for grid-connected PV applications

Elena Villanueva[*], Pablo Correa[*], Jose Rodriguez[†]
[*]University Federico Santa Maria, Valparaiso, Chile, e-mail: *pablo.correa@usm.cl*
[†]University Federico Santa Maria, Valparaiso, Chile, e-mail: *jose.rodriguez@usm.cl*

Abstract—This paper proposes a control scheme for a single phase H-Bridge multilevel converter for photovoltaic applications. This control approach permits each H-Bridge module supply different power levels, allowing an independent maximum power point tracking of each photovoltaic (PV) panel. Since no intermediate stages such boost DC-DC converters are needed, a high efficient solution is obtained compared to the standard PV converter systems. Additionally, this topology offers other advantages such as the operation at lower switching frequency and low current ripple. Simulation results of a two series connected H-bridge rectifier with maximum power point tracking for different conditions are presented.

Keywords— Photovoltaic, solar cell system, multilevel converters.

I. INTRODUCTION

Low power single phase grid connected PV systems are nowadays an important topic of research. In these generation systems issues such as the reliability, the efficiency, the size, weight of the solution and the price are important factors to be considered. The future trend in grid-connected PV applications is to allow an independent control of small group of PV panels to maximize power which is being delivered to mains [1]. Several topologies that fulfils this requirement have been proposed [2][3]. Particularly interesting is the configuration of several series connected H-bridge rectifiers, because it permits an individual tracking of the maximum power point (MPPT) in each panel and do not necessarily requires of an additional stage to boost the voltage [4]. This topology has other advantages such as the generation of a current with reduced ripple and the capacity to operate at a lower switching frequency, increasing in this way the efficiency of the whole system.

Several methods have been proposed for the control of active rectifier with multiple H-bridge converters connected in series. The most ones assumes that the dc-link voltage is constant [5], or are very complex and difficult to implement [4],[6]. In this paper, a simple control scheme based in [7] is applied for the control of PV Multilevel system. This scheme allows an independent MPPT for each panel and delivers current to the mains with unitary power factor. The proposed scheme includes a digital band reject filter in the PV voltage measurement in order to achieve a high quality in the output current without the 100 Hz component proper of the single phase configuration.

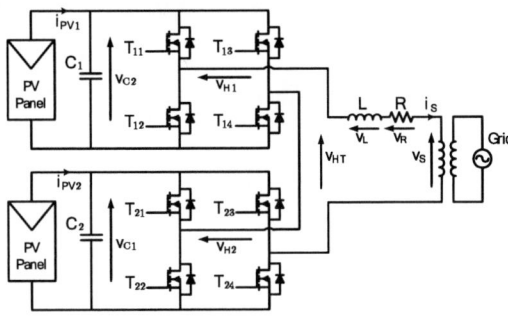

Fig. 1. Topology for grid connection.

The paper is organized in the following manner: first, a description of the topology as well as the basic equations for the model are given. Then, the controller of the converter is described. Simulation results are finally commented in order to validate the proposed control scheme.

II. TOPOLOGY DESCRIPTION

The proposed PV converter system consists of two H-Bridge converters connected in cascade as shown in Fig. 1. This topology can be extended by adding extra cells connected in series, allowing the generation of higher voltage and eventually a connection without transformer to the mains. Since in many PV installations isolation between the panels and the mains is usually required, a transformer will be considered in this work.

Each H-Bridge module can produce 3 levels: $-v_c$, 0 and $+v_c$ where v_c is the DC-Link voltage. The whole system can provide a 5-level voltage if the DC-Links are equal. The output voltage is easily determined out of the following simple relation:

$$\left. \begin{aligned} v_{H1} &= \left(T_{11} - T_{13}\right) \cdot v_{C1} = P_1 \cdot v_{C1} \\ v_{H2} &= \left(T_{21} - T_{23}\right) \cdot v_{C2} = P_2 \cdot v_{C2} \end{aligned} \right\} \tag{1}$$

where T_{XX} represents the state of each switch according to Fig. 1. T_{XX} is defined as follows: 1 for ON, 0 for OFF, therefore P_1 and P_2 can have the discrete values -1, 0 or $+1$ when the output voltage of the H-bridge is $-v_c$, 0 or $+v_c$ respectively. In order to have a complete lineal model of the converter, the functions P_1 and P_2 are replaced by the continuous switching functions S_1 and S_2 [-1,1].

978-1-4244-1741-4/08/$25.00 ©2008 IEEE

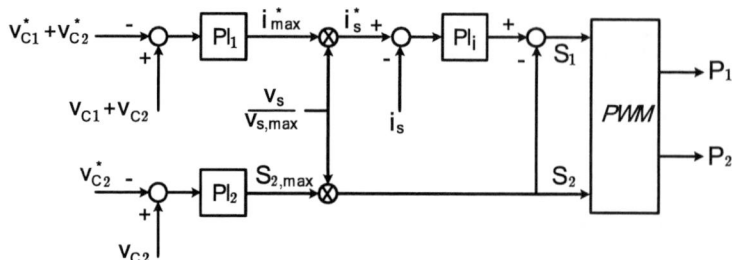

Fig. 2. Proposed control scheme.

Thus, the dynamic behavior of the system can be described by

$$
\left.\begin{array}{l}
\dfrac{di_S}{dt}=\dfrac{1}{L}\left(S_1 v_{C1}+S_2 v_{C2}-Ri_S-v_S\right)\\[2mm]
\dfrac{dv_{C1}}{dt}=\dfrac{1}{C_1}\left(i_{PV1}-S_1 i_S\right)\\[2mm]
\dfrac{dv_{C2}}{dt}=\dfrac{1}{C_2}\left(i_{PV2}-S_2 i_S\right)
\end{array}\right\}
\tag{2}
$$

III. CONTROL SCHEME

The control strategy is based in the classical scheme used for the control of a single H-bridge converter [8]. In [7,9], this idea has been extended for the case of n-cells connected in series for the control of an active rectifier. From these different control schemes, only [7] seems to be suited for this application because it is able to operate with different DC-Link voltages. The use of a similar scheme but including a PI controller in the current control stage, as it is shown in Fig. 2 is proposed.

The control scheme has three control loops: two of them are used to adjust the capacitor voltage in each dc-link, and the other one is necessary for the generation of a sinusoidal input current with unity power factor. As can be seen in Fig. 2, the sum of the dc-link voltages v_{C1} and v_{C2} is controlled through a PI that determines the amplitude of the input current $i_{S,\max}$. By multiplying the output of this controller with a normalized sinusoidal signal in phase with the voltage grid, a suitable reference for the current loop is obtained. On the other hand, the PI current controller PI_i gives the sum of the continuous switching functions S_1+S_2. The control of the voltage v_{C2} is made through another controller that selects the switching function amplitude $S_{2,\max}$ directly. Note that this scheme sets the phase of S_2 equal to grid phase, leading to the phasor diagram of Fig. 3-(b).

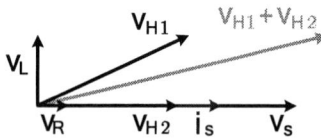

Fig. 3. Phasor diagram.

In order to obtain the maximum power from each PV panel, the Perturb and Observe (P&O) algorithm is used. This algorithm is the most commonly used in practice because of its ease of implementation and has the potential to be very competitive with other methods if it is properly optimized for the given hardware [10]. With this topology, individual MPP tracking of each panel can be made, then the variables to observe are the calculated output power of the PV panels, and the variables to perturb are, in this case, the reference voltages, namely v_{C1} and v_{C2}.

IV. DESIGN OF THE CONTROLLERS

In this section the tuning procedure for the three control loops of Fig. 2 is shown. To facilitate the analysis of the loops, death times due to the calculations and modulation of the converter are neglected. The design of the filters which are used in the PV-voltage measurement to avoid the 100 (Hz) component in the current is also described.

A. Current Loop

Since the dynamic of the control loop is much faster than the dynamic of the voltage loop, the controller has been designed considering only the first equation of the system described in (2). Thus, the following system plant is obtained:

$$
G_i=\frac{I_S(s)}{V_{LR}(s)}=\frac{1}{Ls+R}.
\tag{3}
$$

The scheme of the simplified current control loop is depicted in Fig. 4-(a). The design of the current controller assumes that grid voltage v_S is a variable but slow disturbance for the current loop.

B. Voltage Loop

Two PI controllers are necessary in order to manage the power transfer and the voltage levels on each H-bridge. Adding the last two equations of (2), the for the control of v_{C1} and v_{C2} becomes:

$$
S_1 i_S+S_2 i_S=i_{PV2}+i_{PV1}-C_1\frac{dv_{C1}}{dt}-C_2\frac{dv_{C2}}{dt}.
\tag{4}
$$

Considering only the dc component of the term $S_1\cdot i+S_2\cdot i$, the last equation is equivalent to [7]:

$$
\begin{aligned}
\frac{S_{1,\max}i_{S,\max}+S_{2,\max}i_{S,\max}}{2}&=i_{PV1}+i_{PV2}\\
&\quad-C_1\frac{dv_{C1}}{dt}-C_2\frac{dv_{C2}}{dt}
\end{aligned}
\tag{5}
$$

Considering the currents of the PV panels i_{PV1} and i_{PV2} as disturbances, and that the term $S_{1,\max}+S_{2,\max}$ is almost constant, the following system plant is obtained:

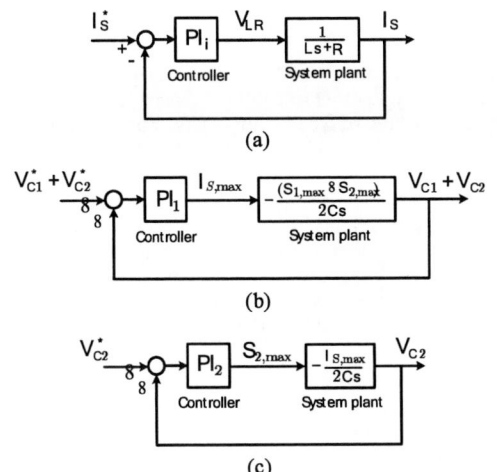

(a)

(b)

(c)

Fig. 4. (a) Current loop; (b) total voltage loop; (c) second cell voltage loop.

$$\frac{V_{C1}(s) + V_{C2}(s)}{I_{S,max}(s)} = -\frac{(S_{1,max} + S_{2,max})}{2Cs} \quad (6)$$

where $C_1 = C_2 = C$ and the term $S_{1,max} + S_{2,max}$ defined for nominal operating conditions as it is indicated the appendix.

The second control loop is responsible for the voltage difference between each dc-link. The last equation in (2), with same considerations of previous loop, provides the basis for the design of the controller.

$$\frac{S_{2,max} i_{S,max}}{2} = -C_2 \frac{dv_{C2}}{dt} \quad (7)$$

In this case, it will be assumed that the current magnitude delivered by the first control loop is constant. Thus, (7) can be expressed in the Laplace domain as:

$$\frac{V_{C2}(s)}{S_{2,max}(s)} = -\frac{I_{S,max}}{2Cs}, \quad (8)$$

where $I_{S,max}$ is defined for nominal steady state conditions as it is shown in the appendix. The schemes for the two voltage loops are shown in Fig. 4-(b) and (c).

The design of the three PI controllers has been carried out with root locus method.

C. Filter design

In order to avoid the 100 Hz component in the current which is generated by the single phase configuration, two digital band reject filters centered in 100 Hz have been placed between the v_{C1} and v_{C2} measurements and the inputs of voltage controllers. Note that filters between PV voltage measurements and MPPT blocks are intentionally avoided because it is necessary to assure the tracking of the optimum power point, and a filter could minimize the effect of environmental perturbations over PV voltages.

The digital filters work as illustrated in Fig. 5. The original signal V is shifted by half cycle and then added to the original waveform to obtain the DC component of the signal. The whole system including the two 100 Hz filters is depicted in Fig. 6.

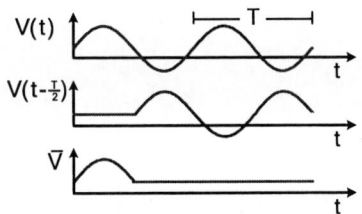

Fig. 5. Filter principle.

Fig. 6. Proposed control scheme with MPPT and band reject filters.

V. RESULTS

In order to validate the proposed ideas a simulation of the PV system was carried out. This simulation considered two H-Bridge rectifiers connected in series to a transformer.

Each H-Bridge is supplied by a PV panel which was modeled according to the specification of commercial PV panel Sharp 208U2. The DC-Link capacitor is 4700 (μF) for each module, and the AC filter has the parameters L=1 (mH) and R=0.1 (Ω). The input transformer provides 30 (V) peak at the terminals. The modulation of each cell is done using unipolar PWM generated by using a triangular carrier signal with a frequency of 5 (kHz).

The operation of the five-level inverter is simulated for three different operation conditions. In the first one, both temperature and solar radiation are equal for the two PV cells with 25°C and 1 (W/m²); then at t=1.5 (s) the solar radiation over the second panel decreases to 0.6 (W/m²); and finally at t=2.5 (s) the temperature in the second panel increases from 25 to 35°C. Fig. 7 shows the DC-Link voltage and reference for each cell. In the three stages, the MPPT gives references around the optimum point in only three levels according to the P&O algorithm best results [10], and the DC-Link voltages follow these references with a short transient even when the changes in solar radiation and temperature are steps, considerably faster than real weather changes. As expected, the first cell MPP voltage doesn't change because the power conditions in the whole simulation are the same. In contrast, the second cell voltage is the same when the solar radiation is lower because only the current decrease proportionally with this variable. Finally, the second cell MPP voltage changes in the third stage because the raise of temperature [11].

453

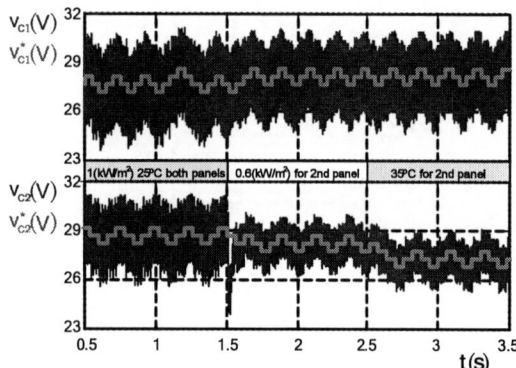

Fig. 7. DC-Links voltages of cells 1 and 2.

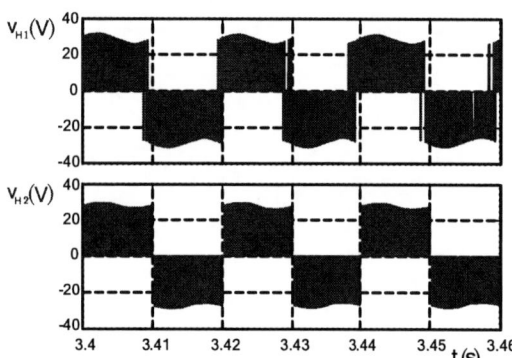

Fig. 8. Output voltage of the each cell.

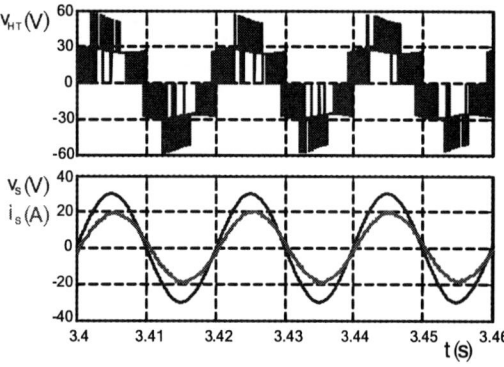

Fig. 9. Output voltage of the multilevel inverter, voltage and current in the terminals of the transformer.

Fig. 8 shows the 3-level voltage output of both cells. These voltages together becomes in the 5-level voltage at the output of the inverter depicted in Fig. 9, along with the voltage and current at the terminals of the transformer. In the upper part of the figure, the five levels in the output voltage of the multilevel inverter is depicted; lower, the current in the terminals of the transformer, which is in phase with the phase voltage.

VI. CONCLUSIONS

In this work a new control scheme is proposed for a multilevel PV system. This control scheme features an independent tracking of the maximum power point in each cell and delivers current to the mains with unitary power factor.

The proposed system features several advantages such as: high quality currents; is able to operate a lower switching frequency than a single converter; the control is simple; it allows the use of small transformer than the configuration with cells connected in parallel, reducing the losses and increasing the global efficiency.

APPENDIX

For $i_{S,\max}$ calculation, the input-output power balance is used. Thus, the following expression defines P_{in} and P_{out}, where the losses in the capacitor and mosfets have been neglected:

$$P_{in} = i_{PV1}v_{C1} + i_{PV2}v_{C2} - \tfrac{1}{2}Ri_{S,\max}^2$$
$$P_{out} = \tfrac{1}{2}v_{S,\max}i_{S,\max} \tag{9}$$

Considering that both PV modules are working in the same operation point, the both equations in (9) are equal, then:

$$P_{in} = P_{out}$$
$$2i_{PV}v_C - \tfrac{1}{2}Ri_{S,\max}^2 = \tfrac{1}{2}i_{S,\max}v_{S,\max} \tag{10}$$

Finally, the following expression for $i_{S,\max}$ is obtained:

$$i_{S,\max} = -\frac{v_{S,\max}}{2R} + \sqrt{\frac{v_{S,\max}^2}{4R^2} + \frac{4i_{PV}v_C}{R}} \tag{11}$$

The value of $S_{1,max} + S_{2,max}$ is derived from phasor relation described in Fig. 3. Considering only the amplitude of the phasors, from the Pythagoras theorem can be obtained:

$$\left(v_{H1,\max} + v_{H2,\max}\right)^2 = \left(v_{S,\max} + i_{S,\max}R\right)^2 + \left(\omega L i_{S,\max}\right)^2 \tag{12}$$

Considering that $v_{C1} = v_{C2} = v_C$, (12) can be written as:

$$\left(S_{1,\max} + S_{2,\max}\right)^2 v_C^2 = \left(v_{S,\max} + i_{S,\max}R\right)^2 + \left(\omega L i_{S,\max}\right)^2 \tag{13}$$

Then, $S_{1,max} + S_{2,max}$ becomes to:

$$S_{1,\max} + S_{2,\max} = \sqrt{\frac{\left(v_{S,\max} + i_{S,\max}R\right)^2}{v_C^2} + \frac{\left(\omega L i_{S,\max}\right)^2}{v_C^2}} \tag{14}$$

When $i_{S,\max}$ and $S_{1,max} + S_{2,max}$ are calculated, the operation point is the MPP of each panel, so MPP values for current and voltage at 1 (kW/m²) are used, i.e. $i_{PV} = i_{MPP}$ and $v_C = v_{MPP}$.

REFERENCES

[1] S. Kjaer, J. Pedersen and F. Blaabjerg, "A review of single-phase grid-connected inverters for photovoltaic modules," *IEEE Trans. Ind. Appl.*, vol.41, no.5, pp. 1292–1306, Sept.–Oct. 2005.

[2] B. Lindgren, "Topology for decentralized solar energy inverters with a low voltage ac-bus," in *Proc. of 8th EPE*, 1999.

[3] M. Meinhardt and G. Cramer, "Multi-string-converter: The next step in evolution of string-converter," in *Proc. of 9th EPE*, 2001.

[4] O. Alonso, P. Sanchis, E. Gubia and L. Marroyo, "Cascaded H-Bridge multilevel converter for grid connected photovoltaic generators with independent maximum power point tracking of each solar array," *Power Electronics Specialist Conference, 2003. PESC '03. 2003 IEEE 34th Annual*, vol.2, pp. 731–735, 15–19 June 2003.

[5] J. A. Barrena, L. Marroyo, M. A. Rodriguez, O. Alonso and J. R. Torrealday, "DC Voltage Balancing for PWM Cascaded H-Bridge Converter Based STATCOM," *IEEE Industrial Electronics, IECON 2006 - 32nd Annual Conference on*, pp.1840-1845, Nov. 2006.

[6] C. Cecati, A. Dell'Aquila, M. Liserre, and V. G. Monopoli, "Passivity-based control of a single phase H-bridge multilevel active rectifier," in *Proc. IEEE IECON'02* Seville, Spain, Nov. 4–10, 2002, pp. 3117–3122.

[7] A. Dell'Aquila, M. Liserre, V. G. Monopoli, and P. Rotondo, "Overview of PI-based solutions for the control of the DC-buses of a single-phase H-bridge multilevel active rectifier," in *Proc. IEEE APEC'04*, Anaheim, CA, 2004, pp. 836–842.

[8] J. Rodriguez, L. Moran, J. Pontt, J. L. Hernández, L. Silva, C. Silva and P. Lezana, "High-voltage multilevel converter with regeneration capability," *IEEE Trans. Ind. Appl.*, vol.49, no.4, pp. 839–846, August 2002.

[9] M. Calais, V. Agelidis, L. Borle and M. Dymond, "A transformerless five level cascaded inverter bases single phase photovoltaic system," *Power Electronics Specialists Conference, 2000. PESC '00. 2000 IEEE 31st Annual*, vol.3, pp.1173-1178, 2000.

[10] D. Hohm and M. Ropp, "Comparative study of maximum power point tracking algorithms," *Pro. Photovoltaics Res. Appl.*, vol.11, no.1, pp.47–62, 2003.

[11] Gow, J.A.; Manning, C.D., "Development of a photovoltaic array model for use in power-electronics simulation studies," *Electric Power Applications, IEE Proceedings*, vol.146, no.2, pp.193-200, Mar 1999.

Switching and Voltage Controls for a Flyback Switch-Mode Rectifier

Yuan-Chih Chang* and Chang-Ming Liaw*, Member IEEE

* Department of Electrical Engineering, National Tsing Hua University, Hsinchu, Taiwan, ROC.,
e-mail: *cmliaw@ee.nthu.edu.tw*

Abstract— **This paper presents the development of a flyback switch-mode rectifier (SMR) and its switching and voltage dynamic controls. In power circuit, the rating derivation and design for its constituted components are made. As to the control affair, a novel charge-regulated varying-frequency current control scheme is first developed. The proposed switching scheme possesses the advantages of without slope compensation, having robust current tracking control and ease of implementation. Then the quantitative and robust voltage regulation controls considering nonlinear behavior are made. The performance of the developed SMR is verified experimentally. Finally, the random switching for the developed SMR is studied.**

Keywords—**Power factor correction (PFC), converter control, robust control, modeling, PWM, random switching.**

I. INTRODUCTION

The application of switch-mode rectifier (SMR) [1,2], or called PFC rectifier, as a front-end converter for power equipments possesses the advantages of having controllable DC output voltage and high line drawn power quality. Generally, a single-stage SMR is formed by inserting a suitable DC/DC converter cell between diode rectifier and output capacitive filter. From input-output voltage magnitude relationship, the buck-boost SMR [3] is perfect in performing power factor correction control. However, the traditional buck-boost SMR possesses the limitations of without isolation and having reverse output voltage polarity. To avoid these limitations, flyback SMR is perhaps the best choice [4]. Till now, there have been a lot of flyback SMRs [4-9]. In [5], the limitations of flyback SMR in PFC characteristics and output voltage dynamic response are discussed. As far as the switching control strategies are concerned, they can be roughly categorized into voltage-follower control [6] and current-mode control [7-9]. The former belongs to open-loop operation under DCM, and thus the current feedback control is not needed. As to the latter, the multiplier-based current control loop is necessary to perform PFC control. The commonly used control approaches include peak current control, average current control, charge control and its modifications [7-9].

For a SMR, the nonlinear behavior and the double-frequency voltage ripple may let the closed-loop controlled SMR encounter undesired nonlinear phenomena [10]. The system may change from the stable operation to the double-period bifurcation, the high-distorted case, and finally enter the chaotic instability case. The key parameters to be observed in nonlinear behavior

of a SMR will be the loading condition, the value of output filtering capacitor and the voltage feedback controller parameters. It follows that suitable design of circuit and controller parameters is essential to avoid the nonlinear instability, and also to yield better control performance.

This paper develops a flyback SMR using a new charge-regulated current mode control. The key ratings of the SMR circuit components are first derived to facilitate their design. In the proposed switching scheme, the turn-on time is fixed and the off time interval of the switching interval is determined by the low-pass filtered switch current and the current command generated from the outer voltage loop. As to the voltage dynamic control, a quantitative designed feedback controller is augmented with a simple robust error cancellation control scheme. Good power factor control and voltage regulation control performances are demonstrated experimentally. Although the developed SMR possesses varying harmonic spectrum, it is not dispersedly distributed but gathers around a center frequency and its multiples. Hence finally, the random switching control [11-14] for the proposed SMR to obtain more uniformly harmonic spectrum is explored in this paper.

II. KEY COMPONENT RATINGS OF A FLYBACK SMR

A. Switching Control Approches

Circuit schematic of a flyback SMR and its sketched key waveforms are shown in Figs. 1(a) and 1(b). Its averaged AC input current within a switching period T_s under DCM can be found as [6]:

$$\langle i_{ac} \rangle_{T_s} = \frac{d^2 T_s}{2L} |v_{ac}| \qquad (1)$$

Hence $\langle i_{ac} \rangle_{T_s}$ is proportional to the input voltage $|v_{ac}|$ if $(d^2 T_s)$ is kept constant. It is naturally acted as an emulated resistor without voltage feedback control. However, the addition of inner loop current-mode control can enhance the performance of a SMR.

B. Voltage Ripple

For an ideal flyback SMR shown in Fig. 1(a), its AC line drawn current i_{ac} is purely sinusoidal and kept in phase with v_{ac}, i.e., $i_{ac} = v_{ac} / R_e$ with R_e being an emulated resistor viewing from v_{ac}. However, in an actual SMR, the double line frequency output voltage ripple always exists for finite value of output filtering capacitor. This ripple may distort the current command and hence

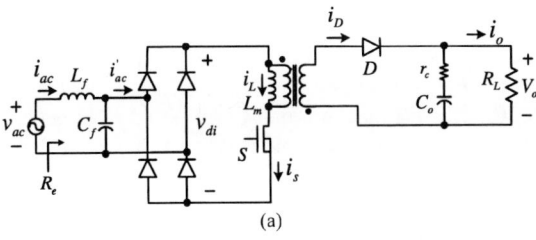

(a)

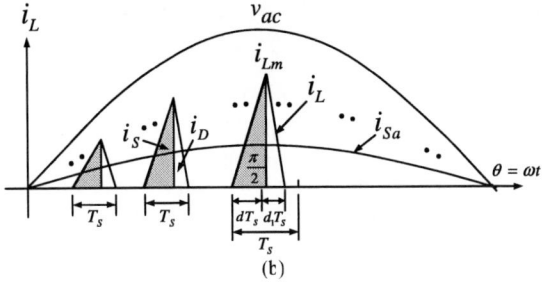

(b)

Fig. 1. A typical flyback SMR: (a) power circuit; (b) typical waveforms of i_L, i_S, i_D, and i_{Sa} under DCM.

worsen the power quality control performance. By neglecting the capacitor ESR r_c, the output ripple voltage can be derived as [6]:

$$\Delta v_{or} = \frac{V_o}{2\omega R_L C_o} \quad (2)$$

C. Component Current Ratings

For the magnetizing current sketched in Fig. 1(b), its peak value at a particular instant θ can be found as:

$$i_{Lm} = \frac{v_{di}}{L_m} dT_s, \; v_{di} = |v_{ac}| = V_m|\sin \omega t|, \; 0 \leq \omega t \leq \frac{\pi}{2} \quad (3)$$

and the instantaneous average value within a switching period of input current or switch current is:

$$i_{Sa} = \frac{v_{di}}{2L_m}(d^2T_s) \quad (4)$$

(a) Estimated RMS Switch Current

From Fig. 1(b) one can find that the maximum peak value of $i_{Sm} = i_{Lm}$ occurs at $\theta = \pi/2$. So the upper limit of the RMS value of i_S can be estimated using the switch current within one switching period at $\theta = \pi/2$ as:

$$I_S < \hat{I}_{S1} = \sqrt{\frac{1}{3}d(\frac{\pi}{2})i_{Lm}(\frac{\pi}{2})} \quad (5)$$

Obviously, this limit $\hat{I}_{S1}$ is too large, since $i_{Lm}(\theta) < i_{Lm}(\pi/2)$, $0 \leq \theta < \pi/2$.

A smaller limit is suggested by considering the RMS value of the instantaneous average value i_{Sa} and the current ripple at $\theta = \pi/2$ as:

$$I_S < \hat{I}_{S2} = \sqrt{\left(\frac{1}{\sqrt{2}}\frac{1}{2}d(\frac{\pi}{2})i_{Lm}(\frac{\pi}{2})\right)^2 + \left(\hat{I}_{S1}^2 - (\frac{1}{2}d(\frac{\pi}{2})i_{Lm}(\frac{\pi}{2}))^2\right)} \quad (6)$$

It can be found that $\hat{I}_{S2} < \hat{I}_{S1}$.

Exact RMS Switch Current:

The exact RMS switch current can be found by:

$$I_S = \sqrt{\frac{1}{T_1}\int_0^{T_1} i_S^2(t)dt} = \sqrt{\frac{1}{T_1}T_s\sum_{k=1}^{T_1/T_s}\left(\frac{1}{T_s}\int_{(k-1)T_s}^{kT_s} i_S^2(t)dt\right)} \quad (7)$$

Using (3) and (5), the RMS values of switch current corresponding to all switching periods between $0 \leq \omega t \leq \pi$ are found, then the exact RMS switch current are obtained by the root mean square value calculation in discrete summation form. Let

$$N \overset{\Delta}{=} T_1 \text{ mod } T_s \quad (8)$$

$$i_{Lm}(n) \overset{\Delta}{=} \frac{v_{di}(n\pi/N)}{2L_m}d(n)T_s = \frac{d(n)T_s}{2L_m}V_m\sin(n\pi/N),$$
$$0 \leq n \leq N \quad (9)$$

Then

$$I_S = \hat{I}_{S3} = \sqrt{\sum_{n=1}^N \frac{1}{3}d(n)i_{Lm}^2(n)} \quad (10)$$

which can be obtained via computer-aided numerical analysis. It is clear that $I_S = \hat{I}_{S3} < \hat{I}_{S2} < \hat{I}_{S1}$.

(b) Estimated RMS Diode Current

Similarly, the upper limit of the RMS value of i_D can be expressed from Fig. 1(b). as:

$$I_D < \hat{I}_{D1} = \sqrt{\frac{d_1(\pi/2)}{3}}i_{Lm}(\frac{\pi}{2})n \quad (11)$$

where d_1 is the off duty time, which can be found as:

$$d_1(\frac{\pi}{2}) = \frac{L_m i_{Lm}(\frac{\pi}{2})f_s}{nV_o} \quad (12)$$

From (12) one can find that the instantaneous average value of i_D is not a sinusoidal function. Hence the limit of i_D like those of (6) is not expressible.

Exact RMS Diode Current:

Similarly, the exact RMS value of diode current can be found following the definition given in (7) and using (11) and (12). The RMS values of diode current corresponding to all switching periods between $0 \leq \omega t \leq \pi$ are found, then the exact RMS diode current are obtained by the root mean square value calculation in discrete summation form. Let N and $i_{Lm}(k)$ be defined as in (9) and (10), then

$$I_D = \hat{I}_{D2} = \sqrt{\sum_{k=1}^N \frac{1}{3}d_1(k)i_{Lm}^2(k)n^2} \quad (13)$$

where $d_1(k)$ is not a constant, which can be found at each PWM switching period as:

$$d_1(k) = \frac{L_m i_{Lm}(\frac{k\pi}{N})f_s}{nV_o} \quad (14)$$

It is also clear that $I_D = \hat{I}_{D2} < \hat{I}_{D1}$.

III. THE FLYBACK SMR WITH THE PROPOSED SWITCHING AND VOLTAGE CONTROL SCHEMES

A. Switching Control Scheme

The developed charge-regulated current-controlled flyback SMR and control mechanism are shown in Fig. 2.

The proposed control methodology possesses the following features:

(i) It belongs to the constant turn-on time current mode control. Hence the switching frequency is not fixed.

(ii) This control is applicable for both DCM and CCM operations, only the DCM is treated in this paper.

(iii) In PWM switching control, the sensed switch current $v_i = K_i i_S$ is used to charge C_1 through R_1 in a constant ON time interval $t_{on} = t_1$ to yield i'_S (or v_s), then it is discharged. As the voltage i'_S (or v_s) across C_1 is smaller than the current command i^*_S (or v_c), the one-shot circuit is triggered again to let the operation be repeated. The use of the low-pass filtered feedback signal to trigger the PWM mechanism can increase the noise immunity. And the stabilized slope compensation is not required.

(iv) The analog circuit realization for the switching scheme shown in Fig. 2 is easily understood. The embedded components in the off-the-shelf voltage-to-frequency controller IC LM331 are employed to accomplish this goal.

The key variables in each instantaneous PWM period of this switching scheme are defined as: $\tau_c = R_1 C_1$ =time constant of the low-pass filter, $t_{on} = t_1$ = PWM ON time, $t_2 \overset{\Delta}{=} T_s - t_1$ =PWM OFF time, $f_s = 1/(t_1 + t_2)$ = switching frequency, $d = t_1/T_s$ =duty ratio, $d_1 T_s$ = output diode conduction time interval. Obviously, the condition of DCM operation is:

$$d_1 T_s < t_2 \qquad (15)$$

Some features of the proposed switching scheme can be found: (i) For the chosen t_1, the operation of the developed SMR is prone to become CCM for the heavier load or the smaller value of time constant $\tau_c = R_1 C_1$. (ii) Within $\theta = 0°$ to $180°$, the instantaneous turn-off time t_2 and hence T_s are shorter at the position closer to $\theta = 0°$. Thus partial CCM operation near $\theta = 0°$ may occur if too small value of t_1 is set. In the developed SMR, overall DCM operation under all load levels is preserved for the chosen t_1 and τ_c.

B. Voltage Control Scheme

The proposed voltage control scheme shown in Fig. 2(a) consists of a basic feedback controller and a robust error cancellation control scheme. The feedback controller is PI-type with:

$$G_{cv}(s) = K_{Pv} + \frac{K_{Iv}}{s} \qquad (16)$$

In the design of $G_{cv}(s)$, the equivalent control block and the desired voltage response due to a step load power change are shown in Fig. 3. The voltage sensing transfer function is set as $K_v(s) = K_v/(1 + \tau_{vf}s) = 0.0235/(1 + 0.001s)$, and $K_v(s) \approx 0.0235$ in the main dynamic frequency range.

At a chosen operating condition ($v_o = 48\,\text{V}$, $R_L = 30.3\Omega$), the voltage response due to a step load power change with an arbitrarily chosen $G_{cv}(s)$ is recorded. From the measured three typical responses points as those sketched in Fig. 3, the plant model parameters are estimated to be a=21.209, b=3.163 and K_{pl} =0.0548. Then the step speed regulation response requirements are specified as: (i) maximum voltage dip Δv_{om} =3.0V, ($\Delta v_o(t = t_m) = v_{om}$); (ii) restore time t_{re} =1.0s ($\Delta v_o(t = t_{re}) = 0.1v_{om}$). Then the parameters of $G_{cv}(s)$ are found analytically to yield:

$$G_{cv}(s) = 6.033 + \frac{30.644}{s} \qquad (17)$$

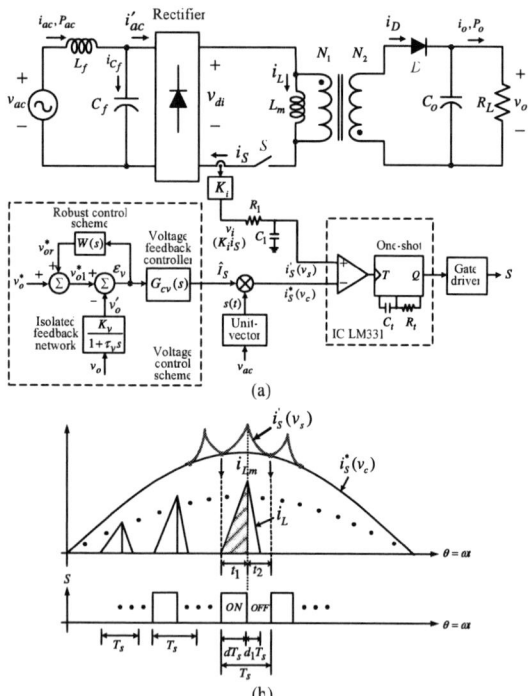

Fig. 2. The proposed flyback SMR: (a) circuit and control scheme; (b) some key waveforms.

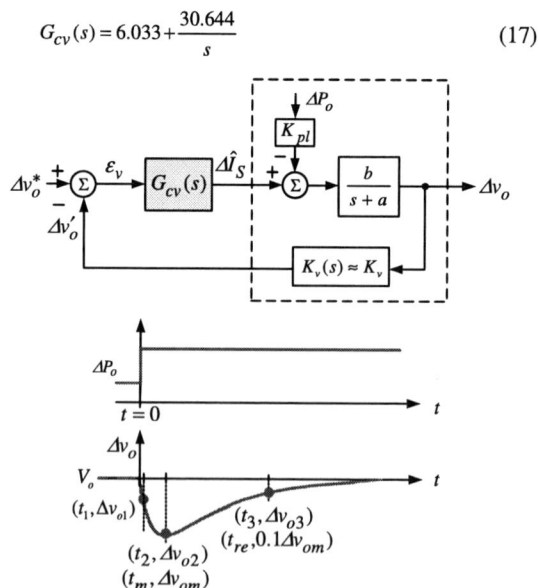

Fig. 3. Voltage feedback control scheme and desired step voltage regulation response.

A simple robust control is added to enhance the voltage regulation dynamic response and the PFC performance. In the proposed robust control scheme, a weighting function

$$W_v(s) = \frac{W_v}{1 + \tau_v s}, \quad 0 \le W_v < 1 \tag{18}$$

is employed to generate a compensation control command v_{or}^* from the tracking error ε_v. The low-passed filter with cut-off frequency $f_c = 1/(2\pi\tau_v)$ is employed to reduce the effect of high-frequency noise on the compensation control behavior. The benefits of adding the proposed robust control lie in: (i) the voltage variation caused by various disturbances can be much reduced to $(1-W_v)\varepsilon_v$ within main dynamic frequency range; (ii) the power factor control performance will also be more robust against the nonlinear behavior and system parameter changes. Here, $W_v = 0.45$ and $\tau_v = 0.00068$ ($f_c = 234\,\text{Hz}$) are chosen from the experimental intuition.

C. Circuit Components

Specifications:

The specifications at nominal case (rated load) are given as: (i) AC input: 110V/60Hz; (ii) DC output: 48V/100W ($R_L = 23.04\,\Omega$), voltage ripple factor $\Delta v_o/V_o < 5\%$; (iii) power factor: $PF > 0.97$, efficiency: $\eta \stackrel{\Delta}{=} P_o/P_{ac} > 70\%$.

The designed and chosen circuit components according to the derived current ratings are summarized as follows:

(a) Flyback transformer: it is made using TDK PC40 EE55 EE core. The primary and secondary coils are wound using 5 AWG #22 wires with $N_1 = 30$ and $N_2 = 10$ turns, respectively. The measured magnetizing inductances are $L_{m1} = 172.55$ H, $L_{m2} = 25.372$ H at 20kHz using Hioki 3532-50 LCR HiTESTER.

(b) Power semiconductor devices: (i) bridge rectifier: GBPC35-10 (700V/35A), Fairchild Semiconductor; (ii) Switch: IRFP460 (500V/20A continuous, 80A pulsed), International Rectifier; (iii) Diode: RURP3060 (600V/30A average, 70A repetitive peak), Fairchild Semiconductor.

(c) Input filter: The input filter cut-off frequency f_c is chosen based on the rule of $(f_1 = 60\text{Hz}) < f_c < (f_s = 20\text{kHz})$. By choosing $f_c = 6\text{kHz}$ and $C_f = 1$ F/250V, one can find the inductance L_f from $f_c = 1/(2\pi\sqrt{L_f C_f})$ to be $L_f = 703.62$ H. The inductor is wound on a toroidal core T130-26 (Micrometals company) using 89 turns of the wire AWG #18. The measured inductance using HIOKI 3532-50 LCR meter are: $L_f = 761.25$ H (60Hz) and 702.39 H (20kHz).

(d) Output filter: The output filtering capacitor is chosen to be $C_o = 1800$ F/450V. By neglecting the ESR, the output voltage ripple is estimated to be:

$$\Delta v_o \approx \frac{V_o}{2\omega_1 R_L C_o} = \frac{48}{2 \times 60 \times 2\pi \times 23.04 \times 1800 \times 10^{-6}} \tag{19}$$
$$= 1.535\,\text{V}$$

It corresponds to 3.20% of $v_o = 48\,\text{V}$. And the hold-up time can be found as:

$$t_h = \frac{C_o(V_o^2 - V_{o,\min}^2)}{2P_o} = \frac{1800 \times 10^{-6}(48^2 - 30^2)}{2 \times 100} \tag{20}$$
$$= 12.64\,\text{ms}$$

where $V_{o,\min} = 30\,\text{V}$ is set.

IV. EXPERIMENTAL RESULTS

Some experimental results are provided to evaluate its static and dynamic performances. Moreover, the improved nonlinear behavior via the developed control approach is also demonstrated.

A. Basic Characteristics

Let the developed SMR be normally operated under only PI feedback control ($W_v = 0$) at the operation condition ($V_{ac} = 110\text{V}/60\text{Hz}$, $v_o = 48\text{V}$, $R_L = 23.2\Omega$, $P_o = 102.28\text{W}$), Fig. 4(a) to Fig. 4(c) respectively show the measured (i_s^*, i_s'), (v_{ac}, i_{ac}) and the spectrum of i_S. Some facts are observed from the results: (i) normal operation is achieved with all key waveforms being similar to the analytically predicted and the simulated ones (not shown here); (ii) the measured switch and diode RMS currents are rather closed to the simulated results; (iii) good AC input power quality ($PF = 0.9936$, $THD_i = 5.92\%$) is obtained; (iv) the measured dominant switch current harmonic frequencies are not fixed and around $f_h = 21\text{kHz}$ and its multiples.

B. Robust Control Effectiveness

Fig. 5(a) ($v_o = 48V$, $R_L = 23.2\Omega$) and Fig. 5(b) ($v_o = 48\text{V}$, $R_L = 230\Omega$) show the measured waveforms ($v_{ac}, \Delta v_o, i_{ac}$) (upper) and phase portrait of ($\Delta v_o, v_{ac}$) (lower) by PI control without (left) and with (right) robust control ($W_v = 0.45$). Table I and Table II list the measured steady-state characteristics at several load powers without and with adding the proposed robust control. As to the voltage control characteristics, Fig. 6 shows the measured voltage responses by PI control without ($W_v = 0$) and with ($W_v = 0.82$) robust control due to a load resistance change $R_L = 30.3\Omega \rightarrow 23.2\Omega$ ($P_o = 77.62\text{W} \rightarrow 102.28\text{W}$) at ($v_o = 48\text{V}$, $R_L = 30.3\Omega$). From the results in Tables I and II, Figs. 5 and 6 one can observe the facts: (i) the designed SMR possesses good static and dynamic performances under wide load range; (ii) at very light load (about 10W), the PI control results in chaotic phenomena. With the developed robust control, the effects of voltage ripple on the PFC control operation are greatly reduced to yield cleaner waveforms and hence the improved SMR control performances; (iii) the faster restoration and smaller dip in voltage step regulation response can be seen from Fig. 6. And the improved PFC control behavior can also be aware from the comparison of the results shown in Fig. 5, Tables I and II.

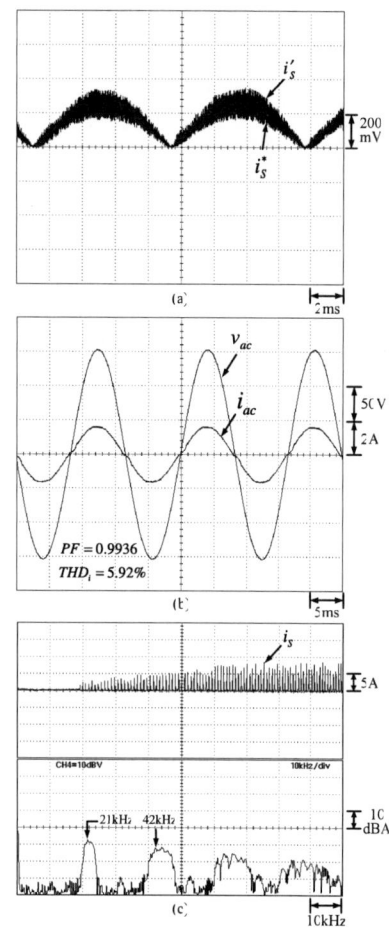

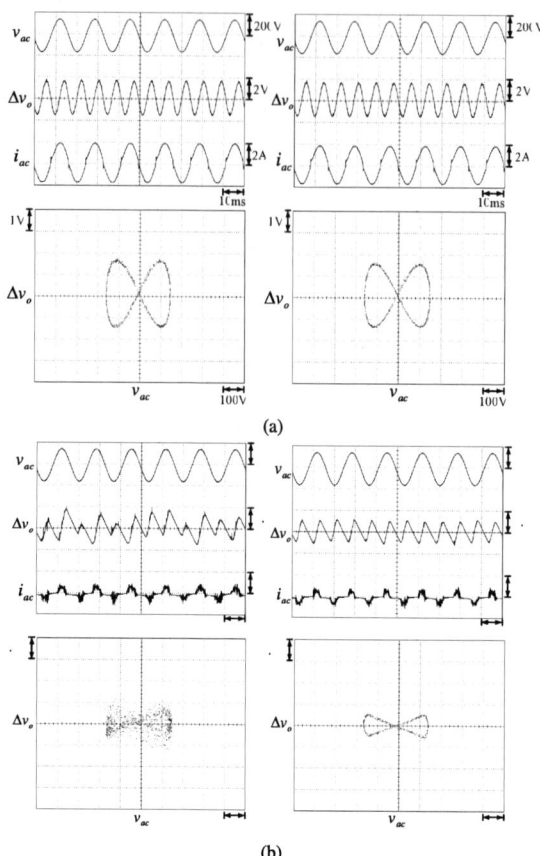

Fig. 5. Measured waveforms (v_{ac}, Δv_o, i_{ac}) (upper) and phase portrait of (Δv_o, v_{ac}) (lower) by PI control without (W_v=0.0)(left) and with (W_v=0.45)(right) robust control: (a) ($v_o = 48V$, R_L =23.2 Ω, P_o =102.28W); (b) ($v_o = 48V$, R_L =230 Ω, P_o=10.08W).

Fig. 4. Measured results of the designed flyback SMR by PI feedback control (W_v=0) at steady-state of ($v_o = 48V$, $R_L = 23.2\Omega$, $P_o = 102.28W$): (a) i_s^* and i_s' ; (b) v_{ac} and i_{ac} ; (c) spectrum of i_s .

TABLE I

MEASURED RESULTS OF THE DEVELOPED SMR UNDER PI CONTROL (W_v=0) AT DIFFERENT OUTPUT POWERS

P_o	102.28W	24.98W	12.04W	11.13W	10.08W
R_L	23.2Ω	92Ω	192Ω	210Ω	230Ω
P_{ac}	138.33W	34.02W	16.32W	15.22W	13.98W
η	73.94%	73.43%	73.77%	73.13%	72.10%
PF	0.9936	0.9849	0.9440	0.9323	0.9215
THD_i	5.92%	16.87%	32.87%	34.47%	42.15%

TABLE II

MEASURED RESULTS OF THE DEVELOPED SMR UNDER PI CONTROL (W_v=0.45) AT DIFFERENT OUTPUT POWERS

P_o	102.12W	25.23W	12.07W	11.18W	10.06W
R_L	23.2Ω	92Ω	192Ω	210Ω	230Ω
P_{ac}	137.89W	34.08W	16.29W	15.18W	13.89W
η	74.06%	74.03%	74.09%	73.65%	72.43%
PF	0.9946	0.9918	0.9820	0.9759	0.9732
THD_i	5.23%	12.62%	20.18%	21.03%	23.52%

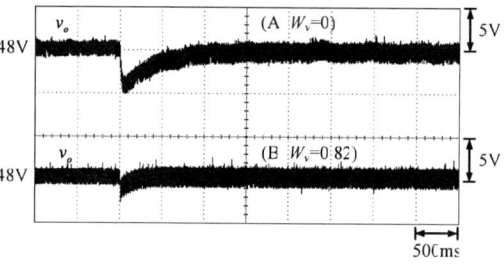

Fig. 6. Measured voltage responses by PI control without (W_v =0) and with (W_v =0.82) robust control due to a load resistance change $R_L = 30.3\Omega \rightarrow 23.2\Omega$ (P_o = 77.62W → 102.28W) at v_o = 48 V, $R_L = 30.3\Omega$).

C. Random Switching

It is known that random switching is an effective means to let a PWM switching controlled converter possess uniformly distributed harmonic spectrum, and there have already been many existing random switch methods [11-14]. In this paper, the feasibility of random switching for the developed flyback SMR and its effectiveness are also studied. The simple approach [14] of fixed turn-on time and repeatedly varying two off-times is adopted here. The proposed random switching scheme is shown in Fig. 7(a) with its key waveforms being sketched in Fig. 7(b). By alternately changing the resistances $R_1 = R_{10}$ and $R_1 = R_{10} // \Delta R_1$ in the timing networks, the off-times will be changed in two values accordingly. It follows that the harmonic spectrum of i_s can be made uniformly distributed.

At the same conditions of Fig. 4 and Fig. 6, the measured steady-state waveforms and voltage dynamic responses are plotted in Fig. 8 and Fig. 9, respectively. The results indicate that without changing the control performances the spectrum of i_s becomes dispersedly distributed with smaller magnitudes at the specific frequencies. However, the more sophisticated RPWM approaches to yield more uniformly harmonic spectrum are worth developing and the research results will be presented in the near future.

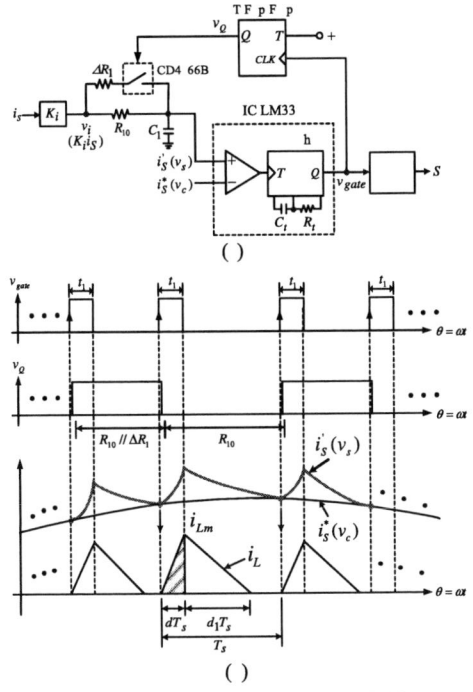

Fig. 7. The proposed flyback SMR with random switching control: (a) implementation of control scheme; (b) some key waveforms.

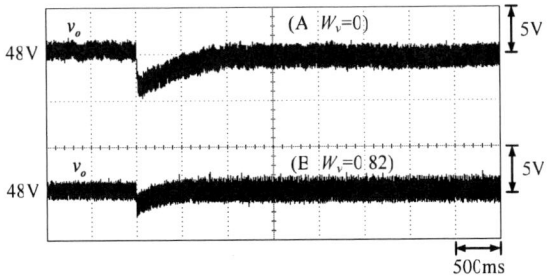

Fig. 8. Measured voltage responses of the flyback SMR with random switching by PI control without ($W_v = 0$) and with ($W_v = 0.82$) robust control due to a load resistance change $R_L = 30.3\Omega \rightarrow 23.2\Omega$ ($P_o = 77.62W \rightarrow 102.28W$) at $v_o = 48V$, $R_L = 30.3\Omega$).

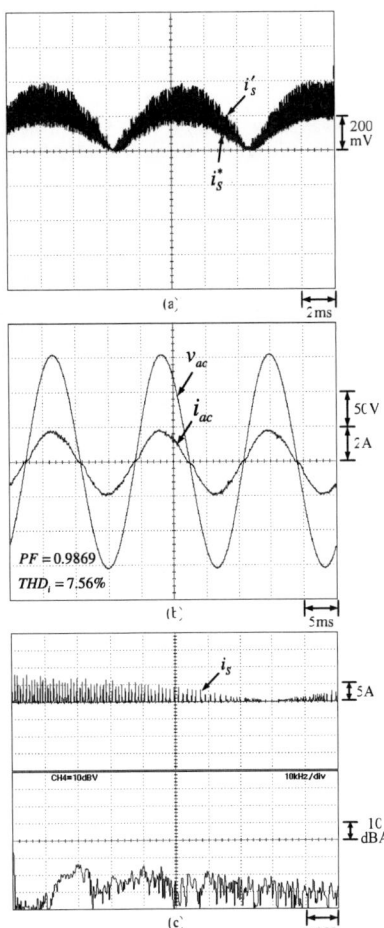

Fig. 9. Measured results of the designed flyback SMR using random switching scheme at steady-state of ($v_o = 48$ V, $R_L = 23.2\Omega$, $P_o = 102.28$W): (a) i_s^* and i_s'; (b) v_{ac} and i_{ac}; (c) spectrum of i_s.

V. CONCLUSIONS

The establishment of a single-stage flyback SMR with charge-regulated varying-frequency current control has been presented. After deriving the ratings of key constituted components, a flyback SMR operating in DCM is designed and implemented. Through the developed switching mechanism, the SMR can be stably operated under wide varying operation conditions. Then the robust voltage control scheme is proposed. The experimental evaluation found that by applying the developed robust control, the cleaner waveforms and hence the improved SMR control performances are obtained. Specifically speaking, the faster restoration and smaller dip in voltage step regulation response are achieved. And meanwhile the nonlinear behavior of a SMR is also improved with much improved SMR performance at lighter loads. Finally, the random switching for the developed SMR is explored. The measured results showed that the switch current harmonic spectrum can be made more dispersedly distributed. However, more sophisticated RPWM methods are needed to develop.

VI. REFERENCES

[1] R. W. Erickson and D. Maksimovic, *Fundamentals of Power Electronics*, 2nd ed., Kluwer Academic Publishers, Norwell Massachusetts, 2001.

[2] O. Garcia, J. A. Cobos, R. Prieto, P. Alou and J. Uceda, "Single phase power factor correction: a survey," *IEEE Trans. Power Electron.*, vol. 18, no.3, pp. 749-755, 2003.

[3] K. Matsui, I. Yamamoto, T. Kishi, M. Hasegawa, H. Mori and F. Ueda, "A comparison of various buck-boost converters and their application to PFC," *Proc. IEEE IECON '02*, 2002, vol. 1, pp. 30-36.

[4] B. Singh, B. P. Singh and S. Dwivedi, "Performance comparison of high frequency isolated AC-DC converters for power quality improvement at input AC mains," *Proc. IEEE PEDES '06*, 2006, pp. 1-6.

[5] D. G. Lamar, A. Fernandez, M. Arias, M. Rodriguez, J. Sebastian and M. M. Hernando, "Limitations of the flyback power factor corrector as a one-stage power supply," *Proc. IEEE PESC'07*, 2007, pp. 1343-1348.

[6] R. Erickson and M. Madigan, "Design of a simple high-power-factor rectifier based on the flyback converter," *Proc. IEEE APEC*, 1990, pp. 792-801.

[7] W. Tang, Y. H. Jiang, G. C. Verghese and F.C Lee, "Power factor correction with flyback converter employing charge control," *Proc. IEEE APEC*, 1993, pp. 293-298.

[8] C. Larouci, J. P. Ferrieux, L. Gerbaud, J. Roudet and J. Barbaroux, "Control of a flyback converter in power factor correction mode: compromise between the current constraints and the transformer volume," *Proc. IEEE APEC*, 2002, vol. 2, pp. 722-727.

[9] S. Buso, G. Spiazzi and D. Tagliavia, "Simplified control technique for high-power-factor flyback Cuk and Sepic rectifiers operating in CCM," *IEEE Trans. Ind. Appl.* .vol. 36, no. 5, pp. 1413-1418, Sept./Oct. 2000.

[10] M. Orabi and T. Ninomiya, "Nonlinear dynamics of power-factor-correction converter," *IEEE Trans. Ind. Electron.*, vol. 50, pp. 1116-1125, 2003.

[11] T. G. Habetler and D. M. Divan, "Acoustic noise reduction in sinusoidal PWM drives using a randomly modulated carrier," *IEEE Trans. Power Electron.*, vol. 6, pp. 356-363, May 1991.

[12] A. M. Trzynadlowski, F. Blaabjerg, J. K. Pedersen, R. L. Kirlin and S. Legowski, "Random pulse width modulation techniques for converter-fed drive systems- A review," *IEEE Trans. Ind. Appl.*, vol. 30, pp. 1166-1175, Oct. 1994.

[13] C. M. Liaw, Y. M. Lin, C. H. Wu and K. I. Hwy, "Analysis, design and implementation of a random frequency PWM inverter," *IEEE Trans. Power Electron.*, vol. 15, no. 5, pp. 843-854, 2000.

[14] J. Y. Chai, Y. H. Ho, Y. C. Chang and C. M. Liaw," On acoustic noise reduction control using random switching technique for switch-mode rectifiers in PMSM drive," *IEEE Trans. Ind. Electron.*, vol. 55, no. 3, pp. 1295-1309, March 2008.

Method Of Designing ZVS Boost Converter

Mirosław Luft[1], Elżbieta Szychta[2], Leszek Szychta[3]

[1] Technical University of Radom, Radom, Poland, e-mail: *m.luft@pr.radom.pl*
[2] Technical University of Radom, Radom, Poland, e-mail: *e.szychta@pr.radom.pl*
[3] Technical University of Radom, Radom, Poland, e-mail: *l.szychta@pr.radom.pl*

Abstract—The article presents a method of designing multiresonant ZVS boost converter including one transistor based on simulation testing. Dependencies are given between parameters of resonant circuit elements and parameters of the control system which condition ZVS operation of the converter. Results of simulation and experimental tests provide grounds for the conclusion that the presented method allows the determination of the values of the resonant circuit elements.

Keywords—Converter circuit, Resonant converter, ZVS converters.

I. INTRODUCTION

Multiresonant ZVS DC/DC converters are resonant circuits where oscillations supporting processes of switching semiconductor elements at zero voltage occur with at least three resonant frequencies in a full operation cycle. High control frequency is the fundamental characteristic of these circuits. Multiresonant ZVS converters are characterised by great energy efficiency ratio, minimum dimensions and minimum electromagnetic and acoustic interference [3]. Power of such converters is usually below 5 kW [1]. These converters are applied, among other uses, in military technology, to supply power to information technology and telematic systems, in transportation systems and many other areas of demand for DC electricity. Interest in the practical potential of these circuits is growing. Designing of multiresonant converters involves necessary application of complex numerical analysis [7], therefore effective methods of designing these circuits, need to be developed. Available research [5] does not cover the problem in full.

The article presents a method of designing multiresonant ZVS boost converter including one transistor. The method is based on simulation testing by means of Simplorer software. It enables to design the circuit without recourse to complex numerical analysis.

II. TOPOLOGY OF ZVS BOOST MRC

Figure 1 presents a simulation model of single-transistor ZVS boost MRC according to Simplorer [2]. The circuit includes models of the following reactive elements: L=7µH, C_S=7nF, C_D=23nF, L_F=600µH C_F=10µF, R_N=0,5 and R_N=1 and models of semiconductor elements: transistor MOSFET IRFP460, diode HFA25TB60. The models of semiconductor elements have parameters of real equipment. Supply voltage is E=50V DC.

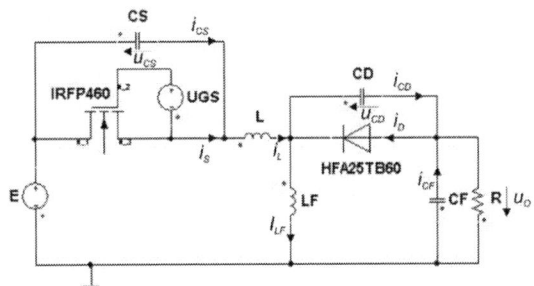

Fig.1. ZVS boost MRC

Essentials notation used in the paper [3]:
ratio of voltage conversion M:

$$M = \frac{U_o}{E} \qquad (1)$$

load current I_O:

$$I_o = \frac{M \cdot E}{R_N \cdot Z_S} \qquad (2)$$

load resistance R_N in relative units:

$$R_N = \frac{R}{Z_S} \qquad (3)$$

characteristic impedance Z_S:

$$Z_S = \sqrt{\frac{L}{(C_S + C_{os})}} \qquad (4)$$

switching frequency in relative units f_N:

$$f_N = \frac{f}{f_S} \qquad (5)$$

where: f – MRC's control frequency
f_S - resonant frequency of L, (C_S+C_{OS}) circuit:

$$f_S = \frac{1}{2\pi\sqrt{L(C_S + C_{os})}} \qquad (6)$$

capacitance factor C_N:

978-1-4244-1741-4/08/$25.00 ©2008 IEEE   463

$$C_N = \frac{C_D + C_{OD}}{C_S + C_{OS}} \qquad (7)$$

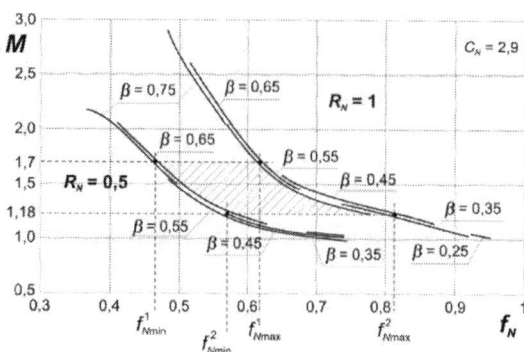

Fig.2. Determination of f_N variation range for $M \in \langle 1,18 \div 1,7 \rangle$, $R_N \in \langle 0,5 \div 1 \rangle$

III. ZVS OPERATING REGION

The control system of ZVS MRC (Fig. 1) is based on the method of frequency control at the constant time of transistor's turn-off t_{off} [4]. The transistor's control modulation ratio β should have such values that MRC's semiconductor elements are switched at zero voltage (ZVS). β is expressed:

$$\beta = \frac{t_{on}}{T} = \frac{T - t_{off}}{T} = 1 - t_{off} \cdot f \qquad (8)$$

where: $f = \dfrac{1}{T} = \dfrac{1}{t_{on} + t_{off}}$

MRC's ZVS operating region is delimited with curves plotted for minimum β_{min} and maximum β_{max} within the acceptable range of f_N variation. Minimum values of β_{min} correspond to maximum time of transistor's turn-off $t_{off\,max}$, while maximum values of β_{max} correspond to minimum time of transistor's turn-off $t_{off\,min}$. MRC's ZVS operating region is defined by such values of β that meet the condition:

$$\beta_{max} \geq \beta \geq \beta_{min} \qquad (9)$$

where minimum $t_{off\,min}$ and maximum $t_{off\,max}$ meet the following condition in the full variation range of control frequency f_N:

$$t_{off\,min} \leq t_{off} \leq t_{off\,max} \qquad (10)$$

IV. METHOD OF SELECTING ELEMENTS OF THE ZVS MRC'S RESONANT CIRCUIT

The following sample input data of the boost MRC's are accepted for selection of the elements:
ratio of voltage conversion $M \in \langle 1,18 \div 1,7 \rangle$,
load resistance R_N in relative units $R_N \in \langle 0,5 \div 1 \rangle$,
supply voltage E=50V,
resonant frequency f_S,
transistor's drain-source voltage V_{ds}=500V,
transistor's drain current I_d=20A.

Selection of the resonant circuit elements is an algorithm.

A. Determination of the switching frequency range f_N

For a given variation range of the voltage conversion ratio $M \in \langle 1.18 \div 1.7 \rangle$, the curves of M (Fig.2) (determined by means of simulation testing) serve to define the range of minimum frequency $f_{N\,min} \in \left(f_{N\,min}^1 \div f_{N\,min}^2 \right)$ at R_N=0.5, and the range of maximum frequency $f_{N\,max} \in \left(f_{N\,max}^1 \div f_{N\,max}^2 \right)$ at $R_N = 1$.

B. Determination of the acceptable ZVS operating region corresponding to the assumptions.

On the basis of boost MRC's ZVS operating region, determined by means of simulation testing (Fig.3 and Fig.4), variation ranges of β for the given $f_{N\,min}$, $f_{N\,max}$ and appropriate $R_N \in \langle 0,5 \div 1 \rangle$ are determined.

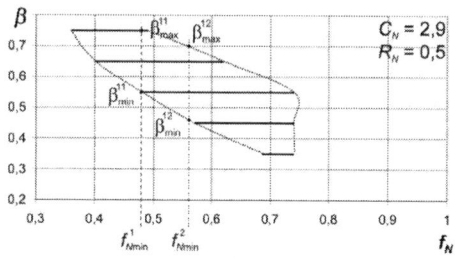

Fig.3. Boost MRC's ZVS operating region for R_N=0.5

Following from the ZVS operating region at R_N=0.5 (Fig.3), β_{min}^{11}, β_{max}^{11} are obtained with respect to $f_{N\,min}^1$, and β_{min}^{12}, β_{max}^{12} are obtained for $f_{N\,min}^2$.

a, b indices in the control modulation ratios of transistor β_{min}^{ab}, β_{max}^{ab} denote:

a=1; for $R_N = 0,5$,   a=2; for $R_N = 1$
b=1; for $f = f_N^1$,   b=2; for $f = f_N^2$

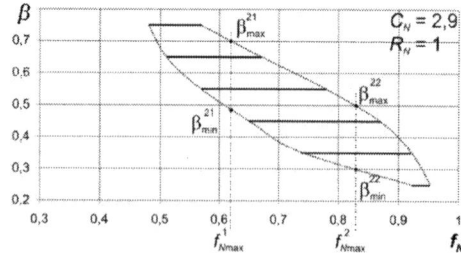

Fig.4. Boost MRC's ZVS operating region for R_N=1

Following from the ZVS operating region at R_N=1 (Fig.4), β_{min}^{21}, β_{max}^{21} result for $f_{N\,max}^{1}$, and β_{min}^{22}, β_{max}^{22} are produced in regard of $f_{N\,max}^{2}$.

C. The determination of t_{off} at a constant value

$t_{off} = \text{const}$ in the full variation range of control frequency f_N at variable values of M and R_N.

MRC's ZVS operation at various f_N (corresponding to various M) and various values of R_N affects t_{CS} of the capacitor's C_S overloading. t_{CS} corresponds to minimum time of transistor's turn-off t_{offmin} (Fig.5). $t_{off} = \text{const}$ in the full variation range of MRC's operation must meet the condition (10). On the basis of assumed variation range of $M \in \langle 1,18 \div 1,7 \rangle$ and load resistance R_N=0.5 and R_N=1, four possible values of t_{off} are obtained (Fig.5):

- $t_{off\,min\,1}$ – determined for M=1,18, R_N=1, (point a),
- $t_{off\,min\,2}$ – determined for M=1,18, R_N=0,5, (point b),
- $t_{off\,min\,3}$ – determined for M=1,7, R_N=1, (point c),
- $t_{off\,min\,4}$ – determined for M=1,7, R_N=0,5, (point d).

$t_{off\,min\,1}$=0,87μs fulfils the condition (10) and determines constant time of transistor's turn-off t_{off} in the assumed range of MRC's operation.

There is a relation between t_{off} and β (8). When (10) is fulfilled by t_{off}, the condition (9) should also be met by β. Variation of β at $t_{off} = \text{const} = t_{off\,min\,1} \div t_{off\,min\,4}$, is illustrated in Figures 6 and 7.

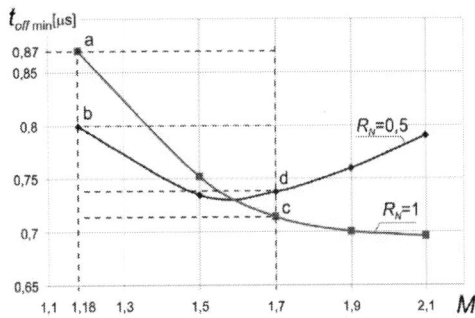

Fig.5. The transistor's turn-off times $t_{off\,min}$ for $M \in \langle 1,18 \div 1,7 \rangle$, R_N=0,5 and R_N=1 (simulation results)

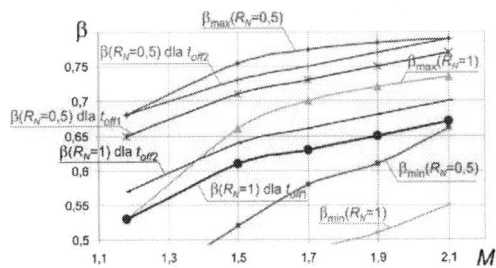

Fig.6. Ratio β at control with constant time of transistor's turn-off $t_{off\,min\,1}$, $t_{off\,min\,2}$, for $M \in \langle 1,18 \div 2,1 \rangle$, R_N=0,5 and R_N=1

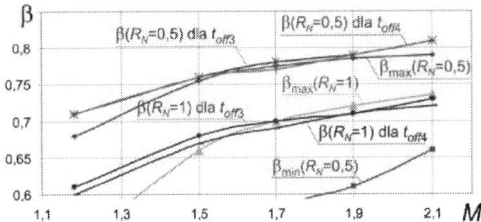

Fig.7. Ratio β at control with constant time of transistor's turn-off $t_{off\,min\,3}$, $t_{off\,min\,4}$ for $M \in \langle 1,18 \div 2,1 \rangle$, R_N=0,5 and R_N=1

Figures 6 and 7 indicate that β is within ZVS operating region in the full variation range of f_N and the conditions (9) and (10) are fulfilled only during control with constant time of transistor's turn-off $t_{off} = t_{off\,min\,1}$. This means that parameters of the resonant circuit L, C_S, C_D elements should be selected with respect to point a (R_N=1) of MRC's operation.

D. Verification of the maximum voltage across the transistor $U_{CS\,max}$.

Maximum voltage across the transistor $U_{CS\,max}$ is defined with the aid of control characteristics of the transistor's maximum voltage U_{CSmax}/E in relative units (Fig.8) for $f_{N\,min}^{1}$ and R_N=0,5, $U_{CSmax}/E \approx 7 \Rightarrow U_{CS\,max} \approx 350$V, up to the acceptable catalogue value Vds=500V.

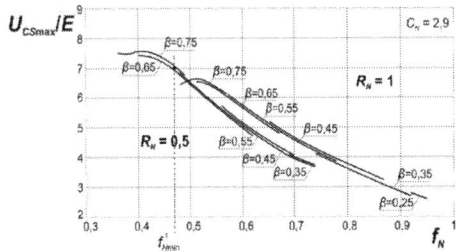

Fig.8. Determination of the transistor's maximum voltage U_{CSmax}/E

E. Verification of the maximum current across the transistor $I_{S\max}$.

The transistor's maximum current $I_{S\max}$ is defined with the aid of control characteristics of the transistor's maximum current $I_{S\max}/I_O$ in relative units (Fig.9) for $f_{N\min}^1$ and $R_N=1$, $I_{S\max}/I_O \approx 3,7 \Rightarrow I_{S\max} \approx 11\,\text{A}$, up to the acceptable catalogue value $Id_{kat}= 20\text{A}$.

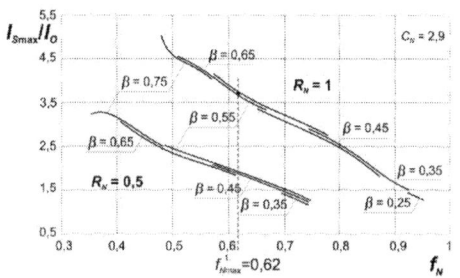

Fig.8. Determination of the transistor's maximum current $I_{S\max}/I_O$

F. Determination of reactive element values

For a known R, on the basis of (3), (4), (6), inductance L is [3]:

$$L = \frac{1}{2\pi}\frac{R}{R_N}\frac{1}{f_s} \qquad (11)$$

C_S is calculated from (6).
For MRC to implement ZVS operation, the following condition must be fulfilled [3]:

$$L \le \left(\frac{1-\beta_{\max}}{\pi \cdot f_{\max}}\right)^2 \cdot \frac{1}{C_S + C_{OS}} \cdot \frac{1+C_N}{C_N} \qquad (12)$$

C_D is calculated from (7).
L_F should be chosen in relation to minimum $f_{\min}$ of the transistor switching and maximum $\beta_{\max}$ within ZVS operating region, thus fulfilling [3]:

$$L_F \ge \left(\frac{1}{\pi \cdot f_{\min}}\right)^2 \cdot \frac{1}{C_S + C_{OS}} \qquad (13)$$

V. RESULTS OF EXPERIMENTAL TESTING OF ZVS BOOST MRC

Based on presented method of designing of ZVS boost MRC, an experimental circuit was designed and executed (Fig.9). Tests were carried out at load resistances $R_N=0.5$ and $R_N=1$ [3].

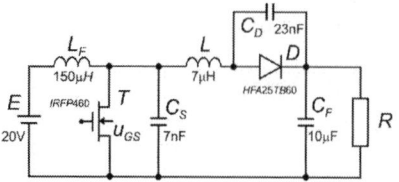

Fig.9. Circuit diagram of the experimental ZVS boost MRC

The control system diagram of the experimental ZVS boost MRC is illustrated in Figure 10 [3].

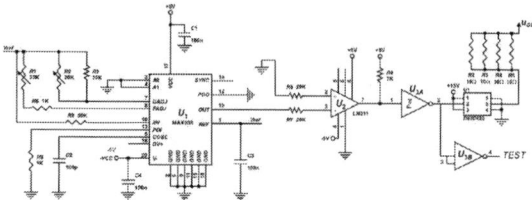

Fig.10. Control system diagram of the experimental ZVS boost MRC

ZVS operation region

On the basis of experimental testing of the system presented in Figure 9, the region of ZVS operation of the multiresonant ZVS boost converter was defined as the dependence $\beta=f(f_N)$, for $R_N=0.5$ and $R_N=1$. Results derived from experimental testing, shown in Figure 11, exhibit conformity with results obtained in the corresponding simulation tests (Fig.3, 4). This conformity is maintained even where different voltages are supplied to the converter.

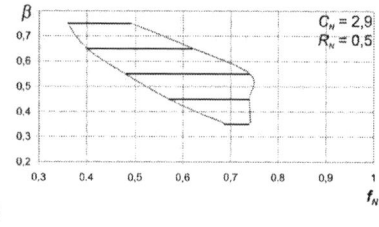

a)

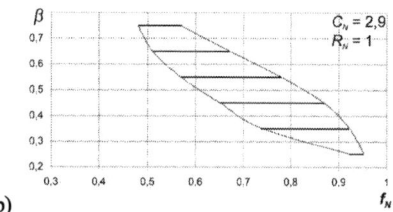

b)

Fig.11. The operating region at zero voltage switching in the experimental ZVS boost MRC.

Selected current and voltage waveforms

Figure 12 presents selected current and voltage waveforms of the resonant circuit elements in the ZVS boost MRC, obtained in simulation (Fig.12a) and experimental (Fig.12b) tests. Notation in regard to the waveforms is shown in Figure 1. To compare the conformity of simulation and experimental results, red-coloured current and voltage waveforms derived from simulation tests were superimposed over i_S, i_L and u_{CS}, u_{CD} waveforms obtained on the basis of experimental testing.

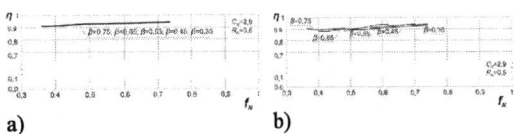

a)   b)

Fig.13. Efficiency ratio η of the ZVS boost MRC; E=20V
a) simulation results, b) experimental results

8. CONCLUSION

1. The paper has presented a selection algorithm of ZVS boost MRC's resonant circuit elements. Determination of the variation range of the transistor's turn-off time t_{off} in the full range of control frequency f_N at varied values of M and R_N is the supreme selection criterion.

2. The design method discussed above enables selection of boost MRC's elements on the basis of ZVS operating region obtained in simulation testing without necessarily resorting to complex numerical calculations.

3. Conformities between results of simulation and experimental tests provide grounds for the conclusion that the designing of a converter based on simulation testing in consideration of ZVS operating region allows the determination of the values of the resonant circuit elements.

4. MRC's ZVS operating region obtained in simulation testing enables to define ranges of the control system parameters.

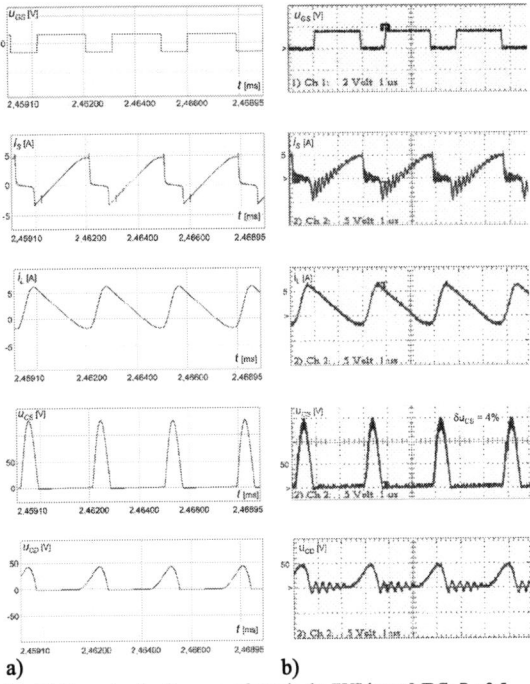

a)   b)

Fig.12. Current and voltage waveforms in the ZVS boost MRC; R_N=0.5, f=345kHz, β=0.65,
a) simulation results; E=20V, I_{LF}=2.92A, U_O=28.48V, η=0.93, P_{in}=58.25W, P_{out}=54.12W
b) experiment results; E=20V, I_{LF}=2.97A, U_O=28.33V, η=0.90, P_{in}=59.40W, P_{out}=53.71W

The result diverges most widely in regard to u_{CS}. With respect to R_N=0.5, the relative error δu_{CS}=4% with:

$$\delta u_{CS} = \frac{\Delta U_{CS\,max}}{U_{CS\,max}} \cdot 100\% \qquad (14)$$

where: $\Delta U_{CS\,max}$ – maximum divergence between simulation and experimental waveforms,

$U_{CS\,max}$ – maximum transistor voltage derived from simulations.

Oscillations in experimental waveforms result from parasitic reactance in the experimental model which is the prototype converter produced in laboratory conditions.

Efficiency ratio

Figure 13 illustrate the results of the simulation and experimental testing of the ZVS boost MRC in the form of efficiency ratio η. The converter's η is similar in both the cases, ranging from 0.91 to 0.97.

REFERENCES

[1] Nowak M., Barlik R., "Poradnik inżyniera energoelektronika" (Handbook of Power Electronic Engineer) WNT 1998.

[2] "Simulation system SIMPLORER 4.0 User Manual", Ansoft Corporation, Pittsburgh, 2002.

[3] Szychta E., "Multirezonansowe przekształtniki ZVS napięcia stałego na napięcie stałe" (Multiresonant DC/DC ZVS converters), Oficyna Wydawnicza Uniwersytetu Zielonogórskiego, Monograph, vol. 6, 2006.

[4] Szychta E., "ZVS operation region of multiresonant DC/DC boost converter", Journal of Advances in Electrical and Electronic Engineering, Faculty of Electrical Engineering, Vol.6, No.2, 2007, Zilina University, pp. 60-62.

[5] Tabisz W.A., Lee F.C., "DC analysis and design of zero-voltage-switched multi-resonant converters", IEEE 20th Annual Power Electronics Specialists Conference, PESC '89, vol. 1, 1989, p. 243 – 251.

[6] Tunia H., Barlik R., "Teoria przekształtników" (Theory of Converters) Oficyna wydawnicza Politechniki Warszawskiej, Warsaw 2003.

[7] Люфт М., Шихта Э., "Математическая модель мультирезонансного инвертора ZVS DC/DC, повышающего напряжение, Вестник МИИТ-а", № 17, сс.74-86, Москва 2007, Россия.

A New DC-DC Converter with Multi Output: Topology and Control Strategies

Arash A Boora, *Student Member, IEEE*, Firuz Zare, *Senior member, IEEE*, Gerard Ledwich, *Senior member, IEEE*, Arindam Ghosh, *Fellow, IEEE*
School of Engineering Systems
Queensland University of Technology
arash.boora@student.qut.edu.au

Abstract -This paper presents a new topology based on a Positive Buck-Boost converter with multi output (MOPBB). A single output positive Buck-Boost converter consists of a Buck and Boost converters in cascade which can be controlled against input voltage fluctuation and load changes. In this paper, the steady state and dynamic analyses of the proposed topology are presented along with simulation results. A control algorithm is presented to control output voltages against input voltage fluctuation and step change in load with a purely logic control system that is based on hysteresis current and voltage control. This topology is suitable for a high power multilevel converter with diode-clamped topology where a series of capacitors are required to generate different voltage levels and capacitors voltage control is an important issue in this application.

Keywords—Multi-output, DC-DC converter, Disturbance robustness,

I. INTRODUCTION

To clarify the advantages of proposed topology in comparison with other multi-output topologies, single output Positive Buck Boost converter (PBB) circuit is shown in (Fig. 1) [1, 2, 3].

PBB has the advantage of an extra freedom degree in comparison with basic DC-DC converters of Buck Boost and Inverting Buck Boost (IBB) [1]. This extra freedom degree can be applied to decouple the inductor current and capacitor voltage.

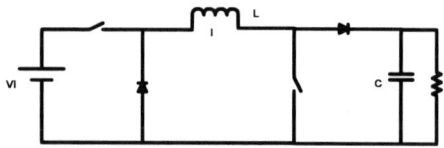

Fig. 1: Positive Buck-Boost Converter

In other words, unlike basic DC-DC converters the inductor current such as PBB is not restricted by voltage conversion ratio and load current.

The relationship between a load current (I_o) and an inductor current (I_L) in basic DC-DC converters and PBB are given in flowing equations.

$$I_L = \frac{I_o}{D'} = \frac{V_C}{RD'} = \frac{V_{in}}{RD'^2} \qquad \text{Boost} \qquad (1)$$

$$I_L = \frac{I_o}{D'} = \frac{V_C}{RD'} = \frac{DV_{in}}{RD'^2} \qquad \text{Inverting Buck-Boost} \qquad (2)$$

$$I_L = I_o = \frac{V_C}{R} = \frac{DV_{in}}{R} \qquad \text{Buck} \qquad (3)$$

$$I_l = \frac{1}{D'_{Boost}} I_o = \frac{1}{D'_{Boost}} \frac{V_o}{R} = \frac{D_{Buck}}{D'^2_{Boost}} \frac{V_i}{R} \qquad \text{PBB} \qquad (4)$$

In this way the inductor in the PBB can be used as an energy storage device as well as energy deliverer while the amount of stored energy is independent from the level of delivered energy by $D`_{Boost}$ (4).

The stored energy is utilized to increase stability of the converter and achieve robustness against input voltage fluctuation and load change.

But an extra current increases switching loss. Calculation to determine how much extra switching loss arises for any situation has been done in [1].

In this paper the Multi-Output Positive Buck Boost (MOPBB) converter is presented.

Here a multi output topology based on PBB is presented. The applications of DC-DC multi output topologies are cited in [4-7].

The main application is in diode clamp multi level inverters. [8] Has developed a multi output Boost converter for diode clamp application. The converter here can be applied for same application of inverter with the advantage of more stability, step down conversion and disturbance rejection heired from PBB.

The sections of this paper cover the new topology, the disturbance rejection theorization, switching frequency increase, the control method of the new topology, and simulation results

978-1-4244-1741-4/08/$25.00 ©2008 IEEE

II. THE NEW MULTI-OUTPUT TOPOLOGY

A Multi Output Positive Buck Boost (MOPBB) converter is shown in Fig.2 where several output voltages are provided by putting capacitors in series. Voltages of the capacitors are controlled by the inductor current and correct switching states to share the energy stored in the inductor with each capacitor. References [4-7] are about some multi output topologies and their control strategies. The main purpose is to supply a multi level inverter.

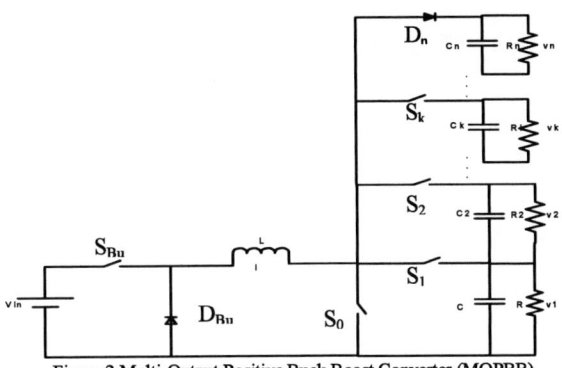

Figure 2 Multi-Output Positive Buck Boost Converter (MOPBB)

In a two output PBB converter, there are eight switching states but only 6 switching states are possible as shown in Fig.3. In this topology, when S_0 is turned on S_1 cannot be turned on as the configuration will be the same when S_1 is turned off. Thus the switching states of (011) and (111) are not allowed in this topology and the converter has six possible switching states as shown in Fig.3.

The advantage of this converter, which is achieved by input voltage switching, is that it can handle a percentage of step change in input voltage and in output current. This capability is identical to Positive Buck-Boost based converters when it comes to decrease in input voltage and increase in output load.

The above mentioned percentage of disturbance which can be dealt by the MOPBB converter without dynamics in output voltage is called disturbance margin in this paper. Disturbance margin depends on the extra current stored in inductor. The inductor current is solely dictated by load in Buck, Boost, and Inverting Buck Boost and the inductor acts only as a deliverer of energy.

Table I shows all switching states and charging and discharging states of the capacitors voltages and the inductor current. There is a switching state (010) which does not exist in the basic DC-DC converters Fig.4 and it has a significant effect on the system performance and dynamic response. Using this switching state a controller can keep the inductor current above the demand level and provide a current source using the buck switch (S_{Bu}). This current source can charge and discharge the capacitors through the boost switch (S_0).

Table I: All possible switching states with charging and discharging states

S_{Buck}	S_0	S_1	V_{c1}	V_{c2}	I_L
0	0	0	Charge	Charge	Discharge
0	0	1	Charge	Discharge	Discharge
0	1	0	Discharge	Discharge	No change
1	0	0	Charge	Charge	Discharge
1	0	1	Charge	Discharge	Discharge
1	1	0	Discharge	Discharge	Charge

Because of 0XX switching states controller can avoid instability of inductor current easily. And because of the switching state of 010 the controller can keep some extra current in the inductor and utilize it to achieve robustness against input voltage fluctuation and load changes.

The switching configurations and states for n output MOPBB can be developed same as Fig. 3 and Fig. 4. The number of switching states for an n output MOPBB is $2(n+1)$. We calculate equations for an n output MOPBB.

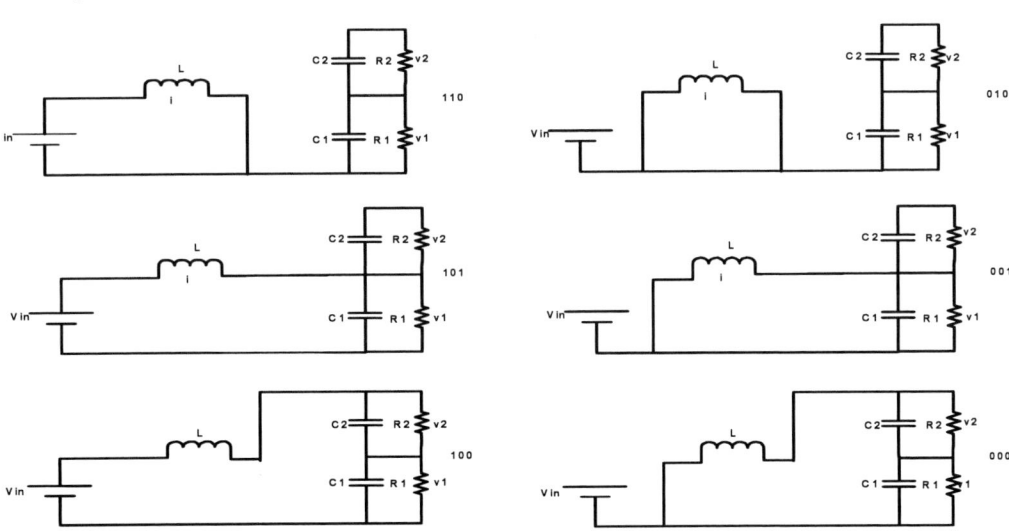

Figure 3: Possible switching configurations for 2 output positive buck-Boost converter

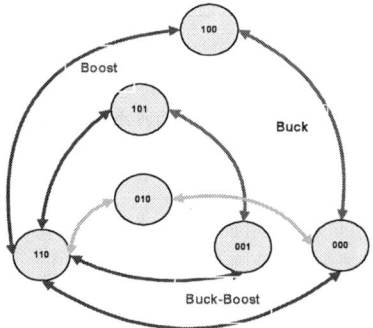

Figure 4: switching configurations and switching between them.

III. STEADY STATE AND DYNAMIC EQUATIONS

To simplify equations we use duty cycle of each switch instead of time intervals of each state. The reason is that low frequency response can be totally explained by duty cycles. D_{Bu} is the duty cycle of the Buck switch. D_0 and D_1 are the duty cycle of the switches S_0 and S_1. D_2 is the duty cycle of the output diode.

Duty cycles should satisfy:

$$0 \le D_J \quad \sum_{j=0}^{n} Dj = 1$$

$$0 \le D_{Bu} \le 1 \tag{5}$$

Using averaging technique, we can find state equations for dynamic analysis such as the inductor current and capacitors voltages in terms of the system variables.

$$L \frac{di_L}{dt} = D_{Bu} v_{in} - \left(\sum_{K=1}^{n} V_k \sum_{j=k}^{n} D_j \right) \tag{6}$$

$$C_k \frac{dv_k}{dt} = i_L \left(\sum_{j=k}^{n} D_j \right) - i_{R2} = i_L \left(\sum_{j=k}^{n} D_j \right) - \frac{v_k}{R_k} \tag{7}$$

Rewriting theses equations as state variable form:

$$\begin{pmatrix} L & 0 & 0 & 0 & 0 & 0 \\ 0 & C_1 & 0 & 0 & 0 & 0 \\ 0 & 0 & ... & 0 & 0 & 0 \\ 0 & 0 & 0 & C_k & 0 & 0 \\ 0 & 0 & 0 & 0 & ... & 0 \\ 0 & 0 & 0 & 0 & 0 & C_n \end{pmatrix} \begin{pmatrix} \dot{i_L} \\ v_1 \\ ... \\ v_k \\ ... \\ v_n \end{pmatrix} = \tag{8}$$

$$\begin{pmatrix} 0 & -\sum_{j=1}^{n} D_j & ... & -\sum_{j=k}^{n} D_j & ... & -D_n \\ \sum_{j=1}^{n} D_j & -1/R_1 & 0 & 0 & 0 & 0 \\ ... & 0 & ... & 0 & 0 & 0 \\ \sum_{j=k}^{n} D_j & 0 & 0 & -1/R_k & 0 & 0 \\ ... & 0 & 0 & 0 & ... & 0 \\ D_n & 0 & 0 & 0 & 0 & -1/R_n \end{pmatrix} \begin{pmatrix} i_L \\ v_1 \\ ... \\ v_k \\ ... \\ v_n \end{pmatrix} + \begin{pmatrix} D_{Bu} \\ 0 \\ ... \\ 0 \\ ... \\ 0 \end{pmatrix} v_i$$

Extracting transfer functions from these equations:

$$v_k(s) = \frac{R_k \sum_{j=k}^{n} D_j}{R_k C_k s + 1} i \tag{9}$$

$$i_L(s) = \frac{D_{Bu}}{Ls + \sum_{k=1}^{n} \left(\frac{R_k \left(\sum_{j=k}^{n} D_j \right)^2}{R_k C_k s + 1} \right)} v_{in} \tag{10}$$

Finally, steady state equations for two output converter will be:

$$V_k = \frac{D_{Bu} R_k \sum_{j=k}^{n} D_j}{\sum_{k=1}^{n} \left(R_k \left(\sum_{j=k}^{n} D_j \right)^2 \right)} V_{in} \tag{11}$$

$$I_L = \frac{D_{Bu}}{\sum_{k=1}^{n} \left(R_k \left(\sum_{j=k}^{n} D_j \right)^2 \right)} V_{in} \tag{12}$$

According to the equations (11) and (12) we can find the effect of the switching state of (010) in dynamic response. We can chose different values for duty cycle in the buck converter, D_{Bu} to have different currents in inductor. Let us assume that we have the series of D_j and a D_{Bu} for a particular series of output voltages. By multiplying D_js and D_{Bu} by a factor of k we can change the inductor current by a factor of $1/k$ while the output voltages are unchanged. Thus, we have same output voltages for different current in inductor (13).

$$I_L = \frac{\sum_{j=1}^{n} R_j I_{Rj}^2}{D_{Bu} V_{in}} \tag{13}$$

IV. MINIMUM INDUCTOR CURRENT

There is a minimum inductor current for any case of input voltage, output voltages, and load currents. It has been shown that the controller can store some extra current in the inductor. To know how much current is stored in the inductor we need to know the minimum inductor current for each case.

The limitation of this topology is on the load current:

$$I_{R1} \ge I_{R2} \ge ... \ge I_{Rn} \tag{14}$$

In step down case:

$$V_{in} \ge \sum_{j=1}^{n} \left(V_j \times \frac{I_{Rj}}{I_{R1}} \right) \quad \Rightarrow \quad I_{min} = I_{R1} \tag{15}$$

In step up case:

$$V_{in} < \sum_{j=1}^{n} \left(V_j \times \frac{I_{Rj}}{I_{R1}} \right) \quad \Rightarrow \quad I_{min} = \sum_{j=1}^{n} \left(\alpha_j \frac{I_{Rj}}{I_{R1}} \right) I_{R1} \tag{16}$$

$$\alpha_j = \frac{V_j}{V_{in}}$$

The extra current stored in the inductor improves the robustness and stability of MOPBB converter against fluctuations in input voltage and load change.

On the other hand MOPBB suffers more switching frequency and switching loss as the current stored in the inductor increases.

In next section we develop the relationship between extra current stored in the inductor and the advantage of robustness (Disturbance rejection) and the disadvantage of extra switching loss. This calculation guides the user of this converter to choose how much extra current is required to be stored in the inductor according to required robustness and acceptable level of switching loss.

V. DISTURBANCE REJECTION

The extra current stored in the inductor lets MOPBB to have a margin of input voltage fluctuation and load change without "low frequency" (lower than switching frequency) effect on output voltage. In other words this extra current lets the MOPBB to block these disturbances from output voltage as far as they are inside the above mentioned margin.

The ratio of actual inductor current in any case to the minimum inductor current (γ) is important because it shows the level of robustness of this converter against input voltage fluctuations and load changes as well as level of extra switching loss arising as a consequence of extra current storage. Here we define the disturbance rejection margin as a function of γ.

$$\gamma = I_L / I_{min} \tag{17}$$

To calculate disturbance rejection margin regarding to input voltage disturbance we need to look at the relationship between D_{Bu} and γ looking at (13) and (15-16) we have:

$$\gamma = \frac{min\left\{1, \sum_{j=1}^{n}\left(\alpha_j \frac{I_{Rj}}{I_{R1}}\right)\right\}}{D_{Bu}} \tag{18}$$

The margin for input voltage rise ($M^+\{V_{in}\}$) is infinite because the controller can reduce D_{Bu} immediately without showing any dynamic at output. This way there is no limit for voltage rise disturbance rejection. When the case is input voltage drop the controller increases the D_{Bu} to let the converter has same average of voltage after Buck switch. The margin in this case depends on γ (20).

$$M^+\{V_{in}\} = \infty \tag{19}$$

$$M^- = 1 - D_{Bu} = 1 - \frac{min\left\{1, \sum_{j=1}^{n}\left(\alpha_j \frac{I_{Rj}}{I_{R1}}\right)\right\}}{\gamma} \tag{20}$$

The load disturbance rejection margin can be calculated according to equation (15, 16, and 18) we can rewrite equation (13) as:

$$I_L = \frac{\left(\sum_{j=1}^{n}\alpha_j I_{Rj}\right)}{min\left\{1, \sum_{j=1}^{n}\left(\alpha_j \frac{I_{Rj}}{I_{R1}}\right)\right\}} \gamma \tag{21}$$

Load change means the change in I_{Rj} if γ can change within one switching cycle sufficient to keep I_L and α_js constant the output will not experience any dynamic with frequencies lower than switching frequency. Because of stability of MOPBB the case of load drop can be handled by reducing the current conducting to load by increasing the duty cycle of boost switch (S_0). So the disturbance rejection margin in case of load drop ($M\{I_{Rj}\}$) is infinite.

$$M^-\{I_{Rj}\} = \infty \tag{22}$$

To calculate this margin for load rise we consider that load current increase by step change to full the margin of load change. To compensate this disturbance γ drops to 1.

$$\frac{\left(\sum_{j=1}^{n}\alpha_j I_{Rj}\right)}{min\left\{1, \sum_{j=1}^{n}\left(\alpha_j \frac{I_{Rj}}{I_{R1}}\right)\right\}} \gamma = \frac{\left(\sum_{j=1}^{n}\alpha_j (1 + M^+\{I_{Rj}\})I_{Rj}\right)}{min\left\{1, \sum_{j=1}^{n}\left(\alpha_j \frac{(1 + M^+\{I_{Rj}\})I_{Rj}}{(1 + M^+\{I_{R1}\})I_{R1}}\right)\right\}} \tag{23}$$

If we assume that same margin for all loads is required,

$$M^+\{I_{Rj}\} = \gamma - 1 \tag{24}$$

Of course if the loads have different sensitivity the controller can devote the stored current to the more sensitive loads or share it asymmetrically which means having wider load change margin for more sensitive loads.

Fig. 5 shows the graph of the relation ship between extra current stored in the inductor and robustness margins.

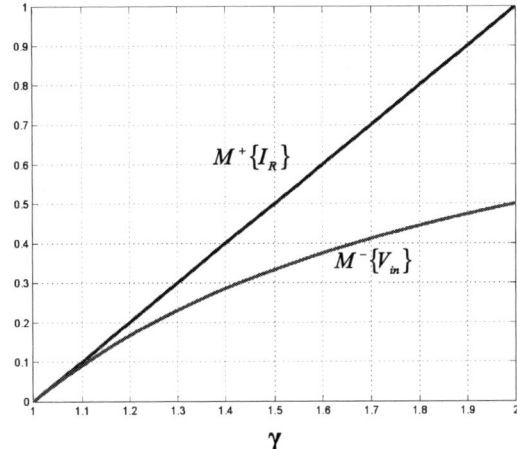

Figure 5 disturbance rejection margins as a function of γ

VI. SWITCHING FREQUENCY

The main disadvantage of extra current stored in the inductor is the increased switching frequency and loss. To have an efficient design for any particular application, the calculation of switching loss is important.

According to equations (11) and (12)

$$\sum_{j=k}^{n} D_j = I_{Rk} / I_L \tag{25}$$

Looking at the switching configuration at Fig. 3, V_2 and V_1 have their rise on the times periods of $D_2 T_{sw}$ and $(D_1 + D_2).Tsw$ respectively. For n output MOPBB we have:

$$C_k \Delta V_k = T_{swBoost} \sum_{j=k}^{n} D_j (I_L - I_{Rk}) \tag{26}$$

According to (26) the ripples of V_ks are dependent. So switching cycle of the Boost switch can be calculated as (27).

$$T_{swBoost} = min\left\{\frac{C_k \Delta V_k I_L}{I_{Rk}(I_L - I_{Rk})}\right\} \qquad k = 1,...,n \qquad (27)$$

To have the switching cycle as a function of γ we apply equation (21):

$$Step\ Down: \quad T_{swBoost} = min\left\{\frac{C_k \Delta V_k \gamma I_{R1}}{I_{Rk}(\gamma I_{R1} - I_{Rk})}\right\}$$

$$stepUp: \quad T_{swBoost} = min\left\{\frac{C_k \Delta V_k \gamma\left(\sum_{j=1}^{n}\alpha_j I_{Rj}\right)}{I_{Rk}\left(\gamma\left(\sum_{j=1}^{n}\alpha_j I_{Rj}\right) - I_{Rk}\right)}\right\} \qquad (28)$$

Inductor current (I_L) is defined by (21). To calculate the Buck switching frequency, the average of positive voltage exposed to the inductor end connecting to the Boost switch is:

$$V_{avgBoost} = \sum_{k=1}^{n}\left(V_k \sum_{j=k}^{n} D_j\right) \qquad (29)$$

So the rise time and fall time of the inductor current will be:

$$-\frac{L\Delta I_L}{T_{fall}} = -\sum_{k=1}^{n}\left(V_k \sum_{j=k}^{n} D_j\right) \Rightarrow T_{fall} = \frac{L\Delta I_L}{\sum_{k=1}^{n}\left(V_k \sum_{j=k}^{n} D_j\right)} \qquad (30)$$

$$\frac{L\Delta I_L}{T_{rise}} = V_{in} - \sum_{k=1}^{n}\left(V_k \sum_{j=k}^{n} D_j\right) \Rightarrow T_{rise} = \frac{L\Delta I_L}{V_{in} - \sum_{k=1}^{n}\left(V_k \sum_{j=k}^{n} D_j\right)}$$

The switching frequency of Buck switch will be:

$$T_{swBuck} = T_{fall} + T_{rise}$$

$$= \frac{L\Delta I_L}{\sum_{k=1}^{n}\left(\alpha_k \sum_{j=k}^{n} D_j\right)\left(1 - \sum_{k=1}^{n}\left(\alpha_k \sum_{j=k}^{n} D_j\right)\right)} \qquad (31)$$

$$= \frac{L\Delta I_L \times I_L}{\sum_{k=1}^{n}(\alpha_k I_{Rk})\left(I_L - \sum_{k=1}^{n}(\alpha_k I_{Rk})\right)}$$

For step up and step down case the equation of Buck switch frequency will be:

$$stepDown: \quad T_{swBuck} = \frac{L\Delta I_L \gamma I_{R1}}{\sum_{k=1}^{n}(\alpha_k I_{Rk})\left(\gamma I_{R1} - \sum_{k=1}^{n}(\alpha_k I_{Rk})\right)}$$

$$stepUp: \quad T_{swBuck} = \frac{L\Delta I_L \gamma \sum_{j=1}^{n}\alpha_j I_{Rj}}{\sum_{k=1}^{n}(\alpha_k I_{Rk})\left(\gamma \sum_{j=1}^{n}\alpha_j I_{Rj} - \sum_{k=1}^{n}(\alpha_k I_{Rk})\right)} \qquad (32)$$

Fig 6 shows the switching frequency of Buck and Boost switches as function of γ for cases of step up and step down when $I_{Rk}=I_{Rj}$.

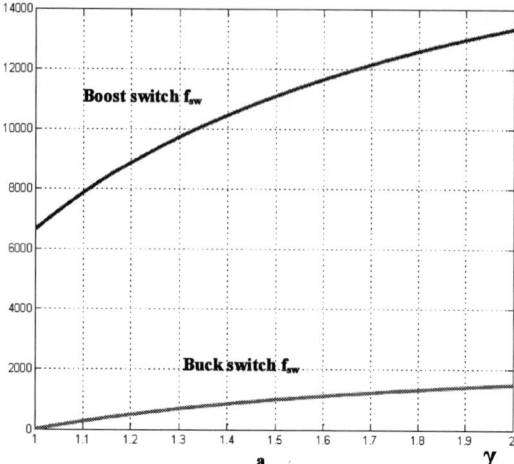

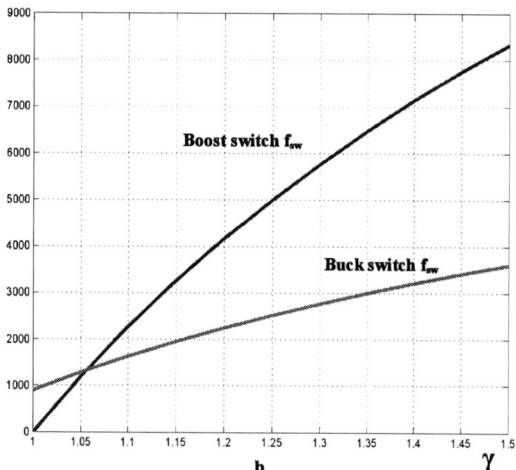

Figure 6 the relationship between switching frequency of Buck switch (purple) and Boost switches (blue) as a function of γ: a) step up b) step down ($I_{Rk}=I_{Rj}$)

As is shown in Fig. 6a the frequency of buck switch is 0 when γ is 1. This means that in step up case if the inductor current be equal to its minimum value the converter is working as a multi output Boost converter (eliminating S_{Bu} in Fig. 2). In Fig. 6b the switching frequency of Boost switch is 0 when γ is 1. This means that in step down case the converter is working as a multi output Buck (eliminating S_0 in Fig. 2) converter when there is no extra current stored in the inductor.

Comparing the Fig. 6a and Fig. 6b, the main switching frequency increase is happening for Boost switches. Particularly, for step up case the switching frequency of the Buck switch is negligible in comparison to the switching frequency of Boost switches.

VII. CONTROL STRATEGY AND SIMULATION RESULT

In this paper the method of Hysteresis control for inductor current and output capacitor voltage are explained.

Because PBB based topologies can decouple inductor current and capacitor voltage by storing some extra current in the capacitor, the control system will have enough freedom to use hysteresis control both for inductor current and capacitor at the same time.

However to have desirable performance the capacitor voltage hysteresis controller needs to consider the inductor as a current source with acceptable fluctuations, so the current loop controller should be faster than the voltage control loop.

The main challenge of this controller is to know the appropriate inductor current. The controller needs to detect the minimum required current to be able to keep the reference voltage at output. On the other hand controller should decide about the level of extra current needed to be stored in the inductor to achieve robustness and stability required by application. In this paper the ratio of actual inductor current to minimum required current is called γ.

Equation 14 suggests to measure I_L, V_{in}, D_{Bu} and estimate load current and extra current storage in the inductor. The controller is responsible to keep enough current stored in the inductor to achieve robustness against disturbances (Eq. 16 and 20) but not too much current should be stored because the switching frequency and loss will be increased (Eq 24 and 27).

Fig. 7 shows the control strategy for a two output MOPBB. The duty cycle of Buck switch is used to control the level of extra current stored in the inductor.

Figure 7 hysteresis control system of MOPBB

The inductor current and both output voltages are controlled by hysteresis method. Fig. 8 shows some simulation results of this control system. The aim of this simulation is to show the robustness of this topology against disturbances in input voltage and load. The level

of extra current is between 25% and 40% ($1.25<\gamma<1.4$) to keep output voltages constant in spite of dramatic changes in load and input voltage.

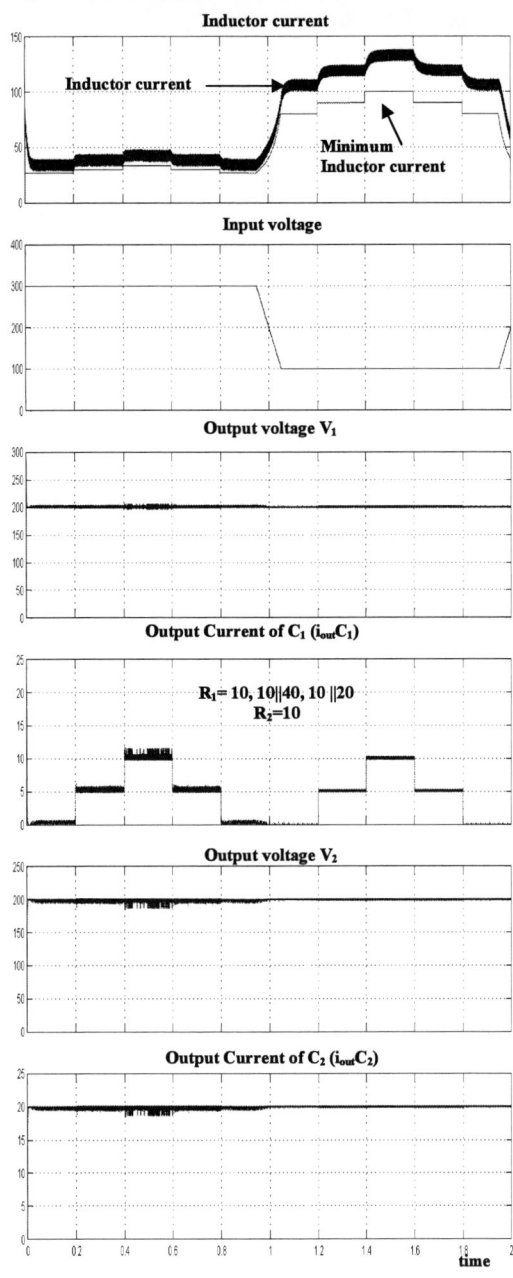

Figure 8 simulation results of the hysteresis control system for MOPBB when input voltage and R_1 change

Fig. 9 shows the same parameters in Fig 8 when input voltage and R_2 change. In both cases all the changes have been in side the disturbance rejection margins. So the output voltages do not endure low frequency dynamics.

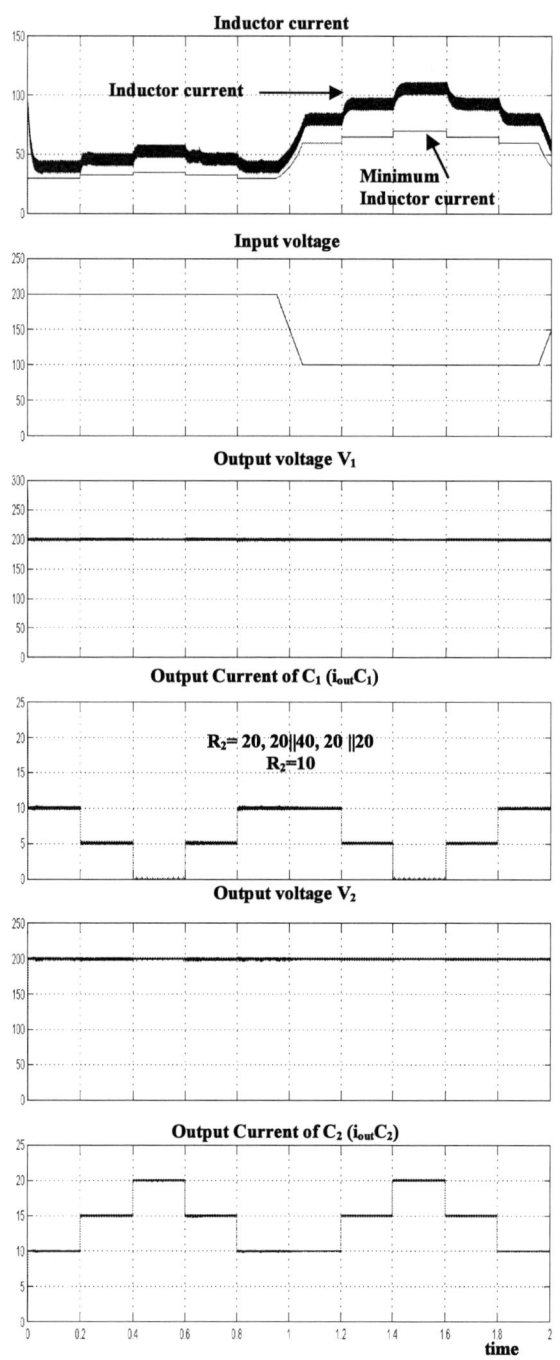

Figure 9 simulation results of the hysteresis control system for MOPBB when input voltage and R_2 change

VIII. CONCLUSION

A multi output DC-DC converter based on positive Buck Boost converter is introduce.

Positive Buck Boost converter also known as noninverting Buck Boost converter has the advantage of an extra freedom degree. This paper has shown the possibility of utilizing this freedom degree to store a level

of extra current in the inductor to achieve robustness against input and output disturbances. The calculation to show the degree of robustness as the advantage of extra current stored in the inductor has been developed. The increase in switching frequency as a disadvantage of storing extra current in the inductor is theorized and formulated. A control strategy has been developed and simulated at last.

The designer of the implementation of this topology and its control strategy should consider the requirements of a particular application, level of input and Load disturbances, and allowed switching frequency and switching loss in that application to decide how much extra current should be stored in the inductor of MOPBB.

IX. ACKNOWLEDGMENT

The authors thank the Australian Research Council (ARC) for the financial support for this project through the ARC Linkage Grant LP0774899.

References

[1] "A General Approach to Control a Positive Buck-Boost Converter to Achieve Robustness against Input Voltage Fluctuations and Load Changes" Arash A Boora, Student member, IEEE, Firuz Zare, Senior member, IEEE, Gerard Ledwich, Senior member, IEEE, Arindam Ghosh, Fellow, IEEE , PESC 2008 (unpublished)

[2] "Combination of Buck and Boost Modes to Minimize Transients in the Output of a Positive Buck-Boost Converter" Chakraborty, Arindam; Khaligh, Alireza; Emadi, Ali; IEEE Industrial Electronics, IECON 2006 - 32nd Annual Conferene on Nov. 2006 Page(s):2372 – 2377

[3] Digital Combination of Buck and Boost Converters to Control a Positive Buck-Boost Converter" Chakraborty, A.; Khaligh, A.; Emadi, A.; Pfaelzer, A.; Power Electronics Specialists Conference, 2006. PESC '06. 37th IEEE 18-22 June 2006 Page(s):1 – 6

[4] "A simple structure of LLC resonant DC-DC converter for multi-output applications" Yilei Gu; Lijun Hang; Huiming Chen; Zhengyu Lu; Zhaoming Qian; Jun Li; Applied Power Electronics Conference and Exposition, 2005. APEC 2005. Twentieth Annual IEEE Volume 3, 6-10 March 2005 Page(s):1485 - 1490 Vol. 3

[5] "Multi-output SC type DC-DC converter using a flexible capacitor ring operation" Harada, I.; Hara, N.; Ueno, F.; Oota, I.; Telecommunications Energy Conference, 1999. INTELEC '99. The 21st International 6-9 June 1999 Page(s):4 pp.

[6] "Programmable Digital Controller for Multi-Output DC-DC Converters with a Time-Shared Inductor" Parayandeh, A.; Stupar, A.; Prodic, A.; Power Electronics Specialists Conference, 2006. PESC '06. 37th IEEE 18-22 June 2006 Page(s):1 – 6

[7] "Behavioral Modeling of Multi-Output DC-DC Converters for Large-Signal Simulation of Distributed Power Systems" Oliver, J.A.; Prieto, R.; Romero, V.; Cobos, J.A.; Power Electronics Specialists Conference, 2006. PESC '06. 37th IEEE 18-22 June 2006 Page(s):1 – 6

[8] "A New Configuration for Multi level converters with diode clamp topology" A. Nami, F. Zare, G. Ledwich, A. Ghosh, IPEC 2007, page. 661-665

Maximum Frequency for Hysteretic Control COT Buck Converters

L.K. Wong* and T.K. Man[†]

* National Semiconductor Corporation, Power Management Design Center, Hong Kong, e-mail: *LK.Wong@nsc.com*

[†] National Semiconductor Corporation, Power Management Design Center, Hong Kong, e-mail: *TK.Man@nsc.com*

Abstract— There are many analyses on the closed-loop performance for COT buck converters. But all of them assume the absence of the multiple pulsing effect, which will increase the output voltage and inductor current swing significantly, and a poor steady state performance is as a result. This paper presents a time domain analysis for COT buck converters to show the relationship between the multiple pulsing effect and the on-time. Results indicate a minimum limit for the on-time that can avoid the multiple pulsing effect, and consequently the maximum frequency with good steady state performance can be found.

Keywords—Converter control, DC power supply, power management.

I. INTRODUCTION

Constant on-time (COT) hysteretic control buck converters have been widely used in industry [1, 2]. By fixing the on-time of the converter, the off-time is determined by detecting the output voltage and comparing it to a reference. Such converter is simple in implementation and design because no compensation network is required. Besides, fast transient response and good robustness can be obtained. With a dedicated circuit for controlling the on-time, the on-time and thus the duty cycle can be made very small, and consequently high converting (step-down) ratio can be achieved.

Previous works on the analysis of COT converters focus on modeling the closed-loop system in the frequency domain [3-6]. They do not show a multiple pulsing effect in the steady state, which is as a result of two or more on-periods without or with minimum off-periods in between, and consequently a large output voltage and inductor current swing are generated. The increase in the output voltage swing will introduce noise and error in the output voltage regulation. A large inductor current swing makes the inductor suffer from larger conduction loss and easier to saturate. Hence, a poor steady state performance is resulted. Figs. 3a and 3b show waveforms of the output voltage and the inductor current of a COT buck converter illustrated in Fig. 1 with parameter values listed in Table 1. The waveforms in Fig. 3a show the multiple pulsing effect. Two on-periods with a minimum off-time in between occur, and the output voltage and inductor current both suffer from large swinging as compared with normal waveforms (no multiple pulsing effect) shown in Fig. 3b.

In this paper, a time domain analysis on the on-period is detailed to show the relationship of the multiple pulsing effect to a number of parameters including the on-time, delay in the control circuit, the input voltage, the reference voltage, the output loading, and component values,

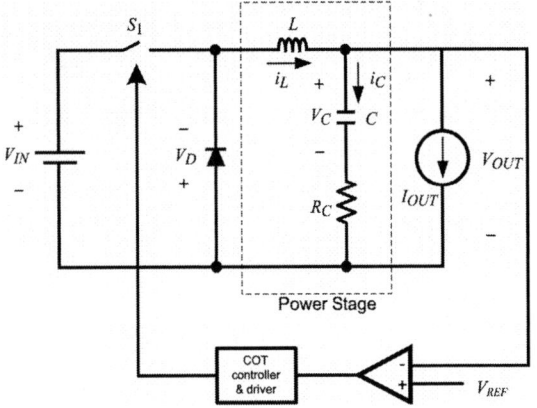

Fig. 1. A constant on-time hysteretic control buck converter.

TABLE I.
PARAMETERS OF THE CONVERTER

Parameter	Value
V_{IN}	18V
V_{REF}	3.3V
V_D	0.85V
L	10μH
C	100μF
R_C	20mΩ
I_{OUT}	2.5A

including the equivalent series resistance (ESR) of the output capacitor. An equation showing the maximum frequency that can avoid the multiple pulsing effect is formulated in Section II. Simulation results will be presented in Section III, and a conclusion will be drawn in Section IV.

II. TIME DOMAIN ANALYSIS

The COT buck converter shown in Fig. 1 consists of a power stage, a comparator for the output voltage, and a constant on-time controller and driver for controlling the switch. The load is assumed to be a constant current source. COT converters require no compensation, so the implementation and design are easy and simple. Fig. 2a shows an illustrated waveform of the output voltage V_{OUT} to address the cause of the multiple pulsing effect. When the comparator detects that V_{OUT} drops below V_{REF}, the switch S_1 will be turned on with a delay at t_1 owing to the offset of the comparator and the unavoidable intrinsic delay of the comparator and the driver. As a result, V_{OUT} will be lower than V_{REF}, say $V_{REF} - V_{OS1}$, at the start of the on-period t_1. If the on-time t_{on} is not long enough for V_{OUT} to rise over a threshold $V_{REF} + V_{OS2}$, where V_{OS2} is the effective threshold that can toggle the comparator output

978-1-4244-1741-4/08/$25.00 ©2008 IEEE

by considering the input offset and the comparator delay, the comparator will generate a control signal to turn-on S_1 at the end of the on-period $t_1 + t_{on}$. Then, the multiple pulsing effect will occur.

A. State Equations and Initial Conditions

To anaylse the multiple pulsing effect in the time domain, the following differential equations of the power stage corresponding to the on-period is considered:

$$V_{IN} = L\frac{di_L}{dt} + V_{OUT}(t) \tag{1}$$

$$i_L(t) = i_C(t) + I_{OUT} \tag{2}$$

The initial conditions at $t = t_1$ are as follows. From Fig. 2a,

$$V_{OUT}(t_1) = V_{REF} - V_{OS1} \tag{3}$$

and as shown in Fig. 2b, let

$$i_L(t_1) = I_1 \tag{4}$$

Also, from Fig. 1,

$$V_{OUT}(t) = V_C(t) + [i_L(t) - I_{OUT}]\,R_C$$

From (3) and (4),

$$V_C(t_1) = V_{REF} - V_{OS1} - [I_1 - I_{OUT}]\,R_C \tag{5}$$

B. Final Conditions of the On-Period

The equations for i_L, V_C and V_{OUT} within the on period are as follows. Since the output ripple is small compared with V_{IN}, in particular in steady state, it can be assumed that

$$V_{IN} - V_{OUT}(t) \approx V_{IN} - V_{REF}.$$

From (1),

$$\frac{di_L}{dt} = \frac{V_{IN} - V_{OUT}(t)}{L}$$

$$\approx \frac{V_{IN} - V_{REF}}{L}$$

$$i_L(t_1 + t) = I_1 + \frac{V_{IN} - V_{REF}}{L}t \tag{6}$$

From (2),

$$i_C(t) = i_L(t) - I_{OUT}$$

$$\frac{dV_C}{dt} = \frac{i_L(t) - I_{OUT}}{C}$$

$$\int_{V_C(t_1)}^{V_C(t_1+t)} dV_C = \int_{t_1}^{t_1+t} \frac{i_L(t) - I_{OUT}}{C}\,dt$$

$$V_C(t_1 + t) = V_C(t_1) + \frac{1}{C}\int_{t_1}^{t_1+t} i_L(t)\,dt - \frac{I_{OUT}t}{C}$$

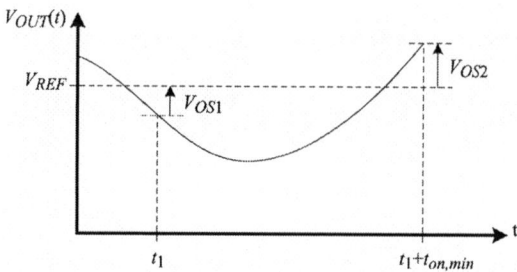

Fig. 2a. Output voltage waveform at the on-period.

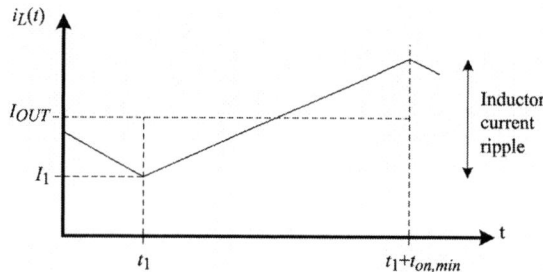

Fig. 2b. Inductor current waveform at the on-period.

The integral term in the above equation corresponding to a trapezoidal area under the curve $i_L(t)$ from t_1 to t_1+t, thus

$$V_C(t_1 + t) = V_C(t_1) + \frac{t}{2C}\left[i_L(t_1+t) + I_1\right] - \frac{I_{OUT}t}{C}$$

From (5),

$$V_C(t_1 + t) = V_{REF} - V_{OS1} - (I_1 - I_{OUT})R_C$$

$$+ \frac{t}{2C}\left[I_1 + \frac{V_{IN} - V_{REF}}{L}t + I_1\right] - \frac{I_{OUT}t}{C} \tag{7}$$

Consider the equation of the output voltage,

$$V_{OUT}(t_1+t) = V_C(t_1+t) + [i_L(t_1+t) - I_{OUT}]\,R_C$$

Substitute (6) and (7),

$$V_{OUT}(t_1 + t) = V_{REF} - V_{OS1} - (I_1 - I_{OUT})R_C$$

$$+ \frac{t}{2C}\left[I_1 + \frac{V_{IN} - V_{REF}}{L}t + I_1\right] - \frac{I_{OUT}t}{C}$$

$$+ \left[I_1 + \frac{V_{IN} - V_{REF}}{L}t - I_{OUT}\right]R_C$$

$$= V_{REF} - V_{OS1}$$

$$+ \frac{t}{2C}\left[I_1 + \frac{V_{IN} - V_{REF}}{L}t + I_1\right] - \frac{I_{OUT}t}{C}$$

$$+ \frac{(V_{IN} - V_{REF})t R_C}{L}$$

$$V_{OUT}(t_1+t) = t^2 \frac{V_{IN}-V_{REF}}{2LC}$$

$$+t\left[\frac{I_1-I_{OUT}}{C}+\frac{(V_{IN}-V_{REF})R_C}{L}\right]$$

$$+V_{REF}-V_{OS1}$$

C. Multiple Pulsing Effect

To avoid the multiple pulsing effect, at t_1+t_{on}, $V_{OUT}(t_1+t_{on})$ should be larger than a threshold $V_{REF}+V_{OS2}$. Hence,

$$V_{OUT}(t_1+t_{on})$$

$$=t_{on}^2\frac{V_{IN}-V_{REF}}{2LC}$$

$$+t_{on}\left[\frac{I_1-I_{OUT}}{C}+\frac{(V_{IN}-V_{REF})R_C}{L}\right]$$

$$+V_{REF}-V_{OS1}$$

$$>V_{REF}+V_{OS2}$$

By re-arranging the terms above,

$$t_{on}^2\frac{V_{IN}-V_{REF}}{2LC}$$

$$+t_{on}\left[\frac{I_1-I_{OUT}}{C}+\frac{(V_{IN}-V_{REF})R_C}{L}\right]$$

$$-V_{OS1}-V_{OS2}$$

$$>0$$

It can be deduced that the above inequality cannot be satisfied if t_{on} is small. This addresses that the multiple pulsing effect will occur for a small t_{on}. Let $t_{on,min}$ be the lower limit of the on-time that can avoid the multiple pulsing effect can be found from the following equation, then

$$t_{on,min}^2\frac{V_{IN}-V_{REF}}{2LC}$$

$$+t_{on,min}\left[\frac{I_1-I_{OUT}}{C}+\frac{(V_{IN}-V_{REF})R_C}{L}\right] \quad (8)$$

$$-V_{OS1}-V_{OS2}$$

$$=0$$

If there is no multiple pulsing effect, $I_{OUT}-I_1$ should be half of the inductor current ripple (Fig. 2b), which is related to the inductor charging rate and t_{on} as indicated from (1) and (6), so

$$I_{OUT}-I_1 = \frac{V_{IN}-V_{REF}}{2L}t_{on,min}$$

Substitute into (8),

$$t_{on,min}\frac{(V_{IN}-V_{REF})R_C}{L}-V_{OS1}-V_{OS2}=0$$

$$t_{on,min}=\frac{(V_{OS1}+V_{OS2})L}{(V_{IN}-V_{REF})R_C} \quad (9)$$

Finally, as the conversion ratio determines the duty cycle at the steady state, the maximum operating frequency $f_{SW,max}$ corresponding to $t_{on,min}$ can be found by considering (1) for the on-period, and the following equation for the off-period:

$$-V_D = L\frac{di_L}{dt}+V_{OUT}(t) \quad (10)$$

where V_D is the forward voltage drop of the diode at the off-period. By averaging (1) and (10),

$$DV_{IN}-(1-D)V_D = L\frac{di_L}{dt}+V_{OUT}(t)$$

where

$$D=t_{on}f_{SW}$$

is the steady-state duty ratio. Also, at steady state,

$$\frac{di_L}{dt}=0$$

Therefore we have

$$DV_{IN}-(1-D)V_D = V_{OUT}(t)$$

$$D(V_{IN}+V_D)-V_D = V_{OUT}(t)$$

$$D=\frac{V_{OUT}(t)+V_D}{V_{IN}+V_D}$$

$$f_{SW,max}=\frac{V_{OUT}(t)+V_D}{(V_{IN}+V_D)t_{on,min}} \quad (11)$$

III. SIMULATION RESULTS

To illustrate the analysis results, a buck converter of Fig. 1 with the parameters listed in Table 1 is used for simulation. The offset values V_{OS1} and V_{OS2} are 6.13mV and 10.01mV respectively, which are obtained from simulation. Let

$$V_{OS1}+V_{OS2} = 16.14\text{mV}$$

Substitute into (9),

$$t_{on,min} = 549\text{ns}.$$

Figs. 3a and 3b show two set of simulated waveforms with t_{on} equals 545ns and 565ns respectively. The multiple pulsing effect occurs in Fig. 3a because its on-time is smaller than $t_{on,min}$, but does not occur in Fig. 3b. From

(11), the maximum operating frequency for this circuit is 401kHz.

IV. CONCLUSION

A time domain analysis has been presented in this paper to find the maximum operating frequency of a COT hysteretic buck converter. Equations have been formulated to show that the minimum on-time $t_{on,min}$ that can avoid the multiple pulsing effect is related to a number of parameters including the input and output voltages, output loading, inductance, the ESR of the output capacitor, and effective offset values of the control circuit. The offset values V_{OS1} and V_{OS2} can varied for different components and operating conditions (for example operating frequency, temperature, input voltage) in practice. But a lower bound can be estimated in the calculation since a larger $t_{on,min}$ is more safe to avoid the multiple pulsing effect. Small ESR of the output capacitor increases $t_{on,min}$. Therefore ceramic capacitors, which have small ESR, are usually not to be used alone as the output capacitors for COT hysteretic converters, unless a small resistor is added in series. However, a large resistance increases the output ripple unnecessarily. The result in this paper help choosing an adequate amount of resistance in series for no multiple pulsing effect once the required operating frequency is determined.

REFERENCES

[1] L.K. Wong and T.K. Man, "LM3102 demonstration board reference design", AN-1646, National Semiconductor, October 2007.

[2] LM2696 3A, Constant On Time Buck Regulator, National Semiconductor, Data Sheet of Chip LM2696, October 2005.

[3] Sung-Soo Hong and Byungcho Choi, "Technique for developing averaged duty ratio model for DC-DC converters employing constant on-time control", *Electronics Letters*, vol. 36. no. 5, pp. 397-199, 2nd March 2000.

[4] R.B. Ridley, "A new continuous-time model for current-mode control with constant frequency, constant on-time, and constant off-time, in CCM and DCM," in *Proc. IEEE PESC '90 record*, 11-14 June 1990, pp. 382-389.

[5] R. Redl, "Small-signal high-frequency analysis of the free-running current-mode-controlled converter", in *Proc. IEEE PESC '91 record*, 24-27 June 1991, pp. 897-906.

[6] Jian Sun, "Small-signal modeling of variable-frequency pulse width modulators", *IEEE Trans. Aerospace and Electronic Syst.*, vol. 38, no. 3, pp. 1104-1108, July 2002.

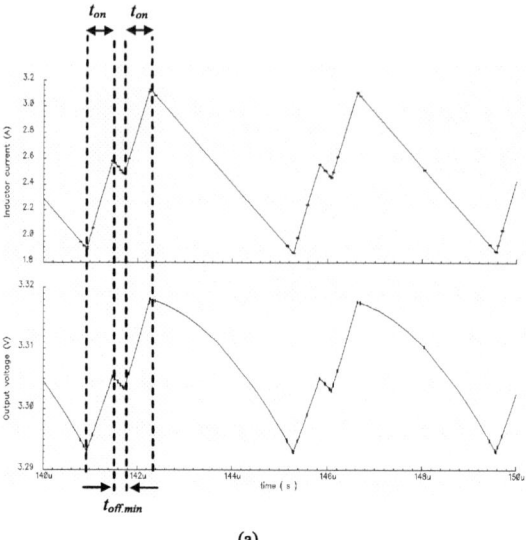

(a)

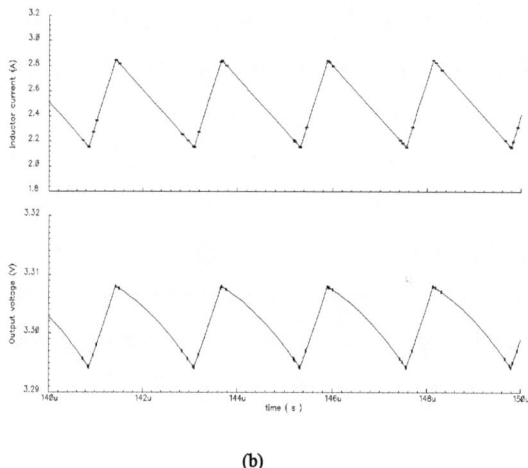

(b)

Fig. 3. Waveforms of a COT buck converter (a) with t_{on} = 545ns, (b) with t_{on} = 565ns; upper trace: inductor current I_L, lower trace: output voltage V_{OUT}, x-axis: time t.

Current Control Method Based on Hysteresis Control Suitable for Single Phase Active Filter with LC Output Filter

Yukinori Kobayashi and Hirohito Funato

Department of Electrical and Electronic Engineering, Utsunomiya University, Utsunomiya city, Japan,
e-mail:mt076215@cc.utsunomiya-u.ac.jp, funato@cc.utsunomiya-u.ac.jp

Abstract—Single phase active power filters are effective to compensate harmonics for home appliance and office equipment. Single phase active power filters, however, have some problems. The one of the problems is current control for high frequency components, because coordinate transform can not be applied to single phase system. In this paper, a new current control method suitable for single phase inverter with LC output filter based on hysteresis control will be discussed with simulations and experiments.

Keywords— Active filter, Harmonics, Single phase system, Current control, Hysteresis control

I. INTRODUCTION

Single phase active power filters (SAF) are effective to compensate harmonics caused by home appliances, office equipments and so on, because harmonics can be compensated near generation point. There are a lot of studies about SAF[1], [2], [3]. Current control of SAF is one of the problems because current reference contains high frequency components but coordinate transform can not be applied. Several methods have been proposed to overcome this problem such as deadbeat control[4], inner model method[5] and so on. Hysteresis control is one of the current control method which can realize robust and fast current controller with simple hardware[6]. On the other hand, LC output filter is required in some cases. In this case, a simple hysteresis control may cause oscillations or current error due to LC filter. In this paper, a new current control method suitable for single phase active filters with LC filter using hysteresis controller with PI controller and state feedback is proposed. The proposed method will be verified through simulations and experiments.

II. PROPOSED CURRENT CONTOROL

Fig. 1 shows power system including SAF (dotted area) applied the proposed current control. In this control, i_1, the current of inductance L_1 is controlled by hysteresis controller then the switching ripple of i_1 is removed by an LC filter (L_2,C). A state feedback of the capacitor voltage is applied to suppress the resonance of the LC filter. PI controller is used to control the output current i_2 of the LC filter. The current reference i_{ref2} is directly added to the i_{ref1} , the reference of hysteresis controller, because the difference between i_{ref1} and i_{ref2} is only the current flowing into the capacitor C. Therefore, the feed forward pass is added to improve the

current response. In this study, a general second order band eliminate filter (BEF) is used to detect harmonic current because the main subject is current control.

III. DESIGN

A. The design of the LC filter

In the first step of design procedure, the parameters of LC filter should be decided. The equivalent circuit of Fig. 1 can be considered as Fig. 2. In this equivalent circuit, the voltage source inverter with L_1 and hysteresis current controller (surrounded by dash-dot line) is represented by ideal current source i_{ref1} and switching ripple components i_{sw}, and the connecting point voltage is V_x. The transfer function $G(s) = \frac{I_2(s)}{I_1(s)}$ can be calculated as equation (1) where switching ripple current i_{ref} and connecting point voltage V_x can be disregarded.

$$G(s) \quad = \quad \frac{\frac{1}{L_2 C}}{s^2 + \frac{1}{L_2 C}} \qquad (1)$$

The filter is designed to obtain 20dB decrease at 10kHz that is the assumed average switching frequency of the inverter. The parameters are derived using Eq.(1). Considering the switching frequency, inductance L_2 was set to 1mH and capacitance C was set to 3.0μF respectively. The bode plot of G(s) is shown in Fig. 3. From Fig.3, the cut-off frequency can be measured as 4.5kHz and the gain is 22dB at 10kHz.

B. The design of hysteresis band and switching frequency

Next, the hysteresis current controller is designed. The output current ripple strongly depends on the hysteresis band. Here, the hysteresis band was set as 0.25[A](0.5[App]) considering switching frequency and filter characteristic. The current ripple is expected to be 50mApp in output by the output filter mentioned above. Because the switching frequency strongly depends on hysteresis band as mentioned previously, the switching frequency in hysteresis control will be calculater at first. The equivalent circuit of main circuit becomes Fig. 4 where the V_i is inverter output voltage and V_x is the connecting point voltage. In this circuit, hysteresis band H.B. determined Fig. 5, where i_1 is current of L_1, as shown is i_{ref} is current reference. Assuming that the current during the switching

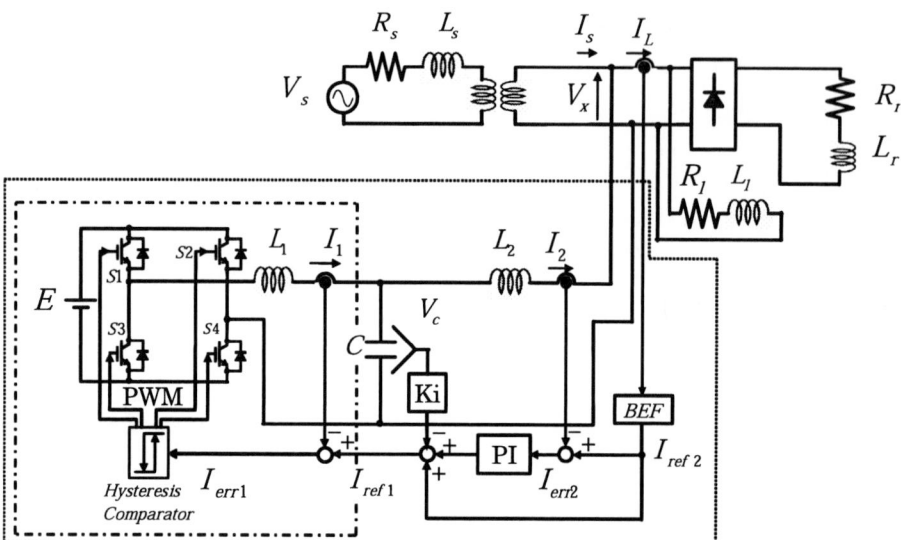

Fig. 1. Single phase active filter using proposed current source

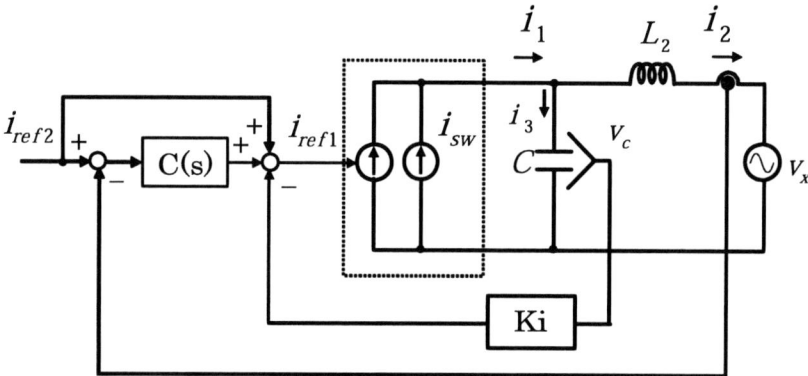

Fig. 2. Equivalent circuit using controlled current source

period changes linearly, the switching frequency of the inverter can be calculated as equation (2),(3).

$$f_{max} = \frac{V_{dc}}{4H.BL_1} \quad (2)$$

$$f_{min} = \frac{V_{dc}^2 - (V_c + \omega L_1 I_{ref})^2}{4H.BL_1 V_{dc}} \quad (3)$$

f_{max} is maximum switching frequency which was calculated under the following assumptions.

1) The gradient of current reference $\frac{di_{ref}}{dt}$ is assumed to be zero.

2) The capacitor voltage v_c is assumed to be zero because the current ripple flowing into the capacitor can be negligible.

f_{min} is the minimum switching frequency. The minimum switching frequency is obtained when the gradient of current

reference becomes maximum. In this case the current reference is assumed to be $i_{ref} = I_{ref} \sin \omega t$ so that maximum gradient of current reference becomes ωI_{ref}. Both f_{max} and f_{min} should be included in the rated switching frequency band of IGBT. This time, actual switching frequency is different from the theoretical one because of the effect of dead-time . Therefore, inductance L_1 is decided to be 1mH from simulations so that the average switching frequency becomes 20kHz.

C. The design of PI controller

Next, the PI controller is designed. Transfer function G(s) from current reference $I_{ref}(s)$ to output current $I_2(s)$ can be calcultated as follow from Fig. 2.

$$G(s) = \frac{(K_p + 1)s + \frac{K_p}{T_i}}{L_2 C s^3 + K_i L_2 s^2 + (K_p + 1)s + \frac{K_p}{T_i}} \quad (4)$$

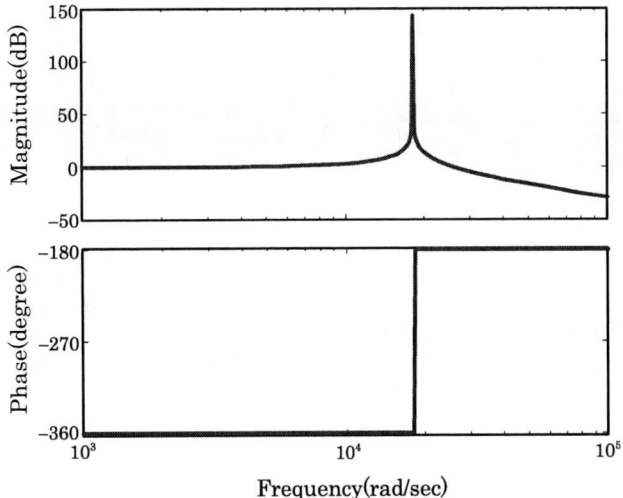

Fig. 3. Bord plot of $I_2(s)/I_1(s)$

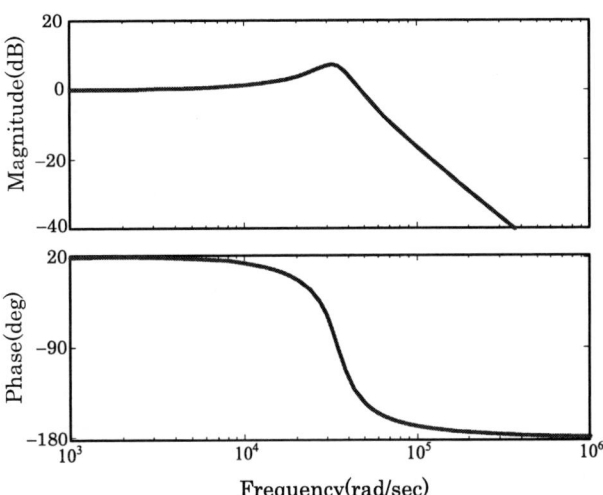

Fig. 6. Bord plot of $I_2(s)/I_ref(s)$

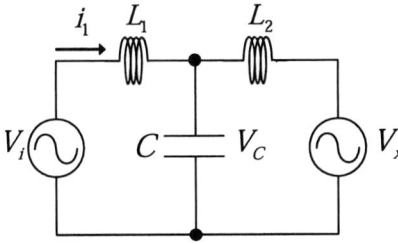

Fig. 4. Equivalent circuit for analysis of switching frequency

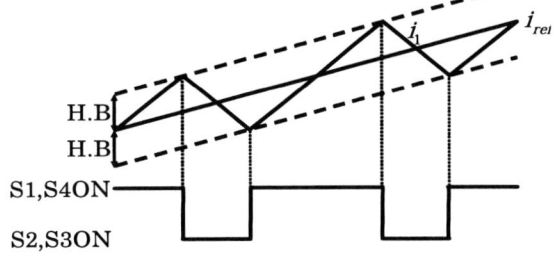

Fig. 5. Current wave form during one switching period

TABLE I
PARAMETERS

V_s	100 [V] in simulation 40 [V] in experiments 50[Hz]	E	200[V] in simulation 120[V] in experiment
R_s	0.16 [Ω]	L_s	0.4[mH]
R_r	10.0[Ω]	L_r	60 [mH]
R_l	100[Ω]	L_l	10[mH]
L_1	2.0[mH]	L_2	1.0[mH]
C	3.0[μF]	K_i	0.0774
PI	K=3,T=0.1[ms]		

In this calculations, the switching ripple current i_{sw} and connecting voltage V_x were considered as disturbance and the transfer function C(s) of PI controller is considered as $C(s) = K_p(1 + \frac{1}{T_i s})$. This time, the controller gain K_p and time constant T_i is decided using trial and error method drawing the bode plot of G(s) . The obtained parameters are K_p=3 and T_i=0.1ms. The bode plot of the obtained transfer function G(s) is shown in Fig.6. From Fig.6, cut-off frequency can be measured as 8.7kHz which is enough wide band for compensating harmonics. In addition resonance can be successfully suppressed using state feedback.

IV. SIMULATION AND EXPERIMENTS

A. Simulation

Fig. 7 shows the simulation result of the single phase active filter with the proposed method. The parameters using in the simulation are shown in Table 1. In Fig.7, Vx is the connecting point voltage V_x, Is is the source current of utility side I_s, IL is the load current IL, iref2 is current reference of the active filter i_{ref2}, i2 is output current of the active filter i_2, ierr2 is current error between i2 and iref2 i_{err2}, ierr1 is the current error in the hysteresis current controller i_{err1}. V and Is becomes almost pure sinsoidal waveform so that it is proven that this system provides good active filter function. In addition, the current error Ierr2 between Iref2 and I2 remained within ±0.01A except zero cross region where current reference has sudden change. From this result, it is confirmed that the proposed method has good current control characteristics.

B. Experiment

Next, the proposed current control method was tested using the circuit of Fig. 1. BEF was composed of a digital filter using DSP and the other part including current control circuit were composed of analog circuit using op-amps. The experimental result is shown in Fig. 8. using parameters in Table 1. Though lower utility voltage was used due to the ratings of experimental setups in experiments, the obtained results were almost similar to the simulations. Therefore, it is verified that the proposed current control method has suitable characteristics for single-phase active filters. Frequency characteristics from current reference i_{ref} to output current i_2 were obtained by the experiment in order to verify the characteristic of the proposal control method as shown in Fig.9. In this figure, the dots are experimental results and the solid line is theoretical curve calculated previously. The gain characteristics in is almost

same while there is phase lag in experimental results. This phase lag may caused by the resistive components of inductor, which is not considered in theoretical calculations. However, the phase lag is 20 degree even at 1.6kHz in experiments. From this analysis, the proposed method has sufficient frequency characteristics for higher harmonics compensation.

V. CONCLUSION

In this paper, a new current control method for an inverter with LC output filter was proposed. The proposed method is composed of hysteresis control and state feedback. The advantages of the proposed method are very low switching ripple in output current and high response suitable for single-phase active filters. The design procedure of the proposed method was explained, then verified by simulations and experiments. The proposed method can be applied not only active filters but also UPS(Uninterrupted Power Supply) and so on. The futre work of this research is digitalization of the proposed method.

REFERENCES

[1] Kenji Hirasaki, Hideaki FujitaF "A Control Method for a Single-Phase Active Filter Capable of Reducing its DC Capacitor" Trans. of IEEJ D, 127, 11, pp.1117-1124 (2007-11) (in Japanese)
[2] Rieko Moriya, Naoki Yamamura, Muneaki Ishida, Takamasa HoriF "A Study of Combined-type Active Filter using Linear Power Amplifier" Trans. of IEEJ D, 124, 5, pp.442-449 (2004-5) (in Japanese)
[3] Toshihiko Tanaka, Kengo Ueda, Kuniaki Sato, Shinji FukudaF "A New Control Method of Single-Phase Shunt Active Filters Using the Correlation and Cross-Correlation Coefficients" Trans. of IEEJ D, 125, 11, pp.1008-1015 (2005-11) (in Japanese)
[4] Tomoki Yokoyama, Kazuya Miyashita, Shinsuke ShimogataF "Multirate Deadbeat Control for PWM Inverter using FPGA based Hardware Controller" Trans. of IEEJ D, 124, 4, pp.380-387 (2004-4) (in Japanese)
[5] Shoji Fukuda, Takehito YodaF "A Current Control Method for Active Filters Using Sinusoidal Internal Model" Trans. of IEEJ D, 120, 12, pp.1440-1446 (2000-12) (in Japanese)
[6] Masahiro Kinoshita, Makoto Nakanishi, Yushin YamamotoF "High efficiency Large Capacity UPS Using Hysteresis Current Controlled PWM" JIAS06, pp.443-444, (2006-8) (in Japanese)
[7] Takeshi Mashiyama, Hirohito Funato, Satoshi OgasawaraF "Consideration on Current Control of Single Phase Active Filter with LC Filter based on Hysteresis Control" IEEJ07,7-057,(2007) (in japanese)

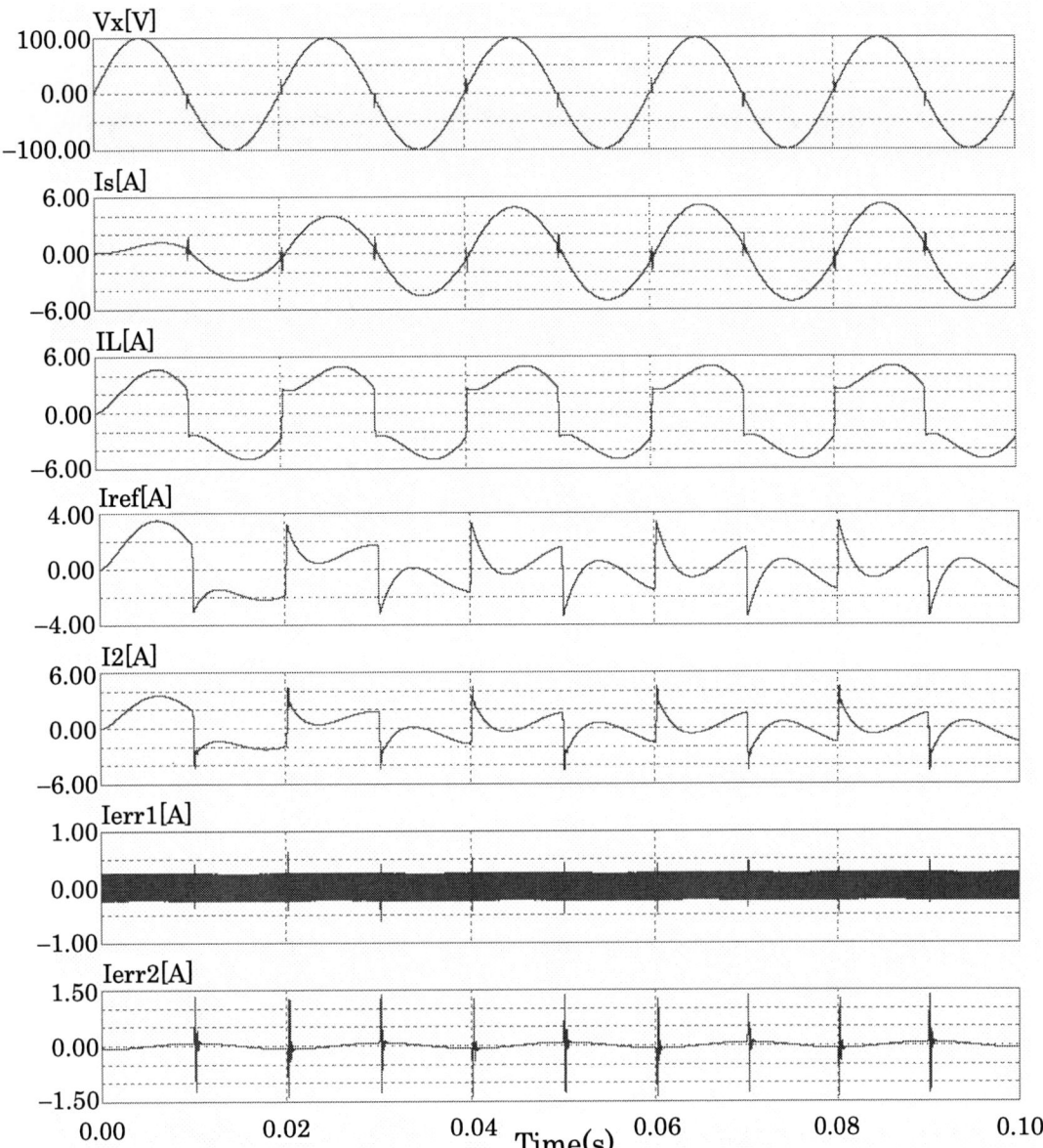

Fig. 7. Simulation results

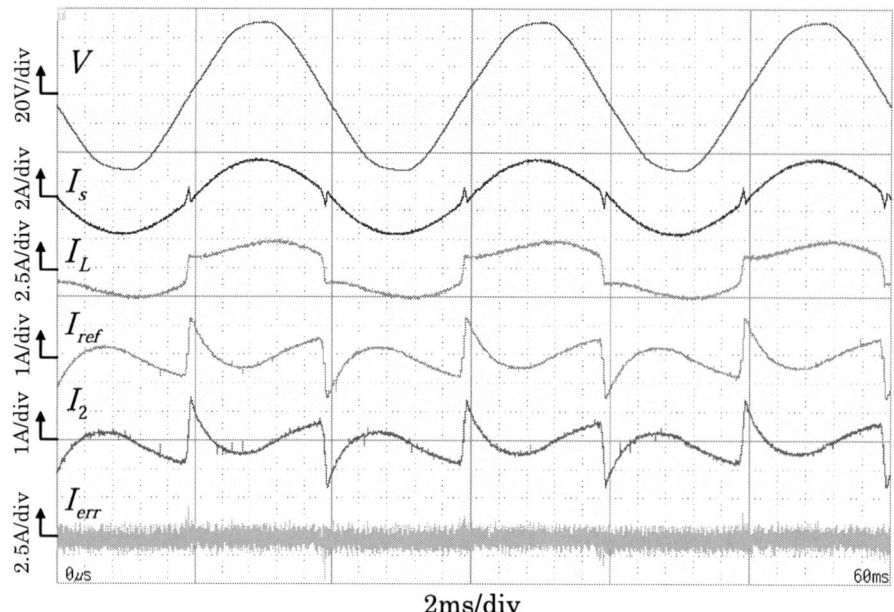

Fig. 8.   Experimental results

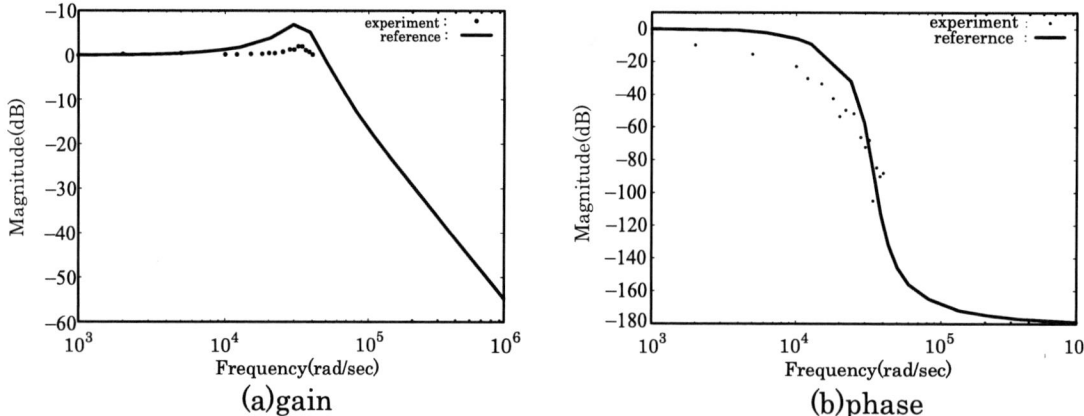

Fig. 9.   Frequency characteristic of the proposed current control

Optimal Slope Compensation for step load in peak current controlled dc-dc Buck Converter

Susovon Samanta[*], Pradipta Patra[†], Siddhartha Mukhopadhyay[*] and Amit Patra[*]

[*] IIT Kharagpur/Electrical Engg., Kharagpur, India, e-mail: (susovon, smukh, amit)@ee.iitkgp.ernet.in
[†] IIT Kharagpur/ATDC, Kharagpur, India, e-mail: pradipta@vlsi.iitkgp.ernet.in

Abstract— **The paper comes up with the variation of slope compensation with step load and duty cycle for one cycle control in a peak current controlled dc-dc buck converter. For an optimal performance under varying load conditions and input voltage the amount of slope compensation desired, vary dynamically. A conventionally used linear slope is definitely not a solution. To meet the dynamic requirement of slope in a current controlled dc-dc buck converter a non-linear ramp generator circuit has been proposed. The primary aim of the circuit is to provide a slope which is very small for low duty cycles and increases steeply as duty cycle increases. Consequently, it proves much more efficient in low duty cycle (less than 0.5) of operation. On the higher side this achieves a similar performance as that by a linear ramp with much less p-p value of the ramp.**

Keywords— **Power management, Voltage Regulator Modules, Automotive application, Switched-mode power supply, Converter control.**

I. INTRODUCTION

With the growing need for consumer applications, the urge for dc-dc converters is also on the rise. Considering high performance, increased efficiency and stability dc-dc switching converters are incorporated.

There are primarily two control methods for PWM DC-DC switching converter, voltage mode and current mode. The voltage mode control has the disadvantage of a small frequency bandwidth. The inductor and the capacitor at the output , which forms a low pass filter, introduce 180° phase shift at the self-resonant frequency; $f_{LC} = 1/2\pi\sqrt{LC}$. This requires a gain of the control loop less than unity at f_{LC} to guarantee stability. The result is a slow response in the output voltage [5-6]. However the current control can realize a large frequency bandwidth. Two loops are required in this configuration: the conventional voltage control loop and the current control loop. Though circuits become complicated and current sensing becomes a challenge it is possible to extend the bandwidth up to half the switching frequency [1].

In order to ensure stability in a current controlled dc-dc switcher with duty cycles more than 0.5, slope compensation (M_c) has to be introduced [5-6]. The amount of slope compensation is important because it affects both on stability and the frequency bandwidth [1]. This value of M_c significantly affects the frequency bandwidth of the system. With a higher value of slope the bandwidth decreases which dampens the dynamic response of the system. For systems which are limited by duty cycles, very common in commercial chips, there is a definite need for high value of ramp [ref]. The value of such a slope also varies depending on the duty cycle of the operating condition, the load step and the gain of the error

amplifier [2] [4]. Thus, a dynamically changing value of M_c is needed which is only possible using a dynamically changing ramp [3].

This paper comes up with results which show the effects on the optimal M_c value for different load steps and different duty cycles. The contribution of the error amplifier gain factor on M_c is also analyzed into and design of a PI based compensator accounted for. Finally, the advantages of using non linear slope compensation over linear slope compensation for varying duty cycle and load currents, has been put forth. The implementation of the non-linear slope compensation has also been discussed.

II. CURRENT MODE CONTROL FOR DC-DC CONVERTERS

In the current-mode control, the duty ratio of the converter is determined by the time taken by the inductor current to reach a threshold value defined by the reference control signal. Fig1 shows a fixed frequency current-mode PWM buck converter. It contains two feedback loops, an outer one which senses the output voltage and develops a control signal to an inner loop which senses the current flowing through the sense resistance series with the inductor, and keeps the output voltage constant on a pulse-by-pulse basis. The major drawback of the current-mode PWM scheme is its instability for duty ratio exceeding 0.5.

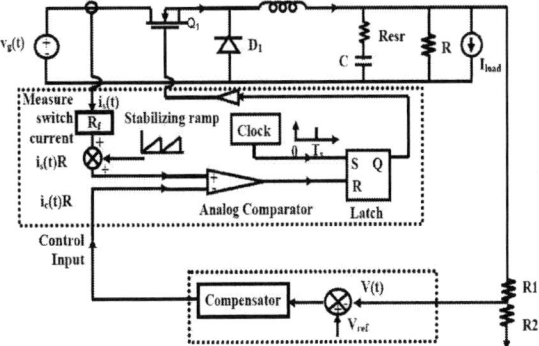

Fig. 1.Current controlled DC-DC Buck Converter

978-1-4244-1741-4/08/$25.00 ©2008 IEEE

III. Optimal Slope compensation for one cycle control of a peak current controlled DC-DC converter

A. Why optimal slope compensation for one cycle control - a discussion

To apply slope compensation for a current controlled buck converter many of the system parameter needs to be concentrated on so as to get the best dynamic performance of the system. The prime consideration in determining the value of M_c is the duty ratio of the operating condition which finally depends on the input and output voltage variation of the system. The following sections would go into the details of such variation of the slope compensation value required for different duty cycles. With different step loads in steady state condition the requirement of the value of the M_c changes quite unlikely to the theoretical $M_c=M_2$ (for one cycle control, where M_2 is the slope of the falling inductor current)or $M_c = M_2/2$ (for stability consideration) [5-6]. Usually, for one cycle control, at high duty ratio, the system requires much more slope compensation than the theoretical values and it depends on the load step as detailed out in Table2. Finally, the gain of the compensator is a serious concern at high frequency as a step change is a high frequency effect. The gain directly affects the control signal (v_c) which effects M_c consequently. So, when it comes to an optimal selection of Mc for one cycle control all the above issues comes into concern. An optimal transient performance with a step load can be characterized by the following effects.

- One cycle control which ensures that the inductor current reaches its initial value in one cycle
- Duty ratio in the transient state should never go beyond 1
- Applying a very high value of Mc the system will more oscillate.

Thus a fixed linear ramp cannot be a solution to all these issues.

B. Effect of control signal on slope compensation

If a step is applied to the load in steady state condition, the control signal (v_c) value too gets a step. This is due to the effect of the ESR of the output capacitance, of the converter, and the amplitude of the load step which is then divided by the voltage divider and amplified by the error amplifier gain. The figure Fig2 illustrates the operation of the outer loop.

From Fig2 the instantaneous change in v_c due to the step load changes can be calculated as;

Change in output voltage

$$Dv_{out} = R_{ESR} \cdot I_{load}$$

Change in feedback voltage

$$Dv_{fb} = Kv_{out} \left[K = \frac{R_2}{R_1 + R_2} \right]$$

$$Dv_C = g_m \cdot (R_{out} \| R_c) \cdot Dv_{fb}$$

$$= g_m \cdot (R_{out} \| R_c) \cdot K \, Dv_{out}$$

$$Dv_c = g_m \cdot (R_{out} \| R_c) \cdot K \cdot R_{ESR} \cdot I_{load}$$

The gain of the error amplifier in high frequency should be nominal, preferably negative. However such a gain depends on the gain of the error amplifier and the parameters of the PI controller and the values which cannot be arbitrarily chosen. Effects of some parameter changes to the system response to achieve a lower value of Mc can be summarized as follows.

1. The value of the R_c of the PI can be made equal to zero so as to achieve a low high frequency gain of the error amplifier which in turns makes the system oscillatory due to double pole.

2. Maintaining the zero within the converter resonant frequency if the R_c and C_c have to be adjusted. Decreasing the R_c value would decrease the gain of the system. This in turn makes the system less sensitive to the change in control signal so the settling time increases. Accordingly the value of R_c has to be chosen such that it incorporates a system gain. This in turn accounts for a greater M_c

TABLE I.
PARAMETER VALUES OF THE CONVERTER

Vg(V)	VO(V)	$R_{load}(\Omega)$	L(H)	C(F)	$R_{esr}(\Omega)$
4 - 10	3.3	10	22µ	60µ	30m
Gm(mho)	Rout(Ω)	Rc(Ω)	Cc(µ)	$R_{sense}(\Omega)$	F_s(Hz)
1300µ	100k	20k	4.7n	0.08	600k

Fig. 2. ESR effects on control signal during load step

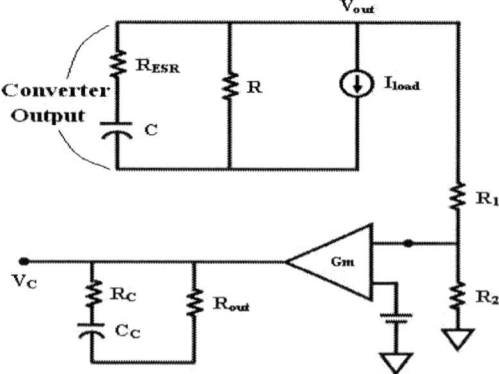

Fig3 shows the ac plots, for a converter with parameters as is tabulated in Table1, of the error amplifier with the zero of PI constant though the high frequency gain is varied simultaneously. But decreasing the gain of the amplifier makes the system response sluggish as shown in Fig4.

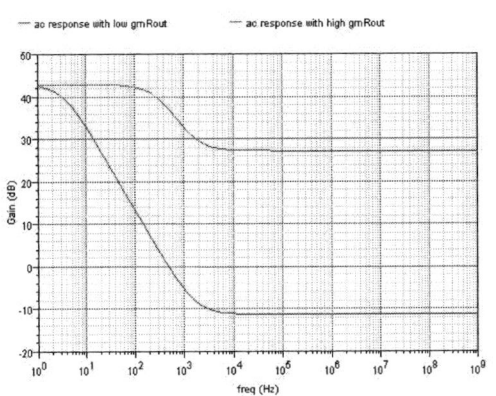

Fig. 3. AC response for different PI's

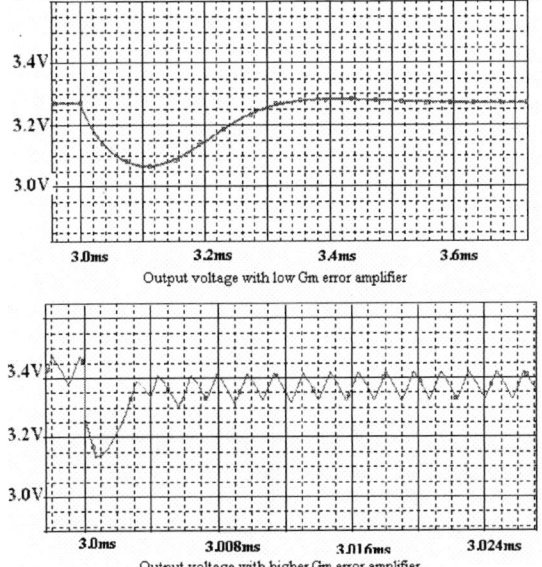

Fig4: Output response at step load for different gain in error amplifier

Consequently to achieve a good dynamic system response we have to compromise with some amount of gain in the compensator which instead makes for higher and varying values of Mc under different operating conditions and different steps as in Table2.

IV. ADVANTAGES OF NON LINEAR SLOPE COMPENSATION OVER LINEAR SLOPE COMPENSATION

When it comes to the design of an on-chip current controlled converter for duty cycles higher than 0.5, selecting the value of the compensating slope becomes a real challenge. Quite contrary to theory Mc=M2 is unable to stabilize the system in one cycle. As has already been discussed apart from the duty cycle the amount of the load step also poses a significant effect on the amount of the compensating slope required for the optimal performance of the converter in the steady state condition. Thus, literally speaking for optimal operation of the converter different values of Mc is needed based on operating condition. This however is not quite achievable with a fixed linear slope which can only be designed to meet the worst case condition for a converter. It not only hinders the optimal performance of the converter but also degrades the system performance at different operating conditions, may be the typical condition also. Based on the above stated analysis on the values of Mc a non-linear ramp can be much worthwhile as compared to its linear counterpart for optimal system performance. Some significant advantages can be

- A low value of Mc for very low duty cycles which does not unnecessarily makes the system sluggish as with a fixed linear slope which offers the same mc for all duty cycles.

- A non-linear ramp can be designed based on the optimal values of the Mc for different operating conditions and the curve obtained is more likely to give an optimal value for its corresponding operating condition. There is no such option for a fixed linear ramp.

- Last but not the least, the p-p value of such a non-linear will definitely be much less than a fixed linear slope.

V. DESIGN OF NON LINEAR SLOPE

Fig5 shows the proposed slope compensation generation circuit for a standard current controlled dc-dc buck converter. The voltage V_{sense} gives the non-linear ramp which is added to the current sense in the similar manner as a linear slope ramp is added before it is compared to the compensated error signal from the outer voltage loop.

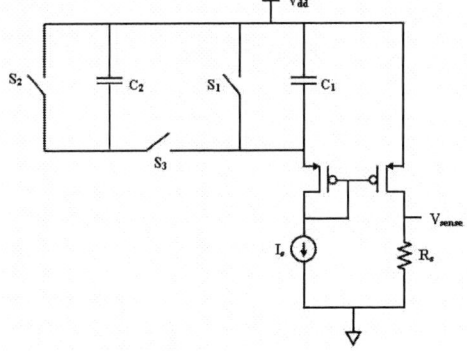

Fig5: Slope Compensation Circuit

The slope compensation circuit described in Fig5 consists of M_1, M_2, a constant bias current I_s, two capacitances C_1 and C_2, and three switches Sw_1, Sw_2 and

Sw$_3$. I$_s$ is a fixed current source and the slope produced by I$_s$ and the combination of C$_1$ and C$_2$ at different time instant is relatively precise. The switching operations of the capacitors must be synchronized with the system clock. With the turning ON of the high side switch S$_3$ turns ON. Consequently the node voltage of N$_s$ is defined by the charging of the capacitors C$_1$ and C$_2$ in parallel. Thus, S$_1$ turns ON to discharge C$_1$. The final part is defined by the switching ON of C$_1$ only. The voltage at the node is due to the charging of C$_1$. Thus maintaining a small value of C$_1$ and a high value of C$_2$ the rate of change of voltage at the node is small when both the capacitors are in parallel than when only C$_1$ is charged. The voltage drop at that node will be linear with respect to supply with two different slopes for the two different charging patterns. The intention of such a slope is to maintain a low slope for low duty cycles and an exponential may be, as the duty cycle increase. This linear voltage change is applied to the gate of M$_2$. As M$_1$ and M$_2$ are mirrored then they would try to maintain the same V$_{gs}$. As the node voltage decreases the Vg tends to decrease to maintain the same V$_{gs}$. Thus the V$_{gs}$ across M$_2$ increases. As a result a current with square relationship with time flows in M$_2$ and is converted to square voltage of the resistor R$_s$. Finally the voltage V$_{sense}$ serves as the ramp for the converter.

VI. SIMULATION RESULTS

Based on the parameters of the current controlled converter tabulated in Table1 the dynamic p-p values of Mc were obtained as in Table2 which give the optimal performance. The corresponding operation of the converter at a constant duty cycle with different load steps has been figured out in Fig6.

TABLE II.
DYNAMIC VALUE OF P-P M$_C$

	Load Step (%)				
	15	30	50	70	90
Duty Cycle 0.4	N.A.	N.A.	80mV	150mV	280mV
0.5	50mV	60mV	100mV	175mV	325mV
0.6	55mV	70mV	130mV	190mV	365mV
0.7	60mV	120mV	170mV	250mV	420mV
0.8	75mV	160mV	310mV	620mV	1.7V

Consequently the transient optimal response of the converter, as from the values of Mc in Table2, with a load step of 50% over a range of duty cycles in plotted in Fig7.

Finally, to account for the advantage of using a non-linear ramp over a linear ramp has been brought forward by Fig8. However for high duty cycles non-linear slope achieves a similar performance even with much smaller p-p value. The transient response at 0.8 duty cycle and 50% load step with a non-linear ramp of about half the peak of linear ramp is shown in Fig9.

However for high duty cycles non-linear slope achieves a similar performance even with much smaller p-p value. The transient response at 0.8 duty cycle and 50% load step with a non-linear ramp of about half the peak of linear ramp is shown in Fig9.

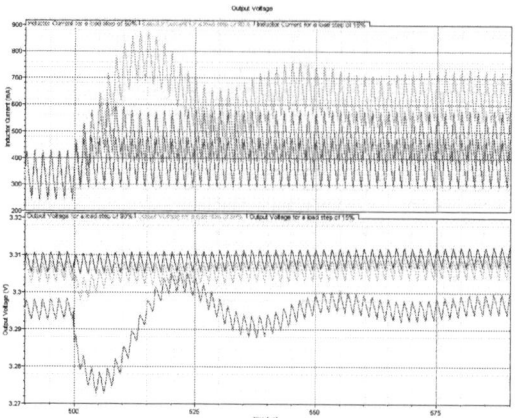

Fig 6. Response at constant d and varying load

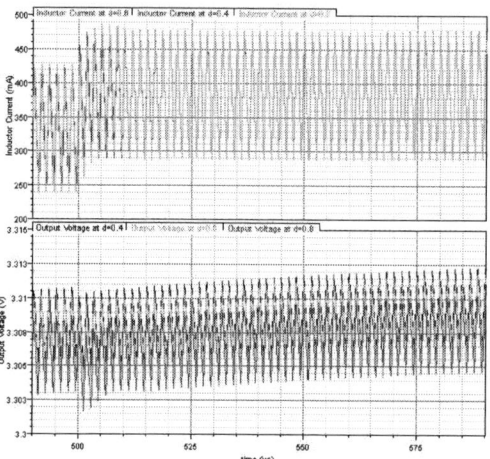

Fig7: Response at constant load and varying d

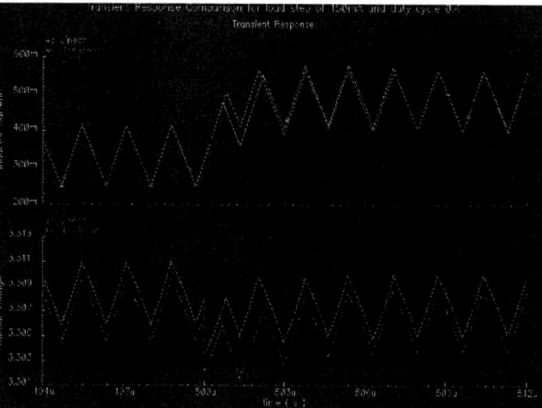

Fig8: Output Comparison with linear and non linear compensation at d=0.4

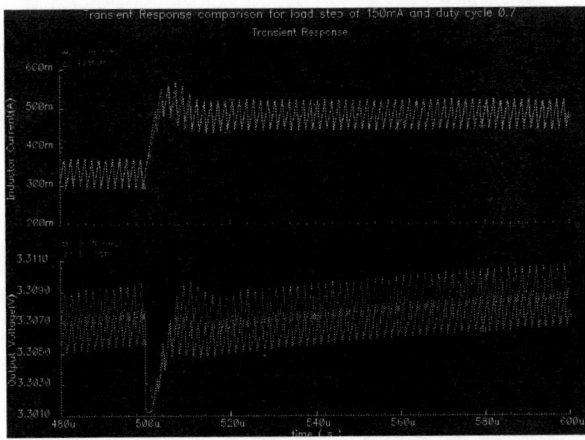

Fig9: Output Comparison with linear and non linear compensation at d=0.8

VII. CONCLUSION

The variation of Mc has been analyzed into with the amount of load disturbance and the changing duty cycle with the change in input and/or output of the system. It can thus be inferred that dynamic slope compensation is of utmost importance to tackle dynamic conditions for a current controlled converter and for one cycle control. Designing a conventional linear ramp is an overdoing it. Based on the values of Mc for optimal performance of the system a non linear ramp generation circuit has been developed and its performance compared with that of the linear one. Such a ramp would deeply improve the performance of peak current controlled converters.

ACKNOWLEDGMENT

The authors would like to express their gratitude to the members of Advanced VLSI Design Lab, IIT Kharagpur who have been always on their toes in times of need

REFERENCES

[1] 1. Tan F.Dong, "Current-Loop Gain with a Nonlinear Compensating Ramp." Power Electronics Specialists Conference, IEEE, Vol. 1., June 1996.

[2] 2. Choi B., "Step Load Response of a Current-Mode-Controlled DC-to-DC Converter." IEEE Transactions on Aerospace and Electronic Systems, Vol.33, No.4, October 1997.

[3] 3. Sakurai H. and Sugimoto Y, "Analysis and Design of a Current-Mode PWM Buck Converter Adopting the Output-Voltage Independent Second-Order Slope Compensation Scheme" IEICE Transaction Fundamentals, Vol.E88-A, No.2 February 2005.

[4] 4. Redl. R, Erisman B and Zansky Z, "Optimizing the Load Transient Response of the Buck Converter." Applied Power Electronics Conference and Exposition, IEEE, 1998.

[5] 5. Erickson Robert W. and Maksimovic Dragan., "Fundamentals of Power Electronics." Kluwer Academic Publishers Group Distribution Centre.

[6] 6. Abraham I. Pressman, "Switching Power Supply Design", McGraw-Hill.

Performances of a PLL Based Digital Filter for double-conversion UPS

Armando Bellini* and Stefano Bifaretti†

*Dept. of Electronic Engineering, University of Rome "Tor Vergata", Rome, Italy, e-mail: *bellini@ing.uniroma2.it*
† Dept. of Electronic Engineering, University of Rome "Tor Vergata", Rome, Italy, e-mail: *bifaretti@ing.uniroma2.it*

Abstract— In uninterruptible power supplies (UPS) using an automatic bypass switch, to reduce the transients due to the switch commutations, it is convenient that the voltage waveforms produced by the inverter are synchronized with the grid ones. In three-phase UPS, the inverter controller employs, as reference signals, the components of the grid voltage phasor, referred to a fix reference frame, suitably multiplied by an amplitude correction factor. Generally, the reference signals are affected by harmonics and amplitude unbalances. Therefore, improper synchronization between the inverter outputs and the grid voltages can arise.

To improve the synchronization, many solutions based on a phase locked loop (PLL) system have been proposed in literature. The paper proposes a different PLL structure employing a Steady-State Linear Kalman Filter (SSLKF), based on a third-order linear and time-invariant observation model. Such filter is able to provide an accurate tracking of the grid voltage phasor also in critical operating conditions. The paper describes, at first, different PLL structures and then the proposed architecture based on a prediction-correction filter and its implementation on a DSP controller. Finally, a comparison, obtained by simulation, with traditional filters and some significant experimental results on a pre-production prototype are carried out.

Keywords—Uninterruptible Power Supply (UPS), Converter control.

I. INTRODUCTION

The three-phase uninterruptible power supplies (UPS) used in industrial applications are usually based on a double stage power conversion system (double-conversion UPS); the first stage (Rectifier) converts the utility AC line to a DC voltage, the second stage (Inverter) supplies the AC output. During normal operation, the rectifier provides the current required to charge the batteries and also to supply the inverter. The inverter feeds the load performing a regulation of the output voltage and frequency. In the event that the AC line fails or deviates significantly from the specified input voltage and frequency tolerances, the inverter uses the batteries as an energy source and operates until the batteries are discharged.

Double-conversion UPS are provided by an automatic bypass switch. This switch usually connects the load to the inverter while, in case of a UPS failure, it commutates the load to the utility power grid. The switch may also be used to help support a temporary overload that the inverter cannot support alone. To reduce the transients connected to the switch commutations, it is convenient that the voltage waveforms produced by the inverter are in phase with the AC source ones. So the voltages synthesized by

the inverted must be synchronized with the grid ones; beside also the inverter phases sequence must coincide with that of the grid. To this aim, the components v_α and v_β of the grid voltage phasor, referred to a fix reference frame α and β, must be determined starting by the measure of two grid voltages.

Reference signals v_α and v_β are contaminated by harmonics, which may have been produced by the power converter itself or generated by another loads. Moreover, due to the mismatch in the measuring chains, the amplitudes of reference voltages v_α and v_β can be different. So, improper synchronization between the inverter and the grid must arise.

To improve the synchronization, many solutions based on a phase locked loop (PLL) system have been proposed in literature [1-6]. Such approaches use a second-order closed-loop transfer function that presents some difficulty on the tuning of its parameters, especially when narrow bandwidth and fast response are required; in addition, PLL may fail to lock the input signals during start-up transient when some adverse conditions occur [6]. The paper proposes a different PLL solution, already employed to reduce the speed measurement noise in drives using an electromagnetic resolver [7, 8], that permits an accurate tracking of input signals, even if critical conditions occur, and an easy tuning of its parameters. The proposed solution employs a prediction-correction filter, whose structure has been derived from a Steady-State Linear Kalman Filter (SSLKF) based on a third order linear and time-invariant observation model; the choice of a SSLKF allows the reduction of the computation complexity, since the correction gain vector can be off-line calculated.

In the paper, after a description of different PLL structures and of the SSLKF algorithm, the architecture of the PLL filter and its implementation on a DSP controller are presented. Finally, a comparison, obtained by simulation, with traditional filters and some experimental results on a pre-production prototype are pointed out.

II. PLL STUCTURES

The simplest PLL structure is based on a phase detector, a low-pass filter and a voltage-controlled oscillator (VCO), as shown in Fig. 1 [4,5]. This structure, widely used in hardware realizations, can be classified as a zero-crossing structure in which the detection of phase and frequency is based on the zero crossing points of the input signal. The dynamic performances of this solution are limited as the zero crossing points are only detected at every half-cycle of the utility frequency [4]. Moreover, the noise around the zero-voltage crossing point makes the output angle oscillate.

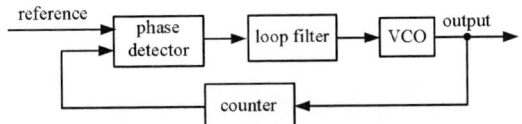

Fig. 1. Block diagram of a PLL with a zero-crossing structure.

A different solution, more convenient for software realizations, makes use of a closed loop system in which the error signal is obtained by the comparison between a measured grid waveform and the obtained one. This solution is rather appropriate in three phase systems allowing the determination of the error signal by a vector product [2, 5].

Fig. 2 shows the basic block diagram of a PLL system based on the vector product between the measured grid voltage phasor and that provided by the PLL. The PLL structure is essentially composed by three blocks.

Block B_1 achieves the vector product and generates signal error e. In particular signal error e is approximated by its sine and it is obtained as:

$$e \cong \sin(\theta_m - \tilde{\theta}) = \frac{v_\beta \cos(\tilde{\theta}) - v_\alpha \sin(\tilde{\theta})}{V_s} = \quad (1)$$
$$= \sin(\theta_m)\cos(\tilde{\theta}) - \cos(\theta_m)\sin(\tilde{\theta}),$$

where v_α and v_β are the components of the grid voltage phasor, referred to a fix reference frame α and β, V_s and θ_m are the module and the phase of the grid voltage phasor and $\sin(\tilde{\theta})$ and $\cos(\tilde{\theta})$ are the sine and the cosine of phase angle $\tilde{\theta}$, provided by the PLL system.

Block B_2 is the loop filter, making the estimation of the phase angle of the grid voltage phasor.

Block B_3 determines the values of the sine and cosine of phase angle $\tilde{\theta}$; this can be achieved employing a numerical procedure or through tables.

The loop filter is generally constituted by a Proportional Integral (PI) regulator, providing the estimated value of angular frequency $\tilde{\omega}$, followed by an integrator, which furnishes the estimation $\tilde{\theta}$ of the phase. A different solution, proposed in the paper, makes use of a suitable deterministic prediction-correction filter, derived by a Steady-State Linear Kalman Filter (SSLKF). Proposed filter has been already employed to reduce the speed measurement noise in drives using an electromagnetic resolver [6, 7] and it is based on a third-order linear and time-invariant observation model.

III. PREDICTION-CORRECTION FILTER

As previously said, the prediction-correction filter, used in the proposed PLL structure is derived by a Steady-State Linear Kalman Filter (SSLKF) fitting the following stochastic discrete-time model:

$$x_n = A x_{n-1} + \xi_n$$
$$y_n = c^T x_n + \eta_n \quad (2)$$

where:

$$x = \begin{bmatrix} \theta \\ \omega \\ a \end{bmatrix} \qquad A = \begin{bmatrix} 1 & T & T^2/2 \\ 0 & 1 & T \\ 0 & 0 & 1 \end{bmatrix}$$

$$c^T = \begin{bmatrix} 1 & 0 & 0 \end{bmatrix},$$

n is the index of the sampling instant,

T is the duration of the sampling interval,

θ is the value of the grid voltage phase angle,

ω is the value of the grid angular frequency,

a is the derivative of the grid angular frequency,

y is the output variable, i.e. the computed value of the phase angle;

ξ is the zero-mean independent Gaussian white-noise vector of the process state, with a covariance matrix Q,

η is the zero-mean independent Gaussian white-noise scalar of the process measurement, with a variance r.

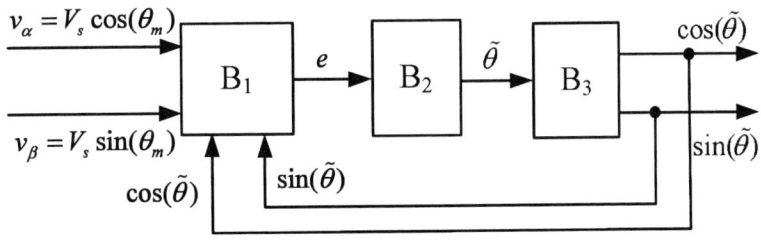

Fig. 2. Block diagram of a PLL system based on the vector product.

The state variables of the previous dynamic model can be estimated by the Kalman filter algorithm, starting from an initial value $\hat{x}_0$ and a matrix $\hat{P}_0$ equal to the covariance matrix of $\hat{x}_0$, and composed by the following steps:

1) prediction of state:
$$\tilde{x}_n = A\hat{x}_{n-1}$$
2) estimation of error covariance matrix:
$$\tilde{P}_n = A\hat{P}_{n-1}A^T + Q$$
3) computation of the Kalman filter gain vector:
$$g_n = \tilde{P}_n c \left(c^T \tilde{P}_n c + r\right)^{-1}$$
4) update of error covariance matrix:
$$\hat{P}_n = \left(I - g_n c^T\right)\tilde{P}_n$$
5) correction of predicted state:
$$\hat{x}_n = \tilde{x}_n + g_n e_n,$$

being $e_n = \theta_n - c^T \tilde{x}_n$ the prediction error.

Since dynamic system (2) is linear and time-invariant, the only variable element in the prediction and correction sections is gain vector g_n. Therefore, it is possible to replace vector's sequence g_n with its limit g when $n \to \infty$; in this way a SSLKF is obtained, taking advantage of a significant reduction in the number of math operations.

Once determined gain vector g, the operations to be effected at every sampling interval are only those related to prediction and correction:
$$\tilde{x}_n = A\hat{x}_{n-1}, \tag{3}$$
$$\hat{x}_n = \tilde{x}_n + g\left(\theta_n - c^T \tilde{x}_n\right). \tag{4}$$

In the SSLKF algorithm, gain vector g depends on measurement noise variance r and on dynamic system covariance matrix Q and can be determined solving a Riccati equation. A different solution, allowing a simple selection of the filter characteristics, consists in imposing the allocation of the filter transfer function poles.

Combining the prediction equation:
$$\tilde{x}_n = A\hat{x}_{n-1}$$
at sampling instant n with the correction equation, at the previous sampling instant:
$$\hat{x}_{n-1} = \tilde{x}_{n-1} + g\left(\theta_{n-1} - c^T \tilde{x}_{n-1}\right),$$
the following dynamic equation of the SSLKF is obtained:
$$\tilde{x}_n = \hat{A}\tilde{x}_{n-1} + Ag\theta_{n-1}, \tag{5}$$
being:
$$\hat{A} = A\left(I - gc^T\right).$$

Defining the following vector:

$$l = Ag = \begin{bmatrix} l_1 \\ l_2 \\ l_3 \end{bmatrix}, \tag{6}$$

dynamic matrix $\hat{A}$ can be expressed as:

$$\hat{A} = A - lc^T = \begin{bmatrix} 1-l_1 & T & T^2/2 \\ -l_2 & 1 & T \\ -l_3 & 0 & 1 \end{bmatrix}.$$

Equating to zero the determinant of matrix $\lambda I - \hat{A}$, the following characteristic equation is obtained:
$$\lambda^3 + \lambda^2 c_2 + \lambda c_1 + c_0 = 0,$$
in which:
$$c_2 = l_1 - 3,$$
$$c_1 = 3 - 2l_1 + l_2 T + l_3 \, T^2/2,$$
$$c_0 = -1 + l_1 - l_2 T + l_3 \, T^2/2. \tag{7}$$

Imposing that the characteristic equation must have a real eigenvalue $\lambda_1 = \rho_0$ and two complex conjugated eigenvalues $\lambda_{2,3} = \rho_1 Exp(\pm j\varphi)$, the characteristic equation coefficients must be:
$$c_2 = -2\rho_1 cos(\varphi) - \rho_0,$$
$$c_1 = \rho_1 \left[2\rho_0 cos(\varphi) + \rho_1\right], \tag{8}$$
$$c_0 = -\rho_0 \rho_1^2.$$

The eigenvalues of matrix $\hat{A}$ can be imposed by discretization of a continuous filter, characterized by a real negative pole of module p_0 and a couple of complex poles $p_{1,2} = -w\left(cos(\phi) \pm j \sin(\phi)\right)$.

The discretization gives:
$$\rho_0 = Exp(-p_0 T), \qquad \rho_1 = Exp(-wT cos(\phi)),$$
$$\varphi = wT \sin(\phi). \tag{9}$$

Different simulations have shown that it is convenient to select the same value for p_0 and w and a value included between 40 and 60 degrees for ϕ, so only the value of w must be selected to modify the behaviour of the filter.

Once chosen the poles of the discrete time transfer function of the filter, coefficients c_2, c_1 and c_0 are determined by (8) and the elements of vector l can be obtained by (7) as:

$$l_1 = c_2 + 3,$$
$$l_2 = \frac{c_1 - c_0 - 4 + 3l_1}{2T},$$
$$l_3 = \frac{c_1 + c_0 - 2 + l_1}{T^2}. \tag{10}$$

Finally, from definition (4), the elements of vector g can be computed as:

$$g_3 = l_3,$$
$$g_2 = l_2 - Tg_3,$$
$$g_1 = l_1 - Tg_2 - g_3 \frac{T^2}{2}.$$
(11)

IV. IMPLEMENTATION ON A FIXED POINT DSP CONTROLLER

The proposed filter has been implemented on the 16-bit fixed point TMS320F2401 DSP controller used to control the UPS. Fig. 3 shows the structure employed for the realization of the PLL-based filter.

Blocks B_1 and B_3 perform the already described operations. Block B_2 is constituted by three blocks; at each sampling interval, firth block, B_{21}, supplies the correction of the prediction provided by block B_{23}, while Block B_{22} represents the delay connected to the discrete time operations of the DSP.

The first implementation problem has been the choice of the scaling factors employed to numerically represent the state variables and the multiplicative constants. Such choice must be performed so to avoid any overflow when performing math operations and to minimize the truncation errors due to the discrete representation of the variables. Moreover, a suitable choice of the scaling factors can allow a reduction of the algorithm complexity and, therefore, of the computation time.

In a scalar form, prediction equation can be expressed as:

$$\theta_n = \hat{\theta}_{n-1} + T\hat{\omega}_{n-1} + \frac{T^2}{2}\hat{a}_{n-1}$$
$$\omega_n = \hat{\omega}_{n-1} + T\hat{a}_{n-1}$$
$$a_n = \hat{a}_{n-1}.$$
(12)

It is clear that its implementation on a fixed point DSP requires to select suitable numeric representations of phase angle θ, grid angular frequency ω and its derivative a. In other words, it is necessary to select the following scaling factors:

$$S_\theta = \frac{\theta}{\theta^*} \quad , \quad S_\omega = \frac{\omega}{\omega^*} \quad , \quad S_a = \frac{a}{a^*}, \quad (13)$$

where θ^*, ω^* and a^* denote the numeric representations as 16-bit integer numbers of θ (rad), ω (rad/s) and a (rad/s^2) respectively.

Taking into account (12) and (13), the prediction equations, referred to the numeric representations of the variables, become:

$$\theta_n^* = \hat{\theta}_{n-1}^* + T\frac{S_\omega}{S_\theta}\hat{\omega}_{n-1}^* + \frac{T^2}{2}\frac{S_a}{S_\theta}\hat{a}_{n-1}^* =$$
$$= \hat{\theta}_{n-1}^* + k_1\hat{\omega}_{n-1}^* + k_2\hat{a}_{n-1}^*$$
$$\omega_n^* = \hat{\omega}_{n-1}^* + T\frac{S_a}{S_\omega}\hat{a}_{n-1}^* = \hat{\omega}_{n-1}^* + k_3\hat{a}_{n-1}^*$$
$$a_n^* = \hat{a}_{n-1}^*.$$
(14)

These equations show that a suitable choice of the scaling factors can allow obtaining a good solution of both the above mentioned implementation problems. In particular, it is clear that the implementation of equations (14) doesn't introduce errors if coefficients:

$$k_1 = T\frac{S_\omega}{S_\theta}, \quad k_2 = \frac{T^2}{2}\frac{S_a}{S_\theta}, \quad k_3 = T\frac{S_a}{S_\omega},$$

are exactly represented by integer numbers; moreover, if they are equal to powers of 2, the prediction values of the angular position and the speed can be simply computed by addition and shift operations.

As regards the phase angle, it is convenient to select a scaling factor S_θ equal to:

$$S_\theta = \frac{\theta}{\theta^*} = \frac{2\pi}{2^{16}}.$$
(15)

In fact, employing this scaling factor, when angle θ changes from 0 to 2π, its representation θ^* assumes values from 0 to 2^{16}; so this choice automatically limits the value of θ inside the round angle.

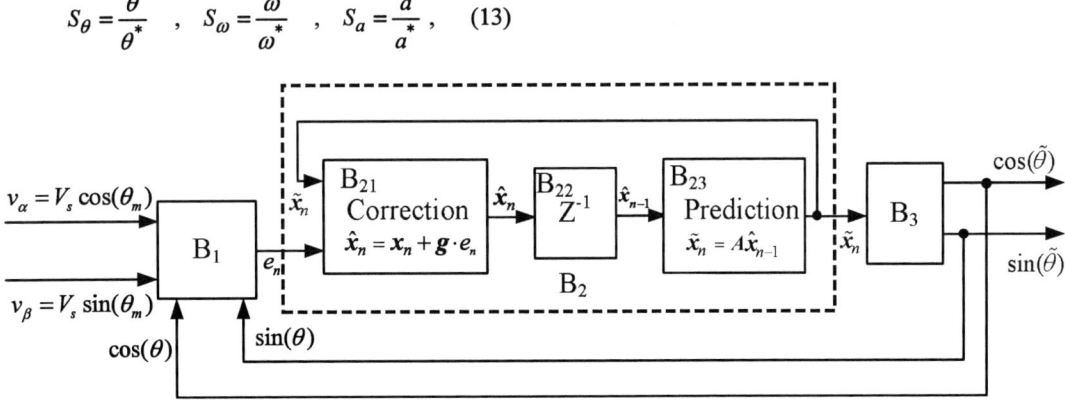

Fig. 3. Structure employed for the PLL-based filter.

Denoting as ω_{min} the angular frequency corresponding to a unitary increment of the angle representation in a sampling period, i.e.:

$$\omega_{\min} = \frac{1}{T} S_\theta = \frac{1}{T} \frac{2\pi}{2^{16}},$$

it's opportune to represent this angular frequency value with an integer number expressed by a power of 2:

$$\omega_{\min}^* = 2^{k_\omega}.$$

In fact, by this choice, angular frequency scaling factor S_ω and coefficient k_1 become:

$$S_\omega = \frac{\omega_{\min}}{\omega_{\min}^*} = \frac{2\pi}{2^{k_\omega} 2^{16} T} = \frac{2\pi}{2^{(16+k_\omega)} T} \qquad k_1 = 2^{-k_\omega}. \quad (16)$$

The exponential value k_ω must be chosen so that the maximum angular frequency representation $\omega_{\max}^*$, corresponding to an angular frequency sufficiently greater than the nominal grid angular frequency, doesn't exceed a value equal to 2^{15}, i.e.:

$$k_\omega = 14 - \mathrm{int}\left(\log_2 \frac{\omega_{\max}}{\omega_{\min}}\right) \quad (17)$$

where $\mathrm{int}(x)$ denotes the integer part of x.

A suitable scaling factor for the derivative of the angular frequency can be achieved accounting that its minimum value a_{min} corresponds to a unitary variation of the angular frequency integer representation that is, on the basis of (10), to an angular frequency variation:

$$\Delta\omega = \omega_{\min} = \frac{2\pi}{2^{(16+k_\omega)} T}.$$

Therefore, considering the above angular frequency variation $\Delta\omega$ during a sampling interval, minimum value a_{min} of the derivative results:

$$a_{\min} = \frac{2\pi}{2^{(16+k_\omega)} T^2}.$$

Imposing $a_{min}^* = 2^{k_a}$, the derivative scaling factor results:

$$S_a = \frac{a_{\min}}{a_{\min}^*} = \frac{2\pi}{2^{(k_\omega + k_a + 16)} T^2}. \quad (18)$$

Consequently, coefficients k_2 and k_3 become:

$$k_2 = 2^{-(1+k_a+k_\omega)} \qquad k_3 = 2^{-k_a}$$

that are powers of 2, as desired.

As for k_ω, the value of k_a must be chosen so that the maximum derivative representation doesn't exceed a value equal to 2^{15}, i.e.:

$$k_a = 14 - \mathrm{int}\left(\log_2 \frac{a_{\max}}{a_{\min}}\right) \quad (19)$$

In order to guarantee a good convergence of the integration procedure, it is convenient to use in prediction and correction equations a double precision representation (32 bit) of its state variables θ, ω and a,

whereas for prediction error e_n^* a 16-bit representation is adequate.

As previously pointed out, the prediction error is computed on the basis of the vector product between the measured grid voltage phasor and that provided by the PLL; more precisely, prediction error e_n is approximated by the sine of its value:

$$e_n \cong \sin(e_n) = \sin(\theta_m - \theta), \quad (20)$$

and it can be computed as:

$$e = \sin(\theta_m - \theta) = \frac{v_\beta \cos(\theta) - v_\alpha \sin(\theta)}{V_s}. \quad (21)$$

Scale coefficient S_p of the sine and cosine of estimated grid voltage phasor angle can be freely selected; to simplify the computation of (15), it is convenient to choose $S_p = 2^{-15}$. On the contrary, the most convenient scale factor S_m of the measured values of v_α and v_β depends on the electronic circuit employed for the measure and on the AD converter resolution.

To keep for the prediction error the same scale employed for the phasor angle, value e_n^* must be computed as:

$$e_n^* = \frac{S_p S_m}{S_\theta} e_n = \frac{S_m}{\pi} e_n. \quad (22)$$

As consequence, the correction equations of the state variables result:

$$\hat{\theta}_n^* = \theta_n^* + g_1^* e_n^*, \quad \hat{\omega}_n^* = \omega_n^* + g_2^* e_n^*, \quad \hat{a}_n^* = a_n^* + g_2^* e_n^*,$$

where coefficients g_1^*, g_2^* and g_3^* correspond to coefficients g_1, g_2 and g_3 of gain vector g, including the scaling factors. Therefore, coefficients g_1^*, g_2^* and g_3^* become:

$$g_1^* = g_1, \quad g_2^* = \frac{S_\theta}{S_\omega} g_2, \quad g_3^* = \frac{S_\theta}{S_a} g_3.$$

Finally, the values of coefficients g_1^*, g_2^* and g_3^* must be further scaled to allow a representation by adequately significant integer numbers; consequently, each of the three multiplication operations requires also a suitable shift.

The described approach was employed in the UPS control system; the PLL filter is characterized by a 50 rad/s bandwidth, an ω_{min} value equal to 1.117 rad/s and an ω_{max} equal to 420 rad/s. With such choice the scaled coefficients of gain vector g become:

$$g_1^* = 18503 * 2^{-6}, g_2^* = 16579 * 2^{-8}, g_3^* = 23466 * 2^{-9}.$$

Using this filter setting and a numerical procedure to determine $\sin(\theta)$ and $\cos(\theta)$, the execution of the whole filtering procedure requires approximately 115 clock cycles, in every sampling interval; 55 clock cycles are required for prediction-correction filter algorithm while 60 clock cycles are necessary, in the worst case, for $\sin(\tilde{\theta})$ and $\cos(\tilde{\theta})$ real-time calculation. So, using a 40 MHz clock frequency, the microcontroller spends for the execution of the PLL based filter algorithm about 2.875 μs that represents about 2% of the UPS control sampling period, imposed to 140 μs.

V. SIMULATION TESTS

The performances of the proposed PLL structure are compared with those obtained by a traditional PLL system, based on the vector product, making use of a PI loop filter. The parameters of the PI regulator have been selected so that the traditional PLL system has two complex poles coincident with the complex poles of the proposed one.

The first tests have considered the effect of unbalanced voltages produced by a three-phase 50 Hz 380V source adding to the positive sequence, a negative sequence with an amplitude equal to the 20% of the positive one. Fig. 4 shows the waveforms, during an half of the period, of positive sequence voltages $v_{\alpha p}$ and $v_{\beta p}$ (in pu) and of signals $\sin(\tilde{\theta})$ and $\cos(\tilde{\theta})$ supplied by the proposed PLL filter. To underline the effect of the negative sequence, Fig. 5 provides the shapes of the phase error and the voltage error module caused by the negative sequence. As it can be observed, the filter action is very effective.

Then, the effects of voltage harmonics are taken into consideration. Fig. 6 shows the shapes of the phase error and of the voltage error module caused, in the three-phase voltage source, by a fifth order harmonic characterized by a negative sequence and an amplitude of 20% of the fundamental. As it can be observed, the harmonic effects are even smaller.

Similar results have been obtained also for the traditional solution. On the contrary, the proposed solution recovers more quickly the errors introduced by frequency variations or by rapid phase changes.

To underline these differences, Fig. 7 illustrates the transient of the estimated angular frequency, assuming that its initial value is chosen for a 60 Hz grid, whilst the UPS is connected to a 50 Hz grid.

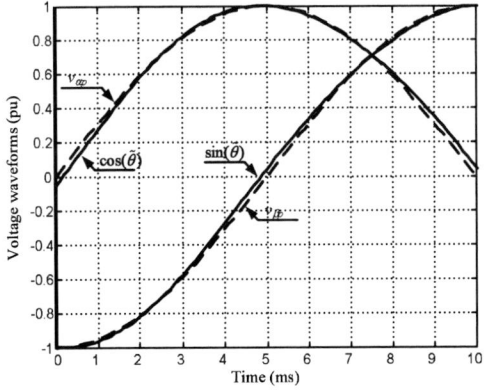

Fig. 4. Waveforms of positive sequence voltages and filter output.

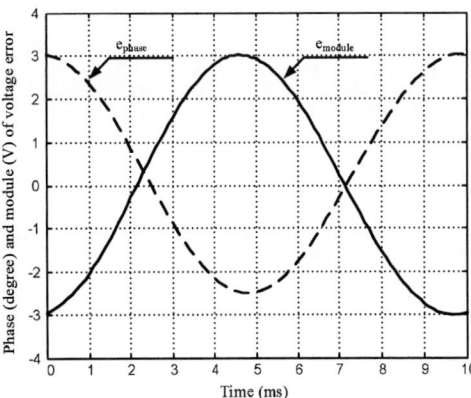

Fig. 5. Shapes of the phase error and of the module of voltage error.

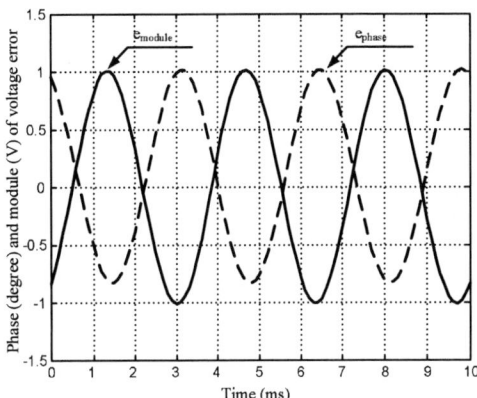

Fig. 6. Shapes of the phase error and of the module of voltage error.

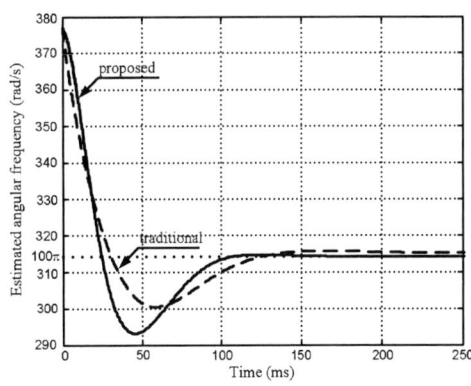

Fig. 7. Estimated angular frequency during a transient.

Finally, Figs. 8 and 9 show, respectively, the shapes of the phase error and of the module of voltage error caused by a phase jump equal to 60 degrees. The figures, in which continuous lines refer to the proposed solution while dotted lines to the traditional one, point out that the transients produced by the proposed filter are quite faster.

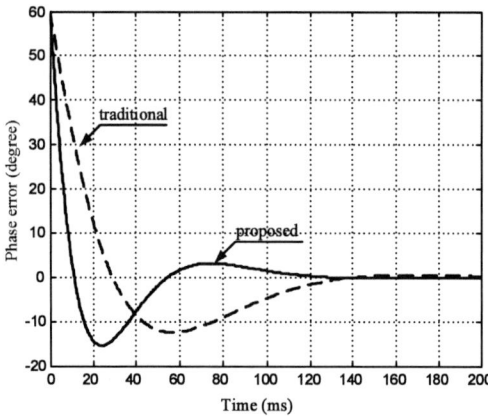

Fig. 8. Phase error due to a 60° phase jump.

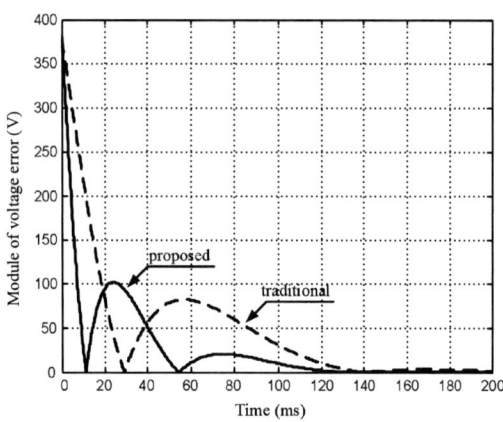

Fig. 9. Module of the voltage error due to a 60° phase jump.

VI. EXPERIMENTAL RESULTS

The performances of the PLL filter has been confirmed by different experimental tests carried out on an electronic board, based on a Texas Instruments TMS320F2401 DSP, connecting the generation system to a three-phase 50 Hz 380V utility grid. As it is possible to notice from Fig. 10, the waveforms of measured phase voltages are generally affected by a significant distortion which, in part, is present on the grid voltages and another part is introduced by the analog measurement circuits that include also transformers.

Fig. 11 illustrates the waveform of phase angle θ_n, used in the algorithm for the correction of predicted state, determined from measured phase voltages v_α and v_β by applying an inverse tangent operation.

In these operational conditions, the proposed filter produces the shapes of the sine and cosine of the phasor angle ($\sin\theta$ and $\cos\theta$) depicted in Fig. 12; the waveform of the filtered phase angle is then shown in Fig. 13. Comparing the waveforms illustrated in Fig. 10 and in Fig. 12, it is possible to notice that the PLL filter is able to track accurately the input signals even if they are strongly distorted.

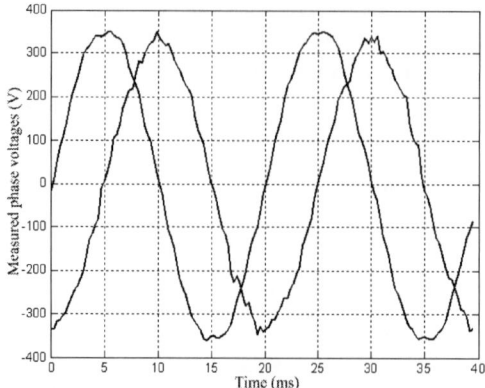

Fig. 10. Waveforms of measured phase voltages v_α and v_β.

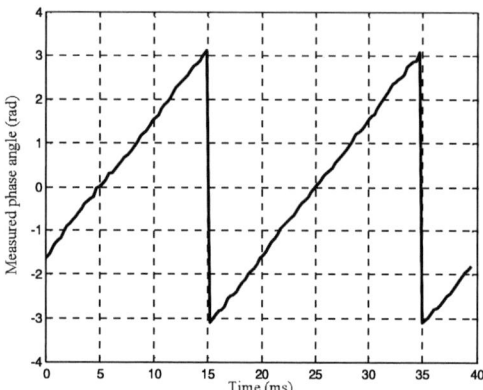

Fig. 11. Waveform of phase angle θ_n calculated from measured phase voltages.

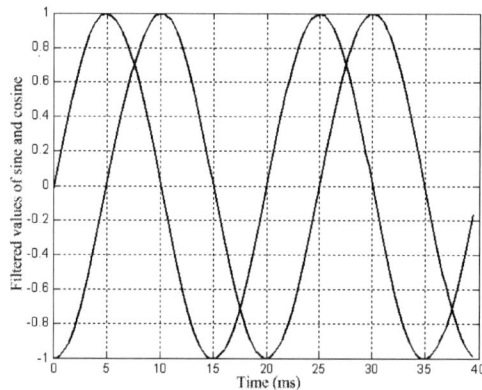

Fig. 12. Shapes of sine and cosine obtained by the filter.

Finally, Figs. 14 and 15 show, respectively, the waveforms of prediction error e_n of the phase angle and the shape of the angular frequency error.

It is possible to highlight that the proposed filter produces a negligible phase angle error and an angular frequency error ranging between ±0.19% of the nominal frequency value.

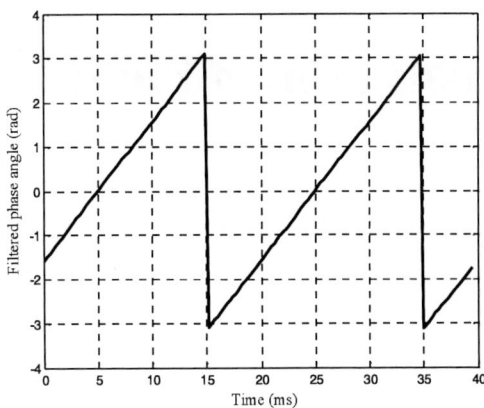

Fig. 13. Waveform of the filtered phase angle.

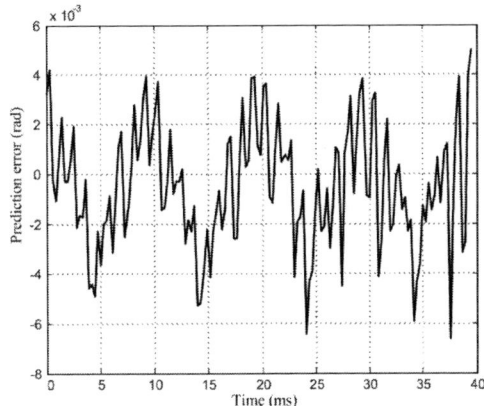

Fig. 14. Waveform of the prediction error of the phase angle.

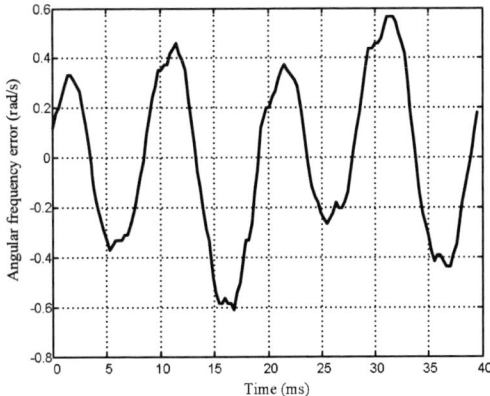

Fig. 15. Waveform of the angular frequency error.

VII. CONCLUSIONS

The paper presents a PLL structure that can be employed in three-phase uninterruptible power supplies (UPS) to synchronize the waveforms, produced by the inverter, with the utility grid ones.

Such synchronization, useful to reduce the transients due to the switch commutations in UPS using an automatic bypass switch, is generally complicated by harmonic distortions and amplitude unbalances on input signals of the filter. Therefore, improper synchronization between the inverter outputs and the grid voltages can arise.

Proposed PLL structure is able to lock the grid voltage frequency in every operating condition. To this aim a suitable prediction-correction filter, whose structure has been derived from a Steady-State Linear Kalman Filter based on a third-order linear and time-invariant observation model, has been employed.

After a short presentation of the SSLKF algorithm and of the proposed prediction-correction filter, the paper describes in detail the architecture of the PLL system and the problems connected with its implementation on the 16-bit fixed point TMS320F2401 DSP controller used to control the UPS.

The performances of the proposed PLL structure are compared, through simulation tests, with those obtained by a traditional PLL system, based on the vector product, making use of a PI loop filter. The tests have shown that both the solution present a good filtering action as concern the effects of unbalanced voltages and harmonics; however, the proposed solution recovers more quickly the errors introduced by frequency variations or by rapid phase changes.

Finally, some experimental results, carried out on a pre-production prototype, have confirmed the simulation results, showing a fully satisfactory behavior of the proposed filter in both transient and steady-state. In particular, the PLL filter presents a considerable robustness against input signals frequency variations and distortions; moreover, the steady-state errors on phase angle and on angular frequency are reduced to negligible values.

REFERENCES

[1] V. Kaura, and V. Blasko, "Operation of a phase locked loop system under distorted utility conditions," *IEEE Trans. Ind. Appl.*, vol. 33, no. 1, pp. 58–63, Jan./Feb. 1997.

[2] G. H. Jung, G. C. Cho, and G. H. Cho, "Improved Control for High PowerStatic Var Compensator Using Novel Vector Product Phase LockedLopp (VP-PLL)," *Intern. Journal of Electronics*, Vol. 86, No.7, 1999, pp. 837-855.

[3] S. Chung, "A phase tracking system for three phase utility interface inverters," *IEEE Trans. on Power Electron.*, vol. 15, no. 3, pp. 431–438, May 2000.

[4] L. N. Arruda, S. M. Silva, and B. J. C. Filho, "PLL structure for utility connected systems," *Conf. Rec. 36th IEEE-IAS Annal. Meeting*, 2001, vol. 4, pp. 2655–2660.

[5] C. Zhan, C. Fitzer, V. K. Ramachandaramurthy, A. Arulampalam, M. Barnes, and N. Jenkins, "Software phase-locked loop applied to dynamic voltage restore (DVR)," *Proc.of IEEE Power Eng. Soc. Winter Meeting*, 2001, vol. 3, pp. 1033–1038.

[6] L.G. Barbosa Rolim, D. Rodrigues da Costa, Jr., and M. Aredes, "Analysis and Software Implementation of a Robust Synchronizing PLL Circuit Based on the pq Theory," *IEEE Trans. Ind. Electron.*, vol. 53, no. 6, pp. 1919–1926, December 2006.

[7] A. Bellini, and S. Bifaretti, "Implementation of a Digital Filter for Speed Noise Reduction in Drives with Electromagnetic Resolver," *Proc. of EPE 2005*, Dresden, Germany, September 2005.

[8] A. Bellini and S. Bifaretti, "A Digital Filter for Speed Noise Reduction in Drives using an Electromagnetic Resolver," *Mathematics and Computers in Simulation*, vol. 71, issues 4-6, June 2006, pp. 476-486.

10A 12V 1 chip digitally-controlled DC/DC converter IC with high resolution and high frequency DPWM

Kazutoshi Nakamura*, Toshiyuki Naka*, Yuki Kamata*, Toyoki Taguchi*, Takaaki Shimizu*, Yoshiko Ikeda*
Akio Nakagawa* and Dragan Maksimovic[†]

* Toshiba Corporation Semiconductor Company
1,Komukai Toshiba-cho,Saiwai-ku,Kawasaki,Japan
, e-mail: *kazutoshi.nakamura@toshiba.co.jp*
[†]Department of Electrical and Compute Engineering, University of Colorado
Boulder, CO 80309-4025, USA

Abstract— **This paper introduces a 10A 12V single chip digitally-controlled DC/DC converter IC based on the low cost 0.6um BiCD process. This IC includes the digital pulse width modulator (DPWM) module with the dead-time programmability. The average time resolution is 1.22ns at the clock frequency 25MHz on 0.6um process. This resolution is as same as that for the counter-based DPWM with the clock frequency 817MHz. The chip adopted low impedance metal bump technology for reducing a parasitic interconnection resistance in the power stage. The fabricated chip achieves a low on resistance 9.7mΩ in the 20V output LDMOS (@drain current=5A, gate voltage=5V). The maximum efficiency is 86.4% at output current 5A when the input voltage, the output voltage and switching frequency and the dead-time are 12V, 1.3V, 780KHz and 15ns, respectively. The maximum voltage deviation and transient response time are 42mV and 8us, respectively in step-load (5A to 10A) transient response.**

Keywords— **High frequency power converter, High voltage IC's, Pulse Width Modulation (PWM) and Regulation.**

I. INTRODUCTION

In recent years, researches in the area of digitally-controlled high-frequency DC-DC converter have attracted much attention. Digital control provides flexibility, programmability and opens up new control algorithm not possible in the converters with traditional analog control.

In pulse-width modulators, analog control system provides very fine resolution for output voltage adjustment. In principle, a voltage can be adjusted to any arbitrary value limited by loop gain, thermal effects and system noise levels. On the other hand, a digital control system has discrete set points resulting from the resolution of quantizing elements in the system. The output voltage resolution corresponds to the time resolution in the digital pulse-width modulators (DPWM). As the output voltage

resolution is higher, the time resolution required in DPWM is higher.

Especially, in the digitally-controlled converter with high input voltage and high switching frequency, high resolution and high frequency DPWM are required to achieve precise voltage regulation.

In the DC/DC converter with synchronous rectifier, it is well known that optimum dead times achieve high converter efficiency [1]. Too long dead times result in additional losses due to the body diode conduction. Too short dead-times may result in simultaneous conduction of the main switch and the synchronous rectifier and the ineffective current from the input voltage source to the ground may flow. The dead-times programmability can make the tuning of the synchronous rectifier commutation timing easy.

With increase in clock speed of microprocessors, high efficiency, high power density, high current slew rate di/dt is strongly demanded for DC/DC converters. Precise voltage regulation under a large di/dt requires high switching frequency.

In order to improve the conversion efficiency at high switching frequency, not only the device structure of switching MOSFET but also the circuit parameters such as parasitic devices on board must be optimized. It has been pointed out that the parasitic inductance between the output MOSFET and the gate driver circuit decreases the conversion efficiency. Multi Chip Module or 1 chip solution are effective to reduce the parasitic wiring inductances [2, 3]. It was also predicted that a low impedance gate drive can reduce the mirror period in the turn-off transient, realizing ideal switching and low turn-off power loss [4]. 1 chip solution is favorable for achieving the ideal switching because the parasitic inductances are small and sufficiently low impedance gate driver circuits are easily integrated with power MOSFETs.

Up to now, the output current of 1 chip converter has been limited to a few or several amperes because the interconnection resistance becomes larger and even exceeds the on-resistance of the device itself if the size of LDMOS increases.

In this paper, we describe a 10A 12V single chip DC/DC converter IC including the DPWM module with

978-1-4244-1741-4/08/$25.00 ©2008 IEEE

the dead-time programmability based on the low cost 0.6um BiCD process.

Section II introduces the overall structure of the developed digital controller for DC/DC converter. Section III introduces the DPWM with the dead-time programmability. In the digitally-controlled converter with high input voltage and high switching frequency, high resolution, high frequency DPWM are required to achieve precise voltage regulation. The technique for realizing the average time resolution 1.22ns at the clock frequency 25MHz on 0.6um process is presented. Section IV introduces the integrated power stage driving high current. In order to reduce the interconnection resistance, the bump technology is applied to the power IC. The implementation and experimental results about 10A 12V single chip DC/DC converter IC are introduced. Conclusions from this research are drawn in Section VI.

where d[n] is the duty cycle command at discrete time n, e[n] is the error signal, a,b and c are constants.

Equation (1) indicates that an implementation of the compensator generally involves the use of digital adders and digital multipliers, which devices increases the size of controller and which tend to increase the clock frequency requirement. Look-up table architecture is applied to the compensator instead of the duty calculation including multiplication [5].

In the controller design, once the coefficients a, b and c are selected (to achieve a desired closed-loop bandwidth and adequate phase margin, for example), the produces $a \cdot e$, $b \cdot e$, and $c \cdot e$ are precomputed. Since digital error signal outputted by six comparators is a small number of values, the number of entries ($a \cdot e$, $b \cdot e$, and $c \cdot e$) in the lookup tables is correspondingly small. The PID compensator can be accomplished in small size and in a small number of system clock cycles.

II. DIGITAL CONTROLLER STRUCTURE

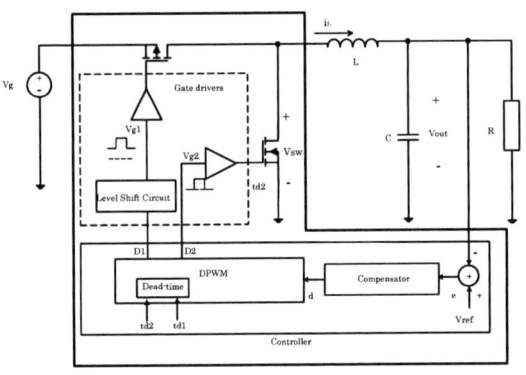

Figure1 Circuit diagram of a synchronous DC-DC buck converter with adjustable commutation dead times: Vg=12V,Vout=1.3V,max(IL)=10A

Figure 1 shows the circuit diagram of the developed synchronous DC-DC buck converter IC. The input voltage is Vg=12V, the reference voltage is Vref=1.3V, the filter parameters are L=0.67uH, C=230uF. It is composed of the flash A/D converter, control circuitry, power switches (the breakdown voltage=25V) and the drivers. The parameters of PID compensator and dead-times are loaded from external EEPROM or PC through a serial interface when the system is started or rebooted. The A/D converter is composed of only six comparators, resulting in a 7-level quantization (the least significant bit value=10mV). The error signal at the output of A/D is processed by a PID compensator, which provides the necessary duty-cycle command for the output voltage regulation.

A typical discrete-time PID control law has the form

$$d[n] = d[n-1] + a \cdot e[n] + b \cdot e[n-1] + c \cdot e[n-2] \ (1)$$

III. PWM MODULE WITH THE DEAD-TIME PROGRAMMABILITY

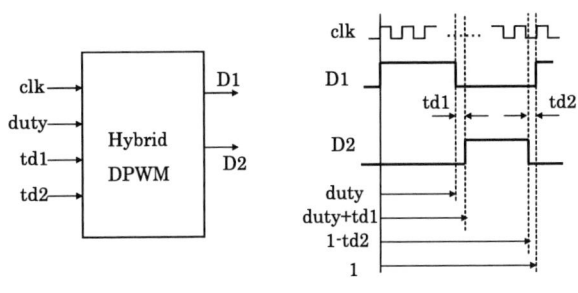

Figure2 Input and output signal entered in the DPWM Block and timing diagram

Figure 2 shows the input and output signal entered in the DPWM Block and timing diagram. In the DPWM, the dead-time programmability is provided to achieve optimum efficiency. The DPWM module with the dead-time programmability suitable for FPGA implementation has been reported [6]. But many input control signals are required to adjust the DPWM. The wiring region between the DPWM and the controlled module must occupy a large space on 0.6um process. The simple analog feedback circuit is applied to the developed DPWM.

The DPWM provides three outputs, clock signal, high side MOSFET ON/OFF signal D1 and low side MOSFET ON/OFF signal D2. The output D1 is a signal with a duty cycle determined by the input duty cycle command. The input dead-time td1 and td2 are applied to the output signal D2.

The output voltage resolution corresponds to the time resolution required in the DPWM. If the desired output voltage tolerance is 1.3 volts +/- 1% in the input voltage 12V and the switching frequency 780KHz, the time resolution and the number of bits required in the DPWM are less than 1.38 ns and more than 10 bits, respectively. In the counter-based DPWM, more than the clock frequency 725MHz is required. Although clock rates of that magnitude, or higher, can be produced in IC devices, they require more advanced process than 0.6um process and driving power that is likely to exceed practical values in most DC/DC conversion products. The hybrid DPWM[7] are introduced to provide the high resolution without the need for a very high clock frequency.

Figure 3 shows the developed hybrid DPWM with the dead-time programmability. The developed 10-bit DPWM is composed of 5-bit counter and 5-bit delay line. The clock period, Tclk is divided into 32 time steps by 5-bit delay line as shown in figure 3. The required clock frequency, 22.7MHz is 32 times smaller than that for the equivalent counter.

lsb(duty),"00001" and "i1" is connected to R. The signal R resets D1. It is important that the required clock frequency is lower as compared to the counter-based DPWM with the same resolution.

The leading and trailing edge of the low side MOSFET ON signal D2 is generated in the same way as the reset signal of the duty cycle command. The setting time resolution for adjusting dead-time is the same as the time resolution of duty cycle command.

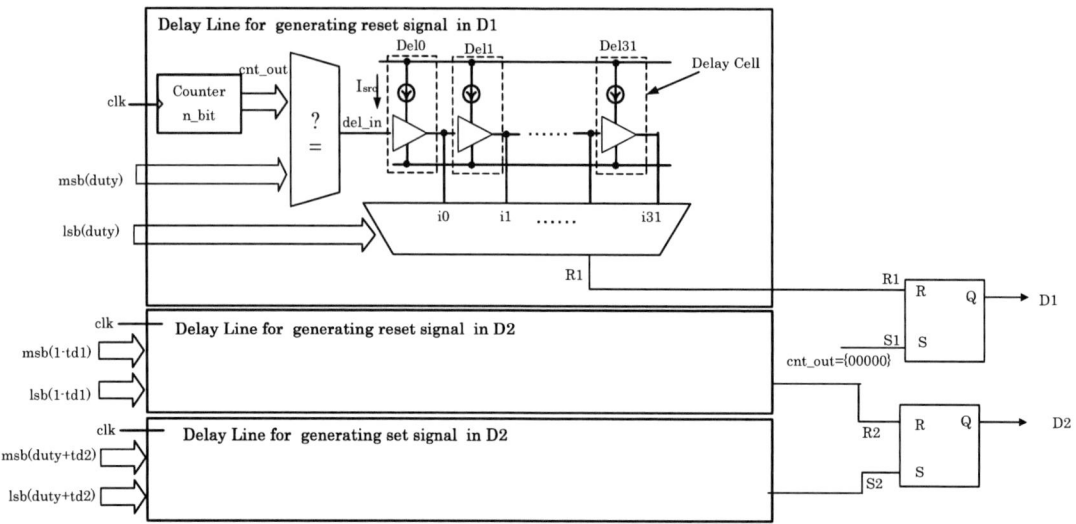

Figure 3 Developed hybrid DPWM with the dead-time programmability

This DPWM is composed of 3 delay lines with 32:1 multiplexer and two SR latches. The delay lines are configured with the same delay cell so that the total propagation delay time of delay cells matches the period of clock. The propagation time in the delay cell can be controlled by the supply current Isrc in figure 3. The supply current Isrc is depended on the clock frequency.

The counter provides the most significant bit(msb) portion of the signal command, for example, the duty cycle command and the delay line provides the least significant bit(lsb) portion. Figure 4 shows a timing diagram implemented with the duty cycle command for the DPWM in case that the duty is 00101_00001. The counter counts at each clock period of input, clk. The output, D1, is set at the zero value of the counter(S1="00000"). The output of the counter is compared with 5 most significant bits of the input duty cycle command, msb(duty), "00101" and the comparator outputs the signal del_in. The signal del_in is propagated through the delay line. The output of each delay cell is tapped out and connected to a 32:1 multiplexer. The propagated signals, i0-i31, in the output of the delay cells, Del0-Del31 are shown in figure 3. The selection of the multiplexer output, "i1" is made by observing the least significant portion of the input duty cycle command,

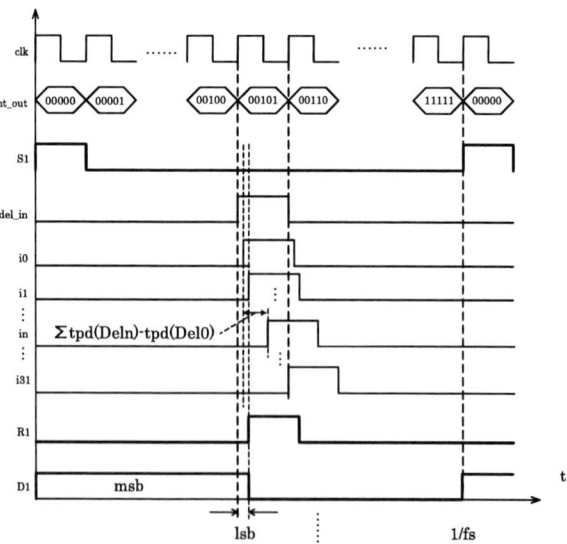

Figure 4 Timing diagram implemented with the duty cycle command for the DPWM in case that the duty is 00101_00001.

500

In the developed DPWM, the digital dither technique is additionally used to improve the time resolution. The LC output filter averages any pulse train that is fed into it. If the output pulse width from the DPWM is increased only once every eight switching periods by the equivalent of the minimum step time, the temporally-smoothed value of the pulse train will be increased by an amount equal to (1/8)*resolution of the minimum step time. The effective time resolution 1.38ns/8=172.5ps is obtained in the DPWM.

Figure 5 shows the measured propagation delay summation of delay cells from Del0 to Deln, for the different clock frequency. The lines are the ideal summation of minimum step time for the each clock frequency (fclk), (Tclk)/32*n. As shown in figure 6, the average resolution is 1.22ns at the clock frequency 25MHz in the developed hybrid DPWM. This resolution is as same as that for the counter-based DPWM with the clock frequency 817MHz.

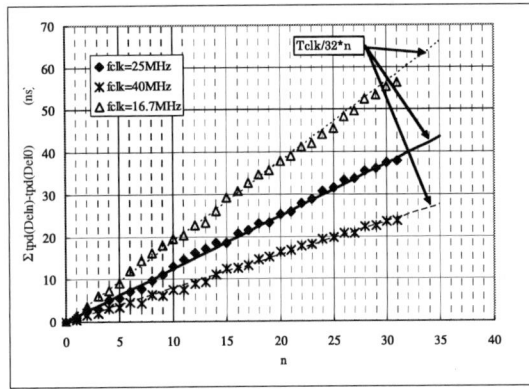

Figure 5 Measured propagation delay summation of delay cells from Del0 to Deln. The lines are the ideal summation of minimum step time for the each clock frequency (fclk), (Tclk)/32*n.

IV. INTEGRATED POWER STAGE DRIVING HIGH OUTPUT CURRENT

A. DEVICE STRUCTURES

Figure 6 shows a cross-sectional view of 20V output LDMOS devices based on the low cost 0.6um BiCD process. The capacitance between gate and drain is depended on the switching loss.

The device is fabricated in the p-well so that gate drain capacitance is minimized. The buried N+ layer is electrically connected to the source electrode to reduce the coupling between the drain and the substrate. Three metal layers with a 3um thick top metal layer are utilized. The drift region is self-aligned to the gate electrode in order to reduce the parasitic gate-drain capacitance which

affects the switching loss. Additionally, the source electrode extended over the gate poly-silicon is effective to reduce the parasitic gate-drain capacitance. In case of the optimized Nch LDMOS, the breakdown voltage, the threshold voltage and the specific on-resistance is 25.0V, 0.85V and 23.1mΩmm2, respectively.

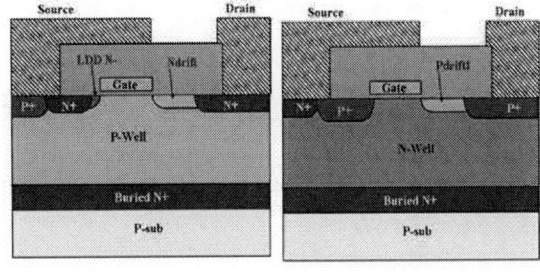

(a)Nch LDMOS   (b)Pch LDMOS

Figure.6 Cross-sectional view of 20V output power devices based on the low cost 0.6um BiCD process.

B. POWER IC USING BUMP TECHNOLOGY

The on resistance of lateral MOSFETs with wire bonding deteriorates considerably with increase in device size due to the parasitic resistance. Interconnection resistance not only increases overall device resistance but also causes debiasing effect in active cells [8]. In order to reduce the interconnection resistance, we have adopted wafer bumping technology [9].

Figure 7 shows the assembled image, the layout of the top metal in IC and the Cu pattern on a printed circuit board (PCB). The chip is attached to the intermediate PCB through bump balls.

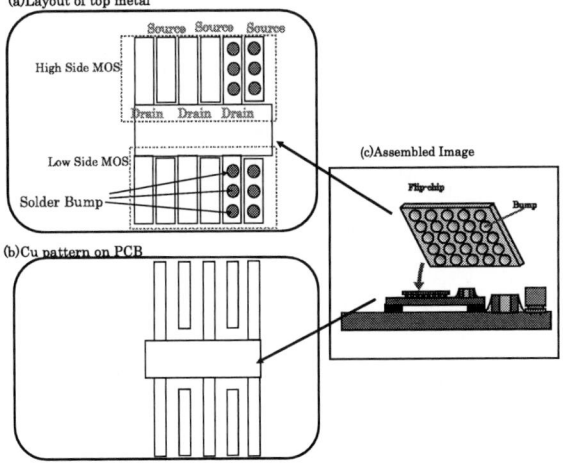

Figure.7 Assembled image, layout of top metal in IC and Cu pattern on printed circuit board (PCB). The drain and the source bumps are electrically connected by parallel running thick Cu metals in the PCB

We have adopted Pch LDMOS for high side switching device in DC/DC converter. The source and drain metals are alternately formed and the drain metals of Pch LDMOS and Nch LDMOS are connected each other. The PCB connects drain and source bumps by parallel running thick Cu metals. The resistance that current laterally flows in the top metal is made as small as possible.

When the input voltage and the output voltage are 12V and 1.3V, respectively, the on time of the low side MOSFET is about 90% in a switching period. It is important to reduce the on resistance of the low side Nch LDMOS. Figure 8 shows the measured output characteristics of a large area Nch LDMOS (the effective area 3.6mm2). The on resistance is 9.7mΩ(@drain current=5A, gate voltage=5V). We have achieved that the value of on resistance is below 10mΩ in 20V LDMOS of Pw IC.

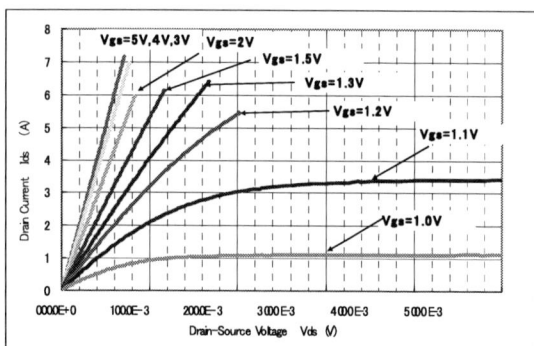

Figure.8 Output characteristics of a large area device (the effective area 3.6mm2).

V. IMPLEMENTATION AND RESULTS

Figure 9 shows a die photo of the fabricated 10A 12V 1chip DC/DC converter IC based on the low cost 0.6um BiCD process. The A/D converter, control circuitry, power switches (the breakdown voltage=25V) and the drivers are integrated on the same die.

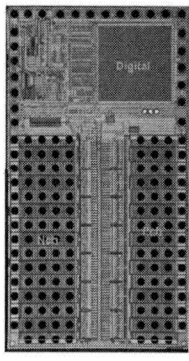

Figure 9 Micrograph of the fabricated 10A 12V 1chip DC/DC converter IC based on the low cost 0.6um BiCD process

Figure 10 shows the measured output voltage ripple, when the input voltage, the output voltage, switching frequency and the load current are 12V, 1.3V, 780KHz and 0A, respectively. As shown in the figure 10, the output voltage ripple is suppressed to below 11mV. The developed DPWM achieves high resolution and high frequency.

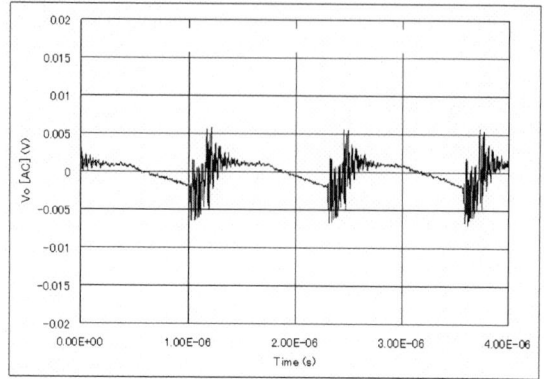

Figure 10 Measured output voltage ripple: Vg=12V, Vout=1.3V, fsw=780 KHz, IL=0A

Figure 11 shows the measured converter efficiency as functions of the dead time td1 in figure 2. It is easy to adjust the dead-time by changing the stored parameter. The clock frequency and switching frequency are 25MHz and 780KHz, respectively. The maximum efficiency is achieved at td1=11.3ns. It is more important to optimize the dead-time in the higher switching frequency.

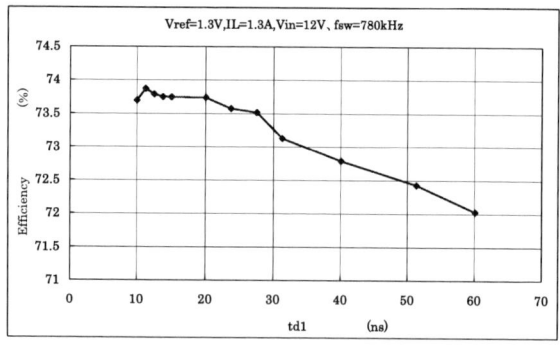

Figure 11 Measured converter efficiency as functions of the dead time td1:Vg=12V,Vout=1.3V,IL=1.3A

Figure 12 shows the measured efficiency of fabricated 1chip DC/DC converter IC when the input voltage, the output voltage and switching frequency and the dead-time(td1,td2) are 12V, 1.3V , 780KHz and 15ns, respectively. The maximum efficiency is 86.4% at output current 5A. The fabricated chip has accomplished the high efficiency.

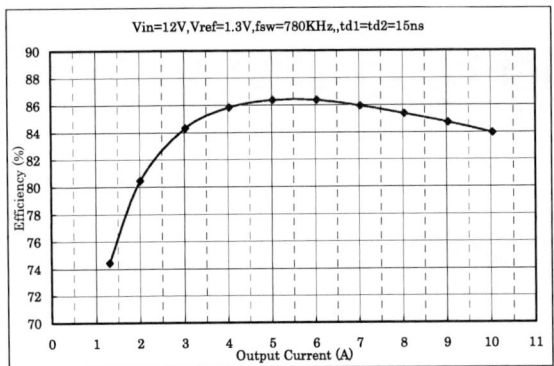

Figure 12 Measured efficiency of fabricated 1chip DC/DC converter IC when the input voltage, the output voltage and switching frequency and the dead-time(td1,td2) are 12V, 1.3V , 780KHz and 15ns, respectively.

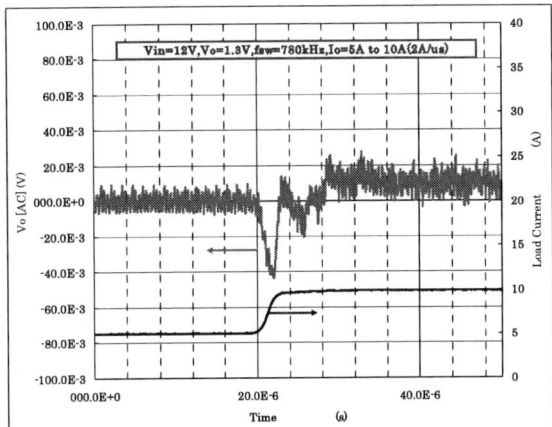

Figure 13 Experimental step-load(5A->10A)transient response. The maximum voltage deviation and transient response time are 42mV and 8us, respectively.

Figure 13 shows the experimental step-load(5A->10A)transient response. The maximum voltage deviation and transient response time are 42mV and 8us, respectively. The transient response of the developed chip is equivalent to that of analog controlled IC.

VI CONCLUSIONS

This paper introduces a 10A 12V single chip digitally-controlled DC/DC converter IC based on the low cost 0.6um BiCD process. This IC includes the digital pulse width modulator (DPWM) module with the dead-time programmability. The average time resolution is 1.22ns at the clock frequency 25MHz on 0.6um process. This resolution is as same as that for the counter-based DPWM with the clock frequency 817MHz. The chip adopted low impedance metal bump technology for reducing a parasitic interconnection resistance in the power stage.

The fabricated chip achieves a low on resistance 9.7mΩ in the 20V output LDMOS (@drain current=5A, gate voltage=5V). The maximum efficiency is 86.4% at output current 5A when the input voltage, the output voltage and switching frequency and the dead-time are 12V, 1.3V, 780KHz and 15ns, respectively. The maximum voltage deviation and transient response time are 42mV and 8us, respectively in step-load (5A to 10A) transient response.

ACKNOWLEDGEMENT

The authors would like to thank K.Morizuka, T.Shimada, Y.Baba, M.Yamaguchi, T.Tsurugai for providing the opportunity of this study and K.Endo, R.Fukamachi, S.Hodama for support of IC fabrication and N.Yasuhara for providing the technical advice.

REFERENCE

[1] Vahid Yousefzadeh,and Dragan Maksimovic,"Sensorless Optimization of Dead Times in DC–DC Converters With Synchronous Rectifiers," IEEE TRANSACTIONS ON POWER ELECTRONICS, VOL. 21, NO. 4, JULY 2006,pp.994-1002.

[2] Y.Kawaguchi et al., "Multi Chip Module with Minimum Parasitic Inductance for New Generation Voltage Regulator," ISPSD'05, pp.371-374

[3] M.Shiraishi et al.,"Low Loss and Small SiP for DC-DC Converters," ISPSD'05, pp.175-178

[4] A. Nakagawa, "Evolution of Silicon Power Devices and Challenges to Material Limit," Proc. of MIEL 2006, pp.167-174

[5] A. Prodic, D. Maksimovic, "Design of a digital PID regulator based on look-up tables," IEEE COMPEL, June 2002, pp. 18-22.

[6] Vahid Yousefzadeh, Toru Takayama, Dragan Maksimovic,"Hybrid DPWM with Digital Delay-Locked Loop,"in IEEE COMPEL Workshop.,2006,pp.142-148.

[7] B. J. Patella, A. Prodic, A. Zirger, and D. Maksimovic, "High-frequency digital PWM controller IC for DC–DC converters," IEEE Trans. Power Electron., vol. 18, no. 1, pp. 438–446, Jan. 2003.

[8] T.Efland et al, " Lateral Thinking About Power Devices (LDMOS),"IEDM'98, pp.679-682

[9] Z.J.Shen et al, " Breaking the Scaling Barrier of Large Area Lateral Power Devices: An 1mΩ Flip-Chip Power MOSFET with Ultra Low Gate Charge," ISPSD'04, pp.387-390

Modelling and Modulation of Voltage Source Converter

Grzegorz Radomski

Kielce University of Technology/Power Electronics Department, Kielce, Poland, e-mail: radomski@tu.kielce.pl

Abstract—The basic operational principle of the three-phase voltage source converter system is the subject of this paper. This type of converter is regarded as well known. Despite this, there are some details worth analysing and describing. In the drawn-out analysis the aggregated transistor state function for one branch of the converter is introduced. Influence of dead time on voltage space vector area is analysed. Unified model of the converter operational principle is presented. Voltage space vector modulation methods with dead time effect compensation for voltage source converter are developed.

Keywords—Voltage Source Converter (VSC), Pulse Width Modulation (PWM), Modulation strategy.

I. INTRODUCTION

Nowadays, bipolar voltage level, transistor-diode voltage source converter is a standard application widely used in industry both as a rectifier and an inverter system. This application is chosen as basic by leading world electrical companies. Voltage source converter may be considered as anti-parallel connection of transistor and diode converter systems. This system may work in rectifier and inverter mode as well as a reactive and distortion power compensator. For this reason it is a very flexible AC/DC system with bidirectional energy flow. Control and modulation methods that are devoted to this structure are still developed [1,2]. Especially, sensor-less control strategies are tested [3]. In voltage sensor-less methods good estimation of input voltage is crucial to method robustness. The transition of the conducting from upper transistor to lower one must always go through dead time state when control signals of both transistors are switched off to avoid short circuit.

Model of voltage source converter that is presented in this paper gives good estimation of input voltage depending on DC side voltage and transistor branches state including system functionality in dead time. For this reason it should be useful for voltage sensor-less control systems (e.g. for virtual flux estimation [3,4]). In high power applications thyristor switches are used. In the case of thyristor switches high frequency modulation are not possible due to the fact that high switching time delays and long dead times are required. On the contrary, in high rotary speed drives modulation frequency should be proportionally high. In both cases voltage deformations introduced into converter system by dead time become important. A robust modulation method compensating dead time effect is required. Methodology of dead time compensation in voltage space vector domain is presented in the paper. Some disadvantages of classic space vector pulse width modulation are illustrated. Modulation

method with four basic voltage space vectors used in modulation step is developed.

II. THREE-PHASE VOLTAGE SOURCE CONVERTER

The three-phase voltage source converter consists of three branches. Figure 1 shows a connection diagram of the three-phase voltage source converter.

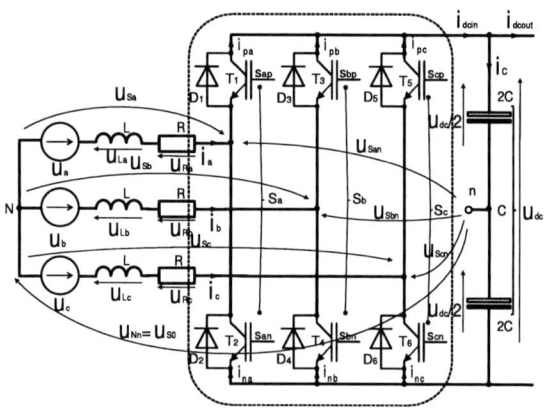

Fig. 1. Three phase voltage source converter

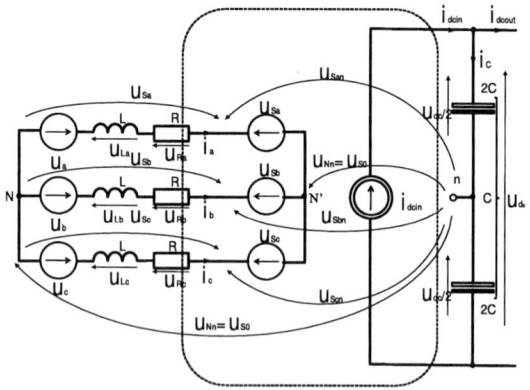

Fig. 2. Three phase voltage source converter equivalent circuit

Voltage of one converter branch is described by (1):

$$u_{Sin} = \left[S_i + \left(1 - |S_i|\right)\mathrm{sign}(i_i) \right]\frac{u_{dc}}{2} \qquad (1)$$

$$r_{Si} = \left(1 - |S_i|\right)\left(1 - |\mathrm{sign}(i_i)|\right)R \qquad (2)$$

978-1-4244-1741-4/08/$25.00 ©2008 IEEE   504

The current that is delivered to the positive output rail is described by (3).While the current is delivered to the negative output rail is described by (4).

$$i_{pi} = \left[\frac{S_i+1}{2}|S_i| + (1-|S_i|)\frac{\text{sign}(i_i)+1}{2}\right]i_i \quad (3)$$

$$i_{ni} = \left[\frac{S_i-1}{2}|S_i| + (1-|S_i|)\frac{\text{sign}(i_i)-1}{2}\right]i_i \quad (4)$$

where $i \in \{a,b,c\}$ and

$$S_i \in \{-1,0,1\}, \quad \begin{cases} S_i = 1 & \Leftrightarrow s_{ip} = 1, \ s_{in} = 0 \\ S_i = 0 & \Leftrightarrow s_{ip} = 0, \ s_{in} = 0 \\ S_i = -1 & \Leftrightarrow s_{ip} = 0, \ s_{in} = 1 \end{cases} \quad (5).$$

Applied sign function and its module are illustrated in Fig. 3.

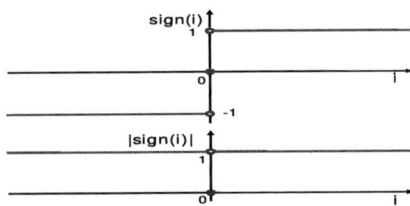

Fig. 3. Sign function and it's module

The model of one converter branch described above will be used for creation of converter structure models. Zero sequence component of converter input voltage is expressed by (6). Converter input voltages referenced to neutral point of the mains are described by (7).

$$u_{Nn} = u_{S0} = \frac{1}{3}\sum_{i\in\{a,b,c\}} u_{Sin} \quad (6) \quad U_{Si} = U_{Sin} - U_{S0} \quad (7)$$

DC side current of transistor-diode matrix may be calculated from (8).

$$i_{dcin} = \sum_{i\in\{a,b,c\}} i_{pi} = \sum_{i\in\{a,b,c\}} i_{ni} =$$

$$= \sum_{i\in\{a,b,c\}}\left[\frac{S_i+1}{2}|S_i|\right]i_i + \sum_{i\in\{a,b,c\}}\left[(1-|S_i|)\frac{\text{sign}(i_i)+1}{2}\right]i_i \quad (8)$$

The DC side current may be decomposed into two components. The first component describes the current delivered by branches with transistor switched on, the second by branches with transistors switched off. Equivalent schematic diagram relevant to equations from (1) to (8) is illustrated in the Fig. 2. While transferring input voltages into stationary two-phase reference frame one can obtain (9). Finally, voltage space vector (9) may be decomposed into two components. One component only depends on transistor state, while the second one defines voltage value in the case of transistors switched off and is dependent on transistor state and current sign. From (9) it may be derived that there is extensive set of basic voltage space vectors. Some of them are stable realisations, while the rest of the vectors are momentary and valid only in the case when one or more current of branch with transistors switched off has zero value.

$$u_{S\alpha\beta} = C_{n\to\alpha\beta}\begin{bmatrix} u_{San} \\ u_{Sbn} \\ u_{Scn} \end{bmatrix} = C_{n\to\alpha\beta}\begin{bmatrix} u_{Sa} \\ u_{Sb} \\ u_{Sc} \end{bmatrix} =$$

$$= \frac{u_{dc}}{2}C_{n\to\alpha\beta}\begin{bmatrix} S_a \\ S_b \\ S_c \end{bmatrix} + \frac{u_{dc}}{2}C_{n\to\alpha\beta}\begin{bmatrix} (1-|S_a|)\cdot\text{sign}(i_a) \\ (1-|S_b|)\cdot\text{sign}(i_b) \\ (1-|S_c|)\cdot\text{sign}(i_c) \end{bmatrix} \quad (9)$$

In this case the input voltage of this phase (1) is equal to zero and output resistance (2) increases. Because input voltage depends on current sign it is convenient to define current sectors regions for control purpose. Possible current sectors are defined by different sets of current signs. Because current sign function (Fig. 3) has three values, current sectors definition (10) reflects this fact.

$$\text{Sectl} = \begin{cases} 0/1/2/3/4/5 & \text{for } \text{sign}(i_a)=0, \ \text{sign}(i_b)=0, \ \text{sign}(i_c)=0 \\ 0 & \text{for } \text{sign}(i_a)=1, \ \text{sign}(i_b)=-1, \ \text{sign}(i_c)=-1 \\ 0/1 & \text{for } \text{sign}(i_a)=1, \ \text{sign}(i_b)=0, \ \text{sign}(i_c)=-1 \\ 1 & \text{for } \text{sign}(i_a)=1, \ \text{sign}(i_b)=1, \ \text{sign}(i_c)=-1 \\ 1/2 & \text{for } \text{sign}(i_a)=0, \ \text{sign}(i_b)=1, \ \text{sign}(i_c)=-1 \\ 2 & \text{for } \text{sign}(i_a)=-1, \ \text{sign}(i_b)=1, \ \text{sign}(i_c)=-1 \\ 2/3 & \text{for } \text{sign}(i_a)=-1, \ \text{sign}(i_b)=1, \ \text{sign}(i_c)=0 \\ 3 & \text{for } \text{sign}(i_a)=-1, \ \text{sign}(i_b)=1, \ \text{sign}(i_c)=1 \\ 3/4 & \text{for } \text{sign}(i_a)=-1, \ \text{sign}(i_b)=0, \ \text{sign}(i_c)=1 \\ 4 & \text{for } \text{sign}(i_a)=-1, \ \text{sign}(i_b)=-1, \ \text{sign}(i_c)=1 \\ 4/5 & \text{for } \text{sign}(i_a)=0, \ \text{sign}(i_b)=-1, \ \text{sign}(i_c)=1 \\ 5 & \text{for } \text{sign}(i_a)=1, \ \text{sign}(i_b)=-1, \ \text{sign}(i_c)=1 \\ 5/0 & \text{for } \text{sign}(i_a)=1, \ \text{sign}(i_b)=-1, \ \text{sign}(i_c)=0 \end{cases} \quad (10)$$

Equation (9) precisely defines converter input voltage also in dead time state that must be always inserted between two basic voltage space vectors. Such description of the converter input voltage may be useful in the case of estimation of voltages supplying converter system [3]. The overall set of basic voltage space vectors is presented in Fig. 4.

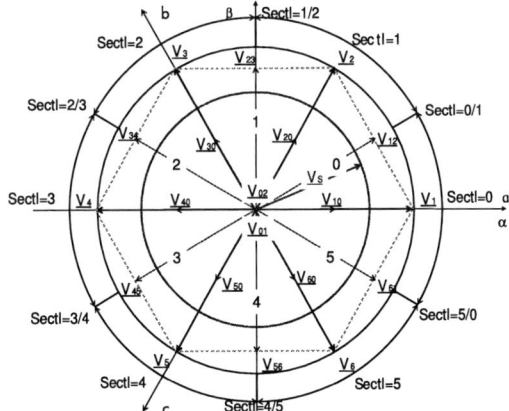

Fig. 4. Basic voltage space vectors of voltage source converter system

Basic voltage space vectors generated for transistor branch state function with non-zero values (11) are accessible in all current sectors. For this reason, the three-phase voltage source converter is a very flexible power electronic system. It may function in rectifier and inverter mode as well as reactive and distortion power compensator.

$$\underline{V}_i = \sqrt{\frac{2}{3}}\,u_{dc}\,e^{j(i-1)\frac{\pi}{3}} \quad (11) \qquad \underline{V}_{i0} = \frac{1}{\sqrt{6}}\,u_{dc}\,e^{j(i-1)\frac{\pi}{3}} \quad (12)$$

$$\underline{V}_{ki} = \frac{1}{\sqrt{2}}\,u_{dc}\,e^{j\left(i-\frac{3}{2}\right)\frac{\pi}{3}} \qquad i,k=1,2,3,4,5. \quad (13)$$

Fig. 5 illustrates the principles of converter functionality with a different input displacement angle for the same input current amplitude. Vectors denoted as (12) and (13) are generated in the case of current space vector crossing current sector boundary, e.g. when at least one phase with current value of zero branch state function has value of zero.

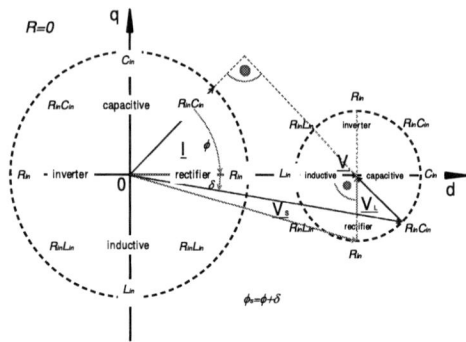

Fig. 5. Vector diagram of voltage source converter system

Table I recapitulates voltage space vectors that are generated for given current sectors and transistor states. Part denoted as 0 contains vectors that are realised when each branch has conducting transistor. These vectors are active in all current sectors. Part 1 depicts vectors realised in the case of one branch having transistors off. These vectors are used during dead times. In part 2 there are vectors generated in the case of two branches off. The

usage of these vectors is generally not advisable. Part 3 replies to pure six diode bridge rectifier.

Dead time is a duration that is always inserted between upper transistor switched on and lower transistor switched on to avoid short circuit. It is important to explain the sense of switching on transistor that is anti-parallel to conducting diode. Generally, there are two reasons:

1. To make converter input voltages independent of current signs. To avoid uncontrolled voltage change in the case of current sign changing.
2. To minimize conducting power losses by paralleling conducting diode by low resistance of MOS transistor canal. This mechanism does not function in the case of IGBT transistors.

It is worth noting that if anti-parallel transistor to conducting diode is not switching on then dead time has no sense [6]. Then converter functions similarly to simple boost converter. If input current is positive then lower transistor and upper diode conduct the current alternately. In the opposite case upper transistor and lower diode conduct current alternately. This kind of control is strongly dependent on current sign and may fail in the case of not exact measurement of low current value. For this reason solutions with diode current sign detection are proposed [6]. Narrow examination of Tab. I and Fig. 4 provides one with conclusion that transition between nearest voltage space vector is possible with changing of state of only one converter branch. This kind of transition is used in voltage space vector modulation algorithms. In this case voltage space vector that is generated during dead time is univocal and depends on only one phase current sign. Also problems with coincidence of transistor switching is omitted.

The effort of this paper is directed to the control of voltage source converter systems with anti-parallel transistor switching on. Strategies of dead time compensation in domain of voltage space vector modulation area are developed in the paper. Improving the quality of electric power that is drawn from the mains by converter system is equivalent to improving the flow of energy stream through converter system.

TABLE I.
INDEXES OF BASIC VOLTAGE SPACE VECTORS THAT ARE REALISED FOR CURRENT SECTOR NUMBER AND TRANSISTOR STATE VECTOR

S_a	-1	1	1	1	-1	-1	-1	1	0	0	0	0	1	-1	-1	1	1	1	-1	-1	0	0	1	-1	0	0	0
S_b	-1	1	-1	-1	1	1	-1	-1	-1	-1	1	1	0	0	-1	1	1	-1	1	-1	0	0	0	0	-1	1	0
Sectl $\;S_c$	-1	1	-1	-1	1	1	1	1	-1	-1	-1	1	-1	1	0	0	0	0	0	0	-1	1	0	0	0	0	0
0/1/2/3/4/5	0	0	1	2	3	4	5	6	10	23	40	56	12	30	45	60	61	20	34	50	20	50	10	40	60	30	0
0	0	0	1	2	3	4	5	6	1	2	0	6	1	0	5	6	1	2	3	0	1	6	1	0	1	2	1
0/1	0	0	1	2	3	4	5	6	1	2	0	6	12	30	45	60	1	2	3	0	12	60	12	30	1	2	12
1	0	0	1	2	3	4	5	6	1	2	0	6	2	3	4	0	1	2	3	0	2	0	2	3	1	2	2
1/2	0	0	1	2	3	4	5	6	10	23	40	56	2	3	4	0	1	2	3	0	23	40	2	3	10	23	23
2	0	0	1	2	3	4	5	6	0	3	4	5	2	3	4	0	1	2	3	0	3	4	2	3	0	3	3
2/3	0	0	1	2	3	4	5	6	0	3	4	5	2	3	4	0	61	20	34	50	3	4	20	34	50	34	34
3	0	0	1	2	3	4	5	6	0	3	4	5	2	3	4	0	6	0	4	5	3	4	0	4	5	4	4
3/4	0	0	1	2	3	4	5	6	0	3	4	5	12	30	45	60	6	0	4	5	30	45	60	45	5	4	45
4	0	0	1	2	3	4	5	6	0	3	4	5	1	0	5	6	6	0	4	5	0	5	6	5	5	4	5
4/5	0	0	1	2	3	4	5	6	10	23	40	56	1	0	5	6	6	0	4	5	10	56	6	5	56	40	56
5	0	0	1	2	3	4	5	6	1	2	0	6	1	0	5	6	6	0	4	5	1	6	6	5	6	0	6
5/0	0	0	1	2	3	4	5	6	1	2	0	6	1	0	5	6	61	20	34	50	1	6	61	50	61	20	61
Part	0								1												2						3

III. ENERGY FLOW OF THREE-PHASE VOLTAGE SOURCE CONVERTER

Voltage source converter is a system with two kinds of energy accumulation. Electric energy is accumulated in the magnetic field of input inductances (dynamic accumulator) and electric field of output capacitor (static accumulator). Energy flow diagram is presented in the Fig. 6.

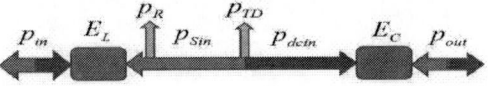

Fig. 6. Energy flow of voltage source converter

In rectifier mode, energy flow from the converter input to the output through magnetic coils, diode-transistor matrix and output capacitance bank. Equ. (14) describes instantaneous input power of voltage source converter. Analogical equation expresses the power absorbed at the input of converter diode-transistor matrix (15).

$$p_{in} = \sum_{i \in \{a,b,c\}} u_i i_i = u^T i = \left(C_{\alpha\beta \to n} u\right)^T \left(C_{\alpha\beta \to n} i\right) = u_{\alpha\beta}^T i_{\alpha\beta}$$

(14)

$$p_{Sin} = \sum_{i \in \{a,b,c\}} u_{Si} i_i = u_S^T i = u_{S\alpha\beta}^T i_{\alpha\beta} \quad (15)$$

$$p_{dcin} = u_{dc} i_{dcin} \quad (16) \quad p_{dcin} = p_{Sin} \quad (17) \quad i_0 = 0 \quad (18)$$

Under the assumption that converter diode-transistor matrix is power losses-free the (17) is true. Difference of the input power (14) and power absorbed at the input of the converter diode-transistor matrix (15) delivers energy that is accumulated in magnetic field and partially lost in the resistances of input magnetic coils. Difference of the DC input power (16) and power absorbed by load delivers energy that is accumulated in the output capacitor electric field. Although it is assigned to be ideal losses free, some amount of energy is lost during conversion in diode-transistor matrix of real converter. For proper operation of the converter system the flow of average active power must be balanced. In the other case, energy accumulator will be overloaded and system will be damaged from short of input circuit (over current case) or output capacitor puncturing (over voltage case). From the analysis of (15) and basic voltage space vectors of voltage source converter system (Fig. 4) one can conclude that the state of energy oscillation between input and output energy accumulators is possible. This state appears when instantaneous power that is transferred by diode-transistor matrix (15), (16) changing its sign during modulation step. To avoid this state only two adjacent voltage sectors with proper phase angle should be used in modulation step.

IV. SPACE VECTOR PULSE WIDTH MODULATION (SVPWM) FOR VOLTAGE SOURCE CONVERTER SYSTEM

Only a subset of overall basic voltage space vectors set is used for modulation. In stable states only vectors with transistors in all legs switched on ($S_i \neq 0$) are used. Voltage space vectors with transistors of one leg switched off are used at the edges between stable voltage space

vectors, during dead time periods. From analysis of (9), it is clear that there is a simple way to incorporate the voltage space vector generated during the dead time into one of the basic voltage space vectors that are used in current modulation step. Because each dead time is incorporated into appropriate voltage space vector thus this voltage space vector has minimal non zero duration. For this reason some areas of average voltage space vector have no realisations. Averaged voltage space vector is a linear combination of basic voltage space vectors used in modulation step (19).

$$\underline{V_S} = \frac{1}{T} \sum_{i \in StepInd} T_i \underline{V_i} = \frac{1}{T} \sum_{i \in StepInd} \left(T_i^* + T_{di}\right) \underline{V_i} =$$

$$= \frac{1}{T} \sum_{i \in StepInd} T_i^* \underline{V_i} + \frac{1}{T} \sum_{i \in StepInd} T_{di} \underline{V_i}$$

(19)

$$T_i = T_i^* + T_{di}$$

(20)

Weight function of each basic voltage space vector is proportional to its duration. Overall duration of basic voltage space vector is a sum (20) of exposition time of appropriate transistor state T_i^* and a sum of dead times for which this basic voltage space vector is generated T_{di}. The overall average voltage space vector may be decomposed into two vectors (21).

$$\underline{V_S} = \underline{V_S^*} + \underline{V_d}$$

(21)

$$\underline{V_S^*} = \frac{1}{T} \sum_{i \in StepInd} T_i^* \underline{V_i} \quad (22) \quad \underline{V_d} = \frac{1}{T} \sum_{i \in StepInd} T_{di} \underline{V_i} \quad (23)$$

The first component (22) is an average voltage space vector that is generated during the time of modulation but without dead times. The second components (23) is an error average voltage space vector generated by basic voltage space vectors during dead times. Average voltage space vector $\underline{V_S^*}$ is modulated in new converted area that is obtained through the input one in the way of shifting by vector $-\underline{V_d}$. Maximal duration of the basic voltage space vector, including generation of this vector during dead times, is expressed by (24).

$$T_{i max} = T - \left(T_{ds} - T_{di}\right) = T - T_{ds} + T_{di}$$

(24)

$$T_{ds} = \sum_{i \in StepInd} T_{di} = nT_d \quad (25) \quad T_{di} \leq T_i \leq T_{i max} \quad (26)$$

$$T_i^* = T_i - T_{di}$$

(27)

$$T_{i max}^* = T_{i max} - T_{di} = T - T_{ds} + T_{di} - T_{di} = T - T_{ds}$$

(28)

$$0 \leq T_i^* \leq T - T_{ds} \quad (29) \quad \gamma = \frac{T - T_{ds}}{T} = 1 - f T_{ds} \quad (30)$$

Time denoted as T_{ds} is a sum of all dead times during modulation step (25) and is equal to multiply of the number of edges between basic voltage space vectors n

that are active in modulation step by dead time duration. Because time T_i^* is calculated according to (27), maximal value of this time T_{imax}^* may be calculated from (28). It is characteristic that time T_{imax}^* has the same value for all basic voltage space vectors. Coefficient γ (30) expresses shortening of linear dimensions of voltage space vector modulation area due to dead time effect in comparison with the area of ideal dead time free system. From the above, it is clear that shortening voltage space vector modulation area has the same shape as the initial dead time free one. For this reason, this shortening is a source of discontinuity of voltage space vector modulation area. Analysing (30), it stays clear that this effect is more important in the case of largest modulation frequency or slower power electronic switches for which longer dead time is required.

A. Modulation algorithm

Two modulation methods: SVPWM3 and SVPWM4 are developed in this paper. They differ from one another in a number of basic space vectors that are used in modulation step. Fundamental principle of voltage space vector modulation algorithm is outlined in Fig. 7, Fig. 8 and Fig. 9, Fig. 10. Modulation algorithm is presented in a number of steps. Step 0 represents calculations that ought to be performed before the start of modulation algorithm.

0) Calculate values of vector V_d for all current and voltage sectors: $V_{dl}(\text{SectI})$, $V_{dV}(\text{SectV})$. This calculation is made according to (36), (37) for SVPWM3 method and according to (38), (39) for SVPWM4. Fig. 9 illustrate components of vector V_d for both methods.

Steps from 1 to 7 describe operations these are realized during time of modulation step (Fig. 8).

1) Shift reference frame to the point of symmetry (Fig. 9):

$$\underline{V}_S^{*I} = \underline{V}_S - \underline{V}_{dl}(\text{SectI}) \qquad (31)$$

2) Determine voltage sector:

$$\text{SectV} = \left\lfloor \arg\left(\underline{V}_S^{*I} e^{-jk\frac{\pi}{6}}\right) / \frac{\pi}{3} \right\rfloor \qquad (32)$$

where $k = \begin{cases} 0 \text{ for method SVPWM3} \\ 1 \text{ for method SVPWM4} \end{cases}$.

3) Translate voltage space vector to the coordinate origin (Fig. 10):

$$\underline{V}_S^* = \underline{V}_S^{*I} - \underline{V}_{dV}(\text{SectV}) \qquad (33)$$

4) Rotate voltage space vector by angle being multiply of voltage sector number by angle of $\pi/3$ (Fig. 10):

$$\underline{V}_S^{*r} = \underline{V}_S^* e^{-j\text{SectV}\frac{\pi}{3}} \qquad (34)$$

5) Calculate constrains of average voltage space vector area.

6) Calculate times T_i^*.

7) Send times T_i^* and corresponding to them transistor state vectors $\underline{S}$ to sequencer unit.

B. Classical and Modified Voltage Space Vector Modulation Method

In classic space vector modulation method (SVPWM3) the average voltage space vector is realised as time superposition of three basic voltage space vectors that create regular triangle. Basic voltage space vectors are sequenced as it is presented in the fig. 7a. The vertexes of the triangle are sequenced in straight and invert direction. In this case transistor branch state function vectors differ from each other only in one component. Particularly, the transition between two null voltage space vectors is omitted. The fig. 8a shows an active area of average voltage space vector for zero current sector. From the analysis of active areas defined for the particular current sectors it can be derived that translation of the voltage space vector active area by vector $-\underline{V}_{dl}$ (36) introduces symmetry into system (Fig. 9a). Equation (37) presents component of shift vector that is dependent on voltage sector (Fig. 9a). The overall shift vector (35) is a sum of component vectors (36) and (37).

$$\underline{V}_d = \underline{V}_{dl} + \underline{V}_{dV} \qquad (35)$$

$$\underline{V}_{dI3} = 2\sqrt{\frac{2}{3}}\frac{T_d}{T} u_{dc} e^{j\text{SectI}\frac{\pi}{3}} \qquad (36)$$

$$\underline{V}_{dV3} = 2\sqrt{2}\frac{T_d}{T} u_{dc} e^{j\left(\text{SectV}+\frac{1}{2}\right)\frac{\pi}{3}} \qquad (37)$$

$$\underline{V}_{dI4} = 2\sqrt{\frac{2}{3}}\frac{T_d}{T} u_{dc} e^{j\text{SectI}\frac{\pi}{3}} \qquad (38)$$

$$\underline{V}_{dV4} = \sqrt{6}\frac{T_d}{T} u_{dc} e^{j\text{SectV}\frac{\pi}{3}} \qquad (39)$$

From the analysis of Fig. 8a it is clear that for instant current space vector there are areas where modulation of voltage space vector is impossible. This effect is a source of converter current deformation. Time plots of branch state functions and related to them converter voltages are illustrated in Fig. 11a.

To avoid disadvantages that are characteristic of classical modulation scheme, modified modulation scheme (SVPWM4) is proposed. In the proposed method, the average voltage space vector is realised as time superposition of four basic voltage space vectors that create regular rhomb. Basic voltage space vectors are sequenced as in the Fig. 7b. This kind of modulation avoids gaps between voltage space vector active areas related to the voltage sectors. This is because introduced forth vector lets compensate the influence of the opposite vector that is realised during dead time. Voltage space vector areas that belong to consecutive voltage sectors overlap one another. In the proposed modulation scheme, the total dead time (26) is equal to that characteristic of classical modulation scheme. The Fig. 8b shows an active area of average voltage space vector for zero current sector. Equation (39) presents component of shift vector that is dependent on voltage sector. The overall shift vector (35) is a sum of component vectors (38) and (39). From the analysis of Fig. 8b it is clear that for instant current space vector there are no gaps between voltage space vector active areas. For this reason the problem of current deformation may be avoided. Time plots of control signals and converter voltages are illustrated in Fig. 11b. In both cases there is equal number of dead time periods. Time plots of converter voltages are symmetrical in the case of SVPWM3 method and asymmetrical in the case of SVPWM4 method.

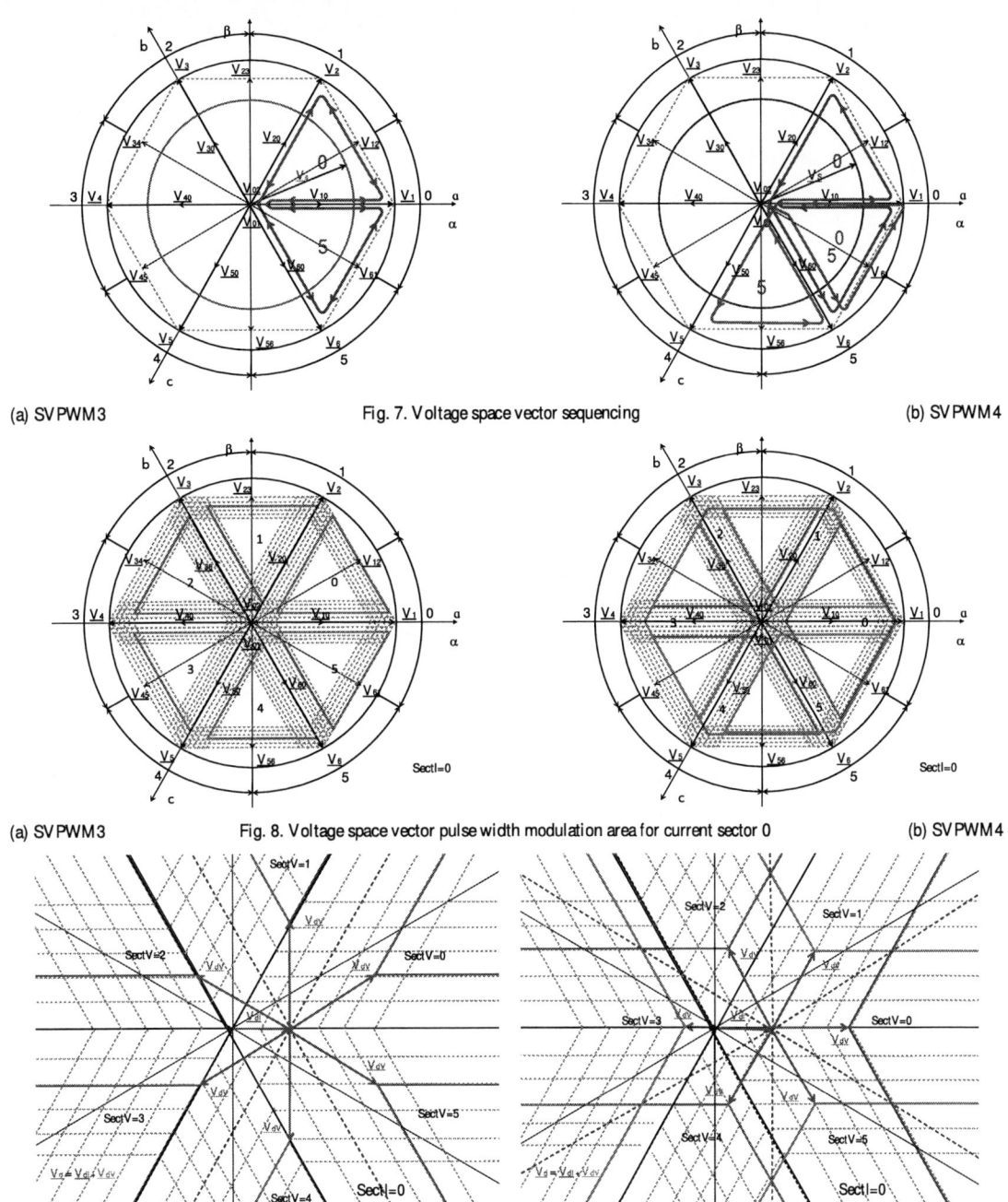

(a) SVPWM3   Fig. 7. Voltage space vector sequencing   (b) SVPWM4

(a) SVPWM3   Fig. 8. Voltage space vector pulse width modulation area for current sector 0   (b) SVPWM4

(a) SVPWM3 Fig. 9. Voltage space vector pulse width modulation area (surroundings of coordinate origin) for current sector 0 (b) SVPWM4

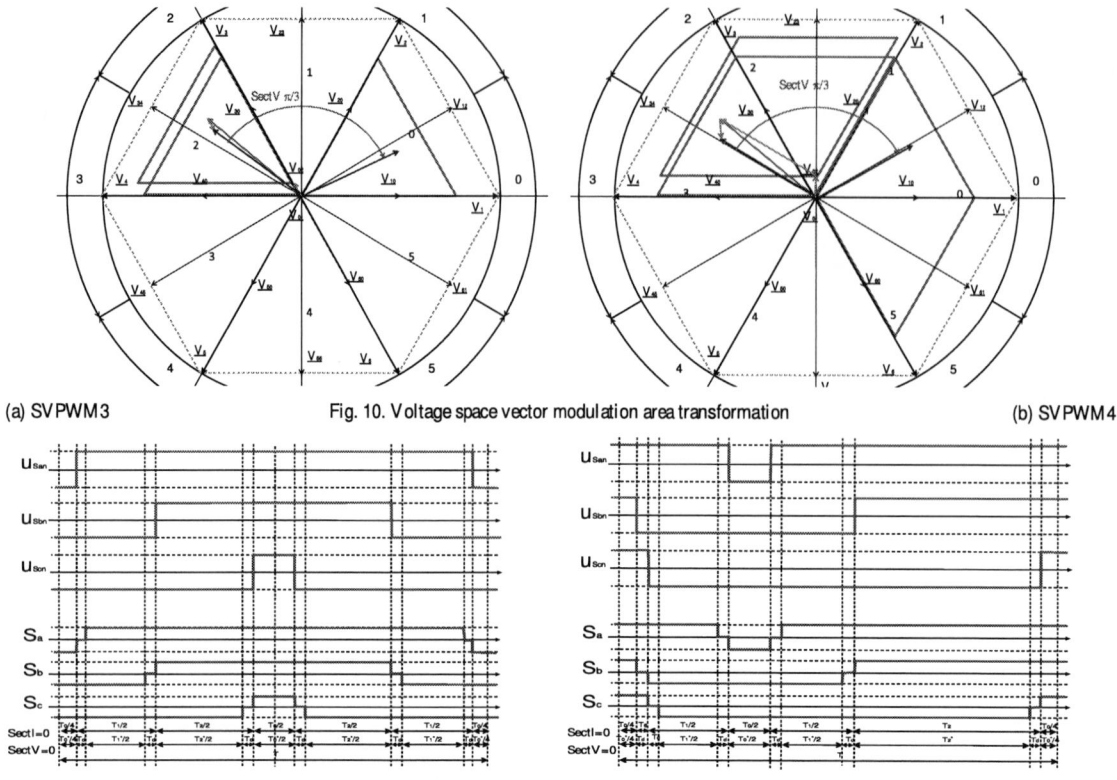

(a) SVPWM3　　　Fig. 10. Voltage space vector modulation area transformation　　　(b) SVPWM4

(a) SVPWM3　Fig. 11. Voltage space vector pulse width modulation time plots of converter voltages and branch state functions　(b) SVPWM4

V. SIMULATIONS

Proprieties of modulation methods SVPWM3 and SVPWM4 were examined by simulation. The simulation results were carried out for the next conditions:

$$L = 10mH, \quad R = 0.1\Omega, \quad C = 200\mu F, \quad T = 100\mu s, \quad T_d = 2\mu s,$$

$$U_{phRMS} = 230V, \quad u_{dc} = 773V, \quad I^*_{RMS} = 5.77A \rightarrow 2.88A,$$

$$S^* = P^* = 3980W \rightarrow 1990W.$$

Fig. 12 illustrates current time plots and current and voltage sector signals from voltage source converter working in rectifier mode and modulated with classical (SVPWM3) and modified (SVPWM4) voltage space vector pulse width modulation method. In both cases dead time compensation function is switched on. Fig. 13 presents time plots of instantaneous powers and state of energy in input magnetic accumulator. Spectra of converter phase currents are presented in Fig. 14 and Fig. 15. The orders of the observed significant parasitic high harmonics of converter current are described by (40) in the case of SVPWM3 method and by (41) in the case of SVPWM4 method:

$$h_{SVPWM3} = n*6 \pm 1 \,(40), \quad h_{SVPWM4} = (n - 1/2)*6 \pm 1 \,(41)$$
$$\text{where } n = 1, 2, 3, \dots.$$

From the analysis of simulation results it is clear that method SVPWM4 has some advantages in comparison to SVPWM3. In the case of SVPWM3 method, voltage sector number value switches many times between two adjacent values in the voltage sector boundary region. This is because some areas of average voltage space vector are not realised in SVPWM3 method. Then significant disturbances of currents are observed. Furthermore, these disturbances are source of multiple current sector transition, which results in further control problems. This situation does not take place in the case of SVPWM4 because this method avoids gaps between voltage space vector active areas related to the voltage sectors. For this reason current generated in the case of SVPWM4 modulation method has slight level of deformation and better spectrum. Moreover, method SVPWM4 has a better controllability than SVPWM3. In the case of SVPWM3 method phase current lags its reference signal and full amplitude is not reached (Fig. 12a). SVPWM4 method is free of these drawbacks (Fig. 12b).

VI. CONCLUSIONS

Theoretical background for modelling and controlling with dead time compensation of bipolar voltage source converter is developed in the paper. Space vector pulse width modulation method capable of reducing the effect of dead time to output voltage distortion is developed. Simulation results are presented. On the basis of theory and simulation results presented in the paper one come to the conclusion that dead time is one of the most important source of current deformation. It has even more importance than converter voltage level. For this reason robust method of dead time compensation should be implemented in the converter system. From the above it may be concluded that bipolar voltage source converter with dead time effect compensation stays on one of the most concurrent converter topology due to its efficiency and simplicity.

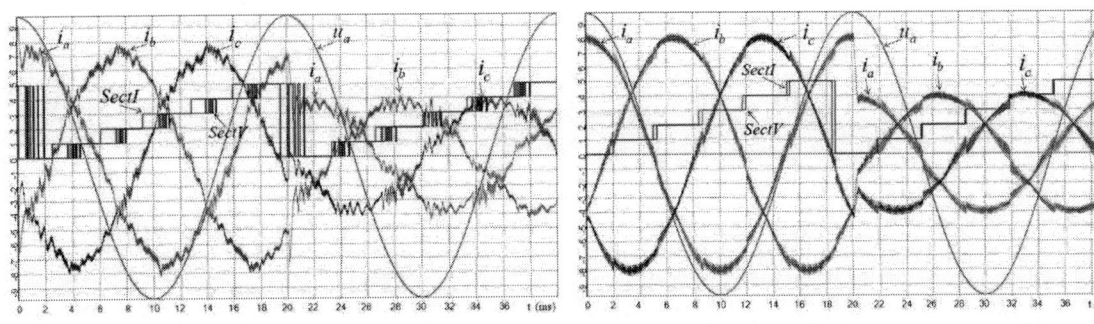

(a) SVPWM3   Fig. 12. Phase currents, phase voltage and current and voltage sector signals of VSC (in rectifier mode)   (b) SVPWM4

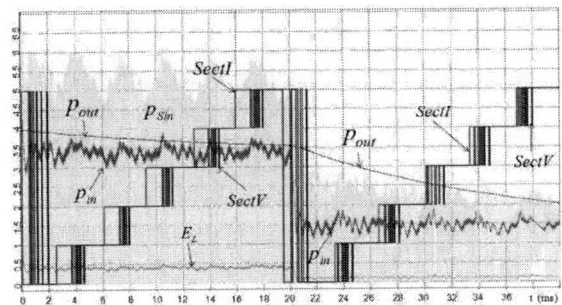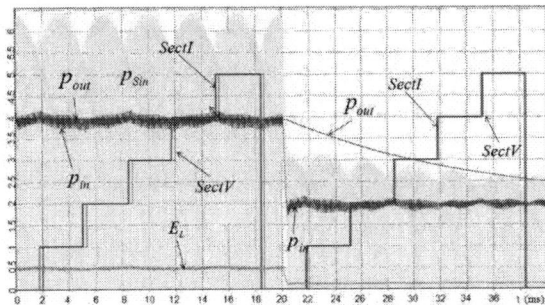

(a) SVPWM3 Fig. 13. Instantaneous powers, induction accumulator energy, current and voltage sector signals of VSC (rectifier mode) (b) SVPWM4

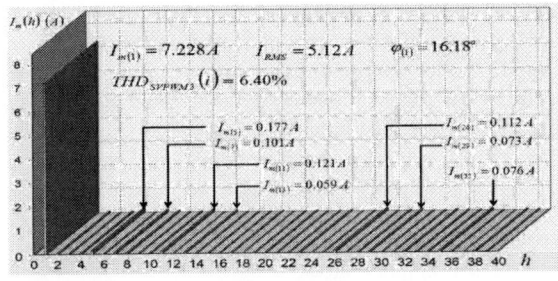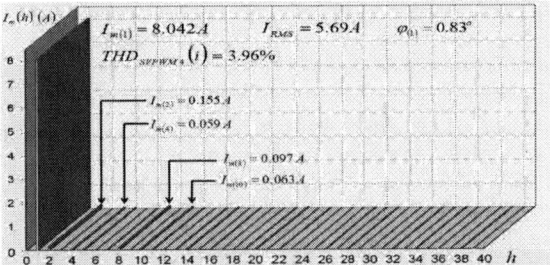

(a) SVPWM3   Fig. 14. Phase current spectrums of VSC working in rectifier mode (harmonics from 0 to 40)   (b) SVPWM4

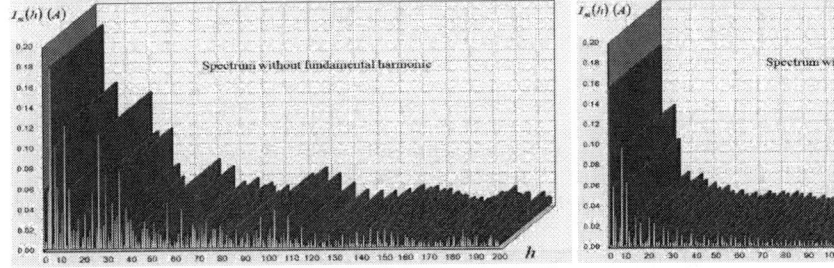

(a) SVPWM3   Fig. 15. Phase current spectrums of VSC working in rectifier mode (harmonics from 2 to 200)   (b) SVPWM4

REFERENCES

[1] J. Holtz: "Pulsewidth modulation – a Survey", IEEE Transactions on Industrial Electronics, Vol. 39, No. 5, Dec. 1992, pp. 410-420.

[2] H. Pinheiro, F. Botterón, C. Reh, L. Schuch, R. F. Camargo, H. L. Hey, H. A. Grundling, R. J. Pinheiro: "Space Vector Modulation for Voltage-Source Inverters: A Unified Approach", IECON 02 Volume 1, Issue, 5-8 November 2002.

[3] M. Malinowski: "Sensorless Control Strategies for Three-Phase PWM Rectifiers", Ph.D. Thesis, Warsaw University of Technology, Warsaw 2001.

[4] M. Jasiński: "Direct Power and Torque Control of AC/DC/AC Converter-Fed Induction Motor Drives", Ph.D. Thesis, Warsaw University of Technology, Warsaw 2005.

[5] R. Strzelecki, H. Supronowicz: "Współczynnik mocy w systemach zasilania prądu przemiennego i metody jego poprawy", Oficyna Wydawnicza Politechniki Warszawskiej, Warszawa 2000, (in Polish).

[6] L. Chen, F. Z. Peng: "Dead-Time Elimination for Voltage Source Inverters", IEEE Transactions on Power Electronics, vol. 23, no. 2, March 2008.

Sliding Mode Control of DC/DC Multiphase Power Converters

Vadim Utkin

Department of Electrical Engineering, Ohio State University, Colmbus, Ohio, USA, e-mail: *utkin.2@osu.edu*

Abstract—The design of a multiphase DC/DC power converter with bounded switching frequency based on the sliding mode control concept is proposed whose main goal is the suppression of chattering. In the framework of this approach: the width of hysteresis loops of switching elements are selected such that oscillation frequency is the same in all phases; the switching command in each phase is chosen as a function of error signals from different phases where these functions are selected to have the best "harmonic cancellation"; and for any range of duty cycles the number of phases can be selected so that the amplitude of chattering is maintained at the desired level.

I. PROBLEM STATEMENT

Traditionally, the control problems of DC/DC converters are solved using PWM techniques. The equivalence between sliding mode control and PWM control in the low frequency range can be easily established. It is recognized that the hardware implementation of sliding mode control can be easier than PWM control. For DC/DC converters, the link current and output voltage are normally selected as state variables which are to be controlled. It has also been established that current dynamics is much faster than that of output voltage. Based on the theory of Singular Perturbation, the control problem can be solved by using two cascaded control loops, namely: an inner current control loop and an outer voltage control loop. The latter is usually implemented using a simple controller, such as that of the PI type, and the current controller is implemented with different types of PWM, for example hysteresis control, which implies enforcing sliding modes.

As the maximal frequencies of the switching elements are tending to be higher and higher, so the sliding mode approach will become more and more popular in the field of converter control.

In general, sliding mode control has long been known as a particularly suitable method for handling nonlinear systems with uncertain dynamics and disturbances due to its order reduction property and low sensitivity to disturbances and plant parameter variations, which thus relaxes the burden of the necessity of exact modeling [1]. Moreover, sliding mode control may reduce the complexity of feedback control design through decoupling of system into independent subsystems of lower dimension. Due of these properties, sliding mode control methodology can be applied in diverse areas such as electric motors, manipulators, power systems, mobile robots, spacecraft, and automotive control.

However, the main drawback of sliding mode control is the occurence of undesirable oscillations of finite amplitude and frequency, either due to the presence of unmodeled dynamics or discrete time implementation. This destructive phenomenon, so-called 'chattering' (the term 'ripple' is used in the literature on power converters), may lower control accuracy or incur unwanted wear of mechanical components. Different methods of chattering suppression have been discussed in literature: use of asymptotic observers, state-dependent gains for discontinuous control, continuous approximation of discontinuous control—all of them lead to either increasing of switching frequency so that the high-frequency component of control is filtered out by the plant dynamics to yield a reduced amplitude of chattering [2]. The asymptotic observer based methodology needs information about the plant model and thus cannot be applied for systems operating under high level of uncertainty. The other approaches are not applicable for the conventional power converters with 'on/off' as the only admissible operation mode (as their outputs are binary time functions).

The natural way to reduce chattering (and as a result an amplitude of the variable to be controlled) is by increasing switching frequency. Unfortunately, this is not always possible due to the limitation of switching frequency to minimise losses in power converters. The idea of chattering suppression subjected to this constraint is oriented toward multiphase converters, consisting of a set of parallel converters [3].

II. SELECTION OF PHASE NUMBER

Suppose that a multiphase converter with m phases is to be designed so that the period of chattering T is the same in each phase, and two subsequent phases have phase shift $\frac{T}{m}$. Since chattering is a periodic time function, it can be represented as a Fourier series with frequencies

$$\omega_k = \omega, \ \omega = \frac{2\pi}{T}, k = (1, 2, \dots, \infty) \tag{1}$$

The effect of the kth harmonic in the output signal, which is the sum of individual outputs from all phases, can be easily calculated from

$$\sum_{i=0}^{m-1} \sin \left[\omega_k \left(t - \frac{2\pi}{\omega m} i \right) \right] = \sum_{i=0}^{m-1} \text{Im} \left[e^{j \left(\omega_k t - \frac{2\pi k}{m} i \right)} \right]$$

$$= \text{Im} \left(e^{j \omega_k t} Z \right), \ Z = \sum_{i=0}^{m-1} e^{-j \frac{2\pi k}{m} i} \tag{2}$$

978-1-4244-1741-4/08/$25.00 ©2008 IEEE

and to find Z, consider the following equation

$$Ze^{-j\frac{2\pi k}{m}} = \sum_{i=0}^{m-1} e^{-j\frac{2\pi k}{m}(i+1)}$$

$$= \sum_{i'=1}^{m} e^{-j\frac{2\pi k}{m}i'} \qquad (3)$$

Thus, from (2)

$$Ze^{-j\frac{2\pi k}{m}} = Z \qquad (4)$$

The function $e^{-j\frac{2\pi k}{m}}$ is equal to 1 only if $\frac{k}{m}$ is an integer or $k = m, 2m\ldots$, which means that $Z = 0$ for all other cases. This analysis shows that all harmonics except for $lm, l = 1, 2\ldots$ are suppressed in the output signal. As a result, the amplitude of chattering can be reduced to the desired level by increasing the number of phases thus providing the desired phase shift between two subsequent phases from the methodology proposed in the previous section.

III. DESIGN METHODOLOGY

This effect can be obtained by providing desired phase shifts for any loads and frequencies to implement the so-called 'ripple cancellation' or 'harmonic cancellation' method. The attempt to apply this idea for PWM strategy has been made so that the phases become interconnected and the phase shift between phases can be controlled correspondingly, by using a transformer with primary and secondary coils in different phases, or by using delays, filters, setd of triangular inputs with selected delays.

The approach in this presentation stems from the nature of sliding mode control and provides a desired phase shift between phases for any frequency without any additional dynamic elements, such as transformers or filters.

The concept 'multidimensional sliding mode' means that control is a vector with each component being a state function undergoing discontinuities at some surface in the system state space, and, after sliding mode occurs, system state trajectories are confined to the intersections of these surfaces. Generally speaking, the oscillation frequencies and phases of control components are not independent. By a proper choice of the surfaces and switching elements all frequencies can be maintained at the desired level and desired phase shifts can provided.

For the problem under study–chattering suppression in DC/DC power converters–even if the plant input is a scalar, we deal with vector control with components as control commands for switching between phases. In the framework of the above approach

- the widths of hysteresis loops of switching elements are selected so that oscillation frequency is the same in all phases for all operation modes; for implementation of the proposed method an average value of the control signal is needed and it can be easily obtained by a low pass filter,
- the switching command in each phase is a function of error signals (difference between reference and real currents) from different phases; these functions are selected such that the desired phase shifts are

provided (their values are found to have the best 'harmonic cancellation'),

- for any range of duty cycles the number of phases is found to reduce chattering amplitude to the desired level.

This methodology is applicable to cases when chattering is caused by unmodeled dynamics and its frequency is lower than th admissible power converter frequency. The generalization is based on the concept of 'equivalent hysteresis'. For control design, unmodeled dynamics are: 1. replaced by a switching element with a hysteresis loop such that the system has the same chattering frequency, and 2. the first phase is implemented with an ideal switching element while the others are designed following the previously described methodology.

The chattering suppression effect is illustrated by a wide range of simulations for the existing power converter and for control of DC motor with unmodeled dynamics. As an example, control of a power converter with low duty cycle is shown in the figures that follow, the 4-phase converter demonstrates a significant reduction of chattering in the output current when compared with that of individual phases, the 8-phase converter is much more efficient in chattering suppression.

IV. SIMULATIONS

The proposed methodology was simulated for a 4- and 8-phase power converter each with the following data: Input voltage $V_s = 12V$, Reference voltage $V_{\text{ref}} = 1.5V$, and load $RL = 1m\Omega$. The results for each are shown in the figures below.

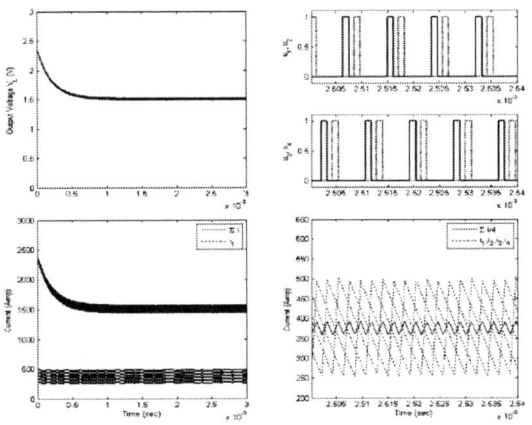

Fig. 1. 4 Phase Converter

REFERENCES

[1] V.I. Utkin, J.G. Guldner, and J. Shi, *Sliding Mode Control in Electromechanical Systems*, Taylor & Francis, 1999.

[2] H. Lee and V.I. Utkin, "The Chattering Analysis," in Advances in Variable Structure and Sliding Mode Control, Lecture Notes in Control and Information Sciences, C. Edwards, E. Fossas Colet and L. Fridman (Eds.), Vol. 334, Springer, Berlin, 2006.

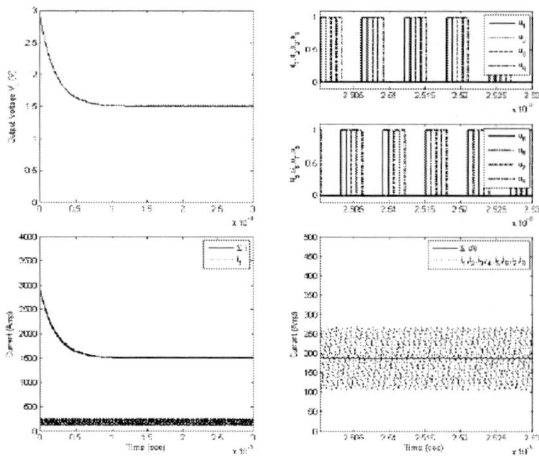

Fig. 2. 8 Phase Converter

[3] M. Lopez, L.G. de Vicuna, M. Castilla, P. Gaya, and O. Lopez, "Current distribution control design for paralleled DC/DC converters using sliding-mode control," in IEEE Transactions on Industrial Electronics, vol. 51, no. 2, 2004, pp. 419–428.

Author Index

A

Abbatelli, L. 61
Abbey, Chad 2178
Abdelhamid, Tamer H. 606
Abdellatif, Meriem 938
Abe, Seiya 393
Abourida, Simon 1077
Abroshan, Mohammad 1117
Abroushan, Mohammad 365
Abuishmais, Ibrahim 867
Abu-Rub, Haithem 1084, 1382
Adabi, Jafar 718, 903
Adamidis, Georgios 1840
Adamowicz, Marek 1729
Adzic, Evgenije 1957
Ahmadi, Muhammad 1847
Ahmed, M.M.R. 1866, 2472
Ahn, Jonng-Bo 2524
Ahn, ong-Bo 2492
Ait-Ahmed, Mourad 1740
Akhondi, Hamidreza 2071
Alarcón, E. 2108
Albert, Laurent 2037
Al-Diab, Ahmad 1710
Alexandrov, Alexandar 787
Al-Khayat, Nazar 2150
Allard, Bruno 2457
Al-Othman, A. K. 606
Amelon, Nicolas 1740
Anaya-Lara, O. 1784, 1941
Andersen, Michael A.E. 127
Ando, Kenji 614
Andrzejewski, Andrzej 1090
Areerak, K-N 2049
Arellano-Padilla, J. 1173
Arellano-Padilla, Jesus 769
Armstrong, S. 1688
Aroudi, A. El 2108
Aroudi, Abdelali El 2115, 2120
Arshad, Waqas M. 867
Asher, G. M. 2261
Asher, G.M. 2049
Asher, Greg 2300
Asiltürk, Ilhan 967
Aurel, Campeanu 893
Averberg, Andreas 213

B

Baalbergen, Freek J.F. 2170
Baghaee, H.R. 313, 629, 750
Bahri, I. 1365
Bailey, Chris 76
Bakas, Panagiotis 1840
Balazovic, Peter 1402

Balouktsis, Anastasios 1840
Baluta, Gh. 2043
Ban, Drago 818
Barai, Mukti 674
Barakat, Georges 1834, 810
Baranowski, Jerzy 1432, 1446
Barbosa, Fabián H. 637
Barlik, Roman 84
Barrero, R. 1512
Bartelt, R. 521
Baskys, Algirdas 1140
Bastiani, A. 1293
Baszynski, Marcin 1779
Bauer, Pavel 422
Bauer, Pavol 2170, 2354, 2368, 2371
Beck, Hans-Peter 1243
Bekbudov, Radiy 337
Bekishev, Anatoly 663
Bélanger, Jean 1077, 1475
Belfkira, Rachid 1834
Belkhodja, I. Slama 1149
Bellini, Armando 490
Bellmunt, Oriol 731
Belter, D. 1044
Benadero, Luis 2115
Bendkowski, Lukas 250
Benecke, Marcel 1280
Benkhoris, Mohamed-Fouad 1740
Bennani, A.Ben Abdelghani 1149
Beran, Leos 782
Bergas-Jane, Joan 731
Bergogne, Dominique 2457
Berthon, A. 1542
Bertoluzzo, Manuele 1491
Bertram, Torsten 1215
Betz, R.E. 1293
Bevilacqua, Pascal 2457
Bifaretti, Stefano 1771, 490, 561
Binder, Andreas 1625, 2385
Binkowski, T. 714
Birolleau, Damien 2037
Biswas, Jayanta 674
Bizon, Nicu 621
Blahník, Vojtech 1535
Blanco, M. 2481
Blazic, B. 2510
Böcker, J. 1598
Böcker, Joachim 159
Bodora, A. 326
Bogalecka, Elzbieta 1975, 804
Bojda, Petr 422
Bolgov, Viktor 154
Bolognani, Silverio 1097
Boora, Arash A 468, 723
Bossche, A. Van den. 1326
Botan, Corneliu 1111

Author Index

Botsali, FatihM. 949
Bouafia, Abdelouahab 703
Boucherit, M.S. 1987
Bouhalli, Nadia 281
Bozhko, S.V. 2049
Brand1tetter, Pavel 1375
Braslavsky, I.Ya. 1050
Breban, Stefan...1896
Brown, Neil L. 2150
Bruno, Francois..2205
Bucher, Alexander 244, 250
Buja, Giuseppe..1491
Bukatov, Alexander 1872
Bulic, Neven 556
Buonomo, S. 61
Buss, Martin..2312

C

C., Ilioudis Vasilios 1105
Caballero, M. 1555
Cabrita, Carlos M. P. 1646
Calado, Maria R. A. 1646
Camara, M.B. 1542
Cambál, Marek 982
Candusso, Denis 734
Cartes, D.A. 793
Case, Michael James..1798
Castaing, Ambroise..2464
Catalão, J. P. S. 1682
Cédl, Marek 1593, 372
Ceglia, Gerardo 268
Cepisca, Costin 1963, 908
Cernat, Mihai 1748
Cernohorský, Josef 1009
Cerovský, Zdenk 982
Cha, Gil-Ro 383
Champenois, Gérard 2015
Chan, Paul K.W. 1688
Chang, Hao-Chi 1652
Chang, Lon-Kou 320
Chang, Yuan-Chih 456
Chante, Jean-Pierre 2457
Charaabi, L. 1365
Chekhet, Eduard 307
Chen, Anyuan 799
Chen, Junling 2000
Chen, Yonggang 1981, 2000, 405, 515
Chen, Zhe..2325
Chen, Zong-Jie 1704
Cheng, K.W.E. 576
Cherif, M. Ghodbane 1149
Cheung, N. C. 1221
Chien, Sywe-Bin 1652
Chillet, Christian 2037
Chimento, F. 61

Chlodnicki, Zdzislaw 2150
Choi, Heung-Kwan 2524
Choi, Jaeho 2498
Choi, Uk-Don 1421
Chou, Ming-Chang 1652
Chrenko, Daniela 2156
Chrzan, Piotr J. 144
Chudzik, Piotr 1568
Chun, Tae-Won 1421
Clare, J. 1326
Clare, Jon C 207
Clare, Jon C. 229, 561
Clare, Jon 1771, 307
Comnac, Vasile 1748
Cook, B.J. 1293
Cook, D. 1326
Coquery, Gérard 2192
Correa, Pablo 451, 699
Courtecuisse, Vincent 1896, 2184
Cousineau, Marc 281
Cuk, Vladimir 1426
Cychowski, Marcin 2241
Czapp, Stanislaw 2059

D

Dabroom, A.M. 1337
Dakyo, B. 1911
Dannehl, J. 444
Darie, Eleonora 1963, 908
Darie, Emanuel 1963, 908
Davey, J. 1918
De Bernardinis, Alexandre 2192
De Castro, M.R. 2126
De Gersem, Herbert 2385
de Kock, H.W. 859
De Souza, Kleber C.A. 1951
Deaconu, Sorin 1409
Debowski, Andrzej 1568, 2289
Degeratu, Sonia 893
Delaney, Kieran 2241
Demenko, Andrzej 2412
Denny, Ernest Edward 1798
Depernet, Daniel 734
Derbel, Nabil 2120
Deskur, Jan 1204, 2227
Deuse, Jacques 2184
Dheilly, Nicolas 2457
Di, Lu 2205
Dianov, Anton 1002
Díaz, Nelson L. 637
Diblík, Martin 1676
Diguet, Marc 1382
Dilevs, Guntis 1811
Dimitrakakis, Georgios S....................................1301
Dinkhauser, Vincenz 1819

Author Index

Dobrucky, Branislav 1402
Dockhorn, Matthias 1734
Dodds, Stephen J. 2551, 2559
Dodds, Stephen James 2543
Doebbelin, Reinhard 1280
Doi, Nobuaki 744
Dong, P. 576
Dong, Wei 1716
Dontchev, Dimitar 787
Drábek, Pavel 1593, 372
Draganov, Denis 1610
Dubowski, Marian Roch 1090
Dudak, Juraj 2368
Dudrik, Jaroslav 295
Duerbaum, Thomas 244, 250
Dufour, Christian 1077, 1475
Duke, Richard 528
Dumur, Guillaume 1475
Durovsky, Frantisek 961
Dybkowski, Mateusz 2211, 2306
Dzieniakowski, Maciej A. 2082

E

Eberhard, Andreas 1371
Eckel, Hans-Guenter 48
Edrington, C.S. 793
Egan, Michael G. 1249
Egorov, Mikhail 1257
Ehsan, Mehdi 1847
Eilenberger, Andreas 945
Elmoctar, Mohamed Y. Ould 810
Empringham, Lee 207, 229, 388
Endo, Tsunehiro 924
Eno, Otu A. 114
Erceg, Gorislav 556
Etxeberria-Otadui, I. 1555

F

Fabianowski, Jan 2082
Fabijanski, Pawel 1040, 2055, 2087
Fahrni, C. 256
Fakham, Hicham 2142
Fan, Yue 1771
Farhangi, Sh. 173
Farshad, Siamak 1575
Fedák, Viliam 2354
Fedak, Viliam 961
Fedyczak, Zbigniew 165, 236
Feki, Moez 2120
Fernández, Herman 1947
Fernandez-Mola, Josep-Maria 731
Ferreira, Jan Abraham 187
Ferreira, Luís António Fialho Marcelino 2076
Fetyko, Jan 961
Filchev, T. 1326

Filho, Braz Jesus Cardoso 1345
Filka, Roman 1402
Fisher, R. 1293
Fleisch, Karl 48
Fodor, D. 2096
Foft, Jiří 1593
Foo, Gilbert 2269
Forster, Stefan 2420
Fotouhi, Reza 1575
Francois, Bruno 2142, 2184
Franke, W. Toke 69
Franko, Marek 2538
Friedli, T. 27
Fröhleke, Norbert 159
Fuchs, F.W. 444
Fuchs, Friedrich W. 1390, 1819, 69
Fujita, Y. 275
Fukushima, Kentaro 148
Funabashi, Toshihisa 2478, 2487
Funato, Hirohito 479
Funayama, Koichi 1020
Futami, Motoo 2337

G

Gabriela, Petropol Serb 893
Gan, W. C. 1221
Gao, Fanqiang 515
Gao, Q. 2261
Gao, Qiang 1058
García-Tabarés, L. 2481
Gardecki, Arkadiusz 1193
Gasiewski, Marcin 1562
Gaubert, Jean-Paul 703
Gavranic, Ivica 818
Gaztañaga, H. 1555
Gelezevicius, Vilius Antanas 1144
Gennadevich, Kiselev Michail 428
Gennadevich, Lepanov Michail 428
Gerada, C. 1173
Gerada, Chris 1058, 388, 769, 887
Ghaedi, Azam 1054
Gharehpetian, G.B. 313, 629, 750
Ghosh, Arindam 468, 723, 903
Gímenez, María Isabel 1947
Giménez, María 268
Giral, Roberto 2115
Gizinski, Zygmunt 1562
Glasberger, Tomá1 1268
Glavin, M.E. 1688
Glushkin, Evgeny 1872
Gnacinski, Piotr 826
Gobis, Vitoldas 1140
Goeldel, C. 2126
Gomis-Bellmunt, Oriol 1670
González-Hernández, S. 1784

Author Index

Gorbounov, Yassen.....................................787
Goto, Hiroki...............................1163, 1168
Grabic, Stevan...1957
Grad, M..714
Grecki, Filip...1440
Grigaitis, Arunas.......................................1144
Grigans, Linards..2066
Grossi, Federica..874
Grzesiak, Lech M.......................................1071
Grzesik, B...956
Gualous, H..1542
Guo, Hai-Jiao..............................1163, 1168
Gustin, F...1542
Gustin, Frederic..734
Guy, Owen J..2464
Guzinski, Jaroslaw.........................1382, 994
Guzmán, Víctor.............................1947, 268
Gwózdz, Michal...728

H

Haan, Sjoerd de...187
Habetler, Thomas G..21
Hadas, Zdenek..1665
Hadjov, Kliment..787
Hájek, Vítezslav...2371
Halasz, S...682
Halgos, Jan...2368
Hamada, Tomoyuki.....................................1884
Hamar, J...1755
Hameyer, K..2393
Hameyer, Kay..2412
Harada, Yosuke..148
Hartansky, Rene..2368
Hartnett, Kevin J.......................................1249
Hasegawa, Masaru.......................................614
Hashimoto, Seiji..932
Hayashi, Kenta..589
Hayashi, Yusuke..2445
Hayes, John G..1249
Heising, C...521
Helmut, Weiss.................................1722, 1934
Henrotte, F..2393
Henze, Olaf...2385
Hercog, Darko...2349
Hicham, Fakham...2205
Himmelstoss, Felix. A....................................331
Hiraki, Eiji..119, 1877
Hirokawa, Masahiko......................................393
Hissel, Daniel..2156
HISSEL, Daniel..734
Hmasic, N..2134
Ho, S.L...576
Hoffmann, Frank..1215
Hõimoja, H..2005
Hõimoja, Hardi..1581

Hojo, Masahide...2487
Holtz, Joachim..1084
Holub, Marcin...195
Horen, Yoram..776
Horga, V..2043
Horga, Vasile..1111
Hrasko, Martin..2538
Hu, Weihao...2325
Hubik, Vladimir...1620
Huiqing, Wen...............................1518, 417
Hurley, W.G..1688
Huttin, N..1523

I

I., Margaris Nikolaos....................................1105
Iannuzzi, Diego...1469
Ibach, Robert..2082
Ibáñez, Fernando...268
Ichinokura, Osamu................1163, 1168, 758
Ichinose, Masaya.......................................2337
Ide, Kazumasa...2337
Igic, P. M..2464
Iida, Takahiko...595
Ikeda, Yoshiko...498
Ikhouane, Faycal.......................................1670
Iman-Eini, H..173
Inoue, Yukinori...1859
Ion, Petropol Serb.......................................893
Iov, Florin....................................1771, 561
Ishikawa, Kazumi.......................................1020
Iskhakov, Albert..663
Ito, Fumio..1309
Itoh, Jun-ichi...581
Itoi, M..275
Ivanovic, Zoran...1957
Iwaji, Yoshitaka..924
Iwanski, Grzegorz...........................1440, 2164
Iwase, Yuta...2487
Izadbakhsh, Alireza....................................2102

J

Jalakas, T...1263
Jalakas, Tanel...1257
Jan, Mucko...1316
Ján, Vittek...2219
Jang, Gil-Soo..2498
Jang, Su-Jin...1924
Jansen, Uwe..88
Janson, Kuno..154
Járdán, Rafael K...916
Jardan, Rafael Kalman.................................2360
Jasim, O..1173
Jasim, Omar...887
Jasinski, Marek...1904
Javurek, Jiri...1465

Author Index

Jedryczka, Cezary 2406
Jennings, Michael R. 2464
Jeon, Jin-Hong 2492, 2524
Jezernik, Karel 2283, 2349, 432
Ji, Young-Hyok 1929
Jian, Xiao 1722
Jin, Zhao 1128
Johnson, C Mark 76
Joós, Géza 2178
Joost, M. 1064
Judek, Slawomir 1497
Jufer, Marcel 1
Jun, Liu 1518, 417
Jung, Doo-Yong 1929
Jung, Yong-Chae 181, 1924, 1929, 383

K

Kalatchikov, P. 837
Kalisiak, Stanislaw 195
Kallaste, Ants 154
Kallenbach, E. 1598
Kalyoncu, Mete 1132, 949, 974
Kamata, Yuki 498
Kamiski, Bartlomiej 2378
Kamper, M.J. 859
Kampisios, Konstantinos 887
Kaneko, Daigo 924
Kanerva, Sami 867
Kaplon, Andrzej 377
Karaffy, Z. 2096
Karsli, Vedat M. 850
Karwowski, Krzysztof 1497
Kasa, Nobuyuki 595
Kasinski, A. 1044
Kasprowicz, Andrzej 1332
Katic, Vladimir 1957
Kato, Koji 581
Katsura, Seiichiro 1187, 1604, 1614
Kawamura, Atsuo 7, 924
Kayhan, Ince 1934
Kazimierz, Jaracz 912
Kazmierkowski, Marian P. 1548, 1904
Kelemen, Franjo 855
Kennel, R.M. 859
Kennel, Ralph 1239
Khaldi, B.S 1987
Kim, Eel-Hwan 2498
Kim, Heung-Gun 1421
Kim, Jae-Hong 2498
Kim, Jae-Hyung 1924, 1929
Kim, Jong-Yul 2492
Kim, Se-Ho 2498
Kim, Seul-Ki 2492, 2524
Kimura, Kensuke 1168
Kimura, Noriyuki 1884

Kinoshita, Hirotaka 2337
Kireev, V. 1598
Klimczak, Pawel 108
Klug, O. 2096
Klyachko, Leonid 663
Klytta, Marius 165
Knop, André 69
Kobayashi, Yukinori 479
Kobougias, Ioannis C. 1274
Koczara, Wlodzimierz 1440, 2150, 2164, 2254
Koda, Noriaki 1877
Kolar, J. W. 27
Kolesnikov, Artem 1872
Kolomeitsev, L. 1598
Kompa, K. 695
Komura, Akiyoshi 2337
Kondo, Masaki 1614
Koneke, Thies 1458
Kong, S.T. 43
Konstantinovich, Rozanov Yurie 428
Korondi, Peter 2360
Korotyeyev, Igor 236
Koskin, Y. 837
Kosmecki, Michal 1975
Kostylev, A.V. 1050
Kotodziejek, Piotr 804
Kouzou, A. 1987
Kowalski, Czeslaw T. 1359
Kraeftner, Wilhelm 331
Kraynov, D. 1598
Krettek, Johannes 1215
Krim, Fateh 703
Krismer, F. 27
Krykowski, K. 326
Krystkowiak, Michal 728
Krzeminski, Zbigniew 1382, 2294
Kubiak, Andrzej 2452
Kubin, Jiri 1815
Kubota, Sachio 1309
Kuchta, Jozef 2538
Kudarauskas, Sigitas 2200
Kuebrich, Daniel 244
Kuhn, Harald 1458
Kuisma, M. 1233
Kulka, Arkadiusz 657
Kumar, Dinesh 207
Kuperman, Alon 776
Kurokawa, Fujio 2434, 2504
Kürschner, Daniel 1696, 1734
Kuß, H. 695
Kusserow, Wolf 1239
Kütt, Lauri 154
kuwata, M. 275
Kyritsis, A.C. 1287

Author Index

L

Laczynski, Tomasz 569, 649
Lafoz, M. 2481
Lagoda, Ryszard 1040, 2055, 2087
Laloya, Eduardo 845
Lange, E. 2393
Lapointe, Vincent 1077
Lastowiecki, Jozef 1440
Latka, M. 714
Latkovskis, Leonards 2066
Laugis, J. 1263
Laugis, Juhan 1017
Laur, R. 1064
Lazar, C. 2043
Lazar, Mihai 2457
Ledwich, Gerard 468, 723, 903
Lee, Joo-Hyuk 1924
Lee, Tzung-Lin 1704
Lehtla, Madis 1581
Lehtla, T. 2011
Leidhold, Roberto 1353
Leszek, Szychta 2091
Leuchter, Jan 422
Levins, Nikolajs 1811
Lewandowski, Daniel 2289, 669
Lewicki, Arka diusz 1382
Lewis, A.W. 1790
Leyva, R. 2108
Li, Kaihang 97
Li, Rongyuan 159
Li, Yaohua 1981, 2000, 405, 515
Li, Zixin 1981, 2000, 405, 515
Liaw, Chang-Ming 1652, 456
Lie, Xu 229
Liffran, Florent 409
Lillo, Liliana de 388
Lindemann, Andreas 1280, 2420
Lingemann, M. 2134
Lis, Jacek D. 1359
Lisik, Zbigniew 2452
Lisowski, Grzegorz 669
Liu, Congwei 405
Liu, Li 793
Lladó, Juan 845
Lodzinski, Michal 2464
Lopez-de-Heredia, A. 1555
Lorenz, Robert D. 903
LU, Di 2142
Lu, Hua 76
Lu, Y. 1221
Luft, Miroslaw 463
Luiz, Alex-Sander Amavel 1345
Luniewski, Piotr 88
Lyons, Brendan J. 1249
Lyskawinski, Wieslaw 2406

M

Macek-Kaminska, Krystyna 1193
Madawala, U. K. 139
Madawala, Udaya K. 1918
Maga, Dusan 2368
Mahmoudi, M.O. 1987
Mahyob, Amin 810
Mailat, Adrian 1748
Majidi, Behrooz 763
Maksimovic, Dragan 498
MAKYS, Pavol 2538
Malekian, Kaveh 1117, 1123, 2071, 365, 763
Milimonfared, Jafar
Malska, W. 714
Man, T.K. 400, 475
Mandache, Lucian 1585
Mandra, Slawomir 1071
Mandrek, Slawomir 144
Marek, Stulrajter 2219
Margaliot, M. 260
Mariano, Sílvio José Pinto Simões 2076
Marouchos, Christos 1967
Martín-del-Brío, Bonifacio 845
Martínez, Abelardo 1947, 845
Martinez, Itziar 437
Martins, Denizar C. 1951
Masada, E. 1755
Mascibrodzki, Ireneusz 1562
Mathis, W. 132
Matsui, Keiju 614
Matsui, Nobumasa 2504
Mawby, P.A. 2472
Mawby, Philip A. 2464
McEachern, Alex 1371
Mecke, Rudolf 1734
Melício, R. 1682
Mendes, V. M. F. 1682
Mertens, A. 132
Mertens, Axel 1458, 213, 569, 649
Meuret, R. 1523
Meynard, Thierry 281
Michalík, Jan 1535, 550
Michalke, N. 695
Mierlo, J. Van 1512
Milanovic, Miro 301
Milimonfared, Jafar 1117, 2071, 365, 763
Mimura, Yasuhiro 2434
Mirsalim, M. 313, 629, 750
Mirsalim, Mojtaba 1123
Mirzaeva, G. 1155
Mishima, Tomokazu 119
Mitani, Tetsuya 2428
Mladenovic, I. 2022
Mohd, A. 2134
Mokrovica, Josipa 855
Mõlder, Heigo 154

Author Index

Molinas, Marta 2318
Möller, T. 2005
Mollov, Stefan V. 350
Molnár, Jan 1535, 550
Mondzik, Andrzej 345
Monmasson, E. 1365
Montesinos-Miracle, Daniel 1670, 731
Morel, Herve 2457
Moreno-Font, Vanessa 2115
Moreno-Goytia, E. 1784, 1941
Morimoto, Shigeo 1859
Morino, Kimio 2478
Morizane, Toshimitsu 1884
Morton, D. 2134
Mouni, Emile 2015
Mukhopadhyay, Siddhartha 485
Munk-Nielsen, Stig 108
Murata, Toshiaki 2337
Musallam, Mahera 76
Mustonen, P. 1233
Musumeci, S. 61
Muszynski, Roman 2227
Mutschler, Peter 1353
Müür, M. 2005
Mysinski, Wojciech 1321

N

Nagy, I. 1755
Nagy, Istvan 2360
Nagy, István 916
Naka, Toshiyuki 498
Nakagawa, Akio 498
Nakamura, Kazutoshi 498
Nakamura, Kenji 758
Nakaoka, M. 275
Nakaoka, Mutsuo 119
Nakayama, Hiroaki 1877
Nanakos, Anastasios Ch. 1827
Naouar, M-W. 1365
Narayanan, E.M. Sankara 43
Narjiss, Abdellah 734
Nasser, Mehdi 1896
Navarro, Daniel 437
Nawaz, Muhammed 2472
Nekoui, Mohammad Ali 1054
Ngwendson, L. 43
Ni, Bingchang 2331
Nichita, C. 1911
Nichita, Cristian 1834
Nicolae, Ileana-Diana 1585
Nicolae, Petre Marian 1585
Nicolae, Petre-Marian 1181
Niechaj, Marek 1890
Niemelä, Markku 1763
Nikolic, Aleksandar 1426

Nilssen, Robert 799
Ninomiya, Tamotsu 148, 393
Nishida, Yasuyuki 2530
Nishikata, Shoji 2343
Nishimiya, Ayumu 1163
Nishioka, Kunihiro 1309
Nitta, Mayumi 932
Noda, Shuji 1877
Norigoe, Isami 148
Novák, Jaroslav 982
Novák, Martin 982
Nowak, Lech 2400
Nowak, Mietek 84
Numata, Shigeo 2478
Nuutinen, Pasi 1763
Nyczkowski, Lukasz 740
Nymand, Morten 127
Nysveen, Arne 799

O

O'Sullivan, D.L. 1790
Ogiwara, H. 275
Ohashi, Hiromichi 2428, 2445, 54
Ohishi, Kiyoshi 1187, 1604, 1614
Ohsaki, H. 1755
Ohyama, Kazuhiro 2300
Okamatsu, Masashi 2434
Oleschuk, Valentin 1548
Omari, O. 2134
OMORI, Hideki 2530
Ondrusek, Cestmir 1665
ONEN, Umit 949
OPROESCU, Mihai 621
Orlik, B. 1064, 830
Orlowska-Kowalska, Teresa 2211, 2306
Ortjohann, E. 2134
Oyarbide, Estanislao 845

P

Pacas, Mario 2248
Pajchrowski, Tomasz 1198, 1204
Pakhomin, S. 1598
Palis, Frank 1610
Palis, Stefan 1660
Panoiu, Caius 1409
Panoiu, Manuela 1409
Papanikolaou, N.P. 1287
Papic, I. 2510
Paquin, Jean-Nicolas 1475
Park, JuneHo 2492
Park, Sang-Hoon 181, 383
Park, So-Ri 181
Parkatti, P. 201
Parker-Allotey, N-A. 2472
Patel, N. D. 139

Author Index

Patra, Pradipta...485
Patra", Amit...485
Pavelka, Jiri..221
Pavelka, Jirí..988
Pavlitov, Constantin.......................................787
Pavlovsky, Martin...7
Pavol, Makys...2219
Pavoni, Alessandro..1491
Peftitsis, Dimosthenis...................................1840
Peltoniemi, Pasi..1763
Peplinski, Marcin..826
Pera, Marie-Cecile..2156
Perez, Francisco...845
Perez-Tomas, Amador....................................2464
Peric, Nedjeljko...2235
Peroutka, Zdenek.............. 1268, 1529, 1535, 550
Peter, Bris...2219
Peter, Zaucher...1722
Petit, Marc..2184
Petrella, Roberto..1097
Petrisor, Anca..893
Piatek, Pawel...1446
Pietrzak-David, Maria......................................938
Piróg, Stanislaw...1779
Pittermann, Martin............................... 1593, 372
Planson, Dominique.......................................2457
Poljugan, Alen..1058
Pollán, Tomás..845
Popa, Anca Sorana..1225
Popa, Mircea..1225
Porada, Ryszard...740
Pospelov, Vladimir..663
Pronin, M...837
Pugachevs, Vladislavs....................................1811
Pyrhönen, Juha...1763

Q

Quiroga, J..793

R

Rabkowski, Jacek...84
Raciti, A..61
Radomski, Grzegorz...504
Raducu, Marian..621
Radulescu, Mircea M.......................................1896
Rafecas-Sabate, Josep......................................731
Rafiei, S.M.R...2102
Rahman, M.F...2269
Rahnamaee, Arash................................. 1117, 365
Rao, Sachit..2312
Rathge, Christian..1696
Ratoi, Marcel...1111
Rawicki, Stanislaw...1481
Raynaud, Christophe.......................................2457
Rednov, F...1598

Reghem, Pascal..................................... 1834, 810
Rerucha, Vladimir...422
Rezaei, Mohammad Mehdi...................................1123
Reznikov, B..260
Ribickis, Leonids..1811
Richter, F..1398
Riipinen, T...1233
Risteiu, Mircea..1243
Riz, A..2096
Roasto, I...2011
Robert, B.G.M..2126
Robert, Bruno Gerard Michel..............................2120
Robinson, Jonathan...2178
Robyns, B...1523
Robyns, Benoît...1896
Robyns, Benoit...2184
Rodic, Miran..2283
Rodriguez, E...2108
Rodriguez, Jose.................................... 451, 699
Rojas, A..1155
Rojko, Andreja...2349
Rolek, Jaroslaw...377
Rompelman, Otto...2354
Ronkowski, Mieczyslaw......................................880
Rosin, A..2005
Rothenhagen, Kai.................................. 1390, 1904
Round, S. D..27
Ru1scin, Vladimír...295
Ruderman, A...260
Ruderman, Michael..1215
Rufer, A..256
Ruger, N. E...132
Rusinov, Radoslav...787
Rylko, Marek S...1249
Ryvkin, Sergey...1505
Rzasa, Janina...357

S

Saadi, S..1987
Sabirin, Chip Rinaldi......................................1625
Saito, Makoto..2439
Saito, Tsuyoshi...744
Sajkowski, M..956
Sakamoto, Kiyoshi...924
Sakamoto, Yosei...288
Salo, M...201
Salonen, Pasi..1763
Samanta, Susovon..485
Samuelsen, Dag...1416
Sanada, Masayuki...1859
Sánchez, Beatriz..845
Sánchez, Carlos...268
Sang-Joon, Lee...1002
Sang-Taek, Lee...1002
Sanjari, M. J..................................... 313, 629, 750

Author Index

San-Sebastian, J. 1555
Santo, António Espírito.................................. 1646
Sarraute, Emmanuel.. 281
Sasaki, Masahiro... 2434
Sato, Muneo.. 1309
Saudemont, C. 1523
Sayed, Mahmoud A. 542
Sayeef, S. 2269
Schallschmidt, Thomas.......................... 1610, 1660
Schanen, JL... 173
Schmelter, A. 2134
Schmid, Markus..244, 250
Schmidt, Istvan 1803
Schmidt-Obermoeller, Richard...................... 1505
Schmitt, Günter... 1239
Schneider, T. 1598
Schnick, O. 132
Schrödl, Manfred 2275
Schroedl, Manfred... 945
Schuffenhauer, U. 695
Scollo, R. 61
Sengupta, Sabyasachi 674
Seppä, L. 1233
Shao, S. 1293
Shapoval, Ivan 307
Sharma, R. 1918
She, X. 710
She, Yun 710
Shieh, Fa-Hwa 1652
Shimaoka, Yoshihiro 1309
Shimizu, Takaaki 498
Shimizu, Toshihisa 2428, 2445, 288, 600
Shimoda, Eisuke 2478
Shiraishi, Keiichi 2504
Shonin, O. 837
Shoyama, Masahito 148, 393
Shyu, Juei Lung 643
Siatkowski, M. 830
Silea, Ioan 1225
Silventoinen, P. 1233
Simetzberger, Christian 2275
Simon, Miklós G. 916
Singule, Vladislav 1620, 1665
Sinsukthavorn, W. 2134
Siostrzonek, Tomasz.. 1779
Sîrbu, Ioana-Gabriela.......................... 1181, 1585
Siroky, Peter 2368
Sitar, Jan 2368
Sivkov, Oleg 221
Skovpen, Sergey 663
Skuta, Ondřej 1375
Slama-Belkhodja, I. 1365
Slama-Belkhodja, Ilhem 938
Smet, Bart 102
Sobczuk, Dariusz 2378
Sobczynski, D. 714

Sochacki, Mariusz 2452
Soltani, Hamid 718
Song, Sang-Hoon 383
Song, Seung-Ho 2498
Soroudi, Alireza 1847
Sosa-Ruiz, J. 1941
Souad, Rafa.. 1209
Sourkounis, C. 1398
Sourkounis, Constantinos 1633, 1710, 2331
Sozanski, Krzysztof Piotr 1995
Stadler, Paul Andreas 2543
Stala, Robert 1852, 345
Stamann, Mario 1660
Stanescu, Dan-Gabriel 1181
Staudt, V. 521
Staudt, Volker 2371
Stefanutti, Fabio 1097
Steimel, A. 521
Steimel, Andreas 1505, 2371
Stenzel, T. 956
Stepanyuk, D.P. 1050
Stepien, P. 1293
Stocco, Piero 1097
Strac, Leonardo 855
Strzelecki, Ryszard Michal 1332
Strzelecki, Ryszard 1729
Stumpf, P. 1755
Stumpf, Péter 916
Sugai, T. 275
Sugimasa, Junji Tamura Masatoshi 2337
Suissa, Uri 776
Sulkowski, Waldemar 1416
Sumida, Yuichi 2434
Sumina, Damir 556
Sumiyoshi, Shinichiro 2530
Summers, T.J. 1293
Sumner, M. 1173, 2261
Sumner, Mark 1058, 2300, 769
Sun, Z. G. 1221
Susluoglu, Berrin 850
Suul, Jon Are 2318
Sveda, Martin 1620
Sweet, M. 43
Sykulski, Jan K. 2383
Szabat, Krzysztof..................................... 2211, 2241
Szamel, Laszlo 1033
Szczeniak, Pawel 165
Szczepankowski, Pawel 1332
Szczesniak, Pawel 236
Szelag, Wojciech 2406
Sziebig, Gabor 2360
Szmidt, Jan 2452
Szubert, Krzysztof 536
Szweda, Mariusz 826
Szychta, Elzbieta 463
Szychta, Leszek 463

Author Index

Szymanski, B. J. 695

T

Tackoen, X. 1512
Tae-Ho, Yoon 1002
Taguchi, Toyoki 498
Takahashi, Nobuo 1877
Takahashi, Rion 2337
Takao, Kazuto 2445, 54
Takeshita, Takaharu 542
Takeuchi, Nobuhito 614
Takeuchi, Toshihiro 924
Tan, Longcheng 1981, 2000, 405, 515
Tanabe, Takayuki 2478
Tanaka, Toshihiko 1877
Taniguchi, Katsunori 1884
Taniguchi, Satoshi 600
Tankari, A.M. 1911
Tao, Zhou 2205
Tapuchi, Saad 776
Tarczewski, Tomasz 1071
Tatakis, E.C. 1287
Tatakis, Emmanuel C. 1274, 1301, 1827
Tatsuta, Fujio 2343
Theodoridis, Michael P. 350
Thomas, D.W.P. 2049
Thomas, David W.P. 1716
Thompson, David S. 114
Tinkir, Mustafa 1132, 949, 974
Tnani, Slim 2015
Tournier, Dominique 2457
Tran, Quang-Vinh 1421
Trentin, Andrew 887
Trujillo, Cesar L. 637
Tsai, Jih-Run 1652
Tseng, K.J. 2516
Tsukakoshi, Kenta 148
Tsuruta, Yukinori 7
Tulbure, Adrian 1243
Turner, Robert W. 528
Tutaj, Andrzej 1432
Tuusa, H. 201

U

Ueda, Yoshinobu 2478, 2487
Ummaneni, Ravindra. B. 799
Undeland, Tore 2318, 657
Ünüvar, Ali 967
Urabe, R. 275
Urbanski, Konrad 1454
Utkin, Vadim 2312, 512

V

Väisänen, V. 1233

Valchev, V. 1326
van Duivenbode, Jeroen 102
Vasak, Mario 2235
Vedrana, Jerkovic 690
Vekic, Marko 1957
Vergnol, Arnaud 1896
Veszpremi, Karoly 1803
Vicuña, Javier 845
Villanueva, Elena 451
Villwock, Sebastian 2248
Vinnikov, D. 1263, 2011
Vinnikov, Dmitri 1257
Viscarret, U. 1555
Vittek, Jan 2551
Vladimír, Vavrus 2219
Vodovozo, Valery 1017
Vorontsov, A. 837
Vrana, Petr 1465

W

Wada, Keiji 2428, 288, 600
Walas, K. 1044
Walter, Julio 268
Walton, Simon 528
Wang, Ping 1981, 2000, 405, 515
Wang, Yi 187
Wang, Yue 2325
Wang, Zhaoan 2325
Weidinger, Thomas 2028
Weiland, Thomas 2385
Weindl, Ch. 2022
Werner, Timur 649
Wheeler, P. 1326
Wheeler, Patrick W. 207
Wheeler, Patrick W. 229
Wheeler, Patrick 388
Wiktor, Hudy 912
Willis, K. 1293
Winternheimer, Stefan 1872
Wisniewski, Janusz 2254
Wlas, Miroslaw 1084
Won, Chung-Yuen 181, 1924, 1929, 383
Wong, L.K. 400, 475
Wu, Dongming 97

X

XiaoyanHuang, ... 388
Xu, Wei 2000, 405
Xuhui, Wen 1518, 417
Xuhui, Zhang 1518, 417

Y

Yaguchi, Hiroyuki 1020
Yamanouchi, Wataru 1187

Author Index

Yang, Lingling..97
Yang, Liu...1128
Yang, Ru-Shiuan...320
Yin, Chunyan..76
Yokokura, Yuki 1187, 1604
Yokoyama, Tomoki 589, 744
Young-Kwan, Kim 1002
Yousefi, Ashkan 1847

Z

Zakrzewski, Zbigniew 1332
Zamma, Toshihiro......................................1020
Zanasi, Roberto...874
Zanchetta, Pericle 1716, 1771, 561, 887
Zare, Firuz 468, 718, 723, 903
Zaring, Carina..2472
Zarko, Damir 855
Zarko, Damirarko 818
Zaskalicka, Maria 899
Zaskalicky, Pavel 899
Zatocil, Heiko 1024
Zawirski, Krzysztof 1198, 1204, 1454
Zdenek, Jiri 1638
Zdravko, Valter..690
Zeljko, Spoljaric 690
Zeman, Karel...1529
Zeroug, Houcine 1209
Zhang, H. 1523
Zhang, S..2516
Zhao, S. W. 1221
Zhou, Tao...2142
Zhu, Haibin 515
Zielinski, K. 1064
Zigic, Aleksandar.......................................1426
Zinoviev, Genady Stepanovic.....................1332
Zlosnikas, Valerijus 1140
Zouhar, Jan 1665
Zulawnik, Marcin 1562
Zych, Michal..1562
Zymmer, Krzysztof.............................1332, 1562

9781424417414

2008 13th International Power Electronics and Motion Control Conference

Poznan, Poland
1-3 September 2008

IEEE Catalog Number: CFP0834A-POD
ISBN: 978-1-42441-741-4